SHORT PROTOCOLS IN MOLECULAR BIOLOGY

Hamed Ghorbani-Moghaddam
Biol 116 TR 9:00-10:15

Third Edition

A Compendium of Methods from
Current Protocols in Molecular Biology

EDITORIAL BOARD

Frederick M. Ausubel
Massachusetts General Hospital & Harvard Medical School

Roger Brent
Massachusetts General Hospital & Harvard Medical School

Robert E. Kingston
Massachusetts General Hospital & Harvard Medical School

David D. Moore
Baylor College of Medicine

J.G. Seidman
Harvard Medical School

John A. Smith
University of Alabama at Birmingham

Kevin Struhl
Harvard Medical School

WILEY

Published by John Wiley & Sons, Inc.

New York • Chichester • Weinheim • Brisbane • Singapore • Toronto

Cover art by Pauline B. Lim

While the authors, editors, and publisher believe that the specification and usage of reagents, equipment, and devices, as set forth in this book, are in accord with current recommendations and practice at the time of publication, they accept no legal responsibility for any errors or omissions, and make no warranty, express or implied, with respect to material contained herein. In view of ongoing research, equipment modifications, changes in governmental regulations, and the constant flow of information relating to the use of experimental reagents, equipment, and devices, the reader is urged to review and evaluate the information provided in the package insert or instructions for each chemical, piece of equipment, reagent, or device for, among other things, any changes in the instructions or indication of usage and for added warnings and precautions. This is particularly important in regard to new or infrequently employed chemicals or experimental reagents.

Library of Congress Cataloging in Publication Data:

Short protocols in molecular biology / edited by Frederick M. Ausubel
...[et al.] —3rd ed.
p. cm.
Includes index.
1. Molecular biology—Technique. 2. Molecular biology—Laboratory manuals. I. Ausubel, Frederick M.

QH506.S54 1992
574.8'8'028—dc20

92-6616
CIP

ISBN 0-471-13781-2

Printed in the United States of America

10 9 8 7 6 5 4

CONTENTS

2 PREPARATION AND ANALYSIS OF DNA

3 ENZYMATIC MANIPULATION OF DNA AND RNA

4 PREPARATION AND ANALYSIS OF RNA

5 RECOMBINANT DNA LIBRARIES

6 SCREENING OF RECOMBINANT DNA LIBRARIES

7 DNA SEQUENCING

10 ANALYSIS OF PROTEINS

11 IMMUNOLOGY

12 DNA-PROTEIN INTERACTIONS

13 *SACCHAROMYCES CEREVISIAE*

15 THE POLYMERASE CHAIN REACTION

16 PROTEIN EXPRESSION

17 ANALYSIS OF PROTEIN PHOSPHORYLATION

APPENDICES

INDEX

Preface

This volume presents shortened versions of the methods published in *Current Protocols in Molecular Biology.* Drawing from both the original "core" manual as well as the quarterly update service, this compendium includes all step-by-step descriptions of the principal methods covered in CPMB. Designed for use at the lab bench, it is intended for graduate students and postdoctorates who are familiar with the detailed explanations found in CPMB. However, sufficient detail is provided to allow the experienced investigator to use it as a standalone bench guide.

Although mastery of the techniques herein will enable the reader to pursue research in molecular genetics, the manual is not intended to be a substitute for graduate-level courses in molecular biology or a comprehensive textbook in the field. In addition, we recommend cross-referencing the commentaries and detailed annotations in *Current Protocols in Molecular Biology.* Finally, we strongly recommend the readers obtain first-hand experience in basic techniques and safety procedures by working in a molecular biology laboratory.

HOW TO USE THIS MANUAL

Organization

This manual is organized by chapters, with individual protocols contained in units. Each unit includes listings of materials, the protocol steps, and references for each technique. Full references for the entire manual can be found in *APPENDIX 5.* The sequence and organization of material in this manual generally follows that of *Current Protocols in Molecular Biology.* Although the unit numbers may not correspond in all cases, the unit titles are identical in both versions. Thus, users who own both manuals will find it easy and convenient to cross-reference CPMB when more explanatory details are required.

Many reagents and procedures are employed repeatedly throughout the manual. Rather than duplicate this information, cross-references among units are used extensively. Early chapters (and *APPENDIX 3A* to *3E*) describe commonly used techniques such as basic microbiology and basic manipulation of enzymes, DNA, and RNA, while later chapters describe more advanced techniques. Thus, whenever a particular enzyme is used in a protocol, the appropriate unit in Chapter 3—describing reaction conditions for that enzyme—is cross-referenced (e.g., *UNIT 3.7* for reverse transcriptase). Similarly, throughout the book readers are referred to *UNIT 1.3* for spreading or streaking a plate, *UNIT 2.1* for phenol extraction/alcohol precipitation, *UNIT 2.5A* for agarose gel electrophoresis, and so on. As a result, protocols in the later chapters of the book are not overburdened with steps describing auxiliary procedures required to prepare, purify, and analyze the sample or molecule of interest.

The appendixes provide recipes for reagents and solutions (*APPENDIX 1*), a list of abbreviations and useful measurements and data (*APPENDIX 2*), commonly used biochemical techniques (*APPENDIX 3*), the names and addresses of suppliers (*APPENDIX 4*), and complete listings for all cited references (*APPENDIX 5*) .

Protocols

Many units contain groups of protocols. The *Basic* Protocol is presented first in each unit and is generally the recommended approach. *Alternate* Protocols are provided where (1) different equipment or reagents can be employed to achieve similar ends, (2) the starting material requires a variation in approach, or (3) requirements for the end product differ from those in the Basic Protocol. *Support* Protocols describe additional steps that are required to perform the Basic or Alternate Protocols; these steps are separated from the core protocol because they might be applicable to other uses in the manual or because they are performed in a time frame separate from the Basic Protocol steps.

Reagents and Solutions

Reagents required for a protocol are listed in the Materials list before the procedure begins. As noted, corresponding recipes are listed in *APPENDIX 1* except for media recipes and buffers for restriction endonucleases—the locations of these recipes are cross-referenced parenthetically in the Materials list. It is important to note that the *names* of some of these special solutions might be similar from unit to unit (e.g., hybridization solution, lysis buffer, etc.), while the *recipes* differ; thus, make certain that reagents are prepared from the proper recipes. To avoid

Special equipment is also itemized in the Materials list of each protocol. We have not attempted to list all items required for each procedure, but rather have noted those items that might not be readily available in the laboratory or that require special preparation. Listed below are standard pieces of equipment in the modern molecular biology laboratory, i.e., items used extensively in this manual and thus not included in the individual materials lists.

Autoclave

Balances analytical and preparative

Bench protectors plastic-backed (including "blue pads")

Centrifuges a low-speed (20,000 rpm) refrigerated centrifuge and an ultracentrifuge (20,000 to 80,000 rpm) are required for many procedures. Vertical ultracentrifuge rotors are very convenient for preparing plasmid DNA. At least one microcentrifuge that holds standard 1.5-ml microcentrifuge tubes is essential. It is also useful to have a large-capacity, low-speed centrifuge for spinning down large bacterial cultures and a tabletop swinging-bucket centrifuge with adapters for spinning 96-well microtiter plates.

Computer and printer

Darkroom and developing tanks or X-Omat automatic X-ray film developer.

Filtration apparatus for collecting acid precipitates on nitrocellulose filters or membrane.

Fraction collector

Freezers and refrigerators for 4°, −20°, and −70°C incubation and storage.

Fume hood

Geiger counter

Gel dryer

Gel electrophoresis equipment at least one full-size horizontal apparatus and one horizontal minigel apparatus, two sequencing gel setups for each person engaged in large-scale sequencing projects, one vertical gel apparatus for polyacrylamide protein gels, and specialized equipment for two-dimensional protein gels as required.

Heating blocks thermostat-controlled metal heating blocks that hold test tubes and/or microcentrifuge tubes are very convenient for carrying out enzymatic reactions.

Ice maker

Incubator (37°C) for growing bacteria. We recommend an incubator large enough to hold a "tissue culture" roller drum that can be used to grow 5-ml cultures in standard 18 × 150–mm test tubes. A convenient and durable tube roller is made by New Brunswick Scientific.

Incubator/shaker(s) an enclosed shaker (such as the New Brunswick Controlled Environment Incubator Shaker) that can spin 4-liter flasks is essential for growing 1-liter *E. coli* cultures. A rotary shaking water bath (New Brunswick R76) is useful for growing smaller cultures in flasks.

Light box for viewing autoradiograms.

Liquid nitrogen

Magnetic stirrers (with heater is useful).

Microcentrifuge Eppendorf-type, maximum speed 12,000 to 14,000 rpm

Microcentrifuge tubes 1.5-ml

Microwave oven to melt agar and agarose.

Mortar and pestle

Paper cutter large size, for 46 × 57–cm Whatman sheets.

pH meter

pH paper

Pipettors that use disposable tips and dispense 1 to 1000 µl. It is best to have a set for each full-time researcher.

Polaroid camera and UV transilluminator for taking photographs of stained gels.

Policemen rubber or plastic

Power supplies 300-volt power supplies are sufficient for agarose gels; 2000-volt power supply required for DNA sequencing.

Radiation shield (Lucite or Plexiglas)

Radioactive ink

Radioactive waste container for liquid and solid waste

Refrigerator 4°C

Safety glasses

Scalpels and blades

Scintillation counter

Seal-A-Meal bag sealer or equivalent

Shakers orbital and platform, room temperature or 37°C

Spectrophotometer UV and visible

Speedvac evaporator

Thermal cycler

Tissue culture equipment CO_2 humidified incubator, phase-contrast microscope, liquid nitrogen storage container, and laminar flow hood.

UV cross-linker

UV light sources long- and short-wave

UV transilluminator

Vacuum desiccator/lyophilizer

Vacuum oven

Vortex mixers

Water baths at least two with 80°C capacity

Water purification equipment or glass distillation apparatus to purify all water used in molecular biology experiments.

X-ray film, cassettes, and intensifying screens

confusion, parenthetical listings of the unit or units in which each recipe is used are provided next to the name of each reagent in *APPENDIX 1*, except in the case of commonly used buffers and solutions—e.g., TE buffer, PBS, and 1 M $CaCl_2$.

NOTE: Deionized, distilled water should be used in all protocols in this manual, and in the preparation of all reagents and solutions.

EQUIPMENT

Standard pieces of equipment in the modern molecular biology laboratory are listed in the accompanying box. These items are used extensively in this manual. The Materials list that precedes each protocol includes only "specialized" items—i.e., items that might not be readily available in the laboratory or that require special preparation.

COMMERCIAL SUPPLIERS

In some instances throughout the manual, we have recommended commercial suppliers of chemicals, biological materials, or equipment. This has been avoided wherever possible because preference for a specific brand is subjective and is generally not based on extensive comparison testing. Our guidelines for recommending a supplier are that (1) the particular brand has actually been found to be of superior quality, or (2) the item is difficult to find in the marketplace. *APPENDIX 4* lists the names, locations, and phone numbers of recommended suppliers, but these are by no means the only vendors of biological supplies. Readers may experiment with substituting their own favorite brands.

SAFETY CONSIDERATIONS

Anyone carrying out these protocols will encounter the following hazardous materials: (1) radioactive substances, (2) toxic chemicals and carcinogenic or teratogenic reagents, (3) pathogens and infectious biological agents, and (4) recombinant DNA. It is essential that these materials be used in strict accordance with local and national regulations. Cautionary notes are included in many instances throughout the manual, but we emphasize that users must proceed with the prudence and precaution associated with good laboratory practice.

ACKNOWLEDGMENTS

Putting this manual together would have been impossible without assistance from the Current Protocols staff at John Wiley & Sons. Among those who helped us, we are extremely grateful to Kaaren Janssen, Kathy Morgan, Kathy Wisch, Janet Blair, Hazel Chan, Rebecca Barr, and Elizabeth Konkle. We are particularly indebted to Sarah Greene, who initially conceived of this project and helped to shape it with skill and patience.

We are especially grateful to our co-workers who have helped with the manual by contributing material to it, by commenting on the chapters, or by field-testing the procedures. To those people—in our own labs in Boston and in academic and industrial labs all over the world—we offer our deepest thanks.

Frederick M. Ausubel, Roger Brent, Robert E. Kingston,
David D. Moore, J.G. Seidman, John A. Smith, and
Kevin Struhl

CONTRIBUTORS

Susan M. Abmayr
Pennsylvania State University
University Park, Pennsylvania

Lisa M. Albright
Reading, Massachusetts

Alejandro Aruffo
Bristol-Myers Squibb
Seattle, Washington

Frederick M. Ausubel
Massachusetts General Hospital
and Harvard Medical School
Boston, Massachusetts

Omar Bagasra
Thomas Jefferson University
Philadelphia, Pennsylvania

Albert S. Baldwin, Jr.
University of North Carolina
Chapel Hill, North Carolina

C.R. Bebbington
Celltech
Slough, England

Daniel M. Becker
Pennie & Edwards
Menlo Park, California

Claude Besmond
Hôpital Robert Debré
Paris, France

Stephen M. Beverley
Harvard Medical School
Boston, Massachusetts

Kenneth Bloch
Massachusetts General Hospital
Boston, Massachusetts

Juan Bonifacino
National Institute of Child Health
and Human Development
Bethesda, Maryland

Ann Boyle
Current Protocols
Madison, Connecticut

Allan R. Brasier
Univerity of Texas
Galveston, Texas

Michael Brenowitz
Albert Einstein College
Bronx, New York

Roger Brent
Massachusetts General Hospital
and Harvard Medical School
Boston, Massachusetts

Terry Brown
University of Manchester
Institute of Science and
Technology
Manchester, England

William Buikema
University of Chicago
Chicago, Illinois

Linda Buonocore
Yale University School
of Medicine
New Haven, Connecticut

Anthony Celeste
Genetics Institute
Cambridge, Massachusetts

Constance Cepko
Harvard Medical School
Boston, Massachusetts

Claudia A. Chen
National Institute of Mental
Health
Bethesda, Maryland

J. Michael Cherry
Stanford University
Palo Alto, California

Lewis Chodosh
Massachusetts Institute of
Technology
Cambridge, Massachusetts

Piotr Chomczynski
University of Cincinnati College
of Medicine
Cincinnati, Ohio

Joanne Chory
The Salk Institute
La Jolla, California

Donald M. Coen
Harvard Medical School
Boston, Massachusetts

Martine A. Collart
Harvard Medical School
Boston, Massachusetts

James F. Collawn
The Salk Institute
La Jolla, California

Helen M. Cooper
Melbourne Hospital
Victoria, Australia

Lynn M. Corcoran
Walter & Eliza Hall Institute
Victoria, Australia

Brendan Cormack
Harvard Medical School
Boston, Massachusetts

Robert L. Dorit
Yale University
New Haven, Connecticut

Allan Duby
Medical City
Dallas, Texas

Barbara Dunn
GenPharm International
Mountain View, California

Christopher D. Earl
Plant Resources Venture Fund
Cambridge, Massachusetts

Patricia L. Earl
National Institute of Allergy &
Infectious Diseases
Bethesda, Maryland

Richard L. Eckert
Case Western Reserve School
of Medicine
Cleveland, Ohio

Elaine Elion
Harvard Medical School
Boston, Massachusetts

Andrew Ellington
Indiana University
Bloomington, Indiana

JoAnne Engebrecht
State University of New York
Stony Brook, New York

Michael J. Evelegh
ADI Diagnostics
Rexdale, Ontario

Michael Finney
MJ Research
Watertown, Massachusetts

John J. Fortin
Tropix, Inc.
Bedford, Massachusetts

Steven A. Fuller
Univax Biologics
Rockville, Maryland

Sean Gallagher
Hoefer Scientific Instruments
San Francisco, California

Subinay Ganguly
SmithKline Beecham
King of Prussia, Pennsylvania

Paul Garrity
University of California
Los Angeles
Los Angeles, California

David H. Gelfand
Roche Molecular Systems
Alameda, California

Michael Gilman
Cold Spring Harbor Laboratory
Cold Spring Harbor, New York

Erica A. Golemis
Fox Chase Cancer Center
Philadelphia, Pennsylvania

Michael E. Greenberg
Harvard Medical School
Boston, Massachusetts

John M. Greene
National Institutes of Health
Bethesda, Maryland

David Greenstein
Massachusetts General Hospital
Boston, Massachusetts

Mitchell S. Gross
SmithKline Beecham
King of Prussia, Pennsylvania

Barbara Grossman
Amersham Life Sciences, Inc.
Cleveland, Ohio

Jeno Gyuris
Mitotix, Inc.
Cambridge, Massachusetts

John Hanson
MJ Research
Watertown, Massachusetts

Joseph S. Heilig
University of Colorado
Boulder, Colorado

Peter Heinrich
Consortium für Elektrochemische
Industrie
Munich, Germany

David E. Hill
Applied Biotechnology
Cambridge, Massachusetts

Timothy Hoey
University of California
Berkeley, California

Charles S. Hoffman
Boston College
Chestnut Hill, Massachusetts

Peter Hornbeck
University of Maryland
Baltimore, Maryland

John G. R. Hurrell
Boehringer Mannheim
Biochemicals
Indianapolis, Indiana

Charles B-C. Hwang
Harvard Medical School
Boston, Massachusetts

Nina Irwin
Cambridge, Massachusetts

Kenneth A. Jacobs
Genetics Institute
Cambridge, Massachusetts

Kaaren Janssen
Current Protocols
Guilford, Connecticut

Mustak A. Kaderbhai
University College of Wales
Penglais, Aberystwyth,
United Kingdom

James T. Kadonaga
University of California
San Diego
La Jolla, California

Steven R. Kain
Clontech Laboratories
Palo Alto, California

Randal J. Kaufman
University of Michigan
Ann Arbor, Michigan

Leslie A. Kerrigan
University of California
San Diego
La Jolla, California

Robert E. Kingston
Massachusetts General Hospital
and Harvard Medical School
Boston, Massachusetts

Carol M. Kissinger
Millipore
Burlington, Massachusetts

Lloyd B. Klickstein
Brigham and Women's Hospital
Boston, Massachusetts

Joan H.M. Knoll
Harvard Medical School
Boston, Massachusetts

Martha F. Kramer
Harvard Medical School
Boston, Massachusetts

Thomas A. Kunkel
National Institute of Environmental Health Sciences
Research Triangle Park, North Carolina

Edward R. LaVallie
Genetics Institute
Cambridge, Massachusetts

Karen Lech
Fish and Richardson
Boston, Massachusetts

Peng Liang
The Vanderbilt Cancer Center
Nashville, Tennessee

Peter Lichter
Deutsches Krebforschungzentrum
Heidelberg, Germany

Victoria Lundblad
Baylor College of Medicine
Houston, Texas

John M. McCoy
Genetics Institute
Cambridge, Massachusetts

David D. Moore
Massachusetts General Hospital and Harvard Medical School
Boston, Massachusetts

Malcolm Moos, Jr.
Center for Biologics Evaluation & Research
Food and Drug Administration
Bethesda, Maryland

Paul R. Mueller
California Institute of Technology
Pasadena, California

Cheryl Isaac Murphy
Cambridge Biotech Corporation
Worcester, Massachusetts

Richard M. Myers
Stanford University School of Medicine
Stanford, California

Rachael L. Neve
McLean Hospital
Belmont, Massachusetts

B. Tracy Nixon
Pennsylvania State University
University Park, Pennsylvania

Marjorie Oettinger
Massachusetts General Hospital
Boston, Massachusetts

Osamu Ohara
Shionogi Research Laboratories
Osaka, Japan

Hiroto Okayama
Osaka University
Osaka, Japan

Salvatore Oliviero
Harvard Medical School
Boston, Massachusetts

Arthur B. Pardee
Dana-Farber Cancer Institute
Boston, Massachusetts

Yvonne Paterson
University of Pennsylvania
Philadelphia, Pennsylvania

Heather Perry-O'Keefe
Perceptive Biosystems
Framingham, Massachusetts

Kevin Petty
University of Texas Southwestern Medical Center
Dallas, Texas

Helen Piwnica-Worms
Washington University School of Medicine
St. Louis, Missouri

Roger Pomterantz
Thomas Jefferson University
Philadelphia, Pennsylvania

Huntington Potter
Harvard Medical School
Boston, Massachusetts

Thomas Quertermous
Massachusetts General Hospital
Boston, Massachusetts

Elisabeth Raleigh
New England Biolabs
Beverly, Massachusetts

K.J. Reddy
State University of New York
Binghamton, New York

Mark Reichardt
Lakeside Biotechnology
Chicago, Illinois

Ann Reynolds
University of Washington
Seattle, Washington

Randall K. Ribaudo
National Institute of Allergy & Infectious Diseases
Bethesda, Maryland

Eric J. Richards
Washington University
St. Louis, Missouri

Paul Riggs
New England Biolabs
Beverly, Massachusetts

Melissa Rogers
Harvard Medical School and Dana-Farber Cancer Institute
Boston, Massachusetts

Sharon Rogers
Lakeside Biotechnology
Chicago, Illinois

M.R. Rolfe
Mitotix, Inc.
Cambridge, Massachusetts

John K. Rose
Yale University School of Medicine
New Haven, Connecticut

Martin Rosenberg
SmithKline Beecham
King of Prussia, Pennsylvania

Nicoletta Sacchi
National Cancer Institute
Frederick, Maryland

Thomas P. St. John
ICOS Corporation
Bothwell, Washington

Joachim Sasse
Shriners Hospital for Crippled Children
Tampa, Florida

Stephen J. Scharf
Cetus Corporation
Emeryville, California

Paul Schendel
Genetics Institute
Cambridge, Massachusetts

Timothy D. Schlabach
Spectra Physics
San Jose, California

Brian Seed
Massachusetts General Hospital
and Harvard Medical School
Boston, Massachusetts

Bartholomew M. Sefton
The Salk Institute
San Diego, California

Christine E. Seidman
Harvard Medical School
Boston, Massachusetts

Richard F Selden
TKT Inc.
Cambridge, Massachusetts

Donald Senear
University of California
Irvine, California

Thikkavarapu Seshamma
Thomas Jefferson University
Philadelphia, Pennsylvania

Raj Shankarappa
Biogenex
San Ramon, California

Allan Shatzman
SmithKline Beecham
King of Prussia, Pennsylvania

Jen Sheen
Massachusetts General Hospital
and Harvard Medical School
Boston, Massachusetts

Harinder Singh
University of Chicago
Chicago, Illinois

Barton E. Slatko
New England Biolabs
Beverly, Massachusetts

Donald B. Smith
University of Edinburgh
Edinburgh, Scotland

John A. Smith
Harvard Medical School
Boston, Massachhusetts

Timothy A. Springer
Center for Blood Research
Boston, Massachusetts

William M. Strauss
Whitehead Institute
Cambridge, Massachusetts

Kevin Struhl
Harvard Medical School
Boston, Massachusetts

Stanley Tabor
Harvard Medical School
Boston, Massachusetts

Miyoko Takahashi
Spectral Diagnostics, Inc.
Toronto, Ontario

Douglas A. Treco
TKT Inc.
Cambridge, Massachusetts

Steven J. Triezenberg
Michigan State University
East Lansing, Michigan

Baruch Velan
Israel Institute of Biological
Research
Ness Ziona, Israel

Daniel Voytas
Iowa State University
Ames, Iowa

Simon Watkins
University of Pittsburgh
Medical School
Pittsburgh, Pennsylvania

John H. Weis
University of Utah School of
Medicine
Salt Lake City, Utah

Michael Whitt
University of Tennessee
Memphis, Tennessee

Kate Wilson
Wye College
Wye, England

Kenneth J. Wilson
Applied Biosystems
Foster City, California

Fred Winston
Harvard Medical School
Boston, Massachusetts

Scott E. Winston
Univax Biologics
Rockville, Maryland

C. Richard Wobbe
Harvard Medical School
Boston, Massachusetts

Barbara Wold
California Institute of
Technology
Pasadena, California

Jerry L. Workman
Pennsylvania State University
University Park, Pennsylvania

Wayne M. Yokoyama
University of California
School of Medicine
San Francisco, California

Rolf Zeller
Harvard Medical School
Boston, Massachusetts

Lou Zumstein
Baylor College of Medicine
Houston, Texas

Escherichia coli, Plasmids, and Bacteriophages

1

Mastery of current DNA technology requires familiarity with a small number of basic concepts and techniques. This chapter provides some of them by introducing the biology and manipulation of *E. coli* and the phage and plasmid vectors used for introducing DNA into the cell.

E. coli is a rod-shaped bacterium with a circular chromosome about 3 million base pairs long. It can grow rapidly (*UNITS 1.1-1.3*) on minimal medium that contains a carbon compound such as glucose (which serves both as a carbon source and an energy source) and salts which supply nitrogen, phosphorus, and trace metals. *E. coli* grows more rapidly, however, on a rich medium that provides the cells with amino acids, nucleotide precursors, vitamins, and other metabolites that the cell would otherwise have to synthesize.

When *E. coli* is placed or inoculated into liquid culture, it first enters a lag period, after which the bacteria begin to divide. In rich medium a culture will double in number every 20 or 30 min. This phase of exponential growth is called log phase. Eventually, the cell density increases to a point at which nutrients or oxygen become depleted from the medium, or at which waste products from the cells have built up to a concentration that inhibits rapid growth. At this point, which under normal laboratory conditions occurs when the culture reaches a density of $1\text{-}2 \times 10^9$ cells/ml, the cells stop dividing rapidly. This is saturation phase and a culture that has just reached this density is said to be freshly saturated.

With very few exceptions, most bacterial strains used in recombinant DNA work are derivatives of *E. coli* strain K-12 (*UNIT 1.4*). Many advances in molecular biology during the 1960s came from studies of this organism and of bacteriophages and plasmids that use it as a host. Much of the cloning technology in current use exploits facts learned during this period.

Bacterial plasmids (*UNIT 1.5*) are frequently used in cloning protocols. These self-replicating, circular, extrachromosomal DNA molecules can be found in natural *E. coli* isolates, specifying important functions including resistance to antibiotics and production of restriction enzymes. During the 1970s, many plasmids were constructed in the laboratory with fragments of DNA from these naturally occurring plasmids. These artificial plasmids, and their derivatives, are the most commonly used vectors in recombinant DNA work. All plasmids used as cloning vectors contain three common features: a replicator, a selectable marker, and a cloning site.

Plasmid DNA can be isolated in small amounts using a miniprep protocol (*UNIT 1.6*). Because minipreps are rapid, DNA from a large number of clones can be prepared and quickly analyzed. Large amounts of highly purified DNA can be isolated using the CsCl/ethidium bromide protocol (*UNIT 1.7*).

Plasmid DNA is introduced into *E. coli* by transformation, either by the calcium chloride method or the electroporation procedure (*UNIT 1.8*). The calcium chloride protocol gives good transformation efficiencies, is simple to complete and requires no special equipment. If considerably higher transformation efficiencies are needed, the electroporation protocol should be followed. Although this method is simple, fast, and reliable, it requires an electroporation apparatus. In both protocols, competent cells can be stored.

A different series of vectors, the M13mp vectors, are derived from filamentous phages (*UNIT 1.9*). These vectors are used because DNA inserted into them can be obtained in two forms—double- and single-stranded circles (*UNIT 1.10*). Foreign DNA is inserted into double-stranded DNA and reintroduced into cells by transformation. Once inside the cells, double-stranded DNA replicates, giving rise to both new double-stranded and single-stranded circles derived from one of the two strands of the vector. Single-stranded circles are packaged into phage particles and are secreted into the medium without lysing the cells. Centrifugation of a culture of infected cells yields a supernatant that is full of particles containing only a single strand of the phage DNA. This availability of single-stranded DNA has made possible new procedures for sequencing DNA (Chapter 7) and mutagenesis (Chapter 8).

Roger Brent

UNIT 1.1

Media Preparation and Bacteriological Tools

MINIMAL MEDIA

To prepare 5× media, add ingredients to water in a 2-liter flask and heat with stirring until dissolved. Pour into bottles with loosened caps and autoclave 15 min at 15 lb/in^2. Cool media to <50°C before adding nutritional supplements and antibiotics (at this temperature the flask still feels hot but can be held continuously in one's hand). Tighten caps and store concentrated media indefinitely at room temperature.

*5× **M9 medium**, per liter*
- 30 g Na_2HPO_4
- 15 g KH_2PO_4
- 5 g NH_4Cl
- 2.5 g NaCl
- 15 mg $CaCl_2$ (optional)

*5× **A medium**, per liter*
- 5 g $(NH_4)_2SO_4$
- 22.5 g KH_2PO_4
- 52.5 g K_2HPO_4
- 2.5 g sodium citrate·$2H_2O$

*5× **M63 medium**, per liter*
- 10 g $(NH_4)_2SO_4$
- 68 g KH_2PO_4
- 2.5 mg $FeSO_4 \cdot 7H_2O$
- Adjust to pH 7 with KOH

Before use, dilute concentrated media to 1× with sterile water and add the following sterile solutions, per liter:

1 ml 1 M $MgSO_4 \cdot 7H_2O$
10 ml 20% carbon source (sugar or glycerol)
and, if required:
0.1 ml 0.5% vitamin B1 (thiamine)
5 ml 20% Casamino Acids *or* L amino acids to 40 µg/ml *or* DL amino acids to 80 µg/ml
Antibiotic (see Table 1.4.1)

RICH MEDIA

Prepare as for minimal media except autoclave 25 min. Do not dilute.

H medium, per liter
10 g tryptone
8 g NaCl

Lambda broth, per liter
10 g tryptone
2.5 g NaCl

LB medium, per liter
10 g tryptone
5 g yeast extract
5 g NaCl
1 ml 1 N NaOH

NZC broth, per liter
10 g NZ Amine A
5 g NaCl
2 g $MgCl_2{\cdot}6H_2O$
Autoclave 30 min
5 ml 20% Casamino Acids

Superbroth, per liter
32 g tryptone
20 g yeast extract
5 g NaCl
5 ml 1 N NaOH

TB (terrific broth)
12 g Bacto tryptone
24 g Bacto yeast extract
4 ml glycerol

Add H_2O to 900 ml and autoclave, then add to above sterile solution 100 ml of a sterile solution of 0.17 KH_2PO_4 and 0.72 M K_2HPO_4.

Tryptone broth, per liter
10 g tryptone
5 g NaCl

TY medium, 2×, per liter
16 g tryptone
10 g yeast extract
5 g NaCl

TYGPN medium, per liter
20 g tryptone
10 g yeast extract
10 ml 80% glycerol
5 g Na_2HPO_4
10 g KNO_3

Tryptone, yeast extract, agar (Bacto-agar), nutrient broth, and Casamino Acids are from Difco. NZ Amine A is from Hunko Sheffield (Kraft).

SOLID MEDIA

Prepare ingredients for minimal and rich plates as described below. Dry plates 2 to 3 days at room temperature, or 30 min with the lids slightly off at 37°C or in a laminar flow hood. Store dried plates wrapped at 4°C.

Minimal Plates

Autoclave 15 g agar in 800 ml water for 15 min. Add 200 ml sterile 5× minimal medium and carbon source. Cool to 50°C and add supplements and antibiotics. Pour 32 to 40 ml medium/plate to obtain 25 to 30 plates/liter.

Rich Plates

Add water to ingredients listed below to 1 liter. Autoclave 25 min. Pour 32 to 40 ml medium for LB and H plates, and 45 ml medium for lambda plates (per plate).

H plates, per liter
10 g tryptone
8 g NaCl
15 g agar

Lambda plates, per liter
10 g tryptone
2.5 g NaCl
10 g agar

LB plates, per liter
10 g tryptone
5 g yeast extract
5 g NaCl
1 ml 1 N NaOH
15 g agar or agarose

Additives

Antibiotics (if required):
Ampicillin to 50 µg/ml
Tetracycline to 12 µg/ml

Other antibiotics, see Table 1.4.1

Galactosides (if required):
Xgal to 20 µg/ml
IPTG to 0.1 mM

Other galactosides, see Table 1.4.2

TOP AGAR

Top agar is used to distribute phage or cells evenly in a thin layer over the surface of a plate. Prepare top agar in 1-liter batches, autoclave for 15 min to melt, cool to 50°C, swirl to mix, pour into separate 100-ml bottles, reautoclave, cool, and store at room temperature. Before use, melt the agar by heating in a water bath or microwave oven then cool to and hold at 45° to 50°C. Top agar contains less agar than plates and stays molten for days when kept at 45° to 50°C.

H top agar, *per liter*
10 g tryptone
8 g NaCl
7 g agar

LB top agar, *per liter*
10 g tryptone
5 g yeast extract
5 g NaCl
7 g agar

Lambda top agar, *per liter*
10 g tryptone
2.5 g NaCl
7 g agar

Top agarose, *per liter*
10 g tryptone
8 g NaCl
6 g agarose

STAB AGAR

Stab agar, *per liter*
10 g nutrient broth
5 g NaCl
6 g agar
10 mg cysteine·Cl
10 mg thymine

Stab agar is used for storing bacterial strains (UNIT 1.3). Cysteine is thought to increase the amount of time bacteria can survive in stabs. Thymine is included so that thy⁻ bacteria can grow.

TOOLS

Inoculating Loops

1. Sterilize the inoculating loop in a bunsen burner flame until it is red hot.
2. Cool by touching to the surface of a sterile agar plate until it stops sizzling.

Sterile Toothpicks

1. Autoclave toothpicks in a beaker covered with foil or in the original box.

   Round wooden toothpicks or the pointed end of flat toothpicks are sometimes used to pick individual colonies or phage plaques.

2. Reautoclave used toothpicks to reuse.

Spreaders

1. Make a spreader by heating and bending a piece of 4-mm glass tubing or a Pasteur pipet (see Sketch 1.1.1).

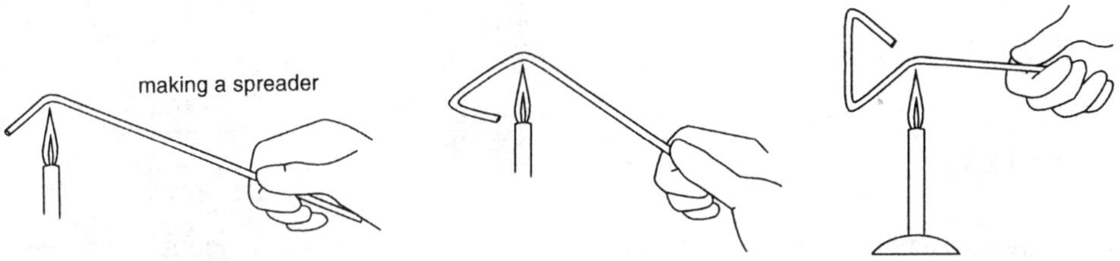

Sketch 1.1.1

2. Sterilize the spreader by dipping the triangular part into a container of ethanol, removing it from the container, and igniting the ethanol on the spreader by passing it through a gas flame.
3. After the flame goes out, cool the spreader by touching it to the surface of a sterile agar plate.

Contributors: Karen Lech and Roger Brent

Growth in Liquid Media

UNIT 1.2

GROWING AN OVERNIGHT CULTURE

BASIC PROTOCOL 1

1. Transfer 5 ml liquid medium into a sterile 16- or 18-mm culture tube.
2. Inoculate with a single bacterial colony on an inoculating loop, by dipping and shaking the loop in the medium.
3. Cap the tube and grow at 37°C to saturation (~6 hr) in a shaker or on a roller drum, 60 rpm.

   A freshly saturated culture can contain 1-2 × 10^9 cells/ml.

GROWING LARGER CULTURES

BASIC PROTOCOL 2

1. Dilute overnight cultures 1:100 in an Erlenmeyer or baffle flask that is ≥5 times the volume of the culture.

   To grow cells without agitation, use a flask that is ≥20 times the volume of the culture.

2. Grow at 37°C with vigorous agitation, ~300 rpm.

MONITORING GROWTH

BASIC PROTOCOL 3

With a Count Slide

1. Cover a clean count slide or Petraff-Hausser chamber with a clean cover slip. See Sketch 1.2.1.
2. Touch a pipet containing a small amount of the culture to the edge of the cover-slip.

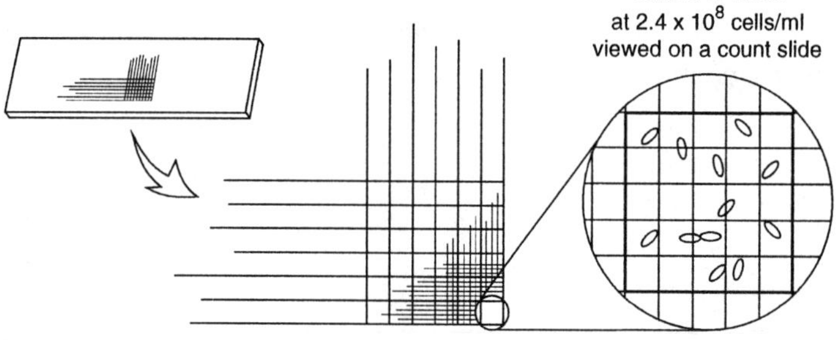

Sketch 1.2.1

3. Place on a phase-contrast microscope at 400× and calculate concentration based on each cell in a small square being approximately equivalent to 2×10^7 cells/ml.

With a Spectrophotometer

1. Dilute culture to achieve an OD_{600} <1.
2. Calculate concentration based on 0.1 OD being roughly equivalent to 10^8 cells/ml.

Contributors: Karen Lech and Roger Brent

UNIT 1.3

Growth on Solid Media

BASIC PROTOCOL 1

TITERING AND ISOLATING BACTERIAL COLONIES BY SERIAL DILUTIONS

1. Add 5 ml LB medium to three sterile culture tubes and label 1 to 3.
2. Transfer 5 µl of cells to the first tube and vortex, transfer 5 µl from the first tube to the second tube and vortex, and transfer 5 µl from the second tube to the third tube and vortex. Use a new pipet tip with each transfer.

   Saturated cultures contain ~10^9 cells/ml.

3. Spread 100 µl from the culture and from each dilution tube onto separate, labeled, dry LB plates, and incubate overnight at 37°C.
4. Calculate cells/ml: multiply number of colonies by 10 times the dilution factor.
5. Save for further use by wrapping plates and storing at 4°C.

BASIC PROTOCOL 2

ISOLATING SINGLE COLONIES BY STREAKING A PLATE

1. Streak an inoculum across one side of a plate using sterile technique (see Sketch 1.3.1).
2. Resterilize an inoculating loop and streak a sample from the first streak across a fresh part of the plate. Repeat until the plate is covered.

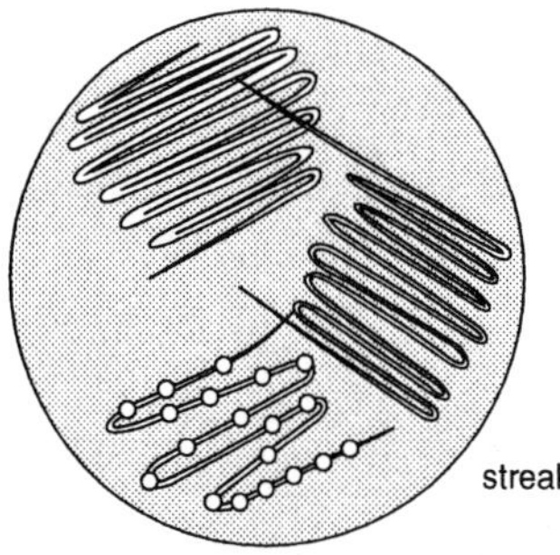

Sketch 1.3.1

3. Incubate at 37°C until colonies appear.

   If colonies must be isolated from many bacteria, divide plate into 4, 6, or 8 sectors and streak for single colonies in each sector.

ISOLATING SINGLE COLONIES BY SPREADING A PLATE

BASIC PROTOCOL 3

1. Pipet 0.05 to 1 ml of the culture onto a dry plate.
2. Spread evenly with a spreader using a circular motion. Alternatively, use the edge of the spreader to make a raster pattern (a series of parallel lines) on the plate's surface; turn plate and repeat at right angle.
3. Incubate at 37°C with plate lid ajar until completely dry.

REPLICA PLATING

SUPPORT PROTOCOL 1

Replica plating is a convenient way to test many colonies for their ability to grow under different conditions. Bacterial colonies are transferred from one plate to another in a way that maintains the original pattern of colonies.

1. Secure a sterile velvet with a metal ring onto a replica block.
2. Press a master plate of well-separated colonies *lightly* onto the velvet.
3. Press new plates, oriented like the master plate, *lightly* onto imprinted velvet to transfer colonies. Up to 10 plates/velvet can be replica plated.

STRAIN STORAGE AND REVIVAL

SUPPORT PROTOCOL 2

Stabs

1. Fill airtight, autoclavable vials with rubber or Teflon caps ⅔ full with stab agar.
2. Inoculate with a single colony, repeatedly poking the inoculating loop deeply into the agar.
3. Incubate at 37°C with the cap loose until cloudy tracks of bacteria become visible (8 to 12 hr).
4. Seal tightly and store in the dark at 15° to 22°C.
5. Revive as follows: using a cool, sterile inoculating loop, remove some bacteria-laden agar. Smear onto an LB plate and streak for single colonies.

Frozen Stocks

1. Add 1 ml of a freshly saturated culture (or 2 ml of a mid-log culture) to a stab vial or Nunc vial containing 1 ml glycerol solution (*APPENDIX 1*) or 1 ml of 7% (v/v) dimethyl sulfoxide (DMSO).

   The only advantage DMSO seems to have over glycerol for frozen stocks is that it is easier to pipet because it is less viscous.

2. Store at –20° to –70°C.

   Most strains remain viable much longer when stored at –70°C.

3. Revive by scraping off splinters of solid ice with a sterile toothpick or pipet (being careful not to allow contents to thaw) and streaking onto an LB plate.

Reference: Lederberg and Tatum, 1953.

Contributors: Karen Lech and Roger Brent

UNIT 1.4

Selected Topics from Classical Bacterial Genetics

Cloning technology exploits facts learned from classical bacterial genetics. Tables 1.4.1-1.4.5 review information critical to this technology.

Table 1.4.1 Antibiotics, Their Modes of Action, and Modes of Bacterial Resistance[a]

Antibiotic[b]	Stock conc. (mg/ml)	Final conc. (μg/ml)	Mode of action	Mode of resistance
Ampicillin[c]	4	50	Bacteriocidal; only kills growing *E. coli*; inhibits cell wall synthesis by inhibiting formation of the peptidoglycan cross-link	β-lactamase hydroylzes ampicillin before it enters the cell
Chloramphenicol, in methanol	10	20	Bacteriostatic; inhibits protein synthesis by interacting with the 50S ribosomal subunit and inhibiting the peptidyltransferase reaction	Chloramphenicol acetyltransferase inactivates chloramphenicol
D-Cycloserine,[d] in 0.1 M sodium phosphate buffer, pH 8	10	200	Bacteriocidal; only kills growing *E. coli;* inhibits cell wall synthesis by preventing formation of D-alanine from L-alanine and formation of peptide bonds involving D-alanine	Mutations destroy the D-alanine transport system
Gentamycin	10	15	Bacteriocidal; inhibits protein synthesis by binding to the L6 protein of the 50S ribosomal subunit	Aminoglycoside acetyltransferase and aminoglycosidenucleotidyltransferase inactivate gentamycin; mutations in *rplF* (encodes the L6 protein) prevent the gentamycin from binding
Kanamycin	10	30	Bacteriocidal; inhibits protein synthesis; inhibits translocation and elicits miscoding	Aminoglycoside phosphotransferase, also known as neomycin phosphotransferase, aminoglycoside acetyltransferase, and aminoglycoside nucleotidyltransferase; inactivates kanamycin
Kasugamycin	10	1000	Bacteriocidal; inhibits protein synthesis by altering the methylation of the 16S RNA and thus an altered 30S ribosomal subunit	Mutations prevent kasugamycin from binding to the ribosome; mutations decrease uptake of kasugamycin
Nalidixic acid, pH to 11 with NaOH	5	15	Bacteriostatic; inhibits DNA synthesis by inhibiting DNA gyrase	Mutations in the host DNA gyrase prevent nalidixic acid from binding
Rifampicin,[e] in methanol	34	150	Bacteriostatic; inhibits RNA synthesis by binding to and inhibiting the β subunit of RNA polymerase; rifampicin sensitivity is dominant.	Mutation in the β subunit of RNA polymerase prevents rifampicin from complexing; rifampicin resistance is recessive
Spectinomycin	10	100	Bacteriostatic; inhibits translocation of peptidyl tRNA from the A site to the P site	Mutations in *rpsE* (encodes the S5 protein) prevent spectinomycin from binding; spectinomycin sensitivity is dominant and resistance is recessive

continued

Table 1.4.1 Antibiotics, Their Modes of Action, and Modes of Bacterial Resistance[a], continued

Antibiotic[b]	Stock conc. (mg/ml)	Final conc. (μg/ml)	Mode of action	Mode of resistance
Streptomycin	50	30	Bacteriocidal; inhibits protein synthesis by binding to the S12 protein of the 30S ribosomal subunit and inhibiting proper translation; streptomycin sensitivity is dominant	Aminoglycoside phosphotransferase inactivates streptomycin; mutations in *rpsL* (encodes the S12 protein) prevent streptomycin from binding; streptomycin resistance is recessive
Tetracycline,[e] in 70% ethanol	12	12	Bacteriostatic; inhibits protein synthesis by preventing binding of aminoacyl tRNA to the ribosome A site	Active efflux of drug from cell

[a]Data assembled from Foster (1983), Gottlieb and Shaw (1967), and Moazed and Noller (1987).

[b]All antibiotics should be stored at 4°C, except tetracycline, which should be stored at −20°C. All antibiotics should be dissolved in sterile distilled H_2O unless otherwise indicated.

[c]Carbenicillin, at the same concentration, can be used in place of ampicillin. Carbenicillin can be stored in 50% ethanol/50% water at −20°C.

[d]D-cycloserine solutions are unstable. They should be made immediately before use.

[e]Light-sensitive; store stock solutions and plates in the dark.

Table 1.4.2 Lactose Analogs Used in DNA Cloning Technology

Galactoside	Stock concentration[a]	Use	Characteristics	Reference
Isopropyl-1-thio-β-D-galactoside (IPTG)	100 mM	Very effective inducer	Nonmetabolizable inducer	Barkley and Bourgeois, 1978 (pp. 177-220)
5-Bromo-4-chloro-3-indolyl-β-D-galactoside (Xgal)	20 mg/ml (dissolved in *N,N* dimethyl formamide)	Identification of *lacZ*+ bacteria, especially useful for detecting β-galactosidase made by recombinant vectors	Noninducing chromogenic substrate of β-galactosidase (cleavage of Xgal results in blue color); production of bluecolor independent of *lacY* gene product	Miller, 1972
Orthonitrophenyl-β-D-galactoside (ONPG)	10 mM	β-galactosidase assays	Chromogenic substrate of β-galactosidase (cleavage of ONPG results in yellow color)	Miller, 1972 (pp. 352-355)
6–O–β-D-Galactopyranosyl D-glucose (allolactose)			Inducer of the lactose operon in vivo; lactose is converted into allolactose by β-galactosidase	Zabin and Fowler, 1978 (pp. 89-121)
Phenyl-β-D-galactoside (Pgal)	2 mg/ml	Selection for *lac* constitutive mutants	Noninducing substrate of β-galactosidase; uptake partly dependent on *lacY* gene product	Miller, 1978 (pp. 31-88)
Orthonitrophenyl-β-D-thiogalactoside (TONPG)	10 mM	Selection for *lac*− mutants	Transported into cells by *lac* permease (the *lacY* gene product); inhibits cell growth at high concentration	Miller, 1978 (pp. 31-88)

[a]Stock solutions should be dissolved in sterile water unless otherwise noted.

Table 1.4.3 Commonly Used Nonsense Suppressors[a]

Suppressor	Map position[b]	Type of suppressor	Amino acid inserted	tRNA gene
supD (sul)	43	Amber	Serine	*serU*
supE (su2)	16	Amber	Glutamine	*glnU*
supF (su3)	27	Amber	Tyrosine	*tyrT*
supB (suB)	16	Ochre/amber	Glutamine	*glnU*
supC (suC)	27	Ochre/amber	Tyrosine	*tyrT*

[a]Data compiled from Bachmann (1983) and Celis and Smith (1979).
[b]Given in minutes; see Bachmann (1983) for description.

Table 1.4.4 Commonly Used Genetic Markers and How to Test Them[a]

Nutritional markers	Streak or replica plate colonies of the strain onto plates with and without the nutrient to be tested, but which contain all other necessary nutrients.
Antibiotic resistance markers	Streak or replica plate colonies of the strain onto plates with and without the antibiotic.
Other markers	
lacZ$^+$	Streak strain on an LB plate with Xgal and IPTG (*UNIT 1.4*). Colonies should turn blue. Colonies of control *lacZ*$^-$ strain should not turn blue.
lacZΔM15[b]	Transform strain with pUC plasmid and with control plasmid such as pBR322. Streak transformants onto LB/ampicillin plate with Xgal and IPTG. Colonies bearing pUC plasmid should turn blue, while colonies bearing pBR322 should not.
F$^+$ or F′	Spot M13 phage onto a lawn of the cells. Small plaques should appear (see *UNIT 1.15*).
recA	Using a toothpick, make a horizontal stripe of cells across an LB plate. Also make a stripe of *recA*$^+$ control cells. Cover half of the plate with a piece of cardboard, and irradiate the plate with 300 ergs/cm^2 of 254 nm UV light from a hand-held UV source (typically 20-sec exposure from a lamp held 50 cm over the plate). *recA*$^-$ cells are very sensitive to killing by UV light, and *recA*$^-$ cells in the unshielded part of the plate should be killed by this level of irradiation.
recBCD	Spot dilutions of λ gam$^-$ on a lawn of cells side by side with dilutions of λ gam$^+$. The *gam*$^-$ plaques should be almost as big as the *gam*$^+$ plaques.
hsdS$^-$	(1) Use the strain and a wild-type strain to plate out serial dilutions of a λ-like phage stock grown on an *hsdS*$^-$ or *hsdR*$^-$ host. If the phage stock came from an *hsdS*$^-$ host, then it should make plaques with 10^4 to 10^6 higher efficiency on the putative *hsdS*$^-$ host than on a wild-type host. If the plate stock came from an *hsdR*$^-$ (*hsdS*$^+$ *hsdM*$^+$) host, it should make plaques with the same efficiency on both strains. (2) Suspend one of the fresh plaques from the putative *hsdS*$^-$ host in 1 ml lambda dilution buffer. Titer this suspension on the putative *hsdS*$^-$ strain and on a wild-type strain. The suspension should make plaques at 10^4 to 10^6 higher efficiency on the *hsdS*$^-$ strain than on the wild-type strain. One plaque contains ~10^7 phage.
hsdR$^-$ (*hsdS*$^+$ *hsdM*$^+$)	(1) Perform step 1 described above, using a plate stock made from an *hsdS*$^-$ host. (2) Suspend one of the fresh plaques in 1 ml lambda dilution buffer. Titer this suspension on the putative *hsdR*$^-$ strain and on a wild-type strain. This suspension should make plaques with the same efficiency on the *hsdR*$^-$ as on a wild-type strain.
dam	Transform the strain and a wild-type strain with a plasmid that contains recognition sites for the enzymes *Mbo*I or *Bcl*I. Prepare plasmid DNA from both strains and verify that plasmid DNA isolated from the *dam*$^-$ strain is sensitive to digestion by the enzyme.
dcm	Transform the strain and a wild-type strain with a plasmid that contains recognition sites for *Scr*FI. Prepare plasmid DNA from both strains to verify that only plasmid DNA from the *dcm* strain is fully sensitive to digestion by the enzyme. Half of the *Scr*FI sites will be cut even when the DNA is *dcm*-methylated.
lon	Streak LB plate for single colonies. Also streak a control plate of a wild-type strain. Incubate at 37°C. Colonies of the *lon*$^-$ strain should be larger, glistening, and mucoidal.

[a]Commonly used protocols in this table are media preparation (*UNIT 1.1*), streaking and replicating a plate (*UNIT 1.3*), and growing lambda-derived vectors.

[b]Encodes omega fragment of β-galactosidase.

Table 1.4.5 Commonly Used *Escherichia coli* Strains

Strain[a]	Genotype
AR58	*sup*^{0} *galK2 galE::*Tn*10* (λcI857 ΔH1 *bio*$^-$ *uvrB kil*$^-$ cIII$^-$) Str^r
AR120	*sup*^{0} *galK2 nad::*Tn*10* (Tet^r) (λcI$^+$ *ind*$^+$ *p*$_L$*-lacZ fusion*) Str^r
AS1[b]	*endA1 thi-1 hsdR17*($r_K^-m_K^+$) *supE44* (λcI$^+$)
BNN102[b]	C600 *hflA150 chr::*Tn*10 mcrA1 mcrB*
BW313[c]	*Hfr lysA*$^-$ *dut ung thi-1 recA spoT1*
C600	*thi-1 thr-1 leuB6 lacY1 tonA21 supE44 mcrA*
CJ236[c]	*dut1 ung1 thi-1 relA1*/pCJ105 (Cm^r)
DH1	*recA1 endA1 thi-1 hsdR17 supE44 gyrA96* (Nal^r) *relA1*
DH5αF′[d]	F′/*endA1 hsdR17*($r_K^-m_K^+$) sup*E44 thi-1 recA1 gyrA* (Nal^r) *relA1* Δ(*lacZYA-argF*)$_{U169}$(*m80lacZ*Δ*M15*)
DK1	*hsdR2 hsdM*$^+$ *hsdS*$^+$ *araD139* Δ(*ara-leu*)$_{7697}$Δ(*lac*)$_{X74}$ *galU galK* rpsL (Str^r) *mcrA mcrB1* Δ(*srl-recA*)$_{306}$
ER1451	F′ *traD36 proAB lacI*^{q} Δ(*lacZ*)*M15*/*endA gyrA96 thi-1 hsdR2* (or *hsdR17*) *supE44* Δ(*lac-proAB*) *mcrB1 mcrA*
HB101[e]	Δ(*gpt-proA*)*62 leuB6 thi-1 lacY1 hsdS*$_B$*20 recA rpsL20* (Str^r) *ara-14 galK2 xyl-5 mtl-1 supE44 mcrB*$_B$
JM101[f]	F′ *traD36 proA*$^+$ *proB*$^+$ *lacI*^{q} *lacZ*Δ*M15*/*supE thi* Δ(*lac-proAB*)
JM105[f]	F′ *traD36 proA*$^+$ *proB*$^+$ *lacI*^{q} *lacZ*Δ*M15*/Δ(*lac-pro*)$_{X111}$ *thi rpsL* (Str^r) *endA sbcB supE hsdR*
JM107[f]	F′ *traD36 proA*$^+$ *proB*$^+$ *lacI*^{q} *lacZ*Δ*M15*/*endA1 gyrA96* (Nal^r) *thi hsdR17 supE44 relA1* Δ(*lac-proAB*) *mcrA*
JM109[g]	F′ *traD36 proA*$^+$ *proB*$^+$ *lacI*^{q} *lacZ*Δ*M15*/*recA1 endA1 gyrA96* (Nal^r) *thi hsdR17 supE44 relA1* Δ(*lac-proAB*) *mcrA*
K38	HfrC (λ)
KM392	*hsdR514*($r_K^-m_K^+$) *supE44 supF58 lacY galK2 galT22 metB1 trp55 mcrA* Δ*lac*$_{U169}$ *proC::*Tn*5*
LE392	*hsdR514*($r_K^-m_K^+$) *supE44 supF58 lacY galK2 galT22 metB1 trp55 mcrA*
MC1061	*hsdR2 hsdM*$^+$ *hsdS*$^+$ *araD139* Δ(*ara-leu*)$_{7697}$Δ(*lac*)$_{X74}$ *galE15 galK16 rpsL* (Str^r) *mcrA mcrB1*
MM294	*endA thiA hsdR17 supE44*
NM539[h]	*supF hsdR* (P2cox3)
P2392	*hsdR514*($r_K^-m_K^+$) *supE44 supF58 lacY galK2 galT22 metB1 trp55 mcrA* (P2)
PR722	F′ Δ(*lacIZ*)$_{E65}$ *pro*$^+$/*proC::*Tn*5* Δ(*lacIZYA*)$_{U169}$ *hsdS20 ara-14 galK2 rpsL20* (Str^r) *xyl-5 mtl-1 supE44 leu*
Q359	*hsdR*$^-$ *hsdM*$^+$ *supE tonA* (φ80^r) (P2)
RR1	Δ(*gpt-proA*)*62 leuB6 thi-1 lacY1 hsdS*$_B$*20 rpsL20* (Str^r) *ara-14 galK2 xyl-5 mtl-1 supE44 mcrB*$_B$
Y1088[h]	*supE supF metB trpR hsdR*$^-$ *hsdM*$^+$ *tonA21 strA* Δ*lac*$_{U169}$ *mcrA proC::*Tn*5*/pMC9
Y1089[h]	Δ*lac*$_{U169}$ *proA*$^+$ Δ(*lon*) *araD139 strA hflA150 chr::*Tn*10*/pMC9
Y1090[h]	Δ*lac*$_{U169}$ *proA*$^+$ Δ(*lon*) *araD139 strA supF trpC22::*Tn*10 mcrA*/pMC9

[a]The original *E. coli* K-12 strain was an F$^+$ λ lysogen, but most K-12 derivatives in common use have been cured of the F factor and prophage and these are indicated only when present. All other genes in these strains are presumed to be wild-type except for the genotype markers noted in the second column.

[b]AS1 is also known as MM294cI$^+$. BNN102 is also known as C600 *hflA*.

[c]Both CJ236 and BW313 are commonly used in oligonucleotide-directed mutagenesis. pCJ105, the plasmid CJ236 carries, is not relevant for this application.

[d]Three strains are in circulation. DH5 is a derivative of DH1 that transforms at slightly higher efficiency. DH5α and DH5αF′ are derivatives that carry a deletion of the lac operon and a Φ80 prophage that directs synthesis of the omega fragment of β-galactosidase. DH5αF′ carries an F′ factor as well. DH5α and DH5αF′ are proprietary strains and the cells are prepared in some way that allows them to be transformed with slightly higher efficiency than DH5.

[e]In this strain, the area of the chromosome that contains the *hsd* genes was derived from the related B strain of *E. coli*.

[f]The continued presence of the F′ factor in JM strains can be insured by starting cultures only from single colonies grown on minimal plates that do not contain proline. These strains encode the omega fragment of *lacZ* and are frequently used with vectors that direct the synthesis of the *lacZ* alpha fragment. These strains are frequently used with M13 vectors for DNA sequencing (*UNITS 1.9, 1.10, & 7.4*).

[g]It is not known whether this strain has markers other than those listed.

[h]pMC9, the plasmid in the Y strains listed here, directs the synthesis of large amounts of *lac* repressor. It also confers resistance to tetracycline and ampicillin (Miller et al., 1984, *EMBO. J.* 3:3117-3121).

References: Miller, 1972; Neidhardt et al., 1987.

Contributors: Elisabeth A. Raleigh, Karen Lech, and Roger Brent

UNIT 1.5

Maps of Plasmids

Table 1.5.1 Characteristics of Commonly Used Plasmid Replicators

Replicator	Prototype plasmid	Size (bp)	Markers on prototype	Copy number	References
pMB1	pBR322	4,362	Amp^r, Tet^r	high; >25	Bolivar et al., 1977
ColE1	pMK16	~4,500	Kan^r, Tet^r, $ColE1^{imm}$	high; >15	Kahn et al., 1979
p15A	pACYC184	~4,000	Eml^r, Tet^r	high; ~15	Chang et al., 1978
pSC101	pLG338	~7,300	Kan^r, Tet^r	low; ~6	Stoker et al., 1982
F	pDF41	~12,800	TrpE	low; 1 to 2	Kahn et al., 1979
R6K	pRK353	~11,100	TrpE	low; <15	Kahn et al., 1979
R1 (R1*drd-17*)	pBEU50	~10,000	Amp^r, Tet^r	low at 30°C; high above 35°C[a]	Uhlin et al., 1983
RK2	pRK2501	~11,100	Kan^r, Tet^r	low; 2 to 4	Kahn et al., 1979
λ *dv*	λ *dvgal*	—[b]	Gal	—	Jackson et al., 1972

[a]Temperature sensitive.
[b]Not known.

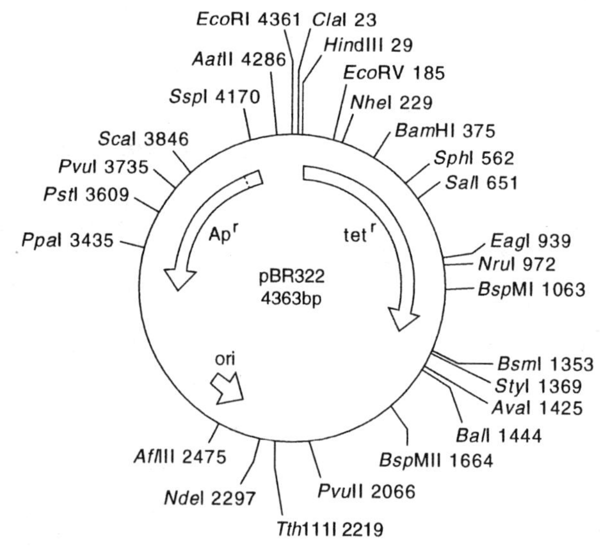

Figure 1.5.1 pBR322. pBR322 is a very commonly used cloning vector. It contains an amplifiable pMB1 replicator and genes encoding resistance to ampicillin and tetracycline. Insertion of DNA into a restriction site in either drug-resistance gene usually inactivates it and allows colonies bearing plasmids with such insertions to be identified by their inability to grow on medium with that antibiotic (Bolivar et al., 1977; sequence in Sutcliffe, 1978).

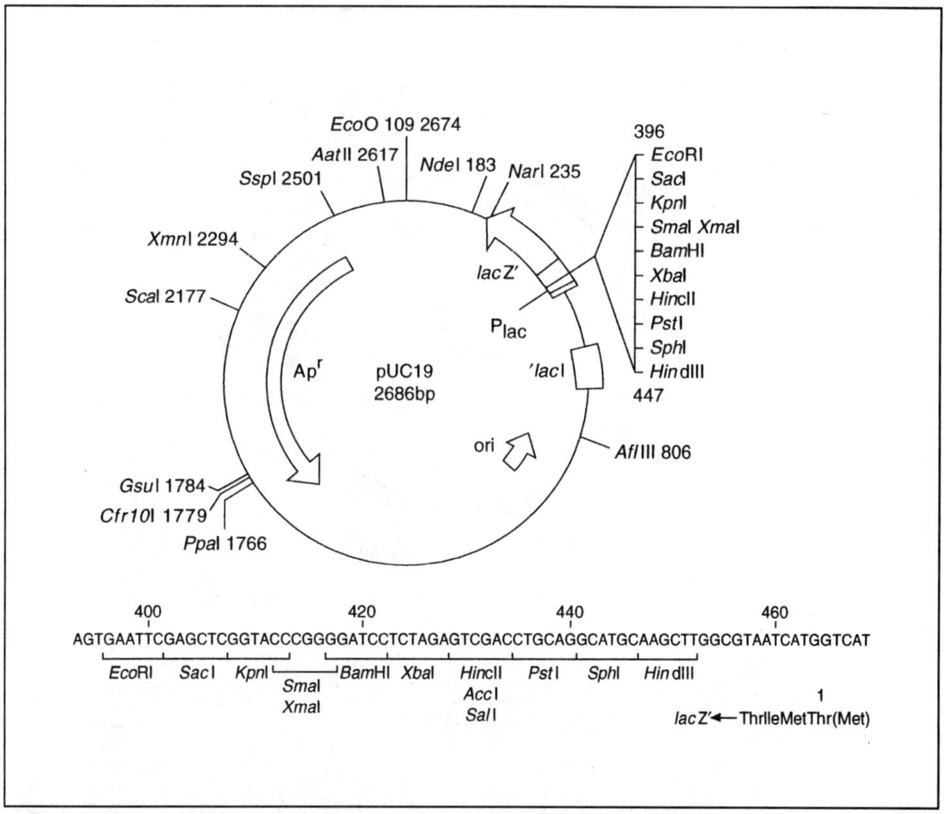

Figure 1.5.2 pUC19. pUC19 belongs to a family of plasmid vectors that contain a polylinker inserted within the alpha region of the *lacZ* gene. The polylinkers are the same as those used in the M13mp series and their sequence is given in Figure 1.14.2. pUC19 and pUC18 have the same polylinker but in opposite orientations. Under appropriate conditions, colonies that bear plasmids containing a fragment inserted into the polylinker form white colonies instead of blue ones. These pMB1-derived plasmids maintain a high copy number because they lack an intact *rop* gene (encoding the Rop protein); moreover, they are thought to bear another mutation in the *ori* region that increases the copy number. Wild-type and recombinant plasmids confer ampicillin resistance and can be amplified with chloramphenicol (Norrander et al., 1983). In additon wild-type plasmids confer a $LacZ^+$ phenotype to appropriate cells (e.g., JM101 cells, Table 1.4.5).

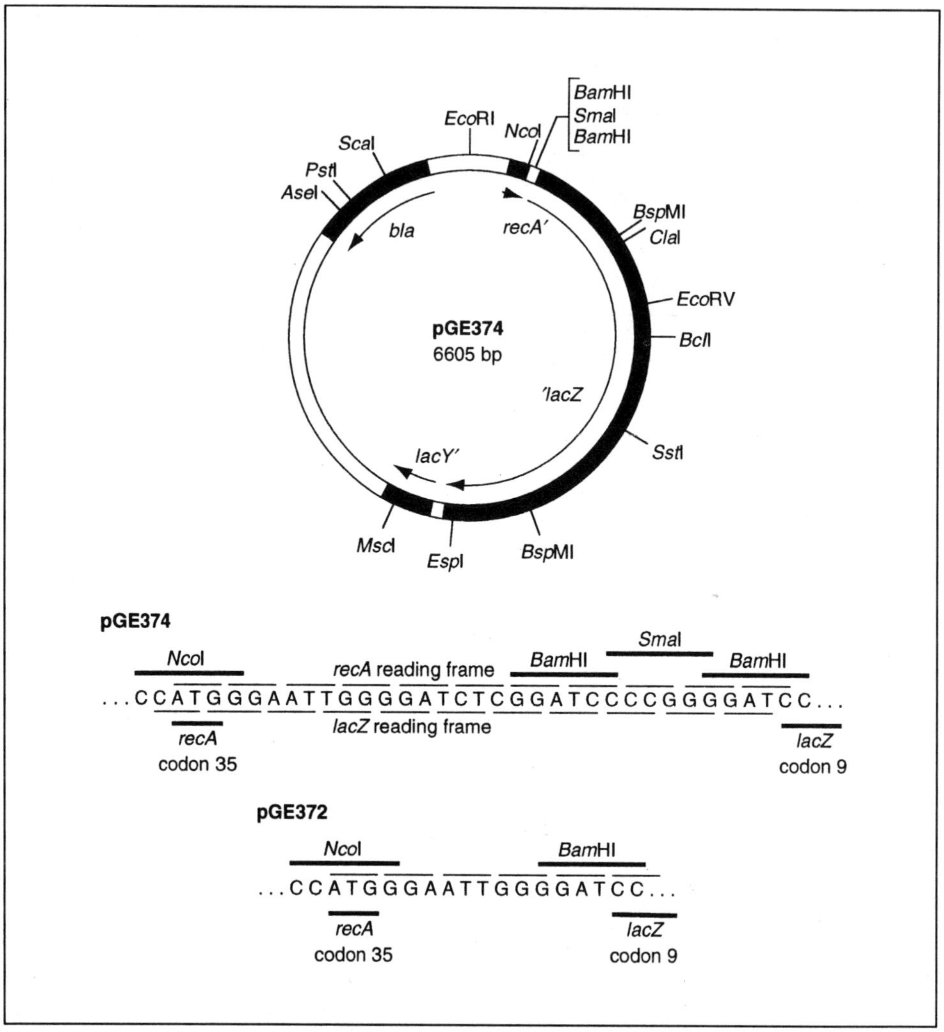

Figure 1.5.3 pGE374. pGE374 is an example of a vector used to clone and express open reading frames (ORFs). This expression system operates on the same principle as vectors that use the *ompF* regulatory region (described by Weinstock et al., 1983); however, the *recA*-carrying vector pGE374 is better behaved and its expression system is more reliable than that of the *ompF* vector system. pGE374 confers a LacZ$^-$ phenotype unless an ORF has been inserted that realigns *recA* and lacZ and creates a LacZ$^+$ phenotype. The basal level of expression of a single-copy *recA* gene is about 1000 molecules per cell, so it is not necessary to induce strains carrying pGE374 with inserts in order to detect their LacZ phenotype. The basal level of expression is sufficient for detection of LacZ in plasmid-carrying strains grown on Xgal indicator plates. The structures of the *recA/lacZ* junctions in pGE374 and its relative pGE372 are shown below the vector. In pGE374 the *recA* and *lacZ* coding sequences are out of frame so that no hybrid protein is produced. Plasmid pGE372 is similar, except that the *recA* and *lacZ* sequences are in frame and hybrid protein is produced.

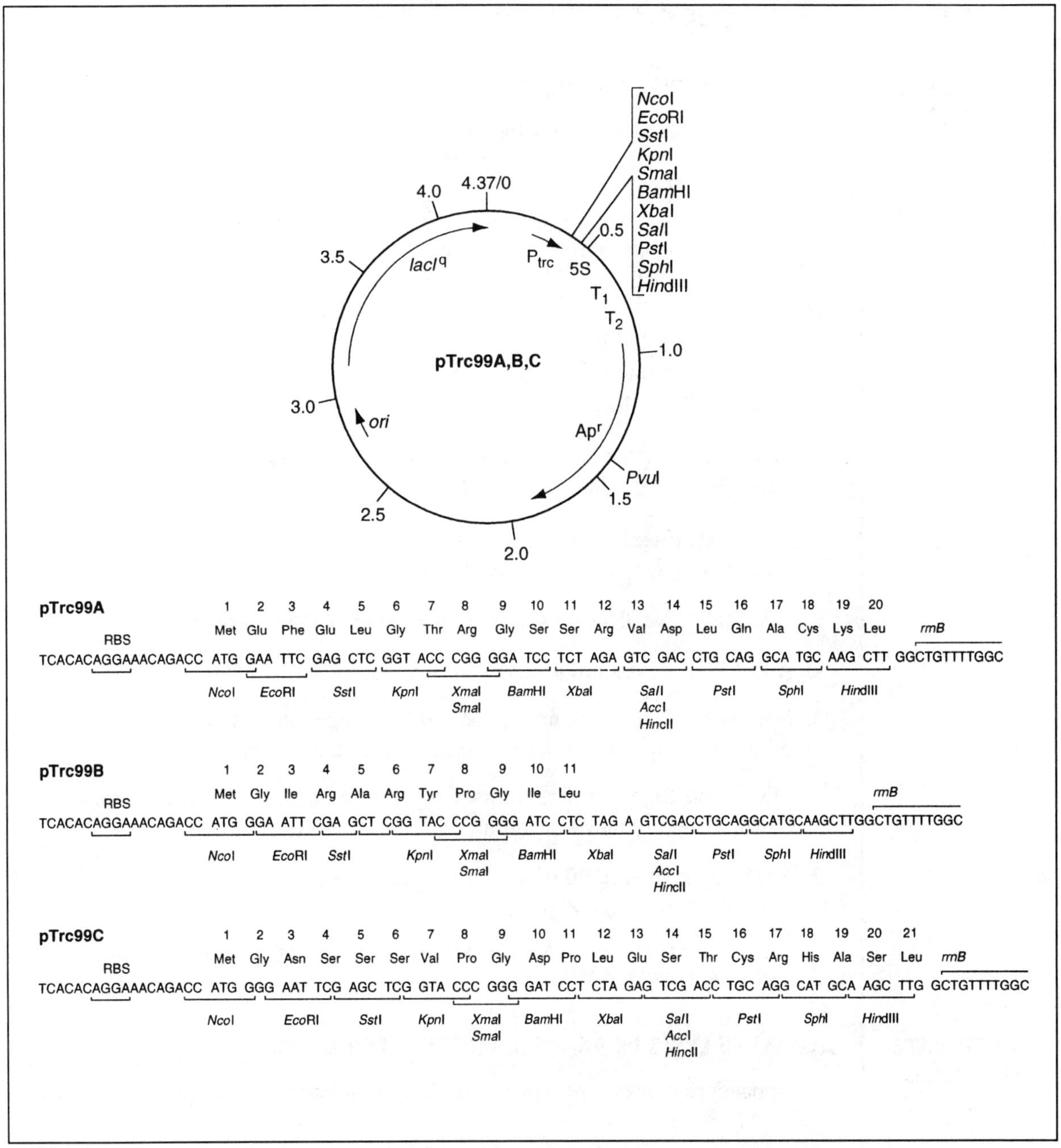

Figure 1.5.4 pTrc 99A,B,C. The pTrc series of plasmid expression vectors facilitates the regulated expression of genes in *E. coli.* These vectors carry the strong hybrid *trp/lac* promoter, the *lacZ* ribosome binding site (RBS), the multiple cloning site of pUC18 that allows insertion in three reading frames, and the *rrnB* transcription terminators (see polylinker sequences given below the vector diagram). These vectors are equally useful for the expression of non-fused proteins (resulting from insertion into the *Nco*I site) or for expression of fusion proteins (using one of the cloning sites in the correct translational frame).The presence of the *lacI*^{q} allele on the plasmid ensures complete repression of the hybrid *trp/lac* promoter during cloning and growth in any host strain (see Amann et al., 1988, for further details).

Reference: Kahn et al., 1979.

Contributors: Roger Brent and Nina Irwin

UNIT 1.6

Minipreps of Plasmid DNA

BASIC PROTOCOL 1

ALKALINE LYSIS MINIPREP

The alkaline lysis procedure is the most commonly used miniprep.

Materials (*see APPENDIX 1 for items with ✓*)

LB medium or enriched medium (e.g., superbroth or terrific broth) (*UNIT 1.1*) containing appropriate antibiotic (Table 1.4.1)
✓ Glucose/Tris/EDTA (GTE) solution
✓ TE buffer
✓ NaOH/SDS solution
✓ Potassium acetate solution, pH 4.8
95% and 70% ethanol

1. Inoculate 5 ml sterile LB medium with a single bacterial colony and grow to saturation (OD_{600} ~4) at 37°C.
2. Microcentrifuge 1.5 ml of cells 20 sec. Resuspend pellet in 100 µl GTE solution and let sit 5 min at room temperature.
3. Add 200 µl NaOH/SDS solution, mix, and place 5 min on ice.
4. Add 150 µl potassium acetate solution, vortex 2 sec, and place 5 min on ice.
5. Microcentrifuge 3 min and transfer 0.4 ml supernatant to a new tube. Add 0.8 ml of 95% ethanol, and let sit 2 min at room temperature.
6. Microcentrifuge 3 min at room temperature, wash pellet with 1 ml of 70% ethanol, and dry under vacuum.
7. Resuspend pellet in 30 µl TE buffer and store as in Support Protocol. Use 2.5 to 5 µl for a restriction digest.

   Destroy contaminating RNA by adding 1 µl of a 10 mg/ml RNase solution (DNase-free; APPENDIX1 & UNIT3.13).

ALTERNATE PROTOCOL

ALKALINE LYSIS IN 96-WELL MICROTITER DISHES

This procedure makes it possible to perform hundreds of rapid plasmid preps in a day.

Additional Materials (*also see Basic Protocol*)

TYGPN medium (*UNIT 1.1*)
Isopropanol

96-well microtiter plates (Dynatech PS plates or equivalent)
Multichannel pipetting device (8-prong Costar; 12-prong Titer Tek)
Multitube vortexer
Sorvall RT-6000 low-speed centrifuge, or equivalent, with microplate carrier in H-1000B rotor

1. Add 0.3 ml TYGPN medium to each well of a 96-well microtiter plate. Inoculate each with a single plasmid-containing colony and grow to saturation at 37°C (~48 hr).
2. Centrifuge plates 10 min at 600 × *g* 4°C. Flick supernatant from wells.

3. Resuspend cells by clamping plate in a multitube vortexer and running it 20 sec at setting 4.

4. Add 50 µl GTE solution and 100 µl NaOH/SDS solution to each well. Wait 2 min and add 50 µl potassium acetate solution to each well.

5. Cover with plate tape, vortex 20 sec at setting 4, and centrifuge 5 min at 600 × *g*, 4°C.

6. Remove 200 µl from each well and transfer to new plate. Add 150 µl isopropanol to each well, cover, agitate, and chill 30 min at –20°C.

7. Centrifuge 25 min at 600 × *g*, 4°C. Wash pellets with cold 70% ethanol, gently decant supernatant, wash with 95% ethanol, and again gently decant supernatant.

8. Air dry pellets 30 min, and resuspend in 50 µl TE buffer. Store as in Support Protocol and use 10-µl aliquots for digestion.

BOILING MINIPREP

BASIC PROTOCOL 2

This procedure is recommended for preparing small amounts of plasmid DNA from 1 to 24 cultures. It is extremely quick, but the quality of DNA produced is lower than that from the alkaline lysis miniprep.

Materials *(see APPENDIX 1 for items with ✓)*

LB medium (*UNIT 1.1*) containing appropriate antibiotic (Table 1.4.1)
✓ STET solution
Hen egg white lysozyme
Isopropanol, ice-cold
✓ TE buffer

1. Inoculate 5 ml LB medium with a single bacterial colony and grow to saturation at 37°C.

2. Microcentrifuge 1.5 ml of culture 20 sec. Resuspend bacteria by vortexing in 300 µl of STET solution containing 200 µg lysozyme. Place tube 30 sec to 10 min on ice.

   Be sure cells are completely resuspended to achieve high yields of DNA.

3. Place tube in boiling water bath (100°C) 1 to 2 min, microcentrifuge 15 to 30 min, and pipet supernatant to a new tube.

4. Mix with 200 µl cold isopropanol and place 15 to 30 min at –20°C.

5. Microcentrifuge 5 min. Remove supernatant and dry under a vacuum.

6. Resuspend in 50 µl TE buffer and store as in the support protocol. Use 5 µl for a restriction digest.

   Destroy contaminating RNA by adding 1 µl of a 10 mg/ml RNase solution (DNase-free; APPENDIX 1 & UNIT 3.13) to the digestion mixture.

SUPPORT PROTOCOL

STORAGE OF PLASMID DNA

1. Store plasmids in bacterial strains on selective plates at 4°C for the short-term, and in a glycerol- or DMSO-based solution (*UNIT 1.3*) at −70°C for the long-term. Grow cells taken from storage on selective plates (*UNIT 1.1*) and check plasmid DNA by restriction analysis (*UNIT 3.1*).

2. Store plasmid DNA in TE buffer at 4°C for short-term storage, and at −20° or −70°C for long-term storage.

References: Birnboim, 1983; He et al., 1991; Holmes and Quigley, 1981.

Contributors: JoAnne Engebrecht, Roger Brent, and Mustak A. Kaderbhai

UNIT 1.7

Large-Scale Preparation of Plasmid DNA

PREPARATION OF CRUDE LYSATES

BASIC PROTOCOL 1

Alkaline Lysis

This is probably the most generally useful plasmid prep procedure. It is fairly rapid, very reliable, and yields reasonably clean, crude DNA that can be further purified by either method described in this unit for purification of plasmid DNA.

Materials *(see APPENDIX 1 for items with ✓)*

Plasmid-bearing *E. coli* strain
LB containing ampicillin or other antibiotic (Table 1.4.1)
✓ Glucose/Tris/EDTA (GTE) solution
Hen egg white lysozyme
✓ NaOH/SDS solution
✓ 3 M potassium acetate solution, pH ~5.5
Isopropanol
70% ethanol

Sorvall GSA, GS-3, or Beckman JA-10 rotor or equivalent
Sorvall SS-34 or Beckman JA-17 rotor or equivalent

1. Inoculate 500 ml LB containing appropriate antibiotic in a 2-liter flask with 5 ml of an overnight culture of *E. coli* containing the desired plasmid. Grow at 37°C until culture is saturated (OD_{600} ≅4).

   To increase yields, maximize aeration using a flask with high surface area and baffles; shake at >400 rpm.

2. Centrifuge 10 min at 6000 × *g*, 4°C. Resuspend bacterial pellet in 4 ml GTE solution and transfer to high-speed centrifuge tube with ≥20 ml capacity.

   Cell pellets can be stored frozen indefinitely at −20°C or −70°C.

3. Add 1 ml GTE solution containing hen egg white lysozyme added fresh to 25 mg/ml. Resuspend pellet and let stand 10 min at room temperature.

4. Add 10 ml freshly prepared NaOH/SDS solution and mix gently with a pipet until solution becomes homogeneous, clear, and viscous. Let stand 10 min on ice.

5. Add 7.5 ml potassium acetate solution and stir gently with a pipet until viscosity is reduced and a large precipitate forms. Leave 10 min on ice.

6. Centrifuge 10 min at 20,000 × *g*, 4°C. Decant supernatant into a clean centrifuge tube. Pour through several layers of cheesecloth if any floating material is visible.

7. Add 0.6 vol isopropanol, mix by inversion, and let stand 5 to 10 min at room temperature.

8. Centrifuge 10 min at 15,000 × *g*, room temperature.

9. Wash pellet with 2 ml of 70% ethanol and centrifuge briefly. Aspirate ethanol and dry pellet under vacuum.

   The pellet can be stored indefinitely at 4°C.

CsCl/Ethidium Bromide Equilibrium Centrifugation

BASIC PROTOCOL 2

This procedure yields high-quality plasmid DNA free of most contaminants but uses ethidium bromide and requires long ultracentrifuge runs to establish the density gradient.

Materials *(see* APPENDIX 1 *for items with ✓)*

✓ TE buffer
Cesium chloride
✓ 10 mg/ml ethidium bromide
CsCl/TE solution: 100 ml TE buffer/100 g CsCl
✓ Dowex AG50W-X8 cation exchange resin
TE buffer/0.2 M NaCl
100% and 70% ethanol
Beckman VTi65 or VTi80 rotor or equivalent

1. Resuspend pellet obtained in final step of crude lysate preparation in 4 ml TE buffer. Add 4.4 g CsCl, dissolve, and add 0.4 ml of 10 mg/ml ethidium bromide.

   CAUTION: *Ethidium bromide is a mutagen and environmental hazard. It should be handled with gloves and disposed of properly.*

2. Transfer solution to 5-ml quick-seal ultracentrifuge tube, fill tube, if necessary, with CsCl/TE solution and seal. Centrifuge 3.5 hr at 500,000 × *g*, 20°C or ≥14 hr at 350,000 × *g*, 20°C.

3. Insert 20-G needle into the top of tube, and recover plasmid band by suction with a 3-ml syringe (inserted ~1 cm below band) with another 20-G needle attached (bevel side facing up).

   CAUTION: *To avoid potentially serious eye injury by UV light wear UV blocking glasses or face shield. Wear gloves when handling ethidium bromide.*

4. Perform a second ultracentrifugation (steps 2 and 3) to eliminate contaminating RNA or chromosomal DNA for a higher purity plasmid.

5. Pour a Dowex AG50W-X8 column, 1.5 to 2 times the volume of the plasmid DNA/ethidium bromide solution. Wash and equilibrate with several volumes of TE buffer/0.2 M NaCl.

6. Load plasmid DNA/ethidium bromide solution directly from syringe to top of resin bed without disturbing the resin.

7. Immediately collect eluate. Wash column with a volume of TE buffer/0.2 M NaCl twice that of the volume loaded. (Final volume collected from column should be three times that in which the plasmid DNA was removed from gradient.)

   The ethidium bromide can also be removed by extraction with an equal volume of TE-saturated n-butanol. Vortex well and remove organic (upper) phase. Repeat until no red color remains. Add 2 vol TE buffer to dilute the CsCl. This procedure generates contaminated organic waste; dispose of properly.

8. Precipitate plasmid DNA with 2 vol of 100% ethanol at room temperature or −20°C. Centrifuge 10 min at 10,000 × *g*, 4°C.

   Do not cool this solution below −20°C as this may cause the cesium chloride to precipitate.

9. Wash pellet with 70% ethanol and dry under vacuum. Resuspend pellet in TE buffer and store at 4°C.

ALTERNATE PROTOCOL

PLASMID DNA PURIFICATION BY ANION-EXCHANGE OR SIZE-EXCLUSION CHROMATOGRAPHY

Chromatographic methods for purifying plasmid DNA take advantage of distinctions between the physical properties of plasmid DNA and those of molecules that copurify with it in the crude lysate. Nucleic acids are negatively charged and can therefore be purified away from contaminants using anion-exchange chromatography (see *UNIT 2.14* for a protocol and discussion of one anion-exchange method). Similarly, the large size of plasmid DNA allows it to be purified away from smaller contaminants by gel-filtration chromatography.

This protocol describes modifications for preparing a crude lysate for chromatographic purification of plasmid DNA.

Preparation of crude lysate. It is unnecessary, and may be futile, to attempt to maximize cell density and plasmid DNA concentration as described in Basic Protocol 1, step 1 if the capacity of the column to be used will be exceeded. The pZ523 column (5 Prime→3 Prime) has a capacity of 4 to 5 mg plasmid DNA; the Qiagen-tip 2500 (Qiagen) and Wizard Maxiprep (Promega) columns have capacities of ~1 to 2 mg.

Bacterial cells should be incompletely lysed for the Qiagen-tip 2500 and Wizard Maxiprep columns, so lysozyme should be omitted from the preparation of crude lysate.

Add 50 µg/ml RNase A (from frozen 1 mg/ml stock, *UNIT 3.13*) to the resuspended cell pellet to remove high-molecular-weight RNA.

To remove floating material from the precipitation (Basic Protocol 1, step 6), decant the supernatant through cheesecloth, add chloroform before centrifugation, or recentrifuge the supernatant. Isopropanol precipitation can be used with either of the alternate protocols for preparation of crude lysate, and the final pellet can be resuspended in the buffer appropriate for the chromatographic matrix to be used.

Column capacity. Plasmid DNA binding capacity is the limiting factor in the use of most popular columns. Overloaded columns will not result in increased yields of plasmid DNA. To optimize recovery, bacterial culture volume, plasmid copy number, and the culture medium must be adjusted to the capacity of the column matrix. The excess DNA will simply run through the column and be discarded. One way to increase the yield is to recover the material that flows through the column when it is initially loaded and apply it to a new or regenerated column.

References: Birnboim, 1983; Clewell and Helinski, 1970, 1972; Lis, 1980; Radloff et al., 1967.

Contributors: J.S. Heilig, Karen Lech, and Roger Brent

Introduction of Plasmid DNA into Cells

UNIT 1.8

TRANSFORMATION USING CALCIUM CHLORIDE

BASIC PROTOCOL 1

Materials *(see APPENDIX 1 for items with ✓)*

Single colony of *E. coli* cells
LB medium (*UNIT 1.1*)
✓ $CaCl_2$ solution, ice-cold
LB plates containing ampicillin (*UNIT 1.1* and Table 1.4.1)
Plasmid DNA (*UNITS 1.6 & 1.7*)
Beckman JS-5.2 rotor or equivalent

NOTE: All materials and reagents coming into contact with bacteria must be sterile.

1. Inoculate a single colony of *E. coli* cells into 50 ml LB medium. Grow overnight at 37°C with shaking (250 rpm; *UNIT 1.2*).

2. Inoculate 4 ml of the culture into 400 ml LB medium in a 2-liter flask. Grow at 37°C, with shaking (250 rpm), to an OD_{590} of 0.375.

   This procedure requires that cells be growing rapidly (early- or mid-log phase).

3. Aliquot culture into eight 50-ml prechilled, sterile polypropylene tubes and leave the tubes on ice 5 to 10 min. Centrifuge cells 7 min at 1600 × *g*, 4°C.

4. Resuspend (gently) each pellet in 10 ml ice-cold $CaCl_2$ solution. Centrifuge 5 min at 1100 × *g*, 4°C.

5. Resuspend each pellet in 10 ml cold $CaCl_2$ solution. Keep resuspended cells on ice for 30 min. Centrifuge 5 min at 1100 × *g*, 4°C.

6. Resuspend each pellet completely in 2 ml of ice-cold $CaCl_2$ solution. Aliquot 250 µl into prechilled, sterile polypropylene tubes. Freeze immediately at −70°C.

7. Use 10 ng of pBR322 to transform 100 µl of competent cells using the steps below. Plate appropriate aliquots (1, 10, and 25 µl) of transformation culture on LB/ampicillin plates and incubate at 37°C overnight.

   The number of transformant colonies per aliquot volume (µl) × 10^5 is equal to the number of transformants per microgram of DNA.

8. Add 10 ng of DNA (10 to 25 µl) into a 15-ml sterile, round-bottom test tube and place on ice. Rapidly thaw competent cells by warming between hands, dispense

100 μl immediately into test tubes containing DNA, swirl, and place on ice 10 min.

9. Heat shock cells by placing tubes 2 min in 42°C water bath. Add 1 ml LB medium to each tube. Incubate 1 hr at 37°C on roller drum (250 rpm).

10. Plate several dilutions on appropriate antibiotic-containing plates, and incubate 12 to 16 hr at 37°C.

Store remainder at 4°C for subsequent platings. Transformation efficiencies range from 10^6 to 10^8 transformants/μg DNA.

ALTERNATE PROTOCOL

ONE-STEP PREPARATION AND TRANSFORMATION OF COMPETENT CELLS

Various strains can be made competent by this procedure and the frequency can be as high as that achieved by the basic protocol. However, frequency is considerably lower than can be obtained by electroporation.

Additional Materials *(also see Basic Protocol 1; see* APPENDIX 1 *for items with ✓)*

✓ 2× transformation and storage solution (TSS), ice-cold
LB medium (*UNIT 1.1*) containing 20 mM glucose

1. Dilute fresh overnight culture of bacteria 1:100 into LB medium and grow at 37°C to an OD_{600} of 0.3 to 0.4.

2. Add a volume of ice-cold 2× TSS equal to the cell suspension, and gently mix on ice.

For long-term storage, freeze small aliquots of the suspension in a dry ice/ethanol bath and store at −70°C. To use frozen cells for transformation, thaw slowly and then use immediately.

3. Add 100 μl competent cells and 1 to 5 μl DNA (0.1 to 100 ng) to an ice cold polypropylene or glass tube. Incubate 5 to 60 min at 4°C.

4. Add 0.9 ml LB medium/20 mM glucose and incubate 30 to 60 min at 37°C with mild shaking. Select transformants on appropriate plates.

The expected transformation frequency is 10^7 and 10^8 colonies/μg DNA.

BASIC PROTOCOL 2

HIGH-EFFICIENCY TRANSFORMATION BY ELECTROPORATION

Electroporation with high voltage is currently the most efficient method for transforming *E. coli* with plasmid DNA. The procedure described may be used to transform freshly prepared cells or to transform cells that have been previously grown and frozen.

Materials *(see* APPENDIX 1 *for items with ✓)*

Single colony of *E. coli* cells
LB medium (*UNIT 1.1*)
Ice-cold H_2O
Ice-cold 10% (v/v) glycerol
✓ SOC medium
LB plates containing antibiotics (*UNIT 1.1* and Table 1.4.1)

Beckman JS-4.2 rotor or equivalent and adapters for 50-ml narrow-bottom tubes
Electroporation apparatus with a pulse controller
Chilled electroporation cuvettes, 0.2-cm electrode gap

NOTE: All materials and reagents coming into contact with bacteria must be sterile.

1. Inoculate a single colony of *E. coli* cells into 5 ml LB medium. Grow 5 hr to overnight at 37°C with moderate shaking (*UNIT 1.2*).

2. Inoculate 2.5 ml of the culture into 500 ml LB medium in a 2-liter flask. Grow at 37°C with shaking (300 rpm) to an OD_{600} of 0.5 to 0.6.

3. Chill cells in an ice-water bath 10 to 15 min and transfer to a 1-liter prechilled centrifuge bottle. Centrifuge 20 min at 5000 × *g*, 2°C. Resuspend pellet in 5 ml ice-cold water.

4. Add 500 ml ice-cold water, mix and centrifuge as in step 3. Repeat once. Pour off supernatant immediately and resuspend pellet by swirling in remaining liquid.

5a. *For fresh cells*: Place suspension in prechilled, narrow-bottom, 50-ml polypropylene tube, and centrifuge 10 min at 5000 × *g*, 2°C. Estimate pellet volume (~500 µl/500-ml culture), add an equal volume of ice-cold water to resuspend cells (2×10^{11} cells/ml), and aliquot 50 to 300 µl cells into chilled microcentrifuge tubes.

Fresh cells work better than frozen cells.

5b. *For frozen cells*: Add 40 ml ice-cold 10% glycerol, mix, and centrifuge as in step 5a. Estimate pellet volume, add an equal volume of ice-cold 10% glycerol to resuspend cells, and aliquot 50 to 300 µl into prechilled microcentrifuge tubes. Freeze on dry ice and store at −80°C.

6. Set electroporation apparatus to 2.5 kV, 25 µF and pulse controller to 200 or 400 ohms.

7. Add 1 µl plasmid DNA (5 pg to 0.5 µg) to tubes containing fresh or thawed cells (on ice) and mix.

8. Transfer mixture to a prechilled cuvette, dry outside of cuvette, and place into sample chamber.

Transformation efficiency is decreased by increasing the DNA volume and salt concentration (which should be <1 mM).

9. Apply pulse. Remove cuvette, and immediately add 1 ml SOC medium and transfer to a sterile culture tube with a Pasteur pipet. Incubate 30 to 60 min with moderate shaking at 37°C.

10. Plate aliquots on LB plates containing antibiotics.

Transformation efficiencies are >10^9 transformants/µg DNA.

References: Dower et al., 1988; Hanahan, 1983.

Contributors: Christine E. Seidman, Kevin Struhl, and Jen Sheen

Introduction to Vectors Derived from Filamentous Phages

Many vectors in current use are derived from filamentous phages. These vectors are used because DNA inserted into them can be recovered in two forms—double-stranded circles and single-stranded circles. Foreign DNA is inserted into double-stranded vector DNA and, then, reintroduced into cells by transformation. Once inside the cells, double-stranded DNA replicates, giving rise both to new double-stranded circles and to single-stranded circles derived from one of the two strands of the vector. Single-stranded circles are packaged into phage particles and secreted from cells (which do not lyse). Single-stranded DNA can be purified from the culture supernatant and used for sequencing (Chapter 7), mutagenesis (Chapter 8), and other techniques described in this book.

DEVELOPMENT AND USE OF FILAMENTOUS PHAGE VECTORS

Prototypes of the filamentous phage vectors are the M13mp derivatives (see Figs. 1.9.1 and 1.9.2). These vectors were developed by Joachim Messing and his co-workers. Foreign DNA is inserted into a *polylinker* located in a non-essential region of the phage genome (Fig. 1.9.3). The polylinker is embedded in-frame within an alpha fragment of the *lacZ* gene. M13mp derivatives form blue plaques on lawns of cells that contain the *lacZ* omega fragment on plates with Xgal and IPTG. Insert-bearing phages resulting from cloning form white plaques.

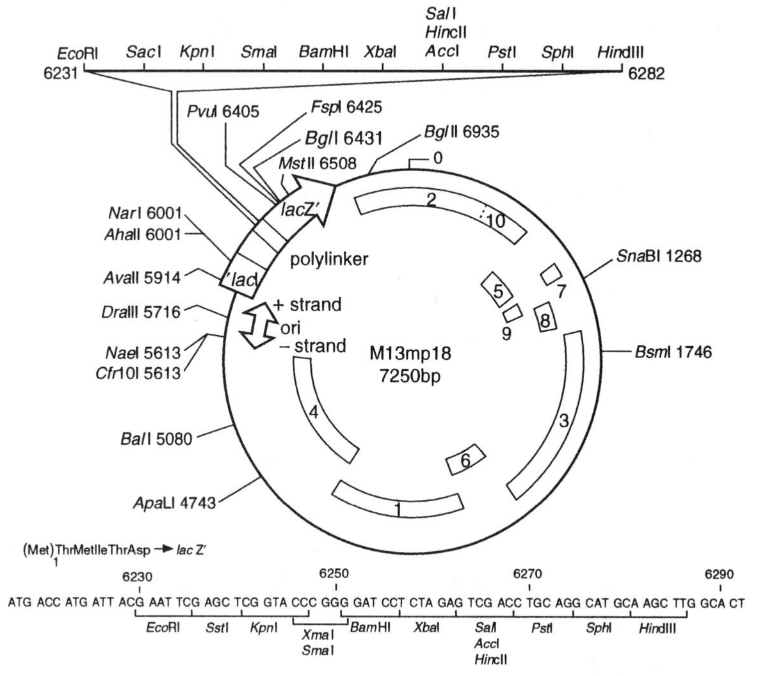

Figure 1.9.1 M13mp18. M13mp18 is one of the M13mp vectors made by Messing and colleagues. Insertion of DNA into the polylinker inactivates the *lacZ* alpha fragment. When insert-containing phages are plated under appropriate conditions (*UNIT 1.10*), they form colorless plaques; vectors that do not contain inserts form blue plaques.

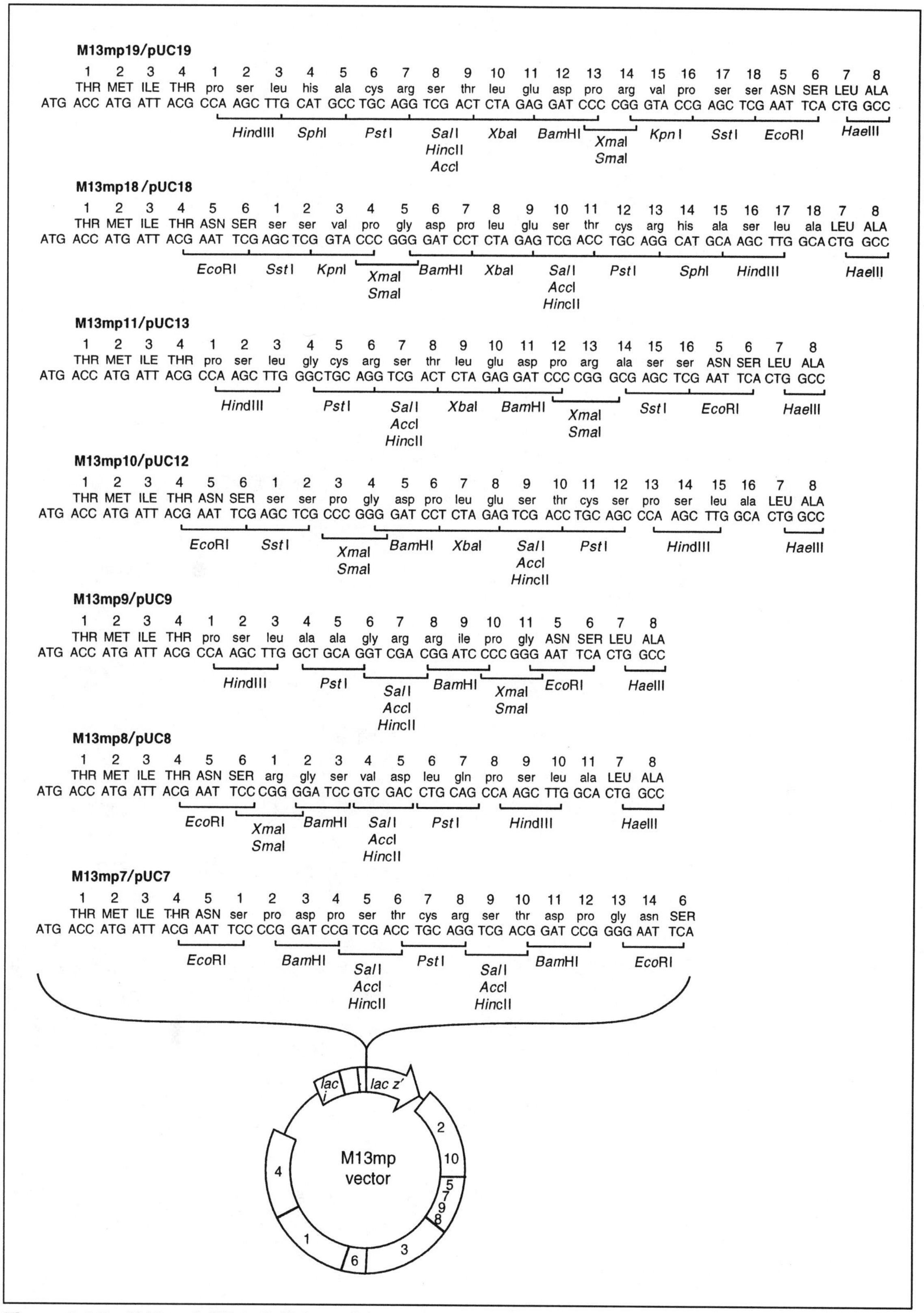

Figure 1.9.2 M13mp/pUC polylinkers. Sequence of polylinkers in the commonly used members of these two series of vectors. Amino acids that have been added to the *lacZ* gene product by insertion of the polylinker are shown in lower case letters. The bracket shows location of polylinkers on vector.

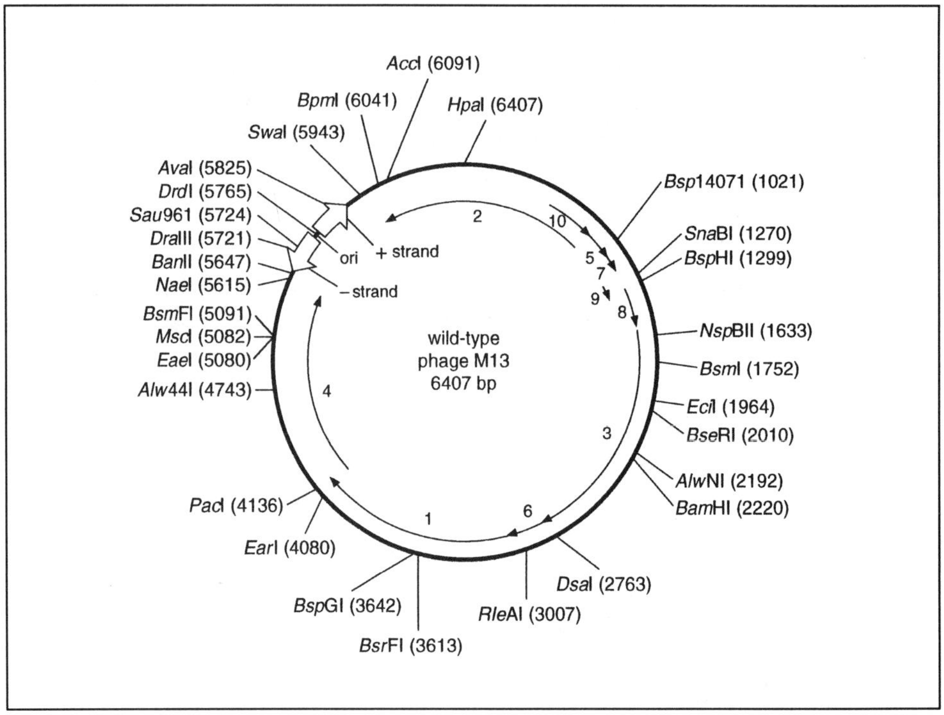

Figure 1.9.3 Wild-type M13.

References: Messing, 1983; Rasched and Oberer, 1986.

Contributors: David Greenstein and Roger Brent

UNIT 1.10

Preparing and Using M13-Derived Vectors

BASIC PROTOCOL

PREPARING SINGLE-STRANDED DNA FROM PLASMIDS USING HELPER PHAGE

Cells containing plasmids with filamentous phage origins (usually the f1 origin) are infected with helper phage. The helper phage provides the gene 2 protein that drives the plasmid into the f1 mode of replication and the DNA packaging and export functions. Single strands of the plasmid are packaged into phage coats and secreted into the supernatant. It is important to remember that only the (+) strand is packaged efficiently. Therefore, only the DNA strand of the insert that is in the same 5′→3′ orientation as the phage (+)-strand origin will be packaged.

Materials

F^+ or Hfr *E. coli* strain (Table 1.4.5) containing a plasmid (pUC118, pBS, or equivalent;

2× TY medium (*UNIT 1.1*)

Sorvall SS-34 rotor or equivalent

1. Dilute a fresh overnight culture started from a single colony 1:50 and grow cells in 2× TY (1 to 5 ml) containing an appropriate antibiotic at 37°C to an OD_{600} of 0.1.

2. Infect cells at MOIs of 1, 10, 20, and 50. Grow cells 4.5 hr at 37°C with vigorous shaking.

After initial determination, use the least amount of phage necessary to give a good yield of plasmid single strands because a portion of the input phage will always be recovered.

3. Centrifuge 10 min at 4000 × *g*. Collect supernatant and heat 15 min at 65°C.
4. Prepare single-stranded DNA from the supernatant. Analyze single-stranded DNA on an agarose gel (*UNIT 2.5*) with the helper phage serving as a control.

References: Dente et al., 1985; Messing, 1983.

Contributors: David Greenstein and Claude Besmond

Preparation and Analysis of DNA

2

2

The application of molecular biology techniques to the analysis of complex genomes depends on the ability to prepare pure, high-molecular-weight DNA. This chapter begins with protocols for purification of genomic DNA from bacteria, plant cells, and mammalian cells. These protocols consist of two parts: a technique to lyse gently the cells and solubilize the DNA, followed by one of several basic enzymatic or chemical methods to remove contaminating proteins, RNA, and other macromolecules. The basic approaches described here are generally applicable to a wide variety of starting materials. A brief collection of general protocols for further purifying and concentrating nucleic acids is also included.

Virtually all protocols in molecular biology require, at some point, fractionation of nucleic acids. Chromatographic techniques are appropriate for some applications—e.g., separation of double- and single-stranded nucleic acids—and may be used for separation of plasmid from genomic DNA as well as separation of genomic DNA from debris in a cell lysate. Gel electrophoresis, however, has much greater resolution than alternative methods and is generally the fractionation method of choice. Gel electrophoretic separations can be either analytical or preparative, and can involve fragments with molecular weights ranging from <1000 Daltons to >10^8.

In general, the use of electrophoresis to separate nucleic acids is simpler than its application to resolve proteins. Nucleic acids are uniformly negatively charged and, for double-stranded DNA, reasonably free of complicating structural effects that affect mobility. A variety of important variables affect migration of nucleic acids on gels. These include the conformation of the nucleic acid, the pore size of the gel, the voltage gradient applied, and the salt concentration of the buffer. The most basic of these variables is the pore size of the gel, which dictates the size of the fragments that can be resolved. In practice, this means that larger-pore agarose gels are used to resolve fragments >500 to 1000 bp and smaller pore acrylamide or sieving agarose gels are used for fragments <1000 bp. This chapter describes analytical and preparative applications and the resolution of very large pieces of DNA on agarose gels using pulsed-field gel electrophoresis (*UNIT 2.5B*).

Frequently it is desirable to identify an individual fragment in a complex mixture that has been resolved by gel electrophoresis. This is accomplished by a technique termed Southern blotting, in which the fragments are transferred from the gel to a nylon or nitrocellulose membrane and the fragment of interest is identified by hybridization with a labeled nucleic acid probe. This chapter gives a complete review of methods and materials required for immobilization of fractionated DNA (*UNIT 2.9*) and associated hybridization techniques (*UNIT 2.10*). These methods have greatly contributed to the mapping and identification of single and multicopy sequences in complex genomes, and facilitated the initial eukaryotic cloning experiments.

Other commonly encountered applications of gel electrophoresis include resolution of single-stranded RNA or DNA. Polyacrylamide gels containing high concentrations of urea as a denaturant provide a very powerful system for resolution of short (<500-nucleotide) fragments of single-stranded DNA or RNA. Such gels can resolve fragments differing by only a single nucleotide in length, and are central to all protocols for DNA sequencing. A detailed description of such denaturing polyacrylamide gels is found in *UNIT 7.6*. Such gels are used for other applications requiring resolution of single-stranded fragments, particularly including the techniques for analyzing mRNA structure by S1 analysis (*UNIT 4.6*), ribonuclease protection (*UNIT 4.7*), or primer extension (*UNIT 4.8*). Denaturing polyacrylamide gels are also useful for preparative applications, such as small-scale purification of radioactive single-stranded probes and large-scale purification of synthetic oligonucleotides (*UNIT 2.11*).

Resolution of relatively large single-stranded fragments (>500 nucleotides) can be accomplished using denaturing agarose gels. This is of particular importance to the analysis of mRNA populations by northern blotting and hybridization. A protocol for use of agarose gels containing formaldehyde in resolution of single-stranded RNA is presented in *UNIT 4.9*. The use of denaturing alkaline agarose gels for purification of labeled single-stranded DNA probes is described in *UNIT 4.6*.

Gels and Electric Circuits

Gel electrophoresis units are almost always simple electric circuits and can be understood using two simple equations. Ohm's law, $V = IR$, states that the electric field, V (measured in volts), is proportional to current, I (measured in milliamps), times resistance, R (measured in ohms). When a given amount of voltage is applied to a simple circuit, a constant amount of current flows through all the elements and the decrease in the total applied voltage that occurs across any element is a direct consequence of its resistance. For a segment of a gel apparatus, resistance is inversely proportional to both the cross-sectional area and the ionic strength of the buffer. Usually the gel itself provides nearly all of the resistance in the circuit, and the voltage applied to the gel will be essentially the same as the total voltage applied to the circuit. For a given current, decreasing either the thickness of the gel (and any overlying buffer) or the ionic strength of the buffer will increase resistance and, consequently, increase the voltage gradient across the gel and the electrophoretic mobility of the sample.

A practical upper limit to the voltage is usually set by the ability of the gel apparatus to dissipate heat. A second useful equation, $P = I^2R$, states that the power produced by the system, P (measured in watts), is proportional to the resistance times the square of the current. The power produced is manifested as heat, and any gel apparatus can dissipate only a particular amount of power without increasing the temperature of the gel. Above this point small increases in voltage can cause significant and potentially disastrous increases in temperature of the gel. It is very important to know how much power a particular gel apparatus can easily dissipate and to carefully monitor the temperature of gels run above that level.

Two practical examples illustrate applications of the two equations. The first involves the fact that the resistance of acrylamide gels increases somewhat during a run as ions related to polymerization are electrophoresed out of the gel. If such a gel is run at constant current, the voltage will increase with time and significant increases in power can occur. If an acrylamide gel is being run at high voltage, the power supply should be set to deliver constant power. The second situation deals with a limitation in number of power supplies, but not gel apparati. A direct application of the first equation shows that the fraction of total voltage applied to each of two gels hooked up in series (one after another) will be proportional to the fraction of total resistance the gel contributes to the circuit. Two identical gels will each get 50% of the total voltage and power indicated on the power supply.

Finally, it should be noted that some electrophoretic systems employ lethally high voltages, and almost all are potentially hazardous. It is very important to use an adequately shielded apparatus, an appropriately grounded and regulated power supply, and most importantly, common sense when carrying out electrophoresis experiments.

David D. Moore

Purification and Concentration of DNA from Aqueous Solutions

This unit presents basic procedures for manipulating solutions of single- or double-stranded DNA through purification and concentration steps.

IMPORTANT NOTE: The smallest amount of contamination of DNA preparations by recombinant phages or plasmids can be disastrous. All materials used for preparation of plasmid or phage DNA should be kept separate from those used for preparation of genomic DNA, and disposable items should be used wherever possible. Particular care should be taken to avoid contamination of commonly used rotors.

PHENOL EXTRACTION AND ETHANOL PRECIPITATION OF DNA

BASIC PROTOCOL

This protocol describes the most commonly used method of purifying and concentrating DNA preparations using phenol extraction and ethanol precipitation; it is appropriate for the purification of DNA from small volumes (<0.4 ml) at concentrations ≤1 mg/ml.

Materials *(see* APPENDIX 1 *for items with ✓)*

DNA to be purified (≤1 mg/ml) in 0.1 to 0.4 ml volume
25:24:1 (v/v/v) phenol/chloroform/isoamyl alcohol (made with *buffered* phenol; Support Protocol 1)
✓ 3 M sodium acetate, pH 5.2
100% ethanol, ice cold
70% ethanol, room temperature
✓ TE buffer, pH 8.0
Speedvac evaporator (Savant)

1. Add an equal volume of phenol/chloroform/isoamyl alcohol to the DNA solution to be purified in a 1.5-ml microcentrifuge tube.
2. Vortex vigorously 10 sec and microcentrifuge 15 sec at room temperature.
3. Carefully remove the top (aqueous) phase containing the DNA using a 200-µl pipettor and transfer to a new tube. If a white precipitate is present at the aqueous/organic interface, reextract the organic phase and pool aqueous phases.
4. Add $\frac{1}{10}$ vol of 3 M sodium acetate, pH 5.2, to the solution of DNA. Mix by vortexing briefly or by flicking the tube several times with a finger.
5. Add 2 to 2.5 vol (calculated *after* salt addition) of ice-cold 100% ethanol. Mix by vortexing and place in crushed dry ice for 5 min or longer.
6. Spin 5 min in a fixed-angle microcentrifuge at high speed and remove the supernatant.
7. Add 1 ml of room-temperature 70% ethanol. Invert the tube several times and microcentrifuge as in step 6.
8. Remove the supernatant. Dry the pellet in a desiccator under vacuum or in a Speedvac evaporator.
9. Dissolve the dry pellet in an appropriate volume of water or TE buffer, pH 8.0.

ALTERNATE PROTOCOL 1

PRECIPITATION OF DNA USING ISOPROPANOL

Isopropanol may also be used to precipitate DNA. Equal volumes of isopropanol and DNA solution are used, allowing precipitation from a large starting volume (e.g., 0.7 ml) in a single microcentrifuge tube.

2.1

SUPPORT PROTOCOL 1

BUFFERING PHENOL AND PREPARING PHENOL/CHLOROFORM/ISOAMYL ALCOHOL

For purification of DNA prior to cloning and other sensitive applications, phenol must be redistilled before use because oxidation products of phenol can damage and introduce breaks into nucleic acid chains. Redistilled phenol is commercially available. Regardless of the source, the phenol must be buffered before use.

CAUTION: Phenol can cause severe burns to skin and damage clothing. Gloves, safety glasses, and a lab coat should be worn whenever working with phenol, and all manipulations should be carried out in a fume hood. A glass receptacle should be available exclusively for disposing of used phenol and chloroform.

Materials *(see APPENDIX 1 for items with ✓)*

8-hydroxyquinoline
Liquefied phenol
50 mM Tris base (unadjusted pH ~10.5)
✓ 50 mM Tris·Cl, pH 8.0
Chloroform
Isoamyl alcohol

1. Add 0.5 g of 8-hydroxyquinoline to a 2-liter glass beaker containing a stir bar.
2. Gently pour in 500 ml of liquefied phenol or melted crystals of redistilled phenol (melted in a water bath at 65°C).

   The phenol will turn yellow due to the 8-hydroxyquinoline, which is added as an antioxidant.

3. Add 500 ml of 50 mM Tris base.
4. Cover the beaker with aluminum foil. Stir 10 min at low speed with magnetic stirrer at room temperature.
5. Let phases separate at room temperature. Gently decant the top (aqueous) phase into a suitable waste receptacle. Remove what cannot be decanted with a 25-ml glass pipet and a suction bulb.
6. Add 500 ml of 50 mM Tris·Cl, pH 8.0. Repeat steps 4 to 6 (i.e., two successive equilibrations with 500 ml of 50 mM Tris·Cl, pH 8.0) as necessary until pH is 8.0.
7. Add 250 ml of 50 mM Tris·Cl, pH 8.0, or TE buffer, pH 8.0, and store ≤2 months at 4°C in brown glass bottles or clear glass bottles wrapped in aluminum foil.
8. For use in DNA purification procedure (Basic Protocol), mix 25 vol phenol (bottom yellow phase of stored solution) with 24 vol chloroform and 1 vol isoamyl alcohol.

CONCENTRATION OF DNA USING BUTANOL

SUPPORT PROTOCOL 2

It is generally inconvenient to handle large volumes or dilute solutions of DNA. Water molecules (but not DNA or solute molecules) can be removed from aqueous solutions by extraction with *sec*-butanol (2-butanol) to reduce volumes or concentrate dilute solutions.

***Additional Materials** (also see Basic Protocol)*

sec-butanol
25:24:1 phenol/chloroform/isoamyl alcohol (made with *buffered* phenol; Support Protocol 1)
Polypropylene tube

1. Add an equal volume of *sec*-butanol to the sample and mix well by vortexing or by gentle inversion if the DNA is of high molecular weight. Perform extraction in a polypropylene tube, as butanol will damage polystyrene.
2. Centrifuge 5 min at 1200 × *g*, room temperature, or 10 sec in a microcentrifuge.
3. Remove and discard the upper (*sec*-butanol) phase.
4. Repeat steps 1 to 3 until the desired volume of aqueous solution is obtained.
5. Extract the lower, aqueous phase with 25:24:1 phenol/chloroform/isoamyl alcohol and ethanol precipitate, or remove *sec*-butanol by two ether extractions (Support Protocol 3).

REMOVAL OF RESIDUAL PHENOL, CHLOROFORM, OR BUTANOL BY ETHER EXTRACTION

SUPPORT PROTOCOL 3

DNA solutions that have been purified by extraction with phenol and chloroform (Basic Protocol) or concentrated with *sec*-butanol (Support Protocol 2) can often be used without ethanol precipitation for enzymatic manipulations or in gel electrophoresis experiments if the organic solvents are removed by extraction with ether. Traces of ether are subsequently removed by evaporation.

CAUTION: Ether is highly flammable and its vapors can cause drowsiness. All manipulations with ether should be carried out in a well-ventilated fume hood.

***Materials** (see APPENDIX 1 for items with ✓)*

Diethyl ether
✓ TE buffer, pH 8.0
Polypropylene tube

1. Mix diethyl ether with an equal volume of water or TE buffer, pH 8.0, in a polypropylene tube. Vortex vigorously for 10 sec and let the phases separate. Collect the top (ether) phase.
2. Add an equal volume of ether to the DNA sample. Mix well by vortexing or by gentle inversion if the DNA is of high molecular weight.
3. Microcentrifuge 5 sec or let the phases separate by setting the tube upright in a test tube rack.
4. Remove and discard the top (ether) layer. Repeat steps 2 and 3.
5. Remove ether by leaving the sample open under a hood for 15 min (for volumes <100 µl), or under vacuum for 15 min (for larger volumes).

ALTERNATE PROTOCOL 2

PURIFICATION OF DNA USING GLASS BEADS

A glass beads suspension can be used for rapid and efficient purification of DNA from contaminating proteins, RNA, or organic solvents.

Additional Materials *(also see Basic Protocol; see* APPENDIX 1 *for items with ✓)*

✓ 6 M sodium iodide (NaI) solution
50 to 200 μl DNA
✓ Wash solution
✓ TE buffer, pH 8.0
✓ Glass beads suspension

NOTE: The above materials are also available as commercial kits (e.g., Glas-Pac, National Scientific Supply; GeneClean, Bio101; and Qiaex Gel Extraction Kit, Qiagen).

1. Add 3 vol 6 M NaI solution to DNA in a 1.5-ml microcentrifuge tube. Add glass beads suspension as follows: for amounts of DNA <5 μg, use 5 μl glass beads suspension; for amounts of DNA >5 μg, use 5 μl plus an additional 1 μl for each 0.5-μg increment above 5 μg. Incubate 5 min at room temperature.
2. Microcentrifuge DNA/glass beads complex 5 sec. Remove and discard supernatant.
3. Wash the DNA/glass beads pellet three times with 500 μl wash solution. Lightly vortex the mixture to resuspend the beads, then microcentrifuge briefly to pellet the beads.
4. Resuspend pellet in TE buffer, pH 8.0, at 0.5 μg/μl. Incubate 2 to 3 min at 45°C to elute DNA from the glass beads.
5. Microcentrifuge 1 min and transfer the DNA-containing supernatant to a clean tube. Store at 4°C until use.

ALTERNATE PROTOCOLS

PURIFICATION AND CONCENTRATION OF RNA AND DILUTE SOLUTIONS OF DNA

Purification and Concentration of RNA

The procedure outlined in the Basic Protocol is identical for purification of RNA, except use 2.5 vol ethanol for the precipitation (step 5). All water used directly or in buffers should be treated with diethylpyrocarbonate (DEPC) to inactivate RNase (see *UNIT 4.1*, Reagents and Solutions, for instructions).

Dilute Solutions of DNA

When DNA solutions are dilute (<10 μg/ml) or when <1 μg of DNA is present, the ratio of ethanol to aqueous volume should be increased to 3:1 and the time on dry ice (step 5) extended to 30 min. Microcentrifugation should be carried out for 15 min in a cold room to ensure the recovery of DNA from these solutions.

Nanogram quantities of labeled or unlabeled DNA can be efficiently precipitated by adding 10 μg of commercially available tRNA from *E. coli*, yeast, or bovine liver as carrier. The DNA will be co-precipitated with the tRNA. Carrier tRNA will be phosphorylated efficiently by polynucleotide kinase and should not be added if this enzyme will be used in subsequent radiolabeling reactions.

Recovery of small quantities of short DNA fragments and oligonucleotides can be enhanced by adding magnesium chloride to a concentration of <10 mM before adding ethanol (step 4).

DNA in Large Aqueous Volumes (>0.4 to 10 ml)

Larger volumes can be accommodated by simply scaling up the amounts used in the Basic Protocol or by using butanol concentration as described in Support Protocol 2. For the phenol extraction (steps 1 through 3), use tightly capped polypropylene tubes. Centrifugation steps should be performed for 5 min at speeds not exceeding 1200 × *g*, room temperature. The ethanol precipitate (step 6) should be centrifuged in sialnized thick-walled Corning glass test tubes (15- or 30-ml capacity) for 15 min in fixed-angle rotors at 8000 × *g*, 4°C.

REMOVAL OF LOW-MOLECULAR-WEIGHT OLIGONUCLEOTIDES AND TRIPHOSPHATES BY ETHANOL PRECIPITATION

ALTERNATE PROTOCOL 4

Small single- or double-stranded oligonucleotides (less than ~30 bp) and unincorporated nucleotides used in radiolabeling or other DNA modification reactions can be effectively removed from DNA solutions by two rounds of ethanol precipitation in the presence of ammonium acetate. This approach is not sufficient to completely remove large quantities of linkers used in cloning procedures (*UNIT 3.16*). T4 polynucleotide kinase is inhibited by ammonium ions so this protocol should be avoided when the nucleic acid is to be phosphorylated.

Additional Materials *(also see Basic Protocol)*

4 M ammonium acetate, pH 4.8

1. Add an equal volume of 4 M ammonium acetate (see recipe), pH 4.8, to the DNA solution. Mix well.
2. Add 2 vol (calculated *after* salt addition) of ice-cold 100% ethanol (67% final). Vortex and set tube in crushed dry ice for 5 min.
3. Microcentrifuge 5 min at high speed, room temperature. Carefully remove supernatant and redissolve pellet in 100 µl TE buffer, pH 8.0.
4. Repeat steps 1 to 3, then proceed to step 5.
5. Add 1 ml of room-temperature 70% ethanol to the tube and invert several times. Microcentrifuge 5 min at high speed, room temperature.
6. Discard ethanol and dry pellet.

Reference: Marmur, J. 1961.

Contributor: David Moore

2.2

UNIT 2.2

Preparation of Genomic DNA from Mammalian Tissue

BASIC PROTOCOL

Materials (*see* APPENDIX 1 *for items with* ✓)

Liquid nitrogen
✓ Digestion buffer
✓ PBS, ice-cold
25:24:1 phenol/chloroform/isoamyl alcohol (UNIT 2.1)
✓ 7.5 M ammonium acetate
100% and 70% ethanol
✓ TE buffer, pH 8
Sorvall H1000B rotor or equivalent

Beginning with whole tissue

1a. Excise and immediately mince tissue quickly and freeze in liquid nitrogen.

If working with liver, remove gall bladder, which contains high levels of digestive enzymes.

2a. Grind 200 mg to 1 g tissue with prechilled mortar and pestle, or crush with hammer to fine powder. Suspend in 1.2 ml digestion buffer per 100 mg of tissue.

Beginning with tissue culture cells

1b. Centrifuge cells 5 min at 500 × *g* and discard supernatant. Trypsinize adherent cells first.

2b. Resuspend cells in 1 to 10 ml ice-cold PBS. Centrifuge 5 min at 500 × *g*, discard supernatant, and repeat. Resuspend cells in 1 vol digestion buffer.

For <3×10^7 cells, use 0.3 ml digestion buffer; for larger numbers of cells, use 1 ml digestion buffer/10^8 cells.

3. Incubate samples, shaking, in tightly capped tubes, 12 to 18 hr at 50°C.

4. Extract samples with an equal volume of phenol/chloroform/isoamyl alcohol. Centrifuge 10 min at 1700 × *g*. If phases do not resolve well, add another volume digestion buffer, omitting proteinase K, and repeat centrifugation. If thick white material appears at interface, repeat organic extraction. Transfer top layer (aqueous) to a new tube.

5. Add ½ vol of 7.5 M ammonium acetate and 2 vol of 100% ethanol. Centrifuge 2 min at 1700 × *g*.

To prevent shearing of high-molecular-weight DNA, remove organic solvents and salt by two dialyses against 100 vol TE buffer for ≥24 hr; omit step 6.

6. Wash with 70% ethanol, air dry, and resuspend in TE buffer at ~1 mg/ml.

Remove residual RNA by adding 0.1% SDS and 1 μg/ml DNase-free RNase, incubating 1 hr at 37°C, and repeating steps 4 and 5.

The expected yield from 1 g cells is ~2 mg DNA.

Reference: Gross-Bellard et al., 1972.

Contributor: William M. Strauss

Preparation of Genomic DNA from Plant Tissue

PREPARATION OF PLANT DNA USING CSCL CENTRIFUGATION

BASIC PROTOCOL

Materials (*see APPENDIX 1 for items with* ✓)

- Cold, sterile H_2O
- Liquid nitrogen
- ✓ Extraction buffer
- 10% (w/v) *N*-lauroylsarcosine (Sarkosyl)
- Isopropanol
- ✓ TE buffer
- Cesium chloride
- 10 mg/ml ethidium bromide
- CsCl-saturated isopropanol
- Ethanol
- ✓ 3 M sodium acetate, pH 5.2
- Beckman JA-14, JA-20 or JA-21, and VTi80 rotors or equivalents

1. Harvest 10 to 50 g fresh plant tissue.

   To reduce starch content, use younger plants and place in dark 1 to 2 days prior to harvest.

2. Rinse with cold, sterile water, blot dry, and freeze with liquid nitrogen. Grind to a fine powder with mortar and pestle.

3. Transfer frozen powder to 250-ml centrifuge bottle, immediately add 5 to 10 ml extraction buffer/g plant tissue, and stir gently to mix. Add 10% (w/v) Sarkosyl to a final concentration of 1%. Incubate 1 to 2 hr at 55°C.

   In all subsequent steps, solutions should not be vortexed or mixed vigorously.

4. Centrifuge 10 min at 5500 × *g* in JA-14 rotor, 4°C. Save the supernatant and repeat this step if necessary to remove debris.

5. Add 0.6 vol isopropanol and mix. If precipitate is not visible, place 30 min at −20°C. Centrifuge 15 min at 7500 × *g* in JA-14 rotor, 4°C. Discard the supernatant.

6. Resuspend pellet in 9 ml TE buffer, add 9.7 g solid CsCl, mix, and incubate 30 min on ice. Centrifuge 10 min at 7500 × *g* in JA-20 rotor, 4°C, and save supernatant.

   The Sarkosyl phase, which may form on the top of the solution, can be removed by filtering the supernatant through 2 layers of cheesecloth.

7. Add 0.5 ml of 10 mg/ml ethidium bromide and incubate 30 min on ice. Centrifuge 10 min at 7500 × *g*, 4°C. Transfer supernatant to two 5-ml quick-seal ultracentrifuge tubes and seal.

8. Centrifuge 4 hr at 525,000 × *g* in VTi80 rotor), 20°C, or overnight at 300,000 × *g*, 20°C. Collect DNA band using a 15-G needle and syringe (*UNIT 1.7*).

9. Remove ethidium bromide by repeatedly extracting DNA with CsCl-saturated isopropanol (*UNIT 1.7*).

10. Add 2 vol water and 6 vol of 100% ethanol, mix, and incubate 1 hr at –20°C. Centrifuge 10 min at 7500 × *g* in JA-20 or JA-21 rotor, 4°C.

11. Resuspend pellet in TE buffer, reprecipitate by adding 1⁄10 vol of 3 M sodium acetate and 2 vol of 100% ethanol. Repeat centrifugation. Air dry and resuspend in TE buffer.

    Yields should be 10 to 40 μg DNA (50-kb length)/g fresh plant tissue.

ALTERNATE PROTOCOL

PREPARATION OF PLANT DNA USING CTAB

The nonionic detergent cetyltrimethylammonium bromide (CTAB) can be used to liberate and complex with total cellular nucleic acids from a wide array of plant genera and tissue types.

Additional Materials *(also see Basic Protocol; see* APPENDIX 1 *for items with ✓)*

2-mercaptoethanol (2-ME)
✓ CTAB extraction solution
✓ CTAB/NaCl solution
24:1 (v/v) chloroform/octanol or chloroform/isoamyl alcohol
✓ CTAB precipitation solution
✓ High-salt TE buffer
80% ethanol
✓ TE buffer

Pulverizer/homogenizer: mortar and pestle, blender, Polytron (Brinkmann), or coffee grinder
Organic solvent–resistant test tube or beaker

1. Add 2-ME to the required amount of CTAB extraction solution to give a final concentration of 2% (v/v). Heat this solution and CTAB/NaCl solution to 65°C.

    Approximately 4 ml of 2-ME/CTAB extraction solution and 0.4 to 0.5 ml CTAB/NaCl solution are required for each gram of fresh leaf tissue.

2. Chill a pulverizer/homogenizer with liquid nitrogen (–196°C) or dry ice (–78°C). Pulverize plant tissue to a fine powder and transfer the frozen tissue to an organic solvent–resistant test tube or beaker.

3. Add warm 2-ME/CTAB extraction solution to the pulverized tissue and mix to wet thoroughly. Incubate 10 to 60 min at 65°C with occasional mixing.

4. Extract the homogenate with an equal volume of 24:1 chloroform/octanol or chloroform/isoamyl alcohol. Mix well by inversion. Centrifuge 5 min at 7500 × *g* (or ~10,000 rpm in a microcentrifuge, for smaller samples), 4°C. Recover the top (aqueous) phase.

5. Add 1⁄10 vol 65°C CTAB/NaCl solution to the recovered aqueous phase and mix well by inversion.

6. Extract with an equal volume of chloroform/octanol. Mix, centrifuge, and recover top (aqueous) phase.

7. Add exactly 1 vol CTAB precipitation solution. Mix well by inversion. If precipitate is visible, proceed to step 8. If not, incubate mixture 30 min at 65°C.

8. Centrifuge 5 min at 500 × *g* (or ~2700 rpm in microcentrifuge), 4°C.

9. Remove but do not discard the supernatant and resuspend pellet in high-salt TE buffer (0.5 to 1 ml per gram of starting material). If the pellet is difficult to resuspend, incubate 30 min at 65°C. Repeat until all or most of pellet is dissolved.

10. Precipitate the nucleic acids by adding 0.6 vol isopropanol. Mix well and centrifuge 15 min at 7500 × *g*, 4°C.

11. Wash the pellet with 80% ethanol, dry, and resuspend in a minimal volume of TE buffer (0.1 to 0.5 ml per gram of starting material).

Reference: Dellaporta et al., 1983.

Contributor: Eric J. Richards

Preparation of Genomic DNA from Bacteria

UNIT 2.4

BASIC PROTOCOL

MINIPREP OF BACTERIAL GENOMIC DNA

Materials *(see* APPENDIX 1 *for items with ✓)*

✓ TE buffer
10% (w/v) sodium dodecyl sulfate (SDS)
20 mg/ml proteinase K (stored in small single-use aliquots at −20°C)
✓ 5 M NaCl
✓ CTAB/NaCl solution
24:1 chloroform/isoamyl alcohol
25:24:1 phenol/chloroform/isoamyl alcohol (*UNIT 2.1*)
Isopropanol
70% ethanol

1. Grow 5-ml bacterial culture to saturation. Microcentrifuge 1.5 ml of the culture 2 min.

2. Resuspend pellet in 567 µl TE buffer by repeated pipetting. Add 30 µl of 10% SDS and 3 µl of 20 mg/ml proteinase K, mix, and incubate 1 hr at 37°C.

3. Add 100 µl of 5 M NaCl and mix thoroughly. Add 80 µl CTAB/NaCl solution, mix, and incubate 10 min at 65°C.

   Polysaccharides and other contaminating macromolecules can be removed starting with this step (see Support Protocol).

4. Add equal volume chloroform/isoamyl alcohol, mix, and microcentrifuge 4 to 5 min. Transfer the supernatant to a fresh tube. If it is difficult to remove the supernatant, remove the interface first with a toothpick.

5. Add equal volume phenol/chloroform/isoamyl alcohol, mix, and microcentrifuge 5 min. Transfer supernatant to a fresh tube.

   With some bacterial strains the interface formed after chloroform extraction is not compact enough to allow easy removal of the supernatant. In such cases, most of the interface can be fished out with a sterile toothpick before removal of the supernatant. Any remaining CTAB precipitate is then removed in the phenol-chloroform extraction.

6. Add 0.6 vol isopropanol and mix gently until DNA precipitates. Transfer precipitate with a sealed Pasteur pipet to 1 ml of 70% ethanol and wash.

   Alternatively, the precipitate can be microcentrifuged briefly and washed with 1 ml of 70% ethanol.

7. Microcentrifuge 5 min, discard supernatant, and dry briefly in a lyophilizer. Resuspend in 100 µl TE buffer. Use 10 to 15 µl per restriction digest.

ALTERNATE PROTOCOL

LARGE-SCALE CsCl PREPARATION OF BACTERIAL GENOMIC DNA

Additional Materials *(also see Basic Protocol)*

Cesium chloride
✓ 10 mg/ml ethidium bromide
CsCl-saturated isopropanol *or* H_2O-saturated butanol
Beckman JA-20 and VTi80 rotors or equivalents

1. Grow 100-ml bacterial culture to saturation. Centrifuge cells 10 min at 4000 × *g* in JA-20 rotor.
2. Resuspend pellet in 9.5 ml TE buffer. Add 0.5 ml of 10% SDS and 50 µl of 20 mg/ml proteinase K, mix, and incubate 1 hr at 37°C.
3. Add 1.8 ml of 5 M NaCl and mix thoroughly. Add 1.5 ml CTAB/NaCl solution, mix, and incubate 20 min at 65°C.
4. Add equal volume chloroform/isoamyl alcohol, extract, and centrifuge 10 min at 6000 × *g*, room temperature. Transfer supernatant with wide-bore pipet to fresh tube.
5. Add 0.6 vol isopropanol and mix gently until DNA precipitates. Transfer precipitate with a sealed Pasteur pipet to 1 ml of 70% ethanol and wash.
6. Centrifuge 5 min at 10,000 × *g* and discard supernatant. Resuspend pellet in 4 ml TE buffer.
7. Measure DNA concentration with a spectrophotometer, and adjust to 50 to 100 µg/ml.
8. Add 4.3 g CsCl/4 ml TE buffer and dissolve. Add 200 µl of 10 mg/ml ethidium bromide. Transfer to 4-ml quick-seal centrifuge tubes, adjust volume, and balance tubes with CsCl in TE buffer (1.05 g/ml). Seal tubes and centrifuge 4 hr at 420,000 × *g* in VTi80 rotor, 15°C, or overnight at 250,000 × *g*.
9. Visualize gradient under longwave UV lamp. Remove band with 15-G needle and 3-ml plastic syringe (*UNIT 1.7*).
10. Remove ethidium bromide by sequential extractions with CsCl-saturated isopropanol or water-saturated butanol (*UNIT 1.7*).
11. Dialyze overnight against 2 liters TE buffer to remove CsCl. If necessary, precipitate DNA and resuspend at desired concentration.

    Yields from both the miniprep and the large-scale prep are 0.5 to 2 mg DNA/100 ml starting culture (10^8 to 10^9 cells/ml).

SUPPORT PROTOCOL

REMOVAL OF POLYSACCHARIDES FROM EXISTING GENOMIC DNA PREPS

1. Adjust NaCl concentration to 0.7 M and add 0.1 vol CTAB/NaCl solution (step 3 of Basic Protocol).
2. Extract with chloroform/isoamyl alcohol (step 4 of Basic Protocol).

   A white interface indicates removal of polysaccharides and other contaminating macromolecules.

3. Repeat CTAB extraction (step 3 of Basic Protocol) and chloroform/isoamyl alcohol extraction (step 4 of Basic Protocol) until no interface is visible.

Reference: Murray and Thompson, 1980.

Contributor: Kate Wilson

Agarose Gel Electrophoresis

UNIT 2.5A

BASIC PROTOCOL

RESOLUTION OF LARGE DNA FRAGMENTS ON AGAROSE GELS

This protocol is used to separate and purify DNA fragments between 0.5 and 25 kb.

Materials *(see* APPENDIX 1 *for items with ✓)*

✓ Electrophoresis buffer (TAE or TBE)
✓ Ethidium bromide solution
✓ Agarose, electrophoresis-grade
✓ 10× loading buffer
DNA molecular weight markers (Fig. 2.5.1)

Horizontal gel electrophoresis apparatus
Gel casting platform
Gel combs (slot formers)
DC power supply

1. Prepare the gel, using electrophoresis buffer and electrophoresis-grade agarose (see Table 2.5.1) by melting in a microwave oven or autoclave, mixing, cooling to 55°C, pouring into a sealed gel casting platform, and inserting the gel comb.

   Ethidium bromide can be added to the gel and electrophoresis buffer at 0.5 μg/ml. CAUTION: *Ethidium bromide is a potential carcinogen. Wear gloves when handling.*

Table 2.5.1 Appropriate Agarose Concentrations for Separating DNA Fragments of Various Sizes

Agarose (%)	Effective range of resolution of linear DNA fragments (kb)
0.5	30 to 1
0.7	12 to 0.8
1.0	10 to 0.5
1.2	7 to 0.4
1.5	3 to 0.2

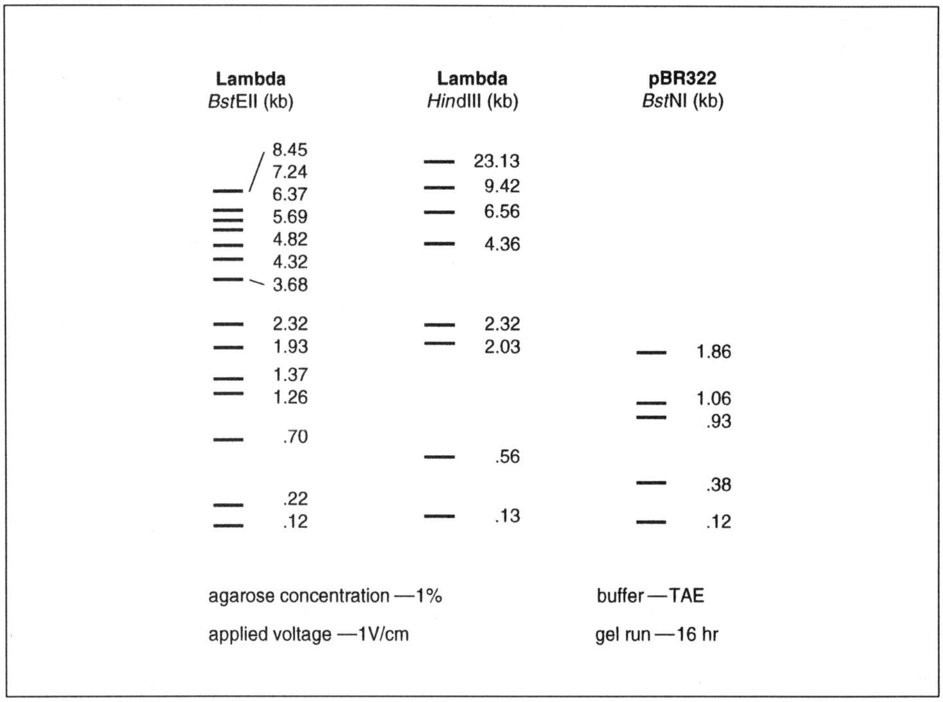

Figure 2.5.1 Migration pattern and fragment sizes for commonly used DNA molecular weight markers.

2. After the gel has hardened, remove the seal from the gel casting platform and withdraw the gel comb. Place into an electrophoresis tank containing sufficient electrophoresis buffer to cover the gel ~1 mm.

3. Prepare DNA samples with an appropriate amount of 10× loading buffer and load samples into wells with a pipettor. Be sure to include appropriate DNA molecular weight markers (Fig. 2.5.1).

4. Attach the leads so that the DNA migrates to the anode or positive lead and electrophorese at 1 to 10 V/cm of gel.

5. Turn off the power supply when the bromphenol blue dye from the loading buffer has migrated a distance judged sufficient for separation of the DNA fragments.

   Bromphenol blue comigrates with ~0.5-kb fragments.

6. Photograph a stained gel directly on a UV transilluminator or first stain with 0.5 µg/ml ethidium bromide 10 to 30 min, destaining 30 min in water, if necessary.

SUPPORT PROTOCOL

MINIGELS AND MIDIGELS

Small gels—minigels and midigels—can generally be run faster than larger gels and are often employed to expedite analytical applications. Because they use narrower wells and thinner gels, minigels and midigels also require smaller amounts of DNA for visualization of the separated fragments. Aside from a scaling down of buffer and gel volumes, the protocol for running minigels or midigels is similar to that for larger gels. Similarly, the parameters affecting the mobility of DNA fragments are the same for both large and small gels.

When selecting a mini- or midigel apparatus consider the volume of buffer held by the gel tank. Smaller gels are typically run at high voltages (>10 V/cm) so electrophoresis buffers are quickly exhausted, therefore, it is advantageous to choose a gel apparatus with a relatively large buffer reservoir. Minigel boxes can also be easily constructed from a few simple materials (Maniatis et al., 1982). A small (e.g., 15 cm long × 8 cm wide × 4 cm high) plastic box can be equipped with male connectors and platinum wire electrodes at both ends to serve as a minimal gel tank.

Although trays for casting small gels are commercially available, gels can also be poured onto glass lantern slides or other small supports without side walls. Such gels are held on the support simply by surface tension. After pouring the gel (e.g., 10 ml for a 5 cm × 8 cm gel), the comb is placed directly onto the support and held up by metal paper-binding clamps placed to the side. There is no agarose at the bottom of such wells, so extra care must be taken to prevent separation of the gel from the support when removing the comb.

Reference: Southern, 1979.

Contributor: Daniel Voytas

Pulsed-Field Gel Electrophoresis

UNIT 2.5B

FIELD-INVERSION ELECTROPHORESIS

BASIC PROTOCOL

Field-inversion gel electrophoresis (sometimes called FIGE) is used to resolve DNA molecules of ~10 to 2000 kb.

Materials *(see* APPENDIX 1 *for items with* ✓*)*

✓ 1% (w/v) agarose gel
✓ GTBE buffer *or* TBE electrophoresis buffer
Peristaltic pump (Cole-Parmer Masterflex or equivalent)
Programmable switching device (MJ Research PPI-200 or equivalent)

1. Prepare a 1% agarose gel for a horizontal apparatus. Place in apparatus and cover with 2 to 3 mm GTBE or TBE buffer.

2. Load liquid samples. Set up a peristaltic pump for buffer circulation. Adjust to 5 to 10 ml/min for minigel and 20 to 50 ml/min for large gel. Connect the tubing ends to the recirculation ports of the gel box or place them directly in the buffer tanks.

   To avoid shearing DNA >100 kb, cut ~5 mm off ends of pipet tips. Samples prepared in agarose blocks (see Support Protocol) should be loaded before gel is placed in apparatus.

3. Connect the programmable switching device to a constant-voltage DC power supply and connect the gel apparatus to the switching device (see Table 2.5.2). Run the gel until the bromphenol blue migrates 1 cm, then start the switching device and the peristaltic pump.

   The most commonly used ratio between forward and reverse times is 3:1.

4. Stain with ethidium bromide and photograph as for a standard agarose gel. The gel may be Southern blotted after acid depurination (UNIT 2.9A).

Table 2.5.2 Empirical Equations for Fragment Resolution and Velocity Using Pulsed-Field Gels[a,b]

Field-inversion gels	
Maximum resolved size[c] (kb)	$0.13 \times (T + 40) \times V^{1.1} \times (3 - A)^{0.6} \times t^{0.875}$
Minimum resolved size (kb)	$0.75 \times$ maximum size
Velocity of 10-kb fragment (cm/hr)	$\frac{0.0016 \times (T + 25) \times V^{1.6} \times (R - 1)}{A \times (R + 1)}$
CHEF or other alternating-angle gels	
Maximum well-resolved size[c] (kb)	$0.034 \times (T + 40) \times V^{1.1} \times (3 - A)^{0.6} \times t^{0.875}$
Minimum well-resolved size (kb)	$0.75 \times$ maximum size
Velocity of 10-kb fragment (cm/hr)	$\frac{0.0012 \times (T + 25) \times V^{1.6} \times \cos(\theta/2)}{A}$

[a]Equations assume use of 0.5× TBE buffer; for GTBE and TAE buffers, sizes separated will be slightly larger and gels will run ~20% and 30% faster, respectively.

[b]Variables: *T*, temperature in °C; *V*, field strength, volts/cm; *A*, % agarose (multiply by 0.8 for pulsed-field grade agarose); *t*, pulse time (reverse time for field-inversion gels) in sec; *R*, forward-to-reverse time ratio; θ, reorientation angle.

[c]Field inversion does not resolve fragments outside this range; alternating-angle gels will resolve fragments outside this range, but not as well.

ALTERNATE PROTOCOL

CHEF ELECTROPHORESIS

It is possible to resolve DNA molecules several million bases in length by periodically changing the angle of the electric field in the gel (Fig. 2.5.2).

Additional Materials (also see Basic Protocol)

CHEF electrophoresis voltage-divider circuitry (Fig. 2.5.3) and gel box

1. Prepare a 1% agarose gel for a CHEF gel apparatus using GTBE or 0.5× TBE buffer.
2. Allow gel to set, then carefully remove comb. Insert into wells any samples that have been prepared in agarose blocks (see Support Protocol).

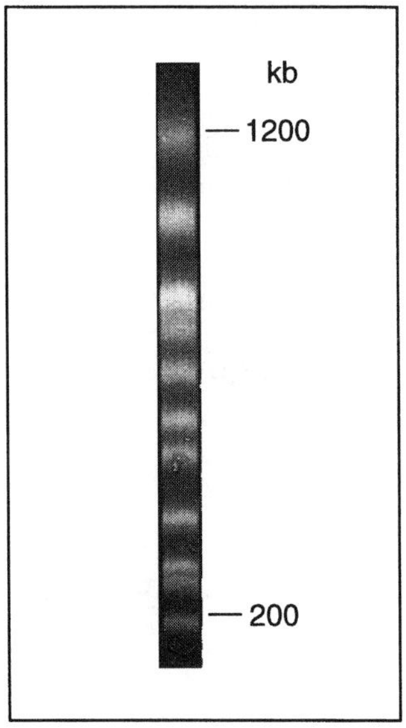

Figure 2.5.2 Chromosomes of *Saccharomyces cerevisiae* separated by field inversion.

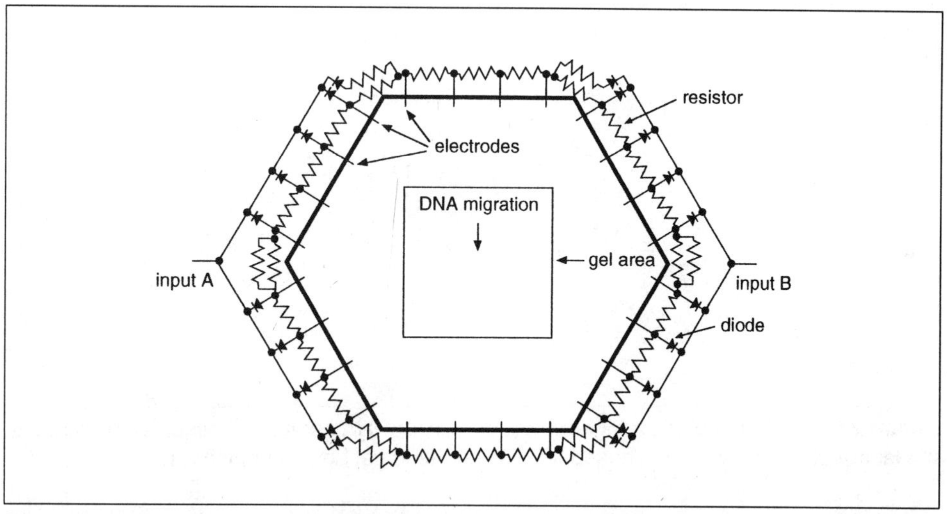

Figure 2.5.3 Circuitry for clamping electrode voltages of a 24-electrode hexagonal CHEF gel (redrawn from Chu, 1989). Positions of the 24 electrodes are shown schematically, as well as the direction of the resulting DNA migration. This circuitry can be driven directly by an inverting gel controller connected to input A and input B. Power dissipation in the resistors limit this circuit to ~250 V input. All resistors are 470 Ω, 1% tolerance, 3 W and diodes are type 1N4004 (1 A, 400 V).

3. Place gel into gel box, cover with buffer to a depth of 2 to 3 mm, adjust recirculation to ≥100 ml/min, and monitor buffer temperature. Wait 15 min after buffer has reached desired running temperature to ensure that the gel has equilibrated at the correct temperature.

4. Load any samples that are in liquid.

5. Paying careful attention to polarity, connect programmable switching device to constant-voltage DC power supply, voltage divider circuitry, and gel apparatus. Set the switching device for an appropriate switching regime and start gel.

6. Complete run and stain gel with ethidium bromide. Photograph as for a standard agarose gel.

PREPARATION OF HIGH-MOLECULAR-WEIGHT DNA SAMPLES AND SIZE MARKERS

SUPPORT PROTOCOL

Materials *(see APPENDIX 1 for items with ✓)*

High-molecular-weight DNA (Table 2.5.3)
1% (w/v) agarose
✓ Lysis buffer
Storage buffer: 10 mM Tris·Cl (pH 8)/10 mM Na_2EDTA (pH 8)
400 mM phenylmethylsulfonyl fluoride (PMSF) in ethanol
✓ 10 mM Tris·Cl, pH 8
Appropriate restriction enzyme and buffer (optional, *UNIT 3.1*)
Block molds or petri plates

1. Seal one end of block mold with tape (see Figure 2.5.4). If block molds are unavailable, prepare samples in agarose poured as a puddle on the bottom of a petri dish and cut blocks to size with a razor blade.

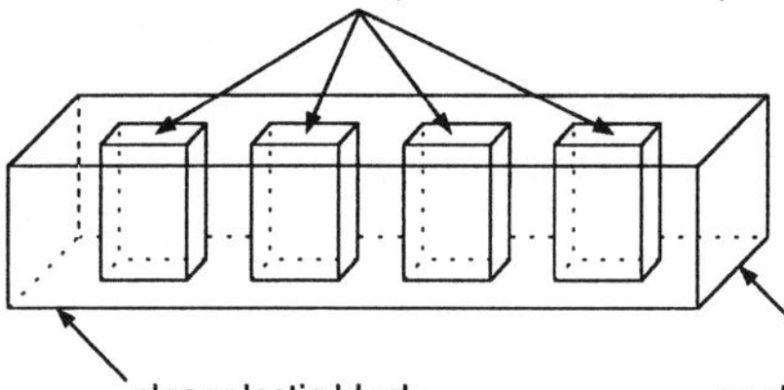

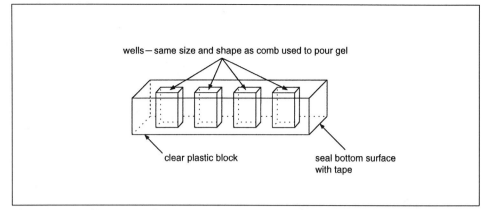

Figure 2.5.4 Block molds for high-molecular-weight DNA samples. These can be made in the laboratory, or may be purchased from pulsed-field gel box manufacturers.

2. Suspend high-molecular-weight DNA at twice the final concentration in room-temperature water or buffer (see Table 2.5.3). Add an equal volume of molten 1% agarose cooled to 50°C, mix quickly, place into block molds, and solidify on ice.

3. Remove tape, push out blocks into a 50-ml conical tube containing ≥20 vol lysis buffer, and incubate overnight at 50°C with gentle shaking.

4. Pour off lysis buffer, add fresh lysis buffer, and incubate overnight at 37°C. Pour off lysis buffer, replace with 20 vol storage buffer, and store at 4°C.

Table 2.5.3 Preparation of High-Molecular-Weight DNA Samples and Size Markers

Starting material	Preparation
Bacteria and phage	Resuspend bacteria (*UNIT 1.2*) or phage particles at a concentration calculated to yield the desired amount of DNA per lane; e.g., 5×10^8 *E. coli* per ml will yield ~100 ng DNA in an average lane.
Lambda ladders for size markers	Start with a concentrated stock of phage λ particles. This procedure does not work well with some lots of commercial λ DNA, possibly because of damaged cohesive ends. Try several dilutions of phage stock to see which works best. The second incubation in lysis buffer should be done at 25° rather than 37°C; during this incubation, the cohesive ends of the molecules will anneal, giving multimers of varying lengths. These concatemers will stay together during electrophoresis provided that the gel is run at <25°C.
Nematodes	Anesthetize worms by resuspending in 10 mM NaN_3, then place in agarose.
Nuclei	Isolate as in *UNIT 4.10*. Approximately 1 µg DNA is contained in 10^5 mammalian nuclei.
Tissue culture cells	Cells should be washed several times in a medium containing no serum, as serum may inhibit proteinase K.
Yeast	*Saccharomyces cerevisiae* cells must have their cell walls removed before being embedded in agarose as described in *UNIT 13.13*, Basic Protocol.

5. For samples to be digested with restriction enzymes, wash three times, 1 hr each, room temperature, with ≥10 vol storage buffer plus 1 mM PMSF.

   CAUTION: *PMSF is toxic and volatile and should be handled in a fume hood. It is unstable in aqueous solution, so always prepare fresh from a stock of 400 mM in ethanol stored at −20°C.*

6. Wash three times, 30 min each, room temperature, with ≥10 vol of 10 mM Tris·Cl, pH 8. Remove excess liquid.

7. Add a volume of 3× restriction buffer containing the enzyme equal to ½ the volume of the block, and incubate at the appropriate temperature until digestion is complete.

Reference: Schwartz and Cantor, 1984.

Contributor: Michael Finney

2.6

Isolation and Purification of Large DNA Restriction Fragments from Agarose Gels

UNIT 2.6

The basic protocol for electroelution is appropriate for purifying large amounts of fragments (>500 ng).

ELECTROELUTION FROM AGAROSE GELS

BASIC PROTOCOL 1

Materials *(see* APPENDIX 1 *for items with ✓)*

Appropriate restriction enzymes and buffers (*UNIT 3.1*)
✓ Ethidium bromide solution
✓ TAE electrophoresis buffer
✓ Elutip high-salt solution
✓ 2.5 M NaCl
✓ Elutip low-salt solution
100% and 70% ethanol
✓ TE buffer
Elutip-d column

1. Digest 0.1 to 25 μg DNA, electrophorese on an agarose gel, stain with ethidium bromide solution, and photograph.

   The gel can also contain ethidium bromide during electrophoresis.

2. Excise DNA band and slide into dialysis tubing (prerinsed with TAE electrophoresis buffer and double-knotted at one end). Fill tubing with TAE electrophoresis buffer, seal, and electroelute at ~2 V/cm. Electroelute 2 hr for a 500- to 2000-bp fragment, 4 hr for a 2000- to 4000-bp fragment, and overnight (at 1 V/cm) for larger fragments.

3. After electroelution is complete, reverse polarity of electrodes for 30 sec at 100 V.

4. Open tubing, collect eluted DNA, and remove gel slice. Rinse tubing with TAE electrophoresis buffer and combine with eluted DNA.

   If gel slice still contains DNA, repeat electroelution (step 2).

5. Wet Elutip-d column by pushing through 2 ml Elutip high-salt solution. Equilibrate by pushing through 5 ml Elutip low-salt solution.

6. Add 2.5 M NaCl to eluted DNA to 0.2 M, and slowly (~20 drops per min) load onto Elutip-d column. Wash slowly with 5 ml Elutip low-salt solution and elute slowly with 400 µl Elutip high-salt solution.

   Yield may be increased by repeating high salt elution and pooling samples.

7. Add 1 ml of 100% ethanol and precipitate at –20°C. Microcentrifuge, wash with 70% ethanol, dry, and resuspend in TE buffer.

   Yields are ~80% to 90% for fragments >1 kb and ~50% to 60% for smaller fragments. See Support Protocol for quantitation.

BASIC PROTOCOL 2

ELECTROPHORESIS ONTO NA-45 PAPER

This procedure is simple and is particularly suitable for small DNA fragments (i.e., <2000 bp).

Materials *(see* APPENDIX 1 *for items with ✓)*

Ultrapure agarose (e.g., SeaKem GTG agarose, FMC Bioproducts)
NA-45 paper
✓ TE buffer, pH 8.0
✓ NA-45 elution buffer
Buffered phenol (*UNIT 2.1*)
25:24:1 (v/v/v) phenol/chloroform/isoamyl alcohol (*UNIT 2.1*)
95% and 70% ethanol, ice cold

1. Digest DNA to completion with appropriate restriction enzymes and load onto gel prepared with ultrapure agarose and 0.5 µg/ml ethidium bromide

2. Electrophorese at appropriate voltage until DNA fragment of interest is well resolved from contaminants, as assessed by visualization with hand-held UV lamp.

3. Stop electrophoresis and, with a clean scalpel or razor blade, cut slits just above and below the fragment.

4. Insert a small piece of NA-45 paper into each of the slits by carefully separating the opening with a flat forceps while placing the paper in the slit with another set of forceps.

5. Gently push on the top and bottom of the gel to ensure that the slits are in contact with the paper. Electrophorese 10 min at the same voltage and current as the prior run, or until the DNA has migrated onto the paper.

6. Carefully remove the paper containing the DNA fragment of interest with forceps. Wash paper gently three times in TE buffer, pH 8.0.

7. Remove the washed paper to a 1.5-ml microcentrifuge tube containing 400 µl NA-45 elution buffer. Heat to 70°C, 15 min for short fragments (<500 bp) or 1 hr for larger fragments (>1500 bp).

8. Extract once with 400 µl buffered phenol. Reextract aqueous phase twice with phenol/chloroform/isoamyl alcohol, then extract twice with chloroform.

9. Add 1 ml of 95% ethanol and precipitate overnight at –20°C.

10. Microcentrifuge at high speed and wash pellet with 70% ethanol chilled to −20°C. Resuspend DNA in TE buffer, pH 8.0.

ISOLATION OF DNA FRAGMENTS USING LOW GELLING/MELTING TEMPERATURE AGAROSE GELS

BASIC PROTOCOL 3

2.6

DNA fragments can be isolated from gel slices from preparative gels made using low gelling/melting temperature agarose. For some applications (including restriction endonuclease digestion, *UNIT 3.1*, and ligation, *UNIT 3.16*) the melted gel slice containing the fragment can be used directly, without further purification by extraction or Elutip-d purification.

Materials *(see APPENDIX 1 for items with ✓)*

DNA to be isolated
Appropriate restriction endonucleases and buffers (*UNIT 3.1*)
Low gelling/melting temperature agarose (SeaPlaque, FMC Bioproducts)
✓ Ethidium bromide solution
✓ TE buffer, pH 8.0
Buffered phenol (*UNIT 2.1*)
✓ Elutip high-salt and low-salt solutions
Elutip-d column (Schleicher & Schuell)

1. Digest DNA sample to completion with appropriate restiction endonucleases and electrophorese on a 1% low gelling/melting temperature agarose gel. Stain the gel using ethidium bromide solution and cut out the target band with a scalpel.

2. Melt the gel slice at 65°C and add enough TE buffer, pH 8.0, to decrease the agarose percentage to ≤0.4%.

3. To remove agarose, add an equal volume of buffered phenol and mix vigorously for 5 to 10 min. Centrifuge 10 min at 15,800 × *g*, room temperature.

4. Collect the aqueous phase and set aside. Reextract the phenol phase and interface with an equal volume of TE buffer, pH 8.0. Centrifuge at 15,000 × *g*. Collect second aqueous phase and reextract a third time if necessary.

5. Ethanol precipitate the combined aqueous phase. Further purify the DNA solution via an Elutip-d column (steps 6 and 7) or use directly after resuspending in the appropriate buffer.

8. Add 10 to 20 vol Elutip low-salt solution.

9. Purify the DNA fragment using an Elutip-d column, maintaining all solutions at 37°C. Ethanol precipitate and quantitate as described in the Support Protocol.

RECOVERY OF DNA FROM LOW GELLING/MELTING TEMPERATURE AGAROSE GELS USING β-AGARASE DIGESTION

ALTERNATE PROTOCOL 1

The following protocol is an alternative method for recovery of DNA from low gelling/melting temperature agarose.

Additional Materials *(also see Basic Protocol; see APPENDIX 1 for items with ✓)*

β-agarase I (New England Biolabs or Calbiochem)
✓ β-agarase buffer

1. Prepare a low gelling/melting temperature agarose gel. Excise the band of interest.
2. Transfer gel slice to a clean tube. Wash twice for 30 min each on ice with 2 vol of 1× β-agarase buffer. Add an approximately equal volume of β-agarase buffer to the washed gel slice. Melt the gel completely by heating 10 min at 65°C.
3. Equilibrate molten agarose to 40°C (allow ~10 min). Add 1 U β-agarase for every 200 μl of 1% agarose and continue incubation for 1 hr.

   Molten β-agarase/DNA solution may be used directly for ligation (UNIT 3.16), transformation (UNIT1.12), or restriction endonuclease digestion (UNIT3.1).

4. To further purify large (>50 kb) DNA fragments, dialyze the DNA solution to remove carbohydrates and β-agarase. To purify smaller fragments, follow steps 5 to 8.
5. To purify small (<50 kb) fragments, adjust the salt concentration of the β-agarase/DNA solution to 0.5 M NaCl and add an equal volume of isopropanol. Chill 15 min on ice.
6. Centrifuge 15 min at 15,000 × *g* to pellet any undigested carbohydrates and transfer supernatant to a clean tube.
7. Add 2 to 3 vol isopropanol, mix thoroughly, and chill 30 min at 0°C.
8. Centrifuge 15 min at 15,000 × *g*, discard supernatant, and dry pellet. Resuspend in TE buffer, pH 8.0, or appropriate buffer.

ALTERNATE PROTOCOL 2

RECOVERY OF DNA FROM LOW GELLING/MELTING TEMPERATURE AGAROSE USING GLASS BEADS

The following rapid approach is based on the glass beads purification method described in UNIT 2.1; it is suitable for fragments >500 bp.

Additional Materials *(also see Basic Protocol 3; see APPENDIX 1 for items with ✓)*

✓ 6 M sodium iodide (NaI) solution
✓ Glass beads suspension
15-ml polypropylene tube

NOTE: The materials required for this procedure are also provided in commercial kits (e.g., GeneClean, Bio101; Glas-Pac, National Scientific Supply; and Qiaex Gel Extraction kit, Qiagen).

1. Digest DNA sample to completion and electrophorese through a 1% low gelling/melting temperature agarose gel. Stain the gel with ethidium bromide and cut out the target band with a scalpel.
2. Transfer gel slice to a 15-ml polypropylene tube and add a volume of NaI solution ~2.5 to 3 times the weight of the gel slice. Incubate 5 min at 45° to 55°C to dissolve agarose.
3. Add glass beads suspension to the DNA/agarose solution as follows: for amounts of DNA ≤5 μg, use 5 μl glass beads suspension; for amounts of DNA >5 μg, use 5 μl plus an additional 1 μl for every 0.5-μg increment above 5 μg. Incubate 5 min at room temperature with mixing every 1 to 2 min.
5. Continue with step 2 of the glass beads purification of DNA protocol (UNIT 2.1).

RAPID ESTIMATION OF DNA CONCENTRATION BY ETHIDIUM BROMIDE DOT QUANTITATION

SUPPORT PROTOCOL

This is a simple method to determine the concentration of isolated DNA fragments or dilute DNA solutions.

1. Prepare the following DNA standards in TE buffer: 0 μg/ml, 1 μg/ml, 2.5 μg/ml, 5 μg/ml, 7.5 μg/ml, 10 μg/ml, and 20 μg/ml.
2. Separately, add 4 μl of each standard DNA solution and 4 μl of each DNA sample of unknown concentration to 4 μl of a 1 μg/ml solution of ethidium bromide. Mix well.
3. Spot the standard and sample DNA/ethidium bromide solutions on plastic wrap placed on a UV transilluminator and photograph. Estimate the unknown DNA concentration by comparison to the fluorescence of the standards.

Reference: Wienand et al., 1978.

Contributors: Richard F Selden and Joanne Chory

Nondenaturing Polyacrylamide Gel Electrophoresis

UNIT 2.7

Large amounts of small, double-stranded DNA fragments (<1000 bp) can be separated and the purified fragments used for cloning, sequencing, and labeling.

BASIC PROTOCOL

Materials *(see APPENDIX 1 for items with ✓)*

✓ 10× and 1× TBE electrophoresis buffer
✓ 29:1 (w/w) acrylamide/bisacrylamide
TEMED (*N,N,N′,N′* tetramethylethlylenediamine)
✓ 10% ammonium persulfate
✓ 10× loading buffer
DNA molecular weight markers (Fig. 2.5.1)
0.5 μg/ml ethidium bromide
✓ Elution buffer, pH 8
100% and 70% ethanol
✓ 3 M sodium acetate, pH 5.2
✓ TE buffer, pH 8

DC power supply
Glass plates, spacers, and combs for pouring gels
Acrylamide gel electrophoresis apparatus
Silanized glass wool (*APPENDIX 3*)
JA-20 rotor or equivalent

1. Assemble glass plates with spacers for casting the gel.
2. See Table 2.7.1 to determine appropriate acrylamide concentration. For a 5% acrylamide gel, mix the following in a side-arm vacuum flask: 5 ml 10× TBE electrophoresis buffer, 8.33 ml 29:1 acrylamide/bisacrylamide, and 36.67 ml water.

Table 2.7.1 Concentrations of Acrylamide Giving Maximum Resolution of DNA Fragments[a]

Acrylamide (%)	Size fragments separated (bp)	Migration of bromphenol blue marker (bp)
3.5	100 to 1000	100
5.0	100 to 500	65
8.0	60 to 400	4
12.0	50 to 200	20
20.0	5 to 100	12

[a]Data are compiled from articles by Maniatis and Ptashne (1973a,b) and Maniatis et al. (1975).

CAUTION: Acrylamide is a neurotoxin. Always wear gloves when working with the unpolymerized monomer.

3. Degas the mixture under vacuum for several minutes. Add 25 µl TEMED and 250 µl of 10% ammonium persulfate and mix.

4. Pour the acrylamide gel mixture between the plates, insert comb, and allow gel to polymerize at room temperature at least 30 min.

5. Remove comb, rinse sample wells with water, and attach plates to electrophoresis tank, which has 1×TBE buffer in lower reservoir. Fill upper buffer chamber with enough buffer to cover wells.

   There should be no bubbles in the wells. Before adding buffer to upper chamber, remove any air bubbles trapped between the plates beneath the bottom of the gel using a syringe with a bent needle.

6. Add 10× loading buffer to DNA samples and size markers to 1× final concentration and load onto gel using a Hamilton syringe for small samples or a micropipet or a pulled plastic capillary for larger volumes (see Support Protocol).

7. Run gel at about 5 V/cm until desired resolution has been obtained.

8. Detach gel plates, stain for 10 to 30 min in 0.5 µg/ml ethidium bromide, and photograph on a UV transilluminator.

9. Excise DNA band and chop up the gel slice into many fine pieces. Transfer pieces to 1.5-ml microcentrifuge tube.

10. Cover gel fragments with elution buffer and incubate at 37°C.

    Small fragments (<250 bp) will be mostly eluted in 2 to 3 hr; larger fragments (>750 bp) will require overnight shaking.

11. Pellet gel fragments at room temperature 10 min in a tabletop centrifuge (~500 × *g*) or 1 min in a microcentrifuge. Remove supernatant, avoiding the polyacrylamide pieces.

12. Rinse polyacrylamide gel fragments with additional elution buffer to recover residual DNA. Recentrifuge if necessary and combine the two supernatants.

    If necessary, remove any remaining acrylamide pieces by filtering the supernatant through a syringe with a silanized glass wool plug or through a 2-µm filter.

13. Precipitate DNA with 2 vol of 100% ethanol and chill 30 min to −20°C. Pellet by centrifugation 10 min at 12,000 × *g* or 5 min in a microcentrifuge.

14. Redissolve DNA pellet in 100 µl of TE buffer, add 10 µl of 3 M sodium acetate, pH 5.2, reprecipitate the DNA with 2 vol of 100% ethanol, and recover by centrifugation.

15. Rinse pellet with 70% ethanol, dry, and resuspend in TE buffer.

PURIFICATION OF FRAGMENTS BY ELECTROELUTION FROM POLYACRYLAMIDE GELS

ALTERNATE PROTOCOL 1

Replace the elution steps (9 to 15) in the Basic Protocol with this electroelution protocol, if you are in a hurry.

1. Excise DNA band and place gel slice into a small dialysis bag containing 0.1× TBE buffer.

2. Place bag in a small horizontal electrophoresis apparatus containing 0.1× TBE buffer and electrophorese 2 hr at ~2 V/cm.

3. Recover DNA in the 0.1× TBE buffer. Rinse gel fragment and inner surface of the dialysis bag. Add 0.1 vol of 3 M sodium acetate and ethanol precipitate.

PURIFICATION OF LABELED FRAGMENTS BY ELECTROELUTION ONTO DEAE MEMBRANE

ALTERNATE PROTOCOL 2

To isolate small quantities of radiolabeled DNA fragments that cannot be detected by staining, labeled fragments are first visualized using autoradiography and the resulting film image is used to locate the desired band on the gel. Then the gel region corresponding to the desired band is excised, immobilized in a pool of agarose, and isolated by electroelution onto DEAE membrane. This method is most typically used for methylation interference assays.

Additional Materials *(also see Basic Protocol; see* APPENDIX 1 *for items with ✓)*

1% (w/v) agarose in 1× TBE electrophoresis buffer
DEAE membrane (Schleicher & Schuell)
✓ DEAE elution buffer
25:24:1 (v/v/v) phenol/chloroform/isoamyl alcohol (*UNIT 2.1*)
10 mg/ml tRNA solution

1. Run radiolabeled DNA fragments on a nondenaturing polyacrylamide gel. At the end of the run, carefully wrap the gel and plate with plastic wrap. Mark the plastic wrap with the outline of the film and autoradiograph the gel 1 to 2 hr at 4°C.

2. Use the developed autoradiogram to line up film with markings on plastic wrap. Excise the bands of interest.

3. Remove the acrylamide strips with forceps.

4. Place the acrylamide strips in the bottom of an agarose minigel apparatus, folding over any strips that are long. Pour 1% agarose in 1× TBE electrophoresis buffer into the minigel without disturbing the acrylamide strip. Allow the agarose to set.

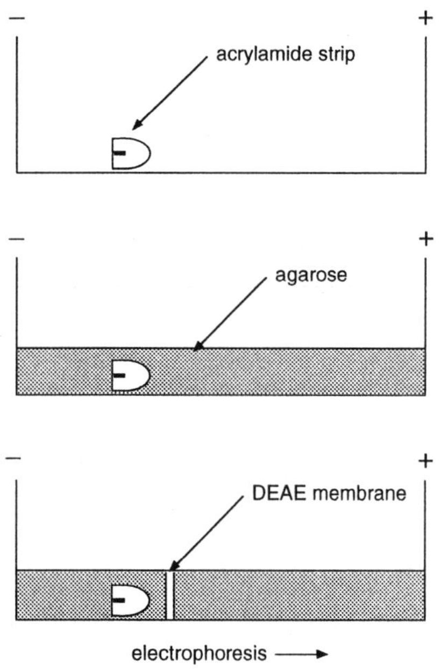

Figure 2.7.1 Agarose gel electrophoresis of labeled DNA from polyacrylamide onto a DEAE membrane. Shown are procedures for pouring gel around acrylamide strip and for inserting DEAE membrane.

5. Cut a slit in the gel with a razor blade ~4 to 5 mm below the strip. Cut a piece of DEAE membrane slightly wider than the acrylamide strip, prewet in 1× TBE, and insert into opening of gel (Fig. 2.7.1). Electrophorese labeled DNA onto membrane at ~60 mA for 10 to 15 min.

6. Remove DEAE membrane, rinse in 1× TBE buffer, and place in a microcentrifuge tube. Cover strip with ~200 µl DEAE elution buffer. Heat 30 min at 65°C.

7. Remove supernatant to a separate tube and rinse the membrane in the original tube with an equal volume of TE buffer, pH 7.5. Add this rinse to tube containing supernatant.

8. Microcentrifuge eluate 5 min at top speed to remove any residual debris. Remove supernatant to a separate tube and extract with an equal volume of phenol/chloroform/isoamyl alcohol.

9. Microcentrifuge 5 min at top speed to separate phases. Remove aqueous phase to a separate tube. Add 1 µl of 10 mg/ml tRNA solution and 2 vol of 100% ethanol. Freeze 10 min in a dry ice/ethanol bath and microcentrifuge 10 min. Rinse pellet carefully with 70% ethanol. Dry inverted for 10 min and resuspend in appropriate buffer or solution (e.g., TE buffer).

REUSABLE PLASTIC CAPILLARIES FOR GEL LOADING

SUPPORT PROTOCOL

1. Break a 1-ml disposable polystyrene pipet in half and heat the center portion of a piece over a Bunsen burner until barely melted.
2. Remove from heat for 1 to 2 sec and—holding vertically—quickly pull to generate a thin capillary. Cut as desired after the piece has cooled.
3. Calibrate by drawing up premeasured volume of water and marking. Load gels using micropipet mouthpiece and tube, or suction apparatus.

Reference: Chrambach and Robard, 1971.

Contributor: Joanne Chory

Sieving Agarose Gel Electrophoresis

UNIT 2.8

BASIC PROTOCOL

Sieving agarose gels are poured and run like conventional agarose gels, but they resolve small DNA fragments like nondenaturing polyacrylamide gels.

Materials *(see APPENDIX 1 for items with ✓)*

Low gelling/melting temperature sieving agarose (FMC Bioproducts)
✓ TAE electrophoresis buffer, pH 8

1. Melt 3-5% low gelling/melting temperature sieving agarose in TAE electrophoresis buffer. Pour gel in ordinary agarose gel apparatus (*UNIT 2.5A*).
2. Load sample and run gel as for an ordinary agarose gel (*UNIT 2.5A*).

   Bromphenol blue migrates at ~50 bp for a 2% gel.

3. Isolate fragment following Basic Protocol 3 or Alternate Protocol 1 for low gelling/melting temperature agarose (*UNIT 2.6*).

   Yields are similar to conventional low gelling/melting temperature agarose (>70%) for fragments <1000 bp. Highest resolution is achieved for fragments <500 bp.

Reference: Literature available from FMC Bioproducts.

Contributor: Joanne Chory

UNIT 2.9A

Southern Blotting

Southern blotting is the transfer of DNA fragments from an electrophoresis gel to a membrane support resulting in immobilization of the DNA fragments, so the membrane carries a semipermanent reproduction of the banding pattern of the gel.

2.9

BASIC PROTOCOL

SOUTHERN BLOTTING ONTO A NYLON OR NITROCELLULOSE MEMBRANE WITH HIGH-SALT BUFFER

The procedure is specifically designed for blotting an agarose gel onto an uncharged or positively charged nylon membrane. With minor modifications, the same protocol can also be used with nitrocellulose membranes.

Materials *(see APPENDIX 1 for items with ✓)*

✓ 0.25 M HCl

Denaturation solution: 1.5 M NaCl/0.5 M NaOH (store at room temperature)

Neutralization solution: 1.5 M NaCl/0.5 M Tris·Cl, pH 7.0 (store at room temperature)

✓ 20× and 2× SSC

Nylon *or* nitrocellulose membrane (see Table 2.9.1 for suppliers)

UV transilluminator (*UNIT 2.5*) or UV light box (e.g., Stratagene Stratalinker) for nylon membranes

CAUTION: Wear gloves from step 2 of the protocol onward to protect your hands from the acid and alkali solutions and to protect the membrane from contamination. Handle the membrane with blunt-end forceps.

1. Digest DNA with appropriate restriction endonuclease(s), run in an agarose gel with appropriate DNA size markers, stain with ethidium bromide, and photograph with a ruler laid alongside the gel so that band positions can later be identified on the membrane.

   The gel should contain the minimum agarose concentration needed to resolve the bands and should be ≤7 mm thick.

2. Treat the gel as follows, with gentle shaking at room temperature. Be sure the gel is covered completely.

   Distilled water
   ~10 gel volumes 0.25 M HCl—30 min
   Distilled water
   ~10 volumes denaturation solution—20 min, twice
   Distilled water
   ~10 volumes neutralization solution—20 min, twice

3. Using Figure 2.9.1 as a guide, assemble the transfer stack consisting of a sponge or solid support with a Whatman 3MM paper wick in a dish filled with enough 20× SSC to half submerge the sponge, three pieces of Whatman 3MM paper, the gel, pieces of plastic wrap on the edges of the gel, a nylon membrane equilibrated in distilled water (or nitrocellulose membrane wet in distilled water and equilibrated 10 min in 20× SSC), five pieces of Whatman 3MM paper and a 4-cm stack of paper towels. As each layer is applied, wet it with 20× SCC and remove any trapped air by carefully rolling a 10-ml glass pipet over the surface.

   Before a sponge is used for the first time, it should be washed thoroughly with distilled water to remove any detergents that may be present.

   For blots of substantial amounts of plasmid or other very-low-complexity DNA, it is important to lay the membrane down precisely the first time.

Table 2.9.1 Properties of Materials used for Immobilization of Nucleic Acids[a]

	Nitrocellulose	Supported nitrocellulose	Uncharged nylon	Positively charged nylon	Activated papers
Application	ssDNA, RNA, protein	ssDNA, RNA, protein	ssDNA, dsDNA, RNA, protein	ssDNA, dsDNA, RNA, protein	ssDNA, RNA
Binding capacity (μg nucleic acid/cm^2)	80-100	80-100	400-600	400-600	2-40
Tensile strength	Poor	Good	Good	Good	Good
Mode of nucleic acid attachment[b]	Noncovalent	Noncovalent	Covalent	Covalent	Covalent
Lower size limit for efficient nucleic acid retention	500 nt	500 nt	50 nt or bp	50 nt or bp	5 nt
Suitability for reprobing	Poor (fragile)	Poor (loss of signal)	Good	Good	Good
Commercial examples	Schleicher & Schuell BA83, BA85; Amersham Hybond-C; PALL Biodyne A	Schleicher & Schuell BA-S; Amersham Hybond-C extra	Amersham Hybond-N; Stratagene Duralon-UV; Du Pont NEN GeneScreen	Schleicher & Schuell Nytran; Amersham Hybond-N$^+$; Bio-Rad ZetaProbe; PALL Biodyne B; Du Pont NEN GeneScreen Plus	Schleicher & Schuell APT papers

[a]This table is based on Brown (1991), with permission from BIOS Scientific Publishers Ltd.

[b]After suitable immobilization procedure (see text).

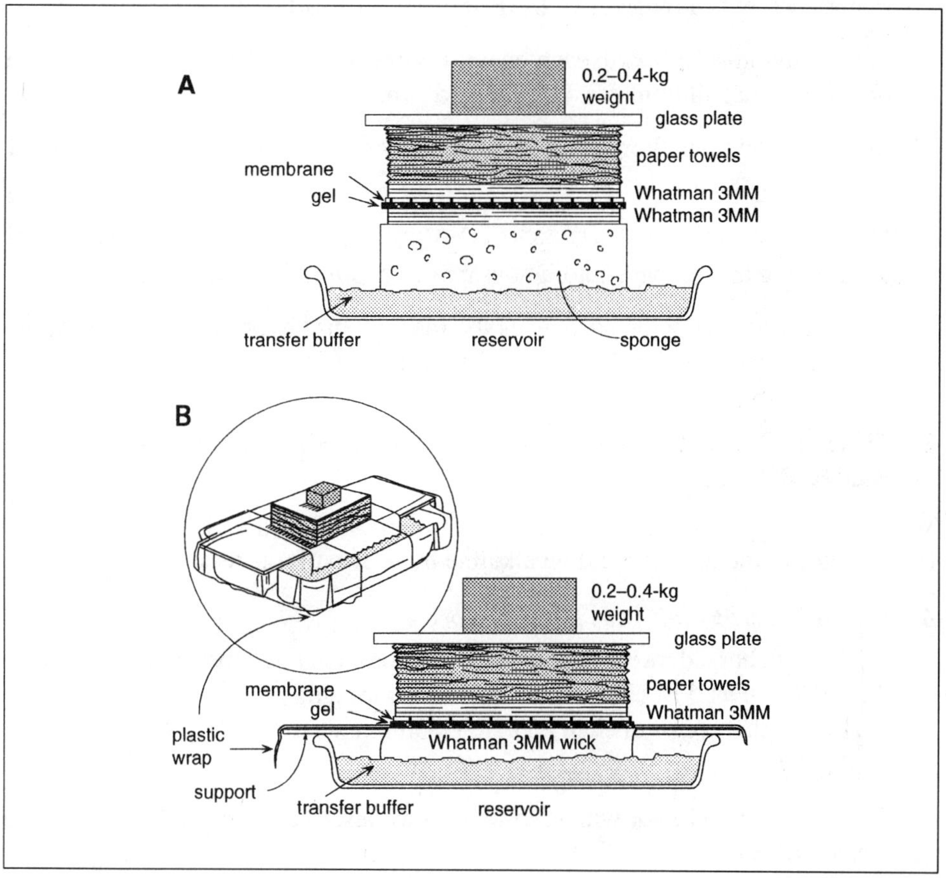

Figure 2.9.1 Transfer stacks for upward capillary transfer. (**A**) Sponge method. (**B**) Filter paper wick method.

4. Lay a glass plate on top of the stack and place a 0.2 to 0.4-kg weight on top to hold everything in place. Leave overnight.
5. Recover the membrane. Mark in pencil the position of the wells on the membrane and cut one corner to mark the orientation.
6. Rinse the membrane in 2× SSC, then place it on a sheet of Whatman 3MM paper and allow to dry completely.
7. Wrap nylon membrane in UV-transparent plastic wrap, place DNA-side-down on a UV transilluminator (254-nm wavelength) and irradiate for the time determined from the Support Protocol.

   Nitrocellulose membranes should be placed between two sheets of Whatman 3MM paper and baked under vacuum for 2 hr at 80°C.
8. Store membranes dry between sheets of Whatman 3MM paper for several months at room temperature.

SUPPORT PROTOCOL

CALIBRATION OF A UV TRANSILLUMINATOR

UV transilluminators must be calibrated to determine their output to ensure the proper intensity of radiation is used for cross-linking.

CAUTION: Exposure to UV irradiation is harmful to the eyes and skin. Wear suitable eye protection and avoid exposure of bare skin.

1. Prepare five identical series of DNA dot blots (*UNIT 2.9B*) on nylon membrane strips, each with a range of DNA quantity from 1 to 100 pg.
2. Dry the nylon strips, wrap each one in UV-transparent plastic wrap, and place on the UV transilluminator, DNA-side-down.
3. Switch on the transilluminator. Remove individual strips after 30 sec, 45 sec, 1 min, 2 min, and 5 min.
4. Hybridize the strips with a suitable DNA probe labeled to a specific activity of 10^8 dpm/μg and prepare autoradiographs (*UNIT 2.10*).
5. Determine which nylon strip gives the most intense hybridization signals; the exposure time used for that strip is the optimal exposure time.

ALTERNATE PROTOCOL 1

SOUTHERN BLOTTING ONTO A NYLON MEMBRANE WITH AN ALKALINE BUFFER

With a positively charged nylon membrane, the transferred DNA becomes covalently linked to the membrane if an alkaline transfer buffer is used.

***Additional Materials** (also see Basic Protocol)*

✓ 0.4 M (for charged membrane) *or* 0.25 M (for uncharged membrane) NaOH
0.25 M NaOH/1.5 M NaCl for uncharged membrane
Positively charged *or* uncharged nylon membrane (see Table 2.9.1 for suppliers)

1. Prepare a gel and treat with 0.25 M HCl as described in steps 1 and 2 of the Basic Protocol.

2. Rinse the gel with distilled water and denature it with 10 gel volumes of 0.4 M NaOH. Shake slowly for 20 min.

   Use 0.25 M NaOH to denature an uncharged nylon membrane.

3. Carry out the transfer, using 0.4 M NaOH as the transfer solution in place of 20× SSC. Leave to transfer ≥2 hr.

   Use 0.25 M NaOH/1.5 M NaCl as the transfer solution for an uncharged nylon membrane.

4. Recover the membrane and rinse in 2× SSC. Place membrane on a sheet of Whatman 3MM filter paper, and allow to air dry. Store at room temperature.

   Baking or UV cross-linking is not necessary with positively charged membranes; however, DNA on uncharged membranes should be cross-linked by UV exposure.

SOUTHERN BLOTTING BY DOWNWARD CAPILLARY TRANSFER

ALTERNATE PROTOCOL 2

One disadvantage with the upward capillary method (Basic Protocol) is that the gel can become crushed by the weighted filter papers and paper towels that are laid on top of it, reducing capillary action. Simple downward capillary transfer that does not cause excessive pressure to be placed on the gel is described in this protocol.

1. Prepare a gel as described in steps 1 to 4 of the Basic Protocol (high-salt transfer) or steps 1 to 2 of Alternate Protocol (alkaline transfer).

2. Using Figure 2.9.2 as a guide, assemble a transfer pyramid consisting of paper towels (2 to 3 cm high) in a glass dish, four pieces of Whatman 3MM filter paper, a fifth filter paper wet with transfer buffer, the membrane wet with distilled water (nylon) or 20× SSC (nitrocellulose), a plastic wrap skirt, the gel, three pieces of Whatman 3MM paper, and two larger pieces of Whatman 3MM paper prewet with transfer buffer and extending from the stack to the transfer buffer reservoir.

3. Place a light plastic cover (e.g., a gel plate) over the top of the stack to reduce evaporation. Let stand 1 hr.

4. Recover the membrane, rinse in 2× SSC, and air dry. Immobilize the DNA and store the membrane at room temperature.

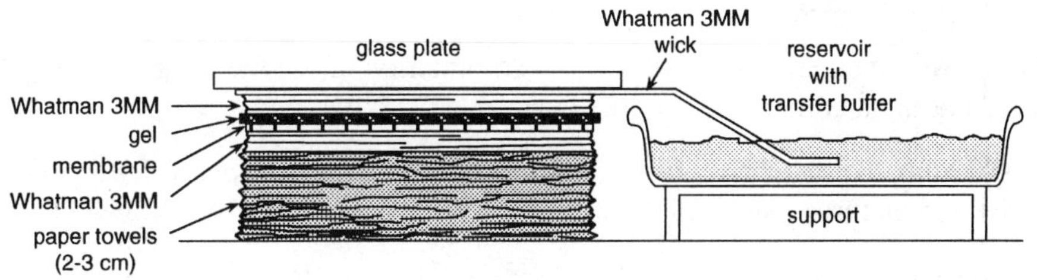

Figure 2.9.2 Transfer pyramid for downward capillary transfer. Adapted with permission from Academic Press.

ELECTROBLOTTING FROM A POLYACRYLAMIDE GEL TO A NYLON MEMBRANE

Capillary transfer does not work with polyacrylamide gels, whose pore sizes are too small for effective transverse diffusion of DNA. Polyacrylamide gels must therefore be blotted by electrophoretic transfer in a buffer of low ionic strength.

Additional Materials *(also see Basic Protocol; see* APPENDIX 1 *for items with ✓)*

✓ 0.5× TBE electrophoresis buffer
Scotch-Brite pads (supplied with Trans-Blot apparatus)
Trans-Blot electroblotting cell (Bio-Rad) with cooling coil, or other electroblotting apparatus (*UNIT 10.8*)

1. Run DNA samples in a nondenaturing or denaturing polyacrylamide gel.
2. When electrophoresis is almost complete, float a piece of nylon membrane sufficient in size to cover the relevant parts of the gel on the surface of distilled water, then submerge it and let stand 5 min.
3. Remove one glass plate from the electrophoresis apparatus and stain and photograph the gel (if nondenaturing). Lay a piece of Whatman 3MM filter paper on the surface of the gel, removing any trapped air bubbles. Lift the gel off the glass plate by peeling the filter paper away.
4. Soak two Scotch-Brite pads in 0.5× TBE electrophoresis buffer and remove air pockets by repeated squeezing and agitation. Cut seven pieces of Whatman 3MM paper to the same size as the gel and soak them 15 to 30 min in 0.5× TBE.
5. Place one saturated Scotch-Brite pad on the inner surface of the gray panel of the gel holder. Place three soaked filter papers on the pad, one at a time, removing trapped air bubbles.
6. Flood the filter paper carrying the gel with 0.5× TBE and place on top of the filter-paper stack. Flood the surface of the gel with 0.5× TBE and place the prewetted membrane onto the gel.
7. Flood the surface of the membrane with 0.5× TBE and place the remaining four sheets of saturated Whatman 3MM paper on top, followed by the second saturated Scotch-Brite pad. Close the gel holder.
8. Half-fill the Trans-Blot cell with 0.5× TBE and place the gel holder in the cell with the grey panel facing towards the cathode. Fill the cell with 0.5× TBE and electroblot at 30 V (~125 mA), with cooling, for 4 hr under the conditions recommended by the manufacturer.
9. Recover the membrane and mark the orientation of the membrane in pencil or by cutting a corner.
10. Denature the membrane from a nondenaturing gel by placing it, DNA-side-up, 10 min on three pieces of Whatman 3MM paper soaked in 0.4 M NaOH.
11. Rinse the membrane in 2× SSC, place on a sheet of Whatman 3MM paper, and allow to dry. Immobilize the DNA and store membrane at room temperature.

Reference: Southern, 1975.

Contributor: Terry Brown

Dot and Slot Blotting of DNA

Dot and slot blotting (Fig. 2.9.3) are simple techniques for immobilizing bulk unfractionated DNA on a nitrocellulose or nylon membrane for hybridization analysis (*UNIT 2.10*) to determine the relative abundance of target sequences in the blotted DNA preparations.

CAUTION: In all of the protocols, wear gloves to protect your hands from the alkali solution and to protect the membrane from contamination. Avoid handling nitrocellulose and nylon membranes even with gloved hands—use clean blunt-ended forceps instead.

DOT AND SLOT BLOTTING OF DNA ONTO UNCHARGED NYLON AND NITROCELLULOSE MEMBRANES USING A MANIFOLD

BASIC PROTOCOL

Dot and slot blots are usually prepared with the aid of a manifold and suction device.

Materials *(see APPENDIX 1 for items with ✓)*

✓ 6× and 20× SSC (*APPENDIX 2*)
Denaturation solution: 1.5 M NaCl/0.5 M NaOH (store at room temperature)
Neutralization solution: 1 M NaCl/0.5 M Tris·Cl, pH 7.0 (store at room temperature)
Uncharged nylon *or* nitrocellulose membrane (see Table 2.9.1, *UNIT 2.9A* for suppliers)
Dot/slot blotting manifold (e.g., Bio-Rad Bio-Dot SF *or* Schleicher and Schuell Minifold II)
UV transilluminator (*UNIT 2.5A*) for nylon membranes

1. Cut a piece of nylon membrane to the size of the manifold. Place the membrane on the surface of 6× SSC and allow to submerge. Let stand 10 min.

   Wet nitrocellulose membrane in 20× SSC.

2. Cut a piece of Whatman 3MM filter paper to the size of the manifold. Wet in 6× SSC.

   Use 20× SSC if transferring onto nitrocellulose.

3. Place the Whatman 3MM paper in the manifold and lay the membrane on top of it. Assemble the manifold according to the manufacturer's instructions, ensuring that there are no air leaks in the assembly.

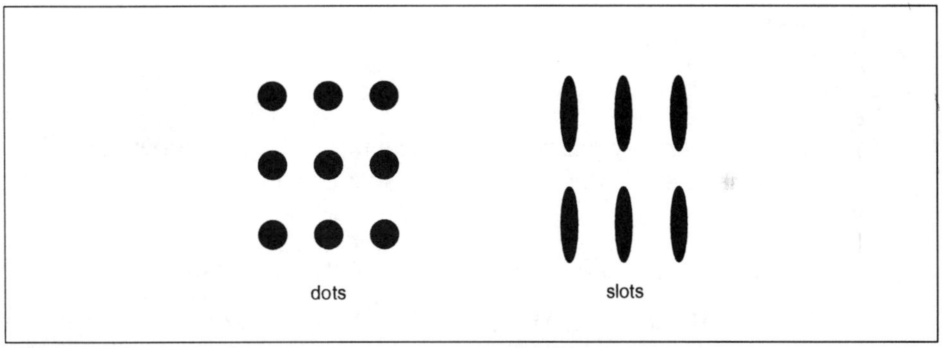

Figure 2.9.3 Dot (left) and slot (right) blot manifold architectures.

4. Add 20× SSC and water to DNA to give a final concentration of 6× SSC in a volume of 200 to 400 μl. Denature 10 min at 100°C, then place in ice.

   For nitrocellulose membrane, add an equal volume of 20× SSC to each sample after placing in ice.

   The amount of DNA that should be blotted will depend on the relative abundance of the target sequence that will subsequently be sought by hybridization probing.

5. Switch on the suction to the manifold device, apply 500 μl of 6× SSC to each well.

   For a nitrocellulose membrane, use 20× SSC.

   Wells that are not being used can be blocked off by placing masking tape over them,

6. Microcentrifuge DNA samples 5 sec. Apply samples to the wells being careful to avoid touching the membrane with the pipet. Allow the samples to filter through.

7. Dismantle the apparatus and place the membrane on a piece of Whatman 3MM paper soaked in denaturation solution. Leave for 10 min.

8. Transfer the membrane to a piece of Whatman 3MM paper soaked in neutralization solution. Leave for 5 min.

9. Place the membrane on a piece of dry Whatman 3MM paper and allow to dry.

10. Wrap the dry membrane in UV-transparent plastic wrap, place DNA-side-down on a UV transilluminator, and immobilize the DNA by irradiating for the appropriate time.

    Nitrocellulose membranes should be placed between two sheets of Whatman 3 MM paper and baked under vacuum for 2 hr at 80°C.

11. Store the membrane dry between sheets of Whatman 3MM filter paper for several months at room temperature.

ALTERNATE PROTOCOL 1

DOT AND SLOT BLOTTING OF DNA ONTO A POSITIVELY CHARGED NYLON MEMBRANE USING A MANIFOLD

Positively charged nylon membranes bind DNA covalently at high pH (see *UNIT 2.9A*). Samples for dot or slot blotting can therefore be applied in an alkaline buffer, which promotes both denaturation of the DNA and binding to the membrane.

Additional Materials *(also see Basic Protocol; see APPENDIX 1 for items with ✓)*

Positively charged nylon membrane (see Table 2.9.1, *UNIT 2.9A*, for suppliers)
✓ 0.4 M and 1 M NaOH
✓ 200 mM EDTA, pH 8.2
✓ 2× SSC

1. Cut a piece of positively charged nylon membrane to the appropriate size. Place the membrane on the surface of distilled water and allow to submerge. Leave for 10 min.

2. Prepare the blotting manifold, using distilled water instead of SSC.

3. Add 1 M NaOH and 200 mM EDTA, pH 8.2, to each sample to give a final concentration of 0.4 M NaOH/10 mM EDTA. Denature 10 min at 100°C. Microcentrifuge each tube 5 sec.

The alkali/heat treatment denatures the DNA.

4. Prewash the membrane with 500 µl distilled water. Apply the samples to the membrane.

5. After applying the samples, rinse each well with 500 µl of 0.4 M NaOH and dismantle the manifold.

6. Rinse the membrane briefly in 2× SSC and air dry.

7. Store membrane at room temperature.

MANUAL PREPARATION OF A DNA DOT BLOT

ALTERNATE PROTOCOL 2

Dot blots can also be set up by hand simply by spotting small aliquots of each sample on to the membrane and waiting for the blot to dry. Repeated applications can be used to apply up to 30 µl of a dilute DNA sample.

1. Cut a strip of uncharged nylon membrane to the desired size and mark out a grid of 0.5-cm × 0.5-cm squares with a blunt pencil. Place membrane on the surface of 6× SSC and allow to submerge. Leave 10 min.

   Wet nitrocellulose membrane in 20× SSC or a positively charged nylon membrane in distilled water.

2. Add ½ vol of 20× SSC to DNA to give a final concentration of 6× SSC in the minimum possible volume. Denature the DNA 10 min at 100°C, then place in ice.

   If using positively charged nylon, add 1 M NaOH and 200 mM EDTA, pH 8.2, to each sample to give a final concentration of 0.4 M NaOH/10 mM EDTA, and denature.

   If using a nitrocellulose membrane, add an equal volume of 20× SSC to each sample after placing in ice.

3. Place the wetted membrane over the top of an open plastic box so that the bulk of the membrane is freely suspended.

4. Microcentrifuge DNA 5 sec, spot onto the membrane using a pipet, and allow to dry.

   Do not touch the membrane with the pipet when applying the samples. Up to 2 µl can be spotted in one application.

5. Denature, neutralize, and immobilize DNA on uncharged nylon or nitrocellulose membrane.

   Rinse and dry a positively charged nylon membrane.

6. Store membrane at room temperature.

Reference: Dyson, N.J. 1991.

Contributor: Terry Brown

Hybridization Analysis of DNA Blots

The principle of hybridization analysis is that a single-stranded DNA or RNA molecule of defined sequence (the "probe" which is usually labeled) can base-pair to a second DNA or RNA molecule that is immobilized and contains a complementary sequence (the "target"), with the stability of the hybrid depending on the extent of base pairing that occurs. The technique permits detection of single-copy genes in complex genomes.

BASIC PROTOCOL

HYBRIDIZATION ANALYSIS OF A DNA BLOT WITH A RADIOLABELED DNA PROBE

This protocol is suitable for hybridization analysis of Southern transfers (*UNIT 2.9A*) and dot and slot blots (*UNIT 2.9B*) with a radioactively labeled DNA probe 100 to 1000 bp in length.

Materials *(see APPENDIX 1 for items with ✓)*

Probe DNA labeled to a specific activity >1 × 10^8 dpm/μg
✓ Aqueous prehybridization/hybridization (APH) solution, room temperature and 68°C
2× SSC/0.1% (w/v) SDS
0.2× SSC/0.1% (w/v) SDS, room temperature and 42°C
0.1× SSC/0.1% (w/v) SDS, 68°C
✓ 2× and 6× SSC

Hybridization oven (e.g., Hybridiser HB-1, Techne) *or* 68°C water bath or incubator
Hybridization tube *or* sealable bag and heat sealer

1. Wet a membrane carrying immobilized DNA in 6× SSC.

2. Place the membrane, DNA-side-up, in a hybridization tube and add ~1 ml APH solution per 10 cm^2 of membrane. Incubate 3 hr in hybridization oven with rotation at 68°C.

   For a nylon membrane, prehybridize 15 min with 68°C, prehybridization/hybridization solution. See Table 2.10.1 for other hybridization solutions and Table 2.10.2 for alternative blocking reagents..

3. Just before the end of the prehybridization incubation, denature probe DNA 10 min at 100°C. Place in ice.

Table 2.10.1 High-Salt Solutions Used in Hybridization Analysis

Stock solution	Composition
20× SSC	3.0 M NaCl/0.3 M trisodium citrate
20× SSPE[a]	3.6 M NaCl/0.2 M NaH_2PO_4/0.02 M EDTA, pH 7.7
Phosphate solution[b]	1 M $NaHPO_4$, pH 7.2[c]

[a]SSC may be replaced with the same concentration of SSPE in all protocols.

[b]Prehybridize and hybridize with 0.5 M $NaHPO_4$ (pH 7.2)/1 mM EDTA/7% SDS [or 50% formamide/0.25 M $NaHPO_4$ (pH 7.2)/0.25 M NaCl/1 mM EDTA/7% SDS]; perform moderate-stringency wash in 40 mM $NaHPO_4$ (pH 7.2)/1 mM EDTA/5% SDS; perform high-stringency wash in 40 mM $NaHPO_4$ (pH 7.2)/1 mM EDTA/1% SDS.

[c]Dissolve 134 g $Na_2HPO_4 \cdot 7H_2O$ in 1 liter water, then add 4 ml 85% H_3PO_4. The resulting solution is 1 M Na^+, pH 7.2.

Table 2.10.2 Alternatives to Denhardt/Denatured Salmon Sperm DNA as Blocking Agents in DNA Hybridization[a]

Blocking agent	Composition	Storage and use
BLOTTO	5% (w/v) nonfat dried milk/0.02% (w/v) NaN_3 in H_2O	Store at 4°C; use at 4% final concentration
Heparin (porcine grade II)	50 mg/ml in 4× SSC	Store at 4°C. Use at 500 µg/ml with dextran sulfate or 50 µg/ml without
Yeast tRNA	10 mg/ml in H_2O	Store at 4°C; use at 100 µg/ml
Homopolymer DNA	1 mg/ml poly(A) or poly(C) in H_2O	Store at 4°C; use at 10 µg/ml in water; appropriate targets: poly(A) for AT-rich DNA, poly(C) for GC-rich DNA

[a]This table is based on Brown (1991) with permission from BIOS Scientific Publishers.

Table 2.10.3 Factors Influencing Hybrid Stability and Hybridization Rate[a]

Factor	Influence
A. Hybrid stability[b]	
Ionic strength	T_m increases 16.6°C for each 10-fold increase in monovalent cations between 0.01 and 0.40 M NaCl
Base composition	AT base pairs are less stable than GC base pairs in aqueous solutions containing NaCl
Destabilizing agents	Each 1% of formamide reduces the T_m by about 0.6°C for a DNA-DNA hybrid. 6 M urea reduces the T_m by about 30°C
Mismatched base pairs	T_m is reduced by 1°C for each 1% of mismatching
Duplex length	Negligible effect with probes >500 bp
B. Hybridization rate[b]	
Temperature	Maximum rate occurs at 20-25°C below T_m for DNA-DNA hybrids, 10-15°C below T_m for DNA-RNA hybrids
Ionic strength	Optimal hybridization rate at 1.5 M Na^+
Destabilizing agents	50% formamide has no effect, but higher or lower concentrations reduce the hybridization rate
Mismatched base pairs	Each 10% of mismatching reduces the hybridization rate by a factor of two
Duplex length	Hybridization rate is directly proportional to duplex length
Viscosity	Increased viscosity increases the rate of membrane hybridization; 10% dextran sulfate increases rate by factor of ten
Probe complexity	Repetitive sequences increase the hybridization rate
Base composition	Little effect
pH	Little effect between pH 5.0 and pH 9.0

[a]This table is based on Brown (1991) with permission from BIOS Scientific Publishers.

[b]There have been relatively few studies of the factors influencing membrane hybridization. In several instances extrapolations are made from what is known about solution hybridization. This is probably reliable for hybrid stability, less so for hybridization rate.

4. Pour the APH solution from the hybridization tube and replace with an equal volume of prewarmed (68°C) APH solution. Add denatured probe and incubate with rotation overnight at 68°C.

If the specific activity of the probe is 10^8 dpm/µg, use 10 ng/ml probe; if it is 1×10^9 dpm/µg, use 2 ng/ml.

See Table 2.10.3 for factors affecting hybrid stability and hybridization rate.

Table 2.10.4 Troubleshooting Guide for DNA Blotting and Hybridization Analysis[a]

Problem	Possible cause[b]	Solution
Poor signal	Probe specific activity too low	Check labeling protocol if specific activity is $<10^8$ dpm/μg.
	Inadequate depurination	Check depurination if transfer of DNA >5 kb is poor.
	Inadequate transfer buffer	1. Check that 20× SSC has been used as the transfer solution if small DNA fragments are retained inefficiently when transferring to nitrocellulose. 2. With some brands of nylon membrane, add 2 mM Sarkosyl to the transfer buffer. 3. Try alkaline blotting to a positively charged nylon membrane.
	Not enough target DNA	Refer to text for recommendations regarding amount of target DNA to load per blot.
	Poor immobilization of DNA	See recommendations in *UNIT 2.9A* commentary.
	Transfer time too short	See recommendations in *UNIT 2.9A* commentary.
	Inefficient transfer system	Consider vacuum blotting as an alternative to capillary transfer.
	Probe concentration too low	1. Check that the correct amount of DNA has been used in the labeling reaction. 2. Check recovery of the probe after removal of unincorporated nucleotides. 3. Use 10% dextran sulfate in the hybridization solution. 4. Change to a single-stranded probe, as reannealing of a double-stranded probe reduces its effective concentration to zero after hybridization for 8 hr.
	Incomplete denaturation of probe	Denature as described in the protocols.
	Incomplete denaturation of target DNA	When dot or slot blotting, use the double denaturation methods described in *UNIT 2.9B*, or blot onto positively charged nylon.
	Blocking agents interfering with the target-probe interaction	If using a nylon membrane, leave the blocking agents out of the hybridization solution.
	Final wash was too stringent	Use a lower temperature or higher salt concentration. If necessary, estimate T_m as described in *UNIT 6.4*.
	Hybridization temperature too low with an RNA probe	Increase hybridization temperature to 65°C in the presence of formamide (see Alternate Protocol).
	Hybridization time too short	If using formamide with a DNA probe, increase the hybridization time to 24 hr.
	Inappropriate membrane	Check the target molecules are not too short to be retained efficiently by the membrane type (see Table 2.9.1).

continued

Table 2.10.4 Troubleshooting Guide for DNA Blotting and Hybridization Analysis[a], continued

Problem	Possible cause[b]	Solution
	Problems with electroblotting	Make sure no bubbles are trapped in the filter-paper stack. Soak the filter papers thoroughly in TBE before assembling the blot. Used uncharged rather than charged nylon.
Spotty background	Unincorporated nucleotides not removed from labeled probe	Follow protocols described in *UNIT 3.4*.
	Particles in the hybridization buffer	Filter the relevant solution(s).
	Agarose dried on the membrane	Rinse membrane in 2× SSC after blotting.
	Baking or UV crosslinking when membrane contains high salt	Rinse membrane in 2× SSC after blotting.
Patchy or generally high background	Insufficient blocking agents	See text for of discussion of extra/alternative blocking agents.
	Part of the membrane was allowed to dry out during hybridization or washing	Avoid by increasing the volume of solutions if necessary.
	Membranes adhered during hybridization or washing	Do not hybridize too many membranes at once (ten minigel blots for a hybridization tube, two for a bag is maximum).
	Bubbles in a hybridization bag	If using a bag, fill completely so there are no bubbles.
	Walls of hybridization bag collapsed on to membrane	Use a stiff plastic bag; increase volume of hybridization solution.
	Not enough wash solution	Increase volume of wash solution to 2 ml/10 cm^2 of membrane.
	Hybridization temperature too low with an RNA probe	Increase hybridization temperature to 65°C in the presence of formamide (see Alternate Protocol).
	Formamide needs to be deionized	Although commercial formamide is usually satisfactory, background may be reduced by deionizing immediately before use.
	Labeled probe molecules are too short	1. Use a ^{32}P-labeled probe as soon as possible after labeling, as radiolysis can result in fragmentation. 2. Reduce amount of DNase I used in nick translation (*UNIT 3.5*).
	Probe concentration too high	Check that the correct amount of DNA has been used in the labeling reaction.
	Inadequate prehybridization	Prehybridize for at least 3 hr with nitrocellulose or 15 min for nylon.
	Probe not denatured	Denature as described in the protocols.
	Inappropriate membrane type	If using a nonradiocative label, check that the membrane is compatible with the detection system.
	Hybridization with dextran sulfate	Dextran sulfate sometimes causes background hybridization. Place the membrane between Schleicher and Schuell no. 589 WH paper during hybridization, and increase volume of hybridization solution (including dextran sulfate) by 2.5%.

continued

Table 2.10.4 Troubleshooting Guide for DNA Blotting and Hybridization Analysis[a], continued

Problem	Possible cause[b]	Solution
	Not enough SDS in wash solutions	Check the solutions are made up correctly.
Extra bands	Final wash was not stringent enough	Use a higher temperature or lower salt concentration. If necessary, estimate T_m as described in *UNIT 6.4*.
	Probe contains nonspecific sequences (e.g., vector DNA)	Purify shortest fragment that contains the desired sequence.
	Target DNA is not completely restriction digested	Check the restriction digestion (*UNIT 3.1*).
	Formamide not used with an RNA probe	RNA-DNA hybrids are relatively strong but are destabilized if formamide is used in the hybridization solution.
Nonspecific background in one or more tracks	Probe is contaminated with genomic DNA	Check purification of probe DNA. The problem is more severe when probes are labeled by random printing. Change to nick translation.
	Insufficient blocking agents	See text for of discussion of extra/alternative blocking agents.
	Final wash did not approach the desired stringency	Use a higher temperature or lower salt concentration. If necessary, estimate T_m as described in *UNIT 6.4*.
	Probe too short	Sometimes a problem with probes labeled by random priming. Change to nick translation.
Cannot remove probe after hybridization	Membrane dried out after hybridization	Make sure the membrane is stored moist between hybridization and stripping.
Decrease in signal intensity when reprobed	Poor retention of target DNA during probe stripping	1. Check calibration of UV source if cross-linking on nylon. 2. Use a less harsh stripping method (support protocol).

[a]Based on Dyson (1991).

[b]Within each category, possible causes are listed in decreasing order of likelihood.

5. Pour out the APH solution and add an equal volume of 2× SSC/0.1% SDS. Incubate with rotation 10 min at room temperature. Change the wash solution after 5 min.

   To reduce background, it may be beneficial to increase the volume of the wash solutions by 100%.

6. Replace the wash solution with an equal volume of 0.2× SSC/0.1% SDS and incubate with rotation 10 min at room temperature. Change the wash solution after 5 min (low-stringency wash).

7. If desired, carry out two 15-min moderate-stringency washes using 42°C 0.2× SSC/0.1% SDS.

8. If desired, carry out two 15-min high-stringency washes using 68°C 0.1× SSC/0.1% SDS.

9. Pour off the final wash solution, rinse the membrane in 2× SSC at room temperature, and blot excess liquid. Wrap in plastic wrap. Autoradiograph.

   Do not allow the membrane to dry out if it is to be reprobed. See Table 2.10.4 for troubleshooting.

HYBRIDIZATION ANALYSIS OF A DNA BLOT WITH A RADIOLABELED RNA PROBE

ALTERNATE PROTOCOL

2.10

Purified RNA polymerases from bacteriophages such as SP6, T3, and T7 (*UNIT 3.8*) are very efficient at synthesizing RNA in vitro from DNA sequences cloned downstream of the appropriate phage promoter (Little and Jackson, 1987). If a radiolabeled ribonucleotide is added to the reaction mixture, the polymerase synthesizes several micrograms of uniformly labeled single-stranded RNA with specific activities $\geq 10^9$ dpm/μg. The hybridization procedure is suitable for both nitrocellulose and nylon membranes, though backgrounds may be higher with nylon.

Additional Materials *(also see Basic Protocol; see APPENDIX 1 for items with ✓)*

- ✓ TE buffer, pH 8.0
- ✓ Labeling buffer
- ✓ Nucleotide mix
- ✓ 200 mM dithiothreitol (DTT), freshly prepared
- 20 U/μl human placental ribonuclease inhibitor
- [α-^{32}P]UTP, 20 mCi/ml (800 Ci/mmol) or 10 mCi/ml (400 Ci/mmol)
- SP6 or T7 RNA polymerase (*UNIT 3.8*)
- ✓ RNase-free DNase I
- ✓ 0.25 M EDTA, pH 8.0
- ✓ Formamide prehybridization/hybridization (FPH) solution
- ✓ 2× SSC containing 25 μg/ml RNase A + 10 U/ml RNase T1 (*UNIT 3.13*)

1. Digest cloned DNA (see Table 2.10.5 for suitable vectors) for the sequence to be transcribed with a restriction endonuclease to linearize and introduce an endpoint for RNA synthesis. Purify the DNA by phenol extraction and ethanol precipitation and resuspend in TE buffer, pH 8.0, at a concentration of 1 mg/ml.

2. Mix the following at room temperature:

   4 μl labeling buffer
   1.5 μl nucleotide mix
   1 μl 200 mM DTT
   1 μl (20 U) human placental ribonuclease inhibitor
   2 μg purified plasmid DNA from step 1
   100 to 200 μCi [α-^{32}P]UTP
   H_2O to a final volume of 20 μl.

Table 2.10.5 Selection of Cloning Vectors Incorporating Promoters for Bacteriophage RNA Polymerases

Vector	Size (bp)	Markers[a]	Promoters
pBluescript	2950	amp, *lacZ′*	T3, T7
pGEM series	2746-3223	amp, *lacZ′*	SP6, T7
pGEMEX-1	4200	amp	SP6, T3, T7
pSELECT-1	3422	tet, *lacZ′*	SP6, T7
pSP18, 19, 64, 65	2999-3010	amp	SP6
pSP70, 71, 72, 73	2417-2464	amp	SP6, T7
pSPORT1	4109	amp, *lacZ′*	SP6, T7
pT3/T7 series	2700, 2950	amp, *lacZ′*	T3, T7
pWE15	8800	amp, neo	T3, T7
pWE16	8800	amp, dhfr	T3, T7

[a]Abbreviations: amp, ampicillin resistance; dhfr, dihydrofolate reductase; *lacZ′*, β-galactosidase α-peptide; neo, neomycin phosphotransferase (kanamycin resistance); tet, tetracycline resistance.

3. Add 5 U of SP6 or T7 RNA polymerase. Incubate 1 hr at 40°C for SP6 polymerase or at 37°C for T7 polymerase.
4. Add 2 U RNase-free DNase I and incubate 10 min at 37°C. Stop the reaction by adding 2 μl of 0.25 M EDTA, pH 8.0.
5. Measure the specific activity of the RNA by acid precipitation and remove unincorporated nucleotides by the spin-column procedure *UNIT 3.4*. Store labeled probe up to 2 days at −20°C.

   The specific activity should be at least 7×10^8 dpm/μg, preferably $>10^9$ dpm/μg.

6. Prehybridize in FPH solution and incubate 3 hr at 42°C.
7. Replace the FPH solution with an equal volume of fresh prewarmed solution. Add the labeled probe to a concentration of 1 to 5 ng/ml and incubate overnight with rotation at 42°C.
8. Carry out two 10-min low-stringency washes in 2× SSC/0.1% at room temperature.
9. Replace the wash solution with an equal volume of 2× SSC containing 25 μg/ml RNase A + 10 U/ml RNase T1; incubate with rotation for 30 min at room temperature.
10. Carry out moderate- and high-stringency washes as desired, rinse the membrane in 2× SSC, and set up autoradiography.

SUPPORT PROTOCOL

REMOVAL OF PROBES FROM HYBRIDIZED MEMBRANES

If the DNA has been immobilized on the membrane by UV crosslinking (for uncharged nylon membranes) or by alkaline transfer (for positively charged nylon), the covalent matrix-target DNA interaction is much stronger than the hydrogen-bonded target-probe interaction, so it is possible to remove (or "strip off") the hybridized probe.

Additional Materials *(also see Basic Protocol; see APPENDIX 1 for items with ✓)*

✓ Mild stripping solution
✓ Moderate stripping solution
✓ 0.4 M NaOH
0.1% (w/v) SDS, 100°C

1a. *Mild treatment:* Wash the membrane in several hundred milliliters of mild stripping solution for 2 hr at 65°C.

1b. *Moderate treatment:* Wash the membrane in 0.4 M NaOH for 30 min at 45°C. Then rinse twice in several hundred milliliters of moderate stripping solution for 10 min at room temperature.

1c. *Harsh treatment:* Pour several hundred milliliters of boiling 0.1% SDS onto the membrane. Cool to room temperature.

If a membrane is to be reprobed, it must not be allowed to dry out between hybridization and stripping.

2. Place membrane on a sheet of dry Whatman 3MM filter paper and blot excess liquid with a second sheet. Wrap the membrane in plastic wrap and set up autoradiography.

   If signal is still seen after autoradiography, rewash using harsher conditions.

3. The membrane can now be rehybridized or dried and stored at room temperature for later use.

Reference: Dyson, N.J. 1991.

Contributor: Terry Brown

Purification of Oligonucleotides Using Denaturing Polyacrylamide Gel Electrophoresis

UNIT 2.11

BASIC PROTOCOL

This method is useful for purifying oligonucleotides because of its speed, simplicity, and high resolution. Although yields tend to be low (<50% of applied sample), the amount of material recovered is usually far in excess of that required for most molecular biology applications. This procedure is also useful for isolating small RNAs or other single-stranded polynucleotides.

Materials *(see APPENDIX 1 for items with ✓)*

Concentrated ammonium hydroxide
✓ 10× and 1× TBE electrophoresis buffer, pH 8
✓ 40% acrylamide/2% bisacrylamide
TEMED
Urea
✓ 10% ammonium persulfate
✓ 2× formamide loading buffer
✓ 0.3 M sodium acetate, pH 7.5
✓ TE buffer (optional)
Thin-layer chromatography (TLC) plate with fluorescent indicator (e.g., Silica Gel F-254 or IB-F)

1. Elute the synthesized oligonucleotides from the controlled-pore glass columns with concentrated ammonium hydroxide. Heat >5 hr at 55°C.

   The elution step is normally performed by the automated DNA synthesizer. The synthesized oligonucleotides used in this procedure should not contain a 5′ trityl group.

2. Transfer the sample to microcentrifuge tubes, lyophilize in a Speedvac evaporator, and resuspend the pellet in 0.2 ml water.

3. Assemble gel casting apparatus (*UNIT 2.7*). Prepare appropriate gel solution (Table 2.11.1). For a gel of 10 cm × 16 cm × 1.6 mm, mix the following in a side-arm flask:

   25.2 g urea (final concentration 7 M)
   6 ml 10 × TBE electrophoresis buffer
   40% acrylamide/2% bisacrylamide solution (desired amount)
   H_2O to 60 ml.

   Degas 10 min and add 200 µl of 10% ammonium persulfate and 30 µl TEMED. Mix and pour the gel. Allow the gel to polymerize >30 min at room temperature.

4. Remove the comb and attach the plates to the electrophoresis tank. Fill the top and bottom troughs with 1× TBE buffer.

2.11

Table 2.11.1 Concentrations of Acrylamide Giving Optimum Resolution of DNA Fragments Using Denaturing PAGE[a]

Acrylamide (%)	Fragment sizes separated (bases)
20 to 30	2 to 8
15 to 20	8 to 25
13.5 to 15	25 to 35
10 to 13.5	35 to 45
8 to 10	45 to 70
6 to 8	70 to 300

[a]Data from Maniatis et al., 1975.

5. Dilute the samples 2-fold with 2× formamide loading buffer prior to loading. Up to 25% of a 0.2 μmol synthesis can be loaded per 2-cm × 1.6-mm well; ~10 μg is required to cast a clear shadow (50 μl of resuspended sample + 50 μl of 2× formamide loading buffer = 100 μl per lane).

   Depending on the degree of secondary structure in the sequence being isolated, it may be useful to heat the sample to 90°C for 3 min prior to loading.

6. Thoroughly rinse the wells by pipetting TBE buffer up and down several times immediately prior to loading.

7. Load the sample. Run the gel at ~5 V/cm^2 until the bromphenol blue dye reaches the bottom of the gel.

   Some heating of the gel is desirable to aid in denaturing the sample. For oligonucleotides ≤30 bases, it may be necessary to omit bromphenol blue from the 2× loading buffer so that the dye will not obscure the bands (i.e., load a separate lane with dye solution).

8. Remove plates from the tank and place them horizontally. Pry off the top plate, cover the gel with plastic wrap, and invert plate. Peel a corner of the gel away from the plate (using a spatula) and onto the plastic wrap atop the gel. Cover the gel with another piece of plastic wrap.

9. Place gel on a TLC plate with a fluorescent indicator. Visualize the bands by briefly shadowing with a shortwave UV lamp; the bands will appear as dark smudges on a greenish background.

   The optimum band is generally the heaviest band on the gel (excluding the material that runs at the dye front); it should also be the slowest-moving band. If deprotection has been inefficient, however, a lighter band containing oligonucleotides with protecting groups will migrate considerably above the major band.

10. Cut out bands directly with a clean scalpel or mark them with ink and cut them out. Peel away plastic wrap and transfer the slices to microcentrifuge tubes.

11. Add 0.5 ml of 0.3 M sodium acetate per 2-cm gel slice. Elute overnight at room temperature on a rotary shaker.

12. Transfer the solution to clean microcentrifuge tubes, avoiding acrylamide fragments. Phenol extract once with an equal volume of phenol, chloroform extract once, and ethanol precipitate (*UNIT 2.1*). Dry and resuspend the samples in TE buffer or water.

Reference: Applied Biosystems, 1984.

Contributor: Andrew Ellington

Enzymatic Manipulation of DNA and RNA

3

Many of the revolutionary changes that have occurred in the biological sciences over the past 15 years can be directly attributed to the ability to manipulate DNA in defined ways. The major tools for this genetic engineering are the enzymes that catalyze specific reactions on DNA molecules. This chapter reviews the properties of the principal enzymes that are critical for carrying out most of the important reactions involved in recombinant DNA technology. In addition, it describes protocols for many of the basic techniques such as restriction mapping, radioactive labeling of nucleic acids, and construction of hybrid DNA molecules (cloning).

From a historical perspective, the discovery of restriction enzymes that cleave DNA at discrete nucleotide sequences was probably the breakthrough that ushered in the rest of the technology. First, the cleavage sites provide specific landmarks for obtaining a physical map of the DNA. Second, the ability to produce specific DNA fragments by cleavage with restriction enzymes makes it possible to purify these fragments by molecular cloning. Third, DNA fragments generated by restriction endonuclease treatment are basic substrates for the wide variety of enzymatic manipulations of DNA that are now possible.

This chapter begins with a protocol for cleaving DNA with restriction endonucleases. In addition to the basic reaction, it includes methods for cleavage with multiple enzymes, for partial digestion of DNA, and for analysis of multiple samples. This section also contains basic information about all the enzymes that are commercially available. For each enzyme the recognition sequence, type of termini that are produced upon cleavage, buffer conditions, and conditions for thermal inactivation are described.

Two different methods for obtaining structural maps of DNA are then presented. The most common method of restriction mapping involves digestion with multiple enzymes. By analyzing the products of different cleavage reactions by gel electrophoresis (see *UNIT 2.5*), the map is deduced. The second method, partial digestion of end-labeled DNA, is valuable for mapping restriction sites that appear frequently in a given DNA region or for mapping small segments of DNA.

The third major section of this chapter describes the properties and reaction conditions for other enzymes that are used to manipulate DNA molecules. These include the following: DNA polymerases which synthesize DNA from double-stranded or primed, single-stranded templates; exonucleases which degrade DNA stepwise from the 5′ and/or 3′ ends; ligases which join double-stranded segments of DNA; RNA polymerases which synthesize RNA from double-stranded DNA templates; phosphatases which remove 5′ terminal phosphate residues from nucleic acids; kinases which phosphorylate 5′ terminal hydroxyl residues; and several other enzymes that have more specialized applications. The major uses of these enzymes are described, as are some specific protocols for radioactively labeling RNA or DNA.

General techniques for constructing hybrid DNA molecules are presented in the fourth section. Because there are numerous tricks of the trade, this section deals with the general principles for designing the best strategy for any particular cloning experiment. Some specific examples are discussed in more detail, e.g., subcloning restriction fragments via cohesive or blunt ends, directional cloning, joining DNA fragments with incompatible ends, and ligations involving oligonucleotide linkers (*UNIT 3.16*). The polymerase chain reaction (PCR) provides additional versatility for constructing recombinant DNA molecules. *UNIT 3.17* gives a general approach and discusses particular scenarios for incorporating specific sequences onto the ends of DNA fragments, for creating in-frame fusion proteins, and for creating deletions and insertions using inverse PCR.

A final section on specialized applications describes the labeling and various means of detecting nonisotopically labeled probes. Biotinylated and digoxygenin-labeled probes are becoming more widely used in place of their radioactive counterparts for the hybridization applications described in *UNITS 2.9, 4.9 & 14.3*. Aside from eliminating the concerns of working with hazardous materials, the use of these probes offers increased stability and reasonable sensitivity to colorimetric (*UNIT 3.18*) or chemiluminescent (*UNIT 3.19*) detection techniques.

Kevin Struhl

UNIT 3.1

Digestion of DNA with Restriction Endonucleases

Restriction endonucleases recognize short DNA sequences and cleave double-stranded DNA at specific sites within or adjacent to the recognition sequences. Restriction endonuclease cleavage of DNA into discrete fragments is one of the most basic procedures in molecular biology. Tables 3.1.1 to 3.1.3 describe restriction endonucleases and their properties.

A number of factors affect restriction endonuclease reactions. Contaminants found in some DNA preparations (e.g., protein, phenol, chloroform, ethanol, EDTA, SDS, high salt concentration) may inhibit restriction endonuclease activity. The effects of contaminants may be overcome by increasing the number of enzyme units added to the reaction mixture (up to 10 to 20 U per microgram DNA), increasing the reaction volume to dilute potential inhibitors, or increasing the duration of incubation. Digestion of genomic DNA (prepared as in *UNITS 2.2-2.4*) can be facilitated by the addition of the polycation spermidine (final concentration 1 to 2.5 mM), which acts by binding negatively charged contaminants. Larger amounts (up to 20-fold more) of some enzymes are necessary to cleave supercoiled plasmid or viral DNA as compared to the amount needed to cleave linear DNA. In addition, some enzymes cleave their defined sites with different efficiencies, presumably due to differences in flanking nucleotides. Some restriction endonucleases are inhibited by methylation of nucleotides within their recognition sequences (Table 3.1.3).

The typical restriction endonuclease buffer contains magnesium chloride, sodium or potassium chloride, Tris·Cl, 2-mercaptoethanol (2-ME) or dithiothreitol (DTT), and bovine serum albumin (BSA). A divalent cation, usually Mg^{2+}, is an absolute requirement for enzyme activity. Some restriction endonucleases are very sensitive to the concentration of sodium or potassium ion, while others are active over a wide range of ionic strengths (Table 3.1.2].

DIGESTING A SINGLE DNA SAMPLE WITH A SINGLE RESTRICTION ENDONUCLEASE

BASIC PROTOCOL

Restriction endonuclease cleavage is accomplished simply by incubating the enzyme(s) with the DNA in appropriate reaction conditions. The amounts of enzyme and DNA, the buffer and ionic concentrations, and the temperature and duration of the reaction will vary depending upon the specific application.

Materials *(see APPENDIX 1 for items with ✓)*

DNA sample in H_2O or TE buffer
✓ 10× restriction endonuclease buffers
Restriction endonucleases (Table 3.1.1)
✓ Loading buffer
✓ 0.5 M EDTA, pH 8.0 (optional)

1. Pipet the following into a clean microcentrifuge tube:

 x µl DNA (0.1 to 4 µg DNA in H_2O or TE buffer)
 2 µl 10× restriction buffer (Table 3.1.2)
 $18 - x$ µl H_2O.

 A 20-µl reaction is convenient for analysis by electrophoresis in polyacrylamide or agarose gels. The amount of DNA to be cleaved and/or the reaction volume can be increased or decreased provided the proportions of the components remain constant.

2. Add restriction endonuclease (1 to 5 U/µg DNA) and incubate the reaction mixture 1 hr at the recommended temperature (in general, 37°C).

 In principle, 1 U restriction endonuclease completely digests 1 µg of purified DNA in 60 min using the recommended assay conditions. The volume of restriction endonuclease added should be less than 1⁄10 the volume of the final reaction mixture, because glycerol in the enzyme storage buffer may interfere with the reaction.

3. Stop the reaction and prepare it for agarose or acrylamide gel electrophoresis (*UNIT 2.5* or *UNIT 2.7*) by adding 5 µl (20% of reaction vol) gel loading buffer.

 The reaction can also be stopped by chelating Mg^{2+} with 0.5 µl of 0.5 M EDTA (12.5 mM final concentration). Many enzymes can be irreversibly inactivated by incubating 10 min at 65°C; some enzymes that are partially or completely resistant to heat inactivation at 65°C may be inactivated by incubating 15 min at 75°C. When the enzyme(s) is completely resistant to heat inactivation, DNA may be purified from the reaction mixture by extraction with phenol and precipitation in ethanol (UNIT 2.1) or by using a silica matrix suspension (UNIT 2.1).

DIGESTING DNA WITH MULTIPLE RESTRICTION ENDONUCLEASES

ALTERNATE PROTOCOL 1

It is often desirable to cleave a given DNA sample with more than one endonuclease. Two or more enzymes may be added to the same reaction mixture if all are relatively active in the same buffer and at the same temperature. Most restriction endonucleases and some DNA-modifying enzymes are active to some extent in potassium glutamate– and potassium acetate–based buffers. If the reaction conditions needed are too dissimilar, follow the procedure below.

1. Digest the DNA with the enzyme(s) that is active at the lower NaCl concentration (Basic Protocol, steps 1 and 2).

 Many enzymes with optimal activity at high temperatures are also active at 37°C.

Table 3.1.1 Cross Index of Recognition Sequences and Restriction Endonucleases[a,b]

	AATT	ACGT	AGCT	ATAT	CATG	CCGG	CGCG	CTAG	GATC	GCGC	GGCC	GTAC	TATA	TCGA	TGCA	TTAA
↓****									MboI[c], Sau3AI[c]							
*↓***		MaeII				HpaII[c], MspI[c]		BfaI		HinPI		Csp6I		TaqI		MseI
↓			AluI				BstUI		DpnI		HaeIII	RsaI				
***↓*										HhaI						
****↓					NlaIII											
A↓****T	ApoI		HindIII		AflIII	AgeI, BsrFI	MluI, AflIII	SpeI	BglII, BstYI							
A*↓***T														ClaI		AseI
A**↓**T				SspI						Eco47III	StuI	ScaI				
A***↓*T																
A****↓T					Nsp7524				HaeII						NsiI	
C↓****G					NcoI, StyI, DsaI	XmaI, AvaI	DsaI	AvrII, StyI			EagI, EaeI, GdiII	BsiWI	SfcI	XhoI, AvaI	AvaI	SfcI
C*↓***G				NdeI												
C**↓**G		PmlI, BsaAI	PvuII, NspBII			SmaI	NspBII									
C***↓*G							SacII		PvuI							
C****↓G									BsiEI		BsiEI				PstI	
G↓****C	EcoRI, ApoI					NgoMI, BsrFI	BssHII	NheI	BamHI, BstYI	KasI, BanI	Bsp120I	Asp718, BanI		SalI	ApaLI	
G*↓***C		BsaHI								NarI, BsaHI			AccI	AccI		
G**↓**C			Ecl136II	EcoRV		NaeI							Bst1107I	HincII		HpaI, HincII
G***↓*C																
G****↓C		AatII	SacI, BanII, HgiAI, Bsp1286		SphI, Nsp7524					BbeI, HaeII	ApaI, BanII, Bsp1286	KpnI			Bsp1286, HgiAI	
T↓****A					BspHI	BspMII		XbaI	BclI		EaeI					
T*↓***A														BstBI		
T**↓**A		SnaBI, BsaAI					NruI			FspI	MscI					DraI
T***↓*A																
T****↓A																

[a]Reprinted by permission of New England Biolabs.

[b]Sequences at the top of each column are written 5′ to 3′. Asterisks at the left of each row are place holders for nucleotides within a recognition sequence, and arrows indicate the point of cleavage. Sequences of complementary strands and their cleavage sites are implied. Enzymes written in bold type recognize only one sequence, while those in light type have multiple recognition sequences.

[c]Sequence cleaved identically by two or more enzymes that are affected differently by DNA modification at that site (see Table 3.1.1).

2. For enzymes active at higher salt concentrations, add 1 M NaCl (1 to 3 µl for a 20-µl reaction) so that the final concentration is suitable for digestion by the next enzyme(s). Add enzyme(s) for the second reaction and incubate appropriately.

3. Stop the reaction for electrophoretic analysis or further enzymatic treatment (Basic Protocol, step 3).

DIGESTING MULTIPLE SAMPLES OF DNA

ALTERNATE PROTOCOL 2

This procedure minimizes the number of pipetting steps when multiple samples are to be digested with the same enzyme(s) and, hence, saves time and reduces the potential for contamination of the enzyme.

1. For each sample to be tested, add a constant *volume* of DNA to a separate microcentrifuge tube.

   Use a different pipet tip for each DNA sample in order to prevent cross-contamination.

2. Prepare a "premix solution" containing sufficient 10× restriction endonuclease buffer and water for digesting all the samples. Place solution on ice.

3. Add sufficient restriction endonuclease(s) for digesting all the samples. Mix quickly by flicking the tube and replace on ice.

   The solution to which the enzyme is added should not be more concentrated than 3× buffer.

4. Add the appropriate amount of solution containing the restriction endonuclease to each tube of DNA and incubate the reactions 1 hr at the appropriate temperature.

5. Stop the reactions for electrophoretic analysis or further enzyme digestion (Basic Protocol, step 3).

PARTIAL DIGESTION OF DNA WITH RESTRICTION ENDONUCLEASES

ALTERNATE PROTOCOL 3

For some purposes, it is useful to produce DNA that has been cleaved at only a subset of the restriction sites. This is particularly important for cloning segments of DNA in which the site(s) used for cloning is also present internally within the segment, and it is also useful for restriction mapping (*UNIT 3.3*).

1. Make up a 100-µl reaction mixture containing DNA in 1× restriction enzyme buffer.

2. Divide up reaction mixture such that tube 1 contains 30 µl, tubes 2 to 4 contain 20 µl, and tube 5 contains 10 µl. Place tubes on ice.

3. Add the selected restriction endonuclease (3 to 10 U/µg DNA) to tube 1, mix quickly by flicking the tube, and place the tube back on ice.

4. Using a different pipet tip, add 10 µl from tube 1 into tube 2, mix quickly, and place back on ice. Continue the serial dilution process by successively pipetting 10 µl from tube 2 to 3, 3 to 4, and 4 to 5. The final volume in each of the tubes is 20 µl.

3.1

Table 3.1.2 Effect of NaCl Concentration on Restriction Endonuclease Activity[a]

Enzyme	0 mM NaCl	50 mM NaCl	100 mM NaCl	150 mM NaCl	Enzyme	0 mM NaCl	50 mM NaCl	100 mM NaCl	150 mM NaCl
*Aat*II	+	++	++	+	*Eco*RI	[b]	+++	+++	+++
*Acc*I	+++	+++	+	+	*Eco*RV	+	+	+	+++
*Aci*I	+	+++	+++	+++	*Fnu*4HI	+++	+++	++	+
*Afl*II	+	+++	++	++	*Fok*I	+++	+++	+++	+++
*Aha*II	+	++	+++	+++	*Fsp*I	+	+++	++	++
*Alu*I	+	+++	+++	++	*Hae*II	+++	+++	+++	++
*Alw*I	+++	+++	++	+	*Hae*III	+++	+++	+++	+++
*Alw*NI	++	+++	+++	+	*Hga*I	+++	+++	+	+
*Apa*I	+++	+++	++	+	*Hgi*AI	+	+	++	+++
*Apa*LI	+++	+++	++	+	*Hha*I	+	+	+	++
*Asc*I	+	++	++	+	*Hinc*II	++	+++	+++	+++
*Ase*I	+	+++	+++	+++	*Hind*III	++	+++	+++	++
*Ava*I	+++	+++	+++	+++	*Hinf*I	++	+++	+++	+++
*Ava*II	+++	+++	++	+	*Hin*PI	+++	+++	+++	+++
*Avr*II	+++	+++	+++	++	*Hpa*I	+	+++	+++	+
*Bal*I	+++	++	++	+	*Hpa*II	+++	+++	++	+
*Bam*HI	+	++	+++	+++	*Hph*I	+++	+++	+++	+
*Ban*I	+++	+++	++	++	*Kas*I	+	+++	+	+
*Ban*II	+++	+++	+++	+++	*Kpn*I	+++	+	+	+
*Bbv*I	+++	+++	+++	+++	*Mbo*I	++	+++	+++	+++
*Bcg*I	+++	+++	+++	+++	*Mbo*II	+++	+++	+++	+++
*Bcl*I	+	+++	+++	+	*Mlu*I	++	+++	+++	++
*Bgl*I	+	+++	+++	+++	*Mnl*I	+++	+++	+++	++
*Bgl*II	++	+++	+++	+++	*Mse*I	+++	+++	++	+
*Bsa*HI	++	+++	+++	+	*Msp*I	+++	+++	+++	+++
*Bsm*I	+++	+++	+++	+++	*Nae*I	+++	+++	+++	+
*Bsp*1286	+++	+++	++	+	*Nar*I	+++	+++	+	+
*Bsp*HI	++	+++	+++	+++	*Nci*I	+++	+++	++	+
*Bsp*MI	++	++	+++	+++	*Nco*I	+	++	+++	+++
*Bsp*MII	++	++	+++	+++	*Nde*I	+	+	++	+++
*Bss*HII	+++	+++	+++	+++	*Nhe*I	+++	+++	+++	++
*Bst*BI	+++	+++	++	+	*Nla*III	+	+	+	+
*Bst*EII	+	++	+++	+++	*Nla*IV	+	+	+	+
*Bst*NI	++	++	+++	+++	*Not*I	+	+++	+++	+++
*Bst*UI	+++	+++	++	+	*Nru*I	+	+++	+++	+++
*Bst*XI	++	+++	+++	+++	*Nsi*I	++	++	++	++
*Bst*YI	+++	+++	++	++	*Pae*R7I	+++	+++	+++	+
*Bsu*36I	+	++	+++	++	*Pfl*MI	++	+++	+++	++
*Cla*I	+++	+++	+++	++	*Ple*I	+++	+	+	+
*Dde*I	++	+++	+++	+++	*Ppu*MI	+++	+++	++	+
*Dpn*I	+	++	+++	+++	*Pst*I	+++	+++	+++	+++
*Dra*I	++	+++	+	+	*Pvu*I	+	++	+++	+++
*Dra*III	++	+++	+++	++	*Pvu*II	+++	+++	+++	+++
*Eae*I	++	+++	++	+	*Rsa*I	+++	+++	+++	+++
*Eag*I	++	+++	+++	+++	*Rsr*II	+++	++	+	+
*Eam*1105I	+	++	+	+	*Sac*I	+++	++	++	+
*Eco*NI	+++	+++	+++	+++	*Sac*II	+++	+	+	+
*Eco*O109	+++	+++	+++	+++	*Sal*I	+	+	++	+++
*Eco*57I	+++	+++	+++	++	*Sau*3AI	+++	+++	+++	+++

continued

Table 3.1.2 Effect of NaCl Concentration on Restriction Endonuclease Activity[a] continued

Enzyme	0 mM NaCl	50 mM NaCl	100 mM NaCl	150 mM NaCl	Enzyme	0 mM NaCl	50 mM NaCl	100 mM NaCl	150 mM NaCl
*Sau*96I	+++	+++	+++	+++	*Stu*I	+++	+++	+++	+++
*Sca*I	+	+++	+++	++	*Sty*I	+	++	+++	+++
*Scr*FI	++	+++	+++	+++	*Taq*I	+++	+++	+++	++
*Sfa*NI	+	+	+++	+++	*Tth*111I	+++	+++	+++	+
*Sma*I	+	+	+	+	*Xba*I	+	+++	+++	+++
*Sna*BI	+++	+++	++	+	*Xca*I	+++	+	+	+
*Spe*I	++	+++	+++	++	*Xho*I	++	+++	+++	+++
*Sph*I	+	+	+++	+++	*Xma*I	+++	+++	++	+
*Ssp*I	++	+++	+++	++	*Xmn*I	+++	+++	+	+

[a]Reprinted by permission of New England Biolabs. The activity of each enzyme listed is compared at specified NaCl concentrations to its activity in recommended assay bufffer. Recommended assay buffers in some cases differ widely in terms of pH and specific ion requirements. The conditions here varied only the NaCl concentration. All buffers contained 10 mM Tris·Cl (pH 7.5), and 100 µg/ml bovine serum albumin. All incubations were done for 60 min at the optimum temperature for each enzyme. Scoring is as follows:

+ <10% of the activity can be obtained using these conditions compared to the recommended conditions.

++ between 100% and 20% of the activity can be obtained using these conditions compared to the recommended conditions.

+++ between 30% and 100% of the activity can be obtained using these conditions compared to the recommended conditions.

[b]Not recommended because of star activity, which refers to cleavage at sites other than the usual recognition sequence. Star activity occurs under nonoptimal reaction conditions, such as low ionic strength, high endonuclease concentrations, high glycerol concentrations, high pH, and when Mn^{2+} is used in place of Mg^{2+}.

5. Incubate all five tubes for 15 min at the appropriate temperature for the restriction endonuclease and stop the reactions for electrophoretic analysis or further enzymatic treatment (Basic Protocol, step 3).

   By virtue of the serial dilution process, the amount of enzyme per microgram of DNA has been varied over a 54-fold range. The extent of digestion is determined by gel electrophoresis of the samples.

METHYLATION OF DNA

SUPPORT PROTOCOL

A number of commercially available methylases covalently join methyl groups to adenine or cytosine residues within specific target sequences (Table 3.1.3). Methylation of these sites renders them resistant to cleavage by the corresponding restriction endonuclease.

Additional Materials *(also see Basic Protocol; see* APPENDIX 1 *for items with ✓)*

✓ 10× methylase buffer
S-adenosylmethionine (SAM)

1. Set up a reaction (typically 20 µl) containing DNA (at a final concentration of ~20 to 200 µg/ml) in 1× methylase buffer.

2. Add SAM (the methyl donor group) to achieve a final concentration of 80 µM.

3. Add a sufficient amount of methylase to completely protect the DNA from cleavage by the corresponding restriction endonuclease.

   A unit of methylase protects 1 µg bacteriophage λ DNA under the recommended conditions.

4. Incubate the reaction mixture 1 hr at the appropriate temperature (usually 37°C).

Table 3.1.3 Recognition Sequences and Reaction Conditions of Commercially Available Methylases[a]

Methylase	Recognition sequence[b]	NaCl	Tris·Cl[c]	EDTA	ME[d]
		(mM)			
*Alu*I	AGC^{m}T0	50	10	5	
*Bam*HI	GGATC^{m}C	50	10	10	5
*Cla*I	ATCGA^{m}T	0	50	10	5
CpG	C^{m}G	50	10	0.1	1[e]
dam	GA^{m}TC	0	50	10	5
*Eco*RI	GAA^{m}TTC	100	100	1	0
*Fnu*DII	C^{m}GCG	0	50	10	5
*Hae*III	GGC^{m}C	50	50	10	1D
*Hha*I	GC^{m}GC	0	50	10	5
*Hpa*II	CC^{m}GG	0	50	10	5
*Msp*I	C^{m}CGG	100	50	10	5
*Pst*I	CTGCA^{m}G	0	50	10	5
*Taq*I	TCGA^m	100	10	0	6[f]

[a]All reaction mixtures contain 80 μM *S*-adenosylmethionine and are incubated at 37°C.
[b]Superscript m signifies methylated nucleotide.
[c]pH 7.5.
[d]2-mercaptoethanol or dithiothreitol (D).
[e]Reaction buffer includes 160 μM *S*-adenosylmethionine.
[f]Reaction buffer includes 6 mM $MgCl_2$.

References: American Chemical Society, 1995; Fuchs and Blakesly, 1983; McClelland et al., 1988; O'Farrell et al/. 1980; Roberts, 1994.

Contributor: Kenneth D. Bloch

3.2

UNIT 3.2

Mapping by Multiple Endonuclease Digestions

BASIC PROTOCOL

Materials *(see APPENDIX 1 for items with ✓)*

✓ Restriction endonucleases and 10× buffers

1. Cleave the DNA (up to 20 kb in length) completely in separate tubes with different restriction endonucleases that cleave infrequently (e.g., *Eco*RI, *Hin*dIII, *Bam*HI, *Pst*I, *Kpn*I, *Xba*I, *Sal*II, and *Xho*I).

2. Fractionate an aliquot of each reaction mixture by gel electrophoresis (*UNITS 2.5A* or *2.7*). Include known DNA size standards on the gel. Store the remainder of the reaction mixture on ice.

3. Calculate the length of the fragments by comparison to DNA molecular weight markers. Determine the number of cleavage sites for each restriction endonuclease.

4. For each sample from step 1, transfer an aliquot into separate tubes and cleave with a second restriction endonuclease. After agarose gel electrophoresis, compare DNA fragments resulting from the digests with (a) two enzymes, (b) the first enzyme alone, and (c) the second enzyme alone.

5. Calculate the length of the DNA fragments. Continue this procedure until an unambiguous, internally consistent map of restriction endonuclease cleavage sites can be determined.

Contributor: Kenneth D. Bloch

Mapping by Partial Endonuclease Digestions

UNIT 3.3

BASIC PROTOCOL

3.3

Restriction maps produced by this method can "miss" a site, because recognition sites may not be cleaved with equal efficiency. Confirm the sites with a standard restriction digest using unlabeled DNA.

Materials *(see APPENDIX 1 for items with ✓)*

✓ Restriction endonucleases and 10× buffers

IMPORTANT NOTE: Enzymes should always be stored at −20°C in freezers that maintain a constant temperature. Do not let enzymes stay for extended periods on ice and do not expose them to temperatures greater than 0°C.

1. Cleave the DNA fragment with a restriction endonuclease that cleaves infrequently (e.g., *Eco*RI, *Hin*dIII, *Bam*HI, *Pst*I, *Kpn*I, *Xba*I, *Sal*I, and *Xho*I).

2. "End label" the products of digestion with ^{32}P. 5′ termini can be labeled by successive treatment with calf intestine alkaline phosphatase and T4 polynucleotide kinase (*UNIT 3.10*), and 3′ termini can be labeled with the Klenow fragment of *E. coli* DNA polymerase I (*UNIT 3.5*) or with T4 DNA polymerase (*UNIT 3.5*).

3. Cleave the fragments with a second restriction endonuclease. Separate the products by gel electrophoresis (*UNITS 2.5A* or *2.7*). Purify by electrophoresis only those DNA fragments that are radiolabeled at one end.

   Only the fragments cleaved with both restriction endonucleases will be labeled at one end. Fragments from step 1 that are not cleaved by the second enzyme retain the label at both ends and are useless for subsequent analysis.

4. Partially digest an isolated DNA fragment with a restriction enzyme that cleaves relatively frequently by serially diluting the enzyme (*UNIT 3.1*). Separate products by gel electrophoresis and visualize by autoradiography.

5. Calculate the length of the fragments by comparison with radiolabeled DNA molecular weight markers.

   MspI-cleaved pBR322 DNA (UNIT 1.5) that has been end labeled is an excellent molecular weight marker for fragments <650 bp.

Reference: Danna, 1980.

Contributor: Kenneth D. Bloch

Reagents and Radioisotopes Used to Manipulate Nucleic Acids

STOCK SOLUTIONS

Solutions should be prepared in deionized water and stored in aliquots. Most of the solutions are stable for years at −20°C or at room temperature if kept sterile. Dithiothreitol should be stored at −20°C and is stable for months. DTT should not be autoclaved, and it should be kept at 0°C during use. (See *APPENDIX 1* for all items with ✓.)

✓ Acid precipitation solution
10 mg/ml autoclaved gelatin
✓ 10 mg/ml bovine serum albumin (BSA), Pentax Fraction V
Buffered phenol (*UNIT 2.1*)
✓ 0.1 M dithiothreitol (DTT)
10 mg/ml *E. coli* tRNA
✓ 0.5 M EDTA, pH 8.0
Ethanol
Glycerol
✓ 1 M KCl
✓ 0.2 M $MgCl_2$
✓ 5 M and 1 M NaCl
✓ 10 mg/ml and 500 μg/ml sonicated salmon or herring sperm DNA
✓ Standard enzyme diluent (SED)
✓ TE buffer
✓ 1 M Tris·Cl, pH 7.5
✓ 1 M Tris·Cl, pH 8.0

10× ENZYME BUFFERS

Nucleoside triphosphates should not be included in the 10× or 5× buffers because the high concentration of Mg^{2+} ions will lead to the formation of insoluble complexes. Bovine serum albumin (BSA) or gelatin is not essential for activity, but it helps stabilize the enzymes. See also Table 3.4.1 for a listing of 1× reaction conditions.

Bal 31 nuclease (5×)
0.25 M Tris·Cl, pH 8.0
50 mM $MgCl_2$
50 mM $CaCl_2$
3 M NaCl
0.25 mg/ml BSA or gelatin

BAP (bacterial alkaline phosphatase)
0.5 M Tris·Cl, pH 8.0
10 mM $ZnCl_2$

CIP (calf intestine alkaline phosphatase)
0.2 M Tris·Cl, pH 8.0
10 mM $MgCl_2$
10 mM $ZnCl_2$
0.5 mg/ml BSA or gelatin

DNase I (deoxyribonuclease I)
0.5 M Tris·Cl, pH 7.5
0.1 M $MgCl_2$ (single-strand breaks; or 0.1 M $MnCl_2$, double-strand breaks)
0.5 mg/ml BSA

E. coli DNA ligase
400 mM Tris·Cl, pH 8
0.1 M $MgCl_2$
50 mM DTT
0.5 mg/ml BSA

E. coli DNA polymerase I or Klenow fragment
0.5 M Tris·Cl, pH 7.5
0.1 M $MgCl_2$
10 mM DTT
0.5 mg/ml BSA or gelatin

E. coli RNA polymerase
0.4 M Tris·Cl, pH 8.0
0.1 M $MgCl_2$
50 mM DTT
0.5 M KCl
0.5 mg/ml BSA or gelatin

Exonuclease III (exo III)
0.5 M Tris·Cl, pH 7.5
50 mM $MgCl_2$
50 mM DTT
0.5 mg/ml BSA or gelatin

Exonuclease VII (exo VII)
- 0.7 M Tris·Cl, pH 8.0
- 80 mM EDTA
- 0.1 M 2-ME
- 0.5 mg/ml BSA or gelatin

Klenow fragment (see E. coli DNA polymerase I)

λ exonuclease
- 0.7 M glycine·KOH, pH 9.4
- 25 mM $MgCl_2$
- 0.5 mg/ml BSA or gelatin

Mung bean nuclease
- 0.3 M sodium acetate, pH 5.0
- 0.5 M NaCl
- 10 mM zinc acetate
- 0.5 mg/ml BSA or gelatin

Poly(A) polymerase
- 0.4 M Tris·Cl, pH 8.0
- 0.1 M $MgCl_2$
- 25 mM $MnCl_2$
- 2.5 M NaCl
- 0.5 mg/ml BSA

Reverse transcriptase
- 0.5 M Tris·Cl, pH 8.2
- 50 mM $MgCl_2$
- 50 mM DTT
- 0.5 M KCl
- 0.5 mg/ml BSA or gelatin

RNase H (ribonuclease H)
- 0.2 M HEPES·KOH, pH 8.0
- 0.5 M KCl
- 40 mM $MgCl_2$
- 10 mM DTT
- 0.5 mg/ml BSA or gelatin

S1 nuclease
- 0.5 M sodium acetate, pH 4.5
- 10 mM zinc acetate
- 2.5 M NaCl
- 0.5 mg/ml BSA or gelatin

Sequenase (see T7 DNA polymerase)

SP6 RNA polymerase
- 0.4 M Tris·Cl, pH 7.5
- 0.1 M $MgCl_2$
- 50 mM DTT
- 0.5 mg/ml BSA or gelatin

T3 RNA polymerase (see SP6 polymerase)

T4 DNA ligase
- 0.5 M Tris·Cl, pH 7.5
- 50 mM $MgCl_2$
- 50 mM DTT
- 0.5 mg/ml BSA or gelatin

T4 DNA polymerase
- 0.5 M Tris·Cl, pH 8.0
- 50 mM $MgCl_2$
- 50 mM DTT
- 0.5 mg/ml BSA or gelatin

T4 polynucleotide kinase
- 0.5 M Tris·Cl, pH 7.5 (forward reaction; or 0.5 M imidazole·Cl, pH 6.6, exchange reaction)
- 0.1 M $MgCl_2$
- 50 mM DTT
- 0.5 mg/ml BSA or gelatin

T4 RNA ligase
- 0.5 M HEPES, pH 8.3
- 0.1 M $MgCl_2$
- 50 mM DTT
- 0.5 mg/ml BSA

T7 DNA polymerase (native and modified)
- 0.4 M Tris·Cl, pH 8.0
- 0.1 M $MgCl_2$ (native or chemically modified; or 50 mM $MgCl_2$, genetically modified)
- 50 mM DTT
- 0.5 M NaCl
- 0.5 mg/ml BSA or gelatin

T7 gene 6 exonuclease (see exonuclease III)

T7 RNA polymerase (see SP6 polymerase)

Taq DNA polymerase
- 0.1 M Tris·Cl, pH 8.4
- x $MgCl_2$ (see *UNIT 15.1*)
- 500 mM KCl
- 1 mg/ml gelatin

Terminal transferase
- 1 M sodium cacodylate, pH 7.0
- 10 mM $CoCl_2$
- 1 mM DTT
- 0.5 mg/ml BSA or gelatin

ENZYME REACTION CONDITIONS AND APPLICATIONS

General guidelines for enzyme reaction conditions are provided in Table 3.4.1. Because many parameters vary depending on the application, *UNITS 3.5-3.15,* which describe specific classes of enzymes, should be consulted for further details.

Some applications of nucleic acid–modifying enzymes are indicated in Table 3.4.2. For some applications, more than one enzyme can be used to carry out the reaction. More detailed information is provided in *UNITS 3.5-3.15.*

3.4

NUCLEOSIDE TRIPHOSPHATES

Nucleoside triphosphates (HPLC-purified) have a limited shelf life in solution and should be stored as aliquots at –20°C, where they are stable for up to 1 year. Deoxyribonucleoside triphosphates (dNTPs) or ribonucleoside triphosphates (NTPs) can be purchased (Pharmacia) as ready-made 100 mM solutions. Alternatively, they can be purchased in lyophilized form and prepared in deionized water as follows:

1. Dissolve in water to ~30 mM and adjust to pH 7 with 1 M NaOH. Determine actual concentration spectrophotometrically, using the extinction coefficients given in Table 3.4.3 (see also *APPENDIX 3*).

   dNTPs and NTPs will undergo acid catalyzed hydrolysis unless they are neutralized.

2. Prepare 5 mM working solutions for each NTP and dNTP from concentrated stocks of nucleoside triphosphates.

3. Prepare NTP and dNTP solutions (mixes) containing equimolar amounts of all four RNA or DNA precursors as follows:

   5 mM 4NTP mix: 5 mM each of ATP, UTP, CTP, GTP
   5 mM 4dNTP mix: 5 mM each of dATP, dTTP, dCTP, dGTP
   0.5 mM 4dNTP mix: 0.5 mM each of dATP, dTTP, dCTP, dGTP

4. For radioactive labeling purposes, prepare stocks lacking one particular NTP or dNTP but containing equimolar amounts of the remaining three precursors such as:

   5 mM 3NTP mix (minus UTP): 5 mM of ATP, GTP, CTP for radiolabeling RNA with radioactive UTP

   5 mM and 0.5 mM 3dNTP mixes (minus dATP): 5 mM and 0.5 mM each of dTTP, dCTP, dGTP for radiolabeling DNA with radioactive ATP.

   Other 3NTP or 3dNTP mixes are made by omitting the precursor corresponding to the selected radiolabel.

RADIOISOTOPES FOR LABELING NUCLEIC ACIDS

^{32}P-labeled NTP or dNTP precursors can be purchased at various specific activities. In general, α-labeled precursors are purchased at 400 to 800 Ci/mmol, whereas γ-labeled precursors are purchased at 3000 to 7000 Ci/mmol. ^{32}P is preferred for preparation of highly radioactive probes and for most autoradiographic procedures.

Table 3.4.1 General Guidelines for Enzyme Reaction Conditions

Many parameters, including reaction volume, dNTP concentration, time, and temperature, will vary depending on application. Dilute appropriate 10× or 5× enzyme buffer (see recipes preceding this section) to 1× and set up reactions as described in table. Stop reactions at 75°C for 10 min, or by adding 2 μl of 0.5 M EDTA (except where indicated otherwise under "Comments"). Enzymes are described in further detail in the units indicated in parentheses.

Enzyme	Rxn. vol. (μl)	Amt. DNA (μg)	Amt. enz. (U)	dNTPs (mM each)	Rxn. temp. (°C)	Time (min)	Comments
DNA Ligases							
E. coli DNA ligase (3.14)	50	1	10	100 NAD	10-25	2-16 hr	
T4 DNA ligase (3.14)	50	1	1	500 ATP	12-30	1-16	
DNA Polymerases							
E. coli DNA polymerase I (3.5)	25	1	3	20	20-37	15-30	*a*
T4 DNA polymerase (3.5)	50	2	5	100	11	20	
T7 DNA polymerase (3.5)	50	2	5	300	37	20	
Modified T7 DNA polymerase (3.5)	50	2	10	300	37	20	
Taq DNA polymerase (3.5 & 3.17)	100	0.1-1	2.5	200	94, 55, 72	1-2	*a, b*
Exonucleases							
Exonuclease II (3.11)	50	2	10	—	37	1-30	
Exonuclease II (3.11)	50	1	0.2	—	37	30	*a*
Lambda exonuclease (3.11)	50	2	10	—	37	1-30	
T7 gene 6 exonuclease (3.11)	50	2	5	—	37	1-30	
Kinases							
T4 polynucleotide kinase (3.10)							
5′ labeling (forward)	30	1-50 pmol[f]	20	50d	37	60	*a*
oligonucleotides	30	1-10[c]	20	1000 ATP	37	60	*a*
5′ labeling (exchange)	30	1-50 pmol[f]	20	60[d]	37	60	*a, e*
Nucleases							
Bal 31 nuclease (3.12)	50	2	10	—	30	1-30	*a*
Deoxyribonuclease I (DNase I) (3.12)	100	2	1	—	37	1-30	*a*
Mung bean nuclease (3.12)	100	1	15	—	37	30	*a*
Sl nuclease (3.12)	100	2	10	—	37	30	*a*
Phosphatases							
BAP (3.10)	50	1-20 pmol[f]	0.1	—	60	30	*a*
CIP (3.10)	50	1-20 pmol[f]	0.1	—	37	30	*a*
RNA Modifying Enzymes							
E. coli RNA polymerase (3.8)	50	2[g]	10	300 NTPs	37	30	*h*
SP6, T7, T3 RNA polymerases (3.8)	50	2[g]	10	400 NTPs	37	30	*h*
Poly(A) polymerase (3.9)	50	12.5 RNA	5	250 ATP	37	30	
Reverse transcriptase (3.7)	50	1 mRNA	40	40	37	30	*i*
RNase H (3.13)	100	2 RNA:DNA	1	—	37	20	*a*
T4 RNA ligase (3.15)	50	2[g]	1	2000 ATP	17	10 hr	*a*
Terminal transferase (3.6)	50	4[f]	10	20	37	30	

[a]Stop reactions as follows: DNase I—5 μl 0.5 M EDTA; *E. coli* DNA polymerase—75°C for 10 min or 1 μl 0.5 M EDTA; Taq DNA polymerase—store at −20°C; Exonuclease VII—phenol extract and ethanol precipitate; T4 polynucleotide kinase—1 μl 0.5 M EDTA, then phenol extract and ethanol precipitate; Bal 31 nuclease—75°C for 10 min or 5 μl 0.5 M EDTA; S1 nuclease, mung bean nuclease, and RNase H—1 μl 0.5 M EDTA; BAP—add SDS to 0.1% and proteinase K to 100 μg/ml, incubate at 37°C for 30 min, then phenol extract twice and ethanol precipitate; CIP—75°C for 10 min or phenol extract and ethanol precipitate; T4 RNA ligase—2 μl 0.5 M EDTA.

[b]DNA is template (genomic); add 0.2-1 μM each oligonucleotide primer.

[c]Forward—dephosphorylated DNA (5′ ends); oligonucleotide—linkers; exchange—phosphorylated DNA (5′ ends).

[d]pmol[γ-^{32}P]ATP, specific activity >3000 Ci/mmol (forward—150 μCi; exchange—180 μCi).

[e]Add 5 mM ADP and use imidazole in the buffer instead of Tris.

[f]BAP and CIP—pmol DNA termini; terminal transferase—pmol DNA as 3′ termini.

[g]RNA polymerases—use DNA with a promoter; T4 RNA ligase—use single-stranded DNA or RNA.

[h]Add 1 mM spermidine for SP6.

[i]Add 100 μCi [α-^{32}P]dNTP, specific activity >400 Ci/mmol, and 1 μg oligo(dT)$_{12\text{-}18}$.

Table 3.4.2 Applications of Nucleic Acid–Modifying Enzymes
Enzymes are described in further detail in the units indicated in parentheses.

Application	Enzyme
Blunt-end generation	
by removal of 3′ protruding ends	T4 and T7 DNA polymerase (3.5)
by filling in 3′ recessed ends	T4 and T7 DNA polymerase (3.5); Klenow fragment (3.5)
cDNA synthesis	
from RNA	Reverse transcriptase (3.7)
of second strand	Klenow fragment (3.5)
Cloning of DNA fragments	CIP (3.10); T4 DNA ligase (3.14); Klenow fragment (3.5); restriction endonucleases (3.1)
Degradation of DNA	
Exonucleases—double strand	
5′→3′	T7 gene 6 and λ exonuclease (3.11)
3′→5′	Exonuclease III (3.11)
5′→3′ and 3′→5′	*Bal* 31 nuclease (3.12)
Exonuclease—single strand (3′ & 5′)	Exonuclease VII (3.11)
Endonucleases—double strand	
nonspecific	DNase I and micrococcal nucleases (3.12)
specific	Restriction endonucleases (3.1)
Endonucleases—single strand	S1, *Bal* 31, and mung bean nucleases (3.12)
Degradation of RNA	Ribonuclease A and/or T1 (3.13); micrococcal nuclease (3.12)
Degradation of RNA in RNA:DNA duplexes	Ribonuclease H (3.13)
	BAP and CPI (3.10)
DNA sequencing	Modified T7 DNA polymerase (3.5); *Taq* polymerase (3.5); Klenow fragment (3.5); T4 polynucleotide kinase (3.10); reverse transcriptase (3.7)
Intron mapping	Exonuclease VII (3.11); S1 nuclease (3.12 & 4.6)
Labeling DNA at 5′ ends	T4 polynucleotide kinase (3.10)
Labeling DNA at 3′ ends	Modified T7 DNA polymerase (3.5); Klenow fragment (3.5); reverse transcriptase (3.7); terminal transferase (3.6)
Labeling RNA at 3′ ends	Poly(A) polymerase (3.9); T4 RNA ligase (3.15)
Ligations of DNA	*E. coli* and T4 DNA ligase (3.14)
Ligations of RNA	T4 RNA ligase (3.15)
Mapping mRNA 5′ ends	S1 or mung bean nuclease (3.12 & 4.6); reverse transcriptase (3.7 & 4.8); ribonuclease A (4.8)
Methylation of DNA	Methylases (3.1)
Nested deletion construction	*Bal* 31 (3.12 & 7.3); exonuclease III (3.11 & 7.3) + S1 or mung bean nuclease (3.12 & 7.3)
Nick translation	DNase I (3.12); *E. coli* DNA polymerase I (3.5)
Oligonucleotide extension	T7 DNA polymerase (3.5); Klenow fragment (3.5)
Phosphorylation	T4 polynucleotide kinase (3.10)
Poly(A) tailing of RNA	Poly(A) polymerase (3.9)
Polymerase chain reaction (PCR)	*Taq* DNA polymerase (3.5 & 3.17)
Primed synthesis	Klenow fragment (3.5)
Replacement synthesis	T4 and T7 DNA polymerase (3.5)
Restriction mapping	Restriction endonucleases (3.1)
Tailing DNA	Terminal transferase (3.6)
Tailing RNA	Poly(A) polymerase (3.9)
Transcription, promoter-specific	*E. coli*, SP6, T3, and T7 RNA polymerase (3.8)

Table 3.4.3 Properties of the Nucleoside Triphosphates

Nucleoside triphosphate	λ max (pH 7.0)	ε max × 10^{-3} (pH 7.0)	Absorbance ratio (280/260) (pH 7.0)	pK_a of base
ATP	259	15.4	0.15	4.0
CTP	271	12.8[a]	0.97	4.8
GTP	252	13.7	0.66	3.3 9.3
UTP	262	10.0	0.38	9.5
dATP	259	15.4	0.15	3.6
dCTP	280	13.1[a]	0.98	4.3
dGTP	253	13.7	0.66	3.5 9.7
dTTP	267	9.6	0.73	9.3

[a]Perform the spectral analysis at pH 2.0.

^{35}S-labeled nucleoside triphosphates have a thio moiety replacing one of the oxygens covalently bound to the phosphate. This is a significant perturbation that inhibits many enzymes. Because the lower energy of ^{35}S does not cause extensive damage to nucleic acids, it is used for the preparation of more stable, but lower specific activity, probes. Recently, ^{35}S has replaced ^{32}P as the preferred isotope for the dideoxy sequencing procedure because its lower energy results in sharper images in autoradiography.

The emission of ^{3}H is too weak for most autoradiographic procedures, although it is used for in situ hybridizations. Its primary use is for quantitative analysis of nucleic acid synthesis and degradation. For some specific purposes, ^{14}C and ^{125}I are used for radiolabeling nucleic acids.

CAUTION: Investigators should wear gloves for all procedures involving radioactivity. All experiments involving ^{32}P should be performed behind lucite screens to minimize exposure. When working with ^{32}P, investigators should frequently check themselves and the working area for radioactivity with a hand-held minimonitor. When finished, radioactive waste should be placed in appropriately designated areas.

BASIC PROTOCOL

Measuring Radioactivity in DNA and RNA by Acid Precipitation

The amount of radioactivity for a known amount of DNA is called the specific activity (usually given in units of cpm/μg of nucleic acid). The most common method for the measurement of radioactivity in nucleic acids is based on the fact that DNA or RNA molecules greater than 20 nucleotides in length are quantitatively precipitated in strong acids, whereas dNTP or NTP precursors remain in solution.

Materials *(see* APPENDIX 1 *for items with ✓)*

✓ 500 μg/ml sonicated salmon sperm DNA in TE buffer
✓ Acid precipitation solution, ice-cold
100% ethanol

Glass microfiber filters (2.4-cm diameter, Whatman GF/A)
Filtration device
Scintillation fluid and vials

1. Add a known volume (typically 1 μl) of a reaction mixture containing radioactive precursors to a disposable glass tube containing 100 μl of 500 μg/ml salmon sperm DNA in TE buffer. Spot 10 μl of the mixture onto a glass microfiber filter.
2. To the remaining 90 μl, add 1 ml of ice-cold acid precipitation solution and incubate 5 to 10 min on ice.

3. Collect precipitate by filtering solution through a second glass microfiber filter. Rinse tube with 3 ml acid precipitation solution and pour through filter. Wash filter four more times with 3 ml acid precipitation solution, followed by 3 ml ethanol.

4. Dry both filters under a heat lamp, and place them in separate vials containing 3 ml of a toluene-based scintillation fluid. Measure radioactivity in a liquid scintillation counter.

   Drying filters is unnecessary with scintillation fluids that can accommodate aqueous samples.

5. Determine incorporation of radioactivity into nucleic acid from ratio of cpm on second filter (which measures radioactivity in nucleic acid) to cpm on first filter (which measures total radioactivity in sample).

3.4

ALTERNATE PROTOCOL

Spin-Column Procedure for Separating Radioactively Labeled DNA from Unincorporated dNTP Precursors

In this method, the packing and running of the column is accomplished by centrifugation rather than by gravity. It is more rapid than conventional chromatography and is useful when many samples are involved; however, removal of the dNTPs may be less quantitative. Spin columns can be purchased commercially at moderate expense.

CAUTION: Use extreme care that centrifuge and work area do not become contaminated with radioactivity. Whenever possible, place samples behind a lucite shield.

1. Plug bottom of a 5-ml disposable syringe with clean, silanized glass wool. Fill syringe with an even suspension of the column resin, and place in a polypropylene tube that is suitable for centrifugation in a tabletop centrifuge (Fig. 3.4.1). Centrifuge 2 to 3 min at a setting of 4 in order to pack the column.

   The resin should not be packed too tightly or too loosely. Therefore, adjust the time and speed in the tabletop centrifuge according to individual laboratory conditions (~1200 rpm).

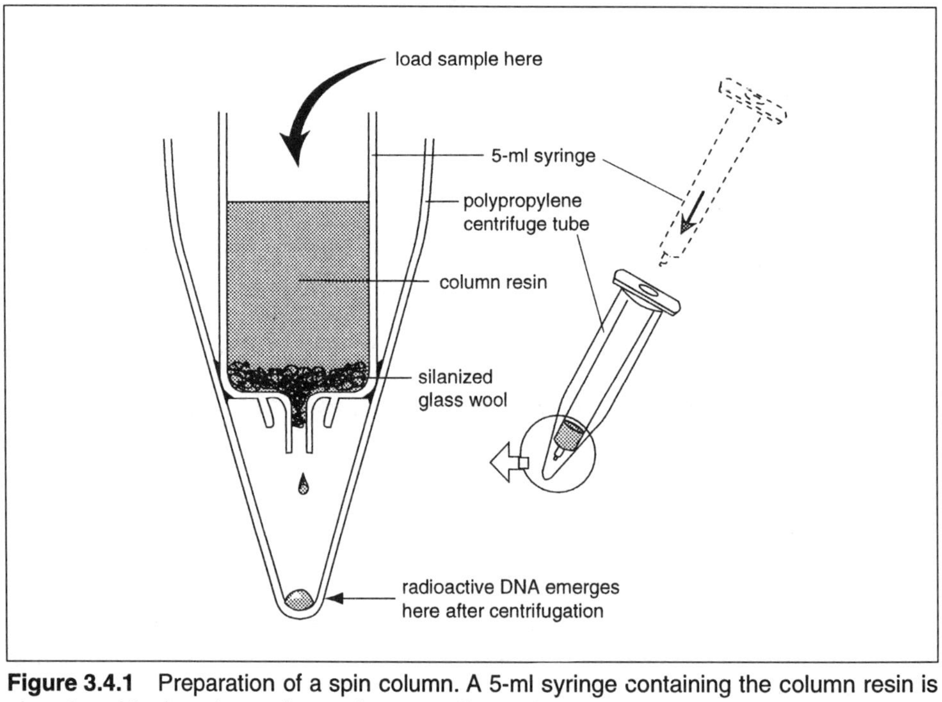

Figure 3.4.1 Preparation of a spin column. A 5-ml syringe containing the column resin is placed upside down in a polypropylene centrifuge tube.

2. Dilute radioactive sample with TE buffer to 100 μl and load in the center of the column. Place syringe in a new polypropylene tube, and centrifuge 5 min at a setting of 5 to 6.

3. Save liquid at bottom of tube containing the labeled DNA. Dispose of all radioactive waste properly.

Contributor: Kevin Struhl

DNA-Dependent DNA Polymerases

UNIT 3.5

All DNA polymerases add deoxyribonucleotides to the 3′-hydroxyl terminus of a primed double-stranded DNA molecule. Synthesis is exclusively in a 5′→3′ direction with respect to the synthesized strand (Fig. 3.5.1). The reaction requires the four deoxyribonucleoside triphosphates (dNTPs) and magnesium ions.

Many DNA polymerases have a 3′→5′ exonuclease activity (Fig. 3.5.2), which (in the absence of dNTPs) degrades both single- and double-stranded DNA from a free 3′-hydroxyl end. In the presence of dNTPs, the exonuclease activity on double-stranded DNA is inhibited by the polymerase activity.

Some DNA polymerases (e.g., *E. coli* DNA polymerase I, *Taq* DNA polymerase) also have an associated 5′→3′ exonuclease activity (Fig. 3.5.3), which degrades double-stranded DNA from a free 5′-hydroxyl end. The 5′→3′ exonuclease activity removes from one to several nucleotides at a time, releasing predominantly nucleoside 5′ phosphates, but also some larger oligonucleotides up to 10 nucleotides in length. See Table 3.5.1 for a summary of properties of DNA polymerases.

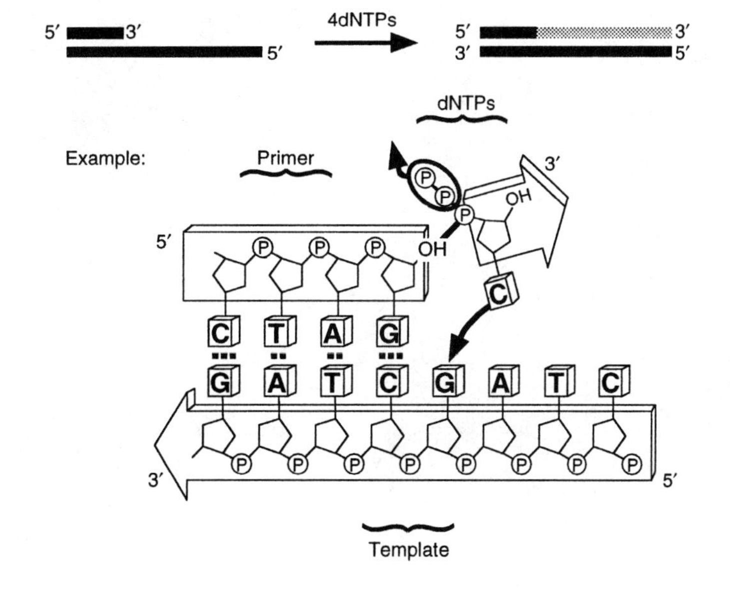

Figure 3.5.1 DNA polymerase 5′→3′ polymerase activity.

Table 3.5.1 Properties of DNA Polymerases

Enzyme	3′→5′ exonuclease	5′→3′ exonuclease	Polymerase rate	Processivity
E. coli DNA polymerase	Low	Present	Intermediate	Low
Klenow fragment	Low	None	Intermediate	Low
Reverse transcriptase	None	None	Slow	Intermediate
T4 DNA polymerase	High	None	Intermediate	Low
Native T7 DNA polymerase	High	None	Fast	High
Chemically modified T7 DNA polymerase	Low	None	Fast	High
Genetically modified T7 DNA polymerase (Δ28)	None	None	Fast	High
Taq DNA polymerase	None	Present	Fast	High

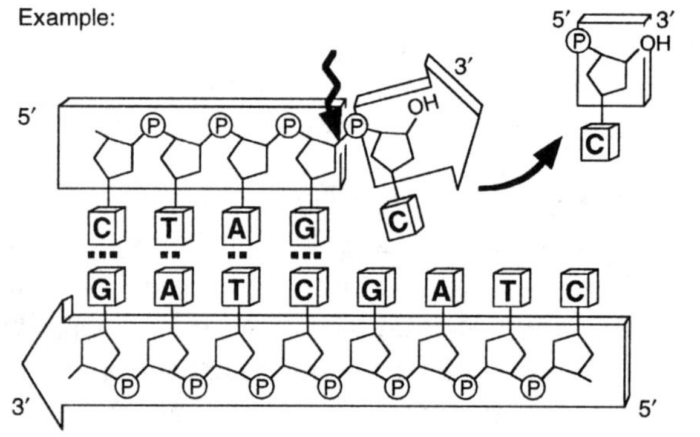

Figure 3.5.2 DNA polymerase 3′→5′ exonuclease activity.

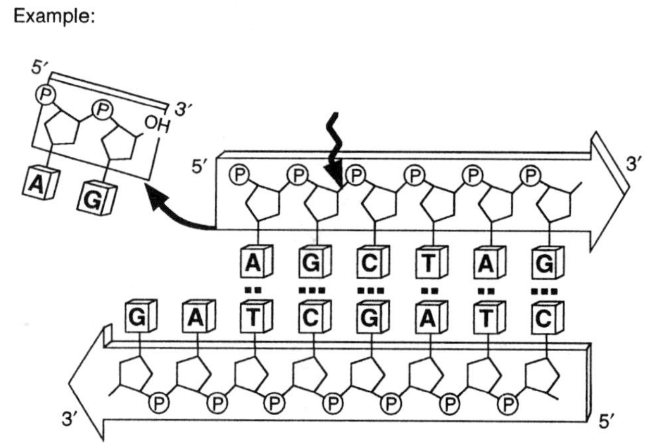

Figure 3.5.3 DNA polymerase 5′→3′ exonuclease activity.

KLENOW FRAGMENT OF *ESCHERICHIA COLI* DNA POLYMERASE I

ENZYME

The Klenow fragment consists of the C-terminal portion (70%) of *E. coli* DNA polymerase I. It retains the DNA polymerase and 3′→5′ exonuclease activity of *E. coli* DNA polymerase I, but lacks the 5′→3′ exonuclease activity. See Table 3.4.1 for detailed reaction conditions.

Labeling the 3′ Ends of DNA

BASIC PROTOCOL 1

This is the method of choice for generating ^{32}P-labeled size standards for autoradiography. Band intensities are independent of fragment length, and end-labeled DNAs are stable for several months. DNA can be labeled selectively at one end by cleaving with two different restriction endonucleases and labeling with a ^{32}P-labeled dNTP that is complementary to only one of the two 5′ extensions.

1. Digest 0.1 to 4 µg DNA with a restriction endonuclease that generates 5′ overhanging ends (*UNIT 3.1*) in a 20-µl reaction mixture.

2. Add 20 µCi of desired [α-^{32}P]dNTP (400 to 800 Ci/mmol) and 1 µl appropriate 5 mM 3dNTP mix (*UNIT 3.4*). If higher specific activities are required, add 80 µCi of the radioactive dNTP at 3000 Ci/mmol. Add 1 U Klenow fragment and incubate 15 min at 30°C.

   Because the Klenow fragment incorporates nucleotides that are complementary to the single-stranded 5′ extensions, the choice of ^{32}P-labeled dNTP depends on the restriction endonuclease used to cleave the DNA.

3. Stop reaction by heating 10 min at 75°C. If desired, remove unincorporated dNTP precursors from labeled DNA (*UNIT 3.4*).

Repairing 3′ or 5′ Overhanging Ends to Generate Blunt Ends

BASIC PROTOCOL 2

For many cloning experiments, it is necessary to convert the ends generated by restriction endonucleases into blunt ends (*UNIT 3.16*).

1. Digest 0.1 to 4 µg DNA with a restriction endonuclease in a 20-µl reaction. Add 1 µl of 0.5 mM each dNTP. Add 1 to 5 U Klenow fragment and incubate 15 min at 30°C.

2. Stop reaction either by heating 10 min at 75°C or by adding 1 µl of 0.5 M EDTA.

Labeling of DNA by Random Oligonucleotide–Primed Synthesis

BASIC PROTOCOL 3

This method is an alternative to nick translation for producing uniformly radioactive DNA of high specific activity.

1. Digest DNA with an appropriate restriction endonuclease (*UNIT 3.1*). Purify by gel electrophoresis (*UNIT 2.6*) or ethanol precipitation (*UNIT 2.1*). Resuspend in TE buffer.

   It is essential to remove the restriction endonuclease buffer from the DNA because the Mg^{2+} ions will make it difficult to denature the DNA. If low gelling/melting temperature agarose is used, it is usually unnecessary to purify the DNA.

2. Prepare the following reaction mix on ice:

   2.5 µl 0.5 mM 3dNTP mix (minus dATP; *UNIT 3.4*)
   2.5 µl 10× Klenow fragment buffer (*UNIT 3.4*)
   5 µl 3000 Ci/mmol [α-^{32}P]dATP (50 µCi)
   1 µl Klenow fragment (3 to 8 U)

   For generating probes with slightly lower specific activity, 2.5 µl (25 µCi) of 800 Ci/mmol [α-^{32}P]dATP can be used (an additional 2.5 µl of water must be added to reaction mixture to maintain a constant volume). Lower-specific-activity probes are useful for most purposes and are less expensive to prepare.

3. Combine DNA (30 to 100 ng) with random hexanucleotides (1 to 5 µg) in 14 µl. Boil 2 to 3 min and place on ice.

   If the DNA has not been purified from gel matrix, place at 37°C.

4. Add 11 µl of the reaction mix (step 2) to the denatured DNA and immediately incubate 2 to 4 hr at room temperature.

5. Stop reaction with 1 µl of 0.5 M EDTA, 3 µl of 10 mg/ml tRNA, and 100 µl TE buffer. Phenol extract (*UNIT 2.1*) and transfer aqueous phase to a fresh tube.

   If DNA was added as a molten gel slice, the stopped reaction mixture must be remelted by incubation at 70°C for 10 min.

6. Separate labeled DNA from unincorporated radioactive precursors by chromatography (*UNIT 3.4*).

7. Remove 1-µl aliquot and determine ^{32}P incorporation by acid precipitation (*UNIT 3.4*). Specific activity should be 10^8 cpm/µg.

Other Applications of Klenow Fragment

1. DNA sequencing by the dideoxy method (*UNIT 7.4*).
2. Synthesis of the second strand for the cloning of cDNAs (*UNIT 5.5*).
3. Extension of oligonucleotide primers on single-stranded templates for synthesis of hybridization probes and for in vitro mutagenesis (*UNITS 8.1 & 8.3*).
4. Converting single-stranded oligonucleotides to double-stranded DNA by mutually primed synthesis (*UNIT 8.2*).

ENZYME

T4 DNA POLYMERASE

T4 DNA polymerase possesses a DNA-dependent DNA polymerase activity and a very active single-stranded and double-stranded 3′→5′ exonuclease. It lacks a 5′→3′ exonuclease activity. See Table 3.4.1 for reaction conditions.

Reaction Conditions

For 50-µl reaction:
50 mM Tris·Cl, pH 8.0
5 mM $MgCl_2$
5 mM DTT
2 µg DNA
100 µM 4dNTP mix (*UNIT 3.4*)
50 µg/ml BSA
0.1 U T4 DNA polymerase

Incubate 20 min at 11°C. Stop reaction with 2 µl of 0.5 M EDTA or by heating 10 min to 75°C. The volume of reaction, concentration of 4 dNTPs, and the temperature of the reaction will vary, depending upon the application.

Effect of Triphosphate Concentration

High concentrations (100 µM) of dNTPs are used to maximize the ratio of polymerase to exonuclease activity in the absence of radioactivity. In labeling experiments, the concentration of the labeled dNTP is reduced to 1 to 2 µM. Levels lower than 1 µM labeled dNTP should not be used because once the dNTPs are exhausted the exonuclease activity will degrade the DNA.

Effect of Temperature

For labeling 3′ termini with T4 DNA polymerase, the temperature of the reaction should be maintained at 11°C. At higher temperatures, if only a subset of the four deoxyribonucleoside triphosphates are added, the exonuclease will degrade the template beyond the nucleotide that the polymerase could replace.

Buffer Compatibility

For many restriction enzymes, cleavage can be carried out in T4 DNA polymerase buffer, and T4 DNA polymerase can be used directly.

Applications

1. Radioactive labeling of the 3′ termini of DNA fragments. DNA fragments containing 5′ protruding ends are incubated with the appropriate [α-^{32}P]dNTPs at 1 to 2 µM 20 min at 11°C.

2. Selective and extensive labeling of the 3′ termini of a linear duplex DNA molecule, known as "replacement synthesis." The duplex DNA fragment is incubated with T4 DNA polymerase in the absence of dNTPs, which degrades DNA selectively from the 3′ ends. The four dNTPs are added (including one radioactively labeled [α-^{32}P]dNTP), to activate the polymerase activity that extends the 3′ ends the length of the template. Optimal labeling is achieved when the 3′→5′ exonuclease activity removes 30 to 40% of the nucleotides from each end.

3. Converting the ends of any duplex DNA fragment to blunt-ended structures suitable for blunt-end ligation for cloning. In the presence of high concentrations of all four dNTPs, the degradation will stop when the enzyme reaches the duplex region. Similarly, if the termini have a 5′ protruding region, the enzyme will simply extend the recessed 3′ termini until the end is blunt. For these applications, the DNA is incubated with T4 polymerase and 100 µM each of the four dNTPs for 20 min at 11°C.

NATIVE T7 DNA POLYMERASE

ENZYME

Purified T7 gene 5 protein has a 3′→5′ exonuclease activity and a nonprocessive DNA polymerase activity. Thioredoxin acts as an accessory protein to increase the affinity of T7 gene 5 protein for the primer template, rendering DNA synthesis processive for thousands of nucleotides.

Native T7 DNA polymerase (T7 gene 5 protein/thioredoxin complex) has a very active single-stranded and double-stranded DNA 3′→5′ exonuclease activity in addition to its polymerase activity. This activity is a detriment to the use of this enzyme for DNA sequence analysis, and must be reduced or eliminated (modified

T7 DNA polymerase) to be used for this application. See Table 3.4.1 for detailed reaction conditions.

Reaction Conditions

For 50-μl reaction:

40 mM Tris·Cl, pH 7.5	2 μg DNA
10 mM $MgCl_2$	300 μM 4dNTP mix (*UNIT 3.4*)
5 mM DTT	50 μg/ml BSA
50 mM NaCl	5 U T7 DNA polymerase

Incubate 20 min at 37°C. Stop with 2 μl of 0.5 M EDTA or by heating 10 min at 75°C. The volume of reaction, concentration of 4 dNTPs, and temperature of the reaction will vary, depending upon the application.

Applications

1. Because T7 DNA polymerase is highly processive, it can be used for extensive synthesis of DNA on long templates (e.g., M13). It will extend thousands of nucleotides from the same primer template without dissociating and is largely unaffected by secondary structures.
2. Native T7 DNA polymerase is the enzyme of choice for the synthesis of the complementary strand during site-directed mutagenesis.
3. Native T7 DNA polymerase can be used analogously to T4 DNA polymerase for labeling 3′ termini either by simple extension or by replacement synthesis.
4. Native T7 DNA polymerase can be used analogously to T4 DNA polymerase to convert the ends of any duplex DNA fragment (either 5′ or 3′ protruding) to blunt-ended structures.

ENZYME

MODIFIED T7 DNA POLYMERASE

Native T7 DNA polymerase has a very active 3′→5′ exonuclease activity in addition to its polymerase activity. In modified T7 DNA polymerase, this exonuclease activity has either been reduced selectively by a chemical reaction or inactivated completely by genetic modification.

Reaction Conditions

For 50-μl reaction:

40 mM Tris·Cl, pH 7.5	300 μM 4dNTP mix (*UNIT 3.4*)
5 or 10 mM $MgCl_2$ (see below)	50 μg/ml BSA
5 mM DTT	10 U modified T7 DNA polymerase
50 mM NaCl	
2 μg DNA	

Incubate 20 min at 37°C. Stop with 2 μl of 0.5 M EDTA or by heating 10 min at 75°C. The volume of reaction, concentration of 4 dNTPs, amount of DNA and enzyme, and temperature of the reaction will vary, depending upon the individual application. Genetically modified T7 DNA polymerase has a lower $MgCl_2$ optimum (5 mM) than the chemically modified polymerase (10 mM).

Applications

1. Modified T7 DNA polymerase has ideal characteristics for a DNA sequencing enzyme: high processivity, lack of 3′→5′ exonuclease, and the lack of discrimination against deoxynucleotide analogs (*UNIT 7.4*).
2. Modified T7 DNA polymerase efficiently labels DNA when dNTPs are present at very low levels (<0.1 µM).
3. Modified T7 DNA polymerase is useful for labeling the 3′ termini of DNA fragments with 5′ protruding ends. It should not be used to make blunt-end fragments because it leaves a one-base overhang at the 3′ end.

TAQ DNA POLYMERASE

ENZYME

Native *Taq* DNA polymerase is a double-stranded DNA polymerase. It has a temperature optimum for polymerization of 75° to 80°C. *Taq* DNA polymerase does not have any 3′→5′ exonuclease activity but does have a 5′→3′ exonuclease activity. For conditions for polymerase chain reaction (PCR) see *UNIT 15.1*.

Applications

1. *Taq* polymerase is active over a broad temperature range and has a high optimum temperature of polymerization, which makes it ideal for use in the polymerase chain reaction (Chapter 15).
2. *Taq* polymerase is useful for DNA sequencing, especially for synthesis at high temperatures to reduce secondary structures in the template.

References: Bebenek and Kunkel, 1989; Challberg and Englund, 1980; Feinberg and Vogelstein, 1983; Innis et al., 1988; Tabor and Richardson, 1987b, 1989.

Contributors: Stanley Tabor, Kevin Struhl, Stephen J. Scharf, and David H. Gelfand

Template-Independent DNA Polymerases

UNIT 3.6

TERMINAL DEOXYNUCLEOTIDYLTRANSFERASE (Terminal Transferase)

ENZYME

Terminal transferase catalyzes the incorporation of deoxynucleotides to the 3′-hydroxyl termini of DNA with the release of inorganic phosphate (Fig. 3.6.1). A template is not required and will not be copied. Divalent cations are required, and the nucleotide preference is determined by the cation used. Single-stranded DNA is the preferred primer. If double-stranded DNA is the primer, the extensions are most efficient when the ends have 3′ protruding termini. In the presence of Co^{2+}, the enzyme will prime any 3′ terminus (although not with uniform efficiency), and will catalyze the limited polymerization of ribonucleotides.

Reaction Conditions

The choice of dNTP and its concentration will depend upon the specific application. Under the conditions in Table 3.4.1, using dCTP, terminal transferase will add ~10 deoxycytidines to each 3′ end in 30 min.

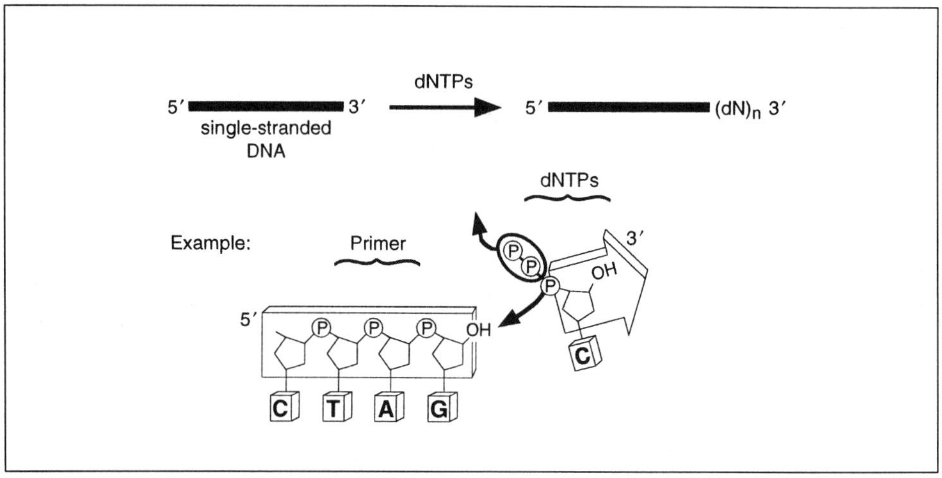

Figure 3.6.1 Terminal transferase activity.

Applications

1. Cloning DNA fragments. It synthesizes a homopolymer "tail" at each end of the DNA and a complementary homopolymer tail at each end of a linearized vector. The vector and insert DNAs are annealed by virtue of their complementary tails.

2. Labeling the 3′ termini of DNA with ^{32}P. For DNA sequence analyses, incorporation can be limited to a single nucleotide by using [α-^{32}P]cordycepin triphosphate (3′-deoxyribonucleoside triphosphate), which is a chain terminator.

3. Incorporating nonradioactive tags onto the 3′ termini of DNA fragments, such as biotin-11-dUTP, which serves as a receptor site for fluorescent dyes or avidin conjugates.

4. Synthesizing model polydeoxynucleotide homopolymers.

Reference: Ratliff, 1981.

Contributor: Stanley Tabor

UNIT 3.7

RNA-Dependent DNA Polymerases

ENZYME

REVERSE TRANSCRIPTASE

Reverse transcriptase makes DNA copies of RNA to synthesize complementary DNA (cDNA). It also has a DNA-directed DNA polymerase activity (Fig. 3.7.1) but dNTP incorporation is very slow. The DNA polymerase activity lacks a 3′→5′ exonuclease activity. A third activity of reverse transcriptase will degrade RNA in an RNA:DNA hybrid from either the 5′ or 3′ terminus (Fig. 3.7.2), which is useful for selectively destroying parts of an RNA molecule to which DNA molecules have been hybridized.

Reaction Conditions

See Table 3.4.1 for detailed reaction conditions. If necessary, the RNA template can be destroyed by adding 10 µl of 5 M NaOH and incubating overnight at 37°C.

Applications

1. Synthesizing cDNA for insertion into bacterial cloning vectors (*UNIT 5.5*). Two types of primers are used: oligo(dT) primers, which bind exclusively at the

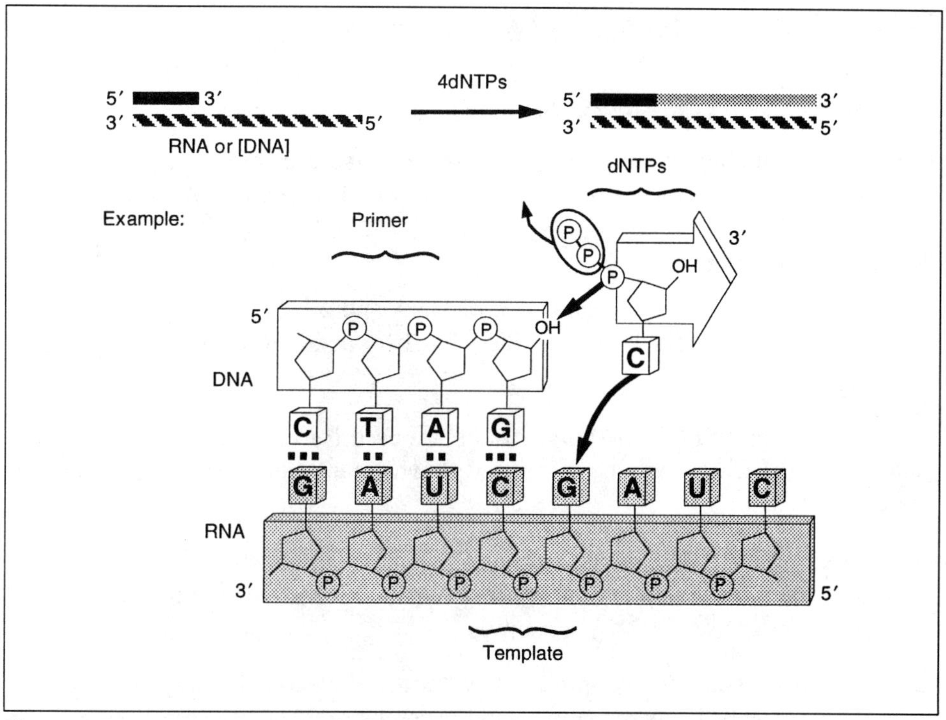

Figure 3.7.1 Reverse transcriptase 5′→3′ DNA polymerase activity.

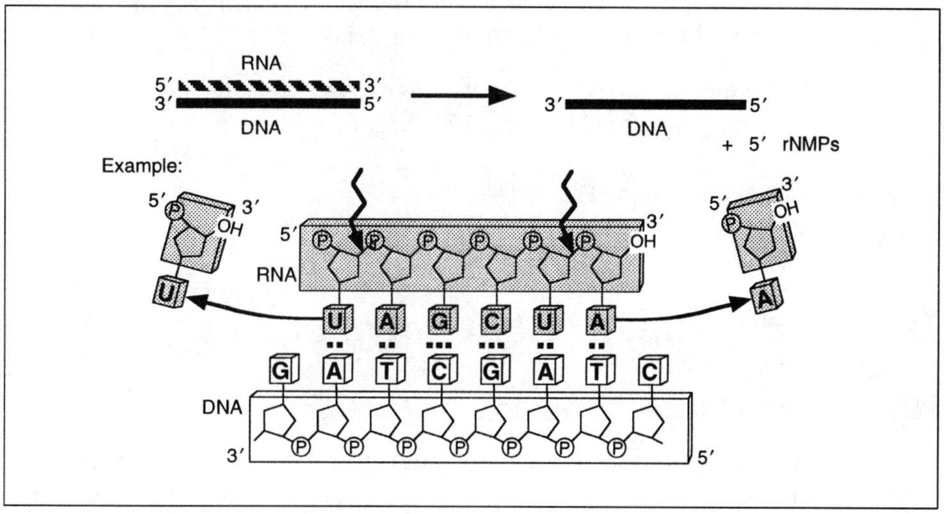

Figure 3.7.2 Reverse transcriptase 5′→3′ and 3′→5′ exoribonuclease activity (RNase H activity).

region corresponding to the 3′ end of an mRNA template, and a population of randomly generated oligodeoxyribonucleotides.

2. Filling in and labeling the 3′ termini of DNA fragments with 5′ protruding ends.

3. For DNA sequencing, in place of Klenow fragment (*UNIT 7.4*). Because it lacks a 3′→5′ exonuclease, reverse transcriptase will often synthesize through regions that impede the Klenow fragment.

Reference: Verma, 1977.

Contributor: Stanley Tabor

UNIT 3.8

DNA-Dependent RNA Polymerases

ENZYME

PHAGE RNA POLYMERASES: SP6, T7, T3

These three homologous bacteriophage RNA polymerases are highly specific for their own promoters. Transcription is both very rapid and extremely processive.

Reaction Conditions

The reaction volume and concentration of DNA and 4 NTPs will vary depending upon the individual application. See Table 3.4.1 for detailed reaction conditions.

3.8

Applications

Phage T7, T3, and SP6 RNA polymerases are used for extensive, highly specific transcription of DNA sequences inserted downstream from the appropriate promoter.

1. Phage polymerases can generate homogeneous single-stranded RNA probes uniformly labeled to a high specific activity, which are useful for detection of homologous DNA or RNA sequences.
2. Uniformly labeled transcripts are used to map the ends of RNA or DNA (*UNIT 4.7*), for genomic DNA sequencing and as precursor RNAs for studies of splicing and processing.
3. The RNA transcripts can be translated in vitro in the presence of radioactively labeled amino acids to generate radiopure proteins.
4. T7 RNA polymerase can be used in vivo to express cloned genes at very high levels exclusively under control of a T7 RNA polymerase promoter.

Reference: Chamberlin and Ryan, 1982.

Contributor: Stanley Tabor

UNIT 3.9

DNA-Independent RNA Polymerases

ENZYME

POLY(A) POLYMERASE

Poly(A) polymerase catalyzes the incorporation of AMP residues onto the free 3′-hydroxyl terminus of RNA, utilizing ATP as a precursor. See Table 3.4.1 for detailed reaction conditions.

Applications

1. Labeling the 3′ ends of RNA with [α-^{32}P]ATP. Labeled RNA prepared by this method can be used for hybridization probes. For cloned genes, RNA probes derived from phage RNA polymerases are much more efficient and produce RNA with a much higher specific activity. cDNA synthesis with reverse transcriptase is the preferred method for labeling cellular RNA.
2. Cloning RNA that lacks a poly(A) tail. A poly(A) tail is synthesized with poly(A) polymerase, and a cDNA copy is made with oligo(dT) as primer.

Reference: Edmonds, 1982.

Contributor: Stanley Tabor

Phosphatases and Kinases

UNIT 3.10

ENZYME

ALKALINE PHOSPHATASES: BACTERIAL ALKALINE PHOSPHATASE (BAP) AND CALF INTESTINE ALKALINE PHOSPHATASE (CIP)

Both enzymes catalyze the hydrolysis of 5′-phosphate residues from DNA, RNA, and ribo- and deoxyribonucleoside triphosphates. For most purposes, CIP is the enzyme of choice because it is inactivated by heating to 70°C for 10 min or by phenol extraction (BAP is more resistant), and has a 10- to 20-fold higher specific activity than BAP. See Table 3.4.1 for detailed reaction conditions.

3.10

Applications

1. Dephosphorylation of 5′ termini of nucleic acids prior to labeling with [γ-^{32}P]ATP and T4 polynucleotide kinase. 5′-^{32}P end-labeled DNA is used for sequencing by the Maxam-Gilbert procedure (*UNIT 7.5*), RNA sequencing by specific RNase digestions and in mapping studies using specific DNA or RNA fragments.

2. Dephosphorylation of 5′ termini of vector DNA to prevent self-ligation of vector termini (*UNIT 3.16*).

ENZYME

T4 POLYNUCLEOTIDE KINASE

The forward reaction catalyzes the transfer of the terminal (γ) phosphate of ATP to the 5′-hydroxyl termini of DNA and RNA (Fig. 3.10.1). Because this reaction is very efficient, it is the preferred method for labeling 5′ ends or for phosphorylating oligonucleotides.

The exchange reaction catalyzes the exchange of 5′-terminal phosphates. In the presence of excess ADP, the 5′-terminal phosphate is transferred to ADP and then rephosphorylated by transfer from the γ phosphate of [γ-^{32}P]. Because it is less efficient than the forward reaction, it is rarely used.

BASIC PROTOCOL 1

Labeling 5′ Ends by the Forward Reaction

The ATP concentration should be ≥1 μM. [γ-^{32}P]ATP can be added to trace the reaction. See Table 3.4.1 for detailed reaction conditions. For linear pBR322, 1 pmol of 5′ ends = 1.6 μg DNA. The enzyme is inhibited by low levels of phosphate buffer or ammonium salts and works very poorly on DNA purified from agarose gels. DNA must be further purified on DEAE cellulose (*UNIT 2.6*).

BASIC PROTOCOL 2

Labeling 5′ Termini by the Exchange Reaction

The ATP concentration should be ≥2 μM. The reaction volume and concentrations of DNA and [γ-^{32}P]ATP will vary, depending upon the application. See Table 3.4.1 for detailed reaction conditions.

Other Applications

1. DNA sequencing by the chemical degradation technique (*UNIT 7.5*).

2. Defining specific protein-DNA interactions by DNase I footprinting or protection from DNA-damaging chemicals (e.g., dimethyl sulfate, methidium propyl EDTA).

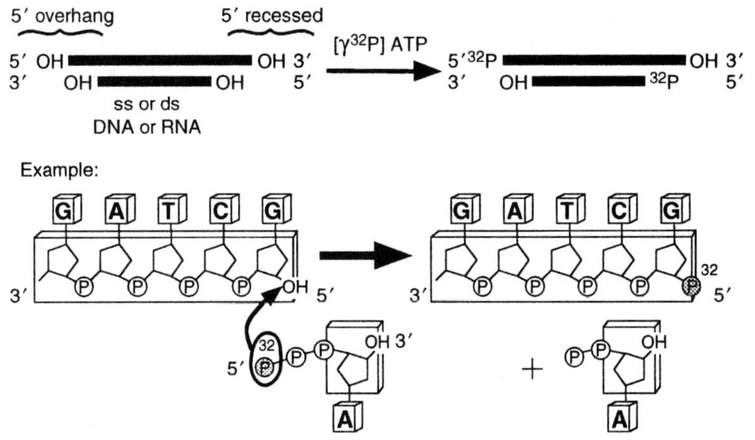

Figure 3.10.1 Kinase activity.

3. Mapping restriction sites by partial digestion of 5′ end-labeled DNA fragments (*UNIT 3.3*), termini of RNA transcripts (*UNIT 4.6*), and positions of introns in DNA.
4. Substrate synthesis for the assay of DNA and RNA ligases.
5. Labeling oligonucleotides for purification by electrophoresis (*UNIT 2.7*).
6. Ligation of oligonucleotides into DNA vectors.

Reference: Chaconas and van de Sande, 1980.

Contributor: Stanley Tabor

UNIT 3.11

Exonucleases

SINGLE-STRANDED 5′→3′ AND 3′→5′ EXONUCLEASES

ENZYME

Exonuclease VII (Exo VII)

Exonuclease VII is a processive single-strand specific exonuclease that acts from both the 3′ and 5′ ends of single-stranded DNA, producing small oligonucleotides. It retains full activity in the presence of 10 mM EDTA. See Table 3.4.1 for detailed reaction conditions.

Applications

1. For mapping positions of introns in genomic DNA (*UNIT 4.6*).
2. To excise segments of DNA inserted into plasmid vectors by the poly(dA-dT) tailing method.

DOUBLE-STRANDED 5′→3′ EXONUCLEASES

Lambda Exonuclease (λ Exo)

ENZYME

This enzyme catalyzes the stepwise and processive hydrolysis of duplex DNA from 5′-phosphoryl termini liberating 5′ mononucleotides. It will not degrade 5′-hydroxyl termini. See Table 3.4.1 for reaction conditions.

Applications

1. Converting double-stranded DNA into single-stranded DNA for use in the dideoxy sequencing method (*UNIT 7.4*).
2. Removing 5′ protruding ends from duplex DNA for tailing with terminal transferase.

3.11

T7 Gene 6 Exonuclease

ENZYME

This enzyme catalyzes the stepwise hydrolysis of duplex DNA from the 5′ termini, liberating 5′ mononucleotides. It has low processivity and will remove both 5′-hydroxyl and 5′-phosphoryl termini. See Table 3.4.1 for detailed reaction conditions.

Applications

T7 gene 6 exonuclease is preferred over λ exo for the controlled and uniform digestion from 5′ ends because it is less processive. It can also degrade from 5′ phosphoryl termini.

DOUBLE-STRANDED 3′→5′ EXONUCLEASES

Exonuclease III (Exo III)

ENZYME

The main application of exo III is as a 3′→5′ double-strand specific exonuclease that catalyzes release of 5′ nucleotides from the 3′-hydroxyl end of double-stranded DNA. The exonuclease activity is nonprocessive, making it ideal for generating uniform single-stranded regions in double-stranded DNA. Its degradation rate is determined by the DNA base composition: C>>A~T>>G.

Reaction Conditions

One unit of exo III will remove 200 nucleotides from each 3′ recessed end of 1 µg of a 5000-bp linear double-stranded DNA template in 10 min at 37°C. See Table 3.4.1 for detailed reaction conditions.

Applications

1. Preparing strand-specific radioactive probes, in conjunction with the Klenow fragment. This method is analogous to replacement synthesis using exonuclease and polymerase functions of T4 DNA polymerase.
2. Preparing single-stranded DNA templates for sequencing by the dideoxy technique (*UNIT 7.4*).
3. Constructing unidirectional deletions from a given position in cloned DNA. These clones are then used for DNA sequencing without prior restriction site mapping.

Reference: Thomas and Olivera, 1978.

Contributor: Stanley Tabor

Endonucleases

ENZYME

BAL 31 NUCLEASE

Bal 31 nuclease is a single strand-specific endodeoxyribonuclease that degrades at nicks or at transient single-stranded regions created by supercoiling on duplex circular DNA. On duplex linear DNA, it degrades from both the 5′ and 3′ termini at both ends, producing a controlled shortening of the DNA. It also acts as a ribonuclease, catalyzing the hydrolysis of ribosomal and tRNA. It is active in SDS and urea.

Reaction Conditions

Analyze the products by agarose gel electrophoresis (*UNIT 2.5A*). For subsequent enzymatic reactions, remove the NaCl by ethanol precipitation (*UNIT 2.1*). See Table 3.4.1 for detailed reaction conditions.

One unit is the amount of enzyme that catalyzes removal of 200 bp from each end of linearized pBR322 in 10 min at 30°C with a DNA concentration of 50 µg/ml in a 50-µl reaction.

Critical Parameters

1. Use DNA that is CsCl-purified (*UNIT 1.7*) or gel-purified (*UNIT 2.6*) because *Bal* 31 activity is inhibited by contaminating RNA.
2. The unit definition for *Bal* 31 should be used only as a rough estimate. For each preparation of DNA, the extent of digestion should be monitored by agarose gel electrophoresis (*UNIT 2.5*).
3. *Bal* 31 activity varies with different DNA templates; AT-rich regions are degraded faster than GC-rich regions.
4. *Bal* 31-digested DNA fragments can be ligated directly at low frequency. To increase efficiency of ligation, digested DNA should be extracted with phenol and precipitated with ethanol (*UNIT 2.1*). The ends of DNA should then be repaired with Klenow fragment or T4 DNA polymerase (*UNIT 3.5*), prior to the ligation reaction with T4 DNA ligase (*UNIT 3.16*).

Applications

1. Cloning for creating deletions of different sizes in a controlled manner.
2. Mapping restriction sites in a DNA fragment.
3. Investigating secondary structure of supercoiled DNA and alterations in the helix structure of duplex DNA caused by treatment with mutagenic agents.

ENZYME

S1 NUCLEASE

S1 nuclease is a highly specific single-stranded endonuclease. It is stable to urea, SDS, and formamide. See Table 3.4.1 for detailed reaction conditions.

Applications

1. Mapping the 5′ and 3′ ends of RNA transcripts by the analysis of S1-resistant RNA:DNA hybrids (*UNIT 4.6*).

2. Mapping the location of introns by digesting a hybrid of mature mRNA with ^{32}P-labeled genomic DNA. S1 cleaves at single-stranded loops created by introns within these hybrid molecules.

3. Digesting the hairpin structures formed during synthesis of cDNA by reverse transcriptase (*UNIT 5.5*).

4. Removing single-stranded termini of DNA fragments to produce blunt ends for ligation (*UNIT 3.16*).

5. Creating small deletions at restriction sites.

MUNG BEAN NUCLEASE

ENZYME

Mung bean nuclease (from sprouts of mung bean) is a highly specific single-stranded endonuclease with properties similar to those of S1 nuclease. See Table 3.4.1 for detailed reaction conditions.

One unit is defined as the amount of enzyme that produces 1 µg of acid-soluble material in 1 min at 37°C using single-stranded salmon sperm DNA as the substrate.

Applications

1. Cleaving immediately adjacent to the last hybridized base pair without removing any of the base-paired nucleotides in transcript mapping experiments (*UNIT 4.6*).

2. Precisely deletes overhanging bases that result from restriction endonuclease cleavage.

MICROCOCCAL NUCLEASE

ENZYME

Micrococcal nuclease from *Staphylococcus aureus* is a relatively nonspecific nuclease that cleaves single- and double-stranded DNA and RNA to oligo- and mononucleotides with 3′ phosphates. Cleavage occurs preferentially at AT- or AU-rich regions. The enzyme is inactivated by Ca^{2+}-specific chelating agents such as EGTA.

One unit is defined as the amount of enzyme that hydrolyzes 1 µmol of acid-soluble oligonucleotides from native DNA per minute at 37°C, pH 8.8.

Reaction Conditions

Typical digestions are performed in 10 mM Tris·Cl, pH 8.0, and 1 mM $CaCl_2$, but the enzyme is somewhat more active at higher pH. Reactions can be stopped by EDTA or the Ca^{2+}-specific chelator EGTA.

Applications

1. Studies of chromatin structure.

2. Removing nucleic acid from crude cell-free extracts without destroying enzyme activities.

ENZYME

DEOXYRIBONUCLEASE I (DNase I)

DNase I is a divalent cation-requiring endonuclease, which degrades double-stranded DNA to produce 3′-hydroxyl oligonucleotides. In the presence of Mg^{2+}, it produces nicks in duplex DNA, while in the presence of Mn^{2+} it produces double-stranded breaks. See Table 3.4.1 for detailed reaction conditions.

Preparation and Storage of DNase I solutions

DNase I is available as a lyophilized powder (2000 to 3000 U/mg protein), which can be prepared and stored as a solution for >1 yr.

3.13

1. Dissolve 1 mg DNase I (without vortexing) in 1 ml of 50% (w/v) glycerol containing 20 mM Tris·Cl, pH 7.5/1 mM $MgCl_2$. Store in liquid form at −20°C.
2. Alternatively, dissolve 1 mg DNase I (without vortexing) in 1 ml of 20 mM Tris·Cl, pH 7.5, containing 1 mM $MgCl_2$. Place in 10-µl aliquots in small microcentrifuge tubes, quick-freeze on dry ice, and store at −80°C. Do not refreeze.

RNase-Free DNase I

For many purposes, DNase I should be free of RNase. High-grade commercial preparations such as Worthington grade DPRF can be satisfactory.

1. Alternatively, dissolve 1 mg/ml DNase I in 0.1 M iodoacetic acid plus 0.15 M sodium acetate at a final pH of 5.3. Heat 40 min at 55°C and cool.
2. Add 1 M $CaCl_2$ to 5 mM and store frozen in small aliquots.

Applications

1. Nick translation (*UNIT 3.5*).
2. Cloning random DNA fragments by catalyzing double-stranded cleavage of DNA in the presence of Mn^{2+}.

References: Alexander et al., 1961; Kroeker et al., 1976; Lau and Gray, 1979; Moore, 1981; Vogt, 1980.

Contributor: Stanley Tabor

UNIT 3.13

Ribonucleases

ENZYME

RIBONUCLEASE A

Ribonuclease A (RNase A) from bovine pancreas is an endoribonuclease that specifically hydrolyzes RNA after C and U residues. Cleavage occurs between the 3′-phosphate group of a pyrimidine ribonucleotide and the 5′-hydroxyl of the adjacent nucleotide. RNase A activity can be inhibited specifically by an RNasin inhibitor, an inhibitor isolated from human placenta.

Reaction Conditions

RNase A is active under an extraordinarily wide range of reaction conditions, and it is extremely difficult to inactivate. At low salt concentrations (0 to 100 mM NaCl), RNase A cleaves single-stranded and double-stranded RNA as well as the RNA strand in RNA:DNA duplexes. However, at NaCl concentrations of 0.3 M or above, RNase A becomes specific for cleavage of single-stranded RNA. Removal of RNase A from a reaction solution generally requires treatment with proteinase K followed by multiple phenol extractions and ethanol precipitation.

DNase-Free RNase A

To prepare RNase A free of DNase, dissolve RNase A in TE buffer at 1 mg/ml, and boil 10 to 30 min. Store aliquots at –20°C.

3.13

Applications

1. Mapping and quantitating RNA species using the ribonuclease protection assay (*UNIT 4.7*). It is used in conjunction with RNase T1.
2. Hydrolyzing RNA that contaminates DNA preparations (*UNITS 1.6 & 1.7*).
3. RNA sequencing.
4. Blunt-ending double-stranded cDNA (*UNIT 5.5*). It is used in conjunction with RNase H.

RIBONUCLEASE H

ENZYME

Ribonuclease H (RNase H) from *E. coli* is an endoribonuclease that specifically hydrolyzes the phosphodiester bonds of RNA in RNA:DNA duplexes to generate products with 3′ hydroxyl and 5′ phosphate ends. It will not degrade single-stranded or double-stranded DNA or RNA. RNase H cleavage can be directed to specific sites by hybridizing short deoxyoligonucleotides to the RNA.

One unit is defined as the amount of enzyme that produces 1 nmol of acid-soluble ribonucleotides from poly(A)·poly(dT) in 20 min at 37°C.

Reaction Conditions

For 100-μl reaction:

20 mM HEPES·KOH, pH 8.0	2 μg RNA:DNA duplex
50 mM KCl	50 μg/ml BSA
4 mM $MgCl_2$	1 U ribonuclease H
1 mM DTT	

Incubate 20 min at 37°C. Stop with 1 μl of 0.5 M EDTA. The volume of reaction, amount of DNA, units of enzyme, temperature, time, and method of stopping the reaction will vary depending on the application.

Applications

1. Facilitating the synthesis of double-stranded cDNA by removing the mRNA strand of the RNA:DNA duplex produced during first strand synthesis of cDNA (*UNIT 5.5*). It is also used later in the protocol to degrade residual RNA in conjunction with RNase A.
2. Creating specific cleavages in RNA molecules by using synthetic deoxyoligonucleotides to create local regions of RNA:DNA duplexes.

ENZYME

RIBONUCLEASE T1

Ribonuclease T1 (RNase T1) from *Aspergillus oryzae* is an endoribonuclease that specifically hydrolyzes RNA after G residues. Cleavage occurs between the 3′-phosphate group of a guanine ribonucleotide and the 5′-hydroxyl of the adjacent nucleotide. The reaction generates a 2′:3′ cyclic phosphate which then is hydrolyzed to the corresponding 3′-nucleoside phosphates.

Reaction Conditions

The enzyme is active under a wide range of reaction conditions, and it is difficult to inactivate. At low salt concentrations (0 to 100 mM NaCl), it cleaves single-stranded and double-stranded RNA as well as the RNA strand in RNA:DNA duplexes. However, at NaCl concentrations of 0.3 M or above, it becomes specific for cleavage of single-stranded RNA. Removal of RNase T1 from a reaction solution generally requires treatment with proteinase K followed by multiple phenol extractions and ethanol precipitation.

Applications

1. Mapping and quantitating RNA species using the ribonuclease protection assay (*UNITS 4.7 & 9.8*). It is used in conjunction with RNase A.
2. RNA sequencing.
3. Determining the level of RNA transcripts synthesized in vitro from DNA templates containing a "G-less cassette."

References: Boehringer Mannheim Biochemicals: Biochemicals for Molecular Biology (catalog); Uchida and Egami, 1971.

Contributor: Kevin Struhl

3.14

UNIT 3.14

DNA Ligases

DNA ligases catalyze the formation of phosphodiester bonds between a juxtaposed 5′ phosphate and a 3′-hydroxyl terminus in duplex DNA.

ENZYME

T4 DNA LIGASE

This is the only DNA ligase that efficiently joins blunt-end termini under normal reaction conditions. See *UNIT 3.16* for a detailed ligation protocol. Cohesive-end ligations are usually carried out at 12° to 15°C. Blunt-end ligations are usually carried out at room temperature (<30°C) with 10 to 100 times more enzyme than cohesive-end ligations. T4 DNA ligase is strongly inhibited by NaCl concentrations >150 mM. Units in Table 3.4.1 are Weiss units, where 1 Weiss unit is equivalent to 60 cohesive-end units.

ENZYME

ESCHERICHIA COLI DNA LIGASE

E. coli DNA ligase can repair single-stranded nicks in duplex DNA and join restriction fragments with homologous cohesive ends, but under normal reaction conditions, it cannot join termini with blunt ends. PEG 8000 increases its rate of cohesive-end joining. Units in Table 3.4.1 are Modrich-Lehman units, where 1 Modrich-Lehman unit is equivalent to 6 Weiss units.

Applications

E. coli DNA ligase can be used instead of T4 DNA ligase when blunt-end ligations are not needed because it produces fewer aberrant ligations and a lower background.

Reference: Engler and Richardson, 1982.

Contributor: Stanley Tabor

RNA Ligases

UNIT 3.15

ENZYME

3.15

T4 RNA LIGASE

T4 RNA ligase catalyzes the ATP-dependent covalent joining of single-stranded 5′-phosphoryl termini of DNA or RNA to single-stranded 3′-hydroxyl termini of DNA or RNA. See Table 3.4.1 for reaction conditions.

Applications

1. Radioactive labeling of 3′ termini of RNA.
2. Circularizing deoxy- and ribo-oligonucleotides.
3. Ligating oligomers for oligonucleotide synthesis to produce internally labeled oligomers at specific residues.
4. Stimulating the blunt-end ligation activity of T4 DNA ligase.

Reference: Uhlenbeck and Gumport, 1982.

Contributor: Stanley Tabor

Subcloning of DNA Fragments

UNIT 3.16

BASIC PROTOCOL

Materials *(see APPENDIX 1 for items with ✓)*

Calf intestine alkaline phosphatase (CIP) and buffer (optional; *UNIT 3.10*)
0.5 mM 4dNTP mix (*UNIT 3.4*)
Klenow fragment of *E. coli* DNA polymerase I *or* T4 DNA polymerase (optional; *UNIT 3.5*)
Oligonucleotide linkers (optional)
10 mM ATP
✓ 0.2 mM dithiothreitol (DTT)
T4 DNA ligase (*UNIT 3.14*) and 2× buffer (*APPENDIX 1*)

1. Cleave the DNA to completion with the appropriate enzyme in a 20-µl reaction. Inactivate the enzymes by heating 15 min at 75°C. If no further enzymatic treatment is required, proceed to step 6.
2. If 5′ phosphates are to be removed, add 2 µl of 10× CIP buffer and 1 U CIP. Incubate 30 to 60 min at 37°C. Stop reaction by heating 15 min at 75°C. If no further enzymatic treatment is required, proceed to step 6.
3. If one or both ends must be converted to blunt ends, add 1 µl of a solution containing all 4 dNTPs (0.5 mM each) and an appropriate amount of Klenow

fragment or T4 DNA polymerase and incubate (Fig. 3.16.1). Stop the reaction by heating 15 min at 75°C. Cleave the reaction products with appropriate restriction enzyme if a DNA fragment with only one blunt end is desired. If no further enzymatic treatment is required, proceed to step 6.

4. If oligonucleotide linkers are to be added, add 0.1 to 1.0 μg of an appropriate oligonucleotide linker, 1 μl of 10 mM ATP, 1 μl of 0.2 M DTT, and 20 to 100 cohesive-end units of T4 DNA ligase. Incubate overnight at 15°C. Stop reaction by heating 15 min at 75°C.

5. Cleave the step 4 products with a restriction endonuclease recognizing the oligonucleotide linker. Cleave the products with an additional restriction endonuclease if only one of the two ends is to contain a linker.

6. Isolate the desired DNA segments by agarose gel electrophoresis (*UNIT 2.5A*). Visualize the DNA by longwave UV light, cut out the desired band, and purify the DNA (*UNIT 2.6*).

   It is critical to use longwave UV light sources to prevent damage to the DNA.

7. Set up the following ligation reaction (Fig. 3.16.2):

   9 μl component DNAs (0.1 to 5 μg)
   10 μl 2× ligase buffer
   1 μl 10 mM ATP
   20 to 500 U (cohesive-end) T4 DNA ligase.

   Incubate at 15°C for 1 to 24 hr.

   Simple ligations with two fragments having 4-bp, 3′ or 5′ overhanging ends require much less ligase than more complex ligations or blunt-end ligations. The quality of DNA will also affect the amount of ligase needed.

8. Introduce 1 to 10 μl of the ligated products into competent *E. coli* cells (*UNIT 1.8*) and select for transformants using genetic markers present on the vector.

9. Purify the plasmid or phage DNAs by miniprep procedures (*UNITS 1.6 & 1.10*) and determine the structure by restriction mapping (*UNITS 3.2 & 3.3*).

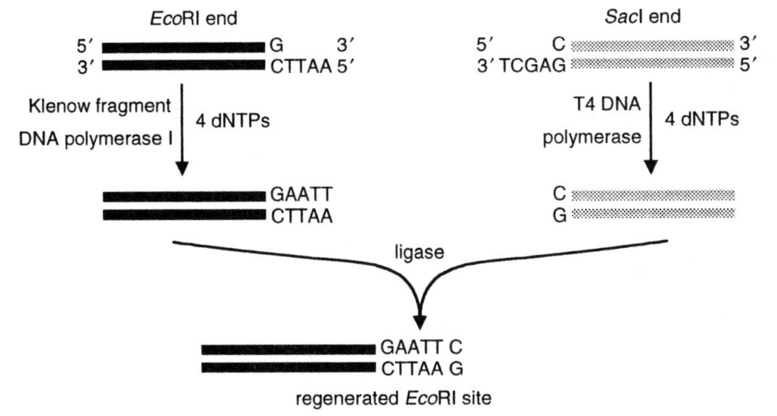

Figure 3.16.1 Joining DNA fragments with incompatible ends.

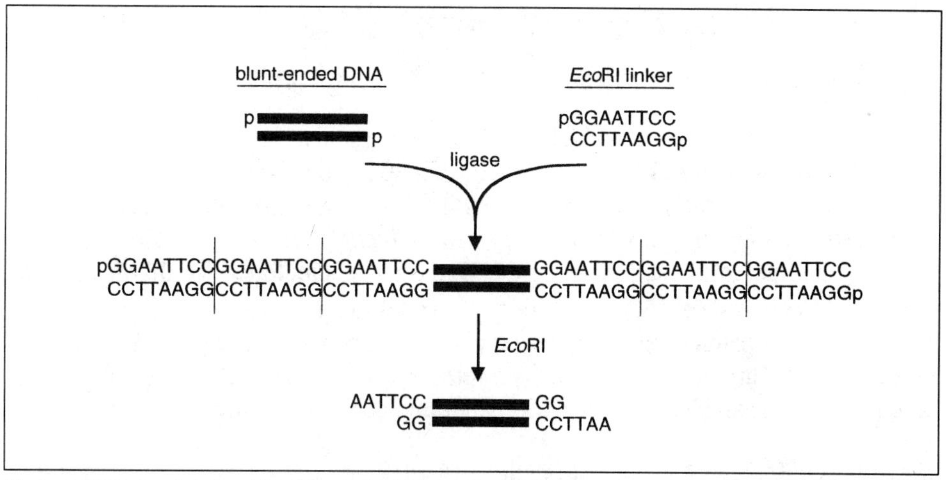

Figure 3.16.2 Joining blunt-ended DNA to *Eco*RI linker.

LIGATION OF DNA FRAGMENTS IN GEL SLICES

ALTERNATE PROTOCOL

This method saves time, but the cloning efficiency may be reduced.

Additional Materials *(also see Basic Protocol; see* APPENDIX 1 *for items with ✓)*

Low gelling/melting temperature agarose (SeaPlaque, FMC Marine Colloids)
✓ TAE electrophoresis buffer

1. Follow steps 1 to 5 of Basic Protocol.
2. Separate the DNAs in high quality low gelling/melting temperature agarose (0.7%) in TAE buffer. Cut the desired band(s) in the smallest possible volume (20 to 50 µl). Place gel slice in a microcentrifuge tube.
3. Melt the gel slices ≥10 min at 70°C. In separate tubes for each ligation reaction, combine gel slices containing appropriate DNAs (and water if necessary) for a total volume of 9 µl. Place for a few minutes at 37°C.
4. Add 11 µl of ice-cold mixture containing 2× buffer, ATP, and T4 DNA ligase. Mix and incubate 1 to 48 hr at 15°C.

   DNA fragments can still be ligated even though the reaction mixture has resolidified into a gel.
5. Remelt the gel slices 5 to 10 min at 73°C. Add 5 µl of ligated products to 200 µl of competent *E. coli* cells as in step 8 of the Basic Protocol.
6. Carry out step 9 of the Basic Protocol

Reference: Struhl, 1985.

Contributor: Kevin Struhl

Constructing Recombinant DNA Molecules by the Polymerase Chain Reaction

Any two segments of DNA can be ligated together into a new recombinant molecule using the polymerase chain reaction (PCR). The DNA can be joined in any configuration, with any desired junction-point reading frame or restriction site, by incorporating extra nonhomologous nucleotides within the PCR primers. It is not necessary to know the nucleotide sequence of the DNA being subcloned by this technique, other than the two short flanking regions (~20 bp) that serve as anchors for the two oligonucleotide primers used in the amplification process. The technique can be used to create an in-frame fusion protein, a recombinant DNA by sequential amplification, or an inserted restriction endonuclease site by inverse PCR.

BASIC PROTOCOL

SUBCLONING DNA FRAGMENTS

Synthetic oligonucleotides incorporating new unique restriction sites are used to amplify a region of DNA to be subcloned into a vector containing compatible restriction sites as summarized in Figures 3.17.1 to 3.17.3.

Materials *(see APPENDIX 1 for items with ✓)*

Template DNA (1 to 10 ng of plasmid or phage DNA; 20 to 300 ng of genomic or cDNA)
Oligonucleotide primers (0.6 to 1.0 mM; *UNIT 8.5*)
Mineral oil
TE-buffered phenol (*UNIT 2.1*) and chloroform
100% ethanol
✓ TE buffer, pH 8.0
Klenow fragment of *E. coli* DNA polymerase I (*UNIT 3.5*)
Vector DNA
Calf intestine alkaline phosphatase (CIP; *UNIT 3.10*)

1. Prepare template DNA. If DNA is not purified by a CsCl gradient, heat sample 10 min at 100°C to inactivate nucleases.

2. Prepare oligonucleotide primers. If the PCR product is to be cloned by blunt-end ligation, phosphorylate the 5′ hydroxyl of the oligonucleotide primers.

   A 5′ phosphate on the ends of the PCR products is needed to form the phosphoester linkage to the 3′OH of the vector during ligation. This step is essential if the vector has been treated with a phosphatase.

3. Set up a standard amplification reaction and overlay with mineral oil (*UNIT 15.1*). Carry out PCR in an automated thermal cycler for 20 to 25 cycles under the following conditions: denature 1 min at 94°C, anneal 1 min at 50°C, and extend 3 min at 72°C. Extend an additional 10 min at 72°C in the last cycle to make products as complete as possible.

   A thermostable DNA polymerase with 3′→5′ exonuclease proofreading activity—e.g. Pfu DNA polymerase (Stratagene) or Vent DNA polymerase (New England Biolabs)— can be used instead of Taq DNA polymerase to reduce the amount of nucleotide misincorporation during amplification.

4. Analyze an aliquot (e.g., 4 to 8 µl) of each reaction mix by agarose or polyacrylamide gel electrophoresis (*UNIT 2.5A or UNIT 2.7*) to verify that the amplification has yielded the expected product.

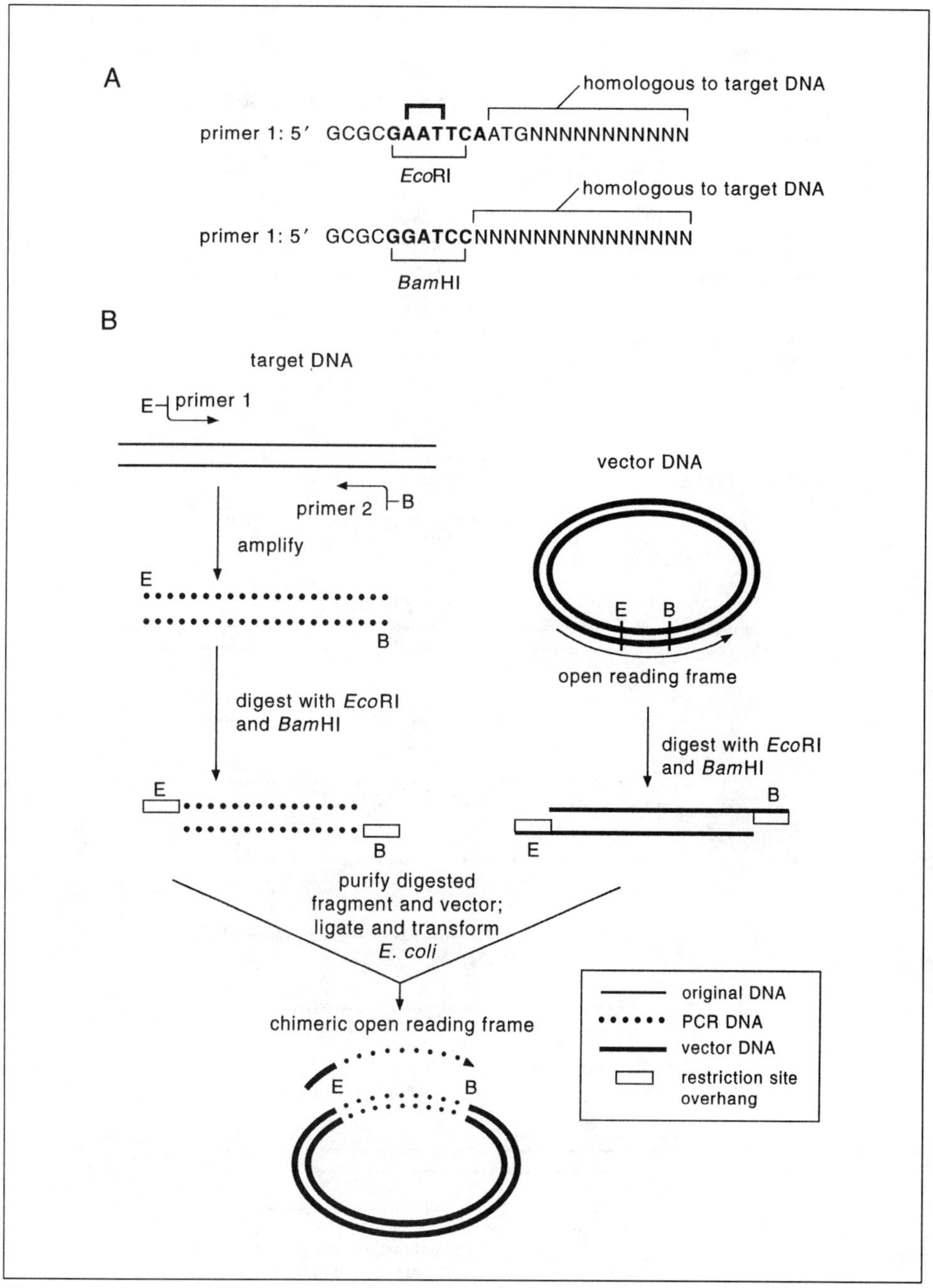

Figure 3.17.1 Introducing unique restriction sites and creating an in-frame fusion protein by PCR. Abbreviations: E, *Eco*RI; B, *Bam*HI.

5. Remove mineral oil overlay from each sample, then extract once with buffered chloroform (*UNIT 2.1*) to remove residual mineral oil. Extract once with buffered phenol and then precipitate DNA with 100% ethanol.

6. Microcentrifuge DNA 10 min at high speed, 4°C. Dissolve pellet in 20 µl TE buffer. Purify desired PCR product from unincorporated nucleotides, oligonucleotide primers, unwanted PCR products, and template DNA using glass beads, electroelution, or phenol extraction of low gelling/melting temperature agarose (*UNIT 2.6*).

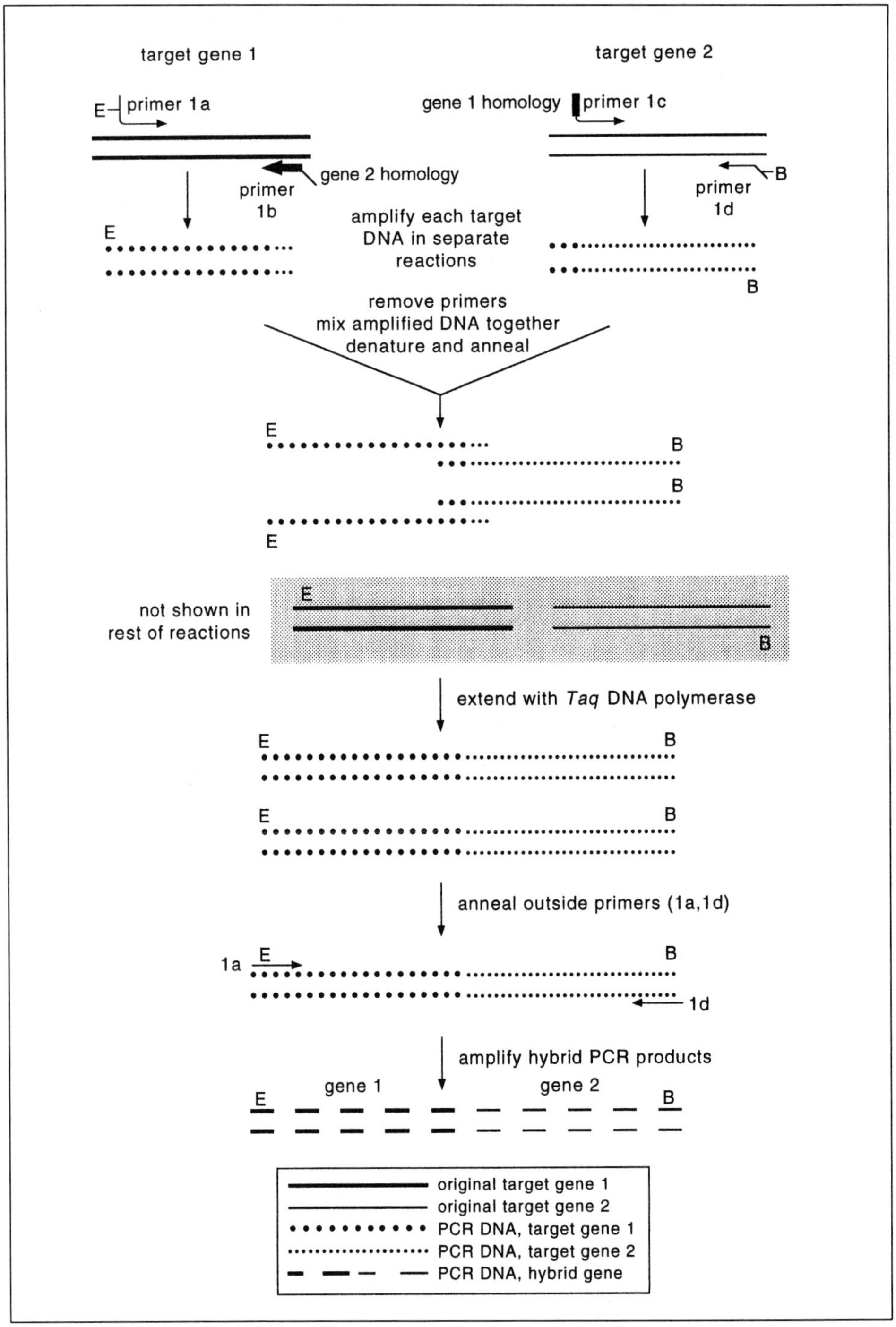

Figure 3.17.2 Creating a recombinant DNA molecule by sequential PCR amplifications. Primer 1b has a region of homology to target gene 2 (open box); primer 1c has a region of homology to target gene 1 (closed box). Abbreviations: E, *Eco*RI; B, *Bam*HI.

7a. *For cloning by blunt-end ligation*: Repair the 3′ ends of the amplified fragment with DNA polymerase I (Klenow fragment).

7b. *For primers that contain unique restriction sites*: Digest half the amplified DNA in 20 μl with the appropriate restriction enzyme(s). Use an excess of enzyme, and digest for several hours.

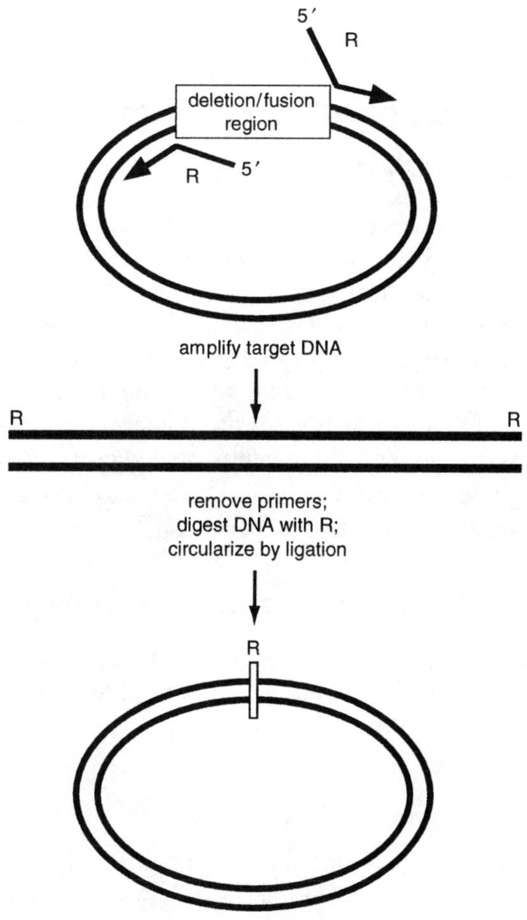

Figure 3.17.3 Inserting a restriction site by inverse PCR. Abbreviation: R, restriction enzyme not found in target plasmid.

8. Digest 0.2 to 2 µg recipient vector in 20 µl with compatible restriction enzymes. If necessary, treat vector DNA with calf intestinal alkaline phosphatase (*UNIT 3.10*) to prevent recircularization during ligation.

9. Separate linearized vector from uncut vector by agarose or low-gelling/melting temperature gel electrophoresis. Recover linearized vector from the gel by adsorption to glass beads, electroelution, or phenol extraction of low-gelling/melting temperature agarose.

10. Ligate the PCR fragment into the digested vector (*UNIT 3.16*).

11. Transform an aliquot of each ligation into *E. coli* (*UNIT 1.8*). Prepare plasmid miniprep DNA from a subset of transformants (*UNIT 1.6*).

12. Digest the plasmid DNA of the selected transformants with the appropriate restriction endonuclease. Analyze the digestions by agarose gel electrophoresis to confirm fragment incorporation.

13. Sequence the amplified fragment portion of the plasmid DNA to check for mutations (*UNIT 7.4*). Alternatively, screen the subset of transformants using a biochemical or genetic functional assay if available.

Reference: Innis et al., 1990.

Contributor: Elaine A. Elion

UNIT 3.18

Labeling and Colorimetric Detection of Nonisotopic Probes

3.18

Biotin and digoxigenin are the nonisotopic labels that are used most frequently and they are commercially available. Either label can be easily incorporated into DNA probes and detected colorimetrically; a number of fluorochromes, as well as alkaline phosphatase and horseradish peroxidase (which produce colored precipitates) are available directly conjugated to anti-digoxigenin antibodies and to avidin. Chemiluminescent detection methods (*UNIT 3.19*) and indirect immunofluorescent techniques (*UNITS 14.6 & 14.7*) also provide sensitive alternatives for many molecular biology applications. Biotin- and digoxigenin-labeled probes have a long shelf life (≥2 years), so many micrograms of DNA can be labeled in one reaction to provide probes of constant quality for multiple experiments.

BASIC PROTOCOL 1

PREPARATION OF BIOTINYLATED PROBES BY NICK TRANSLATION

Biotin-11-dUTP is substituted for dTTP in a standard nick translation reaction mixture (*UNIT 3.5*) and the DNase I concentration is adjusted to ensure a size range of 100 to 500 nucleotides. Other biotinylated nucleotides can be used in place of biotin-11-UTP.

Materials *(see APPENDIX 1 for items with ✓)*

E. coli DNA polymerase I (*UNIT 3.5*) and 10× buffer (*UNIT 3.4*)
0.5 mM 3dNTP mix (minus dTTP; *UNIT 3.4*)
✓ 0.5 mM biotin-11-dUTP stock
100 mM 2-mercaptoethanol (2-ME)
Test DNA
1 mg/ml DNase I stock (*UNIT 3.12*) prepared in 0.15 M NaCl/50% glycerol
DNA molecular weight markers (*UNIT 2.5B*)
✓ 0.5 M EDTA, pH 8.0
10% (w/v) SDS
100% ethanol
✓ SDS column buffer
1-ml syringe

1. Prepare a 100-μl reaction mix as follows:

   10 μl 10× *E. coli* DNA polymerase I buffer
   10 μl 0.5 mM 3dNTP mix
   10 μl 0.5 mM biotin-11-dUTP stock
   10 μl 100 mM 2-ME
   2 μg DNA
   20 U *E. coli* DNA polymerase I
   DNase I stock diluted 1:1000 in cold H_2O immediately before use
   H_2O to 100 μl.

   Incubate reaction 2 to 2.5 hr at 15°C.

2. Place reaction on ice. Remove a 6-μl aliquot, boil it 3 min, and place on ice 2 min.

3. Load aliquot on an agarose minigel, along with suitable size markers (0.1- to 10-kb range). Run gel quickly (15 V/cm) in case additional incubation is necessary.

4. If the digested DNA is between 100 and 500 nucleotides, proceed to step 5. If the probe size is between 500 and 1000 nucleotides (or larger) add a second aliquot of DNase I and incubate further.

   Additional DNase I is added in a more concentrated form than in the initial reaction to avoid significant volume changes. Monitor the additional incubation carefully to avoid complete digestion of the probe.

5. Add 2 µl of 0.5 M EDTA, pH 8.0 and 1 µl of 10% SDS to the reaction. Heat 10 min at 68°C to stop the reaction and inactivate the DNase I.

6. Prepare a Sephadex G-50 spin column in a 1-ml syringe (UNIT 3.4). Wash syringe and silanized glass wool plug with 2 ml of 100% ethanol, then 4 ml water. Pack G-50 resin to the 1-ml mark. Wash column 3 to 4 times with 100 µl SDS column buffer before loading the sample.

7. Separate the biotinylated probe from unincorporated nucleotides. The eluted probe concentration should be ~20 ng/µl (for 2 µg nick-translated DNA) and is ready to use without further treatment. Probe can be stored at −20°C for years without loss of activity.

   Biotinylated DNA should not be subjected to phenol extraction because biotin causes the probe to partition to the phenol/water interface or completely into the phenol if heavily biotinylated.

8. Assess the extent of the biotinylation reaction and the probe quality by colorimetric (Support Protocol) or chemiluminescent (UNIT 3.19) detection.

   Under the standard nick translation conditions provided, ~50 biotin molecules are incorporated per kilobase of DNA.

PREPARATION OF BIOTINYLATED PROBES BY RANDOM OLIGONUCLEOTIDE–PRIMED SYNTHESIS

BASIC PROTOCOL 2

A priming reaction using random octamers that have been biotinylated at the 5′ end ensures that every probe molecule generated contains at least one biotin, but biotin-16-dUTP is also included in the reaction so that additional biotin is incorporated into some of the probe molecules. Other biotinylated nucleotides (e.g., biotin-14-dATP and biotin-11-dUTP) can be used in place of the biotin-16-dUTP.

Materials *(see APPENDIX 1 for items with ✓)*

Linear template DNA ≥200 bp
✓ Biotinylated random octamers
✓ dNTP/biotin mix
5 U/µl Klenow fragment (UNIT 3.5)
✓ TE buffer, pH 7.5
✓ 0.5 M EDTA, pH 8.0
4 M LiCl
100% and 70% ethanol, ice-cold

1. Place 500 ng to 2 µg template DNA in a 1.5-ml microcentrifuge tube. Add nuclease-free water to a total volume of 34 µl.

2. Denature the DNA 5 min in boiling water. Place on ice 5 min and microcentrifuge briefly.

3. Add the following to the sample in the order listed:

   10 µl biotinylated random octamers
   5 µl dNTP/biotin mix
   1 µl (5 U) Klenow fragment.

   Incubate 1 hr at 37°C.

   If desired, an additional 2.5 U of Klenow fragment can be added after 30 min of incubation to "boost" the reaction. Continuing the incubation for up to 6 hr can increase the yield.

4. Terminate reaction by adding 3 µl of 0.5 M EDTA, pH 8.0. Precipitate the probe by adding 5 µl of 4 M LiCl and 150 µl ice-cold 100% ethanol. Place 30 min on dry ice.

5. Microcentrifuge 10 min at top speed, room temperature, and wash DNA pellet with ice-cold 70% ethanol.

6. Resuspend DNA pellet in 20 µl TE buffer, pH 7.5. Assess the quality of the biotinylation reaction by colorimetric (Support Protocol) or chemiluminescent (UNIT 3.19) detection.

SUPPORT PROTOCOL

COLORIMETRIC DETECTION OF BIOTINYLATED PROBES

The detectability of biotinylated probes is a function of the number of biotin molecules per kilobase, rather than the specific activity of a corresponding radiolabeled probe, and the extent of biotinylated deoxynucleotide incorporation is checked by a colorimetric assay (similar assessment can be made by chemiluminescent detection; UNIT 3.19).

Additional Materials *(also see Basic Protocol 1 or 2; see* APPENDIX 1 *for items with ✓)*

Biotinylated standard DNA (Basic Protocols or Life Technologies) and test DNA
✓ DNA dilution buffer
✓ Alkaline phosphatase pH 7.5 (AP 7.5) buffer
Blocking buffer: 3% (w/v) BSA fraction V in AP 7.5 buffer
1 mg/ml streptavidin–alkaline phosphatase (AP) conjugate (Life Technologies)
✓ Alkaline phosphatase pH 9.5 (AP 9.5) buffer
75 mg/ml nitroblue tetrazolium (NBT)
50 mg/ml 5-bromo-4-chloro-3-indoyl phosphate (BCIP)
✓ TE buffer, pH 8.0

Small piece of membrane (e.g., 5 × 3 cm^2): nitrocellulose *or* uncharged nylon
Sealable bags

NOTE: To avoid nonspecific background, wear powder-free gloves when handling the membranes.

1. Prepare biotinylated standard DNA in concentrations of 0, 1, 2, 5, 10, and 20 pg/µl in dilution buffer. Dilute the biotinylated test DNA in a similar fashion.

2a. *For nitrocellulose membrane:* Spot 1 μl of each dilution on a small piece of nitrocellulose. Bake the membrane ~1 hr at 80°C. Proceed directly to step 3.

Nitrocellulose membranes cannot be used in chemiluminescent detection procedures.

2b. *For nylon membrane:* Spot 1 μl of each dilution on the nylon membrane. Air-dry and cross-link DNA to the membrane by UV illumination (*UNIT 3.19*). Proceed to step 3 or develop membrane as described for chemiluminescent detection (*UNIT 3.19*).

If chemiluminescent detection is used, the test DNA should be diluted in a series from 10^{-1} to 10^{-6}, and must be visible at a 10^{-3} dilution. If it is not, the probe is not sufficiently biotinylated.

3. Float membrane in a small volume of AP 7.5 buffer for 1 min to rehydrate. Block membrane with 10 ml blocking buffer in a sealable bag (cut to size) without air bubbles 1 hr at 37°C.

4. Dilute 10 μl streptavidin–AP conjugate with 10 ml AP 7.5 buffer (1 μg/ml final). Cut corner of bag and squeeze out blocking buffer. Replace with the streptavidin–AP solution and reseal. Incubate 10 min at room temperature with agitation on a platform shaker.

5. Remove membrane from bag and transfer to a shallow dish. Wash in 200 ml AP 7.5 buffer twice, 15 min each time, and in 200 ml AP 9.5 buffer once, 10 min with gentle agitation.

6. Add 33 μl of 75 mg/ml NBT to 7.5 ml AP 9.5 buffer and invert to mix. Add 25 μl of 50 mg/ml BCIP and mix gently.

7. Incubate membrane with NBT/BCIP solution in a shallow dish in low light, checking periodically until color development is satisfactory (usually 15 to 60 min).

8. Stop reaction by washing with TE buffer, pH 8.0. Check incorporation of biotinylated dUTP by comparing the intensities of standard and test DNA. If the probe is at least half as intense as the standard DNA at the corresponding dilution, it should be suitable as an in situ hybridization probe.

PREPARATION AND DETECTION OF DIGOXIGENIN-LABELED DNA PROBES

ALTERNATE PROTOCOL

The digoxigenin-based detection system is an alternative nonisotopic labeling method offered by Boehringer Mannheim. Detection is achieved by incubation with antidigoxigenin antibodies coupled directly to one of several fluorochromes or enzymes, or by indirect immunofluorescence (*UNIT 14.6*). Biotin- and digoxigenin-labeled probes can be visualized simultaneously using a different fluorochrome for each probe. Digoxigenin-11-dUTP can be incorporated into DNA by either of the nick translation or random oligonucleotide–primed synthesis protocols.

1. Set up a standard 100-μl reaction (see Basic Protocol 1), but use 10 μl of a 10× digoxigenin-11-dUTP/dTTP stock solution (see recipe) in place of 10 μl of a 10× biotin-11-dUTP stock solution and the same amount of DNase. Incubate 2 hr at 15°C.

2. Electrophorese an aliquot on a minigel to check the probe size.

3. Continue the reaction until the correct probe size is obtained.

4. Stop the reaction by adding 2 µl of 0.5 M EDTA and 1 µl 10% SDS and heating 10 min at 68°C.

5. Separate the probe from unincorporated nucleotides using a G-50 spin column.

   It is advisable to avoid phenol extraction of digoxigenin-labeled probes.

6. Use colorimetric (Support Protocol) or chemiluminescence (*UNIT 3.19*) detection for digoxigenin-labeled probes, using alkaline phospatase–conjugated anti-digoxigenin instead of streptavidin–alkaline phosphatase. To check incorporation, use either digoxigenin-labeled standard DNA (part of the Genius kit, Boehringer Mannheim) or a digoxigenin-labeled DNA that has been used successfully as a control.

Reference: Langer, et al., 1981

Contributors: Ann Boyle, Heather Perry-O'Keefe

UNIT 3.19

Chemiluminescent Detection of Nonisotopic Probes

Chemiluminescent detection is sensitive enough to detect a single-copy gene from 1 µg human genomic DNA in a Southern blot or a moderately repetitive transcript in a northern blot.

BASIC PROTOCOL

CHEMILUMINESCENT DETECTION OF BIOTINYLATED PROBES

The reaction of alkaline phosphatase with a chemiluminescent substrate (Fig. 3.19.1) results in the highly sensitive detection of hybridized target DNA. Many of the reagents used in the protocol are available commercially; Table 3.19.1 lists suppliers of these detection kits.

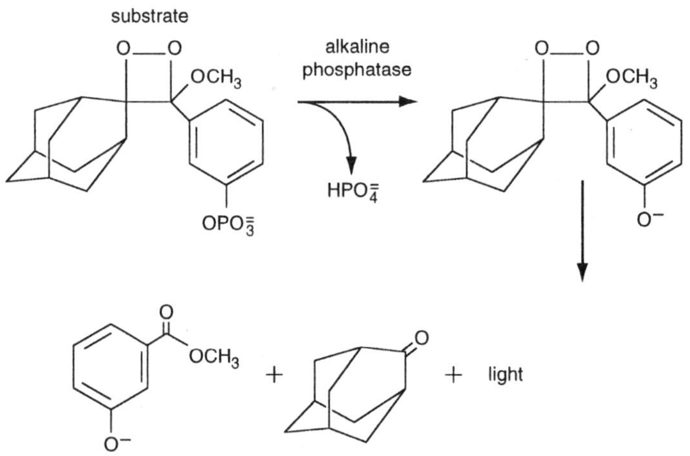

Figure 3.19.1 Chemiluminescence reaction. The enzyme alkaline phosphatase cleaves a phosphate group off the chemiluminescent substrate. The intermediate is unstable and quickly decomposes, emitting light in the process. Reprinted with permission from Millipore.

Table 3.19.1 Chemiluminescent Detection Kits

Kit	Supplier[a]
AmpliProbe	ImClone Systems
ECL	Amersham
Flash	Stratagene
Gene Images	U.S. Biochemical
Genius	Boehringer Mannheim
LightSmith	Promega
PhotoGene	Life Technologies
Phototope	NEB
SouthernLight	Tropix

[a]See suppliers' addresses in APPENDIX 4.

3.19

Materials *(see* APPENDIX 1 *for items with ✓)*

- Uncharged nylon membrane with DNA (UNIT 2.9) blotted via neutral transfer or RNA (UNIT 4.9)
- Biotinylated probes (UNIT 3.18)
- ✓ Blocking solution
- ✓ Wash buffers I and II
- 1 mg/ml streptavidin
- ✓ 0.38 mg/ml biotinylated alkaline phosphatase
- Chemiluminescent dioxetane substrate (Table 3.19.2), room temperature
- Substrate buffer, pH 9.6, room temperature (see Table 3.19.2)

- Blotting paper (Whatman 3MM or equivalent)
- Calibrated UV source (Support Protocol)
- Heat-sealable hybridization bags

CAUTION: Exposure to UV radiation poses a significant health hazard. Wear UV-protected goggles and shield exposed skin.

1. Fasten the corners of the blotted nylon membrane to a dry piece of blotting paper, nucleic acid–side up, with paper clips. Place in an incubator and dry completely, 15 to 30 min at 42° to 80°C or overnight at room temperature.

   Nitrocellulose membrane will quench chemiluminescent reactions and must not be used.

2. Expose membrane, nucleic acid–side up, to the UV source. Cross-link for the calculated optimal time period.

3. Hybridize membrane with biotinylated probes and wash with appropriate stringency.

   The amount of wash and detection solutions needed depend on the size of the membrane. One volume (V) = membrane area in $cm^2 \times 0.05\ ml/cm^2$.

4. Following hybridization, place membrane in a hybridization bag.

5. Seal bag, leaving a small area in one corner that can be made into a spout through which buffers can be easily added and removed.

6. Carefully add 1 vol blocking solution to hybridization bag. Incubate 1 min at room temperature with moderate shaking. Drain and discard solution.

Table 3.19.2 Chemiluminescent Substrates for Detection of Nonisotopic Probes[a]

Dioxetane substrate	Buffer	Source[c]
Lumigen-PPD[b] (0.33 mM)	2-amino-2-methyl-1-propanol (pH 9.6)/0.88 mM $MgCl_2$/750 mM CTAB/1.13 mM fluorescein surfactant	BM, LT, LU, NEB
Lumi-Phos 530[b] (0.33 mM)	2-amino-2-methyl-1-propanol (pH 9.6)/0.88 mM $MgCl_2$/750 mM CTAB/1.13 mM fluorescein surfactant	BM, LT, LU
AMPPD (0.25 mM)	1 mM DEA/1 mM $MgCl_2$, pH 10	TR
CSPD (0.25 mM)	1 mM DEA/1 mM $MgCl_2$, pH 10	TR

[a]Abbreviations: AMPPD, disodium 3-(4-methoxyspiro{1,2-dioxetane-3,2-tricyclo[3.3.1.1^{3,7}] decan}-4-yl)phenyl phosphate; CSPD, AMPPD with substituted chlorine group on adamantine chain; CTAB, cetyltrimethylammonium bromide; DEA, diethanolamine; Lumigen-PPD and Lumi-Phos 530: 4-methoxy-4-(3-phosphate phenyl)-spiro-(1,2,-dioxetane-3,2-adamantine), disodium salt.

[b]Lumi-Phos 530 has a fluorescence enhancer; Lumigen-PPD does not.

[c]Abbreviations: BM, Boehringer Mannheim; LT, Life Technologies; LU, Lumigen; NEB, New England Biolabs; TR, Tropix. Addresses and phone numbers of suppliers are provided in *APPENDIX 4*.

Nonfat dry milk contains biotin and cannot be used as a blocking reagent with biotinylated probes.

7. Add 1 mg/ml streptavidin to 1 vol blocking solution to a final concentration of 1 µg/ml and add to hybridization bag. Incubate 4 min at room temperature with moderate shaking. Drain and discard solution.

8. Add 10 vol wash buffer I to hybridization bag. Wash membrane 4 min at room temperature with moderate shaking. Drain and discard solution. Repeat once and discard solution.

   If background is too high, increase either the number of washes or their duration.

9. Add biotinylated alkaline phosphatase to 1 vol blocking solution to a final concentration of 0.5 µg/ml. Add to hybridization bag. Incubate 4 min at room temperature with moderate shaking. Drain and discard solution.

10. Wash membrane twice as in step 8, using 10 vol of wash buffer II.

11. Using the substrate buffer dilute enough chemiluminescent dioxetane substrate to make 0.5 vol at 1× final concentration. Add diluted substrate to hybridization bag. Incubate 4 min at room temperature with moderate shaking. Open bag and drain as thoroughly as possible.

12. Smooth out any wrinkles in bag and reseal. Place in a film cassette with the nucleic acid side of the membrane facing up and put X-ray film on top. Close cassette and expose 10 to 20 min.

    See Table 3.19.3 for a troubleshooting guide for chemiluminescent detection.

CHEMILUMINESCENT DETECTION OF DIGOXIGENIN-LABELED PROBES

ALTERNATE PROTOCOL

The digoxigenin system utilizes an anti-digoxigenin antibody conjugated to alkaline phosphatase; the antibody recognizes digoxigenin-dUTP that has been incorporated into the probe (*UNIT 3.18*). In this protocol, chemiluminescent detection is based on the use of the Genius kit (Boehringer-Mannheim; see Table 3.19.1), the only kit that works with digoxigenin-labeled probes.

1. Hybridize a positively charged nylon membrane blotted with DNA (*UNIT 2.9*) or RNA (*UNIT 4.9*) with a digoxigenin-labeled probe (*UNIT 3.18*).

2. Block, bind antibody, and develop according to manufacturer's instructions.

3. Expose the membrane to X-ray film for ~20 min.

   The membrane can be used for multiple exposures, and can be stripped and redetected with different probes.

3.19

CALIBRATING AN ULTRAVIOLET LIGHT SOURCE

SUPPORT PROTOCOL

Materials

UV source: transilluminator, hand-held UV lamp, homemade box with germicidal bulbs, or cross-linker (e.g., Stratalinker, Stratagene)
Radiometer and radiometer sensor

The UV source should have sufficient power at 254 nm to generate a minimum of 100 μW/cm^2. Most transilluminators have power outputs in the 500 to 1000 μW/cm^{2}range. Hand-held lamps will have slightly lower outputs.

UV cross-linking of nucleic acids to nylon membrane is a critical step that significantly affects the quality of the data. Therefore, calibration of the UV source is important.

Table 3.19.3 Guide to Troubleshooting Chemiluminescent Detection Techniques

Problem	Possible cause	Solution
Low or insufficient signal	Insufficient biotin labeling—check by comparing sample and control dilution series	If probe is only visible at 10^{-1} and 10^{-2} dilutions, repeat the biotinylation reaction; if visible at dilutions of 10^{-3} or greater, consider other possible causes.
	Inadequate cross-linking	Make sure the membrane is completely dry before cross-linking.
		Calibrate UV source with a radiometer.
	Incorrect pH of the chemiluminescent substrate	Check pH of substrate buffer and adjust to pH 9.6.
Uniform or uneven high background	Inadequate washing	Increase number of washes, ensuring the membrane floats freely in bag during detection steps.
	Contaminated wash solutions	Mold can grow in the wash I solution, leading to spotting on the membrane; filter through a 0.45-μm filter.
	Unbuffered blocking and wash I solutions	Make sure that SDS solutions are made up in phosphate buffer as described.

3.19

UV sources are calibrated with a radiometer, which determines the power output in $\mu W/cm^2$. Radiation safety offices at academic research institutions can often provide radiometers for calibration. The UV light must be turned on and the bulbs warmed up for 1 to 2 min before calibration. The calibration procedure itself takes only a few minutes and should be repeated as the bulbs age. Transilluminator power output is calibrated at the glass surface and hand-held lamp output is calibrated at the distance at which the entire membrane will be evenly exposed to the UV light.

Once the power output is measured, the optimal exposure time for the light source can be calculated as follows:

$$T = \frac{(33{,}000\ \mu W/cm^2)\ (1\ sec)}{P \mu W/cm^2}$$

where T is the exposure time in seconds and P is the power output of the UV source in $\mu W/cm^2$.

Optimal exposure time can also be determined empirically. Transfer duplicate lanes of nucleic acid to a membrane and expose different portions to UV light for increasing amounts of time varying by 1-min increments. Remember to cover previously exposed portions of the membrane with aluminum foil. There should be a marked difference in the amount of detectable and hybridizable nucleic acids.

Reference: Beck and Koster, 1990.

Contributors: Heather Perry-O'Keefe and Carol M. Kissinger

Preparation and Analysis of RNA

4

Isolating clean, intact RNA is important to cloning experiments and is essential to analyzing gene expression. RNA from any cell can be copied into double-stranded DNA and cloned, resulting in the production of a cDNA library specific to the cell type (Chapters 5 and 6). Analysis of RNA structure and synthesis is useful in gene expression studies. This chapter begins by describing several methods commonly used to isolate RNA and concludes with methods used to analyze RNA structure and synthesis.

The critical factor in all RNA experiments is the isolation of full-length RNA. The major source of failure in any attempt to produce RNA is contamination by ribonucleases, enzymes that are very stable and generally require no cofactors. Therefore, a small amount of RNase in an RNA preparation will create a real problem. To avoid contamination problems, all solutions, glassware, and plasticware should be specially treated. Because hands are a major source of contaminating RNase, gloves should be worn during all RNA preparations. Solutions used in RNA preparation should be treated with diethylpyrocarbonate (DEPC), which inactivates ribonuclease. Solutions containing Tris cannot be effectively treated with DEPC because Tris reacts with DEPC to inactivate it. Glassware should be baked at 300°C for 4 hr (autoclaving will not fully inactivate many RNases). Plasticware can be rinsed with chloroform, although if used straight out of the package is generally free from contamination.

The first step in all RNA isolation protocols involves lysing the cell in a chemical environment that results in denaturation of ribonuclease. The RNA is then fractionated from the other cellular macromolecules. The cell type from which the RNA is to be isolated and the eventual use of that RNA will determine which of the procedures described will be appropriate.

Three common methods are presented for preparing total RNA from eukaryotic cells. In *UNIT 4.1*, a gentle detergent is utilized to lyse the cells. In *UNIT 4.2*, cells are lysed using guanidinium isothiocyanate; this requires very few manipulations, gives clean RNA from many sources, and is the method of choice for tissues that have high levels of endogenous RNase. In *UNIT 4.3*, cells are lysed with phenol and SDS. This produces clean, full-length RNA from large quantities of plant cells, and works well with several mammalian cells and tissues.

A protocol for extracting RNA from gram-negative and gram-positive bacteria (*UNIT 4.4*) uses protease digestion and organic extraction to remove protein and nuclease digestion to remove DNA. An alternate protocol provides a relatively simple method for rapidly isolating RNA from *E. coli*, without organic extractions, protease, or nuclease treatment.

The protocols for producing RNA from eukaryotic cells yield total RNA, which contains primarily ribosomal RNA and transfer RNA. Many techniques require messenger RNA that is largely free of contaminating rRNA and tRNA. A protocol for the preparation of poly(A)$^+$ RNA (*UNIT 4.5*), which is highly enriched for messenger RNA, utilizes oligo(dT) cellulose in the purification process.

One of the primary uses of RNA isolation procedures is the analysis of gene expression. In order to elucidate the regulatory properties of a gene, it is necessary to know the structure, amount, level, size, and synthesis rate of the RNA produced from that gene. Protocols commonly used for analyzing in detail RNA structure and amount include S1 analysis (*UNIT 4.6*), ribonuclease protection (*UNIT 4.7*), primer extension (*UNIT 4.8*), and northern blots (*UNIT 4.9*). To determine the number of active RNA polymerase molecules on a given eukaryotic gene, the nuclear runoff technique is used.

S1 analysis, ribonuclease protection, and primer extension can be used to determine both the endpoint and the amount of a specific RNA. The first two protocols use a single-stranded probe that is complementary to the sequence of the measured RNA. The S1 technique uses an end-labeled single-stranded DNA probe, allowing unambiguous determination of the 5′ end of a message, with a low background on the final gel. Ribonuclease protection is a more sensitive technique than S1 analysis because it utilizes a body-labeled probe. However, high background problems may occur. Mapping of RNA across discontinuities, such as splice sites, is possible using primer extension. This protocol employs a labeled oligomer of defined sequence that is extended to the end of any homologous RNA by reverse transcriptase.

Northern blot hybridization can determine the size and amount of any specific RNA. In this protocol, RNA is separated on an agarose gel, transferred to nitrocellulose, and hybridized to a labeled specific probe. This protocol is very sensitive for detection of the level of a specific RNA, but does not allow determination of the precise structure of the RNA.

The nuclear runoff technique is currently the most sensitive procedure for measuring specific gene transcription at the time of lysis. This method can determine the number of active RNA polymerase molecules that are traversing any particular segment of DNA. It analyzes directly how the rate of gene transcription varies when the growth state of a cell changes. It can be used to assess whether changes in the mRNA levels reflect a change in mRNA synthesis or a change in degradation or transport out of the nucleus.

Robert E. Kingston

UNIT 4.1

Preparation of Cytoplasmic RNA from Tissue Culture Cells

BASIC PROTOCOL

Materials *(see APPENDIX 1 for items with ✓)*

✓ Phosphate-buffered saline (PBS), ice-cold
✓ Lysis buffer, ice-cold
20% sodium dodecyl sulfate (SDS)
20 mg/ml proteinase K
25:24:1 phenol/chloroform/isoamyl alcohol (*UNIT 2.1*)
24:1 chloroform/isoamyl alcohol
✓ 3 M sodium acetate, pH 5.2 (DEPC-treated)
100% ethanol
75% ethanol/25% 0.1 M sodium acetate, pH 5.2
✓ DEPC-treated water
Beckman JS-4.2 rotor or equivalent

CAUTION: DEPC is a suspected carcinogen. Handle with care.

1a. *For monolayer cultures*: Wash cultures three times with 1 ml ice-cold PBS for each 10-cm dish. Scrape cells into a small volume of ice-cold PBS and transfer to a centrifuge tube on ice. Centrifuge 5 min at 300 g, discard supernatant, and keep cold.

1b. *For suspension cultures*: Centrifuge suspension cultures 5 min 300 × g, discard supernatant, and resuspend in one-half the original volume ice-cold PBS. Centrifuge and discard supernatant.

Procedure is for 2 × 10^7 cells obtained from two 10-cm dishes or ~20 ml suspension culture.

2. Resuspend cells in 375 µl ice-cold lysis buffer and incubate 5 min on ice. Microcentrifuge 2 min at 4°C. Transfer supernatant to a clean tube containing 4 µl of 20% SDS and vortex immediately.

Resuspend cells with careful but vigorous vortexing and avoid foaming. The suspension should clear rapidly, indicating cell lysis.

3. Add 2.5 µl of 20 mg/ml proteinase K, and incubate 15 min at 37°C.

4. Extract with 400 µl phenol/chloroform/isoamyl alcohol by vortexing 1 min. Microcentrifuge, transfer upper phase to a clean tube, and repeat extraction. Extract with 400 µl chloroform/isoamyl alcohol, and transfer upper phase to a clean tube.

5. Add 40 µl of 3 M sodium acetate, pH 5.2, and 1 ml of 100% ethanol. Mix and precipitate 15 to 30 min on ice or overnight at –20°C.

6. Microcentrifuge 15 min at 4°C and rinse the pellet with 1 ml of 75% ethanol/25% 0.1 M sodium acetate, pH 5.2. Dry and resuspend in 100 µl DEPC-treated water. Dilute 10 µl in 1 ml water and determine A_{260} and A_{280}. Store remainder at –70°C.

Yields are 30 to 100 µg RNA for a confluent 10-cm dish or 10^7 lymphoid cells. RNA at 1 mg/ml has an A_{260} of 25. Ratios of A_{260} to A_{280} should range between 1.7 and 2.0.

4.1

REMOVAL OF CONTAMINATING DNA

SUPPORT PROTOCOL

If RNA is isolated from cells transiently transfected with cloned DNA, contaminating DNA must be removed.

Additional Materials *(also see Basic Protocol; see* APPENDIX 1 *for items with ✓)*

✓ TE buffer, pH 7.4
100 mM $MgCl_2$/10 mM dithiothreitol (DTT)
✓ 2.5 mg/ml DNase I (RNase-free)
Placental ribonuclease inhibitor (e.g., RNAsin from Promega Biotec) *or* vanadyl-ribonucleoside complex
✓ DNase stop mix

1. Redissolve the RNA in 50 µl TE buffer after the microcentrifugation in step 6 of the Basic Protocol.

2. Prepare on ice a cocktail containing (per sample) 10 µl of 100 mM $MgCl_2$/10 mM DTT, 0.2 µl of 2.5 mg/ml RNase-free DNase, 0.1 µl placental ribonuclease inhibitor (25 to 50 U/µl), and 39.7 µl TE buffer. Add 50 µl to each RNA sample, mix, and incubate 15 min at 37°C.

3. Stop the reaction with 25 µl DNase stop mix. Extract once each with phenol/chloroform/ isoamyl alcohol and chloroform/isoamyl alcohol.

4. Add 325 µl ethanol and precipitate 15 to 30 min on ice or overnight at −20°C. Resume the Basic Protocol at step 6.

Reference: Favoloro et al., 1980.

Contributor: Michael Gilman

UNIT 4.2

Guanidinium Methods for Total RNA Preparation

BASIC PROTOCOL

4.2

CsCl PURIFICATION OF RNA FROM CULTURED CELLS

This protocol produces RNA that is clean enough for northern, S1, or SP6 analysis. DEPC treatment of solutions is required to inhibit RNase activity.

Materials *(see APPENDIX 1 for items with ✓)*

✓ Phosphate-buffered saline (PBS)
✓ Guanidinium solution
✓ 5.7 M CsCl, DEPC-treated
✓ TES solution
✓ 3 M sodium acetate, pH 5.2 (DEPC-treated)
100% ethanol
✓ DEPC-treated water

Beckman JS-4.2 and SW-55 rotors or equivalents
6-ml syringe with 20-G needle
13 × 51–mm silanized (*APPENDIX 3*) and autoclaved polyallomer ultracentrifuge tube

CAUTION: DEPC is a suspected carcinogen. Handle with care.

For monolayer cultures:

1a. Wash cells at room temperature with 5 ml PBS per dish, two times.

2a. Add 3.5 ml guanidinium solution for $\leq 10^8$ cells, dividing the solution equally between the dishes. Recover viscous lysate by scraping dishes with a rubber policeman, remove using a 20-G needle fitted on a 6-ml syringe. Combine lysates.

For suspension cultures:

1b. Centrifuge cells ($\leq 10^8$) 5 min at 300 × *g* in JS-4.2 rotor. Resuspend pellet in an amount of PBS equal to half the original volume and centrifuge again.

Carry out steps 1 to 4 at room temperature.

2b. Add 3.5 ml guanidinium solution to tube.

3. Draw the resultant extremely viscous solution up and down four times through a 6-ml syringe with 20-G needle, and transfer to a clean tube.

It is critical that chromosomal DNA is sheared in this step in order to reduce viscosity.

4. Place 1.5 ml of 5.7 M CsCl in a 13 × 51–mm silanized and autoclaved polyallomer ultracentrifuge tube. Layer 3.5 ml of cell lysate on top of CsCl cushion to create a step gradient. The interface should be visible.

5. Centrifuge 12 to 20 hr at 150,000 × *g* in a Beckman SW-55 rotor, 18°C (slow acceleration and deceleration).

6. Remove supernatant by placing the end of Pasteur pipet at top of solution and lowering it as level of solution lowers. Leave ~100 µl in bottom, invert tube carefully, and pour remaining liquid off.

   There should be a white band of DNA at the interface—care must be taken to remove this band completely, as it contains cellular DNA.

7. Drain pellet 5 to 10 min, then resuspend it in 360 µl TES solution by repeatedly drawing solution up and down in a pipet. Allow pellet to resuspend 5 to 10 min at room temperature. Transfer to a clean microcentrifuge tube.

   It is critical to allow ample time for resuspension of this pellet or the yield of RNA will be significantly decreased.

8. Add 40 µl of 3 M sodium acetate, pH 5.2, and 1 ml of 100% ethanol, precipitate 30 min on dry ice/ethanol, and microcentrifuge 10 to 15 min. Resuspend pellet in 360 µl water and repeat precipitation.

9. Drain pellet 10 min and dissolve in ~200 µl water. Quantitate by diluting 10 µl to 1 ml and reading the A_{260} and A_{280} (*APPENDIX 3*). Store RNA at –70°C either as an aqueous solution or as an ethanol precipitate.

CsCl PURIFICATION OF RNA FROM TISSUE

ALTERNATE PROTOCOL 1

Additional precautions must be taken when purifying RNA from tissue, as certain organs (pancreas and spleen) have very high endogenous levels of RNase.

Additional Materials *(also see Basic Protocol; see APPENDIX 1 for items with ✓)*

Liquid nitrogen
✓ Tissue guanidinium solution
20% (w/v) *N*-lauroylsarcosine (Sarkosyl)
Cesium chloride (CsCL)
✓ Tissue resuspension solution
25:24:1 phenol/chloroform/isoamyl alcohol (*UNIT 2.1*)
24:1 chloroform/isoamyl alcohol

Tissuemizer
Sorvall SS-34 and Beckman SW-28 rotors or equivalents
Silanized and autoclaved SW-28 polyallomer tube (*APPENDIX 3*)

1. Rapidly remove tissue from animal, cut pieces of ≤2 g, and quick-freeze in liquid nitrogen. Add 20 ml tissue guanidinium solution (does not contain Sarkosyl) for ~2 g of tissue, and immediately grind in a tissuemizer with two or three 10-sec bursts.

   It is critical to quick-freeze the tissue. Placing the tissue in guanidinium and then waiting to grind it will result in degraded RNA.

2. Centrifuge 10 min at 12,000 × *g* in SS-34 rotor, 12°C. Add 0.1 vol of 20% Sarkosyl to supernatant, and heat 2 min at 65°C.

3. Add 0.1 g CsCl/ml of solution and dissolve. Layer sample over 9 ml of 5.7 M CsCl in silanized and autoclaved SW-28 tube. Centrifuge overnight at 113,000 × *g* in SW-28 rotor, 22°C.

4. Carefully remove supernatant (step 6 of the Basic Protocol) and invert tube to drain. Cut off bottom of tube (containing RNA pellet) and place it in a 50-ml plastic tube. Add 3 ml tissue resuspension buffer and allow pellet to resuspend overnight or longer at 4°C.

5. Extract solution sequentially with 25:24:1 phenol/chloroform/isoamyl alcohol, then with 24:1 chloroform/isoamyl alcohol (*UNIT 2.1*).

6. Add 0.1 vol of 3 M sodium acetate, pH 5.2, and 2.5 vol of 100% ethanol, precipitate, and resuspend RNA in water. Quantitate and store (step 9 of Basic Protocol).

4.2

ALTERNATE PROTOCOL 2

SINGLE-STEP RNA ISOLATION FROM CULTURED CELLS OR TISSUES

Additional Materials *(also see Basic Protocol; see APPENDIX 1 for items with ✓)*

✓ Denaturing solution
✓ 2 M sodium acetate, pH 4
Phenol, water-saturated
49:1 (v/v) chloroform/isoamyl alcohol
100% isopropanol
75% ethanol (prepared with DEPC-treated water)
0.5% (w/v) SDS, DEPC-treated

NOTE: Carry out all steps at room temperature unless otherwise stated.

1a. *For tissue*: Add 1 ml denaturing solution/100 mg tissue and homogenize with a few strokes in a glass Teflon homogenizer.

1b. *For cultured cells*: Either centrifuge suspension cells and discard supernatant, or remove the culture medium from cells grown in monolayer cultures (do not wash cells with saline). Add 1 ml denaturing solution/10^7 cells and pass lysate through a pipet seven to ten times.

The procedure can be carried out in sterile, disposable, round-bottom polypropylene tubes with caps; no additional treatment is necessary. Before using, test if tubes can withstand centrifugation at 10,000 × g with the mixture of denaturing solution and phenol/chloroform.

2. Transfer homogenate into a 5-ml polypropylene tube, add 0.1 ml of 2 M sodium acetate, pH 4, and mix. Add 1 ml water-saturated phenol, mix, and add 0.2 ml of 49:1 chloroform/isoamyl alcohol. Mix and incubate suspension 15 min at 0° to 4°C.

3. Centrifuge 20 min at 10,000 × *g* in SS-34 rotor, 4°C. Transfer upper aqueous phase to a fresh tube. Precipitate RNA with 1 ml (1 vol) of 100% isopropanol, place 30 min at –20°C, and centrifuge 10 min at 10,000 × *g*, 4°C.

For isolation of RNA from tissues with a high glycogen content (e.g., liver), wash out glycogen from RNA pellet by vortexing in 4 M LiCl after the isopropanol precipitation. Sediment insoluble RNA 10 min at 5000 × g. Dissolve pellet in denaturing solution and follow remainder of protocol.

4. Dissolve RNA in 0.3 ml denaturing solution and transfer to a 1.5 ml microcentrifuge tube. Precipitate RNA with 0.3 ml of 100% isopropanol (1 vol) 30 min at −20°C. Centrifuge 10 min at 10,000 × *g*, 4°C.

5. Resuspend RNA in 75% ethanol, vortex, and incubate 10 to 15 min at room temperature. Centrifuge 5 min at 10,000 × *g*.

6. Dry RNA in a vacuum 5 to 15 min, and dissolve in 100 to 200 µl DEPC-treated water or in DEPC-treated 0.5% SDS, depending on the subsequent use of RNA. Quantitate (step 9 of Basic Protocol) and store frozen at −70°C or in ethanol at −20°C.

References: Chirgwin et al., 1979; Chomczynski, 1989; Chomczynski and Sacchi, 1978.

Contributors: Robert E. Kingston, Piotr Chomczynski, and Nicoletta Sacchi

4.3

Phenol/SDS Method for Plant RNA Preparation

UNIT 4.3

BASIC PROTOCOL

This method can be used to prepare RNA from a variety of eukaryotic tissues.

Materials *(see* APPENDIX 1 *for items with ✓)*

Liquid nitrogen
✓ Grinding buffer
✓ Phenol equilibrated with TLE solution
Chloroform
8 M and 2 M LiCl (DEPC-treated)
✓ 3 M sodium acetate (DEPC-treated)
100% ethanol
✓ DEPC-treated water

Polytron (Brinkmann PT 10/35)
Beckman JA-10, JA-20, and JA-14 rotors or equivalents
50-ml Oak Ridge tube
Sarstedt tube

CAUTION: DEPC is a suspected carcinogen. Handle with care.

1. Cool mortar and pestle with liquid nitrogen. Weigh out 15 g frozen plant tissue and grind (keep frozen with liquid nitrogen) in mortar and pestle to a fine powder. Immediately transfer to a 500-ml beaker containing 150 ml grinding buffer plus 50 ml TLE-equilibrated phenol.

   For freshly harvested tissue, quick-freeze in liquid nitrogen.

2. Homogenize with Polytron ~2 min at setting 5 to 6. Add 50 ml chloroform and mix with Polytron at low speed. Transfer mixture to a 500-ml Nalgene centrifuge bottle and heat 20 min at 50°C.

3. Centrifuge 20 min at 17,700 × *g* in JA-10 rotor, 4°C. Transfer aqueous layer to a clean 500-ml Nalgene bottle (and save interface for step 7). Add 50 ml TLE-equilibrated phenol, mix, and add 50 ml chloroform.

4. Transfer remaining aqueous layer together with interface from initial phenol extraction (step 3) to a 50-ml Oak Ridge tube. Centrifuge 20 min at 12,100 × g in JA-20 rotor, 4°C.

5. Remove aqueous layer, combine with aqueous layer from step 3, and mix vigorously.

6. Centrifuge 15 min at 17,700 × *g* in JA-10 rotor, 4°C. Transfer aqueous phase to a fresh 500-ml bottle.

7. Reextract aqueous phase with TLE-equilibrated phenol until no interface is obtained (~3 extractions). Extract aqueous phase with chloroform.

8. Transfer aqueous phase to a clean 250-ml Nalgene bottle and bring to a final concentration of 2 M LiCl (0.33 vol of 8 M LiCl). Precipitate overnight at 4°C.

9. Centrifuge 20 min at 15,300 × *g* in JA-14 rotor, 4°C. Rinse pellet with 2 to 3 ml of 2 M LiCl.

10. Resuspend pellet in 5 ml water and transfer to a 15-ml Sarstedt tube. Bring to 2 M LiCl and precipitate >2 hr at 4°C.

11. Centrifuge 20 min at 12,100 × *g* in JA-20 rotor, 4°C. Rinse pellet with 2 M LiCl, and resuspend in 2 ml water. Add 200 µl of 3 M sodium acetate and 5.5 ml 100% ethanol, and precipitate overnight at −20°C, or 30 min in dry ice/ethanol.

12. Centrifuge 15 min at 17,700 × *g* in JA-10 rotor, 4°C. Resuspend pellet in 1 ml water. Dilute 10 µl to 1 ml with water and measure A_{260} and A_{280}.

Yields are 7 mg total RNA per 15 g starting material for a tissue good for making RNA, such as pea seedlings. Mature Arabidopsis plants yield only 3 mg total RNA per 15 g starting material. 1 A_{260} = 40 µg/ml RNA.

Reference: Palmiter, 1974.

4.4

UNIT 4.4

Preparation of Bacterial RNA

BASIC PROTOCOL 1

ISOLATION OF HIGH-QUALITY RNA FROM GRAM-NEGATIVE BACTERIA

This protocol produces high-quality RNA from *E. coli* or cyanobacteria suitable for northern blotting, S1 mapping, and primer extension.

Materials *(see APPENDIX 1 for items with ✓)*

100-ml *E. coli* culture *or* 500-ml cyanobacteria culture
✓ Stop buffer
✓ STET lysing solution
Buffered phenol (*UNIT 2.1*)
Chloroform (*UNIT 2.1*)
✓ 3 M sodium acetate, pH 6.0
0.2 M and 10 mM vanadyl-ribonucleoside complex (VRC; Life Technologies)
1:1 buffered phenol/chloroform
✓ DEPC-treated water
Cesium chloride, solid
CsCl cushion: 5.7 M CsCl in 100 mM EDTA, pH 7.0

100% and 70% ethanol, ice cold

Beckman JA-14 and JA-17 rotors or equivalents
15-ml polypropylene tube (Sarstedt)
Beckman TL-100 ultracentrifuge with TLA-100.3 rotor and 13 × 51–mm polycarbonate centrifuge tubes, *or* Beckman L5-65 ultracentrifuge with SW-41 rotor and 14 × 89–mm ultraclear centrifuge tubes

NOTE: Water and all other solutions should be treated with DEPC to inhibit RNase activity (*APPENDIX 1*).

CAUTION: DEPC is a suspected carcinogen and should be handled carefully with gloves. Aurintricarboxylic acid (ATA) causes irritation on contact with skin, eyes, and respiratory system. Vanadyl-ribonucleoside complex is harmful if inhaled or swallowed. Use only in a well-ventilated area. Avoid contact with skin.

1. Grow a 100-ml culture of *E. coli* or 500-ml culture of cyanobacteria to log phase, stop growth by adding ½₀ vol stop buffer, and place on ice.

   Stop buffer contains the nuclease inhibitor ATA. This inhibitor can affect certain enzymes and should not be used if RNA will be needed for primer extension or S1 nuclease analysis.

2. Centrifuge cells 5 min at 5500 × *g* in JA-14 rotor, 4°C. Resuspend pellet in 2 ml STET lysing solution, add 100 µl of 0.2 M VRC, and transfer to 15-ml polypropylene tube.

3. Add 1 ml buffered phenol, vortex 1 min, add 1 ml chloroform, and vortex 1 min. Centrifuge 10 min at 10,000 × *g* in JA-17 rotor, 4°C, and collect top aqueous phase (avoid interphase).

4. Add 1/10 vol of 3 M sodium acetate and 2 vol ice-cold 100% ethanol. Centrifuge 10 min at 10,000 × *g*, 4°C, and resuspend pellet in 2 ml of 10 mM VRC.

5. Extract twice with 1:1 phenol/chloroform and reprecipitate as in step 4.

6a. *If using TLA-100.3 rotor:* Resuspend pellet in 2 ml DEPC-treated water. Add 1 g solid CsCl and dissolve completely. Layer 2.25 ml of this solution onto a 0.75-ml CsCl cushion in a 13 × 51–mm TLA-100.3 polycarbonate tube. Centrifuge 1 hr at 280,000 × *g*, 20°C.

6b. *If using SW-41 rotor:* Resuspend pellet in 6 ml DEPC-treated water. Add 4.5 g solid CsCl and adjust volume to 9 ml with DEPC-treated water. Layer this solution onto a 3-ml CsCl cushion in a 14 × 89–mm ultraclear SW-41 tube. Centrifuge 24 hr at 150,000 × *g*, 20°C.

7. Carefully remove DNA at interface and then remove upper CsCl layer with a sterile Pasteur pipet. Pour off remaining supernatant, mark position of RNA pellet, and wipe walls of centrifuge tube with tissue. Resuspend pellet in 0.36 ml DEPC-treated water and transfer to a 1.5-ml microcentrifuge tube.

   Do not let the centrifuge tubes sit after the completion of run.

8. Add 1/10 vol of 3 M sodium acetate and 2.5 vol ice-cold 100% ethanol. Precipitate 20 min at −70°C, and microcentrifuge 5 min at high speed, 4°C. Add 1 ml ice-cold 70% ethanol to RNA pellet and microcentrifuge 5 min at high speed, 4°C.

9. Air dry pellet and dissolve in 200 µl DEPC-treated water. Quantify by measuring A_{260} and A_{280} (*APPENDIX 3*). Adjust to 4 µg/µl final. Place at –70°C for long-term storage or store as an ethanol precipitate.

BASIC PROTOCOL 2

ISOLATION OF RNA FROM GRAM-POSITIVE BACTERIA

Materials *(see APPENDIX 1 for items with ✓)*

10-ml bacteria culture
✓ Lysis buffer
25:24:1 phenol/chloroform/isoamyl alcohol (*UNIT 2.1*)
24:1 chloroform/isoamyl alcohol
✓ 5 M NaCl
100% and 70% ethanol, ice-cold
DNase digestion buffer: 20 mM Tris·Cl (pH 8.0)/10 mM $MgCl_2$
✓ 2.5 mg/ml RNase-free DNase I
✓ TE buffer, pH 8.0
✓ DEPC-treated water

Sorvall SS-34 rotor or equivalent
Microtip sonicator

CAUTION: DEPC is a suspected carcinogen. Handle with care.

1. Centrifuge cells from a 10-ml bacteria culture 10 min at 12,000 × *g*, 4°C. Resuspend in 0.5 ml lysis buffer, transfer to microcentrifuge tube, and freeze on dry ice.
2. Thaw and sonicate (avoid foaming) three times for 10 sec each with a microtip sonicator at 30 W. Incubate 60 min at 37°C.
3. Add an equal volume of 25:24:1 phenol/chloroform/isoamyl alcohol and microcentrifuge 5 min at high speed, room temperature. Remove aqueous (top) layer to a clean microcentrifuge tube.
4. Reextract once with an equal volume of 25:24:1 phenol/chloroform/isoamyl alcohol, then extract once with an equal volume of 24:1 chloroform/isoamyl alcohol.
5. To 400 µl aqueous phase, add 15 µl of 5 M NaCl and fill microcentrifuge tube with ice-cold 100% ethanol. Mix and incubate 15 to 30 min on ice or overnight at –20°C.
6. Microcentrifuge 15 min at 4°C, rinse pellet with 500 µl ice-cold 70% ethanol, and air dry. Dissolve pellet in 95 µl DNase digestion buffer, add 4 µl of 2.5 mg/ml RNase-free DNase I, and incubate 60 min at 37°C.
7. Extract once with 25:24:1 phenol/chloroform/isoamyl alcohol. Add 100 µl TE buffer to remaining organic phase, mix, and microcentrifuge 5 min at high speed, room temperature. Pool the two aqueous phases.
8. Extract once with chloroform/isoamyl alcohol. Add 10 µl of 5 M NaCl to the aqueous phase and 600 µl of 100% ethanol. Precipitate overnight at –20°C or 15 min on dry ice/ethanol. Microcentrifuge 15 to 30 min at high speed, 4°C.
9. Rinse pellet with 500 µl ice-cold 70% ethanol and air dry. Dissolve in 100 µl DEPC-treated water. Dilute 10 µl into 1 ml water and quantify RNA by measuring the A_{260} and A_{280} (*APPENDIX 3*). Store RNA at –70°C or as an ethanol precipitate.

RAPID ISOLATION OF RNA FROM GRAM-NEGATIVE BACTERIA

ALTERNATE PROTOCOL

Additional Materials *(also see Basic Protocol 1 or 2; see* APPENDIX 1 *for items with ✓)*

- 10-ml gram-negative bacteria culture
- ✓ Protoplasting buffer
- 50 mg/ml lysozyme
- ✓ Gram-negative lysing buffer
- Saturated NaCl: 40 g NaCl in 100 ml DEPC-treated H_2O (stir until solution reaches saturation)

1. Centrifuge cells from a 10-ml gram-negative bacteria culture 10 min at 12,000 × *g*, 4°C. Resuspend in 10 ml protoplasting buffer, add 80 µl of 50 mg/ml lysozyme and incubate 15 min on ice.
2. Centrifuge protoplasts 5 min at 5900 × *g*, 4°C. Resuspend in 0.5 ml gram-negative lysing buffer, add 15 µl DEPC, mix, and transfer to a microcentrifuge tube. Incubate 5 min at 37°C, and chill on ice.
3. Add 250 µl saturated NaCl, mix, and incubate 10 min on ice.
4. Microcentrifuge 10 min at high speed, at room temperature or 4°C. Remove supernatant to two clean microcentrifuge tubes. Add to each tube 1 ml ice-cold 100% ethanol and precipitate 30 min on dry ice or overnight at −20°C.
5. Microcentrifuge 15 min at high speed, 4°C, rinse pellet in 500 µl ice-cold 70% ethanol, and air dry. Dissolve in 100 µl DEPC-treated water. Dilute 10 µl into 1 ml water and determine the A_{260} and A_{280} (*APPENDIX 3*). Store RNA at −70°C.

References: Reddy et al., 1990; Summers, 1970.

Contributors: K.J. Reddy and Michael Gilman

Preparation of Poly(A)$^+$ RNA

UNIT 4.5

BASIC PROTOCOL

Most messenger RNAs have a poly(A) tail, while structural RNAs do not. With this protocol, poly(A)$^+$ RNA can be separated from rRNA and tRNA.

Materials *(see* APPENDIX 1 *for items with ✓)*

- ✓ 5 M NaOH
- Oligo(dT) cellulose
- ✓ 0.1 M NaOH
- ✓ DEPC-treated water
- ✓ Poly(A) loading buffer
- 10 M LiCl (DEPC-treated)
- ✓ Middle wash buffer
- 2 mM EDTA/0.1% (w/v) SDS
- ✓ 3 M sodium acetate (DEPC-treated)
- RNase-free TE buffer
- Beckman SW-55 rotor and silanized SW-55 centrifuge tubes

CAUTION: DEPC is a suspected carcinogen. Handle with care.

1. Wash silanized column with 10 ml of 5 M NaOH and rinse with water. Add 0.5 g dry oligo(dT) cellulose powder to 1 ml of 0.1 M NaOH, pour into column, and rinse with ~10 ml water.

   A silanized Pasteur pipet plugged with silanized glass wool or a small disposable column with a 2-ml capacity can be used.

2. Equilibrate the oligo(dT) column with 10 to 20 ml of loading buffer, until pH ≅ 7.5.

3. Heat ~2 mg total RNA in water 10 min at 70°C. Adjust concentration to 0.5 M LiCl with 10 M LiCl.

4. Pass RNA through oligo(dT) column and wash column with 1 ml poly(A) loading buffer. Save eluant from this loading step and pass it through column two more times.

5. Rinse column with 2 ml middle wash buffer. Elute RNA into a clean tube containing 2 ml of 2 mM EDTA/0.1% SDS.

6. Reequilibrate the column, as in step 2. Repeat the poly(A)$^+$ selection with eluted RNA as in steps 2 to 4.

7. Adjust salt concentration of the eluted RNA to 0.3 M sodium acetate and add 2.5 vol ethanol. Transfer to two silanized SW-55 tubes and incubate overnight at −20°C or 30 min on dry ice/ethanol.

8. Centrifuge 30 min at 304,000 × *g*, 4°C, to pellet the (very dilute) RNA. Discard ethanol; air dry pellet. Resuspend in 150 µl RNase-free TE buffer and pool samples. Check quality of RNA by heating 5 µl for 5 min at 70°C and fractionating on a 1% agarose gel.

   Approximately 1% of the input RNA should be recovered as poly(A)$^+$ RNA. It should appear as a smear from ~20 kb down, with greatest intensity in the 5- to 10-kb range.

Reference: Aviv and Leder, 1972.

Contributor: Robert E. Kingston

4.6

UNIT 4.6

S1 Analysis of Messenger RNA Using Single-Stranded DNA Probes

S1 mapping can be used to determine the 5′ ends and intron boundaries of RNA (Fig. 4.6.1). Necessary controls for S1 mapping are described in the Support Protocol.

BASIC PROTOCOL

S1 ANALYSIS OF mRNA USING M13 TEMPLATE

Materials *(see* APPENDIX 1 *for items with ✓)*

Low gelling/melting temperature agarose (*UNIT 2.6*)
✓ Alkaline pour buffer
✓ Alkaline running buffer
[γ-^{32}P]ATP (10 mCi/ml, 6000 Ci/mmol)
100 µg/ml oligonucleotide primer
✓ 10× polynucleotide kinase buffer

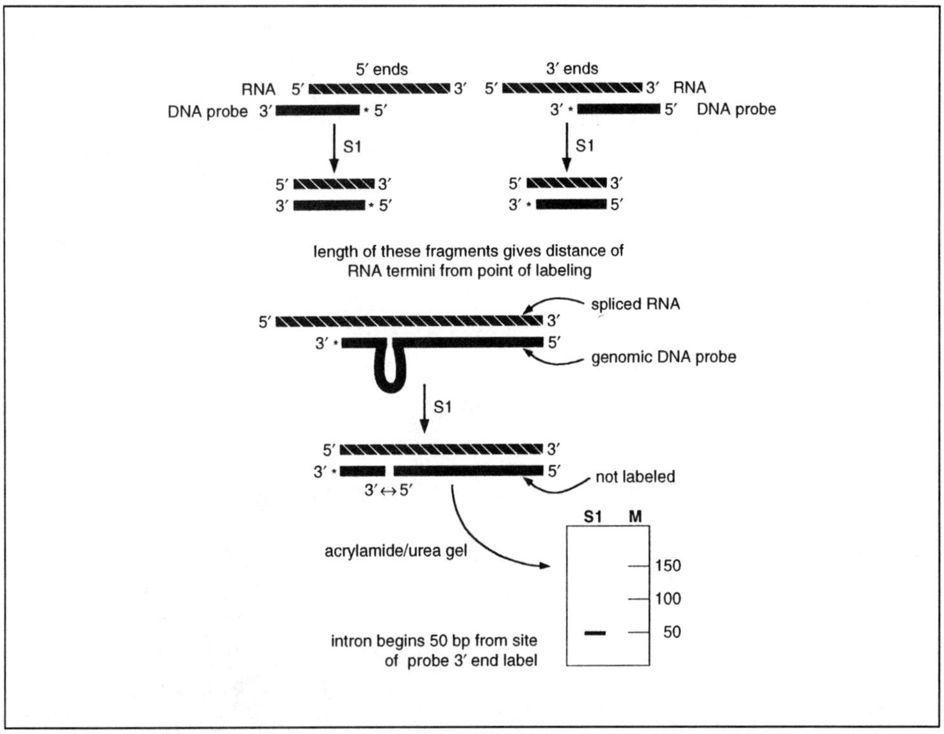

Figure 4.6.1 S1 mapping of RNA 5′ ends and intron boundaries.

T4 polynucleotide kinase (*UNIT 3.10*)
✓ DEPC-treated water
18 μg M13mp template DNA containing sequence of interest
✓ 10× TM buffer
4 mM dNTP mix (*UNIT 3.4*)
Klenow fragment of *E. coli* DNA polymerase I (*UNIT 3.5*)
40 U restriction endonuclease and 10× buffer
✓ 5 M ammonium acetate
100% ethanol
✓ Alkaline loading buffer
✓ TE buffer
10 mg/ml tRNA in H_2O, extracted repeatedly with buffered phenol
Buffered phenol (*UNIT 2.1*)
✓ 3 M and 0.3 M sodium acetate, pH 5.2 (DEPC-treated)
70% ethanol/30% DEPC-treated H_2O
✓ S1 hybridization solution
✓ 2× S1 nuclease buffer
2 mg/ml single-stranded calf thymus DNA
S1 nuclease (*UNIT 3.12*)
✓ S1 stop buffer
✓ Formamide loading buffer

CAUTION: DEPC is a suspected carcinogen. Handle with care.

1. Prepare a 1.2% low gelling/melting temperature agarose gel in 1× alkaline pour buffer. Soak solidified gel overnight in 1× alkaline running buffer.

   Use a comb with 8-mm teeth. Agarose boiled in alkaline buffer does not gel properly and must be soaked overnight. To eliminate soaking, boil the agarose in water, cool to 50° to 60°C, add alkaline running buffer to 1×, and pour the gel.

2. Mix the following to prepare kinased oligonucleotide:

   20 μl [γ-^{32}P]ATP (10 mCi/ml, 200 μCi total)
   1 μl 100 μg/ml oligonucleotide primer (100 ng; ideally a 20- to 30-mer)
   2.5 μl 10× polynucleotide kinase buffer
   4 U T4 polynucleotide kinase

   Incubate 30 min at 37°C. Heat 5 min at 65°C to inactivate kinase.

3. Add the following to the kinased oligonucleotide: 18 μg single-stranded template DNA in 55 μl water and 9 μl of 10× TM buffer. Hybridize 15 min at 40°C.

4. Extend oligonucleotide primer with 9 μl of 4 mM dNTP mix and 2 μl Klenow fragment. Incubate 30 min at 37°C. Heat 5 min at 65°C to inactivate Klenow fragment and place on ice.

5. Cut probe with 10 μl of 10× restriction buffer plus 40 U restriction endonuclease. Incubate 45 min at 37°C. Heat 5 min at 65°C to inactivate enzyme.

   Site should be far enough upstream to clearly distinguish between undigested probe and the protected fragment of interest.

6. Add 100 μl of 5 M ammonium acetate and 500 μl ethanol. Precipitate overnight at −20°C or 15 min in dry ice/ethanol (*UNIT 2.1*). Spin 15 min in microcentrifuge, then dry pellet.

7. Resuspend probe in 25 μl alkaline loading buffer. Load probe onto gel and run 4 hr at 1.8 V/cm maximum (gel will readily melt if the voltage is too high; *UNIT 2.5A*).

8. Expose gel to X-ray film for 3 min and develop the film. Align the gel and film, and excise the probe.

   The upper, darker band is the probe and the lower band is unhybridized oligonucleotide.

9. Transfer the slice to a microcentrifuge tube, melt at 65°C, and determine volume. Add an equal volume of TE buffer, and heat 10 min at 65°C.

10. Add 1 μl of 10 mg/ml tRNA and phenol extract twice (do NOT use phenol/chloroform). Add 0.1 vol of 3 M sodium acetate, pH 5.2, and 2 vol ethanol, and precipitate. Resuspend pellet in 100 μl of 0.3 M sodium acetate, and count 1 μl for Cerenkov counts.

    Yields of 1-2 × 10^7 cpm of single-stranded probe are obtained from 0.1 μg of oligonucleotide.

11. Add 5 × 10^4 Cerenkov counts of probe to ≤50 μg RNA. Adjust the final volume to 100 μl and the salt to 0.3 M sodium acetate. Add 250 μl ethanol and precipitate. Wash with 70% ethanol/30% DEPC-treated water and dry by inverting on Kimwipes 30 min.

12. Resuspend the pellet in 20 µl S1 hybridization solution and denature 10 min at 65°C. Hybridize overnight at 30°C.

13. Add the following S1 nuclease mix for each reaction:

    150 µl 2× S1 nuclease buffer
    3 µl 2 mg/ml single-stranded calf thymus DNA
    147 µl H_2O
    300 U S1 nuclease.

    Incubate 30 min at 60°C. Stop with 80 µl S1 stop buffer.

14. Add 1 ml 100% ethanol and precipitate. Wash the pellet with 70% ethanol and dry 5 min in a Speedvac evaporator. Resuspend in 3 µl TE buffer and 4 µl formamide loading dye. Boil 3 min and place on ice.

15. Prepare an appropriate denaturing polyacrylamide/urea gel from the following (*UNITS 2.11 & 7.6*):

Polyacrylamide urea gel (%)	Size of band (base of pairs)
4	≥200
5	80 to 200
8	40 to 100
12	10 to 50

    Analyze 3 to 5 µl of the sample on the gel. Autoradiograph the gel and determine the size of the labeled DNA fragment (Fig. 4.6.1).

4.6

SYNTHESIS OF SINGLE-STRANDED PROBE FROM DOUBLE-STRANDED PLASMID TEMPLATE

ALTERNATE PROTOCOL 1

***Additional Materials** (also see Basic Protocol; see APPENDIX 1 for items with ✓)*

10× NaOH/EDTA solution: 2 M NaOH/2 mM EDTA, pH 8
✓ 1.5 M ammonium acetate, pH 4.5

1. To 18 µg DNA, add 10× NaOH/EDTA solution to 1×. Incubate 5 min at room temperature.

2. Neutralize with 1.5 vol of 1.5 M ammonium acetate, pH 4.5. Add 2.5 vol ethanol and precipitate 15 min at −70°C. Microcentrifuge, rinse the pellet with 70% ethanol, and dry 5 min in a Speedvac evaporator.

3. Resuspend in 55 µl water. Substitute for single-stranded template in step 3 of Basic Protocol.

ALTERNATE PROTOCOL 2

QUANTITATIVE S1 ANALYSIS OF mRNA USING OLIGONUCLEOTIDE PROBES

Additional Materials (also see Basic Protocol; see APPENDIX 1 for items with ✓)

- 2 pmol each oligonucleotide probe
- ✓ 4 M ammonium acetate
- BioGel P-2 (or equivalent resin; optional)
- ✓ 3× aqueous hybridization solution (optional)
- ✓ 0.5 M EDTA
- ✓ 0.1 M NaOH

1. Design oligonucleotides such that ≤40 residues are complementary to the RNA coding strand. The 5′ end must be complementary to the RNA, and the 5′ terminal nucleotides should contain dG or dC residues. To determine the 5′ termini of the RNA(s), the oligonucleotide 3′ end must extend ≤4 nucleotides beyond the RNA coding sequence (i.e., upstream of the upstream-most RNA initiation site).

2. Set up a T4 polunucleotide kinase reaction in 25 µl:

   2 pmol each oligonucleotide
   150 µCi [γ-^{32}P]ATP (3000 to 7000 Ci/mmol)
   2.5 µl 10× T4 polynucleotide kinase buffer
   10 U T4 polynucleotide kinase.

   Incubate 30 to 60 min at 37°C. Heat 10 min at 75°C to stop the reaction.

   One pmol of a 45-base oligonucleotide = 15 ng. Monitor the extent of phosphorylation by acid precipitation of the oligonucleotide. Assuming that 2 pmol of 2 oligonucleotides are incubated, the theoretical incorporation of labeled ATP into DNA should be 5% to 15%.

3. Add 1 µl of 10 mg/ml tRNA, 26 µl of 4 M ammonium acetate, and 110 µl ethanol, and precipitate (*UNIT 2.1*). Resuspend in 26 µl water, add 26 µl of 4 M ammonium acetate and 110 µl ethanol, and reprecipitate.

   Alternatively, purify the labeled oligonucleotide with BioGel P-2. This method yields enough probe for 50 to 100 hybridization reactions, which can be used >6 weeks (stored at −20°C).

4. Complete the protocol starting at step 11 of the Basic Protocol. Alternatively, set up a 30 µl hybridization reaction for each RNA preparation:

   20 µl RNA (containing up to 50 µg RNA)
   9 µl 3× aqueous hybridization solution
   1 µl probe mixture (0.3 ng each oligonucleotide or ~10^5 cpm).

   Heat 10 min at 75°C. Incubate overnight at 55°C.

5. Microcentrifuge briefly to spin down drops of condensation and place tubes at 37°C. Add 270 µl of S1 nuclease mix with 100 to 300 U of S1 nuclease (step 13, Basic Protocol). Incubate 30 to 60 min at 37°C. Stop the reaction with 3 µl of 0.5 M EDTA, 1 µl of 10 mg/ml tRNA, and 0.7 ml ethanol.

6. Place on dry ice 10 to 15 min, ethanol precipitate, and wash with ethanol. Resuspend in 10 µl of 0.1 M NaOH.

7. Combine 3 µl of resuspended products with 3 µl formamide loading dye, heat 2 min at 90°C, and analyze on a denaturing polyacrylamide gel (step 15, Basic Protocol).

SUPPORT PROTOCOL

CONTROLS FOR QUANTITATIVE S1 ANALYSIS OF mRNA

1. Hybridize varying amounts of RNA to a constant amount of probe to ensure that the band intensity is directly proportional to the amount of RNA added. The probe must be in excess.
2. Optimize reaction by varying the incubation time for the hybridization reaction. The hybridization reaction must go to completion.
3. Optimize temperature such that the RNA:DNA duplexes are equally stable.
4. Vary the level of S1 nuclease.

Reference: Sharp et al., 1980.

Contributors: John M. Greene and Kevin Struhl

4.7

Ribonuclease Protection Assay

UNIT 4.7

BASIC PROTOCOL

This assay uses RNA made from a phage promoter as a probe. It is an alternative to S1 analysis (*UNIT 4.6*) and produces high-specific-activity probes.

Materials *(see APPENDIX 1 for items with ✓)*

✓ 5× transcription buffer
✓ 200 mM dithiothreitol (DTT)
3NTP mix (ATP, UTP, and GTP at 4 mM each; see *UNIT 3.4*)
[α-^{32}P]CTP (10 mCi/ml, 400 to 800 Ci/mmol)
Placental ribonuclease inhibitor (e.g., RNAsin from Promega Biotec)
1 mg/ml template DNA (Support Protocol)
Bacteriophage RNA polymerase (*UNIT 3.8*)
✓ 2.5 mg/ml DNase I (RNase-free)
10 mg/ml tRNA
✓ DEPC-treated water
25:24:1 phenol/chloroform/isoamyl alcohol (*UNIT 2.1*)
✓ 2.5 M ammonium acetate
100% ethanol
75% ethanol/25% 0.1 M sodium acetate, pH 5.2 (DEPC-treated)
✓ Hybridization buffer
✓ Ribonuclease digestion buffer
20% (w/v) sodium dodecyl sulfate (SDS)
20 mg/ml proteinase K (store at −20°C)
40 µg/ml ribonuclease A
2 µg/ml ribonuclease T1
✓ RNA loading buffer

CAUTION: DEPC is a suspected carcinogen. Handle with care.

1. Mix in an autoclaved microcentrifuge tube (20 µl total):

   4 µl 5× transcription buffer
   1 µl 200 mM DTT
   2 µl 3NTP mix
   10 µl [α-^{32}P]CTP (10 mCi/ml, 400 to 800 Ci/mmol)
   1 µl placental ribonuclease inhibitor (20 to 40 U)
   1 µl 1 mg/ml template DNA (50 µg/ml final)
   1 µl bacteriophage RNA polymerase (5 to 10 U).

   Incubate 30 to 60 min at 40°C for SP6 RNA polymerase, or 37°C for T7 and T3 RNA polymerases.

   Add the first four ingredients before adding the template DNA to avoid precipitation of the DNA by the spermidine in the transcription buffer.

   Higher specific activities of labeled CTP yield unstable probes. If labeled GTP or UTP are substituted, use an appropriate mix of the three unlabeled NTPs.

   Use SP6, T3, or T7 RNA polymerase, depending on vector in which probe sequences are cloned. All are functionally equivalent.

2. Add 5 µg or 10 U RNase-free DNase I and incubate 15 min at 37°C.

3. Add 2 µl of 10 mg/ml tRNA and water to 50 µl. Extract with phenol/chloroform/isoamyl alcohol.

4. Add 200 µl of 2.5 M ammonium acetate and 750 µl of 100% ethanol to the aqueous phase. Mix and place 15 min on ice. Microcentrifuge 15 min at 4°C.

5. Dissolve pellet in 50 µl water and reprecipitate twice as in step 4. Rinse final pellet with 75% ethanol/25% 0.1 M sodium acetate, pH 5.2. Dry pellet and dissolve in 100 µl hybridization buffer. Count 1 µl.

   The specific activity is ~10^9 cpm/µg. Use the probe the day it is prepared or within a few days (if stored at 4°C).

6. Ethanol precipitate RNA and dissolve in 30 µl hybridization buffer containing 5×10^5 cpm of the probe RNA. Denature 5 min at 85°C. Transfer to desired hybridization temperature (30° to 60°C) and incubate >8 hr.

7. Add to each reaction 350 µl ribonuclease digestion buffer containing 40 µg/ml ribonuclease A and 2 µg/ml ribonuclease T1. Incubate 30 to 60 min at 30°C.

8. Add 10 µl of 20% SDS and 2.5 µl of 20 mg/ml proteinase K. Incubate 15 min at 37°C.

9. Extract with 400 µl phenol/chloroform/isoamyl alcohol. Transfer the aqueous phase to a clean tube containing 1 µl of 10 mg/ml yeast tRNA and add 1 ml ethanol. Precipitate, dry the pellet, and redissolve in 3 to 5 µl RNA loading buffer.

10. Denature 3 min at 85°C, and analyze on a denaturing polyacrylamide/urea (sequencing) gel (*UNITS 2.11 & 7.6*). Autoradiograph overnight with an intensifying screen.

    RNA has a lower mobility on these gels than DNA. If an RNA species runs with a DNA marker of 100 nucleotides, its actual length is 90 to 95 nucleotides.

GEL PURIFICATION OF RNA PROBES

SUPPORT PROTOCOL 1

Additional Materials (*also see Basic Protocol; see* APPENDIX 1 *for items with* ✓)

✓ TBE electrophoresis buffer
✓ Elution buffer

1. Dry the RNA after the first ethanol precipitation and redissolve the pellet in 10 µl RNA loading buffer.

2. Heat 5 min at 85°C to denature the RNA.

3. Load onto a gel containing 6% polyacrylamide (29:1 acrylamide/bisacrylamide) and TBE electrophoresis buffer. Run at 300 V (higher for longer gels) until the bromphenol blue dye has run one-half to two-thirds down the gel.

   The gel is 0.4 mm thick and 14 cm long. Sequencing-length gels may also be used. Nondenaturing gels are used in the interest of speed, but it is important that the RNA be fully denatured before loading. Denaturing gels containing urea may also be used.

4. Disassemble the gel, leaving it on one plate. Wrap in plastic wrap, mark with radioactive ink, and expose to film for 30 sec. Using the film as a template, excise the full-length RNA band, cutting only a small slice.

5. Elute the RNA in 400 µl elution buffer and shake 2 to 4 hr at 37°C.

6. Remove the eluate to a clean microcentrifuge tube and add 1 ml of 100% ethanol.

7. Incubate 15 min on ice and microcentrifuge 15 min.

8. Redissolve the RNA pellet in 50 µl hybridization buffer and count 1 µl in a liquid scintillation counter.

   Yields will be lower than those obtained without gel purification, but this procedure should yield sufficient probe for >50 hybridizations. Background will be substantially lower.

PREPARATION OF TEMPLATE DNA

SUPPORT PROTOCOL 2

Template DNA is prepared by inserting the sequences of interest into a plasmid vector carrying a bacteriophage promoter. Vectors containing bacteriophage promoters are commercially available.

Digest the DNA with a restriction enzyme that cuts immediately downstream of the probe sequence. This allows the generation of a uniquely sized runoff transcript. The enzyme chosen may cut the plasmid in several places as long as it does not cut within the phage promoter, the probe sequence, or intervening vector DNA. Restriction enzymes that generate 5′ overhangs are best. (Do not cut with enzymes that leave 3′ overhangs because these overhangs serve as initiation sites for the polymerase, leading to synthesis of RNA complementary to the probe.) Cut the DNA to completion but do not grossly overdigest. Extract cut DNA with phenol/chloroform, precipite with ethanol, and redissolve at 0.5 mg/ml in RNase-free TE buffer. As little as 100 ng of template DNA will yield a reasonable amount of probe.

Reference: Melton et al., 1984.

Contributor: Michael Gilman

Primer Extension

BASIC PROTOCOL

Primer extension is used to map the 5′ terminus of an RNA and to quantitate the amount of a given RNA by using reverse transcriptase to extend a primer that is complementary to a portion of the RNA of interest.

Materials *(see APPENDIX 1 for items with ✓)*

✓ Diethylpyrocarbonate (DEPC)
10× T4 polynucleotide kinase buffer (*UNIT 3.4*)
✓ 0.1 M and 1 M dithiothreitol (DTT)
1 mM spermidine
50 to 100 ng/μl oligonucleotide primer (5 to 10 μM)
10 μCi/μl [γ-^{32}P]ATP (3000 Ci/mmol)
20 to 30 U/μl T4 polynucleotide kinase
✓ 0.5 M EDTA, pH 8.0
✓ TE buffer, pH 8.0
Cation-exchange resin (e.g., Bio-Rad AG 50W-X8), equilibrated in 0.1 M Tris·Cl (pH 7.5)/0.5 M NaCl
Anion-exchange resin (e.g., Whatman DE-52), equilibrated in TEN 100
TEN 100 buffer: 100 mM NaCl in TE buffer, pH 7.5
TEN 300 buffer: 300 mM NaCl in TE buffer, pH 7.5
TEN 600 buffer: 600 mM NaCl in TE buffer, pH 7.5
Total cellular RNA (*UNITS 4.1-4.3*)
✓ 10× hybridization buffer
✓ 0.1 M Tris·Cl, pH 8.3
✓ 0.5 M $MgCl_2$
1 mg/ml actinomycin D (store at 4°C protected from light; *UNIT 1.4*)
10 mM 4dNTP mix (*UNIT 3.4*)
25 U/μl AMV reverse transcriptase (*UNIT 3.7*)
✓ RNase reaction mix
✓ 3 M sodium acetate
25:24:1 (v/v/v) phenol/chloroform/isoamyl alcohol (*UNIT 2.1*)
100% and 70% ethanol
✓ Stop/loading dye
9% acrylamide/7 M urea gel (*UNIT 2.11*)

Silanized glass wool and 1000-μl pipet tip (*APPENDIX 3*)
65°C water bath

NOTE: Water should be treated with DEPC to inhibit RNase activity.

CAUTION: DEPC is a suspected carcinogen and should be handled carefully.

1. Mix the following reagents in the order indicated to label the primer (10 μl final):

   2.5 μl H_2O
   1 μl 10× T4 polynucleotide kinase buffer
   1 μl 0.1 M DTT
   1 μl 1 mM spermidine
   1 μl 50-100 ng/μl oligonucleotide primer
   3 μl 10 μCi/μl [γ-^{32}P]ATP
   0.5 μl 20-30 U/μl T4 polynucleotide kinase.

Incubate 1 hr at 37°C.

Oligonucleotides used as primers should be 20 to 40 nucleotides long and selected to yield an extended product of <100 nucleotides.

2. Stop reaction by adding 2 µl of 0.5 M EDTA and 50 µl TE buffer. Incubate 5 min at 65°C.

3. Prepare a small ion-exchange column in a silanized 1000-µl pipet tip with a small plug of silanized glass wool. Add 20 µl of AG 50W-X8 resin and 100 µl of DE-52 resin. Wash the column with 1 ml TEN 100 buffer.

4. Load the labeling reaction from step 2 onto the column. Collect flowthrough and reload it onto column.

5. Wash the column with 1 ml TEN 100, then with 0.5 ml TEN 300 to wash unincorporated nucleotide from the column. Discard eluate as radioactive waste).

6. Elute radiolabeled oligonucleotide using 0.4 ml TEN 600. Collect eluate as a single fraction and store at −20°C in an appropriately shielded container until needed.

7. For each RNA sample, combine the following in a separate microcentrifuge tube (15 µl final):

   10 µl total cellular RNA (10 to 50 µg)
   1.5 µl 10× hybridization buffer
   3.5 µl radiolabeled oligonucleotide (from step 6).

   Seal tubes securely and submerge 90 min in a 65°C water bath. Remove tubes and allow to cool slowly to room temperature.

8. For each sample, prepare the following primer extension reaction mix in a microcentrifuge tube on ice (30.33 µl final per sample):

   0.9 µl 1 M Tris·Cl, pH 8.3
   0.9 µl 0.5 M $MgCl_2$
   0.25 µl 1 M DTT
   6.75 µl 1 mg/ml actinomycin D
   1.33 µl 5 mM 4dNTP mix
   20 µl H_2O
   0.2 µl 25 U/µl AMV reverse transcriptase.

   Do NOT use the 10× reverse transcriptase buffer defined in UNIT 3.4 *because of the salt present in the hybridization reaction.*

9. To each tube containing RNA and oligonucleotide, add 30 µl primer extension reaction mix. Incubate 1 hr at 42°C.

10. Add 105 µl RNase reaction mix to each primer extension reaction tube. Incubate 15 min at 37°C.

11. Add 15 µl of 3 M sodium acetate. Extract with 150 µl phenol/chloroform/isoamyl alcohol, and remove aqueous (top) phase to a clean tube. Precipitate DNA by adding 300 µl of 100% ethanol. Wash the pellet with 100 µl of 70% ethanol. Remove all traces of ethanol using a pipet. Air dry the pellet 5 to 10 min.

12. Resuspend pellet in 5 µl stop/loading dye. Heat tubes 5 min in a 65°C water bath. Load samples on a 9% acrylamide/7 M urea gel and electrophorese until bromphenol blue reaches end of gel.

13. Dry gel and autoradiograph with an intensifying screen.

Reference: Mierendorf and Pfeffer, 1987.

Contributor: Steven J. Triezenberg

UNIT 4.9

4.9

Analysis of RNA by Northern and Slot Blot Hybridization

Blotting and hybridization analysis can be used to detect specific sequences in RNA preparations.

NOTE: To inhibit RNase activity, all solutions for northern blotting should be prepared using sterile deionized water that has been treated with diethylpyrocarbonate (DEPC). For full details on the establishment of an RNase-free environment, see Wilkinson (1991).

CAUTION: DEPC is a suspected carcinogen and should be handled carefully. Because DEPC reacts with ammonium ions to produce ethyl carbamate, a potent carcinogen, special care should be exercised when treating ammonium acetate solution with DEPC.

BASIC PROTOCOL

NORTHERN HYBRIDIZATION OF RNA FRACTIONATED BY AGAROSE-FORMALDEHYDE GEL ELECTROPHORESIS

Materials *(see APPENDIX 1 for items with ✓)*

✓ Diethylpyrocarbonate (DEPC)
✓ 10× and 1× MOPS running buffer
12.3 M (37%) formaldehyde, pH >4.0
RNA sample: total cellular RNA (*UNITS 4.1-4.4*) or poly(A)$^+$ RNA (*UNIT 4.5*)
Formamide
✓ Formaldehyde loading buffer
0.5 M ammonium acetate and 0.5 µg/ml ethidium bromide in 0.5 M ammonium acetate *or* 10 mM sodium phosphate (pH 7.0)/1.1 M formaldehyde with and without 10 µg/ml acridine orange
0.05 M NaOH/1.5 M NaCl (optional)
0.5 M Tris·Cl (pH 7.4)/1.5 M NaCl (optional)
✓ 20×, 2×, and 6× SSC
0.03% (w/v) methylene blue in 0.3 M sodium acetate, pH 5.2 (optional)
DNA suitable for use as probe *or* for in vitro transcription to make RNA probe (Table 2.10.1)
✓ Formamide prehybridization/hybridization solution
2× SSC/0.1% (w/v) SDS
0.2× SSC/0.1% (w/v) SDS, room temperature and 42°C
0.1× SSC/0.1% (w/v) SDS, 68°C

60°C water bath
Oblong sponge slightly larger than the gel being blotted
Whatman 3MM filter paper sheets

Nitrocellulose or nylon membrane (see Table 2.9.1 for list of suppliers)
UV transilluminator, calibrated (*UNIT 3.19*)
UV-transparent plastic wrap (e.g., Saran Wrap or other polyvinylidene wrap)
Hybridization oven (e.g., Hybridiser HB-1, Techne)

1. Dissolve 1.0 g agarose in 72 ml water and cool to 60°C in a water bath. When the flask has cooled to 60°C, place in a fume hood and add 10 ml of 10× MOPS running buffer and 18 ml of 12.3 M formaldehyde (2.2 M final).

   CAUTION: *Formaldehyde is toxic through skin contact and inhalation of vapors. All operations involving formaldehyde should be carried out in a fume hood.*

2. Pour the gel and allow it to set. Remove the comb, place the gel in the gel tank, and add sufficient 1× MOPS running buffer to cover to a depth of ~1 mm.

   A 1.0% gel is suitable for RNA molecules 500 bp to 10 kb in size; a 1.0% to 2.0% gel should be used to resolve smaller molecules or a 0.7% to 1.0% gel for longer molecules. The gel should be 2 to 6 mm thick after it is poured and the wells large enough to hold 60 µl of sample.

3. Adjust the volume of each RNA sample to 11 µl with water, then add:

   5 µl 10× MOPS running buffer
   9 µl 12.3 M formaldehyde
   25 µl formamide.

   Vortex and spin briefly (5 to 10 sec) in a microcentrifuge to collect the liquid. Incubate 15 min at 55°C.

   CAUTION: *Formamide is a teratogen and should be handled with care.*

4. Add 10 µl formaldehyde loading buffer, vortex, spin to collect liquid, and load 0.5 to 10 µg RNA into each of two wells arranged so that half of the gel may be stained. Run the gel at 5 V/cm until the bromphenol blue dye has migrated one-half to two-thirds the length of the gel (~3 hr).

5a. *To stain with ethidium bromide*: Remove the gel and cut off the lanes that are to be stained. Place this portion of the gel in an RNase-free glass dish, add sufficient 0.5 M ammonium acetate to cover, and soak for 20 min. Change solution and soak for an additional 20 min to remove the formaldehyde. Pour off solution, replace with 0.5 µg/ml ethidium bromide in 0.5 M ammonium acetate, and allow to stain for 40 min. If necessary, destain in 0.5 M ammonium acetate for up to 1 hr.

5b. *To stain with acridine orange*: Remove gel, cut off lanes, and stain 2 min in 1.1 M formaldehyde/10 mM sodium phosphate containing 10 µg/ml acridine orange. If necessary, destain 20 min in the same buffer without acridine orange.

6. Examine gel on a UV transilluminator to visualize the RNA and photograph with a ruler laid alongside the gel so that band positions can later be identified on the membrane.

7. Place unstained portion of gel in an RNase-free glass dish and rinse with several changes of sufficient deionized water to cover the gel to remove the formaldehyde.

8. Partially hydrolyze the RNA by soaking the gel 30 min in ~10 gel volumes of 0.05 M NaOH/1.5 M NaCl followed by soaking 30 min in 10 gel volumes of 0.5 M Tris·Cl (pH 7.4)/1.5 M NaCl to neutralize (optional).

9. Replace solution with 10 gel volumes of 20× SSC and soak for 45 min (optional).

10. Place an oblong sponge slightly larger than the gel in a glass or plastic dish (if necessary, use two or more sponges placed side by side). Fill the dish with enough 20× SSC to leave the soaked sponge about half-submerged in buffer.

11. Construct the transfer stack. Cut three pieces of Whatman 3MM paper to the same size as the sponge. Place them on the sponge and wet them with 20× SSC. Place the gel on the filter paper and squeeze out air bubbles by rolling a glass pipet over the surface. Cut four strips of plastic wrap and place over the edges of the gel.

12. Cut a piece of nylon or nitrocellulose membrane just large enough to cover the exposed surface of the gel. Pour distilled water ~0.5 cm deep in an RNase-free glass dish and wet the membrane by placing it on the surface of the water. Allow the membrane to submerge. For nylon membrane, leave for 5 min; for nitrocellulose membrane, replace the water with 20× SSC and leave for 10 min.

13. Place the wetted membrane on the surface of the gel. Remove air bubbles. Flood the surface of the membrane with 20× SSC. Cut five sheets of Whatman 3MM paper to the same size as the membrane and place on top of the membrane. Cut paper towels to the same size as the membrane and stack on top of the Whatman 3MM paper to a height of ~4 cm.

14. Lay a glass plate on top of the stack and add a weight (0.2 to 0.4 kg) to hold everything in place. Leave overnight.

15. Disassemble transfer stack recover the membrane and flattened gel. Mark the position of the wells on the membrane in pencil and mark the orientation of the gel.

16. Rinse the membrane in 2× SSC, then place it on a sheet of Whatman 3MM paper and allow to dry completely.

17a. *For nitrocellulose membranes:* Place between two sheets of Whatman 3MM filter paper and bake under vacuum 2 hr at 80°C.

17b. *For nylon membranes:* Bake as described above *or* wrap the dry membrane in UV-transparent plastic wrap, place RNA-side-down on a UV transilluminator (254-nm wavelength), and irradiate for the appropriate length of time.

18. If desired, check transfer efficiency by either staining the gel in ethidium bromide or acridine orange as in step 5a or 5b or (for a nylon membrane) by staining the membrane in 0.03% (w/v) methylene blue in 0.3 M sodium acetate, pH 5.2, for 45 sec and destaining in water for 2 min.

19. Prepare DNA or RNA probe labeled to a specific activity of $>10^8$ dpm/µg (*UNITS 2.10* or *3.5*) and with unincorporated nucleotides removed (*UNIT 3.4*).

20. Wet the membrane carrying the immobilized RNA in 6× SSC.

21. Place the membrane RNA-side-up in a hybridization tube and add ~1 ml formamide prehybridization/hybridization solution per 10 cm^2 of membrane. Place the tube in the hybridization oven and incubate with rotation 3 hr at 42°C (for DNA probe) or 60°C (for RNA probe).

22. If the probe is double-stranded, denature it by heating in a water bath or incubator 10 min at 100°C. Transfer to ice.

23. Pipet the desired volume of probe (to give 10 ng/ml if the specific activity is 10^8 dpm/μg or 2 ng/ml if the specific activity is 10^9 dpm/μg) into the hybridization tube and continue to incubate with rotation overnight at 42°C (for DNA probe) or 60°C (for RNA probe).

    For denatured probe, add to hybridization tube as soon after denaturation as possible.

24. Pour off hybridization solution and wash in:

    2× SSC/0.1% SDS, 5 min with rotation at room temperature, twice
    0.2× SSC/0.1% SDS, 5 min with rotation at room temperature, twice (low-stringency wash)
    Prewarmed 0.2× SSC/0.1% SDS, 15 min at 42°C, twice (moderate-stringency wash; optional)
    Prewarmed 0.1× SSC/0.1% SDS, 15 min at 68°C, twice (high-stringency wash; optional).

25. Rinse membrane in 2× SSC at room temperature. Blot excess liquid and cover in UV-transparent plastic wrap. Autoradiograph (*APPENDIX 3*).

    Do not allow membrane to dry out if it is to be reprobed.

4.9

NORTHERN HYBRIDIZATION OF RNA DENATURED BY GLYOXAL/DMSO TREATMENT

ALTERNATE PROTOCOL 1

RNA can be denatured by treating samples with a combination of glyoxal and DMSO prior to running in an agarose gel made with phosphate buffer; this produces sharper bands.

Additional Materials *(also see Basic Protocol; see APPENDIX 1 for items with ✓)*

10 mM and 100 mM sodium phosphate, pH 7.0
Dimethyl sulfoxide (DMSO)
✓ 6 M (40%) glyoxal, deionized immediately before use
✓ Glyoxal loading buffer
✓ 20 mM Tris·Cl, pH 8.0
Apparatus for recirculating running buffer during electrophoresis

1. Prepare a 1.0% agarose gel, place gel in gel tank, and add 10 mM sodium phosphate (pH 7.0) until gel is submerged to a depth of ~1 mm.

2. Adjust volume of each RNA sample to 11 μl with water, then add:

    4.5 μl 100 mM sodium phosphate, pH 7.0
    22.5 μl DMSO
    6.6 μl 6 M glyoxal.

    Vortex and spin briefly (5 to 10 sec) in a microcentrifuge to collect the liquid. Incubate 1 hr at 50°C.

3. Cool samples on ice and add 12 μl glyoxal loading buffer to each sample. Load duplicate samples (0.5 to 10 μg RNA) onto gel arranged so that half of the gel may be stained. Run the gel at 4 V/cm with constant recirculation of running buffer for ~3 hr or until bromphenol blue dye has migrated one-half to two-thirds the length of the gel.

   Recirculation is needed to prevent an H^+ gradient forming in the buffer.

4. Remove the gel, cut off lanes, and stain with ethidium bromide as described in the Basic Protocol, step 5a.

   Begin transfer as soon as the gel is cut, before staining the other portion

6. Transfer RNA as in Basic Protocol, steps 7 to 18.

7. Immediately before hybridization, soak the membrane in 20 mM Tris·Cl (pH 8.0) for 5 min at 65°C to remove glyoxal.

8. Continue with hybridization analysis as in Basic Protocol, steps 19 to 25.

4.9

ALTERNATE PROTOCOL 2

NORTHERN HYBRIDIZATION OF UNFRACTIONATED RNA IMMOBILIZED BY SLOT BLOTTING

RNA slot blotting is a simple technique to immobilize unfractionated RNA on a nylon or nitrocellulose membrane. It is followed by hybridization to determine the relative abundance of target mRNA sequences in the blotted samples.

Additional Materials *(also see Basic Protocol; see* APPENDIX 1 *for items with ✓)*

✓ 0.1 M NaOH
✓ 10× SSC
✓ 20× SSC, room temperature and ice-cold
✓ Denaturing solution
✓ 100 mM sodium phosphate, pH 7.0
Dimethyl sulfoxide (DMSO)
✓ 6 M (40%) glyoxal, deionized immediately before use
Manifold apparatus with a filtration template for slot blots (e.g., Bio-Rad Bio-Dot SF, Schleicher and Schuell Minifold II)

1. Clean the manifold with 0.1 M NaOH and rinse with distilled water.

2. Cut a piece of nylon or nitrocellulose membrane to the size of the manifold. Pour 10× SSC (for nylon membrane) or 20× SSC (for nitrocellulose membrane) into a glass dish; place membrane on top of liquid and allow to submerge. Leave for 10 min.

3. Place the membrane in the manifold. Assemble the manifold according to manufacturer's instructions and fill each slot with 10× SSC. Ensure there are no air leaks in the assembly.

4a. Add 3 vol denaturing solution to RNA sample. Incubate 15 min at 65°C, then place on ice.

4b. Alternatively, mix:

   11 μl RNA sample
   4.5 μl 100 mM sodium phosphate, pH 7.0
   22.5 μl DMSO
   6.6 μl 6 M glyoxal.

Vortex and spin briefly in a microcentrifuge to collect liquid. Incubate 1 hr at 50°C.

5. Add 2 vol ice-cold 20× SSC to each sample.
6. Switch on the suction to the manifold device and allow the 10× SSC to filter through. Leave the suction on.
7. Load each sample to the slots and allow to filter through, being careful not to touch the membrane with the pipet tip.
8. Add 1 ml of 10× SSC to each slot and allow to filter through. Repeat.
9. Dismantle the apparatus, place the membrane on a sheet of Whatman 3MM paper, and allow to dry.
10. Immobilize the RNA by baking or UV cross-linking.

    If glyoxal/DMSO denaturation has been used, immediately before hybridization soak the membrane in 20 mM Tris·Cl (pH 8.0) for 5 min at 65°C to remove glyoxal.

11. Carry out hybridization analysis as described in Basic Protocol, steps 19 to 25.

Reference: Thomas, 1980

Contributor: Terry Brown

4.9

Construction of Recombinant DNA Libraries

5

Construction of recombinant DNA molecules by simply ligating vector DNA and a fragment of interest is a straightforward process, discussed in Chapters 1 and 3. Special problems arise, however, when the fragment of interest represents only a very small fraction of the total target DNA. This is the case in two commonly encountered situations: isolation of single-copy genes from a complex genome and isolation of rare cDNA clones derived from a complex mRNA population. This chapter describes techniques to generate recombinant DNA libraries which contain complete representation of genomic or cDNA sequences. Chapter 6 describes strategies and protocols for isolating particular desired sequences from such libraries.

The DNA of higher organisms is remarkably complex: a mammalian haploid genome contains approximately 3×10^9 base pairs. A particular 3000-bp fragment of interest thus comprises only 1 part in 10^6 of a preparation of genomic DNA. Similarly, a particularly rare mRNA species may comprise only 1 part in 10^5 or 10^6 of total poly(A) containing RNA, a ratio that is usually unaffected by the process of copying the RNA into cDNA. Clearly, the main problem in generating a useful recombinant DNA library from either genomic DNA or cDNA is the creation of the huge population of clones necessary to ensure that the library contains at least one version of every sequence of interest. The solutions to this problem are basically similar for genomic and cDNA libraries. As diagrammed in Fig. 5.0.1, the genomic DNA or cDNA is first prepared for insertion into the chosen vector. The vector and target DNA are then ligated together and introduced into *E. coli* either by packaging into phage λ heads in vitro or by direct transformation. In some aspects, however, strategies for isolation of individual genomic or cDNA clones can be quite different.

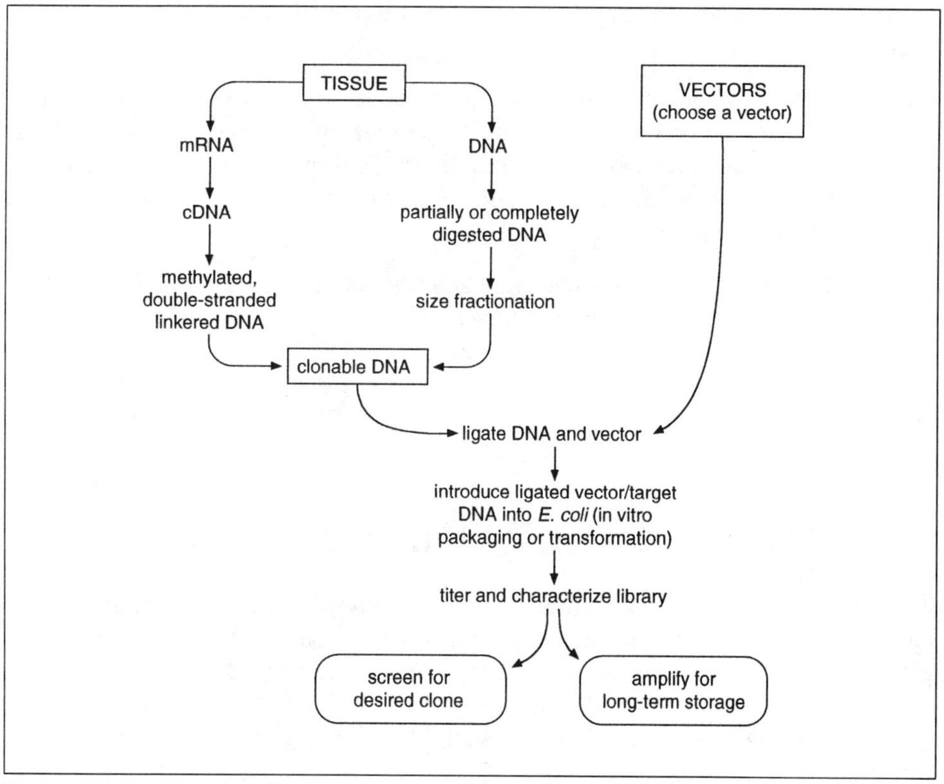

Figure 5.0.1 Steps involved in the construction of cDNA or genomic DNA libraries.

The particular problems of creating these two different types of libraries will be discussed in detail separately (*UNITS 5.1 & 5.2*).

The *E. coli* vectors described in this chapter are limited with regard to the size of insert DNA that can be accommodated (~20 kb for lambda and ~40 kb for cosmid vectors). The ability to clone much larger fragments of DNA, however, has become an essential requirement for many genome analysis projects. Yeast artificial chromosome (YAC) vectors, maintained in yeast hosts, typically carry inserts ranging from 0.3 to 1.2 Mb of genomic DNA. Both the size and complexity of YAC libraries pose special considerations for production, screening, and analysis, and these concerns are addressed in *UNITS 6.9 & 6.10* in the following chapter.

Two important general points pertain to both genomic and cDNA libraries. First, it is essential that both the vector DNA and target DNA used to create the library are not contaminated by exogenous sequences detectable by the probes that will be used to isolate the clones of interest. There are obvious, potentially disastrous effects of contaminated target DNA—for example, by only 1 part in 10^5 of a plasmid containing the cDNA sequences to be used as a probe. (See "Going for the gene," *The Boston Globe Magazine*, Aug. 2, 1987, for an account of such a mistake.) Common sense dictates care and use of absolutely clean and, where possible, disposable materials throughout.

Second, by far the simplest way to generate a library is to "clone by phone" and get one from somebody else. Many useful libraries, including examples of human and other mammalian genomic or cDNA libraries, have been made over the years, and investigators are frequently willing to send out libraries that they have created. In some cases, journals (e.g., *Cell, Science, Proceedings of the National Academy of Sciences U.S.A.*, and the publications of the American Society for Microbiology) require that libraries and individual clones discussed in their pages be freely available to other investigators. In addition, YAC libraries are produced, maintained, and generally available from large "core" academic or institutional laboratories at Washington University (St. Louis), the ICRF (London), and CEPH (Paris); see *UNIT 6.9* for details. Both stock and custom-made libraries are also available from a variety of commercial sources.

Production of recombinant DNA libraries can be a very laborious process. When possible we strongly recommend obtaining previously constructed libraries from other laboratories or purchasing commercially available genomic or cDNA libraries; otherwise, we recommend reagent kits for producing libraries. These kits can save considerable effort and time.

For detailed protocols for constructing a genomic or cDNA library, see Chapter 5 of *Current Protocols in Molecular Biology*.

J.G. Seidman

UNIT 5.1

Overview of Genomic DNA Libraries

Genomic DNA libraries are almost always screened by hybridization using a radioactive nucleic acid probe. The main problem when considering creation of a genomic DNA library is generating a large enough number of recombinant DNA clones. This unit discusses the appropriate numerical considerations for both genomic and subgenomic DNA libraries and describes a limited number of appropriate vectors.

REPRESENTATION AND RANDOMNESS

The size of a library of completely random fragments of genomic DNA that is necessary to ensure representation of a particular sequence of interest is dictated by the size of the cloned fragments and the size of the genome. Specifically, the number of independent clones, *N*, that must be screened to isolate a particular sequence with probability *P* is given by

$$N = \ln(1 - P)/\ln[1 - (I/G)]$$

where *I* is the size of the average cloned fragment, in base pairs, and *G* is the size of the target genome, in base pairs. In general, to have a 99% chance of isolating a desired sequence, the number of clones screened should be such that the total number of base pairs present in the clones screened ($I \times N$) represents a 4.6-fold excess over the total number of base pairs in the genome (*G*; Seed et al., 1982). For an ordinary λ library, this is ~800,000 clones.

When the desired fragment can be purified, the size of the library can be reduced. The library size can then be estimated by: $N \cong 3 \times {}^{1}\!/_{p}$ where p = probability of isolating a particular fragment = 1/total number of fragments in the pool.

It is important to note that this simple analysis assumes that the cloned DNA segments randomly represent the sequences present in the genome. In the strictest sense, this level of randomness can be approached only by the relatively inconvenient means of shearing the target DNA. With common sense and care, however, sufficiently random cleavage of target DNA can generally be obtained using partial digestions with restriction enzymes. One limitation of this approach is that fragments which are larger than the capacity of the vector will be excluded from the library. Clearly, it is best to use an enzyme that cuts the DNA of interest both frequently and without any bias in selection of one site over another (such as *Sau*3A, which recognizes the 4-bp site GATC and generates fragments compatible with several phage λ and cosmid vectors). See Seed et al. (1982) for a detailed discussion of issues affecting representation in genomic libraries.

SUBGENOMIC DNA LIBRARIES

Sometimes only a small and relatively well-characterized fragment such as a 1-kb *Bam*H1 fragment is desired. It can be purified and used to generate a smaller, potentially easier to screen subgenomic DNA library.

Maximizing the fold of purification of target DNA is crucial for subgenomic DNA libraries. Using the equation above, and assuming the genome size is reduced by the amount of purification, it is clear that the fold of purification must exceed the ratio of genomic DNA library insert size to subgenomic insert size.

A simple way to increase the fold of purification is to use multiple, sequential digestion strategies in cases where details of the restriction map of the sequences of interest are known. After initial purification of a given fragment, redigestion with another enzyme that gives a smaller (clonable) fragment will generally yield significant further purification relative to the original DNA.

VECTORS FOR GENOMIC DNA LIBRARIES

Because of their combination of high cloning efficiency and relatively large insert size, either bacteriophage λ vectors or hybrid plasmid vectors called cosmids (which contain particular λ sequences that direct insertion of DNA into phage particles) are generally used to construct genomic DNA libraries.

The choice between phage and cosmid vectors is generally based on the size of the desired genomic DNA segment. Most investigators feel that phage libraries are easier to handle, and choose a phage vector if the desired segment is less than ~20 kb. Larger segments suggest the use of cosmid vectors, which can hold up to 40-kb inserts.

Bacteriophage λ vectors. There are a number of easy-to-use phage λ vectors. These vectors have two basic features in common: ability to accept fragments generated by several restriction enzymes, and biochemical and/or genetic selection against the so-called stuffer sequences present in the original vector in the place of the exogenously added DNA.

One useful vector is λEMBL3, which allows both the genetic and the biochemical strategies to avoid purification of stuffer fragments and includes several useful cloning sites in the polylinker.

Cosmid vectors. Any plasmid cloning vector that contains the λ *cos* site can be used as a cosmid. A number of cosmid vectors designed for particular applications include additional elements that decrease the cloning capacity of the vector and should be avoided unless they are specifically required. One useful, simple cosmid vector is pJB8 (Ish-Horowitz and Burke, 1981), a 5.4-kb plasmid that accepts genomic DNA digested with *Sau*3A and can be used with several cosmid cloning strategies.

Vectors for subgenomic DNA libraries. In general, phage λ vectors designed for direct insertion of foreign DNA rather than substitution for a stuffer fragment are used for subgenomic libraries. It is possible to use simple plasmid vectors for subgenomic DNA libraries if the level of purification and recovery of the target fragment is sufficient to overcome the relative inefficiency inherent in plasmid cloning.

Reference: Seed et al., 1982.

Contributor: David D. Moore

UNIT 5.2

Overview of cDNA Libraries

The most basic step in constructing a cDNA library is the process of generating a double-stranded DNA copy of the mRNA. In general, it should be straightforward to obtain essentially full-length cDNA copies for mRNAs up to the 3- to 4-kb range, and at least feasible for even larger mRNAs. The most important factor affecting quality of cDNA is the quality of the mRNA. Particularly for a large message, it is essential to start with the highest quality RNA available.

Two related issues dominate strategies for constructing cDNA libraries. First is the relative abundance of the clone of interest, which can vary over a wide range. Second is the screening method (see Chapter 6), which can range from simply sequencing several individual isolates until the desired clone is identified, through ordinary hybridization methods, to complex strategies involving expression of identifiable antigens or biological activities.

In general, the relative abundance of the mRNA of interest is not known with precision. It is sensible to aim for a library that contains at least 5 times more recombinants than the total indicated by the lowest abundance estimate. In some cases this number should be multiplied by various factors based on screening efficiency. For example, if it is necessary to fuse a peptide-coding region to a vector in a particular reading frame, the number of identifiable clones is only 1/6 of those present in the library.

If the mRNA of interest is relatively abundant, efficiency of generating clones is not as important, and the choice of cloning strategy and vector should be based on the desired use for the clone. In many cases, however, the mRNA of interest is relatively rare, and high cloning efficiency is of central importance. As with genomic libraries, this has led to development and use of phage λ vectors. In general, there are two types of λ vectors for cDNA cloning adapted for the two most common methods of library screening.

If the library is to be screened by hybridization with a nucleic acid probe (*UNITS 6.1 & 6.3-6.4*), any insertion vector is appropriate. If the library is to be screened by use of antibody probes (*UNIT 6.7*), it is necessary to use an appropriate *E. coli* expression vector. In general, such vectors are based upon expression of a fusion protein in which a segment of the peptide of interest is fused to a highly expressed, stable *E. coli* protein. The most commonly used expression vector is λgt11, in which the cloned peptide coding sequences are fused to coding sequences for β-galactosidase.

Contributor: David D. Moore

Amplification of a Bacteriophage Library

UNIT 5.3

5.3

BASIC PROTOCOL

Amplification should be carried out as soon as possible after the library is packaged. It increases the number of copies of a library, but potential changes in composition may occur due to differences in growth rate during amplification. Changes in composition can be minimized by preadsorbing the library to the bacteria and using a high plating density and a short incubation period.

Materials *(see APPENDIX 1 for items with ✓)*

LB medium containing 0.2% maltose and 10 mM $MgSO_4$ (*UNIT 1.1*)
Suitable host (Table 5.3.1)
In vitro packaged genomic or cDNA library
Top agarose (*UNIT 1.1*), warmed to 47°C
150-mm H plates (*UNIT 1.1*), warmed to 37°C
✓ Suspension medium (SM)
Chloroform
Dimethyl sulfoxide (DMSO)

Table 5.3.1 Suitable *Escherichia coli* Host Strains for Amplifying Lambda-Constructed Libraries

Vector	*E. coli* host	Relevant host genotype
λgt10	C600*hflA*	*hflA*
λgt11	Y1088	*SupF*, *lacI*^{q} *(no antibiotics needed)*
EMBL 3 or 4	P2392, Q359, NM539	P2 lysogen
Charon 4A	LE392	*SupF*

1. Inoculate 250 ml LB medium containing 0.2% maltose and 10 mM $MgSO_4$ with 2.5 ml of a fresh overnight culture of host bacteria. Incubate 2 to 4 hr at 37°C with vigorous shaking to $OD_{600} \cong 0.5$. Prepare cells for plating.

2. Shake tube containing the library, microcentrifuge briefly to separate chloroform, and add packaged, titered phage (without chloroform) to plating bacteria (1×10^5 phage/0.25 ml host cells). Incubate 15 min at 37°C.

3. Add 0.5 ml of the host/phage mixture to 8 ml of 47°C top agarose, mix by inversion, and pour onto a fresh 150-mm H plate warmed to 37°C. Spread evenly and allow to harden 5 min. Invert and incubate 6 to 7 hr at 37°C (39° to 41°C for λgt11 libraries amplified in *E. coli* Y1088).

   Many tiny plaques should be visible at 4 to 5 hr and should expand to near confluence at 6 hr. If plaques are growing slowly, continue incubation up to 8 hr.

4. Remove plates from incubator, cover with 10 ml SM, and let sit 2 to 16 hr at 4°C to elute phage. Combine SM from all plates in a glass or polypropylene tube. Centrifuge 5 min at 2800 × *g*, transfer supernatant to new Teflon-capped glass tube, add 0.5 ml chloroform, and mix. Titer amplified library.

   Expect a titer of 10^{10}-10^{11} pfu/ml. Stored at 4°C, the titer drops only 2- to 3-fold over the years; storage in plastic results in a 100- to 1000-fold drop in titer. For added security, transfer 930-µl aliquots of amplified library without chloroform to screw-cap microcentrifuge tubes, add 70 µl DMSO, mix, and place at −80°C.

Contributor: Lloyd B. Klickstein

5.4

UNIT 5.4

Amplification of Cosmid and Plasmid Libraries

BASIC PROTOCOL

Materials

LB plates containing appropriate antibiotic (*UNIT 1.1*)
LB medium (*UNIT 1.1*)
Sterile glycerol
Nitrocellulose membrane filters (Millipore HATF)

1. Plate drug-resistant bacteria on nitrocellulose filters placed on LB plates containing antibiotic to which colonies are resistant (*UNIT 6.2*). Grow just to confluence.

2. Flood each plate with LB medium using ~2 ml per 10-cm plate and ~4 ml per 15-cm plate. Rub colonies off nitrocellulose filter with a sterile rubber policeman, making a bacterial suspension.

3. Pool suspensions into one 50-ml plastic tube, adding sterile glycerol to 15%. Mix thoroughly, dispense 500 µl into 1-ml tubes, and freeze at −70°C.

   Aliquots should remain viable >1 yr.

4. To screen library, thaw an aliquot, titer bacteria concentration (*UNIT 1.3*), and plate on screening filters.

   NOTE: *The major concern of any amplification step is that each original recombinant be equally represented. Differences in the duplication rate of any recombinant bacteria will result in over- or underpresentation in the amplified library.*

Contributor: John H. Weis

Screening of Recombinant DNA Libraries

6

The usual approach to isolating a recombinant DNA clone encoding a particular gene or mRNA sequence is to screen a recombinant DNA library. As described in Chapter 5, a recombinant DNA library consists of a large number of recombinant DNA clones, each one of which contains a different segment of foreign DNA. Because only a few of the thousands of clones in the library encode the desired nucleic acid sequence, the investigator must devise a procedure for identifying the desired clones. The optimal procedure for isolating the desired clone involves a positive selection for a particular nucleic acid sequence. If the desired gene confers a phenotype that can be selected in bacteria, then the desired clone can be isolated under selective conditions (*UNIT 1.4*). However, most eukaryotic genes and even many bacterial sequences do not encode a gene with a selectable function. Clones encoding nonselectable sequences are identified by screening libraries: the desired clone is identified either because (1) it hybridizes to a nucleic acid probe, (2) it expresses a segment of protein that can be recognized by an antibody, or (3) it promotes amplification of a sequence defined by a particular set of primers.

Screening libraries involves the development of a rapid assay to determine whether a particular clone contains the desired nucleic acid sequence. This assay is used first to identify the recombinant DNA clone in the library and then to purify the clone (see Fig. 6.0.1). Normally, this screening procedure is performed on bacterial colonies containing plasmids or cosmids or on bacteriophage plaques. To test a large number of clones at one time, the library is spread out on agarose plates (*UNIT 6.1*), then the clones are transferred to filter membranes (*UNIT 6.2*). The clones can be simultaneously hybridized to a particular probe (*UNITS 6.3 & 6.4*) or bound to an antibody (*UNIT 6.7*). When the desired clone is first identified, it is usually found among many undesirable clones; an important feature of library screening is the isolation of the desired clones (*UNITS 6.5 & 6.6*). Another method for identifying the desired clone involves hybrid selection (*UNIT 6.8*), a procedure by which the clone is used to select its mRNA. This mRNA is characterized by its translation into the desired protein. Libraries consisting of large genomic DNA fragments (~1 Mb) carried in yeast artificial chromosome (YAC) vectors have proven to be tremendously useful for genome analysis. In general, these libraries (which are usually produced by large "core" laboratories) are intially screened using a locus-specific PCR assay (*UNIT 6.9*); the clone resulting from the initial round of screening is subsequently analyzed by more conventional hybridization methods (*UNIT 6.10*).

To screen a DNA library, one must first devise the screening procedure. The next important choice is the selection of a recombinant DNA library. When choosing which library to screen the investigator should consider whether he or she wants to isolate clones encoding the gene or the mRNA sequence. cDNA clones encode the mRNA sequence and allow prediction of the amino acid sequence, whereas genomic clones may contain regulatory as well as coding (exon) and noncoding (intron) sequences. The differences between genomic and cDNA libraries are discussed in Chapter 5.

Another critical parameter to be determined before proceeding with a library screen is the number of clones in the library that must be screened in order to identify the desired clone. That is, what is the frequency of the desired clone in the library? This frequency is predicted differently for genomic and cDNA libraries, as described below.

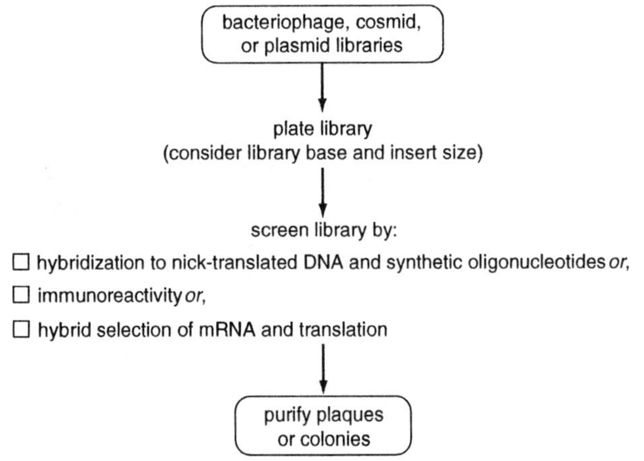

Figure 6.0.1 Flow chart for screening libraries.

Screening a genomic library. In general, genomic libraries can be made from DNA derived from any tissue, because only two copies of the gene are present per cell or per diploid genome. The predicted frequency of any particular sequence should be identical to the predicted frequency for any other sequence in the same genome. The formula for predicting the number of clones that must be screened to have a given probability of success is presented in *UNIT 5.1*. This number is a function of the complexity of the genome and the average size of the inserts in the library clones. For amplified libraries, the base (see *UNIT 5.1*) must exceed this number. Usually about 1 million bacteriophage clones or 500,000 cosmid clones must be screened to identify a genomic clone from a mammalian DNA library. Many of the clones that are screened from an amplified library will be screened more than once; the total number of clones that must be screened is 30 to 40% greater than the number calculated by the formula.

Screening a cDNA library. The optimal cDNA library is one made from a particular tissue or cell that expresses the desired mRNA sequence at high levels. In highly differentiated cells, a particular mRNA may comprise as many as 1 in 20 of the poly(A)$^+$ mRNA molecules, while some mRNAs are either not present at all or comprise as low as 1 molecule in 100,000 poly(A)$^+$ mRNA molecules. When choosing a cDNA library the investigator must make every effort to obtain a library from a cell where the mRNA is being expressed in large amounts. Of course, the number of clones that must be screened is determined by the abundance of the mRNA in the cell. The amount of protein that is found in the cell is frequently a good indicator of the abundance of the mRNA. Thus, proteins that comprise 1% of the total cell protein are made by mRNAs that usually comprise 1% of the total poly(A)$^+$ mRNA, and the desired cDNA clones should comprise about 1% of the clones in the cDNA library.

Screening a YAC library. In the typical genomic libraries maintained in *E. coli* (described in Chapter 5), the size of the insert is limited to 20 to 25 kb for lambda vectors or to 40 to 45 kb for cosmid vectors. Yeast artifical chromosome (YAC) vectors, by contrast, are designed to carry much larger genomic DNA fragments and thereby facilitate genomic analysis, with inserts ranging from 0.3 to ~1 Mb in size. Conventional screening of YAC libraries by hybridization is difficult, both because of the unfavorable signal-to-noise ratio and the sheer numbers of replica films required to represent an entire library.

For example, a standard YAC library representing 5 to 8 genome equivalents requires over 500 microtiter plates (and corresponding filters for screening by hybridization). Thus, most core laboratories screen YAC libraries using a locus-specific PCR assay whose primers define a particular sequence. The PCR screening is initially performed using pools (representing up to 4 microtiter plates or 384 YAC clones) or superpools (representing up to 20 microtiter plates or nearly 2000 clones), followed by subsequent rounds of screening to narrow down the possible candidates.

General considerations. When selecting the library it is critical that the base be larger than the number of clones to be screened. One problem with predicting the number of clones to screen is that most libraries are amplified and in the process of amplifying the library some clones are lost while others may grow more rapidly. Thus, if the desired clone is not found in a particular library, another independent library should be screened.

Having selected the library, the investigator is ready to begin screening for the desired clone. The technologies used to screen libraries are mostly extensions of the techniques that have been described earlier in the manual. Libraries are plated out, transferred to nitrocellulose filters, and hybridized to ^{32}P-labeled probes or bound to antibodies. The major problem associated with this technique is that "false" positives can be identified: the probe may hybridize to clones that do not encode the desired sequence. Approaches to minimize this problem are discussed in *UNIT 6.7*. A second source of undesired clones arises from the power of the screening procedures that are normally used to screen these libraries. The investigator will be screening as many as one million clones. If the library contains any contaminating recombinant DNA clone that has been previously grown in the laboratory, it will be identified in the screening procedure. Thus, extreme care must be exercised to prevent contamination of the library with previously isolated recombinant clones. Despite these problems the ability to screen large DNA libraries to isolate the desired clone provides a powerful tool for molecular biologists.

J.G. Seidman

6.1

Plating and Transferring Bacteriophage Libraries

UNIT 6.1

BASIC PROTOCOL

Materials *(see APPENDIX 1 for items with ✓)*

Host bacteria, selection strain if applicable (Table 1.4.5)
Recombinant phage
0.7% top agarose (*UNIT 1.1*)
82-mm or 150-mm LB plates *or* 245 × 245–mm Nunc Bioassay LB plates (*UNIT 1.1*)
0.2 M NaOH/1.5 M NaCl
0.4 M Tris·Cl, pH 7.6/2× SSC
✓ 2× SSC

Nitrocellulose membrane filters (or nylon filters)
46 × 57–cm Whatman 3MM or equivalent filter paper

1. Determine the titer of a bacteriophage library by serial dilution.

2. Mix recombinant phage and plating bacteria in culture tube (Table 6.1.1) and incubate 20 min at 37°C.

3. Determine the number of plates to pour according to the number of bacteriophage per plate. This number is defined by the number of recombinants in the library and the frequency of the expected clone. Add 0.7% top agarose to culture tubes and pour on LB plates. Incubate 6 to 12 hr, 37°C, until plaques cover the plate but are not confluent. Place at 4°C for >1 hr before applying filters.

   Determine frequency of the clone as follows: cDNA libraries—the expected frequency of the desired RNA among the total cellular RNA, ranging from 1/100 to 1/50,000; genomic libraries—size of insert divided by total genome size; subgenomic libraries—size of insert per total genome size times the fold purification of the DNA fragment (usually 10- to 50-fold).

   The top agarose must be melted and then cooled to 45 to 50°C before use. Top agarose that is too hot will kill the bacteria, while if it is too cold the library will solidify in the tube.

   Because small plaques are preferred, do not incubate unattended overnight; instead, place at 4°C and continue growth the next day.

4. Label nitrocellulose filters with a ballpoint pen and place face down (ink side up) on cold plaque-containing LB plates 1 to 10 min. Mark orientation of the filter with a 20-G needle and remove from plate with blunt, flat forceps. Place face up to paper towels or filter paper and dry ≥10 min.

   Up to five replicas can be made from each plate. Artifacts can be eliminated by hybridizing duplicate filters.

4. In sequence, place filters face up 1 to 2 min on three separate Whatman 3MM filters saturated with:

   0.2 M NaOH/1.5 M NaCl
   0.4 M Tris·C1, pH 7.6/2× SSC
   2× SSC.

5. Dry in a vacuum oven 90 to 120 min at 80°C, or overnight in a regular oven at 42°C. Store at room temperature in absorbent paper until hybridization (*UNITS 6.3 & 6.4*).

Reference: Arber et al., 1983.

Contributor: Thomas Quertermous

6.1

Table 6.1.1 Recommended Mixtures for Plating Bacteriophage Libraries

LB plate ingredient	Plate size		
	82 mm	150 mm	245 × 245 mm[a]
Bacteria[b] (ml)	0.2	0.5	2
Phage, pfu	5,000	20,000-30,000	150,000
Top agarose, ml	3	7	30

[a]Nunc Bioassay plates distributed by Vangard International.
[b]Plating bacteria are prepared as described in Chapter 1.

Plating and Transferring Cosmid and Plasmid Libraries

UNIT 6.2

BASIC PROTOCOL

Materials (*see APPENDIX 1 for items with* ✓)

- LB plates containing antibiotic (*UNIT 1.1*)
- LB medium (*UNIT 1.1*)
- LB plates containing 50 µg/ml chloramphenicol (*UNIT 1.1*)
- 0.5 M NaOH
- ✓ 1 M Tris·Cl, pH 7.5
- 0.5 M Tris·Cl, pH 7.5/1.25 M NaCl

- 10- or 15-cm Whatman 3MM or equivalent filter paper discs
- Nitrocellulose membrane filters (10- or 15-cm, Millipore HATF)
- 20 × 20–cm Whatman 3MM or equivalent filter paper
- 20 × 20–cm glass plate
- 46 × 57–cm Whatman 3MM or equivalent filter paper

1. Determine the titer of a plasmid or cosmid library by serial dilution on plates containing antibiotics. Calculate the optimal amount of the bacterial suspension for plating and dilute to 5 to 10 ml in LB medium (*UNIT 1.3*).

   A 10-cm filter can hold 10,000 to 20,000 colonies and a 15-cm filter can hold up to 50,000.

2. Prepare a layer of sterile 10- or 15-cm Whatman 3MM discs on a sterile sintered-glass Buchner funnel or a porcelain filter funnel, and add 10 to 20 ml LB medium.

3. Place a labeled nitrocellulose filter on an LB/antibiotic plate and transfer to the filtration apparatus (suction off). Pipet bacterial suspension onto the filter, leaving the outer 4 to 5 mm free of solution.

   The antibiotic must be permissive for cosmid- or plasmid-bearing bacterial cells, and usually is tetracycline or ampicillin.

4. Slowly suction the solution, transfer the filter back to the antibiotic plate (making sure there are no air bubbles), and incubate upside down at 37°C until colonies are ~1 mm in diameter.

5. Label and wet another set of nitrocellulose filters. Transfer the initial library filter to several sheets of 20 × 20–cm 3MM paper, bacteria side up. Wearing gloves, position the wetted replica filters on top of the bacterial lawn, offsetting by 2 to 3 mm to facilitate later separation.

6. Lay 3 sheets of 20 × 20–cm 3MM paper on top, followed by a 20 × 20–cm glass plate, and transfer colonies by pressing down on the glass plate.

7. Remove the glass plate and filter paper, and orient by punching holes with a 20-G needle. Separate, place on agar plates (bacteria up), and incubate replica colonies at 37°C and library filters at 25°C, both overnight. Store on agar plates at 4°C, or make additional replicas.

   To make additional replicas, incubate library filters 2 to 4 hr at 37°C or overnight at 25°C to allow regrowth of colonies; repeat steps 5 to 7.

8. Amplify the cosmids or plasmids by transferring to an LB plate containing 50 µg/ml chloramphenicol and incubating 4 to 10 hr at 37°C.

9. In sequence, place filters (bacteria up) 5 min on three separate 46 × 57–cm 3MM filters soaked with:

   0.5 M NaOH
   1 M Tris·Cl, pH 7.5
   0.5 M Tris·Cl, pH 7.5/1.25 M NaCl.

10. Dry on 3MM paper and place in a vacuum oven 90 min at 80°C. Hybridize filters with a nick-translated probe (*UNITS 6.3 & 6.4*).

Reference: Hanahan and Meselson, 1983.

Contributor: John H. Weis

UNIT 6.3

Using DNA Fragments as Probes

BASIC PROTOCOL

HYBRIDIZATION IN FORMAMIDE

Materials *(see* APPENDIX 1 *for items with ✓)*

- Nitrocellulose membrane filters bearing plaques, colonies, or DNA (*UNITS 6.1 & 6.2*)
- ✓ Hybridization solution I
- Radiolabeled probe, 1 to 15 ng/ml (*UNIT 3.5*)
- ✓ 2 mg/ml sonicated herring sperm DNA
- ✓ Low-stringency wash buffer I
- ✓ High-stringency wash buffer I, prewarmed

1. Wet filters in turn with 5 to 20 ml hybridization solution I, producing a stack of 10 to 20 filters. Transfer to a sealable bag and add enough hybridization solution to cover. Seal and prehybridize ≥1 hr at 42°C.

   No more than ten 20 × 20-cm square filters or twenty 82-mm discs should be placed in one stack.

2. Boil radioactive probe (1 to 15 ng/ml hybridization reaction at $>5 \times 10^7$ cpm/µg) with 2 mg (1 ml) sonicated herring sperm DNA for 10 min in a screw-cap tube.

3. Transfer to ice and add 2 ml hybridization solution I. Add probe mixture via syringe with 18-G needle to filters, reseal, mix thoroughly, and incubate overnight at 42°C.

4. Rinse filters three times with 500 ml low-stringency wash buffer I, room temperature, 10 to 15 min for each rinse.

   CAUTION: *The solution is extremely radioactive; handle carefully.*

5. Rinse filters twice with 500 ml high-stringency wash buffer I (prewarmed to wash temperature), 15 to 20 min for each rinse.

   Determine wash temperature empirically. If homology between probe and target approaches 100%, use a high-temperature wash of 65°-75°C. For low homology and short probe lengths, lower temperature to 37°-40°C. Wash very short probes <100 bp at lower temperatures regardless of homology.

6. Mount filters wet wrapped in plastic wrap or dry on plastic backing (e.g., used X-ray film) for autoradiography.

HYBRIDIZATION IN AQUEOUS SOLUTION

ALTERNATE PROTOCOL

Additional Materials (*see* APPENDIX 1 *for items with* ✓)

✓ Hybridization solution II
✓ Low-stringency wash buffer II
✓ High-stringency wash buffer II

1. Prehybridize as in Basic Protocol except incubate in hybridization solution II at 65°C.

2. Prepare probe as in Basic Protocol and dilute with 2 ml hybridization solution II. Hybridize overnight at 65°C. Remove hybridization solution and rinse twice with low-stringency wash buffer II.

3. Wash filters quickly 5 to 8 times with high-stringency wash buffer II at 65°C. Leave in final wash ~20 min. Washed filters should produce a nonspecific signal only a few-fold above background levels.

References: Church and Gilbert, 1984; Denhardt, 1966.

Contributor: William M. Strauss

Using Synthetic Oligonucleotides as Probes

UNIT 6.4

HYBRIDIZATION IN SODIUM CHLORIDE/SODIUM CITRATE (SSC)

BASIC PROTOCOL 1

Materials (*see* APPENDIX 1 *for items with* ✓)

Membrane filters bearing plasmid, bacteriophage, or cosmid libraries (UNITS 6.1 & 6.2)
3× SSC/0.1% SDS
✓ Prehybridization solution
✓ SSC hybridization solution
6× SSC/0.05% sodium pyrophosphate, prewarmed

1. Prepare duplicate nitrocellulose filters (processed and baked) of bacterial colonies or bacteriophage plaques. Wash 82-mm filters 3 to 5 times in 500 ml 3× SSC/0.1% SDS (50 filters) at room temperature. Then wash at 65°C at least 1.5 hr to overnight.

   Use filter forceps (without serrated tips) to handle the membrane filters.

2. Prehybridize in prehybridization solution 1 hr at 37°C.

3. Transfer up to 20 filters into sealable bags containing ≥20 ml SSC hybridization solution, and add 0.125 ng (for bacterial colonies) to 1.0 ng (for bacteriophage plaques) of each ^{32}P-labeled oligonucleotide/ml hybridization solution in one bag. Hybridize oligonucleotides 14 to 48 hr at the temperatures indicated:

   14-base—room temperature
   17-base—37°C
   20-base—42°C
   23-base—48°C.

6.4

4. Remove filters and wash 3 to 5 times in 6× SSC/0.05% pyrophosphate 5 to 15 min at room temperature. Wash 30 min in prewarmed 6× SSC/0.05% pyrophosphate at the temperatures indicated:

   14-base—37°C
   17-base—48°C
   20-base—55°C
   23-base—60°C.

5. If filters are above background radioactivity, increase wash temperature by 2° to 3°C for 15 to 30 min, and recheck. Do not exceed the following temperatures:

   14-base—41°C
   17-base—53°C
   20-base—63°C
   23-base—70°C.

6. Remove, cover with plastic wrap, mount, and expose to X-ray film 14 to 72 hr at −70°C, using an intensifying screen. Spots in the same place on duplicate filters are "positives," and should be processed as in *UNIT 6.3*.

   The intensity of the spot can vary dramatically between the duplicate filters.

BASIC PROTOCOL 2

HYBRIDIZATION IN TETRAMETHYLAMMONIUM CHLORIDE (TMAC)

This protocol facilitates the identification of true positives on the basis of correct nucleotide pairing, allowing hybridization temperatures to be determined on the basis of oligo length alone. False positives resulting from spurious hybridization to G-C rich sequences are eliminated.

6.4

Materials *(see APPENDIX 1 for items with ✓)*

Nitrocellulose or nylon membrane filters bearing plasmid, bacteriophage, or cosmid libraries (*UNITS 6.1 & 6.2*)
150-mm LB agarose plates (*UNIT 1.1*), prewarmed to 37°C
2× SSC/0.5% SDS/50 mM EDTA, pH 8.0, prewarmed to 50°C
✓ TMAC hybridization solution, prewarmed to hybridization temperature
✓ TMAC wash solution
2× SSC/0.1% SDS

1a. Process filters with bacterial colonies as described in *UNIT 6.2*.

1b. Prepare filters with amplified bacteriophage plaques as follows: plate bacteriophage from the library on LB agarose plates at a reduced density (8,000 to 10,000/150-mm plate), incubate until the plaques just become visible, transfer to nitrocellulose or nylon filters, transfer filter to prewarmed LB agarose plate and incubate 5 to 12 hr at 37°C (refrigerate master plates after transfer). Denature and bind DNA to filters (*UNIT 6.1*).

2. Wash filters with bacterial colonies as in step 1 of Basic Protocol 1. Wet bacteriophage-bearing filters in 50°C solution of 2× SSC/0.5% SDS/50 mM EDTA, pH 8.0. Submerge, rub off bacterial debris with gloved hand, and transfer to fresh 2× SSC/0.5% SDS/50 mM EDTA.

   Alternatively, incubate filters in solution for one to several hours, then scrub.

3. Transfer up to 30 filters to a 15-cm glass crystallizing dish containing 5 to 10 ml prewarmed TMAC hybridization solution/filter (see Fig. 6.4.1 for melting

temperatures), seal with plastic wrap, and prehybridize 1 to 2 hr at the hybridization temperature. Hybridization temperature should be 5° to 10°C below the melting temperature, and washing temperature should be 5° to 10°C below the melting temperature.

Alternatively, use a sealable bag with <10 filters per bag (see SSC protocol).

4. Transfer to a hybridization vessel containing fresh, prewarmed TMAC hybridization solution, add 1-2 × 10^6 cpm of ^{32}P-labeled oligonucleotide probe/ml hybridization solution (see Support Protocol), and incubate 40 to 60 hr at the appropriate temperature.

5. Rinse with 5 to 10 ml TMAC wash solution/filter at room temperature, then transfer filters individually to 200 to 250 ml fresh TMAC wash solution and wash 15 min at room temperature with gentle agitation.

6. Replace with a similar volume of prewarmed TMAC wash solution and incubate 1 hr at the appropriate wash temperature. Replace three times with a similar volume of 2× SSC/0.1% SDS, and wash 10 min at room temperature to remove TMAC. Autoradiograph as in Basic Protocol 1.

LABELING THE 5′ ENDS OF MIXED OLIGONUCLEOTIDES

SUPPORT PROTOCOL

Materials

2.5 to 250 pmol mixed oligonucleotides
10× T4 polynucleotide kinase buffer (*UNITS 3.4 & 3.10*)
[γ-^{32}P]ATP (>7000 Ci/mmol)
25 to 50 U T4 polynucleotide kinase (*UNIT 3.10*)
Ice-cold 10% trichloroacetic acid (TCA)

6.4

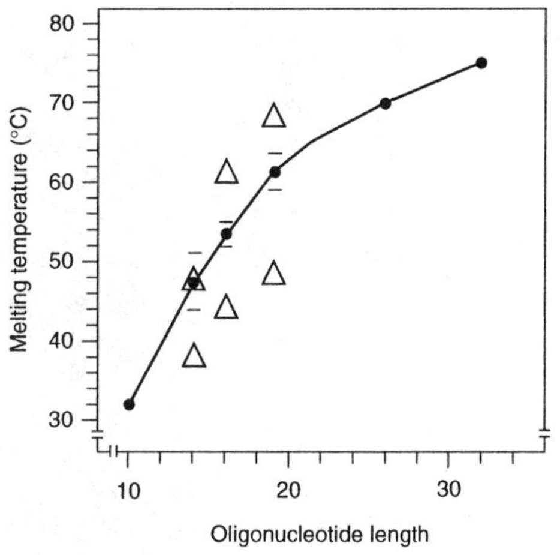

Figure 6.4.1 Melting temperatures of oligonucleotides of different length in TMAC and SSC hybridization solutions. Dots represent the average melting temperature of several different oligonucleotides of length 14, 16, or 19 bases in TMAC; bars represent the high and low melting temperatures for each length. Triangles represent the high and low melting temperatures for the same oligonucleotides in SSC. The melting temperature of only one oligonucleotide of length 10, 26, or 32 bases was determined. Hybridization temperature should be 5° to 10°C below the melting temperature, and washing temperature also should be 5° to 10°C below the melting temperature (Jacobs et al., 1988).

1. Set up reaction mixture on ice in microcentrifuge tube:

    2.5 to 250 pmol mixed oligonucleotides
    7.5 μl 10× T4 polynucleotide kinase buffer
    66 pmol [γ-^{32}P]ATP (200 μCi)
    25 to 50 U T4 polynucleotide kinase
    H_2O to 75 μl.

    Incubate 30 min at 37°C.

    1 mol deoxyribonucleotide ≅ 330 g; 1 A_{260} ≅ 40 μg/ml oligonucleotide.

    The oligonucleotide should be designed based on the amino acid sequence of interest; Table 6.4.1 lists optimal codons for each amino acid.

2. Check for incorporation of label by precipitating 1 μl of a diluted aliquot with ice-cold 10% TCA followed by scintillation counting (*UNIT 3.4*). Store mixture at −20°C.

    Using an excess of oligonucleotide, ~30% to 90% of the counts are incorporated.

References: Jacobs et al., 1988; Wood et al., 1985; Woods et al., 1982.

Contributors: Allan D. Duby, Kenneth A. Jacobs, and Anthony Celeste

Table 6.4.1 Optimum Codon Choice When Deducing a Probe Sequence from Human Amino Acid Sequence Data

Amino Acid	Optimum codon[a] when subsequent codon begins with	
	A or C or T	G
Methionine	ATG	nc[b]
Tryptophan	TGG	nc
Tyrosine	TAC	TAT
Cysteine	TGC	TGT
Glutamine	CAG	nc
Phenylalanine	TTC	TTT
Aspartic acid	GAC	GAT
Asparagine	AAC	AAT
Histidine	CAC[c]	CAT
Glutamic acid	GAG	nc
Lysine	AAG	nc
Alanine	GCC	GCT
Isoleucine	ATC	ATT
Threonine	ACC	ACA[f]
Valine	GTG[d]	nc
Proline	CCC[e]	CCT
Glycine	GGC	nc
Leucine	CTG	nc
Arginine	CGG	nc
Serine	TCC	TCT

[a]The optimum codon is the most frequent codon in all cases except Arg and Ser, where the indicated triplets generate a higher overall homology to all possible codons. Reprinted with permission of *Journal of Molecular Biology.*
[b]No change.
[c]CAT when followed by C.
[d]GTC when followed by T.
[e]CCA when followed by T.
[f]These cases do not follow the "replace C by T" rule applied when the subsequent codon is headed by G.

6.4

Purification of Bacteriophage Clones

UNIT 6.5

BASIC PROTOCOL

Materials (*see APPENDIX 1 for items with* ✓)

- 0.7% top agarose (*UNIT 1.1*)
- Host bacteria (OD_{600} 1.5 to 2 in 10 mM $MgSO_4$)
- LB plates (*UNIT 1.1*)
- ✓ Suspension medium (SM)
- Chloroform
- Nitrocellulose membrane filters

1. Plate 3 ml of 0.7% top agarose containing 200 µl host bacteria on 82-mm LB plates (one plate per clone) and set for 10 min.

2a. Identify potential clone by aligning the primary library filter and its autoradiogram. Insert toothpick first into the primary plate over the hybridization spot, and then into the top agarose of one of the secondary plates prepared in step 1, making 30 to 40 stabs.

2b. Alternatively, remove circular plug of agarose, place in 1 ml SM with one drop of chloroform 1 to 2 hr, and titer. Prepare 3 to 6 plates with <500 phage/plate.

3. Grow secondary plates overnight at 37°C. Transfer plaques to nitrocellulose filters, process, hybridize, wash, and expose (*UNITS 6.1-6.4*). Identify positive plaques on the secondary plates.

4. Insert toothpick into the most strongly hybridizing plaque for each clone and place 5 min in 1 ml SM. Plate 1 µl of phage stock and 1 and 10 µl of 1:100 dilution on tertiary LB plates.

5. Screen tertiary plates as above. Repeat until phage is pure, making a final SM stock.

Expect a recovery of >90% of clones confirmed by duplicate filters.

Reference: Kaiser and Murray, 1984.

Contributor: Thomas Quertermous

Purification of Cosmid and Plasmid Clones

UNIT 6.6

BASIC PROTOCOL

6.6

Materials

- Cold LB medium containing antibiotic (*UNIT 1.1*)
- LB plates containing antibiotic (*UNIT 1.1*)
- Round toothpicks (*UNIT 1.1*)
- Nitrocellulose membrane filters
- Spreader (*UNIT 1.3*)

NOTE: All materials coming into contact with *E. coli* must be sterile.

1. Pick positive clones, as detected by in situ hybridization of nitrocellulose replica filters (*UNITS 6.3* and *6.4*), with a sterile toothpick. Rinse colony into microcentrifuge tube containing 1 ml cold LB medium with appropriate antibiotic. Store at 4°C to inhibit growth.

2. Spread 1 to 25 μl of bacterial suspension onto an LB plate with appropriate antibiotic (*UNIT 1.3*), aiming for 25 to 250 colonies/100-mm plate. Grow overnight at 37°C.

3. Make a replica copy on a nitrocellulose filter, denature, renature, bake, and hybridize (*UNIT 6.2*).

4. From the autoradiograph of the secondary plate (*APPENDIX 3*), select the most isolated, positive colony. Grow the colony and isolate the DNA (*UNIT 1.6*).

Contributor: John H. Weis

UNIT 6.7

Immunoscreening of Fusion Proteins Produced in Lambda Plaques

BASIC PROTOCOL

SCREENING A λgt11 EXPRESSION LIBRARY WITH ANTIBODIES

Materials *(see APPENDIX 1 for items with ✓)*

λgt11 cDNA expression library
E. coli LE392 (Table 1.4.5)
150-mm LB plates (*UNIT 1.1*)
1% LB top agar (*UNIT 1.1*)
India ink containing 0.05% NaN_3 (optional)
✓ Immunoscreening buffer
First-stage antibody
^{125}I-labeled second-stage reagent reactive with first-stage antibody

132-mm nitrocellulose membrane filters

NOTE: All materials coming into contact with *E. coli* must be sterile.

1. Titer and plate a λgt11 cDNA library with *E. coli* LE392 on 150-mm LB plates, using 7 ml of 1% LB top agar/plate (*UNITS 1.11 & 6.1*). Incubate 8 hr at 37°C.

2. Lay a numbered 132-mm nitrocellulose filter on plate (*UNIT 6.1,* step 6), and incubate overnight at 37°C.

3. Mark each filter with needle holes and India ink. Remove filters and block by washing 30 min in immunoscreening buffer at room temperature. Repeat wash 2 to 4 times to remove the bulk of bacteria.

   It is helpful to add sodium azide to 0.05% to India ink to prevent contamination of the plaque plates.

4. Incubate filters with first-stage antibody (0.5 to 10 μg/ml in immunoscreening buffer) in a heat-sealed bag, 2 to 24 hr at 4°C on a horizontal shaker platform.

   Multiple filters may be placed in a single bag if there is sufficient liquid so mixing between filters occurs. In order to stabilize immune complexes, all reactions and washings should be done at 4°C with cold buffer. If reaction to all the plaques is observed, first-stage antibody can be absorbed with commercial E. coli extracts, or reused, as it was absorbed during the first use.

5. Wash filters 4 to 5 times in immunoscreening buffer at 4°C for 5 to 10 min/wash. Incubate with ^{125}I-labeled second-stage antibody (0.5×10^6 cpm/ml) in heat-sealed bags for 2 to 6 hr at 4°C.

6. Wash 4 to 5 times in immunoscreening buffer at 4°C. Blot dry, wrap in plastic wrap, and expose to X-ray film with an intensifying screen at −70°C (*APPENDIX 3*).

   Dispose of isotope and waste appropriately.

7. Purify λ cDNA fusion-protein clones by repeated dilutions until pure, and grow for DNA preparation (*UNIT 6.5*).

INDUCTION OF FUSION PROTEIN EXPRESSION WITH IPTG PRIOR TO SCREENING WITH ANTIBODIES

ALTERNATE PROTOCOL

The probability of success in screening a λgt11 cDNA library with an antibody can sometimes be increased by preventing the expression of the fusion protein until the plaques are well established. The expression of the potential β-galactosidase–cDNA fusion proteins can be induced after 3 to 4 hr of plaque growth by placing a nitrocellulose filter containing the inducer IPTG onto the plate and continuing growth at 37°C. The nitrocellulose filters are then screened with antibodies as in the Basic Protocol.

Additional Materials *(also see Basic Protocol)*

E. coli Y1090 (Table 1.4.5)
10 mM IPTG (Table 1.4.2)
42°C room or incubator

1. Absorb 1–5 × 10^4 cDNA-fusion λ phage with 0.5 to 1.0 ml of a fresh overnight culture of *E. coli* Y1090. Plate on a fresh (2 to 3 days) 150-mm LB plate with 7 ml LB top agar and incubate 3.5 hr at 42°C.

   This higher temperature incubation should make any fusion protein produced as unstable as possible, as well as ensure that the temperature-sensitive λcI857 repressor is completely denatured.

2. Soak a 132-mm nitrocellulose filter in 10 mM IPTG and dry. Lay filters on plates bearing the bacteriophage library and incubate 3.5 hr at 37°C. Mark each filter and remove. Block remaining protein-binding capacity and probe filters (Basic Protocol steps 3 to 7).

Reference: Huynh et al., 1984.

Contributor: Thomas P. St. John

Immunoscreening After Hybrid Selection and Translation

BASIC PROTOCOL

Materials (*see APPENDIX 1 for items with* ✓)

Brain-heart-infusion (BHI) medium (37.5 g/liter, autoclaved), containing appropriate antibiotics
Chloramphenicol (Table 1.4.1)
✓ TE buffer, pH 7.6
1 M NaOH
✓ Neutralization solution
6× SSC
✓ Hybridization solution IV
Poly(A)$^+$ mRNA (*UNIT 4.5*)
TES buffer/0.5% SDS, 65°C
✓ TES buffer, 65°C
10 mg/ml yeast tRNA
Buffered phenol (*UNIT 2.1*)
50:1 chloroform/isoamyl alcohol
✓ 3 M sodium acetate, pH 5.2
Ethanol
Translation mixture
[^{35}S]methionine (800 Ci/mmol)
✓ Immunoprecipitation buffer
Nonimmune serum
✓ Protein A–Sepharose suspension
Polyclonal or monoclonal antibodies
✓ High-salt immunoprecipitation buffer
✓ 2× SDS sample buffer

Beckman JS-4.2 rotor or equivalent
0.45-µm nitrocellulose filters (2.5-cm diameter)
Multifilter washing apparatus
Sterile, silanized 1.5-ml microcentrifuge tubes (*APPENDIX 3*)

1. Pick individual cDNA clones, place into wells of a microtiter dish containing 0.25 ml BHI medium with appropriate selective antibiotics, and grow overnight at 37°C.

2. Inoculate 50 ml BHI/antibiotic medium in a 250-ml flask, with 0.1 ml from each of 10 individual overnight clone cultures, and grow at 37°C until A_{590} = 0.7. Add chloramphenicol (100 µg/ml) and grow overnight at 37°C. Repeat for each set of ten overnight clones.

3. Centrifuge 10 min at 2000 × g, 4°C. Prepare plasmid DNA (*UNIT 1.6*).

4. Dilute ~50 µg crude plasmid DNA in 1.5 ml TE buffer, transfer to a 15-ml sterile, capped glass culture tube, and incubate 10 min in a boiling water bath. Immediately add 1.5 ml of 1 M NaOH and leave 10 min at room temperature. Add 9 ml neutralization solution, mix, place on ice, and check that pH = 6.5 to 7.5.

   Check pH as quickly as possible to avoide renaturation; if necessary, adjust pH with NaOH or HCl.

5. Place 0.45-µm nitrocellulose filter on porous support attached to a vacuum line. Pour denatured DNA solution from step 4 at a flow rate of 1 ml/min. After all liquid has passed through, increase vacuum to maximum for 3 min. Wash each filter with 50 ml of 6× SSC. Bake DNA 2 hr at 80°C in a vacuum oven.

6. Punch out disks of 0.5-cm diameter from the filter (with sterile one-hole paper punch) and mark with a ballpoint pen.

7. Preheat 0.3 ml hybridization solution containing 10 to 50 µg poly(A)$^+$ RNA 10 min at 70°C in a sterile, capped plastic tube. Add up to 10 filter disks and incubate 2 hr at 50°C.

8. Transfer filter disks to a 50-ml test tube (up to 20 filters/tube), wash ten times with 25 ml of 65°C TES/0.5% SDS, hand vortex 30 sec, and suction off supernatant. Wash twice with 25 ml of 65°C TES buffer.

9. Transfer individual filters to sterile, silanized 1.5-ml microcentrifuge tubes. Add 0.3 ml water and 2 µl of 10 mg/ml yeast tRNA to each tube. Boil 60 sec, quick-freeze in dry ice/ethanol, and thaw at room temperature.

10. Remove filters with a sterile needle, add 0.15 ml buffered phenol and 0.15 ml chloroform/isoamyl alcohol to each tube, and extract. Add 30 µl of 3 M sodium acetate and precipitate with 2 vol ethanol. Microcentrifuge 15 min, wash with 0.5 ml ethanol, dry by lyophilization, and resuspend the hybrid-selected RNA in 10 µl water.

11. Add 10 µl translation mixture containing ^{35}S-labeled methionine to 5 µl hybrid-selected RNA, and incubate 60 min at 30°C.

    Any reticulocyte or wheat germ translation mixture system can be used; prepare as recommended by manufacturer.

    The procedure can be interrupted at this stage by freezing the translation mixture at −70°C.

12. Add 15 µl immunoprecipitation buffer and 1 µl nonimmune serum, and incubate 10 min at room temperature.

13. Add 40 µl protein A–Sepharose suspension and leave 30 min at room temperature. Microcentrifuge 2 min and transfer supernatant to a new tube.

14. Add 1 µl polyclonal or monoclonal antibodies directed against the relevant gene product, and incubate 10 min, room temperature.

15. Add 40 µl protein A–Sepharose suspension and incubate 30 min at room temperature. Microcentrifuge and discard supernatant.

16. Wash protein A–Sepharose beads three times with 1 ml immunoprecipitation buffer, once with 1 ml high-salt immunoprecipitation buffer, and once with 1 ml water.

17. Resuspend pellet in 20 µl of 2× SDS sample buffer, and boil 10 min to elute bound polypeptides from protein A–Sepharose beads. Microcentrifuge and fractionate 15- to 20-ml aliquots of supernatant on a denaturing SDS-polyacrylamide gel (*UNIT 10.2*).

    Run a control lane containing the polypeptide of interest.

18. Dry gel and expose to autoradiography.

If any of the lanes contains the required polypeptide band, subject individual clones of that set of ten clones to same procedure.

Reference: Parnes et al., 1981.

Contributor: Baruch Velan

UNIT 6.9

Overview of Strategies for Screening YAC Libraries and Analyzing YAC Clones

Emphasis on identification of disease genes by positional cloning has underscored the need to clone fragments of genomic DNA >100 kb into a vector. The size of genomic inserts that can be carried in traditional cloning vectors has been limited to 20 to 25 kb for λ vectors and 40 to 45 kb for cosmid vectors. These vectors are of limited utility for analyzing very large genes or for "walking" to disease genes from DNA markers that may be 1 to 2 Mb away. Considerable progress has been made in cloning large DNA fragments in *Saccharomyces cerevisiae* using yeast artificial chromosome (YAC) vectors (see Fig. 13.4.6). YACs containing inserts that are >1 Mb have been produced and these are routinely propagated with apparent stability, suggesting that thc major limitation to the size of YAC inserts is the quality of the starting genomic DNA. Large "core" laboratories that generate human YAC libraries prepare human YACs with average insert sizes ranging from 0.3 to 1.2 Mb. Additional high-quality YAC libraries have been constructed using inserts from *Drosophila melanogaster, Caenorhabditis elegans, Schizosaccharomyces pombe*, and mouse.

Anecdotal reports indicate YAC libraries may support the propagation of certain insert sequences that are poorly represented in *Escherichia coli*–based libraries. The YAC cloning system also offers the advantage that large genomic YAC inserts can be easily manipulated in yeast by homologous recombination. Thus, it is relatively simple to truncate a YAC insert or to introduce specific deletions, insertions, or point mutations with high efficiency (*UNIT 13.10*).

This unit provides an introduction to the use of yeast artificial chromosome–bearing yeast clones (hereafter referred to as YAC clones) in genome analysis, criteria for designing a polymerase chain reaction (PCR) assay to screen a YAC core library, and the rationale for verification and characterization of YAC clones obtained from these core laboratories. A summary of the related protocols is presented in Figure 6.9.1.

GENERATING YAC LIBRARIES

Although YAC cloning is one of the methods of choice when insert sizes >100 kb are required, a number of features of the system have interfered with its rapid assimilation for routine cloning. Commonly used YACs are carried as a single copy within the more complex *S. cerevisiae* genome, so the signal-to-noise ratio is less favorable for identifying a cognate clone in a YAC library than in a λ or cosmid library. Moreover, efforts to develop high-density screening methods for YACs have enjoyed only limited success and the effort and resources required to construct YAC libraries and prepare them for screening are enormous. Consequently, it is generally most practical for investigators wishing to obtain YACs carrying a specific DNA sequence to arrange for screening of a preexisting library maintained by a core laboratory.

Initially, YAC libraries were constructed with total genomic DNA (Burke et al., 1987). More recently, there has been interest in generating libraries from targeted DNA using somatic cell hybrids carrying a specific chromosome or portion of a chromosome. This should reduce the cost and effort of screening for loci whose chromosomal location has been established.

YAC LIBRARY SCREENING BY A CORE LABORATORY

Methods used by YAC core laboratories for library screening evolve rapidly. It is possible to screen a library by hybridizing a single-copy probe to nylon filters

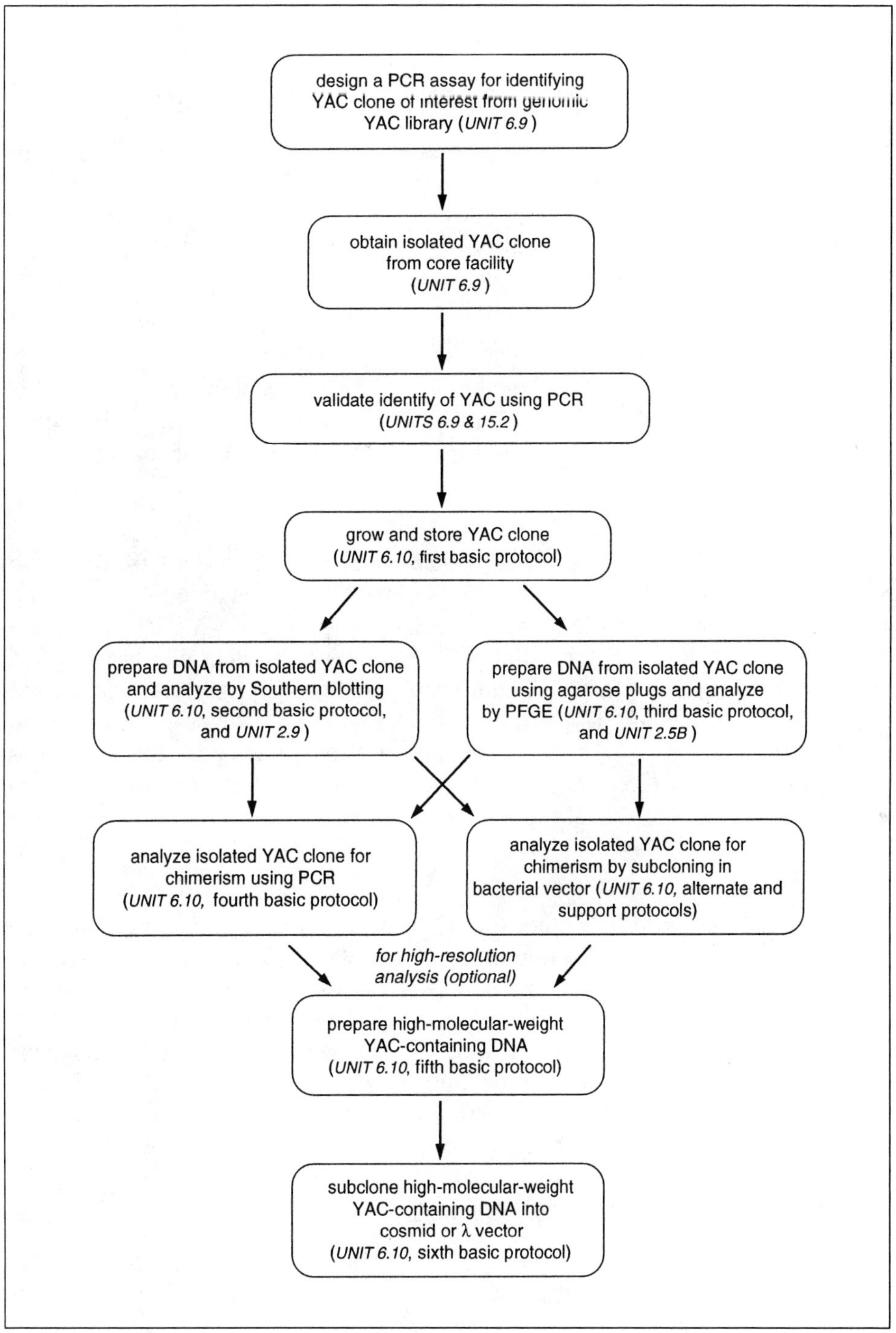

Figure 6.9.1 Flow chart showing protocols used to obtain and analyze YAC clones.

stamped with a replica of one or more microtiter arrays. However, because of the low signal-to-noise ratio for hybridization and the substantial cost required to produce all of the nylon filter replicas, most laboratories perform library screening using PCR. Most core facilities first extract DNA from pools of clones, usually representing 1 to 4 microtiter plates (96 to 384 YACs) per pool, and then combine this pooled DNA into more superpools of ~1500 to 2000 YACs. The pools and superpools are screened by PCR to identify candidate microtiter plates containing at least one amplifying YAC clone. Final identification of the clone is most commonly performed either by colony hybridization using the PCR product as the probe or by screening pools of rows and columns from the same microtiter plate using PCR. The time required for a YAC core laboratory to verify the specificity and parameters of the PCR assay and screen complex clone pools and subpools is usually 2 to 6 weeks.

DESIGNING A LOCUS-SPECIFIC PCR ASSAY FOR SCREENING

An investigator arranging with a core laboratory for library screening is required to design a strategy for detecting the inserted genomic DNA and to provide the appropriate probe(s). In general, any highly specific, sensitive, and robust PCR assay is suitable for screening a YAC library (see Chapter 15). When designing a PCR assay from scratch, it is useful to consider the following:

Fragment Size

The sequence-tagged site (STS) should be 75 to 750 bp in length. Fragments in this range are most efficiently amplified by PCR and are easily detected by either polyacrylamide (*UNIT 2.7*) or standard agarose (*UNIT 2.5A*) gel electrophoresis.

Primer Length

Each primer should ideally be 18 to 30 nucleotides long, be composed of 50% to 55% G + C, and be contained within a single-copy human-genomic-DNA segment. This ensures efficient priming and decreases the probability of false priming, enhancing the sensitivity and specificity of the assay. This also permits the amplified fragment to be used as a hybridization probe in the final hybridization-dependent steps of library screening. If it is not possible to amplify a single-copy fragment, then some other single-copy probe (e.g., a synthetic oligonucleotide 30 nucleotides long) should also be prepared. Oligonucleotide design strategies are discussed further in *UNIT 15.1*.

Primer Affinity

Primers should show little affinity for self-annealing or for annealing with each other. This prevents the production of small, template-independent PCR products that compete for primers in the reaction. A number of academic and commercial DOS-based and Macintosh software programs permit rapid selection of non-self-annealing primers from within a known DNA sequence (see *UNIT 7.7*). Although the use of these programs cannot remove all the uncertainty associated with designing a new PCR assay, it does help eliminate some of the most trivial causes of assay failure.

ANALYZING INDIVIDUAL YAC CLONES

Once library screening has been successfully completed and the isolated YAC clone has been furnished to the investigator, attention should be directed to analyzing its structure. Initial studies should focus on determining whether the genomic insert is chimeric, checking for evidence of rearrangement within the insert, and verifying that the YAC is propagated in stable fashion in the yeast cell. Simply analyzing several isolates of the same YAC in parallel may provide a means of recognizing instability, as each isolate serves as a control for the others.

Chimerism of the YAC Insert

A consistent problem in YAC cloning is chimerism of the YAC insert—i.e., the insert is composed of two or more separate genomic fragments joined in a single YAC. In most existing total genomic YAC libraries, chimeric clones represent from 5% to 50% of the total clones. Preliminary data suggest that targeted, chromosome-specific libraries may contain only 5% to 15% chimeric clones.

The most reliable way to determine if a YAC insert is chimeric is to isolate a small fragment from each end of the insert and determine its chromosome of origin and whether it shares sequences with overlapping YACs derived from the same chromosomal region. It is possible to determine whether a YAC insert is chimeric by preparing probes from the two YAC vector arms and using these to demonstrate that both ends of the YAC map to the same general chromosomal region. This is generally done using hybridization or PCR analysis of a somatic hybrid cell line containing the appropriate human chromosome (or preferably a fragment thereof) as its sole human DNA. These methods are generally favored because of their speed, but they depend on the fortuitous placement of restriction sites close enough to the ends of the genomic insert that a fragment suitably sized for PCR amplification can be generated. Moreover, if highly repetitive sequences are present at the distal portions of the insert, the PCR method may fail to generate useful information.

A reliable but more time-consuming method of generating probes for end-fragment analysis is conventional subcloning of larger YAC-derived restriction fragments into plasmid or λ vectors. Subcloning an end fragment several kilobases in size is time-consuming, but reliably assures identification of nonrepeated sequences for use as probes (*UNIT 6.10*).

Internal Rearrangement or Instability of the YAC Insert

Internal rearrangement of a YAC insert is more difficult to identify than chimerism, and may become apparent only after high-resolution analysis of the clone.

Although YACs are usually stable in culture, deletion or other rearrangements of the insert may occur months after the initial isolation of a clone. Thus, it is wise to verify the size of a YAC following prolonged passage in culture or after it has been thawed from a frozen stock. Several different colonies of the same YAC strain should be analyzed in parallel (*UNIT 6.10*) to confirm that the artificial chromosome is the same size in each of the isolates.

Because cytosine methylation, which is quite frequent in the DNA of higher eukaryotic species, does not occur in yeast, it is not possible to perform direct structural comparisons of the YAC inserts and the corresponding genomic DNA. Direct structural comparisons must be carried out using methylation-insensitive restriction enzymes and frequently spaced probes.

Evidence of internal rearrangement within a YAC clone can be obtained by preparing chromosomes from the clone (*UNIT 6.10*) and analyzing them by pulsed-field gel electrophoresis (PFGE; *UNIT 2.5B*). The CHEF gel system (*UNIT 2.5B*) is particularly useful in that it permits excellent resolution in the size range most common for individual YAC clones. Following electrophoresis, the artificial chromosome can be visualized, using ethidium bromide staining, as an extra chromosome not present in the host yeast strain or by Southern blot hybridization with YAC-specific probes.

CONSTRUCTION AND ANALYSIS OF A YAC-INSERT SUBLIBRARY

Although the large genomic DNA fragments provided by the YAC cloning system are easy to manipulate, it is often convenient to reduce a YAC to smaller fragments by subcloning it into a cosmid or λ vector for high-resolution analysis.

Two general strategies are available for preparing YAC insert DNA in order to create a saturating collection of subclones. The more elegant strategy is to purify the artificial chromosome itself by preparative CHEF gel electrophoresis (*UNIT 2.5B*). This permits isolation and analysis of the resulting recombinant clones without further selection, assuming that only a small amount of contaminating yeast DNA is present in the purified YAC, and that essentially all subclones isolated are derived from the human YAC insert. In practice, however, it is difficult to recover sufficient quantities of purified YAC DNA to permit construction of a cosmid or λ library. An alternate approach is to prepare a library from the total DNA of the YAC-carrying yeast strain. YAC-specific subclones must then be selected by hybridization. An initial round of screening is usually performed with total human genomic DNA (rich in repetitive sequences) as the probe. Additional analysis is performed to identify overlapping sequences and thereby establish an approximate map of the original YAC insert. Ultimately, one or more rounds of chromosome walking may be required to fill in gaps between contiguous groups of subclones.

Reference: Burke et al., 1987

Contributors: David D. Chaplin and Bernard H. Brownstein

6.10

UNIT 6.10

Analysis of Isolated YAC Clones

Methods for analysis of YAC clones involve growing and storing YAC-containing yeast strains and purifying YAC DNA in a form suitable for assessing the size of the artificial chromosome and for conventional Southern blotting.

NOTE: All solutions, media, glassware, and plasticware coming into contact with yeast or bacterial cells must be sterile, and sterile techniques should be followed throughout.

BASIC PROTOCOL 1

PROPAGATION AND STORAGE OF YAC-CONTAINING YEAST STRAINS

YACs prepared using the pYAC4 vector (Fig. 6.10.1) and the *S. cerevisiae* host strain AB1380 (*trp1*$^-$, *ura3*$^-$, *ade2-1*) are grown on AHC plates. They can be stored short-term on AHC plates or stored long-term (after growth in YPD medium) in YPD containing glycerol at −80°C.

Materials *(see APPENDIX 1 for items with ✓)*

S. cerevisiae strain AB1380 containing pYAC4 with insert (from core facility; *UNIT 6.9*)
✓ AHC plates (−Ura −Trp)
YPD medium (*UNIT 13.1*)
80% (v/v) glycerol in YPD medium

30°C orbital shaking incubator (e.g., New Brunswick Scientific)
Cryovials

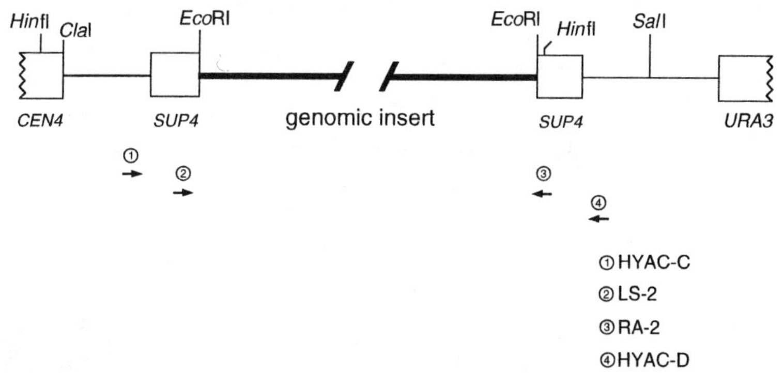

Figure 6.10.1 Structure of a representative pYAC4 clone at the vector/insert junction. Open boxes represent portions of the pYAC4 vector derived from yeast sequences (*CEN4*, the two halves of the *SUP4* element, and *URA3*). Thin lines represent sequences derived from pBR322 and the bold line represents the YAC genomic insert fragment. Sites of annealing of the HYAC-C, LS-2, RA-2 and HYAC-D oligonucleotides are indicated by arrows 1, 2, 3, and 4, respectively. The *Eco*RI cloning site, the *Cla*I and *Sal*I sites (used for the end-fragment subcloning alternate protocol), and the *Hin*fI sites that are utilized in the bubble linker end-fragment isolation protocol are indicated.

1. Streak strain AB1380 containing pYAC4 with insert onto AHC plates. Invert plate and incubate at 30°C until red colonies are 1 to 3 mm in size.

2a. *For short-term storage:* Seal plates with Parafilm and store at 4°C for 4 to 6 weeks.

2b. *For long-term storage:* Inoculate an individual colony into 3.2 ml YPD medium and shake overnight at 30°C. Add 1 ml of 80% glycerol in YPD medium, mix thoroughly, and transfer in 0.2- to 1.0-ml aliquots to cryovials. Store at −80°C.

Strains stored in this fashion are stable for 5 years. Before strains are used in an experiment, they should first be grown on selective medium (e.g., AHC plates) to avoid recovery of a contaminant clone or one that has lost its YAC.

6.10

PREPARATION OF YAC-CONTAINING DNA FROM YEAST CLONES FOR ANALYSIS BY SOUTHERN BLOTTING

BASIC PROTOCOL 2

It is best to confirm the identity of the clone by hybridization analysis. Various methods can be used to prepare DNA suitable for Southern blot analysis using frequently cutting restriction enzymes. This protocol yields substantial quantities of DNA in the size range of 50 to 200 kb.

Materials *(see* APPENDIX 1 *for items with ✓)*

Single colony of *S. cerevisiae* AB1380 containing pYAC4 with insert (Basic Protocol 1)
✓ AHC medium (−Ura −Trp)
✓ SCE buffer
✓ SCEM buffer
50 mM Tris·Cl (pH 7.6)/20 mM EDTA (Tris/EDTA lysis buffer)
10% (w/v) sodium dodecyl sulfate (SDS)
✓ 5 M potassium acetate, pH 4.8, ice-cold

95% (v/v) ethanol, room temperature
✓ TE buffer, pH 8.0
✓ 1 mg/ml DNase-free RNase A
Isopropanol, room temperature
✓ 5 M NaCl
Total genomic DNA of the species or individual from which the library was made (e.g., *UNITS 2.2, 2.3 & 5.3*)
Appropriate single-copy probe designed to hybridize with the YAC insert (see *UNITS 2.9 & 6.9*)

Orbital shaker (e.g., New Brunswick Scientific)
50-ml conical plastic centrifuge tubes
Beckman JS-4.2 rotor or equivalent

1. Inoculate a single red colony of a YAC-containing clone into 20 ml AHC medium in a 250-ml Erlenmeyer flask. Shake 24 hr at 250 rpm, 30°C, on an orbital shaker.

2. Inoculate 1 ml of pink culture from step 1 into 100 ml AHC medium in a 1-liter Erlenmeyer flask. Shake 24 hr at 250 rpm, 30°C.

3. Transfer culture to 50-ml plastic conical centrifuge tubes. Centrifuge 5 min at 2000 × *g*, 4°C. Discard supernatants and resuspend cell pellets in a total of 5 ml SCE buffer. Pool into a single tube.

4. Add 1 ml SCEM buffer. Mix gently 1 to 2 hr at 100 rpm, 37°C, on an orbital shaker.

5. Centrifuge 5 min at 2000 × *g*, 4°C. Discard supernatant and resuspend cell pellet in 5 ml Tris/EDTA lysis buffer.

6. Add 0.5 ml of 10% SDS and invert several times to mix. Incubate 20 min at 65°C.

7. Add 2 ml of ice-cold 5 M potassium acetate, pH 4.8, and invert to mix. Let stand 60 min on ice.

8. Centrifuge 10 min at 2000 × *g*, room temperature. Carefully pour nucleic acid–containing supernatant into a new tube. Add 2 vol room-temperature 95% ethanol and invert to mix.

9. Centrifuge 5 min at 2000 × *g*, room temperature. Discard supernatant and air-dry nucleic acid pellet 10 to 15 min. Add 3 ml TE buffer, pH 8.0, and dissolve overnight at 37°C.

10. Add 0.1 ml of 1 mg/ml DNase-free RNase A and incubate 1 hr at 37°C.

11. Add 6 ml room-temperature isopropanol with swirling, then invert to mix. Spool DNA using a capillary pipet and dissolve in 0.5 ml TE buffer, pH 8.0. Add 50 µl of 5 M NaCl and 2 ml of room-temperature 95% ethanol. Mix by inverting. Spool DNA again and dissolve in 0.5 ml TE buffer. Store at 4°C.

    A yield of 1 to 1.5 µg DNA/10^8 yeast cells can be expected.

12. Analyze 2-µg aliquots of YAC DNA and 15-µg aliquots of total genomic DNA from the species or individual from which the YAC library was made by digesting with several frequently cutting restriction endonucleases. Proceed with Southern blotting and hybridization using a single-copy probe.

13. Once the YAC clone has been verified by Southern blotting, determine its size and obtain a preliminary assessment of its stability by preparing chromosomes in agarose plugs (Basic Protocol 3) and analyzing by pulsed-field gel electrophoresis.

PREPARATION OF YEAST CHROMOSOMES IN AGAROSE PLUGS FOR PULSED-FIELD GEL ELECTROPHORESIS

BASIC PROTOCOL 3

In order to assess size, stability, and possible rearrangements within YACs, and to identify overlapping YACs, it is useful to isolate the YACs by embedding them in agarose plugs for subsequent analysis by pulsed-field gel electrophoresis (PFGE).

Materials *(see APPENDIX 1 for items with ✓)*

✓ AHC medium (–Ura –Trp)
Single colony of *S. cerevisiae* containing pYAC4 with insert (Basic Protocol 1)
✓ 0.05 M EDTA, pH 8.0
✓ SEM buffer
10 mg/ml Lyticase (Sigma or ICN Biomedicals)
2% InCert (w/v) or SeaPlaque agarose (FMC Bioproducts), dissolved in SEM buffer and equilibrated to 37°C
✓ SEMT buffer
✓ Lithium lysis solution
✓ 20% (v/v) NDS solution
✓ 0.5× TBE *or* GTBE buffer

30°C rotary platform shaking incubator
Beckman JS-4.2 rotor or equivalent
Gel sample molds (e.g., CHEF gel molds, Bio-Rad)
60-mm tissue culture plate

1. Inoculate 25 ml AHC medium with a single red colony of a YAC-containing clone. Shake 48 to 60 hr at 250 rpm, 30°C.

   The culture should be pink. It is useful to analyze 4 or 5 individual colonies from the same YAC strain as well as a colony of the untransformed yeast host.

2. Centrifuge 10 min at 2000 × *g*, 4°C. Discard supernatant and resuspend cell pellet in 10 ml of 0.05 M EDTA, pH 8). Centrifuge 10 min at 2000 × *g*, 4°C. Remove all liquid from pellet and resuspend in 150 µl SEM buffer.

3. Warm YAC sample to 37°C and add 25 µl Lyticase. Add 250 µl of 2% InCert or SeaPlaque agarose that has been melted in SEM buffer and equilibrated to 37°C.

4. Mix quickly and pour into CHEF gel sample molds. Chill 10 min at 4°C. Transfer solidified plugs to a 60-mm tissue culture plate. Cover each plug with 4 ml SEMT buffer. Incubate 2 hr with gentle shaking at 37°C.

5. With a pipet, remove SEMT buffer and replace with 4 ml lithium lysis solution. Incubate 1 hr with gentle shaking at 37°C.

6. Remove and replace lithium lysis solution two or three times, shaking 1 hr each time. Shake the last change overnight.

7. Remove lithium lysis solution, replace with 4 ml of 20% NDS solution, and shake 2 hr at room temperature. Repeat once.

8. Cut into plugs of suitable size to fit into wells of a pulsed-field gel. Store plugs individually 4 to 8 weeks in 20% NDS solution at 4°C.

9. Soak each plug 30 min in 1 ml of 0.5× TBE or GTBE buffer. Change buffer three times.

10. Analyze by pulsed-field gel electrophoresis with ethidium bromide visualization. If desired, carry out Southern blotting and hybridization.

BASIC PROTOCOL 4

END-FRAGMENT ANALYSIS USING PCR AMPLIFICATION

This protocol provides a means for recovering end fragments from the YAC insert using PCR amplification of end fragments by ligation of a double-stranded DNA tag containing a "bubble" of noncomplementary sequence flanked by short complementary sequences (Fig. 6.10.2). Primers for selective amplification of the two end-fragment sequences are depicted in Figure 6.10.3.

Materials *(see APPENDIX 1 for items with ✓)*

"Bubble-top" and "bubble-bottom" oligonucleotide primers (Fig. 6.10.2)
YAC-containing DNA (Basic Protocol 2)
*Rsa*I and *Hin*fI restriction endonucleases and appropriate buffers (*UNIT 3.1*)
10× T4 DNA ligase buffer and 1 U/µl T4 DNA ligase (*UNITS 3.4 & 3.14*)
✓ PCR reaction mix
PCR amplification primers HYAC-C, HYAC-D, 224, and RA-2, 4 µM each (Fig. 6.10.2)

Thermal cycling apparatus
65° and 68°C water baths

1. Phosphorylate the bubble-top oligonucleotide. Adjust bubble-top and bubble-bottom oligonucleotide concentrations to 4 nmol/ml with water. Mix together 1 nmol of each, then anneal by heating 15 min at 68°C in a water bath, followed by slow cooling to room temperature over 30 to 60 min.

   Use the bubble vector designed for use with the restriction endonuclease used to digest YAC DNA in step 2.

2. Digest 2.5-µg aliquots of purified YAC-containing DNA to completion with *Rsa*I or *Hin*fI in 20 µl final volume, 37°C. Heat samples 15 min at 65°C to inactivate the restriction enzymes.

3. Prepare the following ligation mix (50 µl total):

   2 µl (250 ng) digested DNA
   1 µl (2 pmol) annealed bubble oligonucleotides (from step 2)
   5 µl 10× ligase buffer
   2 µl (2 U) T4 DNA ligase
   40 µl H_2O.

   Incubate 2 hr at 37°C or overnight at room temperature.

4. Add 200 µl water to bring the DNA concentration to 1 ng/µl final.

6.10

```
A

               GCTGTCTGTCGAAGGTAAGGAACGGACGA
5'-GAAGGAGAGGAC                             GAGAAGGGAGAG-3'      RsaI bubble-top
   ||||||||||||                             ||||||||||||
3'-CTTCCTCTCCTG                             CTCTTCCCTCTC-5'      Universal bubble-bottom
               TCGCTAAGAGCATGCTTGCCAATGCTAAG
               *****************************

            3'-TCGCTAAGAGCATGCTTGCCAATGCTAAGC-5'                 224 primer

B

                    GCTGTCTGTCGAAGGTAAGGAACGGACGA
5'-A(N)TGAAGGAGAGGAC                             GAGAAGGGAGAG-3'  HinfI bubble-top
        ||||||||||||                             ||||||||||||
     3'-CTTCCTCTCCTG                             CTCTTCCCTCTC-5'  Universal bubble-bottom
                    TCGCTAAGAGCATGCTTGCCAATGCTAAG

C

5'-GCTACTTGGAGCCACTATCGACTACGCGAT-3'          HYAC-C primer
5'-GGTGATGTCGGCGATATAGGCGCCAGCAAC-3'          HYAC-D primer
5'-TCGAACGCCCGATCTCAAGATTAC-3'                RA-2 primer
5'-TCTCGGTAGCCAAGTTGGTTTAAGG-3'               LS-2 primer
5'-CGCTGTCCTCTCCTTC-3'                        Bubble sequencing primer
```

Figure 6.10.2 Oligonucleotides for amplification and sequencing of YAC insert end-fragments. (**A**) Annealing of the 53-mer universal "bubble-bottom" oligonucleotide to the 53-mer *Rsa*I bubble oligonucleotide yields a blunt-ended DNA duplex in which 12-bp complementary sequences flank a 29-nucleotide "bubble" of noncomplementary sequence. This bubble linker can be ligated to any blunt-ended fragment (e.g., one generated by digestion with *Rsa*I). The 224 primer does not anneal to either strand of the bubble, but is fully complementary to any DNA strand that is generated during PCR using the universal bubble-bottom strand as a template (see Fig. 6.10.3). (**B**) Annealing of the 53-mer universal bubble-bottom oligonucleotide to the 56-mer *Hin*fI bubble-top oligonucleotide yields a DNA duplex with one blunt end and one cohesive end with the degenerate *Hin*fI site. A mixture of all four nucleotides at a specific position is indicated by (N). (**C**) The HYAC-C, HYAC-D, RA-2, and LS-2 primers anneal to sequences in the pYAC4 vector (see Fig. 6.10.1). The bubble sequencing primer anneals to the *Rsa*I and *Hin*fI bubble-top sequences near their 5′ ends, permitting DNA sequencing from the bubble linker back into the YAC insert end-fragment.

5. Prepare the following PCR on ice (10 µl total):

   8 µl PCR reaction mix
   1 µl (2 µM each) PCR primer pair mix
   1 µl (1 ng) digested, bubble-ligated YAC DNA.

   Carry out 35 cycles of amplification as follows: 1 min at 92°C, 2 min at 65°C, and 2 min at 72°C.

6. Analyze a 1-µl aliquot of PCR product on a 5% polyacrylamide gel. A single, clearly visible amplified fragment should be observed after staining the gel with ethidium bromide.

To amplify the left end (either RsaI- or HinfI-digested DNA):

7a. Amplify 1 µl digested, bubble-ligated YAC DNA (from step 5) with primers 224 and HYAC-C, using 20 cycles of 1 min at 92°C, 2 min at 62°C, and 2 min at 72°C.

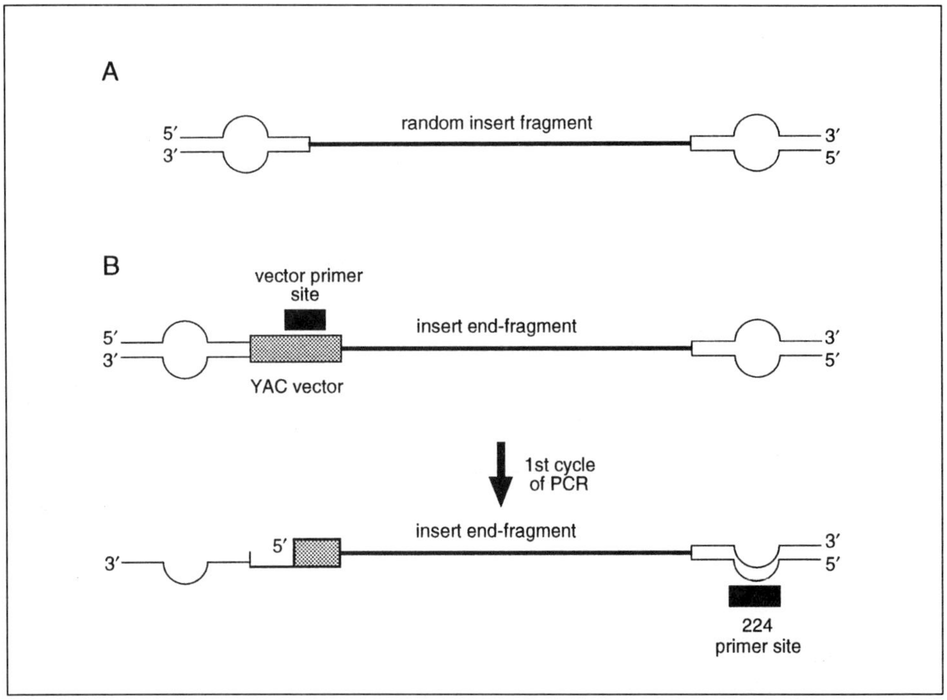

Figure 6.10.3 Selective PCR amplification from the YAC insert end-fragment. (**A**) Result of ligation of the bubble linker to a random fragment from the internal portion of the YAC insert. Because this fragment is not derived from the end of the YAC genomic insert, it contains no sequences from the YAC vector and has no site for annealing any of the HYAC-C, HYAC-D, RA-2, or LS-2 primers or for annealing of the 224 primer. Consequently, no fragment is amplified by PCR. (**B**) Result of ligation of the bubble linker to a fragment derived from the end of the YAC genomic insert and containing its associated YAC vector sequences. During the first cycle of PCR, extension from the YAC vector priming site produces sequences complementary to the universal bubble-bottom primer. This extended fragment provides a template for annealing of the 224 primer, thus permitting successful amplification of the insert end-fragment.

6.10

8a. Dilute the amplification product 1:100 with water and add 1 µl to a new PCR reaction containing primers 224 and LS-2 (specific for the *SUP4* region of pYAC4; see Fig. 6.10.1). Carry out 30 cycles of amplification: 1 min at 92°C, 2 min at 65°C, and 2 min at 72°C.

To amplify the right end using RsaI-digested DNA:

7b. Amplify 1 µl digested, bubble-ligated YAC DNA (from step 5) with primers 224 and HYAC-D, using 20 cycles of 1 min at 92°C, 2 min at 62°C, and 2 min at 72°C.

8b. Dilute amplification product 1:100 with water and add 1 µl to a new PCR reaction (see step 5) containing primers 224 and RA-2. Carry out 30 cycles of amplification: 1 min at 92°C, 2 min at 65°C, and 2 min at 72°C.

Hemi-nesting of the right end cannot be performed with HinfI-digested DNA, because there is a HinfI site only 24 bp from the EcoRI YAC vector cloning site.

9. Analyze a 1-µl aliquot of each final PCR reaction on a 5% polyacrylamide gel.

10. End label amplified fragments with ^{32}P and use as hybridization probes or for nucleotide sequencing to produce an end-specific STS. Alternatively, subclone by blunt-end ligation to a plasmid vector prior to further manipulation.

Blunt-end subcloning of DNA fragments that have been amplified by PCR must be preceded by "polishing" of the ragged PCR ends with S1 nuclease or T4 DNA polymerase (see UNIT 15.7). Alternatively, PCR-amplified fragments may be cloned directly into the TA cloning vector (UNIT 15.7).

END-FRAGMENT ANALYSIS BY SUBCLONING INTO A BACTERIAL PLASMID VECTOR

ALTERNATE PROTOCOL

End fragments are recovered from the YAC insert by double digesting the YAC-containing DNA and subcloning into a pUC19-based vector (Figs. 1.5.2 and 6.10.5). The restriction endonuclease used for digestion of YAC-containing DNA is either *Cla*I (for the left arm) or *Sal*I (for the right arm). Both enzymes cut rarely in human or yeast genomic DNA; therefore, when one is combined with a more frequently cutting enzyme, the resulting doubly digested fragments will represent a minor portion of the total DNA pool.

Additional Materials *(also see Basic Protocol 4)*

*Cla*I, *Sal*I, and other appropriate restriction endonucleases and digestion buffers (*UNIT 3.1*)
Left- and right-vector-arm probes (Fig. 6.10.4)
pUC19-ES and pUC19-HS plasmid vectors (Support Protocol and Fig. 6.10.5)
Transformation-competent *Rec*$^-$ strain of *E. coli* (e.g., DH5; Table 1.4.5)
2× TY or LB agar plates (*UNIT 1.1*) containing 50 to 100 µg/ml ampicillin

1a. *For left arm*: Digest 5-µg aliquots of YAC-containing DNA with *Cla*I and then with each possible second cloning enzyme—*Sac*I, *Kpn*I, *Sma*I, *Bam*HI, *Xba*I, and *Sph*I.

1b. *For right arm:* Digest 5-µg aliquots of YAC-containing DNA with *Sal*I and then with each of the following possible second cloning enzymes—*Sac*I, *Kpn*I, *Sma*I, *Bam*HI, *Xba*I, *Sph*I, or *Hin*dIII.

2. Electrophorese doubly digested DNA on an agarose gel and transfer to a filter for Southern hybridization.

3. Prepare left- and right-vector-arm probes by PCR as described in Fig. 6.10.4. Hybridize each probe to the appropriate filter from step 2.

4. Examine autoradiogram and choose an enzyme combination that yields a hybridizing DNA fragment in the 2- to 7-kb size range. Digest a 50-µg aliquot of YAC-containing DNA with these two enzymes.

Left arm:	5′ ATCGATAAGCTTTAATGCGGTAGT 3′ (pBR322 bases 23-46)
	5′ GATCCACAGGACGGGTGTGGTCGC 3′ (pBR322 bases 379-356)
Right arm:	5′ GATCCTCTACGCCGGACGCATCGT 3′ (pBR322 bases 375-399)
	5′ GTCGACGCTCTCCCTTATGCGACT 3′ (pBR322 bases 656-632)

Figure 6.10.4 Generation of left- and right-vector-arm probes. The 351-bp *Cla*I-*Bam*HI and 276-bp *Bam*HI-*Sal*I fragments of pBR322, which hybridize to sequences immediately flanking the *SUP4* sequences of the YAC vector, are appropriate probes for the YAC left and right vector arms. These probes can be obtained by restriction digestion and gel fractionation of pBR322 plasmid DNA or generated by PCR using 10 ng pBR322 as template for the primers illustrated here. Perform PCR using 25 cycles of 1 min at 92°C, 1 min at 50°C, and 2 min at 72°C. Extract the amplified material once with phenol and once with chloroform, then precipitate with ethanol (*UNIT 2.1*). Label directly by random priming (*UNIT 3.5*) without further purification.

5. Electrophorese doubly digested DNA on an agarose gel. Using a scalpel or razor blade, cut out the segment of gel that should contain the doubly digested DNA fragment.

6. Purify size-fractionated DNA from gel slice and resuspend in a final volume of 20 μl TE buffer, pH 8.0.

7. Ligate 20% of the purified YAC-derived insert DNA with 0.2 μg of gel-purified, compatibly digested pUC19-HS or -ES vector DNA overnight in a total volume of 20 μl.

8. Transform the ligated DNA into a transformation-competent Rec^{-} host strain of *E. coli*. Plate sufficient transformation mix on 2× TY/ampicillin or LB/ampicillin plates to obtain ~200 colonies, a sufficiently low density that individual colonies may be recovered following hybridization. Invert plates and incubate overnight at 37°C.

9. Prepare colony-lift filters and hybridize overnight with ~1–2 × 10^7 cpm of appropriate ^{32}P-labeled left- or right-arm probes. Wash and autoradiograph.

10. Purify plasmid DNA from hybridizing colonies. Verify the structure of the plasmid by comparing its restriction map to the data obtained during the initial analytical double digests of the YAC (steps 1 to 2).

SUPPORT PROTOCOL

DESIGN AND PREPARATION OF pUC19-ES and pUC19-HS SUBCLONING VECTOR

This protocol describes the construction of two vectors for subcloning YACs (Basic Protocol 4). pUC19 is modified by insertion of a "stuffer" fragment in both possible orientations (Fig. 6.10.5).

pUC19-ES: Modify the pUC19 (see Fig. 1.5.2) vector by inserting a stuffer consisting of the 475-bp *Taq*I fragment of pBR322 (positions 653-1128) into the pUC19 polylinker *Acc*I (*Hinc*II) site. In the resulting plasmid, the *Acc*I (and *Sal*I and *Hinc*II) site adjacent to the polylinker *Pst*I site is preserved, but the *Acc*I site previously found next to the polylinker *Xba*I site (which would now be at the other end of the stuffer) is lost (Fig. 6.10.5).

pUC19-HS: Insert the 475-bp *Taq*I fragment stuffer described above into the same pUC19 *Acc*I site but in the opposite orientation. In the resulting plasmid, the polylinker *Acc*I site adjacent to the *Xba*I site is preserved, but the *Acc*I site adjacent to the *Pst*I site is lost (Fig. 6.10.5).

BASIC PROTOCOL 5

PREPARATION OF HIGH-MOLECULAR-WEIGHT YAC-CONTAINING YEAST DNA IN SOLUTION

YAC-containing DNA of sufficiently high molecular weight is purified to provide a source of YAC insert material for subcloning in λ or cosmid vectors. This DNA is also suitable for restriction mapping or other genetic manipulations.

Materials *(see* APPENDIX 1 *for items with ✓)*

Single colony of *S. cerevisiae* containing pYAC4 with insert (Basic Protocol 1)
✓ AHC medium (–Ura –Trp)
✓ SCEM buffer
✓ Lysis buffer

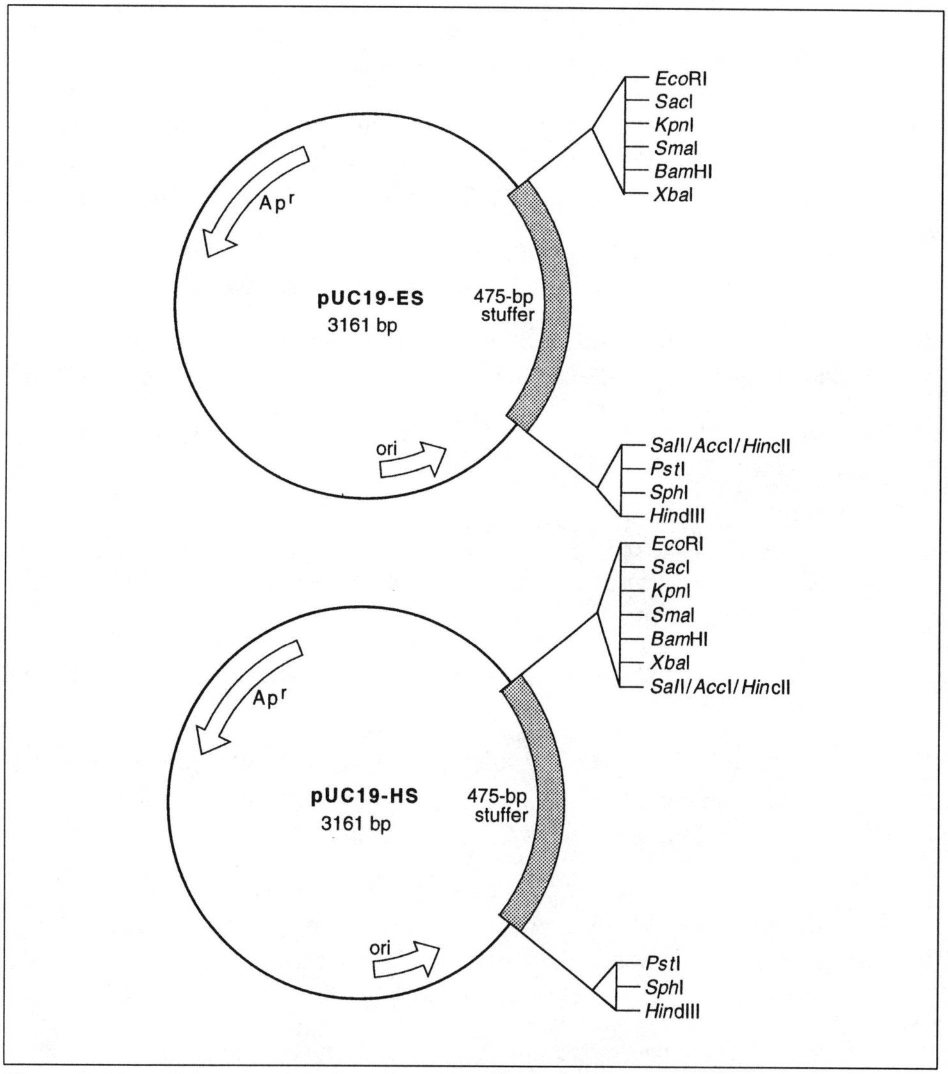

Figure 6.10.5 Structure of the pUC19-ES and pUC19-HS plasmids.

Step-gradient solutions: 50%, 20%, and 15% (w/v) sucrose
✓ TE buffer, pH 8.0
Dry granular sucrose

30°C orbital shaking incubator (e.g., New Brunswick Scientific)
250-ml conical centrifuge bottles (e.g., Corning)
65°C water bath
25 × 89–mm tube (e.g., Beckman)
Beckman JS-4.2 and SW-27 rotors (or equivalents)
Dialysis tubing (*APPENDIX 3*)
Pyrex baking dish
CHEF pulsed-field gel apparatus or equivalent

1. Inoculate a single red colony of a YAC-containing clone into 25 ml AHC medium in a 250-ml flask. Shake at 250 rpm, 30°C, until culture reaches saturation (~3 days).

2. Transfer 1 ml of saturated culture to 100 ml AHC medium in a 1-liter flask. Shake 16 to 18 hr at 250 rpm, 30°C.

3. Harvest yeast cells by centrifuging 10 min at 2000 × g, room temperature, using a 250-ml conical centrifuge bottle. Discard supernatant.

4. Resuspend cells in 50 ml water. Centrifuge 5 min at 2000 × g, room temperature. Discard supernatant. Resuspend cells in 3.5 ml SCEM buffer. Incubate 2 hr at 37°C with occasional gentle mixing. The mixture will become highly viscous.

5. Gradually add cell mixture to 7 ml lysis buffer in a 250-ml Erlenmeyer flask by allowing viscous cell suspension to slide down side of flask.

6. Gently mix by swirling flask until mixture is homogeneous and relatively clear. Incubate 15 min at 65°C, then cool rapidly to room temperature in a water bath.

7. Fractionate on a sucrose step gradient. In a 25 × 89–mm tube, prepare a step gradient consisting of:

   3 ml 50% sucrose
   12 ml 20% sucrose
   12 ml 15% sucrose
   11 ml lysed sample.

   Centrifuge 3 hr at 125,000 × g, room temperature. Discard ~25 ml from the top of the gradient using a 10-ml pipet.

8. Collect viscous DNA-containing solution at the 20% to 50% sucrose interface (~5 ml total volume) and place in dialysis tubing, leaving room for volume to increase 2- to 3-fold. Dialyze overnight against 2 liters TE buffer, pH 8.0, at 4°C.

9. Reconcentrate dialyzed DNA by placing dialysis tubing in an autoclaved Pyrex baking dish and covering with granular sucrose. Recover dialysis tubing when volume of contents has been reduced to ~2 ml.

10. Squeeze DNA solution to one end of dialysis tubing and tie an additional knot to keep DNA in a small volume. Dialyze overnight against 1 liter of TE buffer, pH 8.0, at 4°C.

11. Recover dialyzed DNA and check a small aliquot by electrophoresing in a CHEF pulsed-field gel. Stain with ethidium bromide and estimate DNA content by comparison to a known amount of λ DNA.

    The DNA sample will contain a substantial amount of yeast RNA but should also contain a population of YAC DNA fragments migrating at a size of >100 kb.

BASIC PROTOCOL 6

PREPARATION AND ANALYSIS OF A YAC-INSERT SUBLIBRARY

Construction of a sublibrary of fragments of the YAC insert facilitates high-resolution analysis of the insert sequence.

Materials

High-molecular-weight YAC-containing DNA (Basic Protocol 5)
Vector DNA (e.g., SuperCos 1, Stratagene)
^{32}P-labeled probes (*UNIT 3.10*): total genomic DNA of the individual or species from which the library was made (e.g., *UNITS 2.2, 2.3 & 5.3*), end-specific DNA (*UNIT 3.10*) or RNA (*UNIT 3.8*), and end fragment from YAC (Basic Protocol 4 or Alternate Protocol)

1. Partially digest 1 to 2 μg of YAC-containing DNA with restriction endonuclease(s) appropriate for cosmid vector to be used.

   The quantity of restriction endonuclease should be adjusted to produce digested fragments with an average size of ~40 kb. Only 3000 to 5000 cosmid clones are required to yield 3 yeast genome equivalents. Thus, only 1 to 2 μg of yeast DNA are required to make an adequate library.

2. Perform a series of test ligations. Using optimal conditions, ligate insert DNA to vector DNA.

3. Package cosmid recombinants; dilute packaged extract and determine the titer.

4. Plate and transfer the sublibrary as appropriate for the vector, and prepare resulting filters for hybridization.

5. Perform a preliminary screen of the library using a ^{32}P-labeled probe of total genomic DNA of the individual or species from which the library was made.

6. Organize this first set of cosmid clones into contigs by analyzing shared restriction fragments and by hybridizing with probes contained in the YAC insert or prepared from the ends of individual cosmid inserts.

7. Establish a complete contiguous collection of cosmid clones of the original YAC insert by screening the library with specific YAC-derived probes and cosmid end-specific probes. Repeated hybridization with sequential "walking probes" should reveal new hybridizing colonies at each step.

Contributors: David D. Chaplin and Bernard H. Brownstein

DNA Sequencing

7

OVERVIEW OF DNA SEQUENCING METHODS

For many recombinant DNA experiments, knowledge of a DNA sequence is a prerequisite for its further manipulation. DNA sequencing followed by computer-assisted searching for restriction endonuclease cleavage sites is often the fastest method for obtaining a detailed restriction map (*UNITS 3.1-3.3*). This information is particularly useful when vectors designed to overexpress proteins or to generate protein fusions are utilized for subcloning a gene of interest (Chapter 16). Computer-assisted identification of protein-coding regions (open reading frames or ORFs) within the DNA sequence, followed by computer-assisted similarity searches of DNA and protein data bases, can lead to important insights about the function and structure of a cloned gene and its product (*UNIT 7.7*). In addition, the DNA sequence is a prerequisite for a detailed analysis of the 5′ and 3′ noncoding regulatory regions of a gene. DNA sequence information is essential for site-directed mutagenesis (*UNIT 8.1*). Small amounts of DNA sequence information (sequence tagged sites, or STS, or expressed sequence tags, or EST) are the basis of methods for mapping and ordering large DNA segments cloned into yeast artificial chromosomes or cosmids.

DNA sequencing techniques are based on electrophoretic procedures using high-resolution denaturing polyacrylamide (sequencing) gels. These so-called sequencing gels are capable of resolving single-stranded oligonucleotides up to 500 bases in length which differ in size by a single deoxynucleotide. In practice, for a given region to be sequenced, a set of labeled, single-stranded oligonucleotides is generated, the members of which have one fixed end and which differ at the other end by each successive deoxynucleotide in the sequence. The key to determining the sequence of deoxynucleotides is to generate, in four separate enzymatic or chemical reactions, all oligonucleotides that terminate at the variable end in A, T, G, or C. The oligonucleotide products of the four reactions are then resolved on adjacent lanes of a sequencing gel. Because all possible oligodeoxynucleotides are represented among the four lanes, the DNA sequence can be read directly from the four "ladders" of oligonucleotides as shown in Figure 7.0.1.

The practical limit on the amount of information that can be obtained from a set of sequencing reactions is the resolution of the sequencing gel (see *UNIT 7.6* for protocols on setting up and running sequencing gels). Current technology allows ~300 nucleotides of sequence information to be reliably obtained in one set of sequencing reactions, although more information (up to 500 nucleotides) is often obtained. Thus, if the region of DNA to be sequenced is <300 nucleotides, a single cloning into the appropriate vector is all that is usually necessary to produce a recombinant molecule that can easily be sequenced.

For a larger region of DNA, it is generally necessary to break a large fragment into smaller ones that are then individually sequenced. This can be done in a random or an ordered fashion. *UNIT 7.1* contains a discussion of strategies for sequencing large regions of DNA. Two protocols for subdividing large regions of DNA are provided in *UNIT 7.2*. These protocols are used to create a set of ordered, or nested, deletions for DNA sequencing using exonuclease III or nuclease *Bal*31.

The two methods that are widely used to determine DNA sequences, the enzymatic dideoxy method and the chemical method, differ primarily in the technique used to generate the ladder of oligonucleotides. In the enzymatic dideoxy sequencing method, a DNA polymerase is utilized to synthesize a labeled, complementary copy

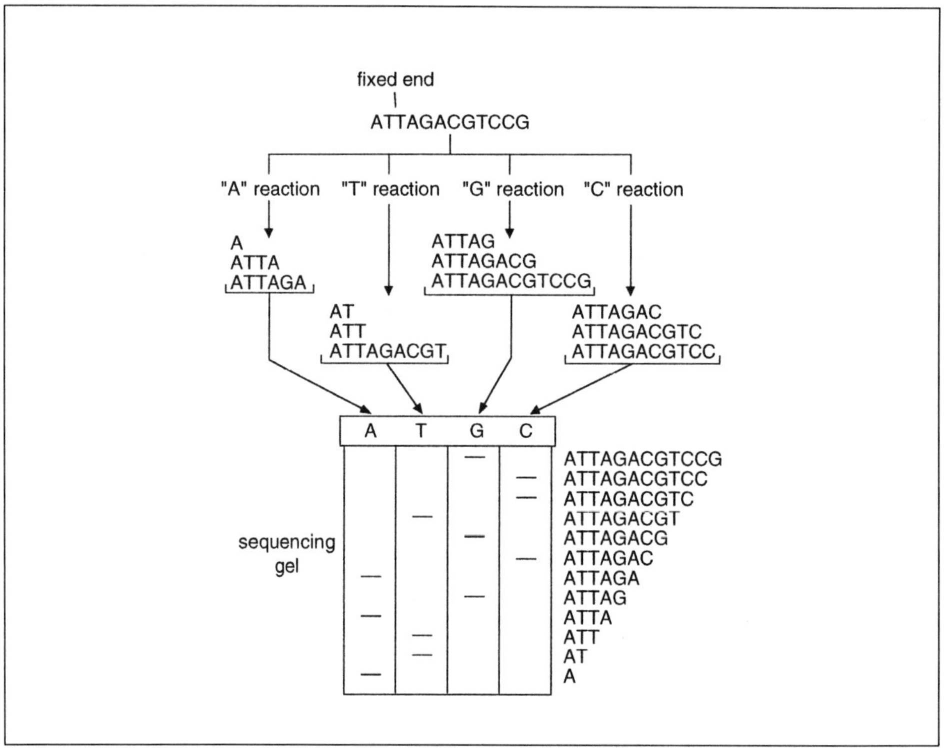

Figure 7.0.1 General strategy for DNA sequencing. To sequence a fragment of DNA, a set of radiolabeled single-stranded oligonucleotides is generated in four separate reactions. In each of the four reactions, the oligonucleotides have one fixed end and one end that terminates sequentially at each A, T, G, or C, respectively. The products of each reaction are fractionated by electrophoresis on adjacent lanes of a high-resolution polyacrylamide gel. After autoradiography, the DNA sequence can be "read" directly from the gel.

of a DNA template. In the chemical sequencing method, a labeled DNA strand is subjected to a set of base-specific chemical reagents. These two techniques are further described below.

DIDEOXY (SANGER) SEQUENCING

Sequencing Method

The dideoxy or enzymatic method, originally developed by F. Sanger and coworkers, utilizes a DNA polymerase to synthesize a complementary copy of a single-stranded DNA template. DNA polymerases cannot initiate DNA chains; rather, chain elongation occurs at the 3′ end of a primer DNA that is annealed to "template" DNA (Fig. 7.0.2). The deoxynucleotide added to the growing primer chain is selected by base-pair matching to the template DNA. Chain growth involves the formation of a phosphodiester bridge between the 3′-hydroxyl group at the growing end of the primer and the 5′-phosphate group of the incoming deoxynucleotide. Thus, overall chain growth is in the 5′→3′ direction.

The dideoxy sequencing method capitalizes on the ability of the DNA polymerase to use 2′,3′-dideoxynucleoside triphosphates (ddNTPs) as substrates. When a ddNMP is incorporated at the 3′ end of the growing primer chain, chain elongation is terminated at G, A, T, or C because the primer chain now lacks a 3′-hydroxyl group. To generate the four sequencing ladders shown in Figure 7.0.1, only one of the four possible ddNTPs is included in each of the four reactions (see below). The ddNTP:dNTP ratio in each reaction is adjusted such that a portion of the elongating primer chains terminates at each occurrence of the base in the template DNA

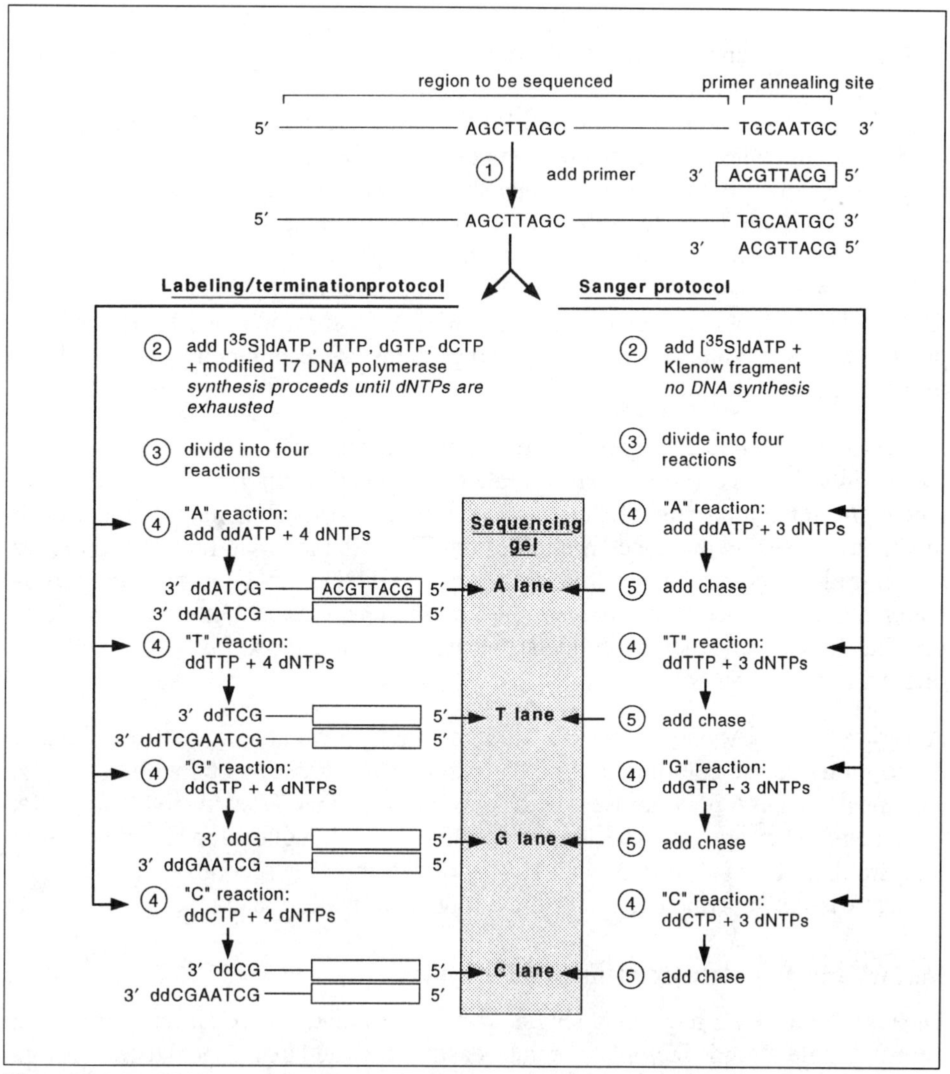

Figure 7.0.2 Dideoxy sequencing methods. In each method, a single-stranded DNA fragment is annealed to an oligonucleotide primer for polymerization (step 1). In the Sanger protocol (right side), the Klenow fragment and radiolabeled dATP are added (step 2). The reaction is divided into four aliquots (step 3) and the other three dNTPs and either ddATP, ddTTP, ddGTP, or ddCTP are added (step 4). DNA synthesis occurs until terminated by the incorporation of a ddNTP. A "chase" of all four dNTPs (step 5) elongates chains not terminated by a ddNMP into higher-molecular-weight DNA. In the labeling/termination protocol (left side), after the first step, a limiting amount of the four dNTPs—one of which is radiolabeled—and Sequenase are added (step 2). DNA synthesis proceeds until the dNTPs are exhausted. The reaction mix is divided into four aliquots (step 3) and all four dNTPs plus either ddATP, ddTTP, ddGTP, or ddCTP are added (step 4). Synthesis resumes, but termination specifically occurs when a ddNMP is incorporated. In each method, after the reactions are terminated, samples are loaded on adjacent lanes of a sequencing gel.

corresponding to the included complementary ddNTP. In this way, each of the four elongation reactions contains a population of extended primer chains, all of which have a fixed 5↔ end determined by the annealed primer and a variable 3′ end terminating at a specific dideoxynucleotide.

Two protocols for dideoxy sequencing are provided in *UNIT 7.4A*. The original dideoxy method, which in this chapter is referred to as the Sanger procedure (Sanger et al., 1977, 1980), was developed for use with the *E. coli* DNA polymerase I large fragment, or the Klenow fragment.

In the “labeling/termination” method, developed for use with modified T7 DNA polymerase (Sequenase; Tabor and Richardson, 1987a), labeling of the primer and termination by incorporation of a dideoxynucleotide occur in two separate reactions (Fig. 7.0.2, left side). After annealing of the primer to the template, the labeling reaction occurs when the primer is elongated and labeled in the presence of a low concentration of all four dNTPs, one of which is radiolabeled. DNA synthesis continues until one or more of the nucleotide pools is exhausted, leading to almost complete incorporation of the labeled nucleotide. The termination step takes place in four separate reactions, each of which contains additional dNTPs and one of the four ddNTPs. In the termination step, a high concentration of dNTPs ensures processive DNA synthesis until the growing chains are terminated by the incorporation of a ddNMP.

In the Sanger procedure, the average length of the sequencing products is controlled by the ddNTP:dNTP ratio, where a higher ratio leads to shorter products. In the labeling/termination protocol, the average length of the sequencing products can be modulated either by the concentration of dNTPs in the labeling reaction (a higher concentration leads to longer products) or by the ddNTP:dNTP ratio in the termination reaction. If Sequenase is used, the labeling/termination method is capable of yielding longer sequencing products, on average, than those obtained using the original Sanger protocol.

A variety of DNA polymerases are commercially available for sequencing. A description of these polymerases and their appropriate uses can be found in *UNIT 7.4*. Thermostable DNA polymerases are the newest class of enzymes available for DNA sequencing. They are useful because they can carry out a sequencing reaction at high temperatures. This property provides a way of destabilizing secondary structures of the DNA template, which can interfere with the elongation reaction.

Vectors and Templates for Dideoxy Sequencing

Dideoxy sequencing requires a single-stranded template to which the primer can anneal. Single-stranded templates can be easily generated using specialized vectors derived from M13, an *E. coli* filamentous phage that contains a single-stranded, circular DNA molecule (Messing, 1983, 1988). Dideoxy sequencing can also be readily carried out using double-stranded DNA if it is first denatured with alkali or heat. Dideoxy sequencing of a double-stranded template is particularly useful when DNA sequencing is the only rapid method available for verifying a particular plasmid construction. However, for large-scale sequencing projects, we recommend using a single-stranded DNA vector system, such as the M13mp series, because the generation of sequencing-quality, single-stranded DNA templates is somewhat more reliable than for plasmid DNA templates.

Radiolabels for Dideoxy Sequencing Reactions

The dideoxy sequencing protocols in *UNIT 7.4* involve radiolabeling nascent DNA chains with [α-^{35}S]dATP rather than with [α-^{32}P]dATP. ^{32}P, however, offers the advantage of short exposure times and is particularly useful in situations, such as verifying plasmid constructions, where maximizing resolution in the higher region of the sequencing gel is not a priority. Recently, the use of a new labeled nucleotide analog, [α-^{33}P]dATP, in sequencing reactions has been described. ^{33}P has a maximum β-emission energy that is 50% stronger than ^{35}S, but 5-fold weaker than ^{32}P. Sequences generated using [α-^{33}P]dATP have short exposure times like ^{32}P, but have band resolution comparable to that of ^{35}S (Zagursky et al., 1991). Another alternative to labeling the nascent oligonucleotide with [α-^{35}S]dATP is to use a 5′-end-labeled primer generated by T4 polynucleotide kinase and [γ-^{32}P]ATP or [γ-^{35}S]ATP (*UNIT 3.10*). Sequencing large double-stranded DNA templates (such as λgt11) with 5′-end-labeled primers has been found to give better results than standard labeling techniques. Protocols for sequencing using end-labeled primers are provided in *UNIT 7.4*.

CHEMICAL (MAXAM-GILBERT) SEQUENCING

Sequencing Method

In the chemical method of DNA sequencing developed by A. Maxam and W. Gilbert (Maxam and Gilbert, 1977, 1980), the four sets of deoxyoligonucleotides are generated by subjecting a purified 3′- or 5′-end-labeled deoxyoligonucleotide to a base-specific chemical reagent that randomly cleaves DNA at one or two specific nucleotides. Because only end-labeled fragments are observed following autoradiography of the sequencing gel, four DNA ladders are observed as shown in Figure 7.0.3.

The Maxam and Gilbert chemical sequencing method is based on the ability of hydrazine, dimethyl sulfate (DMS), or formic acid to specifically modify bases within the DNA molecule. Piperidine is then added to catalyze strand breakage at these modified nucleotides. The specificity of the chemical method resides in the first reaction with hydrazine, DMS, or formic acid, which react with only a few

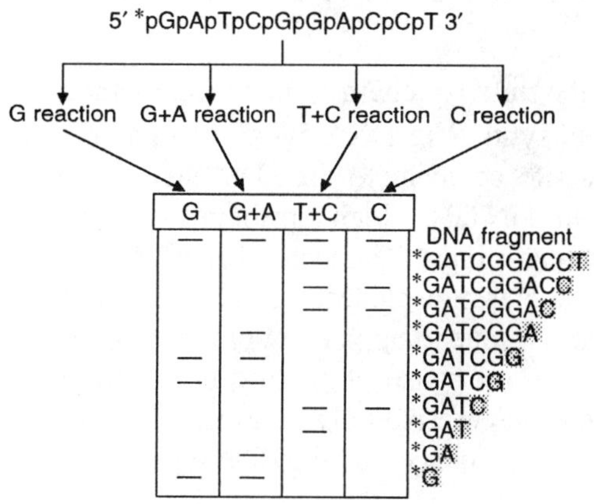

Figure 7.0.3 Chemical sequencing strategy. The ladder of oligonucleotides after gel electrophoresis of the products from the four chemical cleavage reactions is shown. The asterisk (*) indicates the position in the DNA fragment of the ^{32}P label, which is placed on the 5′ end in this example. The direction of fragment migration is downward; smaller DNA oligonucleotides migrate faster in the sequencing gel than larger oligonucleotides. The shaded bases at the 3′ end of the fragments to the right of the gel indicates bases that have been chemically modified and then displaced from the oligonucleotide during piperidine-mediated strand scission. For example, after a limited reaction with dimethyl sulfate (DMS), which is specific for G′s, followed by quantitative release of the modified G residue by piperidine, a set of oligonucleotides are generated that terminate at the base immediately 5′ of each G in the sequence. In this example, the oligonucleotide products are *pGpApTpCpGp and *pGpApTpCp. Each of these products forms a band in the G lane. For *pG, the product is *p, which would most likely run off the gel, making it difficult to determine the identity of the 5′-terminal base. Because formic acid is specific for purines (G′s and A′s), a fragment that terminates in G or A will produce a band in the G + A lane. Hydrazine in the absence of NaCl cleaves T′s and C′s resulting in a band in the T + C lane. Hydrazine in the presence of NaCl cleaves only C′s; thus, a band is observed in the C lane.

7

percent of the bases. The second reaction, piperidine strand cleavage, must be quantitative. The chemical mechanisms of the first reactions are as follows:

> G: DMS methylates nitrogen 7 of G, which then opens between carbon 8 and nitrogen 9. Piperidine then displaces the modified guanine from its sugar.
>
> G+A: Formic acid weakens A and G glycosidic bonds by protonating purine-ring nitrogens. The purines can then be displaced by piperidine.
>
> T+C: Hydrazine splits the rings of T and C. The fragments of these bases can then be displaced by piperidine.
>
> C: In the presence of NaCl, only C reacts with hydrazine. The modified C can then be displaced with piperidine.

In all four reactions, piperidine also catalyzes phosphodiester bond cleavage at the position where the modified base has been displaced by piperidine. Recent improvements in the chemistry of the chemical cleavage rotations utilize reagents that are less hazardous (G. Church, pers. comm.).

Vectors for Chemical Sequencing

Chemical sequencing reactions can be performed on either single- or double-stranded DNA, as long as only one end is labeled. Specialized vectors have been developed that allow labeling of only one strand of the cloned DNA using *Tth*111I or other asymmetric sites adjacent to the cloned DNA (Arnold and Puhler, 1988).

CHOOSING BETWEEN DIDEOXY AND CHEMICAL SEQUENCING METHODS

As described above, the dideoxy chain-termination method is based on the ability of DNA polymerase to synthesize DNA 5′→3′ from a defined primer annealed to the vector DNA at a site adjacent to the DNA being sequenced. Each reaction contains one of the four ddNTPs, which terminates synthesis selectively at G, A, T, or C. Dideoxy sequencing is rapid. The method also offers excellent band resolution if ^{35}S-labeled nucleoside triphosphates are used to label the DNA. The major disadvantage of dideoxy sequencing is that composition or secondary structure of the template can sometimes cause premature termination by DNA polymerase. Despite the existence of alternative polymerases, DNA is sometimes encountered that cannot be accurately sequenced by the dideoxy method.

The second method, chemical cleavage, is based on the ability of various chemicals to cleave DNA with a high specificity. The major advantage of the chemical method is that problems associated with polymerase synthesis of DNA (i.e., premature termination due to DNA sequence or structure) are eliminated, permitting sequencing of stretches of DNA that cannot be sequenced by the enzymatic method. In addition, in contrast to the dideoxy method obtaining the sequence of shorter regions of DNA using the chemical method does not require subcloning into an appropriate sequencing vector. Finally, chemical cleavage is the only sequencing method available for small oligonucleotides. Before the development of specialized cloning vectors, such as pSP64CS, pSP65CS, and others, the major disadvantage of the chemical sequencing method was that preparation of the DNA prior to sequencing was very time-consuming.

ALTERNATIVES TO RADIOLABELED SEQUENCING REACTIONS

Chemiluminescence. Chemiluminescence is a recently developed detection method that is comparable in sensitivity to traditional radiolabeling. Detection of the sequencing products occurs by a chemiluminescent reaction that can be monitored by autoradiography (*UNIT 7.5*).

Multiplex sequencing. Multiplex sequencing uses hybridization to a specific probe to detect an individual sequencing ladder in a mixture of ladders. In this method, the sequencing products derived from a mixture of templates are subjected to electrophoresis on a sequencing gel, transferred to a membrane, and hybridized sequentially with probes specific for different templates. Thus, the amount of sequence information available from one gel is multiplied by the number of times the membrane can be rehybridized (in practice, up to 20 times). A set of sequencing vectors is commercially available (Millipore) for multiplex sequencing using the dideoxy method.

Recent Developments in Sequencing Technology

Commercial kits for sequencing. Commercially available kits eliminate the need to assemble and calibrate numerous mixes and can save a significant amount of startup time, although they are somewhat less flexible and can limit the ability to troubleshoot reactions when necessary.

Automated sequencers and automation of sequencing reactions. Automated sequencing machines automate the gel electrophoresis step, detection of DNA band pattern, and analysis of bands. Currently, all commercially available automated sequencers are designed for enzymatic sequencing reactions producing fluorescently or radioactively labeled products. All automated sequencers possess data-collection capabilities and include either further analysis programs or provide portability to external data-analysis software programs. Fluorescent labels can be incorporated into the sequencing products either through the primer or the ddNTPs. One sequencer detects ^{32}P-reaction products, which are run in four separate lanes. Automation of the sequencing reaction by the use of robotics is also under development (Mardis and Roe, 1989).

Thermal cycle sequencing. Thermal cycle sequencing is a method of dideoxy sequencing in which a small number of template DNA molecules are repetitively utilized to generate a sequencing ladder. A dideoxy sequencing reaction mixture (template, primer, dNTPs, ddNTPs, and a thermostable DNA polymerase) is subjected to repeated rounds of denaturation, annealing, and synthesis steps, similar to PCR (Chapter 15), resulting in linear amplification of the sequencing products from much less. Several thermal cycle sequencing kits are now commercially available utilizing each of the detection methods described in Table 7.4.1.

Solid-phase sequencing. Another recent innovation that is applicable to both manual and automated DNA sequencing is solid-phase DNA sequencing (Hultman et al., 1991). In this approach, one strand of a double-stranded DNA molecule is biotinylated (e.g., by amplification using PCR in which one of the two primers is biotinylated; Chapter 15). The hemibiotinylated DNA molecule is then bound to streptavidin-ferromagnetic beads. The strands are denatured by treating the beads with alkali and the biotinylated strands are separated from the nonbiotinylated strands using a magnet that traps the bead complex to which the biotinylated strands are bound. Sequencing reactions can be performed using either the biotinylated strand-bead complex or the nonbiotinylated strand preparation as the template.

Computer Analysis

Once the gels are run and autoradiograms are obtained, computer software is practically indispensable for analysis of the sequence information. Computer software can assist at three stages. First, DNA sequence data can be entered into a computer database either by "reading" the sequencing gels manually with a digitizer system or by using an automated gel scanner. Second, several software packages are available for detecting overlaps in sequence data and then assembling contiguous DNA sequences (contigs) from individual templates. Third, computer assistance is indispensable for analyzing final sequence data, e.g., in finding open reading frames or finding homologies to other sequences present in the nucleotide or protein data bases. *UNIT 7.7* provides an overview of software and technology currently available.

References: Arnold and Puhler, 1988; Hultman et al., 1991; Mardis and Roe, 1989; Maxam and Gilbert, 1977, 1980; Sanger et al., 1977, 1980; Sears et al., 1992; Tabor and Richardson, 1987a.

Contributors: Frederick M. Ausubel and Lisa M. Albright (contributing editor)

UNIT 7.1

DNA Sequencing Strategies

In general, any sequencing strategy should include plans for sequencing both strands of the DNA fragment, as complementary strand confirmation leads to higher accuracy. The most commonly used methods for generating appropriately sized DNA fragments for dideoxy and chemical sequencing are discussed below.

DIDEOXY SEQUENCING

Planning for Dideoxy Sequencing

Sequencing of a fragment of <300 nucleotides is usually straightforward because this amount of sequence information can be reliably determined from a single set of sequencing reactions. Fragments of this size can usually be subcloned directly into an appropriate single- or double-stranded DNA sequencing vector.

To sequence larger regions of DNA completely, it is generally necessary to subdivide a large fragment into smaller ones that can then be individually sequenced. Three general approaches are currently used. In the first approach, known as "shotgun cloning," random fragments are created from longer DNA. These fragments are combined and the entire pool is ligated into an appropriate sequencing vector. A second subcloning strategy for sequencing large DNA fragments is to generate an ordered set of subclones from a large DNA molecule. This is usually done by making progressive (nested) sets of deletions from a clone containing the entire DNA fragment; two such protocols using exonuclease III and nuclease *Bal* 31 are presented in *UNIT 7.2*. Commercial kits for making nested deletions are currently available from a variety of sources (see Table 7.2.1 for a list of suppliers). A third strategy for sequencing large DNA fragments, known as "primer walking," is uniquely suited to dideoxy DNA sequencing and bypasses the need for subcloning smaller pieces of DNA. Initial sequence data is obtained using a vector-based primer. As new sequence is ascertained, an oligonucleotide is synthesized that hybridizes near the 3′ end of the newly obtained sequence and primes synthesis in a subsequent set of dideoxy reactions.

Vectors for Dideoxy Sequencing

The decision as to vector choice is subject to the following general constraints:

1. Protocols involving construction of nested deletions generally require particular restriction endonuclease sites in the vector that permit generation of deletions or subsequent cloning of deletion products.

2. For deletion strategies that require subcloning of deletion products into a secondary vector, such as the *Bal* 31 procedure (*UNIT 7.2*), it is essential that the secondary vector utilize a positive screening method for inserts.

3. Synthetic oligonucleotides suitable for priming dideoxy sequencing reactions must be available.

4. For large sequencing projects, it is most efficient to use a vector designed to produce single-stranded DNA, such as the phage-based M13mp vectors described below.

5. If the sequencing strategy involves sequencing double-stranded templates, a high-copy-number plasmid is advantageous for sequencing minipreps of plasmid DNA.

M13 vectors. The dideoxy sequencing method has been greatly facilitated by the development of the filamentous *E. coli* phage M13 as a cloning vector. A DNA fragment of interest can be cloned into the RF of M13 and then single-stranded DNA for sequencing can be readily produced. The M13 vectors most widely used for dideoxy sequencing are a series called M13mp constructed by J. Messing and his collaborators (Messing, 1988). The M13mp series contains the *lacZ* promoter and a partial *lacZ* gene, encoding the α-fragment of β-galactosidase (*UNIT 1.9*). After infection of an *E. coli* F′ host containing another partial *lacZ* gene encoding the ω-fragment of β-galactosidase and induction by IPTG, M13mp phage produce blue plaques on Xgal agar (see *UNIT 1.4* for the use of Xgal for assaying β-galactosidase). In addition, each vector in the M13mp series contains a synthetic polylinker inserted into the fifth codon of *lacZ*. Because these polylinkers do not alter the *lacZ* reading frame, M13mp vectors still produce blue plaques on Xgal agar. However, insertion of a DNA fragment into one of the unique polylinker cloning sites has a high probability of disrupting the *lacZ* reading frame, thus generating a recombinant phage that produces colorless plaques.

Plasmid sequencing vectors. Numerous plasmid vectors, most available commercially, can be used for double-stranded dideoxy sequencing (see Table 7.1.1). Each of these plasmids replicates to a high copy number in *E. coli,* each features a blue-versus-white screen for inserts, and for each plasmid, primers are commercially available for sequencing both strands of insert DNA. Several of these vectors also contain an M13, f1, or fd origin of replication that can be activated to produce single-stranded copies of the plasmid by infection with helper phage (*UNITS 1.9 & 1.10*). Several of these hybrid phage/plasmid vectors (designated ± in Table 7.1.1) come as pairs with the phage origin of replication in both orientations such that helper phage directs the synthesis of opposite strands on the hybrid vector.

CHEMICAL SEQUENCING

Planning for Chemical Sequencing

Chemical sequencing requires a DNA fragment that is labeled at only one end (*UNITS 3.5 & 3.10*). Traditionally, chemical sequencing strategies required relatively detailed knowledge of the restriction map for the fragment to be sequenced. This permitted generation of a series of subfragments that could be labeled at both ends and then cut asymmetrically to produce two fragments labeled at each end, which could be purified individually by gel electrophoresis. However, specially designed vectors (described below) circumvent the requirement for purifying and labeling individual restriction fragments and greatly facilitate chemical sequencing projects. These vectors contain asymmetric sites that permit labeling only one end of the target DNA to be sequenced; this strategy can be thought of as the chemical sequencing analog

Table 7.1.1 Dideoxy Sequencing Vectors

Vector[a]	Supplier[b]	Vector[a]	Supplier[b]
M13 phage vectors[c]			
M13mp18/19	AM, ATCC, BR, IBI, ICN, LT, NEB, PH, USB	M13BM20/21	BM
Hybrid phage/plasmid vectors[d]			
BluescriptII series	ST	pSPORT1	LT
pBC series	ST	pSVK3	PH
pBS±	ST	pT3/T7-3	CL
pcDNAII	IN	pT7 T3 18U/19U	PH
pEMBL18/19	ATCC, BM	pT7/T3α-18/19	LT
pfdA/B	ATCC	pT71/2	USB
pGEM9z/11z/13z f±	PR	pTZ18/19	USB
pIBI24(-25)	IBI	pTZ18R	IN
pICEM18/19	ATCC	pTZ18U/19U	BR
pSELECT1	PR	pUC118/119	ATCC
Plasmid vectors[e]			
pAM18/19	AM	pUC18/19	AM, BM, BR, CL, IBI, ICN, LT, NEB, PH, USB
pAT153	ICN		
pT3/T7/lac	BM		
pTTQ	AM	pUCBM20/21	BM
		SP72/73	PR

[a]The most recent version in each vector series is listed (modified from Slatko, 1991). All vectors have commercially available primers, extensive polylinkers, and blue-versus-white screening for inserts using α-complementation of β-galactosidase.

[b]Abbreviations: AM, Amersham; ATCC, American Type Culture Collection; BM, Boehringer Mannheim; BR, Bio-Rad; CL, Clontech; IBI, International Biotechnologies; ICN, ICN Biomedicals; IN, Invitrogen; LT, Life Technologies; NEB, New England Biolabs; PH, Pharmacia LKB; PR, Promega; ST, Stratagene; USB, United States Biochemical. See *APPENDIX 4* for locations and phone numbers. Suppliers' catalogues contain citations of original literature describing these vectors, if available.

[c]These vectors produce both double-stranded RF DNA and packaged single-stranded DNA.

[d]These vectors replicate as plasmids but produce single-stranded DNA upon infection with helper phage.

[e]These vectors produce only double-stranded DNA.

to the universal priming site. These vectors are particularly suited for cloning a set of nested deletions generated by *Bal* 31. We recommend using the nested deletion strategy for large chemical sequencing projects. For small fragments of DNA, the traditional end-labeling strategy is as efficient as subcloning into specialized vectors.

Vectors for Chemical Sequencing

Eckert (1987) has described the construction of specialized vectors that make it possible to rapidly and simultaneously sequence a large number of samples by the chemical cleavage method. These vectors, pSP64CS and pSP65CS (CS refers to chemical sequencing) are high-copy-number plasmids that confer ampicillin resistance and carry a synthetic polylinker containing two *Tth*111I sites flanking a *Sma*I site. The *Bal* 31 procedure (*UNIT 7.2*) can be used to generate a set of nested deletions for subcloning into pSP64CS or pSP65CS. The blunt end generated by *Bal* 31 is ligated to the *Sma*I site in pSP64/65CS. This places the deleted end of each fragment next to a *Tth*111I site for end labeling.

References: Messing, 1988; Eckert, 1987.

Contributors: Barton E. Slatko, Richard L. Eckert, Lisa M. Albright, and Frederick M. Ausubel

Constructing Nested Deletions for Use in DNA Sequencing

Nested deletions are a set of deletions originating at one end of a target DNA fragment and extending various lengths along the target DNA. Each successively longer deletion brings "new" regions of the target DNA into sequencing range (about 300 bp for normal sequencing gels) of the primer site (see also *UNIT 7.1*).

Two protocols for generating nested subclones via enzymatic digestion are included in this unit. The primary advantage of the Basic Protocol, which employs exonuclease III, is that the deletion products generated from the original clone can be recircularized to generate functional plasmids and thus do not require subcloning into another vector. Another advantage of the exo III procedure is that larger DNA fragments (up to 14 kb) can be easily subjected to construction of progressive deletions. On the other hand, the exo III procedure depends on the availability of appropriate restriction sites between the insert and primer site. If the required sites are not available for the exo III procedure, the *Bal* 31 procedure can be used, but this requires subcloning the deletion fragments into a separate vector for subsequent use. Both methods require the presence of unique restriction sites in the vector that are not present in the insert DNA. *Bal* 31 can also be used to generate nested deletions for chemical sequencing in conjunction with specialized chemical sequencing vectors (*UNIT 7.1*). Commercially available kits for making nested deletions using a number of methods are described in *UNIT 7.1* and listed in Table 7.2.1.

USING EXONUCLEASE III TO CONSTRUCT UNIDIRECTIONAL DELETIONS

BASIC PROTOCOL 1

This method is based on the enzymatic properties of exo III, a 3′ exonuclease specific for double-stranded DNA (*UNIT 3.11*). Exonuclease III can initiate digestion at blunt ends or ends with a 5′ overhang, but cannot efficiently initiate digestion at a 3′ overhanging end. This protocol is written for constructing deletions in a ~2.0-kb fragment, but may be scaled up proportionately for longer fragments.

Materials *(see APPENDIX 1 for items with ✓)*

- DNA fragment to be sequenced (insert DNA)
- Appropriate sequencing vector (Table 7.1.1)
- Restriction endonucleases and corresponding buffers (*UNIT 3.1*)
- Buffered phenol (*UNIT 2.1*)
- 25:24:1 phenol/chloroform/isoamyl alcohol (*UNIT 2.1*)
- 100% and 70% ethanol, ice cold
- ✓ 3 M sodium acetate

Table 7.2.1 Commercial Kits for Making Nested Deletions[a]

Deletion method	Supplier[b]
Exo III	NEB, PH, PR, ST
Bal 31	NEB
T4 DNA polymerase	IB
Oligonucleotide-directed	B101

[a]Modified from Slatko, 1991.

[b]Abbreviations: B101, Bio101; IB, International Biotechnologies; NEB, New England Biolabs; PH, Pharmacia LKB; PR, Promega; ST, Stratagene.

Exonuclease III (*UNIT 3.11*) and 1× exonuclease III buffer (15 mM Tris·Cl, pH 8.0/0.66 mM $MgCl_2$)

✓ 1 U/µl S1 nuclease (*UNIT 3.12*) and S1 nuclease buffer

✓ S1 nuclease stop buffer

✓ 10× loading buffer

2 U/µl Klenow fragment of *E. coli* DNA polymerase I (*UNIT 3.5*)

0.25 mM 4dNTP mix (*UNIT 3.4*)

✓ Exo III ligation buffer

10 mM ATP (*UNIT 3.4*)

1 U/µl T4 DNA ligase (measured in Weiss units; *UNIT 3.14*)

If plasmid vector is used:

E. coli DK1 (*recA;* available from BRL) or DH5αF′ (*recA*) made competent for transformation (Table 1.4.5 and *UNIT 1.8*)

LB plates and medium (*UNIT 1.1*) containing appropriate antibiotic

If one of the M13mp plasmids is the cloning vector:

E. coli JM109 (*recA*) or DH5αF′ (*recA;* Life Technologies) made competent for transformation (Table 1.4.5 and *UNIT 1.8*)

LB plates (*UNIT 1.1*), prewarmed

H top agar (*UNIT 1.1*)

Xgal/IPTG mixture [30 µl Xgal stock and 30 µl IPTG stock (Table 1.4.2) for each 3 ml H top agar]

2× TY medium (*UNIT 1.1*)

✓ Phage loading buffer

Preliminary Steps

1. Clone DNA fragment to be sequenced (insert DNA) in polylinker of appropriate sequencing vector (Fig. 7.2.1; *UNIT 3.16*).

2. Prepare DNA by CsCl/EtBr equilibrium centrifugation (*UNIT 1.7*) of recombinant plasmids constructed in step 1 (it is important that DNA be supercoiled; ~5 µg plasmid DNA/experiment).

Exo III Procedure

3. Completely linearize 5 µg vector::insert DNA by double digestion with restriction enzymes that leave a 3′ overhang adjacent to primer site and a 5′ overhang or blunt end adjacent to insert DNA (Fig. 7.2.1).

   The 5′ or blunt restriction site must be positioned between the 3′ restriction site and the insert.

4. Extract with buffered phenol, extract with phenol/chloroform/isoamyl alcohol, add 1/10 vol of 3 M sodium acetate, and precipitate with ethanol (*UNIT 2.1*). Wash pellet in ice-cold 70% ethanol and dry it. Dissolve dried pellet in 1× exo III buffer to 0.1 µg/µl final.

The extent of digestion of exo III is regulated by reaction temperature and incubation time. Rates of 250 bp/min at 37°C and 120 bp/min at 30°C were obtained using the conditions outlined in the following steps.

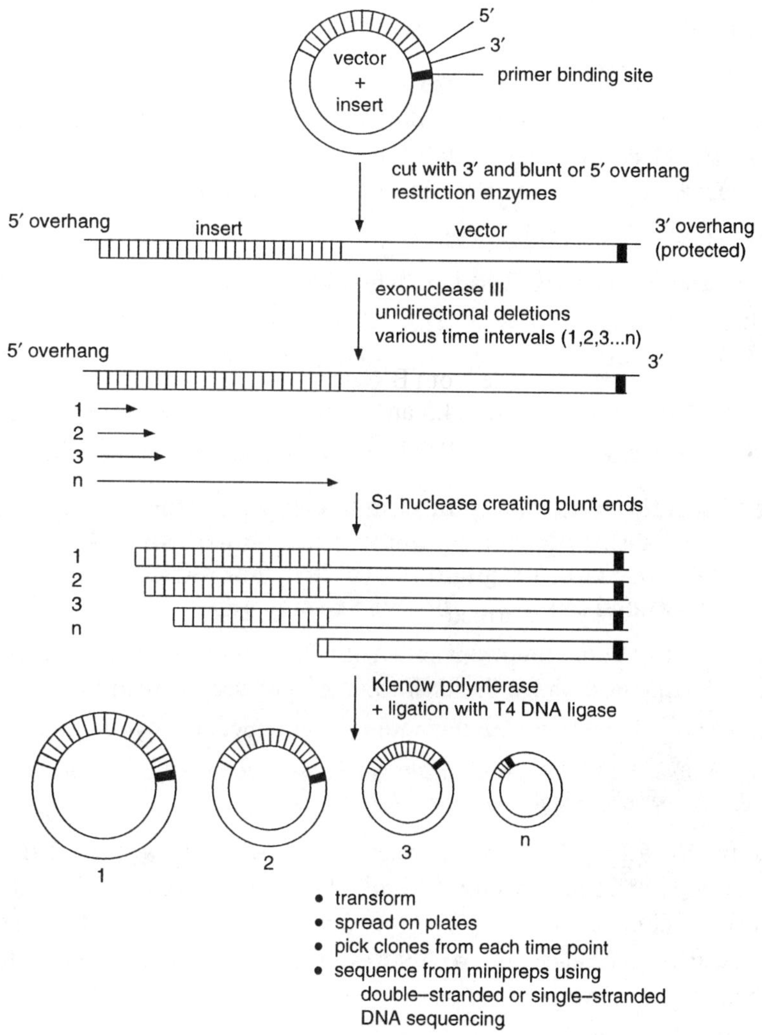

Figure 7.2.1 Construction of unidirectional deletions using exo III. The vector::insert DNA is digested with restriction endonucleases that generate a 3′ overhang next to the sequencing primer site and a 5′ overhang next to the insert DNA. Digestion with exo III generates unidirectional deletions from the end with the 5′ overhang. Treatment with S1 nuclease and repair with Klenow fragment create blunt ends from both the exo III-treated ends and the 3′ overhang that was protected from exo III digestion. The deleted plasmids are treated with DNA ligase and used to transform *E. coli.* Plasmids that have deletions are identified by gel electrophoresis of minipreparations of DNA.

5. Remove 25 µl (2.5 µg) linearized DNA (from step 4) and incubate 2 min at 37°C. Add 150 U exo III per pmol susceptible 3′ end. Continue the 37°C incubation.

   Conversion factor: 1 µg of 1-kb DNA = 1.5 pmol susceptible 3′ ends.

6. Remove 3-µl aliquots at 1-min intervals to individual microcentrifuge tubes (eight samples) and place immediately on dry ice for 5 min. Add 3 µl water and inactivate exo III by incubating 10 min at 70°C. Put samples on ice.

7. Add 15 µl S1 nuclease buffer and 4 µl (4 U) S1 nuclease to each sample. Incubate 20 min at room temperature. Stop reaction (via pH shift) by adding 5 µl S1 nuclease stop buffer to each sample.

8. Add 2 µl of 10× loading buffer to an 8-µl aliquot of each sample and electrophorese on 1% agarose gel, including ethidium bromide in the gel and buffer (*UNIT 2.5A*). Also run a lane of molecular weight markers (e.g., bacteriophage λ DNA, cut with *Hin*dIII) to determine the actual rate of exonuclease degradation. Decide which aliquots are worth saving.

9. Add to each remaining 22 µl of the selected samples 1 µl (2 U) Klenow fragment. Incubate ~2 min at 37°C. Add 1 µl of 0.25 mM 4dNTP mix and continue incubation at 37°C for 10 min to create blunt ends.

10. Add 20 µl exo III ligation buffer, 3 µl of 10 mM ATP, 14 µl water, and 1 µl (1 U) T4 DNA ligase. Incubate 5 hr at room temperature or overnight at 15°C.

If using a plasmid vector and planning to carry out dideoxy sequencing with double-stranded plasmid template, proceed with steps 11 to 15. If using an M13mp vector and planning to carry out dideoxy sequencing with single-stranded templates, proceed to step 16.

11. Transform 100 µl of competent cells with 10 µl of each ligation reaction (*UNIT 1.8*). Following heat-shock in transformation protocol, spread ⅕ of each of the cultures on LB plates containing antibiotic and incubate overnight at 37°C.

    Any competent E. coli recA strain can be used for transformation. About 2000 colonies are expected for each sample.

12. Pick two to four colonies from each plate and inoculate 5 ml LB medium containing antibiotic and grow overnight. Isolate DNA from 1.5 ml of each overnight culture using miniprep procedure in *UNIT 1.6*. Make a frozen glycerol stock with 1 ml of each culture (*UNIT 1.3*) and store remainder at 4°C for DNA sequencing template preparation.

13. Add 2 µl of 10× loading buffer and 10 µl water to each 2.5-µl aliquot of plasmid DNA from step 12 and run on a 1% agarose gel. As size markers, also load 0.1 µg each of undeleted plasmid and parent vector on the gel. It should be possible to discern a ladder of deletions from full insert size to ~300 bp.

14. Choose a set of clones (from step 13) that differ from each other by 250 to 300 bp, spanning the complete region of target DNA. Prepare plasmid DNA from overnight cultures stored at 4°C (step 12).

15. Subject plasmid DNA of deleted clones to sequencing analysis as in *UNIT 7.4* (clones can initially be characterized by T-track analysis, in which each template is used in a dideoxy–T sequencing reaction).

Follow steps 16 to 22 if using an M13mp vector and planning to carry out dideoxy sequencing using single-stranded templates.

16. Transform 100 µl competent *E. coli* cells with 10 µl of each reaction mix (from step 10). Also transform with 0.5 ng of M13mp RF vector DNA and 0.5 ng of M13mp::insert RF DNA that carries nondeleted fragment.

    E. coli DH5αF′ (Table 1.4.5) can be maintained on rich media without loss of F′ factor, is suppressor plus, restriction minus, modification plus, recombination deficient (recA1), and exhibits a higher transformation efficiency than JM109.

17. Following the heat-shock step in the transformation protocol, add 0.2 ml of overnight culture of *E. coli* JM109 or DH5αF′, pipet contents into 3 ml of 45°C H top agar containing 60 µl Xgal/IPTG mixture, mix, and pour on LB plates. Incubate overnight at 37°C.

    Do not leave H top agar containing Xgal too long at 45°C. About 2000 plaques are expected for each sample. All plaques should be white except for those from M13mp RF vector DNA, which should be blue.

18. Picking two to four M13 plaques from each transfection plate (i.e., two to four per exonuclease time point), grow individual M13mp phage stocks as in steps 1 and 2 in Basic Protocol 1 in *UNIT 7.3*. Grow stocks of M13mp vector and M13mp::insert as controls. Store supernatants at 4°C.

19. Place 20 µl of each phage supernatant into 4 µl of phage loading buffer and load on 1% agarose gel. Also load M13mp vector(s) and M13mp::inserts (undeleted) supernatants as size markers. It should be possible to discern a ladder of deletions from full insert size to ~300 bp.

20. Choose clones from step 19 that differ from each other by 250 to 300 bp, spanning complete insert DNA. Respin stored phage supernatants from step 18 (there will be some cell growth, even at 4°C) and transfer supernatants to fresh 1.5-ml microcentrifuge tubes.

21. Proceed immediately to step 3 of Basic Protocol 1 in *UNIT 7.3*.

22. Subject prepared single-stranded DNA templates to dideoxy sequencing as in *UNIT 7.4*. Also perform sequencing reactions on vector containing undeleted fragment.

SUPPORT PROTOCOL 1

PROTECTION OF DNA FROM EXONUCLEASE III DIGESTION USING [α-^{35}S]dNTPs

Creating unidirectional deletions for DNA sequencing using exo III requires that the end of the molecule adjacent to the sequencing primer site be protected from digestion. Restriction sites adjacent to the primer site that leave a 5′ overhang can be filled in with Klenow fragment using the appropriate thio nucleotide analog to cap the 3′ end. This increases flexibility in designing a deletion strategy, as any restriction site that leaves a 5′ overhang can be appropriately modified to provide the necessary protection from digestion by exo III.

Additional Materials *(also see Basic Protocol 1)*

5 mM appropriate dNTPs (*UNIT 3.4*)
5 mM appropriate [α-^{35}S]dNTPs (Pharmacia LKB)

1. In a 20-µl reaction mixture, completely linearize 5 µg vector::insert DNA by digesting with a restriction enzyme that leaves a 5′ overhanging end adjacent to sequencing primer site.

2. Add 1 µl of each appropriate 5 mM [α-^{35}S]dNTPs and 5 mM dNTPs, and 1 U Klenow fragment (*UNIT 3.5*). Incubate 30 min, room temperature.

    The [α-^{35}S]dNTPs and dNTPs required depend upon the sequence of 5′ overhang left by the restriction enzyme. Adequate protection from exo III digestion will be obtained for both partially and completely filled-in restriction sites, providing that the 3′-most nucleotide added by Klenow fragment contains the thio group.

3. Stop Klenow polymerase reaction by incubating 10 min at 75°C. Digest DNA with second restriction enzyme that generates a 5′ overhanging or blunt end adjacent to insert DNA (adjust salt concentration of reaction mix to that appropriate for second enzyme). Do not exceed 20% glycerol in final restriction endonuclease reaction mix.

4. Proceed with step 4 of Basic Protocol 1.

BASIC PROTOCOL 2

USING *BAL* 31 NUCLEASE TO CONSTRUCT NESTED DELETIONS

This method is based on the properties of *Bal* 31 nuclease, a single strand–specific endonuclease that can degrade the ends of duplex linear DNA from both the 5′ and 3′ termini (see *UNIT 3.12*). A nested set of deletions spanning a 1.5- to 2.0-kb region can be reliably generated using this procedure (see Fig. 7.2.2).

Materials *(see APPENDIX 1 for items with ✓)*

DNA fragment to be sequenced (insert DNA)
Appropriate "source" and "sequencing" vectors (Table 7.1.1)
Buffered phenol (*UNIT 2.1*)
Ice-cold 70% ethanol
✓ 3 M sodium acetate
✓ TE buffer
✓ 10× and 1× *Bal* 31 nuclease buffer
Bal 31 nuclease (*UNIT 3.12*)
200 mM EGTA
1 mg/ml yeast tRNA
Restriction endonucleases and corresponding buffers (*UNIT 3.1*)
✓ 10× loading buffer
Low gelling/melting temperature agarose (*UNIT 2.6*)
Ethidium bromide
✓ 50× TAE buffer (optional)
95% ethanol (optional)
10× T4 DNA ligase buffer (*UNIT 3.4*)
400 U/μl T4 DNA ligase (measured in cohesive-end units; *UNIT 3.14*)

If plasmid vector is used:

E. coli JM109, DH5αF′, or equivalent made competent for transformation (Table 1.4.5 and *UNIT 1.8*)
LB plates and medium (*UNIT 1.1*) containing appropriate antibiotic

If one of the M13mp plasmids is the cloning vector:

E. coli MC1061, DH5αF′, JM109, or equivalent made competent for transformation (Table 1.4.5 and *UNIT 1.8*)
Overnight cultures of *E. coli* JM101, JM107, JM109 or equivalent (Table 1.4.5) in LB or 2× TY medium (*UNIT 1.1*)
H top agar (*UNIT 1.1*)
Xgal/IPTG solution [30 μl Xgal stock and 30 μl IPTG (Table 1.4.2) for each 3 ml of H top agar]
LB plates (*UNIT 1.1*), prewarmed
2× TY medium (*UNIT 1.1*)
✓ Phage loading buffer

Preliminary Steps

1. Clone fragment to be sequenced in an appropriate source vector.

   If an M13mp vector is chosen as source vector, clone insert DNA in both orientations with respect to the sequencing priming site so that the sequence at each end of the undeleted insert can be obtained. However, prepare CsCl-purified DNA for only one of the orientations to use for Bal 31 digestion.

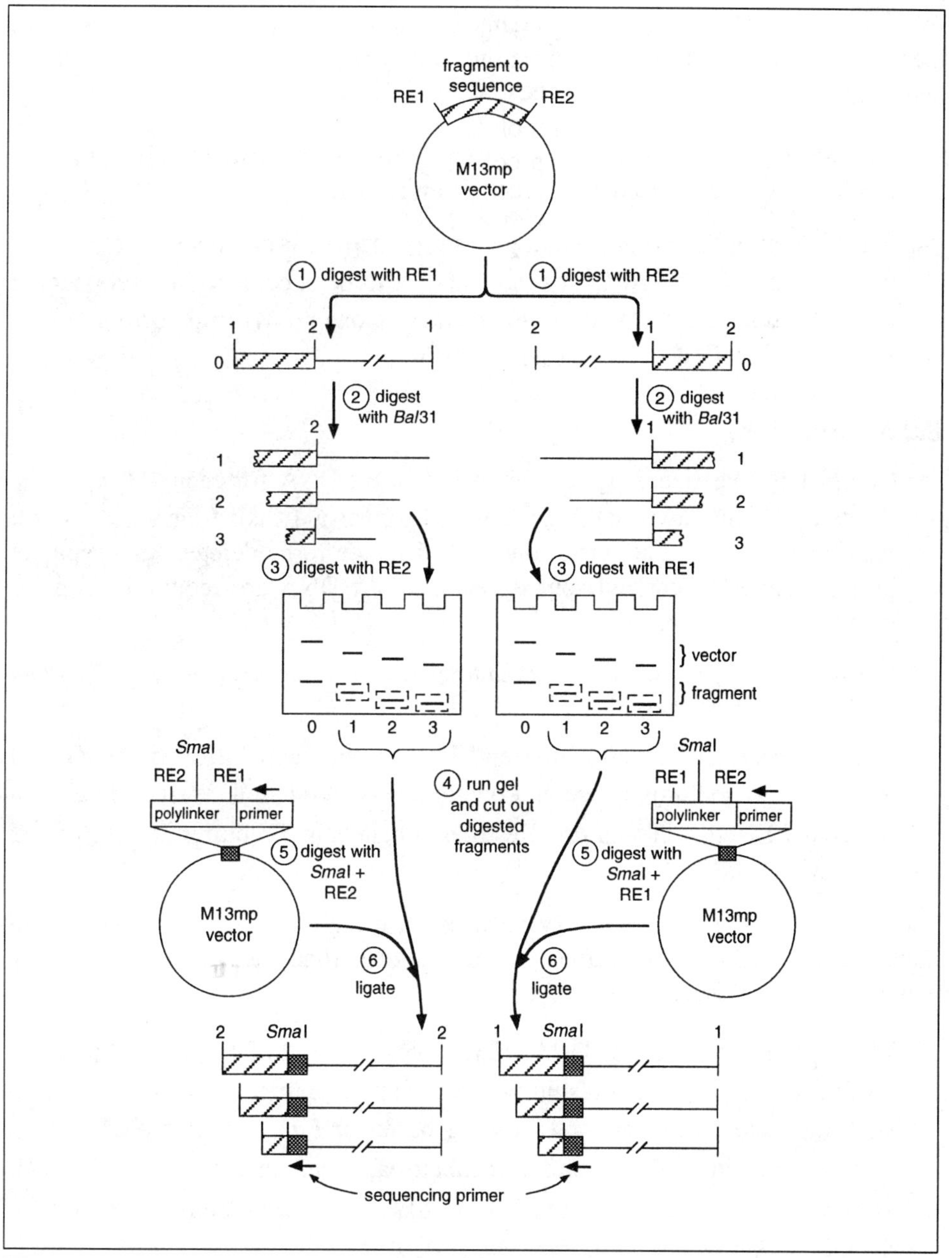

Figure 7.2.2 Strategy for making a nested set of deletions with *Bal*31 nuclease. The source vector containing a fragment to be deleted (in this example RF DNA of an M13mp vector) is linearized at each end of the fragment in separate reactions (step 1) and a nested set of deletions is generated with *Bal* 31 nuclease (step 2). The source vector and insert DNA are separated by restriction endonuclease digestion (step 3) and agarose gel electrophoresis (step 4). The deleted fragments are eluted from the gel and ligated to a sequencing vector (in this example, an M13mp vector) that has been digested with a restriction enzyme generating a blunt end (compatible with ends digested by *Bal* 31; in this example, *Sma*I) and another restriction enzyme corresponding to the site at the other end of the insert (RE1 or RE2). "RE2" and "RE2" refer to two different restriction endonucleases or their respective recognition sites.

2. Prepare 5 µg of CsCl-purified double-stranded DNA (from step 1) for each 150 bp to be deleted (e.g., 50 µg for a 1.5-kb fragment; if RNA contaminates the DNA after one CsCl gradient, purify DNA through a second gradient).

Two separate preparations of sequencing vector DNA are required, each for a set of deletions from one end (Fig. 7.2.2, step 5). For each set of deletions, the sequencing vector should have a blunt end (compatible with *Bal* 31–digested ends) adjacent to a dideoxy sequencing priming site (or chemical sequencing end-labeling site) and a sticky end compatible with the restriction site on the opposite side of the insert from which *Bal* 31 digestion was initiated in that set of deletions (and which was therefore protected from *Bal* 31 digestion). It can be prepared anytime in advance and stored at −20°C.

3. For each set of *Bal* 31 deletions, digest ≥2 µg CsCl-purified double-stranded DNA of sequencing vector with restriction enzymes.

4. For subcloning *Bal* 31–deleted fragments, prepare cells competent for transformation and store at −70°C. If fragments are to be subcloned into sequencing templates that feature *lac* α-complementation or into M13mp vectors, use *E. coli* JM109, DH5αF′, or equivalent host strain.

Bal 31 Procedure

5. Completely linearize 2 µg source vector::insert DNA for each 150 bp to be deleted plus an additional 2 µg (e.g., 22 µg for a 1.5-kb fragment) using a restriction enzyme that cuts at one end of region to be deleted (two separate preparations of linearized source vector::insert DNA are required, each for deletions from one end).

   For each set of deletions in a 2.0-kb fragment, the maximum amount of DNA that should be linearized is 28 to 30 µg.

6. Clean up restriction digest from step 5 by extracting with buffered phenol; add ~1/10 vol 3 M sodium acetate and precipitate with ethanol. Wash pellets with ice-cold 70% ethanol, and dry. Resuspend pellets in TE buffer at 1 µg/µl and store at 4°C.

The following steps are necessary because base composition of the DNA and impurities (such as RNA) can affect the rate of *Bal* 31 digestion and because *Bal* 31 may lose activity during storage.

7. Mix 2 µl (2 µg) linearized DNA, 38 µl water, and 5 µl of 10× *Bal* 31 nuclease buffer, and aliquot 9 µl of resulting mixture into five microcentrifuge tubes. To each tube add 1 µl of either 1× *Bal* 31 buffer or *Bal* 31 nuclease diluted 1/5, 1/10, 1/20, or 1/40 in 1× *Bal* 31 buffer. Incubate each reaction 30 min at 37°C, then stop reactions by adding 1 µl of 200 mM EGTA to each and heating 5 min at 65°C.

8. Analyze samples by agarose gel electrophoresis to determine dilution of *Bal* 31 that completely digests DNA in 30-min incubation.

   DNA samples typically require 0.03 to 0.10 U Bal 31 nuclease per µg linearized DNA. As digestion proceeds, the DNA bands first decrease uniformly in size, but gradually become increasingly diffuse before becoming completely digested.

The number of nucleotides separating the ends of the deletions can be varied by adjusting the amount of Bal 31 used and the time of digestion, and by the use of commercially available preparations of Bal 31 that are characterized as "fast" or "slow." If insufficient digestion occurs with Bal 31, repeat the procedure using a higher concentration of enzyme.

Perform *Bal* 31 digestions on each of the linearized source vector::insert DNA preparations (from step 5). Use 2 μg DNA of each for each 150 bp to be deleted. Use the amount of *Bal* 31 nuclease previously determined (steps 7 and 8) to completely digest the DNA. The protocol assumes that 1-min time points will be taken.

9. For each 150 bp to be deleted, mix 2 μl (2 μg) linearized DNA, 43 μl water, and 5 μl of 10× *Bal* 31 nuclease buffer. Preheat diluted DNA by incubating 10 min at 37°C.

10. For each linearized DNA preparation, set aside a 1.5-ml microcentrifuge tube for each time point (i.e., one tube for each 2 μg DNA to be subjected to *Bal* 31 digestion). Add 5 μl of 200 mM EGTA to each tube and label tubes to indicate the linearized species and the *Bal* 31 digestion time.

11. Add *Bal* 31 nuclease of determined amount to tube of linearized DNA, vortex gently, and incubate at 37°C. Every minute, transfer 45 μl to appropriately labeled tube from step 10 and place at room temperature.

    The digestion described should generate 150- to 200-bp deletions; the time interval or the amount of enzyme can be adjusted. We recommend taking time points for no longer than 10 to 12 min.

    Repeat steps 9 to 11 for each preparation of linearized source vector::insert DNA, one set of deletions from each end of the insert DNA.

12. After all *Bal* 31 digestions have been collected, incubate 5 min at 65°C.

13. Add 2 μl of 1 mg/ml yeast tRNA to each tube and ethanol precipitate samples using 3 M sodium acetate. Dissolve washed and dried pellets in 20 μl of appropriate restriction enzyme buffer for the next step.

    EGTA very effectively stops Bal 31 digestion by chelating Ca^{2+}; however, it also inhibits subsequent digestion with a variety of restriction enzymes. Therefore, it is very important to ethanol precipitate samples carefully after the addition of EGTA. This will allow digestion with minimal amounts of restriction enzymes, thus providing high-quality ends for subsequent ligation reactions.

    Optional: Before digestion with second restriction enzyme, the Bal 31–treated ends can be repaired with Klenow fragment to create blunt ends (UNIT 3.5). This increases the efficiency of ligations for subcloning.

14. Digest each of DNA samples with the restriction enzyme whose recognition site was protected from *Bal* 31 digestion (i.e., flanking the insert opposite the site used for linearization).

15. Add 2 μl of 10× loading buffer to each digested DNA sample and electrophorese on low gelling/melting temperature agarose, including ethidium bromide in gel and buffers. Also run a lane of molecular weight markers spanning the size of fragment to be sequenced. The digested fragments should appear as increasingly diffuse bands with increased digestion time.

16. Using a longwave UV (354-nm) lamp and the molecular weight markers as a guide (and minimizing exposure to UV light), cut out a series of DNA fragments (usually the entire band of digested insert DNA) differing in molecular weight by 150 to 200 bp from each lane and place in individual microcentrifuge tubes.

17. Add TE buffer to ~400 µl total volume for each tube, incubate at 65°C until all the agarose melts (about 5 to 10 min), and extract with 1 ml buffered phenol by vortexing at low speed for 30 sec.

18. Repeat phenol extraction on aqueous phase combined with its milky interface which, after the second extraction, will become much less abundant. If a significant amount of interface remains, repeat again.

19. Precipitate with 95% ethanol. Add 2 µl of 1 mg/ml yeast tRNA, if desired. Dissolve pellets in 30 µl TE buffer and store at 4°C.

20. Add 3 to 8 µl of each DNA sample eluted from agarose gel to appropriate sequencing vector DNA prepared in step 3. Include as a control one sample which contains vector DNA alone. Add 1 µl of 10× T4 DNA ligase buffer and adjust final volume to 10 µl. Add ~0.5 µl T4 DNA ligase and incubate overnight at 4°C.

If planning to sequence double-stranded templates, follow steps 21 and 22. If using M13mp vector, proceed to step 23.

21. Transform 100 µl of competent cells (JM109, DH5αF′ or equivalent from step 4) with 5 to 10 µl of each ligation reaction (from step 20). In addition, transform with sequencing vector DNA untreated with ligase. Following heat-shock of transformation protocol, spread ⅕ of each culture on LB plates containing appropriate antibiotic and incubate overnight at 37°C.

    The hosts described above are appropriate for sequencing vectors that allow a positive screen for inserts using α-complementation of β-galactosidase, where Xgal/IPTG is included in LB plates (UNIT 1.1).

22. If there are many colorless (not blue) colonies, proceed with steps 12 to 15 in Basic Protocol 1. Pick two to four colorless colonies from each time point and grow in LB medium containing antibiotic.

If using an M13mp sequencing vector and planning to carry out dideoxy sequencing using single-stranded templates, proceed with steps 23 to 27.

23. Mix each of ligation reactions with 40 µl of competent cells.

    If the vector was not treated with phosphatase, many more plaques—predominantly blue—may result. Under these circumstances, transform with ≤¼ of each ligation reaction.

24. In addition, transform with M13mp sequencing vector (Fig. 7.2.2, step 6) cut with *Sma*I and additional enzyme, untreated with ligase.

    Other controls: Also transform with 0.5 ng of M13mp RF vector used to subclone the Bal 31–deleted fragments and the source vector::insert clones carrying the undeleted fragment.

25. Following the heat-shock step in the transformation protocol (*UNIT 1.8*), add 0.2 ml of an overnight culture of *E. coli* JM109, DH5αF′, or equivalent in LB medium or 2× TY medium. Pipet contents into 3 ml of 45°C H top agar

containing 60 µl Xgal/IPTG solution, mix, and pour on room temperature LB plates.

26. After top agar hardens, incubate plates upside down overnight at 37°C.

Each transformation should result in ~100 colorless plaques. Depending on vector preparation, these will be accompanied by zero to many blue plaques.

27. If there are many colorless (not blue) plaques, proceed as in steps 18 to 22 in Basic Protocol 1. Picking two plaques for each time point and two plaques from each of control plates is usually sufficient.

PREPARATION OF M13mp SEQUENCING VECTOR DNA FOR SUBCLONING OF *BAL* 31–DIGESTED DNA FRAGMENTS

SUPPORT PROTOCOL

Additional Materials *(also see Basic Protocol 2; see* APPENDIX 1 *for items with ✓)*

M13mp18 or M13mp19
*Sma*I restriction endonuclease and buffer (*UNIT 3.1*)
Buffered phenol (*UNIT 2.1*)
Ice-cold 100% ethanol
Appropriate restriction endonuclease (Fig. 7.2.2) and buffer
Calf intestine alkaline phosphatase (CIP; *UNIT 3.10*; optional)
✓ TE buffer

1. For each 10 µg of DNA that will be subjected to *Bal* 31 digestion, digest 1 µg of mp18 and/or mp19 with *Sma*I.

2. Extract with buffered phenol and precipitate with ethanol as in (*UNIT 2.1*).

   If DNA preparations are really clean, it may be necessary to add 1 to 2 µg yeast tRNA before precipitation to see the pellets.

3. Digest *Sma*I-linearized sequencing vector with restriction endonuclease whose site flanks the insert DNA and which will be protected from *Bal* 31 digestion in that set of deletions (Fig. 7.2.2, step 5). Use the minimum amount of enzyme necessary. To eliminate recircularization of vector during cloning, remove phosphate groups by adding ≤0.2 µl CIP (~5 U) for the last 20 min of digestion (two separate preparations of sequencing vector DNA are required, each for a set of deletions from one end).

   The following restriction buffers work: 10 mM Tris·Cl (pH 7.5), 10 mM $MgCl_2$, and 10 to 150 mM NaCl. A vast excess of phosphatase is used, allowing the use of a suboptimal buffer. Phosphatase treatment sometimes results in a nonfunctional vector. If dephosphorylation is omitted, transform with less DNA in Basic Protocol 2 (step 22 or 26).

4. Extract with buffered phenol, precipitate with ethanol, and dissolve in 50 µl TE buffer (stored at –20°C, such vectors last months).

References: Henikoff, 1987; Hoheisel and Pohl, 1986; Poncz et al., 1982.

Contributors: Barton Slatko, Peter Heinrich, B. Tracy Nixon, and Daniel Voytas

7.2

Preparation of Templates for DNA Sequencing

This unit contains protocols for preparing DNA suitable for use as dideoxy sequencing templates and as material for end labeling and chemical sequencing. Any double-stranded template used in dideoxy sequencing should be denatured before being annealed to the primer. In general practice, alkali denaturation of plasmid DNA works better for sequencing than heat denaturation

BASIC PROTOCOL 1

PREPARATION OF SINGLE-STRANDED M13 PHAGE DNA

Single-stranded M13mp DNA to be sequenced by the dideoxy method is prepared from a single isolated plaque of *E. coli* DH5αF′ (or equivalent *E. coli* F′ host) infected with an M13mp plaque.

Materials *(see APPENDIX 1 for items with ✓)*

- *E. coli* DH5αF′ (or equivalent; Table 1.4.5), competent (*UNIT 1.8*) and overnight cultures in LB medium (*UNITS 1.1 & 1.2*)
- Recombinant M13mp RF DNA or M13mp single-stranded DNA (*UNIT 1.10*)
- H top agar (*UNIT 1.1*)
- LB plates (*UNIT 1.1*), 37°C
- 2× TY medium (*UNIT 1.1*)
- ✓ M13 polyethylene glycol (PEG) solution
- ✓ TE buffer
- Buffered phenol (*UNIT 2.1*)
- ✓ 3 M sodium acetate, pH 5.2
- 100% ethanol and 70% ethanol, ice-cold

- Pasteur pipets, regular and drawn out (sterile)
- 18 × 150–mm tubes

1. If starting with M13 RF DNA or M13 single-stranded DNA, transform 40 μl competent *E. coli* DH5αF′ with ~0.5 ng recombinant M13mp RF DNA or with 5 to 10 ng single-stranded DNA. Following the heat-shock step, add 0.2 ml of an overnight *E. coli* DH5αF′ culture to the transformed cells. Pipet the cells into 3 ml of H top agar, mix, and pour onto 37°C LB plates. Incubate upside-down overnight at 37°C.

2. Dilute an overnight *E. coli* DH5αF′ culture 100-fold with 2× TY medium and dispense 1.5-ml aliquots into loosely capped 18 × 150–mm tubes.

3. Pick one M13mp plaque with a sterile Pasteur pipet and expel it into one of the cultures. Be sure that a plug of agar is transferred. Repeat for each plaque. Grow cells 5 to 6 hr at 37°C (preferably on a roller drum).

4. Transfer cells to a 1.5-ml microcentrifuge tube and microcentrifuge at high speed for 5 min at 4°C to room temperature. Transfer supernatant to a clean 1.5-ml microcentrifuge tube.

5. Add 200 μl of M13 PEG solution to the phage supernatant, mix, and incubate 15 min at room temperature. Microcentrifuge 5 min at high speed and discard the supernatant. Microcentrifuge 30 sec at high speed and remove any remaining supernatant with a drawn-out Pasteur pipet.

   It is critical to remove all traces of the PEG supernatant.

6. Resuspend phage pellet in 100 µl TE buffer. Extract with 100 µl buffered phenol by vortexing 15 to 20 sec at low speed. Microcentrifuge at high speed 5 min at 4°C to room temperature and transfer upper (aqueous) phase to a clean microcentrifuge tube.

7. Add 11 µl of 3 M sodium acetate, pH 5.2, and 250 µl of 100% ethanol. Freeze 20 min at –70°C.

8. Microcentrifuge at high speed 10 min, 4°C, and discard supernatant. Add 100 µl ice-cold 70% ethanol, microcentrifuge, and remove supernatant. Repeat this 70% ethanol wash once. Dry DNA pellet in a Speedvac evaporator.

9. Resuspend pellet (2 to 3 µg) in 20 µl TE buffer. Electrophorese 0.5 µl on a minigel to confirm the DNA concentration. Store the remainder at –20°C until used as template for sequencing reactions (*UNIT 7.4*).

PREPARATION OF λDNA FROM SMALL-SCALE LYSATES

BASIC PROTOCOL 2

Materials *(see APPENDIX 1 for items with ✓)*

Recombinant λ phage
E. coli strain appropriate for λ phage of interest
Lambda top agarose and agarose plates (i.e., substitute agarose for agar; *UNIT 1.1*), freshly prepared on the day of use
1 mg/ml DNase I (*UNIT 3.12*)
✓ 1 mg/ml RNase A, heat-inactivated
✓ Lambda polyethylene glycol (PEG) solution
✓ SM medium
10% (w/v) sodium dodecyl sulfate (SDS)
Buffered phenol (*UNIT 2.1*)
✓ 0.5 M EDTA
1:1 (v/v) phenol/chloroform
Chloroform
✓ 3 M sodium acetate, pH 7.0
Isopropanol
✓ TE buffer, pH 8.0
Pasteur pipets, drawn out (sterile)

1. Prepare the recombinant λ phage stock on an *E. coli* strain appropriate for the λ phage by plate lysis using freshly prepared lambda top agarose and lambda agarose plates. The λ phage stock titer should be >10^9 pfu/ml.

2. Place 700 µl of phage stock in a sterile 1.5-ml microcentrifuge tube. Add 1 µl of 1 mg/ml DNase I and 1 µl of 1 mg/ml heat-inactivated RNase A. Incubate 30 min at 37°C.

3. Add 700 µl lambda PEG solution and incubate 1 hr on ice.

4. Microcentrifuge at high speed 10 min at 4°C and pour off supernatant. Microcentrifuge at high speed 2 min at 4°C and remove any remaining supernatant with a drawn-out Pasteur pipet.

5. Resuspend phage pellet in 500 µl SM medium. Add 5 µl of 10% SDS and 0.5 µl of 0.5 M EDTA to the solution and mix gently. Incubate 15 min at 65°C. Periodically mix gently to dissolve any remaining pellet.

6. Extract with 500 µl buffered phenol by vigorous vortexing. Microcentrifuge at high speed 2 min at room temperature and transfer upper (aqueous) phase to a clean microcentrifuge tube. Repeat using 500 µl of 1:1 phenol/chloroform, then repeat using 500 µl chloroform.

7. Add 50 µl of 3 M sodium acetate, pH 7.0, followed by 550 µl isopropanol. Freeze 30 min at −20°C.

8. Microcentrifuge at high speed 15 min at 4°C and aspirate and discard supernatant. Air dry the DNA pellet. Resuspend pellet (~20 µg) in 50 µl TE buffer, pH 8.0.

9. For sequencing, denature 0.1 to 0.3 pmol (2 to 4 µg) in alkali as described in Basic Protocol 4.

BASIC PROTOCOL 3

MINIPREP OF DOUBLE-STRANDED PLASMID DNA FOR DIDEOXY SEQUENCING

CsCl/ethidium bromide–purified closed-circular DNA (*UNIT 1.7*) can be used directly as a template for double-stranded sequencing.

Materials *(see APPENDIX 1 for items with ✓)*

LB medium containing appropriate antibiotic (*UNIT 1.1*)
E. coli strain carrying recombinant plasmid
✓ Tris/EDTA/glucose buffer
1.0% (w/v) SDS/0.2 N NaOH
✓ 3 M sodium acetate, pH 4.8
Isopropanol
✓ TE buffer, pH 8.0
4 M LiCl
Buffered phenol (*UNIT 2.1*)
Chloroform
Isopropanol
70% ethanol, ice-cold

1. Inoculate 5 ml LB containing appropriate antibiotic with *E. coli* strain carrying the recombinant plasmid. Grow at 37°C until culture is in mid-log phase.

2. Transfer 1.5 ml to a microcentrifuge tube. Microcentrifuge 2 min and discard supernatant. Resuspend pellet in 150 µl Tris/EDTA/glucose buffer and leave 5 min at room temperature.

3. Add 300 µl of 1.0% SDS/0.2 N NaOH and mix by inverting ~15 times (do not vortex). Leave 5 min at room temperature.

4. Add 225 µl of 3 M sodium acetate, pH 4.8, and mix by inverting ~15 times (do not vortex). Leave 45 min on ice.

5. Microcentrifuge at high speed 5 min. Transfer 650 µl of supernatant to a clean microcentrifuge tube. Add 650 µl isopropanol. Mix and let sit 10 min at room temperature.

6. Microcentrifuge at high speed 5 min at room temperature, then decant and discard supernatant. Dry pellet in desiccator. Resuspend pellet in 125 µl TE

buffer, pH 8.0, by vortexing. Add 375 µl of 4 M LiCl and let stand 20 min on ice.

7. Microcentrifuge 5 min at 4°C. Transfer supernatant to a clean microcentrifuge tube. Extract with 500 µl buffered phenol by vigorous vortexing. Microcentrifuge at high speed 2 min at room temperature and transfer upper (aqueous) phase to a clean microcentrifuge tube. Repeat this extraction using 500 µl chloroform.

8. Precipitate DNA with 2 vol isopropanol for 30 min at room temperature. Microcentrifuge 5 min at high speed and decant and discard supernatant.

9. Vortex pellet in ~1 ml ice-cold 70% ethanol. Microcentrifuge at high speed 5 min, then decant and discard supernatant. Dry pellet. Resuspend pellet in 50 µl TE buffer, pH 8.0. Electrophorese 5 to 10 µl on an agarose gel to verify purity and approximate DNA concentration.

10. For sequencing, denature ~0.5 pmol (10 to 45 µl) in alkali as described in Basic Protocol 4.

ALKALI DENATURATION OF DOUBLE-STRANDED PLASMID DNA FOR DIDEOXY SEQUENCING

BASIC PROTOCOL 4

Theoretically, it should be possible to denature double-stranded templates either by treating with alkali or by boiling and achieve equal results in sequencing. In practice, however, alkali denaturation of closed-circular double-stranded templates and λ templates usually gives superior results for dideoxy sequencing. In contrast, when sequencing PCR products heat denaturation appears to yield better results in dideoxy sequencing reactions (see *UNIT 15.2*).

***Materials** (see APPENDIX 1 for items with ✓)*

- Recombinant plasmid DNA (Basic Protocol 3)
- 2 M NaOH/2 mM EDTA
- ✓ 3 M sodium acetate, pH 6.0
- 95% (v/v) and 70% ethanol
- 0.5-ml microcentrifuge tubes

1. Add ~0.5 pmol of recombinant plasmid DNA to a 0.5-ml microcentrifuge tube. If the volume is >20 µl, ethanol precipitate the DNA (*UNIT 2.1*) and redissolve in 20 µl water. If the volume is ≤20 µl, add water to bring the volume to 20 µl.

2. Add 2 µl of 2 M NaOH/2 mM EDTA and gently mix by drawing up and down with a pipet. Incubate 5 min at 25° to 37°C. Place sample on ice, add 7 µl water, and mix thoroughly by drawing solution up and down with the pipettor.

3. Add 7 µl of 3 M sodium acetate, pH 6.0, to neutralize DNA solution. Mix thoroughly by drawing solution up and down with the pipettor. Check the pH of the solution by spotting 1 µl on pH paper. Add 3 M sodium acetate, pH 6.0, in 1-µl increments until the pH is ≤7.0.

4. Add 75 µl of 95% ethanol and place 10 min on dry ice. Microcentrifuge 10 min, 4°C, and carefully remove and discard the supernatant.

5. Add 400 µl of 70% ethanol. Microcentrifuge 10 min at 4°C, then carefully remove and discard the ethanol layer. Dry pellet 10 min in a Speedvac evapo-

rator. Store the pellet at 20°C (up to several weeks) until used as template for dideoxy sequencing reactions (*UNIT 7.4*).

BASIC PROTOCOL 5

PREPARATION OF PLASMID DNA FROM AN *E. COLI* COLONY OR PHAGE DNA FROM A PLAQUE FOR THERMAL CYCLE SEQUENCING

Materials *(see APPENDIX 1 for items with ✓)*

Agar plate containing *E. coli* colonies carrying a recombinant plasmid *or* λ or M13 recombinant phage plaques
✓ Tris/EDTA/proteinase K
Sterile toothpick or glass rod

1a. *For E. coli colony:* add 12 μl of Tris/EDTA/proteinase K mix to a 1.5-ml microcentrifuge tube. With a sterile toothpick or glass rod, transfer a single *E. coli* colony carrying a recombinant plasmid to this solution and vortex 15 sec.

1b. *For phage plaque:* add 12 μl of Tris/EDTA/proteinase K mix to a 1.5-ml microcentrifuge tube. With a sterile toothpick, transfer a single plaque to this solution and vortex 15 sec.

2. Incubate 15 min at 55°C, then 15 min at 80°C. Place 1 min on ice.

3. Microcentrifuge 3 min and immediately transfer supernatant to a clean 1.5-ml microcentrifuge tube. Use 9 μl as the template DNA in the end-labeled thermal cycle sequencing protocols (*UNIT 7.4*). Store remainder at –20°C.

Contributors: Barton E. Slatko, Peter Heinrich, and B. Tracy Nixon

UNIT 7.4

DNA Sequencing by the Dideoxy Method

In the basic dideoxy sequencing reaction, an oligonucleotide primer is annealed to a single-stranded DNA template (prepared by one of the methods described in *UNIT 7.3*) and extended by DNA polymerase in the presence of four deoxyribonucleoside triphosphates (dNTPs), one of which is ^{35}S-labeled. The reaction also contains one of four dideoxyribonucleoside triphosphates (ddNTPs), which terminate elongation when incorporated into the growing DNA chain (see introduction to Chapter 7 for a general overview of dideoxy sequencing). After completion of the sequencing reactions, the products are subjected to electrophoresis on a high-resolution denaturing polyacrylamide gel and then autoradiographed to visualize the DNA sequence.

Commercially available DNA sequencing kits provide most of the reagents required. These kits save a significant amount of startup time, although they may limit flexibility in troubleshooting and are somewhat more expensive than assembling components individually (see Table 7.4.1).

A nonisotopic alternative to this approach (chemiluminescent detection) can be found in *UNIT 7.5*.

Table 7.4.1 Suppliers of DNA Sequencing Kits, Reagents, and Equipment[a]

Product	Supplier[b]
Heat-resistant microtiter plates (round-bottom, 96-well)	BK, CO, PH, SA, ST
Nucleotides, nucleotide analogs	BM, BR, ICN, LT, NEB, PH, PR, USB
Radiolabeled nucleotides	AM, DP, ICN
Sequencing kits	AM, BM, BR, DP, IBI, LT, NEB, PE, PH, PR, ST, USB

[a]Modified from Slatko, 1991.

[b]Abbreviations: AM, Amersham; BK, Beckman; BM, Boehringer-Mannheim; BR, Bio-Rad; CO, Costar; DP, Du Pont NEN; IBI, International Biotechnologies; ICN, ICN Biomedicals; LT, Life Technologies; NEB, New England Biolabs; PE, Perkin-Elmer Cetus; PH, Pharmacia Biotech; PR, Promega; SA, Sarstedt, ST, Stratagene; USB, U.S. Biochemical. Addresses of suppliers are listed in *APPENDIX 4*.

LABELING/TERMINATION SEQUENCING REACTIONS USING SEQUENASE (MODIFIED T7 DNA POLYMERASE)

BASIC PROTOCOL 1

This protocol uses Sequenase. The labeling/termination procedure can also be used with other polymerases (see Table 7.4.2); however, because each polymerase has different buffer and Mg^{2+} concentration optima, and each discriminates to a different extent against ddNTPs, the concentrations of these components must be modified in each case.

Materials *(see APPENDIX 1 for items with ✓)*

0.5 pmol single-stranded *or* denatured double-stranded DNA template (*UNIT 7.3*)

0.5 to 1 pmol/µl oligonucleotide primer in water (store at −20°C; *UNIT 2.11*)

✓ 10× Sequenase buffer

✓ Sequenase termination mixes

✓ Sequenase/pyrophosphatase mix

✓ Sequenase diluent

✓ Labeling mixes

10 mCi/ml [α-^{35}S]dATP (500 to 1200 Ci/mmol)

✓ Stop/loading dye

0.5-ml microcentrifuge tubes

Heat-resistant microtiter plates (optional; Table 7.4.1)

1a. *For each single-stranded DNA template:* Mix the following in a 0.5-ml microcentrifuge tube:

0.5 pmol single-stranded DNA template
0.5 pmol primer
1 µl 10× Sequenase buffer
H_2O to 10 µl.

Mix gently by pipetting up and down (avoid creating bubbles). Incubate 6 min at 65°C, then 20 min at 37°C. Proceed to step 2.

1b. *For each double-stranded DNA template:* Resuspend a dried pellet containing 0.5 pmol denatured double-stranded DNA in the following mixture:

1 pmol primer
1 µl 10× Sequenase buffer
H_2O to 10 µl.

7.4

Table 7.4.2 Characteristics of DNA Sequencing Enzymes[a]

Characteristic	T7	Klenow	*Taq*	Bst	Vent	AMV RT
3′→5′ exonuclease	none	low[b]	none	none	none	none
5′→3′ exonuclease	none	none	present[c]	none	none	none
Processivity	high	low	inter.	inter.	low	inter.
Elongation rate	high	inter.	inter.	high	inter.	low
Use of nucleotide analogs[d]						
dITP	Y	Y	N	Y	Y	ND
7-deaza-dGTP	Y	Y	Y	Y	Y	Y
Band uniformity	very good	poor	good	very good	good	fair
Sequencing reaction temperature (°C)	<55[e]	<50	<70	<65	<75	<42

[a]Modified from U.S. Biochemical. This information is meant to serve as a guideline, not as a reference source. Some characteristics depend strongly on reaction conditions. Enzyme abbreviations: T7, genetically modified T7 DNA polymerase (Sequenase Version 2.0); Klenow, *E. coli* DNA polymerase I, large fragment; *Taq*, *Thermus aquaticus* DNA polymerase; Bst, *Bacillus stearothermophilus* DNA polymerase; Vent, *Thermococcus litoralis* DNA polymerase, exo^-; AMV RT, avian myeloblastosis virus reverse transcriptase. Other abbreviations used in the table are: inter., intermediate; Y, yes; N, no; ND, not determined.

[b]Mutants of Klenow are available that lack a 3′→5′ exonuclease (Derbyshire et al., 1988); however, sequencing results identical to that of normal Klenow fragment are obtained.

[c]Some versions of *Taq* DNA polymerase have the 5′→3′ exonuclease genetically or chemically removed.

[d]This entry refers to whether the enzymes use dITP and dGTP efficiently.

[e]Termination reactions; labeling reactions <37°C.

Mix gently by pipetting up and down (avoid creating of bubbles). Incubate 30 min at 37°C, then keep at this annealing temperature until ready to proceed to step 2.

2. While the primer is being annealed to the template, label four microcentrifuge tubes A, C, G, and T for each template to be sequenced.

3. Add 2.5 µl each of A, C, G, and T Sequenase termination mixes to the bottom of the A, C, G, and T tubes, respectively.

4. Immediately before use, dilute the Sequenase/pyrophosphatase mix in Sequenase diluent to 1 to 2 U Sequenase/µl and keep on ice.

5. Add the following to the annealed primer and template:

   2 µl labeling mix
   0.5 to 1.5 µl 10 mCi/ml [α-^{35}S]dATP
   2 µl diluted Sequenase/pyrophosphatase mix.

   The total volume is 14.5 to 16 µl. Incubate 5 min at 25°C (room temperature).

   Choose the labeling mix appropriate for the lengths of sequencing products that are desired. At least 3 pmol of [α-^{35}S]dATP are required for reactions containing the short labeling mixes whereas 15 pmol must be added to reactions containing the long labeling mixes APPENDIX1).

6. Add 3.5 µl of the labeling reaction mixture to the tube containing Sequenase termination mix A (from step 3). Mix the solution by gently pipetting up and down. Repeat this addition to the C, G, and T tubes, changing pipet tips each time. Incubate 5 to 10 min at 37°C.

7. Add 4 µl stop/loading dye. Heat samples 2 min in a 95°C water bath, then place on ice. Load 2 to 3 µl of each sample on a sequencing gel (*UNIT 7.6*). Electropho-

rese the gel (*UNIT 7.6*) and read the sequence. (See Table 7.4.3 for a troubleshooting guide.)

Excessive boiling of the completed reactions in formamide/dye solution may cause DNA chain breakage and smeared bands on the sequencing gel. If repeated loadings are planned, remove a 3-μl aliquot of each reaction to heat before each loading.

Table 7.4.3 General Troubleshooting Guide for Dideoxy Sequencing

Problem	Possible cause	Solution
Blank autoradiogram	Inactive polymerase Old label Old reagents Incorrect or defective primer Incorrect or defective template	Test using control primer, template, and reagents. Replace defective reagent.
	Incorrect exposure procedure (plastic wrap left on, gel facing backwards, etc.)	Correct error.
Entire autoradiogram too light/poor incorporation of label	Old label	Obtain new label.
	Low enzyme activity	Obtain fresh preparation of enzyme; dilute immediately before use and keep on ice.
	Primer did not anneal well to template	See Critical Parameters for calculating maximum annealing temperature for primers; if template is double-stranded, remember to denature it in alkali before annealing (*UNIT 7.3*).
	Not enough primer	Test using control DNA and primer. Check concentration of primer; use 0.5 pmol/single-stranded template, 1 pmol/ double-stranded template.
	Not enough template	Test using control DNA and primer; check template concentration on gel or by other method. Use ~0.5 pmol template DNA per set of reactions.
	Incorrect exposure or development of autoradiogram	Reexpose gel; change developing reagents, if necessary.
	Failure to remove plastic wrap.	Remove plastic wrap from gel if using ^{35}S.
High background in all lanes	Impure template DNA	See if problem persists with control DNA; if not, make new template DNA. For single-stranded templates, try optional annotations in *UNIT 7.3*. For double-stranded templates, treat with RNase A (*UNIT 3.13*), phenol extract and ethanol precipitate (*UNIT 7.3*); if necessary, CsCl-purify DNA.

continued

7.4

Table 7.4.3 General Troubleshooting Guide for Dideoxy Sequencing, continued

Problem	Possible cause	Solution
Bands throughout the lanes are diffuse or fuzzy	Impure template DNA	See if problem persists with control DNA; if not, make new template DNA; for single-stranded templates, try optional annotations in protocol, *UNIT 7.3*; for double-stranded templates, treat with RNase A, phenol extract and ethanol precipitate as described in *UNIT 7.3*; if necessary, CsCl-purify the DNA.
	Poor quality of acrylamide gel	Prepare fresh acrylamide, bisacrylamide, and buffers using only high-quality reagents; store stock solutions at 4°C in dark (*UNIT 7.6*).
	Gel used too soon after pouring	Let gel set up longer.
	Buffer concentration in gel differs from concentration in reservoirs	Prepare gel and reservoir solutions from same 10× TBE stock.
	Excessive boiling of samples	If samples are to be loaded on multiple gels, remove 3-μl aliquot of each reaction for heating and loading on gel.
	Samples not denatured before running on gel	Heat samples in 95°C water bath 2 to 3 min.
	Gel electrophoresis at too high a temperature	Monitor gel temperature with thermometer, keep temperature below 65°C; if necessary, run gel at lower wattage to keep temperature lower.
Areas on gel where bands are fuzzy	Gel or plastic wrap on top of gel dried with wrinkle	See *UNIT 7.6*, processing sequencing gels, for suggestions to avoid wrinkles.
	Film was not clamped tightly to the gel	Insert filter paper behind gel to take up any extra space in X-ray cassette; use more clamps, if necessary.
Distortion of all bands in 450-550-nucleotide region of the gel	>0.5% glycerol in samples (Tabor and Richardson, 1987b)	Dilute DNA polymerase in diluent without glycerol.
Bands in all four lanes over entire gel	Low enzyme activity	Use fresh enzyme preparation; dilute immediately before use in appropriate diluent or 1× sequencing buffer.
	Incorrect buffer or nucleotide mixes	Prepare new mixes.
	Contaminated reagents	Prepare fresh reagents.
Anomalous spacing of bands, missing bands, or bands at the same position in two or three lanes only at specific regions	Compression (due to secondary structure of newly synthesized DNA strands under conditions of gel electrophoresis)	Increase gel running temperature; include 25% to 40% formamide in gel; chemically modify C residues; substitute dITP for dGTP and include inorganic pyrophosphatase or substitute 7-deaza-dGTP for dGTP.
Bands at the same position in more than one lane throughout the gel	DNA preparation contains two different DNAs that are producing overlapping sequences	Prepare new DNA starting from a single plaque or colony.
	Primer has annealed to secondary sites on template	Adjust primer/template ratio and annealing temperature.

USING Mn^{2+} IN THE LABELING/TERMINATION REACTIONS

ALTERNATE PROTOCOL 1

In the presence of Mn^{2}+, Sequenase incorporates deoxynucleotides and dideoxynucleotides at the same rate To sequence using Mn buffer, carry out Basic Protocol 1, *except* modify step 5 as indicated below:

5. After adding labeling mix and label to the annealed primer and template, add 1 µl Mn (*APPENDIX 1*) buffer to the primer/template/labeling mixture prior to adding Sequenase.

   The 10× Sequenase buffer is still used in this reaction.

USING *TAQ* DNA POLYMERASE IN THE SANGER PROCEDURE

ALTERNATE PROTOCOL 2

Sequencing reactions using *Taq* DNA polymerase can be carried out at temperatures high enough to destabilize many secondary structures that would otherwise inhibit elongation of the polymerase.

To use *Taq* DNA polymerase in the Sanger procedure, modify the steps of Basic Protocol 1 as indicated below. (See Table 7.4.4 for a troubleshooting guide and *APPENDIX 1* for recipes)

1. Use 10× *Taq* sequencing buffer in the annealing reaction.
3. Add A, C, G, and T *Taq* Sanger mixes to the labeled tubes.
4. Immediately before use, dilute *Taq* DNA polymerase to 2.5 U/µl in 1× *Taq* sequencing buffer and keep on ice.
5. Substitute *Taq* DNA polymerase for sequenase in the reaction mixture.
6. Incubate the sequencing reactions 10 min at 50° to 75°C.

   The initial 30 sec of this sequencing reaction should be at a temperature no greater than the maximum annealing temperature of the primer to the template.

7. Incubate the chase reactions 10 min at 50° to 75°C.

Table 7.4.4 Guide to Troubleshooting in the Sanger Procedure

Problem	Possible cause	Solution
High background in "A" track, especially when [^{35}S]dATP is older than one half-life	Old label	Replace with fresh label, or reduce label in reactions
Intensity of bands is too low at top of gel	Ratio of ddNTP to dNTP is too high	If problem is specific to one mix make up new nucleotide mix; otherwise decrease ddNTP/dNTP ratio by lowering ddNTP in each mix
Intensity of bands is too low at bottom of gel	Ratio of ddNTP to dNTP is too low	If problem is specific to one mix, make up new mix; otherwise increase ddNTP/dNTP ratio by increasing ddNTP in each mix

7.4

ALTERNATE PROTOCOL 3

ONE-STEP SEQUENCING REACTIONS USING 5′-END-LABELED PRIMERS

5′-end-labeled primers are used primarily for sequencing very large double-stranded DNA templates (such as λgt11) or for templates that have given less than optimum results with nascent chain labeling. Primers may be 5′ end-labeled with ^{32}P or ^{35}S using T4 polynucleotide kinase.

For end labeling primer using [γ-^{32}P]ATP:

1a. End label ~10.5 pmol primer in a 25-μl reaction as described in *UNIT 3.10* for labeling by the forward reaction. Use ~11.5 pmol [γ-^{32}P]ATP (3000 Ci/mmol) and 10 U of T4 polynucleotide kinase (BSA may be omitted from reaction). Incubate 30 min, 37°C.

2a. Terminate the reaction by incubating 5 min at 95°C. Briefly microcentrifuge at high speed, room temperature. Proceed to step 3.

For end labeling primer using [γ-^{35}S]ATP:

1b. End label ~10.5 pmol primer in a 25-μl reaction as described in *UNIT 3.10* for labeling by the forward reaction. Use ~13 pmol [γ-^{35}S]ATP (1300 Ci/mmol) and 5 U of T4 polynucleotide kinase (BSA may be omitted from the reaction). Incubate 4 hr at 37°C, adding an additional 5 U of T4 polynucleotide kinase every hour.

2b. Terminate the reaction by incubating 5 min at 95°C. Briefly microcentrifuge at high speed, room temperature. Proceed to step 3.

3. Anneal the 5′ end-labeled primer to the DNA template as described in steps 1a or 1b of Basic Protocol 1, using ~2.4 μl of the 25-μl end-labeling reaction (~1 pmol primer) per denatured double-stranded DNA template, and 10× buffer appropriate for the DNA polymerase being utilized.

4. Place 3.5 μl each of the appropriate A, C, G, and T mixes in the bottom of four microcentrifuge tubes labeled A, C, G, and T, respectively—use Sequenase termination mixes or *Taq* termination mixes.

5. Add ~2.5 μl of the annealed labeled primer/DNA mix to each tube from step 4. Incubate 30 sec to 1 min at 37°C.

6. Dilute enzymes in appropriate diluent as follows:

   Sequenase—1 to 2 U/μl final
   Taq DNA polymerase—1 to 2.5 U/μl final.

7. Add 1 to 2 μl diluted DNA polymerase to each of the tubes from step 5. Incubate 5 to 10 min at the temperature appropriate for the DNA polymerase being utilized:

   Sequenase—37°C
   Taq DNA polymerase—50° to 75~°C.

8. Carry out step 7 of Basic Protocol 1.

9. If using Klenow fragment, add 1 μl dNTP chase to each tube after carrying out step 7. Incubate 5 min at 37° to 42°C.

THERMAL CYCLE SEQUENCING REACTIONS USING α-LABELED NUCLEOTIDES

BASIC PROTOCOL 2

Materials *(see APPENDIX 1 for items with ✓)*

✓ $Vent_R$ (exo^-) sequencing mixes
0.04 to 0.1 pmol single-stranded or double-stranded DNA template (*UNIT 7.3*)
0.6 to 1.2 pmol oligonucleotide primer (*UNIT 2.11*)
✓ 10× $Vent_R$ (exo^-) DNA sequencing buffer
3% (v/v) Triton X-100
10 mCi/ml [α-^{35}S]dATP, [α-^{32}P]dATP, *or* [α-^{33}P]dATP (500 to 3000 Ci/mmol)
2000 U/ml $Vent_R$ (exo^-) DNA polymerase
Mineral oil, sterile
✓ TCS stop/loading dye
Thermal cycling apparatus

1. Label four microcentrifuge tubes A,C,G, and T for each template to be sequenced.

2. Add 3 µl each of A, C, G, and T $Vent_R$ (exo^-) sequencing mixes to the bottom of the A, C, G, and T tubes, respectively.

3a. *For each single-stranded DNA template:* Mix the following in a 0.5-ml microcentrifuge tube:

0.04 pmol single-stranded DNA template
0.6 pmol primer
1.5 µl 10× Vent (exo^-) sequencing buffer
1 µl Triton X-100
H_2O to 12 µl.

Mix the solution gently by pipetting up and down.

3b. *For each double-stranded DNA template:* Mix the following in a 0.5-ml microcentrifuge tube:

0.1 pmol double-stranded DNA template
1.2 pmol primer
1.5 µl 10× $Vent_R$ (exo^-) sequencing buffer
1 µl 3% Triton X-100
H_2O to 12 µl.

Mix the solution gently by pipetting up and down.

Individually process each template/primer mixture through step 6. When all sets of reaction mixtures are complete, proceed to step 7.

4. Add to each template/primer mix 1 to 2 µl 10 mCi/ml [α-^{35}S]dATP, [α-^{32}P]dATP, or [α-^{33}P]dATP. Mix the solution gently by pipetting up and down.

5. Add 1 µl of 2000 U/ml $Vent_R$ (exo^-) DNA polymerase to the tube from step 4.

6. Immediately add 3.2 µl of the mix from step 5 to the tube containing $Vent^R$ (exo^-) sequencing mix A from step 2. Mix the solution by gently pipetting up and down. Repeat this addition to the C, G, and T tubes, changing pipet tips each time. Place the reaction mixtures on ice.

7. Overlay each reaction mixture with 1 drop of sterile mineral oil.

8. Set thermal cycle apparatus for the following thermal cycle conditions:

20 cycles:	20 sec	95°C
	20 sec	55°C
	20 sec	72°C

Place the tubes in the thermal cycler and run the program.

9. Add 4 μl TCS stop/loading dye to each tube beneath the mineral oil.

10. Heat samples 3 min at 85°C. Insert pipet tip underneath the oil covering the reaction and load 2 μl on a sequencing gel. Electrophorese the gel (*UNIT 7.6*) and read the sequence (see Figure 7.4.1 for an example of typical results and Table 7.4.5 for a troubleshooting guide).

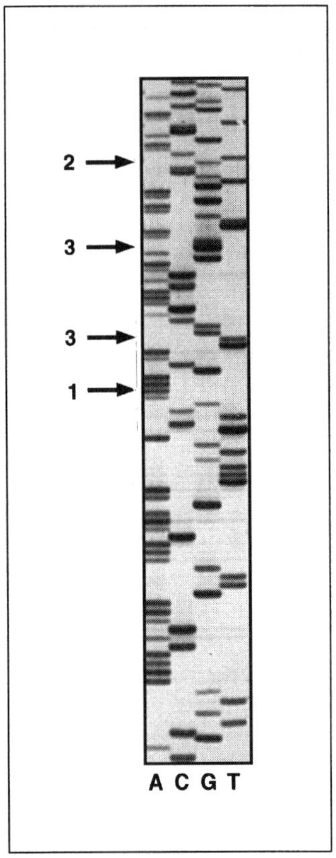

Figure 7.4.1 The results of thermal cycle sequencing of a double-stranded PCR DNA template utilizing the $Vent_R$ (exo^-) DNA polymerase followed by chemiluminescent detection. The lane order is A, C, G, T. The effects of sequence-dependent discrimination against ddNTPs are shown: (1) in the A lane, position marker shows an example of the second A in a run of A's darker than the preceding (and/or following) A's; (2) in the C lane, position marker shows that the first C in a run of C's is darker than the following C's; (3) in the G, C, or T lanes, position markers show that a nucleotide following an A tends to be darker than other nucleotides. (Photo courtesy of Millipore Corp./New England Biolabs, Inc.)

Table 7.4.5 Guide to Troubleshooting in the Thermal Cycle Sequencing Protocols

Problem	Possible cause	Solution
Bands are seen at same position in all four lanes, especially near primer.	Impure template DNA	See if problem exists with control DNA. If not, repurify DNA.
	Incorrect primer annealing temperature	Reduce primer annealing temperature; use longer (more stable) primer; increase annealing and extension steps to 1 min each.
	Impure or old reagents; ddNTP amount too high or dNTP amount too low	Prepare fresh reagents and readjust dNTP:ddNTP ratios.
Bands below a certain site on gel are very dark, and bands above that are very faint.	Secondary structure of template impeding extensions	Use higher reaction temperatures.
A dark band is present across all 4 lanes of gel; bands below and above it are equally dark.	Secondary structure of reaction products causing anomalous electrophoresis position of products in gel	Use higher gel temperatures or formamide gel; use base analogs or TdT in reaction; for PCR products, check purification procedure.
Intensity of bands at bottom of gel is too low.	Proportion of reaction products terminating near bottom of gel too low	Increase ddNTP:dNTP ratio in sequencing mixes.
Intensity of bands at top of gel is too low.	Proportion of reaction products terminating near top of gel too low	Decrease ddNTP:dNTP ratio in sequencing mixes.
One lane of reaction failed or is weak or smeary.	Possible cycler fault in one reaction slot	Check cycler performance.
	Pipetting error	Review chemistry procedure.
All lanes are smeary or show high backgrounds.	Template impurity	Review template preparation; run control DNA template.
	Primer impurity	Check primer purity.
	Cycler problem	Check cycler function.
	Reagent problem	Review chemistry procedure; make fresh reagents.
	Primer/template ratio too high	Review chemistry procedure.
	Incorrect priming temperature	Calculate T_m as guideline for reaction.
Entire sequence is light.	Insufficient template or primer in reaction	Double DNA and primer quantities in reaction.
	Cycler error; incorrect or inadequate cycling conditions	Check cycler performance; make steps 1 min instead of 30 sec; increase cycles to 30.
Oil covering reaction makes it difficult to load gel.	Excess oil present	Remove oil or use hot top-apparatus; microcentrifuge reaction before loading gel.

7.4

ALTERNATE PROTOCOL 4

THERMAL CYCLE SEQUENCING REACTIONS USING 5′-END-LABELED PRIMERS

This method allows detection of a few nanograms of M13-derived or plasmid DNA or 300 ng of λ DNA.

Additional Materials *(also see Basic Protocol 2)*

10 mCi/ml [γ-^{32}P]ATP or [γ-^{33}P]ATP (3000 Ci/mmol), *or* biotin *or* fluorescent conjugates

1. End label primer using [γ-^{32}P]ATP or [γ-^{33}P]ATP (3000 Ci/mmol) as described in steps 1a and 2a of Alternate Protocol 3.

2. Set up tubes and sequencing mixes as described in steps 1 and 2 of Basic Protocol 2.

Individually process tubes as described in Basic Protocol 2.

3a. *For each single-stranded DNA template:* Mix the following in a 0.5-ml microcentrifuge tube:

0.004 pmol single-stranded DNA template
0.6 pmol 5′-end-labeled primer
1.5 µl 10× $Vent_R$ (exo^-) sequencing buffer
1 µl 3% Triton X-100 solution
H_2O to 14 µl.

Mix gently by pipetting up and down.

3b. *For each double-stranded DNA template:* Mix the following in a 0.5-ml microcentrifuge tube:

0.01 pmol double-stranded DNA template
1.2 pmol 5′-end-labeled primer
1.5 µl 10× $Vent_R$ (exo^-) sequencing buffer
1 µl 3% Triton X-100 solution
H_2O to 14 µl.

Mix gently by pipetting up and down.

4. Add 1 µl of 2000 U/ml $Vent_R$ (exo^-) DNA polymerase to the tube from step 3a or 3b. Mix gently by pipetting up and down.

5. Complete the sequencing reactions as described in steps 6 to 10 of Basic Protocol 2.

References: Sanger et al., 1977; Sears et al., 1992; Tabor and Richardson, 1987b, 1990.

Contributors: Barton E. Slatko, Lisa M. Albright, snd Stanley Tabor

7.4

Dideoxy DNA Sequencing with Chemiluminescent Detection

Standard dideoxy DNA sequencing (*UNIT 7.4*) can be performed easily and efficiently with nonisotopic, chemiluminescent detection (Fig. 7.5.1) by utilizing primers labeled with biotin in the sequencing reactions. Primers derivatized with nonisotopic labels such as biotin can be utilized in protocols for DNA sequencing reactions designed for 5′-end-labeled primers. Chemiluminescent DNA sequencing kits incorporating nonisotopically labeled primers and chemiluminescent detection reagents are commercially available (Table 7.5.1). Suppliers for specialized equipment are listed in Table 7.5.2.

BASIC PROTOCOL

DNA SEQUENCING USING BIOTINYLATED PRIMERS WITH CHEMILUMINESCENT DETECTION

Materials *(see APPENDIX 1 for items with ✓)*

1 μg single-stranded DNA (~0.5 pmol) *or* 1 to 3 μg denatured double-stranded DNA template
Biotinylated DNA sequencing primer
5× *Bst* reaction buffer: 100 mM Tris·Cl, pH 8.5/100 mM $MgCl_2$
✓ Primer termination mixes (A, C, G, T)
1 U/μl *Bst* polymerase (Bio-Rad)
20× chase solution: 10 mM each of dATP, dCTP, dGTP, dTTP in H_2O
✓ Stop solution
0.2- to 0.4-mm-thick sequencing gel containing 8 M urea (*UNIT 7.6*)
✓ TBE electrophoresis buffer
✓ Blocking buffer I
✓ Conjugate solution
✓ Wash buffer I
✓ Assay buffer
✓ Dioxetane detection solution

65°, 70°, and 80°C water baths or heating blocks *or* thermal cycler
Filter paper (Whatman 3MM or equivalent)
Nylon membrane (Table 7.5.2)
UV light source: UV cross-linking apparatus, UV transilluminator, *or* hand-held UV lamp
Large heat-sealable bags (~40 × 53–cm; Table 7.5.1)
X-ray film (Kodak XAR-5 or equivalent)

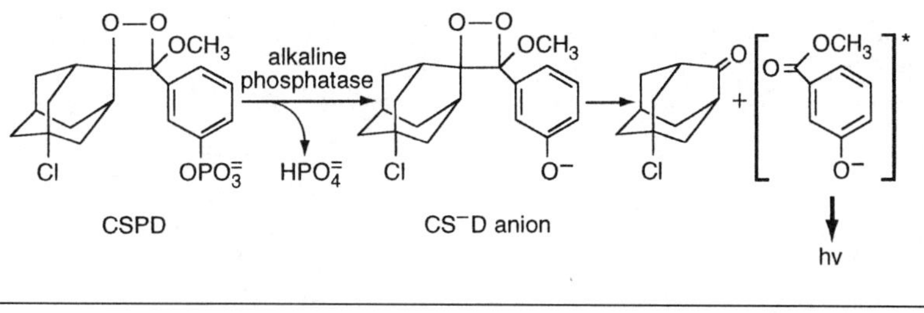

Figure 7.5.1 Pathway of decomposition of CSPD chemiluminescent substrate. Asterisk indicates compound in its excited state. Reprinted with permission of Tropix.

1. Prepare DNA template mix by combining the following in a microcentrifuge tube:

   1 μg single-stranded DNA (~0.5 pmol) *or* 1 to 3 μg denatured double-stranded DNA template
   0.5 pmol biotinylated DNA sequencing primer
   2 μl 5× *Bst* reaction buffer
   H_2O to 11 μl.

   Heat to 70°C, then cool slowly over 30 min to 30°C.

Table 7.5.1 Suppliers of Nonisotopic DNA Sequencing Reaction Kits, Chemiluminescent Detection Reagents, and Supplies

Product	Supplier[a]
Biotin DNA sequencing reaction kit	MI, NEB, TR, USB
Digoxigenin DNA sequencing reaction kit	BM
Bst polymerase	BR
Biotin DNA sequencing chemiluminescent detection kit	MI, NEB, TR, USB
Digoxigenin chemiluminescent detection kit	BM
Chemiluminescent 1,2-dioxetane substrate[b]	
CSPD	BM, BR, ST, TR
AMPPD	TR
Lumigen-PPD	LU
Membrane-blocking reagent	
I-Block	TR
Genius blocking agent	BM
Streptavidin–alkaline phosphatase conjugate	LT, TR, USB
Antibody–alkaline phosphatase conjugate	
anti-digoxigenin	BM
anti-fluorescein	DK, DP, TR
anti-DNP	DK
Large heat-sealable bags	NEB, MI, TR, USB
Nylon membranes	LT, MI, MS, NEB, PA, ST, TR, USB

[a]Abbreviations: BM, Boehringer Mannheim; BR, Bio-Rad; DP, Du Pont NEN; DK, Dako; LT, Life Technologies; LU, Lumigen; MI, Millipore; MS, Micron Separations; NEB, New England Biolabs; PA, PALL; PR, Promega; ST, Stratagene; TR, Tropix; USB, U.S. Biochemical. Addresses of suppliers are listed in *APPENDIX 4*.

[b]Abbreviations: AMPPD, disodium 3-(4-methoxyspiro[1,2-dioxetane-3,2′-tricyclo[3.3.1.1^{3,7}]decan]-4-yl)phenyl phosphate; CSPD, disodium 3-(4-methoxyspiro[1,2-dioxetane-3,2′-(5′-chloro)tricyclo [3.3.1.1^{3,7}]decan]-4-yl)phenyl phosphate; Lumigen-PPD, 4-methoxy-4-(3-phosphate phenyl)-spiro-(1,2-dioxetane-3,2′-adamantane), disodium salt.

Table 7.5.2 Suppliers of Specialized Instruments for DNA Sequencing With Chemiluminescent Detection

Instrument	Supplier
Large bag sealer, 60 cm (24 in.)	National Bag Company
Horizontal electroblotter	Panther HEP-3, Owl Scientific Plastics TE 90, Hoefer
Direct transfer electrophoresis (DTE) apparatus	TwoStep, Hoefer
Rotating-bottle apparatus	Navigator, Biocomp Instruments
Automated blot development apparatus	PR 1000, Hoefer

2. While the template mix is cooling, aliquot 2 µl each of the A, C, G, and T primer termination mixes into four separate microcentrifuge tubes or microtiter plate wells, and label them A, C, G, and T. Warm the tubes or plate containing termination mixes to 65°C.

3. Add 1 µl *Bst* polymerase to the cooled template mixture from step 1.

4. Add 2.5 µl polymerase/template mixture to each primer termination mix set up in step 2 and incubate 5 min at 65°C.

5. Prepare a 1× chase solution by diluting the 20× chase solution stock with water. Add 2 µl of 1× chase solution to each sample tube or well and incubate 5 min at 65°C.

6. Add 4 µl stop solution to each tube or well.

7. Set up a standard 0.2- to 0.4-mm-thick sequencing gel (*UNIT 7.6*) containing 8 M urea.

8. Immediately before loading, heat reactions 2 min at 80°C and place on ice. Load 2 to 3 µl. Run gels as in *UNIT 7.6*, monitoring the migration of the marker dyes (see Table 7.6.2).

9. Cut three pieces of Whatman 3MM filter paper to size of gel or slightly larger. Cut a piece of nylon membrane that will cover the desired region of gel.

10. Disassemble gel apparatus and separate glass plates. Place one piece of dry Whatman 3MM filter paper on the gel, gently rub the paper, and carefully lift the gel up by peeling back the paper. Place the paper with gel attached, gel-side-up, on the glass plate.

11. Wet nylon membrane thoroughly with TBE buffer, either by immersion or by squirting the membrane with buffer from a squeeze bottle. Carefully place wet membrane on gel. Roll a pipet or smooth rod over membrane to remove air bubbles.

12. Place two pieces dry Whatman 3MM filter paper on top of membrane. Place another glass plate on top of filter paper and cover with ~2 to 4 kg of weight. Allow transfer to proceed 1 hr.

13. Separate gel/membrane sandwich. Carefully peel membrane back from gel. Mark side of membrane that was in contact with gel with a pencil. Place membrane on filter paper, DNA-side-up, and air dry ~10 min.

14. Using a UV cross-linking apparatus, UV transilluminator, or hand-held UV lamp, achieve a total UV exposure of the membrane of 120 millijoules/cm^2. Proceed with step 15 or store membrane between two pieces of plastic wrap ≤6 months at 4°C.

Perform the following steps at room temperature with moderate shaking (120 to 140 rpm). Reagent volumes are for a single 1000- to 1200-cm^2 membrane. Only one membrane per bag is recommended. Solutions should not be reused.

15. Place the membrane in large heat-sealable bag. Add 500 ml blocking buffer I, carefully remove air bubbles and seal the bag, then incubate 10 min.

16. Incubate membrane in the following:

    20 min in 200 ml conjugate solution
    5 min in 500 ml blocking buffer I
    10 min in 500 ml wash buffer I, three times
    2 min in 500 ml assay buffer, two times
    5 min in 50 ml dioxetane detection solution.

17. Drain excess detection solution from bag, smooth out any wrinkles, and reseal.

18. Place wrapped membranes in direct contact with X-ray film at room temperature.

    An initial 60-min exposure is normally adequate. Additional shorter or longer exposures can be performed to optimize signal intensity and resolution. Table 7.5.3 is a troubleshooting guide for DNA sequencing with chemiluminescence detection.

ALTERNATE PROTOCOL 1

TWO-STEP (INDIRECT) DETECTION USING STREPTAVIDIN AND BIOTINYLATED ALKALINE PHOSPHATASE

Additional Materials *(also see Basic Protocol; see* APPENDIX 1 *for items with ✓)*

✓ Two-step blocking solution
✓ Streptavidin solution
✓ Two-step wash solution I
✓ Biotinylated alkaline phosphatase solution
✓ Two-step wash solution II

1. Carry out DNA sequencing reactions, electrophoresis, DNA transfer, and UV cross-linking as in steps 1 to 14 of the Basic Protocol.

2. Place membrane in a bag or container and incubate in the following with moderate shaking (120 to 140 rpm) at room temperature:

    5 min in 100 ml of two-step blocking solution
    5 min in 50 ml streptavidin solution
    5 min in 500 ml two-step wash solution I, two times
    5 min in 50 ml biotinylated alkaline phosphatase solution
    5 min in 500 ml two-step wash solution II, two times
    5 min in 500 ml assay buffer
    5 min in 50 ml dioxetane detection solution.

3. Drain dioxetane detection solution, process membrane, and expose to X-ray film as described in steps 17 and 18 of the Basic Protocol.

ALTERNATE PROTOCOL 2

SEQUENCING WITH HAPTEN-LABELED PRIMERS AND DETECTION WITH ANTIBODY-ALKALINE PHOSPHATASE CONJUGATES

As an alternative to biotin, other haptens such as digoxigenin, fluorescein, and 2,4-dinitrophenyl (DNP) may be used to label the products of the DNA sequencing reactions by incorporating primers labeled with these haptens.

Additional Materials *(also see Basic Protocol; see* APPENDIX 1 *for items with ✓)*

5′-hapten-end-labeled primer
✓ Blocking buffer II
✓ Wash buffer II
✓ Antibody-conjugate solution

Table 7.5.3 Troubleshooting Guide for DNA Sequencing with Chemiluminescent Detection[a]

Problem	Possible cause	Solution
High background	Contamination of reagents with alkaline phosphatase	Prepare all buffers daily; use ultrapure deionized water and other reagents free of alkaline phosphatase contamination; avoid cross-contamination of wash solutions by conjugate solution—i.e., cover solutions during storage; wash funnel used for liquid transfers; clean outside of bag after each solution addition.
	Glove dust	Use powder-free gloves.
	Insufficient agitation during incubations	Increase shaker speed; increase volume of liquid; remove air bubbles.
	Nonspecific binding of conjugate to membrane	Increase incubation time with blocking buffer or increase number of washes after enzyme/conjugate incubation.
Splotchy images	Bacterial or alkaline phosphatase contamination of membrane	Determine that all buffers are free of contamination prior to use and that membrane, blotting paper, and hybridization bags are clean and fingerprint-free.
Spots or lines	Mechanical damage to the membrane	Repeat with new membrane.
Poor band resolution	Exposure begun too long after substrate removal (depending on specific dioxetane substrate and membrane used)	Perform film exposures soon after substrate addition.
	Membrane not in good contact with gel during transfer	Carefully press membrane onto gel with pipet; add heavier weight during capillary transfer procedure.
	X-ray film not in good contact with membrane	Reseal membrane with plastic wrap and avoid wrinkles; use exposure cassette with clamps.
Low signal	Incorrect assay buffer pH	Adjust assay buffer to pH 10.
	Inefficient DNA transfer	Review transfer procedure and repeat experiment; when using capillary transfer method, it is very important to have gel sandwich on an absolutely flat surface; increase amount of weight to provide more uniform transfer.
Low-intensity signal from low-molecular-weight fragments	Low-molecular-weight fragments bind poorly, especially to neutral nylon	Use charged nylon membrane; always cross-link DNA to membrane.

[a]General troubleshooting information on DNA sequencing reactions and gel electrophoresis is described in Critical Parameters and *UNIT 7.4*.

7.5

1. Prepare the DNA template mix as in step 1 of the Basic Protocol, using a 5′-hapten-end-labeled primer in place of the biotinylated sequencing primer.

2. Perform the DNA sequencing reactions, run the sequencing gel, and transfer the DNA to a nylon membrane as in steps 2 through 13 of the Basic Protocol.

3. Place the membrane in a bag or container and incubate in the following with moderate shaking (120 to 140 rpm) at room temperature:

   30 min in 200 ml antibody-conjugate solution
   5 min in 500 ml blocking buffer II
   10 min in 500 ml wash buffer II, three times
   2 min in 500 ml assay buffer, two times
   5 min in 50 ml dioxetane detection solution.

4. Drain dioxetane detection solution and expose to X-ray film as described in steps 17 and 18 of the Basic Protocol.

References: Creasey et al., 1991; Martin et al., 1991; Olesen et al., 1993.

Contributor: Chris S. Martin

UNIT 7.6

Denaturing Gel Electrophoresis for Sequencing

The accuracy of DNA sequence determination depends largely upon resolution of the sequencing products in denaturing polyacrylamide gels. In general, the gels required for DNA sequencing are 40-cm long, of uniform thickness, and contain 4% to 8% acrylamide and 7 M urea (Basic Protocol). Modifications of the Basic Protocol increase the length of readable sequence information which can be obtained from a single gel.

We recommend running 6% acrylamide, nongradient gels as a starting point. Two gels can be run, one for a relatively short time to retain the shorter oligonucleotides, the second for a longer time to maximize separation of the longer oligonucleotides. This two-gel procedure, using 6% acrylamide, allows 300 to 350 bases of sequence information to be obtained from a single set of sequencing reactions. If fewer than five sets of sequencing reactions have been performed, one 6% gel can be used for both the short and the long runs (the typical gel format has room for ten sets of sequencing reactions, or 40 samples). An alternative practice is to use an 8% gel for the shorter oligonucleotides and a 6% gel for the longer oligonucleotides.

To ensure even conduction of the heat generated during electrophoresis, an aluminum plate (0.4 cm thick, 34 × 22 cm) can be clamped onto the front glass plate with the same book-binder clamps used to hold the gel sandwich to the apparatus. The aluminum plate must be positioned so that it does not touch any buffer during electrophoresis.

CAUTION: Acrylamide and bisacrylamide are neurotoxins and should be handled with gloves. Handle dimethyldichlorosilane with gloves also and carry out plate treatment with this solution in a fume hood (see silanization, *APPENDIX 3*). Also handle TEMED and formamide with care.

POURING, RUNNING, AND PROCESSING SEQUENCING GELS

BASIC PROTOCOL

Materials *(see* APPENDIX 1 *for items with ✓)*

- 70% (v/v) ethanol *or* isopropanol in squirt bottle
- 5% (v/v) dimethyldichlorosilane (diluted in $CHCl_3$; Sigma)
- ✓ Denaturing acrylamide gel solution
- TEMED
- 10% (w/v) ammonium persulfate (made fresh weekly and stored at 4°C)
- ✓ TBE electrophoresis buffer, pH 8.3-8.9
- Sequencing samples in formamide/dye solution (*UNITS 7.4* or *7.5*)
- 5% acetic acid/5% methanol (v/v) fixer solution

- 30 × 40–cm front and back gel plates
- 0.2- to 0.4-mm uniform-thickness spacers
- Large book-binder clamps
- 60-ml syringe
- Pipet tip rack or stopper
- 0.2- to 0.4-mm sharkstooth or preformed-well combs
- Sequencing gel electrophoresis apparatus
- Pasteur pipet *or* Beral thin stem
- Power supply with leads
- Sequencing pipet tip
- Gel dryer
- Shallow fixer tray
- 46 × 57–cm gel blotting paper (e.g., Whatman 3MM)
- Kodak XAR-5 X-ray film

NOTE: Many companies provide equipment needed for sequencing experiments; a list of suppliers is provided in Table 7.6.1.

1. For each gel, meticulously wash a pair of 30 × 40–cm front and back gel plates with soap and water; rinse well with deionized water and dry. Wet plates with 70% ethanol or isopropanol in a squirt bottle and wipe dry with a Kimwipe or other lint-free paper towel.

2. Wearing gloves and working inside a fume hood, apply a film of 5% dimethyldichlorosilane in $CHCl_3$ to one side of each plate by wetting a Kimwipe with the solution and wiping the whole plate carefully. After the film dries, wipe with 70% ethanol or isopropanol and dry with a Kimwipe. Check plates a final time for dust and other particulates.

3. Assemble gel plates according to manufacturer's instructions with 0.2- to 0.4-mm uniform-thickness spacers and large book-binder clamps, making certain the side and bottom spacers fit tightly together.

4. For each gel, prepare 60 ml of desired denaturing acrylamide gel solution in a 100-ml beaker. Thoroughly mix 60 µl TEMED, then 0.6 ml of 10% ammonium persulfate, into each acrylamide solution immediately before pouring each gel (step 5).

   Before adding TEMED and ammonium persulfate, the gel mix can be heated; however, to prevent degradation of acrylamide, do not heat over 55°C. Allow to cool to room temperature (≤25°C) to prevent polymerization while pouring gel. If particulate matter remains, filter through Whatman No. 1 filter paper in a funnel. To

Table 7.6.1 Suppliers of Sequencing Gel Electrophoresis Equipment[a]

Equipment	Supplier[b]
Sequencing gel apparatus, including necessary clamps, combs and spacers	AAP, ABA, BR, HO, IBI, JS, LT, OSP, PH, SS
Gel tape	ABA, HS, IBI, LT, OSP
Power supplies	ABA, ACS, BR, EC, FD, HO, IBI, IS, LT, OSP, PH, SS, ST
Beral thin stem	BE
Ultrathin pipet tips for loading gels	ABA, BR, CO, DR, DY, HO, IBI, IS, MB, PH, SS, ST
Gel thermometer	BR
Shallow fixer tray	OSP
Transfer paper	ABA, SS, WH
Gel dryers	ATR, BR, HO, SV

[a]Modified from Slatko, 1991.

[b]Abbreviations: AAP, Ann Arbor Plastics; ABA, American Bioanalytical; ACS, Accurate Chemical and Scientific; ATR, ATR; BE, Beral Enterprises; BR, Bio-Rad; CO, Costar; DR, Drummond; DY, Dynalab; EC, EC Apparatus; FD, Fotodyne; HO, Hoefer; IBI, International Biotechnologies; IS, Integrated Separation Systems; JS, Jordan Scientific; LT, Life Technologies; MB, Marsh Biomedical; OSP, Owl Scientific Plastics; PH, Pharmacia LKB; SS, Schleicher & Schuell; ST, Stratagene; SV, Savant; WH, Whatman. See *APPENDIX 4* for addresses of suppliers.

achieve slower polymerization, reduce amounts of TEMED and ammonium persulfate to 40 µl and 0.4 ml, respectively.

5. Pour gel immediately. Gently pull acrylamide solution into a 60-ml syringe, avoiding bubbles. With short plate on top, raise top of each gel sandwich to a 45° angle from the benchtop and slowly expel acrylamide between plates along one side. Adjust angle of plates so gel solution flows slowly down one side.

6. When solution reaches the top of the short plate, lay gel sandwich down so the top edge is ~5 cm above the benchtop; place an empty disposable pipet tip rack or stopper underneath the sandwich to maintain the low angle. Insert flat side of a 0.2- to 0.4-mm sharkstooth comb into solution 2 to 3 mm below top of short plate, being very careful to avoid bubbles. Use book-binder clamps to pinch combs between plates so no solidified gel forms between combs and plates. Layer extra acrylamide gel solution onto comb to ensure full coverage. Rinse syringe with water to remove acrylamide.

7. Remove bottom spacer or tape at the bottom of each gel sandwich. Remove extraneous polyacrylamide from around combs with razor blade. Clean spilled urea and acrylamide solution from plate surfaces with water. Remove sharkstooth combs gently from each gel sandwich, avoiding pulling or stretching top of gel. Clean combs with water so they will be ready to be reinserted in step 10. If a preformed-well comb was used, take care to prevent tearing of polyacrylamide wells.

8. Fill bottom reservoir of each gel apparatus with 1× TBE buffer such that gel plates will be submerged 2 to 3 cm in buffer. Place each gel sandwich in a sequencing electrophoresis apparatus and clamp plates to support.

9. Pour 1× TBE buffer into top reservoir to ~3 cm above top of gel. Rinse top of gel with 1× TBE buffer using a Pasteur pipet or Beral thin stem.

10. Reinsert teeth of cleaned sharkstooth combs into gel sandwich with points just barely sticking into gel. Using a Pasteur pipet or Beral thin stem, rinse wells thoroughly with 1× TBE buffer to remove stray fragments of polyacrylamide.

11. Preheat gels by turning on power supplies to 45 V/cm, 1700 V, 70 W constant power ~30 min before loading sequencing samples.

12. Rinse wells just prior to loading gels to remove urea that has leached into them.

13. Cover and heat completed sequencing samples in formamide/dye solution for 2 min at 95°C, then place on ice. Load 2 to 3 μl of each sample per well on each gel to be run. Rinse sequencing pipet tip twice in lower reservoir after dispensing from each reaction tube. Run gels at 45 to 70 W constant power. Maintain a gel temperature of ~65°C. Temperatures higher than this can result in cracked plates or smeared bands, while too low a temperature can lead to incomplete denaturation of sequencing products. Observe migration of marker dyes (Table 7.6.2) to determine length of electrophoresis.

14. Fill dry ice traps attached to gel dryer (if required) and preheat to 80°C.

15. After electrophoresis of each gel is complete, drain buffer from upper and lower reservoirs of sequencing apparatus and discard liquid as radioactive waste.

16. Remove gel sandwich from electrophoresis apparatus and place under cold running tap water until surfaces of both glass plates are cool. Lay sandwich flat on paper towels with notched (short) plate up. Remove excess liquid and remaining clamps or tape. Remove one side spacer and insert a long metal spatula between glass plates where spacer had been. Pry the plates apart with a gentle rocking motion of spatula. The gel should stick to bottom plate. If it sticks to top plate, flip sandwich over. Slowly lift top plate from the side with inserted spatula, gradually increasing the angle until top plate is completely separated from gel.

17. Once plates are separated, remove second side spacer and any extraneous bits of polyacrylamide around gel.

18. If samples contain ^{32}P or ^{33}P the fixing steps are optional. If samples contain ^{35}S, transfer gel on glass plate to a shallow fixer tray and gently cover to 2 cm with 5% acetic acid/5% methanol fixer solution. Soak gel 10 to 15 min. Gently rock tray periodically to gradually loosen gel from glass plate.

Urea quenches the ^{35}S signal and must be removed from ^{35}S-containing gels. Although not required, removing urea will also increase the clarity of ^{32}P-containing gels.

19. Reposition gel over plate and remove fixer solution by aspiration or gravity. Take care to keep gel centered over glass plate. Carefully lift plate with gel on top from tray. Place gel on benchtop. Hold two pieces of dry blotting paper

Table 7.6.2 Migration of Oligodeoxynucleotides (Bases) in Denaturing Polyacrylamide Gels in Relation to Dye Markers

Polyacrylamide (%)	Bromphenol blue	Xylene cyanol
5	35 b	130 b
6	26 b	106 b
8	19 b	75 b
10	12 b	55 b

together as one piece. Beginning at one end of gel and working slowly towards the other, lay paper on top of gel. Take care to prevent air bubbles from forming between the paper and gel.

20. Peel blotting paper up off the plate—the gel should come with it. Gradually curl paper (Whatman 3MM) and gel away from plate as it is being pulled away.

21. Place paper and gel on preheated gel dryer. Cover with plastic wrap. Dry gel thoroughly 20 min to 1 hr at 80°C. Peel plastic wrap away.

    Remove any bubbles between plastic wrap and gel by gently rubbing covered surface of gel from middle toward edges with a Kimwipe. When gel is completely dry, the plastic will easily peel off without sticking.

22. Remove plastic wrap on the dried gel and place each dried gel in a separate X-ray cassette with Kodak XAR-5 film in direct contact with gel and autoradiograph at room temperature (*APPENDIX 3*). After sufficient exposure time (usually overnight), remove X-ray film and process.

ALTERNATE PROTOCOL 1

BUFFER-GRADIENT SEQUENCING GELS

In this protocol, the sequencing gel contains a higher concentration of buffer at the bottom of the gel than at the top and as a result, the migration of shorter oligodeoxynucleotides (at the bottom) are slowed down relative to the longer oligodeoxynucleotides (at the top). This allows the gel to be run longer without losing the shorter oligodeoxynucleotides off the bottom and improves the resolution between longer oligodeoxynucleotides.

Additional Materials *(see APPENDIX 1 for items with ✓)*

✓ Buffer-gradient gel solutions (containing 0.5× and 2.5× TBE)
25-ml pipet equipped with a rubber pipet-filler bulb

1. Assemble one gel plate sandwich as in steps 1 to 3 of Basic Protocol.

2. In two 100-ml beakers, prepare buffer-gradient gel solutions: 50 ml using 0.5× TBE (clear solution) and 25 ml using 2.5× TBE (blue solution). Heat gently at ≤50°C while stirring until all solids have dissolved. Cool solutions until they are room temperature (≤25°C). Add 20 µl TEMED and 200 µl of 10% ammonium persulfate to 0.5× TBE/gel solution and mix gently. Add 10 µl TEMED and 100 µl of 10% ammonium persulfate to 2.5× TBE/gel solution and mix gently.

3. Using a 25-ml pipet equipped with a rubber pipet-filler bulb, pull up 12.5 ml of clear 0.5× TBE/gel solution followed gently by 12.5 ml of blue 2.5× TBE/gel solution. Allow three or four air bubbles to be taken up into pipet by gently squeezing the suction inlet on rubber bulb (to cause two layers to gently mix at interface).

4. Release solution down glass plate sandwich in a gentle, even manner. Using the same 25-ml pipet, fill the rest of gel with 0.5× TBE/gel solution. Insert combs, position clamps, and observe polymerization as in step 6 of Basic Protocol.

6. Run and process the gel as in steps 7 to 22 of Basic Protocol.

ELECTROLYTE-GRADIENT SEQUENCING GELS

ALTERNATE PROTOCOL 2

In this protocol, the ionic strength of the bottom of the gel is increased simply by increasing the salt concentration in the bottom buffer chamber. During the run, the salt is electrophoresed into the gel and generates a reproducible and effective gradient.

Additional Materials *(see* APPENDIX 1 *for items with ✓)*

✓ 0.5× and 1× TBE electrophoresis buffer
✓ 3 M sodium acetate (unbuffered)

1. Pour and prerun sequencing gel as in steps 1 to 11 of Basic Protocol. Use 0.5× TBE buffer in top and 1× TBE buffer in bottom reservoir.

2. Prepare and load sequencing samples as in steps 12 and 13 of the Basic Protocol.

3a. If ≤400 bases of sequence information is needed, add 3 M sodium acetate to bottom reservoir to 1 M final. Proceed to step 4.

3b. If ≥400 bases of sequence information is desired, proceed to step 4 and wait 2 to 3 hr after beginning electrophoresis to add 3 M sodium acetate to the bottom reservoir to 1 M final.

4. Run gels at 60 W constant power.

   It will take ~75% longer than usual for dye markers to migrate to the same location on gel. The temperature of gel should be monitored carefully. If gel becomes significantly hotter at top than bottom, reduce the power.

5. Process gel as in steps 14 to 22 of Basic Protocol.

FORMAMIDE-CONTAINING SEQUENCING GELS

ALTERNATE PROTOCOL 3

Formamide is added to the acrylamide gel solution to destabilize the secondary structures of sequencing products that can cause compressions.

Additional Materials *(see* APPENDIX 1 *for items with ✓)*

✓ Formamide gel solution
5% acetic acid/20% methanol (v/v) fixer solution

1. Assemble gel sandwich as in steps 1 to 3 of Basic Protocol.

2. Prepare formamide gel solution according to acrylamide concentration desired. Heat gently with stirring until dissolved, then cool until ≤30°C.

3. Add 0.15 ml TEMED and 1 ml of 10% ammonium persulfate. Immediately pour solution into gel sandwich. Observe polymerization.

   The gel solution is viscous. Hold plates at a nearly vertical angle while pouring solution—this is easiest if solution is in a beaker. The gel should polymerize within 30 min.

4. Run gel 45 to 70 W, constant power.

   This requires a higher voltage (60% higher) than nonformamide-containing gels. The DNA migrates about half as fast.

5. Process and dry gel as in steps 13 to 22 of Basic Protocol, except fix gel 15 min in 5% acetic acid/20% methanol fixer solution in step 18.

Reference: Slatko, 1991.

Contributors: Barton E. Slatko and Lisa M. Albright

Computer Manipulation of DNA and Protein Sequences

The ability to determine DNA sequences is now commonplace in many molecular biology laboratories. As the amount of DNA sequence data available to researchers has increased, the use of computers to manipulate, compare, and analyze this data has grown accordingly. This unit outlines a variety of methods by which DNA sequences can be manipulated by computers.

The Appendix to this unit, located at the end of the unit, lists the addresses and phone numbers of all software and hardware vendors discussed below. Additional information about relevant journals and databases is also provided in the Appendix.

SEQUENCE DATA ENTRY

To begin analysis of a DNA sequence, the information must be in a form that computers can understand. Generally this requires that the DNA sequence information be contained in a file (a file on the computer is simply a collection of information in computer-readable form). For small sequencing projects, DNA sequence data can be easily entered and manipulated using a word processor or text editor on a microcomputer such as an Apple Macintosh or IBM-compatible. As the size of a project increases, however, a specialized editor for data entry becomes more and more desirable. All software packages for DNA sequence analysis provide some form of sequence editor—a program that acts like a word processor for sequence data.

Currently there is no standard format for DNA or peptide sequence files. In the early days of sequence analysis, most software programs had their own unique formats for storing sequence information. Although these format differences still exist today (see Fig. 7.7.1), most available software can import several different sequence formats.

Manual Entry Using Word Processors and Sequence Editors

Word processors. A DNA or peptide sequence can be entered into a computer using a word processor or text editor program, simply by creating a new document and then typing in the sequence data. It is imperative that DNA or peptide sequence documents be saved in "text" or ASCII format because sequence analysis programs cannot translate the default files used by word processor programs; however, most packages provide an easy one-step method of reformatting text files containing DNA or peptide sequences into a format appropriate to the computer at hand.

Sequence files often contain more information than just the DNA or peptide sequence data. Comments and reference information about the sequence and the history and dates of changes made to the sequence are commonly included. Such comments must be distinguishable by the program from the sequence data in order for the sequence analysis software to function properly. The most commonly used file formats are shown in Figure 7.7.1. Note that some—e.g., the Intelligenetics and DNA Strider text formats—use specific characters to identify the comments before the sequence file; other formats simply require a string of characters that is used to denote the end of the comments and the beginning of the sequence—e.g., the GenBank flatfile format uses a line beginning with "ORIGIN" and GCG uses ".." (two periods). The NBRF software package (also known as PIR) allows only a single line of comments. Some formats (e.g., NBRF and Intelligenetics) require a specific terminator character for the sequence entry.

```
A) EMBL:
   ID hummycc.primate
   DE hummycc.primate
   SQ            100 BP
   AGCTTGTTTGGCCGTTTTAGGGTTTGTTGGAATTTTTTTTTCGTCTATGT
   ACTTGTGAATTATTTCACGTTTGCCATTACCGGTTCTCCATAGGGTGATG
   //
B) GenBank:
   LOCUS           HUMMYCC     10996 bp ds-DNA      PRI      25-JUL-1994
   DEFINITION      Human (Lawn) c-myc proto-oncogene, complete coding ...
   ACCESSION       J00120 K01908 M23541 V00501 X00364
   KEYWORDS        Alu repeat; c-myc proto-oncogene; myc oncogene; ...
   SOURCE          Human DNA (genomic library of Lawn et al.), clones ...
      ORGANISM     Homo sapiens ...
   REFERENCE       1 (bases 3507 to 7559)
      AUTHORS      Colby, W.W., Chen, E.Y., Smith, D.H., and Levinson, A.D.
      TITLE        Identification and nucleotide sequence of a human ...
      JOURNAL      Nature 301 (5902), 722-725 (1983)
      MEDLINE      83141777
   COMMENT         The myc gene is the cellular homologue of the ...
   FEATURES          Location/Qualifiers
   BASE COUNT      2747 a   2723 c   2733 g   2793 t
   ORIGIN          198 bp upstream of Sau96A site, on chrmosome 8 (q24)
                   1 agcttgtttg gccgttttag ggtttgttgg aattttttttt tcgtc ...
   //
C) GCG:
   Human (Lawn) c-myc proto-oncogene, complete coding sequence and flanks.
      hummycc.primate  Length: 100  Sun, Mar 17, 1991  9:48 PM  Check: 6864  ..
          1  AGCTTGTTTG GCCGTTTTAG GGTTTGTTGG AATTTTTTTT TCGTCTATGT
         51  ACTTGTGAAT TATTTCACGT TTGCCATTAC CGGTTCTCCA TAGGGTGATG

D) Intelligenetics:
   ; hummycc.primate, 100 bases.
   ;Human (Lawn) c-myc proto-oncogene, complete coding sequence and flanks.
   ;
   hummycc.primate
   AGCTTGTTTGGCCGTTTTAGGGTTTGTTGGAATTTTTTTTTCGTCTATGT
   ACTTGTGAATTATTTCACGTTTGCCATTACCGGTTCTCCATAGGGTGATG1

E) NBRF:
   >DL;hummycc.primate
   hummycc.primate, 100 bases.
          1  AGCTTGTTTG GCCGTTTTAG GGTTTGTTGG AATTTTTTTT TCGTCTATGT
         51  ACTTGTGAAT TATTTCACGT TTGCCATTAC CGGTTCTCCA TAGGGTGATG*

F) DNA Strider Text:
   ; DNA sequence  hummycc.primate, 100 b.p.
   ;
   AGCTTGTTTGGCCGTTTTAGGGTTTGTTGGAATTTTTTTTTCGTCTATGT
   ACTTGTGAATTATTTCACGTTTGCCATTACCGGTTCTCCATAGGGTGATG
   //
G) FASTA:
   >hummycc Human (Lawn) c-myc proto-oncogene
   AGCTTGTTTG GCCGTTTTAG GGTTTGTTGG AATTTTTTTT TCGTCTATGT
   ACTTGTGAAT TATTTCACGT TTGCCATTAC CGGTTCTCCA TAGGGTGATG
```

Figure 7.7.1 Commonly used sequence file formats. Different formats have specific defined elements and defining codes. (**A**) EMBL comment lines begin with two-letter codes: ID, denoting short sequence name; DE, denoting description; and SQ, denoting sequence length. DNA or protein sequence follows; sequence end is denoted by two slashes (//) on a separate line. (**B**) GenBank comments and sequence are separated from it by the code "ORIGIN"; material before "ORIGIN" is comment and material after "ORIGIN" is sequence. Sequence end is denoted by two slashes on a separate line. The actual text ot this entry has been abbreviated due to space limitations. (**C**) GCG comments and sequence are separated from it by two dots (..); material before dots is considered comments and material after dots is sequence. (**D**) Intelligenetics comment lines begin with semicolons (;). A single description line follows, and then the sequence begins on a separate line. Sequence end is denoted by a numeral one (1). (**E**) NBRF (sometimes called PIR format) first line starts with four required characters: a greater-than sign (>); either "D" for DNA or "P" for protein; either "L" for linear or a "C" for circular; and a semicolon. The short sequence name follows immediately on the same line. The next line is a description line. Sequence starts on a new line and end of sequence is denoted by an asterisk (*). (**F**) DNA Strider Text is similar to the Intelligenetics format, but lacks the description line. (**G**) FASTA (sometimes called Pearson format) first line begins with a greater-than sign (>), followed by the sequence name and a short description. Sequence data then starts on a separate line. *Note:* Some formats (including GenBank, GCG, and NBRF) allow numbers to be included within the sequence for ease of reading (the numbers are ignored during sequence analysis).

It is a good idea to test the file format before investing a large amount of time typing in data with the word processor. This can be done by creating a small DNA sequence, then experimenting with the comment and sequence formats as well as the word processor's Save option. The resulting file can then be checked in the analysis program.

Sequence editors. All analysis programs include some form of sequence editor. This is the most effective tool for entering DNA or peptide sequence data because it produces a sequence file in the correct format with the comments separate from the sequence as they should be. DNA Strider (Macintosh) is built around a very easy-to-use sequence editor. It also makes restriction maps, finds open reading frames, and translates DNA sequences into amino acid sequences.

Recently, several programs have become available for microcomputers that "speak" the sequence as it is entered. This is quite useful for entering data by hand from an autoradiogram. The audio feature permits verification of the sequence. SeqSpeak is designed for Macintosh equipped with HyperCard version 2.0 and distributed free of charge (see Table 7.7.2).

Semiautomated Entry Using Digitizing Hardware and Software

Several sequence analysis software packages can accommodate the connection of a relatively inexpensive digitizing pad to the computer; a digitizing pad permits DNA sequences to be read directly from an autoradiogram. The result of the digitized entry process is a file containing the newly determined DNA sequence. In some cases the file must be reformatted before it can be used with an analysis program—this can be determined by comparing the format defined by the entry software with the list of acceptable formats in the sequence analysis software.

DNA sequence entry using a digitizing program can be very quick and accurate when the gel is of good quality. However, if the lanes are irregular due to curving or excessive smiling, the program may incorrectly identify which lane contains a selected band. Commercial gel-reading programs are designed to handle some types of common gel problems. The SEQED program of the GCG package for UNIX and VMS and the DNAStar package for IBM-compatibles and Macintosh can both be used with a digitizing pad for semiautomatic DNA sequence determination.

Automated Entry Using Gel Readers and Automated Sequencers

Automated gel readers. Several automated autoradiogram readers are now available that include a scanner and a computer. Automated readers use a high-resolution gray-scale scanner or a digital video camera to digitize an autoradiogram of a sequencing gel produced by conventional methods. The digitized image of the film is then analyzed by the computer using image-analysis techniques. The software is designed to identify first the locations of lanes and then bands within the lanes. This process can be confounded by smudges and random spots on the original film, as well as by gel smiling and curved lanes. Indeed, a major challenge to designers of automatic gel-reading programs is the elimination of such noise in scanned autoradiograms. Automated gel-reader systems (sold by Life Technologies and Milligen, among others) are generally priced for use by large laboratories and core facilities.

When using automated image analysis, it may be advantageous to employ the multiplex method of DNA sequencing (introduction to Chapter 7 and *UNIT 15.2*), because it is then possible to incorporate a known sequence of DNA into each gel as an internal standard.

Automated DNA sequencers. Automated DNA sequencing machines (discussed in *UNIT 7.0*) determine the sequence of a DNA fragment and then place the DNA sequence into a file. The most common automated sequencers use special fluorescence-tagged oligonucleotide primers in a dideoxy primer-extension reaction. The output from the sequencer is a trace of the amount of fluorescence observed as the tagged reaction products pass off the gel. An attached computer analyzes the trace and determines the DNA sequence it represents. Thus, entry of sequence data is completely automated; the user merely has to define or identify the particular sequencing run. Automated sequencers are quick and currently provide about the same level of accuracy as manual sequencing.

Editing automated sequence data. Because automated autoradiogram scanners are not absolutely accurate, the software packages that accompany them typically permit the user to edit the automatically entered sequence. A sequence-assembly program can then be used to align the sequences based on overlapping regions. Automatic sequencing machines (such as those produced by Pharmacia Biotech, Li-Cor, and Applied Biosystems) provide a similar editing feature whereby the user can view the fluorescent trace on a computer screen and override the computer's choice of base if it appears to be incorrect. Some vendors of automated sequencing machines use a statistical means of correcting for discrepancies between regions that have been sequenced multiple times. That is, if a particular region is sequenced several times and the number of differences is low, the machine will automatically adopt the majority consensus sequence.

SEQUENCE DATA VERIFICATION AND ASSEMBLY

As with any experimental result, it is important to verify DNA sequence data. This is typically accomplished by sequencing a given region more than once and by sequencing both strands. The computer can be used to compare sequences, find overlapping regions, and highlight differences.

Comparison of Multiple Entries

To identify errors introduced during manual or automated data entry, each gel can be read more than once and the independent readings compared. The comparison can be carried out automatically using an alignment program (see Homology Searching) and any differences can then be investigated. Similarly, several of the digitizing gel readers provide a confirmation option in which each segment of a sequence can be checked automatically, with the machine simply re-entering the band locations. These programs, such as SEQED (included in the GCG package; see Table 7.7.1), alert the user as differences occur. In this way the entry process can be quite fast without sacrificing accuracy.

For large sequencing projects that involve sequencing a number of random clones from a library covering the region of interest, it is usually possible to achieve a sufficient level of redundancy that a majority consensus sequence can be determined.

Comparison to Known Restriction Maps

If a restriction map for the DNA region of interest already exists, a restriction map generated from the DNA sequence (as described below) can serve as a check on the accuracy of the sequence.

Test Translations to Detect Shifts in Reading Frame

One of the most widespread features of sequencing programs is the capacity to translate a DNA sequence into amino acids, thereby generating a putative protein sequence from an open reading frame (ORF); many restriction mapping programs also include a translation feature. Translating the sequence in this fashion provides a simple check for deletions and insertions. Some software is able to take into account variant genetic codes; for instance, DNA Strider allows the user to select from a list of variant codes, and the GCG package allows the user to reset individual codons. Although this technique will detect only a subset of possible errors, it can often quickly identify common problems such as simple typing mistakes or miscounting of the number of the same nucleotides in a run of identical nucleotides.

Most DNA sequence packages allow the user to specify a range and reading frame to be used in translating a DNA sequence. DNA Strider (Table 7.7.1) has a useful feature that hunts for ORFs in a DNA sequence and highlights the putative coding sequence, thus doing all the necessary work.

Detecting Overlap with Other Sequenced Fragments

With increasing worldwide interest in genome sequencing projects, sequence assembly packages now provide very effective automatic DNA sequence assembly, connecting shorter pieces of DNA to build the longest continuous sequence possible. Once the sequences to be assembled are identified, they are compared, overlaps identified, and contiguous sequences (contigs) constructed. Typically, parameters are available to allow adjustment of the alignment process. A good commercial assembly program is LaserGene, available from DNAStar for both Macintosh and IBM-compatible computers (Table 7.7.1). Once a sequence contig has been assembled, the LaserGene program creates a graphic overview of the sequencing project, highlighting regions that need further verification. Another commercial program is Sequencher (Gene Codes). This program provides expanded features for the assembly, processing, and editing of DNA sequences determined with the ABI sequencer; however, it is only available for the Macintosh computer.

If a sophisticated assembly program is not available, contigs can be constructed "by hand" using a comparison program and a multiple sequence editor. Both the input DNA sequence and its complement should be considered in assembling contigs. When using a program that does not automatically consider the complement sequence, conduct a separate search of that sequence.

When conducting sequence comparisons to identify overlaps, it is important to take into account the fact that different programs take slightly different approaches to this process. Some comparison programs that use the Needleman and Wunsch algorithm (e.g., the GCG GAP program) are designed to find the maximum number of matches between two sequences over the entire length of the two sequences, with the minimum number of gaps. This type of program is best suited for aligning two sequences along their entire lengths. If the two sequences differ greatly in size, however, the result may not satisfactorily represent the similarity between them. For identifying regions of similarity between two pieces of DNA that are largely different in sequence or of significantly different lengths, programs employing algorithms such as those described by Smith and Waterman or Wilbur and Lipman—e.g., the GCG BESTFIT program—are more suitable. These algorithms do not try to align the two sequences being compared in their entirety, but instead search for short matches within the sequences.

Editing a Contig and Verifying the Sequence

Generally, software packages that provide sequence assembly programs include multiple sequence editors that display the individual sequences of the aligned contig together one on top of the other, one sequence per line (see Fig. 7.7.2), and can generate the consensus sequence automatically. The most significant feature of a multiple sequence editor is its ability to produce a consensus sequence for an aligned set of sequences. If the consensus sequence is not satisfactory, the alignment can instantly be changed and a new consensus sequence generated.

Another use of a multiple sequence editor is for manually comparing a group of overlapping sequences by aligning common regions vertically on the screen (Fig. 7.7.2), which makes it easy to identify differences in the aligned sequences. It is generally useful to go back to the original gel to determine whether these differences result from misreading, sequence compression, or gel defect. Multiple sequence editors will often have the ability to display the reverse complement of a particular sequence, irrespective of which strand was sequenced.

Some very useful programs are available free of charge; these programs provide automatic multiple sequence alignment functions as well as other types of analysis on a set of sequences. Examples of these are the MACAW software for Microsoft Windows or Macintosh, developed by Greg Schuler of the NCBI, and GDE for X-Windows, developed by Steven Smith of the Harvard Genome Laboratory (Table 7.7.2).

RESTRICTION MAPPING

Once a sequence contig is generated, a map of restriction endonuclease cleavage sites can be a useful aid in further analysis of the region. As stated above, construction of a restriction map based on newly determined sequence information allows rapid visual comparison of the sequence with a known restriction map. Restriction maps also identify sites for subcloning or other molecular genetic manipulations and can provide useful summaries of newly determined DNA sequences.

Mapping All Known Commercial Restriction Enzyme Recognition Sites

Programs are available that identify on a strand of DNA the sites of all known restriction enzymes, according to either the first nucleotide of the recognition site or the location of cleavage. Most of these programs produce one list of the sizes of fragments that would be created by cutting with each restriction enzyme and a separate list of the recognition sites identified. They can also display the cleavage sites of specified sets of restriction enzyme of interest, such as those that generate

```
    TACCTCAGCCAGCATGGCAGCCTCTTTCCCACCCACCTTGGGACTCAGTTCTGCCCCAGATGAAATTCAGCACCC
AAGGTACCTCAGCCAGCATGGCAGCCTCTTTCCCACCCACCTTGGGACTCAGTTC
AAGGTACCTCAGCCAGCATGGCAGCCTCTTTCCCACCCACCTTGGGACTCAGTTC
AAGGTACCTCAGCCAGCATGGCAGCCTCTTTCCCACCCACCTTGGGACTCAGTTCTGCCCCAGATGAAATTCAGCACCC
....|.........|.........|.........|.........|.........|.........|.........|....
   100       110       120       130       140       150       160       170
 7
 6
 5A                               +--------------------->
 4A                   +--------------------->
 3A   +---------------*---->
 2A   +--------------------->
 C    +------------------------------------------------->
      |......|......|......|......|......|......|......|......|......|......|
      0     50     100    150    200    250    300    350    400    450    500
```

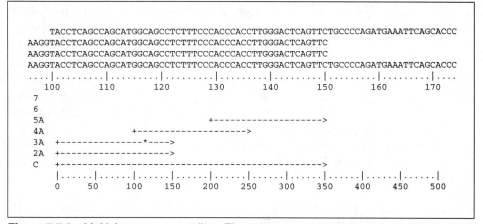

Figure 7.7.2 Multiple sequence editor. The GCG program GELASSEMBLE displays the aligned sequences on the top of the screen and a schematic of the sequenced fragments on the bottom. Arrows indicate the direction of sequencing; the asterisk in the lower part of the display indicates the position of the cursor in the sequence alignment as the user edits the sequence.

3′- or 5′-strand overhangs or blunt ends. All restriction mapping programs contain a file or files in which restriction site data are stored; this information can be updated as new restriction enzymes and their cleavage sites are identified. Restriction mapping programs are one of the most common types of molecular biology software; virtually all commercial software provides excellent restriction mapping features.

An up-to-date list of restriction enzymes is maintained by Richard Roberts (New England Biolabs). It is accessible in the form of a text file called rebase and contains all known type II restriction enzyme recognition sites, sites of cutting, a complete cross-reference of isoschizomers, reference citations, and commercial sources for the restriction enzyme with addresses and phone numbers. The latest version of rebase can be obtained free from a variety of electronic mail (e-mail) and network servers.

Mapping to Predict Band Sizes

Some software programs can model double digests, predicting the band sizes that would result from cleavage of an entered sequence with a given pair of restriction enzymes. Such double-digest band-size analyses simplify the reading of restriction fragment patterns on gels and can also be useful for planning subsequent cloning strategies. A few programs can also predict the results of partial digests. Software packages that provide restriction fragment pattern analysis features are available from Textco's Gene Construction Kit (for Macintosh only) and Pro-RFLP from DNA ProScan (for Macintosh and IBM-compatible computers).

Graphical Restriction Mapping

Several very elegant programs are available that produce graphical restriction-site maps that are useful for searching for possible sites to be used in further recombinant DNA manipulations. Graphical restriction maps are often represented as collections of horizontal lines, one for each restriction enzyme, with the recognition sites represented by short vertical marks at the appropriate locations (Fig. 7.7.3). A limited number of programs produce pictures of sequences that look like standard circular plasmid maps, including coding regions and other features of interest within the sequence, as well as the vector DNA and polylinker. These graphical maps can be saved and then manipulated in a graphics editor or drawing program, greatly simplifying the task of preparing figures for presentation; an excellent example of such a program is the Textco Gene Construction Kit. This program reads a variety of sequence-file formats and generates a picture of the DNA sequence in standard restriction map form. The pictures can be edited to include arrows and information boxes that are of publishable quality. In contrast to most graphical mapping programs, the Gene Construction Kit and other commercially available programs such as MacVector and LaserGene use DNA sequence data directly to generate the pictorial maps. Many shareware mapping programs, such as Jingdong Liu's MacPlasMap, do not work with the sequence data directly; rather, recognition site locations must be input by the user. One shareware program that does have the capacity to create graphical maps directly from sequence data, and provides many analysis features as well, is DNA Strider, developed by Christian Marck (Table 7.7.2).

Recent versions of many mapping programs can actually simulate the electrophoretic-gel banding patterns that will be observed following particular restriction digests. The programs allow the user to specify the type of gel medium being used, making it possible to determine visually whether the banding pattern observed on a gel could be produced by a given sequence. Some programs will even simulate

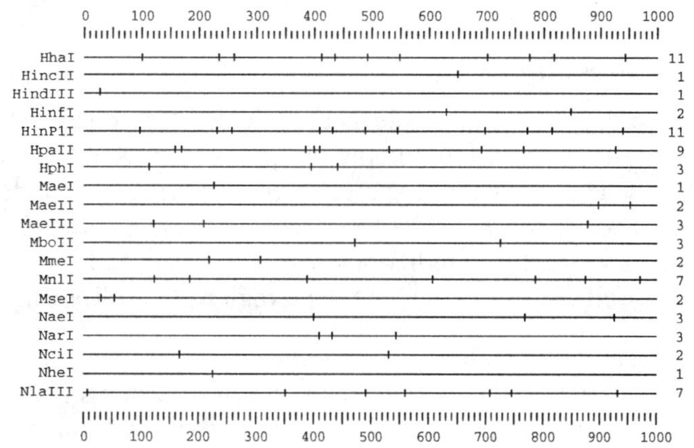

Figure 7.7.3 One type of graphical restriction map. This figure was produced by the free PlotZ program; GCG MAPPLOT produces similar output.

partial digests. The Gene Construction Kit provides the gel simulation feature as does the ACEDB genome database software (see Genetic Sequence Databases); the latter is limited in that it can simulate the gel banding patterns only for sequences already contained within the database being used.

PREDICTION OF NUCLEIC ACID STRUCTURE

Once the DNA sequence of a region has been identified, a number of analyses can be performed to identify interesting features such as repeats, areas of atypical base composition, and RNA secondary structure. These in turn can help to define functional regions within the sequence of interest.

Base Repeats

Direct and inverted repeats are often part of transcriptional or translational control regions. Most sequence analysis software can identify repeats and provide an optional graphic display of their location. The standard "dot matrix" plot is a simple and effective method of identifying repeated regions as diagonal lines. Some dot matrix programs show inverted repeats—which may indicate potential stem-loop structures in the corresponding DNA or RNA—as lines with negative slopes relative to direct repeats. To identify repeats within a single sequence, the same sequence is used for both the *x* and *y* axes.

Analysis packages that do not produce a graphical presentation of the repeated regions usually include a program that lists the repeats and their locations. This involves searching a sequence against itself to find direct repeats or against its complement to detect inverted repeats. The repeats must then be plotted by hand or the distances between them calculated to determine if a periodicity is present.

GC and AT Content

The GC/AT content of a sequence may provide some insight into structural features such as Z DNA and bent DNA. GC/AT content can also serve as a reasonable indicator of a coding region in many invertebrate and plant species. A few sequence-analysis packages contain specialized programs, such as GCG STATPLOT, that show the GC and AT content of a sequence in graphical form—allowing AT- or GC-rich regions to be readily identified. However, many packages provide little more than a tally of nucleotide composition, i.e., the program only lists the number of A, G, C, and T residues in a sequence.

RNA Secondary Structure

Although folding programs are available that predict RNA secondary structure, this type of analysis is still an art. RNA-folding programs help identify possible stable stems in an RNA molecule, but a trial-and-error process is required to determine the biological significance of these results for a given RNA molecule. Even with this limitation, secondary structure predictions can be useful for identifying mRNA control regions as well as possible stable folded regions of an RNA molecule.

The greatest problem with predicting secondary structure is modeling the interactions present in a tertiary structure and then relating those back to the primary sequence for use in a folding program. Indeed, current RNA-folding programs do not take into account possible tertiary structures of a nucleic acid molecule. These programs determine the energetics of a limited number of two-dimensional folded structures. The most stable structure predicted by the program in a two-dimensional world may be far from the most stable structure in three dimensions where loops can interact with loops, helical regions can stack, and various non-Watson-Crick base-pairing structures can occur.

Currently, the most sophisticated RNA-folding program is MFOLD (for Multifold, an extension of an earlier program known as RNAFold or FOLD in the GCG package), designed by Michael Zuker of the National Research Council of Canada (Table 7.7.2). In addition to standard analysis of base-pairing energetics, MFOLD takes into account base-pair-stacking energies and single-base-stacking (dangle) enthalpies. Another major feature of MFOLD is that it also depicts many suboptimal structures. VMS, UNIX, DOS, and Macintosh versions of this program are available from many of the software archives (see Table 7.7.2). Although the output of MFOLD is text-based (Fig. 7.7.4A), several programs are available that generate graphic representations of the predicted structure (e.g., LoopViewer, developed by Don Gilbert; see Fig. 7.7.4B).

OLIGONUCLEOTIDE DESIGN STRATEGY

Increased use of polymerase chain reaction (PCR) methods has stimulated the development of many programs to aid in the design or selection of oligonucleotides used as primers for PCR. Four such programs that are freely available via the Internet (see Table 7.7.2) are: PRIMER by Mark Daly and Steve Lincoln of the Whitehead Institute (UNIX, VMS, DOS, and Macintosh), Oligonucleotide Selection Program (OSP) by Phil Green and LaDeana Hiller of Washington University in St. Louis (UNIX, VMS, DOS, and Macintosh), PGEN by Yoshi (DOS only), and Amplify by Bill Engels of the University of Wisconsin (Macintosh only). Generally these programs help in the design of PCR primers by searching for bits of known repeated-sequence elements and then optimizing the T_m by analyzing the length and GC content of a putative primer. Commercial software is also available and primer selection procedures are rapidly being included in most general sequence analysis packages.

Sequencing and PCR Primers

Designing oligonucleotides for use as either sequencing or PCR primers requires selection of an appropriate sequence that specifically recognizes the target, and then testing the sequence to eliminate the possibility that the oligonucleotide will have a stable secondary structure. Inverted repeats in the sequence can be identified using

A

```
                              10                      20
---------------G        --C     ---------  UUC      G
               AGUUGU     UGCC            GU    GGGUC C
               UCGACA     ACGG            CA    CCCAG U
AUUAUGCUGAGUGAUA      UUU    GUUGCUGAU  --U       U
 90          80        70          60          30

                                            40
                                             UACA
                                        UCCCU     U
                                        AGGGA     U
                                             CAUC
                                           50
```

B

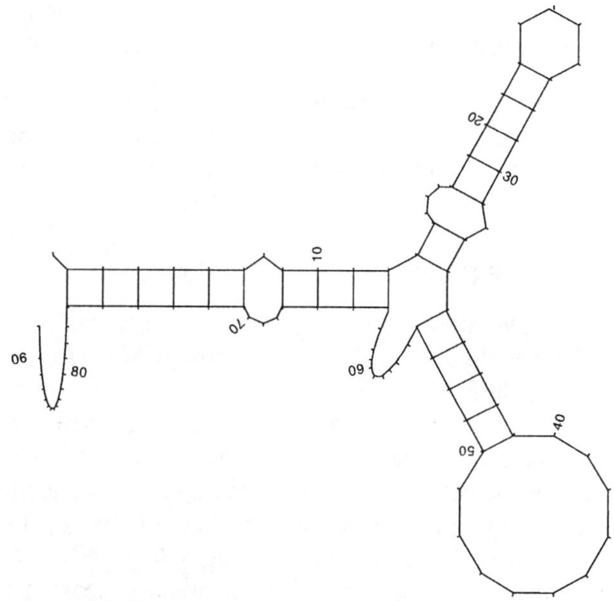

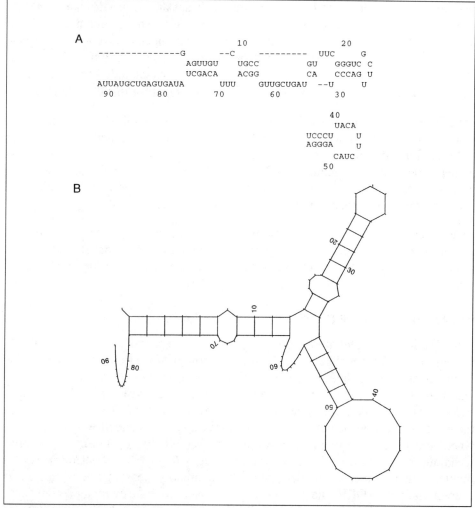

Figure 7.7.4 (**A**) Text-based output from Zuker's RNA-folding program, available from GCG under the name of FOLD. This type of representation is difficult to visualize, but acceptable when only a quick view of the possible folded structures is desired. (**B**) Graphic representation of the structure shown in part A, produced by the GCG Squiggles program. The free LoopViewer program (for Macintosh) produces similar representations.

a repeat-identification or RNA-folding program such as those described above (see Prediction of Nucleic Acid Structure). If a possible stem structure is observed, the sequence of the primer can be shifted a few nucleotides in either direction to minimize the predicted secondary structure. The sequence of the oligonucleotide should also be compared with the sequences of both strands of the appropriate vector and insert DNA. Obviously, a sequencing primer should only have a single match to the target DNA. It is also advisable to exclude primers that have only a single mismatch with an undesired target DNA sequence. For PCR primers used to amplify genomic DNA, the primer sequence should be compared to the sequences in the GenBank database to determine if any significant matches occur. If the oligonucleotide sequence is present in any known DNA sequence or, more importantly, in any known repetitive elements, the primer sequence should be changed.

Degenerate Probes for Detecting Related Genes

Once a conserved protein sequence has been identified, a degenerate oligonucleotide can be designed for use as a hybridization probe to screen a library to identify additional members of the protein family (see *UNIT 6.4*). To design this oligonucleotide, the conserved protein sequence must be translated into a degenerate DNA sequence. Most software packages provide this feature; their output is a DNA sequence that is produced using the IUPAC degenerate-nucleotide codes. A degenerate oligonucleotide is then synthesized to correspond to the back-translated protein sequence. Most DNA synthesizers will create an oligonucleotide with more than one nucleotide at any position within the sequence except for the 3′ nucleotide.

For efficiency of synthesis and hybridization, the following guidelines, designed to yield oligonucleotides with the lowest possible level of degeneracy, should be followed. First, it is not necessary to incorporate both G and A to match a consensus sequence position containing C and T, because G pairs with both C and T. Second, inosine (I) pairs with G, C, and A. Third, a pyrimidine-pyrimidine mismatch does not disrupt base pairing, but a purine-purine mismatch is destabilizing.

As a general rule, create the minimum length of sequence that hybridizes to the consensus sequence. For each species (unique sequence) of oligonucleotide in the synthesis, the concentration of oligo used in hybridizations must be increased to achieve an equivalent C_0t value.

IDENTIFICATION OF PROTEIN-CODING REGIONS

Identification of potential protein-coding regions, especially in genomic DNA sequences from higher eukaryotes, is still not a completely automated process. It is helpful to simply translate the region in all six reading frames (three on each strand) and then identify all possible exon regions as uninterrupted open reading frames (ORFs). Although identifying the AUG initiation codon is a simple task, determining the location of introns is not as straightforward (unless the sequence is from yeast, where the rules for splicing are simple and seemingly absolute). One useful technique available with some software packages (e.g., GCG TestCode) uses a purely statistical method to determine the nonrandomness of the triplet code characteristic of an ORF. The basic principle is that introns are evolving without any restraint and are thus more random in sequence than exons, which are subject to stabilizing selection.

Prediction of ORFs is a very active research area. The rules used by the cellular machinery to define splice sites for higher eukaryotic sequences are still illusive. Four projects are trying to develop truly automatic gene-identification methods. The first, Gene Recognition and Analysis Internet Link (GRAIL), is available via an electronic mail (e-mail) server and as a UNIX application; it is being developed by a group led by Edward Uberbacher of the Oak Ridge National Laboratory. GRAIL utilizes an artificial intelligence technique called a neural network that learns by example. The GRAIL software is given a set of well-characterized human sequences for which the locations of exons and introns have been experimentally identified. The neural network is programmed to search for particular types of simple features and then to correlate these features with the input set's exon-intron boundaries.

In contrast to GRAIL, the second program, GeneID, developed by Steen Knudsen and Kathleen Klose of Temple Smith's group at Boston University, utilizes many features of coding regions, including exon/intron consensus sequences and codon preferences. Because its rules are generally based on human sequences, GeneID's usefulness with nonmammalian sequences may be limited. GeneID is available from BMERC at Boston University Bio Molecular and Engineering Research Center via an e-mail server (geneid@darwin.bu.edu).

The third gene-finding resource, BCM Gene Finder, is provided by the Baylor College of Medicine via e-mail and the World Wide Web (see Appendix). A set of analysis programs is available to aid in the identification of genes in human DNA sequences. The analysis method involves predicting all possible internal exons based on the combination of characteristics describing a potential splice site. Then the set of potential exons is analyzed to determine the optimal combination and a model for the putative gene is constructed.

The final service was announced in 1992 and like G uses a neural network program to identify coding regions. The NetGene service is available through an e-mail server and is provided by the Department of Physical Chemistry at the Technical University of Denmark. This server appears to be changing the slowest, but because no one has yet produced the definitive gene-finding software, it is suggested that all of these servers should be tried to determine which resource is the most useful.

HOMOLOGY SEARCHING

Searching for homology between a newly obtained sequence and a sequence already listed in one of the DNA or protein databases can be very informative. Similarity to a known sequence can suggest the function of the new protein or indicate that no similar sequence has yet been deposited in the database. Because of the size of the databases and the speed with which they are expanding, the task of searching the database is not always easy to accomplish using an isolated laboratory microcomputer. Searching sequence databases for similarities is one of the few sequence-analysis tasks that is still best performed on a larger computer system. To search a single sequence against the entire GenBank, European Molecular Biology Laboratory (EMBL), or DNA Data Bank of Japan (DDBJ) databases requires about an hour on a smaller VAX or microcomputer. For this reason, many laboratories obtain an account on a large computer system or modern workstation that provides access to the large genetic sequence databases. Several sources, including the European Bioinformatics Institute (EBI), the National Center for Biotechnology Information (NCBI), and the University of Houston, provide free databases and homology searches over the Internet to anyone with access to e-mail.

Comparison of Two Sequences

Many programs are able to align two DNA or protein sequences. Such programs are often used to format an alignment for publication or simply to identify regions of similarity between two input sequences. In addition, these programs also introduce gaps into a sequence to optimize the alignment. Because alignment programs assign numerical penalties to gaps and mismatches, the alignment can be influenced by varying gap and mismatch parameters.

Protein sequences are aligned using a scoring matrix developed by Margaret Dayhoff known as PAM250, which represents the evolutionary change that takes place in a protein sequence over time. A PAM (rearranged acronym for Accepted Point Mutations) is a measure of the number of individual amino acid changes occurring per 100 amino acid residues as a result of evolution. The so-called PAM250 log odds matrix, the log of the probability that a given amino acid could mutate in the evolutionary time equal to 250 PAM, was created by analyzing many families of protein sequences that were available in the late 1970s. The PAM250 matrix is used to determine the score of an aligned pair of sequences by summing the matrix values corresponding to each aligned pair of amino acids. Some amino acid changes have positive effects on the alignment score, and others have negative effects. The larger the number, the better the match. The Gonnet tables are the first recalculation of the PAM tables since their initial formulation. Subsequently a new amino acid substitution matrix was derived directly from sequence or three-dimensional structural alignments of distantly related proteins. This matrix, called BLOSUM62, is reported to perform better than the previous matrix at detecting distant relationships using either BLAST or FASTA. The BLAST servers at the NCBI use the BLOSUM62 matrix by default for protein sequence comparisons.

Comparison of Multiple Sequences

Programs for simultaneously aligning multiple sequences require more computer power and are less common than the alignment programs discussed above. Three standard multiple-alignment algorithms have been developed for UNIX and VMS-based systems: Des Higgins' Clustal (see Table 7.7.2), the Feng and Doolittle algorithm implemented in the GCG PILEUP program (see Table 7.7.1), and PIRAlign from NBRF, which is based on a variation of the Needleman-Wunsch algorithm (see Table 7.7.1). A powerful multiple-alignment tool for Microsoft Windows and Macintosh is the MACAW program created by Greg Schuler. These programs identify common regions (or segments) in all the sequences that have been input, and then use the common regions as a starting point for building an overall alignment. They generally work best if the extent of the sequences being aligned is limited to the regions that are conserved among them.

Database Searches

Most researchers currently carry out both DNA and protein database searches over the Internet using the program Basic Local Alignment Search Tool (BLAST), developed by NCBI. The BLAST family of programs allows rapid similarity searching of nucleic acid or protein databases. The basic BLAST algorithm is used in several different programs that are each specific for a particular database and a particular type of input sequence. BLASTN is used to search a nucleic acid sequence against a nucleic acid database, BLASTP is used to search an amino acid sequence against a protein database, and TBLASTN is used to search an amino acid sequence against a nucleic acid database. In the latter program, the database is translated in all six reading frames prior to the search. The converse analysis is performed by BLASTX, which takes a nucleic acid input sequence and translates it in all six reading frames before searching against a protein-sequence database. Finally, if BLASTP does not identify significant sequence similarities, a more extensive program, BLAST3, is available that also compares an amino acid input sequence against a protein database. Like BLASTP, BLAST3 identifies regions of similarity between the input sequence and sequences present in the database, but the initial search is at lower stringency. This search produces a collection of pairwise matching sequences that are then compared to each other, resulting in three-way matches where the component two-way matches were not significant. BLAST3 can be useful in identifying divergent members of a common gene family.

If the sequence of interest contains a protein-coding region, it is more informative to search the predicted protein sequence against one of the protein databases than to search the DNA sequence against a nucleic acid database. Because protein sequences evolve at a slower rate than DNA sequences, a distant homology between protein sequences may be missed at the DNA level. If no coding region has been defined, BLASTX can be used to translate the DNA sequences in all six reading frames before searching against a protein database. Because the protein-sequence databases only contain identified proteins, it is also important to check a newly defined amino acid sequence, or translated DNA sequence, against the current GenBank, EMBL, or DDBJ DNA sequence databases using TBLASTN. Such a search may identify a significant similarity to a DNA sequence that was not previously known to encode a protein.

An important feature of BLAST is a statistical significance score of the reported match. This statistical significance score is determined using an implementation of Karlin's significance formula, which calculates the Poisson probability that the observed sequence similarity will occur by chance based on the size and composition of the sequence database as well as on the size and quality of the match.

Another frequently used program for searching both protein and DNA sequence databases is FASTA, an updated version of the FASTN and FASTP programs. FASTA, which has been included in many commercial analysis packages, uses an initial fast search through the database to identify sequences with a high degree of identity to the test sequence. This fast search is performed by limiting the search to short regions of identity between the test sequence and the database.

FASTA builds a list or dictionary of all possible sequences of the size specified by the k-tuple value. The word size (or k-tuple) parameter used in this program is the size of the initial match (sequence segment) that is used. The size of the k-tuple can be varied and indirectly affects the speed and sensitivity of the initial search through the database. The test sequence and all sequences in the database are then processed to find the locations of all segments in the sequence of a length equal to the k-tuple value that are present in the dictionary. The dictionaries of the two sequences can be more quickly compared than the sequences themselves, allowing for efficient identification of regions that contain small similarities. Once a list of the highest-scoring sequences is produced using the initial fast search, a second comparison is performed on just the top-scoring sequences. This secondary alignment uses the algorithm of Needleham and Wunsch (1970) to produce an alignment with gaps and is the output at the conclusion of the analysis. If no good homologies are observed after running FASTA, it is sometimes helpful to repeat the analysis using a smaller k-tuple value or an alternate scoring matrix.

A convenient method of performing a FASTA search on a local computer is to use a free e-mail server. A computerized service provided by several institutions automatically accepts requests for FASTA searches via e-mail, searches the sequence contained within the mail message against a variety of databases, and then returns the results via e-mail. A mail server for FASTA is available from the University of Houston and the European Bioinformatics Institute. A sample mail message is shown in Figure 7.7.5. The FASTA program is currently available for most computers in common use via the Internet or from Bill Pearson at the University of Virginia. The DOS and Macintosh versions of FASTA can search the CD-ROM version of GenBank or EMBL databases in a few hours.

Both FASTA and TFASTA (which, like TBLAST, checks an amino acid sequence against DNA sequence databases) produce a score representing the quality of the

```
TITLE A test search of the EMBL Other Mammalian DNA sequences
LIB EMAM
WORD 4
LIST 100
ALIGN 20
SEQ
tgcttggctgaggagccataggacgagagcttcctggtgaagtgtgtttcttgaaatcat
caccaccatggacagcaaa
END
```

Figure 7.7.5 Text of a message sent to the EBI FASTA mail server (e-mail address: fasta@ebi.ac.uk). This message requests that the sequence be searched against the Other Mammalian section of the EMBL database. The answer will include the top 100 matching sequences and alignments of the top 20 matching sequences.

match for each matching pair of sequences generated using the PAM250 matrix described earlier. FASTA does not give a significance value like that provided by BLAST programs. With either FASTA or BLAST, determining whether a match is biologically significant is up to the investigator. In making this determination, the known or presumed function of the protein, the consensus match to known active sites or sequence motifs, and the number of distinct sequences that fit the proposed homologous group must be taken into account.

Because BLAST and FASTA use different algorithms, it is advisable to perform searches on a given sequence using both of these programs. If a significant match is not identified using one of the programs, use the other.

The BLAST programs are available from e-mail servers and as an Internet service using either a command-line interface or graphical interface (NetBLAST). UNIX and VMS versions are available via network file servers (see Table 7.7.2). A public domain program called MAILFASTA is now available; this program will take a DNA or protein sequence, reformat it, and e-mail it to any or all of the following e-mail servers: BLAST, BLITZ, BLOCKS, FASTA, GeneID, GRAIL, PredictProtein, and Pythia.

GENETIC SEQUENCE DATABASES AND OTHER ELECTRONIC RESOURCES AVAILABLE TO MOLECULAR BIOLOGISTS

Different gene sequence databases can be searched via the Internet or are available to anyone free of charge either via network file transfer protocol (FTP) or for the price of the distribution media. Some of the larger sequence analysis packages (e.g., GCG) provide tools that reformat the database data files as they are received from the database distributors. Reformatting is generally performed by a computer systems manager. Many users currently access the protein and DNA databases on the Internet.

Acknowledgement: We wish to thank Rose Marie Woodsmall and her colleagues at the National Center for Biotechnology Information for their assistance in updating this unit.

Contributor: J. Michael Cherry

7.7

APPENDIX

This section lists products and other resources designed for DNA and protein sequence analysis—e.g., databases, software, and journals—and how to obtain them. Some but not all of these products have been described or cross-referenced earlier in this unit.

Commercially Available Software

Table 7.7.1 is a brief description of sequence analysis software packages and some comments concerning their notable features.

Shareware and Free Software

Table 7.7.2 is a brief description of software that is distributed without charge. See also FTP archive and electronic mail listings below.

Electronic Mail Servers

Electronic mail is available at most research institutions, and these servers provide useful information, software, and analysis at no charge. With most servers, sending the simple message "HELP" will cause the latest documentation to be transmitted via e-mail (see *UNIT 19.1*).

FTP Archives Available on Internet

A vast amount of software and databases applicable to molecular biology is available at no charge from anonymous FTP servers as described in *UNIT 19.1*.

Free Databases

A brief description of databases that are available without charge is in *UNIT 19.1*.

Addresses of Databases for Retrieval and Submission

DNA Data Bank of Japan (DDBJ) Submissions
Laboratory of Genetic Information Analysis
Center for Genetic Information Research
National Institute of Genetics (NIG)
Mishima, Shizuoka 411 Japan
E-mail: ddbjsub@ddbj.nig.ac.jp
WWW URL: http://www.nig.ac.ip

EMBL Data Library
EMBL Nucleotide Sequence Submissions
European Bioinformatics Institute
Hinxton Hall, Hinxton
Cambridge CB10 1RQUK
E-mail (retrieval): datalib@ebi.ac.uk
E-mail (submission): datasub@ebi.ac.uk
WWW URL: http://www.ebi.ac.uk

GenBank
National Center for Biotechnology Information
National Library of Medicine
National Institutes of Health
Building 38A, Room 8N-803
8600 Rockville Pike
Bethesda, MD 20894
(301) 496-2475
E-mail: info@ncbi.nlm.nih.gov
E-mail (submission): gb-sub@ncbi.nlm.nih.gov

International Protein Information Database in Japan
Research Institute for Biosciences
Science University of Tokyo
2669 Yamazaki
Noda 278 Japan

Martinsried Institute for Protein Sequences
Am MPI für Biochemie
8033 Martinsried, Germany
49-89-8578-2656
E-mail: datasub@mips.embnet.org

Protein Identification Resource (PIR)
National Biomedical Research Foundation
Georgetown University Medical Center
3900 Reservoir Road, N.W.
Washington, DC 20007
E-mail: pirmail@nbrf.georgetown.edu

Publications

The Applications of Computers to Research on Nucleic Acids—published annually by IRL Press, P.O. Box Q, McLean, VA 22101. (703) 356-4031; FAX: (703) 356-4303.

BioTechniques—published monthly by Eaton Publishing, 154 E. Central Street, Natick, MA 01760. (508) 655-8282; FAX (508) 655-9910.

BioTechnology Software—published bimonthly by Mary Ann Liebert, 1651 Third Avenue, New York, NY 10128. (212) 289-2300; FAX (212) 289-4697.

Computer Applications in the Biosciences (CABIOS)—published monthly by Oxford University Press, 2001 Evans Road, Cary, NC 27513. (919) 677-0977; FAX: (919) 677-8877.

Table 7.7.1 Commercially Available Sequence-Analysis Software

Name	Description	Source
Ball & Stick	Molecular graphics display, printing, and manipulation for the Macintosh; a helpful demo is available via ftp.bio.indiana.edu.	Cherwell Scientific Publishing 27 Park End Street Oxford, OX1 1HU, UK (44) 865 774 800 FAX (44) 865 794 664
ChemDraw **Chem3D**	Desktop publishing for chemical structures in two and three dimensions; very useful for creating journal figures or instructional illustrations of molecular structures.	Cambridge Scientific Computing 875 Massachusetts Ave., Suite 61 Cambridge, MA 02139 (617) 491-6862 FAX (617) 491-8208
DNA Strider	A simple and very useful Macintosh sequence analysis program; includes restriction mapping and circular plasmid maps generated from DNA sequence.	Christian Marck Service de Biochimie—Bat 142 Centre d'Etudes Nucléaires de Saclay 91191 Gif-sur-Yvette Cedex France
EUGENE & SAM	Extensive nucleic and amino acid analysis package for Sun Microsystems SPARCstations; includes FASTA for sequence searches and very quick keyword searching on DNA and protein database.	Lark Sequencing Technologies 9545 Katy Freeway, Suite 200 Houston, TX 77024 (713) 464-7488; (800) 288-3720 FAX (713) 464-7492
GCG	Package including nucleic and amino acid sequence analysis, sequencing project management, database searching, RNA folding, protein secondary structure prediction, and sequence motif generation and searching for VMS and UNIX multi-user systems.	Genetics Computer Group University Research Park 575 Science Drive, Suite B Madison, WI 53711 (608) 231-5200 FAX (608) 231-5202 E-mail: help@gcg.com
Gene Construction Kit **DNA Inspector**	The ultimate plasmid database, design, and presentation tool for the Macintosh; DNA Inspector is a basic analysis program for the Macintosh; demo disks available.	Textco 27 Gilson Road West Lebanon, NH 03784 (603) 643-1471
GENEPRO	Complete sequence-analysis software for DOS, including nucleic and amino acid analysis; GenBank and EMBL DNA databases or PIR and SWISS-PROT protein databases can be searched on floppy disks; demo disk available.	Riverside Scientific Enterprises 15705 Point Monroe Drive N.E. Bainbridge Island, WA 98110 (206) 842-9498 FAX (206) 842-9534
HIBIO DNASIS **PROSIS**	Complete DNA and protein analysis packages for DOS; includes restriction analysis, secondary structure prediction, sequencing project management, digitizer and speech synthesizer support, and database searches from CD-ROM drive.	Hitachi Software Engineering America Computer Division 1111 Bayhill Drive, Suite 395 San Bruno, CA 94066 (800) 624-6176 In CA (800) 225-9925 FAX (415) 615-7699
Intelligenetics Suite **PC Gene** **GeneWorks**	Multifunction sequence analysis package—Intelligentics Suite is for VMS and Sun Microsystems, PC Gene is for DOS, and GeneWorks is for Macintosh.	Intelligenetics 700 East El Camino Real Mountain View, CA 94040 (415) 962-7300 FAX (415) 962-7302

continued

Table 7.7.1 Commercially Available Sequence-Analysis Software, continued

Name	Description	Source
LaserGene **Protean**	Excellent nucleic and amino acid sequence analysis packages for DOS and Macintosh computers; programs for sequence comparison and alignment, editing and analysis, sequencing project management, restriction analysis and mapping, and database searching; demo available.	DNAStar, Inc. 1228 South Park Street Madison, WI 53715 (608) 258-7420 FAX (608) 258-7439
MacMolly	Good basic sequence analysis package for Macintosh.	Soft Gene Berlin Offenbacher Str 5, D-100 Berlin, 33 Germany 030-821-1407
MacVector	Complete package for the Macintosh including restriction map presentation, sequence comparison and database searching from hard disk or CD-ROM drive, and protein secondary structure analysis.	International Biotechnologies P.O. Box 9558 New Haven, CT 06535 (203) 786-5600; (800) 243-2555 FAX (203) 786-5694
Pearson Sequence Analysis package **FASTA**	UNIX, VMS, Macintosh, and DOS-based sequence analysis software and fast database searching; may be stored on either hard disk or CD-ROM.	William R. Pearson Department of Biochemistry Box 440 Jordan Hall University of Virginia Charlottesville, VA 22908
Plasmid Artist	Publication-quality restriction map diagram production program for the Macintosh.	Clontech Laboratory 4030 Fabian Way Palo Alto, CA 94303 (415) 424-8222 FAX (415) 424-1352
Pro-RFLP	Macintosh and DOS gel analysis software for use with a scanner or TV camera; provides a number of features including calibration, database for unknowns, searching for matches to unknowns, complete printer output.	DNA ProScan, Inc. P.O. Box 12185 Nashville, TN 37212 (800) 841-4362
Rodger Staden programs	Several sequence assembly and analysis packages and complete UNIX analysis package; includes features that can analyze the fluorescence trace output files from ABI automatic sequencer machines.	Rodger Staden MRC Lab of Molecular Biology Hills Road Cambridge, England CB2 2QH
Sequencher	Macintosh sequencing support software including vector screening, ORF reports, translations, restriction maps, and a good interface with fluorescent sequencers.	Gene Codes Corporation 2901 Hubbard Road Ann Arbor, MI 48105 (313) 769-7249; (800) 497-4939 FAX (313) 930-0145
XQS **ALIGN** **PSQ** **NAQ** **ATLAS**	Database management and searching software featuring very fast and flexible keyword searches for VMS.	National Biomedical Research Foundation. Georgetown Univ. Medical Center 3900 Reservoir Road, N.W. Washington, DC 20007 (202) 687-2121 FAX (202) 687-1662

Table 7.7.2 Free Sequence-Analysis Software Programs

Name	Description	Source[a]
Amplify	Utility to aid in designing, analyzing, and even simulating experiments for PCR reactions.	FTP: sumex-aim.stanford.edu
BinHex 4.0	Essential utility required to decode files retrieved from FTP archives or received from mail servers for the Macintosh.	Available from most FTP archives; however, it is generally easiest to ask a colleague for a copy
BLAST	Ultra-fast database searching program for UNIX and VMS C source.	FTP: ncbi.nlm.nih.gov EM: netserv@ebi.ac.uk
BioSCAN	Forms-based search and retrieval resource from the University of North Carolina, Chapel Hill.	WWW URL: http://genome.cs.unc.edu/
BLOCKS Search	Forms-based submission tool via WWW for comparisons to BLOCKS database of highly conserved regions in protein sequences.	WWW URL: http://www.blocks.fcrc.org/
BoxShade	Prepares publication-quality figures of aligned sequences; resulting PostScript output contains shading over regions of similiarity.	EM: netserv@ebi.ac.uk
Clustal	Multiple sequence alignment and production of phylogenetic dendrograms; available for DOS, Macintosh, UNIX, and VMS.	FTP: ftp.bio.indiana.edu FTP: ch.embnet.org
Covariation	HyperCard stack to analyze an aligned set of sequences to aid in the identification of covariations (two or more sites in an RNA or DNA molecule that are evolving together).	FTP: ftp.bio.indiana.edu
Disinfectant	Excellent Macintosh virus detection and removal application.	FTP: ftp.acns.nwu.edu FTP: sumex-aim.stanford.edu
Entrez	Application for access to genetics and biomolecular subset of MEDLINE bibliographic database; available for Macintosh, Windows, UNIX, and VMS; computer must be registered with NCBI to obtain access; registration is free.	FTP: ncbi.nlm.nih.gov WWW URL: http://atlas.nlm.nih/gov:5700/Entrez/index.html/
Enzyme Kinetics	HyperCard stack to analyze and plot enzyme kinetic experimental data.	FTP: ftp.bio.indiana.edu
FASTA	Fast sequence database searching program for UNIX, and VMS.	FTP: uvaarpa.virginia.edu
GDE	Superb analysis package for Sun Microsystems SPARCstations; includes multiple sequence alignment, sequencing project management, database searching, and more.	FTP: golgi.harvard.edu
GenBank Search	HyperCard stack that prepares a mail message for the GenBank FASTA and sequence retrieval mail servers; requires MacTCP software on Macintosh computers.	FTP: ftp.bio.indiana.edu
GeneMapper	Full-featured coding region identification and more for Sun Microsystems SPARCstations.	FTP: haywire.nmsu.edu
GenoBase	An experimental WWW resource provided by NIH; provides tables and query capabilities to an object-oriented molecular biology database.	WWW URL: http://specter.dcrt.nig.gov:8004/

continued

Table 7.7.2 Free Sequence-Analysis Software Programs, continued

Name	Description	Source[a]
GenQuest	Forms-based database searching via WWW from Genome Data Base; provides BLAST, FASTA, and Smith-Waterman searching SWISS-PROT, Genome Sequence Database (GSDB), Protein Databank (PDB), and Prosite.	WWW URL: http://www.gdb.org
Gopher	Client application to the growing number of Internet Gopher servers.	FTP: ftp.bio.indiana.edu FTP: boombox.micro.umn.edu
Linkage	Performs linkage analysis on genetic markers; available for UNIX, DOS, and VMS computers.	FTP: corona.med.utah
LoopViewer	Displays results of Zuker RNA folding programs on a Macintosh; produces output similar to that shown in Figure 7.7.4B.	FTP: ftp.bio.indiana.edu
MACAW	Excellent multiple sequence alignment and editing tool for Microsoft Windows and Macintosh.	FTP: ncbi.nlm.nih.gov
MacMolecule	Displays and rotates molecular structures of nucleic acids and proteins; very useful as a teaching aid.	FTP: ftp.bio.indiana.edu
MacPattern	Macintosh application that searches for patterns in protein sequences utilitizing the PROSITE motif database; allows documentation from PROSITE to be directly accessed as well as entry of user-defined patterns.	FTP: ftp.bio.indiana.edu EM: netserv@ebi.ac.uk
MacPlasMap	Plasmid map drawing program for the Macintosh; the cutting sites of restriction enzymes must be entered by the user.	FTP: ftp.bio.indiana.edu EM: netserv@ebi.ac.uk
MandM	Materials and methods HyperCard stack for organizing laboratory protocols.	EM: netserv@ebi.ac.uk
MapMaker	Linkage analysis of F_2 and CEPH populations; available for Sun Microsystems and VMS.	EM: mapm@genome.wi.mit.edu FTP: genome.wi.mit.edu
Mase	Multiple sequence alignment editor for UNIX.	FTP: mbcrr.harvard.edu
MFOLD (or LRNA & CRNA)	VMS version for the prediction of Zuker RNA secondary structures.	FTP: amber.mgh.harvard.edu FTP: ftp.bio.indiana.edu
MulFold	Macintosh version of the MFOLD for prediction of Zuker RNA secondary structures, created by Don Gilbert.	FTP: ftp.bio.indiana.edu
NCSA GelReader	Automates the measurement of DNA fragment lengths using a scanned image of an autoradiogram or EtBr-stained gel on a Macintosh; identifies lanes and bands, then exports the predicted band sizes as a text file for use with a spreadsheet.	FTP: ftp.ncsa.uiuc.edu
NCSA Mosaic	First WWW browser for Macintosh, Windows, and UNIX; requires an Internet connection.	FTP: ftp.ncsa.uiuc.edu
Netscape	Commercial WWW browser, available to educational users for free; provides fast displays and inline images in either GIF or JPEG encoding formats; available for Macintosh, Windows, and UNIX.	FTP: ftp.netscape.com

continued

Name	Description	Source[a]
NIH Image	General image-analysis program for biologists that uses scanned images on a Macintosh.	FTP: alw.nih.gov
OSP	Oligonucleotide selection program to create appropriate PCR primers; selects primers using the sequence of the target region determining GC content, possible secondary structure, primer and applified product length.	FAX: (314) 362-2985 c/o Paula Kassos (Phil Green, Washington University, St. Louis) EM: pg@genome.wustl.edu
PAUP	Phylogenetic analysis software for DOS and Macintosh.	EM: swofford@uxh.cso.uiuc.edu (David Swofford, Illinois Natural History Survey, Champaign, IL)
PGEN	Oligonucleotide design and analysis software for DOS; requires EGA or VGA graphics adaptors; will work from a 360K floppy drive.	FTP: ftp.bio.indiana.edu
Phylip	Phylogenetic analysis software for UNIX.	FTP: ftp.bio.indiana.edu
Plot/A	Protein analysis software with a variety of secondary structure predictions.	FTP: ftp.bio.indiana.edu
PlotZ	Graphical restriction mapping for UNIX and VMS.	FTP: amber.mgh.harvard.edu
PLSearch	Database of primary protein sequence patterns determined by analyzing the SWISS-PROT database, for UNIX.	FTP: mbcrr.harvard.edu
Primer	Primer selection software for UNIX, VMS, DOS, and Macintosh; selects primers using the sequence of the target region by determining the GC content, possible secondary structure, primer, and amplified product length.	FTP: genome.wi.mit.edu
SeqApp	Analyzes DNA and protein sequences while also allowing easy access to WAIS and Internet Gopher resources; developed by Don Gilbert; requires a Macintosh and MacTCP.	FTP: ftp.bio.indiana.edu
SeqSpeak	Simple DNA sequence editor; developed by Keith Conover for the Macintosh; provides audio feedback using a digitized human voice.	FTP: ftp.bio.indiana.edu
SLINK and FASTSLINK	Linkage analysis software for UNIX computers	FTP: watson.hgen.pitt.edu
Virtual Library of BioSciences	Expanding collection of Internet resources available for biology; continually updated and an excellent starting point for exploration of information available on the Internet.	WWW URL: http://golgi.harvard.edu/biopages .html
WAIS	Wide area information service, providing effective access to large collections of information.	FTP: think.com (NeXT, Motif, and Macintosh software) FTP: samba.oit.unc.edu (DOS, SunView, and VMS software)

[a]Source abbreviations: FTP, file transfer protocol (all are anonymous); EM, electronic mail; WWW URL, World Wide Web Uniform Resource Locator. See *UNIT 19.1* for descriptions of these sources. Contact person is listed in parentheses.

Mutagenesis of Cloned DNA

8

Genes and other genetic elements are frequently characterized by correlating specific changes in DNA sequence with effects on function. The classical approach to this problem has been to obtain mutations by selecting for organisms having new properties. The relevant wild-type and mutant genes can then be cloned and subjected to DNA sequence analysis. This approach, however, suffers from several disadvantages. First, the methods of mutagenesis severely constrain the kinds of mutations that are obtained. Second, since the entire organism is subjected to mutagenesis, mutations occurring in the gene of interest are relatively rare. Third, since mutations are identified by virtue of their phenotype, it is essentially impossible to obtain "mutants" that behave indistinguishably from the wild-type gene; these are particularly valuable for determining which parts of the gene are not important for function.

The advent of recombinant DNA technology has made it possible to reverse the procedure of classical mutagenesis. Mutations are first generated in cloned segments of DNA by using a variety of chemical and enzymatic methods. These methods can produce mutations at an extremely high frequency (approaching 100% in some cases), and essentially all possible mutations can be generated. Once generated, the mutant DNAs are subjected to DNA sequence analysis and then analyzed for the specific function of interest. In this way, mutations can be obtained in a systematic manner without regard to their phenotype. The end result is that the functions of a given region of DNA can be investigated in much more detail.

This chapter describes several protocols for altering the nucleotide sequence of cloned DNA segments. Oligonucleotide-directed mutagenesis (*UNIT 8.1*) uses a mutagenic oligonucleotide primer to alter the DNA sequence in a defined way. Oligonucleotides can also be used to create a large number of mutations within either a small region of DNA, or in larger regions, using mutually primed synthesis (*UNIT 8.2*). By using specific chemicals, it is possible to generate many random mutations in cloned DNA fragments as large as 3 kb (*UNIT 8.3*). Linker-scanning mutagenesis (*UNIT 8.4*) is used to introduce clusters of point mutations throughout a sequence of interest. The polymerase chain reaction (*UNIT 8.5*) can be used to introduce restriction endonuclease sites or point mutations into specific DNA sequences.

In oligonucleotide-directed mutagenesis, it is necessary to synthesize an oligonucleotide whose sequence contains the mutation of interest. This oligonucleotide is then hybridized to a template containing the wild-type sequence and extended with T4 DNA polymerase. The resulting product is a heteroduplex molecule containing a mismatch due to the mutation in the oligonucleotide. The mutation is "fixed" upon repair of the mismatch in *E. coli* cells. This method is extremely valuable for situations in which it is desired to determine the effects of particular changes in the DNA, and it is also useful for introducing restriction sites at specific positions within a given stretch of DNA. However, it is relatively expensive (one oligonucleotide per mutation) and hence is limited to circumstances where one or a few specific mutations are desired.

Synthetically derived mixtures of oligonucleotides, obtained by adding small, defined amounts of "incorrect" precursors at each step of DNA synthesis, can be used to generate a large number of mutations within a small region of DNA. Each oligonucleotide in the mixture thus has a defined probability of being altered from the wild-type sequence. This degenerate oligonucleotide mixture is converted to double-stranded DNA, whereupon individual oligonucleotide molecules are isolated by molecular cloning. In principle, mutations occur at the frequency that was programmed into the DNA synthesis, and they occur at random positions throughout the region of interest. The major limitation is that only oligonucleotides ≤80 bases can be mutagenized. By using a set of adjacent degenerate oligonucleotides, it is possible to mutagenize larger regions.

A variety of chemicals can be used to generate a large number of randomly distributed nucleotide substitution mutations over several kilobases. In this method, single-stranded DNA containing the region of interest is treated with these chemicals. By using an appropriate oligonucleotide primer, the mutated region is copied and then cloned. The mutation frequency is set by the severity of the chemical treatment, and essentially all possible base substitutions are obtained. This method is valuable when mutagenizing regions of DNA that are larger than can be accommodated in a single or a few oligonucleotides. However, since there are many possible mutations in such relatively large regions, this method is less useful for saturating a region with mutations. Instead, it is best suited for obtaining mutations that confer phenotypes of interest.

By introducing clusters of point mutations using linker-scanning mutagenesis, important sequence elements can be located and analyzed. The mutations are introduced by first generating a nested series of deletion mutations in a specific region. Next, synthetic linkers are added to the deletion endpoints. A pair of complementary oligonucleotides is then added to fill in the gap in the sequence of interest between the linker at the deletion endpoint and a nearby restriction site. By its position at the varied endpoints of the deletion mutation series, the linker sequence actually provides the desired clusters of point mutations as it is moved, or "scanned," across the region.

Kevin Struhl

UNIT 8.1

Oligonucleotide-Directed Mutagenesis Without Phenotypic Selection

The DNA sequence is altered with a mutated oligonucleotide, which primes synthesis on a template (Fig. 8.1.1). Mutants are recovered at 50% to 80% efficiency.

BASIC PROTOCOL

8.1

Materials *(see APPENDIX 1 for items with ✓)*

- Single-stranded bacteriophage vector with insert
- TY medium containing 0.25 µg/ml uridine (*UNIT 1.1*)
- *E. coli* CJ236 (or alternative *dut*⁻ *ung*⁻ F′ strain)
- 5× PEG/NaCl solution: 15% (w/v) PEG 8000/2.5 M NaCl
- ✓ TE buffer
- 10× polynucleotide kinase buffer (*UNIT 3.4*)
- 10 mM ATP (*UNIT 3.4*)
- Mutagenic oligonucleotide primer
- T4 polynucleotide kinase (*UNIT 3.10*)
- ✓ 100 and 500 mM EDTA, pH 8
- ✓ 20× SSC
- ✓ 5× polymerase mix
- T7 DNA polymerase (*not* Sequenase) or T4 DNA polymerase (*UNIT 3.5*)
- T4 DNA ligase (measured in Weiss units; *UNIT 3.14*)

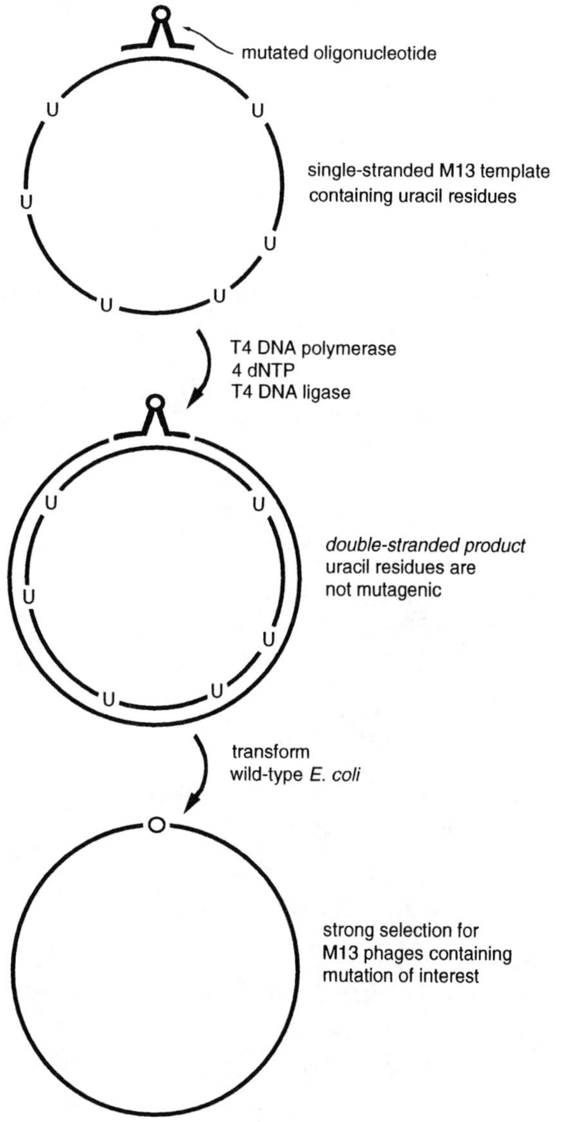

Figure 8.1.1 Oligonucleotide-directed mutagenesis without phenotypic selection. Single-stranded DNA containing a small number of uracil (U) residues in place of thymine is prepared from a *dut*$^-$ *ung*$^-$ strain. A synthetic oligonucleotide containing the mutation of interest is annealed to the template (the mismatch is shown as a discontinuity in the oligonucleotide) and treated with T4 DNA polymerase and T4 DNA ligase to produce a double-stranded circular molecule. Introduction of this heteroduplex molecule into a wild-type (*dut*$^+$ *ung*$^+$) strain allows for the efficient recovery of mutant DNA.

1. Place one plaque produced by a single-stranded phage in a 1.5-ml microcentrifuge tube containing 1 ml sterile TY medium. Incubate 5 min at 60°C to kill cells. Vortex vigorously to release phage from the agar and then microcentrifuge 2 min.

2. Place 100 µl of the supernatant in a 1-liter flask containing 100 ml TY medium supplemented with uridine to 0.25 µg/ml. Add 5 ml of a midlog culture of *E. coli* CJ236 and incubate 6 to 18 hr at 37°C with vigorous shaking.

3. Centrifuge 30 min at 5000 × *g* and save the supernatant. Titer the phage on any *E. coli ung*⁻ (e.g., CJ236) versus *ung*⁺ strain (e.g., JM105, JM107, or JM109; *UNIT 1.11*).

   The supernatant should contain phage at 10^{10} to 10^{11} pfu/ml. Phage containing uracil in the DNA have normal biological activity in the ung^- host, but >100,000-fold lower survival in the ung^+ host.

4. Add 1 vol of 5× PEG/NaCl solution to 4 vol supernatant to precipitate the phage. Mix and incubate 1 hr at 0°C.

5. Centrifuge 15 min at 5000 × *g* and drain the pellet well. Resuspend the pellet in 5 ml TE buffer in a 15-ml Corex tube and vortex vigorously.

6. Place the phage solution on ice for 1 hr and centrifuge as above to remove debris. Purify the single-stranded phage DNA by phenol extraction and ethanol precipitation (*UNIT 2.1*). Determine the DNA concentration by reading absorbance at 260 nm (1 A_{260} = 36 µg/ml).

7. To a 1.5-ml microcentrifuge tube, add the following:

   2 µl 10× T4 polynucleotide kinase buffer
   2 µl 10 mM ATP
   Mutated oligonucleotide (15 to 50 nucleotides long)
   H_2O to 20 µl
   2 U T4 polynucleotide kinase.

   Incubate 60 min at 37°C. Add 3 µl of 100 mM EDTA and heat to 70°C to terminate the reaction.

   Ratio of oligonucleotide to single-stranded template ranges from 2:1 to 10:1; for short primers (15-20 bases), this corresponds to 4 to 30 ng/µg single-stranded M13 template.

8. Add the single-stranded circular uracil-containing DNA template (usually 1 µg in 1 µl) to the phosphorylated oligonucleotide. Add 1.25 µl of 20× SSC, mix thoroughly, and microcentrifuge 5 sec. Place the tube in a 500-ml beaker of water at 70°C. Allow to cool to room temperature. Microcentrifuge 5 sec and place the tube on ice.

9. To the hybridization mixture add the following:

   20 µl 5× polymerase mix
   2.5 U T7 or T4 DNA polymerase
   2 U T4 DNA ligase
   H_2O to 100 µl.

   Mix thoroughly, then incubate 5 min at 0°C, 5 min at room temperature, and 2 hr at 37°C. Add 3 µl of 500 mM EDTA to terminate the reaction.

   T4 DNA polymerase activity in commercial preparations can vary with the source and age of the enzyme, so it may be necessary to add more than 2.5 U. Add the enzymes last. T7 and T4 DNA polymerase are preferred over the Klenow fragment because they do not "strand displace" the mutagenic oligonucleotide after synthesis is completed. T7 DNA polymerase (not Sequenase) is ideal for complete synthesis due to its high processivity.

10. Electrophorese 20 µl (200 ng DNA) in a 0.8% agarose gel (*UNIT 2.5*). Control lanes should contain single-stranded circular viral DNA, double-stranded closed

circular DNA (RFI), and nicked double-stranded circular DNA (RFII). Based upon an estimate from the gel analysis, use 1 to 100 ng of the double-stranded DNA product to transfect any desired *ung*$^+$ strain of *E. coli* cells (*UNIT 1.8*).

11. The resulting clones (as phage plaques or colonies) can be selected or chosen randomly for isolation of pure genetic stocks. Analyze the clones by DNA sequencing (*UNIT 7.4*).

Reference: Kunkel, 1985; Kunkel et al., 1987.

Contributor: Thomas A. Kunkel

Mutagenesis with Degenerate Oligonucleotides: Creating Numerous Mutations in a Small DNA Sequence

UNIT 8.2A

BASIC PROTOCOL

An important feature of this method is that a single-stranded degenerate oligonucleotide is converted to double-stranded homoduplex molecules that can be cloned directly into standard vectors (see Fig. 8.2.1). The palindromic character of the 3′ end of the oligonucletide allows nonidentical DNAs to hybridize; thus, the oligonucleotides serve as mutual primers for extension by the Klenow fragment of *E. coli* DNA polymerase I.

By cloning homoduplex DNAs, mismatch repair in vivo is avoided, potential bias against particular mismatched nucleotides is prevented, and mutations located in all possible positions of a sequence can be obtained. The main critical parameters are the structures at the 5′ and 3′ ends. The only absolute requirement of the method is that the 3′ end be a palindromic sequence that can be cleaved by a restriction

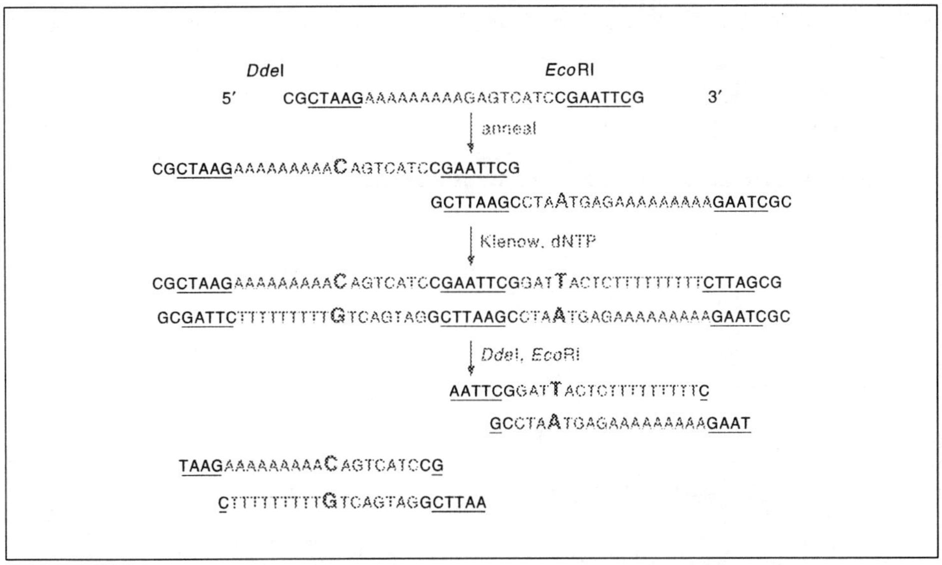

Figure 8.2.1 Mutually primed synthesis for cloning degenerate oligonucleotides. A degenerate oligonucleotide (top line; see Fig. 8.2.2) is self-annealed via the 8-nucleotide palindrome on the 3′ end that encompasses the *Eco*RI site, and then treated with the Klenow fragment. The resulting double-stranded DNA contains two oligonucleotide units each derived from an original oligonucleotide molecule (nucleotides that deviate from the "wild-type" sequence are depicted as large letters). Upon cleavage by *Eco*RI and *Dde*I, the double-stranded oligonucleotides are cloned into an appropriate vector.

endonuclease. The palindrome is necessary for the mutual priming reaction, and the restriction site is necessary for cleavage of the initial product, an oligonucleotide dimer, to oligonucleotide units suitable for cloning. It is preferable if the 5′ end also contains a sequence that is recognized by a restriction endonuclease. However, the 5′ terminal sequences can be anything, since the oligonucleotides can be cloned via the blunt ends generated by the mutually primed synthesis procedure.

When restriction sites are to be used at both the 5′ and 3′ end of the oligonucleotide, the restriction site having the highest GC composition should be used at the 3′ site to facilitate the annealing reaction. In addition, the palindrome at the 3′ end should usually be extended to 8 bases by flanking the site with additional G or C residues. If the 5′ end contains a sequence that is to be cleaved with a restriction endonuclease, it is useful to include one to three extra nucleotides beyond the recognition sequence at the 5′ end to facilitate cleavage. The length of the palindrome at the 5′ end should be minimized in order to disfavor hybridizations that might block the extension reaction.

Materials *(see* APPENDIX 1 *for items with ✓)*

- 10× *E. coli* DNA polymerase I buffer (*UNIT 3.4*)
- 4dNTP mix (2.5 mM each dNTP; *UNIT 3.4*)
- [α-^{32}P]dNTP (400 to 800 Ci/mmol; *UNIT 3.4*)
- Klenow fragment of *E. coli* DNA polymerase I (*UNIT 3.5*)
- ✓ 0.5 M EDTA, pH 8
- ✓ TE buffer, pH 7.5
- ✓ 3 M sodium acetate, pH 5.2
- Buffered phenol (*UNIT 2.1*)
- 100% ethanol
- ✓ Elution buffer
- T4 DNA ligase (measured in cohesive-end units; *UNIT 3.14*)

1. Design the oligonucleotide such that the 3′ end contains an 8-nucleotide palindromic sequence encompassing a restriction endonuclease cleavage site. The 5′ end should consist of sequences encompassing a restriction endonuclease site (if possible). The central region should contain the mutagenized sequence of interest (see Fig. 8.2.2).

   The length and mutation frequency of the central region will vary, depending on the specific experiment.

2. Synthesize the oligonucleotide. Where mutations are not desired, program the synthesis to use homogeneous solutions of individual nucleotide precursors.

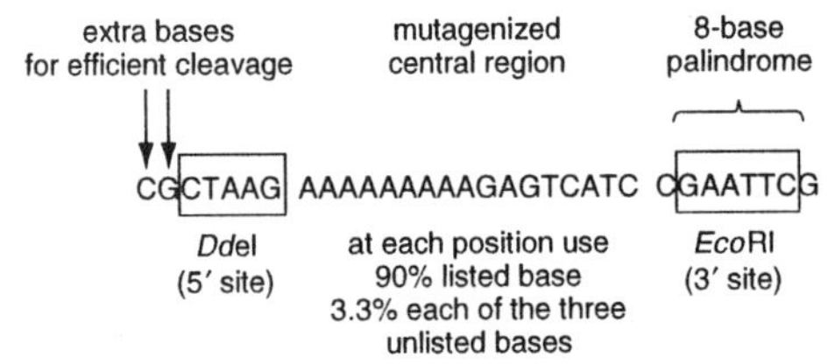

Figure 8.2.2 Design of degenerate oligonucleotide. A 17 base-pair region located between *Eco*RI and *Dde*I sites is mutagenized at a frequency of 10% per position.

Where mutations are desired, program the synthesis to use defined mixtures of nucleotide precursors.

The frequency of mutation at a given position is determined by the relative molarities of the precursors present in the solution. For example, a 10% mutation rate would be achieved with a mixture of 90% wild-type nucleotide and 3.3% each of the three "incorrect" nucleotides.

3. Purify the oligonucleotide by high-performance liquid chromatography and/or electrophoresis in denaturing polyacrylamide gels containing 7 M urea (*UNIT 2.7*). Adjust the concentration to 1 mg/ml in water (~50 to 100 μM).

4. Transfer 200 pmol (1 to 2 μg) of the oligonucleotide to a 500-μl microcentrifuge tube. Adjust the volume to 7 μl with water. Incubate 5 min at 70°C.

   The oligonucleotide does not need to be phosphorylated if there will be a restriction site at the 5′ end after conversion to double-stranded form. If the 5′ end will not contain a restriction site, the oligonucleotide should be phosphorylated by T4 polynucleotide kinase (UNIT 3.10).

5. Add 1 μl of 10× DNA polymerase I buffer, cool to a temperature that will allow the 3′-end palindromic sequences to hybridize (typically 23°C), and incubate ≥60 min.

6. Add 2 μl of 2.5 mM 4dNTP mix, 5 U Klenow fragment, and 10 μCi of any one [α-^{32}P]dNTP. Incubate 1 hr at 23°C. Add another 5 U Klenow fragment and continue incubation 2 hr to overnight.

   A low level of radioactively labeled dNTP is included in the reaction to facilitate purification of the DNA.

7. Add 1 μl of 0.5 M EDTA to stop the reaction. Adjust the volume to 50 μl with TE buffer and add sodium acetate to 0.3 M. Extract with buffered phenol and ethanol precipitate the DNA (*UNIT 2.1*). Resuspend the DNA in 20 μl TE buffer. Reserve 2 μl for analysis on denaturing polyacrylamide gel (step 12).

8. In a 30-μl reaction, digest the now double-stranded oligonucleotide with the restriction endonuclease recognizing the outside sites for ≥2 hr with 10 to 40 U enzyme/original μg oligonucleotide (*UNIT 3.1*). If there is no 5′ restriction site, cleave the DNA with the enzyme recognizing the internal restriction site.

   The cleavage occurs at what was originally the 5′ site of the single-stranded oligonucleotide.

9. Reserve 2 μl for analysis on denaturing polyacrylamide gel (step 12). Extract the remainder with buffered phenol, and ethanol precipitate. Resuspend in ≥10 μl TE buffer and purify by electrophoresis on nondenaturing polyacrylamide gel (*UNIT 2.7*).

   To prevent denaturation of the DNA, do not allow the gel to heat above room temperature. This step effectively removes the small single-stranded DNA fragments from the restriction digest, and the unreacted single-stranded oligonucleotides which could interfere with the subsequent ligation and transformation steps. The double-stranded oligonucleotide is a heterogeneous mixture of different sequences, so the DNAs might not migrate as sharp band, even though they are of identical length.

10. Excise and elute the DNA in gel elution buffer. Resuspend the double-stranded oligonucleotide in 20 μl TE buffer and store at −20°C.

11. Digest the double-stranded oligonucleotide with the enzyme recognizing the internal restriction site (original 3′ site) to produce a double-stranded, homoduplex version of the oligonucleotide mixture with 5′ and 3′ ends suitable for ligation into standard vectors. Analyze 2 µl on denaturing polyacrylamide gel (step 12). Phenol extract the remainder, ethanol precipitate, and resuspend in 20 µl TE buffer.

12. Electrophorese the 2-µl aliquots (steps 7, 9, and 11) on denaturing polyacrylamide gel to confirm that the reactions have produced the desired products (*UNIT 7.6*).

13. Serially dilute the double-stranded oligonucleotide in TE buffer by successive factors of 10 until a 10,000-fold dilution is reached. Set up a series of ligation reactions, each containing a constant amount of vector and a portion of each dilution of the oligonucleotide (*UNIT 3.16*).

14. Introduce the ligation mixtures into an appropriate *E. coli* strain by standard transformation procedures (*UNIT 1.8*). Analyze the DNA obtained from the transformants by restriction endonuclease digestion and by DNA sequence analysis (*UNITS 7.4 & 7.5*).

In order to avoid DNAs containing multiple oligonucleotide insertions, it is best to analyze clones with the lowest amount of oligonucleotides. Numerous different mutants are generated that saturate a small region of DNA with all possible single-base changes. Sequencing is necessary to identify the specific mutation(s) introduced.

Reference: Hill et al., 1986.

Contributor: David E. Hill

UNIT 8.2B

Gene Synthesis: Assembly of Target Sequences Using Mutually Priming Long Oligonucleotides

BASIC PROTOCOL

The most straightforward way to generate a desired sequence is simply to synthesize it. This protocol uses pairs of oligonucleotides annealed at a short duplex segment at their 3′ ends as both templates and primers (mutually primed synthesis) to generate desired sequences up to 400 bp in a single step.

The strategy for large synthetic projects is diagrammed in Figure 8.2.3. Overall, this approach is simpler and substantially less expensive than some previous approaches, which relied on annealing and ligating large numbers of shorter oligonucleotides. The use of pairs of relatively long oligonucleotides (>100 nucleotides) for mutually primed synthesis minimizes the number of base pairs synthesized chemically, thus minimizing cost. The simple priming reaction is easier to set up and control than the multi-oligonucleotide annealing reaction. In general, it is also more convenient to synthesize and purify two relatively long oligonucleotides than a larger number of shorter ones.

The actual length and design of the oligonucleotides will vary with each experiment. A 15-bp duplex is a convenient length for mutual priming, and duplexes <10 base pairs should be avoided to ensure stability, particularly if they are A·T rich. Some common mutations that are introduced into the oligonucleotides include changes in

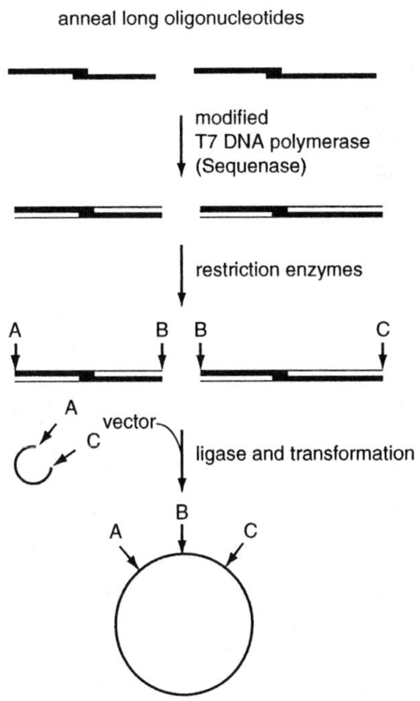

Figure 8.2.3 Strategy for constructing a synthetic gene using two pairs of oligonucleotides. Lines represent two pairs of oligonucleotides. Note the region of duplex in each pair which makes mutually primed synthesis possible. A, B, and C represent restriction endonuclease sites needed for cloning the annealed, extended oligonucleotide pairs into the vector.

amino acid sequence, changes to facilitate expression in specific cell types, and addition or removal of restriction endonuclease sites. At various stages of a project involving mutually primed oligonucleotides, sequences should be checked carefully to determine that the correct sequence is present.

Materials *(see* APPENDIX 1 *for items with ✓)*

- 10× Sequenase buffer (*UNIT 3.5*)
- 4dNTP mix (2.5 mM each dNTP; *UNIT 3.4*)
- Modified T7 DNA polymerase (Sequenase, U.S. Biochemical; *UNIT 3.5*)
- ✓ 10× restriction endonuclease buffer
- 100% and 95% ethanol
- ✓ TE buffer

The following steps replace steps 4 to 14 of the Basic Protocol in *UNIT 8.2A*.

1. Add 1 μg of each of the two oligonucleotides to a microcentrifuge tube. Adjust to 17 μl with water and add 2 μl of 10× Sequenase buffer. Heat 5 min at 70°C, followed by 5 min at the appropriate annealing temperature. Remove 2 μl for later analysis.

   The annealing temperature may be crudely estimated by adding 2°C for each A·T base pair plus 4°C for each G·C base pair in the duplex segment.

2. Add 2 μl 4dNTP mix and 10 U Sequenase. Incubate 30 min at 30°C.

3. Heat the reaction 10 min at 70°C to inactivate the DNA polymerase. Remove 2 μl for later analysis.

8.2

4. Increase the volume to 100 µl with 10× restriction endonuclease buffer (final concentration 1×), water, and 20 to 100 U restriction endonucleases appropriate for cloning (*UNIT 3.1*). Digest ≥2 hr at the appropriate temperature.

5. Phenol extract (*UNIT 2.1*). Add 100 µl of 4 M ammonium acetate and 400 µl of 100% ethanol, chill 15 min at −70°C, and microcentrifuge 5 min. Redissolve the pellet in 100 µl TE buffer and repeat the ethanol precipitation. Rinse the pellet with 95% ethanol, dry, and resuspend in 20 µl TE buffer. Remove 2 µl for later analysis.

   This ammonium acetate precipitation is only appropriate if the extended product is ≥50 bp. Shorter extended duplexes can be purified by electrophoresis on nondenaturing polyacrylamide gels (UNIT2.7) or used directly for cloning after recovery using two ethanol precipitations with 0.3 M sodium acetate.

6. Check starting material, extended product, and digested product by electrophoresis on a sieving agarose gel (*UNIT 2.8*). Estimate the amount of extended oligonucleotide and subclone using the desired vector.

7. Sequence several appropriate subclones (*UNITS 7.4 & 7.5*).

Reference: Uhlmann, 1988.

Contributor: David D. Moore

UNIT 8.3

Region-Specific Mutagenesis

BASIC PROTOCOL

This method makes it possible to generate a large number of randomly distributed nucleotide substitution mutants in cloned DNA fragments as large as 3 kb. After the DNA is cloned into a vector that permits isolation of single-stranded DNA, it is treated with a variety of chemicals that cause damage to specific bases (Fig. 8.3.1).

The incubation times suggested here are roughly calibrated to generate mutations in 10-20% of target fragments 150 bp in length. For larger target fragments, the chemical treatment times should be decreased in a linear fashion. Some variation in mutagenesis efficiency occurs with different batches of chemicals, so some modification in incubation times may be required.

Materials *(see APPENDIX 1 for items with ✓)*

Restriction endonucleases and buffers (*UNIT 3.1 & APPENDIX 1*)
Single-stranded DNA containing the target DNA sequences
2.5 M sodium acetate, pH 4.3
2 M sodium nitrite
18 M formic acid
12 M hydrazine
2.5 M sodium acetate, pH 5.5
10 mg/ml tRNA in TE buffer
100% ethanol
✓ TE buffer
Oligonucleotide primer
4dNTP mix (2.5 mM each dNTP; *UNIT 3.4*)
AMV reverse transcriptase (Life Sciences; *UNIT 3.7*) and 10× buffer (*UNIT 3.4*)
Buffered phenol (*UNIT 2.1*)
2 mg/ml RNase A in H_2O (*UNIT 3.13*)

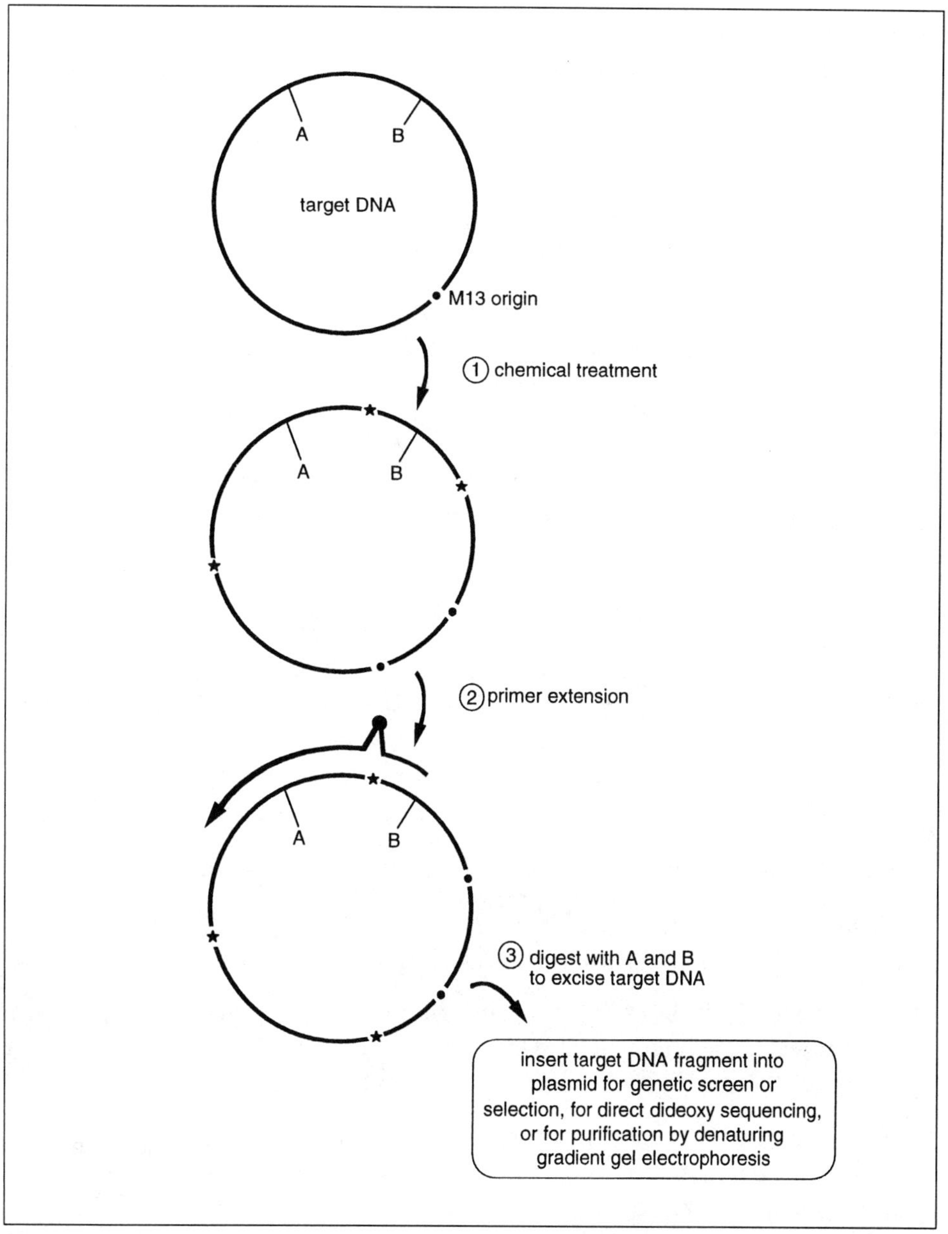

Figure 8.3.1 Region-specific mutagenesis by chemical treatment, primer extension, and cloning.

✓ Elution buffer
Ethidium bromide
Double-stranded cloning vector
T4 DNA ligase (measured in cohesive-end units; *UNIT 3.14*) and 10× buffer (*UNIT 3.4*)

1. Cleave DNA with restriction endonucleases to generate the DNA fragment of interest (*UNIT 3.1*). Clone the target fragment in both orientations into an M13 phage vector or into a plasmid vector containing the M13 origin of replication (*UNIT 3.16*). Prepare single-stranded DNA from a 100-ml culture of infected cells (*UNITS 1.9 & 1.10*).

2. To each of three microcentrifuge tubes, add 40 µg of 1 mg/ml single-stranded DNA. If both strands of the target sequence are used, there will be six reactions.

   If possible, steps 2 to 10 should be performed in one day.

3. Nitrous acid reaction: To one tube, add 10 µl of 2.5 M sodium acetate, pH 4.3, and 50 µl of 2 M sodium nitrite. Mix and incubate 60 min at room temperature.

4. Formic acid reaction: To another tube, add 60 µl concentrated (18 M) formic acid. Mix and incubate 10 min at room temperature.

5. Hydrazine reaction: To the last tube, add 60 µl concentrated (12 M) hydrazine. Mix and incubate 10 min at room temperature.

6. Add 100 µl of 2.5 M sodium acetate, pH 5.5, 30 µg carrier tRNA (10 mg/ml in TE), and 1 ml of 100% ethanol to each tube. Chill to −80°C in dry ice for 10 min. Microcentrifuge 10 min and remove supernatant. Repeat the ethanol precipitation twice and resuspend the DNA in 80 µl TE buffer.

7. To the resuspended DNA add 10 µl of 4dNTP mix, 10 µl of 10× reverse transcriptase buffer, and 100 pmol oligonucleotide primer. Heat 5 min at 85° C, then 15 min at 40°C.

8. Add 10 µl of 4dNTP mix and 30 to 40 U AMV reverse transcriptase. Incubate 1 hr at 37°C.

9. Extract with buffered phenol and ethanol precipitate the DNA. Resuspend the pellets in 100 µl of 1× restriction enzyme buffer and cleave with the appropriate restriction endonucleases to excise the target fragment from the vector.

10. Phenol extract and ethanol precipitate the DNA (*UNIT 2.1*). Resuspend the pellets in 10 to 15 µl elution buffer and add 0.5 µl of 2 mg/ml RNase A. Incubate 15 min at 37°C to degrade carrier tRNA.

    RNase treatment can be omitted if the target fragment is >150 bp.

11. Electrophorese the DNA on a nondenaturing, 6% polyacrylamide gel to separate the excised target fragment from the vector (*UNIT 2.7*). Stain the gel with ethidium bromide (0.5 µg/ml), and excise and elute the DNA.

12. Ligate the purified insert DNA into a plasmid or bacteriophage vector containing ends compatible with those of the target fragments. Introduce ligation mixture into competent bacteria by standard transformation procedures (*UNIT 1.8*).

    Thousands of colonies containing mutagenized DNA will be obtained, with 10% to 20% having mutations in the region of interest.

8.3

SUPPORT PROTOCOL

ENRICHMENT OF MUTANT CLONES

There are several methods for enrichment and identification of mutant clones:

1. Phenotypic selection, where mutants are selected or screened on the basis of a biological property.

2. Direct DNA sequencing, which is most useful for short DNA fragments or for small numbers of mutants. Eliminate clones that do not contain the desired mutant sequence.

3. Isolation of a large number of mutants without regard to phenotype, where the resulting transformants are pooled together and amplified as a population. DNA is isolated and cleaved with appropriate restriction endonucleases to liberate the

target DNA fragments. The wild-type and mutant target fragments are separated by denaturing PAGE, and recloned into an appropriate double-stranded vector.

Reference: Myers et al., 1985, 1987.

Contributor: Richard M. Myers

Linker-Scanning Mutagenesis of DNA

UNIT 8.4

LINKER SCANNING USING NESTED DELETIONS AND COMPLEMENTARY OLIGONUCLEOTIDES

BASIC PROTOCOL

This protocol can be used to generate clusters of point mutations throughout a sequence of interest (Fig. 8.4.1).

Materials *(see APPENDIX 1 for items with ✓)*

High-copy-number plasmid vector
Linkers
Bal 31 nuclease (*UNIT 3.12*) *or* exonuclease III and S1 nuclease (*UNITS 3.11* and *3.12*, respectively)
Restriction endonucleases and buffers (*UNIT 3.1 & APPENDIX 1*)
✓ 0.5 M EDTA, pH 8 (optional)
100% ethanol
Synthetic oligonucleotides for the gene of interest
TE buffer containing 150 mM NaCl
T4 DNA ligase (*UNIT 3.14*) and 10× buffer (*UNIT 3.4*)
Selective plates (*UNIT 1.1*)

1. Generate a nested set of 5′ or 3′ deletion mutations in the region of interest in the plasmid with either *Bal* 31 or exo III and S1 nuclease (*UNIT 7.2*). First, linearize the plasmid with a restriction enzyme that cuts near the sequence of interest. For *Bal* 31 digestion, determine the rate of digestion so that the time points contain deletion endpoints every six to eight base pairs and span the region to be scanned.

2. Ligate a synthetic linker to the ends of the DNA from the appropriate time points (*UNIT 3.16*). Digest with the appropriate restriction endonucleases to produce fragments with the linker at the deleted end and a vector site at the other (*UNIT 3.1*).

3. Isolate the desired fragments on low gelling/melting temperature agarose gels (*UNIT 2.6*). Ligate the fragments to a vector fragment to create intact plasmids with the deletion mutations.

4. Transform competent *E. coli* cells with the ligation mix, pick colonies, and prepare minipreps of plasmid DNA (*UNIT 1.8*).

5. Digest the miniprep DNA with restriction enzymes to produce a small DNA fragment with one terminus at the deletion endpoint (*UNIT 1.6*). Electrophorese on a nondenaturing polyacrylamide gel or sieving agarose gel to determine the fragment size that will allow determination of the deletion endpoints (*UNIT 2.7* or *2.8*). Choose enzymes so that small fragments of 200-300 bp are sized.

6. Sequence the set of deletions that span the region of interest to determine the exact deletion endpoint.

7. Design complementary oligonucleotides to restore the wild-type sequence between the linker present at each deletion endpoint and a nearby restriction site upstream of the region for a 5′ deletion series, or downstream of the region for a 3′ deletion series (*Eco*RI was chosen in Fig. 8.4.1).

8. Make the following fragments for ligation to the oligonucleotides: (a) a fragment spanning the linker site and a convenient site in the plasmid's drug maker ("promoter fragment"), cut from each deletion mutation; and (b) a fragment spanning the site in the drug marker and the restriction site near the region to be scanned ("backbone fragment"), cut from the wild-type parental plasmid. Isolate the fragments.

9. Resuspend each oligonucleotide at ~100 μg/ml in TE buffer containing 150 mM NaCl. Mix equimolar amounts of each strand and heat 10 min at 65°C.

10. Cool slowly to room temperature for 20 min. Check the A_{260} to determine the concentration. Monitor hybridization by running an aliquot on a 4% sieving agarose gel.

11. Set up the three-part ligation (Fig. 8.4.1) as follows: Mix equimolar amounts of the promoter and backbone fragments with a 50-fold molar excess of the

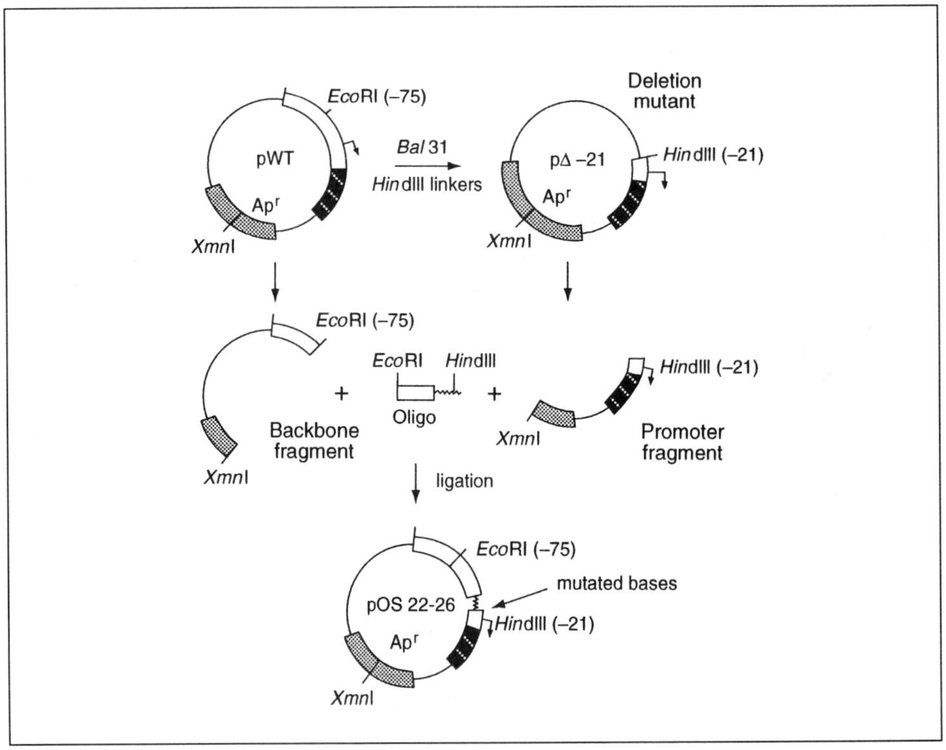

Figure 8.4.1 Assembly of linker-scan mutations using Basic Protocol. pWT is linearized near the region of interest at a site such as the *Eco*RI site at −75. *Bal* 31 is used to create deletions and *Hin*dIII linkers are added to the deletion endpoints. The DNA is digested with *Hin*dIII and *Xmn*I and then is ligated to a complementary *Xmn*I-*Hin*dIII vector fragment from pWT to form the deletion mutation pΔ −21. The backbone fragment is prepared from pWT by an *Xmn*I-*Eco*RI digest and the promoter fragment is prepared from pΔ −21 by a *Hin*dIII-*Xmn*I digest. After isolation, the two fragments are ligated together with the hybridized oligonucleotides to create the linker-scan mutation pOS 22-26. This plasmid is identical to pWT except for the mutated bases at the site of the *Hin*dIII linker at −21.

hybridized oligonucleotides relative to the plasmid fragments; use ~400 ng total DNA in 10 μl. Add 0.4 μl (160 cohesive-end units) T4 DNA ligase and incubate 2 hr at 15°C. If a blunt site is being used at the promoter–backbone fragment junction, add 1 μl (400 U) ligase and incubate an additional 2 hr at 30°C. Dilute 40-fold with TE buffer to 1 μg/ml.

12. Heat the diluted ligation reaction 10 min at 65°C and cool slowly to room temperature.

    Because the hybridized oligos do not have 5′ terminal phosphates, each end of the backbone and promoter fragment will be covalently attached to one strand of a double-stranded oligonucleotide. To circularize the plasmid, it is necessary to melt the noncovalently bound strand of each oligonucleotide away from the ligated fragments.

13. Transform competent *E. coli* cells with 20 to 25 μl of the diluted ligation reaction and plate on selective media. If 5% of the ligation is plated, 20 to 30 colonies are obtained, where virtually all colonies have the correct structure.

14. Pick colonies and prepare minipreps of plasmid DNA. Check the DNA by restriction digests and DNA sequencing (*UNITS 7.4 & 7.5*).

LINKER SCANNING USING OLIGONUCLEOTIDE-DIRECTED MUTAGENESIS

ALTERNATE PROTOCOL

Additional Materials

M13 template (*UNIT 1.15*)
E. coli dut⁻ ung⁻ strain (*UNIT 8.1*)
T4 DNA polymerase (*UNIT 3.5*)

1. Design a set of oligonucleotides, with several adjacent bases mutated as a group flanked by wild-type sequences to allow hybridization to the appropriate region of the gene. A set of exact linker-scan mutations will be produced.

2. Hybridize the oligonucleotides to the M13 template, extend with T4 DNA polymerase, and transform wild-type *E. coli* (*UNIT 8.1*). Plate and isolate double-stranded RF DNA from colonies of each transformation. Sequence the DNA purified from the transformants.

    Mutations are introduced at 50% to 80% efficiency.

Reference: McKnight and Kingsbury, 1982.

Contributor: John M. Greene

UNIT 8.5

Site-Directed Mutagenesis by the Polymerase Chain Reaction

PCR can be used as a quick and efficient method for introducing any desired sequence change into the DNA of interest.

BASIC PROTOCOL 1

INTRODUCTION OF RESTRICTION ENDONUCLEASE SITES BY PCR

The strategy is outlined in Figure 8.5.1.

Materials *(see* APPENDIX 1 *for items with ✓)*

DNA sample to be mutagenized
pUC19 plasmid vector (*UNIT 1.5*) or similar high-copy-number plasmid having M13 flanking primer sequences
500 ng/µl (100 pM/µl) oligonucleotide primers incorporating the restriction enzyme site
✓ TE buffer
✓ 10× amplification buffer
2 mM 4dNTP mix (*UNIT 3.4*)
500 ng/µl (100 pM/µl) M13 flanking sequence primers: forward (NEB) and reverse (NEB)
5 U/µl *Taq* DNA polymerase (*UNITS 15.1 & 3.5*)
Mineral oil
Chloroform (*UNIT 2.1*)
Buffered phenol (*UNIT 2.1*)
100% ethanol
Appropriate restriction endonucleases (Table 8.5.1)

500-µl microcentrifuge tube
Automated thermal cycler

1. Subclone DNA to be mutagenized into high-copy-number vector using restriction sites flanking the area to be mutated (*UNIT 3.16*).

   A different restriction site should be used for each end of the fragment to facilitate cloning at various steps in the procedure.

2. Prepare template DNA by plasmid miniprep (*UNIT 1.6*). Resuspend 100 ng in TE buffer to 1 ng/µl final.

3. Synthesize oligonucleotide primers (1 and 2; Fig. 8.5.1B) and purify by denaturing polyacrylamide gel electrophoresis (*UNIT 2.11*). Resuspend oligonucleotides in 500 µl TE buffer. Determine absorbance at A_{260} and adjust to 500 ng/µl.

   The 5′ end of each primer consists of the restriction site to be introduced, preceded by three or four extra nucleotides (a "clamp" sequence). The restriction site sequences are followed by ≥15 bases that are homologous to the template DNA.

4. Combine the following in each of two 500-µl microcentrifuge tubes, adding oligonucleotides 1 and 2 to separate tubes:

   10 µl (10 ng) template DNA
   10 µl 10× amplification buffer
   10 µl 2 mM 4dNTP mix
   1 µl (500 ng) oligonucleotide 1 *or* 2 (100 pM final)

continued

8.5

1 μl (500 ng) appropriate M13 flanking sequence primer, forward *or* reverse (100 pM final)

H_2O to 99.5 μl

0.5 μl *Taq* DNA polymerase (5 U/μl).

Overlay reaction with 100 μl mineral oil.

5. Carry out PCR in an automated thermal cycler for 20 to 25 cycles under the following conditions (*UNIT 15.1*):

45 sec	93°C	(denaturation)
2 min	50°C	(hybridization)
2 min	72°C	(extension)

After last cycle, extend for an additional 10 min at 72°C.

6. Analyze 4 μl by nondenaturing agarose or acrylamide gel electrophoresis to verify that the amplification has yielded the predicted product (*UNITS 2.5 & 2.7*).

7. Remove mineral oil and extract once with chloroform to remove remaining oil (*UNIT 2.1*). Extract with buffered phenol and concentrate by precipitation with 100% ethanol.

8. Digest half the amplified DNA with the restriction endonucleases for the flanking and introduced sites (*UNIT 3.1*). Purify digested fragments on a low gelling/melting agarose gel (*UNIT 2.6*).

9. Ligate and subclone both fragments into an appropriately digested vector to obtain a recombinant plasmid containing a single DNA fragment incorporating the new restriction site.

10. Transform plasmid into *E. coli* (*UNIT 1.8*). Prepare DNA by plasmid miniprep (*UNIT 1.6*).

11. Analyze amplified fragment portion of plasmid by DNA sequencing to confirm the addition of the mutation (*UNITS 7.4 & 7.5*).

INTRODUCTION OF POINT MUTATIONS BY PCR

BASIC PROTOCOL 2

The strategy is outlined in Figure 8.5.2.

Materials

DNA sample to be mutagenized
Oligonucleotide primers incorporating the point mutation
Klenow fragment of *E. coli* DNA polymerase I (*UNIT 3.5*)
Appropriate restriction endonuclease (Table 8.5.1)

1. Prepare template DNA (steps 1 and 2 of Basic Protocol 1).

2. Synthesize and purify oligonucleotide primers (3 and 4; Fig. 8.5.2B), and phosphorylate 5′ ends (*UNIT 3.10*).

   The oligonucleotide primers must be homologous to the template DNA for ≥15 bases. A 4-base "clamp" sequence is not added to these primers.

3. Amplify template DNA (steps 4 and 5 of Basic Protocol 1). After final extension step, add 5 U Klenow fragment and incubate 15 min at 30°C.

4. Analyze and process reaction (steps 6 and 7 of Basic Protocol 1).

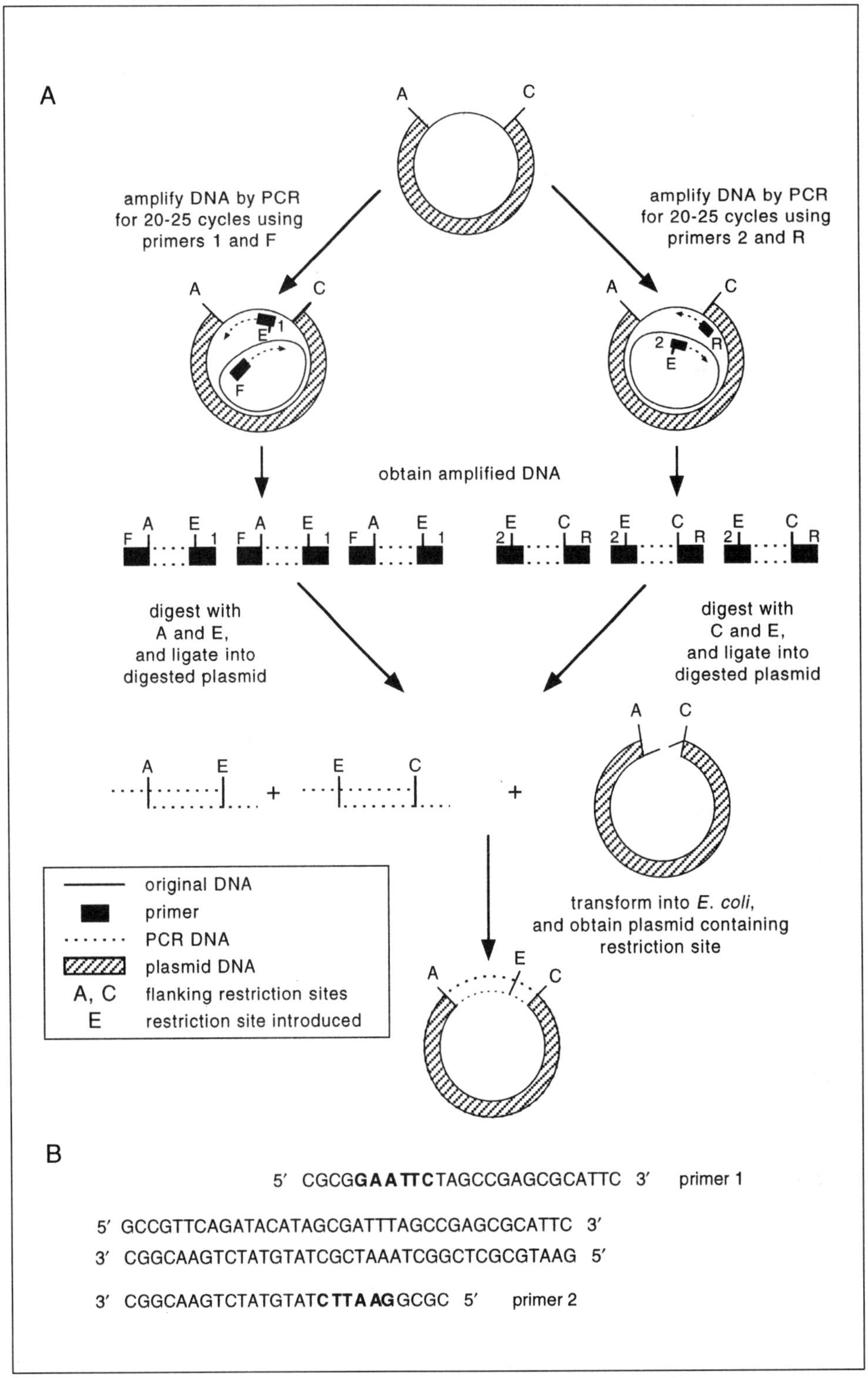

Figure 8.5.1 Introduction of a restriction site into a specific DNA fragment. (**A**) The fragment of interest is cloned into a high-copy-number vector. Sites for two oligonucleotide primers, such as the M13 forward and reverse primers (F and R), flank the cloning site. In two separate reactions, fragments upstream (A—E) and downstream (E—C) of the introduced site are PCR-amplified using flanking primers and oligonucleotides containing site to be introduced (primers F and 1, R and 2). The amplified fragments are digested with the appropriate restriction endonucleases, then ligated and subcloned into an appropriately cut vector and transformed into *E. coli*. The resultant plasmid contains an inserted fragment identical to the original DNA except for the introduced restriction site. (**B**) The oligonucleotides needed to change the primary sequence to an *Eco*RI restriction site are indicated. Note the four-base "clamp" sequence at 5′ end of primers.

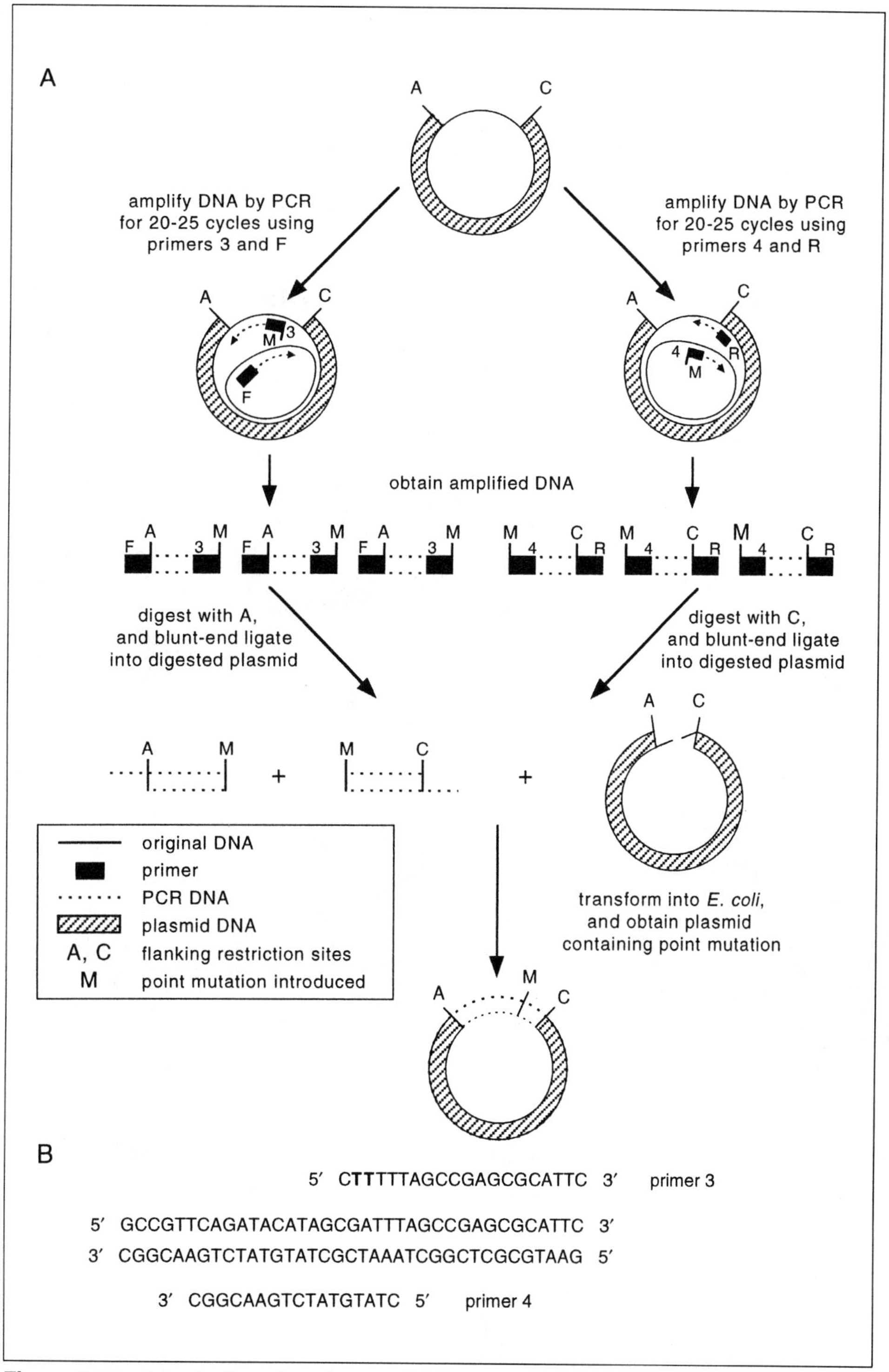

Figure 8.5.2 Introduction of a point mutation into a specific DNA fragment. (**A**) The fragment of interest is cloned into a high-copy-number plasmid vector. Sites for two oligonucleotide primers, such as the M13 forward and reverse primers (F and R), flank the cloning site. In two separate reactions, fragments upstream (A—M) and downstream (M—C) of the point mutation are PCR amplified using the flanking primers and oligonucleotides containing the point mutation to be introduced (primers F and 3, R and 4). The fragments are digested with the appropriate restriction endonucleases and made blunt-ended. These fragments are ligated and subcloned into an appropriately cut vector and transformed into *E. coli*. The resultant plasmid contains an inserted fragment identical to the original DNA except for the introduced point mutation. (**B**) The oligonucleotides needed to generate a point mutation in the primary sequence are shown. Note that there are no "clamp" sequences at the 5′ ends of the primers.

Table 8.5.1 Relative Efficiencies of Restriction Enzyme Cleavage When Restriction Site is Near End of DNA Fragment[a]

Enzyme	Oligo sequence	Chain length	% cleavage	
			2 hr	20 hr
*Bam*HI	CGGATCCG	8	10	25
	CGGGATCCCG	10	>90	>90
	CGCGGATCCGCG	12	>90	>90
*Bgl*II	CAGATCTG	8	0	0
	GAAGATCTTC	10	75	>90
	GGAAGATCTTCC	12	25	>90
*Cla*I	CATCGATG	8	0	0
	GATCGATC	8	0	0
	CCATCGATGG	10	>90	>90
	CCCATCGATGGG	12	50	50
*Eco*RI	GGAATTCC	8	>90	>90
	CGGAATTCCG	10	>90	>90
	CCGGAATTCCGG	12	>90	>90
*Hin*dIII	CAAGCTTG	8	0	0
	CCAAGCTTGG	10	0	0
	CCCAAGCTTGGG	12	10	75
*Nhe*I	GGCTAGCC	8	0	0
	CGGCTAGCCG	10	10	25
	CTAGCTAGCTAG	12	10	50
*Not*I	TTGCGGCCGCAA	12	0	0
	ATTTGCGGCCGCTTTA	16	10	10
	AAATATGCGGCCGCTATAAA	20	10	10
	ATAAGAATGCGGCCGCTAAACTAT	24	25	90
	AAGGAAAAAAGCGGCCGCAAAAGGAAAA	28	25	>90
*Pst*I	GCTGCAGC	8	0	0
	TGCACTGCAGTGCA	14	10	10
*Pvu*I	CCGATCGG	8	0	0
	ATCGATCGAT	10	10	25
	TCGCGATCGCGA	12	0	10
*Sac*I	CGAGCTCG	8	10	10
*Sac*II	GCCGCGGC	8	0	0
	TCCCCGCGGGGA	12	50	>90
*Sma*I	CCCGGG	6	0	10
	CCCCGGGG	8	0	10
	CCCCCGGGGG	10	10	50
	TCCCCCGGGGGA	12	>90	>90
*Spe*I	GACTAGTC	8	10	>90
	GGACTAGTCC	10	10	>90
	CGGACTAGTCCG	12	0	50
	CTAGACTAGTCTAG	14	0	50
*Sph*I	GGCATGCC	8	0	0
	CATGCATGCATG	12	0	25
	ACATGCATGCATGT	14	10	50
*Xba*I	CTCTAGAG	8	0	0
	GCTCTAGAGC	10	>90	>90
	TGCTCTAGAGCA	12	75	>90
	CTAGTCTAGACTAG	14	75	>90
*Xho*I	CCTCGAGG	8	0	0
	CCCTCGAGGG	10	10	25
	CCGCTCGAGCGG	12	10	75

[a]Reprinted with permission from New England Biolabs.

A

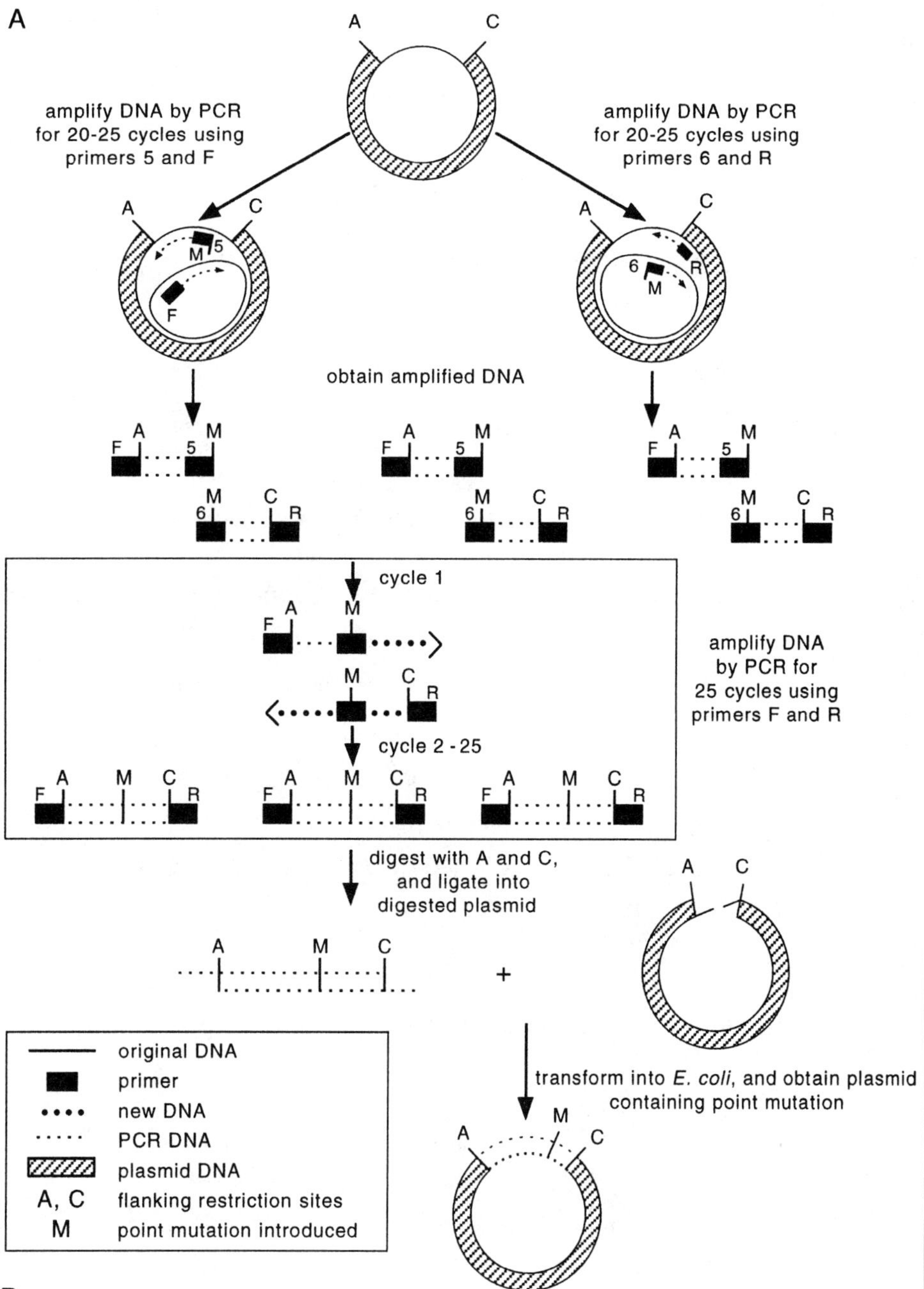

B

5′ ATAGC**TT**TTTAGCCGAGCGCAT 3′ primer 5

5′ GCCGTTCAGATACATAGCGATTTAGCCGAGCGCATTC 3′
3′ CGGCAAGTCTATGTATCGCTAAATCGGCTCGCGTAAG 5′

3′ GTCTATGTATCG**AA**AAATC 5′ primer 6

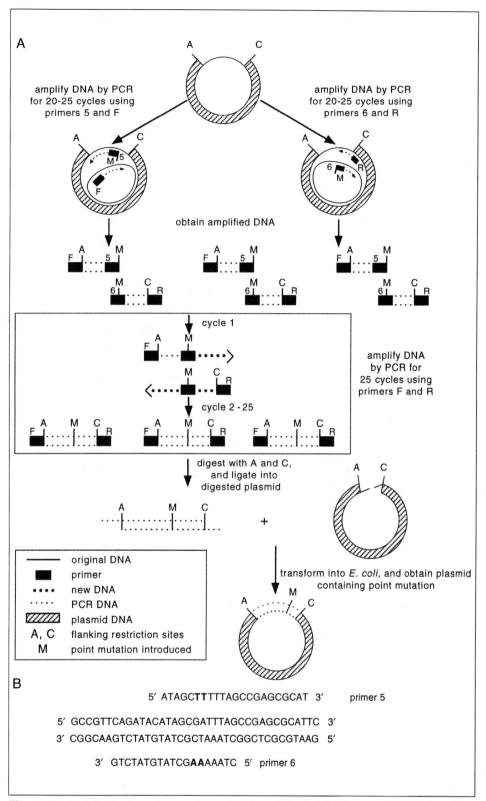

Figure 8.5.3 Introduction of a point mutation by sequential PCR steps. (**A**) The first steps of this protocol are as described in Figure 8.5.2A except primers 5 and 6 are used. The amplified fragments are then placed in the same tube and amplified in a second PCR step (using primers F and R only). This second PCR step obviates the need for the blunt-end ligation. The full-length fragment generated here is digested with the appropriate restriction endonucleases, subcloned into an appropriately cut vector, and transformed into *E. coli.* The resultant plasmid contains an inserted fragment identical to the original DNA except for the point mutation. (**B**) The oligonucleotide primers needed to generate a point mutation in this procedure are shown. Note that there are 12 bases of overlap between the two primers.

5. Digest half the amplified fragments with the restriction endonucleases for the flanking sequences. Purify digested fragments on a low gelling/melting agarose gel.

6. Subclone the two amplified fragments into an appropriately digested vector by blunt-end ligation.

7. Carry out steps 10 and 11 of Basic Protocol 1.

ALTERNATE PROTOCOL

INTRODUCTION OF A POINT MUTATION BY SEQUENTIAL PCR STEPS

The strategy is outlined in Figure 8.5.3.

1. Prepare the template DNA (steps 1 and 2 of Basic Protocol 1).

2. Synthesize and purify the oligonucleotide primers (5 and 6; Fig. 8.5.3B).

   The oligonucleotides must be homologous to the template for 15 to 20 bases and must overlap with one another by ≥10 bases. The 5′ end does not have a "clamp" sequence.

3. Amplify the template and generate blunt-end fragments (step 3 of Basic Protocol 2).

4. Purify fragments by nondenaturing agarose gel electrophoresis. Resuspend in TE buffer at 1 ng/µl.

5. Combine the following in a 500-µl microcentrifuge tube:

   10 µl (10 ng) each amplified fragment
   1 µl (500 ng) each flanking sequence primer (each 100 pM final)
   10 µl 10× amplification buffer
   10 µl 2 mM 4dNTP mix
   H_2O to 99.5 µl
   0.5 µl *Taq* DNA polymerase (5 U/µl).

   Overlay with 100 µl mineral oil.

6. Carry out PCR for 20 to 25 cycles (step 5 of Basic Protocol 1). Analyze and process the reaction mix (steps 6 and 7 of Basic Protocol 1).

7. Digest DNA fragment with appropriate restriction endonuclease for the flanking sites. Purify fragment on a low gelling/melting agarose gel. Subclone into an appropriately digested vector.

8. Carry out steps 10 and 11, Basic Protocol 1.

Contributor: Brendan Cormack

Introduction of DNA into Mammalian Cells

9

After cloning a gene, many researchers wish to analyze its characteristics by reintroducing it into various cell types. There are many reasons to do this. The genetic elements responsible for regulation of expression can be determined by mutating the gene and subsequently examining its activity under a variety of physiological conditions. The effects that the gene has on cellular growth can be ascertained by producing cell lines that express the gene and determining their phenotypes. Cell lines can be produced that overexpress the gene, allowing purification of the product for biochemical characterization, or large-scale production of a product for use as a drug. All these goals require the ability to introduce DNA into a cell efficiently. The protocols in this chapter cover methods that are used to introduce genes into mammalian cells, as well as techniques used to investigate their regulation and to overexpress the gene product.

Perhaps the most frequent reason for introducing DNA into mammalian cells is to analyze gene expression. It is critical that an appropriate cell line be used in these studies. For example, the β-globin gene is expressed at a low level that is not inducible in HeLa cells, but is inducible upon differentiation of MEL cells. Clearly, MEL cells should be used to investigate the mechanism of the induction. Many researchers must therefore transfect DNA into a cell line that is not chosen for ease of transfection, but instead is chosen for its biological properties.

The first section of the chapter presents four techniques for introducing DNA into mammalian cells: calcium phosphate transfection (*UNIT 9.1*), DEAE-dextran transfection (*UNIT 9.2*), electroporation (*UNIT 9.3*), and liposome-mediated transfection (*UNIT 9.4*). The first two procedures produce a chemical environment that results in DNA attaching to the cell surface; the DNA is then endocytosed by as yet uncharacterized pathways. The parameters for transfecting cells by these techniques vary for each different cell type and therefore need to be carefully optimized (*UNIT 9.9*).

Electroporation uses an electric field to open up pores in the cell. The DNA presumably diffuses into the cell through the pores. This technique therefore is not dependent upon special characteristics of the cell and can be used with virtually any cell type. The optimal amplitude and length of pulse will vary for each cell type, so this procedure too requires fine-tuning.

The simplest way to optimize parameters for transfection is to use an easily assayed reporter gene as the transfected DNA during the optimization. Reporter systems are covered in the second section of this chapter (*UNITS 9.6 & 9.7*), with descriptions of how these genes can be used to optimize transfection efficiency, and how they can be used to analyze promoter function. Reporter genes allow an indirect measure of promoter activity, so the researcher will need to analyze RNA levels and the structure of the RNA produced from the transfected gene. *UNIT 9.8* describes how the techniques of RNA isolation and analysis (Chapter 4) can be used in conjunction with transfection procedures to analyze directly the regulation of a gene.

CHOICE OF TRANSFECTION METHOD

There are two types of transfections that are routinely done in mammalian systems—*transient* and *stable* or *permanent*. In a transient transfection, transcription or replication of the transfected gene can be analyzed between 1 and 4 days after introduction of the DNA, generally by harvesting the transfected cell. Alternatively, many experiments require formation of cell lines that contain gene(s) that are integrated into chromosomal DNA resulting in a stable or permanent transfection; *UNIT 9.5* describes selectable markers that are used in this type of transfection.

All four protocols described above can be used for transient transfections. Three of these—calcium phosphate transfection, electroporation, and liposome-mediated transfection—can be used to efficiently produce cell lines containing stably integrated DNA. Electroporation is most easily done using suspension cultures, as cells must be transferred into cuvettes, while calcium phosphate and liposome-mediated transfections are most easily done using adherent cells. Two methods for calcium phosphate–mediated transfection are presented in the first unit. The HEPES-buffered system can be used for producing either transiently or stably transformed lines. The second method, based on a BES buffer system, is now widely used for the stable transformation of most common fibroblast and epithelial cells, and for these applications is at least 10-fold more efficient than other methods. DEAE-dextran transfection does not work well when producing stable cells lines, but is more reproducible than calcium phosphate transfection when used in transient protocols. Electroporation is also very reproducible, but requires more cells than either of the chemical procedures do. While these considerations are important in choosing a transfection protocol, the most critical parameter may be the efficiency with which DNA is introduced into the recipient cell, and this might have to be determined experimentally.

There are numerous techniques used in mammalian cell culture that are not discussed in this chapter. In particular, it is assumed that the conditions needed for optimal growth and passage of the cell lines used will already have been determined. In order for transfection to work efficiently, it is necessary that the recipient cells be healthy. A brief discussion below describes the reagents used in mammalian cell culture.

VIRAL VECTORS

DNA can also be introduced into eukaryotic cells by using viral vectors. Viral vectors can allow introduction of DNA into virtually all cells in a given experiment. Certain viral systems will result in overexpression of proteins (e.g., baculovirus and vaccinia virus; see Chapter 16) or will allow single-copy stable integrants of expressed genes (e.g., retroviral vectors, *UNITS 9.10-9.14*). Viral systems therefore can be much more powerful than transfection systems—however, constructing a recombinant virus and packaging that virus involves a tremendous effort. It is therefore prudent to plan experiments extremely carefully when using viral vectors to ensure that a recombinant virus is really needed and that the optimal virus is constructed.

Retroviruses are not generally used for overproducing a protein, but are used instead to introduce a gene in a stable fashion into a cell line or into cells in an animal. These vectors can therefore be used either to express a gene product in a set of cells or to express a marker gene that will allow identification of an infected cell and its progeny. An overview of the retroviral life cycle and of retroviral vectors is provided in *UNIT 9.10*. Standard cloning procedures are used to create a retroviral vector, after which a packaging line must be made to produce infectious particles (*UNIT 9.11*). These cell lines can then be used to produce large amounts of recombinant retroviruses (*UNIT 9.12*). The resultant retrovirus stocks may then be used in a variety of experiments; however, it is critical to ensure that they are free of wild-type "helper" retroviruses that may arise from the packaging cell lines. These helper viruses are replication-competent, and therefore can create serious complications in certain experiments. Assays that can be used to detect helper virus in a preparation are described in *UNIT 9.13*. Detailed protocols for introducing recombinant retroviruses into animals are beyond the scope of this chapter, although *UNIT 9.14* provides a beginning point for these in vivo experiments.

MAMMALIAN CELL CULTURE

Unlike *E. coli,* mammalian cells are extremely sensitive to growth conditions. In addition, because their doubling time is usually 12 to 48 hr, mammalian cultures are susceptible to contamination by bacteria and fungi, and the necessity to maintain sterile conditions during experimental procedures involving cell growth cannot be overemphasized. The four transfection protocols described in this chapter are stressful to the cell and thus can cause extensive cell death if the transfected cells are not healthy prior to transfection. This means that cell culture conditions that are marginal for normal growth and passage of cells will probably not suffice for transfection purposes. The following precautions and ingredients are crucial to success in culturing mammalian cells:

1. ***Glassware.*** Numerous solutions commonly used in the laboratory are cytotoxic when present even in trace amounts. Many laboratories therefore keep separate glassware for tissue culture purposes. This glassware is never mixed with glassware used in the laboratory. Tissue culture glassware should be left soaking in water after use in order to prevent drying of residual chemicals onto the glass. For these reasons, plasticware is frequently used for pipets and for storage of reagents used in tissue culture.

2. ***Water.*** Only distilled and deionized water should be used for preparation of solutions used in tissue culture. Some laboratories use a glass still to produce this water, while in some newer facilities, house purification systems are sufficient.

3. ***Media.*** It is possible to purchase sterile media in solution. Alternatively, packaged, premixed powders can be purchased. These can be made in 10- to 20-liter batches in the laboratory. The resulting solution must then be filter sterilized. This is most conveniently done using a high-capacity pump (500 ml/min) and filter apparatus.

Media are supplemented shortly before use with serum. Frequently, fetal bovine serum (FBS; also known as fetal calf serum) is used, but for some uses, less expensive sera (see below) may be used. Other supplements can include glutamine, nonessential amino acids, 2-mercaptoethanol (2-ME), penicillin, and streptomycin sulfate to formulate "complete" media; however, the exact formulation will vary with different cell types. Nomenclature of media employed in this manual indicates the base medium as well as the percentage of serum. For example, "complete DMEM-10" means that DMEM is supplemented with 10% serum plus the other ingredients noted above (see recipe, *UNIT 9.4*).

4. ***Serum.*** Horse, calf, and fetal bovine serum can be purchased. In using serum, it is accepted that a partially undefined material is being used, which may show considerable variation in the ability to support growth of particular cells. Thus, for every culture application, serum should be "screened" to determine if, in fact, it supports the growth of the cells of interest. Such screening can also be done to determine if the serum is unduly stimulatory to the cells of interest, thereby leading to high background responses. There is no easy way to screen lots of serum because the suitability of a given lot can only be determined in the application at hand. It is recommended that for extensive studies of a given cell population, small amounts of different lots of serum be obtained from one or more suppliers with the proviso that the supplier put aside a much larger amount of the lot(s) for possible delivery. The various lots are then tested in an assay of interest (a proliferation assay, a

transient transfection assay, etc.) to determine which lot provides the most vigorous response and the least background response (response in the absence of specific stimulus). The lot with this characteristic is then bought in large amounts and the other lots are "released" (allowed to be sold to other purchasers). Reliable suppliers of serum are Armour and Hyclone.

Generally, heat-inactivated (1 hr at 56°C) serum is required to support long-term cultures. The heat treatment is thought by some labs to reduce the number of viral and other adventitious contaminants and by others to be required to inactivate complement. It therefore has developed into a standard procedure before the use of a certain batch of serum, despite alleged disadvantages (i.e., inactivation of certain growth factors and production of heat-labile serum components that are toxic to cells). If desired, test whether the growth-supporting function of a certain lot of serum can be improved by not heating it, particularly when culturing fastidious cells.

Given the high cost of serum, a lower percentage (5%) can be used for short-term procedures such as washing cells. In addition, calf serum, which is about ten times less expensive than fetal bovine serum, can sometimes be substituted in short-term procedures. Store all serum at −20°C.

5. ***Supplements.*** Many researchers use antibiotics in mammalian culture media to prevent contamination. The most effective addition is gentamicin sulfate which is purchased as a 50 mg/ml solution and used at 20 to 50 µg/ml. A somewhat less expensive (and less stable) antibiotic combination that is also somewhat less effective is a cocktail of penicillin (50 to 100 U/ml) and streptomycin sulfate (100 µg/ml). Note that use of antibiotics in the medium can mask sloppy tissue culture technique. Sloppy technique can result in contamination of a culture by agents that are not susceptible to the antibiotics, such as mycoplasma. Some laboratories therefore do not use antibiotics in order to ensure that strict sterility is being maintained. Some experiments will require other additions to the medium, such as nucleosides or drugs. These solutions should be made up using tissue culture glassware or plasticware. Care should be taken to avoid the use of stir bars or other items that might have trace amounts of laboratory chemicals on them. These solutions should be filter sterilized through Nalgene filters.

2-mercaptoethanol is often included in medium but the precise mode(s) of action of 2-ME has not been defined. The presence of 2-ME during growth of established cell lines appears not to be required, but it has proven to be a critical ingredient in primary cultures. 2-ME must be added to medium immediately before use because its activity rapidly declines in diluted form. Thus, concentrated 2-ME (14.3 M) is diluted to 50 mM in HBSS or PBS and this solution is added to medium at a final concentration of 50 µM. The 50 mM stock solution is stored at 4°C and should be replaced after ~4 months.

Other reagents are added to medium immediately before use; these are typically made in large batches at 100× or 200× concentrations and stored in aliquots of convenient size until use. Stock solutions containing L-glutamine, nonessential amino acids, and antibiotics are stored at −20°C. Stock solutions containing HEPES are stored at room temperature.

6. ***Trypsin/EDTA.*** A buffered salt solution containing 0.5% (w/v) trypsin and 0.2% (w/v) EDTA is used to dissociate adherent cells from tissue culture dishes. This solution can be purchased.

Robert E. Kingston

Calcium Phosphate Transfection

UNIT 9.1

This unit contains two methods for calcium phosphate–based eukaryotic cell transfection that can be used for both transient and stable (*UNIT 9.5*) transfections. Both methods of transfection require very high-quality plasmid DNA.

TRANSFECTION USING CALCIUM PHOSPHATE–DNA PRECIPITATE FORMED IN HEPES

BASIC PROTOCOL

A precipitate containing calcium phosphate and DNA is formed by slowly mixing a HEPES-buffered saline solution with a solution containing calcium chloride and DNA. Depending on the cell type, up to 10% of the cells on a dish will take up the DNA precipitate. Glycerol or dimethyl sulfoxide shock increases the amount of DNA absorbed in some cell types (Support Protocol).

Materials *(see APPENDIX 1 for items with ✓)*

Exponentially growing eukaryotic cells (e.g., HeLa, BALB/c 3T3, NIH 3T3, CHO, or rat embryo fibroblasts)
Complete medium (depending on cell line used)
CsCl-purified plasmid DNA (10-50 µg/transfection, twice-purified; support protocol)
✓ 2.5 M $CaCl_2$
✓ 2× HEPES-buffered saline (HeBS)
✓ Phosphate-buffered saline (PBS)

37°C, 5% CO_2 humidified incubator
10-cm tissue culture plates
15-ml conical tube

1. Split cells into 10-cm tissue culture plates the day before transfection. When transfecting adherent cells that double every 18 to 24 hr, a 1:15 split from a confluent dish generally works well. On the day of transfection, cells should be well separated on the dish. Feed cells with 9.0 ml complete medium 2 to 4 hr prior to addition of precipitate.

2. Ethanol precipitate DNA (10-50 µg/10-cm plate) to be transfected and air dry pellet (*UNIT 2.1*). Resuspend pellet in 450 µl sterile water and add 50 µl of 2.5 M $CaCl_2$.

   Supercoiled DNA works well in transfections. See UNIT 9.9 for optimization parameters.

   Ethanol precipitation sterilizes the DNA to be transfected. For transient transfections, this is not necessary. Instead, make a 450-µl aqueous solution containing the DNA. Keep amount of Tris in the solution to a minimum.

3. Place 500 µl of 2× HeBS in a sterile 15-ml conical tube. Use a mechanical pipettor attached to a plugged 1-ml pipet to bubble the 2× HeBS. Add DNA/$CaCl_2$ solution dropwise with a Pasteur pipet (see Sketch 9.1.A) and immediately vortex 5 sec. Allow precipitate to sit 20 min at room temperature.

4. Add precipitate evenly over a 10-cm plate of cells and gently agitate.

5. Incubate cells 4 to 16 hr under standard growth conditions. Remove medium, wash cells twice with 5 ml of 1× PBS, and feed cells with 10 ml complete medium.

formation of calcium phosphate precipitate

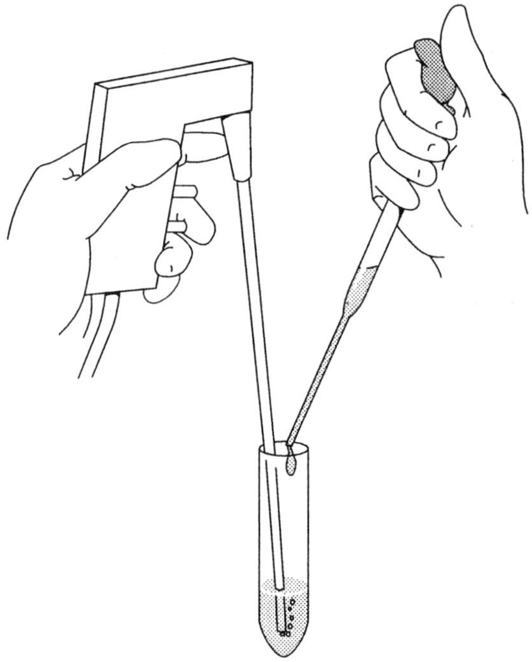

Sketch 9.1A

For hardy cells such as HeLa, NIH 3T3, and BALB/c 3T3, the precipitate can be left on for 16 hr.

6. For transient analysis, harvest cells at desired time point (*UNITS 9.6-9.8, 14.6, & 16.12*). For stable transformation, allow cells to double twice before plating in selective medium (*UNITS 9.5 & 16.12*).

SUPPORT PROTOCOL 1

GLYCEROL/DMSO SHOCK OF MAMMALIAN CELLS

Additional Materials *(also see Basic Protocol)*

Sterile 10% (v/v) glycerol solution or DMSO in complete medium

Transfection efficiency in some cell lines, such as CHO DUKX, is dramatically increased by "shocking" the cells with glycerol or DMSO. Precipitates are left on the cell for only 4 to 6 hr, and the cells are shocked immediately after removal of the precipitate. If desired, replace step 5 with the following:

5a. Incubate cells 4 to 6 hr and remove medium. Add 2 ml of a sterile 10% glycerol solution and let cells sit 3 min at room temperature.

Alternatively, 10% or 20% DMSO can be used. DMSO tends to be somewhat less harmful to the cells, but also may not work as well.

5b. Add 5 ml of 1× PBS to glycerol solution on the cells, mix, and remove solution. Wash cells twice with 5 ml of 1× PBS and feed.

Excessive exposure to glycerol will kill cells.

HIGH-EFFICIENCY TRANSFECTION USING CALCIUM PHOSPHATE–DNA PRECIPITATE FORMED IN BES

ALTERNATE PROTOCOL

With this method, 10% to 50% of the cells stably integrate and express the DNA transfected. Transient expression under these conditions is comparable to that obtained with the basic protocol. Glycerol or DMSO shock does not increase the number of cells transformed.

Additional Materials *(also see Basic Protocol; see* APPENDIX 1 *for items with ✓)*

Complete medium
✓ TE buffer
✓ 2× BES-buffered solution (BBS)

10-cm tissue culture plates
Humidified 35°C, 3% CO_2 incubator
Humidified 35°-37°C, 5% CO_2 incubator

1. Seed exponentially growing cells at 5×10^5 cells/10-cm plate in 10 ml complete medium the day prior to transfection. There should be $<10^6$ cells/plate just prior to infection with enough surface area remaining for at least two more doublings.
2. Dilute plasmid DNA with TE buffer to 1 µg/µl. Store at 4°C.

   The optimum amount of plasmid to use can be determined by transfecting three plates of cells with 10, 20, and 30 µg of plasmid DNA and incubating overnight. The plates are examined with a microscope at 100× magnification. An even, granular precipitate will form at the optimal DNA concentrations.

3. Mix optimum amount of plasmid DNA with 500 µl of 0.25 M $CaCl_2$. Add 500 µl 2× BBS, mix well, and incubate 10 to 20 min at room temperature.
4. Add calcium phosphate–DNA solution dropwise onto plate while swirling the plate. Incubate 15 to 24 hr in a 35°C, 3% CO_2 incubator.

   Level of carbon dioxide is critical. Use a Fyrite gas analyzer to measure percent CO_2 prior to incubation.

5. Wash cells twice with 5 ml PBS, and add 10 ml complete medium. For stable transformation, incubate overnight in a 35°-37°C, 5% CO_2 incubator. For transient expression, incubate cells for 48 to 72 hr (*UNITS 9.6-9.8, 14.6, & 16.12*).
6. Split cells before beginning selection of stable transformants—1:10 to 1:30 depending on growth rate. Incubate overnight in a 35°-37°C, 5% CO_2 incubator.
7. Start selection by changing to selection medium or by incubating cells under appropriate selection conditions (*UNITS 9.5 & 16.12*).

PLASMID DNA PURIFICATION

SUPPORT PROTOCOL 2

Plasmid DNA prepared by the modified lysozyme-Triton method described below gives consistently good transfection efficiencies in the Basic and Alternate Protocols. This procedure yields ~1 mg plasmid DNA/liter culture.

Additional Materials *(also see Basic Protocol; see* APPENDIX 1 *for items with ✓)*

Plasmid-bearing *E. coli* strain
10 mg/ml chloramphenicol (*UNIT 1.4*)
✓ Sucrose/Tris/EDTA solution
10 mg/ml chicken egg white lysozyme (Sigma; prepare fresh in sucrose/Tris/EDTA solution)

✓ Triton lytic mix
Cesium chloride
✓ 10 mg/ml ethidium bromide
CsCl/TE solution: 100 ml TE buffer/100 g CsCl, with and without 0.2 mg/ml ethidium bromide (optional)
1:1 (v/v) phenol/chloroform
0.5% (v/v) SDS in TE buffer
100% ethanol

Beckman J-6B low-speed centrifuge with JS-4.2 rotor and 1-liter centrifuge bottles
Beckman type 35, VTi65, and Sorvall SS-34 rotors or equivalents
5-ml quick-seal ultracentrifuge tubes

1. Grow 1 liter of a plasmid-bearing *E. coli* strain to an OD_{600} of 1.0, add solid chloramphenicol to 150 µg/ml, and grow overnight. Centrifuge cells 30 min at 1300 × *g*, 4°C, in 1-liter bottle.

2. Resuspend pellet in 10 ml sucrose/Tris/EDTA solution. Add 2 ml of freshly prepared 10 mg/ml lysozyme, gently mix, and incubate ≤30 min at room temperature.

3. Add 4 ml Triton lytic mix and incubate to ≤30 min at 37°C, swirling occasionally. Centrifuge 30 min at 69,000 × *g*. Decant and measure supernatant, leaving any gelatinous material behind.

4. Add 0.95 g cesium chloride/ml supernatant and dissolve completely and 0.1 ml of 10 mg/ml ethidium bromide solution/ml supernatant. Centrifuge 20 min at 7000 rpm in a Sorvall SS-34 rotor.

5. Transfer supernatant to 5-ml ultracentrifuge tube. Top the tube, if necessary, with CsCl/TE solution, and seal. Band plasmid by centrifuging overnight at 260,000 × *g*.

6. Remove (lower) plasmid band as described in *UNIT 1.7*. Add plasmid band to another ultracentrifuge tube, top with CsCl/TE solution containing 0.2 mg/ml ethidium bromide, and seal. Centrifuge overnight at 260,000 × *g*.

7. Remove plasmid (lower) band. Extract with 1:1 (v/v) phenol/chloroform two times. Remove and discard upper phase, being careful not to remove white layer containing DNA at the interface. Measure volume of lower phase and transfer to clean tube. Add 2 vol water, mix, then add 6 vol 100% ethanol (66% final). Precipitate overnight at −20°C.

8. Centrifuge 20 min at 3000 × *g*, 4°C. Discard supernatant and dry pellet.

9. Dissolve pellet in 1 ml 0.5% SDS in TE buffer. Repeat phenol/chloroform extraction, saving upper phase and leaving behind white layer at interface this time.

10. Ethanol precipitate twice and dissolve pellet in TE buffer. Store at 4°C and use as described in the Basic and Alternate Protocols.

References: Chen and Okayama, 1987, 1988; Ishiura et al., 1982.

Contributors: Robert E. Kingston, Claudia A. Chen, and Hiroto Okayama

Transfection Using DEAE-Dextran

UNIT 9.2

BASIC PROTOCOL

This procedure is very simple and is more reproducible than calcium phosphate–mediated transfection. It works very well in transient expression experiments (UNITS 16.12 & 16.13), but is not useful for production of stably transfected cell lines. This protocol has been optimized for use with mouse L cell fibroblasts but minor modifications can be made so that it is appropriate for essentially any cell type. It is critical that the procedure be optimized for the cell type being studied (UNIT 9.9).

Materials *(see APPENDIX 1 for items with ✓)*

Mouse L cell fibroblasts
Plasmid DNA
✓ Tris-buffered saline (TBS)
✓ Complete Dulbeccos minimum essential medium containing 10% NuSerum (complete DMEM-10 NS)
✓ 10 mg/ml DEAE-dextran in TBS, 37°C
10% (v/v) dimethyl sulfoxide (DMSO) in PBS (filter sterilized and stored at room temperature)
✓ Phosphate-buffered saline (PBS)
Humidified 37°C, 5% CO_2 incubator

1. Plate ~5×10^5 mouse L cell fibroblasts/10-cm dish. Grow 3 days to 30% to 50% confluence.

    Some cell types that are sensitive to DEAE-dextran should be plated at higher densities, e.g., mouse AtT-20 cells (1×10^6 cells/10-cm dish) and certain primary cells (up to 5×10^6 cells/10-cm dish).

2. Ethanol precipitate 4 µg of desired plasmid DNA for each plate to be transfected (UNIT 2.1). Air dry pellet in a tissue culture hood and resuspend in 40 µl TBS.

3. Aspirate medium, wash plates with 10 ml of 1× PBS, and 4 ml complete DMEM-10 NS.

    NuSerum allows cells to withstand the DEAE-dextran mixture for a greater length of time than does calf or fetal bovine serum, resulting in increased transfection efficiencies. The volume of 10% NS used varies with the size of the plate as follows (regardless of container size, the cells should be 30% to 50% confluent when transfected; see step 1):

Container	10% NuSerum (ml)
150-cm^2 flask	10
10-cm dish	4
60-mm dish	2
35-mm dish	1.5

4. Add DNA slowly (while constantly shaking the tube) to 80 µl of warm 10 mg/ml DEAE-dextran in TBS. Using a 200-µl pipettor, add 120 µl DNA/DEAE-dextran dropwise to each plate, distributing it evenly over the entire plate. Swirl gently until a uniform red color is attained. Incubate 4 hr (less for some cell types) in a humidified CO_2 incubator.

    The final concentration of DEAE-dextran in the medium is 200 µg/ml; however, the optimal concentration for transfecting a given cell type must be determined experi-

mentally. To modify DEAE-dextran concentration (assuming 4 ml of NuSerum on a 10-cm dish), scale the volumes as in the following examples:

Final DEAE-dextran in medium (μg/ml)	DNA in TBS (μl)	10 mg/ml DEAE-dextran (μl)	Total volume (μl)
400	80	160	240
200	40	80	120
100	20	40	60
50	10	20	30

5. Aspirate DNA/DEAE-dextran/medium from plates. Shock cells by adding 5 ml of 10% DMSO in PBS. Incubate 1 min at room temperature. Aspirate DMSO, wash with 5 ml of 1× PBS, and aspirate. Add 10 ml complete medium to each plate.

6. Incubate cells and analyze at appropriate time points (*UNITS 9.6-9.8, 14.6, 16.12, & 16.13*).

ALTERNATE PROTOCOL 1

DEAE-DEXTRAN TRANSFECTION IN BATCH

1. Plate 3×10^6 mouse L cell fibroblasts on a 25 × 25–cm dish in 60 ml of complete medium. Grow 3 days to 30% to 50% confluence.

2. Ethanol precipitate DNA and resuspend in 360 μl TBS. Add DNA to 720 μl of warm 10 mg/ml DEAE-dextran in TBS. Add DNA/DEAE-dextran to 36 ml complete DMEM-10 NS.

3. Aspirate medium from cells and add DNA/DEAE-dextran/medium mixture. Incubate 4 hr in a humidified 37°C, 5% CO_2 incubator.

4. Aspirate medium and shock cells by adding 30 ml of 10% DMSO in PBS for 1 min. Aspirate DMSO, wash dish with 30 ml of 1× PBS, aspirate, and add 60 ml of complete medium. Incubate overnight at 37°C.

5. Aspirate medium. Wash twice with 35 ml of 1× PBS. Split cells into six 10-cm dishes, harvest, and analyze (*UNITS 9.6-9.8, 14.6, 16.12, & 16.13*).

ALTERNATE PROTOCOL 2

TRANSFECTION OF CELLS IN SUSPENSION

This procedure has been used successfully for transfecting lymphoid cell lines in suspension.

Additional Materials *(also see Basic Protocol; see* APPENDIX 1 *for items with ✓)*

- ✓ Suspension TBS (STBS) solution
- ✓ 10 mg/ml DEAE-dextran in STBS solution
- Medium without serum
- Beckman JS-4.2 rotor or equivalent

1. Ethanol precipitate DNA (~10 μg/2×10^7 cells) and air dry. Resuspend in 0.5 ml STBS solution.

2. Centrifuge 2×10^7 cells in a sterile 50-ml polypropylene tube 5 min at 640 × *g*, room temperature. Wash cells with 5 ml warm STBS solution and centrifuge again.

3. Prepare 0.5 ml of 2× DEAE-dextran in STBS solution. Add this solution to 0.5 ml DNA solution (step 1) and mix. Resuspend cells in 1.0 ml DEAE-dextran/DNA solution. Incubate 30 to 90 min, tapping cells every 30 min to keep them from clumping.

   The optimal DEAE-dextran concentration varies between 100 and 500 μg/ml. Precise time for incubation depends on the cell line and should be optimized (UNIT 9.9).

4. Add DMSO to cells, dropwise to 10% final, swirling constantly to ensure rapid mixing. Incubate 2 to 3 min at room temperature.

5. Add 15 ml STBS solution and centrifuge 5 min at 300 × *g*.

6. Resuspend cells in 10 ml STBS solution and centrifuge again. Wash cells with 10 ml medium without serum.

7. Resuspend cells in complete medium at $2\text{-}10 \times 10^5$ cells/ml and incubate until ready to harvest and analyze (*UNITS 9.6-9.8, 14.6, 16.12, & 16.13*), generally 48 hr posttransfection.

CHLOROQUINE TREATMENT OF CELLS

ALTERNATE PROTOCOL 3

Some investigators claim chloroquine disphosphate significantly increases transfection efficiency, but others believe it is more trouble than it is worth. The cells should be closely monitored while they are in chloroquine diphosphate because it is extremely cytotoxic, and the medium should be changed immediately if the health of the cells is deteriorating. Modify the Basic Protocol by adding chloroquine to the medium immediately after the DNA/DEAE-dextran is added at step 4:

4a. Add chloroquine diphosphate to 100 μM final in medium and mix immediately. Incubate ≤4 hr. Proceed as in Basic Protocol.

References: Lopata et al., 1984; Sussman and Milman, 1984.

Contributor: Richard F Selden

Transfection by Electroporation

UNIT 9.3

Electroporation—the use of high-voltage electric shocks to transiently or stably introduce DNA into cells—is a procedure that is gaining in popularity. It can be used with most cell types, yields a high frequency of both stable transformation and transient gene expression, and, because it requires fewer steps, can be easier than alternate techniques (*UNITS 9.1, 9.2, 9.4 & 9.9*).

ELECTROPORATION INTO MAMMALIAN CELLS

BASIC PROTOCOL

Materials *(see APPENDIX 1 for items with ✓)*

Mammalian cells to be transfected
✓ Complete medium without and with appropriate selective agents
✓ Electroporation buffer, ice-cold
Linear or supercoiled, purified DNA preparation (see step 5)

Beckman JS-4.2 rotor or equivalent
Electroporation cuvettes (Bio-Rad) and power source

1. Grow cells to be transfected to late-log phase in complete medium. Harvest cells by centrifuging 5 min at 640 × *g*, 4°C.

   Adherent cells are first trypsinized (introduction to Chapter 9) and the trypsin inactivated with serum.

2. Wash the cells by resuspending cell pellet in half its original volume of ice-cold electroporation buffer and centrifuging 5 min at 640 × *g*, 4°C.

3. Resuspend cells at 1 × 10^7/ml in electroporation buffer at 0°C for stable transfection. Higher concentrations of cells (up to 8 × 10^7/ml) may be used for transient expression.

4. Transfer 0.5-ml aliquots of the cell suspension into desired number of electroporation cuvettes set on ice.

5. Add DNA to cell suspension in the cuvettes on ice. Mix DNA/cell suspension by holding the cuvette on the two "window sides" and flicking the bottom. Incubate 5 min on ice.

   For stable transformation, 1 to 10 μg DNA should be linearized by cleavage with a restriction enzyme (UNIT 3.1) that cuts in a nonessential region and purified by phenol extraction and ethanol precipitation (UNIT 2.1). For transient expression, the DNA, 10 to 40 μg, may be left supercoiled. In either case, the DNA should be purified through two preparative CsCl/ethidium bromide equilibrium gradients (UNIT 1.7) followed by phenol extraction and ethanol precipitation. The DNA stock may be sterilized by one ether extraction (UNIT 2.1); the (top) ether phase is removed and the DNA solution allowed to dry for a few minutes to evaporate any remaining ether.

6. Place cuvette in the holder in the electroporation apparatus (at room temperature) and shock one or more times at the desired voltage and capacitance settings.

   The number of shocks and the voltage and capacitance settings will vary depending on the cell type and should be optimized (UNIT 9.9).

7. Place cuvette on ice for 10 min.

8. Dilute transfected cells 20-fold in nonselective complete medium and rinse cuvette with this same medium to remove all transfected cells.

9a. *For stable transformation:* Grow cells 48 hr (about two generations) in nonselective medium, then transfer to antibiotic-containing medium.

   Selection conditions will vary with cell type. For example, neo selection generally requires ~400 μg/ml G418 in the medium. XGPRT selection requires 1 μg/ml mycophenolic acid, 250 μg/ml xanthine, and 15 μg/ml hypoxanthine in the medium.

9b. *For transient expression:* Incubate cells 50 to 60 hr, then harvest cells for transient expression assays.

9.3

ALTERNATE PROTOCOL

ELECTROPORATION INTO PLANT PROTOPLASTS

Additional Materials *(also see Basic Protocol; see APPENDIX 1 for items with ✓)*

5-mm strips (1 g dry weight) sterile plant material
✓ Protoplast solution
✓ Plant electroporation buffer

80-μm-mesh nylon screen
Sterile 15-ml conical centrifuge tube

1. Obtain protoplasts from carefully sliced 5-mm strips of sterile plant material by incubating in 8 ml protoplast solution for 3 to 6 hr at 30°C on a rotary shaker.

2. Remove debris by filtration through an 80-µm-mesh nylon screen. Rinse screen with 4 ml plant electroporation buffer. Combine protoplasts in a sterile 15-ml conical centrifuge tube.

3. Centrifuge 5 min at 300 × *g*. Discard supernatant, add 5 ml plant electroporation buffer, and repeat wash step. Resuspend in plant electroporation buffer at 1.5–2 × 10^6 protoplasts/ml.

4. Carry out electroporation as described for mammalian cells (Basic Protocol, steps 4 to 8). Use one or several shocks at 1 to 2 kV with a 3- to 25-µF capacitance as a starting point for optimizing the system.

   Alternatively, use 200 to 300 V with 500 to 1000 µF capacitance if the phosphate in the electroporation buffer is reduced to 10 mM final.

5. Harvest cells after 48 hr growth and isolate RNA, assay for transient gene expression, or select for stable transformants.

Reference: Potter et al., 1984.

Contributor: Huntington Potter

Liposome-Mediated Transfection

UNIT 9.4

Using liposomes to deliver DNA into different eukaryotic cell types results in higher efficiency and greater reproducibility than other transfection methods. Optimization procedures are described in *UNIT 9.9*, and protein expression strategies are discussed in *UNIT 16.12*.

TRANSIENT EXPRESSION USING LIPOSOMES

BASIC PROTOCOL

Materials *(see APPENDIX 1 for items with ✓)*

Exponentially growing mammalian cells (Table 9.4.1)
Plasmid DNA (miniprep or CsCl purified; Table 9.4.1)
✓ Complete Dulbeccos minimum essential medium, serum free (DMEM; or appropriate growth medium)
✓ Complete DMEM containing 10% and 20% fetal bovine serum (DMEM-10 and -20; or appropriate complete growth medium)
Liposome suspension (Table 9.4.1; Lipofectin or TransfectACE; Life Technologies)

6-well, 35-mm tissue culture dishes
Humidified 5% CO_2, 37°C incubator
Polystyrene tubes (Falcon or Corning)

1. Plate exponentially growing cells in 6-well tissue culture dishes at 5 × 10^5 cells/well and grow overnight in a CO_2 incubator at 37°C to 80% confluency.

   If 100-mm dishes are used in place of 6-well dishes, grow cells to 80% confluency and scale up all amounts by a factor of 8.

2. Prepare DNA/liposome complex in a polystyrene tube as follows: dilute plasmid DNA into 1 ml complete DMEM-SF, vortex 1 sec, then add liposome suspension

Table 9.4.1 Amount of DNA and Liposome Required for Liposome-Mediated Transfection[a]

Cell type	Plasmid DNA[b] (μg)	Liposome suspension (μl)
BHK-21	0.5	5
COS-7	0.5	5
CV-1	1.0	10
HeLa	2.0	10

[a]Amounts of DNA and liposome suspension are recommended for transfection of each cell type in a total volume of 1 to 1.5 ml DMEM-SF in a 6-well, 35-mm dish.

[b]Plasmid DNA can be prepared using a miniprep protocol (*UNIT 1.6*) or purified by CsCl/ethidium bromide equilibrium centrifugation (*UNIT 1.7*).

and vortex again. Incubate 5 to 10 min at room temperature to allow binding of DNA to cationic liposomes.

See Table 9.4.1 for amounts of DNA and liposome suspension, according to cell type. For other cell types, see optimization of transfection, UNIT 9.9. It is important to use polystyrene rather than polypropylene tubes because the DNA/liposome complex apparently sticks to polypropylene.

3. Aspirate complete DMEM-10 from cells, wash cells once with 1 ml complete DMEM-SF, and aspirate DMEM-SF. To each 35-mm well, add 1 ml DNA/liposome complex directly to the cells. Incubate 3 to 5 hr in a CO_2 incubator at 37°C.

4. To each well of cells, add 1 ml complete DMEM-20 and incubate an additional 16 to 24 hr in a CO_2 incubator at 37°C.

5. Aspirate complete DMEM/DNA/liposome complex and add 2 ml fresh, complete DMEM-10 to each well. Incubate an additional 24 to 48 hr.

6. Harvest cells by scraping, trypsinization, or freeze-thaw lysis (*UNITS 9.7A & 9.7B*). Perform appropriate expression assay (*UNITS 9.5 & 9.6*).

ALTERNATE PROTOCOL

STABLE TRANSFORMATION USING LIPOSOMES

1. Plate cells (Basic Protocol, step 1); grow to 50% confluency.

2. Prepare DNA/liposome complex and transfect cells (Basic Protocol, steps 2 and 3).

3. To each well of cells, add 1 ml complete DMEM-20 and incubate 48 hr in a CO_2 incubator at 37°C.

4. Aspirate DMEM and dilute cells into selective medium. Grow cells for appropriate length of time to select true transfected colonies (*UNIT 9.5*).

Reference: Felgner et al., 1987.

Contributors: Michael Whitt, Linda Buonocore, and John K. Rose

Stable Transfer of Genes into Mammalian Cells

BASIC PROTOCOL

Analysis of gene function frequently requires formation of mammalian cell lines that contain the gene in a stably integrated form (*UNIT 16.12*). Approximately one in 10^4 cells in a transfection will stably integrate DNA so a dominant selectable marker is used to permit isolation of stable transfectants. Highly transfectable cell lines, such as HeLa, CHO DUKX, and NIH3T3 should be used to ensure success. In addition, appropriate controls should be included to be certain that the gene of interest is not cytotoxic.

Materials

Complete medium
Selective medium (see below)
Cloning cylinders

1. Ensure that cell line to be transfected can grow as an isolated colony. For adherent cells, plate ~100 cells on a 10-cm tissue culture dish and feed every 4 days for 10 to 12 days. Count the number of viable colonies using methylene blue staining.

2. Determine selection conditions for parental cell line by splitting a confluent dish of cells ≥1:15 into medium containing various levels of the drug. Incubate cells for 10 days, feeding with appropriate selective medium every 4 days and examining for viable cells.

3. Determine the most efficient means of transfecting parental cell type (*UNIT 9.9*); either calcium phosphate transfection or electroporation should be used. For calcium phosphate transfection, split parental cell line 1:15 into complete medium the day before the DNA is applied to the cell. For electroporation, refer to *UNIT 9.3*.

4. Transfect desired gene into parental cell line, using ≥5:1 molar ratio of the plasmid containing the gene of interest to the plasmid containing the selective marker. Include a control where carrier DNA (e.g., pUC13) is used instead of the plasmid containing the gene to be studied and do several separate transfections.

   It is not necessary to physically link the selective marker to the gene of interest prior to transfection if a 5:1 ratio is used. If selection requires transfection of a large amount of the selection plasmid, then a 5:1 ratio will not be achievable in some cases, and a single plasmid should be constructed that contains both the selective marker and the gene of interest. If no colonies result from the transfection containing the gene of interest, but colonies do appear in the control containing pUC13, then the gene of interest may be cytotoxic.

5. Allow cells to double twice under nonselective conditions. Split cells 1:15 into selective medium.

6. Place ≥5 dishes into selective medium from each transfected dish to maximize the number of colonies that can be picked and expanded into cell lines. Feed cells with selective medium every 4 days for 10 to 12 days. Inspect the plates for colonies by holding the plates up to a light at an angle. One of the plates can be stained to facilitate counting the colonies. Pick large, healthy colonies.

   Colonies should contain ~500 to 1000 cells when they are picked.

ADENOSINE DEAMINASE (ADA)

Selection conditions. Medium supplemented with 10 μg/ml thymidine, 15 μg/ml hypoxanthine, 4 μM 9-β-D-xylofuranosyl adenine (Xyl-A), and 0.01 to 0.3 μM 2′-deoxycoformycin (dCF). Fetal bovine serum contains ADA which will detoxify the medium, so serum should be added to the medium immediately prior to use.

Basis for selection. Xyl-A can be converted to Xyl-ATP and incorporated into nucleic acids, resulting in cell death. Xyl-A is detoxified to its inosine derivative by ADA. dCF is a transition state analogue inhibitor of ADA and thus inactivates ADA endogenous to the parental cell type. The level of endogenous ADA varies with cell type, so the appropriate concentration of dCF for selection will vary.

Comments. High levels of expression of the transfected ADA gene will be necessary to achieve selection in cells with high endogenous ADA levels. ADA-deficient CHO cells are available. ADA can be used in amplification systems (Kaufman et al., 1986) by increasing the level of dCF.

Reference: Kaufman et al., 1986.

AMINOGLYCOSIDE PHOSPHOTRANSFERASE (neo, G418, APH)

Selection conditions. 100 to 800 μg/ml G418 in complete medium. G418 should be prepared in a highly buffered solution (e.g., 100 mM HEPES, pH 7.3) so that addition of the drug does not alter the pH of the medium.

Basis for selection. G418 blocks protein synthesis in mammalian cells by interfering with ribosomal function. It is an aminoglycoside, similar in structure to neomycin, gentamycin, and kanamycin. Expression of the bacterial APH gene (derived from tn5) in mammalian cells therefore results in detoxification of G418.

Comments. Varying concentrations of G418 should be tested, as cells differ in their susceptibility to killing by G418. Many investigators buy a large amount of one lot of G418, as different lots can differ in potency. Cells will divide once or twice in the presence of lethal doses of G418, so the effects of the drug take several days to become apparent.

Reference: Southern and Berg, 1982.

DIHYDROFOLATE REDUCTASE (DHFR)

Selection conditions. α– medium supplemented with 0.01 to 300 μM methotrexate (MTX) and dialyzed fetal bovine serum (see *UNIT9.9*).

Basis for selection. DHFR is necessary for purine biosynthesis and is thus required for cell growth in the absence of exogenous purines. Dialysis of serum to remove endogenous nucleosides and use of nucleoside-free media are necessary for selection. MTX is a potent competitive inhibitor of DHFR, so increasing MTX concentrations can select for increased DHFR expression.

Comments. Extremely high levels of expression of the transfected normal DHFR gene are needed for selection in cell lines with high endogenous DHFR levels. A mutant DHFR gene is available that encodes an enzyme resistant to MTX (Simonsen et al., 1983) and can be used for dominant selection in most cell types. DHFR can be used to amplify transfected genes using a DHFR-deficient CHO cell line (*UNIT9.9*).

Reference: Simonsen and Levinson, 1983.

HYGROMYCIN-B-PHOSPHOTRANSFERASE (HPH)

Selection conditions. Complete medium supplemented with 10 to 400 μg/ml hygromycin-B.

Basis for selection. Hygromycin-B is an aminocyclitol that inhibits protein synthesis by disrupting translocation and promoting mistranslation. The HPH gene (isolated from *E. coli* plasmid pJR225; Gritz and Davies, 1983) detoxifies hygromycin-B by phosphorylation.

Comments. The HPH gene has only recently been used in mammalian systems, and vectors that efficiently express the gene are not widespread. While the level of hygromycin-B needed for selection can vary from 10 to 400 μg/ml, many cell lines require 200 μg/ml.

References: Gritz and Davies, 1983; Palmer et al., 1987.

THYMIDINE KINASE (TK)

Selection conditions. Forward (TK^- to TK^+): Complete medium supplemented with 100 μM hypoxanthine, 0.4 μM aminopterin, 16 μM thymidine, and 3 μM glycine (HAT medium). Reverse (TK^+ to TK^-): Complete medium supplemented with 30 μg/ml 5-bromodeoxyuridine (BUdr).

Basis for selection. Under normal growth conditions, cells do not need thymidine kinase, as the usual means for synthesizing dTTP is through dCDP. Addition of BUdr to the medium will kill TK^+ cells, as BUdr is phosphorylated by TK and then incorporated into DNA. Selection of TK^+ cells in HAT medium is primarily due to the presence of aminopterin, which blocks the formation of dTDP from dCDP. Cells thus require TK to synthesize dTTP from thymidine.

Comments. Thymidine kinase is widely used because both forward and reverse selection conditions exist. Unlike ADA and DHFR, it is not possible to select for variable levels of TK, so the gene cannot be used for amplification. Like ADA and DHFR, most mammalian cell lines express TK, removing the possibility of using the marker in those lines unless BUdr is used to select a TK^- mutant.

Reference: Littlefield, 1964.

XANTHINE–GUANINE PHOSPHORIBOSYLTRANSFERASE (XGPRT, gpt)

Selection conditions. Medium containing dialyzed fetal bovine serum and 250 μg/ml xanthine, 15 μg/ml hypoxanthine, 10 μg/ml thymidine, 2 μg/ml aminopterin, 25 μg/ml mycophenolic acid, and 150 μg/ml L-glutamine.

Basis for selection. Aminopterin and mycophenolic acid both block the de novo pathway for synthesis of GMP. Expression of XGPRT allows cells to produce GMP from xanthine, allowing growth on medium that contains xanthine but not guanine. It is therefore necessary for selection to use dialyzed fetal bovine serum and a guanine-free medium.

Comments. XGPRT is a bacterial enzyme that does not have a mammalian homologue, allowing it to function as a dominant selectable marker in mammalian cells. The amount of mycophenolic acid necessary for selection varies with cell type and can be determined by titration in the absence/presence of guanine.

Reference: Mulligan and Berg, 1981.

References: Perucho et al., 1980; Robins et al., 1981; Wigler et al., 1977.

Contributor: Robert E. Kingston

UNIT 9.6

Overview of Genetic Reporter Systems

Transfection is the most commonly used procedure for analyzing mammalian gene expression in vivo (*UNITS 9.1-9.4, 9.9, 16.12 & 16.13*). Many experimenters use transient assay systems that are based on the use of fusion genes and assess gene expression within 48 hr after introduction of the DNA. The fusion gene usually consists of the promoter activator binding site or enhancer sequence under study attached to a gene directing the synthesis of a reporter molecule; the amount of reporter protein synthesized under various conditions is presumed to reflect the ability of the inserted sequence to direct and/or promote transcription. It follows that a useful transient expression system should be based on the synthesis of an easily assayed reporter protein that has minimal or no effects on the physiology of the transfected cell. Ideally, the assay for this reporter molecule would be extremely sensitive. It is also important to verify that any regulatory event observed with a fusion gene also affects expression of the gene in vivo, both in terms of kinetics and magnitude of the response.

A central question in molecular and cell biology is how *cis*-acting DNA sequences and *trans*-acting factors act in unison to control eukaryotic gene expression. These interactions are mediated by specific binding of a transcription factor or a complex of factors to enhancer and promoter elements generally found upstream of the transcription start site. Direct quantitation of changes in gene expression requires the measurement of specific mRNAs using techniques such as northern blot hybridizations (*UNIT 4.9*) or nuclease protection assays (*UNIT 4.7*). These procedures can be time-consuming and are not always practical for analysis of many different gene constructs. Furthermore, such techniques require that the analyzed gene be modified in some manner to distinguish it from the native gene in transfected cells.

An alternate approach to gauge changes in transcription is to link the presumed *cis*-acting sequence(s) from the gene of interest to the coding sequence for an unrelated reporter gene. To test for complete promoters, the DNA fragment is placed directly upstream of the reporter gene in a vector lacking endogenous promoter activity (Fig. 9.6.1). Similarly, for activator binding sites, the heterologous sequence is placed in a vector proximal to a basal promoter that contains sequences required for recognition by the basic transcription machinery (e.g., RNA polymerase II). Following introduction of the chimeric reporter construct into an appropriate cell type or animal, measurement of reporter-gene product provides an indirect estimate of the induction in gene expression directed by the regulatory sequences. It is important to recognize that these assays use reporter constructs to measure protein level or activity and not RNA level. The two levels are frequently, but not always, correlated with one another. Any treatment of the cells or extracts that may affect message translation will alter the accuracy of measurements of the reporter construct. Furthermore, when dealing with an inducible system, the stability of the reporter must be taken into account. For example, if the reporter-gene product is very stable, high levels of protein can build up in the cell prior to induction. Upon induction following addition of the effector, the protein level may well increase, but the observed fold increase may be less dramatic because of the high basal level of reporter protein that has already accumulated in the cell.

With these considerations, the reporter can be used to characterize the cellular, tissue, or temporal specificity of the *cis*-acting element or its responsiveness to external stimuli such as hormones. Furthermore, *cis*-regulatory elements can be functionally characterized by constructing successive deletion mutants (Fig. 9.6.1), by altering sequence orientations, or by site-directed mutagenesis. Moreover, the requirements for *trans*-acting factor(s) can be investigated by expression of the reporter construct in different cell types and organisms or by biochemical manipulation of the source of the factors. For example, involvement of specific signaling pathways in the response to external stimuli can be investigated by treating cells

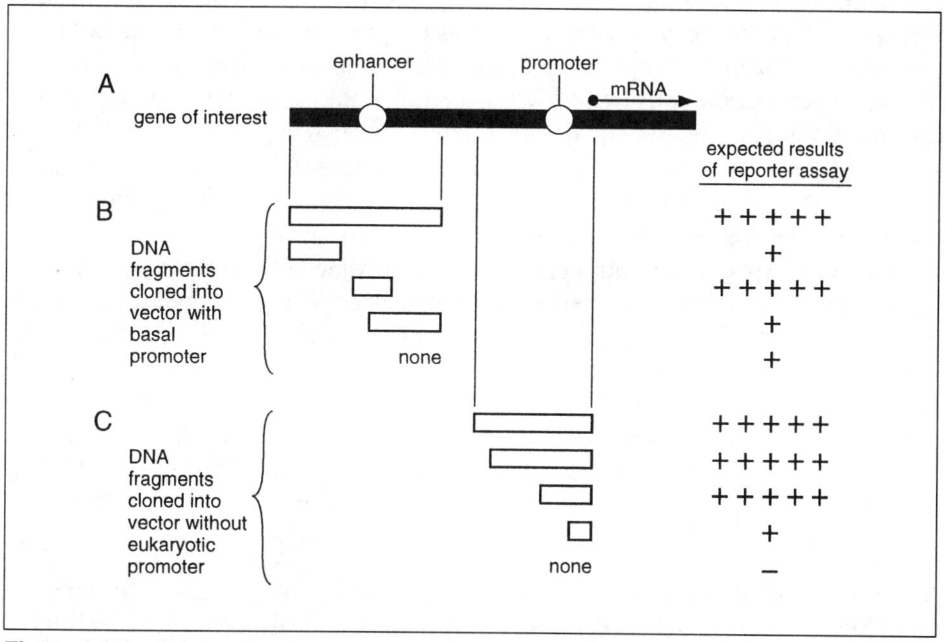

Figure 9.6.1 Schematic of promoter and enhancer definition studies. (**A**) A hypothetical eukaryotic gene with promoter and enhancer elements indicated by open circles. (**B**) A series of DNA fragments that could be cloned into the MCS of a reporter vector containing a basal promoter for the purpose of identifying the position of an enhancer. (**C**) Cloning of DNA into a promoterless reporter vector to define the minimal functional promoter. The expected results of the subsequent reporter assays for these constructs are shown to the right of the corresponding DNA fragments. + and +++++ indicate lesser and greater expression, and – indicates no expression.

with inhibitors and/or activators of protein phosphorylation, glycosylation, or fatty acylation. Kinetic studies with protein synthesis inhibitors may be employed to correlate the requirement for new protein synthesis with induction of the reporter gene.

There are several criteria for selection of a transcription reporter gene. (1) The reporter protein should be absent from the host, or easily distinguished from endogenous versions. (2) A simple, rapid, sensitive, and cost-effective assay should be available to detect the reporter protein. (3) The assay for the reporter protein should have a broad linear range to facilitate analysis of both large and small changes in promoter activity. (4) Expression of the gene must not alter the physiology of the recipient cells or organism. These criteria are met to varying degrees by the reporter systems outlined in Table 9.6.1. Each of these reporter genes and the corresponding assay systems have specific features and limitations that must be considered in choosing a system tailored to the particular question being studied. For example, the reporter protein may be expressed intracellularly or secreted from the cell and assayed in the culture medium. There may be limitations in the cell types suitable for a particular reporter system due to low expression levels or the presence of endogenous activities that contribute to high background. The reporter proteins that are commonly used vary widely in their relative stabilities, which is an important consideration in the study of inducible reporter constructs. Finally, the reporter protein may be measured by an in vitro activity or immunological assay or detected in vivo via histochemical procedures. These and other issues will be addressed for each reporter gene in the sections that follow.

DESIGN OF REPORTER VECTORS

The general design of a reporter vector referred to as "pGENERIC," is shown in Figure 9.6.2. The vector backbone typically contains a bacteriophage origin of replication (f1 *ori*) for single-stranded DNA production to facilitate sequencing and mutagenesis studies, and a bacterial origin of replication (often from pBR322 or pUC-derived plasmids) for propagation of the vector in *Escherichia coli*. Additional origins of replication are often added to permit stable propagation in other host species. Production of the vector in bacteria is also facilitated by an antibiotic resistance gene, in most cases coding for β-lactamase, which imparts ampicillin resistance to *E. coli* cells transformed with the vector. A polyadenylation signal downstream of the reporter gene ensures proper and efficient processing of the reporter transcript in eukaryotic cells. A second polyadenylation site may be placed upstream of the reporter to prevent background transcription from cryptic promoters located in the flanking DNA. Lastly, a multiple cloning site (MCS) upstream of the reporter gene allows for the insertion of foreign DNA containing putative promoter and/or enhancer elements. For cloning purposes, the MCS typically contains five to seven closely aligned unique restriction endonuclease sites. In addition to the MCS, reporter vectors often contain one or more unique restriction sites downstream of the reporter gene for testing putative enhancer elements or for excision of the reporter gene.

Some vectors already contain a basal promoter immediately upstream of the reporter gene. Such promoters are frequently of viral origin, and ideally provide a constitutive level of transcription in a broad range of cell types. These reporter vectors are used to identify and characterize activator binding sites and enhancer sequences, and also to study the process of transcription activation. A reporter vector containing a basal promoter, and possibly an upstream enhancer, may also be used as a positive control to normalize transfection efficiencies of experimental reporter constructs.

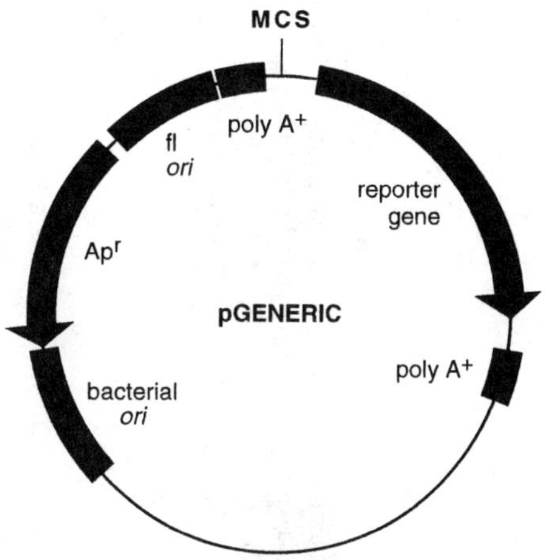

Figure 9.6.2 Plasmid map of a representative reporter vector. MCS denotes the multiple cloning site used for insertion of foreign DNA into the reporter vector. Poly A^+ indicates the positions of polyadenylation signals. *Ori* refers to the bacteriophage (f1) and bacterial plasmid origins of replication. Arrows indicate the direction of transcription for the reporter gene and an antibiotic resistance gene (Ap^r).

IN VITRO REPORTER ASSAYS

In vitro reporter assays refer to procedures in which the reporter protein is quantified using either cell or tissue lysates containing the reporter, or (for secreted reporter proteins) the culture medium from transfected cells. These assays utilize direct quantitation of the reporter protein by enzymatic or immunological means to provide an indirect estimate of the transcriptional activity of the reporter vector encoding the protein. Such assays are distinguished from the in vivo reporter systems described below in that numerical data suitable for comparative studies of different promoters, enhancers, and cell-type requirements are more readily obtained.

Chloramphenicol Acetyltransferase (CAT)

Many of the above criteria for genetic reporters are satisfied by the enzyme chloramphenicol acetyltransferase (CAT; *UNIT 9.7A*). This enzyme catalyzes the transfer of acetyl groups from acetyl-coenzyme A to chloramphenicol. Because CAT is a prokaryotic enzyme, there are minimal competing activities in most eukaryotic cells, resulting in high signal-to-background ratios in these reporter assays. However, most formats of the CAT assay require a relatively expensive radioactive substrate, the assays are very time-consuming to perform, and the sensitivity of CAT assays is inferior to that of recently developed nonisotopic reporter systems. CAT is quite stable (half-life 50 hr in mammalian cells), making it well-suited as a reporter in transient transfection experiments, but less desirable for stable expression studies. Moreover, the stability exhibited by CAT can be a disadvantage in inducible systems (e.g., inducible enhancers), where accumulation of the reporter protein may mask the induction brought about by addition of the effector. For applications of this type, a reporter protein with a relatively short half-life, such as firefly luciferase, is more desirable.

The most common CAT assay is based on incubation of cell lysates prepared from transfected cells with ^{14}C-labeled chloramphenicol. Acetylated and unacetylated forms of this compound are separated by thin-layer chromatography on siliconized glass plates. The plates are exposed to X-ray film for a qualitative estimate of CAT

Table 9.6.1 Genetic Reporter Systems

Reporter gene	In vitro assay	In vivo assay	Strengths	Weaknesses	Limit of detection (molecules)[a]	Cost/assay (U.S. dollars)[b]	Available reporter vectors[c]
Chloramphenicol acetyltransferase (CAT)	Chromatography, differential extraction, fluorescence, or immunoassay	None commonly available	Minimal endogenous activity in mammalian cells; stable protein; various assay formats available	Assays are time-consuming and laborious; most formats require expensive radioactive substrate; relatively low sensitivity and narrow linear range (2 orders of magnitude); fluorescent assays offer better sensitivity, but require a fluorometer for quantitation; short half-life of CAT mRNA	$5\text{-}10 \times 10^7$	2.00-7.00	pSV0CAT and pSV2CAT (Gorman et al., 1982); pA_{10}CAT2 (Laiminis et al., 1984); pCAT-Basic, pCAT-Enhancer, pCAT-Promoter, and pCAT-Control (Promega)
Firefly luciferase	Bioluminescence (luminometer or scintillation counter)	Bioluminescence assay in live cells with luciferin esters as substrate	Nonisotopic; good sensitivity and broad linear range (4 orders of magnitude); minimal endogenous activity in mammalian cells; relatively inexpensive	Short half-life of protein; assay requires a luminometer or scintillation counter; conventional assay lacks reproducibility	$1\text{-}2.5 \times 10^5$	0.10-0.25	pT8lluc, pSluc2, pXP1/pXP2, and pOLUC (Nordeen, 1988) pMAMneo-LUC (Clontech) pGL2-Basic, pGL2-Enhancer, pGL2-Promoter, and pGL2-Control (Promega)
β-galactosidase	Colorimetric, fluorescence, chemiluminescence (luminometer or scintillation counter)	Histochemical staining with X-gal substrate; bioluminescence assay in live cells with fluorescein di-β-D-galactopyranoside (FDG)	Nonisotopic; various assay formats available for different applications; chemiluminescent assay is very sensitive, and has an extremely broad linear range (5-6 orders of magnitude); well suited as internal control for normalizing other reporters	Many cell types have high endogenous β-galactosidase activity; assays require fluorometer or luminometer	$10^4\text{-}10^5$ (chemiluminescence assay)	0.01-0.02	pβgal-Basic, pβgal-Enhancer, pβgal-Promoter, and pβgal-Control (Clontech) pSV-β-gal (Promega)
Secreted alkaline phosphatase (SEAP)	Colorimetric, bioluminescence, or chemiluminescence (luminometer or scintillation counter)	None commonly available	Nonisotopic; secreted reporter protein; various assay formats available; chemiluminescent assay is very sensitive, and has a broad linear range (4 orders of magnitude); useful for high throughput assays performed in 96-well microtiter plates	May not be suitable for cells expressing low levels of placental-type alkaline phosphatase (lung, testes, and cervix); may not be suitable if experimental design affects secretory capacity of target cells	$10^4\text{-}10^5$ (chemiluminescence assay)	0.01-0.04	pSEAP-Basic, pSEAP-Enhancer, pSEAP-Promoter, and pSEAP-Control; (Clontech) pBC12/RSV/SEAP and pBC12/HIV/SEAP (Berger et al., 1988)

continued

Table 9.6.1 Genetic Reporter Systems, continued

Reporter gene	In vitro assay	In vivo assay	Strengths	Weaknesses	Limit of detection (molecules)[a]	Cost/assay (U.S. dollars)[b]	Available reporter vectors[c]
Human growth hormone (hGH)	Radioimmuno-assay	None commonly available	Secreted reporter protein; direct detection of protein levels; simple assay, easy to perform; high-throughput assay performed in 96-well microtiter plates; useful as internal control for normalizing other reporters (e.g., SEAP)	Radioactive assay requiring ^{125}I-labeled antibody; no signal amplification as with activity assays; relatively low sensitivity and narrow linear range; expensive	3×10^8	1.00-2.00	pXGH, pOGH, and pTKGH (Selden et al., 1986)
β-glucuronidase (GUS)	Colorimetric, fluorescence, or chemiluminescence (luminometer or scintillation counter)	Histochemical staining with X-Gluc substrate	Nonisotopic; various assay formats available for different applications; GUS protein is stable; chemiluminescent assay is very sensitive, and has an extremely broad linear range (6 orders of magnitude); predominant reporter used in plant genetic research	Assays with best sensitivity require fluorometer or luminometer (scintillation counter can substitute)	N/A	N/A	pBI101, pBI101.2, pBI101.3, pBI121, pBI221, and pGUSN358-S (Clontech)
Green fluorescent protein (GFP)	Fluorescence (UV light box or fluorescence imaging device)	Fluorescence (UV light box, fluorescence microscopy, or FACS analysis)	Reporter of gene expression and protein localization in live cells; fluorescence occurs in a species-independent fashion; fluorescence is an intrinsic property of the protein and does not require additional gene products, substrates, or cofactors; fluorescent signal is highly resistant to photobleaching; no apparent toxic effects of GFP expression in bacteria or eukaryotes	Signal intensity may be too weak for some applications	N/A	Essentially free, no additional reagents necessary	pGFP, pGFP1, pGFP-N1,2,3, and pGFP-C1,2,3 (Clontech)

[a,b,c]Much of this information was taken from Alam and Cook, 1990. The lists of available reporter vectors may not be complete. N/A: information not available.

activity. For more quantitative data, the spots corresponding to the positions of the mono- and diacetylated forms of chloramphenicol can be scraped from the plate and counted in a liquid scintillation counter. This assay can be particularly tedious with several samples and is difficult to interpret due to the presence of two reaction products. Differential extraction techniques have been developed for the quantitative separation of chloramphenicol from its acetylated derivatives, eliminating the need for time-consuming chromatography (*UNIT 9.7A*). In this format, the radiolabel resides on acetyl-coenzyme A, and ^{3}H can be substituted for ^{14}C, making the assay somewhat less hazardous and less expensive. Use of an alternate solvent system results in enhanced sensitivity by improving the signal-to-background ratio.

There are two nonisotopic detection methods for quantifying CAT. A CAT ELISA procedure performed in 96-well microtiter plates is available to directly quantify CAT using an anti-CAT antibody (from Boehringer Mannheim). This assay has the advantage of directly measuring CAT protein levels rather than CAT activity. A nonisotopic CAT activity assay has also been developed; this uses a fluorescent derivative of chloramphenicol (from Molecular Probes) that undergoes a single acetylation reaction. Because only a single reaction product is produced from this substrate, quantitative analysis of CAT activity is more reliable than with chloramphenicol. Moreover, this assay is more sensitive than those employing a radioactive substrate and has a broader linear range.

Firefly Luciferase

Cloning of the *luc* gene from the firefly *Photinus pyralis* provided the first nonisotopic genetic reporter system with widespread utility in mammalian cells. The bioluminescent reaction catalyzed by luciferase requires luciferin (the substrate), ATP, Mg^{2+}, and molecular O_2. Mixing these reagents with cell lysates containing luciferase results in a flash of light that decays rapidly (in <1 sec). The light signals are detected using a luminometer equipped with an autoinjection device to facilitate rapid mixing of reaction components. Alternatively, the light signal can be recorded using a liquid scintillation counter. The total light emission is proportional to the luciferase activity of the sample, which in turn provides an indirect estimate of the transcription of the luciferase reporter gene. Luciferase is sensitive to degradation by proteases, and therefore has a half-life of ~3 hr in transfected mammalian cells. The rapid turnover of luciferase protein makes this reporter a good candidate for studying inducible systems where the increase above basal expression levels needs to be maximized.

The original luciferase assay provides good sensitivity (10 to 1000 times as sensitive as the radioactive CAT assay), but is somewhat intricate to perform and lacks reproducibility between samples, largely due to the rapid "flash" kinetics of the photon emission. An improved assay for luciferase that includes coenzyme A in the reaction mixture produces a more sustained light signal because of the favorable reaction of luciferase with luciferyl-CoA (luciferyl-coenzyme A). Moreover, the longer duration of the light signal increases luminescence intensity by tenfold, thereby making the luciferase assay more sensitive. Future prospects for improving the assay further include the production of modified luciferase enzymes with shifted emission wavelengths which would permit simultaneous analysis of two or more genetic reporters in the same population of cells via detection of different colors of light.

β-Galactosidase

The *lacZ* gene from *E. coli*, which encodes the enzyme β-galactosidase, is among the most versatile genetic reporters, having both in vitro and in vivo assay formats employing a variety of different substrates. The enzyme catalyzes the hydrolysis of various β-galactosides, including several specialized substrates tailored to different assay formats. In addition to its use as a reporter for uncharacterized *cis*-regulatory sequences, expression of β-galactosidase under the control of a constitutive promoter is frequently used as an internal control to normalize the variability of other reporter assays. β-galactosidase is particularly useful for normalizing CAT and firefly luciferase, as cell lysates prepared for these reporter assays are also suitable for measurement of β-galactosidase activity. Both a colorimetric assay using *o*-nitrophenyl-β-D-galactopyranoside (ONPG) and a fluorometric assay using 4-methylumbelliferyl-β-D-galactoside (MUG) have been developed for β-galactosidase, but these have received limited use, due primarily to their poor sensitivity. The development of chemiluminescent 1,2-dioxetane substrates for β-galactosidase has greatly improved the utility of *lacZ* as a transcriptional reporter by increasing the sensitivity of the assay and extending the linear dynamic range of detection (*UNIT 9.7B*). When a luminometer is used to detect the chemiluminescent signal, the assay is 50,000-fold more sensitive than the colorimetric assay. Sensitivity can also be enhanced by using assay conditions that minimize endogenous enzyme activity contributed by eukaryotic β-galactosidase.

Secreted Alkaline Phosphatase (SEAP)

Secreted alkaline phosphatase (SEAP) differs from the abovementioned reporter proteins in that SEAP is secreted from transfected cells and can thus be assayed using a small aliquot of the culture medium. The SEAP gene encodes a truncated form of human placental alkaline phosphatase (PLAP), which lacks a critical membrane-anchoring domain, thereby allowing the protein to be efficiently secreted from transfected cells. Levels of SEAP activity detected in culture medium are directly proportional to changes in intracellular concentrations of SEAP mRNA and protein (Cullen and Malim, 1992). SEAP has the unusual properties of being extremely heat-stable and resistant to the phosphatase inhibitor L-homoarginine. Therefore, endogenous alkaline phosphatase activity can be eliminated by pretreatment of samples at 65°C and incubation with the inhibitor. The secreted nature of SEAP provides several advantages for its use as a genetic reporter: (1) preparation of cell lysates is not required; (2) the kinetics of gene expression can be easily studied by repeated collection of medium from the same cultures; (3) transfected cells are not disturbed during measurement of SEAP activity and remain intact for further investigations; (4) background from endogenous alkaline phosphatase activity in the culture medium is almost absent; and (5) sample collection and assay can be automated using 96-well microtiter plates.

The original SEAP assay was a colorimetric procedure using the alkaline phosphatase substrate *p*-nitrophenyl phosphate. This procedure is fast, simple to perform, and very inexpensive. However, the assay's sensitivity is poor, and it has a narrow linear dynamic range. Sensitivity can be improved by using a two-step bioluminescent assay for SEAP based on hydrolysis of D-luciferin-*O*-phosphate. The dephosphorylation reaction catalyzed by SEAP yields free luciferin, which in turn serves as the substrate for firefly luciferase. The sensitivity of this assay is roughly equivalent to that of the conventional bioluminescent assay for luciferase. The most sensitive SEAP assays use chemiluminescent alkaline phosphatase substrates such as 1,2-dioxetane CSPD (see Table 10.7.1). Dephosphorylation of CSPD results in a sustained "glow"-type luminescence that remains constant up to 60 minutes and is readily detected using a luminometer or scintillation counter. In addition to enhancing sensitivity, the chemiluminescence assay for SEAP greatly increases the linear dynamic range of detection.

Human Growth Hormone (hGH)

Human growth hormone (hGH) is normally secreted exclusively from the somatotropic cells of the anterior pituitary gland. The restricted pattern of hGH expression makes this protein an attractive choice as a genetic reporter for most mammalian cell types. hGH reporter assays have many of the same advantages as SEAP (*UNIT 9.7A*), but are less desirable due to the relatively low sensitivity of the procedure and the need for a hazardous radioimmunoassay to quantitate hGH. Measurement of hGH has been used as an internal control to normalize transfection efficiency and is well suited in this regard to normalize expression from experimental SEAP reporter constructs.

β-Glucuronidase (GUS)

The bacterial β-glucuronidase (GUS) gene is the predominant reporter used to study gene expression in plants. GUS is used as a reporter in both plant and mammalian cells, but is particularly useful in higher plants due to the absence of endogenous GUS activity in most species. Higher plants transformed with GUS are healthy, develop normally, and are fertile. As with β-galactosidase, one of the principal advantages of GUS as a reporter is the wide diversity of assays available for the enzyme. Several different colorimetric assays for GUS have been developed using a variety of β-glucuronides as substrates. The most popular of these uses X-Gluc, a substrate that can also be used for histochemical staining of tissues and cells expressing GUS activity. A more sensitive fluorescence assay for GUS uses the substrate 4-MUG. A chemiluminescent assay for GUS has also been developed; this is very similar to the procedure used to quantify β-galactosidase, and employs an adamantyl 1,2-dioxetane aryl glucuronide substrate. This assay is approximately 100-fold more sensitive than the fluorometric assay using 4-MUG, and has a wider linear dynamic range. In addition to the uses of GUS as a genetic reporter, GUS fusion proteins have been used extensively to study protein localization and transport in both plant and animal species.

IN VIVO REPORTER ASSAYS

In vivo reporter assays may be defined as those in which the reporter protein is detected in either live cells or tissues (e.g., of transgenic animals), or in cells or tissues that have been fixed for histochemical staining. The data provided by this approach is less quantitative than in vitro reporter systems, but provides important information regarding the cell-type specificity of promoter/enhancers and the tissue distribution of specific transcription factors. Because this method focuses on localization of reporter proteins to the site of their production, it is well suited to reporter assays that employ either precipitating substrates or substrates that generate a luminescent or fluorescent signal. Conversely, secreted proteins such as SEAP and hGH are of less use as in vivo reporter genes.

Green Fluorescent Protein (GFP)

Light is produced in several bioluminescent jellyfish as a result of energy transfer to green fluorescent proteins (GFPs). One such GFP, from *Aequorea victoria*, fluoresces in vivo after receiving energy from the Ca^{2+}-activated photoprotein aequorin. This process occurs without requirement for a substrate or cofactor and proceeds via direct transfer of energy between the two proteins. Purified GFP has similar spectral properties to the protein expressed in vivo, absorbing blue light and emitting green light that is detectable using a fluorescence microscope or a UV light box or by fluorescence-activated cell sorting (FACS). The absorption and emission wavelengths of GFP are similar to those of fluorescein, and conditions used for visualizing fluorescein are also suitable for GFP. Full-length GFP appears to be required for fluorescence; however, the minimal chromophore responsible for light emission can be assigned to a hexapeptide within the protein. This region of the protein contains a Ser-dehydroTyr-Gly trimer, which cyclizes to yield a chromophore that emits light by an as yet unknown mechanism.

Cloning and sequencing of the *A. victoria* GFP gene allowed expression of the protein in heterologous systems. GFP expressed in either prokaryotic or eukaryotic cells yields a bright green fluorescence when the cells are excited by blue or UV light. GFP fluorescence does not require additional gene products from *A. victoria* and occurs in a species-independent fashion. Recent studies using the nematode *Caenorhabditis elegans*, an organism often used for developmental studies, have demonstrated the utility of GFP as a reporter of gene expression in vivo. These researchers expressed GFP in *C. elegans* under the control of a neuron-specific promoter and used GFP fluorescence to monitor the formation of neuronal processes in real time as the worms developed. GFP has also been expressed in yeast, mammalian cells, and *Drosophila.* Moreover, both N- and C-terminal protein fusions with GFP have been constructed and shown to maintain the fluorescence properties of native GFP.

A critical aspect of GFP-based reporter systems is the ability to monitor transcriptional changes in real time. Future directions for improvement of GFP as a genetic reporter include efforts to increase the intensity of the fluorescent signal and thereby enhance detection sensitivity, and to reduce the variability of the signal between replicate samples. As with firefly luciferase, variant GFPs with shifted emission wavelengths may allow the use of this reporter to study two or more transcriptional events in the same cell population. The availability of multiple fluorescent proteins derived from wild-type GFP will not only facilitate the simultaneous detection of several transcription reporters, but also permit a similar analysis of protein trafficking events using GFP fusion proteins.

Firefly Luciferase

Recent studies have shown that firefly luciferase activity can be detected in live cells by means of soluble luciferase substrates capable of crossing the plasma membrane. These compounds are generally uncharged luciferin esters that readily penetrate and cross lipid bilayers. Once inside the cell, these substrates are converted by endogenous esterases into firefly luciferin, which in turn serves as the substrate for the firefly luciferase encoded by the reporter gene. A photolyzable "caged" luciferin compound has been described that appears to enter cells more efficiently. This compound can be hydrolyzed to yield free luciferin either by endogenous esterases or by visible light. In its present form, the in vivo luciferase assay is somewhat insensitive and lacks the capability for real-time analysis provided by GFP. Through the development of alternate luciferase substrates having a higher quantum yield and more sophisticated detection instruments, it is possible that luciferase may develop into a useful reporter system for in vivo analyses.

β-Galactosidase

In vivo levels of the β-galactosidase reporter protein can be determined in prokaryotic and eukaryotic cells, tissue sections, and intact embryos using the precipitating substrate Xgal. The reaction with Xgal produces a rich blue color that can easily be scored against background in most applications. Histochemical staining with Xgal requires fixation of the cells or tissue prior to the enzymatic assay. Detection of *lacZ* expression in whole embryos has been particularly useful for characterizing tissue-specific gene expression during early development. Detection of β-galactosidase activity in live cultured cells is achieved with the substrate fluorescein di-β-D-galactopyranoside (FDG). FDG is delivered into live cells by hypotonic loading, and after cleavage by β-galactosidase is trapped inside the cell due to the hydrophobic structure of the reaction product. Cells expressing the reporter protein are detected via fluorescence from the fluorescein moiety of the metabolized substrate.

Another important application of in vivo expression of β-galactosidase are so-called "enhancer trap" studies, in which positional effects of reporter integration into the chromosomes of transgenic organisms is used as a means of locating enhancer sequences. In one approach, a vector containing a weak basal promoter cloned directly upstream of the *lacZ* gene is introduced into mouse eggs and integrates into the mouse genome. Stable transgenic animals are bred from the progeny of the resulting adult animals. A sufficiently weak promoter is chosen such that expression of β-galactosidase occurs only when the promoter is proximal to an enhancer sequence near the site of integration. The expression pattern of β-galactosidase in transgenic lines indirectly reflects both the tissue specificity and strength of nearby enhancer(s). Experiments of this type utilize expression of a reporter to identify novel genes and regulatory flanking sequences that have potentially interesting patterns of expression during early development.

References: Alam and Cook, 1990; Bronstein et al., 1994.

Contributors: Steven R. Kain and Subinay Ganguly

UNIT 9.7A

Isotopic Assays for Reporter Gene Activity

BASIC PROTOCOL 1

CHROMATOGRAPHIC ASSAY FOR CAT ACTIVITY

Materials *(see APPENDIX 1 for items with ✓)*

Cells transfected with CAT expression plasmid (*UNITS 9.1-9.4 & UNIT 16.13*; Fig. 9.7.1), in 100-mm petri plates
✓ Phosphate-buffered saline
✓ TEN solution
✓ 1 M and ice-cold 0.25 M Tris·Cl, pH 7.5
200 μCi/ml [^{14}C]chloramphenicol (35 to 55 mCi/mmol)
4 mM acetyl CoA (store ≤2 weeks at −20°C)
Ethyl acetate
19:1 (v/v) chloroform/methanol

Rubber policeman or equivalent
Thin-layer chromatography (TLC) tank
Whatman 3MM filter paper
Thin-layer chromatography (TLC) sheets (plastic-backed, silica gel 1B; J.T. Baker)
Pen or marker with radioactive ink

1. Wash 100-mm plate of adherent cells transfected with CAT expression plasmid twice with 5 ml PBS per wash. Add 1 ml TEN solution to each plate. Let cells sit 5 min on ice.

   If transfected cells grow in suspension, wash by centrifugation and proceed to step 3.

2. Scrape the cells off the plate with a rubber policeman. Transfer them to a 1.5-ml microcentrifuge tube on ice.

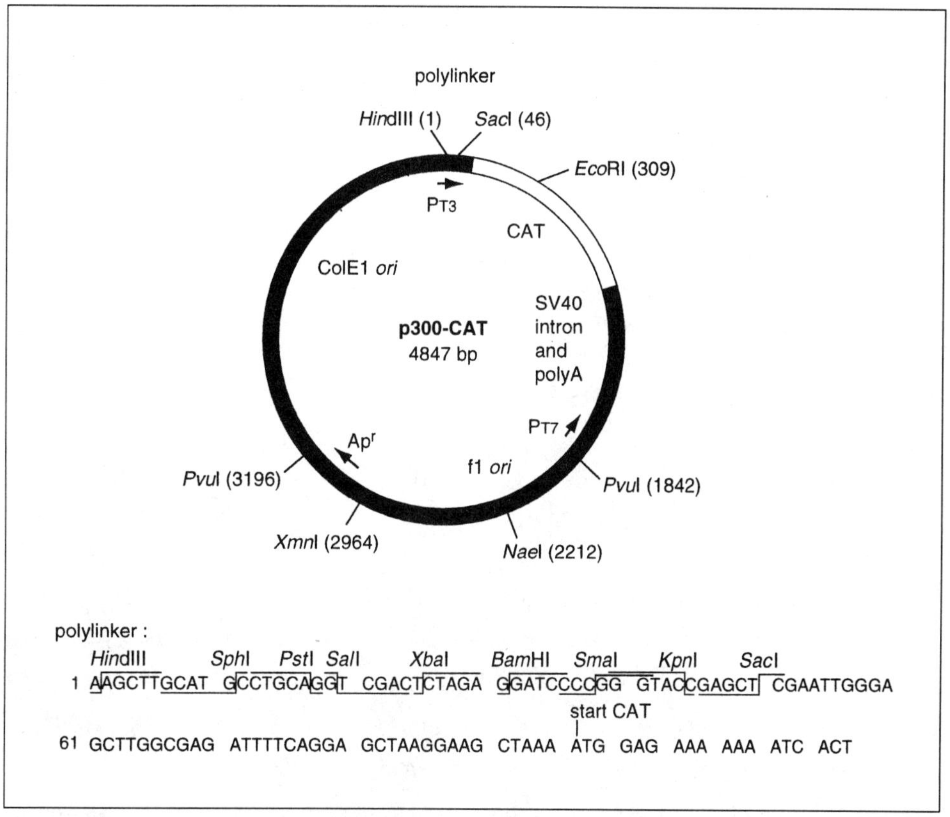

Figure 9.7.1 Expression vector p300-CAT. This vector is designed for insertion of the desired control elements. It was constructed by cutting the CAT cassette out of pUC-CAT (Gilman et al., 1986) and blunt-end coning it into the *Eco*RI site of pBluescript M13⁻. Every restriction endonuclease site in the polylinker, which is pUC-like, is unique and therefore available for cloning. The vector is a phagemid that produces the strand that permits sequencing of promoter inserts in the 5′ to 3′ direction using the reverse primer. In addition, the design of mutagenic oligonucleotides is facilitated by use of the sequence of the top strand of a double-stranded sequence.

3. Microcentrifuge cells 1 min at maximum speed, 4°C. Resuspend cell pellet in 100 µl ice-cold 0.25 M Tris·Cl, pH 7.5.

4. Freeze cells 5 min in dry ice/ethanol. Transfer to 37°C and thaw 5 min. Repeat this freeze-thaw process twice more.

5. Cool cell lysate on ice, then microcentrifuge 5 min at maximum speed, 4°C. Remove and save the supernatant (cytoplasmic extract). Freeze and store at −20°C.

6. Assay 20 µl of cell extract in the following cocktail (130 µl per reaction):

   2 µl 200 µCi/ml [^{14}C]chloramphenicol (35 to 55 mCi/mmol)
   20 µl 4 mM acetyl CoA
   32.5 µl 1 M Tris·Cl, pH 7.5
   75.5 µl H_2O.

   To assay different amounts of extract (up to 50 µl), adjust the reaction to give 0.25 M Tris·Cl in a final volume of 150 µl.

7. For each assay, add 130 µl cocktail and 20 µl extract to a microcentrifuge tube and mix gently. Incubate 1 hr at 37°C.

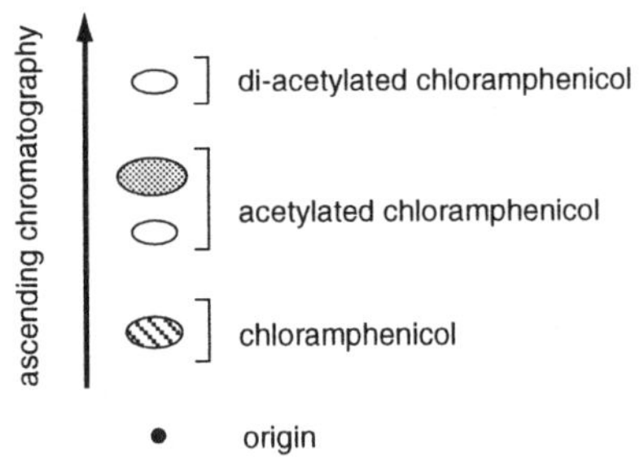

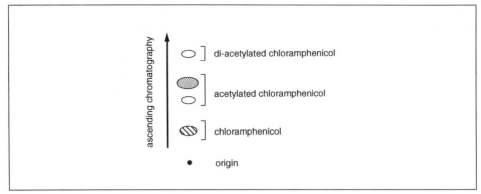

Figure 9.7.2 Schematic depiction of results of a CAT assay.

8. Add 1 ml ethyl acetate to the reaction and vortex. Microcentrifuge 1 min and remove the top (ethyl acetate) layer. Dry the ethyl acetate 45 min in a Speedvac evaporator.

9. Equilibrate the thin-layer chromatography tank: add 190 ml chloroform, 10 ml methanol, and a piece of Whatmann 3MM filter paper approximately the size of the TLC sheet to the tank and let stand 2 hr.

10. Resuspend each sample in 30 µl ethyl acetate. Spot sample, 5 µl at a time, 2 cm above the edge of a plastic-backed TLC sheet.

11. Develop the chromatogram in an equilibrated chromatography tank containing 200 ml of 19:1 chloroform/methanol. Allow the chromatography to run 2 hr or until the solvent front is close to the top of the sheet. Remove the TLC sheet and air dry. Mark the TLC sheet with radioactive ink, cover with plastic wrap, and place on film for autoradiography (*APPENDIX 3*).

The final autoradiogram will have up to five spots for each sample. They are, in ascending order, a weak spot at the origin, nonacetylated chloramphenicol, the two forms of acetylated chloramphenicol, and diacetylated chloramphenicol (Fig. 9.7.2). If the diacetylated spot is present, the assay is out of the linear range (conversion to acetyl chloramphenicol ≥20% to 30%). If this occurs, dilute sample or reduce assay time.

12. Calculate the activity of the extract by first determining the percentage of counts that are in the monoacetylated chloramphenicol species by cutting out each spot and adding it to scintillation fluid and counting. Calculate CAT activity as follows:

$$\%\text{ acetylated} = \frac{\text{counts in acetylated species}}{\text{counts in acetylated species} + \text{counts in nonacetylated chloramphenicol.}}$$

ALTERNATE PROTOCOL 1

IN SITU LYSIS OF CELLS FOR CAT ASSAY

Additional Materials *(also see Basic Protocol 1; see APPENDIX 1 for items with ✓)*

Cells transfected with CAT expression plasmid (*UNITS 9.1-9.4 & 16.13*), in 60-mm petri plates
✓ Hypotonic buffer
✓ Triton lysis buffer

1. Wash 60-mm plate of cells transfected with CAT expression plasmid once with 2 ml PBS. Add 2 ml hypotonic buffer per plate. Incubate at room temperature 2 to 5 min.

2. Aspirate hypotonic buffer and add 400 µl Triton lysis buffer. Scrape the plate with a rubber policeman and transfer lysate to a 1.5-ml microcentrifuge tube using a 1000-µl pipettor.

3. Microcentrifuge 1 min to remove nuclei and insoluble proteins. Transfer supernatant to a clean microcentrifuge tube and proceed with the assay for CAT activity (Basic Protocol 1, step 6, or Alternate Protocol 2, step 3).

PHASE-EXTRACTION ASSAY FOR CAT ACTIVITY

ALTERNATE PROTOCOL 2

Additional Materials *(also see Basic Protocol 1; see* APPENDIX 1 *for items with ✓)*

- Mammalian cells transfected with CAT expression plasmid (*UNITS 9.1-9.4 & 16.13*) *or* protoplasts transfected with CAT expression plasmid (*UNIT 9.3*)
- ✓ 0.01 µCi/µl [^{3}H]chloramphenicol solution
- 5 mg/ml butyryl CoA (store ≤4 months at −20°C)
- 100 mg/ml unlabeled chloramphenicol
- ✓ 2 M Tris·Cl, pH 8.0
- 2:1 (v/v) tetramethylpentadecane (TMPD)/xylenes

From mammalian cells:

1a. To prepare cell extracts from mammalian cells, follow either the freeze-thaw method in steps 1 to 5 of Basic Protocol 1 or the in situ lysis method in Alternate Protocol 1. If cells are harvested using trypsin or by scraping, replace steps 1 to 4 of the in situ lysis method with steps 2a to 4 below.

2a. Spin cells in a clinical centrifuge 5 min at 300 × *g*, room temperature, and discard supernatant. Add 2 ml hypotonic buffer per 10^7 cells, centrifuge, and discard supernatant. Add 200 µl Triton lysis buffer per 10^7 cells.

From plant protoplasts:

1b. Harvest plant protoplasts (10^5 cells) in 1.5-ml microcentrifuge tubes by microcentrifuging 5 min at 2000 to 4000 rpm, room temperature. Discard supernatant.

2b. Add 50 µl hypotonic buffer to the pellet and vortex vigorously. Freeze the sample by incubating the tube ≥15 min at −70°C or 5 min in a dry ice/ethanol bath. Thaw and microcentrifuge 2 min at 13,000 rpm, room temperature. Transfer supernatant (containing the CAT extract) to a clean tube.

3. To assay 50 µl cell extract, make the following CAT assay mix (50 µl per reaction):

   20 µl 0.01 µCi/µl [^{3}H]chloramphenicol solution
   5 µl 5 mg/ml butyryl CoA
   5 µl 2 M Tris·Cl, pH 8.0
   20 µl H_2O.

4. For each assay, add 50 µl CAT assay mix to 50 µl cell extract. Incubate 30 to 90 min at 37°C.

   The assay must be in the linear range for both time and activity (<20% to 30% conversion). For extract with greater activity, dilute the sample or assay for a shorter time.

5. Extract acylated chloramphenicol with 200 µl of 2:1 TMPD/xylenes by vigorous shaking. Centrifuge and remove the top (organic) phase to a scintillation vial.

6. Add 3 to 5 ml scintillation fluid to the sample and count to determine CAT activity.

BASIC PROTOCOL 2

RADIOIMMUNOASSAY FOR HUMAN GROWTH HORMONE

Materials

Cells transfected with hGH expression plasmid (*UNITS 9.1-9.4 & UNIT 16.13*; see Fig. 9.7.3)
Human Growth Hormone Radioimmunoassay Kit (Allegro Human Growth Hormone, Nichols Institute Diagnostics) containing:
- ^{125}I-labeled antibody solution
- Wash solution
- Human growth hormone (hGH) standards
- Avidin-coated beads

12 × 75–mm round-bottom test tube
γ counter

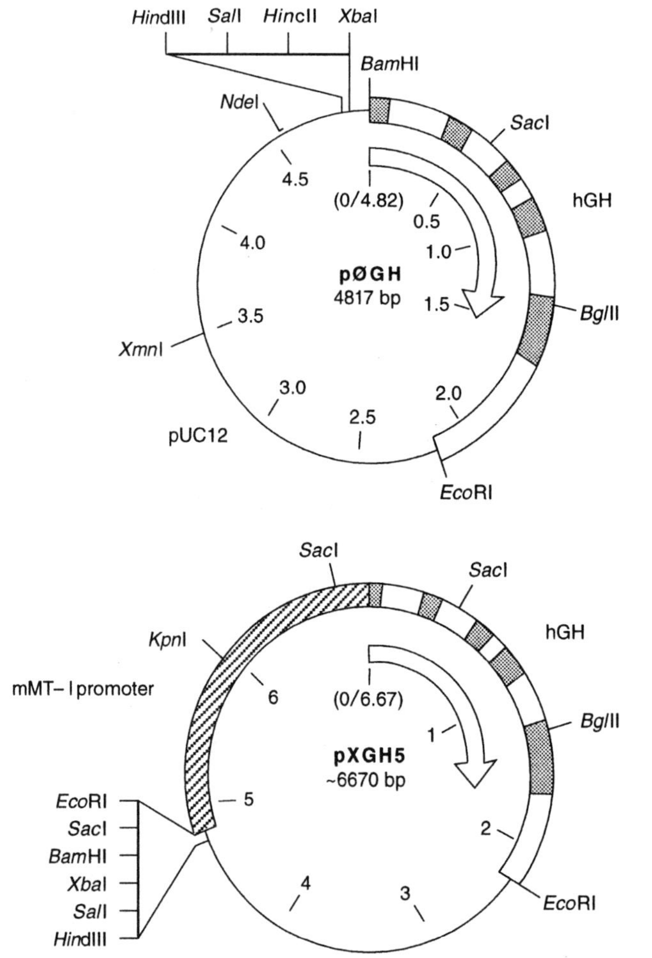

Figure 9.7.3 Expression vectors for human growth hormone (hGH; Selden et al., 1986).

1. Remove 100 to 500 μl medium from mammalian cells transfected with hGH expression plasmid (Fig. 9.7.3).

2. Pipet 100 μl medium or standards into a 12 × 75–mm round-bottom test tube. Add 100 μl of ^{125}I-labeled antibody solution to the tube and mix. Add one avidin-coated bead and cap the tube or cover it with Parafilm. Incubate 90 min at room temperature on a horizontal rotating shaker set at ~170 rpm.

3. Wash the bead twice by adding 2 ml wash solution and aspirating completely.

4. Count the tube in a γ counter 1 min.

5. Plot a standard curve with the data from the hGH standards supplied with growth hormone assay kit. Calculate the values of the unknowns using this standard curve.

   The assay is not linear for values >50 ng/ml. Samples with values >50 ng/ml should be diluted and reassayed.

References: Gorman et al., 1982; Seed and Sheen, 1988; Selden et al., 1986.

Contributors: Robert E. Kingston, Jen Sheen, and David Moore

Nonisotopic Assays for Reporter Gene Activity

UNIT 9.7B

FIREFLY LUCIFERASE REPORTER GENE ASSAY

BASIC PROTOCOL 1

Materials *(see* APPENDIX 1 *for items with* ✓*)*

Cells transfected with luciferase expression plasmid (*UNITS 9.1-9.4 & 16.13*; Fig. 9.7.4), in 60-mm petri plates
✓ PBS, ice-cold
✓ Triton/glycylglycine lysis buffer
✓ Luciferase assay buffer
✓ Luciferin stock solution
25 mM glycylglycine, pH 7.8 (free base, crystalline; Sigma)
Firefly luciferase of known activity (Sigma) for use as standard

Rubber policeman or equivalent
Luminometer (manual or automated) with printer or chart recorder and cuvettes

1. Wash three 60-mm plates of cells transfected with luciferase expression plasmid three times with 4 ml ice-cold PBS per wash, aspirating medium between each wash.

   The luciferase enzymatic reaction is inhibited by traces of calcium.

2. Add 350 μl Triton/glycylglycine lysis buffer to each plate. Scrape with a rubber policeman. Transfer solubilized cells to a 1.5-ml microcentrifuge tube. Microcentrifuge 5 min at maximum speed, 4°C. Transfer supernatant (cell lysate) to a clean microcentrifuge tube and store on ice for assay.

3. Gently vortex cell lysate and place 100 μl in a luminometer cuvette. Add 360 μl luciferase assay buffer. Place cuvette in luminometer chamber.

4. Dilute luciferin stock solution to 200 μM in 25 mM glycylglycine, pH 7.8.

5. Inject 200 µl diluted luciferin solution into the sample in the luminometer and measure light output for 20 sec at 25°C. Quantitate by subtracting machine background (determined by measuring light output in a cuvette that has no cellular lysate) and by comparing bioluminescence of each sample with luciferase standards obtained commercially.

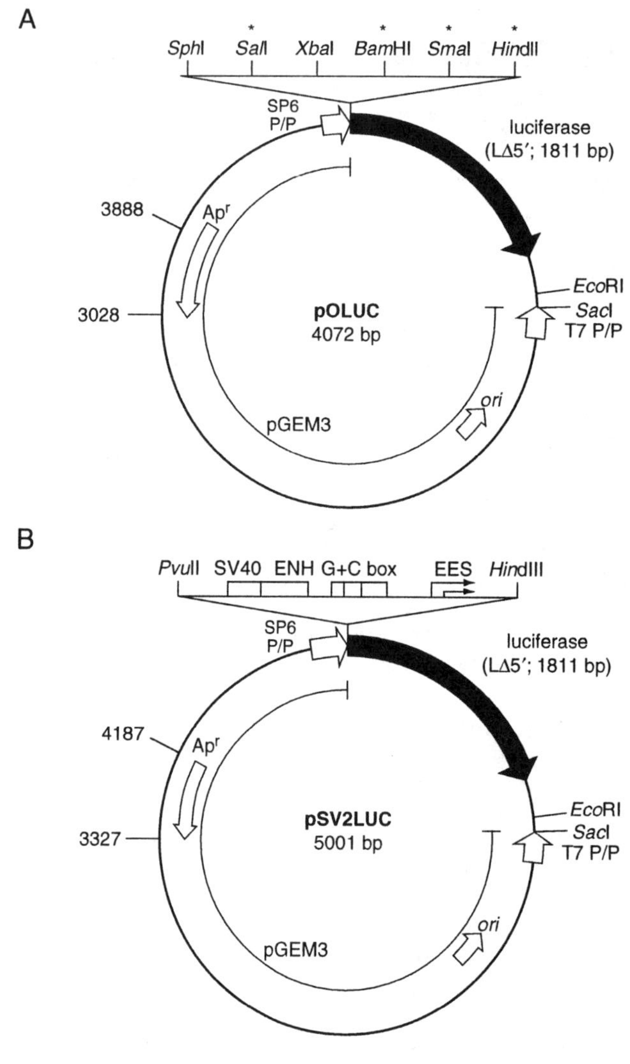

Figure 9.7.4 Luciferase expression vectors. (**A**) pOLUC is a promoterless vector containing a luciferase gene. It is derived from pGEM3, a high-copy ampicillin-resistance plasmid whose multiple cloning sites are flanked by SP6 promoter/primer sequences (SP6 P/P) and T7 promoter/primer sequences (T7 P/P). Unique restriction sites for insertion of sequences of interest are indicated by asterisks (*). (**B**) pSV2LUC contains the strong eukaryotic simian virus-40 (SV40) promoter/enhancer sequences inserted into pOLUC for use as a positive control for luciferase expression or as a standard for normalization between transfections. SV40 ENH designates the 72-base-pair repeats of the SV40 enhancer, G+C box(es) are the SP1-binding sites of the SV40 promoter, and EES designates the early-early transcription start sites.

LUCIFERASE ASSAY IN FREEZE-THAW-LYSED CELLS

ALTERNATE PROTOCOL

Additional Materials (*also see Basic Protocol 1; see APPENDIX 1 for items with ✓*)

✓ Extraction buffer

1. Wash three 60-mm plates of cells transfected with luciferase plasmid in PBS (Basic Protocol 1, step 1).
2. Add 1 ml extraction buffer to each plate. Scrape cells immediately with a rubber policeman. Transfer cells to a 1.5-ml microcentrifuge tube. Microcentrifuge 15 to 30 sec. Remove and discard supernatant.
3. Add 100 µl extraction buffer to the cell pellet. Lyse cell membranes by three cycles of freeze-thaw.
4. Microcentrifuge lysed cells 5 min at maximum speed, 4°C. Remove and assay the supernatant for luciferase activity (Basic Protocol 1, steps 3 to 5).

CHEMILUMINESCENT β-GALACTOSIDASE REPORTER GENE ASSAY

BASIC PROTOCOL 2

Materials (*see APPENDIX 1 for items with ✓*)

Cells transfected with a β-galactosidase expression plasmid (*UNITS 9.1-9.4 & 16.13*; Fig. 9.7.5), in 60-mm petri plates
Mock-transfected control cells, in 60-mm petri plates
✓ PBS
✓ Triton lysis solution
✓ β-galactosidase reaction buffer
✓ Light-emission accelerator solution

Rubber policeman (or equivalent)
Luminometer with chart recorder and tubes *or* scintillation counter

1. Rinse one 60-mm plate of cells transfected with β-galactosidase expression plasmid and one 60-mm plate of mock-transfected cells twice with PBS.
2. Add 250 µl Triton lysis solution. Detach cells from plate using a rubber policeman. Transfer cell extract to a 1.5-ml microcentrifuge tube and microcentrifuge 2 min at maximum speed, 4°C. Transfer supernatant to a clean microcentrifuge tube.

   Cell extracts may be used immediately or frozen several months at −70°C before continuing with the procedure.

3. Place 2 to 20 µl cell extract into a luminometer tube. Add 200 µl β-galactosidase reaction buffer to the luminometer tube and incubate 60 min at room temperature.

   Analyze the activity in the samples one at a time, delaying the addition of reaction buffer by the time it takes to complete one measurement.

4. Place tube into luminometer and inject 300 µl light-emission accelerator solution. Wait 2 to 5 sec after injection and take the measurement for 5 sec.

References: Gould and Subramani, 1988; Jain and Magrath, 1991.

Contributors: Allan R. Brasier and John J. Fortin

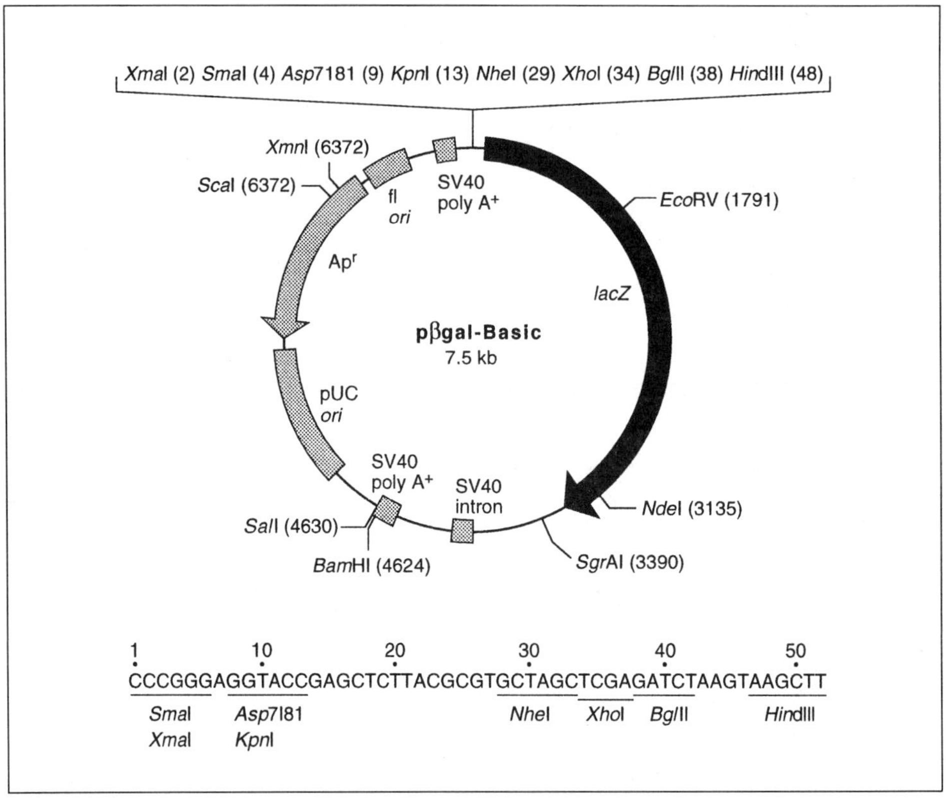

Figure 9.7.5 β-galactosidase expression vector. pβ-gal-Basic lacks eukaryotic promoter and enhancer sequences and allows insertion of promoter-containing DNA fragments into the polylinker upstream of the *lacZ* gene. Enhancer sequences can be cloned into the polylinker or into unique sites downstream of *lacZ*. All four vectors in this series contain an SV40 intron and SV40 polyadenylation signal downstream of *lacZ* to ensure proper and efficient processing of the transcript. The vector backbone contains a f1 origin for single-stranded DNA production and a pUC-19 origin of replication and ampicillin-resistance gene for propagation in *Escherichia coli*. The multiple cloning site (polylinker) region is identical in all four vectors in the series; in pβ-gal-Promoter and in pβ-gal-Control a 202-bp promoter-containing fragment has been inserted between *BglII* and *HindIII*. The sequence of pβ-gal-Basic has been deposited in GenBank (Accession #U13184).

UNIT 9.8

Direct Analysis of RNA After Transfection

It is possible to transfect mammalian cells with a gene of interest and directly detect the RNA made from that gene 24 to 48 hr later (*UNITS 9.1-9.4, 16.12, & 16.13*). This type of analysis circumvents the need for fusion genes, but takes many more hours to do. The ability to analyze the RNA directly allows mutational and functional analysis of an intact gene, as the promoter does not have to be subcloned. This technique is also essential when fusion genes are used. To be sure that the level of the reporter protein produced by a fusion gene provides an accurate measure of appropriately initiated RNA from the promoter under study, it is necessary to determine the amount and 5′ end of the fusion message. An investigator who verifies this using direct RNA analysis can proceed with some confidence that the level of the reporter protein is a measure of promoter activity.

The difficulty in directly analyzing RNA is that the sensitivity of detection is at the limits of present technology. Consequently, the transfection protocol, RNA preparation, and RNA analysis techniques all have to be working near optimum for an experiment to work. The cell type used must be readily transfectable, and the promoter being studied must be relatively efficiently expressed or the amount of RNA produced will be below the limits of detection. Direct analysis of RNA is several orders of magnitude less sensitive than the commonly used reporter gene systems.

The following parameters should be considered when directly analyzing RNA after a transfection.

TRANSFECTION EFFICIENCY

It is imperative that the cell type and transfection protocol used result in efficient transfection. The efficiencies that result from $CaPO_4$-mediated transfection of HeLa, BALB/c 3T3, or NIH 3T3 cells are sufficient to do direct analysis of RNA expression after transient transfection. Optimization of transfection efficiency in the cell type used (*UNIT 9.9*) should be done prior to attempting to analyze RNA from transfected genes. One way to avoid the extensive workup of an experiment that is doomed to failure due to inefficient transfection is to include an easily assayed reporter gene in the transfection as an indicator of transfection efficiency.

RNA PREPARATION

RNA must be efficiently made in an intact form. The total cytoplasmic (*UNIT 4.1*) and guanadinium isothiocyanate (*UNIT 4.2*) protocols are both well suited to this purpose. Two 10-cm dishes of adherent cells should produce 100 µg total RNA (this number will be lower for particularly flat cell types). Expression from a transfected gene can be detected by analyzing 40 µg total RNA. Sensitivity will be increased by using poly(A)-selected RNA (*UNIT 4.5*), but this is time-consuming.

ANALYSIS OF RNA

Both S1 analysis using single-stranded probes (*UNIT 4.6*) and ribonuclease protection analysis (*UNIT 4.7*) have the sensitivity necessary to detect gene expression at transient times after transfection. S1 analysis tends to have a lower background, allowing longer exposure times to be used. Ribonuclease protection analysis uses a probe of higher specific activity and thus should be significantly more sensitive if optimized.

PROMOTER STRENGTH

If the promoter strength is more than an order of magnitude below that of, for example, the SV40 early promoter in HeLa cells, heroic efforts may be necessary in order to analyze the RNA. In such cases, poly(A) selection may be necessary as well as the use of body-labeled probes.

References: Kingston et al., 1986; Treisman, 1986.

Contributor: Robert E. Kingston

Optimization of Transfection

Every cell type has a characteristic set of requirements for optimal introduction of foreign DNA. There is a tremendous degree of variability in transfection conditions that work for various cell types. It is thus helpful to have a straightforward, systematic approach to optimizing transfection efficiency. Transient assay systems are particularly useful for this purpose. The human growth hormone (hGH; *UNIT 9.7A*) assay system is particularly useful for this purpose because both harvest and assay take very little time.

The single most important factor in optimizing transfection efficiency is selecting the proper transfection protocol. This usually comes down to a choice among DEAE-dextran-mediated gene transfer (*UNITS 9.2, 16.12, & 16.13*), calcium phosphate–mediated gene transfer (*UNIT 9.1*), electroporation (*UNIT 9.3*), and liposome-mediated transfection (*UNIT 9.4*). It is recommended that any adherent cell line under investigation be tested for transfection ability with DEAE-dextran, calcium phosphate, and liposome-mediated transfection. Nonadherent cell lines can be transfected by electroporation and liposome-mediated transfection.

DEAE-DEXTRAN TRANSFECTION

There are several factors that can be varied in DEAE-dextran transfection. The number of cells, concentration of DNA, and concentration of DEAE-dextran added to the dish are the most important to optimize. To a first approximation, most cell types that can be transfected using DEAE-dextran will have a preference for 1 to 10 µg DNA/10-cm dish and for 100 to 400 µg DEAE-dextran/ml of medium. Table 9.9.1 shows how the dishes in an optimization might be chosen. If an hGH expression vector such as pXGH5 is used, a time course of expression under each condition can be determined by removing 100-µl aliquots of the medium 2, 4, and 7 days posttransfection (with a medium change after the day 4 aliquot is removed).

With the results of this pilot experiment in hand, a second experiment using a narrower range of DEAE-dextran concentrations and a wider range of DNA doses should be undertaken. For example, if the cells appear to express more hGH at 100 µg/ml DEAE-dextran than at higher concentrations in the pilot experiment, the second experiment should cover from 25 to 150 µg/ml DEAE-dextran. Because DEAE-dextran is toxic to some cells, a brief exposure to small concentrations may be optimal. The wide range of added DNA in this experiment is crucial in two respects. First, it is valuable to know the smallest amount of the transfected reporter

Table 9.9.1 Optimization of DEAE-Dextran Transfection

5×10^5 cells/10-cm dish:			2×10^6 cells/10-cm dish:		
Dish	pXGH5 (µg)	DEAE-dextran (µg/ml)	Dish	pXGH5 (µg)	DEAE-dextran (µg/ml)
1	1	400	11	1	400
2	1	200	12	1	200
3	1	100	13	1	100
4	4	400	14	4	400
5	4	200	15	4	200
6	4	100	16	4	100
7	10	400	17	10	400
8	10	200	18	10	200
9	10	100	19	10	100
10	0	200	20	0	200

Table 9.9.2 Optimization of Calcium Phosphate Transfection

Dish (10-cm)	pXGH5 (μg)	pUC13 (μg)	Exposure to precipitate (hr)	Glycerol shock (min)
1	5	5	6	—
2	5	15	6	—
3	5	35	6	—
4	5	5	16	—
5	5	15	16	—
6	5	35	16	—
7	5	5	6	3
8	5	15	6	3
9	5	35	6	3
10	5	5	16	3
11	5	15	16	3
12	5	35	16	3

gene that can give a readily detectable signal. Second, the linearity of the dose of DNA with the amount of reporter gene expression generally decays for large amounts of input DNA. When excessive (i.e., nonlinear) amounts of DNA are used in transfection experiments, it is possible that the effects observed are dose-response effects rather than the phenomenon intended for study. This serious and common problem can be eliminated by doing a careful DNA dose–response curve as above.

CALCIUM PHOSPHATE TRANSFECTION

The primary factors that influence efficiency of calcium phosphate transfection are the amount of DNA in the precipitate, the length of time the precipitate is left on the cell, and the use and duration of glycerol or DMSO shock. A calcium phosphate optimization is shown in Table 9.9.2. Generally, higher concentrations of DNA (10-50 μg) are used in calcium phosphate transfection. Total DNA concentration in the precipitate can have a dramatic effect on efficiency of uptake of DNA with calcium phosphate–mediated transfection. With some cell lines, more than 10 to 15 μg of DNA added to a 10-cm dish results in excessive cell death and very little uptake of DNA. With other cell types, such as primary cells, a high concentration of DNA in the precipitate is necessary to get any DNA at all into the cell.

The optimal length of time that the precipitate is left on cells varies with cell type. Transfection efficiency of some cell types is dramatically increased by glycerol or DMSO shock (*UNIT 9.1*). The pilot experiment listed will indicate whether the cell type is tolerant to long exposure to a calcium phosphate precipitate and whether glycerol shock should be used.

Once optimal conditions for transfection are found, extensive DNA curves varying the amount of reporter plasmid can be prepared, as for the DEAE-dextran technique. The total amount of DNA that is transfected should be held constant, with the difference made up using carrier plasmid.

ELECTROPORATION

Perhaps because it is not a chemically based protocol, electroporation tends to be less affected by DNA concentration than either DEAE-dextran- or calcium phosphate–mediated gene transfer. Generally, DNA amounts in the range of 10-40 μg/10^7 cells work well, and there is a good linear correlation between the amount of DNA present and the amount taken up. The parameter that can be varied to optimize

Table 9.9.3 Optimization of Liposome-Mediated Transfection

Dish (35-mm)	pSV2CAT (μg)	Liposomes (μl)
1	0.1	1
2	0.1	2
3	0.1	4
4	0.1	8
5	0.1	12
6	0.5	1
7	0.5	2
8	0.5	4
9	0.5	8
10	0.5	12
11	5	5
12	5	10
13	5	15
14	5	20
15	5	30
16	10	5
17	10	10
18	10	15
19	10	20
20	10	30

electroporation is the amplitude and length of the electric pulse, the latter being determined by the capacitance of the power source. The objective is to find a pulse that kills between 20% and 60% of the cells (generally 1.5 kV at 25 μF). If excessive cell death occurs, the length of the pulse can be lowered by lowering the capacitance (3 and 25 μF).

LIPOSOME-MEDIATED TRANSFECTION

Three primary parameters—the concentrations of lipid and DNA and incubation time of the liposome-DNA complex—affect the success of DNA transfection by cationic liposomes. These should be systematically examined to obtain optimal transfection frequencies.

Concentration of lipid. In general, increasing the concentrations of lipid improves transfection; however, at high levels (>100 μg), the lipid can be toxic. For each particular liposome mixture tested, it is important to vary the amount as indicated in Table 9.9.3.

Concentration of DNA. In many of the cell types tested, relatively small amounts of DNA are effectively taken up and expressed. In fact, higher levels of DNA can be inhibitory in some cell types with certain liposome preparations. In the optimization protocol outlined in Table 9.9.3, the standard reporter vector pSV2CAT is used; however, any plasmid DNA whose expression can be easily monitored would be suitable.

Time of incubation. When the optimal amounts of lipid and DNA have been established, it is desirable to determine the length of time required for exposure of the liposome-DNA complex to the cells. In general, transfection efficiency increases with time of exposure to the liposome-DNA complex, although after 8 hr, toxic conditions can develop.

Contributors: Richard F Selden and John K. Rose

Overview of the Retrovirus Transduction System

A retrovirus vector is an infectious virus used to introduce a nonviral gene into mitotic cells in vivo or in vitro. The vectors are modified in various ways and originate from replication-competent viruses isolated from rodents or chickens. The transduction process produces a single copy of the viral genome stably integrated into a host chromosome.

Retrovirus vectors are useful in achieving stable and efficient transduction of a gene or genes into cells that are not easily transfected, such as primary cells and cells in

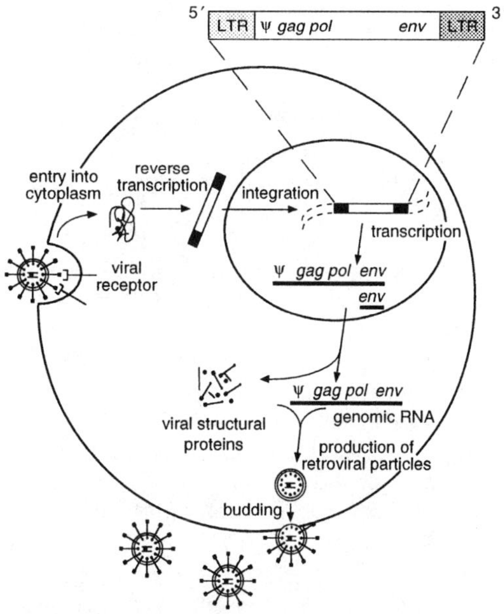

Figure 9.10.1 Replication-competent retrovirus life cycle. All retroviruses infect host cells through an interaction with a specific viral receptor on the host-cell surface. They thereby enter the host-cell cytoplasm and initiate reverse transcription of the RNA viral genome using reverse transcriptase contained within the viral particle. After synthesis of a linear, duplex DNA copy of the viral genome, the viral DNA integrates into the host genome. Integration is mediated by the viral *int* gene product (encoded at the 3′ end of the *pol* region) and usually occurs at the ends of the viral genome. Within the host genome, integration can occur at a very large number of chromosomal locations with little or no apparent specificity. One infectious virion produces one integrated copy of the viral genome. This "provirus" is subsequently replicated with the host DNA and is passed on to all progeny cells. After integration, the viral LTR promoter at the 5′ end of the genome is usually active and directs synthesis of a full-length (i.e., unspliced) copy of the viral genome, which terminates in the 3′ LTR.

In replication-competent viruses, the viral genome encodes the viral proteins *gag, pol,* and *env* required to make viral particles. The full-length viral transcript serves several functions. It is both the mRNA for the *gag* proteins (which comprise the bulk of the viral particle), reverse transcriptase, and the *int* gene product and is the precursor RNA that is used to generate a spliced mRNA for the *env* protein, found on the surface of the viral particle. Finally, in its unspliced form, it is also the RNA that encodes Ψ, the recognition sequence for encapsidation. Because it contains the Ψ sequence, it can be recognized by the packaging proteins and become packaged into a viral particle. Due to this role, the unspliced full-length transcript is also referred to as the "genomic RNA." Infection by a replication-competent virus thus leads to production of infectious retroviral particles identical to the original infecting particle. These particles are budded from the host cell and can go on to initiate more rounds of infection in other host cells (by C. Cepko, D. Altshuler, D. Fekete, D. Wu, unpub. results).

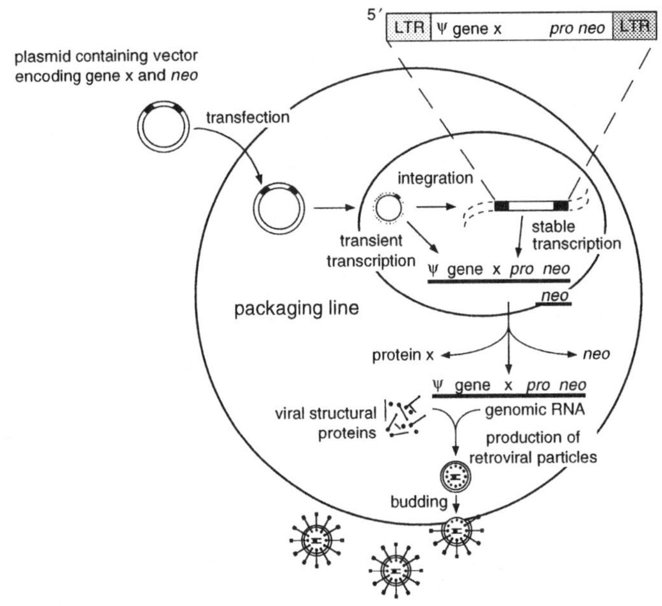

Figure 9.10.2 Production of an infectious retroviral particle from a bacterial plasmid that encodes a retroviral genome requires the introduction of the bacterial plasmid into mammalian or avian cells. This is true for both replication-competent and replication-incompetent vectors. In the example shown here, a replication-incompetent vector encoding *neo* and a gene of interest (*x*) are introduced, via $CaPO_4$ transfection, into a packaging line. The packaging line is a mouse or quail cell line (see Table 9.10.1) that encodes the proteins necessary for production of viral particles (i.e., *gag, pol,* and *env*). The bacterial plasmid can be transiently (for a few days after transfection) transcribed from nonintegrated plasmid molecules or can be stably transcribed from integrated plasmid molecules. The viral transcript is initiated in the 5′ LTR and is terminated in the 3′ LTR, and is thus a full-length viral transcript. It contains the packaging sequence, Ψ, which is recognized by the capsid proteins and allows it to be packaged into viral particles. A fully infectious viral particle containing the vector genome is thus budded from the packaging cell. The culture supernatant is removed from cells and used as the source of virus for future experiments (e.g., see Figure 9.10.3).

Also shown in this figure is the production of another transcript, initiated at an internal promoter (*pro*) and encoding the *neo* gene. The *neo* transcript produced by the internal promoter does not contain Ψ and thus does not become encapsidated. It will, however, serve as an mRNA for *neo*, thereby allowing for growth of these cells in the selective drug G418. As described in *UNIT 9.11,* this drug-resistance feature can be exploited when one wishes to select cells that have been stably transfected with the retroviral plasmid (by C. Cepko, D. Altshuler, D. Fekete, D. Wu, unpub. results).

vivo. However, the retrovirus transduction system is not the method of choice when transient expression studies can be performed using transfected cells in vitro (for transient expression protocols, see *UNITS 9.1-9.4 & 16.13*) or when large (>8 kb) fragments of DNA are to be transduced. Moreover, postmitotic cells cannot be transduced, because a cellular S phase seems to be a requirement for complete reverse transcription and possibly viral integration.

This unit presents an overview of the retrovirus life cycle and a description of vector designs and packaging cell lines (for review, see Weiss et al., 1984, 1985). Transfection of a vector into a packaging line is then detailed (*UNIT 9.11*), with protocols for titering and assaying the resultant virus stock. For various applications it is desirable to produce large quantities of the virus, to concentrate the virus stock (*UNIT*

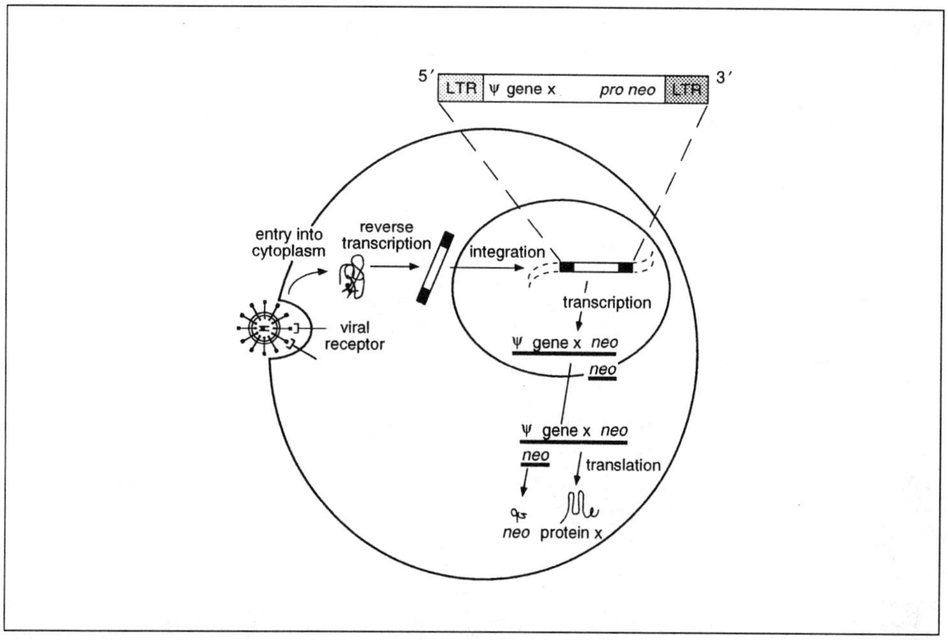

Figure 9.10.3 Infection and expression of a target cell by a replication-incompetent retrovirus vector occurs through interaction with a host-cell receptor. The virus then reverse transcribes and integrates, just as a replication-competent virus, as described in Figure 9.10.1. A full-length viral transcript can also be initiated in the 5′ LTR and end in the 3′ LTR, exactly as in the case of the replication-competent virus. However, the viral transcript typically does not encode any of the proteins required to make a viral capsid. It usually encodes a gene of interest (*x*). In addition, many replication-incompetent vectors encode another promoter (*pro*), within the viral genome for synthesis of an mRNA for another gene (e.g., *neo*). Because such a vector does not encode viral capsid proteins, cells infected with a replication-incompetent vector cannot make more viral particles. Thus, there is not spread of the viral genome from an infected cell to other cells. However, the viral genome does pass to all progeny cells through the normal processes used by the host to pass on all host genes (by C. Cepko, D. Altshuler, D. Fekete, D. Wu, unpub. results).

9.12), and/or to check the stock for contamination by helper viruses (particularly when animals are to be infected; *UNIT 9.13*). Finally, outlines for infection of cells in vivo and in vitro are suggested (*UNIT 9.14*).

RETROVIRUS LIFE CYCLE

Although retroviral genomes are made of RNA, retroviruses produce a DNA copy of their genome immediately after infection of the host cell (see Fig. 9.10.1). This linear, double-stranded DNA copy is made by reverse transcriptase, a product of the viral *pol* gene that is present in the viral particle. The viral DNA then integrates (mediated by the viral *int* gene product) into the host genome where it is referred to as a provirus. Most experimental retrovirus vectors were originally derived from cloned proviral sequences.

An infectious virus stock is made from cloned vectors encoded by bacterial plasmids by transfecting a retrovirus vector plasmid into cell lines called packaging lines, for replication-incompetent vectors (Figure 9.10.2), or into host lines that support replication, for replication-competent vectors. Once in a cell line, transcription proceeds from the viral LTR promoter encoded by the plasmid, generating an RNA viral genome. The viral genome is then encapsidated by viral structural proteins and infectious viral particles are produced by budding from the surface of cells (Figure 9.10.2). The supernatant produced by such cells comprises a virus stock. The number

Table 9.10.1 Cell Lines for Packaging and Titering Retroviral Stocks

Cell line	Source	Helper virus produced	Features
Cell Lines for Packaging			
Murine-Ecotropic			
Ψ2	Mann et al., 1983	Yes	Most reliable for high titers
ΨCRE	Danos and Mulligan, 1988	No	
GP+E-86	Markowitz et al., 1988	No	
ΩE	Morgenstern and Land, 1990	No	
Murine-Amphotropic			
Ψam	Cone and Mulligan, 1984	Yes	
PA12	Miller et al., 1985	Yes	
PA317	Miller and Buttimore, 1986	Rarely	Reliable for high titers
ΨCRIP	Danos and Mulligan, 1988	No	Reliable for high titers
Avian			
Q2bn	Stoker and Bissel, 1988	Yes	Works well for transient expression but does not give high titers from stably transfected lines; uses subgroup A *env*
Isolde	Cossett et al., 1990	No	Gives good titers from stably transfected lines; uses subgroup A *env*
Cell Lines for Titering			
Murine			
NIH 3T3 (mouse fibroblast)	Generally available	NA	Established cell line; easy to grow
Avian			
QT6 (Japanese quail line)	Moscovici et al., 1977	NA	Established cell line; preferable for titration over CEF
CEF (chicken embryo fibroblast)	Hunter, 1979	NA	Not a cell line; primary cells must be prepared from embryos; more difficult to grow than QT6

of infectious viral particles per ml in a stock, referred to as colony forming units/ml (CFU/ml), is determined by a bioassay (*UNIT 9.11*). The bioassay, or "titration," is carried out on a convenient target cell line that is easy to grow and is very susceptible to infection by a particular virus (Table 9.10.1). The titered stock is then used to infect cells in vivo or in vitro. Once integrated into the host cell the vector provirus makes mRNA(s) for the gene(s) of interest.

REPLICATION-INCOMPETENT VECTORS

Replication-incompetent retrovirus vectors were derived from proviruses that were cloned from cells infected with a naturally occurring retrovirus. The vectors are deleted for viral products necessary to produce an infectious particle but retain the *cis*-acting viral sequences necessary for transmission. These sequences include the Ψ packaging sequence (necessary for recognition of the viral RNA for encapsidation into the virus particle), reverse transcription signals, integration signals, and viral promoter, enhancer, and polyadenylation sequences. Using the transcription regulatory sequences provided by the virus, any cDNA can be expressed in the vector. Alternate constructs that lack the viral promoter and/or enhancer have also been made, which are used for studies of alternative promoter regulation, or when trying to avoid repression of transcription that sometimes can occur when using the viral LTR.

REPLICATION-COMPETENT VECTORS

Vectors that retain all of the genes encoding the virion structural proteins, as well as all of the *cis*-acting viral elements necessary for transmission, are "replication-competent." Expression of additional, nonviral sequences has frequently been successful for genes ≤2 kb using avian replication-competent vectors (Hughes et al., 1987), but few successes have been reported for the murine replication-competent vectors.

PACKAGING LINES AND VIRUS PRODUCTION

Replication-incompetent retrovirus vectors do not encode the structural genes whose products comprise the viral particle (Fig. 9.10.3). To produce infectious viral particles from a retrovirus plasmid, the viral structural proteins encoded by *gag, pol,* and *env* are supplied by packaging cell lines. These cells are derived from either stable mouse-fibroblast lines or the QT6 quail line and contain the viral *gag, pol,* and *env* genes as a result of introduction by transfection. However, the viral RNA that encodes the structural proteins does not contain the packaging sequence Ψ. Thus, the packaging lines do not efficiently package the RNA that encodes the genes *gag, pol,* or *env.* Prior to the introduction of vector DNA into these lines, cellular RNAs are randomly encapsidated and budded as normal viral particles. To produce viral particles that contain the vector genome, the vector DNA is introduced via transfection or infection. The vector RNA is preferentially packaged because it contains the packaging sequence, Ψ (Fig. 9.10.2).

As with naturally occurring retroviruses, packaged retrovirus vectors enter the host cell via interaction of a viral envelope glycoprotein (a product of the viral *env* gene) with a host-cell receptor. Packaging lines that express various viral envelope proteins are listed in Table 9.10.1, along with a summary of some of their properties. Cell lines for titering are also listed and will be discussed in *UNIT 9.11*.

SAFETY ISSUES

Biological safety issues must be carefully considered when working with retrovirus vectors. These issues will vary with any given experiment, so detailed protocols should be discussed with the appropriate biological safety committee at the experimenter's company or institution. In particular, extreme precautions and containment levels may be necessary when working with amphotropic viruses.

References: Danos and Mulligan, 1988; Morgenstern and Land, 1990.

Contributor: Constance Cepko

UNIT 9.11

Preparation of a Specific Retrovirus Producer Cell Line

Establishing a cell line that produces high levels of a specific retrovirus construct is a lengthy process. First, the retroviral construct must be stably introduced into an appropriate packaging cell line. After stable lines are produced, they must be characterized to identify lines that produce high titers of virus with an appropriate structure.

BASIC PROTOCOL 1

INTRODUCTION OF A RETROVIRUS VECTOR INTO A PACKAGING CELL LINE

Producing an infectious virus from a bacterial plasmid involves introduction of the plasmid into a packaging line. The transfection protocol most commonly used is a modification of the $CaPO_4$ transfection protocol (*UNIT 9.1*). This protocol has been successfully used with the packaging lines Ψ2, ΨCRE, ΨCRIP, PA317, Q2bn, and Isolde. Stable producers can be selected among the stably transfected cells, or by cross-infection using transiently produced virus from one class of packaging line to infect a packaging line bearing a different class of envelope (Fig. 9.11.1).

Materials *(see APPENDIX 1 for items with ✓)*

Appropriate packaging cell line(s) (*UNIT 9.10*) with appropriate medium
Retrovirus plasmid DNA
✓ HEPES-buffered saline (HeBS)
2 M $CaCl_2$
Medium, without and with serum
HeBS containing 15% glycerol (HeBS/glycerol)
800 µg/ml (100×) polybrene in dH_2O, filter sterilized (store at −20°C)
10% to 15% dimethylsulfoxide (DMSO)

10-cm or 6-cm dishes
24-well or 6-well tissue culture dishes
Cloning cylinders

NOTE: All incubations involving tissue culture cells should be performed in a humidified 5% CO_2 incubator at 37°C unless otherwise noted.

1. Plate packaging cells to ~10% to 20% the density of confluent cells (a 1:10 or a 1:5 split) in a 10-cm dish the day before transfection.

2. Place 10 µg of retrovirus plasmid DNA containing drug resistance gene into 0.5 ml HeBS (Falcon 5-ml tube). Add 32 µl of 2 M $CaCl_2$ while gently shaking. Tap tube for ~30 sec and incubate 45 min at room temperature, until a fine, hazy blue precipitate develops.

3. Remove medium from packaging cells and gently pipet HeBS-DNA precipitate onto center of dish. Expose cells to DNA 20 min in the tissue-culture hood; after ~10 min, gently rock dish to evenly redistribute solution. Add 10 ml medium and place cells 4 hr at 37°C.

4. Aspirate medium completely, and gently add 2.5 ml of room-temperature HeBS/glycerol. Return dish to incubator for 3.5 min (Ψ2, ΨCRE, ΨCRIP, and PA317), 90 sec (Q2bn), or a time determined to be optimal for the cells. Quickly remove HeBS/glycerol and gently rinse with 10 ml medium. Repeat medium rinse and add 5 ml medium plus serum. Incubate 18 to 24 hr.

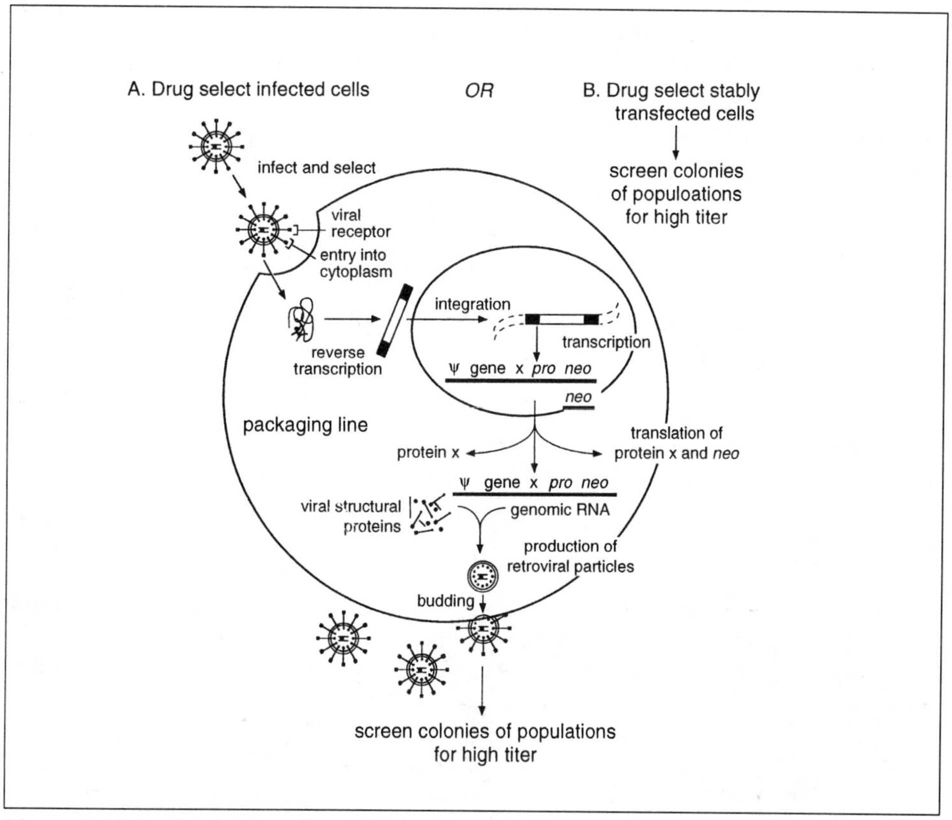

Figure 9.11.1 Production of a stable producer line for a replication-incompetent vector. To have a continual source of high-titer virus stocks, make a stable producer line by stable transduction of the viral genome into packaging cells. One of two basic methods for transduction is used. The starting point for both methods is to introduce a bacterial plasmid encoding the vector into packaging cells as shown in Figure 9.10.2. The virus produced transiently by the transfected cells is then harvested and used to infect a packaging line bearing a different *env* protein (see text) as shown in (**A**). Alternatively, selection of the transfected packaging cells in which integration of the bacterial plasmid has occurred can be carried out as shown in (**B**). In either method, drugs (e.g., G418 selection for the virus shown here) are used to select those cells that have stably integrated the viral sequences. Drug-resistant colonies can then be screened for production of high-titer virus stocks (by C. Cepko, D. Altshuler, D. Fekete, D. Wu, unpub. results).

5. Remove medium and filter through a 0.45-µm filter to obtain supernatant containing the transiently produced virus. Store at −70°C or −80°C or use immediately for an infection of another packaging line bearing a different env class in order to make an infected producer line (steps 7 to 9 below).

If the purpose of the transfection is to make stable producer lines, there are two options. The first option (step 6 followed by steps 10 to 13) is to select among the transfected cells for those that have stably integrated the vector plasmid. They can be placed under drug selection and the resulting drug-resistant cells screened for virus production (Support Protocol). The second option is to use "cross-infected" packaging cells. The transiently produced virus harvested in step 5 above can be used to infect another packaging line, as in steps 7 to 9. The *infected* packaging cells can then be placed under drug selection (steps 10 to 13). Regardless of the method used, the goal is to isolate packaging cells that have stably integrated the viral genome and that make the highest possible titer.

6. Add 10 ml medium to transfected cells, incubate 2 to 3 days, and then proceed to step 10.

This method may be preferable over using transfected cells, as the provirus that results from infection is usually stably associated with the host genome and does not rearrange or delete after integration. To use this method, two packaging lines expressing different classes of viral glycoproteins must be used. When a packaging line produces a particular env glycoprotein (e.g., murine amphotropic env), the receptors on the surface of the packaging cells become blocked by the binding of that viral glycoprotein. When using cross-infection, the two packaging lines must be homologous (i.e., both must be murine or both must be avian).

7. Split a packaging line (different envelope class than the cell line used in step 1) 1:10 to 1:20 the day before the infection.

8. Remove medium from packaging line to be infected. Add virus (from step 5) as follows: for a 10-cm dish, dilute 0.1 to 1.0 ml virus stock to 3 to 5 ml final volume and add 800 µg/ml polybrene to 8 µg/ml; for a 6-cm dish, dilute 0.1 to 1 ml virus stock to 2 ml final and add 800 µg/ml polybrene to 8 µg/ml final. Incubate ≥1 hr.

   Use several amounts of virus stock (e.g., 0.1 ml; 0.4 ml; 1.0 ml) to infect several dishes. Include polybrene for murine viruses and avian subgroup E. Other avian subgroups do not require polybrene. Longer incubation times do not hurt the cells if they are insensitive to polybrene (as most packaging cells are).

9. Add medium to 10 ml final for a 10-cm dish, and to 4 ml for a 6-cm dish. Incubate 2 to 3 days.

   If cells are insensitive to polybrene, add medium to increase volume.

10. Split transfected (step 6) or infected cells (step 9) 1:10 or 1:20 (2 to 3 days after transfection or infection), plate in selective medium, and incubate 3 days. Replace medium with fresh selective medium and incubate an additional 4 to 7 days until colonies are visible.

11. Pick well-isolated colonies using cloning cylinders and transfer into two wells for each clone in 24-well or 6-well tissue culture dishes. Grow until 50% to 90% confluent.

12. Remove medium and replace with one-half the normal volume. Incubate 1 to 3 days depending on packaging line. Remove medium and either titer it immediately (Support Protocol) or store at −70°C or −80°C.

13. Continue to passage producer clones until they can be frozen, and/or until the best producers are identified. When a good producer clone is identified, store 20 to 25 vials (1- to 2-ml aliquots) in 10% to 15% DMSO in liquid nitrogen.

BASIC PROTOCOL 2

DETERMINATION OF VIRAL TITER: IDENTIFICATION OF PRODUCER CLONES MAKING HIGH-TITER VIRUS

After isolating clones of virus producers or after generating pools of producers, those making high titers need to be identified. The usual method is to perform a bioassay using the virus to infect target cells as described here, although rough quantitative analyses can alternatively be performed on the producer cells to quantitate the presence of vector RNA or protein (see Alternate and Support Protocols). It is also necessary to ensure that the titered virus has not rearranged or deleted.

Additional Materials *(also see Basic Protocol)*

Virus stock (first basic protocol; e.g., Ψ2 BAG supernatant)
Target cell line (e.g., NIH 3T3 fibroblasts)
G418 (*UNIT 9.5*) or other selection drug

1. Split NIH 3T3 target cells (Table 9.10.1) 1:10 to 1:20 into 6-cm dishes the day before infection.

    If using avian viruses, substitute chicken embryo fibroblasts split 1:4 or 1:5, or use quail cell line QT6 split 1:6.

2. On the day of infection, remove medium from target cells and add medium containing virus stock. Use 1 to 2 ml medium containing from 0.01 µl to 0.1 ml of virus stock to infect a 6-cm dish of cells, or 3 to 5 ml for a 10-cm dish. Add 800 µg/ml polybrene to 8 µg/ml final for murine or avian subgroup E stocks and incubate cells 1 to 3 hr at 37°C.

    Titer at least two dilutions that differ by 10-fold to obtain a countable number of colonies (20 to 100).

3. Add medium to dilute polybrene to 2 µg/ml and incubate at least 2 or 3 times the length of a cell cycle (2 to 3 days for NIH 3T3 and QT6 cells).

4a. If virus carries a histochemical marker gene such as *lacZ*, stain infected cells with Xgal (Support Protocol). No drug selection is needed. Proceed to step 6 to calculate titer by counting number of blue colonies.

4b. If virus carries a drug resistance gene, split infected cells into selection conditions. If the resistance gene is *neo*, split the cells 1:10 to 1:20, set up two 10-cm dishes containing G418 for each, and incubate 3 days.

    The amount of G418 to use depends upon the cells being selected.

5. Change medium, including appropriate selection drug(s). Incubate a total of 7 to 10 days, at which time colonies should be obvious. Count colonies before they spread (usually before 12 days under selection).

6. Calculate titer as follows:

$$\text{G418} - \text{RCFU/ml} = \frac{\text{no. of colonies}}{\text{virusvolume(ml)} \times \text{replicationfactor} \times \text{fraction of infection cells plated}}$$

or, if the virus carries *lacZ*,

$$\text{Xgal CFU/ml} = \frac{\text{no. of Xgal}^{+}\text{ colonies}}{\text{virus volume (ml)}}$$

A rule of thumb is that titers varying >3-fold are significantly different.

7. Run an assay to determine that the producer clone is making a virus that is not rearranged or deleted.

    If such an assay is not straightforward, at least the structure of the viral genome, and the presence of viral RNAs, should be investigated. A Southern (UNIT 2.9) or northern (UNIT4.9) blot analysis can be performed on cells infected with virus from the producer clone.

ALTERNATE PROTOCOL

RAPID EVALUATION OF PRODUCER COLONIES

Rather than titrate the virus stock for activity, one can quantitatively analyze the producer cells for the presence of vector RNA or protein. There is often (though not always) a correlation between amount of viral product within the packaging cell and the titer of virus vector in the supernatant. Alternatively, RNA dot blots can be used to quickly assess the amount of vector RNA, which has also been correlated with titer. Because Xgal staining, as well as RNA analysis methods, are invasive and lethal, duplicate wells containing the colonies picked from the drug-selection dishes are made. Direct titration of the culture supernatants (Basic Protocol 2) is the more typical method to screen producer clones, but takes longer.

SUPPORT PROTOCOL

XGAL STAINING OF CULTURED CELLS

This protocol can be used to stain infected cells to obtain a direct measure of titer or to stain producer lines themselves (Alternate Protocol) to determine which producer clones make high levels of β-galactosidase.

Additional Materials *(also see Basic Protocol; see* APPENDIX 1 *for items with ✓)*

Cells infected or transfected with a *lacZ*-encoding virus (Basic Protocol 1)
✓ Phosphate-buffered saline (PBS)
✓ Fixative solution: 0.05% glutaraldehyde *or* 2% paraformaldehyde solution
✓ Xgal solution

1. Discard medium and add fixative solution (2 ml for 6-cm dish and 5 ml for 10-cm dish). Incubate at room temperature in a fume hood with 2% paraformaldehyde for 60 min, or 5 to 15 min for 0.05% glutaraldehyde.
2. Discard fixative solution using proper chemical disposal protocol for your institution and rinse cells thoroughly three times in PBS at room temperature. Leave the second rinse on for ~10 min and perform first and third washes quickly.
3. Add a minimal volume Xgal solution to cover cells. Incubate 1 hr to overnight at 37°C. Positive cells will stain blue.

References: Cepko, 1989; Cossett et al., 1990.

Contributor: Constance Cepko

Large-Scale Preparation and Concentration of Retrovirus Stocks

UNIT 9.12

For some applications, (i.e., infection of cells in vivo) it is necessary to concentrate retrovirus stocks in order to increase their titer. In addition, if one wishes to infect a population of cells uniformly with a virus that does not encode a selectable marker, high-titered stocks are necessary. The protocol requires several hundred milliliters to a few liters of producer cell supernatant.

NOTE: Retroviral particles are fragile, with short half-lives even under optimum conditions. Resuspension of pellets and other procedures must be carried out gently, and materials should be kept cold.

BASIC PROTOCOL

PREPARATION OF VIRUS STOCK AND CONCENTRATION BY CENTRIFUGATION

Materials *(see* APPENDIX 1 *for items with ✓)*

Identified high-titer Ψ2 producer cells, 50% to 90% confluent (*UNIT 9.11*)
✓ Complete DMEM containing 10% calf serum (DMEM-10, prepared with calf serum instead of FBS)
NIH 3T3 cells
800 μg/ml polybrene
0.45-μm filters for large volumes (e.g., Nalgene 115-ml or 500-ml filters)
Beckman JA-14 rotor with 250-ml centrifuge bottles if concentrating large volumes *or* Beckman SW-27 or SW-41Ti rotor or equivalents if concentrating small volumes

NOTE: All incubations involving tissue culture cells should be performed in a humidified 5% CO_2 incubator at 37°C unless otherwise noted.

1. Split producer cells 1:10 or 1:20 from newly confluent cultures into twenty to fifty 10-cm dishes and incubate cells to 50% to 90% confluency. Discard medium, gently add half the normal volume of medium, and make sure that the medium is not alkaline. Incubate 2 to 3 days.

   For different packaging lines, optimum conditions may vary—e.g., the supernatant may be harvested just as the cells reach confluency, or the medium may be changed to half volume the day before confluency and harvested the following day.

2. Harvest the supernatant, filter through a 0.45-μm filter, and store at −70 or −80°C, or titer and concentrate immediately (steps 3 to 5).

   The cells will be extremely densely packed when the harvest is made and the medium will be yellow.

3. Titer the virus stock (*UNIT 9.11*; it is useful to know the starting titer prior to concentration of the virus).

4. For large volumes, centrifuge virus stock 20 min at 25,000 × *g* in JA-14 rotor at 4°C in sterile 250-ml bottles. For smaller volumes, centrifuge 10 min at 20,000 rpm in SW-27 or SW-41Ti rotor. Rinse tubes with 70% ethanol prior to use to avoid bacterial contamination.

5. Pour supernatant directly into sterile bottles and centrifuge for 5 to 16 hr at 14,000 rpm in a JA-14 rotor, or 2 hr at 20,000 rpm in an SW-27 or SW-41Ti

rotor, 4°C. Remove supernatant carefully and gently resuspend pellet in DMEM-10 using 0.1% to 1% of the original volume.

A small volume of supernatant may be saved for titering to determine whether virus was pelleted. Resuspension may take 2 hr using a pipet if pellet is fairly sticky. Leave centrifuge bottle in an ice bucket in the tissue culture hood and pipet suspension (with same pipet to avoid losses) approximately every 15 min. Avoid making bubbles as this can denature the viral proteins.

6. Store virus at −70° or −80°C in 5- to 50-µl aliquots. Titer an aliquot of concentrated and unconcentrated virus (*UNIT 9.11*).

ALTERNATE PROTOCOL

CONCENTRATION BY PEG PRECIPITATION AND CHROMATOGRAPHY

Additional Materials *(also see Basic Protocol; see APPENDIX 1 for items with ✓)*

✓ 5 M NaCl, filter sterilized
Polyethylene glycol (PEG) 6000, filter sterilized
✓ NTE buffer
Sepharose CL-4B or CL-2B (Pharmacia)
Savant high-speed centrifuge *or* Beckman SW-41Ti rotor

1. Prepare a virus stock (Basic Protocol, steps 1 and 2). Add 5 M NaCl to virus stock to 0.4 M final while stirring at 4°C. Slowly add PEG 6000 to 8.5% final (w/v) and continue stirring 1 to 1.5 hr at 4°C.

2. Centrifuge 10 min at 7000 × *g* (7500 rpm in SW-41Ti rotor or 10,000 rpm in Savant high-speed centrifuge). Dissolve pellet in NTE buffer in 1% of original volume. Use directly or store at −70° or −80°C.

3. Prepare column of Sepharose CL-4B or CL-2B, equilibrating in NTE buffer. Use 10-ml Econo-Columns (Bio-Rad) for pellets prepared from 100 ml of virus stock.

4. Apply virus and chromatograph at 1 ml/min with NTE buffer, collecting 0.3- to 0.5-ml fractions. Assay fractions by measuring absorbance at 280 or 260 nm or by reverse transcriptase (RT) assay (*UNIT 9.13*). Pool appropriate fractions and titer (*UNIT 9.11*).

ALTERNATE PROTOCOL

CONCENTRATION USING MOLECULAR-WEIGHT-CUTOFF FILTERS

This method is the simplest means of concentrating small to medium volumes of retrovirus stocks. Supernatants prepared as in steps 1 and 2 of the Basic Protocol are centrifuged through filters, either the Centricon-30 microconcentrator from Amicon or the CentriCell 60 from Polysciences, essentially following the manufacturers' instructions. These filters allow passage of molecules that are much smaller than the virus (e.g., Centricon-30 filters allow passage of molecules of <30 kDa). For example, supernatants can be centrifuged through one of these filters at 5000 × *g* in a Sorvall SS-34 centrifuge (Stoker et al., 1990); the material that does not flow through the filter is removed and used as the concentrated stock (the longer the centrifugation, the more concentrated the stock).

Reference: Aboud et al., 1982.

Contributor: Constance Cepko

Detection of Helper Virus in Retrovirus Stocks

UNIT 9.13

Helper virus is a replication-competent virus that is sometimes present in stocks of replication-incompetent virus. The presence of helper virus can be problematic in animal infections and lineage analysis.

DETECTION OF HELPER VIRUS THROUGH HORIZONTAL SPREAD OF DRUG RESISTANCE

BASIC PROTOCOL

The following is a very sensitive method of examining the ability of a virus supernatant to promote the horizontal spread of a viral genome from an infected cell to neighboring nonsibling cells. The presence of virus that can transmit a marker (neomycin resistance [*neo*]) indicates that the original stock contained helper virus. For controls in this assay, a helper virus–free stock (previously verified) and a helper-containing stock are used.

Materials *(see APPENDIX 1 for items with ✓)*

NIH 3T3 cells (used only with murine viruses)
Three titered virus stocks: test virus containing *neo* marker, helper virus–free control containing *neo* marker (negative control), and wild-type helper virus
800 µg/ml polybrene, filter sterilized (stored at −20°C)
✓ Complete DMEM containing 10% calf serum (DMEM-10 prepared with calf serum instead of FBS)
6-cm dishes
0.45-µm filter

NOTE: All incubations involving tissue culture cells should be performed in a humidified, 5% CO_2 incubator at 37°C unless otherwise noted.

1. Split NIH 3T3 cells 1:50 into three 6-cm dishes the day before the assay is initiated. Label one dish as a positive control, one dish as a negative control, and one dish for the stock to be tested.

2. *Prepare test virus stock containing selectable marker:* Add 0.01 ml of 800 µg/ml polybrene (8 µg/ml final) to 1 ml of test virus supernatant or 1 to 30 µl of the concentrated stock diluted in 1 ml medium and filter through a 0.45-µm filter.

3. *Prepare a positive control stock containing selectable marker:* Prepare a mixture of 1 ml helper-free stock and 10 to 100 CFU of helper stock. Add 0.01 ml of 800 µg/ml polybrene (8 µg/ml final) and filter through a 0.45-µm filter.

   The amount (CFU) of helper-free virus should equal the amount (CFU) of the test virus.

4. *Prepare a negative control stock*: Prepare 1 ml of helper-free stock alone by adding 0.01 ml of 800 µg/ml polybrene (8 µg/ml final) and filtering through a 0.45-µm filter.

5. Infect one dish (step 1) with each virus stock. Place all three infected dishes 1 to 3 hr in a CO_2 incubator at 37°C to permit absorption. Add 4 ml DMEM-10 and incubate until cells reach confluency (~3 to 4 days).

   Polybrene concentration must be maintained at 2 µg/ml during the assay to allow spread of both helper and marker-containing viruses.

6. Split cells 1:50 into new 6-cm dishes. Set up only one new dish for each of the three dishes, and discard unused portion of originally infected cells. Include polybrene at 2 µg/ml final. Incubate until cells reach 50% to 90% confluency.

   Save some of the initial supernatants (e.g., remove ~2 ml and store at –80°C) prior to splitting. If helper virus is present, the relative titers of these initial supernatants and the final supernatants (step 8) may be informative concerning the amount of helper present in the stock.

7. Discard medium and replace with half the volume of fresh DMEM-10. Incubate an additional 2 to 3 days.

8. Harvest final supernatant from confluent cells, filter through a 0.45-µm filter, and add polybrene to 8 µg/ml final. Store at –70° or –80°C or use immediately to titer and assay virus for marker resistance.

9. Split fresh, uninfected NIH 3T3 cells 1:10 or 1:20 into three 6-cm dishes the day before titering supernatants. Use 1 ml of each supernatant (step 8) to infect NIH 3T3 cells and carry out titration steps (*UNIT 9.11*, Basic Protocol 2, steps 1 to 5). The presence of several thousand *neo* resistant colonies indicates that the original test stock was contaminated with helper virus.

   Use CEF cells for avian replication-incompetent virus stock. In steps 1 and 6 split cells 1:10. Polybrene is not required for avian viruses unless a subgroup E virus is used.

ALTERNATE PROTOCOL

REVERSE TRANSCRIPTASE ASSAY TO DETECT HELPER VIRUS

Virus supernatants generated in the basic protocol (step 8) can be monitored for incorporation of radioactive dTTP. This serves as an assay for reverse transcriptase activity that only is observed when helper virus is in the original stock. This assay also can be used for other purposes—e.g., to test packaging lines for maintenance of virus production.

Additional Materials *(also see Basic Protocol; see APPENDIX 1 for items with ✓)*

✓ Reverse transcriptase (RT) reaction cocktail
Virus supernatants: test and control samples (Basic Protocol, step 8)
✓ 2× SSC
95% ethanol

96-well microtiter dish
Whatman DE52 or DE81 paper, precut to 2.5-cm circles or sheets

1. Add 50 µl RT reaction cocktail to separate wells of a 96-well microtiter dish, one well for each test or control sample. Add 10 µl virus supernatant, cover dish, and place 1 to 2 hr in a CO_2 incubator at 37°C.

2. Spot 10 µl of each reaction onto a 2.5-cm circle of DE52 or DE81 paper, set up on a piece of plastic wrap. Label paper with a No. 2 lead pencil.

3. Place paper in a tray and cover with 2× SSC. Wash 20 min with gentle shaking on a shaker at room temperature, discard SSC, and repeat wash step twice. Soak 1 min in 95% ethanol and air dry ~10 min.

4. Count in scintillation counter or expose to film.

References: Goff et al., 1981; Omer and Faras, 1982; Stoker and Bissell, 1987.

Contributor: Constance Cepko

Retrovirus Infection of Cells In Vitro and In Vivo

There are many applications in which retrovirus vectors are used as transduction agents. For each application, the infection protocol may vary and must often be optimized. Guidelines for infection of cells in some typical in vivo and in vitro experiments are presented here. To optimize infection of a particular type of cell, it is often advantageous to use vectors that are easy to score.

CAUTION: When working with human blood, cells, or infecting agents, strict biosafety practices must be followed.

INFECTION OF CELLS IN VITRO

Infection of target cells in vitro is accomplished by simply incubating the virus with the cells (e.g., *UNIT 9.11*, Basic Protocol 2). For most in vitro applications using a murine virus, a polycation such as polybrene is used to aid viral infection. When incubating target cells with virus in vitro, problems may occur if undiluted, high-titer virus is used directly on some cell types—i.e., fusion of target cells can occur with subsequent death of most fused cells within a few days of infection.

Alternatively, cells can be incubated with the packaging line that produces the desired vector (cocultivation method). This method is used to infect hematopoetic cells and appears to greatly increase the infection efficiency. In some cases, it may be desirable to prevent cell division of the producer cells during or after cocultivation. This can be achieved by using protocols that allow virus production, but that prevent further cell division of the producer cells. For example, prior to cocultivation, confluent or nearly confluent producer cells can be killed by irradiation (2800 rad) or a 3-hr treatment with mitomycin C (10 µg/ml in medium) followed by several rinses with medium. After this treatment, the cells to be infected are plated onto producer cells. Target cells should continue to grow, but producer cells will die as they can no longer divide. If the target cells are nonadherent, they can be removed by simply washing the producer monolayer gently after ~48 hr of cocultivation.

It is difficult to generalize about the efficiency of infection, although it can approach 100%. For the most part, the variables that influence infectability of a given cell are unknown. However, it is clear that for optimal results, the cells should be as mitotically active as possible.

Expression of viral gene products is usually assayed two or three cell cycles after infection. When assaying *lacZ* expression in NIH 3T3 cells via Xgal staining (*UNIT 9.11*), we have found that the number of clones peaks at ~48 hr postinfection (three cell cycles). However, Xgal staining is visible within 24 hr in a subset of the infected cells.

Infections of tissue explants or cultured embryos can also be performed essentially as described above. A tissue explant can be bathed in as much virus stock as is desired for a few hours (when using murine viruses or avian subgroup E viruses, include polybrene at 8 µg/ml), or it can be cultured over a monolayer of producer cells.

INFECTION OF RODENTS IN VIVO

The following paragraphs describe in vivo infection of rodents, the laboratory animal most often used in experimental procedures; however, the techniques apply to other species (e.g., avian) as well.

NOTE: Detailed protocols for the care and handling of laboratory animals are beyond the scope of this unit. The reader is referred to *Current Protocols in Immunology*, Chapter 1 (Donovan and Brown, 1991) for instructions on proper animal restraint, anesthesia, injections, and euthanasia techniques that are essential to the infection methods described here.

Infection of Postnatal Animals In Situ

Volume and titer of virus stock. The volume that can safely be delivered to a tissue in vivo is generally quite small—0.1 to 1 µl. It is therefore important to prepare high-titer (usually concentrated) virus stock (*UNIT 9.12*), at 10^6 to 10^8 CFU/ml. Because of these limitations, it is quite important to deliver the virus directly to the area containing the highest percentage of mitotic target cells. It is unclear if there are factors in tissue fluids that inhibit or destroy viral infectivity.

Delivery of virus. Virus delivery to postnatal animals is fairly straightforward. It is possible to use a hand-held Hamilton syringe with a 33-G needle or use a drawn-out glass pipet. The skin, and even the skull, are soft enough on the first few days after birth for direct injection into the tissue. For tougher injection sites on older animals, it may be necessary to make a hole using a stronger needle prior to inserting a more delicate needle. Co-injection with a dye such as 0.05% (w/v) trypan blue or 0.025% fast green aids in detecting the accuracy of injections and does not impair viral infectivity. Rodents ≤7 days can be anesthetized by simply cooling on ice for a few minutes. Landmarks (e.g., sutures or blood vessels) near the area to be injected can be visualized using a fiber-optics light source. It is best to practice a series of injections with dye alone and then immediately dissect the animal for examination of the injection site. Animals can be examined for evidence of viral infection at any time.

Setup of pilot experiments to optimize injection and expression efficiency. When performing infections of tissue for the first time, or when infection and expression efficiency of the target cells is unknown, it is useful to perform pilot experiments using a retrovirus vector carrying a histochemical marker gene such as *lacZ*, which allows for examination of the injected animal a few days after infection. For such experiments, it is necessary to optimize conditions of the Xgal staining reaction (*UNIT 9.11*) for the particular tissue under study.

An excellent way to simultaneously determine the accuracy of injection, and whether the Xgal histochemistry is working, is to inject cells that contain the *lacZ* gene, such as the Ψ2 BAG producer cells, *UNIT 9.11*. The cells can be prelabeled with a fluorescent dye such as carboxyfluorescein diacetate succinimyl ester (CFSE; Molecular Probes; prepare 10 mM stock in DMSO and store in foil at 4°C); see Bronner-Fraser, 1985, for details. If the Xgal histochemistry is working properly, no fluorescent cells should be visible because Xgal-stained cells are usually so full of indigo dye that all fluorescence is absorbed. Once the injection method and Xgal histochemistry have been optimized, BAG virus is injected into the site and the tissue is examined for viral infection several days later. If the Xgal histochemistry is *not* working well, fluorescent cells will be visible. In this case, vary fixation and tissue preparation methods until good Xgal staining is achieved (Cepko, 1989).

Infection of Prenatal Rodents In Utero

Injections made in utero are performed with drawn-out glass pipets (Austin and Cepko, 1990). The actual diameter and shape of the tip should be determined empirically. Animals are anesthetized with a mixture of ketamine (20 to 40 mg/kg) and xylazine (3 to 5 mg/kg) and are opened via an incision along the midline. The orientation of an embryo can be determined by visualizing the head with fiber optics. Injections of 0.1 µl are made through the uterine wall into the area of interest. Practice injections followed by immediate dissection and examination are again recommended. The mother is then closed with suture. The injected animals can be delivered prenatally or allowed to finish gestation, depending upon the experiment.

Infection of Prenatal Rodents Exo Utero

It is difficult to make precisely directed injections into many of the mitotic zones of prenatal mammals through the uterine wall. The exo utero surgical procedure (Muneoka et al., 1986) at least partially circumvents this problem. Mouse embryos at embryonic days 11 to 19 (E11 to E19) are released from the uterus by cutting the uterine wall, but remain attached via the placenta. The abdominal cavity of the mother is filled with a buffered saline solution to protect the embryos. An incision can be made in the extra-embryonic membranes that surround the embryo so that the embryo can be directly manipulated or injected. Subsequently, the extra-embryonic membranes are closed with fine suture. The embryos can be brought to term in the abdominal cavity and delivered by Caesarean section.

Additional factors that influence the success of this method are choice of mouse strain and health of the mouse colony. Outbred mouse strains such as CD-1 or Swiss Webster appear to be best.

References: Austin and Cepko, 1990; Bronner-Fraser, 1985; Cepko, 1989; Donovan and Brown, 1995, Lemischka et al., 1986; Muneoka et al., 1986.

Contributor: Constance Cepko

Analysis of Proteins

The isolation and analysis of proteins is integral to designing oligodeoxynucleotide probes and gene cloning, confirming DNA sequence data, and synthesizing peptides for eliciting antipeptide antibodies.

This chapter consists of protocols that provide answers to the following questions: How much protein is there? Is the protein pure? Does the protein have subunits? How many protein subunits are there? How is the protein isolated? How can the protein be synthesized in vitro? How can a protein be biosynthetically labeled? How can amino-terminal sequences of scarce proteins be determined? How can internal sequences be derived from N-terminally blocked proteins?

As shown in Figure 10.0.1, the answer to the first question, "How much protein is there?" is provided by colorimetric methods (*UNIT 10.1*). The amount of protein can also be determined by comparing the staining intensity of an unknown protein to the staining intensity of protein standards separated by either one- or two-dimensional electrophoresis (*UNITS 10.2 & 10.3*). The proteins can be stained while they are still within the polyacrylamide gel (*UNIT 10.5*) or after the proteins are transferred to a blot transfer membrane (*UNITS 10.6 &10.18*). In addition, protein-specific monoclonal or polyclonal antibodies can be used for detection by immunoblotting (*UNIT 10.7*).

The answers to the next three questions come from an analysis of the data derived from a combination of electrophoresis and chromatography (i.e., conventional gel filtration or size-exclusion high-performance liquid chromatography; see Fig. 10.0.2). For a protein without subunits or a protein with identical subunits, detection of a single protein band after one-dimensional gel electrophoresis under denaturing conditions or a single spot after two-dimensional gel electrophoresis indicates that the protein is pure. If the protein consists of multiple subunits of different molecular sizes, purity is confirmed by detecting a single, stainable band after gel electrophoresis under nondenaturing conditions. Once the protein is demonstrated to be pure,

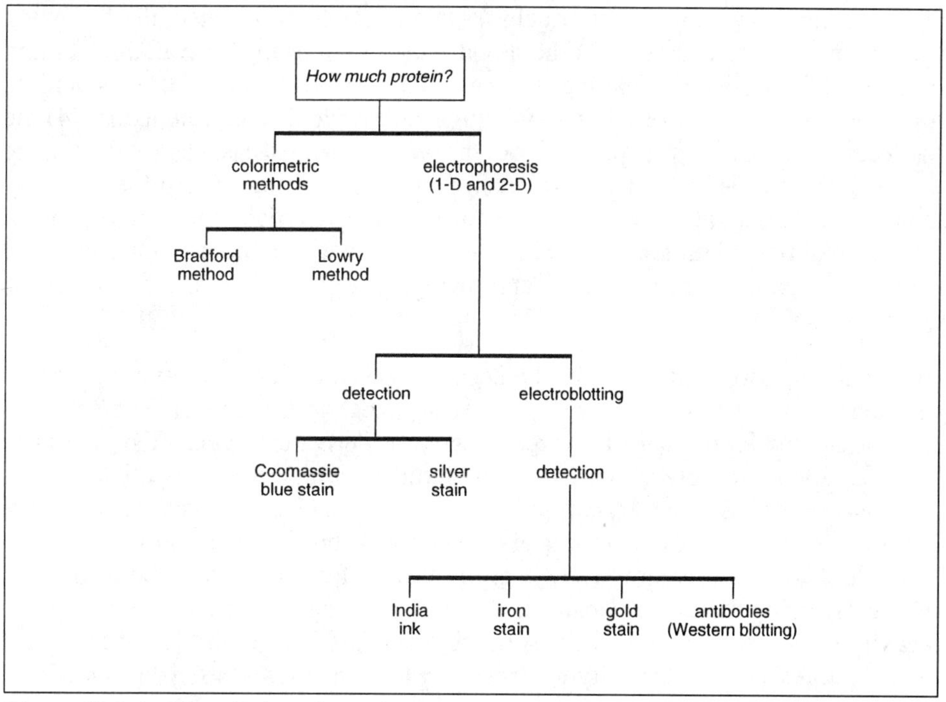

Figure 10.0.1 Quantitation of proteins.

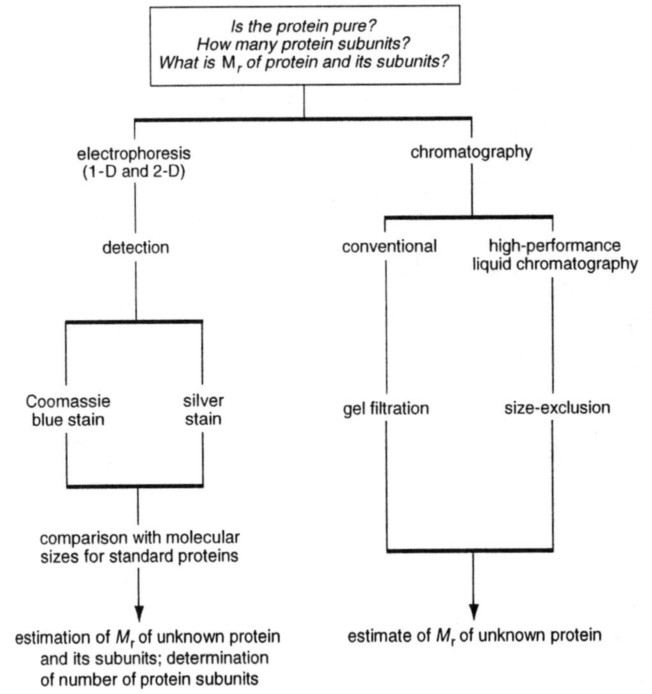

Figure 10.0.2 Analysis of protein purity, molecular weight, and subunit structure.

an estimate of molecular size of the protein is made by comparing the elution volume of the protein from a conventional gel-filtration or high-performance size-exclusion column to the elution volumes of standard proteins. An estimate of the size of the subunits can be determined by subsequent electrophoresis under denaturing conditions. The number of each subunit is then deduced by comparing the molecular size of "native" (i.e., nondenatured) protein and the molecular size(s) of the subunit(s).

To determine how a protein or protein fragment should be isolated, the following factors must be considered: (1) the amount of a protein in the available starting material, (2) the cost of preparing starting material (e.g., cell culture, fermentation, or organs) and the cost of labor, (3) the molecular size of the protein, and (4) the molecular size and physical properties of the protein. In most cases a protein is being isolated and purified in order to ascertain partial protein sequence information by automated Edman degradation using a commercially available protein sequencer. However, almost all proteins are isolated by a combination of conventional chromatography, high-performance liquid chromatography (HPLC), and electrophoresis (Fig. 10.0.3).

Both one- and two-dimensional gel electrophoresis are high-resolution separation methods, yielding protein whose sequence can be determined after either electroelution or electroblotting onto Polybrene-coated or derivatized glass fiber sheets or polyvinylidene difluoride membrane filter (all of which are compatible with a gas-phase protein sequencer; *UNIT 10.18*). In most cases, electrophoretic methods are used after several successive modes of conventional chromatography or HPLC have been used to purify progressively a given protein from a crude protein mixture. However, if the protein is separated under denaturing conditions, the biological activity of a desired protein will likely be lost. This is a reason for utilizing gel electrophoresis last when purifying a protein whose identity is based on a functional assay.

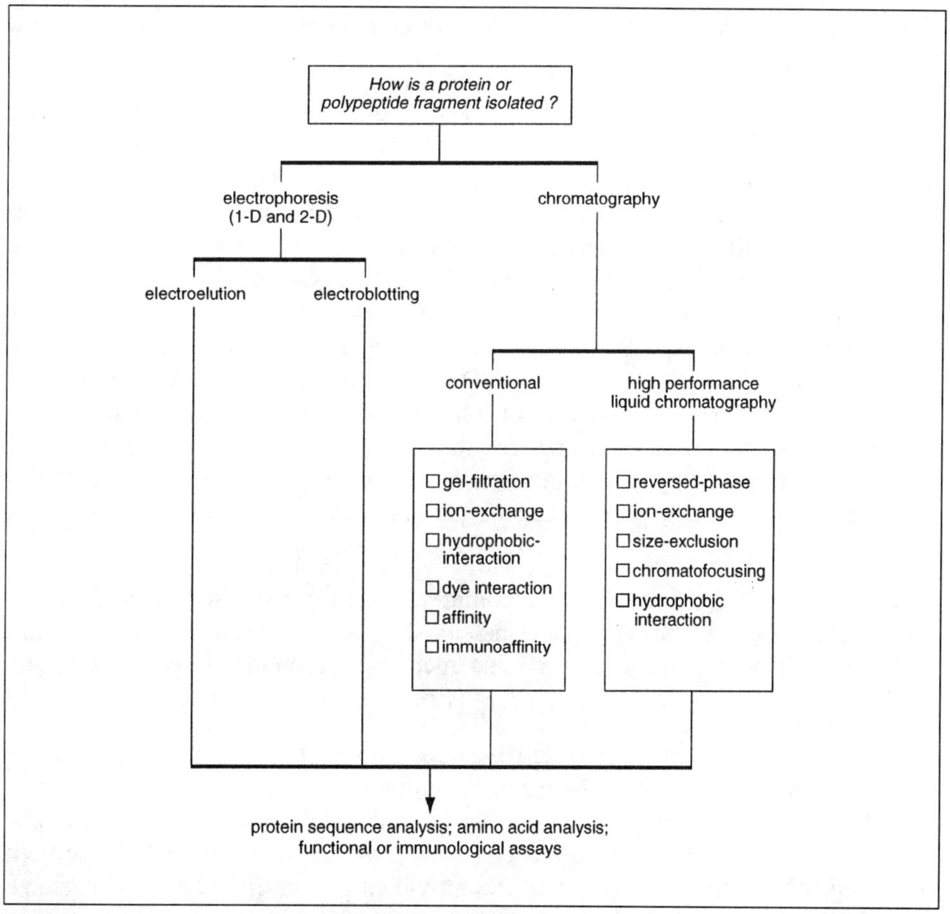

Figure 10.0.3 solation of proteins and polypeptide fragments.

Conventional chromatography includes gel filtration (*UNIT 10.8*), ion exchange (*UNIT 10.9*), immunoaffinity (*UNIT 10.10*), affinity on immobilized dyes, affinity on immobilized ligands, and hydrophobic interaction. Only the first three of these conventional chromatographic modes are discussed in this chapter. Metal-chelate affinity chromatography (MCAC; *UNIT 10.11*) can be used to isolate proteins that contain amino acid residues with affinity for immobilized metal ions. All of these methods can accommodate large amounts of crude starting material. Because gel-filtration chromatography has a greater separating range than comparable size-exclusion HPLC, gel filtration is preferred for separating proteins of similar molecular sizes. None of these chromatographic modes will cause protein denaturation, although contaminating proteases in crude protein mixtures can always lead to degradation of a protein during purification.

Three HPLC methods are provided in this chapter: reversed phase (*UNIT 10.12*), ion exchange (*UNIT 10.13*), and size exclusion (*UNIT 10.14*). Ion-exchange and size-exclusion HPLC are equivalent to their counterparts in conventional chromatography. However, the advantages of HPLC versus conventional chromatography are that small amounts (<1 nmol) of a given protein can be handled and purified, less time is required for a separation (<3 hr), and reduced chromatographic peak volume is achieved. Reversed-phase HPLC is primarily useful for purifying small proteins (molecular weight <20,000) and protein fragments (discussed below).

Immunoprecipitation is another method of protein purification (*UNIT 10.15*) by which a specific protein can be selectively precipitated from a complex protein mixture, provided that specific antibodies directed against the protein are available. Both conventional and monoclonal antibodies may be employed for immunoprecipitation.

An approach that differs from these methods is the in vitro synthesis of proteins by transcription and translation of cloned genes (*UNIT 10.16*). Such in vitro-synthesized proteins are extremely useful for a variety of purposes including analysis of DNA-protein interactions and for studies of mutant proteins obtained from mutagenesis of cloned DNA. In this approach, the protein-coding sequences are cloned into a vector containing a promoter for SP6 or T7 RNA polymerase. Messenger RNA encoding the protein is generated by transcribing the DNA template with the appropriate bacteriophage RNA polymerase. The essentially pure mRNA is then translated in vitro using wheat germ extracts or reticulocyte lysates. By using [^{35}S]methionine during the translation reaction, the protein is synthesized as a radiolabeled species. For most applications, in vitro-synthesized proteins can be used directly without any further purification. A major advantage of this method is that any desired mutant protein can be generated simply by altering the DNA template.

Biosynthetic labeling techniques are commonly used in the study of biochemical properties, processing, intracellular transport, secretion and degradation of proteins. Methods for labeling many secreted and membrane proteins are presented (*UNIT 10.17*).

Many proteins, however, are normally synthesized with the α-N_2 group blocked by an acyl moiety (e.g., acetyl) and are consequently refractory to Edman degradation. Therefore, it is wise to fragment a protein (by chemical and/or enzymic methods) after it has been purified to homogeneity by a combination of separation methods including SDS-PAGE, and to isolate the individual protein fragments by reversed-phase HPLC (*UNIT 10.18*). The partial sequence analysis of several peptides may then be completed. These data may be used collectively to confirm the identity of an unknown gene or to form the basis for the synthesis of oligodeoxynucleotides (i.e., primers or probes) or peptides.

John A. Smith

UNIT 10.1

Colorimetric Methods

The Bradford method, which is faster than the Lowry method, is the method of choice for determining protein concentration. Certain substances interfere with these assays. Substances that interfere with the Bradford method include glycerol, detergents, 2-mercaptoethanol, acetic acid, ammonium sulfate, Tris, and certain alkaline buffers. Precipitation of a protein with deoxycholate/trichloroacetic acid will eliminate many of these interfering substances; alternatively, appropriate controls can be used.

BASIC PROTOCOL

BRADFORD METHOD

This method quantitates the binding of Coomassie brilliant blue to an unknown protein and compares this binding to that of different amounts of a standard protein, usually bovine serum albumin. This protocol quantifies 1 to 10 µg protein using a standard curve. For 10 to 100 µg protein, increase the volume of the dye solution 5-fold and use larger tubes.

Materials (*see* APPENDIX 1 *for items with* ✓)

✓ 0.5 mg/ml bovine serum albumin (BSA)
0.15 M NaCl
✓ Coomassie brilliant blue solution

1. Add duplicate aliquots of 0.5 mg/ml BSA (5, 10, 15, and 20 µl) into microcentrifuge tubes. Bring the volume in each tube to 100 µl with 0.15 M NaCl. Prepare 2 blank tubes containing only 100 µl of 0.15 M NaCl.
2. Add 1 ml Coomassie brilliant blue solution to each tube and vortex. Leave 2 min at room temperature.
3. Measure the A_{595} using a 1-cm pathlength microcuvette (1 ml). Make a standard curve by plotting absorbance versus protein concentration. Measure the A_{595} of the unknown and determine the protein concentration in the unknown from the BSA standard curve.

   If the unknown protein concentration is too high, dilute the protein, assay a smaller aliquot, or generate another standard curve in a higher concentration range (e.g., 10-100 µg).

Reference: Darbre, 1986.

Contributor: John A. Smith

One-Dimensional SDS Gel Electrophoresis of Proteins

UNIT 10.2

Electrophoresis is used to separate complex mixtures of proteins, to investigate subunit compositions and to verify homogeneity of protein samples. It can also serve to purify proteins for use in further applications. In polyacrylamide gel electrophoresis, the combination of gel pore size and protein charge, size, and shape determines the migration rate of the protein.

CAUTION: Before any protocols are used, it is extremely important to read the following section about electricity and electrophoresis.

ELECTRICITY AND ELECTROPHORESIS

Many researchers are poorly informed concerning the electrical parameters of running a gel. It is important to note that the voltages and currents used during electrophoresis are dangerous and potentially lethal. Thus, safety should be an overriding concern. A working knowledge of electricity is an asset in determining what conditions to use and in troubleshooting the electrophoretic separation, if necessary.

Safety Considerations

1. Never remove or insert high-voltage leads unless the power supply voltage is turned down to zero and the power supply is turned off. Always grasp high-voltage leads one at a time with one hand only. Never insert or remove high-voltage leads with both hands. This can shunt potentially lethal electricity through the chest and heart should electrical contact be made between a hand and a bare wire. On older or homemade instruments, the banana plugs may not be shielded and can still be connected to the power supply at the same time they make

contact with a hand. Carefully inspect all cables and connections and replace frayed or exposed wires immediately.

2. Always start with the power supply turned off. Have the power supply controls turned all the way down to zero. Then hook up the gel apparatus: generally, connect the red high-voltage lead to the red outlet and the black high-voltage lead to the black outlet. Turn the power supply on with the controls set at zero and the high-voltage leads connected. Then, turn up the voltage, current, or power to the desired level. Reverse the process when the power supply is turned off: i.e., turn the power supply down to zero, wait for the meters to read zero, turn off the power supply, and then disconnect the gel apparatus one lead at a time.

   CAUTION: *If the gel is first disconnected and then the power supply turned off, a considerable amount of electrical charge is stored internally. The charge will stay in the power supply over a long time. This will discharge through the outlets even though the power supply is turned off and can deliver an electrical shock.*

Ohm's Law and Electrophoresis

Understanding how a gel apparatus is connected to the power supply requires a basic understanding of Ohm's law: voltage = current × resistance, or $V = IR$. A gel can be viewed as a resistor and the power supply as the voltage and current source. Most power supplies deliver constant current or constant voltage. Some will also deliver constant power: power = voltage × current, or $VI = I^2R$.

Most modern commercial equipment is color-coded so that the red or positive terminal of the power supply can simply be connected to the red lead of the gel apparatus, which goes to the lower buffer chamber. The black lead is connected to the black or negative terminal and goes to the upper buffer chamber. This configuration is designed to work with vertical slab gel electrophoreses in which negatively charged proteins or nucleic acids move to the positive electrode in the lower buffer chamber (an anionic system).

When a single gel is attached to a power supply in an anode system, the negative charges flow from the negative cathode (black) terminal into the upper buffer chamber, through the gel, and into the lower buffer chamber which is connected to the positive anode (red) terminal to complete the circuit. Occasionally, proteins are separated in cationic systems. In these gels, the proteins are positively charged because of the very low pH of the gel buffers or the presence of a cationic detergent (e.g., cetyltrimethylammonium bromide, CTAB). Proteins move toward the negative electrode (cathode) in cationic gel systems, and the polarity is reversed compared to SDS-PAGE.

Most SDS-PAGE separations are performed under constant current. The resistance of the gel will increase during SDS-PAGE in the standard Laemmli system. If the current is constant, then the voltage will increase during the run. If more than one gel is connected directly to the outlets of a power supply, then these gels are connected in parallel. In a parallel circuit, the voltage is the same across each gel. The total current, however, is the sum of the individual currents going through each gel. Therefore, under constant current it is necessary to increase the current for each additional gel that is connected to the power supply. Gel thickness affects the above relationships. If a gel thickness is doubled, then the current must also be doubled. There are limits to the amount of current that can be applied. Unless temperature control is available in the gel unit, a thick gel should be run more slowly than a thin gel.

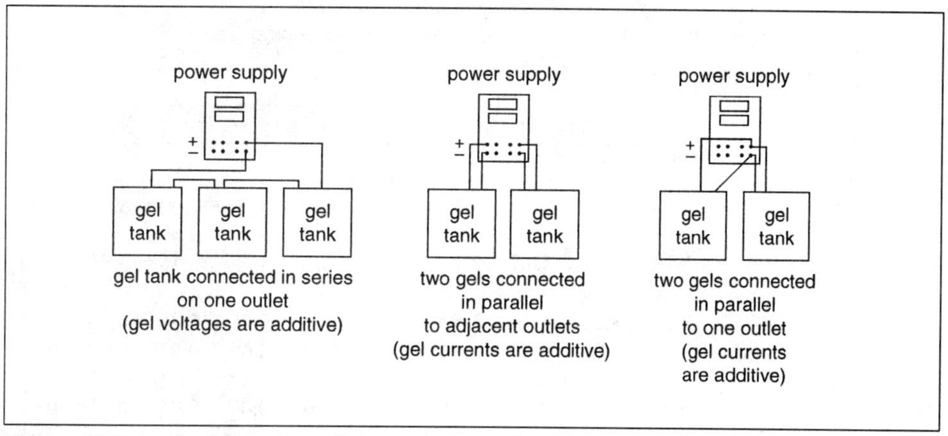

Figure 10.2.1 Series and parallel connections of gel tanks to power supply.

Multiple gel apparatuses can also be connected to one pair of outlets on a power supply. The gels can be connected in parallel or in series (Fig. 10.2.1). In the case of two or more gels running off the same outlet in series, the current is the same for every gel. The voltage, however, is additive for each gel.

NOTE: Milli-Q-purified water or equivalent should be used throughout the protocols.

DENATURING (SDS) DISCONTINUOUS GEL ELECTROPHORESIS: LAEMMLI GEL METHOD

BASIC PROTOCOL 1

One-dimensional gel electrophoresis under denaturing conditions (i.e., in the presence of 0.1% SDS) separates proteins based on molecular size as they move through a polyacrylamide gel matrix toward the anode. The polyacrylamide gel is cast as a separating gel topped by a stacking gel and secured in an electrophoresis apparatus. Sample proteins are solubilized by boiling in the presence of SDS. 2-Mercaptoethanol (2-ME) or dithiothreitol (DTT) is added during solubilization to reduce disulfide bonds.

This protocol is designed for a vertical slab gel with a maximum size of 0.75 mm × 14 cm × 14 cm. For thicker gels, or minigels (see Basic Protocol 2 and Support Protocol 5), the volumes of stacking and separating gels and the operating current must be adjusted.

Materials *(see* APPENDIX 1 *for items with ✓)*

Separating and stacking gel solutions (Table 10.2.1)
H_2O-saturated isobutyl alcohol
✓ 1× Tris·Cl/SDS, pH 8.8 (dilute 4× Tris·Cl/SDS, pH 8.8; Table 10.2.1)
Protein sample to be analyzed
✓ 2× and 1× SDS sample buffer
Protein molecular-weight-standards mixture (Table 10.2.2)
✓ 6× SDS sample buffer (optional)
✓ 1× SDS electrophoresis buffer

Electrophoresis apparatus with clamps, glass plates, casting stand, and buffer chambers (Bio-Rad, Hoefer Pharmacia Biotech)
0.75-mm spacers
0.45-µm filters (used in stock solution preparation)
25-ml Erlenmeyer side-arm flask
Vacuum pump with cold trap

Table 10.2.1 Recipes for Polyacrylamide Separating and Stacking Gels[a]

SEPARATING GEL

Stock solution[b]	Final acrylamide concentration in separating gel (%)[c]									
	5	6	7	7.5	8	9	10	12	13	15
30% acrylamide/ 0.8% bisacrylamide	2.50	3.00	3.50	3.75	4.00	4.50	5.00	6.00	6.50	7.50
4× Tris·Cl/SDS, pH 8.8	3.75	3.75	3.75	3.75	3.75	3.75	3.75	3.75	3.75	3.75
H_2O	8.75	8.25	7.75	7.50	7.25	6.75	6.25	5.25	4.75	3.75
10% (w/v) ammonium persulfate[d]	0.05	0.05	0.05	0.05	0.05	0.05	0.05	0.05	0.05	0.05
TEMED	0.01	0.01	0.01	0.01	0.01	0.01	0.01	0.01	0.01	0.01

Preparation of separating gel

In a 25-ml side-arm flask, mix 30% acrylamide/0.8% bisacrylamide solution, 4× Tris·Cl/SDS, pH 8.8 (see *APPENDIX 1*), and H_2O. Degas under vacuum about 5 min. Add 10% ammonium persulfate and TEMED. Swirl gently to mix. Use immediately.

STACKING GEL (*3.9% acrylamide*)

In a 25-ml side-arm flask, mix 0.65 ml of 30% acrylamide/0.8% bisacrylamide, 1.25 ml of 4× Tris·Cl/SDS, pH 6.8 (see *APPENDIX 1*), and 3.05 ml H_2O. Degas under vacuum 10 to 15 min. Add 25 µl of 10% ammonium persulfate and 5 µl TEMED. Swirl gently to mix. Use immediately. Failure to form a firm gel usually indicates a problem with the persulfate, TEMED, or both.

[a]The recipes produce 15 ml of separating gel and 5 ml of stacking gel, which are adequate for a gel of dimensions 0.75 mm × 14 cm × 14 cm. The recipes are based on the SDS (denaturing) discontinuous buffer system of Laemmli (1970).
[b]All reagents and solutions used in the protocol must be prepared with Milli-Q-purified water or equivalent. See *APPENDIX 1* for recipes.
[c]Units of numbers in table body are milliliters. The desired percentage of acrylamide in the separating gel depends on the molecular size of the protein being separated. See annotation to step 3, Basic Protocol 1.
[d]Best to prepare fresh.

0.75-mm Teflon comb with 1, 3, 5, 10, 15, or 20 teeth
25- or 100-µl syringe with flat-tipped needle
Constant-current power supply

1. Assemble the glass-plate sandwich of the electrophoresis apparatus according to manufacturer's instructions using two clean glass plates and two 0.75-mm spacers. Lock the sandwich to the casting stand.

2. Prepare the separating gel solution as directed in Table 10.2.1 and degas. After adding the specified amount of 10% ammonium persulfate and TEMED to the degassed solution, stir gently to mix.

   The desired percentage of acrylamide in the separating gel depends on the molecular size of the protein being separated. Generally, use 5% gels for SDS-denatured proteins of 60 to 200 kDa, 10% gels for SDS-denatured proteins of 16 to 70 kDa, and 15% gels for SDS-denatured proteins of 12 to 45 kDa (Table 10.2.1).

3. Using a Pasteur pipet, immediately apply the separating gel solution to the sandwich along an edge of one of the spacers until the height of the solution between the glass plates is ~11 cm.

   Sample volumes <10 µl do not require a stacking gel.

Table 10.2.2 Molecular Weights of Protein Standards for Polyacrylamide Gel Electrophoresis[a]

Protein	Molecular weight
Cytochrome *c*	11,700
α-Lactalbumin	14,200
Lysozyme (hen egg white)	14,300
Myoglobin (sperm whale)	16,800
β-Lactoglobulin	18,400
Trypsin inhibitor (soybean)	20,100
Trypsinogen, PMSF treated	24,000
Carbonic anhydrase (bovine erythrocytes)	29,000
Glyceraldehyde-3-phosphate dehydrogenase (rabbit muscle)	36,000
Lactate dehydrogenase (porcine heart)	36,000
Aldolase	40,000
Ovalbumin	45,000
Catalase	57,000
Bovine serum albumin	66,000
Phosphorylase *b* (rabbit muscle)	97,400
β-Galactosidase	116,000
RNA polymerase, *E. coli*	160,000
Myosin, heavy chain (rabbit muscle)	205,000

[a]Protein standards are commercially available in kits (e.g., Hoefer Pharmacia Biotech, Life Technologies, Bio-Rad, or Sigma).

4. Using another Pasteur pipet, slowly cover the top of the gel with a layer (~1 cm thick) of H_2O-saturated isobutyl alcohol, by gently layering the isobutyl alcohol against the edge of one and then the other of the spacers. Allow the gel to polymerize 30 to 60 min at room temperature.

   A sharp optical discontinuity at the overlay/gel interface will be visible on polymerization. Failure to form a firm gel usually indicates a problem with the ammonium persulfate, TEMED (N,N,N′,N′-tetramethylethylenediamine), or both.

5. Pour off the layer of H_2O-saturated isobutyl alcohol and rinse with 1× Tris·Cl/SDS, pH 8.8.

6. Prepare the stacking gel solution as directed in Table 10.2.1. Using a Pasteur pipet, slowly allow the stacking gel solution to trickle into the center of the sandwich along an edge of one of the spacers until the height of the solution in the sandwich is ~1 cm from the top of the plates.

7. Insert a 0.75-mm Teflon comb into the layer of stacking gel solution. If necessary, add additional stacking gel to fill the spaces in the comb completely. Allow the stacking gel solution to polymerize 30 to 45 min at room temperature.

8. Dilute a portion of the protein sample to be analyzed 1:1 (v/v) with 2× SDS sample buffer and heat 3 to 5 min at 100°C in a sealed screw-cap microcentrifuge tube. If the sample is a precipitated protein pellet, dissolve the protein in 50 to 100 µl of 1× SDS sample buffer and boil 3 to 5 min at 100°C. Dissolve

protein-molecular-weight standards mixture in 1× SDS sample buffer according to supplier's instructions as a control (Table 10.2.2).

For dilute protein solutions, consider adding 5:1 protein solution/6× SDS sample buffer to increase the amount of protein loaded. DO NOT leave the sample in SDS sample buffer at room temperature without first heating to 100°C to inactivate proteases.

For a 0.8-cm-wide well, 25 to 50 μg total protein in <20 μl is recommended for a complex mixture when staining with Coomassie blue, and 1 to 10 μg total protein is needed for samples containing one or a few proteins. If silver staining is used, 10- to 100-fold less protein can be applied (0.01 to 5 μg in <20 μl depending on sample complexity).

9. Carefully remove the Teflon comb without tearing the edges of the polyacrylamide wells. After the comb is removed, rinse wells with 1× SDS electrophoresis buffer and fill the wells with 1× SDS electrophoresis buffer.

10. Attach gel sandwich to upper buffer chamber using manufacturer's instructions. Fill lower buffer chamber with the recommended amount of 1× SDS electrophoresis buffer.

11. Place sandwich attached to upper buffer chamber into lower buffer chamber. Partially fill the upper buffer chamber with 1× SDS electrophoresis buffer so that the sample wells of the stacking gel are filled with buffer.

12. Using a 25- or 100-μl syringe with a flat-tipped needle, load equal volumes of the protein sample(s) at the same concentration into one or more wells by carefully applying the sample as a thin layer at the bottom of the wells. Load control wells with molecular weight standards. Add an equal volume of 1× SDS sample buffer to any empty wells to prevent spreading of adjoining lanes.

13. Fill the remainder of the upper buffer chamber with additional 1× SDS electrophoresis buffer so that the upper platinum electrode is completely covered. Do this slowly so that samples are not swept into adjacent wells.

14. Connect the power supply to the cell and run at 10 mA of constant current for a slab gel 0.75 mm thick, until the bromphenol blue tracking dye enters the separating gel. Then increase the current to 15 mA and run until the bromphenol blue tracking dye has reached the bottom of the separating gel.

For a standard 16-cm gel sandwich, 4 mA per 0.75-mm-thick gel will run ~15 hr (i.e., overnight); 15 mA per 0.75-mm gel will take 4 to 5 hr.

15. Turn off the power supply and disconnect the gel. Discard electrode buffer and remove the upper buffer chamber with the attached gel sandwich.

16. Orient the gel so that the order of the sample wells is known, remove the sandwich from the upper buffer chamber, and lay the sandwich on a sheet of absorbent paper or paper towels.

17. Carefully slide one of the spacers halfway from the edge of the sandwich along its entire length. Use the exposed spacer as a lever to pry open the glass plate, exposing the gel.

18. Carefully remove the gel from the lower plate. Cut a small triangle off one corner of the gel so the lane orientation is not lost during staining and drying. Proceed with protein detection.

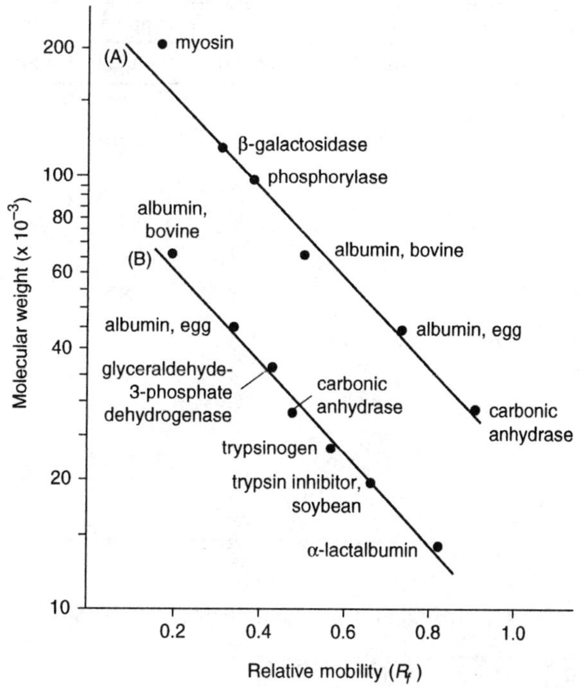

Figure 10.2.2 Typical calibration curves obtained with standard proteins separated by nongradient denaturing (SDS) discontinuous gel electrophoresis based on the method of Laemmli (1970). **(A)** Gel with 7% polyacrylamide. **(B)** Gel with 11% polyacrylamide. (Redrawn with permission from Sigma.)

The gel can be stained with Coomassie blue or silver (UNIT 10.5), or proteins can be electroeluted (UNIT 10.4), electroblotted onto a polyvinylidene difluoride (PVDF) membrane for subsequent staining or sequence analysis (UNIT 10.18), or transferred to a membrane for immunoblotting (UNIT 10.7). If the proteins are radiolabeled (UNIT 10.17), they can be detected by autoradiography (APPENDIX 3A). See Figure 10.2.2 for typical calibration curves for 7% and 11% gels.

ELECTROPHORESIS IN TRIS-TRICINE BUFFER SYSTEMS

ALTERNATE PROTOCOL 1

Separation of peptides and proteins under 10 to 15 kDa is not possible in the traditional Laemmli discontinuous gel system (Basic Protocol 1) because the comigration of SDS and smaller proteins obscures the resolution. The Tris-tricine method uses a modified buffer to separate the SDS and peptides, thus improving resolution. Several precast gels are available for use with the tricine formulations (Table 10.2.3).

Additional Materials *(also see Basic Protocol 1; see APPENDIX 1 for items with ✓)*

- Separating and stacking gel solutions (Table 10.2.4)
- ✓ 2× tricine sample buffer
- Peptide molecular-weight-standards mixture (Table 10.2.5)
- ✓ Cathode buffer
- ✓ Anode buffer
- ✓ Coomassie blue G-250 staining solution
- 10% (v/v) acetic acid

Table 10.2.3 Vertical Format Precast Gel Compatibility

Gel type and compatibility	Gel supplier				
	Bio-Rad	ISS/Daiichi	Jule	Millipore	Novex
SDS-PAGE gel type offered					
Peptide (tricine)	×	×	×	×	×
Single concentration	×	×	×	×	×
Gradient	×	×	×	×	×
Minigel size	×	×	×	×	×
Standard gel size		×	×		
Compatibility of gel with equipment manufactured by					
Hoefer Pharmacia Biotech		×	×	×	×
Bio-Rad	×	×	×	×	
Life Technologies	×	×	×	×	×
Novex		×		×	×
ISS/Daiichi		×	×	×	

Table 10.2.4 Recipes for Tricine Peptide Separation Gels[a]

SEPARATING AND STACKING GELS

Stock solution[b]	Separating gel	Stacking gel
30% acrylamide/0.8% bisacrylamide	9.80 ml	1.62 ml
Tris·Cl/SDS, pH 8.45	10.00 ml	3.10 ml
H_2O	7.03 ml	7.78 ml
Glycerol	4.00 g (3.17 ml)	—
10% (w/v) ammonium persulfate[c]	50 µl	25 µl
TEMED	10 µl	5 µl

Prepare separating and stacking gel solutions separately.

In a 25-ml side-arm flask, mix 30% acrylamide/0.8% bisacrylamide solution (Table 10.2.1), Tris·Cl/SDS, pH 8.45 (see *APPENDIX 1*), and H_2O. Add glycerol to separating gel only. Degas under vacuum 10 to 15 min. Add 10% ammonium persulfate and TEMED. Swirl gently to mix, use immediately. Failure to form a firm gel usually indicates a problem with the persulfate, TEMED, or both.

[a]The recipes produce 30 ml of separating gel and 12.5 ml of stacking gel, which are adequate for two gels of dimensions 0.75 mm × 14 cm × 14 cm. The recipes are based on the Tris-tricine buffer system of Schagger and von Jagow (1987).

[b]All reagents and solutions used in the protocol must be prepared with Milli-Q-purified water or equivalent. See *APPENDIX 1* for recipes.

[c]Best to prepare fresh.

1. Prepare and pour the separating and stacking gels (Basic Protocol 1, steps 1 to 7), using Table 10.2.4 in place of Table 10.2.1.

2. Prepare the sample (see Basic Protocol 1, step 8), *except* substitute 2× tricine sample buffer for the 2× SDS sample buffer and treat the sample at 40°C for 30 to 60 min prior to loading. Use the peptide molecular-weight-standards mixture for peptide separations (Table 10.2.5).

3. Load the gel and set up the electrophoresis apparatus (see Basic Protocol 1, steps 9 to 13) *except* use the tricine-containing cathode buffer or water to rinse and fill wells. Fill the lower buffer chamber with anode buffer, assemble the

Table 10.2.5 Molecular Weights of Peptide Standards for Polyacrylamide Gel Electrophoresis[a]

Peptide	Molecular weight
Myoglobin (polypeptide backbone)	16,950
Myoglobin 1-131	14,440
Myoglobin 56-153	10,600
Myoglobin 56-131	8,160
Myoglobin 1-55	6,210
Glucagon	3,480
Myoglobin 132-153	2,510

[a]Peptide standards are commercially available from Sigma.

unit, and attach the upper buffer chamber. Fill the upper buffer chamber with cathode buffer and load the samples.

4. Connect the power supply to the cell and run 1 hr at 30 V (constant voltage) followed by 4 to 5 hr at 150 V (constant voltage). Use heat exchanger to keep the electrophoresis chamber at room temperature.

5. After the Coomassie blue G-250 tracking dye has reached the bottom of the separating gel, turn the power supply to zero and disconnect the power supply.

6. Disassemble the gel (see Basic Protocol 1, steps 15 to 18). Stain proteins in the gel for 1 to 2 hr in Coomassie blue G-250 staining solution. Follow by destaining with 10% acetic acid, changing the solution every 30 min until background is clear (3 to 5 changes). For higher sensitivity, use silver staining as a recommended alternative (*UNIT 10.5*).

NONUREA PEPTIDE SEPARATIONS WITH TRIS BUFFERS

ALTERNATE PROTOCOL 2

A simple modification of the traditional Laemmli buffer system presented in Basic Protocol 1, in which the increased concentration of buffers provides better separation between the stacked peptides and the SDS micelles, permits reasonable separation of peptides as small as 5 kDa.

Additional Materials *(also see Basic Protocol 1; see APPENDIX 1 for items with ✓)*

Separating and stacking gel solutions (Table 10.2.6)
✓ 2× SDS electrophoresis buffer
✓ 2× Tris·Cl/SDS, pH 8.8 (dilute 4× Tris·Cl/SDS, pH 8.8; Table 10.2.1)

1. Prepare and pour the separating gel (Basic Protocol 1, steps 1 to 4), using Table 10.2.6 in place of Table 10.2.1.

2. Prepare and pour the stacking gel (Basic Protocol 1, steps 5 to 7), using 2× Tris·Cl/SDS, pH 8.8, rather than 1× Tris·Cl/SDS buffer, for rinsing the separating gel after removing the isobutyl alcohol overlay.

3. Prepare the sample and load the gel (Basic Protocol 1, steps 8 to 13) and substitute 2× SDS electrophoresis buffer for the 1× SDS electrophoresis buffer. Use peptide molecular weight standards (Table 10.2.5).

4. Run the gel (Basic Protocol 1, step 14).

The separations will take ~25% longer than those using Basic Protocol 1.

Table 10.2.6 Recipes for Modified Laemmli Peptide Separation Gels[a]

SEPARATING AND STACKING GELS

Stock solution[b]	Separating gel	Stacking gel
30% acrylamide/0.8% bisacrylamide	10.00 ml	0.65 ml
8× Tris·Cl, pH 8.8	3.75 ml	—
4× Tris·Cl, pH 6.8	—	1.25 ml
10% (w/v) SDS[c]	0.15 ml	50 µl
H_2O	1.00 ml	3.00 ml
10% (w/v) ammonium persulfate[d]	50 µl	25 µl
TEMED	10 µl	5 µl

Prepare separating and stacking gel solutions separately.

In a 25-ml side-arm flask, mix 30% acrylamide/0.8% bisacrylamide solution (see Table 10.2.1), 8× Tris·Cl, pH 8.8 (separating gel) or 4× Tris·Cl, pH 6.8 (stacking gel), 10% SDS (see *APPENDIX 1*), and H_2O. Degas under vacuum 10 to 15 min. Add 10% ammonium persulfate and TEMED. Swirl gently to mix; use immediately. Failure to form a firm gel usually indicates a problem with the persulfate, TEMED, or both.

[a]The recipes produce 15 ml of separating gel and 5 ml of stacking gel, which are adequate for one gel of dimensions 0.75 mm × 14 cm × 14 cm. The recipes are based on the modified Laemmli peptide separation system of Okajima et al. (1993).

[b]All reagents and solutions used in the protocol must be prepared with Milli-Q-purified water or equivalent. See *APPENDIX 1* for recipes.

[c] Prepare fresh.

[d]Best to prepare fresh.

5. Disassemble the gel (Basic Protocol 1, steps 15 to 18).

   If Coomassie blue staining is used, stain using the solutions outlined in UNIT 10.5 for 1 hr followed by destaining with several changes of 25% methanol/7% acetic acid.

ALTERNATE PROTOCOL 3

CONTINUOUS SDS-PAGE

With continuous SDS-PAGE, the same buffer is used for both the gel and electrode solutions. Although continuous gels lack the resolution of the discontinuous systems, they are extremely versatile, less prone to mobility artifacts, and much easier to prepare. The stacking gel is omitted.

Additional Materials *(also see Basic Protocol 1; see APPENDIX 1 for items with ✓)*

Separating gel solution (Table 10.2.7)
✓ 2× and 1× phosphate/SDS sample buffer
✓ 1× phosphate/SDS electrophoresis buffer

1. Prepare and pour a single separating gel (Basic Protocol 1, steps 1 to 4), *except* use solutions in Table 10.2.7 and fill the gel sandwich to the top. Insert the comb (Basic Protocol 1, step 7) and allow the gel to polymerize 30 to 60 min at room temperature.

2. Mix the protein sample 1:1 with 2× phosphate/SDS sample buffer and heat to 100°C for 2 min.

3. Assemble the electrophoresis apparatus and load the sample (Basic Protocol 1, steps 9 to 13), using the phosphate/SDS electrophoresis buffer and loading empty wells with 1× phosphate/SDS sample buffer.

Table 10.2.7 Recipes for Separating Gels for Continuous SDS-PAGE[a]

SEPARATING GEL

Stock solution[b]	Final acrylamide concentration in separating gel (%)[c]										
	5	6	7	8	9	10	11	12	13	14	15
30% acrylamide/ 0.8% bisacrylamide	2.50	3.00	3.50	4.00	4.50	5.00	5.50	6.00	6.50	7.00	7.50
4× phosphate/SDS, pH 7.2	3.75	3.75	3.75	3.75	3.75	3.75	3.75	3.75	3.75	3.75	3.75
H_2O	8.75	8.25	7.75	7.25	6.75	6.25	5.75	5.25	4.75	4.25	3.75
10% (w/v) ammonium persulfate[d]	0.05	0.05	0.05	0.05	0.05	0.05	0.05	0.05	0.05	0.05	0.05
TEMED	0.01	0.01	0.01	0.01	0.01	0.01	0.01	0.01	0.01	0.01	0.01

Preparation of separating gel

In a 25-ml side-arm flask, mix 30% acrylamide/0.8% bisacrylamide solution (see Table 10.2.1), 4× phosphate/SDS, pH 7.2, and H_2O. Degas under vacuum about 5 min. Add 10% ammonium persulfate and TEMED. Swirl gently to mix. Use immediately.

[a]The recipes produce 15 ml of separating gel, which is adequate for one gel of dimensions 0.75 mm × 14 cm × 14 cm. The recipes are based on the original continuous phosphate buffer system of Weber et al. (1972). The stacking gel is omitted.

[b]All reagents and solutions used in the protocol must be prepared with Milli-Q-purified water or equivalent. See *APPENDIX 1* for recipes.

[c]Units of numbers in table body are milliliters. The desired percentage of acrylamide in the separating gel depends on the molecular size of the protein being separated. See Basic Protocol 1, annotation to step 3.

[d]Best to prepare fresh.

4. Connect the power supply and start the run with 15 mA per 0.75-mm-thick gel until the tracking dye has entered the gel. Continue electrophoresis at 30 mA for 3 hr (5% gel), 5 hr (10% gel), 8 hr (15% gel), or until the dye reaches the bottom of the gel. Use temperature control if available to maintain the gel at 15° to 20°C.

5. Disassemble the gel (Basic Protocol 1, steps 15 to 18).

   Proteins in the gel may be stained using the protocols given in UNIT10.5.

CASTING AND RUNNING ULTRATHIN GELS

ALTERNATE PROTOCOL 4

Ultrathin gels provide superb resolution but are difficult to handle. In this application, gels are cast on Gel Bond, a Mylar support material. Silver staining is recommended for the best resolution. Combs and spacers can be adapted from combs and spacers used for DNA sequencing to cast gels from 0.2 to 0.5 mm thick.

Additional Materials *(also see Basic Protocol 1)*

95% (v/v) ethanol
Gel Bond (FMC) cut to a size slightly smaller than the gel plate dimensions
Glue stick
Ink roller (available from art supply stores)
Combs and spacers (0.19- to 0.5-mm; sequencing gel spacers and combs can be cut to fit)

1. Wash gel plates thoroughly with water-based laboratory detergent followed by successive rinses with hot tap water, deionized water, and finally 95% ethanol. Allow to air dry.

   Gloves should be worn throughout these procedures to prevent contamination by proteins on the surface of skin.

2. Apply a streak of adhesive from a glue stick to the bottom edge of the glass plate. Quickly position the Gel Bond with the hydrophobic side down (a drop of water will bead up on the hydrophobic surface). Apply pressure with Kimwipe tissue to attach the Gel Bond firmly to the plate. Pull the top portion of the Gel Bond back, place a few drops of water underneath, and roll flat with an ink roller.

   Make sure the Gel Bond does not extend beyond the edges of the upper and lower sealing surface of the plate.

3. Assemble the gel cassette according to the manufacturer's instructions (Basic Protocol 1, step 1). Just prior to assembly, blow air over the surface of both the Gel Bond and the opposing glass surface to remove any particulate material (e.g., dust).

   After the plates are positioned in the clamps, use a razor blade to trim any excess spacer at top and bottom to get a reusable spacer exactly the size of the plate.

4. Prepare and pour the separating and stacking gels (Basic Protocol 1, steps 2 to 7). In place of the Teflon comb, insert a square well sequencing comb cut to fit within the gel sandwich. Allow the stacking gel to polymerize 30 to 45 min at room temperature.

5. Prepare the sample and load the gel (Basic Protocol 1, steps 8 to 13).

   When preparing protein samples for ultrathin gels, 3 to 4 μl at 5 μg protein/μl is required for Coomassie blue R-250 staining, whereas 10-fold less is needed for silver staining.

6. Run the gel (Basic Protocol 1, step 14), *except* conduct the electrophoresis at 7 mA/gel (0.25-mm-thick gels) or 14 mA/gel (0.5-mm-thick gels) for 4 to 5 hr.

7. When the separation is complete, disassemble the unit and remove the gel (Basic Protocol 1, steps 15 to 18). With a gloved hand, wash away the adhesive material under a stream of water before proceeding to protein detection.

   Either Coomassie blue or silver staining (UNIT 10.5) may be used, but silver staining produces particularly fine resolution with thin Gel Bond–backed gels.

SUPPORT PROTOCOL 1

CASTING MULTIPLE SINGLE-CONCENTRATION GELS

Casting multiple gels at one time has several advantages. All the gels are identical, so sample separation is not affected by gel-to-gel variation. Furthermore, casting ten gels is only slightly more difficult than casting two gels. Once cast, gels can be stored for several days in a refrigerator.

Additional Materials *(also see Basic Protocol 1)*

Separating and stacking gels for single-concentration gels (Table 10.2.8)
H_2O-saturated isobutyl alcohol

Multiple gel caster (Bio-Rad, Hoefer Pharmacia Biotech)
Extra plates and spacers
14 × 14–cm acrylic blocks or polycarbonate sheets
250- and 500-ml side-arm flasks (used in gel preparation)

continued

Table 10.2.8 Recipes for Multiple Single-Concentration Polyacrylamide Gels[a]

SEPARATING GEL

Stock solution[b]	Final acrylamide concentration in separating gel (%)[c]										
	5	6	7	8	9	10	11	12	13	14	15
30% acrylamide/0.8% bisacrylamide	52	62	72	83	93	103	114	124	134	145	155
4× Tris·Cl/SDS, pH 8.8	78	78	78	78	78	78	78	78	78	78	78
H_2O	181	171	160	150	140	129	119	109	98	88	78
10% (w/v) ammonium persulfate[d]	1.0	1.0	1.0	1.0	1.0	1.0	1.0	1.0	1.0	1.0	1.0
TEMED	0.21	0.21	0.21	0.21	0.21	0.21	0.21	0.21	0.21	0.21	0.21

Preparation of separating gel

In a 500-ml side-arm flask, mix 30% acrylamide/0.8% bisacrylamide solution (see Table 10.2.1), 4× Tris·Cl/SDS, pH 8.8 (Table 10.2.1), and H_2O. Degas under vacuum about 5 min. Add 10% ammonium persulfate and TEMED. Swirl gently to mix; use immediately.

STACKING GEL

In a 250-ml side-arm flask, mix 13.0 ml 30% acrylamide/0.8% bisacrylamide solution, 25 ml 4× Tris·Cl/SDS, pH 6.8 (Table 10.2.1), and 61 ml H_2O. Degas under vacuum about 5 min. Add 0.25 ml 10% ammonium persulfate and 50 µl TEMED. Swirl gently to mix. Use immediately. Failure to form a firm gel usually indicates a problem with the persulfate, TEMED, or both.

[a]The recipes produce about 300 ml of separating gel and 100 ml of stacking gel, which are adequate for ten gels of dimensions 1.5 mm × 14 cm × 14 cm. Volumes were measured using 1.5-mm spacers (or fewer gels). For thinner spacers or fewer gels, calculate volumes using the equation in the annotation to step 4. The recipes are based on the SDS (denaturing) discontinuous buffer system of Laemmli (1970).

[b]All reagents and solutions used in the protocol must be prepared with Milli-Q-purified water or equivalent. See *APPENDIX 1* for recipes.

[c]Units of numbers in table body are milliliters. The desired percentage of acrylamide in separating gel depends on the molecular size of the protein being separated. See Basic Protocol 1, annotation to step 3.

[d]Best to prepare fresh.

Long razor blade *or* plastic wedge (Wonder Wedge, Hoefer Pharmacia Biotech)
Resealable plastic bags

1. Assemble the multiple gel caster according to the manufacturer's instructions.

2. Assemble glass sandwiches and stack them in the casting chamber. Stack up to ten 1.5-mm gels and fill in extra space with acrylic blocks or polycarbonate sheets to hold the sandwiches tightly in place. Make sure the spacers are straight along the top, right, and left edges of the glass plates and that all edges of the stack are flush.

3. Place the front sealing plate on the casting chamber, making sure the stack fits snugly. Secure the plate with four spring clamps and tighten the bottom thumb screws.

4. Prepare the separating (resolving) gel solution (Table 10.2.8).

   A 12-cm separating gel with a 4-cm stacking gel is recommended.

   If fewer than ten gels are prepared (Table 10.2.8), use the following formula to estimate the amount of separating gel volume needed:

Volume = gel number × height (cm) × width (cm) × thickness (cm) + 4 × gel number + 10 ml.

5. Using a 100-ml disposable syringe with flat-tipped needle, inject the resolving gel solution down the side of one spacer into the multiple caster. A channel in the silicone plug distributes the solution throughout the whole caster. Avoid introducing bubbles by giving the caster a quick tap on the benchtop once the caster is filled.

6. Overlay the center of each gel with 100 µl H_2O-saturated isobutyl alcohol and let polymerize for 1 to 2 hr.

7. Drain off the overlay and rinse the surface with 1× Tris·Cl/SDS, pH 8.8. If the gels will not be used immediately, skip to step 11.

8. Immediately before use, prepare the stacking gel solution either singly (Basic Protocol 1, step 6) or for all the gels at once (Table 10.2.8). Fill each sandwich in the caster with stacking gel solution.

9. Insert a comb into each sandwich and let the gel polymerize for 2 hr.

10. Remove the combs and rinse wells with 1× SDS electrophoresis buffer.

11. Remove the gels from the caster and separate by carefully inserting a long razor blade or knife between each gel sandwich. A plastic wedge (Hoefer Pharmacia Biotech's Wonder Wedge) also works well. Clean the outside of each gel plate with running water to remove the residual polymerized and unpolymerized acrylamide.

12. Overlay gels to be stored with 1× Tris·Cl/SDS, pH 8.8, place in a resealable plastic bag, and store at 4°C until needed (up to 1 week).

ALTERNATE PROTOCOL 5

SEPARATION OF PROTEINS ON GRADIENT GELS

Gels that consist of a gradient of increasing polyacrylamide concentration resolve a much wider size range of proteins than standard uniform concentration gels (see Figure 10.2.3). The protein bands, particularly in the low-molecular-weight range, are also much sharper.

Additional Materials *(also see Basic Protocol 1)*

Light and heavy acrylamide gel solutions (Table 10.2.9 and Table 10.2.10)
Bromphenol blue (optional; for checking practice gradient)
TEMED

Gradient maker (30 to 50 ml, Hoefer Pharmacia *or* 30 to 100 ml, Bio-Rad)
Tygon tubing with micropipet tip
Peristaltic pump (optional; Markson)
Whatman 3MM filter paper

1. Assemble the magnetic stirrer and gradient maker on a ring stand as shown in Figure 10.2.4. Connect the outlet valve of the gradient maker to Tygon tubing attached to a micropipet tip that is placed over the vertical gel sandwich. If desired, place a peristaltic pump in line between the gradient maker and the gel sandwich.

2. Place a small stir-bar into the mixing chamber of the gradient maker (i.e., the chamber connected to the outlet).

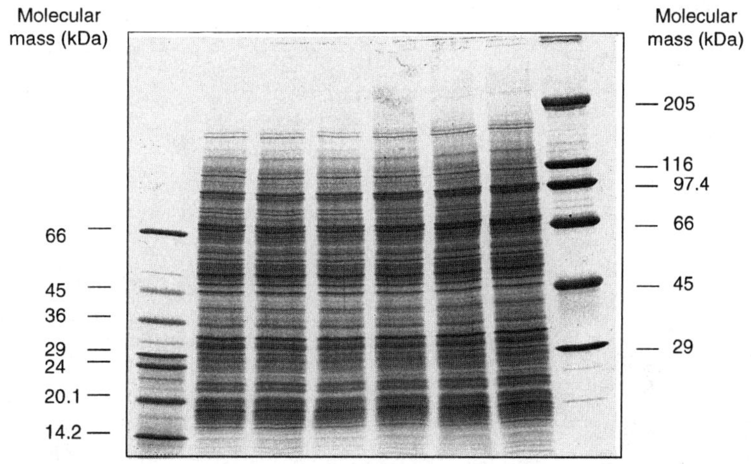

Figure 10.2.3 Separation of membrane proteins by 5.1% to 20.5% T polyacrylamide gradient SDS-PAGE. Approximately 30 µl of 1× SDS sample buffer containing 30 µg of Alaskan pea (*Pisum sativum*) membrane proteins was loaded in wells of a 14 × 14–cm, 0.75-mm-thick gel. Standard proteins were included in the outside lanes. The gel was run at 4 mA for ~15 hr.

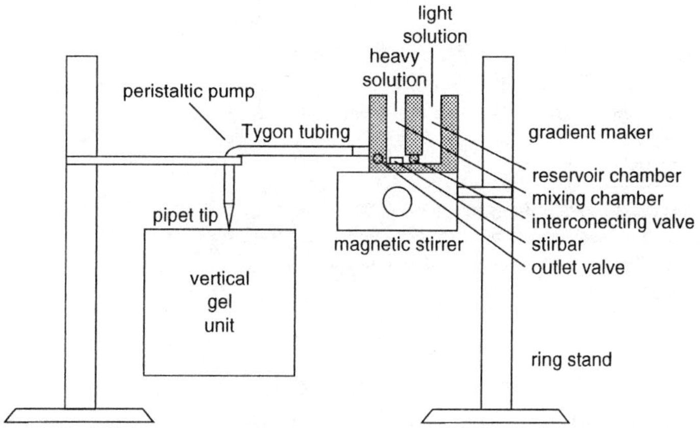

Figure 10.2.4 Gradient gel setup. A peristaltic pump, though not required, will provide better control.

3. Using the recipes in Table 10.2.9 and Table 10.2.10, prepare light and heavy acrylamide gel solutions. Do not add ammonium persulfate until just before use (step 7). Keep the heavy acrylamide gel solution on ice until used to prevent polymerization when the ammonium persulfate is added.

   Deaeration is not recommended for either the light or heavy solution. Omitting the deaeration will allow polymerization to proceed more slowly, letting the gradient establish itself in the gel sandwich before polymerization takes place.

4. With the outlet port and interconnecting valve between the two chambers closed, pipet 7 ml of light (low-concentration) acrylamide gel solution into the reservoir chamber for one 0.75-mm-thick gradient gel.

Table 10.2.9 Light Acrylamide Gel Solutions for Gradient Gels[a]

Stock solution	Acrylamide concentration of light gel solution (%)[b]									
	5	6	7	8	9	10	11	12	13	14
30% acrylamide/ 0.8% bisacrylamide[c]	2.5	3.0	3.5	4.0	4.5	5.0	5.5	6.0	6.5	7.0
4× Tris·Cl/SDS, pH 8.8[c]	3.75	3.75	3.75	3.75	3.75	3.75	3.75	3.75	3.75	3.75
H_2O	8.75	8.25	7.75	7.25	6.75	6.25	5.75	5.25	4.75	4.25
10% (w/v) ammonium persulfate[d]	0.05	0.05	0.05	0.05	0.05	0.05	0.05	0.05	0.05	0.05

[a]To survey proteins ≥10 kDa, 5-20% gradient gels are recommended. To expand the range between 10 and 200 kD, a 10-20% gel is recommended.

[b]Numbers in body of table are milliliters of stock solution. Deaeration is not required. Keep solution at room temperature prior to adding TEMED no longer than 1 hr.

[c]See *APPENDIX 1* for preparation.

[d]Best to prepare fresh.

Table 10.2.10 Heavy Acrylamide Gel Solutions for Gradient Gels[a]

Stock solution	Acrylamide concentration of heavy gel solution (%)[b]										
	10	11	12	13	14	15	16	17	18	19	20
30% acrylamide/ 0.8% bisacrylamide[c]	5.0	5.5	6.0	6.5	7.0	7.5	8.0	8.5	9.0	9.5	10.0
4× Tris·Cl/SDS, pH 8.8[c]	3.75	3.75	3.75	3.75	3.75	3.75	3.75	3.75	3.75	3.75	3.75
H_2O	5.0	4.5	4.0	3.5	3.0	2.5	2.0	1.5	1.0	0.5	0
Sucrose (g)	2.25	2.25	2.25	2.25	2.25	2.25	2.25	2.25	2.25	2.25	2.25
10% (w/v) ammonium persulfate	0.05	0.05	0.05	0.05	0.05	0.05	0.05	0.05	0.05	0.05	0.05

[a]Deaeration is not recommended for gradient gels.

[b]Numbers in body of table are milliliters of stock solution (except sucrose). Do not add the ammonium persulfate until just before use. The heavy acrylamide will polymerize, albeit more slowly, without the addition of TEMED. Keep the heavy solution on ice after adding ammonium persulfate.

[c]See *APPENDIX 1* for preparation.

5. Open the interconnecting valve briefly to allow a small amount (~200 µl) of light solution to flow through the valve and into the mixing chamber.

6. Add 7 ml of heavy (high-concentration) acrylamide gel solution to the mixing chamber.

7. Add the specified amount of ammonium persulfate and ~2.3 µl TEMED per 7 ml acrylamide solution to each chamber. Mix the solutions in each chamber with a disposable pipet. Open the interconnecting valve completely.

8. Turn on the magnetic stirrer and adjust the rate to produce a slight vortex in the mixing chamber. Open the outlet of the gradient maker slowly. Adjust the outlet valve to a flow rate of 2 ml/min.

9. Fill the gel sandwich from the top. Place the pipet tip against one side of the sandwich so the solution flows down one plate only. The heavy solution will flow into the sandwich first, followed by progressively lighter solution.

10. Watch as the last of the light solution drains into the outlet tube and adjust the flow rate to ensure that the last few milliliters of solution do not flow quickly into the gel sandwich and disturb the gradient.

11. Overlay the gradient gel with H_2O-saturated isobutyl alcohol. Allow the gel to polymerize ~1 hr.

12. Remove the H_2O-saturated isobutyl alcohol and rinse with 1× Tris·Cl/SDS, pH 8.8. Cast the stacking gel (Basic Protocol 1, steps 5 to 7).

13. Prepare the protein sample and protein molecular-weight-standards mixture. Load and run the gel (Basic Protocol 1, steps 8 to 18).

    The gel can be stained with Coomassie blue or silver (UNIT 10.5).

14. After staining, dry the gels onto Whatman 3MM or equivalent filter paper.

    Gradient gels >0.75 mm thick with ≤20% acrylamide solutions will dry without cracking as long as the vacuum pump is working properly and the cold trap is dry at the onset of drying. For gradient gels >0.75 mm thick, add 3% (w/v) glycerol to the final destaining solution to help prevent cracking. Another method is to dehydrate and shrink the gel in 30% methanol for up to 3 hr prior to drying. Then place the gel in distilled water for 5 min before drying.

CASTING MULTIPLE GRADIENT GELS

SUPPORT PROTOCOL 2

Casting gradient gels in a multiple gel caster has several advantages. In addition to the time savings, batch casting gives gels that are essentially identical. This is particularly important for gradient gels, where slight variations in casting technique can cause variations in protein mobility. The gels may be stored for up to 1 week after casting to ensure internal consistency from run to run during the week.

Additional Materials *(also see Alternate Protocol 5; see APPENDIX 1 for items with ✓)*

✓ Plug solution
Light and heavy acrylamide gel solutions for multiple gradient gels (Table 10.2.11 and Table 10.2.12)
TEMED
H_2O-saturated isobutyl alcohol

Multiple gel caster (Bio-Rad, Hoefer Pharmacia Biotech)
Peristaltic pump (25 ml/min)
500- or 1000-ml gradient maker (Bio-Rad, Hoefer Pharmacia Biotech)
Tygon tubing

1. Assemble the multiple caster as in casting multiple single-concentration gels (Support Protocol 1, steps 1 to 3), making sure to remove the triangular space filler plugs in the bottom of the caster.

2. Set up the peristaltic pump (Fig. 10.2.5). Using a graduated cylinder and water, adjust the flow rate so that the volume of the gradient solution plus volume of plug solution is poured in ~15 to 18 min (~25 ml/min).

3. Set up a gradient maker that holds no more than four times the total volume of solution to be poured. Close all valves and place a stir-bar in the mixing chamber, which is the one with the outlet port. Attach one end of a piece of Tygon tubing to the outlet of the gradient maker. Run the other end of the tubing through the peristaltic pump and attach it to the red inlet port at the bottom of the caster.

4. Prepare solutions for the gradient maker (Table 10.2.11 and Table 10.2.12). Keep the heavy acrylamide solution on ice until used.

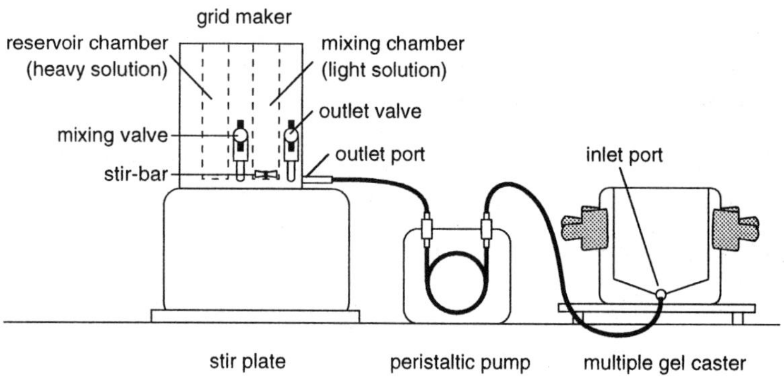

Figure 10.2.5 Setup for casting multiple gradient gels. Casting multiple gradient gels requires a peristaltic pump and a multiple gel caster. Gel solution is introduced through the bottom of the multiple caster.

5. Add the TEMED to both heavy and light solutions (54 µl/165 ml) and immediately pour the light (low-concentration) solution into the mixing chamber (the one with the port). Open the mixing valve slightly to allow the tunnel to fill and to avoid air bubbles. Close the valve again and pour the heavy (high-concentration) acrylamide solution into the reservoir chamber.

6. Start the magnetic stirrer and open the outlet valve; then start the pump and open the mixing valve.

   In units for casting multiple gels, acrylamide solution flows in from the bottom. To use a multiple casting unit, the light solution is placed in the mixing chamber and the heavy in the reservoir.

7. When almost all the acrylamide solution is gone from the gradient maker, stop the pump and close the mixing valve. Tilt the gradient maker toward the outlet side and remove the last milliliters of the mix. Do not allow air bubbles to enter the tubing.

8. Add the plug solution to the mixing chamber and start the pump. Make sure that no bubbles are introduced. Continue pumping until the bottom of the caster is filled with plug solution to just below the glass plates; then turn off the pump. Clamp the tubing close to the red port of the casting chamber.

9. Quickly overlay each separate gel sandwich with 100 µl H_2O-saturated isobutyl alcohol. Use the same amount on each sandwich. Allow the gel to polymerize.

10. Drain off the overlay and rinse the surface of the gels with 1× Tris·Cl/SDS, pH 8.8.

11. Prepare and cast the stacking gel as in casting multiple single-concentration gels (Support Protocol 1, steps 7 to 10).

12. Remove gels from the caster and clean the gel sandwiches (Support Protocol 1, step 11). Store gels, if necessary, according to the instructions for multiple single-concentration gels (Support Protocol 1, step 12).

Table 10.2.11 Light Acrylamide Gel Solutions for Multiple Gradient Gels[a, b]

Stock solution	Acrylamide concentration of light separating gel solution (%)[c]									
	5	6	7	8	9	10	11	12	13	14
30% acrylamide/ 0.8% bisacrylamide[d]	28	33	39	44	50	55	61	66	72	77
4× Tris·Cl/SDS, pH 8.8[d]	41	41	41	41	41	41	41	41	41	41
H_2O	96	91	85	80	74	69	63	58	52	47
10% (w/v) ammonium persulfate	0.55	0.55	0.55	0.55	0.55	0.55	0.55	0.55	0.55	0.55

[a]To survey proteins ≥10 kDa, 5-20% gradient gels are recommended. To expand the range between 10 and 200 kDa, a 10-20% gel is recommended.

[b]Recipes produce ten 1.5-mm-thick gradient gels with 10 ml extra solution to account for losses in tubing.

[c]Numbers in body of table are milliliters of stock solution. Deaeration is not required. Keep solution at room temperature prior to adding TEMED no longer than 1 hr.

[d]See *APPENDIX 1* for preparation.

Table 10.2.12 Heavy Acrylamide Gel Solutions for Multiple Gradient Gels[a,b]

Stock solution	Acrylamide concentration of heavy gel solution (%)[c]										
	10	11	12	13	14	15	16	17	18	19	20
30% acrylamide/ 0.8% bisacrylamide[d]	55	61	66	72	77	83	88	94	99	105	110
4× Tris·Cl/SDS pH 8.8[d]	41	41	41	41	41	41	41	41	41	41	41
H_2O	55	50	44	39	33	28	22	17	11	5.5	0
Sucrose (g)	25	25	25	25	25	25	25	25	25	25	25
10% (w/v) ammonium persulfate	0.55	0.55	0.55	0.55	0.55	0.55	0.55	0.55	0.55	0.55	0.55

[a]Deaeration is not recommended for gradient gels.

[b]Recipes produce 10 ml extra solution to account for losses in tubing.

[c]Numbers in body of table are milliliters of stock solution (except sucrose). Do not add the ammonium persulfate until just before use. The heavy acrylamide will polymerize, albeit more slowly, without the addition of TEMED. Keep the heavy solution on ice after adding ammonium persulfate.

[d]See *APPENDIX 1* for preparation.

ELECTROPHORESIS IN SINGLE-CONCENTRATION MINIGELS

BASIC PROTOCOL 2

Separation of proteins in a small-gel format is becoming increasingly popular for applications that range from isolating material for peptide sequencing to performing routine protein separations. The unique combination of speed and high resolution is the foremost advantage of small gels. Small gels are easily adapted to single-concentration, gradient, and two-dimensional SDS-PAGE procedures. A multiple gel caster is the only practical way to produce small linear polyacrylamide gradient gels (see Support Protocol 3).

Materials

- Minigel vertical gel unit with glass plates, clamps, and buffer chambers (Hoefer Pharmacia Biotech, Bio-Rad)
- 0.75-mm spacers
- Single or multiple gel caster (Hoefer Pharmacia Biotech, Bio-Rad)
- Acrylic plate (Hoefer Pharmacia Biotech, Bio-Rad) or polycarbonate separation sheet (Hoefer Pharmacia Biotech, Bio-Rad)

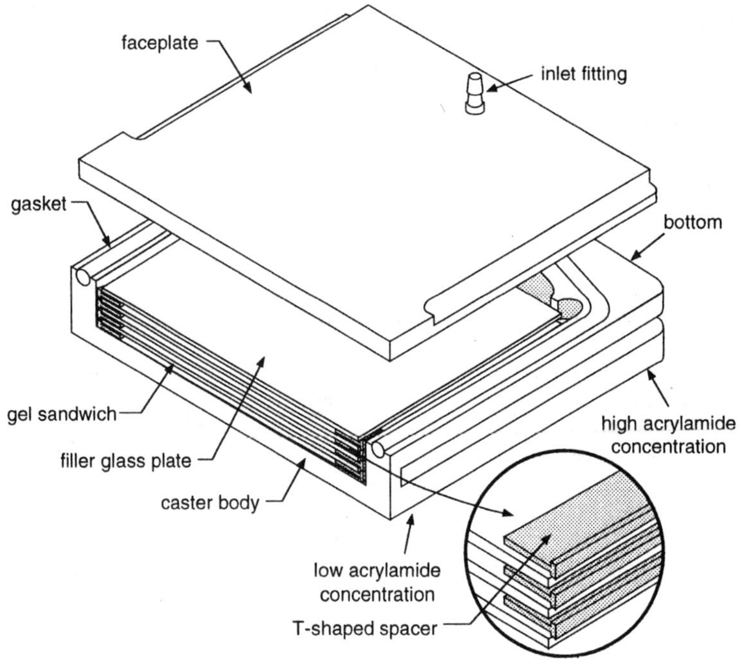

Figure 10.2.6 Minigel sandwiches positioned in the multiple gel caster. Extra glass or acrylic plates or polycarbonate sheets are used to fill any free space in the caster and to ensure that the gel sandwiches are held firmly in place.

10- and 50-ml syringes
Combs (Teflon; Hoefer Pharmacia Biotech, Bio-Rad)
Long razor blade
Micropipet

1. Assemble each gel sandwich by stacking, in order, the notched (Hoefer Pharmacia Biotech) or small rectangular (Bio-Rad) plate, 0.75-mm spacers, and the larger rectangular plate. Be sure to align the spacers properly with the ends flush with the top and bottom edge of the two plates when positioning the sandwiches in the multiple gel caster (Fig. 10.2.6).

   Alternatively, precast minigels can be purchased from a number of suppliers (see Table 10.2.3).

2. Fit the gel sandwiches tightly in the multiple gel caster. Use an acrylic plate or polycarbonate separation sheet to eliminate any slack in the chamber.

   Alternatively, gels can be cast singly with a stand-alone easter.

3. Place the front faceplate on the caster, clamp it in place against the silicone gasket, and verify alignment of the glass plates and spacers.

4. Prepare the separating gel solution as directed in Table 10.2.1. For five 0.75-mm-thick gels, prepare ~30 ml solution (i.e., double the volumes listed). Do not add TEMED and ammonium persulfate until just before use.

   To compute the total gel volume needed, multiply the area of the gel (e.g., 7.3 × 8.3 cm) by the thickness of the gel (e.g., 0.75 mm) and then by the number of gels in the caster. If needed, add about 4 to 5 ml of extra gel solution to account for the space around the outside of the gel sandwiches.

5. Fill a 50-ml syringe with the separating gel solution and slowly inject it into the caster until the gels are 6 cm high, allowing 1.5 cm for the stacking gel. Overlay each gel with 100 µl H_2O-saturated isobutyl alcohol. Allow the gels to polymerize for ~1 hr.

6. Remove the isobutyl alcohol and rinse with 1× Tris·Cl/SDS, pH 8.8.

   Stacking gels can be cast one at a time with the gel mounted on the electrophoresis unit, or all at once in the multiple caster.

7. Practice placing a comb in the gel sandwiches before preparing the stacking gel solution. Press the comb against the rectangular or taller plate so that all teeth of the comb are aligned with the opening in the gel sandwich, then insert into the sandwich. Remove combs after practicing.

8. Prepare the stacking gel solution (2 ml per gel) as directed in Table 10.2.1. Fill a 10-ml syringe with stacking gel solution and inject the solution into each gel sandwich. Insert combs, taking care not to trap bubbles. Allow gels to polymerize 1 hr.

9. Remove the front faceplate. Carefully pull the gels out of the caster, using a long razor blade to separate the sandwiches.

   The gels can be stored tightly wrapped in plastic wrap with the combs left in place inside a sealable bag to prevent drying for ~1 week. Without the stacking gel, the separating gel can be stored for 2 to 3 weeks. Keep gels moist with 1× Tris·Cl/SDS, pH 8.8, at 4°C. Do not store gels in the multiple caster.

10. Remove the combs and rinse the sample wells with 1× SDS electrophoresis buffer. Place a line indicating the bottom of each well on the front glass plate with a marker.

11. Fill the upper and lower buffer chambers with 1× SDS electrophoresis buffer. The upper chamber should be filled to 1 to 2 cm over the notched plate.

12. Prepare the protein sample and protein-standards mixture (Basic Protocol 1, step 8). Load the sample using a micropipet. Insert the pipet tip through the upper buffer and into the well. The mark on the glass plate will act as a guide. Dispense the sample into the well.

    For a complex mixture, 20 to 25 µg protein in 10 µl SDS sample buffer will give a strongly stained Coomassie blue pattern. Much smaller amounts (1 to 5 µg) are required for highly purified proteins, and 10- to 100-fold less protein amount in the same volume (e.g., 10 µl) is required for silver staining.

13. Electrophorese samples at 10 to 15 mA per 0.75-mm gel until the dye front reaches the bottom of the gel (~1 to 1.5 hr).

14. Disassemble the gel (Basic Protocol 1, steps 15 to 18). Proceed with detection of proteins.

SUPPORT PROTOCOL 3

PREPARING MULTIPLE GRADIENT MINIGELS

Polyacrylamide gradients not only enhance the resolution of larger format gels but also greatly improve protein separation in the small format. Casting gradient minigels one at a time is not generally feasible because of the small volumes used, but multiple gel casters make it easy to cast several small gradient gels at one time. The gels are cast from the bottom in multiple casters, with the light acrylamide solution entering first.

Additional Materials *(also see Basic Protocol 2; see* APPENDIX 1 *for items with ✓)*

✓ Plug solution

1. Assemble minigel sandwiches in the multiple gel caster as described for single-concentration minigels (Basic Protocol 2, steps 1 to 3).
2. Set up the 30-ml gradient maker, magnetic stirrer, peristaltic pump (optional), and Tygon tubing as in Figure 10.2.5. Connect the outlet of the 30-ml gradient maker to the inlet at the base of the front faceplate of the caster.
3. Prepare light (Table 10.2.9) and heavy (Table 10.2.10) acrylamide gel solutions, omit degassing. Add ammonium persulfate just before use. Use ~12 ml of each solution for five 0.75-mm-thick minigels. Keep heavy acrylamide solution on ice until used.
4. With the outlet and interconnecting valve closed, add the heavy solution to the reservoir chamber. Briefly open the interconnecting valve to let a small amount of heavy solution through to the mixing chamber, clearing the valve of air.
5. Fill the mixing chamber with light solution. Add ammonium persulfate and 4 µl TEMED per 12 ml acrylamide solution to each chamber and mix with a disposable pipet.
6. Turn on the magnetic stirrer. Open the interconnecting valve and allow the chambers to equilibrate. Then slowly open the outlet port to allow the solution to flow from the gradient maker to the multiple caster by gravity (a peristaltic pump may be used for better control). Adjust the flow rate to 3 to 4 ml/min.
7. Close the outlet port as the last of the gradient solution leaves the mixing chamber, just before air enters the outlet tube. Fill the two chambers with plug solution and slowly open the outlet once again.
8. Allow the plug solution to push the acrylamide in the caster up into the plates. Close the outlet when the plug solution reaches the bottom of the plates.
9. Quickly add 100 µl H_2O-saturated isobutyl alcohol to each gel sandwich. Let the gels polymerize undisturbed for ~1 hr.
10. Prepare and pour the stacking gel (Basic Protocol 2, step 8).
11. Disconnect the gradient maker, place the caster in a sink, and remove the front faceplate. The plug solution will drain out from bottom of the caster. Remove the gels (Basic Protocol 2, step 9).

    Gradient minigels can be stored as described for single-concentration minigels (Basic Protocol 2, step 9 annotation). For instructions on preparing, loading, and running the gels, see Basic Protocol 2, steps 10 to 14.

Reference: Hames and Rickwood, 1990.

Contributor: Sean R. Gallagher

Two-Dimensional Gel Electrophoresis Using the O'Farrell System

Two-dimensional gel electrophoresis is the combination of two high-resolution electrophoretic procedures (isoelectric focusing and SDS–polyacrylamide gel electrophoresis) to provide much greater resolution than either procedure alone. The resulting two-dimensional gel contains numerous round or elliptical protein spots well separated from each other; depending on the sample, as many as 1500 protein spots may be detected by silver staining (*UNIT 10.5*) or autoradiography (*APPENDIX 3*).

FIRST-DIMENSION (ISOELECTRIC-FOCUSING) GELS

BASIC PROTOCOL 1

Materials *(see APPENDIX 1 for items with ✓)*

Urea (ultrapure)
✓ 30% acrylamide/1.8% bisacrylamide
Ampholytes, pH 4 to 8 (ISO-DALT-grade Resolytes from BDH Chemicals are available from Hoefer; suppliers for other ampholytes are Bio-Rad, Pharmacia LKB, and Serva)
Nonidet P-40 (NP-40)
TEMED (*N,N,N′,N′* tetramethylethylenediamine)
✓ 10% (w/v) ammonium persulfate
✓ 0.085% (v/v) phosphoric acid
✓ 0.02 M NaOH (prepare from 10 M NaOH stock and freshly deaerated water just before use)
Protein samples (Support Protocol)
✓ Concentrated bromphenol blue
✓ Chromic acid cleaning solution

1.0- to 3.0-mm-inner-diameter glass gel tubes (~1.5 in. longer than the width of the second-dimension gel; 4- to 6-mm outer diameter)
2.5- to 3.0-cm-inner-diameter gel-casting glass tube, ~2 cm shorter than gel tubes
Small vacuum flask
50-μl, 1-ml, and 20-ml syringes
0.2- or 0.45-μm filter capsule (Acrodisk; Gelman)
Single-edge razor blade
Rubber grommets
Tube cell (Bio-Rad, Hoefer)
22-G hypodermic needle (2-in. long)
200-μl pipettor tip
1-dram gel vials

NOTE: Distilled, deionized water should be used throughout this protocol.

1. Mark clean, dry 1.5-mm-i.d. gel tubes to indicate the desired height of the gel (usually the same as the width of the second-dimension gel). Place a rubber band around the gel tubes so that they form a tight bundle (~12 tubes fit into a bundle). Hold the bundle vertically on a flat surface and push down on the tops of the tubes so that the bottoms are even.

2. Carefully seal one end of the 2.5- to 3.0-cm-i.d. gel-casting glass tube with three or four layers of Parafilm to form a strong, water-tight seal.

3. Place the bundle of gel tubes inside the gel casting tube and support the glass tube in a vertical position with a ring stand and clamp to allow the sealed end of the glass tube to rest on a solid surface.

4. Add 8.25 g urea, 6.0 ml water, 2.0 ml of 30% acrylamide/1.8% bisacrylamide, and 0.75 ml ampholytes, pH 4 to 8, to a small vacuum flask. Add a small stir-bar to the flask. Place the flask in a warm water bath on a magnetic stirrer and stir just until the urea is in solution; do not heat the solution to >30°C.

   If increased resolution in a narrow pH range is needed, narrow-range ampholytes can be added to the broad-range ampholytes in a 2:1 ratio.

5. Deaerate the solution by applying a strong vacuum for 2 to 3 min. Add 0.3 ml NP-40 and swirl until dissolved.

6. Pour solution into a 20-ml syringe fitted with a 0.2- or 0.45-μm filter capsule and force through the filter. Add 10 ml TEMED and swirl. Add 70 ml of 10% ammonium persulfate and swirl. Immediately pipet the gel solution into the space between the gel tubes and the large glass tube.

7. Gently run water down the outside of the gel tubes using a wash bottle. Add water until the level of the acrylamide solution inside the tubes reaches the desired height.

8. Remove the Parafilm from the bottom of the gel casting tube and push the gel tubes, containing the polymerized gel, out the bottom. Cut across the gel-tube bottoms with a single-edge razor blade to remove excess acrylamide. Rinse the bottom of the gel tubes under running deionized water to remove residual acrylamide.

9. Place rubber grommet on the top of each tube, making sure that the top surface of the gel is visible below the grommet; ~5 mm of the gel tube should be visible above the grommet.

10. Seat the tube and grommet assemblies in the holes of the upper buffer reservoir of the tube cell. Plug any unused holes with rubber stoppers.

11. Fill the lower reservoir with ~3 liters of 0.085% phosphoric acid.

12. Place the upper reservoir into lower reservoir and adjust lower buffer level to cover the entire gel.

13. Fill the upper buffer reservoir with 250 ml of 0.02 M NaOH. Fill the gel tubes to the top with 0.02 M NaOH using a 1-ml syringe equipped with a 22-G hypodermic needle. Be careful to eliminate any air bubbles in the gel tubes.

14. Connect the tube cell to the power supply. The black (–) lead goes to the upper reservoir. Prefocus the gel 1 hr at 200 V constant voltage (see *UNIT 10.2* Introduction for a discussion of electricity and electrophoresis). Disconnect the tube cell from the power supply.

15. Layer 10 to 30 μl of protein samples (100 to 150 μg) on top of the gels through the upper buffer with a 50-μl syringe. Fill the remainder of the tube with 0.02 M NaOH to eliminate any bubbles.

16. Place the lid on the upper reservoir and attach the electrical leads to a power supply. Turn on the power supply and adjust to the desired settings at constant voltage (700 to 800 V). Run the gel 16 hr (11,000 to 13,000 V hr).

17. Reduce the voltage setting to zero and turn off the power supply to end the run. Add ~1 µl concentrated bromphenol blue to the top of each gel with a 50-µl syringe.

18. Extrude the gels from the tubes using water pressure from a 1-ml syringe fitted with a 200-µl pipettor tip (cut off ~1 cm of the large end of the tip so it fits on the syringe).

19. Place each gel in a labeled gel vial. The gels can be stored at –70°C for many weeks or used immediately.

20. Soak gel tubes overnight in chromic acid cleaning solution, then rinse thoroughly under running deionized tap water for 15 min. Remove excess water from the gel tubes with suction and allow them to dry.

SECOND-DIMENSION GELS

BASIC PROTOCOL 2

Materials *(see* APPENDIX 1 *for items with ✓)*

✓ 30% acrylamide/0.8% bisacrylamide
✓ Gel buffer
10% (w/v) SDS
TEMED
✓ 10% (w/v) ammonium persulfate
Isobutyl alcohol, H_2O-saturated
✓ Stacking gel buffer (optional)
First-dimension gel (Basic Protocol 1)
✓ Equilibration buffer
✓ Hot 0.5% and 1% (w/v) agarose (keep in boiling water bath)
Protein molecular weight standards (*UNIT 10.2*, Table 10.2.2; available as kits from Bio-Rad or Pharmacia LKB)
✓ SDS/solubilization buffer
✓ Reservoir buffer, prechilled to 10° to 20°C
Coolant (from running tap water or circulating refrigerated water bath)

Gel plates, one long and one short
1.5-mm spacers (~14 cm × 14 cm × 0.75 mm)
Casting stand
Gel identification tag (e.g., typed consecutive numbers on filter paper)
Nylon screen
5 × 15–cm glass plate
Protean II electrophoresis cell (Bio-Rad)

1. Assemble the gel plates with 1.5-mm spacers. Position clamps on each side of the gel sandwich over the spacers and place on the casting stand. Be sure the plates and spacers are properly aligned, then tighten the clamps and cams to get a leak-proof seal. Make adjustments so that plates are level and vertical. Place the gel identification tag between the glass plates so that it rests in the lower right hand corner.

2. Prepare the gel solution by combining 30% acrylamide/0.8% bisacrylamide, gel buffer, and water in a vacuum flask. Deaerate the solution by applying vacuum for 5 min.

3. Add 10% SDS and TEMED and swirl, then add 10% ammonium persulfate and swirl.

4. Fill the gel sandwich to 5 mm below the top of the short plate and overlay with H_2O-saturated isobutyl alcohol or water. Allow the gel to polymerize 1.5 hr.

   If a stacking gel is desired, stop the gel 1.5 cm from the top of the short plate. After the gel polymerizes, remove the overlay and fill the remaining space with stacking gel solution: 4.8 ml stacking gel buffer, 3.8 ml water, and 1.4 ml 30% acrylamide/0.8% bisacrylamide (all deaerated). Add 200 µl of 10% ammonium persulfate and 8 µl TEMED. Pipet onto the separating gel, overlay with H_2O-saturated isobutyl alcohol, and allow to polymerize 30 min.

5. Add equilibration buffer to completely cover the first-dimension gel (thaw at room temperature, if necessary).

   The time the gel is in equilibration buffer containing SDS can vary from seconds to several minutes depending upon the sample.

6. Pour the gel and equilibration buffer onto a nylon screen placed over a beaker and transfer the first-dimension gel to a 5 × 15–cm glass plate (Parafilm is not rigid enough). Using a spatula, lay the gel out straight along one edge of the glass plate.

7. Pipet a very thin layer of hot 0.5% agarose on the top of the slab gel to be loaded.

8. Using a spatula, carefully slide the first-dimension gel off the glass plate and place it across the top of the slab gel. Orient the first dimension gel with the blue (basic) end to the right.

9. Pipet a thin layer of hot 0.5% agarose over the first-dimension gel to seal it in place. Allow the agarose to solidify.

   Protein molecular-weight markers may be run as one-dimensional separations on the sides of the second-dimension gel. Solubilize marker proteins in SDS/solubilization buffer by boiling for 5 min, then dilute 1:1 with hot 1% agarose solution and draw up the hot solution in a glass tube the same diameter as the first-dimension gel. A short piece of the solidified agarose can be applied to one or both sides of the second-dimension gel and held in place with 0.5% agarose.

10. Mount the gels on the electrophoresis cell. Fill the upper and lower reservoirs with prechilled reservoir buffer.

11. Attach tubing for coolant to the in and out ports and start the flow of coolant to maintain the temperature of the tank buffer at 10° to 20°C during the run to ensure that the gels are adequately cooled.

12. Attach the electrical leads to the power supply (the upper reservoir is connected to the negative lead). Electrophorese at 15 to 20 mA/gel until the tracking dye reaches the end of the gel (or 3 to 5 mA/gel overnight).

13. Reduce the voltage setting to zero and turn off the power supply at the end of the run. Remove the gel from the electrophoresis unit and take off the clamps. Pry the glass plates apart with a spatula.

14. Stain the gel (*UNIT 10.5*) or process for immunoblotting (*UNIT 10.7*) or autoradiograph (*APPENDIX 3*).

ISOELECTRIC FOCUSING OF VERY BASIC AND VERY ACIDIC PROTEINS

ALTERNATE PROTOCOLS 1 AND 2 FOR FIRST-DIMENSION GELS

Nonequilibrium pH gradient electrophoresis (NEPHGE) can be used for first-dimension gels to resolve basic (pI > 8) or acidic (pI < 3.8) proteins.

NEPHGE for Very Basic Proteins

Additional Materials *(also see Basic Protocol 1)*

Ampholytes, pH 2 to 11 (Serva)
0.01 M phosphoric acid, deaerated
4 M urea, deaerated

To analyze very basic proteins, the procedure is the same as described in Basic Protocol 1 with the following exceptions in the indicated steps:

4a. Use 0.75 ml ampholytes, pH 2 to 11, in gel solution.

11a. Fill the lower reservoir with ~3 liters deaerated 0.02 M NaOH.

13a. Fill the upper reservoir with 250 ml deaerated 0.01 M phosphoric acid.

14a. Do not prefocus gels.

15a. Overlay samples with deaerated 4 M urea to protect proteins from phosphoric acid.

16a. Attach the negative lead to the lower tank and the positive to the upper tank (this is the reverse of the usual setup). Focus 1 hr at 400 V and then 4 to 5 hr at 800 V for a total of 4000 V-hr.

Because the bromphenol blue applied to the top of the gel at the end of the run now marks the acidic end, load first-dimension gels onto the second-dimension gel with the blue end to the left.

NEPHGE for Very Acidic Proteins

Additional Materials *(also see Basic Protocol 1)*

Ampholytes, pH 2.5 to 4 (Pharmacia LKB)
Concentrated sulfuric acid
Water, deaerated

To analyze very acidic proteins, the procedure is the same as described in Basic Protocol 1 with the following exceptions in the indicated steps:

4b. Prepare the gel solution as follows:

8.25 g urea
5.5 ml H_2O
2.0 ml 30% acrylamide/1.8% bisacrylamide
1.0 ml ampholytes, pH 2.5 to 4 (Pharmacia LKB)
0.3 ml ampholytes, pH 2 to 11 (Serva).

6b. Add 90 µl of 10% ammonium persulfate and swirl, then add 10 µl TEMED and swirl.

11b. Prepare lower buffer by adding 4.5 ml concentrated sulfuric acid to 3 liters of water and fill the lower reservoir.

13b. Prepare upper buffer by adding 3 ml ampholyte, pH 2 to 11, to 120 ml deaerated water and place in the upper reservoir.

16b. Run at 800 V for 4.5 to 5.0 hr or 250 V for 16 hr for a total of 3600 to 4000 V-hr.

ALTERNATE PROTOCOL 3

TWO-DIMENSIONAL MINIGELS

Though limited in resolving area, small two-dimensional SDS-PAGE is a quick way to separate proteins for a variety of applications. These include peptide sequencing, purity checks, and protocol development. A minigel separation (*UNIT 10.2*) takes 4 to 5 hr.

To perform two-dimensional SDS-PAGE in the small format, a few simple changes from Basic Protocols 1 and 2 are needed. Tube-gel adaptors for IEF, available from Hoefer and Bio-Rad, fit in the minigel unit (*UNIT 10.2*). Because the tube gels are much shorter (6 to 8 cm), the isoelectric focusing time is less; 2 to 4 hr at 500 V is usually adequate. Furthermore, less protein is generally required. Stacking gels are not normally required. The second-dimension gel is processed in the same way as a one-dimensional minigel (*UNIT 10.2*)

SUPPORT PROTOCOL

SOLUBILIZATION AND PREPARATION OF PROTEINS IN TISSUE SAMPLES

Materials *(see APPENDIX 1 for items with ✓)*

Tissue samples
✓ SDS/ or urea/solubilization buffer

Dounce homogenizer with pestles A and B
200-μl centrifuge tubes
Beckman 42.2-Ti rotor (or equivalent)

1. Weigh and place tissue samples in a Dounce homogenizer. Add 1.5 to 2.0 ml SDS/ or urea/solubilization buffer per 100 mg tissue. Homogenize using 50 strokes with pestle B, then 50 strokes with pestle A.

2. Let stand a few minutes, then transfer an aliquot to a 200-μl centrifuge tube. Centrifuge ≥2 hr at 100,000 × *g*, or 1 hr at >200,000 × *g*, 20°C. Save the supernatant (protein sample) to load onto the first-dimension gel in Basic Protocol 1.

   Boil samples in SDS/solubilization buffer 5 min before centrifugation. Never heat samples in urea/solubilization buffer. Centrifuge sample just prior to loading the first-dimension gel.

References: Anderson, 1988; Celis and Bravo, 1984; Dunbar, 1987; O'Farrell, 1975.

Contributors: Lonnie D. Adams and Sean R. Gallagher

Electroelution of Proteins from Stained Gels

BASIC PROTOCOL

Materials *(see APPENDIX 1 for items with ✓)*

SDS-polyacrylamide gel containing electrophoresed proteins (UNIT 10.2)
✓ Staining solution
✓ Destaining solution
✓ 10% dithiothreitol (DTT)
✓ Soaking buffer
✓ Elution buffer
✓ Dialysis buffer
50:50 (v/v) methanol/acetone

Electroelution apparatus (CBS Scientific)
✓ Spectrapor 6 dialysis membrane (Spectrum Medical)
Cartridge filter (Applied Biosystems)

NOTE: Distilled, deionized water should be used throughout this protocol.

1. Soak polyacrylamide gel in staining solution 15 to 20 min at room temperature with gentle shaking. Transfer to destaining solution 2 to 3 hr at 4°C with gentle shaking.

   For proteins containing acid-cleavable Asp-Pro sequences, stain with 4 M sodium acetate 1 hr at room temperature. No destaining in acidic solution is needed.

2. Cut out stained gel bands and soak in water 2 to 3 hr at 4°C, with several changes of water. Transfer each gel slice to a separate petri dish and cover with 10 ml water. Mince the gel into ~1 mm^3 pieces. Aspirate water with a Pasteur pipet.

3. Cover the ends of the elution cell wells with a disc of Spectrapor 6 dialysis membrane and tighten the caps (Fig. 10.4.1).

   Use a pore size less than or equal to the molecular weight of the protein being eluted. Carry out electroelution at room temperature, or at 4°C using Tris/acetate buffers if the protein is susceptible to digestion by trace levels of contaminating proteases.

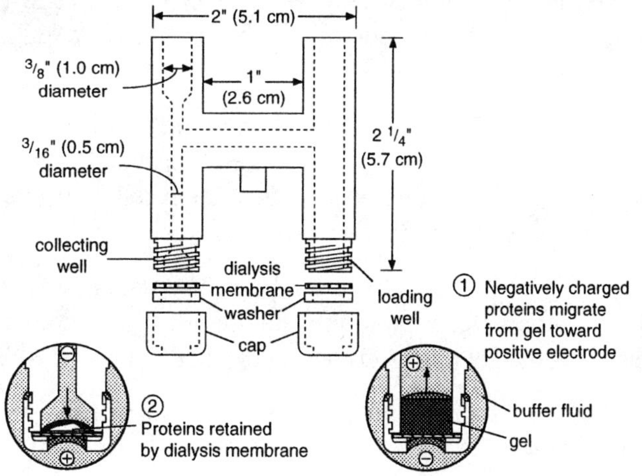

Figure 10.4.1 Elution cell.

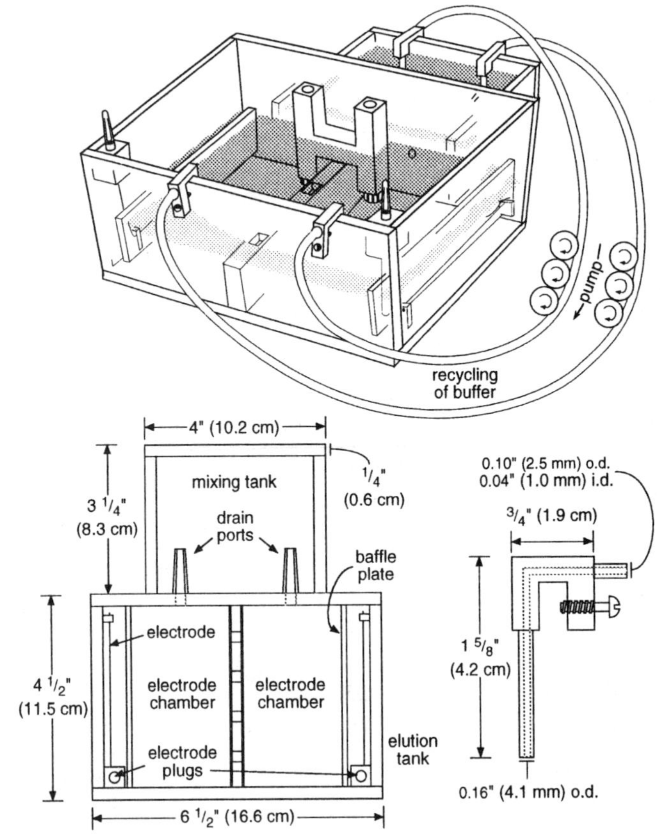

Figure 10.4.2 Elution apparatus and cell. Clockwise from top: operating schematic, port connector, and top view of assembly.

4. Transfer the minced gel pieces to the large side of an elution cell and cover with soaking buffer (if the protein was separated under reducing conditions, adjust solution to 0.1% DTT). Overlay the soaking buffer with elution buffer to just above the horizontal channel. Eliminate all air bubbles.

5. Place the cell in the elution tank and add elution buffer to a level slightly above the drain ports in each electrode chamber (Fig. 10.4.2). Add additional buffer to fill the smaller mixing tank.

6. Attach a two-channel peristaltic pump and adjust the flow rate so the buffer drips gently from the mixing tank into the elution tank. Remove any air bubbles from under the dialysis membranes of the elution cell with a bent Pasteur pipet, taking care not to puncture the dialysis membrane.

7. Soak the gel pieces 3 to 5 hr. Connect the electrodes, taking care to connect the cathodic (negative) electrode to the chamber of the elution tank with the cell side containing the gel pieces. Run at a constant voltage of 50 V for 12 to 16 hr.

8. Turn off the power supply and disconnect the electrodes. Replace the elution buffer in the tank with dialysis buffer. Reconnect the electrodes at a constant voltage of 80 V for 12 to 24 hr.

9. Turn off the power supply, disconnect the electrodes, and turn off the peristaltic pump. Remove the elution cell from the tank, and remove all buffer in the well

containing the eluted protein from above the blue dye-protein layer. Mix the remaining solution in the well using a 50-µl syringe with a bent, flat-tipped needle.

10. Transfer protein solution to a microcentrifuge tube. Rinse the collection well with 50 µl dialysis buffer and pool with the protein solution (final volume ~200 µl).

11. Lyophilize the solution in a Speedvac evaporator attached to a lyophilizer. The sample can be stored at −20°C.

12. Redissolve the pellet in 100 µl water and precipitate with 400 µl of 50:50 methanol/acetone. Store overnight at −20°C. Microcentrifuge to pellet the protein (save the supernatant in case the protein did not pellet).

13. Analyze 5% to 10% of the precipitated protein on a minislab gel using the Laemmli gel system (*UNIT 10.2*) to determine extent of protein elution, molecular weight, and recovery. Protein can be detected with Coomassie brilliant blue or silver stain (*UNIT 10.5*).

14. Apply the remainder of the sample to a cartridge filter in preparation for protein sequence analysis.

    Protein sequence analysis requires an automated protein sequencer (e.g., Applied Biosystems 476A or equivalent).

Reference: Hunkapiller and Lujan, 1986.

Contributor: John A. Smith

Staining Proteins in Gels

UNIT 10.5

The location of a protein in a gel can be determined by either Coomassie blue staining or silver staining. The former is easier and more rapid; however, silver staining methods are considerably more sensitive and thus can be used to detect smaller amounts of protein.

COOMASSIE BLUE STAINING

BASIC PROTOCOL 1

Detection of protein bands in a gel by Coomassie blue staining depends on nonspecific binding of the dye to the proteins. The detection limit is 0.3-1 µg/protein band.

Materials *(see APPENDIX 1 for items with ✓)*

Polyacrylamide gel (*UNIT 10.2*)
✓ Fixing solution
✓ Coomassie staining solution
✓ Destaining solution
7% (v/v) aqueous acetic acid

Cellophane membrane backing sheets (optional; Hoefer or Bio-Rad)
Whatman 3MM filter paper (optional)

NOTE: Deionized, distilled water should be used throughout this protocol.

1. Place polyacrylamide gel in a plastic container and cover with 3 to 5 gel volumes of fixing solution. Agitate slowly 2 hr on an orbital shaker.

2. Pour out fixing solution, cover gel with Coomassie staining solution and agitate slowly 4 hr.

3. Pour out staining solution. Rinse gel briefly with ~50 ml fixing solution.

4. Pour out fixing solution. Cover gel with destaining solution and agitate slowly 2 hr. Pour out destaining solution. Add fresh destaining solution and continue destaining until blue bands and a clear background are obtained. Store the gel in 7% acetic acid or water.

5. If desired, photograph gel (Support Protocol).

   Use a yellow-orange filter, such as a Wratten #8 or #9 filter series A with a fine-grained panchromatic halftone film, such as Kodak T-Max 100, for photography.

6. Alternatively, place gel on two sheets of Whatman 3MM filter paper and cover top with plastic wrap. Dry in a conventional gel dryer 1 to 2 hr at 80°C.

BASIC PROTOCOL 2

SILVER STAINING

Detection of protein bands in a gel by silver staining depends on binding of silver to various chemical groups (e.g., sulfhydryl and carboxyl moieties) in proteins. The detection limit is 2 to 5 ng/protein band.

Materials *(see* APPENDIX 1 *for items with ✓)*

✓ Fixing and destaining solutions
10% (v/v) glutaraldehyde (freshly prepared from 50% stock; Kodak)
✓ Silver nitrate solution (ammoniacal)
✓ Developing solution
Kodak Rapid Fix Solution A

NOTE: Wear gloves at all times to avoid fingerprint contamination.

1. Place polyacrylamide gel in a plastic container and add 5 gel volumes of fixing solution. Agitate slowly ≥30 min on an orbital shaker.

2. Pour out fixing solution. Add 5 gel volumes of destaining solution for ≥60 min, agitating slowly.

3. Pour out destaining solution. Add 5 gel volumes of 10% glutaraldehyde and agitate slowly 30 min in a fume hood.

   CAUTION: *Wear gloves and work only in a fume hood.*

4. Pour out glutaraldehyde solution. Wash gel ≥4 times with water, ≥30 min for each and preferably overnight for last wash. Agitate slowly.

5. Pour out water. Equilibrate gel with ~5 gel volumes of silver nitrate solution (to cover the gel) for 15 min with vigorous shaking.

   CAUTION: *Dispose of ammoniacal silver solution immediately by flushing with copious amounts of water, as it becomes explosive upon drying.*

6. Transfer gel to another plastic box and wash 5 times with deionized water, exactly 1 min for each wash. Agitate slowly.

7. Dilute 25 ml developing solution with 500 ml water. Transfer gel to another plastic box, add enough diluted developer to cover gel and shake vigorously

until bands reach desired intensity. If developer turns brown, change to fresh developer.

8. Transfer to Kodak Rapid Fix Solution A for 5 min. If necessary, swab gel surface with soaked cotton to remove residual silver deposits. Pour off Rapid Fix Solution and wash gel exhaustively in water (4 to 5 times).

9. Photograph gel (Support Protocol). Dry gel (Basic Protocol 1, step 6) or store in sealable plastic bag (6 to 12 months).

   Use a blue-green filter such as a Wratten #58 filter with Kodak T-Max 100. Photograph gels as soon as possible because there may be slight changes in color intensity and increases in nonspecific background.

NONAMMONIACAL SILVER STAINING

ALTERNATE PROTOCOL

This nonammoniacal silver staining procedure uses more stable solutions and detects certain proteins not stained by using the preceding protocol.

Additional Materials *(also see Basic Protocol 2; see* APPENDIX 1 *for items with ✓)*

5 µg/ml dithiothreitol (DTT)
0.1% silver nitrate (store in brown bottle at room temperature ~1 month)
✓ Carbonate developing solution
2.3 M citric acid

1. Place gel in a glass or polyethylene container and add 100 ml fixing solution. Agitate slowly 30 min on an orbital shaker.

2. Pour out fixing solution. Immerse gel in destaining solution and agitate slowly 30 min.

3. Pour out destaining solution. Cover gel with 50 ml of 10% glutaraldehyde and agitate slowly 10 min in a fume hood.

   CAUTION: *Wear gloves and work only in a fume hood.*

4. Pour out glutaraldehyde solution. Wash the gel *thoroughly* in several changes of water for 2 hr to ensure low background levels.

5. Pour out water. Soak gel in 100 ml of 5 µg/ml DTT for 30 min.

6. Pour out DTT solution. Without rinsing, add 100 ml of 0.1% silver nitrate and agitate slowly 30 min.

7. Pour out silver nitrate solution. Wash gel once quickly with a small amount of water, then twice rapidly with a small amount of carbonate developing solution.

8. Soak gel in 100 ml of carbonate developing solution and agitate slowly until desired level of staining is achieved.

9. Stop staining by adding 5 ml of 2.3 M citric acid per 100 ml of carbonate developing solution for 10 min and agitate slowly.

10. Pour off solution. Wash several times in water, agitating slowly 30 min.

11. Photograph gel (Support Protocol). Store by soaking 10 min in 0.03% sodium carbonate. Wrap in plastic wrap or seal in a heat-sealable bag.

BASIC PROTOCOL 3

RAPID SILVER STAINING

This protocol is rapid and gives low background but may not be quite as sensitive in detecting very small proteins because there is no glutaraldehyde fixation.

Additional Materials *(also see Basic Protocol 2; see* APPENDIX 1 *for items with ✓)*

✓ Formaldehyde fixing solution
0.2 g/liter sodium thiosulfate ($Na_2S_2O_3$)
✓ Thiosulfate developing solution
✓ Drying solution
Dialysis membrane soaked in 50% methanol

1. Place gel in a plastic container and add 50 ml formaldehyde fixing solution. Agitate slowly 10 min on an orbital shaker.

   Times indicated are flexible and are appropriate for a 0.75-mm × 5.5-cm × 8-cm, 12.5% acrylamide slab gel. Each gel is placed in an 8 × 14–cm plastic container.

2. Pour out fixing solution. Wash gel twice with water, 5 min for each wash. Agitate slowly for each wash.
3. Pour out water. Soak gel 1 min in 50 ml of 0.2 g/liter $Na_2S_2O_3$, agitating slowly.
4. Pour out $Na_2S_2O_3$ solution. Wash gel twice with water, 20 sec/wash.
5. Pour out water. Soak gel 10 min in 50 ml of 0.1% silver nitrate, agitating slowly.
6. Pour out silver nitrate solution. Wash gel with water and then with a small volume of thiosulfate developing solution.
7. Soak gel in 50 ml fresh thiosulfate developing solution and agitate slowly until band intensities are adequate (~1 min). Development continues a little after stopping (next step), so do not overdevelop here.
8. Add 5 ml of 2.3 M citric acid per 100 ml thiosulfate developing solution and agitate slowly 10 min.
9. Pour off solution. Wash gel in water, agitating slowly 10 min.
10. Pour off water. Soak gel 10 min in 50 ml drying solution.
11. Sandwich gel between two pieces of wet dialysis membrane on a glass plate. Clamp edges of plate with notebook clamps and dry overnight at room temperature.

SUPPORT PROTOCOL

GEL PHOTOGRAPHY

Any good single-lens reflex camera attached to a copy stand will give good results. If larger-format cameras are available, sheet film will give spectacular resolution. The light box must produce relatively even lighting.

Kodak T-Max 100 and 400 are extremely high-resolution films that work well for gel photography. For contrast enhancement of Coomassie blue–stained gels, photographing the gels through a deep-yellow to yellow-orange filter (Wratten #8 or #9 or a yellow-orange Cokin filter) is recommended. Silver-stained gels are photographed with a blue-green filter (Wratten #58 or a blue Cokin filter). Develop according to manufacturer's instructions. Medium-contrast resin-coated paper is recommended for printing.

The instant films from Polaroid are ideal for fast, high-quality photographs of gels for laboratory notebooks and publications. Typically, Type 57 (4 × 5–in., MP4 camera) or Type 667 (3¼ × 4¼–in., DS 34 camera) black and white film is used for documentation at a shutter speed of 1/60 to 1/30 sec with an aperture of F16. Type 55 and 665 positive/negative films not only provide a high-quality print, but also a fine-grain, medium-format negative that can be used to produce multiple photographs of the gel in any size; for these films, use a shutter speed of ¼ to ½ sec and an aperture of F16. Aperture settings of F11 or higher should be used for sharp photographs. If the picture is too light, increase the shutter speed or go to a higher F number. If the picture is too dark, decrease shutter speed or go to a lower F number (adapted from *Polaroid Guide to Instant Imaging*).

References: Bloom et al., 1987; Merril et al., 1984; Morrissey, 1981.

Contributors: Joachim Sasse and Sean R. Gallagher

Detection of Proteins on Blot Transfer Membranes

UNIT 10.6

Proteins are stained after electroblotting from polyacrylamide gels to blot transfer membranes. If duplicate samples are prepared, half the membrane can be stained to determine the efficiency of transfer to the membrane and the other half can be used for immunoblotting (*UNIT 10.7*). Table 10.6.1 provides detection limits and compatible blot transfer membranes and gels.

INDIA INK STAINING

BASIC PROTOCOL

The transferred bands will appear as black bands on a gray background.

Materials

Tween 20 solution: 0.3% (v/v) Tween 20 in PBS, pH 7.4
India ink solution: 0.1% India ink (e.g., Pelikan 17 black) in Tween 20 solution

NOTE: Deionized, distilled water should be used throughout this protocol.

1. Place blot transfer membrane(s) in a plastic box on an orbital shaker and wash in Tween 20 solution three times, 30 min each, at 37°C. Continue to wash the membrane in Tween 20 solution two times, 30 min each, at room temperature.

2. Stain the membrane in India Ink solution 3 hr or overnight.

Table 10.6.1 Properties and Compatibilities of "On Blot" Membrane Stains[a]

Stain	Detection limit (ng)	Membrane types			Gel types	
		Nitrocellulose	Nylon	PVDF	SDS-PAGE	Native PAGE
India ink	50	+	–	+	+	+
Gold	3	+	–	+	+	n.d.

[a]Abbreviations: n.d., not determined; PVDF, polyvinylidene difluoride; SDS-PAGE, sodium dodecyl sulfate–polyacrylamide gel electrophoresis.

3. Rinse the membrane twice in Tween 20 solution. Destain in Tween 20 solution until the protein bands appear black against a gray background. Air dry.

ALTERNATE PROTOCOL

GOLD STAINING

The transferred proteins will appear as red bands on an almost white background.

Additional Materials *(also see Basic Protocol)*

AuroDye colloidal gold solution (Janssen Pharmaceutica)

1. Wash blot transfer membrane(s) (Basic Protocol, step 1).

   Do not attempt to stain nylon membranes.

2. Stain the membrane in AuroDye colloidal gold sol 3 hr or overnight with continuous shaking.
3. Rinse the membrane briefly in water and air dry.

SUPPORT PROTOCOL

ALKALI ENHANCEMENT OF PROTEIN STAINING

A brief pretreatment of nitrocellulose-bound protein with alkali enhances subsequent staining with either India ink or colloidal gold.

Materials *(also see Basic Protocol; see APPENDIX 1 for items with ✓)*

1% (w/v) KOH
PBS
Glass or pyrex dish

1. Place nitrocellulose blot or dot membrane with transferred proteins in a glass dish containing enough 1% KOH to cover the membranes. Soak 5 min at 20°C.
2. Rinse 10 min in PBS twice.

References: Hancock and Tsang, 1983; Moeremans et al., 1985; Sutherland and Skerritt, 1986.

Contributors: Joachim Sasse and Sean Gallagher

UNIT 10.7

Immunoblotting and Immunodetection

Immunoblotting (often referred to as western blotting) is used to identify specific antigens recognized by polyclonal or monoclonal antibodies.

BASIC PROTOCOL 1

PROTEIN BLOTTING WITH TANK TRANSFER SYSTEMS

Materials *(see APPENDIX 1 for items with ✓)*

Samples for analysis
Protein molecular weight standards (*UNIT 10.2*), prestained (Sigma or Bio-Rad) or biotinylated (Vector Laboratories or Sigma)
✓ Transfer buffer appropriate for membrane
Powder-free gloves
Scotch-Brite pads (3M) or equivalent sponge
Whatman 3MM filter paper or equivalent

Transfer membrane: 0.45-µm nitrocellulose (Millipore or Schleicher & Schuell), PVDF (Millipore), neutral nylon (Pall Biodyne), *or* positively charged nylon (Pall Biodyne; Bio-Rad) membrane
Electroblotting apparatus (EC Apparatus, Bio-Rad, or Hoefer)
Paper-Mate pen

NOTE: Deionized, distilled water should be used throughout this protocol.

NOTE: Use gloves when manipulating filter papers, gels, and membranes. Oil from hands blocks the transfer.

1. Prepare antigenic samples and separate proteins using small or standard-sized one- or two-dimensional gels. Include prestained or biotinylated protein molecular weight standards in one or more gel lanes.

2. When electrophoresis is complete, disassemble gel sandwich, and remove stacking gel. Equilibrate gel 30 min at room temperature in appropriate transfer buffer.

3. Fill a tray large enough to hold the plastic transfer cassette with transfer buffer so that cassette is covered (Fig. 10.7.1).

4. On bottom half of plastic transfer cassette, assemble part of transfer sandwich:

    Scotch-Brite pad or sponge
    A sheet of filter paper cut to same size as gel and prewet with transfer buffer
    Gel.

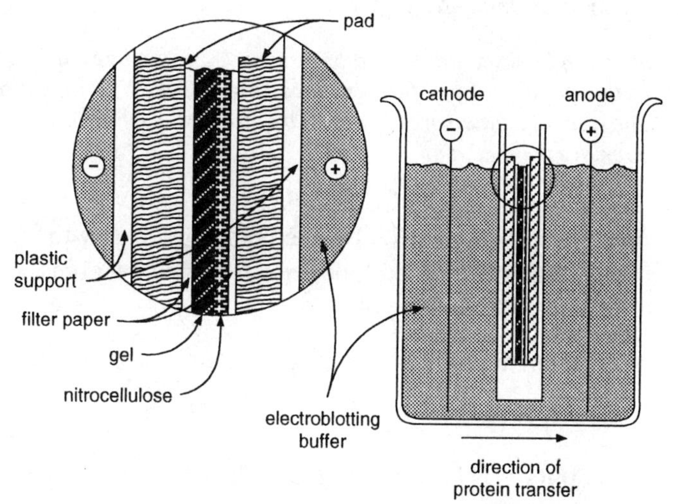

Figure 10.7.1 Immunoblotting with a tank blotting unit. The polyacrylamide gel containing the protein is laid on a sheet of filter paper. The uncovered side of the gel is overlaid with a sheet of membrane precut to the size of the gel plus 1 to 2 mm on each edge, then this membrane is overlaid with another sheet of filter paper. The filter paper containing the gel and membrane is sandwiched between Scotch-Brite pads. This sandwich is placed in a plastic support, and the entire assembly is placed in a tank containing transfer buffer. For transfer of negatively charged protein, the membrane is positioned on the anode side of the gel. For transfer of positively charged protein, the membrane is placed on the cathode side of the gel. Charged proteins are transferred electrophoretically from the gel onto the membrane. Transfer is achieved by applying a voltage of 100 V for 1 to 2 hr (with cooling) or 14 V overnight.

Remove any air bubbles between gel and filter paper by gently rolling a test tube or glass rod over surface of gel.

The side of the gel touching the paper arbitrarily becomes the cathode side of the gel (i.e., ultimately toward the negative electrode when positioned in the tank).

5. Prepare transfer membrane. Cut membrane to same size as gel plus 1 to 2 mm on each edge. Place nylon or nitrocellulose membrane into distilled water slowly, with one edge at a 45° angle. The water will wick up into the membrane, wetting the entire surface. Equilibrate 10 to 15 min in transfer buffer. Immerse PVDF membrane briefly in 100% methanol, then equilibrate in transfer buffer 10 to 15 min.

6. Moisten surface of gel with transfer buffer. Place prewetted membrane directly on top side of gel (i.e., anode side) and remove all air bubbles.

7. Wet another piece of Whatman 3MM filter paper, place on anode side of membrane, and remove all air bubbles. Place another Scotch-Brite pad or sponge on top of this filter paper.

8. Complete assembly by locking top half of the transfer cassette into place (Fig. 10.7.1).

9. Fill tank with transfer buffer and place transfer cassette containing sandwich into electroblotting apparatus in correct orientation. Connect leads of power supply to corresponding anode and cathode sides of electroblotting apparatus.

10. Electrophoretically transfer proteins from gel to membrane for 30 min to 1 hr at 100 V with cooling or overnight at 14 V (constant voltage), in a cold room.

11. Turn off the power supply and disassemble the apparatus. Remove membrane from blotting apparatus and note orientation by cutting a corner or marking with a soft lead pencil or Paper-Mate pen.

    Membranes can be dried and stored in resealable plastic bags at 4°C for 1 year or longer at this point. Prior to further processing, dried PVDF membranes must be placed into a small amount of 100% methanol to wet the membrane, then in distilled water to remove the methanol.

12. Stain gel for total protein with Coomassie blue to verify transfer efficiency. If desired, stain nitrocellulose or PVDF membrane reversibly to visualize transferred proteins (Support Protocol), or irreversibly with Coomassie blue, India ink, naphthol blue, or colloidal gold.

13. Proceed with immunoprobing and visual detection of proteins (Basic and Alternate Protocols 2 and 3).

ALTERNATE PROTOCOL 1

PROTEIN BLOTTING WITH SEMIDRY SYSTEMS

Even and efficient transfer of most proteins is also possible with semidry blotting (Fig. 10.7.2), a convenient alternative to tank transfer systems.

Additional Materials *(also see Basic Protocol 1)*

Six sheets of Whatman 3MM filter paper or equivalent, cut to size of gel and saturated with transfer buffer

Semidry transfer unit (Hoefer, Bio-Rad, or Sartorius)

1. Prepare samples and separate proteins using small or standard-sized one- or two-dimensional gels.

Because transfer efficiency depends on many factors (e.g., gel concentration and thickness, protein size, shape, and net charge), results may vary. The table below is a guideline for 0.75-mm-thick SDS-PAGE gels transferred by semidry blotting.

Percent acrylamide (resolving gel)	Size range transferred (~100% efficiency)
5–7	29–150 kD
8–10	14–66 kD
13–15	<36 kD
18–20	<20 kD

2. Prepare transfer membrane (Basic Protocol 1, step 5).
3. Disassemble gel sandwich. Remove and discard stacking gel.
4. Assemble the transfer stack:

    Mylar mask (optional for some equipment)
    3 sheets of filter paper saturated with transfer buffer
    Equilibrated transfer membrane
    Gel
    3 sheets of filter paper.

    Roll out bubbles as each compnent is added to the stack.

    Multiple gels can be transferred using semidry blotting. Simply put a sheet of porous cellophane (Hoefer) or dialysis membrane (Bio-Rad or Sartorius) equilibrated with transfer buffer between each transfer stack (Fig. 10.7.2).

5. Place top electrode onto transfer stack.

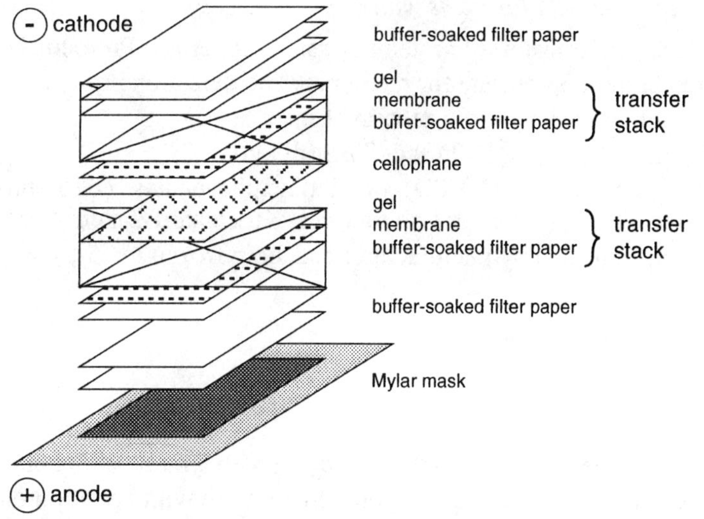

Figure 10.7.2 Immunoblotting with a semidry transfer unit. Generally, the lower electrode is the anode, and one gel is transferred at a time. A Mylar mask (optional in some units) is put in place on the anode. This is followed by three sheets of transfer buffer–soaked filter paper, the membrane, the gel, and finally, three more sheets of buffer-soaked filter paper. To transfer multiple gels, construct transfer stacks as illustrated, and separate each with a sheet of porous cellophane. For transfer of negatively charged protein, the membrane is positioned on the anode side of the gel. For transfer of positively charged protein, the membrane is placed on the cathode side of the gel. Transfer is achieved by applying a maximum current of 0.8 mA/cm^2 of gel area. For a typical minigel (8 × 10 cm) and standard-sized gel (14 × 14 cm), this means 60 and 200 mA, respectively.

6. Carefully connect high-voltage leads to the power supply (see UNIT 10.2 for safety precautions). Apply constant current to initiate protein transfer. Transfers of 1 hr are generally sufficient.

   In general, do not exceed 0.8 mA/cm² of gel area. For a typical minigel (8 × 10 cm) and standard-sized gel (14 × 14 cm) this means ~60 and 200 mA, respectively.

   Monitor the temperature of the transfer unit directly above the gel by touch. The unit should not exceed 45°C. If the unit is hot, lower the current.

7. After transfer, turn off power supply and disassemble unit. Remove membrane from transfer stack, marking orientation. Proceed with staining and immunoprobing.

SUPPORT PROTOCOL

REVERSIBLE STAINING OF TRANSFERRED PROTEINS

Additional Materials *(also see Basic Protocol 1; see APPENDIX 1 for items with ✓)*

✓ Ponceau S solution

1. Place nitrocellulose or PVDF membrane in Ponceau S solution 5 min at room temperature.
2. Destain 2 min in water. Photograph membrane if required (UNIT 10.5) and mark the molecular-weight-standard band locations with indelible ink (e.g., Paper-Mate).
3. Completely destain membrane by soaking an additional 10 min in water.

BASIC PROTOCOL 2

IMMUNOPROBING WITH DIRECTLY CONJUGATED SECONDARY ANTIBODY

Materials *(see APPENDIX 1 for items with ✓)*

Membrane with transferred proteins (Basic or Alternate Protocol 1)
✓ Blocking buffer appropriate for membrane and detection protocol
Primary antibody specific for protein of interest
✓ TTBS (nitrocellulose or PVDF) or TBS (nylon)
Horseradish peroxidase (HRPO)- or alkaline phosphatase (AP)-anti-Ig conjugate (Cappel, Vector Laboratories, Kirkegaard & Perry, or Sigma; dilute as indicated by manufacturer and store frozen in 25-µl aliquots until used)

Heat-sealable plastic bag
Powder-free gloves
Plastic box

1. Place membrane in heat-sealable plastic bag with 5 ml blocking buffer and seal bag. Incubate 30 min to 1 hr at room temperature with agitation on an orbital shaker or rocking platform.
2. Dilute primary antibody in blocking buffer (1:10 to 1:100,000 depending on the antibody and detection system; Fig. 10.7.3).

   To determine the appropriate concentration of the primary antibody, a dilution series is easily performed with membrane strips. Separate antigens on a preparative gel and immunoblot the entire gel. Cut 2- to 4-mm strips by hand or with a membrane cutter (Schleicher and Schuell, Inotech) and incubate individual strips in a serial

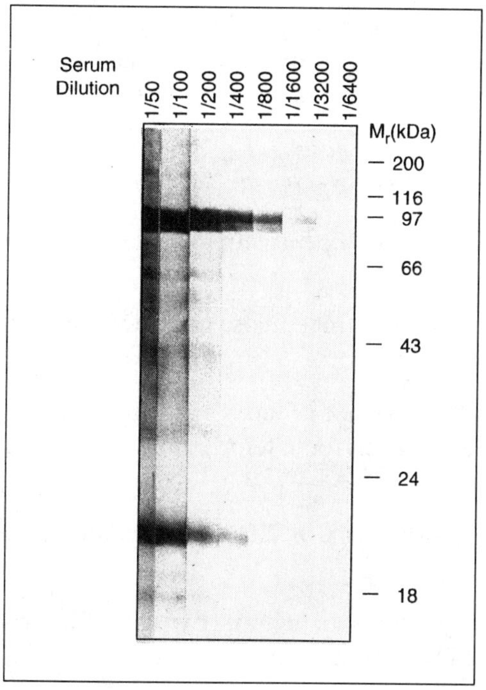

Figure 10.7.3 Serial dilution of primary antibody directed against the 97-kDa catalytic subunit of the plant plasma membrane ATPase. Blot was developed with HRPO-coupled avidin-biotin reagents according to the second alternate protocol and visualized with 4-chloro-1-naphthol (4CN). Note how background improves with dilution.

dilution of primary antibody. The correct dilution should give low background and high specificity (Fig. 10.7.3).

3. Open bag and pour out blocking buffer. Replace with diluted primary antibody and incubate 30 min to 1 hr at room temperature with constant agitation.
4. Remove membrane from plastic bag with gloved hand. Place in plastic box and wash four times by agitating with 200 ml TTBS (nitrocellulose or PVDF) or TBS (nylon), 10 to 15 min each time.
5. Dilute HRPO- or AP-anti-Ig conjugate in blocking buffer (1:200 to 1:2000, see supplier's instructions).
6. Place membrane in new heat-sealable plastic bag, add diluted HRPO- or AP-anti-Ig conjugate, and incubate 30 min to 1 hr at room temperature with constant agitation.
7. Remove membrane from bag and wash as in step 4. Develop according to appropriate visualization protocol (Basic Protocol 3 or Alternate Protocol 2 or 3).

IMMUNOPROBING WITH AVIDIN-BIOTIN COUPLING TO THE SECONDARY ANTIBODY

ALTERNATE PROTOCOL 2

Additional Materials *(also see Basic Protocol 2; see* APPENDIX 1 *for items with ✓)*

✓ Blocking buffer appropriate for membrane and detection protocol
✓ TTBS (nitrocellulose or PVDF) *or* TBS (neutral or positively charged nylon)
Vectastain ABC (peroxidase) or ABC-AP (alkaline phosphatase) kit (Vector Laboratories) containing:
- Reagent A (avidin)
- Reagent B (biotinylated HRPO or AP)
- Biotinylated secondary antibody

1. Equilibrate membrane in appropriate blocking buffer with constant agitation using an orbital shaker or rocking platform. For nitrocellulose and PVDF, incubate 30 to 60 min at room temperature. For nylon, incubate ≥2 hr at 37°C.

   Tween 20/TTBS is well suited for avidin-biotin systems. For nylon, protein-binding agents are recommended. Casein, nonfat dry milk, serum, and some grades of BSA may interfere with the formation of the avidin-biotin complex and should not be used in the presence of avidin or biotin reagents. Azide is a peroxidase inhibitor and should not be used as a preservative.

2. Prepare primary antibody solution in TTBS (nitrocellulose or PVDF) or TBS (nylon).
3. Remove membrane from blocking buffer and place in enough primary antibody solution to cover membrane. Incubate 30 min at room temperature with gentle rocking.
4. Wash membrane three times over a 15-min span in TTBS (nitrocellulose or PVDF) or TBS (nylon).
5. Prepare biotinylated secondary antibody solution by diluting two drops biotinylated antibody with 50 to 100 ml TTBS (nitrocellulose or PVDF) or TBS (nylon).
6. Transfer membrane to secondary antibody solution. Incubate 30 min at room temperature with slow rocking, then wash as in step 4.
7. While membrane is being incubated with secondary antibody, prepare avidin-biotin-HRPO or -AP complex. Mix two drops Vectastain reagent A and two drops reagent B into 10 ml TTBS (nitrocellulose or PVDF) or TBS (nylon). Incubate 30 min at room temperature, then further dilute to 50 ml with TTBS or TBS.
8. Transfer membrane to avidin-biotin-enzyme solution. Incubate 30 min at room temperature with slow rocking, then wash over a 30-min span as in step 4.
9. Develop membrane according to the appropriate visualization protocol (Basic or Alternate Protocol 3).

BASIC PROTOCOL 3

VISUALIZATION WITH CHROMOGENIC SUBSTRATES

Materials *(see APPENDIX 1 for items with ✓)*

Membrane with transferred proteins and probed with antibody-enzyme complex (Basic or Alternate Protocol 2)
✓ Tris-buffered saline (TBS)
Chromogenic visualization solution (Table 10.7.1)

1. Wash membrane 15 min at room temperature in 50 ml TBS to remove excess phosphate and Tween 20.
2. Place membrane into chromogenic visualization solution. Bands should appear in 10 to 30 min.
3. Terminate reaction by washing membrane in distilled water. Air dry and photograph (*UNIT 10.5*) for a permanent record.

Table 10.7.1 Chromogenic and Luminescent Visualization Systems[a]

System	Reagent[b]	Reaction/Detection	Comments[c]
Chromogenic			
HRPO-based	4CN	Oxidized products form purple precipitate	Not very sensitive (Tween 20 inhibit reaction); fades rapidly upon exposure to light
	$DAB/NiCl_2$[d]	Forms dark brown precipitate	More sensitive than 4CN but potentially carcinogenic; resulting membrane easily scanned
	TMB[e]	Forms dark purple stain	More stable, less toxic than $DAB/NiCl_2$; may be somewhat more sensitive[e]; can be used with all membrane types; kits available from Kirkegaard & Perry, TSI, Moss, and Vector Laboratories
AP-based	BCIP/NBT	BCIP hydrolysis produces indigo precipitate after oxidation with NBT; reduced NBT precipitates; dark blue-gray stain results	More sensitive and reliable than other AP-precipitating substrates; note that phosphate inhibits AP activity.
Luminescent			
HRPO-based	Luminol/H_2O_2/ *p*-iodophenol	Oxidized luminol substrate gives off blue light; *p*-iodophenol increases light output	Very convenient, sensitive system; reaction detected within a few seconds to an hour
AP-based	Substituted 1,2-dioxetane-phosphates (e.g., AMPPD, CSPD, Lumigen-PPD, Lumi-Phos 530[f])	Dephosphorylated substrate gives off light	Protocol described gives reasonable sensitivity on all membrane types; consult instructions of reagent manufacturer for maximum sensitivity and minimum background

[a]Abbreviations: AMPPD or Lumigen-PPD, disodium 3-(4-methoxyspiro{1,2-dioxetane-3,2′-tricyclo[3.3.1.1^{3,7}]decan}-4-yl)phenyl phosphate; AP, alkaline phosphatase; BCIP, 5-bromo-4-chloro-3-indolyl phosphate; 4CN, 4-chloro-1-napthol; CSPD, AMPPD with substituted chlorine moiety on adamantine ring; DAB, 3,3′-diaminobenzidine; HRPO, horseradish peroxidase; NBT, nitroblue tetrazolium; TMB, 3,3′,5,5′-tetramethylbenzidine.

[b]Recipes and suppliers are listed in APPENDIX 1 and APPENDIX 4 unless indicated under "comments" in this table.

[c]See commentary for further details.

[d]$DAB/NiCl_2$ can be used without the nickel enhancement, but it is much less sensitive.

[e]McKimm-Breschkin (1990) reported that by first treating nitrocellulose filters with 1% dextran sulfate for 10 min in 10 mM citrate-EDTA (pH 5.0), TMB precipitates onto the membrane with a sensitivity much greater than 4CN or DAB, and equal to or better than that of BCIP/NBT.

[f]Lumi-Phos 530 contains dioxetane phosphate, $MgCl_2$, CTAB, and fluorescent enhancer in a pH 9.6 buffer.

VISUALIZATION WITH LUMINESCENT SUBSTRATES

ALTERNATE PROTOCOL 3

Additional Materials *(also see Basic Protocol 3; see* APPENDIX 1 *for items with ✓)*

- Luminescent substrate buffer: 50 mM Tris·Cl, pH 7.5 (HRPO) *or* dioxetane phosphate substrate buffer (alkaline phosphatase)
- Nitro-Block solution (AP reactions only): 5% (v/v) Nitro-Block (Tropix) in dioxetane phosphate substrate buffer, prepared just before use
- Luminescent visualization solution (Table 10.7.1)
- Clear plastic wrap

1. Equilibrate membrane in two 15-min washes with 50 ml substrate buffer.

2. *For alkaline phosphatase reactions using nitrocellulose or PVDF membranes:* Incubate 5 min in Nitro-Block solution, followed by 5 min in 50 ml substrate buffer.

   Nitro-Block enhances light output from the dioxetane substrate in reactions using AMPPD, CSPD, or Lumigen-PPD concentrate. It is required for nitrocellulose and recommended for PVDF membranes. It is not needed for Lumi-Phos 530, alkaline phosphate reactions on nylon membranes, or peroxidase–based reactions on any type of membrane. Lumi-Phos 530 is not recommended for nitrocellulose membranes.

3. Transfer membrane to visualization solution. Soak 30 sec (HRPO reactions) to 5 min (alkaline phosphatase reactions).

4. Remove membrane, drain, and place face down on a sheet of clear plastic wrap. Fold wrap back onto membrane to form a liquid-tight enclosure.

5. In a darkroom, place membrane face down onto film and autoradiograph (*APPENDIX 3*) for a few seconds to overnight depending upon the sample.

6. If desired, wash membrane in two 15-min washes of 50 ml TBS and process for chromogenic development (Basic Protocol 3).

References: Bejurrum and Schafer-Nielsen, 1986; Gillespie and Hudspeth, 1991; Harlow and Lane, 1988; Schneppenheim et al., 1991.

Contributors: Sean Gallagher, Scott E. Winston and Steven A. Fuller, and John G.R. Hurrell

UNIT 10.8

Gel-Filtration Chromatography

This method separates proteins in order of large to small molecules. Table 10.8.1 describes common problems observed during gel-filtration chromatography.

BASIC PROTOCOL

GEL FILTRATION TO SEPARATE PROTEINS

Materials *(see APPENDIX 1 for items with ✓)*

GF gel matrix (Support Protocol and Table 10.8.2)
✓ GF buffer
GF protein standards (Table 10.8.3)

Buchner or sintered-glass funnel
GF chromatography column (Support Protocol)
Column extension or gel reservoir
Buffer reservoir
Peristaltic pump
UV detector
Fraction collector

1. If the GF gel matrix is a dry powder, swell completely in GF buffer. DO NOT stir with a magnetic stirrer. Allow the gel to settle and aspirate or decant the fine particles that have not settled to the gel bed. Alternatively, follow manufacturer's instructions. For a preswollen GF gel matrix, wash with excess buffer on a Buchner or sintered-glass funnel to remove preservatives. Resuspend the settled gel in an equal volume of GF buffer, pour the slurry into a filtration flask, and degas the suspension before use.

Table 10.8.1 Troubleshooting Guide for Conventional Chromatography[a]

Problem	Cause	Solution
Air bubbles in gel	Matrix not degassed before pouring gel; column moved from cold to warm location	Repack column
Cracks in gel	Column ran dry	Repack column and check tubing and end fittings for air leaks
No column flow	Bed support, end fittings, or tubing is blocked	Check for gel particle clogs or an air lock
Reduced column flow	See "no column flow" above;	
	Insoluble particles in sample	Filter or centrifuge sample prior to application
	Compressed column bed	Repack column at reduced hydrostatic pressure or after prolonged use
	Microbial growth	Store column in 0.02% sodium azide
	Plugging by fine gel particles	Resuspend gel matrix and aspirate the "fines" from a settling gel slurry
Gel particles in fractions	Bed support is loose or broken	Dismantle column and check bed support; repack column
Poor recovery	Specific or nonadsorption	Use different chromatographic matrix
	Microbial growth	Store column in 0.02% sodium azide
	Precipitation of protein; loss of activity due to removal of an interacting subunit or cofactor	Use buffer in which protein is soluble during the separation
Poor resolution	Microbial growth	Store column in 0.02% sodium azide
	Proteins or lipids precipitated	Remove 1 cm of gel from top of gel bed and discard
	High sample viscosity	Sample viscosity should be ≤2× the buffer viscosity
Poor resolution	Sample volume too large	Sample volume should be <5% of total bed volume
	Column too short	Optimize based on manufacturer specifications
	Flow rate too high	Maintain at a linear flow rate of 2 to 10 cm/hr; column flow rate (ml/hr) = linear flow rate × column cross-sectional area (cm^2)
	Poor column packing; column not vertical	Repack column using correct hydrostic pressure to maintain correct linear flow rate (see above)
	Gel particles too large	Use very fine or fine particle sizes

[a]Based on information contained in *Gel Filtration: Theory and Practice* from Pharmacia.

Table 10.8.2 Fractionation Ranges of Commonly Used Gel Filtration Matrices[a]

Gel type[b]	Fractionation range (molecular weight)
Sephadex G-10	≤700
Bio-Gel P-4	800–4,000
Sephadex G-25	1,000–5,000
Bio-Gel P-6	1,000–6,000
Bio-Gel P-10	1,500–20,000
Sephadex G-50	1,500–30,000
Bio-Gel P-30	2,500–40,000
Bio-Gel P-60	3,000–60,000
Sephadex G-75	3,000–70,000
Ultrogel AcA 54	5,000–70,000
Sephadex G-100	$4,000–1 \times 10^5$
Bio-Gel P-100	$5,000–1 \times 10^5$
Ultrogel AcA 44	$10,000–1.3 \times 10^5$
Sephadex G-200	$5,000–2.5 \times 10^5$
Sephacryl S-200 Superfine	$5,000–2.5 \times 10^5$
Bio-Gel A-0.5m	$<10,000–5 \times 10^5$
Ultrogel AcA 34	$20,000–3.5 \times 10^5$
Bio-Gel P-200	$30,000–2 \times 10^5$
Bio-Gel A-1.5m	$<10,000–1.5 \times 10^6$
Sephacryl S-300	$10,000–1.5 \times 10^6$
Sepharose 6B	$10,000–4 \times 10^6$
Ultrogel AcA 22	$1 \times 10^5–1.2 \times 10^6$
Bio-Gel A-1.5m	$40,000–15 \times 10^6$
Sepharose 4B	$60,000–20 \times 10^6$
Sepharose 2B	$70,000–40 \times 10^6$
Bio-Gel A-50m	$100 \times 10^5–50 \times 10^6$

[a]The fractionation range indicates the molecular size of a protein expected to elute at an elution volume equal to the bed volume (number on left) and the molecular size of a protein expected to be totally excluded from the column and to elute at the void volume (number on right). The fractionation ranges given are for dilute aqueous buffers and will differ for separations carried out under denaturing conditions (e.g., 6 M guanidine·Cl, 8 M urea, or detergents).

[b]Sephadex, Sephacryl and Sepharose are available from Pharmacia. Bio-Gel P and Bio-Gel A are available from Bio-Rad. Ultrogel AcA is available from Sepracor.

2. Mount the column vertically, as determined with a level, on a stable laboratory stand in a constant-temperature environment away from traffic or direct sunlight. Inject GF buffer with a syringe into the column outlet tubing until it is just above the bed support screen. Leave the syringe in place to block the end of the outlet tubing.

    Good resolution is achieved with columns ≥50 cm in length with an inner diameter of 1-2.5 cm.

3. Fill the column with the gel suspension to the desired bed height by pouring it onto a glass rod placed against the inner column wall. After the column is packed to the desired bed height, carefully pipet a 1-cm layer of buffer onto the top of

Table 10.8.3 Molecular Weights of Protein Standards for Gel Filtration[a]

Protein	Molecular weight
Cytochrome *c*	11,700
Myoglobin (sperm whale)	16,800
Trypsinogen (PMSF-treated)	24,000
Carbonic anhydrase	29,000
Ovalbumin	45,000
Hemoglobin	64,500
Bovine serum albumin	66,000
Transferrin	74,000
Immunoglobulin G	158,000
Fibrinogen	341,000
Ferritin	470,000
Thyroglobulin	670,000

[a]Protein standards are commercially available in kits.

the gel. Connect the buffer reservoir, remove the syringe from the column outlet tubing, and wash the column with 2 to 3 bed volumes of buffer.

Poorly packed columns result from packing the column in stages and from using a thin gel suspension.

4. Drain the buffer down to the top of the gel and close the outlet tubing. Apply protein sample (0.2 to 0.5 mg/ml) in a buffer volume equal to 1% to 5% of the total column bed volume. Open the outlet tubing and allow the sample to penetrate into the bed. Rinse the column wall above the gel with buffer. Pipet a 1-cm layer of buffer onto the top of the gel, reconnect the buffer reservoir, and begin the elution.

5. Collect fractions, each equal to 1% of the total bed volume. Measure the A_{280} of the individual fractions. Determine the biological activity using aliquots from individual fractions. Use SDS-PAGE (*UNIT 10.2*) to monitor the number of proteins and their quantity in selected fractions. Pool the fractions containing the desired protein(s).

   Proteins containing tyrosine and tryptophan residues absorb at 280 nm.

6. Wash the column with 2 to 3 additional bed volumes of GF buffer and store in GF buffer containing 0.02% sodium azide. The column can be used indefinitely.

GEL FILTRATION TO DESALT PROTEINS

ALTERNATE PROTOCOL

Additional Materials *(also see Basic Protocol; see APPENDIX 1 for items with ✓)*

✓ Desalting solution

1. Repeat the Basic Protocol using Sephadex G-25 or Bio-Gel P-6 for desalting proteins. Apply the sample in a volume that is ≤25% of the total bed volume.

2. The protein will elute in the void volume and can be concentrated by lyophilization.

   Void volume equals ~⅓ of bed volume.

SUPPORT PROTOCOL

SELECTING A GF GEL MATRIX AND COLUMN

For a protein of known molecular weight, select a gel matrix on which the protein will elute midway in the fractionation range (Table 10.8.2). For a protein of unknown molecular weight, a broad fractionation range should first be selected. If this protein elutes in the void volume, select another matrix with a higher exclusion limit. If this protein elutes near the bed volume, select another matrix that will separate lower-molecular-weight proteins.

References: Fischer, 1980; Pharmacia, Gel Filtration; Bio-Rad Price List L.

Contributor: John A. Smith

UNIT 10.9

Ion-Exchange Chromatography

BASIC PROTOCOL

Materials

- IEX gel matrix (Support Protocols and Table 10.9.1)
- IEX buffers (Table 10.9.2)
- IEX chromatography column (see Support Protocols)
- Gradient mixer (Fig. 10.9.1; Pharmacia GM-1 or equivalent)
- Conductivity meter (e.g., Radiometer CDM3 or Bio-Rad conductivity monitor)

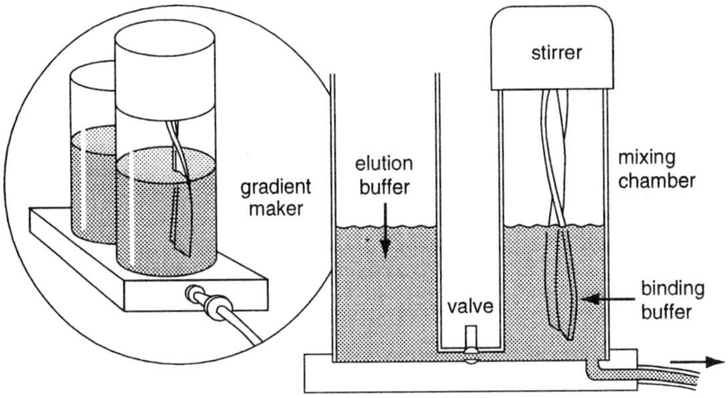

Figure 10.9.1 Gradient mixer for forming salt gradients and pH gradients used during ion-exchange (IEX) chromatography of proteins. The apparatus shown is a Pharmacia Gradient Mixer GM-1, which can be used for preparing gradients of 500 ml or less. If a gradient elution is used for a given IEX separation, a gradient apparatus should be substituted for the buffer reservoir in one of the column configurations. A *salt gradient* is produced by filling the reservoirs with buffers at the same pH and different ionic strength. The mixing chamber contains the lower ionic strength buffer, and the other chamber contains the higher ionic strength buffer. A linear salt gradient can be produced with this apparatus. A *pH gradient* is produced by filling the reservoirs with buffers at the same ionic strength and different pH. A gradient of increasing pH is used for cationic separation, and a gradient of decreasing pH is used for anionic separations. A continuous, nonlinear pH gradient can be produced with this apparatus. The buffer chambers are joined by a channel controlled by a valve. The outflow from the mixing chamber is also controlled by a valve.

Table 10.9.1 Ion-Exchange Matrices Suitable for Protein Separations[a,b]

Anion exchange matrix	Anion exchange type	Cation exchange matrix	Cation exchange type
Polystyrene			
AG 1	Strong	AG 50W	Strong
AG 2	Strong	Bio-Rex 70	Weak
Bio-Rex 5	Intermediate		
AG 3-X4A	Strong		
Bio-Rex MSZ 1-X8	Strong		
Cellulose			
Aminoethyl-cellulose	Weak	CM-cellulose	Weak
DEAE-cellulose	Weak		
Benzyl DEAE-cellulose	Weak		
ECTEOLA-cellulose	Weak		
PEI-cellulose	Weak		
QAE-cellulose	Strong		
TEAE-cellulose	Weak		
DEAE-Sephacel	Weak		
Sephadex			
DEAE-Sephadex (A-25 or A-50)	Weak	CM-Sephadex (C-25 or C-50)	Weak
QAE-Sephadex (A-25 or A-50)	Strong	SP-Sephadex (C-25 or C-50)	Strong
Sepharose			
DEAE-Sepharose CL-6B	Weak	CM-Sepharose CL-6B	Weak
DEAE-Sepharose (fast flow)	Weak	CM-Sepharose (fast flow)	Weak
Q Sepharose (fast flow)	Strong	S Sepharose (fast flow)	Strong

[a]Abbreviations: CM, carboxymethyl; DEAE, diethylaminoethyl; ECTEOLA, epichlorohydrin triethanolamine; PEI, polyethylenimine; Q, quaternary amine; QAE, diethyl-[2-hydroxypropyl]-aminoethyl; S, sulphonate; SP, sulphopropyl; TEAE, triethylaminoethyl.

[b]Sephadex, Sephacel, and Sepharose are available from Pharmacia. Various derivatized forms of cellulose are available from Whatman, Sigma, and other suppliers. AG and Bio-Rex are polystyrene-based matrices available from Bio-Rad. Similar matrices are available from other suppliers, e.g., Dow ("Dowex") and Rohm & Haas ("Amberlite").

1. Prepare the IEX gel and set up the column according to the Basic Protocol in UNIT 10.8, using IEX buffer in place of GF buffer. See Support Protocols and Tables 10.9.1 and 10.9.2 for selection of IEX gel matrix, column, and buffer.

   Volatile or nonvolatile buffers may be used depending on the next separation procedure or method of analysis for the protein of interest. If a chaotropic agent (e.g., urea), organic solvent (e.g., alcohol or acetonitrile), or detergent (e.g., nonionic forms) is added to solubilize the protein of interest, it is important to demonstrate that the additive does not cause precipitation of buffer salts.

2. Dissolve the sample containing a mixture of proteins in the starting gradient buffer. Remove most of the buffer from the column bed and apply to the top of the gel. Any volume may be used as long as the ionic strength does not exceed the ionic strength of the starting buffer. Allow the sample to penetrate the gel and rinse the inner wall with buffer. Pipet a 1-cm layer of starting buffer onto the top of the gel.

   Do not let the center of the gel become dry.

Table 10.9.2 Buffers Commonly Used for Gradient Elution of Proteins During Ion Exchange Chromatography[a]

Anion exchange		Cation exchange	
pH	Buffer and gradient	pH	Buffer and gradient
Volatile[b]			
2.3–3.5	Pyridine/formic acid	2.0	Formic acid
3.0–5.0	Trimethylamine/formic acid		
3.0–6.0	Pyridine/acetic acid		
7.9	Ammonium bicarbonate		
7.0–8.5	Ammonia/formic acid		
8.9	Ammonium carbonate		
8.0–9.5	Ammonia/ammonium carbonate		
8.5–10.0	Ammonia/acetic acid		
7.0–12.0	Trimethylamine/CO_2		
7.0–12.0	Triethylamine/CO_2		
Nonvolatile[c,d]			
6.0–7.5	0.01-0.025 M Bis-Tris ± 0.3-1 M NaCl	3.5	0.06 M sodium formate ± 0.95 M NaCl
6.5	0.1 sodium or potassium phosphate ± 0.3 M sodium or potassium phosphate	5.7	0.01 M malonate ± 0.3 M LiCl
7.6–8.6	0.01-0.05 M Tris·Cl ± 0.75-1.0 M NaCl or KCl	6.0	0.025-0.05 M MES ± 0.35-1.0 M NaCl
7.8	0.05 M triethanolamine + 7.2 M urea ± 0.5-1.0 M NaCl or KCl		
10.4	0.1 M CAPS ± 0.5 M sodium acetate		

[a]Abbreviations: Bis-Tris, 2-bis[2-hydroxyethyl]amino-2-[hydroxymethyl]-1,3-propanediol; CAPS, [cyclohexyl-amino]-1-propanesulphonic acid; MES, 2-N-morpholinoethane sulphonic acid-NaOH.

[b]Data for volatile buffers systems compiled from "FPLC Ion Exchange and Chromatofocusing: Principles & Methods" available from Pharmacia. Linear gradients from 0.01 M to 0.5 M are most frequently used with volatile buffers.

[c]Nonionic detergents are compatible with ion-exchange chromatography. Deoxycholate cannot be used with anion exchange matrices. Zwitterionic detergents may be used, but there is a pH dependence. The chromatographic column must be equilibrated with a detergent solution before beginning to run a gradient containing detergent. Organic solvents including acetonitrile and alcohols may be used at concentrations <90%. Urea is also compatible with ion-exchange chromatography.

[d]± refers to buffer at same pH containing the concentration of the salt listed (+) or not containing additional salt (–). See Figure 10.9.1 for description of apparatus used to form a salt gradient.

3. Attach the gradient mixer and collect 100 fractions, each containing ~1% of the total gradient volume. Measure the A_{280} to identify the protein peaks and the ionic strength with a conductivity meter to determine the gradient shape.

SUPPORT PROTOCOL

SELECTING AN IEX GEL MATRIX AND COLUMN

1. Choice of an IEX gel depends upon:

   a. pH range where the protein is stable. Use an anion-exchange gel matrix for a protein stable at a pH above its pI; use a cation-exchange gel matrix for a protein stable at a pH below its pI.

   b. Molecular size of the protein being separated. For proteins of molecular weight 10,000 to 100,000, use DEAE-Sephacel and DEAE-Sepharose (Pharmacia); for larger proteins, use Sephadex A-25 or C-25.

2. Column size depends on the binding capacity of the gel matrix. The most frequently used diameters are 1, 2, and 2.5 cm; the length is usually <20 cm. For complex mixtures of protein, use a longer column.

3. Commmercially available prepacked, ion-exchange FPLC columns are available. They are Mono-Q (anion) and Mono-S (cation) from Pharmacia in three sizes as 0.5 mm × 5 cm (25 mg sample), 1 × 10 cm (200 mg sample), and 1.6 × 10 cm (500 mg sample).

Reference: Bio-Rad Price List L; Pharmacia, Ion-Exchange Chromatography.

Contributor: John A. Smith

Immunoaffinity Chromatography

UNIT 10.10

ISOLATION OF SOLUBLE OR MEMBRANE-BOUND ANTIGENS

BASIC PROTOCOL

Materials (*see* APPENDIX 1 *for items with* ✓)

Antibody (Ab)-Sepharose (*UNIT 10.15*, Support Protocol)
Activated, quenched (control) Sepharose, prepared as for Ab-Sepharose but eliminating Ab or substituting irrelevant Ab during coupling
Cells or homogenized tissue
✓ Tris/saline/azide (TSA) solution, ice-cold
✓ Lysis buffer, ice-cold
5% (w/v) sodium deoxycholate (Na-DOC; filter sterilize and store at room temperature)
✓ Wash buffer
✓ Tris buffers, pH 8.0 and 9.0, ice-cold
✓ Triethanolamine solution, ice-cold
✓ 1 M Tris·Cl, pH 6.7, ice-cold
✓ Column storage solutions, ice-cold

Columns
Quick-seal centrifuge tubes (Beckman)

NOTE: Carry out all procedures involving antigen in a 4°C cold room or on ice.

1. Prepare an activated, quenched (control) Sepharose precolumn (5 ml packed bed volume) and an Ab-Sepharose immunoaffinity column (5 ml; 5 mg/ml antibody per milliliter packed Sepharose) linked in series (Fig. 10.10.1).

2. Suspend 50 g of cells at 1-5 × 10^8 cells/ml in ice-cold TSA solution, or add 1 to 5 volume of ice-cold TSA per volume packed cells or homogenized tissue. Add an equal volume of ice-cold lysis buffer and stir 1 hr at 4°C.

3. Centrifuge 10 min at 4000 × *g* to remove nuclei. Decant supernatant and save.

4. For purification of membrane antigens, add 0.2 vol of 5% Na-DOC to the postnuclear supernatant, and leave 10 min at 4°C or on ice. Transfer to quick-seal centrifuge tubes and centrifuge 1 hr at 100,000 × *g*. Carefully remove supernatant and save.

5. Attach Sepharose precolumn to immunoaffinity column (Fig. 10.10.1).

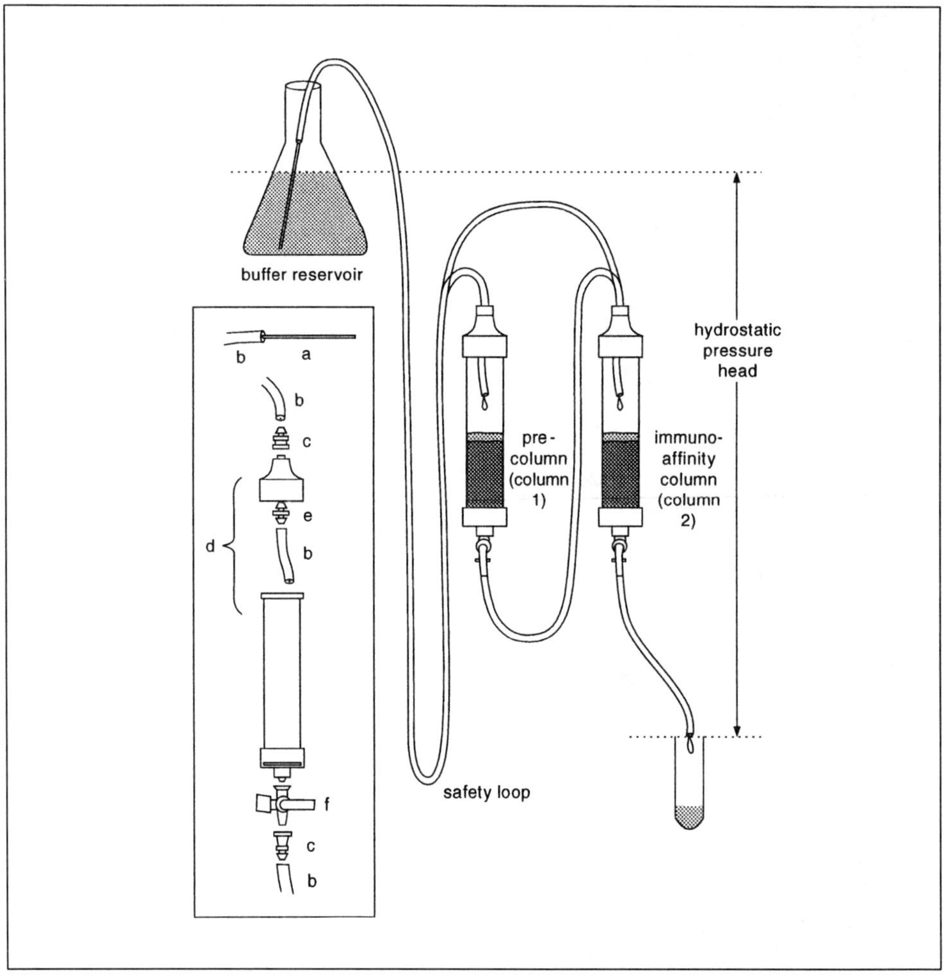

Figure 10.10.1 Immunoaffinity chromatography. During the application of the supernatant or sample, two columns, an immunoaffinity column (with covalently bound antibody) (1), and a Sepharose precolumn (without covalently bound antibody) (2), are attached in series to a buffer reservoir containing the supernatant or sample. After the sample is applied the precolumn is removed, and the tubing of the safety loop is connected to the immunoaffinity column. The hydrostatic pressure head is the distance between the top of the solution in the buffer reservoir and the tip of the tube at the bottom of the immunoaffinity column. When the sample reservoir is emptied, the hydrostatic head becomes zero when the fluid level reaches the safety loop, preventing columns from running dry. Fluid remaining above the column beds can be removed by raising the safety loop. After rinsing the tubing, next elution is begun by placing the end of the safety loop in another reservoir containing next elution buffer.

6. Wash both columns with the following:

   10 column volumes of wash buffer
   5 column volumes of Tris buffer, pH 8.0
   5 column volumes of Tris buffer, pH 9.0
   5 column volumes of triethanolamine solution
   5 column volumes of wash buffer.

7. Apply the supernatant (reserving a sample for analysis) to the precolumn and allow it to flow through the precolumn and immunoaffinity column linked in series at a flow rate of 5 column volumes/hr. Collect the flowthrough as fractions of ¹⁄₁₀ to ¹⁄₁₀₀ the volume of the applied supernatant.

8. Wash with 5 column volumes of wash buffer, then close the stopcocks on both columns and disconnect the precolumn from the immunoaffinity column. Open

the stopcock of the immunoaffinity column and allow fluid above the top of the column to drain out to bed level.

NOTE: Between each change of buffers, wash the immunoaffinity column as follows. Close the stopcock and remove the end cap of the column. With a syringe connected to the outlet of the tubing from the buffer reservoir, aspirate all buffer from the tubing. Place tubing into the next buffer contained in another reservoir. Aspirating with a syringe, fill the tubing from the reservoir and remove the syringe. Crimp the tubing to regulate flow and rinse the inside wall of the column with the buffer. Open the column stopcock and drain the buffer to bed level. Put end cap loosely on column and allow buffer to drain into the column to a level several centimeters above the bed. Secure end cap and commence washes or elution.

9. Wash with the following and collect fractions:

   5 column volumes of wash buffer
   5 column volumes of Tris buffer, pH 8.0
   5 column volumes of Tris buffer, pH 9.0.

10. Elute the antigen with 5 column volumes of triethanolamine solution. Collect fractions of one column volume into tubes containing 0.2 vol of 1 M Tris·Cl, pH 6.7, to neutralize the fractions as they are collected.

    In some cases it may be desirable to lower the pH of the triethanolamine solution to preserve the functional activity of the ligand. The ideal pH gives complete release of the ligand, as verified by SDS-PAGE evaluation of a sample (~20 μl) of the eluted column bed (Ab-Sepharose) and eluate (50 μl).

11. Wash the column with 5 column volumes of TSA solution.

    A column may be reused many times and remain active for several years after storage at 4°C in TSA solution.

12. Analyze fractions for the presence of antigen—50-μl aliquots of each eluate fraction should be analyzed by SDS-PAGE and silver staining. Analyze 0.5- to 1-ml aliquots of the sample applied to the column and representative flow-through and wash fractions by immunoprecipitation with Ab-Sepharose followed by SDS-PAGE and detection by silver staining to determine whether the column was saturated.

    If antibody leaches off the column during elution, it may be removed from eluate by passage through Protein A–Sepharose (Ey et al., 1978).

BATCH PURIFICATION OF ANTIGENS

ALTERNATE PROTOCOL 1

1. Obtain the post-nuclear supernatant from 50 g of cells (Basic Protocol, steps 2 to 4).

2. Suspend Ab-Sepharose in the supernatant in a flask. Shake gently on a rotary shaker for 3 hr. Stop shaking and allow the Sepharose to settle. Decant most of the supernatant. Pour the Ab-Sepharose and the remainder of the supernatant into a column and open the stopcock. Continue draining the column until all the Sepharose has been added. Allow the fluid to drain to bed level and close the stopcock.

3. Wash the column and elute the antigen (Basic Protocol, steps 9 to 11). Analyze the fractions (Basic Protocol, step 12).

ALTERNATE PROTOCOL 2

LOW-pH ELUTION OF ANTIGENS

Additional Materials (*also see Basic Protocol; see* APPENDIX 1 *for items with* ✓)

✓ Sodium phosphate buffer, pH 6.3
✓ Glycine buffer
✓ 1 M Tris·Cl, pH 9.0

1. Prepare the sample and the column (Basic Protocol, steps 1 to 9) *except* do not wash with Tris buffer, pH 9.0.
2. Wash with 5 column volumes of sodium phosphate buffer.
3. Elute with 5 column volumes of glycine buffer. Collect fractions into tubes containing 0.2 vol of 1 M Tris·Cl, pH 9.0 and mix each fraction immediately after collection.
4. Analyze fractions for antigen (Basic Protocol, step 12).

References: Harlow and Lane, 1988; Wilchek et al., 1984.

Contributor: Timothy A. Springer

UNIT 10.11

Metal-Chelate Affinity Chromatography

Recombinant native or denatured proteins engineered to have six consecutive histidine residues on either the amino or carboxyl terminus (see Fig. 10.11.1 for primers) can be purified using a resin containing nickel ions (Ni^{2+}) that have been immobilized by covalently attached nitrilotriacetic acid (NTA), metal-chelate affinity chromatography (MCAC).

NOTE: All solutions and equipment coming into contact with bacteria must be sterile, and proper sterile technique should be used accordingly.

A 5′-GGGNNNNNNATGCATCATCATCATCATCAT . . . $N_{15\text{-}30}$-3′
RE site – MetHisHisHisHisHisHis . . .

B 5′-GGGNNNNNNTTAATGATGATGATGATGATG . . . $N_{15\text{-}30}$-3′
RE site – ENDHisHisHisHisHisHis . . .

Figure 10.11.1 Sequences of primers required to create histidine tails at protein termini, with functions (e.g., protein sequence) marked below. Three guanines are included at the 5′ ends of each primer to facilitate restriction enzyme digestion of the PCR product prior to subcloning. NNNNNN represents a unique restriction enzyme site compatible with the selected vector. $N_{15\text{-}30}$ represents 15 to 30 additional nucleotides specific to the cDNA beginning with the second codon. Met represents an initiator methionine and END represents a termination codon. (**A**) Sequence of 5′ primer used to create a histidine tail at the amino terminus. The 3′ primer should include a second unique restriction enzyme site and the final 5 to 10 codons (including a stop codon) of the cDNA sequence. (**B**) Sequence of 3′ (antisense) primer used to create a histidine tail at the carboxyl terminus. The 5′ primer should contain a second unique restriction site and the first 5 to 10 codons of the cDNA.

NATIVE MCAC FOR PURIFICATION OF SOLUBLE HISTIDINE-TAIL FUSION PROTEINS

BASIC PROTOCOL

Materials *(see APPENDIX 1 for items with ✓)*

✓ M9ZB medium containing 50 µg/ml ampicillin and 25 µg/ml chloramphenicol
E. coli BL21(DE3)pLysS or other suitable strain (Novagen) containing a pET vector (Fig. 10.11.2) expressing a histidine-tail fusion protein
0.1 M IPTG, filter sterilized (Table 1.4.2)
NTA resin slurry: 50% (w/v) suspension in 20% (w/v) ethanol (Qiagen)
50 mM $NiSO_4 \cdot 6H_2O$
✓ MCAC-0, MCAC-20, MCAC-40, MCAC-60, MCAC-80, MCAC-100, MCAC-200, and MCAC-1000 buffers
✓ 150× protease inhibitor cocktail
10% (v/v) Triton X-100
✓ MCAC-EDTA buffer

1.5 × 10–cm glass or polypropylene column
Beckman JA-10 and Ti-70 rotors or equivalent

1. Inoculate 10 ml M9ZB/ampicillin/chloramphenicol with *E. coli* BL21(DE3)pLysS containing a pET vector expressing a desired histidine-tail fusion protein. Grow overnight with shaking at 37°C.

2. Inoculate 100 ml M9ZB/ampicillin/chloramphenicol with 1 ml of the overnight culture and grow with shaking at 37°C to OD_{600} = 0.7 to 1.0.

3. Add 1 ml of 0.1 M IPTG (to 1 mM final) and continue shaking incubation 1 to 3 hr at 37°C.

4. Centrifuge 10 min at 4400 × *g*, 4°C. Discard supernatant. Store pellet at –70°C.

   Alternatively, extract precipitation may be carried out immediately and the column prepared during centrifugation.

5. Add 1 ml NTA resin slurry to a 1.5 × 10–cm column and allow liquid to drain just to the top of the packed resin. Wash the column with the following:

   0.5 ml (3 bed volumes) deionized water
   0.5 ml (5 bed volumes) of 50 mM $NiSO_4 \cdot 6H_2O$
   0.5 ml MCAC-0 buffer.

   Charged resin can be stored at 4°C. If column is to be stored >1 day, wash with 10 bed volumes of 20% ethanol and add 1 bed volume of 20% ethanol prior to storage. Keep column sealed to prevent evaporation.

6. Thaw cell pellet on ice. Add 5 ml MCAC-0 buffer and 33 µl of 150× protease inhibitor cocktail and resuspend by pipetting, sonication, or homogenization.

   Beginning with this step, all procedures should be performed on ice or in a cold room.

7. Add 0.05 ml of 10% (w/v) Triton X-100 (0.1% final). Mix thoroughly and freeze 10 min at –70°C.

8. Thaw cell suspension on ice. Spin 60 min at 260,000 × *g*, 4°C. Decant supernatant into a clean container on ice and discard pellet. Set aside and freeze a 10-µl aliquot at –70°C for later analysis.

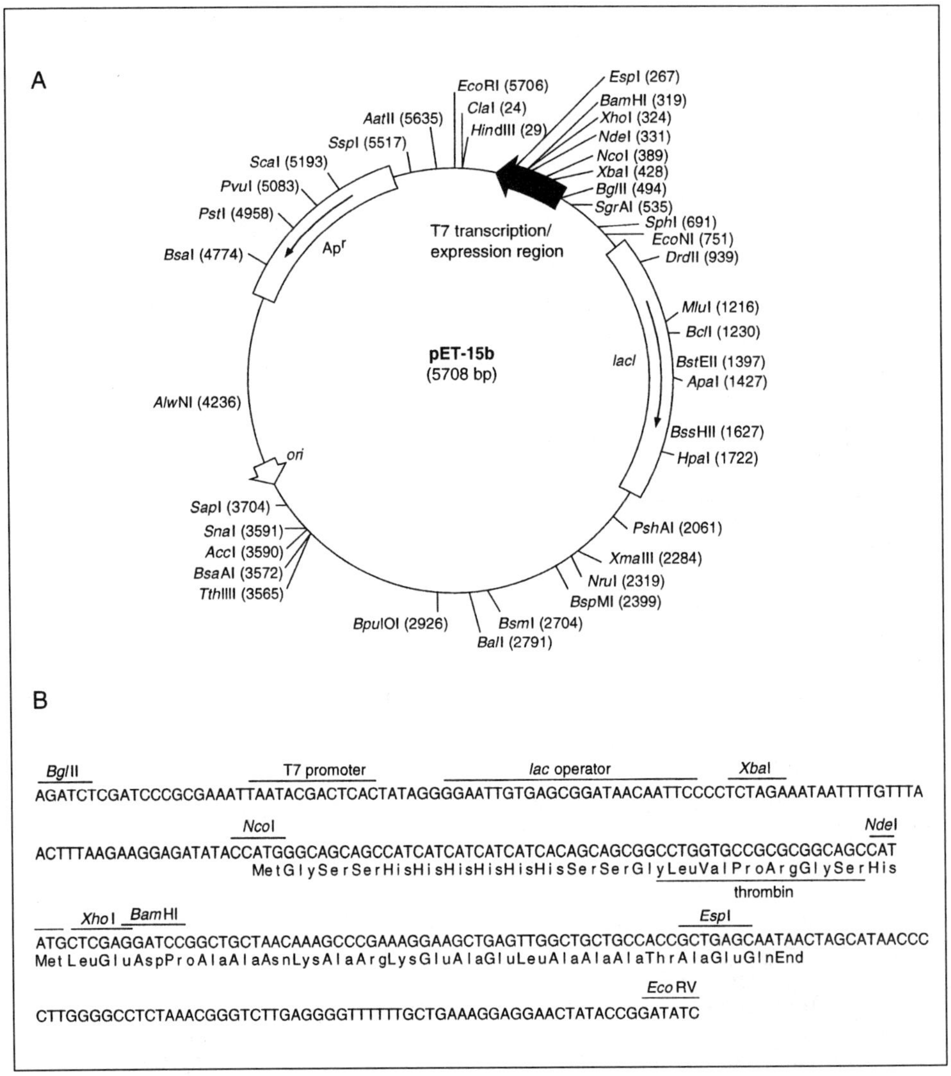

Figure 10.11.2 pET-15b, one of a series of pET vectors designed for cloning, expression, and purification of recombinant proteins (reprinted with the permission of Novagen). (**A**) pET-15b vector; (**B**) sequence of pET-15b cloning/expression region. The target gene is cloned into the pET plasmid such that its expression is under the control of bacteriophage T7 transcription and translation signals (see *UNIT 16.2*). pET-15b encodes an amino-terminal His·Tag leader that allows purification of the resulting recombinant protein over Ni^{2+}-NTA resin. Following purification, the His tag can be removed by thrombin cleavage (*UNIT 16.7*). The pET series are derivatives of vectors originally described by Studier et al. (1990) and are available from Novagen.

9. If extract is frozen, thaw on ice. Load onto Ni^{2+}-NTA column and allow to flow through at a rate of 10 to 15 ml/hr. Collect column flowthrough and save for SDS-PAGE (Support Protocol 1).

10. Wash column with 5 ml MCAC-0 buffer at a flow rate of 20 to 30 ml/hr. Discard flowthrough.

11. Wash column in stepwise fashion with 5 ml each of MCAC-20, MCAC-40, MCAC-60, MCAC-80, MCAC-100, MCAC-200, and MCAC-1000 buffers. Collect 1-ml fractions and save on ice for SDS-PAGE.

Alternatively, the column can be eluted with a 20-ml linear gradient (UNIT10.10) of 0 to 400 mM imidazole in MCAC.

The second and third fractions of each wash will contain most of the eluted proteins. Most proteins with hexahistidine tails will elute with 100 to 200 mM imidazole (MCAC-100 or -200).

12. Elute column with 5 ml MCAC-EDTA buffer at a flow rate of 10 to 15 ml/hr, collecting 1-ml fractions.

    The same column may be used three to five times before EDTA stripping and nickel recharging are necessary, but only one protein should be purified on any given column.

13. Analyze fractions for the presence of eluted protein.

    A quick and easy method to determine which fractions contain eluted protein is to place 2 μl of undiluted Protein Assay Dye Reagent Concentrate (Bio-Rad) on a piece of Parafilm, add 8 μl from fraction to be tested, and mix by pipetting up and down. Immediate appearance of blue color indicates that the fraction contains protein. This does not work in the presence of Triton X-100 because the detergent itself produces an intense blue color; for this reason, Triton X-100 is excluded from the washing and elution buffers.

14. Combine the fractions containing eluted protein and remove a 10-μl aliquot for SDS-PAGE. If desired, remove MCAC buffer by dialysis. Freeze all samples in smaller aliquots at −70°C or in liquid nitrogen.

DENATURING MCAC FOR PURIFICATION OF INSOLUBLE HISTIDINE-TAIL FUSION PROTEINS

ALTERNATE PROTOCOL 1

Additional Materials *(also see Basic Protocol; see* APPENDIX 1 *for items with ✓)*

✓ GuMCAC-0, GuMCAC-20, GuMCAC-40, GuMCAC-60, GuMCAC-100, and GuMCAC-500 buffers
✓ GuMCAC-EDTA buffer
Appropriate final buffer for protein (e.g., for proteolytic cleavage or long-term storage)
Guanidine·HCl
Beckman JA-20 rotor or equivalent

1. Prepare the pellet of *E. coli* expressing a histidine-tail fusion protein (Basic Protocol, steps 1 to 4).

2. Prepare NTA resin column (Basic Protocol, step 5) *except* wash the charged column with 0.5 ml GuMCAC-0 buffer.

3. Thaw cell pellet on ice. Resuspend in 5 ml GuMCAC-0 buffer by pipetting, sonication, or homogenization. Freeze 10 min at −70°C and thaw at room temperature.

   Subsequent steps can be performed at room temperature. However, if solid-phase renaturation (Alternate Protocol 2) is used, it is probably better to maintain lower temperatures throughout the process.

4. Gently mix samples for 60 min using a rocker, rotating mixer, or magnetic stirrer. Spin 30 min at 27,000 × *g*, 4°C. Decant supernatant into a clean container and discard pellet. Set aside a 10-μl aliquot for analysis by SDS-PAGE (Support Protocol 1).

5. If extract is frozen, thaw at room temperature. Load onto Ni^{2+}-NTA column and allow to flow through at a rate of 10 to 15 ml/hour. Collect flowthrough and save a 10-μl aliquot for SDS-PAGE.

6. Wash column with 5 ml GuMCAC-0 buffer at a rate of 20 to 30 ml/hour. Discard the flowthrough.

7. Wash column in stepwise fashion with 5 ml GuMCAC-20, -40, -60, -100, and -500 buffers. Collect 1-ml fractions and save for SDS-PAGE.

   The second and third fractions from each wash will contain most of the unbound protein. The histidine tail binds slightly less avidly under denaturing conditions.

8. Elute with 5 ml GuMCAC-EDTA buffer at a rate of 10 to 15 ml/hr, collecting 1-ml fractions.

9. Identify fractions containing the protein and pool together. Prepare samples for dialysis or store them at −70°C.

   Guanidine precipitates in the presence of SDS and must be removed by dialysis before SDS-PAGE.

10. Prepare appropriate final buffer for protein (e.g., for proteolytic cleavage or long-term storage) and add sufficient guanidine to bring final concentration to 4 M. Dialyze purified protein ≥ 2 hr at 4°C against 1000 ml of buffer/4 M guanidine.

11. Remove 500 ml buffer/guanidine and add 500 ml buffer without guanidine. Continue dialysis ≥2 hr. Repeat.

   With some proteins, renaturation by dialysis may require longer dialysis periods and more gradual decrements in the guanidine concentration of the buffer. Conditions for each protein must be determined empirically.

12. Remove dialysis bag to a container containing 1000 ml of fresh buffer without guanidine at 4°C. Continue dialysis 2 hr to overnight.

13. Remove sample from dialysis bag, divide into aliquots, and freeze at −70°C or in liquid nitrogen.

17. Analyze fractions and process protein.

ALTERNATE PROTOCOL 2

SOLID-PHASE RENATURATION OF MCAC-PURIFIED PROTEINS

Additional Materials *(also see Basic Protocol)*

1:1 (v/v) MCAC-20/GuMCAC-20 buffer
3:1 (v/v) MCAC-20/GuMCAC-20 buffer
7:1 (v/v) MCAC-20/GuMCAC-20 buffer

1. Perform steps 1 to 6 of Alternate Protocol 1.

2. Wash column with:

   5 ml 1:1 MCAC-20/GuMCAC-20 buffer
   5 ml 3:1 MCAC-20/GuMCAC-20 buffer
   5 ml 7:1 MCAC-20/GuMCAC-20 buffer
   5 ml MCAC-20 buffer.

   Slow elution (between 1 to 2 hr) with a 50-ml linear gradient (UNIT10.10) from 100% GuMCAC-20 to 100% MCAC-20 may also yield efficient renaturation.

6. Proceed with steps 12 to 14 of Basic Protocol.

ANALYSIS AND PROCESSING OF PURIFIED PROTEINS

SUPPORT PROTOCOL 1

Materials *(see* APPENDIX 1 *for items with* ✓*)*

Fractions from MCAC column purification (crude extract, flowthroughs, and purified protein; Basic or Alternate Protocols)
✓ MCAC-0 buffer
✓ 2× SDS sample buffer

1. Thaw aliquots of fractions to be analyzed on ice. Mix 5 µl from crude extract and crude flowthrough fractions and 10 µl from the second and third fractions from each washing step with an equal volume of 2× SDS sample buffer.

2. Load samples onto a standard (e.g., 1 mm × 14 cm × 14 cm) SDS-PAGE gel. Run gel and visualize to identify the fractions containing purified protein.

3. Thaw the remaining aliquots of fractions containing purified protein, dialyze against the appropriate proteolysis buffer, and carry out cleavage procedure if desired. If necessary, after cleavage dialyze the protein against an appropriate storage buffer and freeze in aliquots.

   The size of the cleaved histidine tail will generally be <3 kDa, depending on the design of the fusion protein. This fragment will usually be removed by the dialysis. It may also be removed by ultrafiltration (which will also concentrate the protein), size-exclusion chromatography, or a second MCAC chromatography.

NTA RESIN REGENERATION

SUPPORT PROTOCOL

Materials *(see* APPENDIX 1 *for items with* ✓*)*

Stripping solution: 0.2 M acetic acid/6 M guanidine·HCl
2% (w/v) SDS
20%, 25%, 50%, 75%, and 100% ethanol
✓ 0.1 M EDTA, pH 8.0

1. Wash 1 ml of the resin with the following:

   5 ml stripping solution
   5 ml water
   2.5 ml 2% SDS
   2.5 ml 25% ethanol
   2.5 ml 50% ethanol
   2.5 ml 75% ethanol
   12.5 ml 100% ethanol
   2.5 ml 75% ethanol
   2.5 ml 50% ethanol
   2.5 ml 25% ethanol
   2.5 ml water
   12.5 ml of 0.1 M EDTA, pH 8.0
   7.5 ml water.

2. Either charge the column with nickel (Basic Protocol, step 5) or add 2.5 ml of 20% ethanol to resin and store at 4°C.

Reference: Hochuli, 1990.

Contributor: Kevin J. Petty

UNIT 10.12

Reversed-Phase High-Performance Liquid Chromatography

This method is used in the fractionation of shorter peptides and water-soluble (hydrophilic) proteins. Tables 10.12.1 and 10.12.2 provide column specifications, separation conditions, and buffers used in this protocol.

BASIC PROTOCOL

PEPTIDE AND PROTEIN ISOLATION

Materials *(see* APPENDIX 1 *for items with ✓)*

Degassed, HPLC-grade H_2O (Support Protocol)
0.1% (v/v) trifluoroacetic acid (TFA) in H_2O, both HPLC-grade
TFA/acetonitrile buffer: 0.085% (v/v) TFA/70% (v/v) acetonitrile in H_2O
TFA/guanidine buffer: 0.1% (v/v) TFA/6M guanidine·HCl
✓ RP peptide standards
✓ RP protein standards
RP-HPLC column (Table 10.12.1)
Gradient HPLC instrument, equipped with 210-nm detector
HPLC syringe with appropriate needle

NOTE: See APPENDIX 1 entries for "Buffer preparation" and "Protein and peptide sample preparation." To minimize contaminants that absorb at 210 to 220 nm, use the highest purity solvents and buffers.

1. Remove the organic solvent from the RP column with degassed, HPLC-grade water, using a gradient from 100% organic solvent to 100% water over 15 min at 1 ml/min.

   4-mm-i.d. columns make isolations less efficient and microisolations impossible. 2-mm-i.d. columns are compatible with most commercial HPLCs.

2. Equilibrate the RP column with 100% TFA/acetonitrile buffer at 1 ml/min until the pressure and absorbance become constant. Gradually switch to 0.1% TFA with a 10 to 15 min linear gradient and equilibrate at 0.1% TFA for 20 min.

3. Check the column with a blank run. With a flow rate of 1 ml/min, run a linear gradient from 0 to 100% TFA/acetonitrile buffer over 45 min, maintain for 5

Table 10.12.1 Column Specifications and Separation Conditions for Reversed-Phase HPLC of Peptides and Proteins[a]

Variable	Peptides	Proteins
Solid support matrix		
Bond phase	C_3–C_{18}, phenyl	C_1–C_{18}, phenyl
Particle size, µm	3–10	3–10
Pore size, Å	100–300	300
Column dimensions		
Length, cm	1–25	1–25
Internal diameter, mm	1–5	1–25
Separation conditions		
Temperature	Ambient	5–40°C
Flow rate, ml/min	0.025–2	0.025–2
Buffers and organic modifiers	See Table 10.12.2	

[a]It is not possible to describe all available column matrices, column dimensions, separation conditions, and manufacturers, because there are many alternatives available. In addition to the key reference, the following annual product information sources are useful: Labguide (*Analytical Chemistry*), Buyer's Guide (*LC·GC*), and Guide to Biotechnology Products and Instruments (*Science*).

Table 10.12.2 Buffers for Reversed-Phase HPLC of Peptides and Proteins[a,b]

Acidic (pH 2 to 3)	0.1% TFA*
	0.1% H_3PO_4
	0.1% HFBA*
	10 mM HCl*
	5 to 60% formic acid*
	20 mM TEA-phosphate, pH 3
Acidic (pH 4 to 6)	10 mM NH_4-acetate, pH 4 to 6*
	10 mM TEA-acetate, pH 4.5*
	100 mM $NH_4H_2PO_4$, pH 4.5
Neutral (pH >6)	100 mM Na-acetate, pH 7.5
	10 mM KH_2PO_4, pH 6 to 8
	20 mM Tris·Cl, pH 7 to 8
	50 mM NaH_2PO_4, pH 7
	50 mM NH_4HCO_3, pH 8*

[a]Abbreviations: TFA, trifluoroacetic acid; HFBA, heptafluorobutyric acid; TEA, triethylammonium.

[b]Only volatile (*) and nonvolatile aqueous buffer components are listed. TFA is the most frequently used buffer for peptide and protein separations. The organic phase most commonly used is acetonitrile. For hydrophobic peptides or proteins, 1- or 2-propanol should be used as a 50% to 70% (v/v) solution in water containing the same concentration of buffer salts as the aqueous buffer described in the protocol. There are instances where mixed organic phases (e.g., acetonitrile/propanol, propanol/butanol, or aceto- nitrile/butanol) have been used.

min, return to 0.1% TFA with a linear gradient over 15 min, and maintain at 0.1% TFA for 15 min. Detection settings (chart speed of 0.5 cm/min) should be 0.1 absorption units full scale (AUFS) at ~210 to 220 nm for 50 to 200 pmol peptide, 0.3 AUFS for 500 pmol, 0.5 AUFS for ~1 nmol, and 1.0 AUFS for ~2 nmol. The corresponding settings for proteins are 2 to 3 times higher.

4. Centrifuge the RP peptide standards or a digestion mixture of peptides 5 min at 5000 × *g*. Withdraw an aliquot of the supernatant into an HPLC syringe (rinsed with 0.1% TFA) and load the injection loop (without air bubbles).

5. Run the gradient as in step 3. Collect each chromatographic peak into a separate polypropylene tube. Wash and store the column in organic solvent, when not used for >2 days.

   To ensure the purity of a sample, rechromatograph a peak at a different pH using either 20 mM ammonium acetate or 20 mM triethylammonium acetate (pH 7) and a linear gradient of 70% acetonitrile containing the corresponding buffer salt. Alternatively, rechromatograph the peak isocratically at a lower percentage of TFA/acetonitrile buffer or with a gradient on another type of RP-HPLC column.

6. Dissolve the collected peptides in TFA/guanidine buffer and vortex. Centrifuge 5 min at 5000 × *g* and complete the second HPLC separation. For amino acid sequence analysis, apply sample directly to glass filter of the protein sequencer. For amino acid analysis, dry an aliquot in an appropriate container for subsequent hydrolysis.

ALTERNATE PROTOCOL

PROTEIN ISOLATION

Repeat steps 1 to 5 above for proteins in the RP protein standards. For an unknown mixture, repeat steps 1 to 6.

SUPPORT PROTOCOL

DEGASSING WATER, BUFFERS, AND SOLVENTS

Degas water until it stops bubbling by evacuation with a trapped vacuum pump. Add volatile buffers after degassing. Nonvolatile buffer solutions should be filtered through a 0.45-µm nylon filter prior to being degassed.

Reference: Hancock, 1984.

Contributors: Kenneth J. Wilson and Timothy D. Schlabach

UNIT 10.13

Ion-Exchange High-Performance Liquid Chromatography

Ion-exchange (IEX) high-performance liquid chromatography (HPLC) can be used to separate proteins based on molecular charge. Tables 10.13.1 and 10.13.2 provide column specifications, separation conditions, and buffers used in these protocols.

Materials *(see APPENDIX 1 for items with ✓)*

Degassed, HPLC-grade H_2O (*UNIT 10.12*)
✓ IEX protein standards
0.2 M Tris·Cl, pH 7.5
Tris/NaCl buffer: 0.02 M Tris·Cl (pH 7.5)/0.5 M NaCl
0.02 M sodium phosphate, pH 6.0
Sodium phosphate/NaCl buffer: 0.02 M sodium phosphate (pH 6.0)/0.5 M NaCl
Anion-exchange (AEX) HPLC column (Table 10.13.1)
Cation-exchange (CEX) HPLC column (Table 10.13.1)

NOTE: See *APPENDIX 1* entries for "Buffer preparation" and "Protein and peptide sample preparation." All stainless-steel parts will deteriorate after prolonged contact with halides. Flush columns with water immediately after use and store in 50% methanol.

BASIC PROTOCOL 1

ANION-EXCHANGE HPLC

1. Before using the IEX column, remove adsorbed protein by flushing sequentially with 10 column volumes degassed water and 10 column volumes of a strong salt buffer (~0.5 to 1 M) at pH 8 to 10 for polymeric IEX columns, or a strong salt buffer at pH 7 to 8 for silica-based IEX columns.
2. Equilibrate the AEX column at room temperature (higher temperatures have been used for some separations) with a 50% mixture of 0.2 M Tris·Cl, pH 7.5, and Tris/NaCl buffer for 20 min at 1.0 ml/min for 4 mm i.d. Flush column with >10 column volumes of 0.2 M Tris·Cl, pH 7.5.
3. Make a blank run at 210 to 220 nm with 0.1 to 1.0 absorbance units full scale (AUFS) as follows: a linear gradient of 0 to 75% Tris/NaCl buffer over 20 min, hold at 75% buffer for 5 min, return to 0% buffer over 5 min, and hold at 0% buffer for 15 min prior to next run.

   Presence of peaks >5% full scale indicates contaminants left on the column or present in buffers. To eliminate contaminants, wash the column sequentially with 10 column

Table 10.13.1 Column Specifications and Separation Conditions for Ion-Exchange HPLC of Proteins[a,b]

Support	Weak[c]	Strong[c]
Polymeric anion exchange	WAX	SAX
Bonded phase	DEA	QAE
Particle size, µm	5–15	5–15
Pore size, Å	100–500	100–500
Column dimensions		
Inner diameter, mm	4–8	4–8
Length, cm	5–30	5–30
Flow rate, ml/min	0.2–2.0	0.2–2.0
Silica anion exchange	WAX	SAX
Bonded phase	PEI	Me-PEI
Particle size, µm	5–10	5–10
Pore sizes, Å	100–300	100–300
Column dimensions		
Inner diameter, mm	1–10	1–10
Length, cm	5–25	5–25
Flow rate, ml/min	0.02–2.0	0.02–2.0
Polymeric cation exchange	WCX	SCX
Bonded phase	CM	SP
Particle size, µm	5–15	5–15
Pore size, Å	100–500	100–500
Column dimensions		
Inner diameter, mm	4–8	4–8
Length, cm	5–30	5–30
Flow rate, ml/min	0.2–2.0	0.2–2.0
Silica cation exchange	WCX	SCX
Bonded phase	CM	SP
Particle size, µm	5–10	5–10
Pore size, Å	100–300	100–300
Column dimensions		
Inner diameter, mm	1–10	1–10
Length, cm	5–25	5–25
Flow rate, ml/min	0.02–2.0	0.02–2.0
Temperature range	5°–40°C	5°– 40°C

[a]Abbreviations: WAX, Weak anion exchange; SAX, strong anion exchange; DEA, diethylamine; QAE, quaternary ammonium ethyl; PEI, polyethyleneimine; Me-PEI, methyl polyethyleneimine; WCX, weak cation exchange; SCX, strong cation-exchange; CM, carboxymethyl; SP, sulfopropyl.
[b]It is not possible to describe all available column matrices, column dimensions, separation conditions, and manufacturers, because there are many alternatives available. In addition to the key reference, the following annual product information sources are useful: Labguide (*Analytical Chemistry*), Buyer's Guide (*LC·GC*), and Guide to Biotechnology Products and Instruments (*Science*).
[c]The terms *strong* and *weak* refer to the degree of ionization of the exchange material, not to the strength of retention.

volumes distilled water, 10 column volumes 50% methanol, 10 column volumes distilled water, and 10 column volumes 0.2 M Tris·Cl, pH 7.5.

4. Do a sham run by injecting 10 µl of 0.2 M Tris·Cl, pH 7.5.

5. Centrifuge IEX protein standards and peptide mixture for 5 min at 5000 × *g*. Withdraw 5 µl supernatant into an HPLC syringe rinsed with 0.2 M Tris·Cl, pH 7.5. Load the injection loop, inject the sample onto the column, and start the gradient program as in step 3.

6. Collect each chromatographic peak into a separate polypropylene tube. Desalt the sample with reversed-phase (RP) HPLC (UNIT 10.12) and remove the solvent on a Speedvac evaporator.

Table 10.13.2 Buffers and Buffer Components for Ion-Exchange (IEX) HPLC of Proteins

IEX mode	Buffers (10-50 mM)	Na salts (0.2-0.8 M)	Denaturants/ detergents[a]	Organic solvents[b]
Anion exchange	Tris·Cl Bis-Tris	Chloride Acetate Phosphate	Urea (4-7 M) CHAPS (0.05%)	Methanol (40%) Acetonitrile (30%) Propanol (20%)
Cation exchange	Sodium phosphate	Chloride Acetate Phosphate	Urea (4-7 M) CHAPS (0.05%) SDS (0.02%)	Methanol (40%) Acetonitrile (30%) Propanol (20%)

[a]Abbreviations: SDS, sodium dodecyl sulfate; CHAPS, 3-[(3-cholamidopropyl)-dimethylammonio]-1-propane-sulfonate; Bis-Tris, 2-bis[2-hydroxyethyl]amino-2-[hydroxymethyl]-1,3-propanediol.

[b]Percentages indicate maximum strength of solvent.

BASIC PROTOCOL 2

CATION-EXCHANGE HPLC

1. Remove the storage solvent with 10 column volumes degassed water, and flush with 10 column volumes of a strong salt buffer (~0.5 M) at low pH (pH 3 to 5) for silica-based or polymeric IEX columns.

2. Follow the steps as for anion-exchange HPLC (described above) using sodium phosphate and sodium phosphate/NaCl buffers to separate the protein standards. Substitute the following buffers (steps 3 and 5): use a linear gradient of 0 to 80% sodium phosphate/NaCl buffer over 20 min and use a complex linear gradient of 10% sodium phosphate/NaCl buffer at 0 min, 60% buffer at 15 min, 80% buffer at 20 min, holding at 80% buffer for 5 min, returning to 10% buffer over 5 min, and holding at 10% buffer for 15 min prior to the next injection.

Reference: Regnier, 1984.

Contributors: Kenneth J. Wilson and Timothy D. Schlabach

UNIT 10.14

Size-Exclusion High-Performance Liquid Chromatography

BASIC PROTOCOL

Materials *(see APPENDIX 1 for items with ✓)*

Degassed, HPLC-grade H_2O (*UNIT 10.12*)
✓ SE buffer
✓ SE protein standards
0.75 × 30–cm SE column (Toyo Soda; Table 10.14.1)

1. Remove storage solvent from the SE column with degassed, HPLC-grade water at a flow rate of 1 ml/min. Equilibrate the SE column in degassed SE buffer at 1 ml/min for 30 min at room temperature or until the baseline is stable (*UNIT 10.12*).

2. Carry out a sham run with the injection of 100 µl SE buffer. The detector at 210 to 220 nm should be set at 0.1 to 1.0 absorbance units full scale (AUFS). Convenient AUFS settings are 0.1 for 100 pmol, 0.3 for 500 pmol, 0.5 for 1 nmol, and 1.0 for ~2 nmol. Chart speed should be 0.5 cm/min.

Table 10.14.1 Column Specifications and Separation Conditions for Size-Exclusion HPLC of Proteins[a]

Variable	Options
Support matrix	
Material	Silica, polymeric
Bonding	Hydrophilic (–OH)
Particle size, µm	5–25
Pore size, Å	100–500
Column dimensions	
Length, cm	25–90
Diameter, mm	4.6–21.5
Separation conditions	
Temperature	Ambient to manufacturer's limit
Flow rate, ml/min	0.10–2
Buffers	20–100 mM sodium acetate or sodium phosphate (pH 5 to 8) containing 0.1–0.4 M NaCl; addition of denaturants or detergents (i.e., 8 M urea, 6 M guanidine·Cl, or 0.1% SDS) for determining denatured molecular size. SDS should be used with membrane proteins, since these frequently will not dissolve in aqueous buffers.

[a]It is not possible to describe all available column matrices, separation conditions, and manufacturers, since there are so many alternatives. In addition to the key references, the following annual product information sources are useful: Labguide (*Analytical Chemistry*), Buyer's Guide (*LC·GC*), and Guide to Biotechnology Products and Instruments (*Science*).

3. Centrifuge the SE protein standard or protein mixture for 5 min at 2000 × *g*. Inject 100 µl of the supernatant, and repeat step 2. Collect each chromatographic peak into a separate polypropylene tube and determine the elution volume for each peak.

   The molecular weight of the unknown can be estimated from a standard curve generated by plotting the log_{10} of the protein standard molecular weight versus the respective elution volume.

4. Desalt each fraction and remove the solvent in a Speedvac evaporator.

5. The retention time for the last peak should be no more than twice that for the first peak. If this is not the case, add 20% methanol to the mobile phase to inhibit proteins from binding to the column.

References: Kato, 1984; Unger, 1984.

Contributors: Kenneth J. Wilson and Timothy D. Schlabach

UNIT 10.15

Immunoprecipitation

BASIC PROTOCOL

IMMUNOPRECIPITATION OF RADIOLABELED ANTIGEN WITH ANTIBODY-SEPHAROSE

This protocol follows the steps presented in Figure 10.15.1. It relies on the formation of an insoluble immune complex between a protein antigen and an antigen-specific monoclonal (or polyclonal) antibody bound to Sepharose.

Materials *(see* APPENDIX 1 *for items with ✓)*

Surface-labeled cells (with ^{125}I) *or* biosynthetically ^{35}S-, ^{3}H-, or ^{14}C-labeled cells (*UNIT 10.17*)
✓ Lysis buffer
✓ Dilution buffer
Antibody (Ab)-Sepharose (Support Protocol)
Activated, quenched (control) Sepharose, prepared as for Ab-Sepharose (Support Protocol) but eliminating Ab or substituting irrelevant Ab during coupling
✓ Tris/saline/azide (TSA) solution
✓ 50 mM Tris·Cl, pH 6.8
✓ SDS/sample buffer

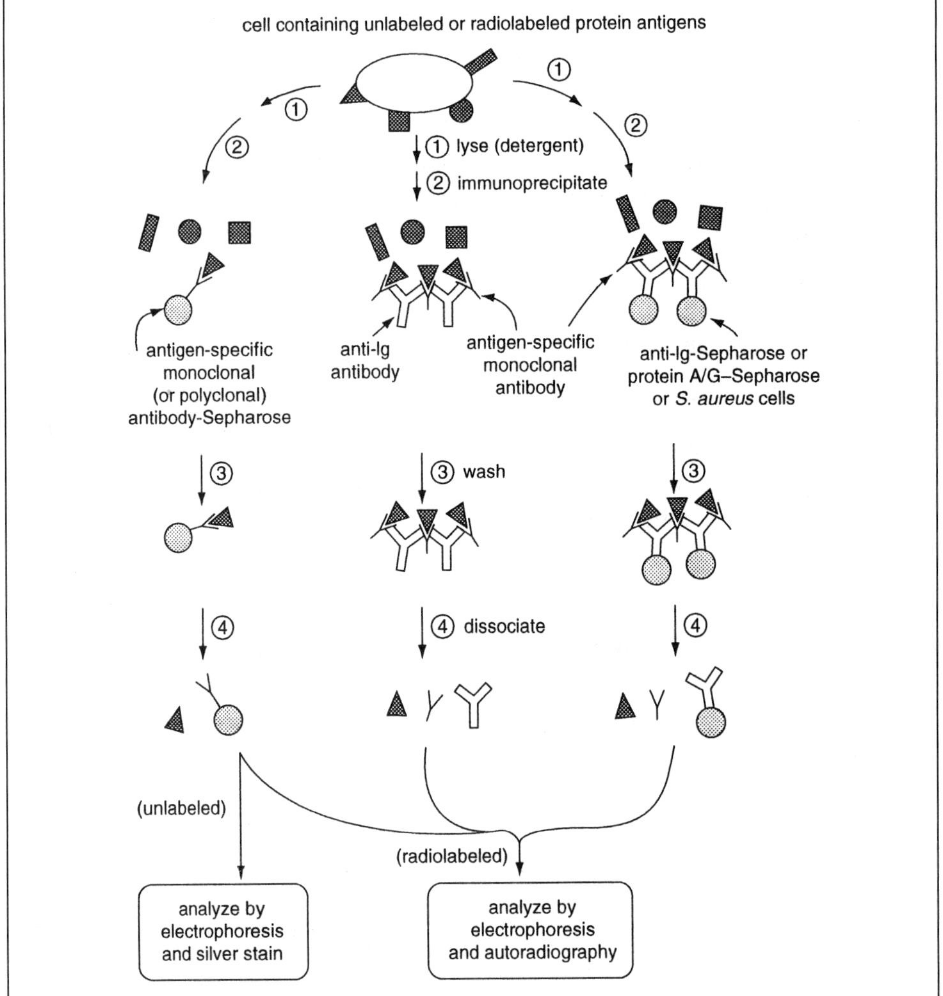

Figure 10.15.1 Immunoprecipitation for the isolation of protein antigens.

NOTE: Carry out all procedures in a 4°C cold room or on ice.

1. Incubate surface- or biosynthetically labeled cells in lysis buffer (5×10^7 cells/ml) for 1 hr at 4°C. Centrifuge lysate 10 min at $3000 \times g$ to remove nuclei, then centrifuge resulting supernatant 1 hr at $100,000 \times g$.

   The supernatant must be used within several days or stored at −70°C. The length of storage is limited by autoradiolysis and the half-life of the isotope. Repeated freezing and thawing may disrupt antigenic determinants and dissociate some protein complexes, especially those that are noncovalently associated.

2. Preclear supernatant in one batch by adding 10 µl activated, quenched (control) Sepharose/200 µl supernatant. Shake on an orbital shaker 2 hr at room temperature or overnight at 4°C. Centrifuge 1 min at $200 \times g$ and save supernatant.

   Control Sepharose can be prepared without antibody or coupled with irrelevant (nonspecific) antibody.

3. Precoat 1.5-ml microcentrifuge tubes by filling with lysis buffer 10 min at room temperature. Remove solution by aspiration, and add 10^5 to 10^6 cpm of radiolabeled (^{125}I or ^{35}S) supernatant containing antigen (step 2). Bring volume to 200 µl with dilution buffer.

   The recommended amount of radioactivity is appropriate for eukaryotic cells with >1000 molecules of antigen/cell.

4. Add ~10 µl of a 1:1 slurry of Ab-Sepharose/dilution buffer (Support Protocol) and shake 1.5-3 hr at 4°C on an orbital shaker, keeping Sepharose suspended.

5. Wash Ab-Sepharose with 1 ml of the buffers listed below. After each wash, centrifuge 1 min at $200 \times g$ or microcentrifuge 5 sec. Carefully aspirate supernatant with fine-tipped Pasteur pipet and leave 10 µl fluid above pellet. After fourth wash, centrifuge to bring down any residual drops on side of the tube, aspirate, and leave 10 µl over pellet.

   First wash: dilution buffer
   Second wash: dilution buffer
   Third wash: TSA solution
   Fourth wash: 50 mM Tris·Cl, pH 6.8.

6. Add 20 to 50 µl of SDS/sample buffer. Because sample buffer has a higher density than the wash solution, it will sink into the Sepharose. Do not vortex because Sepharose may stick to side of tube above buffer level. Cap tube securely and incubate 5 min at 100°C.

7. Vortex and centrifuge 1 min at $200 \times g$ or microcentrifuge 5 sec. Load supernatant into a gel lane and analyze by SDS-PAGE (*UNIT 10.2*). Detect labeled proteins by autoradiography (*APPENDIX 3*) with an enhancing screen (^{125}I) or by fluorography (^{35}S, ^{14}C, and ^{3}H).

SUPPORT PROTOCOL

PREPARATION OF ANTIBODY-SEPHAROSE

This protocol describes covalently linking an antibody to Sepharose using the cyanogen bromide activation method.

Additional Materials *(also see Basic Protocol; see* APPENDIX 1 *for items with ✓)*

1 to 30 mg/ml antigen-specific monoclonal or polyclonal antibody
0.1 M $NaHCO_3$/0.5 M NaCl
Sepharose CL-4B (or Sepharose CL-2B for high-molecular-weight antigens; Pharmacia)
0.2 M Na_2CO_3
✓ Cyanogen bromide (CNBr)/acetonitrile
1 mM and 0.1 mM HCl, ice-cold
50 mM glycine (*or* ethanolamine), pH 8.0

Dialysis tubing (MWCO >10,000)
Whatman No. 1 filter paper

1. Dialyze 1 to 30 mg/ml antibody against 0.1 M $NaHCO_3$/0.5 M NaCl at 4°C with three buffer changes during 24 hr. Use a volume of dialysis solution that is 500 times the volume of antibody solution.
2. Centrifuge 1 hr at 100,000 × *g*, 4°C, to remove aggregates.
3. Measure the A_{280} of an aliquot of the supernatant and determine concentration of the antibody (mg/ml IgG = A_{280}/1.44). Dilute with 0.1 M $NaHCO_3$/0.5 M NaCl to 5 mg/ml (or to same concentration as desired for Ab-Sepharose) and keep at 4°C. Measure A_{280} of this solution for later use in step 10.
4. Allow Sepharose slurry to settle in a beaker and decant and discard the supernatant. Weigh out amount of Sepharose equal to volume of antibody (assume density = 1.0).
5. Set up a filter apparatus using Whatman No. 1 filter paper in a Buchner funnel and an Erlenmeyer filtration flask attached to a water aspirator. Wash Sepharose with 10 vol water.
6. Transfer Sepharose to 50-ml beaker and add equal volume 0.2 M Na_2CO_3.
7. Activate Sepharose at room temperature using 3.2 ml CNBr/acetonitrile per 100 ml Sepharose. Add CNBr/acetonitrile dropwise with a Pasteur pipet over 1 min, while slowly stirring the slurry with a magnetic stirrer. Continue stirring slowly for 5 min.

   CAUTION: *Activation should be carried out in a fume hood.*

8. Rapidly filter the CNBr-activated Sepharose as in step 5. Aspirate to semidryness (i.e., until the Sepharose cake cracks and loses its sheen). Wash with 10 vol ice-cold 1 mM HCl. Wash with 2 vol of ice-cold 0.1 mM HCl. Hydrate cake with enough ice-cold 0.1 mM HCl so the cake regains its sheen, but so there is no excess liquid above the cake.

   CNBr-activated Sepharose can be purchased from Pharmacia, but the coupling capacity will be lower.

9. Immediately transfer a weighed amount of Sepharose (assume density = 1.0) to a beaker. Add an equal volume of a solution of antibody dissolved in 0.1 M

$NaHCO_3$/0.5 M NaCl (step 2). Stir gently with a magnetic stirrer 2 hr at room temperature or overnight at 4°C.

10. Add 50 mM glycine (or ethanolamine), pH 8.0, to saturate the remaining reactive groups on the Sepharose and allow slurry to settle. Remove aliquot of the supernatant, centrifuge to remove any residual Sepharose, and measure A_{280}. Compare absorbance to that of the A_{280} of the antibody solution from step 2 to determine the percentage coupling.

11. Store Ab-Sepharose in TSA solution.

IMMUNOPRECIPITATION OF RADIOLABELED ANTIGEN WITH ANTI-Ig SERUM

ALTERNATE PROTOCOL 1

This protocol relies on the formation of soluble immune complexes between a protein and an antigen-specific antibody, followed by immunoprecipitation of the immune complexes by antibodies contained in anti-immunoglobulin (Ig) serum.

Additional Materials *(also see Basic Protocol)*

Normal serum
Anti-Ig serum (Zymed Laboratories)
Antigen-specific antiserum *or* antigen-specific purified MAb *or* antigen-specific hybridoma culture supernatant

Follow Basic Protocol, with the following modifications at indicated steps:

2a. Preclear by adding normal serum at a concentration of 2 µl/ml radiolabeled antigen. Add proper amount of anti-Ig serum and let stand 12 to 18 hr at 4°C. Centrifuge 10 min at 1000 × *g* and reserve supernatant.

Normal serum is the source of carrier Ig. The proper amount of anti-Ig serum must be determined by titration with radiolabeled antigen or Ig. For high-titered anti-Ig serum, this amount would be 20× to 40× the volume of antigen-specific antiserum, 2 to 4 µl/µg purified MAb, or ⅓ the volume of hybridoma culture supernatant.

4a. Add 1 µl antigen-specific antiserum, 3 µg antigen-specific purified MAb, or antigen-specific hybridoma culture supernatant (30 µl cloned line or 100 µl uncloned line). Vortex and allow to stand 2 hr at 4°C. Then add the proper amount of anti-Ig serum, vortex, and allow to stand 12 to 18 hr at 4°C.

5a. Wash immunoprecipitate as in Basic Protocol, except centrifuge 7 min at 1000 × *g*.

6a. Add 20 to 50 µl of SDS/sample buffer. Do not vortex, as immunoprecipitates may stick to side of tube above buffer level. Cap tube securely. For immunoprecipitates, first incubate 1 hr at 56°C and then 5 min at 100°C. Load mixture into a gel lane and analyze as in step 7 of the basic protocol.

IMMUNOPRECIPITATION OF RADIOLABELED ANTIGEN WITH ANTI-Ig-SEPHAROSE, PROTEIN A– OR G–SEPHAROSE, OR *S. AUREUS* CELLS

ALTERNATE PROTOCOL 2

Additional Materials *(also see Basic Protocol)*

1:1 (v/v) anti-Ig-Sepharose/dilution buffer (coupled at 10 mg/ml Sepharose as in support protocol), 1:1 (v/v) protein A– or G–Sepharose (Pharmacia

LKB, Calbiochem, or Sigma)/dilution buffer, *or* 10% suspension *S. aureus* Cowan strain II bacteria

Follow Basic Protocol (Fig. 10.15.1), with the following modifications at indicated steps:

2b. Preclear as described below in one of the alternatives to step 4b (i, ii, or iii) as appropriate.

4b. Add 1 µl antigen-specific antiserum, 3 µg antigen-specific MAb, or antigen-specific hybridoma culture supernatant (30 µl cloned line, 100 µl uncloned line). Then perform (i), (ii), or (iii) below.

(i) Add a 1:1 slurry of anti-Ig-Sepharose/dilution buffer.

(ii) Add a 1:1 slurry of protein A– or G–Sepharose/dilution buffer.

For (i) and (ii) above, use an amount 20× to 40× the volume of antiserum, 2 to 4 µl/µg MAb, or ⅓ the volume of hybridoma culture supernatants. Shake 1.5 hr at 4°C.

(iii) Wash a 10% suspension *S. aureus* Cowan II bacteria in lysis buffer in a low-speed centrifuge and resuspend at 10% in dilution buffer. Add 50 µl of the 10% suspension. Shake 10 min at 4°C.

5b. Wash immunoprecipitates as in basic protocol, except centrifuge slurries from (i) or (ii) (step 4b above) 1 min at 200 × *g* or slurry from (iii) 7 min at 1000 × *g*.

6b. Add 20 to 50 µl of SDS/sample buffer. Do not vortex, as Sepharose and bacteria stick to side of tube above buffer level. Cap tube securely and heat 5 min at 100°C. Vortex, microcentrifuge, and load supernatant into a gel lane and analyze as in step 7 of basic protocol.

ALTERNATE PROTOCOL 3

IMMUNOPRECIPITATION USING MORE STRONGLY DISSOCIATING LYSIS AND WASH BUFFERS

This protocol should be used when protein antigens are suspected of being nonspecifically associated with other proteins after immunoprecipitation by the basic or alternate protocols.

Additional Materials *(also see Basic Protocol; see* APPENDIX 1 *for items with ✓)*

10% sodium deoxycholate (Na-DOC)
10% SDS
✓ RIPA buffer

1. To supernatant obtained after step 2 of Basic Protocol, add $\frac{1}{10}$ vol of 10% Na-DOC and $\frac{1}{100}$ vol of 10% SDS to the lysate.

2. Proceed with steps 3 and 4 of Basic Protocol.

3. Proceed with step 5 of Basic Protocol, except use RIPA buffer for first and second washes.

4. Proceed with steps 6 and 7 of the Basic Protocol.

IMMUNOPRECIPITATION OF UNLABELED ANTIGEN WITH ANTIBODY-SEPHAROSE

ALTERNATE PROTOCOL 4

Immunoprecipitation of unlabeled antigen followed by visualization with silver staining eliminates radiolabeling, one of the most tedious and expensive steps in immunoprecipitation protocols. Protein antigens present in greater than ~10^4 copies per eukaryotic cell may be detected by immunoprecipitation of unlabeled (i.e., not radiolabeled as in the Basic Protocol) cell lysates with Ab-Sepharose followed by SDS-PAGE (*UNIT 10.2*) and silver staining (*UNIT 10.6*). If antigen is eluted from beads in SDS lacking reducing agents, little antibody is coeluted.

Additional Materials *(also see Basic Protocol; see APPENDIX 1 for items with ✓)*

✓ Modified lysis buffer
0.1% Triton X-100 in TSA solution
✓ SDS sample buffer *without* 2-mercaptoethanol (2-ME)

1. Incubate 5×10^7 cells/ml in modified lysis buffer 1 hr at 4°C. Centrifuge lysate 15 min at 3000 × *g* to remove nuclei, and then centrifuge supernatant 1 hr at 100,000 × *g*. Save supernatant.

2. Preclear lysate with 50 µl activated, quenched Sepharose (Basic Protocol, step 4) or irrelevant Ab–Sepharose/ml antigen by gently shaking 1 hr at 4°C. Centrifuge 5 min at 200 × *g* and save supernatant.

3. Mix 25 µl of a 1:1 slurry of Ab-Sepharose/TSA solution/ml lysate and gently shake on an orbital shaker 1 hr at 4°C.

   To control for antibody eluting from the beads, a control of Ab-Sepharose incubated with mock lysate should be run simultaneously. A control of lysate incubated with irrelevant Ab-Sepharose should also be run.

4. Wash as in step 5 of Basic Protocol, except use the following buffers:

   First and second washes: 0.1% Triton X-100 in TSA solution
   Third wash: TSA solution
   Fourth wash: 50 mM Tris·Cl, pH 6.8.

5. Add 20 to 50 µl of SDS sample buffer without 2-ME. Do not vortex because Sepharose sticks to side of tube above buffer level. Cap tube securely and heat 5 min at 100°C. Vortex and microcentrifuge 5 sec to pellet Sepharose. Save supernatant and apply to SDS-PAGE directly (nonreducing) or after incubation 1 hr at 37°C with 5% 2-ME (reducing). Load mixture into a gel lane and analyze by SDS-PAGE (*UNIT 10.2*). Detect antigen by silver staining (*UNIT 10.6*).

   Elution of antigen from Sepharose should be done under nonreducing conditions because antibody eluted from the beads under reducing conditions gives background staining.

References: Dustin et al., 1986; Harlow and Lane, 1988; Hjelmeland and Chrambach, 1984; Springer, 1981.

Contributor: Timothy A. Springer

UNIT 10.16

Synthesizing Proteins In Vitro by Transcription and Translation of Cloned Genes

BASIC PROTOCOL

The basis of the method is to clone the protein-coding sequences into a vector containing a promoter for SP6 or T7 RNA polymerase (*UNIT 1.5*), to produce messenger RNA by transcribing the DNA template with this enzyme (*UNIT 3.8*), and to synthesize the desired protein (often as a ^{35}S-labeled species) by translation of this mRNA in vitro. A major advantage is that any desired mutant protein can be generated simply by altering the DNA template.

Materials *(see APPENDIX 1 for items with ✓)*

Plasmid DNA containing SP6 or T7 promoter
DNA containing cloned gene *or* cDNA encoding protein of interest
Appropriate restriction endonucleases (*UNIT 3.1*)
✓ TE buffer
✓ 5× ribonucleoside triphosphate mix
10× SP6/T7 RNA polymerase buffer (*UNIT 3.4*)
10 mM spermidine (for SP6 RNA polymerase only)
Pancreatic ribonuclease inhibitor (e.g., RNasin from Promega)
SP6 *or* T7 RNA polymerase (*UNIT 3.8*)
Buffered phenol (*UNIT 2.1*)
Isobutanol
✓ 10 M ammonium acetate
100% ethanol
In vitro translation kit (wheat germ extract or reticulocyte lysate)
^{35}S-labeled methionine (1400 Ci/mmol)
0.1 M NaOH
10% (v/v) trichloroacetic acid (TCA)
EN^{3}HANCE (Du Pont NEN)

1. Subclone protein-coding DNA sequences of interest into a plasmid vector that contains a promoter for SP6 or T7 RNA polymerase (e.g., pSP64) at a site downstream of the promoter (*UNIT 3.16*).

   The protein-coding sequences must be contiguous (uninterrupted by introns) and in the correct orientation downstream of the bacteriophage promoter such that the correct initiation codon is the first AUG in the RNA to be synthesized and relatively close (25 to 100 bases) to the 5′-end of the RNA.

2. Prepare plasmid DNA (step 1) by CsCl/ethidium bromide centrifugation or PEG precipitation (*UNIT 1.7*).

   The DNA must be high quality and devoid of ribonuclease activity.

3. Cleave 10 µg DNA with a restriction endonuclease that cuts just downstream of termination codon (ideally 50 to 200 bp) and does not cut within the protein-coding region (*UNIT 3.1*). Remove a small aliquot and check by agarose gel electrophoresis (*UNIT 2.5A*).

4. Purify DNA by phenol extraction and ethanol precipitation (*UNIT 2.1*). Resuspend in 50 µl TE buffer.

   This amount of DNA is enough for ten separate in vitro transcription and translation reactions.

5. Set up the following 25-µl reaction mixture at room temperature (to avoid precipitation of the DNA template by spermidine):

   8 µl H_2O
   5 µl DNA (total 1 µg)
   5 µl 5× ribonucleoside triphosphate mix
   2.5 µl 10× SP6/T7 RNA polymerase buffer
   2.5 µl 10 mM spermidine (for SP6 RNA polymerase only)
   1 µl RNasin (30 to 60 U)
   1 µl SP6 or T7 RNA polymerase (5 to 20 U).

   Incubate 60 min at 40°C. For reactions with T7 RNA polymerase, omit the spermidine and add an additional 2.5 µl water.

6. Add 25 µl buffered phenol, vortex, and extract immediately. Transfer aqueous phase to a new microcentrifuge tube, extract twice with isobutanol, and add 6 µl of 10 M ammonium acetate and 70 µl ethanol. Ethanol precipitate with RNA and wash once with ethanol.

7. Resuspend RNA in 24 µl TE buffer, add 6 µl of 10 M ammonium acetate and 70 µl ethanol, reprecipitate, and wash once with ethanol. Resuspend RNA in 10 µl TE buffer. The RNA should be translated immediately or quick frozen and stored at −70°C.

8. Add 1 to 10 µl RNA to an in vitro translation kit and follow directions of manufacturer. Add 15 µCi of $[^{35}S]$methionine (1400 Ci/mmol) to radioactively label the protein. Reactions are typically performed in 30-µl volumes at room temperature for 30 to 60 min. After reaction is complete, store at 0° to 4°C for ≤ 1 week.

9. Remove 1 µl of the translation products and add to 50 µl of 0.1 M NaOH; incubate 15 min at 37°C. Add 1 ml of 10% TCA and incubate 15 min on ice. Collect precipitated protein on glass fiber filters and quantitate incorporated $[^{35}S]$methionine by scintillation counting (*UNIT 3.4*).

10. Remove 3 µl of the translation reaction and analyze by one-dimensional SDS–polyacrylamide gel electrophoresis, including lanes for molecular weight standards. Visualize $[^{35}S]$proteins by fluorography with EN^3HANCE and autoradiograph for 1 to 4 hr (*APPENDIX 3*).

Reference: Melton et al., 1984.

Contributor: Kevin Struhl

Biosynthetic Labeling of Proteins

UNIT 10.17

Biosynthetic labeling techniques are commonly used in the study of biochemical properties, synthesis, processing, intracellular transport, secretion, and degradation of proteins. In this unit, protocols are described for biosynthetically labeling many secreted and membrane proteins. They have been optimized for suspension cultures or from single-cell suspensions of cells from spleen or thymus.

BASIC PROTOCOL

SHORT-TERM LABELING OF CELLS IN SUSPENSION WITH [^{35}S]METHIONINE

Because of its high specific activity (>800 Ci/mmol) and ease of detection, [^{35}S]methionine is the amino acid of choice for biosynthetic labeling of proteins. Its main disadvantage is the relatively low abundance of methionine in proteins (~1.8% of the average amino acid composition; Table 10.17.1). For proteins that contain little or no methionine, other radiolabeled amino acids should be used.

Materials *(see* APPENDIX 1 *for items with* ✓*)*

✓ Short-term labeling medium, warmed to 37°C in a water bath
Cells in suspension grown in a humidified 37°C, 5% CO_2 incubator (e.g., HeLa, BALB/c 3T3, NIH 3T3, CHO, rat embryo fibroblasts, or mouse L)
[^{35}S]L-methionine (800 to 1200 Ci/mmol)
✓ PBS, ice-cold
15- or 50-ml conical tubes with screwcaps, sterile

1. Grow suspension culture to exponential phase. Harvest 10^7 to 10^8 cells by centrifuging 5 min at 300 × *g*, room temperature.

2. Wash cells in conical tube(s) with ~10 ml of 37°C short-term labeling medium/2 × 10^7 cells. Centrifuge 5 min at 300 × *g*, at room temperature. Resuspend cells gently and repeat wash.

Table 10.17.1 Radioactive Amino Acids Used in Biosynthetic Protein Labeling

Amino acid	Frequency (%)[a]	Isotope	Specific activity (Ci/mmol)	Comments[b]
Leucine	10.4	^{3}H	25-190	E
Serine	8.1	^{3}H	5-30	T, I
Glutamate	7.3	^{3}H	15-50	T, I
Lysine	7.0	^{3}H	15-110	E
Alanine	7.0	^{3}H	30-120	T
Valine	6.2	^{3}H	5-65	E
Glycine	5.7	^{3}H	10-60	T, I
Threonine	5.6	^{3}H	5-25	E
Arginine	5.0	^{3}H	15-70	E
Aspartate	4.9	^{3}H	15-40	T, I
Proline	4.9	^{3}H	1-130	—
Glutamine	4.5	^{3}H	20-60	T, I
Phenylalanine	4.5	^{3}H	15-130	E
Tyrosine	3.6	3H	25-100	—
Asparagine[c]	3.5	—	—	—
Cysteine	3.4	^{35}S	>600	—
Isoleucine	2.9	^{3}H	30-120	E
Histidine	2.5	^{3}H	10-60	E
Methionine	1.8	^{35}S	>800	E
Tryptophan	1.3	^{3}H	2-60	E

[a]The frequency of amino acid residues in proteins was taken from Lathe (1985).

[b]E, essential amino acids; T, amino acids whose incorporation into proteins is most decreased by transamination (Coligan et al., 1983); I, amino acids that are converted to other amino acids.

[c]Asparagine is difficult to label (Coligan et al., 1983).

3. Resuspend cells at 5×10^6 cells/ml in 37°C short-term labeling medium and incubate 15 min at 37°C to deplete intracellular pools of methionine. Swirl periodically.

4. Thaw [^{35}S]methionine at room temperature and prepare working solution of 0.1 to 0.2 mCi/ml in 37°C short-term labeling medium.

   CAUTION: *Volatile ^{35}S compounds can be released during labeling. Keep [^{35}S]methionine medium in a tightly capped tube in a 37°C water bath until use. Do not let it sit >30 min at 37°C.*

5. Centrifuge cells 5 min at $300 \times g$ at room temperature. Resuspend cells in conical tubes, using 4 ml of [^{35}S]methionine working solution/2×10^7 cells (5×10^6 cells/ml). Incubate cells 30 min to 3 hr in a 37°C water bath, resuspending frequently by inversion.

6. Centrifuge cells 5 min at $300 \times g$, 4°C. Resuspend with gentle swirling in 10 ml ice-cold PBS and repeat centrifugation.

   CAUTION: *The medium and wash are radioactive—follow safety regulations for use and disposal.*

7. If desired, determine amount of label incorporation by TCA precipitation (Support Protocol). Process and analyze cells as described in immunoaffinity chromatography (*UNIT 10.10*), immunoprecipitation (*UNIT 10.16*), and one- and two-dimensional gel electrophoresis (*UNITS 10.2 & 10.4*).

   If cell pellets cannot be processed immediately, they can be kept on ice for a few hours or frozen at −80°C for several days. Thaw frozen cell pellets on ice before analysis.

SHORT-TERM LABELING OF ADHERENT CELLS WITH [^{35}S]METHIONINE

ALTERNATE PROTOCOL 1

Additional Materials *(also see Basic Protocol)*

Adherent cells, grown in a humidified 37°C, 5% CO_2 incubator

1. Grow adherent cells in 100-mm-diameter tissue culture dishes ($0.5\text{-}2 \times 10^7$ cells) to 70% to 90% confluency. Aspirate culture medium and wash twice by gently swirling with 10 ml of 37°C short-term labeling medium.

2. Add 5 ml of 37°C short-term labeling medium and incubate 15 min in a humidified 37°C, 5% CO_2 incubator to deplete intracellular pools of methionine.

3. Prepare working solutions of [^{35}S]methionine (Basic Protocol, step 4).

4. Remove medium from cells. Add 2 to 4 ml [^{35}S]methionine working solution (from step 3). Incubate 30 min to 3 hr in a humidified 37°C, 5% CO_2 incubator.

5. Remove medium from cells, wash once with 10 ml ice-cold PBS, and remove. Add 10 ml ice-cold PBS and scrape cells from plate.

   CAUTION: *The medium and wash are radioactive—follow safety regulations for use and disposal.*

6. Transfer suspension to a 15-ml conical tube, centrifuge 5 min at $300 \times g$, 4°C, and discard supernatant.

7. Process and analyze cells (Basic Protocol, step 7).

ALTERNATE PROTOCOL 2

LONG-TERM LABELING OF CELLS WITH [^{35}S]METHIONINE

It may be necessary to label cells for a longer period of time when studying proteins that are synthesized at a relatively low rate. Long-term labeling is also used to label proteins in steady state. In this procedure, unlabeled methionine is added to the medium to maintain cell viability and to sustain incorporation of label during the experiment. The amount of unlabeled methionine added will depend on factors such as the length of the labeling period and the cell density. Media containing between 5% and 20% normal methionine concentration are used. The conditions described below are suitable for overnight labeling.

Additional Materials *(also see Basic Protocol; see* APPENDIX 1 *for items with ✓)*

✓ Long-term labeling medium, warmed to 37°C in a water bath

1. Prepare working solution of [^{35}S]methionine (Basic Protocol, step 4).
2. Wash cells in suspension (Basic Protocol, step 2) or adherent cells (Alternate Protocol 1, step 1) twice with long-term labeling medium.
3. For cell suspensions resuspend 1-2.5 × 10^{7} cells in 50 ml [^{35}S]methionine working solution (from step 1) and transfer to a 150-mm^{2} flask. For adherent cells, add 50 ml [^{35}S]methionine working solution to each 150-mm^{2} flask containing 1-2.5 × 10^{7} scraped cells. Incubate 16 hr in a humidified 37°C, 5% CO_2 incubator.
4. Complete step 6 of Basic Protocol or steps 5 to 7 of Alternate Protocol 1.

ALTERNATE PROTOCOL 3

PULSE-CHASE LABELING OF CELLS WITH [^{35}S]METHIONINE

Pulse-chase experiments are employed to analyze time-dependent processes, such as posttranslational modification, transport, secretion, or degradation of newly synthesized proteins. These methods can label biosynthetic precursors; therefore labeling times (pulses) are shorter than in the previous protocols. Labeling is followed by incubation in complete medium containing excess unlabeled amino acid (chase).

Additional Materials *(also see Basic Protocol; see* APPENDIX 1 *for items with ✓)*

✓ Chase medium, 37°C

1. Prepare and label cells with [^{35}S]methionine (Basic Protocol, steps 1 to 5 or Alternate Protocol 1, steps 1 to 4. Pulse-label cells for 5 to 30 min using 0.2 to 1.0 mCi/ml [^{35}S]methionine.

   If necessary, the concentration of cells in suspension can be raised to 1-2 × 10^{7} cells/ml. Use 1-2 × 10^{7} cells per sample (per time-point).

2. After pulse-labeling, remove [^{35}S]methionine medium, wash once with 10 ml 37°C chase medium, and add 10 ml 37°C chase medium.

   Rapid termination of the labeling reaction can be achieved by adding two times the volume of chase medium containing excess unlabeled methionine (15 mg/ml) directly to the labeling mixture.

3. Incubate for desired time at 37°C. Incubate cell suspensions in tightly capped tubes with rotation. Incubate adherent cells in a humidified 37°C, 5% CO_2 incubator.

   Final cell concentration should be 10^6 cells/ml for chases ≤2 hr, or 2 × 10^6 cells/ml for chases >2 hr. For adherent cells, add 10 ml/100-mm-diameter dish.

4a. Collect suspended cells by centrifuging 5 min at 300 × *g*, 4°C, and discard supernatant

4b. Scrape off adherent cells and transfer to 15-ml conical tubes, centrifuge 5 min at 300 × *g*, and discard supernatant. Wash once with 10 ml ice-cold PBS and repeat centrifugation.

5. Process and analyze cells (Basic Protocol, step 7).

BIOSYNTHETIC LABELING WITH OTHER AMINO ACIDS

ALTERNATE PROTOCOL 4

It may be desirable or necessary to label proteins with amino acids other than [^{35}S]methionine—e.g., when the proteins have a low content of methionine or no methionine residues at all. In these cases, the best choices are [^{35}S]cysteine or [^{3}H]leucine.

Cysteine residues are more abundant than methionine residues in proteins (3.4% versus 1.8%; Table 10.17.1). Formulations of [^{35}S]cysteine of high specific activity can be obtained (>600 Ci/mmol). Leucine has the advantage of being the most frequent amino acid in proteins (10.4%; Table 10.17.1) and specific activities of [^{3}H]leucine are the highest among ^{3}H-labeled amino acids (25 to 190 Ci/mmol). If proteins are known to be rich in a particular amino acid or for special methods, such as radiochemical sequencing or multiple labeling, other ^{3}H- or ^{14}C-labeled amino acids can be used. Label cells with these amino acids as described for [^{35}S]methionine (see previous protocols). Substitute [^{35}S]cysteine or [^{3}H]leucine in the labeling medium appropriately; a higher concentration of labeled amino acid is needed to achieve a similar level of labeling with these amino acids.

CAUTION: Solutions containing [^{35}S]cysteine release volatile ^{35}S compounds. The same precautions described above for handling of [^{35}S]methionine should be observed.

TCA PRECIPITATION TO DETERMINE LABEL INCORPORATION

SUPPORT PROTOCOL

In biosynthetic labeling experiments, it is often useful to monitor the incorporation of radioactivity into total cellular proteins. This can be easily achieved by precipitation with trichloroacetic acid (TCA) using bovine serum albumin (BSA) as a carrier protein. This protocol can be used at the end of the labeling procedures in each protocol.

Additional Materials *(also see Basic Protocol)*

0.1 mg/ml bovine serum albumin (BSA) containing 0.02% NaN_3
20% and 10% (v/v) TCA, ice-cold
100% ethanol
Glass microfiber filters, 2.5-cm diameter (Whatman GF/C)

CAUTION: TCA is extremely caustic. Protect eyes and avoid contact with skin when preparing and handling TCA solutions

1. Add 10 to 50 µl of a labeled cell suspension to 0.5 ml of 0.1 mg/ml BSA containing 0.02% NaN_3 and place on ice. Add 0.5 ml ice-cold 20% TCA, vortex vigorously, and incubate 30 min on ice.

2. Filter suspension through a filtration apparatus onto glass microfiber disks. Wash disks twice with 5 ml ice-cold 10% TCA and twice with 100% ethanol. Air dry 30 min.

3. Transfer disks to scintillation vials, add scintillation fluid, and measure radioactivity in a scintillation counter.

Reference: Coligan et al., 1983.

Contributor: Juan S. Bonifacino

UNIT 10.18

Isolation of Proteins for Microsequence Analysis

BASIC PROTOCOL 1

DETERMINATION OF AMINO ACID SEQUENCE OF SAMPLES ON PVDF MEMBRANES

Materials *(see* APPENDIX 1 *for items with ✓)*

Separating and stacking gel solutions (Table 10.18.1)
✓ 4× gel buffer
Glutathione, reduced powder (Sigma)
✓ 10× lower reservoir buffer
✓ 10× upper reservoir buffer
Mercaptoacetic acid, sodium salt
Protein sample in sample buffer (~50 pmol; Support Protocol)
Methanol
✓ Transfer buffer
0.1% Coomassie blue in 50% methanol (v/v)
10% acetic acid in 50% methanol (v/v)

Vertical minigel unit (Bio-Rad)
Power supply (constant voltage and constant current)
Microvolume syringe or gel-loading pipet tip
Powder-free plastic gloves
PVDF membranes (Millipore, Applied Biosystems)
Small-format transfer apparatus (Hoefer, Pharmacia LKB, Bio-Rad)
Automated protein sequencer (Applied Biosystems)

1. Pour denaturing minigels as in UNIT 10.2, substituting the separating and stacking gel solutions listed in Table 10.18.1.

   After removal of combs, the wells should be rectangular and firm. If they are not, prepare a fresh gel; poorly polymerized stacking gels are the most common cause of low sequencing yields.

2. Assemble vertical minigel unit.

3. Dilute 80 ml of 4× gel buffer to 320 ml (to 1× gel buffer). Pour 200 ml of 1× gel buffer into lower buffer reservoir. Add reduced glutathione (powder) to remaining 1× gel buffer to 1.0 mM final. Pour this into upper buffer reservoir.

Table 10.18.1 Recipes for Polyacrylamide Separating and Stacking Gels

SEPARATING GEL

Stock solutions	Final acrylamide concentration in the separating gels (%)										
	5	6	7	8	9	10	11	12	13	14	15
30% acrylamide monomer[a]	3.33	4.00	4.67	5.33	6.00	6.67	7.33	8.00	8.67	9.33	10.00
H_2O	11.49	10.83	10.16	9.50	8.84	8.17	7.51	6.84	6.18	5.51	4.85
TEMED[b]	0.040	0.033	0.029	0.025	0.025	0.020	0.018	0.017	0.015	0.014	0.013

Mix the above ingredients (listed in milliliters of stock solution) with 5 ml of 4× gel buffer and 0.14 ml of 5% potassium persulfate or 70 µl of 10% ammonium persulfate.

STACKING GEL

0.666 ml 30% acrylamide monomer[a]
3.033 ml H_2O
0.025 ml 10% ammonium persulfate *or* 0.05 ml 5% potassium persulfate
0.025 ml TEMED[b]

[a]Gas-stabilized monomer solution (containing 37.5:1 acrylamide/*N,N′*-methylene-bisacrylamide) from which acrylic acid and carbonyl-containing compounds have been removed (Protogel, National Diagnostics; or PAGE1 protein gel mix, Boehringer Mannheim)

[b]TEMED may have to be altered to facilitate proper polymerization. Values given are reasonable approximations.

Attach gel to a constant voltage/constant current power supply (see *UNIT 10.2* for a discussion of electricity and electrophoresis).

4. Preelectrophorese by applying 10 mA per minigel for 45 min.
5. Turn off power supply. Allow gel to stand overnight.
6. Pour off gel buffer and blot wells with tissue or filter paper.
7. Dilute 10× lower and upper reservoir buffers to 1×. Pour 200 ml of 1× lower reservoir buffer into lower reservoir. Add 0.1 g mercaptoacetic acid (sodium salt) to 150 ml of 1× upper reservoir buffer and pour into upper reservoir.
8. Load protein sample (<30 µl for an 0.75-mm gel with 5 wells) dissolved in sample buffer with a microvolume syringe or gel-loading pipet tip.
9. Electrophorese sample by applying 10 mA until pink tracking dye (pyronin Y; see Support Protocol) reaches bottom of gel (60 to 80 min). Turn off power supply.

NOTE: Wear powder-free plastic gloves for all subsequent steps.

10. Wet two PVDF membranes with methanol and immerse them in transfer buffer.
11. Disassemble minigel unit and gently separate glass plates.
12. Assemble blotting sandwich supplied with the small-format transfer apparatus in the following sequence: plastic frame, sponge, filter paper, gel, two PVDF membranes, filter paper, sponge, plastic frame (see *UNIT 10.7*, Fig. 10.7.1; substitute PVDF membrane for nitrocellulose).
13. Insert sandwich into transfer apparatus, placing membranes closest to the anode (red or positive electrode). Fill apparatus with transfer buffer.

14. Apply ~6 V/cm (e.g., 50 V in the LKB Midget MultiBlot apparatus) across electrodes of transfer apparatus for a period of time appropriate to protein of interest and gel concentration. Turn off power supply.

    At the specified voltage, transfer takes 30 to 40 min for 20- to 50-kDa proteins in a 15% gel, 60 to 70 min for 50- to 100-kDa proteins in a 10% to 12% gel, and 90 min for 150- to 200-kDa proteins in a 7% gel. For proteins >60 kDa, reduce the amount of methanol in transfer buffer to 1%.

15. Disassemble transfer apparatus. Immerse PVDF membrane blots in 0.1% Coomassie blue in 50% methanol and agitate 5 min.

16. Destain blots in 10% acetic acid prepared in 50% methanol by agitating until bands become clearly visible (5 to 10 min).

17. Transfer to water and photograph (optional; *UNIT 10.6*).

18. Excise band of interest with a razor blade (use a new razor blade for each band). Place each band in a microcentrifuge tube and allow to air dry at room temperature (do not heat). Store excised bands at –20°C.

    A sample for sequencing must be a purified protein free of contaminants.

19. Insert excised band into sequencer reaction cartridge, protein side facing the solvent delivery (see manufacturer's instructions).

20. Sequence the sample on an automated protein sequencer.

SUPPORT PROTOCOL

PREPARATION OF PROTEIN SAMPLES FOR SDS-PAGE

***Additional Materials** (also see Basic Protocol 1; see APPENDIX 1 for items with ✓)*

Protein samples
1 M $NaHCO_3$ (optional)
100% ethanol, ice-cold (containing no denaturants; USP grade)
✓ Sample buffer
0.1% (w/v) pyronin Y
Ultrafiltration concentrator (Amicon) or Speedvac evaporator (Savant)
Drawn-out Pasteur pipet *or* gel-loading pipet tip
Boiling water bath

1. Adjust the salt concentration of the protein sample to >100 mmol/liter with 1 M $NaHCO_3$, if necessary.

2. Concentrate samples to 50 to 100 µl by ultrafiltration (follow manufacturer's instructions explicitly) or vacuum centrifugation (it is crucial to avoid introduction of airborne debris when vacuum is released). Transfer to 1.5-ml microcentrifuge tubes.

3. Add 9 vol ice-cold 100% ethanol to the samples. Incubate 1 hr on dry ice or overnight at –20°C.

   Most samples may be kept indefinitely at this stage.

4. Microcentrifuge 15 min at maximum speed. Aspirate the supernatant with a drawn-out Pasteur pipet or a gel-loading pipet tip and save the pellet.

5. Dissolve the pellet in 10 µl sample buffer by drawing the sample buffer up and down with a pipettor. Boil 3 min.

6. Add 1 µl of 0.1% pyronin Y (tracking dye) to the sample. Load on the minigel.

DETERMINATION OF INTERNAL AMINO ACID SEQUENCE FROM N-TERMINALLY BLOCKED PROTEINS

BASIC PROTOCOL 2

Materials *(see APPENDIX 1 for items with ✓)*

Protein sample (~200 pmol)
✓ 0.1% (w/v) Ponceau S (Sigma) in 1% (v/v) acetic acid
1% (v/v) acetic acid
✓ 0.2 mM NaOH
0.5% (w/v) polyvinylpyrrolidone (Sigma) in 0.1 M acetic acid
✓ Digestion buffer
1 mg/ml sequencing-grade trypsin (Promega)
Chromatography solvent A: 5% (v/v) acetonitrile in 0.1% (v/v) trifluoroacetic acid (TFA)
Chromatography solvent B: 70% (v/v) acetonitrile in 0.085% (v/v) TFA

0.22–µm nitrocellulose membrane (e.g., Schleicher & Schuell)
Acid-washed glass plate or petri dish
Powder-free gloves
Fine-tipped forceps
0.5-ml microcentrifuge tube
Bath sonicator (Bransonic)
Centrifugal filter device, 0.22-µm membrane, low-protein-binding (e.g., Millipore)
Reversed-phase HPLC column (e.g., Vydac), UV column monitor, and chart recorder
Column oven (optional)

1. Resolve protein(s) of interest by electrophoresis and transfer to nitrocellulose membrane as described in *UNIT 10.7*.

   Although PVDF membranes may be used, nitrocellulose seems to give better yields.

2. Place nitrocellulose membrane in an acid-washed glass petri dish (or similar vessel) containing 50 ml (for an 8 × 10–cm minigel) of 0.1% Ponceau S prepared in 1% aqueous acetic acid. Agitate gently 1 min.

3. Transfer to 1% acetic acid for 1 min, changing the solution as necessary to allow easy visualization of band(s).

4. Using a new razor blade and wearing powder-free gloves, cut out the band(s) of interest, carefully removing all excess nitrocellulose. Place in 1.5-ml microcentrifuge tube(s). Use a piece of blank nitrocellulose as a control.

   Meticulous technique is crucial at this step to eliminate contamination of the sample by adventitious proteins. A clean work area, scrupulously clean glassware and instruments, and powder-free gloves are minimum precautions.

   The bands may be transferred to microcentrifuge tubes containing 0.5 ml water and stored at −20°C at this step.

5. Transfer membrane pieces to 1 ml of 0.2 mM NaOH and vortex 1 min to destain. Aspirate the NaOH.

6. Immediately add 1 ml of 0.5% PVP-40 in 0.1 M acetic acid and agitate tube gently 30 min at room temperature.

7. Aspirate the PVP-40 and wash membrane five times with 1 ml water. Be sure to remove any liquid droplets caught under tube cap.

8. Using clean fine-tipped forceps, transfer excised band to a clean glass surface (e.g., acid-washed glass plate or petri dish) and cut it into 1- to 2-mm pieces. Use forceps to collect the pieces and squeeze out excess liquid.

9. Immediately transfer pieces to a 0.5-ml microcentrifuge tube containing 25 µl digestion buffer. Add 1 µl 1 mg/ml trypsin. Mix so that membrane pieces are evenly coated with solution. Incubate overnight at room temperature.

10. Microcentrifuge sample 1 sec at high speed, room temperature, to recover liquid that may have condensed on tube walls and cap.

11. Sonicate sample 5 min at room temperature.

12. Microcentrifuge 1 min at top speed, room temperature. Transfer supernatant to a centrifugal filter device. Rinse membrane pieces with 100 µl digestion buffer and add to supernatant. Microcentrifuge sample 20 to 30 sec at top speed.

    Samples may be stored frozen until they are fractionated.

13. Equilibrate a 2.1-mm-i.d. × 250-mm reversed-phase HPLC column with 95% chromatography solvent A/5% solvent B at a flow rate of 0.15 ml/min. For optimal resolution, perform the separation at 60°C if a column oven is available.

14. Inject the sample. Wash column 10 to 15 min with 95% solvent A/5% solvent B.

15. Elute the peptides with a gradient between chromatography solvents A and B as follows: 5% to 40% solvent B over 1 hr; 40% to 75% solvent B over 30 min; and 75% to 100% solvent B over 15 min. Monitor elution of the peptides at 215 nm.

    The TFA concentration should be adjusted to equalize the UV absorbance (215 nm) of the eluants. Alternatives to TFA, such as phosphoric acid or hydrochloric acid, are compatible with the procedure. Buffers containing ammonia or UV-containing impurities should not be used.

    Trypsin elutes at 60% solvent B and can be used as an internal standard. Characteristically, it elutes in a broader peak than the majority of peptides.

16. Monitor the appearance of peptide peaks with a chart recorder adjusted so that 0.05 to 0.1 AUFS corresponds to a full-scale deflection.

    Minimize the length and capacity of the capillary tubing between the column and the point of collection as much as possible. Polyethylether ketone (PEEK) tubing (0.005-in. i.d.) is helpful for this purpose.

17. When chart-recorder pen begins a deflection indicative of a peak, wipe tip of capillary tubing against a clean surface (e.g., a Kimwipe or the previous collection tube) and collect this fraction in a microcentrifuge tube.

    Clean gloves should be worn at this point. Commercial microcentrifuge tubes are sufficiently clean as supplied by the manufacturer if they are stored in the original container and handled with clean gloves.

18. Immediately cap tube and place it on dry ice or store at −80°C.

19. Sequence samples using glass support disks precycled with polybrene as prescribed by Applied Biosystems.

References: Moos et al., 1988; Tempst et al., 1990.

Contributor: Malcolm Moos, Jr.

Immunology

Certain technological advances in the field of molecular biology were made possible in part by earlier progress in the field of immunology. A review of the earlier chapters in this book documents the importance of immunological methods to the purification of proteins as well as to the identification of specific cDNA clones. Specific antibodies have greatly facilitated the purification of proteins by immunoaffinity chromatography (*UNIT 10.10*) and immunoprecipitation (*UNIT 10.15*). When pure protein has been unavailable for deducing the complementary oligonucleotide sequence, specific antibodies have been utilized to screen recombinant DNA libraries for the desired cDNA clones (*UNIT 6.7*) and select mRNA for the translation of desired protein (*UNIT 6.8*). Specific antibodies have also been utilized to identify antigens by immunoblotting (*UNIT 10.7*).

Just as immunology has facilitated the advances made in the field of molecular biology, the latter in turn has contributed to a better understanding of the basis for antibody diversity. The clonal selection theory proposed by Sir Macfarlane Burnet in 1959 is now an accepted concept: each B cell differentiates into a plasma cell committed to the production of antibodies specific for one antigen—i.e., the antibodies are monoclonal in nature. "Clonal selection" refers to the fact that when an antigen binds to one of these antibodies on the membrane of the B cell, the cell is stimulated to proliferate (at which point some variation may be introduced in the "monoclonal" cell line). Generally, many clones respond to a single antigen, as most proteins carry multiple antigenic sites (called epitopes). The overall immune response is polyclonal, with specific recognition of multiple, discrete epitopes.

An understanding of the genetic mechanisms responsible for antibody (or immunoglobulin) diversity requires some knowledge of antibody structure. Man has five major immunoglobulin classes: IgG, IgA, IgD, IgE, and IgM, which share the same type of combining site for antigen. The immunoglobulin molecule is similar for the first four classes; it consists of four polypeptides—two heavy chains and two light chains—arranged in the shape of the letter "Y," with a molecular weight of ~150,000. The IgM class, with a molecular weight of ~800,000, consists of five Y-shaped molecules arranged in a cyclic pentamer, with the antigen-binding sites facing outward. Although the different immunoglobulin classes can share the same κ or λ light chains, they are each distinguished by their unique heavy chains, designated γ (IgG), α (IgA), δ (IgD), ε (IgE), and μ (IgM). The heavy and light chains are each composed of constant and variable regions. The antigen-binding site, a cleft of about 15 Å × 20 Å × 10 Å deep formed by interactions of hypervariable regions of the heavy- and light-chain variable regions, is unique for each antibody.

For many years it was assumed that the mammalian germ line must include a separate gene for every polypeptide that ultimately appears in an antibody; this model presupposes a vast number of immunoglobulin genes. In the past decade, however, recombinant DNA technology has shown that diversity in antigen-binding sites arises through genetic recombination in somatic cells—i.e., while B lymphocytes are maturing and differentiating in the bone marrow. Located on different chromosomes are approximately 50 genes coding for the "constant" C regions, the "variable" V regions, the "joining" J segments (which combine with the C and V regions to make up the antibody's light chain) and the "diversity" D segments (which combine with C, J, and V regions to comprise the antibody's heavy chain). Mouse germ cells have a few hundred V segments, approximately 20 D segments, and 4 J segments, which can be assembled in >10,000 combinations. Subsequent assemblage of heavy and light chains could yield >10 million specific antigen-binding sites. (For an excellent review of the molecular biology of the immune system, see Tonegawa, 1985.)

This chapter presents the methodologies for the preparation of both monoclonal and polyclonal antibodies. The first part describes the enzyme-linked immunosorbent assay (ELISA), a highly sensitive, versatile, and quantitative technique that requires little equipment and for which critical reagents are readily available. The preparation of enzyme-antibody conjugates, which forms the basis of this assay, is described in *UNIT 11.1*. The versatility of ELISAs is demonstrated by the six distinct ELISA protocols presented in *UNIT 11.2*. These provide general methods for the detection of specific antibodies, soluble antigens, or cell-surface antigens.

The pioneering studies of Kohler and Milstein (1975) enable investigators to obtain milligram quantities of specific monoclonal antibodies after immunizing mice with relatively impure antigen. The spleen is removed from a previously immunized mouse that has a sufficient antibody titer. After separation into individual cells, B cells from the spleen are fused with myeloma cells of B cell origin to produce immortal antibody-secreting hybridoma cells of predetermined specificity. Each hybridoma cell is capable of producing an unlimited supply of a single, antigen-specific monoclonal antibody. Production of cell culture supernatants of monoclonal antibodies in ascites fluid is described in *UNIT 11.3* and purification of these monoclonal antibodies by affinity chromatography is presented in *UNIT 11.4*. Detection of antibody in serum, hybridoma supernatants (micrograms per milliliter), and ascites fluid (milligrams per milliliter) by ELISA is described in *UNIT 11.2*.

Although monoclonal antibodies can be made available in unlimited quantities and without the need to purify the antigen to homogeneity, the reliance upon only monoclonal antibodies for detection and identification of antigen and cDNA clones can produce equivocal results. Because monoclonal antibodies may be specific for short peptide sequences, there is a possibility of obtaining false positives, since unrelated proteins can share small regions of homology. One way in which this uncertainty can be minimized is to utilize several different monoclonal antibodies specific for different sites on the antigen. Another disadvantage of using a monoclonal antibody is that it may have a relatively low affinity for a given antigenic site.

These problems caused by the use of monoclonal antibodies may be circumvented by generating polyclonal antibodies, which consist essentially of numerous monoclonal antibodies with different epitope specificities. When a purified antigen is available in sufficient amount for immunization, it is possible to obtain specific polyclonal antibodies with high affinity after repeated immunizations (*UNIT 11.5*; Klinman and Press, 1975). Choice of animal is determined by the amount of antiserum required for subsequent experiments. Although animals such as goats, sheep, or horses can provide larger volumes of antiserum, few institutions have adequate facilities for their care and maintenance. Mice, rats, and guinea pigs, on the other hand, may not yield sufficient volumes of antiserum. For these reasons, rabbits have become the animal of choice for the generation of polyclonal antibodies. *UNIT 11.5* describes the proper preparation of antigen as well as various routes of immunization in rabbits to optimize the antibody response. Although a schedule for immunization and boosting is provided, this procedure is only a recommendation of what has worked for the author; optimal conditions should be determined empirically. *UNIT 11.6* discusses the purification from serum, ascites fluid, or hybridoma supernatant of the immunoglobulin G fraction, which becomes the predominant antibody class after the booster injection.

If purified antigen is in limited supply, polyclonal (as well as monoclonal) antibodies can still be raised by immunization with synthetic peptides whose sequences are based on that of the protein which it is designed to mimic. In this case, the selection of an immunogenic peptide is vital for obtaining a good antibody response. *UNIT 11.7* discusses the necessary parameters to consider in the selection of a particular peptide sequence that will elicit an antibody that recognizes the native form of the protein. To enhance the immunogenicity of the peptide, it can be chemically cross-linked to a carrier molecule (*UNIT 11.8*). Such cross-linking of the peptide has been demonstrated to be helpful in generating an antibody response to peptides that might not otherwise elicit antibody production.

References: Burnet, 1959; Klinman and Press, 1975; Kohler and Milstein, 1975; Tonegawa, 1985.

John A. Smith

11.1

Conjugation of Enzymes to Antibodies

UNIT 11.1

Enzymes are conjugated with antigen-specific antibodies to permit detection of antigen using an ELISA (*UNIT 11.2*) or immunoblotting (*UNIT 10.7*). Horseradish peroxidase (HRPO) and alkaline phosphatase conjugates can be used in both of these assays; both conjugates can detect 1 to 10 ng/ml of antigen, although the latter are more stable.

CONJUGATION OF HRPO TO ANTIBODIES

BASIC PROTOCOL

Materials *(see APPENDIX 1 for items with ✓)*

1 mg/ml antibody solution (affinity-purified polyclonal or monoclonal IgG)
✓ 0.1 M sodium phosphate buffer, pH 6.8
Horseradish peroxidase (HRPO; Sigma Type VI)
✓ 0.1 M carbonate buffer, pH 9.2
Sodium periodate ($NaIO_4$) solution: 1.71 mg $NaIO_4$/ml H_2O (prepare fresh)
Sodium borohydride ($NaBH_4$) solution: 5 mg $NaBH_4$/ml 0.1 mM NaOH (prepare fresh)
✓ Saturated ammonium sulfate (SAS) solution
✓ Tris/EDTA/NaCl (TEN) buffer, pH 7.2
Bovine serum albumin (BSA)
Glycerol

Dialysis membrane
Sephadex G-25, medium (Pharmacia)

1. Dialyze ≥1 mg/ml antibody solution (A_{280}/1.44 = 1 mg IgG/ml) against 2 liters of 0.1 M sodium phosphate buffer, pH 6.8, overnight at 4°C with gentle stirring.

2. Dissolve 10 mg HRPO in 1 ml of 0.1 M carbonate buffer, pH 9.2. Mix 0.25 ml with 0.25 ml freshly prepared $NaIO_4$ solution, cap tightly, and incubate 2 hr at room temperature in the dark.

3. Into a Pasteur pipet fitted with glass wool and blocked at the tip with Parafilm, add 1 ml of the dialyzed antibody solution (step 1) to the entire mixture from

step 2. Add 0.25 g Sephadex G-25 and incubate 3 hr at room temperature in the dark.

4. Wash column with 0.75 ml carbonate buffer to elute conjugate. To the eluate, add 38 µl freshly prepared $NaBH_4$ solution and incubate 30 min at room temperature in the dark.

5. Add 112 µl freshly prepared $NaBH_4$ and incubate 60 min in the dark. Add 0.9 ml saturated (SAS) solution and stir gently 30 min at 4°C.

6. Centrifuge 15 min at 10,000 × *g*, 4°C. Discard supernatant, resuspend pellet in 0.75 ml TEN buffer, and dialyze against 2 liters TEN buffer overnight at 4°C. Change TEN solution and continue dialysis 4 hr.

7. Remove conjugate from dialysis membrane and add BSA to 20 mg/ml. Add an equal volume of glycerol and store at −20°C.

ALTERNATE PROTOCOL

CONJUGATION OF ALKALINE PHOSPHATASE TO ANTIBODIES

Additional Materials *(also see Basic Protocol; see* APPENDIX 1 *for items with ✓)*

5 mg/ml antibody solution (*UNIT 11.11*)
10 mg/ml alkaline phosphatase (enzyme immunoassay grade; Boehringer Mannheim; source is important)
25% (v/v) glutaraldehye in H_2O
✓ Tris/ovalbumin buffer
Sodium azide

1. Dialyze 5 mg/ml antibody solution as in step 1 of Basic Protocol except use PBS. When dialysis is complete, place antibody solution in a tube and read A_{280}. Dilute with PBS to 3 mg/ml.

2. Add 100 µl of 3 mg/ml antibody solution to 90 µl of 10 mg/ml alkaline phosphatase in a 1.5-ml microcentrifuge tube. Add 5 µl of 25% glutaraldehyde, mix gently, and let stand at room temperature.

3. Remove 25-µl aliquots at 0, 5, 10, 15, 30, 60, and 120 min to individual 1.5-ml microcentrifuge tubes (store on ice). Add 125 µl PBS to each sample, then add 1.1 ml Tris/ovalbumin buffer.

4. Dialyze samples as in step 1. Test samples for alkaline phosphatase activity using a direct ELISA (*UNIT 11.2*) to determine optimal conjugation time. Repeat steps 1 to 3 using optimal time.

5. Add sodium azide to 0.1% and store protected from light ≤1 year at 4°C, or add equal volume glycerol and store ≤1 year at −20°C.

Reference: Van Vunakis and Langone, 1980.

Contributors: Scott E. Winston, Steven A. Fuller, Michael J. Evelegh, and John G.R. Hurrell

Enzyme-Linked Immunosorbent Assay (ELISA)

This unit describes six ELISA sytems for detecting antigen and antibodies (Figs. 11.2.1-11.2.6). In all protocols, the solid-phase reagents are incubated with secondary or tertiary reactants covalently coupled to an enzyme. Unbound conjugates are washed out and a chromogenic or fluorogenic substrate is added. As the substrate is hydrolyzed by the bound enzyme conjugate, a colored or fluorescent product—proportional to the amount of analysate in the test mixture—is generated and detected visually or with a microtiter plate reader. Antibody-sandwich ELISAs are generally the most sensitive and can detect 100 pg/ml to 1 ng/ml protein antigen (direct ELISAs are often an order of magnitude less sensitive).

INDIRECT ELISA TO DETECT SPECIFIC ANTIBODIES

BASIC PROTOCOL

This assay is useful for screening antisera or hybridoma supernatants for specific antibodies when milligram quantities of purified or semipurified antigen are available (1 mg of purified antigen will permit screening of 80 to 800 microtiter plates; Fig. 11.2.1).

Materials *(see* APPENDIX 1 *for items with ✓)*

Developing reagent: protein A–alkaline phosphatase conjugate (Sigma), protein G–alkaline phosphatase conjugate (Calbiochem), *or* anti-Ig-alkaline phosphatase conjugate (*UNIT 11.1*; Southern Biotechnology or *Linscott's Directory*)

Test antigen solution (coating reagent)

PBS containing 0.05% NaN_3 (PBSN)

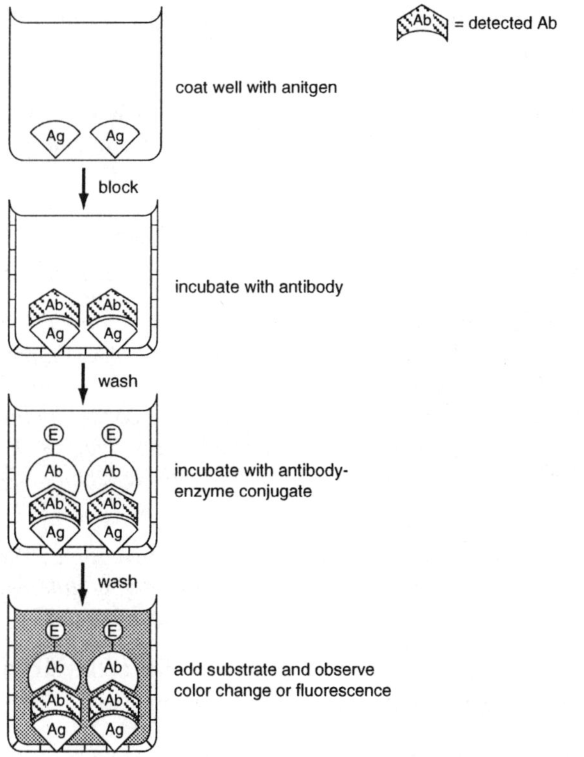

Figure 11.2.1 Indirect ELISA to detect specific antibodies. Ag = antigen; Ab = antibody; E = enzyme.

✓ Blocking buffer
Antibody samples to be tested
✓ 4-methylumbelliferyl phosphate (MUP) *or* *p*-nitrophenyl phosphate (NPP) substrate solution
0.5 M NaOH (optional)

Immulon 2 (Dynatech), Immulon 4 (Dynatech), or equivalent microtiter plates
Microtiter plate reader (optional): spectrophotometer with 405-nm filter *or* spectrofluorometer (Dynatech) with 365-nm excitation filter and 450-nm emission filter

1. Determine optimal concentrations of developing reagent (conjugate) and antigen coating reagent by criss-cross serial dilution analysis (Support Protocol 1). Prepare antigen solution in PBSN at this final concentration (0.2 to 10.0 µg/ml; ~6 ml/plate; can be stored at 4°C).

   Pure antigen solution concentrations are usually ≤2 µg/ml. Although pure antigen preparations are not essential, >3% of the total protein should be antigen. Do not raise the antigen concentration to >10 µg/ml.

2. Using a multichannel pipet, dispense 50 µl antigen solution into each well of an Immulon microtiter plate and tap or shake the plate to evenly distribute antigen. Wrap coated plates in plastic wrap to seal and incubate overnight at room temperature or 2 hr at 37°C.

   Sealed plates can be stored at 4°C with antigen solution for months.

3. Fill well of plate with deionized or distilled water from a plastic squirt bottle and flick into sink. Repeat rinse two more times.

4. Fill each well with blocking buffer using a squirt bottle and incubate 30 min at room temperature.

5. Rinse plate three times with water as in step 3. After last rinse, remove residual liquid by wrapping each plate in a large paper tissue and gently flicking it face down onto several paper towels.

6. Add 50 µl antibody samples diluted in blocking buffer to each coated well, wrap plate in plastic wrap, and incubate ≥2 hr at room temperature.

   Hybridoma supernatants (UNIT 11.3) can usually be diluted 1:5 and ascites fluid (UNIT 11.3) and antisera (UNIT 11.5) diluted 1:500 in blocking buffer and still generate a strong positive signal. Dilutions of nonimmune ascites or sera should be assayed as a negative control. Prepare antibody dilutions in cone- or round-bottom microtiter plates before adding them to antigen-coated plates. Sources of appropriate antibodies and conjugates can be found in Linscott's Directory of Immunological and Biological Reagents.

   The specific signal may be increased by longer incubations. The same pipet tips can be reused for hundreds of separate transfers involving different solutions if the tips are washed 5 times in blocking buffer between transfers.

7. Rinse plate three times in water as in step 3. Fill each well with blocking buffer, vortex, and incubate 10 min at room temperature.

8. Rinse three times in water as in step 3. After final rinse, remove residual liquid as in step 5.

9. Add 50 µl developing reagent in blocking buffer (at optimal concentration determined in step 1) to each well, wrap in plastic wrap, and incubate ≥2 hr at room temperature. Wash plates as in steps 7 and 8.

   The strength of the signal may be increased by longer incubations.

   After final rinsing, plates may be wrapped in plastic wrap and stored for months at 4°C prior to adding substrate.

10. Add 75 µl MUP or NPP substrate solution to each well and incubate 1 hr at room temperature. Monitor hydrolysis qualitatively by visual inspection or quantitatively with a microtiter plate reader (see below). Hydrolysis can be stopped by adding 25 µl of 0.5 M NaOH.

    a. Hydrolysis of NPP appears yellow. For spectrophotometer measurement of NPP hydrolysis, use a 405-nm filter.

    b. Hydrolysis of MUP can be monitored in a darkened room with a long-wavelength UV lamp. For spectrofluorometer measurement, use a 365-nm excitation filter and a 450-nm emission filter.

    The fluorogenic system using MUP is 10 to 100 times faster than the chromogenic system using NPP.

DIRECT COMPETITIVE ELISA TO DETECT SOLUBLE ANTIGENS

ALTERNATE PROTOCOL 1

This assay is most useful when both a specific antibody and milligram quantities of purified or semipurified antigen are available (Fig. 11.2.2).

Additional Materials *(also see Basic Protocol)*

Specific antibody–alkaline phosphatase (AP) conjugate (*UNIT 11.1*)
Standard antigen solution

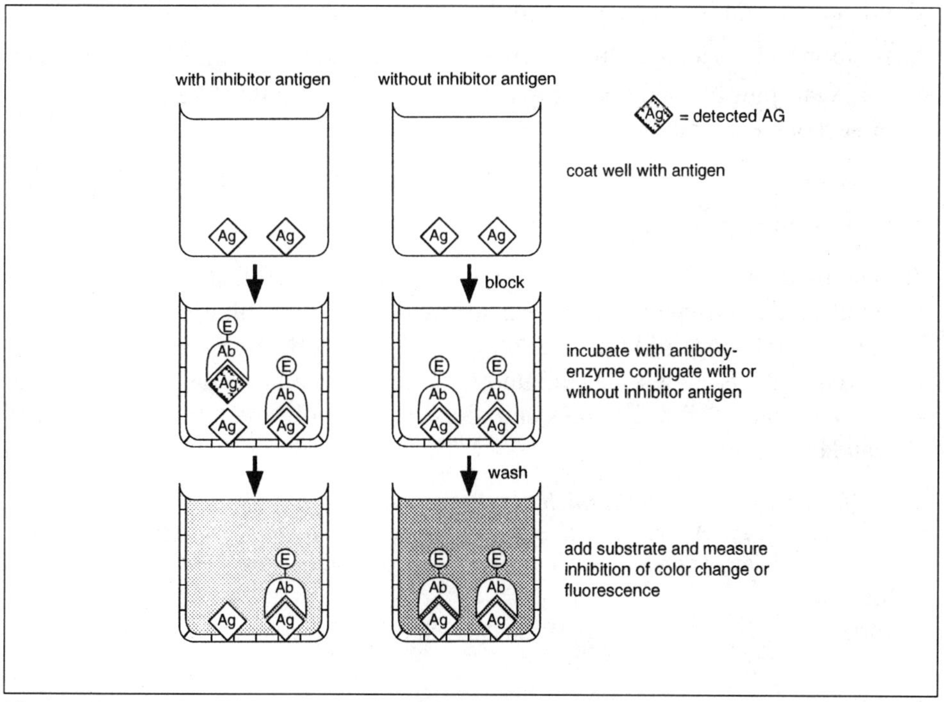

Figure 11.2.2 Direct competitive ELISA to detect soluble antigens. Ag = antigen; Ab = antibody; E = enzyme.

1. Determine optimal concentrations of coating reagent (antigen) and developing reagent (antibody-AP conjugate) by criss-cross serial dilution analysis (Support Protocol 1). Dilute the conjugate in blocking buffer to twice the optimal concentration.

   The final concentration is usually 25 to 500 ng antibody/ml. Prepare 3 ml antibody-AP conjugate for each plate.

2. Coat, block, and rinse wells of an Immulon microtiter plate with 50 µl antigen solution (Basic Protocol, steps 2 to 5).

3. Prepare six 1:3 serial dilutions of standard antigen solution in blocking buffer (Support Protocol 1)—these antigen concentrations will be used in preparing a standard inhibition curve (step 7).

   Antigen concentrations should span the dynamic range of inhibition, i.e., the range of inhibitor concentrations that produces detectable changes in the amount of inhibition. This must be determined empirically in an initial assay in which antigen concentration is varied from 10^{-6} M to 10^{-12} M. If possible, initial concentration should be ~100 µg/ml, followed by nine 1:4 serial dilutions in blocking buffer. These are assayed for their ability to inhibit binding of conjugate to antigen-coated plates. From this initial assay, six 1:3 antigen dilutions spanning the central segment of the dynamic range of inhibition are used as standard antigen-inhibitor dilutions. Prepare ≥75 µl of each dilution for each plate to be assayed. Inhibitor curves are most sensitive in the region of the curve where small changes in inhibitor concentrations produce maximal changes in the amount of inhibition (normally 15% to 85% inhibition). In most systems, this range of inhibition is produced by concentrations of inhibitor between 1 and 250 ng/ml.

4. Add 75 µl of 2× conjugate solution (from step 1) to each well of a round- or cone-bottom microtiter plate, followed by 75 µl inhibitor—either test antigen solution or standard antigen solution (from step 3). Mix solutions by pipetting up and down three times and incubate ≥30 min at room temperature.

5. Prepare uninhibited control samples by mixing equal volumes of 2× conjugate solution and blocking buffer. Transfer 50 µl conjugate plus inhibitor (step 4) or conjugate plus blocking buffer to antigen-coated plate (step 2) and incubate ≥2 hr at room temperature.

6. Wash plate (Basic Protocol, steps 7 and 8). Add 75 µl of MUP or NPP substrate solution to each well and incubate ≥1 hr at room temperature.

7. Quantitate on the microtiter plate reader (Basic Protocol, step 10). Prepare a standard antigen-inhibition curve constructed from the inhibitions produced by the standard antigen-inhibitor dilutions (step 3). Plot antigen concentration on x axis, which is a log scale, and fluorescence or absorbance on y axis, which is a linear scale. Interpolate concentration of antigen in the test solutions from standard curve.

   The dynamic range of the inhibition curve may deviate from linearity if the specific antibodies are heterogeneous and possess significantly different affinities or if the standard antigen preparation contains heterogeneous forms of the antigen.

ANTIBODY-SANDWICH ELISA TO DETECT SOLUBLE ANTIGENS

Antibody-sandwich ELISAs are frequently 2 to 5 times more sensitive in detecting antigen than those ELISAs in which antigen is directly bound to the solid phase (Fig. 11.2.3).

Additional Materials *(also see Basic Protocol)*

Specific antibody or Ig fraction from antiserum (*UNIT 11.5*), or ascites fluid (*UNIT 11.3*), *or* hybridoma supernatant (*UNIT 11.5*), *or* bacterial lysate (Support Protocol 2)

1. Determine concentrations of capture antibody and conjugate necessary to detect desired concentration of antigen by criss-cross serial dilution analysis (Support Protocol 1). Prepare capture antibody solution in PBSN at this concentration, usually between 0.2 and 10µg/ml.

2. Coat wells of an Immulon plate with capture-antibody solution (Basic Protocol, steps 2 and 3), and block wells (Basic Protocol, steps 4 and 5).

3. Prepare a standard antigen-dilution series by successive 1:3 dilutions of homologous antigen stock in blocking buffer (Support Protocol 1).

   The standard antigen-dilution series should span most of the dynamic range of binding (usually between 0.1 to 1000 ng antigen/ml).

4. Prepare dilutions of antigen test solutions in blocking buffer. Add 50-µl aliquots of antigen test solutions and standard antigen dilutions (from step 3) to antibody-coated wells and incubate ≥2 hr at room temperature.

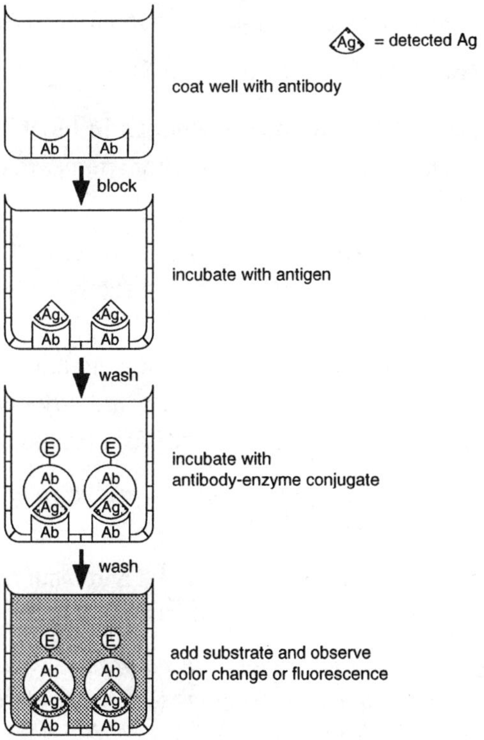

Figure 11.2.3 Double antibody–sandwich ELISA to detect specific antibodies. Ag = antigen; Ab = antibody; E = enzyme.

5. Wash plate (Basic Protocol, steps 7 and 8). Add 50 µl specific antibody-AP conjugate and incubate 2 hr at room temperature.

   The antibody concentration is typically 25 to 400 ng/ml.

6. Wash plate (Basic Protocol, steps 7 and 8). Add 75 µl of MUP or NPP substrate solution to each well and incubate ≥1 hr at room temperature.

7. Quantitate on a microtiter plate reader (Basic Protocol, step 10). Prepare standard curve constructed from data produced by serial dilutions of standard antigen (step 4). Plot antigen concentration on x axis which is a log scale, and fluorescence or absorbance on y axis which is a linear scale. Interpolate concentration of antigen in test solutions from standard curve.

ALTERNATE PROTOCOL 3

DOUBLE ANTIBODY–SANDWICH ELISA TO DETECT SPECIFIC ANTIBODIES

This assay is especially useful for detecting specific antibodies when small amounts are available and purified antigen is unavailable. Additionally, this method can be used for epitope mapping of different monoclonal antibodies that are directed against the same antigen (Fig. 11.2.4).

Additional Materials *(also see Basic Protocol)*

Capture antibodies specific for Ig from the immunized species
Specific antibody–alkaline phosphatase (AP) conjugate (*UNIT 11.1*)

1. Coat wells of an Immulon microtiter plate with 50 µl of 2 to 10 µg/ml capture antibodies (Basic Protocol, steps 2 and 3).

   Capture antibodies must not bind the antigen or conjugate antibodies. When analyzing hybridoma supernatants or ascites fluid, coat plates with 2 µg/ml capture antibody. When analyzing antisera, coat plates with 10 µg/ml capture antibody.

2. Block wells (Basic Protocol, steps 4 and 5).

3. Prepare dilutions of test antibody solutions in blocking buffer. Add 50 µl to coated wells and incubate ≥2 hr at room temperature. Wash plate (Basic Protocol, steps 7 and 8).

   Hybridoma supernatants (UNIT 11.3), antisera (UNIT 11.5), or ascites fluid (UNIT 11.3) can be used as the test samples. Dilute hybridoma supernatants 1:5 and antisera or ascites fluid 1:200.

4. Prepare antigen solution in blocking buffer containing 20 to 200 ng/ml antigen. Add 50-µl aliquots of antigen solution to antibody-coated wells and incubate ≥2 hr at room temperature. Wash plate (Basic Protocol, steps 7 and 8).

   Although purified antigen preparations are not essential, the limit of detection for most protein antigens is 2 to 20 ng/ml.

5. Add 50 µl antibody-AP conjugate to wells and incubate 2 hr at room temperature. Wash plate as in steps 7 and 8 of basic protocol.

   The conjugate concentration is typically between 25 to 500 ng specific antibody/ml, and should be high enough to result in ~0.50 absorbance units/hr at 405 nm when using NPP as a substrate or a signal of 1000 to 1500 fluorescence units/hr when using MUP. If no specific antibodies from the appropriate species are available to serve as a positive control, then a positive control system should be constructed out

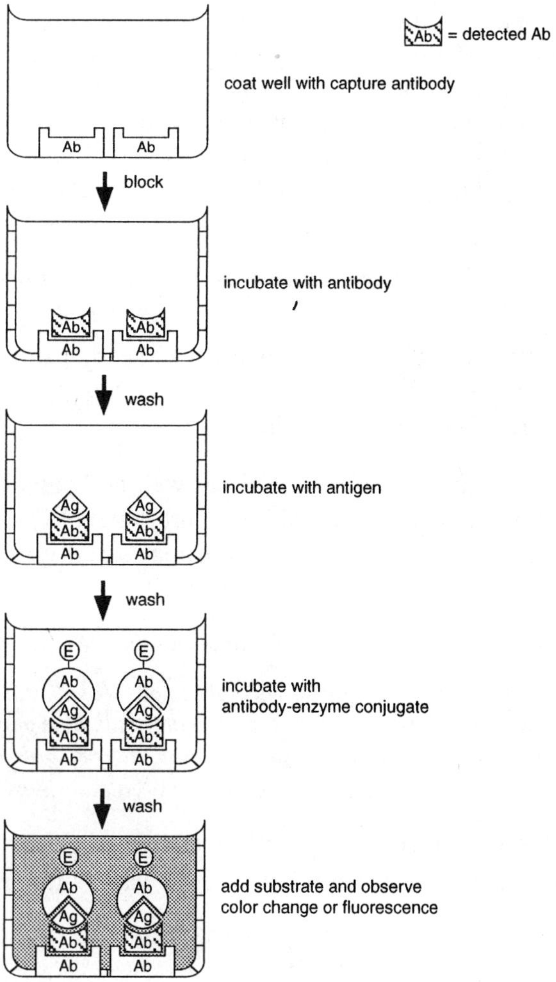

Figure 11.2.4 Double antibody–sandwich ELISA to detect specific antibodies. Ag = antigen; Ab = antibody; E = enzyme.

of available antibodies and antigens. Such reagents can be found in Linscott's Directory of Immunological and Biological Reagents.

6. Add 75 µl of MUP or NPP substrate solution to each well and incubate 1 hr at room temperature. Examine hydrolysis visually or spectrophotometrically (Basic Protocol, step 10).

   In order to detect low-level reactions, the plate can be read again after several hours or days.

7. Check for false positives by rescreening samples that test positive for antigen-specific antibody. For each positive sample, coat four wells with capture antibody and arm the capture antibody with test antibody (steps 1 to 3). Incubate two wells with antigen (step 4) and two wells with blocking buffer. Add conjugate and wash (step 5). Add substrate and measure hydrolysis (step 6).

DIRECT CELLULAR ELISA TO DETECT CELL-SURFACE ANTIGENS

Expression of cell-surface antigens or receptors is measured using existing antibodies or other ligands specific for cell-surface molecules (Fig. 11.2.5). This procedure can be as sensitive as flow cytometry analysis in quantitating the level of antigen expression on a population of cells (Coligan et al., 1995). Unlike flow cytometry analysis, however, this method is not sensitive for mixed populations. This assay can be converted to an indirect assay by substituting biotinylated antibody for the enzyme-antibody conjugate, followed by a second incubation with avidin–alkaline phosphatase.

Additional Materials *(also see Basic Protocol; see APPENDIX 1 for items with ✓)*

Cell samples
Specific antibody–alkaline phosphatase (AP) conjugate (*UNIT 11.1*)
✓ Wash buffer, ice-cold
Sorvall H-1000B rotor or equivalent

1. Determine optimal number of cells per well and conjugate concentration by criss-cross serial dilution analysis (Support Protocol 1) using variable numbers of positive- and negative-control cell samples and varying concentrations of antibody alkaline phosphatase conjugate.

   Titrate cells initially at 1-5 × 10^5/well and conjugate at 0.5 to 10 μg/ml.

   Because eukaryotic cells express variable amounts of alkaline phosphatase, test cells must be assayed in a preliminary experiment for alkaline phosphatase by incubation with substrate alone. If the test cells express unacceptable levels of alkaline phosphatase, another enzyme conjugate such as β-galactosidase should be used.

2. Place cell sample in a 15- to 50-ml centrifuge tube and spin 5 min at 450 × *g*, 4°C. Count cells and resuspend in ice-cold wash buffer at 1-5 × 10^6 cells/ml.

   If the surface antigen retains its antigenicity after fixation, cells may be fixed at the beginning of the experiment. Fix cells by suspending in glutaraldehyde (0.5% final; from a 25% stock, EM grade, Sigma) and incubating 30 min at room temperature. Centrifuge cells, resuspend in PBS containing 100 mM lysine and 100 mM ethanolamine (PBSLE), and incubate 30 min at 37°C. Wash twice in PBSLE and resuspend in wash buffer. Cells can be kept for months at 4°C after fixation.

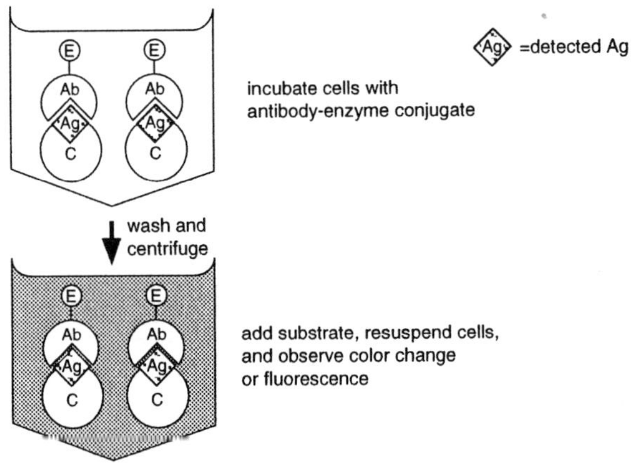

Figure 11.2.5 Direct cellular ELISA to detect cell-surface antigens. Ab = antibody; E = enzyme; C = cell.

3. Dispense 100 µl cell suspension (1-5 × 10^5 cells) into wells of cone- or round-bottom microtiter plates, and centrifuge 1 min at 450 × *g*, 4°C. Remove supernatant by vacuum aspiration, and disrupt pellet by gently shaking microtiter plate on a vortex mixer or microtiter plate shaker.

4. Resuspend pellet in 100 µl conjugate in ice-cold wash buffer at optimal concentration (step 1). Incubate 1.5 hr at 4°C, resuspending cells by gently shaking at 15-min intervals.

5. Centrifuge cells 1 min at 450 × *g*, 4°C, remove supernatant by vacuum aspiration, briefly vortex pellet, and resuspend in 200 µl ice-cold wash buffer. Repeat three times.

6. Add 100 µl MUP or NPP substrate solution. Incubate ≥1 hr at room temperature, resuspending cells by gently shaking at 15-min intervals during hydrolysis. Determine hydrolysis by visual inspection or using a microtiter plate reader (Basic Protocol, step 10).

INDIRECT CELLULAR ELISA TO DETECT ANTIBODIES SPECIFIC FOR SURFACE ANTIGENS

ALTERNATE PROTOCOL 5

Figure 11.2.6 illustrates the basis of this ELISA.

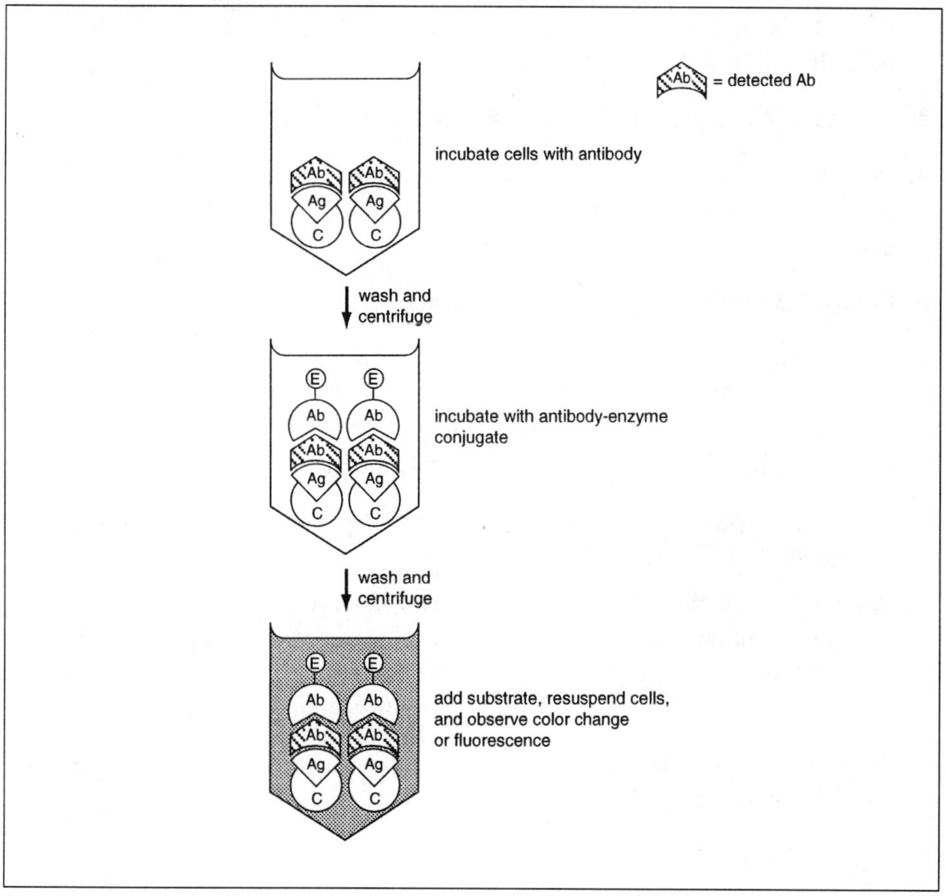

Figure 11.2.6 Indirect cellular ELISA to detect antibodies specific for surface antigens. Ab = antibody; E = enzyme; C = cell.

Additional Materials *(also see Basic Protocol)*

Positive-control antibodies (i.e., those that react with the experimental cells and are from the immunized species)
Negative-control antibodies (i.e., those that do not react with the experimental cells)
Test antibody solution
Antibody or F(ab′)$_2$ (against Ig from the immunized species) conjugated to alkaline phosphatase

1. Centrifuge and resuspend cell sample as in step 2 of Alternate Protocol 4 at 1–5 × 10^6 cells/ml.

   All steps must be performed at 4°C in physiological buffers containing NaN_3. Because eukaryotic cells express variable amounts of alkaline phosphatase, test cells must be assayed for alkaline phophatase activity.

   Because this technique detects antibodies against uncharacterized epitopes, fixation prior to analysis is not recommended.

2. In preliminary assays, determine optimal number of cells per well and conjugate concentration by criss-cross serial dilution analysis using positive- and negative-control antibodies instead of test antibodies (Support Protocol 1). In adapting the criss-cross serial dilution analysis, whole cells replace the solid-phase coating reagent; techniques for handling cells are outlined in steps 3 to 7 below. Set up titrations by varying the number of cells between 1 × 10^5 and 5 × 10^5/well, the concentration of positive- and negative-control antibodies between 0.1 and 10 µg/ml, and the concentration of antibody-enzyme conjugate between 0.1 and 10 µg/ml.

3. Dispense cells (Alternate Protocol 4, step 3).

4. Resuspend cells in 100 µl of ice-cold wash buffer solutions containing 1 to 10 µg/ml test antibody or control antibodies. Incubate 1.5 hr at 4°C, resuspending cells by gently shaking at 15-min intervals.

5. Centrifuge and wash cells (Alternate Protocol 4, step 5).

6. Resuspend pellet in 100 µl enzyme-antibody conjugate or F(ab′)$_2$-enzyme conjugate diluted in ice-cold wash buffer. The optimal concentration of antibody, determined in step 2, is usually 100 to 500 ng/ml. Incubate 1.5 hr at 4°C, resuspending cells by gently shaking at 15-min intervals.

   When working with cells that may express Fc receptors, it is best to use enzyme conjugated to F(ab′)$_2$ fragments.

7. Wash cells three times (Alternate Protocol 4, step 5). Add 100 µl MUP or NPP substrate solution. Allow hydrolysis to proceed until the signal has reached the desired levels; resuspend cells by gently shaking at 15 min intervals during hydrolysis. If desired, stop hydrolysis by adding 25 µl of 0.5 M NaOH.

8. Determine hydrolysis by visual inspection or using microtiter plate reader (Basic Protocol, step 10).

CRISS-CROSS SERIAL DILUTION ANALYSIS TO DETERMINE OPTIMAL REAGENT CONCENTRATIONS

SUPPORT PROTOCOL 1

Serial dilution titration analyses are performed to determine optimal concentrations of reagents to be used in ELISAs. In this protocol, all three reactants in a three-step ELISA—a primary solid-phase coating reagent, a secondary reagent that binds the primary reagent, and an enzyme-conjugated tertiary developing reagent that binds to the secondary reagent—are serially diluted and analyzed by a criss-cross matrix analysis.

Additional Materials *(also see Basic Protocol)*

Coating reagent
Secondary reagent
Developing reagent

1. Place four 17 × 100–mm test tubes in a rack and add 6 ml PBSN to the last three tubes. In tube 1, prepare a 12-ml solution of coating reagent at 10 μg/ml in PBSN. Transfer 6 ml of tube 1 solution to tube 2. Mix by pipetting up and down five times. Repeat this transfer and mix for tubes 3 and 4; the tubes now contain coating reagent at 10, 5, 2.5, and 1.25 μg/ml.

2. Using a multichannel pipettor, dispense 50 μl of coating reagent dilutions into wells of four Immulon microtiter plates (i.e., each plate is filled with one of the four dilutions). Wrap plates in plastic wrap and incubate overnight at room temperature or 2 hr at 37°C.

3. Rinse and block plates with blocking buffer as in steps 3 to 5 of Basic Protocol.

4. Place five 12 × 75–mm test tubes in a rack and add 3 ml blocking buffer to the last four tubes. In tube 1, prepare a 4-ml solution of secondary reagent at 200 ng/ml in PBSN. Transfer 1 ml of tube 1 solution to tube 2. Pipet up and down five times. Repeat this transfer and mix for tubes 3 to 5; the tubes now contain secondary reactant at 200, 50, 12.5, 3.125, and 0.78 ng/ml. If possible, prepare and test serial dilutions of a nonreactive heterologous form of the secondary reactant in parallel.

   If the assay is especially insensitive, it may be necessary to increase the secondary reactant concentrations so the tube-1 solution is 1000 ng/ml.

5. Dispense 50 μl of secondary reagent dilutions into first five columns of all four coated plates. The most dilute solution is dispensed into column 5, while solutions of increasing concentration are added successively into columns 4, 3, 2, and 1. Incubate 2 hr at room temperature.

6. Wash plates (Basic Protocol, steps 8 and 9). Place five 17 × 100–mm test tubes in a rack and add 3 ml blocking buffer to the last four tubes. In tube 1, prepare a 6-ml solution of developing reagent at 500 ng/ml in blocking buffer. Transfer 3 ml of tube 1 solution into tube 2 and mix. Repeat this transfer and mixing for tubes 3 and 4; the tubes now contain developing reagent at 500, 250, 125, 62.5, and 31.25 ng/ml.

7. Dispense 50 μl developing reagent solutions into wells of rows 2 to 6 of each plate, dispensing the most dilute solution into row 6 and solutions of increasing concentration successively into rows 5, 4, 3, and 2. Incubate 2 hr at room temperature.

8. Wash plates (Basic Protocol, steps 8 and 9). Add 75 µl MUP or NPP substrate solution to each well, incubate 1 hr at room temperature, and measure hydrolysis visually or with a microtiter plate reader (Basic Protocol, step 11).

   Assay should result in 0.50 absorbance units/hr at 405 nm when using NPP or 1000 to 1500 fluorescence units/hr when using MUP.

SUPPORT PROTOCOL 2

PREPARATION OF BACTERIAL-CELL-LYSATE ANTIGENS

A culture of *E. coli* containing proteins expressed from cloned genes is lysed for use as test antigen in any of the first three protocols of this unit.

Materials *(see* APPENDIX 1 *for items with ✓)*

Escherichia coli culture in broth or agar
Cell resuspension buffer: 10 mM HEPES
✓ Lysozyme solution
✓ Tris/EDTA/NaCl (TEN) buffer
10% sodium dodecyl sulfate (SDS)
8 M urea (optional)

Nylon-tipped applicator (Falcon, Becton Dickinson)

1. For liquid culture, centrifuge 5 ml cells 10 min at 2500 rpm in tabletop centrifuge. Decant supernatant and resuspend pellet in 5 ml cell resuspension buffer by vortexing gently. For agar culture, remove about 10 colonies from plate using a nylon-tipped applicator and resuspend in 2 ml cell resuspension buffer. Press swab against side of tube to remove as much liquid as possible.

   Yield of expressed protein may vary with growth phase. Samples should be taken for analysis at various periods of growth (e.g., mid-log and stationary phases). If samples are taken from agar plates, the culture should be grown overnight at 37°C.

2. Place 1 ml resuspended cells in a microcentrifuge tube on ice. Add 0.2 ml lysozyme solution to tube and leave 5 min on ice.

3. Microcentrifuge 5 min. Decant supernatant and save. Resuspend pellet in 1.2 ml TEN buffer.

   Because many expressed proteins are insoluble, it is worthwhile to assay both pellet and supernatant.

4. Add 65 µl of 10% SDS solution to each sample. Incubate 10 min at 37°C. Samples are ready for ELISA. Store frozen if not used within several hours.

   To avoid denaturation, add Triton X-100 to a final concentration four times greater than the SDS concentration.

References: Coligan et al., 1995; Engvall and Perlman, 1971; *Linscott's Directory of Immunological and Biological Reagents.*

Contributors: Peter Hornbeck, Scott E. Winston, and Steven A. Fuller

Production of Monoclonal Antibody Supernatant and Ascites Fluid

A major advantage of using monoclonal antibodies over polyclonal antisera is the potential availability of large quantities of the specific monoclonal antibody (MAb). Preparations containing the MAb generally include a hybridoma supernatant, ascites fluid from a mouse inoculated with the hybridoma, and purified MAb. Hybridoma supernatants are easy to produce, especially for large numbers of different MAb, but are relatively low in MAb concentration. Ascites fluid contains a high concentration of MAb, but the fluid is not a pure MAb preparation. To obtain a purified preparation, affinity chromatography (*UNITS 10.8 & 10.9*) of culture supernatants or ascites fluid can be performed.

Most culture supernatants will have saturating MAb titers of ≥1:10 when tested at 100 µl for 10^6 cells. Following affinity chromatography, 1 to 10 mg of purified MAb/liter can be anticipated. Most hybridomas can be grown as ascites tumors. The saturating concentration of the MAb in such fluids should be detected at dilutions of 1:500.

PRODUCTION OF A MONOCLONAL ANTIBODY SUPERNATANT

BASIC PROTOCOL 1

The hybridoma is grown and split 1:10. The cells are then overgrown until cell death occurs. The supernatant is harvested and the titer determined. If the titer is high, the hybridoma can be used for large-scale production (Alternate Protocol) in anticipation of purification or ascites production.

Materials *(see APPENDIX 1 for items with ✓)*

Hybridoma of interest
✓ Complete DMEM-10

175-cm² tissue culture flasks
Humidified 37°C, 5% CO_2 incubator
50-ml conical centrifuge tubes, sterile
Beckman TH-4 rotor or equivalent

1. Place hybridoma in a 175-cm² tissue culture flask in complete DMEM-10 medium. Grow in CO_2 incubator at 37°C until vigorously growing and ready to split.

   Most cell lines need to be split into new medium or new flasks when cell density reaches 1-2 × 10^2 cells/ml. Tissue culture flasks can be inspected with an inverted microscope and cell viability and density determined. In addition, the culture can be monitored for contamination. Cells should never be allowed to become so crowded that cell death occurs, because this crisis phase increases the likelihood of phenotypic change.

2. Split cells 1:10 in a new 175-cm² flask. Fill the flask with complete DMEM-10 to 100 ml total and place in CO_2 incubator until cells are overgrown, medium becomes acidic (yellow), and cells die (~5 days).

3. Transfer flask contents to sterile 50-ml conical centrifuge tubes. Centrifuge 10 min at 1500 × *g*, room temperature. Collect the supernatant and discard the pellet.

4. Assay the titer of MAb supernatant by ELISA (*UNIT 11.2*) or flow cytometry (Holmes and Fowlkes, 1995; Yokoyama, 1995).

5. Store the supernatant under sterile conditions; it is generally stable at 4°C for weeks to months, at –20°C for months to years, and indefinitely at –70°C. Minimize thawing and refreezing by storing several aliquots.

11.3

ALTERNATE PROTOCOL 1

LARGE-SCALE PRODUCTION OF MONOCLONAL ANTIBODY SUPERNATANT

The first step in monoclonal antibody purification by affinity chromatography is production of large amounts of culture supernatant. Cells are first grown in smaller flasks, then gradually expanded into large-volume roller flasks. The supernatant is harvested and stored until needed.

Additional Materials *(also see Basic Protocol; see* APPENDIX 1 *for items with ✓)*

✓ Complete DMEM-10 containing 5 to 10 mM HEPES, pH 7.2 to 7.4
70% ethanol

850-cm² roller flask and roller apparatus
250-ml conical centrifuge tubes, sterile
Beckman JS-5.2 rotor or equivalent

1. Repeat step 1 of Basic Protocol 1 and split 1:10 in complete DMEM-10/HEPES to 100 ml total.

   This procedure can be scaled up as desired—each 175-cm² flask will ultimately seed 2.35 to 2.5 liters of culture medium. It is usually not necessary to adapt and grow cells in serum-free medium (which frequently decreases yield); however, if the supernatant will be used for MAb purification by protein A–affinity chromatography, test the culture medium with FBS alone for contaminants (i.e., other proteins) that may copurify with the MAb. Bovine newborn serum frequently contains significant amounts of Ig that will bind to protein A and should not be used as a medium supplement.

2. When cells are ready to split, transfer contents of the 175-cm² flask (100 ml) to an 850-cm² roller flask. Add an additional 150 ml complete DMEM-10/HEPES. Cap tightly, place on roller apparatus at 37°C and grow 1 to 2 days.

3. Wipe cap and neck of roller flask with a sterile gauze sponge soaked in 70% ethanol. Open roller flask and add an additional 250 ml complete DMEM-10/HEPES (500 ml total volume). Cap tightly and incubate on roller apparatus another 1 to 2 days.

4. Repeat wiping as in step 3. Open roller flask and add ~2 liters complete DMEM-10/HEPES until flask is almost full (~2.5 liters total volume depending on capacity of flask). Cap tightly and incubate on roller apparatus at 37°C until the medium turns yellow (~5 days).

   Avoid foaming by first pouring in growth medium without FBS followed by FBS to 10% final. Avoid prolonged rolling, as cells will fragment and the debris will be difficult to pellet with large centrifuge tubes.

5. Pour the culture into sterile 250-ml conical centrifuge tubes. Harvest the supernatant by centrifuging 20 min at 250 × *g*, room temperature. Collect the supernatant and discard the pellet. Freeze the supernatant in aliquots.

   If the supernatant will not be used in a bioassay, add 10% sodium azide to 0.02% final. If the supernatant will undergo affinity chromatography or salt fractionation, sterile filtration through a 0.45-μm filter is recommended to eliminate debris.

LARGE-SCALE PRODUCTION OF HYBRIDOMAS OR CELL LINES

ALTERNATE PROTOCOL 2

11.3

The following procedure is used to produce large amounts of cells which can be used to isolate cellular components such as membrane proteins. Individual small flasks are grown, then each is used to inoculate a larger roller flask. The cells are gradually expanded by addition of fresh medium, and are harvested when the cells are near saturation densities.

1. Follow steps 1 to 4 for large-scale production of MAb supernatants (Alternate Protocol 1) but harvest when the density is appropriate or if the cell growth plateaus.

   Estimate the amount of cells needed. This procedure will yield ~10^6 cells/ml. Seed (introduce cells into) the number of 175-cm^2 flasks necessary to produce the amount of cells required (one 175-cm^2 flask for every 2.4 liters of medium). Each flask should be treated as an independent culture. Estimate time of harvesting by macroscopic inspection for medium color and turbidity and by taking daily cell counts and checking viability on several flasks. When cell concentration reaches a plateau, harvesting is indicated. Do not allow cells to overgrow or cell viability will drop precipitously.

2. Pour cells into sterile 250-ml conical centrifuge tubes, centrifuge 15 min at 250 × *g*, 4°C, and discard supernatant.

   Each tube can be used for two spins. Harvesting 80 liters of cells requires at least three centrifuges to spin four to six tubes (1 to 1.5 liters) each, two or three people, and nearly one day. Conical centrifuge tubes are recommended because resulting pellets are easier to work with.

3. Place cell pellets on ice. Pool 10 cell pellets into one tube by resuspending cells in 250 ml of 4°C PBS. Centrifuge at 250 × *g*, 4°C, and discard supernatant.

4. Repeat and further consolidate tubes into one tube. After three washes, the cells are ready for further processing (e.g., cell lysis, radiolabeling).

PRODUCTION OF ASCITES FLUID CONTAINING MONOCLONAL ANTIBODY

BASIC PROTOCOL 2

High-titer monoclonal antibody preparations can be obtained from the ascites fluid of mice inoculated intraperitoneally with MAb–producing hybridoma cells.

Materials *(see APPENDIX 1 for items with ✓)*

Nude mice, 6 to 8 weeks old and specific-pathogen free, *or* syngeneic host if mouse-mouse hybridomas are injected
Pristane (2,6,10,14-tetramethylpentadecane; Aldrich)
Hybridoma of interest
✓ Complete DMEM-10 medium containing 10 mM HEPES and 1 mM sodium pyruvate
✓ PBS *or* HBSS, sterile and *without* FBS

20- or 22-G needle and 18-G needle
Beckman TH-4 rotor or equivalent
50- and 15-ml polypropylene conical centrifuge tubes, sterile

1. Using a 20- or 22-G needle, inject mice intraperitoneally with 0.5 to 1 ml Pristane per mouse 1 week prior to inoculation with cells (see Donovan and Brown, 1995).

   Mice should be maintained in specific-pathogen-free (spf) facility.

2. Grow hybridoma cells in a 175-cm^2 flask in complete DMEM-10/HEPES/pyruvate under conditions that promote log-phase growth.

   Before injecting mice in step 7, test supernatants for MAb activity by ELISA (UNIT 11.2) or appropriate assay, preferably before the cells are expanded. To minimize the risk of introducing a pathogen into the rodent colony, screen cells for pathogens by antibody-production assay (Donovan and Brown, 1995).

3. Transfer culture to 50-ml conical centrifuge tubes. Centrifuge 5 min at 500 × *g*, room temperature.

4. Wash cells by resuspending in 50 ml sterile PBS or HBSS without FBS, then centrifuging 5 min at 500 × *g*, room temperature, and discarding supernatant. Repeat twice and resuspend cells in 5 ml PBS or HBSS.

   Avoid washing in FBS-containing medium because the mouse will produce antibodies to the FBS.

5. Count cells and determine viability by trypan blue exclusion (UNIT 11.5; cells should be nearly 100% viable).

6. Adjust cell concentration to 2.5×10^2 cells/ml using PBS or HBSS without FBS.

7. Using a 10-ml sterile syringe with a 22-G needle, inject nude mouse intraperitoneally with 2 ml cells. Wait for ascites to form (1 to 2 weeks).

   Typically, three mice are injected at one time. In most cases, at least one and frequently all of the mice develop ascites.

8. Harvest ascites by grasping and immobilizing the mouse in one hand in such a way as to stretch the abdominal skin taut. With the other hand, insert an 18-G needle 1 to 2 cm into the abdominal cavity. Enter either the left or right lower quadrants to avoid the vital organs in the upper quadrants and the major vessels in the midline. Allow the ascites to drip into a sterile 15-ml polypropylene conical centrifuge.

   If the mouse has a large amount of ascites and the fluid stops dripping from the 18-G needle, it may be necessary to reposition the needle tip by withdrawing it slowly and reinserting it in a different plane. If no ascites fluid accumulates, the mouse may be reinjected. Occasionally, the ascites is under such high pressure that a large amount squirts out as soon as the needle is inserted; be sure that the hub of the needle is pointed into a tube before inserting the needle into the peritoneal cavity.

   Rather than tapping the mouse as soon as the ascites is apparent, allow fluid to build up (3 to 7 days) to obtain highest yield. Frequently, 5 to 10 ml (sometimes >40 ml) can be collected from each mouse.

   If the mice die without any ascites forming, particularly within 2 weeks of inoculation, try fewer cells. If the mice do not form ascites after 2 weeks and they appear healthy, inject those mice—as well as naive, Pristane-primed mice—with more cells. If solid tumors form, tease cells into suspension and inject tumor cells into another Pristane-primed mouse. Even if a little ascites forms, the fluid can be transferred to another mouse (~0.5 ml/mouse), and large amounts of ascites should accumulate. Once the ascites is formed, the mouse-adapted cells can be frozen and used to reinoculate mice in the future.

9. Centrifuge ascites 10 min at 1500 × *g*, room temperature. Harvest supernatants and discard pellet. Store ascites fluid at 4°C until all collection is completed (<1 week).

   If the fluid clots, "rim" the clot by passing with a wooden applicator stick around its edge (between clot and tube) before centrifugation.

10. Allow the mouse to reaccumulate ascites (2 to 3 days) before reharvesting as in step 8. Process the ascites as in step 9. Repeat this process until no further ascites accumulates, the fluid cannot be collected, or the mouse becomes ill. The mouse should be euthanized at this point (Donovan and Brown, 1995).

11. Pool ascites fluid collected on different days and heat-inactivate 45 min in a 56°C water bath. If a clot reforms, remove it by rimming and centrifuge as in step 9.

12. Assay the titer of MAb-containing ascites by the appropriate method.

    Saturating concentration (maximal activity) of the MAb should be apparent at 0.5% or higher dilutions. Lower titers usually are the result of unstable hybridomas that stop producing MAb or too many (>2) in vivo serial passes of the hybridomas.

13. Dilute >1:10 and filter sterilize through a 0.45-μm filter. Aliquot and freeze at −70°C, avoiding repeated freezing and thawing. Shelf life should be several years.

    Add sodium azide to 0.02% if ascites will not be used for bioassay.

References: Andrew and Titus, 1995; Donovan and Brown, 1995; Holmes and Fowlkes, 1995; Yokoyama 1995.

Contributor: Wayne M. Yokoyama

Purification of Monoclonal Antibodies

UNIT 11.4

PURIFICATION USING PROTEIN A–SEPHAROSE

BASIC PROTOCOL

Staphylococcal protein A binds immunoglobulins of several mammalian species. In this affinity chromatography method, murine antibodies are fractionated by employing a protein A–Sepharose column matrix and a stepwise pH elution. The matrix will bind some IgM, most IgG1, and nearly all IgG2a, IgG2b, and IgG3 MAb.

Materials *(see* APPENDIX 1 *for items with ✓)*

Protein A–Sepharose CL-4B (Pharmacia LKB)
✓ Tris buffer, pH 8.6
Ascites fluid (*UNIT 11.10*)
✓ Citrate buffer, pH 5.5
✓ Acetate buffer, pH 4.3
✓ Glycine·Cl buffer, pH 2.3
✓ Neutralizing buffer, pH 7.7
Ultrafiltration cells and XM50 membrane (Amicon)

1. Fully swell protein A–Sepharose in Tris buffer (50-fold excess, v/v). Pour into a 2.5-cm glass chromatography column and equilibrate with Tris buffer at room temperature (*UNIT 10.10*).

    Antibody from 10 to 20 ml of ascites fluid can be purified with ~3 g dry weight protein A–Sepharose.

2. Dilute ascites fluid in 3 vol Tris buffer and apply to column at 1 to 5 ml/min. Wash column with Tris buffer until all unbound proteins are eluted, as monitored at A_{280}.

3. Elute bound proteins successively with 2 to 3 column volumes of citrate, acetate, and glycine·Cl buffers directly into tubes containing neutralizing buffer equal to one-quarter of collected volume.

4. Assay for antigen-specific MAb by ELISA (*UNIT 11.2*).

5. Concentrate monoclonal antibody fractions to 1-5 mg/ml in ultrafiltration cell with XM50 ultrafiltration membrane. Store at −20°C.

   Measure A_{280} to calculate concentration. A mouse IgG solution at 1 mg/ml = 1.44 absorbance units.

ALTERNATE PROTOCOL 1

ALTERNATIVE BUFFER SYSTEM FOR PROTEIN A–SEPHAROSE

An alternative buffer system (suggested by Pharmacia) results in higher binding capacity for certain monoclonal antibodies. Proceed with the basic protocol but replace the buffers as follows:

Additional Materials *(also see Basic Protocol; see* APPENDIX 1 *for items with ✓)*

✓ Glycine·OH buffer replaces Tris buffer
✓ Elution buffers pH 6.0 and pH 5.0 replace citrate buffer (carry out elution with pH 6.0 buffer first, then use pH 5.0 buffer)
✓ Elution buffer pH 4.0 replaces acetate buffer
✓ Elution buffer pH 3.2 replaces glycine·Cl buffer.

ALTERNATE PROTOCOL 2

PURIFICATION BY ANTIGEN-SEPHAROSE AND ANTI-MOUSE-Ig-SEPHAROSE

When the antibody of interest does not bind quantitatively to staphylococcal protein A (e.g., IgM and some IgG1 antibodies), affinity columns can be prepared using a matrix comprised of protein antigens or anti-mouse-Ig coupled to CNBr-activated Sepharose (see also *UNIT 10.15*). The antigen-Sepharose matrix will isolate specific antibody while the anti-mouse-Ig-Sepharose will isolate all mouse immunoglobulins.

Additional Materials *(also see Basic Protocol; see* APPENDIX 1 *for items with ✓)*

CNBr-activated Sepharose 4B (Pharmacia)
1 mM HCl
✓ Coupling buffer
Protein antigen (previously purified; Chapter 10) or anti-mouse-Ig antibody (commercially available)
✓ 1 M ethanolamine, pH 8.0
✓ PBS
✓ Washing buffer
60- or 150-ml sintered glass funnel, medium porosity

1. Swell 10 ml CNBr-activated Sepharose 4B in 1 mM HCl. Transfer to sintered glass funnel and wash successively over 30 min with (a) 200 to 500 ml of 1 mM HCl and (b) 200 to 500 ml coupling buffer.

   One gram of dry gel swells to ~3 ml. Use 5 to 10 mg ligand/ml gel.

2. Transfer gel to 50-ml conical plastic centrifuge tube. Add 50 to 100 mg ligand (2 to 5 mg/ml), and mix overnight at 4°C on a rocker platform.

3. Centrifuge 10 min at 250 × *g*, and discard supernatant. Fill tube with coupling buffer, centrifuge, and discard supernatant.

4. Add 20 to 40 ml of 1 M ethanolamine, pH 8.0, and incubate on a rocker platform 4 to 5 hr at 4°C. Centrifuge and discard supernatant. Wash gel successively by filling the tube with PBS and washing buffer followed by centrifugation.
5. Transfer gel to a 2.5-cm glass column and equilibrate with 100 ml PBS. Dilute ascites fluid with 3 vol PBS and apply to the column at a flow rate of 1 to 5 ml/min at room temperature.
6. Wash unbound protein from the column with 30 ml PBS and elute bound protein with glycine·Cl buffer (Basic Protocol).
7. Proceed with step 5 of Basic Protocol.

Reference: Langone and Van Vunakis, 1986.

Contributors: Steven A. Fuller, Miyoko Takahashi, and John G.R. Hurrell

Production of Polyclonal Antisera in Rabbits

UNIT 11.5

Polyclonal antipeptide antisera are generated using native protein antigen or a peptide–carrier protein conjugate (*UNITS 11.7 & 11.8* describe selecting an immunogenic peptide sequence and preparing conjugates, respectively). The following protocols detail multiple immunization routes, which are often required for obtaining high-titer serum. Titers of 5-10 mg/ml result after repeated boosts using large or nonevolutionarily related proteins, while 1-2 mg/ml titers result when using small or highly conserved protein species. Figure 11.5.1 shows the kinetics of an antibody response.

INTRAMUSCULAR IMMUNIZATION

BASIC PROTOCOL

Materials

Complete and incomplete Freunds adjuvant
1 to 2 mg/ml antigen in PBS
Adult New Zealand red or white rabbit (2 to 5 kg body weight)
70% ethanol
Petroleum jelly

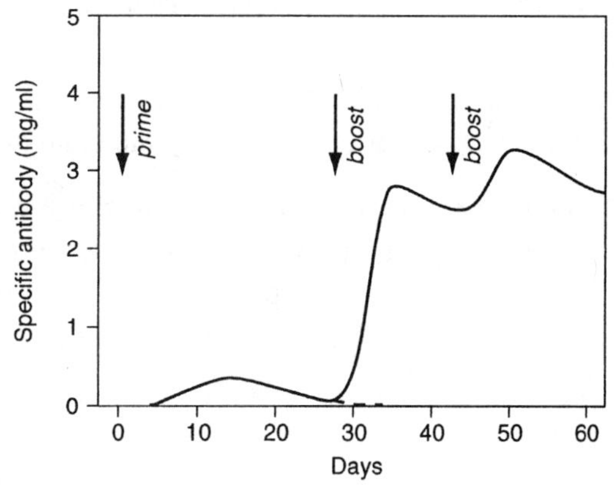

Figure 11.5.1 Kinetics of the development of the specific antibody response. Arrows indicate when the priming and boosting immunizations were administered. Actual amounts of specific antibody produced will vary considerably depending on the immunogenicity of the protein.

22-G needle
3-ml syringes with locking hubs (Luer-Lok, Becton Dickinson)
Double-ended locking hub connector (Luer-Lok, Becton Dickinson)
Rabbit restrainer
Heat lamp

1. Add 2 ml complete Freunds adjuvant to 2 ml purified antigen in PBS and emulsify. Transfer emulsion into 3-ml syringe, attach 22-G needle, and remove air bubbles.

   As the amount of immunizing antigen is decreased, the strength of the response—as evidenced by the quantity and affinity of antibodies produced—may diminish.

2. Place rabbit onto flat surface. Grasp by scruff of neck with one hand and place other hand under rabbit's hindquarters. Bunch rabbit so hind leg muscles cannot extend. Clean area to be injected with 70% ethanol. Insert needle ~1 cm into the heavy thigh muscle of each hind leg, and inject 0.5 ml emulsion (maximum) into each thigh muscle (intramuscular injection; be careful not to contact thigh bone).

   Do not support rabbit from its underside. The powerful hind legs and sharp claws can cause deep scratches. Recruit another person to restrain animal while injection is administered.

3. Boost rabbit intramuscularly 4 weeks later with 1 mg antigen emulsified in incomplete Freunds adjuvant (1:1) as in steps 1 and 2. Repeat booster immunization 2 weeks after initial boost.

4. Bleed rabbit from marginal vein of the ear 10 days after second booster immunization by placing rabbit into restrainer (restrain tightly) with its ear extended. Swab ear with 70% ethanol and place it under heat lamp. Gently tap ear with your finger; the blood vessels should be fully dilated within 1 to 2 min. Make a small, clean, quick cut, ~0.5-cm long, through marginal vein (Fig. 11.5.2) using the point of a sterile scalpel blade. Collect blood in 50-ml plastic centrifuge tube until the natural clotting reaction stops the flow. Allow blood to flow down the side of the tube to avoid hemolysis.

   From a 2-kg rabbit, 30 to 40 ml of blood should flow before clotting becomes inhibitory. Remove no more blood than 20 ml/kg, once a week. If 50 ml of blood have been collected but clotting has not occurred, stop bleeding by wiping cut area with petroleum jelly and applying gentle pressure to wound with a piece of gauze or cotton for 2 to 3 min.

5. Wipe ear clean and swab it with 70% ethanol. Apply a layer of petroleum jelly to the ear to minimize skin irritation induced by the ethanol.

6. Allow blood to stand several hours at room temperature before placing overnight at 4°C. Gently loosen and remove clot from sides of tube with a wooden applicator. Transfer serum into appropriate centrifuge tube and pellet any remaining red blood cells and debris by centrifuging 10 min at 5000 × *g*.

   If a clot does not form, initiate clotting by placing wooden applicator into tube containing collected blood.

7. Administer further booster immunizations every 2 weeks. Bleed rabbit 10 days after each boost. Alternate ear from which blood is obtained.

8. Determine specific antibody titer of antiserum by ELISA (*UNIT 11.2*) or RIA. If desired, purify the specific antibody population using procedures in *UNIT 11.4* or *11.6*.

INTRADERMAL IMMUNIZATION

ALTERNATE PROTOCOL 1

Additional Materials *(also see Basic Protocol)*

Electric animal shaver
24-G needle

1. Prepare antigen for immunization of rabbit as in step 1 of Basic Protocol.
2. Hold rabbit firmly and shave back. Swab exposed skin with 70% ethanol. Stretch area of skin to be immunized between your thumb and index finger. Hold 3-ml syringe equipped with 24-G needle at an angle of 30° to skin surface (with bevel of needle facing up) and insert needle just under outer skin layer. Lower syringe until it is laying along rabbit's back. Inject ≤100 µl emulsified antigen between dermal layers. Similarly, inject ≤100 µl antigen into four other sites on rabbit's back.

   Small bumps should appear at injection sites; two weeks after immunization, the skin will become ulcerated at the injection sites.

3. Boost rabbit intramuscularly using incomplete Freunds adjuvant as in Basic Protocol. If a high specific antibody titer is not achieved after two intramuscular boosts, boost rabbit intradermally, but at only two sites.

SUBCUTANEOUS IMMUNIZATION

ALTERNATE PROTOCOL 2

1. Prepare antigen for injection into rabbit, and prepare rabbit as for intradermal immunization.

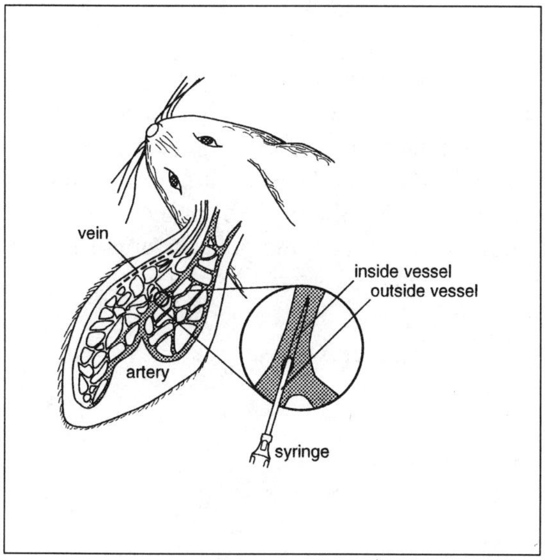

Figure 11.5.2 Circulatory system in the ear of a rabbit. The site of the incision along the marginal vein is indicated by the dotted line. The site for the insertion of the syringe needle into the artery is indicated by the magnified segment. The tip of the needle should point toward the head of the rabbit.

2. Pull skin of rabbit's back away from underlying muscle. Insert 24-G needle all the way through skin, being careful not to pierce muscle tissue. Inject 0.5 ml at four different sites.

3. Boost rabbit intramuscularly using incomplete Freunds adjuvant (Basic Protocol), or subcutaneously as above.

ALTERNATE PROTOCOL 3

BLEEDING FROM THE EAR ARTERY

When rabbits are bled from the ear artery, 50 to 100 ml of blood can be readily obtained. However, with recoveries of >50 to 60 ml in a single bleed, the rabbit must be rested longer to allow its hematocrit to return to normal (36% to 48%).

1. Prepare rabbit for bleeding (Basic Protocol).

2. Support the half of the ear closest to rabbit's head with the first and second fingers, using the third and fourth fingers to bend the other half of the ear downward, to allow blood to flow with gravity. Lay an 18-G needle (without a syringe attached) along the line of the central artery (Fig. 11.5.2), with needle tip pointing in direction of base of ear.

3. Insert needle into vessel, then push needle along 5 to 10 mm inside artery without again piercing vessel walls. Collect blood in a 50-ml plastic centrifuge tube (Basic Protocol), stop blood flow, and dress ear.

   The total blood volume of a rabbit is 57 to 65 ml/kg; no more than 20 ml blood can be drawn per kilogram of body weight in 1 week.

Reference: Klinman and Press, 1975.

Contributors: Helen M. Cooper and Yvonne Paterson

UNIT 11.6

Purification of IgG Antibodies from Antiserum, Ascites Fluid, or Hybridoma Supernatant

Although affinity chromatography (*UNIT 11.4*) is the method of choice for purifying specific antibody, such purification procedures may sometimes be unnecessary or not applicable (e.g., when there is insufficient antigen). This unit describes the isolation of the IgG fraction (containing antibodies of all specificities) in a complex mixture by precipitation with saturated ammonium sulfate (SAS) or by chromatography on DEAE resin. In many cases, transferrin will coprecipitate or copurify and may be removed by gel-filtration chromatography (*UNIT 10.8*).

BASIC PROTOCOL

PRECIPITATION OF IgG WITH AMMONIUM SULFATE

Materials *(see APPENDIX 1 for items with ✓)*

✓ Saturated ammonium sulfate (SAS) solution
33% (v/v) SAS solution: 33 ml SAS solution + 67 ml PBS, pH 7.0
IgG-containing antiserum, ascites, or tissue culture supernatant
Dialysis membrane (MWCO 12,000 to 14,000; Spectrapor 2)

1. With constant mixing at 4°C, add 1 vol SAS solution, pH 7.0, dropwise to 2 vol IgG-containing mixture. Allow precipitate to form over 2 to 4 hr at 4°C with constant mixing. Centrifuge 20 min at 12,000 × *g*.
2. Wash pellet by vortexing in a volume of cold 33% SAS solution equivalent to original volume of IgG-containing mixture.
3. Dissolve pellet in appropriate cold buffer (5% to 10% of original volume) by gentle vortexing. Dialyze IgG solution over 48 hr at 4°C against three changes of desired buffer (4 liters/change; *APPENDIX 3*).

FRACTIONATION OF IgG BY CHROMATOGRAPHY ON DEAE-AFFI-GEL BLUE

ALTERNATE PROTOCOL

Additional Materials *(also see Basic Protocol)*

IgG-containing antiserum or ascites
Loading buffer: 20 mM Tris·Cl/30 mM NaCl, pH 8.0
Elution buffer: 20 mM Tris·Cl/50 mM NaCl, pH 8.0
NaN_3, crystalline form
DEAE-Affi-Gel Blue (Bio-Rad)

1. Prepare a column of DEAE–Affi-Gel Blue and equilibrate with five bed volumes of loading buffer (7 ml bed volume/ml antiserum or ascites). Perform this and subsequent steps 4°C.
2. Dialyze sample against two changes of loading buffer (~4 liters/change) 40 hr at ~4°C. Apply to column at 10 ml/hr. Elute unbound protein with three bed volumes of loading buffer at 10 ml/hr.

   Alternatively, perform elution overnight; however, run at least three bed volumes of loading buffer before specific elution of IgG fraction.
3. Elute bound IgG fraction with elution buffer at 10 ml/hr. Collect 10-ml fractions and store at 4°C until they are pooled (transferrin, M_r 76,000, usually coelutes with IgG fraction).
4. Identify fractions containing IgG by analyzing 30 to 50 µl of each fraction on a 10% SDS-PAGE under reducing conditions (*UNIT 10.2*). Pool fractions containing IgG and dialyze over ~40 hr at 4°C against two changes of the desired buffer (4 liters/change).

   Under reducing conditions, the IgG heavy and light chains run separately with molecular weights of ~50,000 and 25,000, respectively.
5. Re-equilibrate DEAE–Affi-Gel Blue column with five bed volumes of loading buffer, add a few crystals of NaN_3, and store at 4°C.

Reference: Cooper, 1977.

Contributors: Helen M. Cooper and Yvonne Paterson

UNIT 11.7 Selection of an Immunogenic Peptide

Until recently the complete amino acid sequence was known for just a few proteins, obtained after years of laborious effort. Today investigators are able to deduce a complete protein sequence from the corresponding cDNA sequence without purifying the protein (Chapter 5). Based on an amino-terminal protein sequence or a variety of predictive methods, it is now possible to deduce a specific peptide sequence to be used, which then will cross-react with the intact native protein (*UNIT 11.8*). Figure 11.7.1 illustrates the steps involved in selecting an immunogenic peptide and its carrier and in producing an antipeptide antibody used to detect the protein from which the peptide sequence was derived.

Choosing an appropriate peptide is critical in obtaining an antibody that cross-reacts with the native antigen. The most common practice is to choose a 10- to 15-residue peptide corresponding to the carboxyl- or amino-terminal sequence of a protein antigen and to chemically cross-link it to a carrier molecule such as keyhole limpet hemocyanin (KLH) or bovine serum albumin (BSA). However, if an internal-sequence peptide is desired, selection of the peptide is based on the use of algorithms that predict potential antigenic sites. These predictive methods are, in turn, based on predictions of hydrophilicity (Kyte and Doolittle, 1982; Hopp and Woods, 1983) or secondary structure (Chou and Fasman, 1978). The objective is to choose a region of the protein that is (1) surface exposed, e.g., a hydrophilic region or (2) conformationally flexible relative to the rest of the structure, e.g., a loop region or a region predicted to form a β-turn. The selection process is also limited by constraints imposed by the chemistry of the coupling procedures used to attach peptide to carrier protein.

Selection of the carboxyl-terminal peptide. Reasons for choosing a carboxyl-terminal peptide are that peptides from the termini of proteins are often more mobile than the rest of the molecule, and the peptide can be coupled to the carrier in a

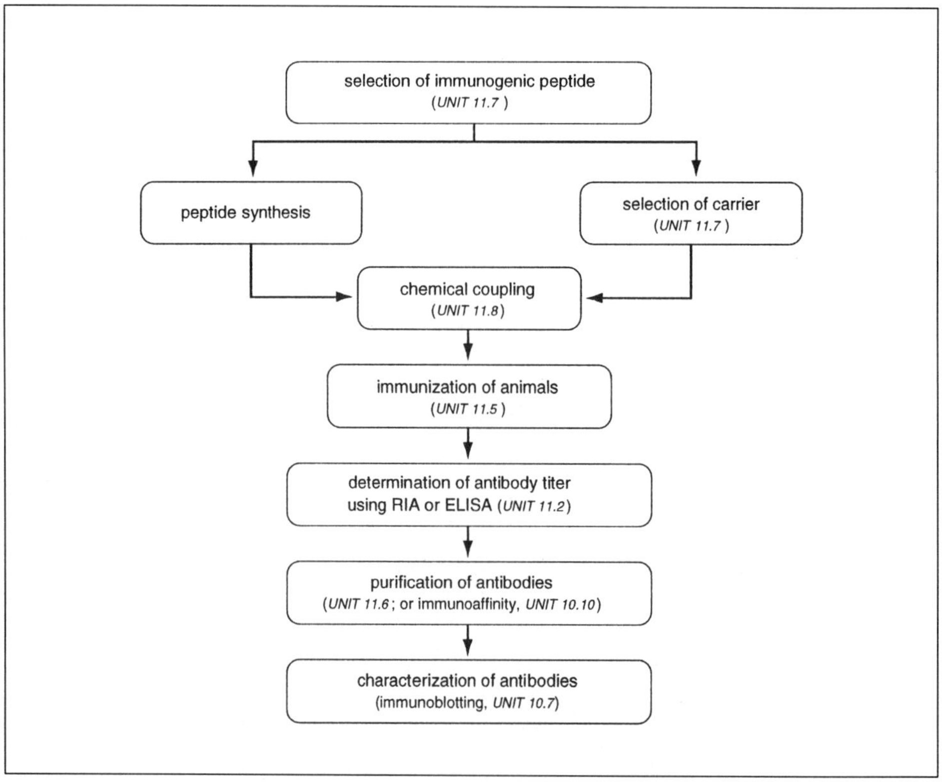

Figure 11.7.1 Flow chart for preparation of antipeptide antibodies.

straightforward manner using glutaraldehyde (*UNIT 11.8*). Coupling the peptide via its amino terminus to the carrier will bind it in a fashion similar to that found in the native antigen. However, this method precludes the presence of basic amino acids within the peptide (*UNIT 11.8*). For example, if the carboxyl-terminal sequence is

```
1 2 3 4 5 6 7 8 9 10
S Y G R N Q A E K Q,
```

coupling via glutaraldehyde is not useful because of the lysine (K) at position 9. In this case, synthesize

```
-1 1                 10
C  S Y G R N D A E K Q,
```

inserting a cysteine (C) to provide a "linker" and couple using *m*-maleimidobenzoyl-*N*-hydroxysuccinimide ester (MBS; *UNIT 11.8*). The cysteine/residue will be positioned next to the carrier molecule, which minimizes interference with reactivity toward the native sequence in the peptide.

A carboxyl-terminal sequence that is part of a transmembrane region will not be a suitable immunogen. This can be determined by examining a hydropathy plot of the sequence (Kyte and Doolittle, 1982).

Selection of the amino-terminal peptide. The amino terminus has the same advantages as described for the carboxyl terminus, but it has the disadvantage that it may be modified post-translationally by acetylation or by the removal of a leader sequence. Coupling an N-terminal peptide containing a C-terminal cysteine to a carrier is best accomplished using MBS (*UNIT 11.8*).

Selection of an internal peptide sequence. Selection of a peptide from an internal part of the sequence requires the use of algorithms to predict those regions most likely to be antigenic. The only information needed is the primary sequence. Two methods, those of Hopp and Woods (1983) and Kyte and Doolittle (1982), are based upon the fact that antigenic sites are located in surface-exposed, hydrophilic regions of the antigen. The algorithm of Chou and Fasman (1978) is an empirical method that utilizes a library of known structures to determine the distribution of each amino acid in the various conformational states. A computer program for performing this analysis can be found in Corrigan and Huang (1982). Regions predicted to form turns should be selected as peptide immunogens.

Self proteins. If the antigen in question is of mammalian origin, comparison of its sequence with the protein sequence from the animal to be immunized can yield important information. Only those regions with sequence differences are likely to be immunogenic. If the immunogenicity of homologous proteins is known, then select similar regions as the basis for synthesizing synthetic peptide immunogens in the protein being studied.

Selection of a carrier. The most important requirements of a carrier molecule are that it remain soluble after chemical coupling and that it have a sufficient number of lysines for coupling. The carriers most often used are hemocyanin and BSA. Hemocyanin has the advantage of being very immunogenic, and BSA offers good solubility.

Peptide synthesis. The most critical parameter in peptide synthesis is ensuring that the peptide mimics the native antigen as closely as possible. In synthesizing an N-terminal peptide of an N-terminally blocked antigen, acetylation of the α-amino group may be required. Acetylation also offers the opportunity to incorporate a trace radioactive label into the peptide, which may be important for quantitating the amount of peptide cross-linked to the carrier molecule (Miller et al., 1984).

If an internal peptide sequence is desired, the free α-amino and α-carboxyl groups should be modified to mimic the native antigen. One end will be blocked in the coupling procedure, but the other end must be modified.

References: Chou and Fasman, 1978; Corrigan and Huang, 1982; Hopp and Woods, 1983; Kyte and Doolittle, 1982; Miller et al., 1984.

Contributors: James F. Collawn and Yvonne Paterson

UNIT 11.8

Production of Antipeptide Antibodies

This unit describes chemical cross-linking of synthetic peptides to carrier proteins (see also *UNIT 11.7*). The Basic Protocol can be used to conjugate peptides through an amino- or carboxyl-terminal cysteine residue. An Alternate Protocol for coupling through α-amino groups is also given.

BASIC PROTOCOL

CHEMICAL COUPLING OF SYNTHETIC PEPTIDE TO CARRIER PROTEIN USING MBS

MBS (*m*-maleimidobenzoyl-*N*-hydroxysuccinimide ester) cross-links amino- and carboxyl-terminal thiol groups in the peptide with lysine side chains present in the carrier protein. An N-terminal cysteine is useful for carboxyl-terminal peptides, whereas a C-terminal cysteine is useful for amino-terminal peptides. Tris buffers should not be used as they compete with the carrier protein for the MBS-derivatized peptide.

Materials *(see APPENDIX 1 for items with ✓)*

Carrier protein—e.g., bovine serum albumin (BSA)
✓ 0.1 M sodium phosphate buffer, pH 6.8
MBS/DMF solution: 25 mg MBS (Pierce; store aliquots at −20°C under dessication) in final volume of 1 ml dimethylformamide (use within 1 hr)
Cysteine-containing peptide (*UNIT 11.7*)
20 mM EDTA prepared in 0.1 M sodium phosphate buffer, pH 6.8
✓ PBS
PD-10 column (Pharmacia)

1. Dissolve 10 mg BSA in 0.5 ml of 0.1 M sodium phosphate buffer, pH 6.8, in a 2-ml glass tube and add 50 µl MBS/DMF solution (MBS/carrier protein molar ratio, 27:1). Stir gently 30 min at room temperature.

2. Load onto PD-10 column (*UNIT 10.8*) pre-equilibrated with 10 ml of 0.1 M phosphate buffer, pH 6.8, and collect 0.6-ml fractions. Read A_{280} and collect first protein peak (first peak contains MBS/BSA complex and second peak contains free MBS). Pool first peak and place in 15-ml glass tube.

3. Dissolve 10 µmol peptide in 20 mM EDTA to ~1 mg/ml final and add to MBS/BSA complex in glass tube. Stir under N_2 for 4 hr at room temperature. Dialyze overnight at 4°C against 3 liters PBS. Change PBS and dialyze 4 hr.

   For peptides insoluble in 20 mM EDTA, dissolve (heating if necessary) in 0.5 to 1 ml DMSO and dilute to 2 ml with 20 mM EDTA.

4. Divide sample into aliquots and store at −20°C. Use 1 mg of total protein for each immunization as in *UNIT 11.5*.

CHEMICAL COUPLING OF SYNTHETIC PEPTIDE TO CARRIER PROTEIN USING GLUTARALDEHYDE

ALTERNATE PROTOCOL

Glutaraldehyde cross-links peptide and carrier molecules through their amino groups. Peptides having lysines at positions other than the amino terminus are best avoided. Tris buffers should not be used.

***Additional Materials** (also see Basic Protocol; see APPENDIX 1 for items with ✓)*

Carrier protein—e.g., *Limulus polyphemus* (horseshoe crab) or *Megathura crenulata* (keyhole limpet) hemocyanin (Worthington or Sigma)
✓ Borate buffers, pH 10 and pH 8.5
0.3% glutaraldehyde (Sigma) in borate buffer, pH 10 (prepare just before use)
1 M glycine

1. Dissolve 10 mg hemocyanin in 2 ml borate buffer, pH 10, in 15-ml glass tube by gentle stirring. Add 10 µmol synthetic peptide.
2. Slowly add 1 ml of 0.3% glutaraldehyde solution, stirring at room temperature. Allow to react 2 hr (solution will turn yellow). Add 0.25 ml of 1 M glycine to block unreacted glutaraldehyde. Allow to react for 30 min.

   CAUTION: *Glutaraldehyde is a carcinogen and may cause allergic reaction. Use in a fume hood and discard according to institutional regulations.*

3. Dialyze conjugate against 3 liters borate buffer, pH 8.5, overnight at 4°C. Change borate buffer and dialyze 4 hr. Store at 4°C.

Reference: Doolittle, 1986.

Contributors: James F. Collawn and Yvonne Paterson

DNA-Protein Interactions

12

For several decades, DNA-binding proteins have been studied because of their involvement in cellular processes such as replication, recombination, viral integration, and transcription. In recent years, the number of workers interested in the study of these proteins has greatly increased as the advent of recombinant DNA technology has led to the isolation of numerous biologically important genes. Many investigators are interested in how transcription of these genes is controlled in response to environmental or developmental signals, and have started to characterize the DNA sequences responsible for this regulation. This analysis has naturally led to the detection, isolation, and characterization of the proteins that bind to these regulatory sequence elements. This chapter summarizes the techniques currently used to characterize DNA-protein interactions and to isolate DNA-binding proteins.

The preparation of cell extracts from either the nucleus or cytoplasm is often the first step in studying these proteins and their interactions (*UNIT 12.1*). The resulting preparations can be used directly in a variety of functional studies, or as the starting point for the purification of proteins involved in gene regulation. In the past three years many advances have been made in the technologies used for detecting DNA-binding proteins and for purifying those proteins. Perhaps the most widely applied technology has been the mobility shift assay (*UNIT 12.2*), whereby proteins are detected by their ability to retard mobility of a labeled DNA fragment through a nondenaturing gel. This technique—originally developed to characterize the interaction of purified prokaryotic proteins with DNA—has been refined to allow detection of numerous DNA-binding proteins in crude extracts from a wide variety of cells. Because the mobility shift assay is simple and rapid, it is typically the method of choice when purifying DNA-binding proteins.

After detection of a DNA-binding protein, it is often necessary to determine its binding specificity. Guesses about the binding site can be verified by using mutated DNA fragments as either cold competitor or as labeled probe in the mobility shift assay. Alternatively, one can chemically modify the DNA template, and ask how these alterations affect binding of a specific protein. The most widely used of these modification, or "interference," techniques is methylation interference (*UNIT 12.3*). This technique allows direct detection of nucleotides that are in close contact with the binding protein. Here, guanine and adenine residues are methylated at an average of one modification per DNA molecule. Modified DNA molecules bound to a specific protein are then separated from those molecules that are not bound via the mobility shift assay. Molecules that are methylated at nucleotides which are important for binding will be underrepresented in bound DNA.

While methylation interference tends to be more informative than most protection techniques, DNase I protection mapping (or "footprinting") can provide another level of information (*UNIT 12.4*). In these procedures, protein is first bound to DNA and then the DNA is cleaved either by DNase or chemical agents. Footprinting typically reveals, without perturbing, the general region(s) of DNA to which a protein binds. Footprinting is rapid and sensitive once it is optimized for a particular interaction and can be used as a routine assay in purification. The footprinting assay can also be used as a quantitative technique to determine both the binding curves for individual proteins, as well as cooperative interactions among proteins bound to multiple sites along DNA.

After detection and preliminary characterization of a DNA-binding protein, investigators frequently wish to purify the protein. This can be accomplished using standard chromatography techniques (Chapter 10) and assaying the fractions for presence of the protein by mobility shift gel electrophoresis or footprinting. Another procedure that has proved to be very powerful in these purification schemes is the use of an affinity column containing large amounts of DNA specifying the binding site for the factor (*UNIT 12.9*; Kadonaga and Tjian, 1986).

These affinity columns take advantage of the extraordinary specificity that a protein has for its cognate binding site. When such a column is used, protein is applied to it in the presence of high levels of competing, nonspecific DNA, and proteins recognizing specific sequence motifs on the affinity column partition onto those sequences as the column is loaded. In an alternative procedure, protein is first fractionated on a standard, nonspecific DNA column (e.g., DNA-cellulose), and then applied in high salt to the specific affinity column. Specific protein-DNA complexes are frequently stable to moderate salt concentrations, while many nonspecific protein-DNA complexes are disrupted in high salt. A high degree of purification can therefore be achieved using a standard DNA-cellulose column and a DNA affinity column in tandem.

Similar degrees of purification can be accomplished through the use of biotinylated DNA fragments that contain the binding site for a protein and column matrices that are coated with streptavidin (*UNIT 12.6*). In this procedure, a biotinylated DNA fragment containing a specific sequence motif is mixed with crude or partially purified protein and an excess of competitor DNA. As with the affinity column described above, the specific protein will partition onto the DNA fragment containing the binding site. The specific DNA fragment—as well as the attached protein—is then fished out of this mixture using streptavidin and a column matrix, and the specific protein is eluted with high salt. This protocol is relatively easy to optimize, as the success or failure of several of the steps can be monitored using mobility shift gels. It is also extremely flexible, as one single type of column matrix is compatible with any biotinylated DNA probe, and thus can be used to purify numerous different DNA-binding proteins.

To help ensure that the appropriate protein has been purified, it is useful to know the size of the protein that interacts with a specific DNA sequence. In addition, regulatory proteins can be modified in the cell in response to environmental or developmental signals. Therefore, knowing the apparent size of a DNA-binding protein under various conditions is also important as this information can provide insight into regulatory events concerning that protein.

The size of a DNA-binding protein can be determined by covalently crosslinking the protein to its regulatory sequence using UV light and resolving the protein complexes on an SDS polyacrylamide gel (*UNIT 12.5*). This procedure works even for impure proteins in crude extracts. Crosslinking can be accomplished by irradiating a protein solution containing a specific labeled DNA probe. Alternatively, protein-DNA complexes can first be separated on a mobility shift gel and then irradiated. This latter protocol increases confidence that the identified protein is actually part of an appropriate complex. In both protocols, it is critical to verify that the crosslinked protein interacts specifically with a given sequence motif, for example by performing competition studies with unlabeled DNA fragments.

In many instances, an important goal is to clone the gene that encodes a DNA-binding protein, thus allowing detailed genetic characterization of the protein. This can be accomplished by purifying the binding protein to homogeneity, sequencing it, and using that sequence to identify a cDNA clone (see Chapter 6). However, despite the recent advances in purification techniques, the above approach can be extremely time-consuming, particularly if the DNA-binding protein is present at very low levels. An alternative approach is to use the sequence known to be recognized by the protein to directly identify a cDNA clone (*UNIT 12.7*). In this protocol, a library that expresses inserted cDNAs in *E. coli* is plated out, and proteins expressed in plaques produced by recombinant phages are transferred to nitrocellulose filters. These filters are then probed with a specific labeled sequence in order to detect clones that express a given DNA-binding protein. Success of this protocol relies on the assumption that the protein as expressed in *E. coli* will be capable of specifically binding DNA. For example, if the protein binds DNA as a heterodimer, or requires a particular covalent modification to bind to its site, it will not be detectable using this approach.

Once a cloned gene encoding a DNA-binding protein is identified, it is possible to synthesize radiolabeled protein by in vitro transcription and translation (*UNIT 10.6*). The resulting labeled protein can then be used to detect and analyze DNA-protein interaction (*UNIT 12.8*). The striking advantage of this approach is the possibility for constructing mutations in the cloned gene which can affect the DNA binding properties of the expressed gene.

The protocols described in this chapter can therefore be used to detect and characterize specific DNA-protein interactions, to purify specific DNA-binding proteins, and to clone the genes for these proteins. Using these protocols, one can start with a defined regulatory sequence motif and isolate the gene that encodes the factor responsible for regulation. The characteristics of the regulatory protein can be determined, and the clone of the regulatory gene can be dissected to define functional domains of the regulatory protein. These procedures have been widely applied in mammalian systems where there is very little that can be done to dissect gene regulation by classical genetic means, and have developed to the point where it is feasible, though certainly still difficult, to dissect complex regulatory loops at the molecular level.

Robert E. Kingston

Preparation of Nuclear and Cytoplasmic Extracts from Mammalian Cells

UNIT 12.1

Extracts prepared from the isolated nuclei of cultured cells (Fig.12.1.1) are functional in accurate in vitro transcription and mRNA processing and can be used as the starting material for purification of the proteins involved in these processes.

Protein content of the extracts is generally 8 to 12 mg/ml.

PREPARATION OF NUCLEAR EXTRACTS

BASIC PROTOCOL

Materials *(see APPENDIX 1 for items with ✓)*

Mammalian (e.g., HeLa) cells from spinner or monolayer cultures
✓ PBS
✓ Hypotonic buffer
✓ Low-salt and high-salt buffer (the latter with 1.2 M KCl)
✓ Dialysis buffer
Liquid nitrogen

Beckman JS-4.2 and JA-20 rotors or equivalents, 4°C

Preparation of Nuclear and Cytoplasmic Extracts

Date ______________________ Cell type ________________ Source ________________

Cell count ____________________ Culture volume ____________ Total cell number __________

Starting time: ______________________

Collect and centrifuge spinner cultures 20 min at 1850 x *g* in conical tubes.
Wash monolayer cultures with PBS and collect in conical tubes.
Centrifuge 10 min at 1850 x *g*.
Packed cell volume (pcv): ________________________
Wash spinner cultures with 5 pcv PBS and centrifuge 10 min at 1850 x *g*.
Resuspend in 5 pcv hypotonic buffer. ________________ *(vol hypotonic buffer)*
Centrifuge 5 min at 1850 x *g*.
Resuspend in hypotonic buffer to 3 pcv. ________________ *(vol hypotonic buffer)*
Swell on ice 10 min.
Homogenize swollen cells 10 strokes.
Pellet nuclei 15 min at 3300 x *g* ; remove cytoplasmic extract.

Nuclear Extract

Packed nuclei volume (pnv): ________________________
Resuspend nuclei in 1/2 pnv low-salt buffer. ________________ *(vol low-salt buffer)*
Add 1/2 pnv high-salt buffer. ________________ *(vol high-salt buffer)*
Homogenization? ________________ *(no. strokes)*
Extract 30 min: start: ____________ stop: ____________
Centrifuge 30 min at 25,000 x *g*.
Conductivity of supernatant: __________ = ____________ M
Dialysis: start: ____________ finish: ____________
Conductivity of nuclear extract: __________ = ____________ M
Centrifuge 20 min at 25,000 x *g*. ________________ *(vol nuclear extract)*
Aliquots: no. ____________ vol. ____________
Freeze in liquid nitrogen; store at –80°C.
Ending time: ________________________
Protein concentration: ________________________

Cytoplasmic (S-100) Extract

Cytoplasmic extract volume: ________________________
Add 0.11 vol 10x cytoplasmic extract buffer ________________ *(vol buffer)*
Centrifuge 1 hour at 100,000 x *g*.
Conductivity of supernatant: __________ = ____________ M
Dialysis: start: ____________ finish: ____________
Centrifuge 20 min at 25,000 x *g*. ________________ *(vol S-100 fraction)*
Conductivity of S-100 fraction: __________ = ____________ M
Aliquots: no. ____________ vol. ____________
Freeze in liquid nitrogen; store at –80°C.
Ending time: ________________________
Protein concentration: ________________________

Figure 12.1.1 Flow sheet for recording data when preparing nuclear and cytoplasmic extracts.

50-ml graduated conical polypropylene centrifuge tubes (or 15-ml tubes for smaller extract volumes)
Glass Dounce homogenizer (type B pestle)
Dialysis membrane tubing (≤14,000 MWCO; *APPENDIX 3*)
Conductivity meter

NOTE: Perform all protocols at 0° to 4°C, preferably in a cold room, using precooled buffers and equipment. Carry out all centrifugations at 4°C.

1a. *Collect cells from spinner cultures.* Centrifuge 5-10 × 10^8 cells/liter in 1-liter plastic bottles 20 min at 1850 × *g*. Pool cells in 50-ml graduated conical centrifuge tubes (one tube for every 2 to 3 liters of cells). Proceed to step 2.

1b. *Collect cells from monolayer cultures.* Remove culture medium from confluent monolayer cultures. Wash with PBS. Scrape cells into fresh PBS and pool in a graduated conical centrifuge tube. Proceed to step 2.

2. Centrifuge cells 10 min at 1850 × *g*.

3a. *Spinner cultures.* Measure packed cell volume (pcv). Resuspend cells in a volume of PBS ~5 times the pcv. Centrifuge cells 10 min at 1850 × *g*. Proceed to step 4.

3b. *Monolayer cultures.* Measure the pcv and proceed to step 4.

4. Rapidly resuspend cell pellets in a volume of hypotonic buffer ~5 times the pcv. Centrifuge cells 5 min at 1850 × *g*. Resuspend cells in hypotonic buffer to a final volume of 3 times the original pcv (step 3) and allow to swell on ice 10 min (cells should swell ≥2-fold).

5. Homogenize cells with ten or more slow up-and-down strokes in a glass Dounce homogenizer using a type B pestle. Check for cell lysis in a microscope using trypan blue exclusion (lysis should be >80% to 90%).

6. Centrifuge 15 min at 3300 × *g*. Save supernatant for S-100 cytoplasmic extract prep (Support Protocol).

7. Measure packed nuclear volume (pnv) from step 6. Resuspend nuclei (homogenize one stroke in Dounce if necessary) with a volume of low-salt buffer equal to ½ pnv.

8. While stirring gently, add dropwise a volume of high-salt buffer equal to ½ the pnv (step 7); KCl concentration should be ~300 mM final. Homogenize one stroke in Dounce if necessary. Allow nuclei to extract for 30 min with continuous gentle stirring.

9. Centrifuge nuclei 30 min at 25,000 × *g*. Measure conductivity of the nuclear extract (supernatant; step 10).

10. Place nuclear extract in dialysis tubing. Dialyze against 50 vol of dialysis buffer until the conductivities of extract and buffer are equal (when the extract reaches 100 mM KCl). Minimize dialysis time.

    Conductivity is checked by diluting 5 to 10 µl of the extract to 1 ml water. Read the conductivity of this dilution directly with a conductivity meter and compare it to that of an equivalent dilution of dialysis buffer.

11. Remove extract from dialysis bag; centrifuge 20 min at 25,000 × *g*.

12. Determine protein concentration of supernatant by Bradford assay (*UNIT 10.1*). Aliquot into tubes, freeze by submerging in liquid nitrogen, and store at −80°C.

SUPPORT PROTOCOL 1

OPTIMIZATION OF NUCLEAR EXTRACTION

This protocol describes a simple method to optimize the salt concentration during nuclear extraction for specific cell types or applications including transcription, splicing, and gel-shift analysis (*UNIT 12.2*).

Additional Materials *(also see Basic Protocol 2; see APPENDIX 1 for items with ✓)*

✓ High-salt buffer with 0.8, 1.0, 1.2, 1.4, and 1.6 M KCl

1. Perform steps 1 to 7 of Basic Protocol.
2. Divide nuclear suspension into five equal aliquots. With continuous mixing in cold room, add ⅓ aliquot vol of high-salt buffer to each aliquot as follows: add high-salt buffer with lowest KCl concentration (0.8 M) to first aliquot and add high-salt buffer with increasing KCl concentrations (up to 1.6 M) to subsequent aliquots. Gently mix 30 min.
3. Centrifuge 30 min at 25,000 × *g*. Dialyze supernatants (Basic Protocol, step 10) until the one with the highest KCl concentration approaches the conductivity of the dialysis buffer (100 mM KCl). Centrifuge the extracts 20 min at 25,000 × *g*.
4. Check conductivity of each extract and determine protein concentration (Basic Protocol, step 12). Analyze the extracts by the intended procedure to determine optimum extraction conditions.

SUPPORT PROTOCOL 2

PREPARATION OF THE CYTOPLASMIC (S-100) FRACTION

Additional Materials *(also see Basic Protocol; see APPENDIX 1 for items with ✓)*

Cytoplasmic extract (Basic Protocol, step 6)
✓ 10× cytoplasmic extract buffer
Beckman Type 50 fixed-angle rotor or equivalent

NOTE: Perform all steps at 0° to 4°C and use precooled buffers and equipment.

1. Measure volume of cytoplasmic extract. Add 0.11 vol of 10× cytoplasmic extract buffer and mix thoroughly. Centrifuge 1 hr at 100,000 × *g*.
2. Place supernatant (S-100) in dialysis tubing. Dialyze fraction until conductivity of S-100 reaches that of dialysis buffer (100 mM KCl). Centrifuge 20 min at 25,000 × *g*.
3. Check conductivity of supernatant and determine protein concentration. Aliquot into tubes, freeze in liquid nitrogen, and store at −80°C.

Reference: Dignam et al., 1983.

Contributors: Susan B. Abmayr and Jerry Workman

Mobility Shift DNA-Binding Assay Using Gel Electrophoresis

This assay is a simple, rapid, and extremely sensitive method for the detection of sequence-specific DNA-binding proteins in crude extracts. Proteins that bind specifically to an end-labeled DNA fragment retard the mobility of the fragment during electrophoresis, resulting in discrete bands corresponding to the protein-DNA complexes. This assay also permits the quantitative determination of the affinity, abundance, association and dissociation rate constants, and binding specificity of DNA-binding proteins.

MOBILITY SHIFT ASSAY USING LOW-IONIC-STRENGTH PAGE

BASIC PROTOCOL

Materials *(see APPENDIX 1 for items with ✓)*

Plasmid DNA with desired binding site
Appropriate restriction endonucleases (*UNIT 3.1*)
3000 to 6000 Ci/mmol [α-^{32}P]dNTP (*UNIT 3.4*)
5 mM 3dNTP mix (excluding labeled dNTP; *UNIT 3.4*)
Klenow fragment of *E. coli* DNA polymerase I (*UNIT 3.5*)
✓ TE buffer
✓ 10× loading buffer
✓ TAE or TBE buffer
✓ Ethidium bromide solution
✓ DEAE elution solution
✓ 1 M $MgCl_2$
100% ethanol
✓ Low-ionic-strength electrophoresis buffer and gel mix
30% (w/v) ammonium persulfate
TEMED (*N,N,N′,N′*-tetramethylethylenediamine)
Bulk carrier DNA, e.g., poly(dI-dC)·poly(dI-dC) (with minimal resemblance to the specific binding site)
Bovine serum albumin (BSA)
Buffered crude protein extract (containing DNA-binding protein)
DEAE membrane (Schleicher & Schuell NA45)

1. Digest 10 µg plasmid DNA in 100 µl with one or more restriction endonucleases to produce a 25- to 100-bp fragment containing the binding site and at least one 5′ overhang (*UNIT 3.1*).

   Gel-purified, double-stranded synthetic oligonucleotides may also be used as probes. Klenow fragment–labeled probes are preferable to kinased probes because some extracts contain active phosphatases.

2. Add 100 µCi of the desired [α-^{32}P]dNTP, 4 µl of 5 mM 3dNTP mix, and 2.5 U Klenow fragment. Incubate 20 min at room temperature.

3. Add 4 µl of a solution containing 5 mM of the dNTP corresponding to the radioactive dNTP. Incubate 5 min at room temperature.

4. Precipitate DNA (*UNIT 2.1*) and resuspend in TE buffer.

5. Add 10× loading buffer and electrophorese in TAE or TBE buffer on a 2% agarose minigel containing 5 µg/ml ethidium bromide (*UNIT 2.5A*).

Use a 3% to 4% sieving agarose gel for smaller fragments (UNIT 2.8). If nondenaturing polyacrylamide gel electrophoresis is used, detect the fragment by autoradiography, excise the fragment, embed the gel slice in an agaorse gel, and proceed as outlined below.

6. Visualize bands on a long-wavelength UV transilluminator. Cut a horizontal slit below band and slide a piece of DEAE membrane (wet in gel buffer) into slit. Squeeze gel firmly against paper and resume electrophoresis until the DNA has run onto membrane.

7. Wash membrane in electrophoresis buffer and blot on filter paper. Place membrane in bottom of 1.5-ml, round-bottom screw-cap vial containing 400 µl DEAE elution solution. Incubate 30 min at 68°C.

8. Transfer elution solution to 1.5-ml microcentrifuge tube and microcentrifuge 15 min at 4°C. Transfer supernatant to clean tube, leaving 10 µl at the bottom of tube.

9. Add 4 µl of 1 M $MgCl_2$ to supernatant. Precipitate DNA with 1.0 ml of 100% ethanol and resuspend in 100 µl TE buffer.

10. Count 1 µl to determine cpm/µl. Estimate DNA concentration by ethidium bromide dot quantitation (*UNIT 2.6*, Support Protocol).

    Efficient recovery of DNA yields 2-10 µg/ml that can be used for 4 to 6 weeks.

11. Assemble 16-cm-long glass plates and 1.5-mm spacers. Add 100 µl of 30% ammonium persulfate and 34 µl TEMED to 40 ml low-ionic-strength gel mix and pour between plates, inserting a comb with teeth ≥7 mm wide. Allow gel to polymerize 20 min.

12. Remove comb and bottom spacer and place gel in electrophoresis tank with low-ionic-strength electrophoresis buffer. Recirculate buffer with a two-head pump at 5-30 ml/min. Pre-run gel ≥90 min at 100 V (~22 mA).

13. In a microcentrifuge tube, combine the following in 10-15 µl final: 10,000 cpm DNA probe (0.1 to 0.5 ng), 2 µg poly(dI-dC)·poly(dI-dC), 300 µg/ml BSA final, and ~15 µg protein from a buffered crude protein extract. Mix and incubate mix 15 min at 30°C.

    Many of the salt, buffer, and glycerol components needed in the binding reaction are present in the buffer containing the protein extract. Optimum levels must be titrated for each protein-DNA interaction.

14. Load a small volume of 10× loading buffer into one well and run dyes into gel. Load each binding reaction into a well and electrophorese at ~30 to 35 mA until bromphenol blue (≅ 70 bp) approaches the bottom of the gel (~1.5 to 2 hr).

15. Disassemble gel apparatus and pry glass plates apart. Lay plate with gel attached on bench and place 3 sheets of Whatman 3 MM filter paper on top of gel. Invert and peel paper with gel attached from plate.

16. Cover gel with plastic wrap and dry under vacuum. Autoradiograph dried filter overnight without an intensifying screen or 2 to 3 hr with an intensifying screen (*APPENDIX 3*).

    Femptomole quantities, as well as the number and type of DNA-binding proteins, can be detected on the autoradiogram.

MOBILITY SHIFT ASSAY USING HIGH-IONIC-STRENGTH PAGE

ALTERNATE PROTOCOL

Higher ionic strength of the buffer results in sharper bands than those detected in the low-ionic-strength system. However, some DNA-protein interactions do not survive the high-ionic-strength buffer. Therefore, both buffer systems should be tried for each protein-DNA interaction.

Additional Materials *(also see Basic Protocol; see APPENDIX 1 for items with ✓)*

✓ High-ionic-strength gel mix and electrophoresis buffer

Follow Basic Protocol with the following exceptions:

1. Pour gel using high-ionic-strength gel mix (step 11).
2. Run gel using high-ionic-strength electrophoresis buffer, without recirculating the buffer (step 12).
3. Prepare binding reactions with 2- to 10-fold less carrier DNA (step 13).

Reference: Chodosh et al., 1986.

Contributor: Lewis A. Chodosh

Methylation and Uracil Interference Assays for Analysis of Protein-DNA Interactions

UNIT 12.3

Interference assays identify specific residues in the DNA binding site that, when modified, interfere with binding of the protein.

METHYLATION INTERFERENCE ASSAY

BASIC PROTOCOL 1

In methylation interference, probes are generated by methylating guanines at the N-7 position and adenines at the N-3 position with DMS; these methylated bases are cleaved specifically by piperidine.

Materials *(see APPENDIX 1 for items with ✓)*

✓ TE buffer, pH 7.5 to 8.0
Dimethyl sulfate (DMS)
✓ DMS reaction buffer
✓ DMS stop buffer
10 mg/ml tRNA solution
0.3 M sodium acetate/1 mM EDTA, pH 5.2
1 M piperidine (dilute from 10 M piperidine stock)
✓ Stop/loading dye
90° to 95°C water bath

1. Prepare a DNA probe labeled at one end (as for mobility-shift assay, *UNIT 12.2*).
2. Suspend ~10^6 cpm of probe in 5 to 10 µl TE buffer. Add 200 µl DMS reaction buffer and 1 µl DMS. Mix well by vortexing. Incubate 5 min at room temperature.

   CAUTION: *DMS is a powerful poison and must be used in a fume hood with careful handling. Liquids containing DMS should be disposed of in a designated DMS waste bottle and pipet tips that come into contact with DMS should be placed in a separate DMS solid-waste bottle for disposal by institutional safety officials.*

3. Add the following to the probe mixture:

   40 µl DMS stop buffer
   1 µl 10 mg/ml tRNA solution
   600 µl 100% ethanol.

   Mix and incubate ~10 min in a dry ice/ethanol bath. Microcentrifuge 10 min at maximum speed, 4°C. Carefully remove supernatant with a drawn-out Pasteur pipet and dispose in liquid DMS waste bottle.

4. Resuspend pellet in 250 µl of 0.3 M sodium acetate/1 mM EDTA. Keep the tube on ice. Add 750 µl of 100% ethanol, mix, and ethanol precipitate as in step 3.

5. Repeat ethanol precipitation once exactly as in step 4. Wash pellet once in 70% ethanol and microcentrifuge 10 min again. Carefully remove supernatant, invert tube on tissue, and air dry 10 min.

6. Measure the pellet for Cerenkov counts in a scintillation counter to determine cpm. Resuspend pellet in TE buffer at ~20,000 cpm/µl.

   Labeled probe should have an average of one modified residue per molecule.

7. Using optimized conditions for DNA binding, set up a reaction as described in step 13 of the *UNIT 12.2* Basic Protocol, but scaled up ~5-fold (use 10^5 cpm of probe in 50 µl).

8. Load binding reaction on three lanes of a nondenaturing polyacrylamide gel. Electrophorese the binding reaction using the procedure for the mobility-shift assay.

9. Autoradiograph gel, cut out bands corresponding to the protein-DNA complex and free probe, and purify DNA from gel by electroelution onto a DEAE membrane (*UNIT 2.7*).

10. Resuspend pellet in 100 µl of 1 M piperidine. Place in a 90° to 95°C water bath for 30 min. Put a glass plate over tubes to keep tops from popping open. Carefully remove from water bath and place on dry ice.

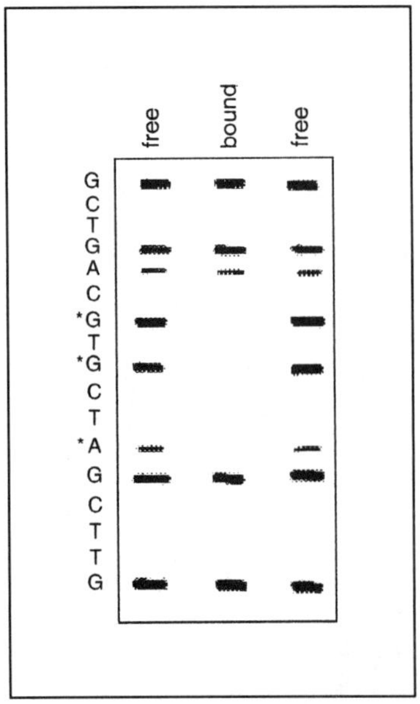

Figure 12.3.1 Hypothetical autoradiogram of a methylation interference experiment, as described in Anticipated Results. Guanines and adenines that interfere are indicated by asterisks.

11. Make holes in the tops of the tubes with a large needle and lyophilize samples in a vacuum evaporator (e.g., Speedvac) for ~1 hr or until dry. Add 100 µl of distilled water. Freeze and lyophilize again. Repeat addition of water, freezing, and lyophilizing. Measure sample for Cerenkov counts to determine cpm.

12. Add sufficient stop/loading dye to pellet so that 1 to 2 µl will contain the sample to be loaded (~3000 cpm for an overnight exposure). Heat 5 min at 95°C. Quickly chill on ice.

13. Load the samples from the free probe and bound complex on a 6% or 8% polyacrylamide/urea sequencing gel. Electrophorese samples as for a sequencing gel and expose gel for autoradiography (Fig. 12.3.1).

    It is critical to equalize the number of counts applied from the DNA-protein complex and from the free probe to allow accurate comparison between samples.

URACIL INTERFERENCE ASSAY

BASIC PROTOCOL 2

In uracil interference, probes are generated by PCR amplification (*UNIT 15.1*) in the presence of a mixture of TTP and dUTP, thereby producing products in which thymine residues are replaced by deoxyuracil residues (which contain hydrogen in place of the thymine 5-methyl group). Uracil bases are specifically cleaved by uracil-*N*-glycosylase to generate apyrimidinic sites that are susceptible to piperidine (Fig. 12.3.2).

Materials *(see APPENDIX 1 for items with ✓)*

DNA containing protein-binding site
Oligonucleotide primers specific for sequences flanking the binding site on the two complementary DNA strands
2 mM 4dNTP mix (*UNIT 3.4*)
0.5 mM dUTP
Taq polymerase buffer (*UNIT 3.4*) and *Taq* polymerase (*UNIT 3.5*)
✓ TE buffer, pH 7.5 to 8.0
Uracil-*N*-glycosylase (Perkin Elmer-Cetus)
1 M piperidine (diluted from 10 M piperidine stock)
✓ Stop/loading dye
90° to 95°C water bath

1. ^{32}P-label the 5′ ends of oligonucleotide primers using T4 polynucleotide kinase.

   Primers should be designed for PCR amplification (UNIT 15.1) and should contain sequences that flank the protein-binding site and are on opposite strands (Fig. 12.3.1).

2. Set up two parallel 50-µl PCR reactions containing the following:

   5 µl DNA fragment containing protein-binding site (0.2 pmol)
   5 µl of one ^{32}P-labeled oligonucleotide primer (20 pmol)
   5 µl of other oligonucleotide primer (unlabeled; 20 pmol)
   5 µl 2 mM 4dNTP mix
   5 µl 0.5 mM dUTP
   5 µl 10× *Taq* DNA polymerase buffer
   19 µl H_2O
   1 µl *Taq* DNA polymerase (5 U).

   Carry out 8 cycles of optimized PCR amplification (*UNIT 15.1*).

3. Purify PCR products by electrophoresis on a nondenaturing polyacrylamide gel followed by phenol extraction and ethanol precipitation.

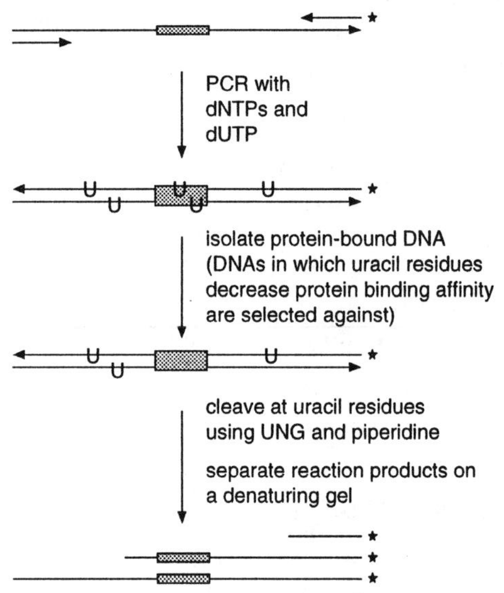

Figure 12.3.2 Uracil interference. An oligonucleotide or restriction fragment (long arrow) containing a protein-binding site (hatched box) is amplified by PCR using one unlabeled primer (short arrow) and one 5′-labeled primer (asterisk) in the presence of dGTP, dATP, dCTP, dTTP, and dUTP, producing reaction products in which deoxyuracil is randomly substituted for thymine on both strands. This collection of DNA molecules is incubated with the protein of interest and DNA molecules containing deoxyuracil substitutions that do not interfere with protein binding are selected by purifying the DNA-protein complex away from unbound DNA. The resulting DNA is cleaved at uracil residues using uracil-*N*-glycosylase followed by piperidine, and the reaction products are separated on a denaturing polyacrylamide gel. (Reprinted from Pu and Struhl, 1992, by permission of Oxford University Press.)

4. Measure pellet for Cerenkov counts in a scintillation counter to determine cpm. Resuspend in TE buffer at ~20,000 cpm/μl.

5. Carry out a DNA-binding reaction (*UNIT 12.2* Basic Protocol, step 13) using 10^5 cpm of probe in 50 μl. Electrophorese the binding reaction on a nondenaturing polyacrylamide gel.

6. Autoradiograph gel (1 to 12 hr), cut out bands corresponding to the protein-DNA complex and free probe, and purify DNA from gel slices. Resuspend each DNA sample in 25 μl TE buffer.

7. Set up the following uracil-*N*-glycosylase reaction (50 μl final):

   19 μl H_2O
   5 μl 10× *Taq* DNA polymerase buffer
   25 μl DNA (from step 6)
   1 μl uracil-*N*-glycosylase (1 U).

   Incubate 60 min at 37°C. Stop the reaction by ethanol precipitation.

8. Resuspend DNA pellets in 100 μl of 1 M piperidine. Incubate 30 min in a 90° to 95°C water bath. Place a glass plate over tubes to keep tops from popping open. After piperidine cleavage, place tubes on dry ice.

9. Make holes in tops of tubes with a large needle and lyophilize samples in a vacuum evaporator (e.g., Speedvac) for 1 hr or until dry. Add 100 µl water, freeze, and lyophilize again. Repeat addition of water, freezing, and lyophilizing. Measure the samples for Cerenkov counts.

10. Add stop/loading dye to the pellet. Heat 5 min at 95°C and quickly chill on ice. Electrophorese samples from free probe and bound complex on a 6% or 8% sequencing gel and autoradiograph.

    To easily compare the samples, it is critical to equalize the number of counts prior to electrophoresis. For an overnight exposure with an intensifying screen, 3000 cpm is sufficient.

References: Hendrickson and Schleif, 1985; Maxam and Gilbert, 1980; Pu and Struhl, 1992.

Contributors: Albert S. Baldwin, Jr., Marjorie Oettinger, and Kevin Struhl

DNase I Footprint Analysis of Protein-DNA Binding

UNIT 12.4

In this assay—designed to determine the specific binding sites of protein on DNA—bound protein protects the phosphodiester backbone of DNA from DNase I–catalyzed hydrolysis. Binding sites are visualized by autoradiography of the DNA fragments that result from hydrolysis, following separation by electrophoresis on denaturing DNA sequencing gels. Footprinting has been developed further as a quantitative technique to determine separate binding curves for individual protein-binding sites on the DNA. For each binding site, the total energy of binding is determined directly from that site's binding curve. For sites that interact cooperatively, simultaneous numerical analysis of all binding curves can be used to resolve both the intrinsic binding and cooperative components of these energies.

DNASE I FOOTPRINT TITRATION

BASIC PROTOCOL

Materials *(see* APPENDIX 1 *for items with ✓)*

Plasmid DNA containing protein-binding sites (*UNIT 1.6*)
Appropriate restriction endonucleases (*UNIT 3.1*)
100% and ice-cold 70% ethanol
✓ TE buffer
Aqueous [α-^{32}P]dNTP (3000 to 6000 Ci/mmol; *UNIT 3.4*)
10× Klenow fragment buffer (*UNIT 3.4*) and Klenow fragment (*UNIT 3.5*)
5 mM 4dNTP mix (*UNIT 3.4*)
Deoxyribonuclease I (DNase I; *UNIT 3.12*)
✓ Assay buffer
✓ DNase I stop solution
✓ DNase I storage buffer
✓ Formamide loading buffer

1. Digest ~5 pmol plasmid with a restriction enzyme that generates a 3′-recessed end (*UNIT 3.1*), 25 to 100 bp from first protein-binding site (Fig. 12.4.1).

2. Ethanol precipitate DNA (*UNIT 2.1*), wash once with 1 ml cold 70% ethanol, and dry pellet in Speedvac evaporator. Dissolve pellet in 5 µl TE buffer.

3. Add 50 µCi each of appropriate aqueous [α-^{32}P]dNTPs, 5 µl of 10× Klenow fragment buffer, and water to 49 µl. Add 1 µl Klenow fragment (5 to 10 U), mix gently, microcentrifuge, and incubate 25 min at room temperature.

4. Add 2 µl of 5 mM 4dNTP mix, mix, and incubate 5 min. Remove unincorporated nucleotides with spin column procedure (*UNIT 3.4*). Ethanol precipitate DNA as in step 2.

5. Digest with second enzyme to generate a restriction fragment labeled on one strand and only at one end, >150 bp beyond the protein-binding site most distal to labeled end of fragment (Fig. 12.4.1).

6. Purify labeled binding-site-containing DNA using agarose gel electrophoresis (*UNIT 2.5A*), followed by electroelution (*UNIT 2.6*) and reversed-phase chromatography (*UNIT 2.6*).

7. Ethanol precipitate DNA as in step 2. Dissolve DNA in 100 µl TE buffer, pH 8.0 and determine counts. Store at 4°C ≤2 weeks. Do not freeze.

8. Prepare $(n + 2) \times 180$ µl assay buffer in 10-ml disposable plastic tube (n is number of binding reactions in experiment). Add volume of ^{32}P-labeled DNA to yield 10,000 to 15,000 cpm/lane and vortex gently.

   The assay buffer used depends on the nature of the protein, the protein-DNA system, and the questions being addressed. Millimolar concentrations of Mg^{2+} and Ca^{2+} are required for DNase I activity.

9. Aliquot 180 µl of assay buffer containing ^{32}P-labeled DNA into each of $n + 1$ silanized 1.5-ml microcentrifuge tubes (extra tube is control to which no DNase I is added).

10. Prepare serial protein dilutions that cover the range of concentrations to be analyzed.

    The ligand concentrations should span a range from 0% to ≥99% saturation of all of the protein-binding sites and should define an evenly spaced, logarithmic series with at least several points to define each asymptote of the titration curves (Fig. 12.4.2).

11. Pipet protein dilution (2 to 20 µl) into each tube. Add assay buffer (without ^{32}P-labeled DNA) to a final total volume of 200 µl per tube. Gently mix and microcentrifuge briefly. Equilibrate in a regulated water bath 30 to 45 min (determined empirically) at the desired temperature.

12. Prepare an excess of DNase I stop solution (700 µl/tube) and place in dry ice/ethanol bath.

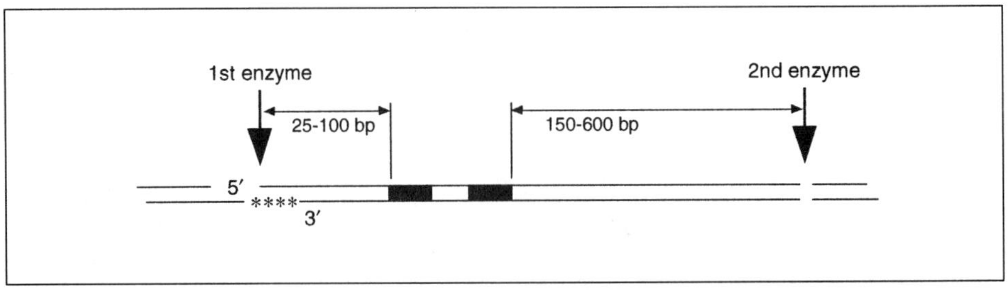

Figure 12.4.1 Correct positioning of protein-binding sites from the restriction cuts used to generate a linear, singly end-labeled fragment. Black boxes represent protein-binding sites. Asterisks (*) represent [^{32}P]dNTP incorporated in the Klenow labeling reaction.

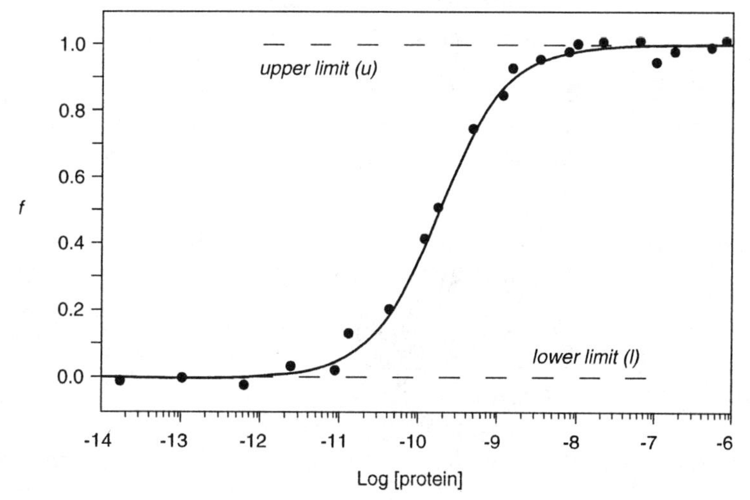

Figure 12.4.2 Footprint titration curve showing the proper range and spacing of protein concentrations. The upper and lower limits of the titration curve are indicated by dashed lines. See Equation 1 for a definition of *f*.

13. Prepare ≥500 µl dilute DNase I solution (determine concentration empirically) in assay buffer *without* BSA and calf thymus DNA. Place solution in regulated water bath with samples and allow it to equilibrate.

14. To start DNase I exposure, pipet exactly 5 µl dilute DNase I into first tube, mix quickly, and immediately return tube to regulated bath. After precisely 2 min, rapidly add 700 µl DNase I stop solution to tube. Vortex vigorouly. Place tube in dry ice/ethanol bath.

    The exposure time and DNase I concentration can be varied (between, but never within, experiments) as long as the product of the two (and hence, the number of DNA nicks produced) remains constant. Overall exposure time is not an issue.

15. Repeat step 14 for each tube.

16. Add stop solution to control tube prepared in step 9 (without DNase I).

17. Precipitate DNA in dry ice/ethanol bath for ≥15 min. Start timing when last tube is placed in bath. Microcentrifuge 15 min and carefully remove supernatant.

18. Add 1 ml cold 70% ethanol and microcentrifuge 5 min. Repeat wash step. Dry pellets in Speedvac evaporator (10 to 15 min).

19. Resuspend DNA in 5 µl formamide loading buffer and vortex vigorously. Store samples overnight at −70°C if not analyzed immediately.

20. Prepare and pre-electrophorese a polyacrylamide (usually 6% to 8%) DNA sequencing gel (*UNIT 2.12*).

21. Heat samples at 90°C in a dry-block heater 5 to 10 min, followed by immediate quenching in wet ice. When the temperature of the gel reaches 50° to 55°C during pre-electrophoresis, load samples.

22. Electrophorese until protein-binding sites migrate to, or just below, the middle of gel. Dry gel and autoradiograph with preflashed Kodak X-Omat AR film and a single calcium tungstate intensifying screen at −70° to −85°C (*APPENDIX 3*).

23. Develop films. Make two or three autoradiograms of each gel (at different exposure levels) to ensure a correct exposure. Store films carefully to avoid marks that will interfere with quantitation.

SUPPORT PROTOCOL

12.4

QUANTITATION OF PROTEIN-BINDING EQUILIBRIA BY DENSITOMETRIC AND NUMERICAL ANALYSES

This protocol outlines the resolution of binding curves from autoradiograms and the numerical analyses of those curves to obtain equilibrium binding constants. Accurate densitometric analysis is critical to obtaining thermodynamically valid individual site–binding curves. Required hardware includes a two-dimensional optical scanning device and a micro- or minicomputer with a high-resolution graphics display.

The digital image consists of an array of values corresponding to the diffuse optical density of picture elements (pixels). The pixels must be small enough to resolve closely spaced bands clearly. Spatial resolution of 250 to 200 μm is optimal. The scanner must resolve at least 256 levels of gray. This optical resolution is adequate to quantitate 0 to 1.6 OD when the levels represent equal optical density increments, such as those produced by flying spot or laser scanning devices. Camera devices, however, measure transmitted intensity (I) rather than optical density. The transformation to optical density (OD = log I/I_0) must be computed. I_0, the incident intensity, is determined by scanning the light source with no film present. For analysis, the film image should be displayed on a graphic display monitor with a minimum of 16 gray or color levels. Display devices with 256 levels are preferred. Spatial resolution ≥1024 × 1024 pixels is preferred, although lower-resolution display devices can be used.

Computer software to expedite both film scanning and analysis is available for PC/DOS and VAX/VMS computers. Information concerning these programs can be obtained by contacting the authors. Also, some commercially available image-analysis software packages have functions that can be used for the analysis of titration autoradiograms. The protocol described can be implemented using such systems.

1. Use a film-scanning device to construct a two-dimensional digital image of the film.

2. Integrate optical density in each lane (i.e., at each ligand concentration) for protected DNA bands in each binding site (Fig. 12.4.3).

   It is best to consider all protected bands by analyzing contiguous groups of bands as single blocks (Fig. 12.4.3). This simplification is appropriate whenever all the bands included titrate equivalently. Choose the ends of blocks by drawing boundaries at the minimum density level between bands of similar maximum density that can be visualized in all lanes (i.e., even at high protection). It is crucial that each block include precisely the same bands in every lane of the film.

3. Correct integrated optical density value of each block in each lane for local film background optical density. This can be calculated using image-analysis software as follows:

   a. Define small rectangles in the center of the interlane space and at the position of each block (Fig. 12.4.3).

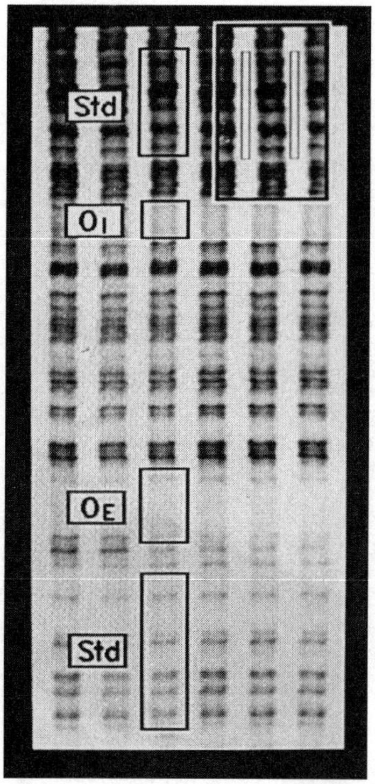

Figure 12.4.3 Portion of a footprint titration autoradiogram showing the different blocks that need to be defined for analysis. "Std." denotes a standard block used to correct for variations in the DNA loaded onto each lane. O_I and O_E denote protein-binding sites 1 and 2, respectively. The insert shows regions for a standard block for which background determinations will be made.

b. Calculate a histogram plot of the pixel optical density values in each rectangle and define local background as the most probable pixel optical density.

c. For each block and lane, average the background values on either side of the lane. Multiply this average value by the number of pixels in the block.

d. Subtract the product from the integrated optical density to give corrected integrated optical density for that block.

4. Standardize corrected, integrated optical density values for binding-site blocks to total DNA in lane.

a. Choose one or more blocks, excluding titrating sites or very hypersensitive bands, to represent the total DNA concentration in a lane.

b. Follow steps 2 and 3 to determine corrected, integrated optical density for these standard blocks.

c. Calculate optical density ratio (D_{site}/D_{std}) for each binding-site block. If more than one standard block is chosen, take the sum of their values as D_{std}.

It is practical and sufficient to choose two large standard blocks (Fig. 12.4.3), one nearer the origin of electrophoresis than the binding site(s) and one farther away. Carefully check each standard for systematic, ligand concentration–dependent variations that invalidate its use.

5. Convert OD ratios to fractional protection (f), according to

$$f = 1 - \left\{ \frac{(D_{n,site}/D_{n,std})}{(D_{r,site}/D_{r,std})} \right\} \quad (1)$$

where n refers to any lane with finite protein ligand concentration and r refers to reference lane (that must be included in every experiment) in which no protein ligand has been added to reaction mixture. Plot data, f versus [ligand], for each binding-site block to define binding curves.

12.4

Table 12.4.1 Calculated Distribution of DNase I Nicks Among Fragments[a]

Uncut[b] (%)	Singly nicked[c] (%)	Multiply nicked[c] (%)	Ave. no. nicks/ fragment[d]
10	23.1	67.0	2.3
20	32.2	47.8	1.6
30	36.1	34.0	1.2
50	34.7	15.3	0.7
75	21.6	3.4	0.3
90	9.5	0.5	0.1

[a]Calculated for a 682-bp fragment, before DNase I exposure.
[b]Experimentally measured percentage of fragments left after exposure to DNase I.
[c]Percentages were calculated assuming a Poisson distribution of nicks.
[d]Average number of nicks per labeled strand calculated assuming a Poisson distribution of nicks.

If outlined protocols are followed and if control experiments (see Brenowitz et al., 1986a,b) demonstrate no effect of exposure to DNase I on the protein–DNA binding equilibria, then f yields the fractional saturation ($\bar{Y}$). Observed values of f frequently do not span the entire range, 0 to 1. One reason for this is the failure to account for nicking of the DNA prior to DNase I exposure (i.e., $D_{n,\text{site}} > 0$, even at saturating ligand concentration; see Table 12.4.1). Therefore, in analyzing data, it is best to treat f as defining a transition curve, proportional to the binding curve, and to include both curve endpoints (u and l for upper and lower endpoints, respectively) as adjustable parameters.

For a single binding site or multiple sites that do not interact cooperatively, the binding curve is the familar Langmuir isotherm given by

$$\bar{Y} = \frac{k[P]}{1 + k[P]} \qquad (2)$$

where $[P]$ is the free protein concentration and k is the microscopic equilibrium association constant (the more complex case of multiple, interacting binding sites is discussed in the commentary). Thus, the equation used to analyze the data is

$$f = m \times \left\{\frac{k[P]}{1 + k[P]}\right\} + b \qquad (3)$$

with $m = 1/(u - l)$ and $b = 1/(l - u)$ and other units as defined above. Methods of nonlinear least-squares parameter estimation (Johnson and Frasier, 1985) should be used to estimate k, l, and u, along with their confidence limits and the variance of the fit. Pay particular attention to the pattern of residuals (i.e., difference between fitted and observed f) in evaluating the goodness of fit of the nonlinear least-squares parameter estimation.

ALTERNATE PROTOCOL

DNASE FOOTPRINTING IN CRUDE FRACTIONS

DNase footprinting is frequently used to locate proteins in crude fractions, thus providing an assay for use during purification. This is accomplished by varying conditions such that nonspecific DNA-binding proteins are inhibited from binding to the labeled DNA containing the site of interest—i.e., by including substantial amounts of competitor DNA in the reaction, or by increasing the salt concentration of the reaction.

1. Prepare singly ^{32}P end-labeled DNA as in steps 1 to 7 of Basic Protocol.

Table 12.4.2 Assay Conditions for Footprinting in Crude Fractions

Reaction no.[a]	10× assay buffer (μl)	Probe (cpm)	Crude fraction (μl)	Poly(dI-dC) (μg)	DNase (μl)[b]
1	20	10-20k	0	0.4	5 (1×)
2	20	10-20k	12	0.1	5 (1×)
3	20	10-20k	12	0.4	5 (1×)
4	20	10-20k	12	2.0	5 (1×)
5	20	10-20k	12	10.0	5 (1×)
6	20	10-20k	0	0.4	5 (4×)
7	20	10-20k	12	0.1	5 (4×)
8	20	10-20k	12	0.4	5 (4×)
9	20	10-20k	12	2.0	5 (4×)
10	20	10-20k	12	10.0	5 (4×)
11	20	10-20k	0	0.4	5 (16×)
12	20	10-20k	12	0.1	5 (16×)
13	20	10-20k	12	0.4	5 (16×)
14	20	10-20k	12	2.0	5 (16×)
15	20	10-20k	12	10.0	5 (16×)

[a]Reaction volume is 200 μl.

[b]Dilutions are shown in parentheses. 1× refers to the dilution that leaves 50% unnicked DNA (step 2 of Alternate Protocol). 4× and 16× refer to higher concentrations (as indicated) of DNase I.

2. Characterize dilutions of stock DNase I for a concentration of DNase I that will produce ~50% non-nicked DNA (~0.1 mg/ml). Set up a 200-μl reaction containing probe and assay buffer, but no added protein, as described in step 8 of Basic Protocol. Add 5 μl of diluted DNase I, mix gently, and microcentrifuge briefly.

3. Incubate 2 min and stop as described in step 14 of Basic Protocol. Ethanol precipitate and analyze products on a sequencing gel.

4. Perform footprinting reaction as in Basic Protocol. Instead of using varying amounts of protein, use a constant amount of crude fraction that is known to contain activity of interest. Perform a set of assay conditions as in Table 12.4.2, to find appropriate condition.

   In the experiment described in the table, poly(dI-dC) concentration is varied in an attempt to find an optimal concentration that will "soak up" contaminating nonspecific proteins. A similar protocol can be used to vary salt concentration (e.g., try 200 mM, 300 mM, 500 mM, and 800 mM) or calf thymus DNA concentration.

References: Brenowitz et al., 1986a,b; Dabrowiak and Goodisman, 1989; Johnson and Frasier, 1985.

Contributors: Michael Brenowitz, Donald F. Senear, and Robert E. Kingston

UV Cross-Linking of Proteins to Nucleic Acids

Irradiation of protein-nucleic acid complexes with ultraviolet light causes covalent bonds to form, selectively labeling DNA-binding proteins based on their specific interaction with a DNA recognition site. As a consequence of label transfer, the molecular weight of a DNA-binding protein in a crude mixture can be rapidly and reliably determined (Fig. 12.5.1).

BASIC PROTOCOL

UV CROSS-LINKING USING A BrdU-SUBSTITUTED PROBE

DNA molecules containing halogenated analogs of thymidine, such as bromodeoxyuridine (BrdU), are considerably more sensitive to UV-induced cross-linking compared to unsubstituted DNA. The efficiency of UV cross-linking in this protocol is 0.1% to 10%, and it is exceedingly rare to see more than one cross-linking event in a single complex. Because BrdU cannot be incorporated into RNA, a method for cross-linking to RNA is described in the Alternate Protocol.

Materials *(see APPENDIX 1 for items with ✓)*

Single-stranded M13 vector with desired binding site (*UNIT 1.10*)
17-bp M13 universal primer (Pharmacia)
✓ 1× and 10× restriction endonuclease buffers (500 μM and 500 mM NaCl, respectively)
3000 Ci/mmol [α-^{32}P]dCTP (*UNIT 3.4*)
✓ 50× dNTP/BrdU solution
✓ 0.1 M dithiothreitol (DTT)
25 U Klenow fragment of *E. coli* DNA polymerase I (*UNIT 3.5*)
Ammonium acetate
100% ethanol
✓ TE buffer
Buffered extract containing DNA-binding protein
Bulk carrier DNA, e.g., poly(dI-dC)·poly(dI-dC)
0.5 M $CaCl_2$
Deoxyribonuclease I (Worthington; *UNIT 3.12*)
1 U micrococcal nuclease (Worthington; *UNIT 3.12*)
✓ 2× SDS sample buffer
Fluor (Du Pont NEN EN^{3}HANCE or equivalent)
^{14}C-labeled protein markers (*UNIT 10.2*)

DEAE membrane (Schleicher & Schuell NA45)
UV transilluminator (305 nm, 7000 μW/cm^2; Fotodyne)

1. Mix 10 μg single-stranded M13 vector (with high-affinity protein-binding site) with equimolar amount of 17-bp M13 universal primer. Adjust final volume to 100 μl in 1× restriction buffer; heat 5 min at 90°C, and cool overnight at room temperature.
2. Add the following to hybridized mixture:

   50 μl 3000 Ci/mmol [α-^{32}P]dCTP
   3.5 μl 50× dNTP/BrdU solution
   1.75 μl 0.1 M DTT
   7.5 μl 10× restriction buffer (50 mM NaCl final)
   7 μl H_2O
   5 μl Klenow fragment (25 U).

   Incubate 90 min at 16°C. Heat-inactivate Klenow fragment 10 min at 68°C.

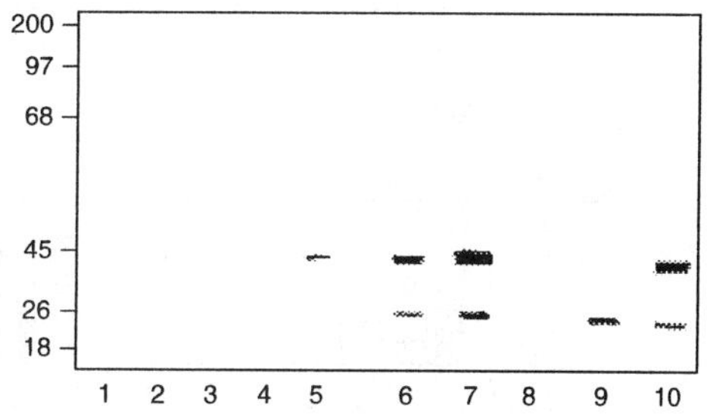

Figure 12.5.1 Representation of autoradiogram of a typical UV crosslinking experiment. As described in commentary, procedure involves UV crosslinking of protein to DNA probe, digestion of free DNA with DNase, and electrophoresis by SDS-PAGE of the treated binding reactions. Molecular weights of ^{14}C-labeled protein markers in Lane 1 are indicated to the left. Lanes 2 through 10 each contained a ^{14}C-labeled DNA probe with a site for a binding protein X. Samples applied to the lanes differed as follows (see also anticipated results): Lane 2—no protein; Lane 3—no UV irradiation; Lane 4—5 min irradiation of complete binding reaction; Lane 5—15 min irradiation of complete binding reaction; Lane 6—30 min irradiation of complete binding reaction; Lane 7—60 min irradiation of complete binding reaction; Lane 8—60 min irradiation of complete binding reaction then proteinase K treatment; Lane 9—60 min irradiation of complete binding reaction plus 100-fold molar excess of unlabeled probe containing a binding site for protein X; Lane 10—60 min irradiation of complete binding reaction plus 100-fold molar excess of unlabeled mutant probe lacking a binding site for protein X.

3. Add 40 U restriction endonuclease(s) to generate DNA fragment between 20 and 600 bp. Digest under appropriate conditions (*UNIT 3.1*).

4. Add ammonium acetate to 0.3 M and precipitate with 2 vol of 100% ethanol. Resuspend in TE buffer.

5. Electrophorese on agarose gel containing 0.5 µg/ml ethidium bromide. Isolate desired fragment using DEAE membrane (*UNIT 12.2*).

6. Determine specific activity of BrdU-substituted fragment and estimate DNA concentration by ethidium bromide dot quantitation (*UNIT 2.6*).

   Integrity and function of probe can be tested in a mobility shift DNA-binding assay (UNIT 12.2).

7. Set up the following binding reaction in 1.5-ml round-botton vial: 10^5 cpm of uniformly labeled probe, buffered extract containing binding protein, and 10-20 µg DNA carrier such as poly(dI-dC)·poly(dI-dC). Adjust volume to 50 µl. Seal with plastic wrap held in place with Parafilm.

8. Set vial in test-tube rack (Fig. 12.5.2). Irradiate from 5 cm directly above, 5 to 60 min by inverting a UV transilluminator, 305 nm, 7000 $\mu W/cm^2$.

9. Add the following to each binding reaction: 1 µl 0.5 M $CaCl_2$, 4 µg DNase I, and 1 U micrococcal nuclease. Digest 30 min at 37°C.

10. Add an equal vol of 2× SDS sample buffer and boil 5 min at 100°C.

11. Electrophorese sample through discontinuous SDS-polyacrylamide gel of appropriate percentage (*UNIT 10.2*; include ^{14}C-labeled protein markers). Cut away dye front and dry gel. Autoradiograph 1 to 3 days with intensifying screen (*APPENDIX 3*).

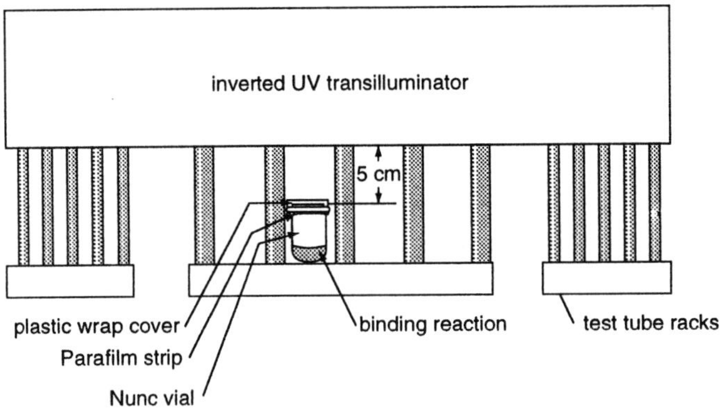

Figure 12.5.2 Experimental setup for UV crosslinking DNA-binding proteins to uniformly labeled DNA.

ALTERNATE PROTOCOL 1

UV CROSS-LINKING USING A NON–BrdU-SUBSTITUTED PROBE

This protocol can be used to UV-cross-link proteins to RNA and to sequences that do not contain thymidine residues. The procedure is similar to the Basic Protocol except for steps noted below.

Additional Materials *(also see Basic Protocol; see* APPENDIX 1 *for items with ✓)*

✓ 50× dNTP/TTP solution
UV transilluminator (254 nm)

2. Substitute 50× dNTP/TTP solution for 50× dNTP/BrdU solution.

   *If cross-linking to RNA, construct template using SP6 RNA polymerase (*UNIT 4.7*).*

8. Irradiate 5 min to 3 hr at 254 nm.

ALTERNATE PROTOCOL 2

UV CROSS-LINKING IN SITU

The advantage of using this protocol is that when multiple proteins bind to the probe, proteins present in individual protein–nucleic acid complexes can be visualized.

Additional Materials *(also see Basic Protocol; see* APPENDIX 1 *for items with ✓)*

Low gelling/melting temperature agarose (*UNIT 2.6*)
✓ TBE electrophoresis buffer
✓ SDS gel solution

1. Prepare 50-bp DNA probes, using M13 system with appropriate restriction enzymes or a small oligonucleotide primer complementary to the 3′ end of a larger synthetic oligonucleotide. Follow steps 1 to 6 of Basic Protocol.

2. Set up standard 50-μl binding reaction (Basic Protocol, step 7 or as in mobility shift assay, *UNIT 12.2*, scaled up 5×). Electrophorese in a cold room on a 1% low gelling/melting temperature agarose gel in 1× TBE buffer 2 to 3 hr at 4 V/cm.

3. Place gel on plastic wrap on surface of 305-nm UV transilluminator. Irradiate in cold room 5 to 30 min. Autoradiograph 1 to 3 hr at 4°C.

4. Excise regions corresponding to specific protein-DNA complexes. Add 10 µl of SDS gel solution to each 50 µl of gel slice and boil 2 min.

5. Electrophorese on discontinuous SDS-polyacrylamide gel. Autoradiograph as in Basic Protocol.

Reference: Chodosh et al., 1986.

Contributor: Lewis A. Chodosh

Purification of DNA-Binding Proteins Using Biotin/Streptavidin Affinity Systems

UNIT 12.6

12.6

This purification system is based on the tight and essentially irreversible complex that biotin forms with streptavidin (Fig. 12.6.1).

BASIC PROTOCOL

Materials *(see APPENDIX 1 for items with ✓)*

Plasmid DNA with desired binding site
Appropriate restriction endonucleases (*UNIT 3.1*)
0.5 mM biotin-11-dUTP (Enzo)
Labeled (3000-6000 Ci/mmol) and unlabeled 2 mM dNTPs (*UNIT 3.4*)
Klenow fragment of *E. coli* DNA polymerase I (*UNIT 3.5*)
✓ TE buffer
Biotin-cellulose (Pierce)
✓ Biotin-cellulose binding and elution buffers
20 mg/ml bovine serum albumin (BSA)
2 mg/ml bulk carrier DNA [e.g., poly(dI-dC)·poly(dI-dC), salmon sperm DNA, or *E. coli* DNA]
Protein solution
Streptavidin (Celltech; may be stored as 5 mg/ml stock ~2 months)
DEAE membrane (Schleicher & Schuell NA45)
0.025-µm filter discs (Millipore VS)

1. Digest 50 µg of plasmid DNA that contains a binding site for the protein with one or more restriction endonucleases (*UNIT 3.1*) in 100 µl.

2. Add the following to probe mixture: 0.5 mM biotin-11-dUTP to 20 µM final, 0.5 µM of 3000-6000 Ci/mmol radioactive dNTP for incorporation into 5′ overhang, 100-fold molar excess of corresponding 2 mM unlabeled dNTP (50 µM final), remaining two unlabeled dNTPs to 200 µM final, and 5 U Klenow fragment.

3. Ethanol precipitate (*UNIT 2.1*) and isolate probe by agarose gel electrophoresis using DEAE membrane (*UNIT 12.2*).

4. Resuspend in TE buffer, count an aliquot, and estimate DNA concentration by ethidium bromide dot quantitation (*UNIT 2.6*). Check ability of biotinylated probe to bind the protein as in *UNIT 12.2*.

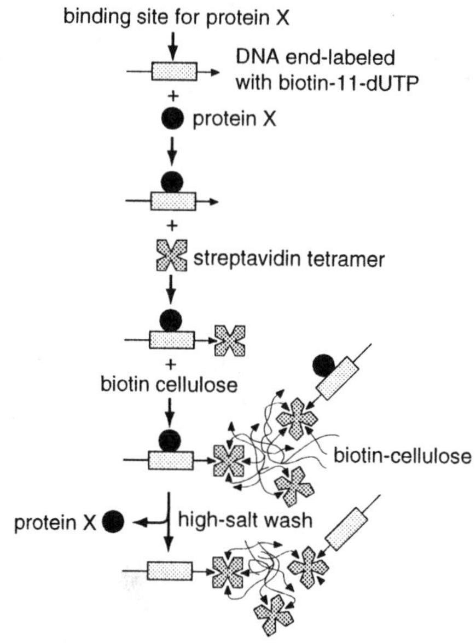

Figure 12.6.1 Purification of DNA-binding proteins using the biotin/streptavidin affinity technique. The protocol involves the following steps: (1) a biotinylated, labeled DNA fragment is prepared containing a binding site for the protein to be purified; (2) the biotin-cellulose resin is prepared; (3) a binding reaction containing a crude protein fraction and the biotinylated probe is set up; (4) free streptavidin is added to the binding reaction; (5) the protein/biotinylated DNA fragment/streptavidin complex is bound to the biotin-cellulose resin; (6) unbound protein is removed by extensive resin washing; and (7) the protein is eluted from the resin with high-ionic-strength buffer. Each of these steps can be monitored and optimized in solution, using the mobility shift DNA-binding assay (see *UNIT 12.2*).

5. Microcentrifuge 200 µl biotin-cellulose for 30 sec. Add the following to pellet: 1.0 ml biotin-cellulose binding buffer, 20 mg/ml BSA to 500 µg/ml final, and 2 mg/ml carrier DNA to 200 µg final. Gently mix 5 min on rotating wheel.

6. Microcentrifuge, and resuspend pellet in 1.0 ml biotin-cellulose elution buffer. Gently mix 5 min on rotating wheel. Repeat wash, and rinse (without rotating) two times with biotin-cellulose elution buffer.

   This 1:6 dilution is ready for use and can be stored several months.

7. Use mobility-shift assay (*UNIT 12.2*) to determine concentration of protein to be purified.

8. Set up standard binding reaction as follows (using conditions that optimize protein binding to its recognition site): protein to be purified, carrier DNA, and 10-fold molar excess of biotinylated fragment relative to protein. Incubate 15 min at appropriate temperature. Add 5-fold molar excess of streptavidin relative to biotinylated fragment. Incubate 5 min at 30°C.

9. In separate tube, place 2 µl pretreated biotin-cellulose (12 µl of 1:6 dilution) per picomole of biotinylated DNA fragment in binding reaction. Microcentrifuge resin and discard supernatant.

10. Add binding reaction mix to the biotin-cellulose resin. Resuspend resin and incubate 30 min at 4°C or room temperature on a rotating wheel.

11. Microcentrifuge and save supernatant. Resuspend pellet in 500 µl biotin-cellulose binding buffer. Invert 1 to 2 min, microcentrifuge, and wash resin twice. Transfer to clean tube at last wash.

    Use mobility shift assay to determine percentage of biotinylated fragment and protein that has been removed from supernatant.

12. Resuspend pellet in equal volume biotin-cellulose elution buffer. Mix 20 min on rotating wheel. Microcentrifuge and assay supernatant for binding activity (e.g., dialyze on 0.025-µm filter discs).

PURIFICATION USING A MICROCOLUMN

ALTERNATE PROTOCOL 1

This method is useful for large volumes of biotin-cellulose resin. Protein can also be eluted in as small a volume and as high a concentration as possible.

Additional Materials *(also see Basic Protocol; see* APPENDIX 1 *for items with ✓)*

✓ Silanized glass wool

1. Prepare biotinylated DNA fragment and biotin-cellulose resin, and set up binding reaction (Basic Protocol, steps 1 to 8).
2. Plug bottom of 1.0-ml pipet tip with silanized glass wool (prewet in biotin-cellulose binding buffer), and attach to a ring stand. Add 500 µl binding buffer to microcolumn and maintain a steady flow.
3. Add ≥40 µl of 1:1 biotin-cellulose slurry to microcolumn and equilibrate pretreated resin (Basic Protocol, steps 5 and 6) with 3 column-volumes of biotin-cellulose binding buffer. For untreated resin, wash sequentially with 3 column-volumes each of:

    Biotin-cellulose binding buffer
    Biotin-cellulose binding buffer with 500 µg/ml BSA and 200 µg/ml poly(dI-dC)·poly(dI-dC)
    Biotin-cellulose elution buffer.

    Equilibrate with biotin-cellulose binding buffer.

4. Load binding reaction mix (Basic Protocol, step 8) on the microcolumn and collect flowthrough.
5. Wash with 4 column-volumes of biotin-cellulose binding buffer and discard flowthrough.
6. Wash with 3 column-volumes of biotin-cellulose elution buffer. Collect 2-drop fractions and assay.

PURIFICATION USING STREPTAVIDIN-AGAROSE

ALTERNATE PROTOCOL 2

This protocol can be used when high-quality streptavidin is unavailable or cellulose is an inappropriate resin. The biotinylated DNA fragment is removed from solution directly by streptavidin-agarose (Fig. 12.6.2).

Additional Materials *(also see Basic Protocol)*

Streptavidin-agarose

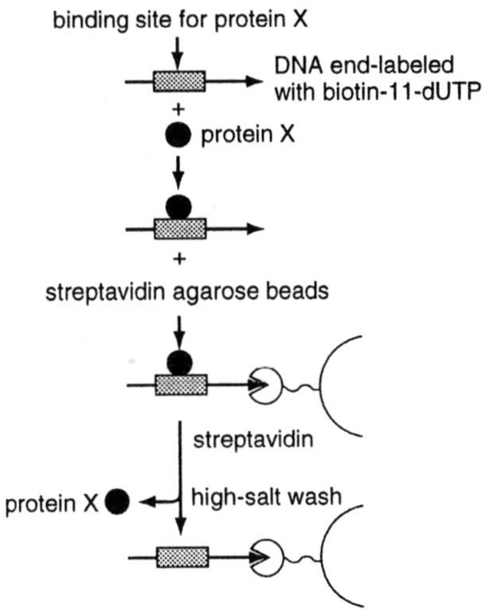

Figure 12.6.2 Purification of DNA-binding proteins using streptavidin-agarose.

1. Prepare biotinylated DNA fragment and streptavidin-agarose resin, and set up binding reaction (Basic Protocol, steps 1 to 8).
2. In separate tube, add 50 µl pretreated streptavidin-agarose (300 µl of 1:6 dilution) for each picomole of biotinylated DNA fragment in binding reaction. Microcentrifuge and discard supernatant.
3. Add binding-reaction mix to streptavidin-agarose. Resuspend the resin and incubate 30 min to 2 hr on rotating wheel (streptativin-agarose may also be used in a microcolumn; see Alternate Protocol 1).
4. Wash and elute protein (Basic Protocol, steps 11 and 12).

Reference: Kasher et al., 1986.

Contributor: Lewis A. Chodosh

UNIT 12.7

Detection, Purification, and Characterization of cDNA Clones Encoding DNA-Binding Proteins

In this unit, an appropriate recombinant clone is detected in an expression library with a DNA recognition-site probe, purified, and shown to encode a DNA-binding domain of defined sequence specificity. This strategy obviates purification of a sequence-specific DNA-binding protein for the purpose of isolating its gene; it simply requires a suitable cDNA library constructed in the expression vector λgt11 and a DNA recognition-site probe.

SCREENING A λgt11 EXPRESSION LIBRARY WITH RECOGNITION-SITE DNA

BASIC PROTOCOL

The success of the screening strategy is critically dependent on the quality of the cDNA library and the recognition-site probe used. Recombinant cDNA libraries made by random priming and multisite recognition probes are the preferred reagents. The synthetic probe should be tested for binding by the desired protein (*UNIT 12.2*) before use in screens. If possible, the highest affinity site among a set of related sites should be chosen for the construction of the multisite probe.

Materials *(see APPENDIX 1 for items with ✓)*

pUC recombinant plasmid DNA containing multiple tandem copies of the wild-type or mutant recognition site (*UNITS 1.5 & 3.16*)
*Eco*RI and *Hin*dIII restriction endonucleases
✓ 1 M Tris·Cl, pH 7.5
5 mM each dCTP, dGTP, dTTP, and dATP (*UNIT 3.4*)
5000 Ci/mmol [α-^{32}P]dATP
Klenow fragment of *E. coli* DNA polymerase I (*UNIT 3.5*)
500 mM EDTA, pH 8.0
25:24:1 phenol/chloroform/isoamyl alcohol (*UNIT 2.1*)
24:1 chloroform/isoamyl alcohol
✓ 3 M sodium acetate, pH 5.2
100% ethanol
✓ Elution buffer
E. coli Y1090 (Table 1.4.5)
λgt11 cDNA library
LB medium containing 0.2% maltose and 50 μg/ml ampicillin (*UNIT 1.1*)
0.7% top agarose, melted and equilibrated to 47°C (*UNIT 1.1*)
100- and 150-mm LB plates containing 50 μg/ml ampicillin, dry and prewarmed to 37°C (*UNIT 1.1* and Table 1.4.1)
10 mM IPTG (prepare 1 M stock with sterile H_2O and store at −20°C; Table 1.4.2)
✓ BLOTTO
✓ Binding buffer
1 mg/ml sonicated (to ~1 kbp) calf thymus DNA, denatured 10 min at 100°C
Chloroform

Elutip-d columns (Schleicher & Schuell)
132- and 82-mm nitrocellulose membrane filters (Schleicher & Schuell)

1. Digest 20 μg of an appropriate pUC recombinant plasmid DNA with *Eco*RI and *Hin*dIII in 100 μl to release insert containing 2 to 10 protein-binding sites.

2. Add the following components to the restriction digest:

   1 M Tris·Cl, pH 7.5 (50 mM final)
   5 mM each dCTP, dGTP, dTTP (100 μM final)
   200 μCi [α-^{32}P]dATP (5000 Ci/mmol)
   10 U Klenow fragment.

   Incubate 30 min at room temperature. Add cold 5 mM dATP to 100 μM and continue incubation 30 min.

3. Add EDTA to 20 mM and extract sequentially with 25:24:1 phenol/chloroform/isoamyl alcohol and 24:1 chloroform/ isoamyl alcohol.

4. Adjust to 0.3 M sodium acetate, and precipitate with 100% ethanol (*UNIT 2.1*). Resuspend pellet in 200 µl water, add 22 µl of 3 M sodium acetate, pH 5.2, and reprecipitate with 100% ethanol.

5. Separate labeled fragments by nondenaturing polyacrylamide gel electrophoresis using a 0.75-mm, 5% gel (*UNIT 2.7*).

6. Visualize fragments by autoradiography (*APPENDIX 3*). Cut out a gel slice containing binding-site fragment and elute by incubating gel slice in 1.5 ml elution buffer at 4°C overnight with shaking.

7. Purify labeled fragment from supernatant by Elutip-d chromatography (*UNIT 2.6*). Count probe and store at 4°C.

   These reaction conditions yield DNA probes with specific activities of 2-4 × 10^7 cpm/pmol, which is sufficient to screen 20 large filters representing ~10^6 plaques.

8. Grow *E. coli* Y1090 to saturation in LB/maltose/ampicillin medium at 37°C. For each plating, mix 0.5-ml of overnight culture with λgt11 library (containing 3-5 × 10^4 pfu) in a tube (*UNIT 6.1*). Incubate 15 min at 37°C.

9. Add 9 ml top agarose to each tube, inverting a few times, and plate onto a 150-mm LB/ampicillin plate. Incubate plates ~3 hr at 42°C until tiny plaques are visible. Move plates to 37°C incubator.

10. Overlay each plate with a dry nitrocellulose filter impregnated with IPTG (immerse clean filter in 10 mM IPTG 30 min and air dry at room temperature 60 min) and continue incubation 6 hr at 37°C.

11. Cool plates 10 min at 4°C. Mark position of each filter with needle holes. Lift filters and immerse in BLOTTO. Incubate 60 min at room temperature with gentle swirling.

    Seal master plates with Parafilm and store at 4°C.

12. Immerse all filters in binding buffer (500 ml/10 filters) and incubate 5 min at room temperature with shaking. Repeat twice with fresh binding buffer (filters can be stored in binding buffer up to 24 hr at 4°C prior to screening).

13. Incubate filters 60 min at room temperature in binding buffer (50 ml/10 filters) containing ^{32}P-labeled binding-site probe (10^6 cpm/ml, ~10^{-10} M) and 5 µg/ml denatured, sonicated calf thymus DNA.

14. Wash filters in batches, 4 times (7.5 min per wash, 30 min total) at room temperature with 500 ml binding buffer/10 filters. Blot filters dry, cover with plastic wrap, and expose 12 to 24 hr to X-ray film with an intensifying screen at –70°C.

15. Align phage plates with autoradiographs, isolate agarose plugs corresponding to positive signals, and generate secondary phage stocks.

    True positives have a diffuse halo-like appearance. Generate multiple autoradiographs to eliminate spots that do not reproduce.

16. Plate secondary phage stocks (~5000 pfu/100-mm plate; step 8) using a 0.2-ml aliquot of Y1090 culture and 3 ml top agarose.

17. Prepare and screen secondary nitrocellulose filter replicas (steps 10 to 14) using 82-mm filters.

18. Screen secondary stocks of positive phage with control DNA probe (lacking mutant recognition sequence). Plaque purify and store positive phage over a drop of chloroform at 4°C (UNIT 6.5).

ALTERNATE PROTOCOL

DENATURATION/RENATURATION OF DRIED FILTERS

Processing protein replica filters through a denaturation/renaturation cycle using 6 M guanidine·HCl may enhance the binding to DNA-recognition-site probes (Vinson et al., 1988).

12.7

Additional Materials (*also see Basic Protocol; see* APPENDIX 1 *for items with* ✓)

✓ HEPES binding buffers (with and without guanidine·HCl)

1. Follow steps 1 to 10 of Basic Protocol.
2. Cool plates 10 min at 4°C. Mark position of each filter with needle holes before removing. Air dry 15 min at room temperature.
3. Immerse filters in HEPES binding buffer/6 M guanidine·HCl and incubate 10 min at 4°C with gentle shaking. Repeat once.
4. Immerse filters in HEPES binding buffer/3 M guanidine·HCl (50 ml/10 filters) 5 min at 4°C. Repeat four times, each time using HEPES binding buffer with a 2-fold dilution of guanidine·HCl from previous wash.
5. Incubate filters in HEPES binding buffer (50 ml/10 filters) 5 min at 4°C. Repeat once with fresh buffer.
6. Incubate in BLOTTO to block (Basic Protocol, step 11) and process for screening (Basic Protocol, steps 12 to 18).

SUPPORT PROTOCOL

PREPARATION OF A CRUDE EXTRACT FROM A RECOMBINANT LYSOGEN TO CHARACTERIZE DNA-BINDING ACTIVITY

The specific detection of a recombinant phage with a wild-type recognition-site probe suggests that the phage expresses a β-galactosidase fusion protein that binds specifically to the recognition site. Direct evidence is obtained by analyzing a crude extract from a lysogen harboring the recombinant phage for appropriate DNA-binding activity.

Additional Materials (*also see Basic Protocol; see* APPENDIX 1 *for items with* ✓)

E. coli Y1089 *hflA*150 (Table 1.4.5)
LB medium (UNIT 1.1) with and without 10 mM $MgCl_2$
10^{10} pfu/ml λgt11 recombinant phage stock (Basic Protocol)
1 M IPTG (Table 1.4.2)
✓ Extract buffer
5 mg/ml lysozyme in extract buffer (store at −20°C)
4 M NaCl
0.025-μm filter disks (Millipore VS)

1. Grow *E. coli* Y1089 *hflA*150 to saturation in LB/maltose/ampicillin medium at 37°C. Dilute cell suspension 100-fold in LB/$MgCl_2$ medium. Infect 100 μl of cell suspension with 5 μl of 10^{10} pfu/ml λgt11 recombinant phage stock and incubate 20 min at 32°C.

2. Dilute infected cell suspension 1000-fold in LB medium. Spread 100 μl on LB/ampicillin plates and incubate overnight at 32°C to obtain ~100 colonies/plate.

3. Test single colonies for temperature-sensitive growth. Spot single colonies onto two LB/ampicillin plates. Incubate one plate at 42°C and the other at 32°C.

   Clones that grow at 32°C but not at 42°C represent lysogens. Lysogens should arise at a frequency of 10% to 80%.

4. Grow overnight cultures of recombinant phage lysogens in LB/ampicillin medium at 32°C. Inoculate 2 ml LB/ampicillin medium with 20 μl of overnight cultures. Incubate at 32°C with good aeration until OD_{600} = 0.5 (~3 hr). Shift temperature rapidly to 44°C and incubate 20 min.

5. Add 1 M IPTG to 10 mM final and reduce temperature to 37°C. Continue incubation for 1 hr.

6. Microcentrifuge 1 ml of induced cultures ~45 sec at room temperature. Resuspend in 100 μl extract buffer and quick freeze with dry ice/ethanol. Store cell suspensions at –70°C.

7. Rapidly thaw frozen cell suspension. Add 5 mg/ml lysozyme to 0.5 mg/ml final. Incubate 15 min on ice (cell suspension will be viscous). Add 4 M NaCl to 1 M final and mix thoroughly. Incubate on rotator for 15 min at 4°C.

8. Microcentrifuge 30 min at 4°C. Remove supernatants carefully and dialyze on 0.025-μm filter disks against 100 ml extract buffer 60 min at 4°C (*APPENDIX 3*). Freeze in dry ice/ethanol bath and store at –70°C.

References: Singh et al., 1989; Vinson et al., 1988.

Contributor: Harinder Singh

UNIT 12.8

Analysis of DNA-Protein Interactions Using Proteins Synthesized In Vitro from Cloned Genes

BASIC PROTOCOL

In vitro synthesized proteins are extremely useful for determining whether a cloned gene encodes a specific DNA-binding protein and for analyzing DNA-protein interactions. To detect DNA-binding activity, the labeled protein is incubated with specific DNA fragments, and protein-DNA complexes are separated from free protein by electrophoresis in native acrylamide gels (*UNIT 12.2*). Unlike the more conventional mobility shift assay that utilizes ^{32}P-labeled DNA and unlabeled protein, the assay described here generally utilizes ^{35}S-labeled protein and unlabeled DNA. Any desired mutant protein can be tested for its DNA-binding properties simply by altering the DNA template, and the subunit structure (e.g., dimer, tetramer) can be determined.

Materials *(see APPENDIX 1 for items with ✓)*

Plasmid DNA containing desired binding sites
Appropriate restriction endonucleases (*UNIT 3.1*)
^{35}S-labeled protein (*UNIT 10.16*)

✓ 5× binding buffer
10 mg/ml poly(dI-dC)·poly(dI-dC) or other bulk carrier DNA
✓ Loading buffer
45% methanol/10% acetic acid
EN^{3}HANCE (Du Pont NEN)

1. Cleave 0.5 μg plasmid DNA containing binding site (assuming a plasmid of 5 kb) with appropriate restriction endonucleases, generating a well-isolated fragment whose length is between 150 and 700 bp. Purify by phenol extraction and ethanol precipitation (*UNIT 2.1*).

2. Set up the following 15-μl binding reaction:

   5 μl H_2O
   3 μl 5× binding buffer
   5 μl DNA (DNA fragments each at 9 nM)
   1 μl 10 mg/ml poly(dI-dC)·poly(dI-dC)
   1 μl ^{35}S-labeled protein.

   Incubate 20 min at room temperature.

   Perform control reactions that contain no DNA and nonspecific DNA (i.e., lacking a binding site).

3. Add 5 μl loading buffer. Carry out electrophoresis on a 5% nondenaturing polyacrylamide gel until bromphenol blue is near bottom of gel.

   Gel composition (UNIT 2.7 or UNIT 12.2) and ionic strength of buffer can be varied and may be important for detecting a protein-DNA interaction.

4. Fix gel in 45% methanol/10% acetic acid for 1 hr at room temperature. Treat the gel with EN^{3}HANCE for 1 hr, dry, and autoradiograph.

Reference: Hope and Struhl, 1987.

Contributor: Kevin Struhl

12.9

Purification of Sequence-Specific DNA-Binding Proteins by Affinity Chromatography

UNIT 12.9

Many biological processes, such as recombination, replication, and transcription, involve the action of sequence-specific DNA-binding proteins that typically make up <0.01% of the total cellular protein. Affinity chromatography is a very effective means of purifying a protein based on its sequence-specific DNA-binding properties.

PREPARATION OF DNA AFFINITY RESIN

BASIC PROTOCOL 1

Correct choice of oligonucleotide sequence and preparation of the affinity resin are probably the most important parts of the affinity chromatography procedure.

NOTE: Glass-distilled or other high-quality water should be used throughout these procedures.

Materials *(see* APPENDIX 1 *for items with ✓)*

440 μg each of two synthetic oligonucleotides with desired binding site (Support Protocol 1 or commercial HPLC-purified)
✓ TE buffer, pH 7.8
✓ 10× T4 polynucleotide kinase buffer
20 mM ATP (Na^+ salt), pH 7.0
150 mCi/ml [γ-^{32}P]ATP (6000 Ci/mmol)
10 U/μl T4 polynucleotide kinase (New England Biolabs; *UNIT 3.10*)
✓ 10 M ammonium acetate
25:24:1 (v/v/v) phenol/chloroform/isoamyl alcohol (*UNIT 2.1*)
24:1 (v/v) chloroform/isoamyl alcohol
✓ 3 M sodium acetate
100% and 75% (v/v) ethanol
✓ 10× linker/kinase buffer
6000 U/ml T4 DNA ligase (measured in Weiss units; New England Biolabs; *UNIT 3.14*)
Buffered phenol (*UNIT 2.1*)
Isopropanol (2-propanol)
Sepharose CL-2B (Pharmacia Biotech)
Cyanogen bromide (CNBr; Aldrich)
N,N-dimethylformamide
✓ 5 N NaOH
✓ 10 mM and 1 M potassium phosphate buffer, pH 8.0
✓ 1 M ethanolamine hydrochloride, pH 8.0
NaOH, solid
Glycine
✓ 1 M KCl
✓ Column storage buffer

15-ml screw-cap polypropylene tubes
15°, 37°, 65°, and 88°C heating blocks or water baths
60-ml coarse-sintered glass funnel
Rotating wheel

1. In a 1.5-ml microcentrifuge tube, prepare a mixture containing 440 μg of each oligonucleotide in TE buffer in a total volume of 130 μl. Add 20 μl of 10× T4 polynucleotide kinase buffer. Incubate 2 min at 88°C, 10 min at 65°C, 10 min at 37°C, and 5 min at room temperature.

   A 1-μmol synthesis of oligonucleotide should yield enough purified DNA for ~20 ml of affinity resin. The oligonucleotide sequence should contain the binding site, flanking nucleotides, and a single-stranded overhang for ligation, e.g., GATC-XXX-BINDING SITE-XXX where XXX is flanking sequence.

2. Divide mixture in half in separate microcentrifuge tubes. To each 75-μl aliquot, add 15 μl of 20 mM ATP (pH 7.0), ~5 μCi [γ-^{32}P]ATP, and 10 μl of 10 U/μl T4 polynucleotide kinase (100 U total). Incubate 2 hr at 37°C.

   To add the labeled ATP, do not thaw, but simply touch (do not jab) the top of the frozen [γ-^{32}P]ATP with a yellow pipet tip and transfer the resulting tiny amount to the reaction tube.

3. Inactivate the kinase by adding 50 µl of 10 M ammonium acetate and 100 µl water to each tube and heating 15 min at 65°C. Allow to cool to room temperature.

4. Add 750 µl of 100% ethanol and mix by inversion. Microcentrifuge at high speed 15 min at room temperature to pellet DNA. Discard supernatant.

5. Resuspend each pellet in 225 µl TE buffer.

6. Add 250 µl of 25:24:1 phenol/chloroform/isoamyl alcohol to each tube. Vortex 1 min. Microcentrifuge at high speed 5 min to separate phases. Transfer the aqueous phase (upper layer) to a new tube.

7. Add 250 µl of 24:1 chloroform/isoamyl alcohol to aqueous phase. Vortex 1 min. Microcentrifuge at high speed 5 min to separate phases. Transfer the aqueous phase (upper layer) to a new tube.

8. Add 25 µl of 3 M sodium acetate to aqueous phase and mix by vortexing. Then add 750 µl of 100% ethanol and mix by inversion. Microcentrifuge at high speed 15 min to pellet DNA. Discard supernatant.

9. Wash pellet with 800 µl of 75% ethanol. Mix by vortexing. Microcentrifuge at high speed 5 min. Discard supernatant.

10. Dry pellet in vacuum evaporator (e.g., Speedvac).

11. Add 65 µl water and 10 µl of 10× linker/kinase buffer to each pellet. Dissolve DNA by vortexing. Add 20 µl of 20 mM ATP (pH 7.0) and 5 µl of 6000 U/ml T4 DNA ligase (30 Weiss units). Incubate ≥2 hr at room temperature or overnight at 15°C.

    Depending upon the size and sequence of the oligonucleotides used, the optimal temperature for ligation will vary from 4° to 30°C.

12. Monitor the ligation reaction by agarose gel electrophoresis, using 0.5 µl of ligation reaction per gel lane and including lanes containing size markers. Visualize DNA by ethidium bromide staining and UV photography.

    If ligation has not occurred, extract the DNA once with 25:24:1 phenol/chloroform/isoamyl alcohol and once with 24:1 chloroform/isoamyl alcohol, then ethanol precipitate using sodium acetate as the precipitating salt. Wash with 75% ethanol, dry in Speedvac, and repeat ligation.

13. Add 100 µl buffered phenol to the 100-µl ligation reactions. Vortex 1 min. Microcentrifuge at high speed 5 min at room temperature. Transfer aqueous phase (upper layer) to a new tube.

14. Add 100 µl of 24:1 chloroform/isoamyl alcohol to aqueous phase. Vortex 1 min. Microcentrifuge at high speed 5 min, room temperature. Transfer aqueous phase (upper layer) to a new tube.

15. Add 33 µl of 10 M ammonium acetate to the aqueous phase. Mix by vortexing.

16. Add 133 µl isopropanol. Mix by inversion. Incubate 20 min at −20°C. Microcentrifuge at high speed 15 min to pellet DNA. Discard supernatant.

    The ammonium acetate/isopropanol precipitation removes residual ATP, which would otherwise interfere with coupling of the ligated DNA to the CNBr-activated Sepharose.

17. Add 225 µl TE buffer. Vortex to dissolve pellet. Add 25 µl of 3 M sodium acetate. Mix by vortexing. Add 750 µl of 100% ethanol. Mix by inversion. Microcentrifuge at high speed 15 min to pellet DNA. Discard supernatant.

18. Wash DNA twice with 75% ethanol. Dry pellet in vacuum evaporator.

19. Dissolve DNA in 50 µl water. Store at −20°C.

    Do not dissolve the DNA in TE buffer, as the Tris buffer in TE will interfere with the coupling reaction.

NOTE: It is best to assemble all equipment and reagents required for the activation and coupling reactions *before* proceeding with the following steps. Do not allow the resin to dry.

20. Place 10 to 15 ml (settled bed volume) of Sepharose CL-2B in a 60-ml coarse-sintered glass funnel and wash extensively with 500 ml water.

21. Transfer moist Sepharose resin to a 25-ml graduated cylinder, estimating 10 ml of resin. Add water to 20 ml final volume. Transfer the resulting slurry to a 150-ml glass beaker containing a magnetic stir bar. Place beaker in a water bath equilibrated to 15°C and set up over a magnetic stirrer in a fume hood. Turn on stirrer to slow medium speed.

22. In the fume hood, measure 1.1 g CNBr into a 25-ml Erlenmeyer flask, keeping the mouth of the flask covered with Parafilm or a ground glass stopper as much as possible (it is better to have slightly more than 1.1 g than slightly less). Add 2 ml *N,N*-dimethylformamide (the CNBr will dissolve instantly). Over the course of 1 min, add the resulting CNBr solution dropwise to the stirring Sepharose slurry.

    CAUTION: *CNBr is highly toxic and volatile. Use only in a fume hood with extreme caution. It is important to clean up all CNBr waste carefully. In the fume hood, add solid NaOH and glycine (~10 to 20 mg/ml) to inactivate the CNBr. Soak contaminated instruments in a similar solution. Let sit overnight in the fume hood, then discard.*

23. Immediately add 5 N NaOH as follows: add 30 µl to the stirring mixture every 10 sec for 10 min until 1.8 ml NaOH has been added.

24. Immediately add 100 ml ice-cold water to the beaker and pour the mixture into a 60-ml coarse-sintered glass funnel.

25. Still working in the fume hood, wash the resin in the funnel with four 100-ml washes of ice-cold (≤4°C) water followed by two 100-ml washes of ice-cold 10 mM potassium phosphate, pH 8.0.

26. Immediately transfer the resin to a 15-ml polypropylene screw-cap tube and add ~4 ml of 10 mM potassium phosphate (pH 8.0) until the resin has the consistency of a thick slurry.

27. Immediately add the two 50-µl aliquots of DNA from step 19. Incubate on a rotating wheel overnight (≥8 hr) at room temperature.

28. In the fume hood, transfer the resin to a 60-ml coarse-sintered glass funnel and wash with two 100-ml washes of water and one 100-ml wash of 1 M ethanolamine hydrochloride, pH 8.0.

Using a Geiger counter, compare the level of radioactivity in the filtrate with that of the washed resin to estimate the efficiency of incorporation of DNA to the resin. Usually, all detectable radioactivity is present in the resin.

29. In the fume hood, transfer the resin to a 15-ml polypropylene screw-cap tube and add 1 M ethanolamine hydrochloride (pH 8.0) until the mixture is a smooth slurry. Incubate the tube on a rotating wheel 2 to 4 hr at room temperature.

30. Wash the resin in a 60-ml coarse-sintered glass funnel with 100 ml of 10 mM potassium phosphate (pH 8.0), 100 ml of 1 M potassium phosphate (pH 8.0), 100 ml of 1 M KCl, 100 ml water, and 100 ml column storage buffer.

31. Store the resin at 4°C (stable at least one year; do not freeze).

COUPLING THE DNA TO COMMERCIALLY AVAILABLE CNBr-ACTIVATED SEPHAROSE

ALTERNATE PROTOCOL

Additional Materials *(also see Basic Protocol 1)*

1 mM HCl, prepared fresh before use
CNBr-activated Sepharose 4B (Pharmacia Biotech)

1. Prepare labeled, ligated oligonucleotides (Basic Protocol 1, steps 1 to 19).

2. Weigh out 3 g CNBr-activated Sepharose 4B (1 g freeze-dried resin gives ~3.5 ml final gel volume).

3. Place the dry resin in a 15-ml conical polypropylene tube. Hydrate resin with 10 ml of 1 mM HCl and mix gently by flicking and inverting the tube. After 1 min, transfer slurry to a 60-ml coarse-sintered glass funnel. Wash and swell the beads by gradually pouring 500 ml of 1 mM HCl through the funnel (this will take ~15 min).

4. Wash the resin with 100 ml water and then with 100 ml of 10 mM potassium phosphate, pH 8.0.

5. Couple oligonucleotides to CNBr-activated Sepharose 4B (Basic Protocol 1, steps 26 to 31).

PURIFICATION OF OLIGONUCLEOTIDES BY PREPARATIVE GEL ELECTROPHORESIS

SUPPORT PROTOCOL 1

Additional Materials *(also see Basic Protocol 1; see APPENDIX 1 for items with ✓)*

40% (w/v) 19:1 acrylamide/bisacrylamide
✓ 10× TBE buffer
Urea
✓ 10% (w/v) ammonium persulfate
TEMED
✓ Formamide loading buffer
sec-butanol (2-butanol)
Diethyl ether
✓ 1 M $MgCl_2$

Saran wrap or other UV-transparent plastic wrap
Intensifying screen (e.g., Lightning Plus, DuPont NEN)
Hand-held short-wavelength UV light source

Silanized glass wool (*APPENDIX 3*)
Dry ice/ethanol bath (−78°C)

1. Prepare a 20 cm × 40–cm × 1.5–mm denaturing gel with four wells 3 cm in width, using 16% polyacrylamide/urea (see recipe) for separating oligonucleotides ~10 to ~45 bases long (or 8% or 6% polyacrylamide/urea for longer oligonucleotides). Let gel polymerize for ≥30 min, then prerun at 30 W for ≥1 hr.

2. Dissolve each oligonucleotide in formamide loading buffer to 200 μl final. Heat 15 min at 65°C to remove any secondary structure in the DNA. Load 50 μl of samples into separate wells, and run the gel at 30 W for ~4 hr, until bromphenol blue migrates three-quarters of the way down the gel.

   The amount of oligonucleotide that is prepared in a 1-μmol synthesis (~1 to 2 mg) can be loaded on one gel (0.25 μmol per 3-cm well). This is the maximum amount that can be applied to the gel without overloading it.

   In a 16% gel, bromphenol blue comigrates with ~10-base oligonucleotides and xylene cyanol comigrates with ~30-base oligonucleotides. If the oligonucleotide is 25 to 35 bases long, it is recommended that the formamide loading buffer be made without xylene cyanol.

3. Remove one of the glass gel plates and cover the gel with Saran wrap. Flip gel over and remove the other plate so that the gel is lying on the Saran wrap. Cover the other side of the gel with Saran wrap.

4. In a darkroom, lay the gel on an intensifying screen and hold a hand-held short-wavelength UV light source directly over it to visualize the DNA. Identify the major oligonucleotide band and mark its position directly on the Saran wrap using a marker, making sure that the light is directly over the band.

5. Carefully cut out the band with a razor blade, trying to avoid shredding the gel material or pulverizing the gel slice into small pieces.

6. Soak the gel piece in 5 ml TE buffer in a 15-ml polypropylene tube overnight at 37°C with shaking.

7. Place a silanized glass wool plug in a Pasteur pipet and prerinse with ~5 ml water. Filter the supernatant containing the DNA through the glass wool.

8. Concentrate the DNA to ≤180 μl by repeated extractions with *sec*-butanol.

9. Extract the DNA once with diethyl ether and place in a vacuum evaporator (e.g., Speedvac) until all traces of ether are removed.

10. Adjust the volume of the liquid to 180 μl with TE buffer. Add 20 μl of 3 M sodium acetate and 2 μl of 1 M $MgCl_2$. Mix by vortexing.

11. Add 600 μl of 100% ethanol. Mix by inversion. Chill 10 min in dry ice/ethanol. Let stand 5 min at room temperature. Microcentrifuge at high speed 15 min, room temperature, to pellet DNA. Discard supernatant.

12. Add 180 μl TE buffer. Dissolve pellet by vortexing. Add 20 μl of 3 M sodium acetate and 2 μl of 1 M $MgCl_2$. Mix by vortexing.

13. Add 600 μl of 100% ethanol. Mix by inversion. Chill 10 min in dry ice/ethanol. Let stand 5 min, room temperature. Microcentrifuge at high speed 15 min, room temperature. Discard supernatant.

14. Add 800 μl of 75% ethanol. Mix by vortexing. Microcentrifuge at high speed 5 min, room temperature. Discard supernatant.

15. Carefully dry pellet in vacuum evaporator (sometimes static electricity will cause the pellet to jump out of the tube).

16. Dissolve DNA in 200 μl TE buffer and measure A_{260} and A_{280}. Store at −20°C indefinitely.

For sequence-specific DNA affinity resins, assume that 1 A_{260} unit = 40 μg/ml DNA.

12.9

DNA AFFINITY CHROMATOGRAPHY

BASIC PROTOCOL 2

Affinity chromatography is used to purify proteins that bind specifically to a defined DNA sequence (Fig. 12.9.1).

Materials *(see APPENDIX 1 for items with ✓)*

Prepared DNA affinity resin (Basic Protocol 1 or Alternate Protocol)
✓ Buffer Z or other column buffer (e.g., buffers Z^e or TM) made with varying KCl concentrations (buffers Z/0.1 M KCl through Z/1 M KCl)
Partially purified protein fraction dialyzed against buffer Z/0.1 M KCl
Nonspecific competitor DNA (Support Protocol 2)
✓ Column regeneration buffer
✓ Column storage buffer

Disposable chromatography column (Bio-Rad)
Sorvall SS-34 rotor or equivalent
Liquid nitrogen
Narrow glass rod, silanized (*APPENDIX 3*)

1. Equilibrate 1 ml settled bed volume of the DNA affinity resin in a disposable chromatography column with two 10-ml washes of buffer Z/0.1 M KCl.

2. Combine the partially purified protein fraction in buffer Z/0.1 M KCl with nonspecific competitor DNA as determined by DNA-binding studies. Incubate mixture 10 min on ice.

3. Centrifuge mixture 10 min at 12,000 × *g*, 4°C, to pellet insoluble protein-DNA complexes.

4. Load the supernatant onto the column at gravity flow (e.g., 15 ml/hr per column for Sepharose CL-2B).

A single 1-ml column is sufficient for a standard nuclear extract (e.g., 12 liters of HeLa cells or 150 g of Drosophila embryos). When purifying a larger quantity of material, it is preferable to use multiple 1-ml columns. It is common practice to apply as much as 50 ml of protein sample onto a single 1-ml column.

Typically, 1 ml of affinity resin contains ~80 to 90 μg DNA, which corresponds to a protein-binding capacity of 7 nmol/ml of resin, assuming one recognition site per 20 bp. Typically such a column will yield >100-fold purification of a binding protein (and any contaminants that may be present).

5. After loading the starting material, wash the column four times with 2-ml aliquots of buffer Z/0.1 M KCl, rinsing the sides of the column each time.

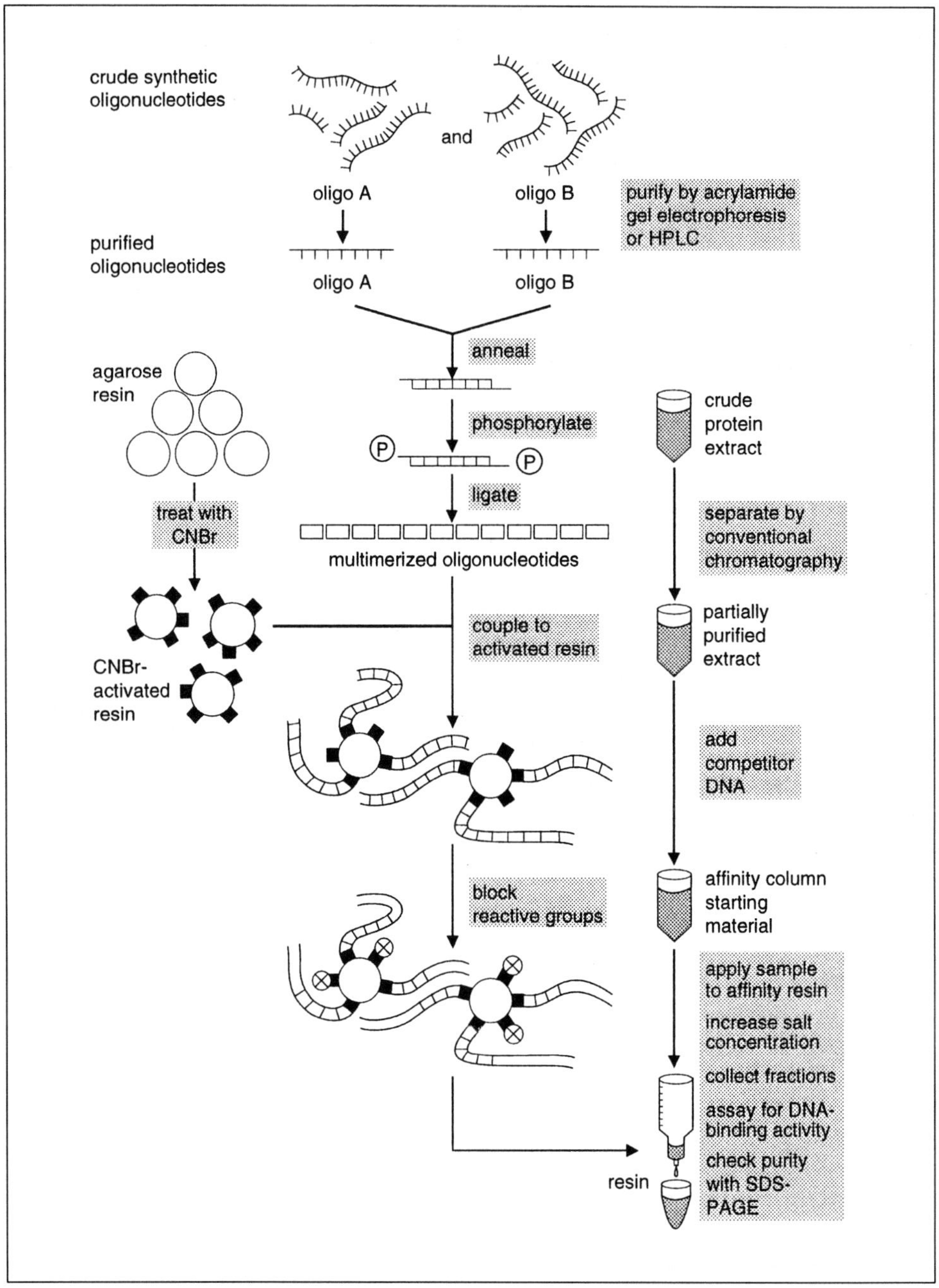

Figure 12.9.1 Purification of sequence-specific DNA-binding proteins with DNA affinity chromatography. Shown are the steps required to perform an affinity chromatography experiment using the methods described in this unit.

It is very important to wash the affinity column thoroughly at this step by rinsing down the sides of the column with the 2-ml wash buffer aliquots; a single 8-ml wash is not effective.

6. Elute the protein from the column by successive addition of 1-ml portions of buffer Z/0.2 M KCl, buffer Z/0.3 M KCl, buffer Z/0.4 M KCl, buffer Z/0.5 M KCl, buffer Z/0.6 M KCl, buffer Z/0.7 M KCl, buffer Z/0.8 M KCl, and buffer Z/0.9 M KCl, followed by three 1-ml aliquots of buffer Z/1 M KCl. Collect 1-ml fractions that correspond to the addition of the 1-ml portions of buffer. Quick-freeze the protein samples in liquid nitrogen and store at –80°C. Samples can generally be stored at least 2 years.

It is convenient to save separate aliquots of each fraction for DNA-binding assays (20 µl each) and SDS-PAGE analysis (50 µl each).

7. Assay the protein fractions for the sequence-specific DNA-binding activity using a DNA-binding assay. Estimate the purity of the protein fractions by SDS-PAGE followed by silver staining to visualize the protein.

   If further purification is desired, combine the fractions that contain the activity and, depending on the KCl concentration, either dilute (using buffer Z without KCl) or dialyze (against buffer Z/0.1 M KCl) to 0.1 M KCl. Then combine the protein fraction with nonspecific competitor DNA and reapply to either fresh or regenerated DNA affinity resin.

   If the desired protein is not detected in either the column fractions or flowthrough, it may still be on the column, and a wash at a higher salt concentration (e.g., 2 M KCl) may be needed to elute the protein from the resin.

8. Regenerate the affinity resin as follows: At room temperature, stop the column flow and add 5 ml column regeneration buffer to the column. Stir the resin with a silanized narrow glass rod to mix the resin with the regeneration buffer. Let the buffer flow out of the column. Repeat step.

9. To store column, add 10 ml column storage buffer and allow to flow through. Repeat this wash, then close the bottom of the column and add another 5 ml buffer. Cover top of column and store up to 1 year at 4°C.

SELECTION AND PREPARATION OF NONSPECIFIC COMPETITOR DNA

SUPPORT PROTOCOL 2

The amount and type of competitor to use in a given experiment must be determined experimentally. A typical method is to mix aliquots of a protein sample with varying amounts of different competitor DNAs and evaluate DNA binding using assays such as DNase I footprinting (*UNIT 12.4*) or gel mobility shifts (*UNIT 12.2*). The competitor DNA that inhibits DNA binding the least should be used for DNA affinity chromatography.

First, determine the highest amount of competitor DNA that can be added to a DNA-binding reaction that does not interfere with the binding of the sequence-specific factor. The optimal amount of competitor DNA will vary with the purity of the partially purified protein sample. Fractions that contain more nonspecific DNA-binding proteins will normally require more competitor DNA and thus it is necessary to determine experimentally the optimal amount of DNA to use with each protein fraction. In addition, the amount of competitor DNA is estimated as the mass of DNA (in micrograms) to add per volume (in milliliters) of protein fraction; this is because it is likely that competitor DNA acts by forming a complex with high-affinity, nonspecific DNA-binding proteins in the crude extract. To move up to a full-scale experiment from the pilot study, use one-fifth of the amount that would be required if the binding reaction were directly scaled up. Different competitor DNAs can be used in a single experiment.

To prepare poly(dI-dC), poly(dG-dC), and poly(dA-dT), dissolve the desired amount of competitor DNA to a final concentration of 10 A_{260} units in TE buffer per 100 mM NaCl. Heat the sample to 90°C and slowly cool to room temperature over 30 to 60 min. If the average length of the DNA is >1 kb, degrade it by sonication. Estimate the length of the DNA by agarose gel electrophoresis (*UNIT 2.5*).

References: Kadonaga, J.T. 1991; Kadonaga, J.T. and Tjian, R. 1986.

Contributors: Leslie A. Kerrigan and James T. Kadonaga

Saccharomyces Cerevisiae

13

Saccharomyces cerevisiae, or baker's yeast, has been called the *Escherichia coli* of eukaryotic organisms. Yeast has been extensively characterized genetically, with more than 900 genes currently mapped; this genetic map will soon be correlated with a complete physical map of the genome. It is as easy to grow as other microorganisms and has a haploid nuclear DNA content only 3.5 times that of *E. coli.* However, despite this small genome size, yeast displays most of the features of higher eukaryotes. The increasing evidence which suggests that many cellular processes are mechanistically conserved among different eukaryotic species—combined with the powerful genetic and molecular tools that are available for *S. cerevisiae*—has made yeast the organism of choice for a variety of basic problems in eukaryotic molecular biology.

Culturing yeast is simple, economical, and rapid, characterized by a doubling time of about 90 min on rich media. In addition, yeast has been well adapted to both aerobic and anaerobic large-scale culture. Cells divide mitotically by forming a bud, which pinches off to form a daughter cell. The progression through the cell cycle can be monitored by the size of the bud; this has been used to isolate a large collection of mutants (called *cdc* mutants) that are blocked at various stages of the cell cycle. Because yeast can be grown on a completely defined medium (see *UNIT 13.1*), many nutritional auxotrophs have been isolated. This has not only permitted the analysis of complex metabolic pathways but has also provided a large number of mutations useful for genetic analysis.

13

Yeast can exist stably in either haploid or diploid states. The haploid cell can be either of two mating types, called **a** and α. Diploid **a**/α cells—formed by fusion of an α cell and an **a** cell (*UNIT 13.2*)—can grow mitotically indefinitely, but under conditions of carbon and nitrogen starvation will undergo meiosis. The meiotic products, called spores, are contained in a structure called an *ascus*. After gentle enzymatic digestion of the thick cell wall of the ascus, the haploid spore products can be individually isolated and analyzed (*UNIT 13.2*). This ability to recover all four products of meiosis has allowed detailed genetic studies of recombination and gene conversion that are not possible in most other eukaryotic organisms. The existence of stable haploid and diploid states also facilitates classical mutational analysis, such as complementation tests and identification of both dominant and recessive mutations.

The haploid yeast cell has a genome size of about 15 megabases and contains 16 linear chromosomes, ranging in size from 200 to 2200 kb. Thus, the largest yeast chromosome is still 100 times smaller than the average mammalian chromosome. This small chromosome size, combined with the advent of techniques for cloning yeast genes as well as manipulating yeast chromosomes, has allowed detailed studies of chromosome structure. Three types of structural elements required for yeast chromosome function have been identified and cloned: origins of replication (*ARS elements*), centromeres (*CEN elements*), and *telomeres.* The cloning of these elements has led to the construction of artificial chromosomes that can be used to study various aspects of chromosome behavior, such as how chromosomes pair and segregate from each other during mitosis and meiosis. In addition, systems using artificial chromosomes have been designed that allow cloning of larger contiguous segments of DNA (up to 400 kb) than are obtainable in other cloning systems. These structural elements, as well as cloned selectable yeast genes, have permitted the construction of yeast/*E. coli* shuttle vectors which can be maintained in yeast as well as in *E. coli* (*UNITS 13.4 & 13.6*).

Procedures for high-efficiency transformation of yeast (*UNIT 13.7*) have been available for over a decade, allowing cloning of genes by genetic complementation (*UNITS 13.8 & 13.9*). Because yeast has a highly efficient recombination system, DNAs with alterations in cloned genes can be reintroduced into the chromosome at the corresponding homologous site (*UNIT 13.10*). This has permitted the rapid identification of the phenotypic consequences of a mutation in any cloned gene, a technique generally unavailable in higher eukaryotes. The recent development of methods that permit the resolution of individual yeast chromosomes on agarose gels (*UNIT 2.5B*) has also allowed any cloned gene to be easily mapped to its chromosome location.

Despite its small genome size, yeast is a characteristic eukaryote, containing all the major membrane-bound subcellular organelles found in higher eukaryotes, as well as a cytoskeleton. Yeast DNA is found within a nucleus and nucleosome organization of chromosomal DNA is similar to that of higher eukaryotes, although no histone H1 is present. Three different RNA polymerases transcribe yeast DNA, and yeast mRNAs (transcribed by polymerase II) show characteristic modifications of eukaryotic mRNAs [such as a 5′ methyl-G cap and a 3′ poly(A) tail], although only a few *S. cerevisiae* genes contain introns. Transcriptional regulation has been extensively studied and at least one yeast transcriptional activator has been shown to function in higher eukaryotes as well. High-molecular-weight yeast DNA and RNA can be prepared fairly quickly (*UNITS 13.11 & 13.12*). Another characteristic of eukaryotes is the proteolytic processing of precursor proteins to yield functional products, which is often coupled to secretion. Yeast has several well-studied examples of secreted proteins and pheromones, and the large number of genes that have been identified as involved in protease processing and secretion suggests a highly complex pathway. Yeast protein extracts can be prepared using three different protocols (*UNIT 13.13*) where the best choice will depend on the particular application. The ease and power of genetic manipulation in yeast facilitate the use of this organism to detect novel interacting proteins using the two-hybrid system or interaction trap (*UNIT 13.14*).

References: Strathern et al., 1981, 1982; Watson et al., 1987.

Contributor: Victoria Lundblad

UNIT 13.1

Preparation of Yeast Media

Like *Escherichia coli*, yeast can be grown in either liquid media or on the surface of (or embedded in) solid agar plates. Yeast cells grow well on a minimal medium containing dextrose (glucose) as a carbon source and salts which supply nitrogen, phosphorus, and trace metals. Yeast cells grow much more rapidly, however, in the presence of protein and yeast cell extract hydrolysates, which provide amino acids, nucleotide precursors, vitamins, and other metabolites which the cells would normally synthesize *de novo*. During exponential or log-phase growth, yeast cells divide every 90 min when grown in such media.

The rich medium, YPD (yeast extract, **p**eptone, **d**extrose; also called YEPD medium), is most commonly used for growing *S. cerevisiae*. Additional recipes are provided for minimal and complete minimal dropout media, which are routinely used for testing mating type, selecting diploids, determining auxotrophic requirements (*UNIT 13.2*), and selecting for transformants (*UNIT 13.7*). Finally, recipes for media required in the more advanced techniques described in *UNITS 13.6* and *13.10* are presented.

Preparation of sterile media of consistently high quality is essential for the genetic manipulation of yeast. It is recommended that ingredients always be purchased from the same manufacturer. The following sources are recommended for specific ingredients (complete addresses and phone numbers are provided in *APPENDIX 4*):

J.T. Baker
dextrose
ammonium sulfate
potassium acetate
glycerol

Difco
agar (Bacto-agar)
peptone (Bacto-peptone)
yeast extract (Bacto-yeast extract)
yeast nitrogen base (YNB, *without* amino acids or ammonium sulfate)

Sigma
amino acids
nucleotide bases
canavanine
cycloheximide
L-α-aminoadipic acid
galactose (with 0.01% contaminating glucose)
potato starch

PCR, Inc
5-fluoroorotic acid

Autoclaving is usually carried out for 15 min at 15 lb/in^2, but times should be increased when large amounts of media are being prepared (20 min for 4 to 6 liters and 25 min for 6 to 12 liters).

LIQUID MEDIA

Ingredients for liquid media are dissolved in water to 1 liter, mixed until completely dissolved, and autoclaved in 100- or 500-ml media bottles. Alternatively, liquid media can be filter sterilized, resulting in faster preparation, less carmelization (of dextrose), and faster growth of cells. Recipes for "premixes" are provided to minimize the number of materials that must be weighed each time medium is prepared. When preparing premixes, break up any large chunks of dextrose before mixing with other components, then shake the container vigorously until contents are homogenized. It is convenient to make the premix in the empty plastic containers in which 2.5 kg of dextrose is packaged. Throughout this chapter, YNB –AA/AS refers to yeast nitrogen base without amino acids or ammonium sulfate. (See listing of recommended suppliers of ingredients in unit introduction above.)

Rich Medium

YPD medium (YEPD medium)

Per liter:	*Premix:*	*Final concentration:*
10 g yeast extract	250 g yeast extract	1% yeast extract
20 g peptone	500 g peptone	2% peptone
20 g dextrose	500 g dextrose	2% dextrose
	Use 50 g/liter	

It is preferable to use a 20% (10×) solution of dextrose that has been filter sterilized or autoclaved separately (and added to the other ingredients after autoclaving) to prevent darkening of the media and to promote optimal growth.

Minimal Media

Minimal medium [synthetic dextrose (SD) medium]

Per liter:	*Premix:*	*Final concentration:*
1.7 g YNB –AA/AS	68 g YNB –AA/AS	0.17% YNB –AA/AS
5 g $(NH_4)_2SO_4$	200 g $(NH_4)_2SO_4$	0.5% $(NH_4)_2SO_4$
20 g dextrose	800 g dextrose	2% dextrose
	Use 27 g/liter	

This minimal medium can support the growth of yeast which have no nutritional requirements. However, it is used most often as a basal medium to which other supplements are added (see CM dropout medium below).

Complete minimal (CM) dropout medium, *per liter*

1.3 g dropout powder (Table 13.1.1)
1.7 g YNB –AA/AS
5 g $(NH_4)_2SO_4$
20 g dextrose
(Alternatively, replace last three ingredients with 27 g minimal medium premix)

CM dropout powder, also known as minus or omission powder, lacks a single nutrient but contains the other nutrients listed in Table 13.1.1. Complete minimal (CM) dropout medium is used to test for genes involved in biosynthetic pathways and to select for gene function in transformation experiments. To test for a gene involved in histidine biosynthesis, determine if the yeast strain in question can grow on CM minus histidine (–His) or "histidine dropout" plates. It is convenient to make several dropout powders, each lacking a single nutrient, to avoid weighing each component separately for all the different dropout plates required in the laboratory.

It may be preferable to use a 10× solution of dropout powder (i.e., 13 g of dropout powder in 100 ml water) that has been "sterilized" separately (and added to the other ingredients after autoclaving) to improve the growth rate in this medium.

Sporulation medium, *per liter*

10 g potassium acetate (1% final)
1 g yeast extract (0.1% final)
0.5 g dextrose (0.05% final)

This nitrogen-deficient "starvation" medium contains acetate as a carbon source to promote high levels of respiration, which induces diploid yeast strains to sporulate. Sporulation can be carried out in liquid media or on plates (see below and UNIT 13.2). If nutrients are required, add them at the concentrations listed in Table 13.1.1.

Table 13.1.1 Nutrient Concentrations for Dropout Powders[a]

Nutrient[b]	Amount in dropout powder(g)[c]	Final conc. in prepared media (µg/ml)	Liquid stock conc. (mg/100 ml)[d]
Adenine (hemisulfate salt)	2.5	40	500
L-arginine (HCl)	1.2	20	240
L-aspartic acid[e]	6.0	100	1200
L-glutamic acid (monosodium salt)	6.0	100	1200
L-histidine	1.2	20	240
L-leucine[f]	3.6	60	720
L-lysine (mono-HCl)	1.8	30	360
L-methionine	1.2	20	240
L-phenylalanine	3.0	50	600
L-serine	22.5	375	4500
L-threonine[e]	12.0	200	2400
L-tryptophan	2.4	40	480
L-tyrosine	1.8	30	180[f]
L-valine	9.0	150	1800
Uracil	1.2	20	240

[a]CM dropout powder lacks a single nutrient but contains the other nutrients listed in this table. Nomenclature in this manual refers to, e.g., a preparation that omits histidine as histidine dropout powder. Conditions from Sherman et al., 1979.

[b]Amino acids not listed here can be added to a final concentration of 40 µg/ml (40 mg/liter).

[c]Grind powders into a homogeneous mixture with a clean, dry mortar and pestle. Store in a clean, dry bottle or a covered flask.

[d]Use 8.3 ml/liter of each stock for special nutritional requirements. Store adenine, aspartic acid, glutamic acid, leucine, phenylalanine, tyrosine, and uracil solutions at room temperature. All others should be stored at 4°C.

[e]Although these amino acids can be used reliably when included in autoclaved media, they supplement growth better when added after autoclaving.

[f]Use 16.6 ml/liter for L-tyrosine nutritional requirement.

Alternative Carbon Sources

Wild-type yeast can use a variety of carbon sources other than glucose to support growth. These include galactose, maltose, fructose, and raffinose. In particular, galactose is often used to induce transcription of sequences fused to the *GAL10* promoter (*UNIT 13.6*). All are used at a concentration of 2% w/v (20 g/liter) and should be made by replacing dextrose in the recipe for minimal or complete minimal (CM) dropout media. For a nonfermentable carbon source—which will not support the growth of *petites* (cells lacking functional mitochondria)—2% potassium acetate (w/v), 3% glycerol, 3% ethanol, or 2% glycerol and 2% ethanol (v/v each) can be used. YPA medium (2% potassium acetate, 2% peptone, and 1% yeast extract) is excellent for inducing high levels of respiration in cells prior to sporulation (*UNIT 13.2*).

SOLID MEDIA

Making solid media for yeast is—for the most part—no different from preparing plates for bacteriological work (see *UNIT 1.1*). For all plates, agar is added at a concentration of 2% (20 g/liter). A pellet of sodium hydroxide (~0.1 g) should be added per liter to raise the pH enough to prevent agar breakdown during autoclaving. In addition, add a stir bar to facilitate mixing after autoclaving. After autoclaving, flasks are left for 45 to 60 min at room temperature until cooled to 50° to 60°C. (Drugs and other nutrients are added after 30 min at room temperature.) Just prior to pouring, put the flask on a stir plate at medium to high speed and mix until contents are homogeneous (~5 min). After pouring, a few bubbles can be removed from the agar surface by passing the flame of a Bunsen burner lightly over the surface of the molten agar ("flaming" the plates). One liter of medium will yield 30 to 35 plates.

While a specific brand of petri plate is not required, we recommend Fisher plates (100 × 15 mm), which have ridges around the tops of the covers to allow easy stacking, making plate pouring less cumbersome. Most plates can be stored at room temperature for ≤4 months. Plates containing drugs (cycloheximide, 5-fluoroorotic, canavanine, and L-α-aminoadipic acid) or Xgal are stable for 2 to 3 months when stored at 4°C.

Minimal Plates and Rich Plates

YPD, minimal, and CM dropout plates

Per liter: Follow recipes for liquid medium above, adding 20 g agar and a pellet of NaOH.

Premixes: Follow recipes for liquid medium premixes above, adding 500 g agar for YPD premix and 800 g agar for minimal premix. To prepare plates, add one NaOH pellet and the following amounts of premix (per liter):

YPD plates—70 g YPD plate premix
Minimal plates—47 g minimal plate premix
CM dropout plates—47 g minimal plate premix + 1.3 g dropout powder

Specialty Plates

The recipes for α-aminoadipate, canavanine, and cycloheximide plates are included even though no specific use for them is described in this chapter. They are commonly used in negative selection experiments in the same way that 5-FOA plates are used (see below). It is possible to select against the wild-type *LYS2*, *CAN1*, and *CYH2* genes by growth on plates that select for cycloheximide, α-aminoadipate, or canavanine resistance, respectively (for review, see Brown and Szostak, 1983).

5-fluoroorotic acid (5-FOA) plates

To a 2-liter flask (containing a stir bar), add:
1 g 5-FOA powder
500 ml H_2O
5 ml 2.4 mg/ml uracil solution
Stir with low heat ~1 hr until completely dissolved; filter sterilize

To a separate 2-liter flask, add:
1.7 g YNB –AA/AS
5 g $(NH_4)_2SO_4$
20 g dextrose
20 g agar
1.3 g uracil dropout premix (Table 13.1.1)
H_2O to 500 ml and autoclave
(Alternatively, replace first four ingredients with 47 g minimal plate premix)

When the molten agar cools to ~65°C, gently add the sterile 5-FOA/uracil solution to the uracil dropout medium by pouring it down the inside wall of the flask containing the latter. Swirl gently to mix and pour the plates.

URA3$^+$ *strains are unable to grow on media containing the pyrimidine analog 5-fluoroorotic acid (Boeke et al., 1984). This observation has led to methods that use 5-FOA to select against the functional URA3 gene (see* UNIT 13.10*). This type of selection—termed "negative selection" (Brown and Szostak, 1983)—can also be used to select against the wild-type LYS2, CAN1, and CYH2 genes.*

5-FOA is quite expensive, and plates should be used sparingly. The material is prepared in bulk and is substantially discounted for members of the Genetics Society of America (Bethesda, Md.).

Xgal plates, *per liter*

1.7 g YNB –AA/AS
5 g $(NH_4)_2SO_4$
20 g dextrose
20 g agar
0.8 g dropout powder (omitting appropriate amino acids; see Table 13.1.1)
NaOH pellet
(Alternatively, replace first four ingredients with 47 g minimal plate premix)

Add H_2O to 900 ml and autoclave. Add 100 ml of 0.7 M potassium phosphate, pH 7.0, and 2 ml of 20 mg/ml Xgal prepared in 100% *N,N*-dimethylformamide (stored as frozen stock; see Table 1.4.2).

Dissection plates

Follow the recipe for YPD plates, keeping in mind that dissection plates used for tetrad analysis should be of uniform thickness and free of imperfections. After the plates have been poured they should be flamed. Stack the plates on a level surface in piles of six and move gently in a circular motion to "even out" the agar. Certain batches of agar produce plates that have microscopic precipitates embedded in the agar, often looking much like yeast spores. If this occurs, an agar of higher purity can be used, such as Noble agar (Difco) or agarose.

α-aminoadipate plates

1.7 g YNB –AA/AS
20 g dextrose
20 g agar
H_2O to 1 liter

Add ingredients to a 2-liter flask and autoclave. When the molten agar cools to ~65°C, add 34 ml of a solution of 6% L-α-aminoadipic acid (prepared by dissolving α-aminoadipate in 100 ml water and adjusting the pH to ~6.0 with 1 M KOH). The final concentration of α-aminoadipic acid in the plates should be 0.2%. Swirl gently to mix and pour the plates.

Lys2⁻ yeast use α-aminoadipic acid as an alternate nitrogen source. Because yeast can use certain amino acids as nitrogen sources, only those amino acids which are required by the particular strain being used should be added to these plates. These can be added from liquid stocks prior to autoclaving. When using this medium for isolating lys2⁻ yeast, lysine must be added at 30 μg/ml.

Canavanine plates

Follow recipe for complete minimal (CM) dropout plates, omitting the nutrient arginine. When the 1-liter autoclaved solution has cooled to ~65°C, add 10 ml of 6 mg/ml filter-sterilized canavanine sulfate solution. The final concentration of canavanine should be 60 μg/ml.

The sterile canavanine sulfate solution can be stored frozen.

Cycloheximide plates

Follow the recipe for YPD plates. When the agar cools to ~65°C, add 1 ml of 10 mg/ml filter-sterilized cycloheximide solution. The final concentration of cycloheximide in the plates should be 10 μg/ml.

The sterile cycloheximide stock solution can be stored frozen.

STRAIN STORAGE AND REVIVAL

Yeast strains can be stored at –70°C in 15% glycerol (viable for >5 years), or at 4°C on slants consisting of rich medium supplemented with potato starch (viable for 1 to 2 years).

Preparation and Inoculation of Frozen Stocks

BASIC PROTOCOL

Make a solution of 30% (w/v) glycerol. Pipet 1 ml into 15 × 45–mm, 4-ml screwcap vials. Loosely cap the vials and autoclave 15 min.

To inoculate vials for storage, add 1 ml of a late-log or early-stationary phase culture, mix, and set on dry ice. Store at –70°C. Revive by scraping some of the cells off the frozen surface and streak onto plates. Do not thaw the entire vial. Cells can also be stored in the same way by adding 80 μl dimethyl sulfoxide (DMSO) to 1 ml cells (8% v/v) and storing at –70°C.

References: Boeke et al., 1984; Brown and Szostak, 1983; Sherman et al., 1979.

Contributors: Douglas A. Treco and Victoria Lundblad

Growth and Manipulation of Yeast

Aside from different media requirements, yeast cells are physically manipulated essentially as described for bacterial cells—i.e., they are grown in liquid culture (in tubes or flasks) or on the surface of agar plates and are manipulated using the basic equipment described in *UNITS 1.1-1.3*. In addition, a well-equipped yeast laboratory requires static and shaking incubators dedicated to 30°C and a microscope with magnification up to 400×. A second microscope adapted for dissecting yeast tetrads is extremely valuable for the genetic analyses and strain constructions described in this unit. A small electric clothes dryer is indispensable when replica plating is done frequently and large numbers of velvets are regularly used.

BASIC PROTOCOL 1

GROWTH IN LIQUID MEDIA

Wild-type *S. cerevisiae* grows well at 30°C with good aeration and with glucose as a carbon source. When using culture tubes, vortex the contents briefly after inoculation to disperse the cells. Erlenmeyer flasks work well for growing larger liquid cultures, and baffled-bottom flasks to increase aeration are especially good. It is important that all glassware be detergent-free. For good aeration, the medium should constitute no more than one-fifth of the total flask volume, and growth should be carried out in a shaking incubator at 300 rpm. For small-scale preparations of DNA and RNA, yeast can be grown in glass or plastic culture tubes filled one-third full with medium and shaken at 350 rpm in a rack firmly attached to a shaking incubator platform.

BASIC PROTOCOL 2

GROWTH ON SOLID MEDIA

Yeast cells can be streaked or spread on plates as shown for bacteria in the sketches in *UNIT 1.3*. When a dilute suspension of wild-type haploid yeast cells is spread over the surface of a YPD plate and incubated at 30°C, single colonies may be seen after ~24 hr but require ≥48 hr before they can be picked or replica plated (see below). Growth on dropout media (*UNIT 13.1*) is ~50% slower.

BASIC PROTOCOL 3

DETERMINATION OF CELL DENSITY

The density of cells in a culture can be determined spectrophotometrically by measuring its optical density (OD) at 600 nm. For reliable measurements, cultures should be diluted such that the OD_{600} is <1. In this range, each 0.1 OD_{600} unit corresponds to ~3×10^6 cells/ml. Thus, an OD_{600} of 1 is equal to ~3×10^7 cells/ml. It is advisable to calibrate the spectrophotometer by graphing the OD_{600} as a function of the cell density that has been determined by some other means, such as direct counting in a hemacytometer chamber (*UNIT 1.2*) or titering for viable colonies (*UNIT 1.3*).

Log phase growth can be divided into three stages based on the rate of cell division (or the proportion of budded cells within a culture), which is in turn a function of the cell density of the culture. As cell density increases, nutrient supplies drop and the rate of cell division slows. Early-log phase is the period when cell densities are $<10^7$ cells/ml. Mid-log phase cultures have densities between 1 and 5×10^7 cells/ml. Late-log phase occurs when cell densities are between 5×10^7 and 2×10^8 cells/ml. At a density of 2×10^8 cells/ml yeast cultures are said to be saturated and the cells enter stationary, or G_0, phase.

DETERMINATION OF PHENOTYPE BY REPLICA PLATING

BASIC PROTOCOL 4

Cells from yeast colonies grown on any medium can be tested for their nutritional requirements by replica plating (*UNIT 1.3*). An inexpensive replica plating block can be constructed by gluing a circular plexiglass disk (8-cm diameter, 1 cm thick) onto the end of a hollow plexiglass tube (8 cm long with an 8-cm outer diameter). Sterile velveteen squares (velvets) are held in place by a large adjustable tube clamp (available in any automotive supply outlet) set to fit snugly around the outside of the tube.

A master plate containing the strain or strains of interest is first printed onto a velvet. A copy of this impression is transferred to plates made with all the relevant selective media, which may include various dropout and drug media, as well as alternative carbon sources (*UNIT 13.1*). For analysis of temperature-sensitive mutations (*UNIT 13.8*), a copy of the master plate is made on a plate that will provide all the nutritional requirements of the strain. This plate is then incubated at 37°C.

DIPLOID CONSTRUCTION

BASIC PROTOCOL 5

Diploids are constructed by mating strains of opposite mating types on the surface of agar plates (patch mating). Mix cells from freshly grown colonies of each haploid parent with a toothpick in a circle ~0.5 cm in diameter on an agar plate (the plate should allow growth of both haploid strains). Allow mating to proceed ≥4 hr at 30°C, then streak the mating mixture onto a plate that will select for the diploid genotype.

When there is no selection specific for the diploid genotype, isolate diploids by physically "pulling zygotes" out of the mating mixture using a dissecting microscope. After mating for 4 hr, streak the mix in several parallel lines on an agar plate that will support the growth of the diploid. Using the dissecting microscope, identify zygotes by their characteristic shape (large 2- and 3-lobed structures in which the lobes are connected by large smooth "necks"), pick them up with the dissecting needle, move them away from the streak of cells, and set them down. To ensure that the selected cells are actually diploids, patch them onto sporulation plates (*UNIT 13.1*) and examine microscopically for tetrad formation after appropriate incubation. Alternatively, attempt to mate selected cells with a pair of mating type tester strains and examine microscopically for zygote formation with each tester. Zygotes should not form if a diploid was correctly selected.

SPORULATION OF DIPLOID CELLS

BASIC PROTOCOL 6

Starvation of diploid yeast cells for nitrogen and carbon sources induces meiosis and spore formation, during which chromosomes replicate and proceed through two divisions to produce haploid nuclei. These nuclei (along with surrounding cytoplasm) are individually packaged into spores, and the four spore products (tetrad) of a single meiosis are held together in a thick-walled sac (ascus). The sporulation process can be induced in cells growing either on solid or in liquid medium. Because some strains do not sporulate well on plates, and other strains do not sporulate well in liquid, both methods should be used, and one of the two methods should result in reasonably good spore formation for any given diploid.

Materials

Yeast cells
Sporulation plates *or* sporulation medium, with appropriate nutrients (*UNIT 13.1*)
YPD medium (*UNIT 13.1*)

Sporulation on plates

1. Patch cells that have been grown on YPD or selective plates onto a sporulation plate.

   If no selective conditions are required, grow cells several days on YPD plates prior to transfer to sporulation medium. Allow single colonies to grow for 3 to 4 days on YPD; for patches of cells, allow 2 days growth on YPD. Pregrowth is not essential, but it results in much more efficient sporulation. A small dab of cells should be smeared over a relatively large area (~1 cm²) of the sporulation plate, such that no thick patches of the inoculum are visible.

2. Incubate 4 days at 25°C.

   Sporulation is generally less efficient at higher temperatures. While sporulation can occur in the absence of amino acids or other nutrients that are required by the strain for mitotic growth, sporulation is much more complete when those (and only those) nutrients that the particular strain requires are added (see Table 13.1.1).

3. Visualize tetrads by suspending a small dab of cells in a drop of water on a microscope slide and examining at a magnification of 250× to 400×.

   Tetrads will appear as clusters of four small spheres (the spores), all held within a tight-fitting sac. The four spores can be in either a diamond or tetrahedral configuration. Not all asci will contain four spores.

Sporulation in liquid medium

For unknown reasons, some strains do not sporulate well on plates. Even for strains that do, the efficiency can often be increased by sporulation in liquid. In fact, sporulation in liquid is so efficient that higher incubation temperatures (30°C) do not appear to inhibit this process. Under these conditions, sporulation is effectively complete within 48 hr. In addition, the yield of asci with four spores is usually higher using this protocol.

1. In any suitable culture vessel, grow the diploid to be sporulated to an OD_{600} of 2.5 to 3.0 (~8×10^7 cells/ml) in YPD medium.

2. Transfer 1 ml culture to a sterile, disposable 15-ml polypropylene tube and centrifuge 5 min at 1200 × *g*.

3. Pour off supernatant and resuspend cells in 5 ml sterile water. Vortex to resuspend cells and spin as in step 2.

4. Pour off supernatant and resuspend cells in 1 ml liquid sporulation medium supplemented with nutritional requirements of the particular diploid.

5. Shake for 2 to 3 days at ≥350 rpm, 30°C, and examine the culture microscopically for spore formation.

   If a strain appears refractory to induction of sporulation, try pregrowing cells in YPA medium (UNIT 13.1). The use of acetate as a carbon source requires respiration, which is a requirement for sporulation.

BASIC PROTOCOL 7

PREPARATION AND DISSECTION OF TETRADS

Analysis of yeast tetrads requires a standard light microscope with a stage that is movable along both the *x* and *y* axes in precisely measurable intervals, but that does not move up and down (focusing is accomplished by moving the objectives). The microscope must be modified with an assembly for mounting an inverted petri dish on its stage and a micromanipulator for holding and moving a fine glass needle (see

Support Protocol). Plans for the construction of micromanipulators for tetrad analysis have been published (Sherman, 1987). An assembled instrument is available from Rainin Instrument.

Materials

Glusulase (Du Pont NEN), diluted 1:10 in H_2O *or* Zymolyase-100T solution: 0.5 mg/ml Zymolyase-100T (ICN Immunobiologicals) in 1 M sorbitol

Dissecting microscope (Rainin Instrument)
Dissecting needle (Support Protocol)

1a. *For glusulase treatment:* Wash spores three times in sterile water. Resuspend in 5 ml sterile water and dilute 30 µl in 270 µl sterile water. Examine cells under a light microscope to detect intact asci. Add 3 µl diluted glusulase to spore preparation. Mix by gently vortexing. Incubate 2 min at room temperature, then examine the cells microscopically. Continue incubation at room temperature until glusulase treatment is complete.

The glusulase treatment is complete when approximately half of the tetrads show disruption of their ascus walls (usually 2 to 6 min).

1b. *For Zymolyase-100T treatment:* Suspend pellet in 50 µl Zymolyase-100T solution. Examine cells under a light microscope to detect intact asci. Incubate 10 min at 30°C. Gently add 0.8 ml sterile water by slowly running it down side of tube.

The Zymolyase treatment is complete when approximately half of the tetrads show disruption of their ascus walls. Although Zymolase-100T is more expensive, it generally gives more uniform and reproducible digestion.

3. Set tubes on ice and leave them there. Streak treated spores (using an inoculating loop) in two parallel lines across the surface of a YPD dissection plate (*UNIT 13.1*), as shown in Figure 13.2.1.

4. Examine plate, inverted with the lid off, on the dissecting microscope. Individual tetrads, grouped into tetrahedral- or diamond-shaped clusters of spores, should be visible.

The following procedure is specifically adapted for the Rainin Instrument dissecting microscope (Fig. 13.2.1).

5. Position the plate so that streaks are parallel to the *x* axis of the stage. Focus on spores and position a tetrad that is away from other tetrads in center of field. Adjust dissecting needle upward using coarse adjustment so that its tip is touching the plate surface when the joystick is depressed halfway.

6. Gently touch needle to surface of plate *immediately adjacent to the tetrad* to pick up the tetrad.

7. Move the plate so that the needle is positioned ~1 cm away from and perpendicular to the streak of treated spores. Gently touch the needle to the agar surface to deposit the tetrad on the agar. The following variations are possible:

   a. No spores were deposited. In this case repeat step 7, using more force when touching the needle to the agar.

   b. Only a single spore is on the plate (the best result). Move the *y*-axis adjustment ~0.5 cm and repeat step 7.

   c. If two or three spores are on the plate, move the *y*-axis adjustment ~1 cm and

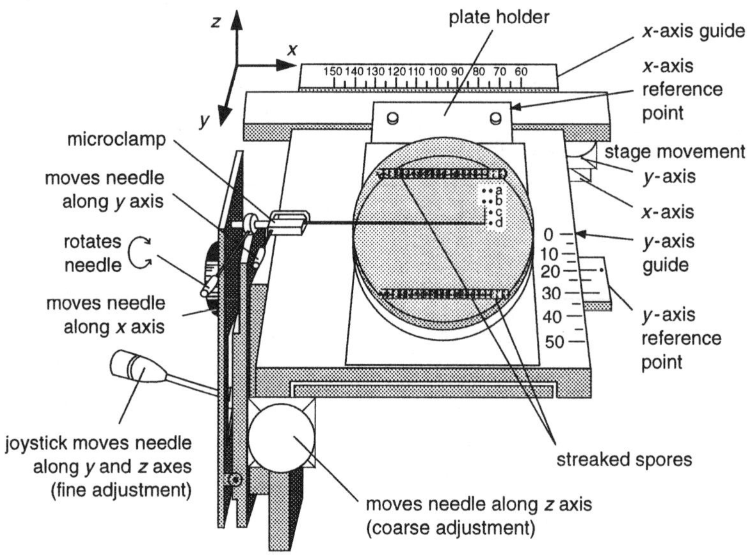

Figure 13.2.1 The dissection microscope. Not all parts are shown, including the light source and condenser (below) and the objective and eyepieces (above). The micromanipulator is attached to the base of the microscope. The mark with the large dot on the *y*-axis guide is aligned with demarcations on the right side of the stage to identify positions *a, b, c,* and *d* (corresponding to positions 10, 15, 20, and 25, respectively, in this figure). The upper right-hand corner of the plate holder is aligned with demarcations along the rear of the stage to identify positions 1, 2... 13 (corresponding to positions 60, 65... 120, respectively, in this figure).

repeat step 7 to deposit one or more spores. It may be necessary to break the spores apart by moving the joystick in circular motion, so that the tip of the needle hits the plate and drags over the spores. Pick up and deposit spores until they are arranged in a line perpendicular to the original streak with 0.5-cm intervals between spores.

d. If all four spores are present, use the needle to break them apart. Once this is done, pick up one or more of the spores and deposit them in a line perpendicular to the original streak with 0.5-cm intervals between spores.

If the four spores cannot be separated after repeatedly dragging with the needle, the digestion was probably not sufficient. Let the treated culture (which should have been on ice) sit at room temperature for a few minutes and then streak onto a new YPD dissection plate.

8. Use *x*- and *y*-axis controls to move the plate back to a position where the needle is directly below the streak of treated spores. Repeat steps 6 and 7, with spores of each successive tetrad set down in a line 1 cm to right of and parallel to that of the preceding tetrad. Placement of these lines is determined by aligning an arbitrary marking on the movable stage (upper right-hand corner of the plate in Fig. 13.2.1) with a fixed *x*-axis scale that is even with back edge of stage. Alternatively, a strip of paper marked in 0.5-cm intervals can be taped along front edge of stage. In this case the lower right-hand corner of the plate holder is used as the arbitrary alignment point.

9. Continue steps 6 through 8 until all positions along the *x* axis are occupied by dissected tetrads. Rotate the plate 180° for dissection of additional tetrads using

the streak on the opposite side of plate as a source. Thirteen tetrads can be dissected on each side of the plate.

PREPARATION OF DISSECTING NEEDLES

SUPPORT PROTOCOL

Making needles requires patience and practice. The general strategy is to first produce a long fine thread of glass, which is broken into multiple short segments. Individual segments are then glued to the ends of capillary tubes. The finished product is L-shaped, with a short arm of ~1.3 cm, culminating in a needle "tip" that is 40 to 150 μm in diameter.

1. Heat a piece of glass tubing until it is flexible and, in one motion, remove it from the heat source and pull the ends apart very quickly to draw out the molten glass into a very fine thread (40 to 150 μm in diameter—a human hair ranges from 40 to 100 μm) before it (nearly instantly) hardens. Either capillary tubes or pyrex glass pipets can be used as sources of glass tubing.
2. Break the very fine thread into segments ~1.3 cm long. To do this, simply begin at one end, pulling the segments straight away from the rest of the thread using your fingers. Wear latex gloves for a better grip. Alternatively, 1.3-cm segments can be cut using a fresh razor blade.
3. Examine these segments microscopically. Look for chips and protrusions in the glass. The best needles (those that easily pick up and release cells and do not cut the surface of the agar) have an absolutely flat surface that is perpendicular to the shaft. Save any with an acceptable end, noting which end is which, and discard those with chipped or uneven ends.
4. Create a dissecting needle by attaching a segment with an acceptable tip to the end of a capillary tube. Introduce a right-angle bend in the capillary tube 0.5 to 1.0 cm from the end by briefly heating the tube in a Bunsen burner flame and bending the tip using forceps. Glass threads can be glued in place inside, or against the outside edge, of the short arm of the L-shaped capillary tube, using any fast-acting glue.
5. Examine the finished product after mounting the needle on a dissection microscope. Place a plate on the microscope and press the needle firmly against the surface of the agar to make sure the needle leaves an even, circular impression.

   Even with all these precautions, needles will perform differently when it is time to dissect, so several candidates should be made at the same time and tested to determine which function best.

RANDOM SPORE ANALYSIS

ALTERNATE PROTOCOL

As an alternative to separating spores by tetrad dissection, meiotic products can be released from their asci, dispersed by sonication, and plated directly onto agar plates. The spore colonies can then be screened for the desired genotypes by replica plating.

Materials

Zymolyase-20T solution: 1 mg/ml Zymolyase-20T (ICN Immunobiologicals) in H_2O and filter sterilized
2-mercaptoethanol (2-ME)
1.5% (v/v) Nonidet P-40 (NP-40)
Ethanol

Sonicator and probe

1. Prepare cells for tetrad dissection (Basic Protocol 7). If sporulation was done on plates, resuspend several toothpicks-full in 5 ml water in a 50-ml flask. If sporulation was done in liquid, resuspend all the cells from a 1-ml sporulation culture in 5 ml water. Add 0.5 ml Zymolyase-20T solution and 10 µl of 2-ME. Incubate overnight at 30°C with gentle shaking.

   Treatment of the sporulated culture with Zymolyase-20T in a hypotonic solution results in lysis of unsporulated diploid cells. The preparation should be examined microscopically after the Zymolyase treatment to evaluate its effectiveness.

2. Add 5 ml of 1.5% Nonidet P-40 (NP-40). Transfer the suspension to a 15-ml disposable tube and set 15 min on ice.

3. Before sonicating, clean the sonicator probe with water followed by a wipe-down with ethanol. Hold the tube in one hand and insert the sonicator probe as far into the liquid as possible, but without touching the bottom or the sides of the tube. Sonicate 30 sec at 50% to 75% full power, then set on ice 2 min. Repeat twice.

4. Centrifuge spores 10 min at 1200 × *g*. Aspirate or pour off supernatant and resuspend in 5 ml of 1.5% NP-40. Vortex vigorously. Repeat twice.

5. Sonicate as in step 3 (with repeats). Examine the spores after the last sonication.

   If spores remain stuck to each other, add 2 ml glass beads (Type I, Sigma) and shake 30 min at 300 rpm in an Erlenmyer flask at 30°C. Let the beads settle and remove the supernatant containing the spores.

6. Centrifuge spores 10 min at 1200 × *g*. Aspirate or pour off supernatant and resuspend in 5 ml water. Vortex vigorously. Repeat.

7. Count a 10-fold dilution of the treated spores using a hemacytometer. Dilute the spores with water to get 10^3 spores/ml. Plate 100 µl on several YPD plates and incubate 3 days at 30°C.

8. Screen spore colonies for markers of interest by replica plating (Basic Protocol 4 and UNIT 1.3).

Reference: Sherman et al., 1979.

Contributors: Douglas A. Treco and Fred Winston

13.3

UNIT 13.3

Mutagenesis of Yeast Cells

Two common mutagens of yeast cells are ethyl methanesulfonate (EMS) and ultraviolet (UV) light. Mutagenesis can increase the frequency of mutation up to 100-fold per gene without excessive killing of the cells and without a significant frequency of double mutants. Mutagenized cells can be screened for any phenotype, including auxotrophics, cold sensitivity, and radiation sensitivity. Once a gene of interest has been identified by such procedures, it can be cloned, mutagenized, and manipulated in other ways to study its function in greater detail (UNITS 13.8 & 13.10).

Mutagenesis can be monitored by a control, measuring the frequency of canavanine-resistant mutants. Under optimal conditions of mutagenesis, the frequency of canavanine-resistant mutants should be at least 100-fold greater after mutagenesis. This control experiment is important in order to know that the mutagenesis has worked properly as the frequency of obtaining the desired class of mutant is likely to be unknown. The frequency of finding a particular class of mutant will depend entirely upon the nature of the mutation sought.

MUTAGENESIS USING ETHYL METHANESULFONATE

BASIC PROTOCOL

Materials (*see* APPENDIX 1 *for items with* ✓)

YPD medium and plates (*UNIT 13.1*)
✓ 0.1 M sodium phosphate buffer, pH 7.0
Ethyl methanesulfonate (EMS; Kodak)
5% (w/v) sodium thiosulfate (Sigma), autoclaved for sterility
Canavanine plates (*UNIT 13.1*)

13 × 100–mm culture tube
30°C incubator with rotating platform

CAUTION: EMS is a dangerous mutagen. All solutions, plasticware, and glassware that come into contact with EMS should be rinsed with 5% sodium thiosulfate to inactivate the EMS.

NOTE: All solutions, plasticware, glassware, and velveteens coming into contact with yeast cells must be sterile.

1. Grow an overnight culture of desired yeast strain in 5 ml YPD medium at 30°C (*UNIT 13.2*). Determine cell density, record, and adjust to ~2×10^8 cells/ml.

2. Transfer 1 ml to sterile microcentrifuge tube. Microcentrifuge 5 to 10 sec at maximum speed. Resuspend cells in 1 ml sterile water. Repeat wash. Resuspend cells in 1.5 ml sterile 0.1 M sodium phosphate buffer, pH 7.0.

3. Add 0.7 ml cell suspension to 1 ml buffer in 13 × 100–mm culture tube. Save remaining cells on ice as control. Add 50 µl EMS to cells and disperse by vortexing. Place on rotating platform 1 hr at 30°C.

   EMS treatment should cause ~40% of cells to be killed.

4. Transfer 0.2 ml cell suspension to culture tube containing 8 ml sterile 5% sodium thiosulfate, which will stop mutagenesis by inactivation of EMS. If cells are to be stored before plating, centrifuge in tabletop centrifuge (5 min, 3000 rpm), resuspend in equal volume sterile water, and store at 4°C.

5. Plate 0.1 ml mutagenized cells directly on each of two canavanine plates. As controls, plate 0.1 ml nonmutagenized cells (step 3) on duplicate canavanine plates. Incubate at 30°C until colonies form (~2 to 4 days). Calculate relative levels of canavanine-resistant mutants in mutagenized (step 5) and nonmutagenized (step 1) cultures.

6. Dilute EMS-treated cells (from step 3) 1:1000 with sterile water to obtain 100 to 200 viable cells/plate. Plate 0.1 and 0.2 ml of diluted cells on separate sets of YPD plates, using ten plates in each set (or more depending on frequency of desired mutant class). Incubate all plates 3 to 4 days at room temperature (23°C). For screening of auxotrophs, incubate plates at 30°C.

7. Choose at least ten YPD plates from step 6 that contain 100 to 200 colonies per plate. Replica-plate each to two fresh YPD plates. For temperature-sensitive mutants, incubate one YPD plate per set at 37°C and one at room temperature (23°C) overnight. In general, one of two replica plates should be permissive for

mutant growth (typically a YPD plate). The other plate should represent conditions that do not permit growth of desired class of mutants.

8. Compare 37°C plate with 23°C plate. Any colonies that failed to grow at 37°C are candidates for temperature-sensitive mutants.

9. To recheck, pick corresponding colonies from YPD plates incubated at 23°C and streak for single colonies on fresh YPD plates. On each plate, also streak parental control strain for single colonies. Incubate plates at 23°C until colonies form (2 to 4 days). Six to eight strains can be purified on each YPD plate.

10. After single colonies form, replica-plate YPD plates with temperature-sensitive candidates to two YPD plates, as before. Incubate one plate at 37°C and other at room temperature for 1 day. Record growth response. For each candidate that is reproducibly temperature-sensitive, place into permanent storage (*UNIT 13.1*).

13.3

ALTERNATE PROTOCOL

MUTAGENESIS USING UV IRRADIATION

Additional Materials *(also see Basic Protocol)*

YPD medium and plates (*UNIT 13.1*)
Canavanine plates (*UNIT 13.1*)
UV germicidal light bulb (Sylvania G15T8; 254-nm wavelength)
UV dosimeter (optional)

CAUTION: Wear safety glasses to protect eyes from UV light.

NOTE: All solutions, plasticware, glassware, and velveteens coming into contact with yeast cells must be sterile.

1. Grow overnight culture of desired yeast strain in 5 ml YPD medium at 30°C. Microcentrifuge 5 to 10 sec. Resuspend pellet in 1 ml sterile water and repeat wash. Resuspend in 1 ml sterile water.

2. Determine cell density and record this number. Adjust to 2×10^8 cells/ml. Make serial dilutions of culture in sterile water so that each plate has 100 to 200 viable cells. Plate 0.1 and 0.2 ml on YPD plates (Basic Protocol, step 6).

3. Warm UV light ≥20 minutes before use. Remove petri plate lid prior to irradiation. Irradiate all but two plates from each set with UV light using 300 ergs/mm^2 to obtain ~40% to 70% survival (nonirradiated plates serve as controls to determine degree of killing).

   Light from a UV germicidal bulb can be measured using a UV dosimeter. From this measurement, the proper length of time can be calculated for irradiation to attain 300 ergs/mm^2. Alternatively, the proper time of irradiation can be empirically determined by measuring the time that results in 40% to 70% survival.

4. As a control, plate 0.1 ml of original culture on duplicate canavanine plates and irradiate one plate for time determined in step 3. From the number of canavanine-resistant colonies, calculate frequency of mutations with and without UV irradiation.

5. Incubate plates from step 2 at 23°C until colonies form (~2 to 4 days). Screen colonies for temperature-sensitive mutants (Basic Protocol, steps 6 to 10).

Reference: Rose et al., 1990.

Contributor: Fred Winston

Yeast Cloning Vectors and Genes

This unit describes some of the most commonly used yeast vectors, as well as the cloned yeast genes that form the basis for these plasmids. Yeast vectors can be grouped into five general classes, based on their mode of replication in yeast: YIp, YRp, YCp, YEp, and YLp plasmids. With the exception of the YLp plasmids (yeast linear plasmids), all of these plasmids can be maintained in *E. coli* as well as in *S. cerevisiae* and thus are referred to as shuttle vectors.

Table 13.4.1 summarizes the general features of a number of these vectors, including the phenotypes that allow selection in either *E. coli* or yeast (or both). These plasmids contain two types of selectable genes, both of which can confer a dominant phenotype: plasmid-encoded drug-resistance genes (*UNIT 1.5*) and cloned yeast genes. The drug resistance marker is dominant because the recipient (bacterial) cell does not encode a gene for drug resistance. In contrast, the cloned yeast gene present on the plasmid is a copy of a gene that is present in the yeast genome as well. Thus, this gene functions as a dominant selectable marker only when the recipient yeast cell has a recessive mutation in the corresponding chromosomal copy of the cloned gene. Many of these cloned yeast genes encode functions involved in biosynthetic pathways of yeast and are capable of complementing certain mutations in similar biosynthetic pathways of *E. coli* (e.g., the cloned *URA3* gene of yeast can complement *ura3*$^-$ mutations of yeast as well as mutations in the *pyrF* gene of *E. coli*). The availability of bacterial mutations that can be complemented by these yeast genes can greatly simplify plasmid constructions, allowing genetic screening for the expected recombinant plasmid.

Table 13.4.2 presents characteristics of a number of cloned yeast genes, along with mutants of bacteria and yeast in which the wild-type cloned gene can be selected. For two yeast genes, a positive selection for mutant alleles also exists: *ura3*$^-$ cells

Table 13.4.1 Yeast Plasmids and their Selectable Markers

Plasmid	Size	*E. coli* replicon	Yeast replicon	Phenotypes selectable in *E. coli*	Phenotypes selectable in yeast	Reference
YIp5	5541 bp	pMB1	none	Ap^r, Tet^r, PyrF$^+$	Ura$^+$	Struhl et al., 1979
YRp7	5816 bp	pMB1	*ARS1*	Ap^r, Tet^r, TrpC$^+$	Trp$^+$	Struhl et al., 1979
YRp17	7002 bp	pMB1	*ARS1*	Ap^r, Tet^r, PyrF$^+$, TrpC$^+$	Ura$^+$, Trp$^+$	Stinchcomb et al., 1982
YEp13	10.7 kb	pMB1	2μm	Ap^r, Tet^r, LeuB$^+$	Leu$^+$	Broach et al., 1979
YEp24	7769 bp	pMB1	25μm	Ap^r; Tet^r in some constructions; PyrF$^+$	Ura+	Botstein et al., 1979
YCp19	10.1 kb	pMB1	*ARS1*	Ap^r, PyrF$^+$, TrpC$^+$	Ura$^+$, Trp$^+$	Stinchcomb et al., 1982
YCp50	7.95 kb	pMB1	*ARS1*	Ap^r, Tet^r, PyrF$^+$	Ura$^+$	Rose et al., 1987
YLp21	55 kb	none	*ARS1*	n.a.	Trp$^+$, His$^+$	Murray and Szostak, 1983
pYAC3	11.4 kb	pMB1	*ARS1*	Ap^r, Tet^r, TrpC$^+$, PyrF$^+$, HisB$^+$	Trp$^+$, Ura$^+$, His$^+$	Burke et al., 1987
2μm	6318 bp	none	2μm	n.a.	none	Hartley and Donelson, 1980

Table 13.4.2 Cloned Yeast Genes

Yeast gene	Length of sequenced fragment (bp)	Map position in yeast genome	Selectable phenotype in *E. coli*	*E. coli* strain[a]	Selectable phenotype of wild type in yeast	Common nonreverting mutant alleles	Selectable phenotype of mutant alleles in yeast[b]	GenBank file name
ARG4	2296	8R	$ArgH^+$	JA209	Arg^+	none	none	YSCARG4
ARS1	1453[c]	4R	none	n.a.	none	n.a.	none	YSCTRP1
CAN1[d]	n.s.	5L	none	n.a.	Can^s	*can1*-100 *can1*-11	canavanine sulfate^r	YSCCAN1
CEN3	627	3	none	n.a.	none	n.a.	none	YSCCEN3
CEN4	2095	4	none	n.a.	none	n.a.	none	YSCCEN4
CYH2[e]	1393	7L	none	n.a.	viability[f]		cycloheximide^r	YSCRPL29
GAL1-GAL10 regulatory region	907	2R	none	n.a.	none	n.a.	n.a.	YSCGAL
HIS3	1822	15R	$HisB^+$	BA1	His^+	*his3*-Δ1 *his3*-200	none	SCHIS3Y
LEU2	2230	3L	$LeuB^+$	JA300	Leu^+	*leu2*-3,112	none	YSCLEU2
LYS2	n.s.	2R	none	n.a.	Lys^+	*lys2*-Δ1	α-aminoadipate^r	n.a.
TCM1[g]	1529	15R	none	n.a.	viability[f]	n.a.	tricodermin^r	YSCRP13
TRP1	1453[c]	4R	$TrpC^+$	JA300	Trp^+	*trp1*-289 *trp1*-Δ901	none	YSCTRP1
URA3	1170	5L	$PyrF^+$	DB6656	Ura^+	*ura3*-52 *ura3*-Δ1	5-fluoroorotic acid^r	YSCODCD

n.s. = not sequenced; n.a. = not applicable.

[a]The genotypes and references for these strains can be found in the legends to Figures 13.4.1, 13.4.2, and 13.4.6, except for JA209 (*argH1 metE xyl trpA36 recA56 str^r*; Clarke and Carbon, 1978).

[b]Selection for α-aminoadipate resistance produces both *lys2*$^-$ and *lys5*$^-$ mutants. Selection for 5-FOA resistance generates both *ura3*$^-$ and *ura5*$^-$ mutants.

[c]This 1453-bp fragment contains both the *TRP1* and the *ARS1* genes.

[d]The wild-type *CAN1* gene encodes dominant sensitivity to the arginine analog canavanine sulfate.

[e]The sensitive (wild-type) and resistant alleles of the *CYH2* gene are codominant.

[f]The *CYH2* and *TCM1* genes encode ribosomal proteins, which are required for viability, but which can be mutated to confer resistance to cycloheximide or tricodermin, respectively.

[g]The resistant allele of *TCM1* is dominant to the wild-type sensitive allele.

can be selected on plates containing the drug 5-fluoroorotic acid (5-FOA), whereas *lys2*$^-$ cells can be selected on α-aminoadipate plates (see UNIT 13.2 for recipes for these two types of plates).

The nomenclature of different classes of yeast vectors, as well as details about their mode of replication in yeast, are described below.

PLASMID NOMENCLATURE

YIp plasmids (yeast integrating plasmids) contain selectable yeast genes but lack sequences that allow autonomous replication of the plasmid in yeast. Instead, transformation of yeast occurs by integration of the YIp plasmid into the yeast genome by recombination between yeast sequences carried on the plasmid and homologous sequences in the yeast genome. This recombination event results in a tandem duplication of the yeast sequences that bracket the rest of the plasmid DNA. If the YIp plasmid contains an incomplete portion of a cloned gene, this technique can be used to create a gene disruption (see UNIT 13.10). The reversal of the integration process occurs at a low frequency (about 0.1% to 1% per generation), with excision of the integrated plasmid occurring by recombination between duplicated yeast sequences. The frequency of transformation of YIp plamids is only 1 to 10 transformants/μg DNA, but transformation frequency can be increased 10- to 1000-fold by

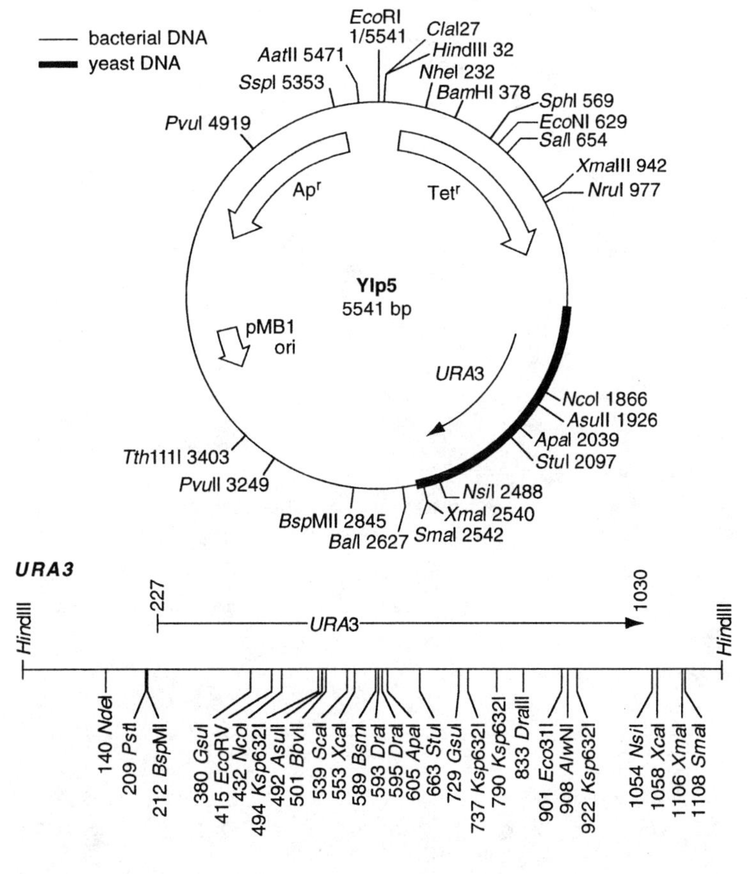

Figure 13.4.1 YIp5. YIp5 contains the 1.1-kb *Hin*dIII *URA3* gene cloned into the *Ava*I site of pBR322 via the addition of poly(dG-dC) tails (Struhl et al., 1979). As this plasmid does not contain a yeast origin of replication, transformants occur by integration into the yeast genome at the *URA3* locus; the frequency of transformation can be increased by linearization of the plasmid within the *URA3* insert. The complete nucleotide sequence is available from the Vecbase database (file name: Vecbase.YIp5) and a detailed restriction map can be found in the New England Biolabs catalog.

The *URA3* gene encodes orotidine-5′-phosphate (OMP) decarboxylase, a 267-amino-acid protein required for uracil biosynthesis. The map shown is the gene from strain +D4, which can be expressed in *E. coli* without an external bacterial promoter (Rose et al., 1984). Loss of $URA3^+$ function can be directly selected using 5-FOA: $ura3^-$ cells are resistant to 5-FOA, whereas 5-FOA is toxic to cells synthesizing the *URA3* gene product (Boeke et al., 1984). This negative selection has been exploited in a variety of gene replacement schemes, discussed in *UNIT 13.10*. The *URA3* gene can also complement mutations in the *pyrF* gene in *E. coli* using strain DB6656 (*pyrF*::Mu, $lacZ_{am}$ trp_{am} hsr_k^- hsm_k^+; Bach et al., 1979).

linearizing the plasmid within yeast sequences that are homologous to the intended site of integration on the yeast chromosome. Linearization also directs the integration event to the site of the cleavage, which is useful when several different homologous yeast sequences are present on the plasmid.

Three classes of yeast vectors are circular plasmids capable of extrachromosomal replication in yeast. YRp plasmids (yeast replicating plasmids) contain sequences from the yeast genome which confer the ability to replicate autonomously. These autonomous replication sequences (*ARS*) have been shown to be chromosomal origins of replication. YRp plasmids have high frequencies of transformation (10^3 to 10^4 transformants/µg DNA), but transformants are very unstable both mitotically

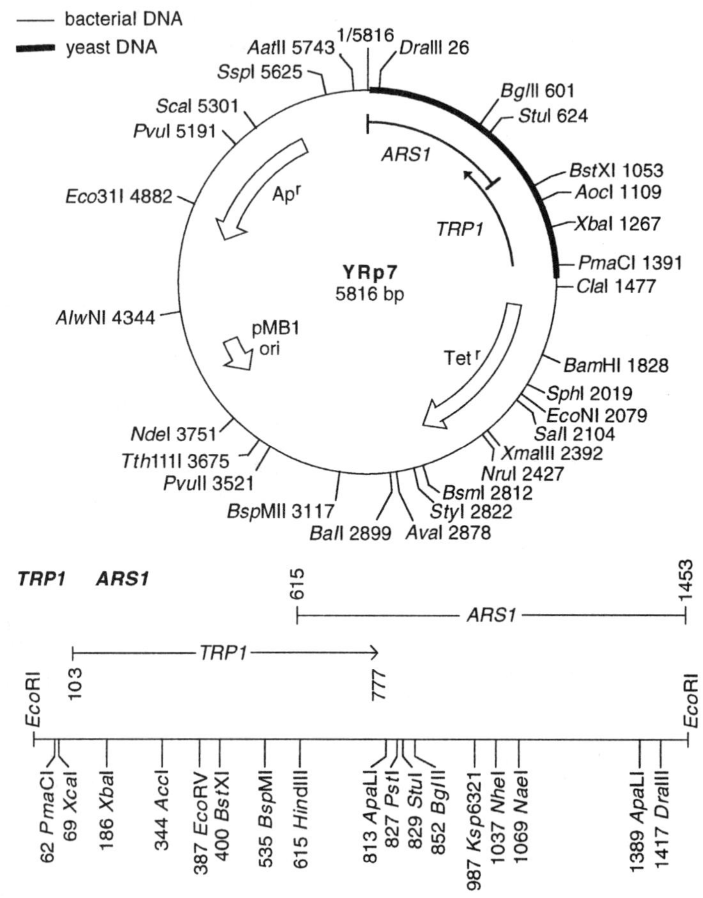

Figure 13.4.2 YRp7. This plasmid contains the 1453-bp *Eco*RI *TRP1 ARS1* fragment from *S. cerevisiae* inserted into the *Eco*RI site of pBR322 (Struhl et al., 1979).

The *TRP1* RI circle is a derivative of YRp7 containing only the 1453-bp *TRP1 ARS1 Eco*RI fragment. This plasmid is mitotically and meiotically unstable, but is present in 100 to 200 copies per plasmid-bearing cell in both cir$^+$ and cir^o strains.

A genomic plasmid bank has been constructed by inserting size-selected *Sau*3A partial fragments into the *Bam*HI site of YRp7 (Nasmyth and Reed, 1980).

The *TRP1 ARS1* 1453-bp *Eco*RI fragment contains both the *TRP1* gene, encoding N-(5′-phosphoribosyl)-anthranilate isomerase, and the autonomous replication sequence *ARS1*. The intact *TRP1* gene can complement mutations in the *trpC* gene of *E. coli*, using *E. coli* JA300 (*thr1 leuB6 thi1 thyA trpC*1117 *hsr*$_k^-$ *hsm*$_k^-$ *str*^{r}; Tschumper and Carbon, 1982). The chromosomal replicator *ARS1* lies between positions 615 and 1453 (on a *Hin*dIII-*Eco*RI fragment) and is composed of three domains. Domain A contains an 11-bp core sequence (position 857 to 867) consisting of a consensus sequence found in many other *ARS* elements and which is essential for *ARS1* function. Domains B and C flank Domain A and are relatively AT-rich regions that contribute to, but are not essential, for *ARS* function.

and meiotically. Despite the fact that *ARS*-containing plasmids replicate only once during the cell cycle, YRp plasmids can be present in high copy number (up to 100 copies per plasmid-bearing cell, although the average copy number per cell is 1 to 10). During mitosis, both instability and high copy number are due to a strong bias to segregate to the mother cell, to the extent that as few as 5% to 10% of cells grown selectively still retain the plasmid.

Incorporation of DNA segments from yeast centromeres (*CEN* elements) into YRp plasmids, to generate vectors called YCp plasmids (yeast centromeric plasmids),

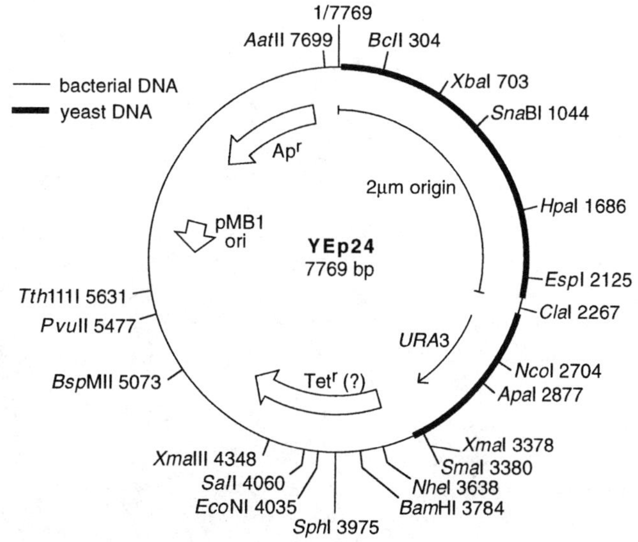

Figure 13.4.3 YEp24. YEp24 has the 2.2-kb *Eco*RI fragment of the B form of the 2μm plasmid and the 1.1-kb *Hin*dIII *URA3* gene inserted into the *Eco*RI and *Hin*dIII sites, respectively, of pBR322 (Botstein et al., 1979). The expression of the tet^r gene is variable among different isolates of this plasmid. YEp24 is mitotically stable in cir$^+$ strains at a copy number of about 20 but is unstable in cir^o strains. The complete sequence of YEp24 is available from the Vecbase database (file name: Vecbase.Yep24) and a detailed restriction map can be found in the New England Biolabs catalog.

greatly increases plasmid stability during mitosis and meiosis. Such plasmids—present in 1 to 2 copies per cell—have a loss rate of approximately 1% per generation and show virtually no segregation bias. During meiosis, *CEN* plasmids behave like natural chromosomes, generally segregating in a $2^+:2^-$ ratio.

The last class of circular replicating plasmids, YEp vectors (yeast episomal plasmids), contain sequences from a naturally occurring yeast plasmid called the 2μm circle. These 2μm sequences allow extrachromosomal replication and confer high transformation frequencies (~10^4 to 10^5 transformants/μg DNA). These plasmids are commonly used for high-level gene expression in yeast, due to their ability to be propagated relatively stably through mitosis and meiosis in high copy number. YEp vectors vary in the portion of 2μm DNA that they carry, although most carry only the sequences essential for autonomous replication. If the 2μm-encoded *REP1* and *REP2* functions are present (either on the YEp plasmid or due to the presence of endogenous 2μm circles), transformants are relatively stable and present in high copy number (20 to 50 copies). In the absence of *REP* functions, transformants are much more unstable, with a segregation bias and copy number similar to those observed with YRp plasmids. In the 2μm circle, a highly efficient site-specific recombination event occurs between two perfect 599-bp inverted repeats, mediated by the 2μm–encoded *FLP* gene. Most YEp plasmids carry at least one copy of this repeat; thus, FLP-mediated recombination between YEp vectors and other plasmids carrying one or more of these repeats (either the endogenous 2μm plasmid or other shuttle vectors) can result in a variety of recombinant plasmid multimers.

YLp plasmids (yeast linear plasmids) contain certain G-rich repeated sequences at their termini which function as telomeres and allow the plasmid to replicate as a linear molecule. In yeast, the telomeric sequence consists of tandem repeats of the

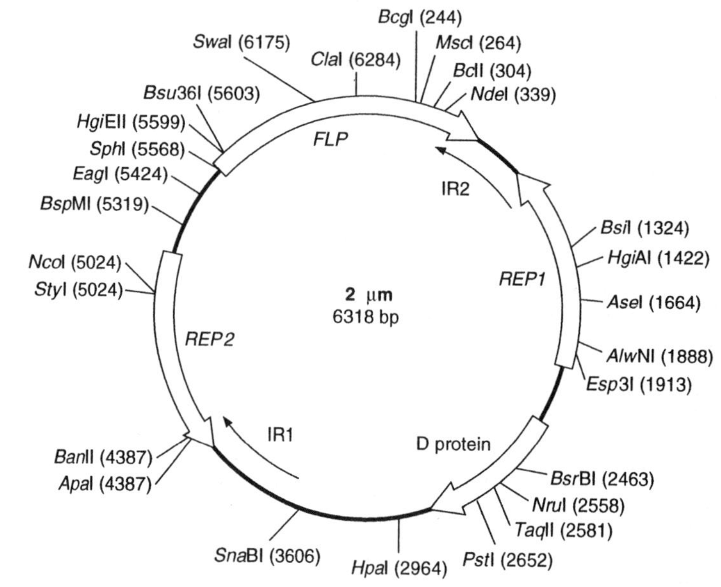

Figure 13.4.4 2μm plasmid. The 2μm circle is a naturally occurring DNA plasmid found in almost all strains of *S. cerevisiae*, with a copy number of ~20 to 80. The plasmid exists in two different forms, A and B (the former is shown above), due to intra-molecular recombination between two perfect 599-bp inverted repeats. Strains that carry this plasmid are called cir^+; strains missing the plasmid cir^0 have been identified or isolated (*see UNIT 13.9*). It is extremely stable mitotically, with a spontaneous loss rate in haploid cells of 10^{-4} per generation; during meiosis the plasmid is transmitted to all four spore products. The plasmid has been completely sequenced (Hartley and Donelson, 1980) and the sequence is available from GenBank (Plant: yscplasm).

sequence 5′($dG_{1\text{-}3}dT$)3′. Very short *CEN*-containing YLp plasmids (10 to 15 kb) are unstable and present in high copy number due to random segregation during mitosis. Increasing the size to 50 to 100 kb produces YLp vectors that disjoin from each other in a manner similar to that of natural chromosomes, resulting in a copy number of about one per cell. However, these artificial chromosomes—which are lost at a rate of 10^{-2} to 10^{-3} per cell division—are still ~100-fold less stable than a natural yeast chromosome.

MAPS OF SELECTED PLASMIDS AND GENES

Restriction maps and a brief description of selected plasmid vectors from several of the five general classes are presented in Figures 13.4.1, 13.4.2, 13.4.3, 13.4.4, 13.4.5, 13.4.6, and 13.4.7. These plasmids were chosen because they are used by many investigators and are generally applicable for a wide variety of purposes. However, where different selectable markers or unique restriction sites may be required, the reader is referred to two reviews on vector systems used in yeast (Parent et al., 1985; Pouwels et al., 1985). In addition, a method for constructing new plasmids in vivo in yeast has been described and employed to construct an extended series of new YRp, YCp, and YEp plasmids. Yeast shuttle vectors have also been constructed that are derived from either pUC18 or the Bluescript plasmids, providing a greater variety of unique cloning sites and allowing both identification of recombinant plasmids by screening for alpha-complementation of the *lacZ–M15* mutation of *E. coli* (see UNIT 1.4) and the ability to produce single-stranded DNA.

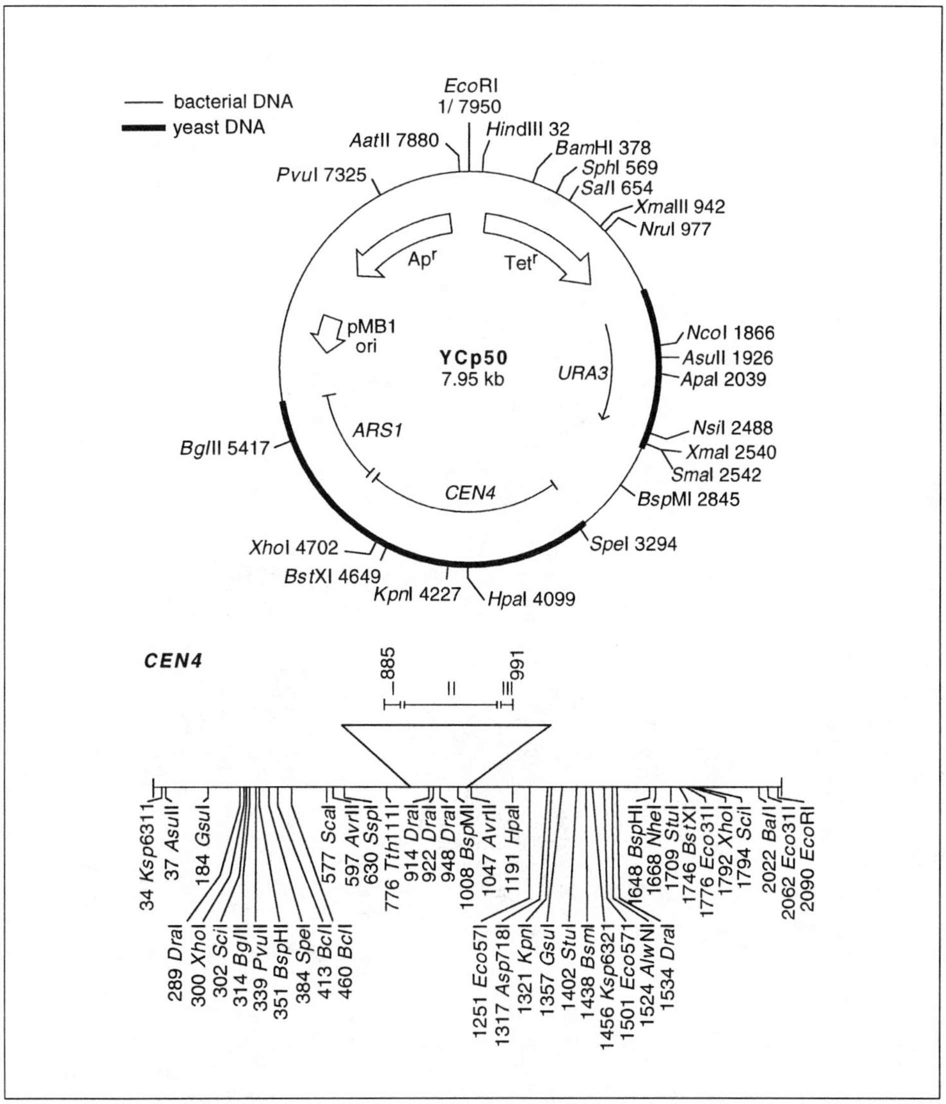

Figure 13.4.5 **YCp50.** This vector is a derivative of YIp5 and YCp19. The *Eco*RI site of YCpl9 was removed (producing an unsequenced deletion of about 190 bp) and a *Pvu*II-*Hin*dIII fragment (containing *CEN4* and *ARS1*) from this derivative was cloned into the *Pvu*II site of YIp5, with loss of the *Pvu*II site (Rose et al., 1987). Due to the presence of the *CEN* element, this plasmid exists in low copy in yeast (1 to 2 copies/cell) and is mitotically stable (<1%loss per cell per generation). This plasmid has not been completely sequenced; a more complete restriction map is available in Rose et al., 1987.

A set of genomic plasmid banks using YCp50 and size-selected DNA fragments has been constructed (Rose et al., 1987). These plasmid banks provide an alternative to genomic libraries constructed in high-copy-number vectors, useful when isolating genes that would be lethal in yeast when present in high copy.

Both the *CEN3* (Fig. 13.4.7) and *CEN4* (above) sequences were identified based on their ability to confer mitotic stability and proper meiotic segregation to autonomously replicating plasmids (Fitzgerald-Hayes et al., 1982; Mann and Davis, 1986). Nucleotide sequence comparison combined with functional analysis has shown that centromeres contain three conserved structural elements. Elements I and III show the highest degree of sequence conservation between different centromeres, and are separated by an extremely AT-rich region of about 90 bp, designated Element II. Full *CEN4* activity is contained within the 850-bp *Pvu*II-*Hpa*I fragment (which contains Elements I, II, and III), although the adjacent 905-bp *Hpa*I-*Eco*RI fragment also confers some mitotic stability to unstable *ARS*-containing plasmids (Mann and Davis, 1986).

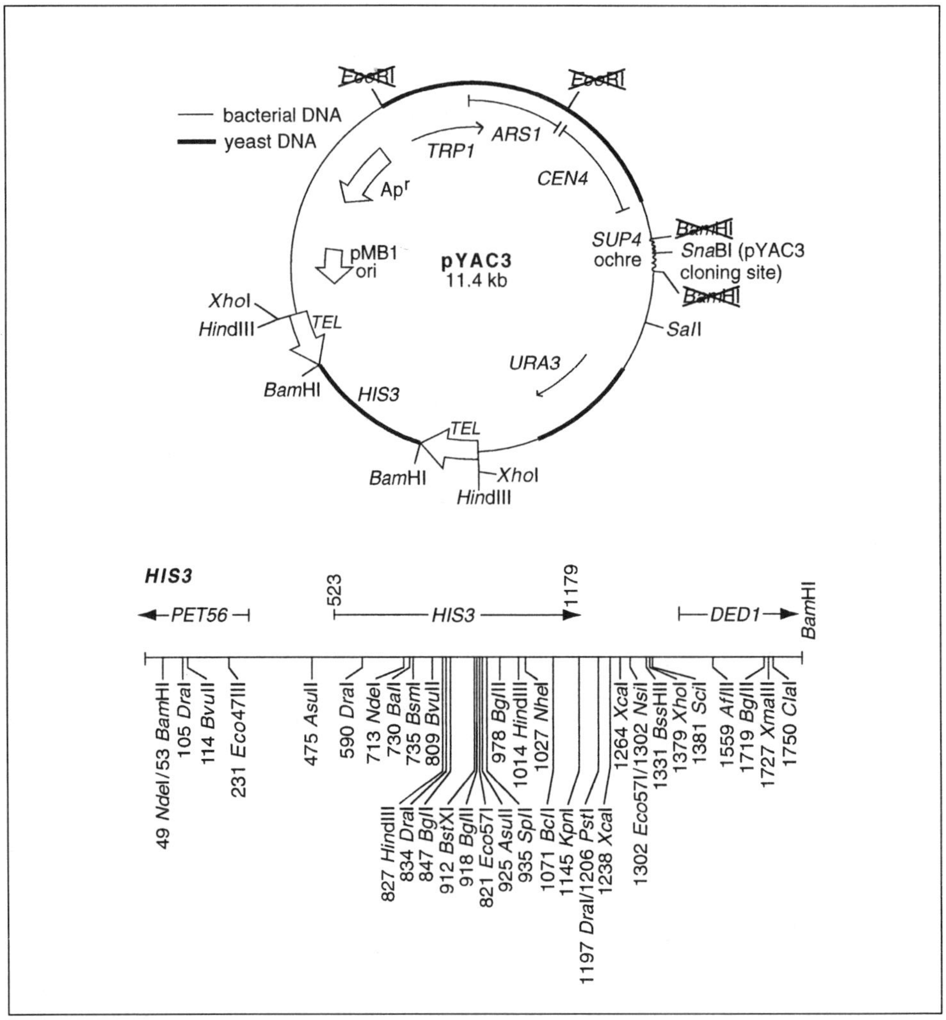

Figure 13.4.6 pYAC3. pYAC vectors are used to clone very large fragments of exogenous DNA onto artificial linear chromosomes, which can be stably maintained in yeast. This vector, which can be propagated as a circular plasmid in *E. coli*, contains a unique cloning site in the *SUP4* gene (an ochre-suppressing allele of a tyrosine tRNA), as well as *ARS1* and *CEN4* elements, required for stable single-copy propagation of the artificial chromosome. The *TEL* sequences are derived from *Tetrahymena* telomeres and have been shown to function as telomeres in yeast. To clone an insert, pYAC3 is digested with *Bam*HI (which cuts adjacent to the telomere sequences) and *Sna*BI; the resulting vector arms (containing either *TRP1*, *ARS1*, and *CEN4* or *URA3*) are ligated to insert fragments with *Sna*BI-compatible ends. The resulting ligation products are transformed into a $ura3^-$ $trp1^-$ ade2-1 yeast strain, using the spheroplast protocol, selecting for Ura^+ and subsequently screening for Trp^+ (to insure that both vector arms are present). Transformants can be further screened for the presence of inserts in the middle of the *SUP4* gene by using a color assay: colonies in which the *ade2-1* ochre mutation is suppressed by *SUP4* are white, whereas inactivation of the suppressor results in red colonies.

The pYAC vector shown above is one of a collection of three plasmids, each with a different cloning site inserted into the *SUP4* gene: pYAC4 and pYAC5 contain *Eco*RI and *Not*I sites, respectively, in place of the *Sna*BI site found in pYAC3. Selected restriction sites (not necessarily unique) are shown for pYAC3, as well as sites that have been destroyed in the process of plasmid construction. The *SUP4* gene is shown as a wavy line. For more detailed discussion of the cloning protocol, as well as details of the construction of this vector, see Burke et al., 1987.

The *HIS3* gene encodes imidazoleglycerolphosphate (IGP) dehydratase, which catalyzes a step in the histidine biosynthetic pathway. This 1822-bp fragment also contains a portion of two other genes: *pet56*, required for mitochondrial function, and *ded1*, required for cell viability (Struhl, 1985). Mutations in the *hisB* gene of *E. coli* can be complemented by the cloned *HIS3* yeast gene, using *E. coli* BA1 (*thr1 leuB6 trpC*1117 *hisB463*Tn*10*::near *hisB thi1 thyA* hsr_k^- hsm_k^- str^r; Murray et al., 1986).

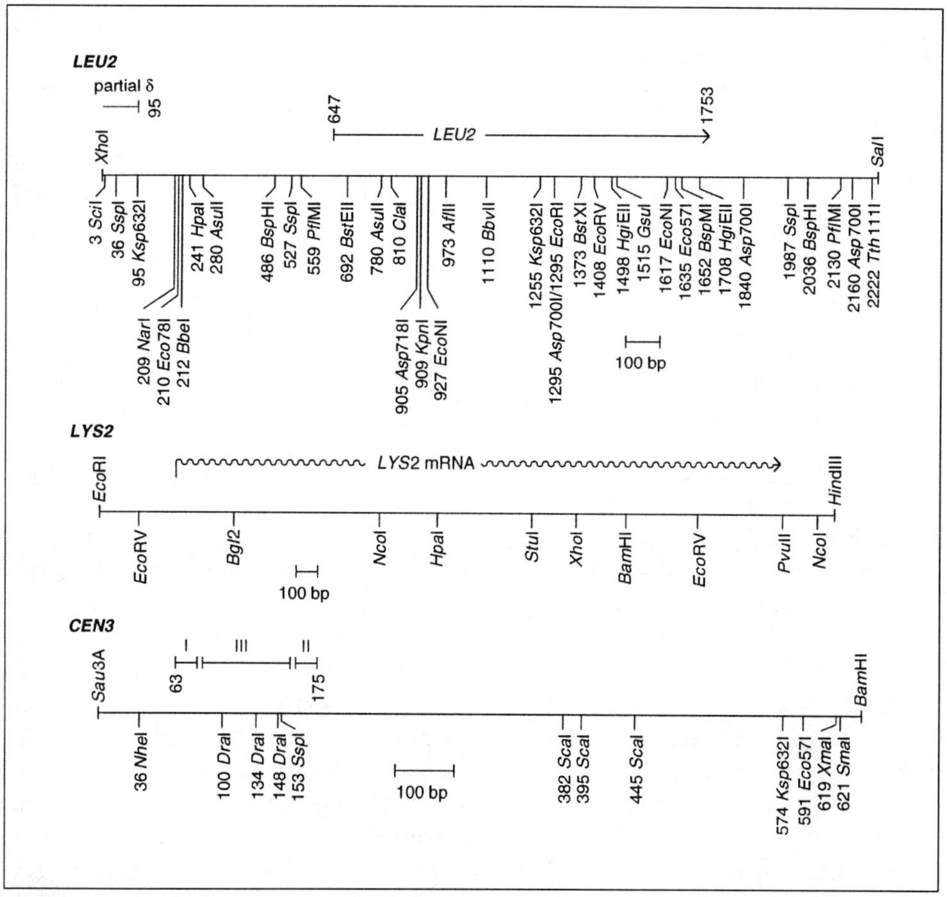

Figure 13.4.7 The *LEU2* gene encodes β-isopropylmalate (β-IPM) dehydrogenase, which catalyzes the third step in leucine biosynthesis (Andreadis et al., 1982). Unlike several other yeast genes involved in amino acid biosynthesis, *LEU2* (and *LEU1*, which is coordinately regulated with *LEU2*) is under specific amino acid control: gene expression is repressed by elevated concentrations of leucine. The *leu2-d* allele is a deletion of the 5′-flanking region of the *LEU2* message which leaves only 29 bp preceding the *LEU2* intiation codon; this derivative of the *LEU2* gene, when present on a YEp plasmid, requires a very high plasmid copy number to give a Leu^+ phenotype and has been used to cure cir^+ strains of the endogenous 2μm plasmid (see *UNIT 13.9*). Also contained in this 2230-bp *XhoI*-*SalI* fragment are 95 bases of the 330-nucleotide δ element. Although this δ element diverges in sequence from other δ elements, when the entire 2230-bp fragment is used as a probe of genomic yeast DNA δ elements present elsewhere in the genome will be detected at a low level. The cloned *LEU2* gene can complement mutations in the *leuB6* gene of *E. coli* using the strain JA300 (*thr1 leuB6 thi1 thyA trpC*1117 hsr_K^- hsm_K^- str^r; Tschumper and Carbon, 1982).

The *LYS2* gene is the structural gene for α-aminoadipate reductase, which catalyzes an essential step in lysine biosynthesis. The gene, which has not yet been sequenced, is present on a 4.6-kb *Eco*RI-*Hin*dIII genomic fragment, and gives rise to a 4.2-kb *LYS2* transcript, which is under general amino acid control (Eibel and Philippsen, 1983; Barnes and Thorner, 1986). Much larger genomic fragments (up to 15.7 kb) containing the *LYS2* gene have been isolated, providing a large variety of restriction sites flanking the gene for cloning purposes. As with *URA3*, a positive selection for $lys2^-$ mutants exists: such mutants can be selected on medium containing α-aminoadipatic acid and lysine, with a spontaneous frequency of 10^{-5} to 10^{-6}. Because the pathways for lysine biosynthesis in bacteria and fungi are not the same, no *E. coli* mutations can be complemented by the cloned *LYS2* gene.

See the legend to Figure 13.4.5 for a discussion of *CEN3*.

Finally, a vector system has been designed using yeast artificial chromosome (pYAC; Fig. 13.4.6) plasmids, allowing direct cloning into yeast of contiguous stretches of DNA up to 400 kb. The circular pYAC plasmids (without inserts) can replicate in *E. coli.* In vitro digestion of the pYAC vector, ligation to exogenous DNA, and direct transformation of the subsequent linear molecules (with telomeric sequences at each terminus) into yeast generate a library that can then be screened by standard techniques.

References: Burke et al., 1987; Ma et al., 1987; Parent et al., 1985; Pouwels et al., 1985.

Contributor: Victoria Lundblad

13.5

UNIT 13.5

Yeast Vectors For Expression of Cloned Genes

This unit describes vectors used for expression of proteins in *S. cerevisiae*. Yeast is often used industrially to produce large amounts of desired proteins because the cells can be grown (fermented) cheaply in large amounts. Because so many good (perhaps better) alternative ways to synthesize working quantities of recombinant gene products are available, this use is not common in academic labs. Most yeast expression studies are done for another reason. A number of insights in molecular biology have come from studies in which proteins from other eukaryotic and prokaryotic species have been expressed in yeast. Frequently, expression of the foreign protein perturbs the cell in a way that illuminates some aspect of eukaryotic biology. Such studies have, for example, made major contributions to our understanding of transcription activation and oncoprotein function. The expression vectors described in this unit are all plasmids, and they are all shuttle vectors—i.e., they contain replicators and genetic markers that allow their selection and maintenance either in *S. cerevisiae* and *E. coli*. The drug resistance markers and replicators that allow selection and maintenance in *E. coli* are standard, whereas those that allow use in yeast are more specialized (Table 13.5.1).

Selectable markers. The selectable markers in common use are wild-type genes such as *URA3, LEU2, HIS3,* and *TRP1.* These genes complement a particular metabolic defect (nutritional auxotrophy) in the yeast host (*UNIT 13.4* and Table 13.4.2). Thus, in contrast to the antibiotic resistance markers commonly used in bacterials plasmids, these markers must be used in conjunction with special host strains that carry the appropriate complementable mutations. Dominant markers that confer resistance to fungicides such as benomyl, or eukaryotic poisons such as G418, have been reported (and patented), but have not been incorporated into widely used expression vectors.

Replicators. Most expression vectors use a replicator derived from the 2μm circle, a naturally occurring plasmid native to most lab strains of *S. cerevisiae* (see *UNIT 13.4*). The native 2μm circle contains DNA sites and genes whose products ensure proper copy number control and proper segregation (partitioning) into daughter cells, but these sequences are typically missing from 2μm-derived vectors. As a result, 2μm based plasmids are typically maintained at a copy number of 10 to 40 per cell. Such 2μm derivatives do not exhibit plasmid incompatibility, allowing several different plasmids with different selectable markers to be maintained in the same cell. The high copy number results in relatively high gene dosage that favors efficient expression, and 2μm-based plasmids can attain an even higher copy number if they carry the *leu2-def* allele (see *UNIT 13.9*) as a selectable marker. This allele is a reduction-in-function mutant version of *LEU2* that is so severe that 100 to 200

Table 13.5.1 Yeast expression vectors[a]

UAS	Vector	Cloning sites	Marker	Origin	ATG	Terminator sequence
Inducible						
GAL1-*GAL10*	pBM150	*Eco*RI, *Bam*HI	*URA3*	*CEN4* *ARS1*	No	—
	pYEp51	*Sal*I to *Bam*HI, *Hin*dIII or *Bcl*I	*LEU2*	2μm	No	—
	pLGSD5	*Bam*HI	*URA3*	2μm	Yes	—
	YEp62	*Sal*I, *Bam*HI or *Sma*I	*LEU*	2μm	Yes	—
PHO5	pAM82	*Xho*I	*LEU2*	2μm	No	—
	pYE4	*Eco*RI	*TRP1*	2μm	No	*TRP1*
Constitutive						
ADHI	pAAh5	*Hin*dIII	*LEU2*	2μm	No	*ADHI*
	pMA56	*Eco*RI	*TRP1*	2μm	No	—
	pAH9/10/21	*Hin*dIII	*LEU2*	2μm	Yes	—
PGK	pMA230	*Bam*HI	*LEU2*	2μm	Yes	—
	pMA91	*Bgl*II	*LEU2*	2μm	No	*PGK*
GPD	pG-1/2	Polylinker	*TRP1*	2μm	No	*PGK*

[a]Adapted with permission from Academic Press (Schneider and Guarente, 1991).

copies per cell are needed to complement a *leu2* auxotrophy and allow growth in the absence of leucine.

Expression cassettes. In addition to the common elements described above, the vectors described in this unit also contain regions of DNA known as expression cassettes. These regions include a yeast promoter and one or more downstream restriction sites for insertion of desired protein-coding sequences, and may also contain translation-initiation sequences, transcription terminators, and coding sequences specifying useful protein moieties.

Promoters. The vectors described here carry strong promoters utilized by RNA polymerase II. The promoters can be either inducible (e.g., *GAL, GAL10, PHO5*) or constitutive (see Table 13.5.1). As in other eukaryotes, transcription from these promoters depends on activator proteins bound to sites upstream of the transcription start site (in yeast, termed upstream activation sites or UAS). As always, use of these promoters for protein expression often benefits from some knowledge of the molecular genetics of their regulation.

Inducible promoters are used to give conditional expression of the plasmid-encoded protein—e.g., if the gene product is toxic to the cell. The *GAL* and *PHO5* promoters are regulated by analogous sets of proteins: activators (GAL4 and PHO4, respectively) that bind to UAS upstream of the transcription start, and negative regulators (GAL80 and PHO80 products) that suppress activation by the activator. Transcription from the *GAL1* (galactokinase) promoter, for example, occurs when cells are grown in glucose-containing medium (<1 transcript/cell), because GAL80 forms a complex with GAL4 and thereby obscures a transcription-activating region, and probably also because GAL4 is not able to bind its site when the cells are grown in glucose. *GAL1* transcription is massively induced [~1% poly(A)$^+$ mRNA] when cells are grown in medium that contains galactose as the sole source of carbon; under these conditions GAL80 dissociates from GAL4, and GAL4 is bound to the *GAL1* UAS. Similarly, transcription from the *PHO5* (alkaline phosphatase) promoter is low (repressed) when cells are grown in medium that contains sufficient inorganic phosphate, but is induced when cells are shifted to a medium that lacks phosphate.

Genetic manipulations can allow induction of these promoters without changes in the growth medium. For example, *PHO5* transcription can be induced in strains grown in the presence of sufficient phosphate if the strains carry loss-of-function mutations in *PHO80*, and conditional (temperature-sensitive) loss-of-function mutations in *PHO4*. In such strains, growth at high temperature does not allow transcription, because the activator is not active. A shift to the permissive temperature restores PHO4 protein function, and the *PHO80* mutation ensures that the *PHO4* protein is functional even if the medium contains phosphate. Thus, such strains can be grown initially at high temperature without expression of the cloned gene, then shifted to low temperature to allow production.

Constitutive promoters are also often employed. The *ADH1* (alcohol dehydrogenase I) promoter is probably the most popular, but the *TPI* (triose phosphate isomerase) and *PGK* (3-phosphoglycerate kinase) promoters are also commonly used. Transcription from any of these typically results in mRNA that constitutes >1% of total mRNA in the cell. As for the inducible promoters, transcription of these promoters depends on activator proteins bound to upstream sites, and the activity of these activators typically varies with changes in the metabolic state of the cells, although less drastically than that of the inducible promoters. Thus, despite the fact that these promoters are said to be constitutive, their activity varies with changes in the growth medium; in particular, all of the above promoters are less active when the yeast is growing on a carbon source other than glucose.

Leaders. As in other organisms, the translation efficiency in yeast is influenced by DNA sequences upstream of the gene coding sequence. However, the relationship between the sequence of untranslated mRNA leader sequences and translation efficiency in yeast is not very well understood, and a body of largely anecdotal evidence suggests that sequence requirements are not very restrictive. Typically, the only step taken to ensure proper translation is to make sure that the ATG that begins the gene's coding sequence encodes the first AUG in the mRNA. However, ATG fusion vectors (like those described in *UNIT 16.4*) containing untranslated sequences from well-expressed genes are now available in some systems (Table 13.5.1).

Terminators. By contrast, most yeast expression vectors do carry "transcription terminators" downstream from the promoter and the cloning sites. As their name implies, these sequences are thought to cause RNA polymerase II to cease transcribing. While this has not been completely established, it is clear that their presence in transcribed regions causes formation of 3′-ended, polyadenylated transcripts. More importantly, there is some evidence that the presence of transcription terminators downstream of a promoter increases the total amount of message (perhaps by stabilizing it) as well as the total amount of protein expression.

Useful protein moieties. Many expression vectors also carry, downstream of the promoter, DNA that contains an ATG and that encodes a useful protein sequence. Examples of such moieties include the following: signal sequences, to direct secretion of the expressed protein into the extracellular medium, nuclear localization sequences, to direct its transport into the nucleus; and epitope or other protein tags such as those from Myc, influenza virus hemaglutinin, or the *E. coli TrpE* gene product, to facilitate its immunological detection and purification. Recently, vectors have been described that allow the expression of the protein fused to a eukaryotic transcription activation region; such fusion proteins are used to detect and study protein-protein interactions.

Reference: Schneider and Guarente, 1991.

Contributors: Roger Brent and Kaaren Janssen

Yeast Vectors and Expression Assays

UNIT 13.6

LACZ FUSION VECTORS FOR STUDYING GENE REGULATION

BASIC PROTOCOL 1

Because of the ease and sensitivity of the β-galactosidase assay, yeast genes are often "tagged" with a functional portion of the *lacZ* gene, to monitor the regulation of expression of the yeast gene in question. These fusions are constructed such that the promoter region of the yeast gene—plus several amino acids from the N terminus of the protein encoded by this gene—is fused to the carboxy-terminal region of the *lacZ* gene, which encodes a protein fragment that still retains β-galactosidase activity. When constructing *lacZ* fusions, it is crucial that the translational reading frame across the fusion junction is maintained. A more detailed discussion of construction of in-frame fusions can be found in Guarente, 1983.

The plasmid pLG670-Z (Fig. 13.6.1) can be used for constructing *lacZ* fusions (Guarente, 1983). Two other plasmids (pLG200 and pLG400)—containing different translational reading frames and/or different unique restriction sites at the 5′ end of the *lacZ* fragment—have also been constructed (Guarente et al., 1980). These two plasmids do not contain yeast selectable genes or yeast replication origins but can be used to first construct an in-frame fusion, which is then transferred onto a yeast shuttle vector (*UNIT 13.4*).

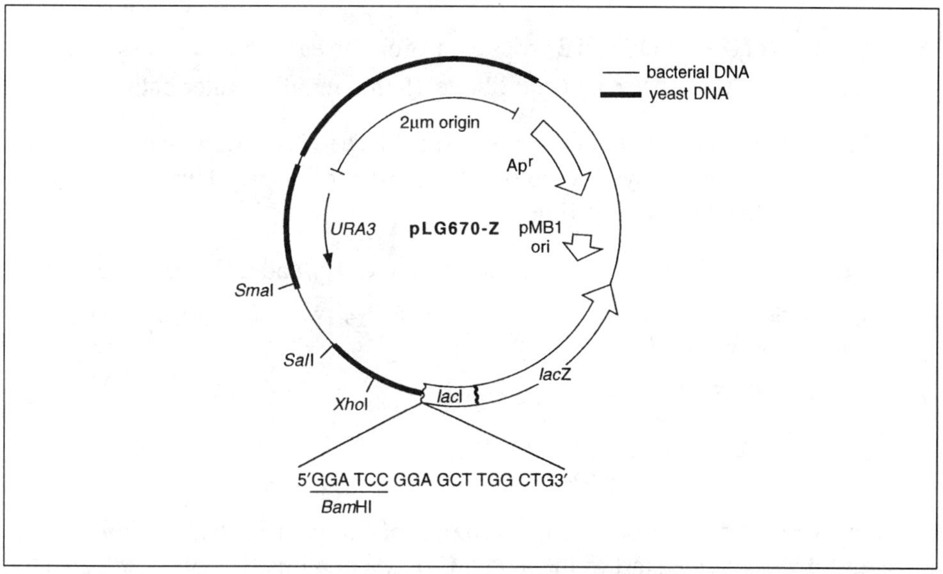

Figure 13.6.1 1pLG670-Z (derived from YEp24) contains the 2μm origin of replication, *URA3* as a selectable yeast gene, and a 3′ fragment of the *lacZ* gene with a unique *Bam*HI site at the 5′ end of the *lacZ* fragment. The sequence just after the *Bam*HI, as well as the translational reading frame, is shown.

BASIC PROTOCOL 2

ASSAY FOR β-GALACTOSIDASE IN LIQUID CULTURES

Materials (*see APPENDIX 1 for items with ✓*)

YPD medium (*UNIT 13.1*)
✓ Z buffer
0.1% sodium dodecyl sulfate (SDS)
Chloroform
4 mg/ml ONPG in 0.1 M KPO_4, pH 7.0 (filter sterilize and store frozen)
1 M Na_2CO_3
30°C water bath

1. Inoculate 5 ml YPD (or appropriate) medium with single yeast colony. Grow 2 to 3 independent single-colony cultures of a strain containing *lacZ* fusion protein overnight at 30°C. If fusion is on a plasmid, grow cells in medium that selects for plasmid.

2. Inoculate 5 ml YPD medium (or appropriate selective medium and/or inducing conditions) with 20 to 50 µl of each overnight culture. Grow to mid- or late-log phase: 0.5-1 × 10^8 cells/ml (OD_{600} = 2.0) for rich medium or 2-5 × 10^7 cells/ml (OD_{600} = 0.5 to 1.0) for minimal medium.

   If goal is to investigate activity of yeast promoter under conditions that induce its expression, grow cells in parallel under inducing and noninducing conditions.

3. Centrifuge cells 5 min at 1100 × *g*. Resuspend in equal volume of Z buffer and place on ice. Determine OD_{600} for each sample.

   If anticipated level of β-galactosidase activity is low, it may be necessary to concentrate cells. For cells demonstrating 100 to 1000 U of activity within 30 min to 4 hr, concentration is not necessary.

4. Set up the following two reaction tubes for each sample (1 ml each), with mixing: (a) 100 µl cells with 900 µl Z buffer and (b) 50 µl cells with 950 µl Z buffer.

5. Add 1 drop of 0.1% SDS and 2 drops chloroform to each sample using a Pasteur pipet. Vortex 10 to 15 sec and equilibrate 15 min in 30°C water bath.

6. Add 0.2 ml of 4 mg/ml ONPG; vortex 5 sec. Place in 30°C water bath and begin timing. When medium-yellow color (OD_{420} = 0.3 to 0.7) has developed, add 0.5 ml of 1 M Na_2CO_3 and note time.

7. Centrifuge cells 5 min at 1100 × *g*. Determine OD_{420} and OD_{550} of supernatant.

   If the cell debris has been well pelleted, the OD_{550} is usually zero and therefore is not necessary to read.

8. Calculate units with the following equation:

$$U = \frac{1000 \times [(OD_{420}) - (1.75 \times OD_{550})]}{(t) \times (v) \times (OD_{600})}$$

   where t = time of reaction (min); v = volume of culture used in assay (ml); OD_{600} = cell density at the start of the assay; OD_{420} = combination of absorbance by *o*-nitrophenol and light scattering by cell debris; and OD_{550} = light scattering by cell debris.

References: Guarente, 1983; Guarente et al., 1980.

Contributors: Ann Reynolds and Victoria Lundblad

Introduction of DNA into Yeast Cells

NOTE: All solutions and glassware coming into contact with yeast cells must be sterile. Traces of soap on glassware may decrease the transformation efficiency. In addition, the water used for washes and in solution preparation must be of the highest quality.

TRANSFORMATION USING LITHIUM ACETATE

BASIC PROTOCOL

This is the most commonly used method for yeast transformation. It is reasonably fast and provides transformation efficiencies of 10^5 to 10^6 transformations/µg.

Materials *(see APPENDIX 1 for items with ✓)*

YPD medium (*UNIT 13.1*)
Yeast strain to be transformed
YPAD medium: YPD medium supplemented with 30 mg/liter adenine hemisulfate
✓ Highest-quality sterile H_2O
✓ 10× TE buffer, pH 7.5, sterile
10× lithium acetate stock solution: 1 M lithium acetate, pH 7.5 (adjust pH with dilute acetic acid), filter sterilized
DNA: high-molecular-weight, single-stranded carrier DNA (Support Protocol) and transforming DNA
50% (w/v) PEG 4000 or 3350 (do not use PEG 8000), filter sterilized
CM dropout plates (*UNIT 13.1*) prepared with Difco agar

30°C incubator with shaker
Sorvall GSA and SS-34 rotors (or equivalents)
42°C water bath

1. Two days before the experiment, inoculate 5 ml YPD medium with a single yeast colony of the strain to be transformed. Grow overnight to saturation at 30°C.

2. The night before transformation, inoculate a 1-liter sterile flask containing 300 ml YPAD medium with an appropriate amount of the saturated culture and grow overnight at 30°C to 1×10^7 cells/ml ($OD_{600} \cong 0.3$ to 0.5, depending on strain). For 2- to 3-fold higher efficiency, dilute at this point to 2×10^6 cells/ml in fresh YPAD medium and grow for another 1 to 2 generations (2 to 4 hr).

   Because growth phase and cell density are important in achieving the highest transformation efficiency, the easiest approach is to inoculate three independent flasks with varying amounts of the saturated overnight culture (try 1, 5, and 25 µl for a 12-hr growth period), then check the OD_{600} of each flask the next day.

3. Harvest cells by centrifuging 5 min at 4000 × *g*, room temperature. Resuspend in 10 ml highest-quality sterile water. Transfer to a smaller centrifuge tube and pellet cells by centrifuging 5 min at 5000 to 6000 × *g*, room temperature.

4. Resuspend in 1.5 ml buffered lithium solution, freshly prepared as follows:

   1 vol 10× TE buffer, pH 7.5
   1 vol 10× lithium acetate stock solution
   8 vol sterile water.

5. For each transformation, mix 200 µg carrier DNA with ≤5 µg transforming DNA in a sterile 1.5-ml microcentrifuge tube. Keep total volume of DNA ≤20 µl.

Maximal transformation efficiency is achieved by repeating the denaturation cycle (boiling and chilling) of carrier DNA immediately prior to use.

6. Add 200 µl yeast suspension to each microcentrifuge tube. Add 1.2 ml PEG solution, freshly prepared as follows:

   8 vol 50% PEG
   1 vol 10× TE buffer, pH 7.5
   1 vol 10× lithium acetate stock solution.

   Shake 30 min at 30°C.

7. Heat shock exactly 15 min at 42°C. Microcentrifuge 5 sec at room temperature. Resuspend yeast in 200 µl to 1 ml of 1× TE buffer (freshly prepared from 10× stock) and spread up to 200 µl onto CM dropout plates made with Difco agar. Incubate 2 to 5 days at 30°C until transformants appear.

   Difco agar gives 3-fold higher transformation efficiency.

13.7

ALTERNATE PROTOCOL 1

SPHEROPLAST TRANSFORMATION

This method is more time-consuming than the lithium procedure, but can result in a higher efficiency of transformation per input DNA.

Materials *(see APPENDIX 1 for items with ✓)*

YPD medium and plates (*UNIT 13.1*)
Yeast strain to be transformed
1 M sorbitol
2-mercaptoethanol (2-ME)
Glusulase (Du Pont NEN)
✓ $CaCl_2$ solution
✓ Sorbitol/$CaCl_2$ solution
DNA: transforming DNA and 5 mg/ml carrier DNA (sheared calf thymus or salmon sperm DNA)
✓ PEG/$CaCl_2$ solution
✓ Selective regeneration agar
CM drop-out plates (*UNIT 13.1*)

30°C incubator with rotating platform
55°C water bath

NOTE: Before starting the transformation, melt selective regeneration agar (microwave at low setting), aliquot 10-ml samples into sterile glass test tubes, and place in a 55°C water bath.

1. Two days before the experiment, inoculate 5 ml YPD medium with a single yeast colony of the strain to be transformed. Grow overnight to saturation at 30°C (or lower temperature, if strain is temperature sensitive).

2. The night before transformation, inoculate a 250-ml sterile flask containing 50 ml YPD with an appropriate amount of the 5-ml culture and grow overnight to 1–2 × 10^7 cells/ml ($OD_{600} \cong 0.5$ to 1.0, depending on strain).

   The frequency of transformation drops with cell densities higher than this. If the exact inoculum for a given strain is not known, it is best to inoculate three independent flasks with varying amounts of the saturated 5-ml overnight culture (Basic Protocol, step 2).

3. Pellet cells 5 min at 1100 × *g*, room temperature, and resuspend in 10 ml of 1 M sorbitol. Repeat. Pellet cells once more and resuspend in 5 ml of 1 M sorbitol. Make 10^{-5} dilution in sterile water and plate 0.1 ml onto a YPD plate. Incubate 2 days at 30°C.

4. Transfer resuspended cells to a 50-ml sterile flask and add 5 µl 2-ME and 150 µl glusulase. Incubate ~30 to 60 min at 30°C with very gentle shaking. Plate 0.1 ml of a 10^{-3} dilution (made in sterile water) onto a YPD plate and incubate 2 days at 30°C.

   The number of colonies that grow up on this plate can be compared to the number that grew on the plate from step 3 to calculate the percentage of spheroplast formation. Each new lot of glusulase and each new strain should be assayed for the percentage of spheroplast formation.

   Spheroplast formation can also be monitored during the 30- to 60-min incubation with glusulase by observing the degree of lysis in water under a microscope (spheroplasts will lyse in a nonisotonic medium such as water). At 15-min intervals, remove 10 µl of the cells, place on a microscope slide, and cover with a cover slip. After focusing (at 240× to 400× magnification), gently touch a drop of water to the edge of the cover slip. As the water leaks under the cover slip, observe how many cells lyse. Spheroplasts should swell slightly before lysis; upon lysis, a "ghost" (membrane) should still be visible.

5. Transfer spheroplasts to a sterile 50-ml, round-bottom centrifuge tube and pellet 4 min at 400 × *g*, room temperature.

6. Gently decant supernatant without dislodging pellet. Add 2 ml of 1 M sorbitol; resuspend pellet by gently swirling liquid across surface of pellet (which should easily come off the side of the tube).

   DO NOT vortex or otherwise vigorously agitate spheroplasts. If pellet does not easily resuspend, reduce either the time or the rpm of subsequent spins.

7. Add 8 ml of 1 M sorbitol and pellet at 400 × *g*. Repeat steps 6 and 7.

8. Repeat step 6, add 7 ml of 1 M sorbitol and 1 ml $CaCl_2$ solution, mix gently by swirling and centrifuge 4 min at 400 × *g*. Resuspend in 1 ml sorbitol/$CaCl_2$ solution.

9. Mix 150 to 200 µl cells with the DNA to be transformed (up to 10 µg) plus 10 µl carrier DNA. Incubate 10 min at room temperature.

   The total volume of added DNA should be no greater than 1⁄10 the volume of the cells. For each transformation, it is advisable to mix an aliquot of cells with only carrier DNA, to monitor the frequency of reversion of the relevant mutation.

   The amount of transforming DNA to be added depends on several parameters, including the frequency with which the strain can be transformed (determined experimentally) and the state of the DNA (fragments of DNA without replication origins will transform several orders of magnitude less efficiently than intact YEp, YRp or YCp plasmids—see UNIT 13.4).

10. Add a 10-fold volume of PEG/$CaCl_2$ solution and thoroughly resuspend. Let sit 10 min at room temperature.

11. Pellet 4 min at 400 × *g* and decant the PEG-containing supernatant. Gently resuspend pellet in 0.5 ml of sorbitol/$CaCl_2$ solution and pipet into 10 ml melted regeneration agar tempered to 55°C. Vortex briefly to mix and immediately pour

onto the appropriate CM dropout (selective) plate, swirling the plate to further mix cells and agar together.

The 3% agar will quickly solidify at temperatures <55°C, so it is important to plate each 10-ml aliquot as rapidly as possible.

12. Incubate at 30°C (or other appropriate temperature) until colonies appear, both on the surface of the regeneration agar overlay and embedded in the agar. Pick individual colonies and streak for single colonies on the same selective plates. If the purpose of this transformation was to disrupt genomic sequences (*UNIT 13.10*), make DNA from several transformants in order to analyze the relevant chromosomal region by Southern blot analysis (*UNIT 13.11*).

ALTERNATE PROTOCOL 2

TRANSFORMATION BY ELECTROPORATION

Additional Materials *((also see Basic Protocol; see APPENDIX 1 for items with ✓)*

✓ 1 M dithiothreitol (DTT; filter sterilize and store at −20°C)
1 M sorbitol
✓ Sorbitol selection plates
Gene Pulser with Pulse Controller (Bio-Rad) *or* Cell-Porator (Life Technologies)
0.2-cm-gap disposable electroporation cuvettes (Bio-Rad) *or* 0.15-cm-gap microelectroporation chambers (Life Technologies); ice-cold

1. Two days before the experiment, inoculate 5 ml of YPD medium with a single yeast colony of the strain to be transformed. Grow overnight to saturation at 30°C.

2. The night before transformation, inoculate a 2-liter sterile flask containing 500 ml YPD with an appropriate amount of the saturated culture and grow overnight with vigorous shaking at 30°C to 1×10^8 cells/ml ($OD_{600} \cong 1.3$ to 1.5, depending on strain).

This cell density is achieved in mid- to late-log phase. If the exact inoculum for a given strain is not known, inoculate three independent flasks with varying amounts of the saturated culture (Basic Protocol, step 2).

3. Harvest culture by centrifuging at 4000 × *g*, 4°C, and resuspend vigorously in 80 ml sterile H_2O. To increase electrocompetence of the cells, proceed to step 4. If this treatment is not required, proceed to step 6.

Treatment with lithium acetate is optional. It increases the handling time in preparing the yeast and should not be performed if the cells are to be frozen for subsequent use.

4. Add 10 ml of 10× TE buffer, pH 7.5. Swirl to mix. Add 10 ml of 10× lithium acetate stock solution. Swirl to mix. Shake gently 45 min at 30°C.

5. Add 2.5 ml of 1 M DTT while swirling. Shake gently 15 min at 30°C.

This treatment increases the transformation efficiency >5-fold.

6. Dilute yeast suspension to 500 ml with water. Wash and concentrate the cells three times by centrifuging at 4000 to 6000 × *g*, 4°C, and resuspend the successive pellets as follows:

   First pellet—250 ml ice-cold water
   Second pellet—20 to 30 ml ice-cold 1 M sorbitol
   Third pellet—0.5 ml ice-cold 1 M sorbitol.

   Resuspension should be vigorous enough to completely dissociate each pellet. The final volume of resuspended yeast should be 1.0 to 1.5 ml and the final OD_{600} should be ~200.

Using the Bio-Rad Gene Pulser:

9a. In a sterile, ice-cold 1.5-ml microcentrifuge tube, mix 40 µl concentrated yeast cells with ≤100 ng transforming DNA contained in ≤5 µl.

Transforming DNA should be in a low-ionic-strength buffer such as TE or in sterile high-quality water. There is no required length of incubation; this time can be varied to convenience. Do not include carrier DNA in this procedure; it drastically reduces transformation efficiency. Maximal efficiency (transformants/µg) will be obtained with <10 ng of transforming DNA, while the largest number of transformants will be obtained with 100 ng.

10a. Transfer to an ice-cold 0.2-cm-gap disposable electroporation cuvette.

11a. Pulse at 1.5 kV, 25 µF, 200 Ω. It is crucial that the Bio-Rad Pulse Controller be included in the circuit; failure to do so will result in damage to the Gene Pulser.

The time constant reported by the Gene Pulser will vary from 4.2 to 4.9 msec. Times <4 msec or the presence of a current arc (evidenced by a spark and smoke) indicate that the conductance of the yeast/DNA mixture is too high.

12a. Add 1 ml ice-cold 1 M sorbitol to the cuvette and recover the yeast, with gentle mixing, using a sterile 9-inch Pasteur pipet.

13a. Spread aliquots of the yeast suspension directly on sorbitol selection plates. Incubate 3 to 6 days at 30°C until colonies appear.

Using the Life Technologies Cell-Porator:

9b. In a sterile, ice-cold 1.5-ml microcentrifuge tube, mix 20 µl concentrated yeast with ≤100 ng transforming DNA contained in ≤5 µl.

Incubation time can be varied for convenience (step 9a).

10b. Transfer to an ice-cold, 0.15-cm-gap micro-electroporation chamber.

11b. Pulse at 400V, 10 µF, low resistance.

12b. Remove 10 µl of electroporated mixture to a sterile 1.5-ml microcentrifuge tube containing 0.5 ml ice-cold 1 M sorbitol.

13b. Spread aliquots of the yeast suspension directly on sorbitol selection plates. Incubate 3 to 6 days at 30°C until colonies appear.

SUPPORT PROTOCOL

PREPARATION OF SINGLE-STRANDED HIGH-MOLECULAR-WEIGHT CARRIER DNA

Addition of denatured high-molecular-weight carrier DNA is critical to achieving high transformation efficiency using the lithium acetate ~protocol. Carrier DNA prepared according to this Support Protocol may also be used for spheroplast transformations, although single-stranded DNA has been reported to give a slightly lower transformation frequency than the more easily prepared double-stranded calf thymus or salmon sperm carrier DNA.

Materials *(see APPENDIX 1 for items with ✓)*

DNA (type III sodium salt from salmon testes; Sigma)
✓ 1× TE buffer, pH 8.0
Buffered phenol (*UNIT 2.1*)
1:1 (v/v) phenol/chloroform
Chloroform
✓ 3 M sodium acetate, pH 5.2
100% ethanol, ice-cold
Probe sonicator

13.7

1. Dissolve DNA in 1× TE buffer, pH 8.0, to a final concentration of 10 mg/ml. Stir overnight at 4°C.

   Each individual transformation requires 200 μg; it is useful to prepare a large amount (e.g., 1 g) and freeze aliquots at −20°C.

2. Sonicate using a large probe ~30 sec at 75% power until viscosity appears to decrease slightly. Run 1 μg on a 0.8% agarose gel to determine size distribution of sonicated fragments. Repeat sonication as necessary to achieve the appropriate size distribution—fragments that range from 2 to 15 kb, with a mean size of ~7 kb.

3. Extract once with buffered phenol, once with phenol/chloroform, and once with chloroform.

4. Precipitate DNA with ¹⁄₁₀ vol of 3 M sodium acetate, pH 5.2, and 2.5 vol ice-cold 100% ethanol.

5. Resuspend pellet in 1× TE buffer to 10 mg/ml final concentration.

   Resuspension may require stirring overnight at 4°C.

6. Transfer carrier DNA to a Pyrex flask. Microwave to a rolling boil and continue to boil 2 to 3 min.

7. Chill flask rapidly in ice water. Aliquot and freeze DNA in sterile tubes at −20°C.

References: Becker and Guarente, 1991; Hinnen et al., 1978

Contributors: Daniel M. Becker and Victoria Lundblad

Cloning Yeast Genes by Complementation

UNIT 13.8

BASIC PROTOCOL

This unit presents a generalized protocol using a hypothetical mutation of yeast for cloning yeast genes by complementation. The mutation, *cdc101-1*is both recessive and temperature-sensitive for growth: it can grow relatively normally at 23°C but is unable to make a colony at 37°C. A genomic DNA clone that complements this mutation will be isolated by transforming the *cdc101-1* strain with a yeast genomic library and subsequently screening for temperature-resistant colonies. Once isolated, two steps are necessary to prove that the insert present on the plasmid contains the wild-type *CDC101* gene. First, segregation of the complementing plasmid must result in co-loss of both the plasmid-borne selectable marker and the complementing phenotype, demonstrating that the observed complementation is plasmid-specific and is not due to reversion of the *cdc101-1* mutation. Second, it must be ruled out whether the cloned gene encodes a phenotypic suppressor of the mutation, rather than the wild-type gene. This is done via a complementation test, which demonstrates whether or not a disruption of the cloned gene that is integrated into the genome can complement the original mutation.

13.8

Cloning the Gene

1. Transform *leu2*$^-$ *cdc101-1* yeast strain with yeast genomic DNA library containing *LEU2* as selectable marker (*UNIT 13.7*) and selecting, in this case, for Leu$^+$ at 23°C. Depending on insert size, screen between 2,000 and 20,000 transformants (~4 to 8 genome equivalents).

   If mutation reverts at a very low frequency (10^8), transformants that complement the mutant phenotype (in this case, temperature sensitivity) can be selected directly after transformation. However, if mutation reverts at higher frequencies (or if complementing phenotype cannot be directly selected), the mutation must be present in a strain that also contains a low or nonreverting mutation in the same selectable gene present on library vector.

2. Replica plate transformants onto prewarmed selective plates (leucine dropout plates; *UNIT 13.1*) and incubate at 37°C to screen for complementation of temperature-sensitive phenotype. With overnight incubation, colonies containing plasmids with inserts that complement the *cdc101-1* mutation, as well as potential *CDC101*$^+$ revertants that contain random noncomplementing plasmids, will grow up. Restreak each single colony that appears on this plate and save for further analysis.

   Many mutant phenotypes, including ts phenotypes, cannot be distinguished by replica plating (presumably due to leakiness of the mutation). If this is the case, transformants must be recovered from the original transformation plate and replated directly on a second plate that screens for the mutant phenotype.

   Recover transformants by pipetting 0.5 ml sterile water onto each plate and resuspending individual colonies in liquid by mushing about plate with sterile spreader (UNIT 1.3). Pipet the 0.5 ml back into sterile tube and replate for single colonies onto prewarmed selective plates. If mutation displays more than one phenotype, complementation of additional phenotypes should also be tested. Complementation of some but not all of mutant phenotypes indicates that the wild-type gene has not been cloned. However, such a complementation pattern suggests that a gene which performs a related function has been isolated.

Proof That the Correct Gene Has Been Cloned

Determine whether segregation of complementing plasmid results in co-loss of both the plasmid-borne selectable marker and the complementing phenotype

3. Using protocol in *UNIT 13.9*, isolate Leu^- segregants for each candidate and test their ability to grow at 37°C. Save transformants that are only able to grow at 37°C in presence of plasmid.

   If plasmid segregants can be isolated that still display "complementation," this suggests the complementation pattern is due to either reversion of the original mutation or acquisition of a suppressor mutation.

4. Isolate plasmid DNA from each transformant identified in step 3 by transforming total yeast DNA into *E. coli* (*UNIT 13.11*). Retransform each plasmid into mutant yeast strain and confirm that correct plasmid has been isolated (in this example, demonstrate that retransformation into a *cdc101-1* strain results in a temperature-resistant phenotype).

5. Analyze plasmids by restriction mapping (*UNIT 3.2*) to determine whether genomic DNA inserts present in different plasmids have overlapping segments, which can help define the boundaries of the complementing region. Construction of various deletion and insertion mutations, and tests of their ability to complement the mutation in yeast, will provide more precise information about location of gene within cloned insert.

Determine whether disruption of cloned gene that is integrated into the genome can complement original mutation (complementation test)

The following steps are designed to test whether the cloned gene encodes a phenotypic suppressor of the mutation, rather than the wild-type gene.

6. Introduce a yeast selectable marker (see *UNIT 13.4*) into a site in the middle of the complementing region (based on the information gained in step 5).

7. Transform this disrupted plasmid back into original mutant strain and screen for mutant phenotype to test whether complementation has been abolished. Using techniques in *UNIT 13.10*, construct diploid strain containing one wild-type copy of the gene and one disrupted copy of the gene. After sporulation and dissection (*UNIT 13.2*), cross haploid spore product containing disruption (usually identified by selectable marker associated with disruption) to strain carrying original mutation (*UNIT 13.2*). If this diploid has the same mutant phenotype as the strain with the original mutation, this demonstrates lack of complementation and indicates that wild-type gene corresponding to this mutation has been cloned. As a control, cross strain with gene disruption (which may itself display a non-wild-type phenotype) to a wild-type strain, to demonstrate that the phenotype of this diploid is wild-type (showing that disruption is recessive mutation).

References: Rose, 1987; Rothstein, 1985.

Contributor: Victoria Lundblad

Segregation of Plasmids from Yeast Cells

UNIT 13.9

Most *E. coli*/yeast shuttle vectors are lost at a frequency of ~1% when grown nonselectively. One plasmid that will not easily segregate by this protocol is the endogenous 2μm plasmid; spontaneous loss of 2μm DNA occurs with a frequency of $\sim 10^{-4}$ per generation.

BASIC PROTOCOL

Materials

Plasmid-bearing yeast strain
YPD or other nonselective liquid medium and plates (*UNIT 13.1*)
CM dropout plates (*UNIT 13.1*)
Sterile velvets and a replica block (*UNITS 13.2 & 1.3*)

1. Inoculate several single colonies of plasmid-bearing yeast strain into individual 10-ml aliquots of nonselective medium and grow overnight at 30°C.

   If no other selection is necessary, YPD medium can be used. If, other unstable genetic markers or plasmids are to be retained, use defined medium supplemented with nutrient corresponding to selectable marker to be lost.

2. Plate for single colonies on nonselective plates and incubate for 2 days at 30°C. In most cases, ~200 to 300 single colonies should be examined (~100/plate).

3. Replica plate onto selective plates that identify colonies that have lost plasmid-borne selectable marker. Incubate plates overnight at 30°C. Those colonies that are present on master plate but fail to grow on selective plate have lost the plasmid.

Reference: Toh-e and Wickner, 1981.

Contributor: Victoria Lundblad

Manipulation of Cloned Yeast DNA

UNIT 13.10

INTEGRATIVE TRANSFORMATION (Hinnen et al., 1978)

BASIC PROTOCOL 1

In the steps described below, a YIp plasmid (*UNIT 13.4*) harboring both a selectable marker and a cloned gene of interest is integrated at the chromosomal location of the cloned gene via homologous recombination. Although the frequency of integrative transformation is low, the transformation frequency can be increased by linearizing the plasmid at a restriction site within the cloned gene that is homologous to the intended site of integration. The resulting integrant contains the entire plasmid, bracketed by intact copies of the gene. In the case of cloned genes for which no mutations have been identified (or where identified mutations have a phenotype that is difficult to score), the integrated plasmid can be used as a genetic marker—by virtue of the presence of the selectable gene—for mapping and other genetic studies. It is important to note that this method introduces a selectable marker at the genomic site of the cloned gene but does not disrupt the copy of the duplicated gene. Often, multiple tandem integrations can occur. In addition, these integrants are unstable and are lost at a frequency of ~1%/generation when grown nonselectively.

1. Subclone (*UNIT 3.16*) gene to be studied onto YIp plasmid (*UNIT 13.4*).

2. Linearize plasmid with restriction enzyme that cuts within cloned gene.

   If a unique restriction site is not available, digestion with one or more enzymes to produce a gap in this region is also acceptable, as long as sufficient homology is present on either side of the gap (>250 bp).

3. Transform appropriately marked strain with 1 to 10 μg DNA plus carrier DNA (*UNIT 13.7*) selecting for marker present on plasmid. Purify several transformants on selective medium (*UNIT 13.2*) and make DNA (*UNIT 13.11*). Confirm by Southern blot hybridization (*UNIT 13.11*) that integration has occurred at the desired site and determine whether multiple integrations have occurred.

GENE REPLACEMENT TECHNIQUES

The two methods described below provide a means of constructing a mutation in vitro in a cloned gene and reintroducing this mutation at the correct chromosomal site. This allows assessment of the genetic consequences of a mutation, and is often used to determine whether or not a gene is essential (by determining if a complete gene deletion is viable). The one-step gene disruption technique generates either insertion or deletion mutations. Transplacement can be used to introduce insertion or deletion mutations containing a selectable marker, as well as to introduce nonselectable mutations, such as conditional lethal mutations in an essential gene.

In each of these procedures, if the goal is to examine whether or not a null mutation is viable, a diploid strain should be transformed with the appropriate DNA construct. This allows a potentially lethal mutation to be complemented by the wild-type allele on the other chromosome.

BASIC PROTOCOL 2

Integrative Disruption (Shortle et al., 1982)

This technique generates a deletion in the chromosomal copy of a gene. In this method, an internal fragment of a cloned gene is introduced into the chromosome on an integrating plasmid. This generates a gene duplication (which brackets the integrated plasmid), but neither copy of the gene consists of an intact copy: one copy is missing the 3′ end of the gene and one copy is missing the 5′ end. The procedure is the same as for integrative transformation, with two exceptions: (1) the starting YIp plasmid, instead of containing an intact gene, contains a completely internal fragment of the cloned gene, and (2) the strain to be transformed should be diploid.

Two limitations of this technique are that a knowledge of the 5′ and 3′ boundaries of the coding region of the gene is required and the size of the subcloned fragment must be at least 250 bp to promote efficient recombination. However, the disruption can be constructed in one step, without requiring an insertion or insertion/deletion mutation in the cloned gene, and a selectable phenotype (from the selectable marker on the integrated YIp plasmid) is associated with the disruption. Like integrating transformation, the integrants are unstable when grown nonselectively, and multiple integration events can occur. In addition, linearization of the plasmid increases the frequency of transformation and targets the integrants to the desired site.

1. Subclone (*UNIT 3.16*) an internal fragment of the gene onto a YIp vector (*UNIT 13.4*).
2. Linearize the plasmid within this internal fragment.
3. Follow step 3 of integrative transformation (Basic Protocol 1).

One-Step Gene Disruption (Rothstein, 1983)

BASIC PROTOCOL 3

This method generates a gene disruption in one step via transformation, using a fragment of DNA containing a cloned gene that is disrupted by a selectable genetic marker. Homologous recombination between the free DNA ends, which are highly recombinogenic, and homologous sequences in the yeast genome result in replacement of the wild-type gene by the disrupted copy. The disrupted gene can contain either a simple insertion (of the selectable marker) or a deletion/insertion mutation. Introduction of these disruptions into the genome can be achieved in a single step, resulting in stable, nonreverting mutations.

1. Subclone (*UNIT 3.16*) into gene of interest a suitable selectable gene, creating in the process of subcloning a deletion as well, if desired.

2. Using appropriate restriction sites, excise a linear fragment that contains disrupted gene from plasmid constructed in step 1 and gel purify (*UNIT 2.6*). Transform with 1 to 10 µg of gel-purified fragment selecting for inserted marker.

   Small amounts of vector sequences can be retained on this fragment without deleterious effects. Ideally, ≥250 bp of the cloned gene should bracket either side of inserted selectable gene, to promote recombination at the chromosomal locus of cloned gene, rather than at site of selectable marker.

3. Confirm structure of disruption by Southern blot hybridization (*UNIT 13.11*). If a diploid was transformed, sporulate and dissect (*UNIT 13.2*) to obtain haploid spore products with disruption (or to observe inviability).

Transplacement (Scherer and Davis, 1979)

BASIC PROTOCOL 4

This method, also called allele replacement, provides a general means of introducing any type of mutation constructed in vitro that does not have a selectable phenotype into its corresponding chromosomal location (Fig. 13.10.1). It can be used to introduce either a single defined mutation or to screen a mutagenized collection of plasmids. The mutated gene is introduced into yeast on a YIp plasmid, which usually contains the *URA3* gene as a selectable marker. After transformation, selection for Ura^+ results in an integration event, such that mutant and wild-type alleles bracket the *URA3*-containing plasmid sequences (Fig. 13.10.1). Subsequent eviction of the plasmid (containing one copy of the gene) is monitored by screening for colonies that are Ura^- (resistant to 5-fluoroorotic acid; 5-FOA). The 5-FOA^r colonies are then screened for the desired mutant phenotype; the proportion of 5-FOA^r colonies

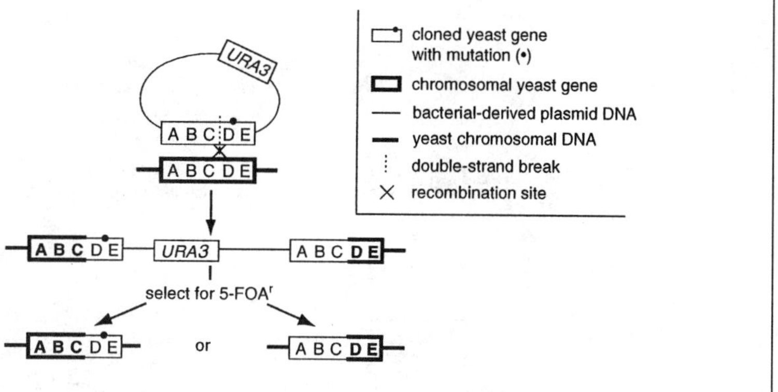

Figure 13.10.1 Transplacement allows the introduction of any mutation in a cloned gene that does not have a selectable phenotype (see accompanying text).

containing the mutant allele will depend on the position of the mutation relative to the length of the flanking homologous DNA.

1. Subclone gene of interest into YIp5. Mutagenize plasmid either by techniques presented in Chapter 8 or by introducing defined deletion or insertion mutations.

2. Follow protocol for integrative transformation (Basic Protocol 1), starting with plasmid linearization.

3. Grow transformants overnight in liquid medium without selection (either in YPD or uracil-containing minimal media; *UNIT 13.1*). Plate for single colonies on nonselective medium (~100 single colonies) and grow two days at 30°C.

   If the goal is to identify conditional lethal mutants, a large number of transformants should be examined (some proportion of transformants will have received unmutagenized plasmid and conditional lethal mutations can be rare events). In addition, this nonselective period of growth should be at permissive temperature (usually 23°C).

4. Replica plate onto 5-FOA plates and grow overnight to identify Ura^- segregants. Colonies that have evicted the plasmid during the period of growth in liquid medium will give completely resistant replicas on 5-FOA plates; these colonies should be recovered from a nonselective plate (rather than from a 5-FOA plate) and used in step 5. Other colonies that have evicted the plasmid during the growth of the colony will appear "patchy" on a 5-FOA replica (often referred to as papillation).

5. Screen 5-FOA^r colonies for mutant phenotype. If mutation has an anticipated alteration in restriction pattern of wild-type gene, confirm genotype of mutant colonies by Southern hybridization (*UNIT 13.11*)

BASIC PROTOCOL 5

PLASMID GAP REPAIR (Orr-Weaver et al., 1983)

The gap repair technique is used to rescue genomic mutations onto a plasmid-borne copy of the gene. This method can also be used to map multiple alleles of a gene (Fig. 13.10.2). Although several variations on this technique have been reported, the simplest method (technically) rescues a chromosomal mutant allele onto an autonomously replicating plasmid. Introduction of a gapped plasmid into a yeast strain carrying the mutation—followed by a gene conversion event between the chromosomal copy and the gap—rescues the chromosomal DNA covered by the gapped

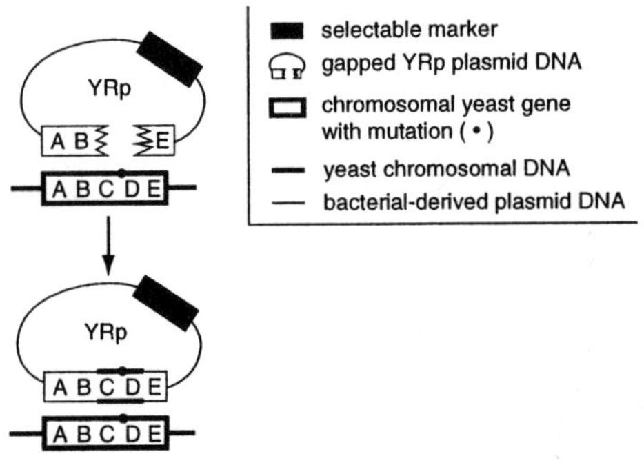

Figure 13.10.2 Plasmid gap repair rescues genomic DNA mutations onto a plasmid-borne copy of the gene (see accompanying text).

region (Fig. 13.10.2). This plasmid can be isolated by transformation of total yeast genomic DNA into *E. coli* (*UNIT 13.11*) and subsequently analyzed.

1. Subclone (*UNIT 3.16*) gene of interest onto YRp plasmid (*UNIT 13.4*). Create a gapped region in this gene by cutting at two restriction sites within the gene, leaving ≥100 to 250 bp of homology on either side of gap. Gel purify fragment containing plasmid backbone (*UNIT 2.6*).

2. Transform 1 to 10 µg of gapped plasmid DNA into yeast strain containing mutant allele to be rescued, selecting for selectable gene on YRp plasmid. Identify transformants that are unstable for selectable marker using plasmid segregation technique (*UNIT 13.9*).

   Transformants resulting from integration of the plasmid into the chromosome (which are relatively stable for the selectable marker) and transformants that contain the repaired autonomously replicating plasmid (unstable for this marker) will be recovered with approximately equal frequencies, so relatively few colonies need to be screened.

3. Among unstable transformants, screen for those that still display mutant phenotype.

   If mutation is located within gapped region, all transformants will contain plasmids with the mutant allele. If, however, mutation is adjacent to a gapped region, some transformants will contain plasmids with mutation and some will contain plasmids with wild-type gene (which will now complement chromosomal mutation), depending on where crossover event occurs. The farther the mutation is from the gapped region, the lower will be the proportion of plasmids containing the mutation.

4. Make DNA from transformants that display mutant phenotype and transform into *E. coli* (*UNIT 13.11*). Analyze miniprep DNA (*UNIT 1.6*).

PLASMID SHUFFLING (Boeke et al., 1987)

BASIC PROTOCOL 6

This technique provides a rapid means for identifiying conditional lethal mutations in an essential cloned gene carried on a plasmid (Fig. 13.10.3). In this system, the chromosomal copy of the essential gene is deleted or otherwise disrupted. Viability is maintained by the presence of a YEp plasmid carrying an intact copy of this essential gene, as well as the *URA3* gene. Introduction of a second plasmid carrying a temperature-sensitive copy of this gene can relieve selective presssure on the YEp plasmid at the permissive temperature, generating Ura^- derivatives due to loss of the YEp plasmid. However, at the nonpermissive temperature, the YEp plasmid cannot be lost and no Ura^- segregants are generated. Loss of this plasmid is assayed by replica plating single colonies onto 5-FOA plates grown at the permissive and nonpermissive temperatures. On these plates, Ura^- segregants appear as 5-FOA resistant papillae. This replica-plating technique allows a large number of mutagenized clones to be screened rapidly.

1. Construct haploid strain that contains a chromosomal disruption of essential gene and YEp plasmid carrying both intact essential gene and *URA3* as selectable marker (*UNIT 13.8*).

2. Subclone essential gene onto YCp vector (bearing selectable marker other than *URA3*) and mutagenize ~10 to 20 µg using hydroxylamine or other techniques (Chapter 8).

3. Transform (*UNIT 13.7*) into strain from step 1, selecting for YCp plasmid on dropout plates that are supplemented with uracil. Incubate at permissive temperature (usually 25°C).

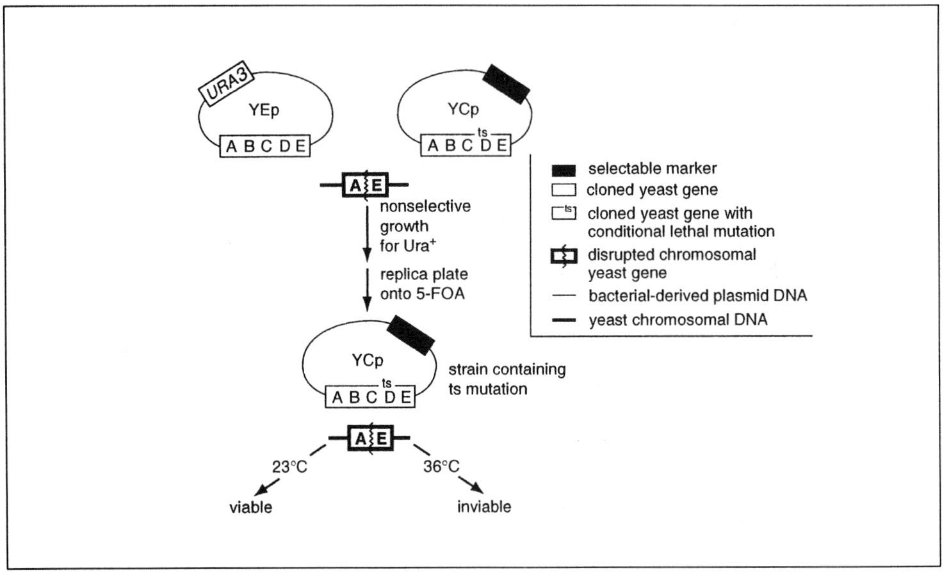

Figure 13.10.3 Plasmid shuffling is a rapid means of identifying conditional lethal mutations in an essential cloned gene carried on a plasmid (see accompanying text).

4. Replica plate each transformation plate onto two plates that contain 5-FOA and that also maintain selection for YCp vector. Incubate one plate at permissive temperature and one plate at nonpermissive temperature (usually 36°C).

   Replica plating distinguishes between different categories of transformants by determining whether a portion of the colony has lost the Ura$^+$ YEp plasmid: such Ura$^-$ cells appear as 5-FOA^r papillae growing out of the background of the mostly Ura$^+$ replica. Transformants that carry an unmutagenized copy of the essential gene on the YCp vector will give Ura$^-$ papillae at both temperatures, whereas transformants that contain a temperature-sensitive mutation in the YCp-borne essential gene will only produce Ura$^-$ papillae at the permissive temperature. Those transformants containing a null mutation in this gene will not generate papillae at either temperature.

5. Recover from transformation plate those colonies that gave Ura$^-$ papillae at permissive temperature but not at nonpermissive temperature and streak for single colonies.

6. Retest ~6 single colonies from each candidate for Ura$^-$ papillation at two temperatures. As a control, test for growth on YPD plates at both temperatures, to rule out an unlinked chromosomal temperature-sensitive mutation. For each candidate that retests as a temperature-sensitive mutation, recover Ura$^-$ papillae from permissive temperature 5-FOA plate (by streaking out papillae on separate plate that maintains selection for YCp vector). Recover plasmid from these strains (which now contain only YCp vector) by transformation into *E. coli* (*UNIT 13.11*).

7. Subclone gene (now carrying ts mutation) onto YIp5 vector and introduce this mutagenized gene into its chromosomal site using transplacement technique.

References: Boeke et al., 1987; Hinnen et al., 1978; Orr-Weaver et al., 1983; Rothstein, 1983; Scherer and Davis, 1979.

Contributor: Victoria Lundblad

Preparation of Yeast DNA

UNIT 13.11

Molecular studies in yeast often require the isolation of both plasmid and chromosomal yeast DNA. Plasmid DNA is used in the transformation of *E. coli*, whereas chromosomal DNA is used for Southern hybridization analysis (UNIT 2.9A), in vitro amplification by the polymerase chain reaction (PCR; UNIT 15.1), or cloning of integrated plasmids.

RAPID ISOLATION OF PLASMID DNA FROM YEAST

BASIC PROTOCOL

Materials *(see* APPENDIX 1 *for items with ✓)*

- YPD or selective medium (UNIT 13.1)
- Yeast colony containing the plasmid of interest
- ✓ Breaking buffer
- 25:24:1 (v/v/v) phenol/chloroform/isoamyl alcohol (with buffered phenol; UNIT 2.1)
- ✓ 0.45- to 0.52-mm acid-washed glass beads (Thomas Scientific)
- Competent *E. coli* cells HB101 or MH1 (UNIT 1.8)
- LB plates (UNIT 1.1) containing appropriate antibiotic (UNIT 1.4)

- 13 × 100–mm glass tubes, sterile
- 30°C incubator with shaker or roller drum

1. Inoculate 2 ml medium in a 13 × 100–mm sterile glass tube with a single yeast colony containing the plasmid of interest. Grow overnight to stationary phase at 30°C in either a roller drum or a shaking incubator.
2. Transfer 1.5 ml of the overnight culture to a microcentrifuge tube and spin 5 sec at high speed, room temperature. Pour off supernatant and disrupt pellet by vortexing briefly.
3. Resuspend cells in 200 µl breaking buffer. Add 0.3 g glass beads (~200 µl volume) and 200 µl phenol/chloroform/isoamyl alcohol. Vortex 2 min at highest speed. Microcentrifuge 5 min at high speed, room temperature.
4. Transform competent *E. coli* HB101 or MH1 with 1 to 2 µl of the aqueous layer. Plate on LB plates containing the appropriate antibiotic to select for the drug-resistance marker on the plasmid.

   Save 50 µl of the aqueous layer and store at –20°C in case additional transformations are required.

RAPID ISOLATION OF YEAST CHROMOSOMAL DNA

ALTERNATE PROTOCOL

Additional Materials *(also see Basic Protocol; see* APPENDIX 1 *for items with ✓)*

- ✓ TE buffer
- 1 mg/ml RNase A
- ✓ 4 M ammonium acetate solution
- 100% ethanol

- 18 × 150–mm glass culture tubes *or* 17 × 100–mm disposable polypropylene tubes, sterile
- Tabletop centrifuge

1. Grow a 10-ml culture of yeast in YPD overnight to stationary phase in either 18 × 150–mm sterile glass culture tubes or in 17 × 100–mm sterile, polypropylene tubes (Basic Protocol, step 1).

13.11

Saccharomyces cerevisiae

2. Spin culture 5 min in a tabletop centrifuge at 1200 × *g*, room temperature. Aspirate or pour off supernatant, and resuspend cells in 0.5 ml water.
3. Transfer the resuspended cells to a microcentrifuge tube and spin 5 sec at room temperature. Pour off supernatant and disrupt pellet by vortexing briefly.
4. Resuspend cells in 200 µl breaking buffer. Add 0.3 g glass beads (~200 µl volume) and 200 µl phenol/chloroform/isoamyl alcohol and vortex at highest speed for 3 min.
5. Add 200 µl TE buffer and vortex briefly. Microcentrifuge 5 min at high speed, room temperature and transfer aqueous layer to a clean microcentrifuge tube. Add 1 ml of 100% ethanol and mix by inversion.
6. Microcentrifuge 3 min at high speed, room temperature. Remove supernatant and resuspend pellet in 0.4 ml TE buffer.
7. Add 30 µl of 1 mg/ml RNase A, mix, and incubate 5 min at 37°C.
8. Add 10 µl of 4 M ammonium acetate and 1 ml of 100% ethanol. Mix by inversion. Microcentrifuge 3 min at high speed, room temperature. Discard supernatant and dry pellet. Resuspend DNA in 100 µl TE buffer.

   Yields of ~20 µg of chromosomal DNA should be obtained. This DNA is ready to use for restriction digestion (UNIT 3.1), in vitro PCR amplification (UNIT 15.1), or Southern blot analysis (UNIT 2.9). For Southern blots, best results are obtained when 5 µl DNA (~1 µg) is digested in a total volume of 20 µl. To amplify by PCR, 2 µl of DNA should be used in a 50-µl reaction.

Reference: Hoffman and Winston, 1987.

Contributor: Charles S. Hoffman

UNIT 13.12

Preparation of Yeast RNA

NOTE: Take precautions to prevent contamination by RNases; use DEPC-treated water for all solutions.

BASIC PROTOCOL

PREPARATION OF YEAST RNA BY EXTRACTION WITH HOT ACIDIC PHENOL

This procedure yields RNA that is relatively free of contaminating DNA, is convenient to perform with multiple samples, and gives little or no sample-to-sample variation.

Materials *(see APPENDIX 1 for items with ✓)*

Yeast cells and desired medium (*UNITS 13.1 & 13.2*)
✓ TES solution
✓ Acid phenol
Chloroform
✓ 3 M sodium acetate, pH 5.3
100% and 70% ethanol, ice-cold

50-ml centrifuge tube (Falcon)
Centrifuge: tabletop *or* Sorvall equipped with an SS-34 rotor

1. Grow yeast cells in 10 ml of desired medium to mid-exponential phase (OD_{600} = 1.0).

   It is not advisable to prepare RNA from cells that have reached a higher density because as the stationary phase is approached, the results are less consistent and RNA yields will vary.

2. Transfer culture to 50-ml centrifuge tube and centrifuge cells 3 min at 1500 × *g*, 4°C.

3. Discard supernatant, resuspend pellet in 1 ml ice-cold water. Transfer to a clean 1.5-ml microcentrifuge tube. Microcentrifuge 10 sec at 4°C, and remove supernatant. Proceed to step 4 or if desired, immediately freeze pellet by placing tube in dry ice.

4. Resuspend cell pellet in 400 µl TES solution. Add 400 µl acid phenol and vortex vigorously 10 sec. Incubate 30 to 60 min at 65°C with occasional, brief vortexing.

5. Place on ice 5 min. Microcentrifuge 5 min at top speed, 4°C.

6. Transfer aqueous (top) phase to a clean 1.5-ml microcentrifuge tube, add 400 µl acid phenol, and vortex vigorously. Repeat step 5.

7. Transfer aqueous phase to a clean 1.5-ml microcentrifuge tube and add 400 µl chloroform. Vortex vigorously and microcentrifuge 5 min at top speed, 4°C.

8. Transfer aqueous phase to a new tube, add 40 µl of 3 M sodium acetate, pH 5.3, and 1 ml of ice-cold 100% ethanol and precipitate. Microcentrifuge 5 min at top speed, 4°C. Wash RNA pellet by vortexing briefly in ice-cold 70% ethanol. Microcentrifuge as before to pellet RNA.

9. Resuspend pellet in 50 µl H_2O. Determine the concentration spectrophotometrically by measuring the A_{260} and A_{280} (*UNIT 4.1*). Store at −70°C, or at −20°C if it is to be used within 1 year.

   Make sure that the RNA is well dissolved; if necessary, heat the resuspended pellet at 65°C for 10 to 20 min and/or dilute further with more water. The yield from 10 ml of cells grown in YPD medium is ~300 µg. Cells grown in less optimal medium will yield less RNA per ml culture.

PREPARATION OF RNA USING GLASS BEADS

ALTERNATE PROTOCOL 1

This preparation is suitable for S1, northern hybridization, or primer extension analyses (*UNITS 4.6, 4.9,* and *4.8,* respectively) and can be prepared quickly and easily from a relatively small quantity of yeast cells. Although the RNA isolated by this procedure is contaminated with DNA, the DNA component does not interfere with most analytical studies.

Additional Materials *(also see Basic Protocol; see APPENDIX 1 for items with ✓)*

✓ RNA buffer

25:24:1 phenol/chloroform/isoamyl alcohol (equilibrated with RNA buffer; see *UNIT 2.1*)

✓ 0.45- to 0.55-mm, chilled, acid-washed glass beads (Sigma)

1. Grow and process yeast cells and freeze cell pellet (Basic Protocol, steps 1 to 3).

2. Resuspend pellet in 300 µl RNA buffer.

3. Add a volume of chilled acid-washed glass beads equivalent to ~200 µl water. Add 300 µl of 25:24:1 phenol/chloroform/isoamyl alcohol equilibrated with RNA buffer. Close the cap, then invert and shake up and down to ensure that the beads are suspended. Vortex vigorously for 2 min at highest speed.

4. Microcentrifuge 1 min at room temperature. Transfer 200 to 250 µl of the aqueous (top) layer to a clean microcentrifuge tube.

5. Add an equal volume of 25:24:1 phenol/chloroform/isoamyl alcohol. Vortex vigorously 10 sec.

6. Repeat step 4. Add 3 vol~ of ice-cold 100% ethanol. Mix well and place at –20°C for ≥30 min or on dry ice for 5 min.

7. Microcentrifuge 2 min at 4°C. Aspirate or pour off the supernatant and wash pellet with ice-cold 70% ethanol.

8. Microcentrifuge 1 min at 4°C. Aspirate or pour off supernatant and dry pellet. Resuspend pellet in 50 µl H_2O. Determine the concentration spectrophotometrically by measuring the A_{260} and A_{280} (*UNIT 4.1*). Store the RNA at –70°C.

If 2×10^8 cells are used, the RNA concentration of the final solution will be ~2 mg/ml.

ALTERNATE PROTOCOL 2

PREPARATION OF POLY(A)$^+$ RNA

Larger quantities of RNA can be isolated by simply scaling-up the procedures. For 10^{10} *S. cerevisiae* cells, use the following guidelines:

For the hot acidic phenol protocol (Basic Protocol): Increase volumes of TES and acid phenol solutions to 4 ml each; use 50-ml polypropylene centrifuge tubes for all manipulations. The yield will be ~10 mg of RNA.

For the glass beads protocol (Alternate Protocol 1): Increase volumes to 15 ml RNA buffer, 10-ml volume glass beads, 15 ml phenol/chloroform/isoamyl alcohol, and 30 ml of 100% ethanol. Perform the procedure in a 50-ml disposable polypropylene tube and centrifuge 10 min, 1200 × *g*, 4°C in a fixed-angle or swinging-bucket type rotor at each phenol extraction step. Good recovery of the precipitated nucleic acid can be accomplished by centrifugation under these same conditions. The yield will be ~5 mg total RNA.

Prepare poly(A)$^+$ RNA from either method using the protocol presented in *UNIT 4.5*.

Contributors: Martine A. Collart and Salvatore Oliviero

Preparation of Protein Extracts from Yeast

Three protocols are presented for preparing protein extracts; they differ primarily in the way the cells are broken. If possible, it is advantageous to use protease-deficient strains such as BJ926 or EJ101. Cell extracts produced by any of these procedures contain ~10 to 30 mg/ml of the total cellular protein. These extracts can be used directly for some purposes, such as mobility-shift DNA-binding assays (Chapter 12), or can serve as the first step in protein purification (Chapter 10). The extracts will also contain large amounts of nucleic acid, especially RNA. Nuclear extracts will contain significant concentrations of chromosomal DNA.

SPHEROPLAST PREPARATION AND LYSIS

BASIC PROTOCOL

Materials *(see APPENDIX 1 for items with ✓)*

- Protease-deficient yeast cells (BJ926, EJ101, or equivalent)
- YPD medium (*UNIT 13.1*)
- ✓ Zymolyase buffer, room temperature and ice-cold
- Zymolyase-100T (ICN Immunobiologicals)
- 1 M sorbitol (optional)
- ✓ Lysis buffer
- ✓ Extraction buffer
- ✓ Storage buffer
- Liquid nitrogen

- Sorvall GS-3 or GSA, SS-34 or SA-600, and Beckman 45Ti rotors or equivalents
- 30°C shaker platform

1. Grow cells to mid-log phase ($OD_{600} \cong 1$ to 5) in YPD medium (100 ml to 20 liters) with vigorous shaking or forced aeration (*UNITS 13.1 & 13.2*). Centrifuge 5 min at 1500 × *g*, 4°C, in preweighed centrifuge bottles.
2. Determine wet weight (in grams) of yeast cells. This is approximately equal to packed cell volume (in milliliters), and for all subsequent steps will be considered 1 vol (one liter of BJ926, a diploid strain, at $OD_{600} = 1.0$, yields packed cell volume of 2 to 3 ml).
3. Resuspend cells in 2 to 4 vol ice-cold water and immediately centrifuge 5 min at 1500 × *g*, 4°C. Resuspend cells by adding 1 vol Zymolyase buffer containing 30 mM DTT and incubate 15 min at room temperature.
4. Centrifuge 5 min at 1500 × *g*, 4°C, and resuspend in 3 vol Zymolyase buffer.
5. Add 2 mg (200 U) Zymolyase-100T/ml original packed cell volume to resuspended cells. Incubate 40 min at 30°C on shaker platform at ~50 rpm. Determine if conversion to spheroplasts has been completed by lysis in water technique (*UNIT 13.7*). If spheroplasting is incomplete, continue incubation until complete.

Perform all procedures from this point on at 4°C:

6. Centrifuge spheroplasts 5 min at 1500 × *g*. Decant supernatant carefully—spheroplast pellet will not be as tight as previous cell pellets.
7. Wash spheroplasts by gently resuspending pellet in 2 vol ice-cold Zymolyase buffer. Centrifuge 5 min at 1500 × *g*. Repeat two more times.

8. Gently resuspend pellet in 2 vol lysis buffer. Do not try to achieve a homogeneous suspension; simply dislodge pellet from side of tube and swirl 10 to 20 times. Centrifuge spheroplasts 10 min at 1500 × *g*.

9. Thoroughly resuspend spheroplast pellet with 1 vol lysis buffer using glass rod.

   Resuspended spheroplasts can be quick-frozen at this point in liquid nitrogen and stored at –80°C; thaw frozen spheroplasts overnight on ice before proceeding.

10. Lyse spheroplasts with 15 to 20 strokes of tight-fitting pestle (clearance 1 to 3 µm) in Dounce homogenizer.

11. Half-fill ultracentrifuge tubes with lysate. Add equal volume extraction buffer and seal tubes. Gently invert tubes on rotating wheel or rocker 15 to 30 min at 4°C.

12. Centrifuge 90 min at 100,000 × *g* in 45Ti rotor, 4°C. Collect supernatant and dialyze 2 to 4 hr against 100 vol storage buffer (*APPENDIX 3*). Transfer dialysis bag to 100 vol fresh storage buffer; dialyze an additional 2 to 4 hr.

   A flocculent precipitate may form during dialysis. These precipitates usually contain negligible amounts of most protein factors and can be discarded.

13. Remove a few microliters of dialysate, dilute 1:1000 with water, and determine conductivity (*UNIT 10.10*). If it is equal to that of similarly diluted storage buffer or below some acceptable value (usually 100 to 250 mM NaCl), proceed to step 14. If not, continue dialysis.

14. Centrifuge dialysate 10 min at 10,000 × *g*, 4°C. Collect supernatant, freeze in small aliquots in liquid nitrogen, and store at –80°C.

   This crude extract contains most DNA-binding proteins as well as transcription and replication factors. The pellet contains proteins that can be "salted in" by resuspending in storage buffer containing 0.5 to 1.0 M KCl, if desired.

SUPPORT PROTOCOL

NUCLEI PREPARATION BY DIFFERENTIAL CENTRIFUGATION

Additional Materials *(also see Basic Protocol; see APPENDIX 1 for items with ✓)*

✓ Nuclei buffer
✓ Ficoll buffer, ice-cold
Teflon pestle tissue homogenizer, motor-driven (optional; Thomas)

1. Perform steps 1 to 7 of Basic Protocol. Resuspend cells in 0.5 vol nuclei buffer.

2. Pipet cells drop by drop into beaker containing 15 to 25 vol ice-cold Ficoll buffer with continuous stirring in ice bath or cold room.

   Alternatively, resuspend cells in several volumes Ficoll buffer and homogenize with 5 to 10 strokes using a motor-driven tissue homogenizer and moderately tight Teflon pestle (clearance 0.15 to 0.23 mm) at medium speed.

3. Transfer suspension to centrifuge tubes and centrifuge 5 min at 3000 × *g*, 4°C. Repeat several times if it is important to completely remove cell debris.

4. Transfer supernatant to new centrifuge tubes and spin 20 min at 20,000 × *g*, 4°C. Resuspend pellet in 5 to 10 vol nuclei buffer and centrifuge at 12,000 × *g*, 4°C.

5. To obtain nuclear lysates, resuspend nuclear pellet in 1 vol lysis buffer and carry out steps 8 to 14 of Basic Protocol.

   For chromatin studies, nuclei can be suspended in appropriate buffer and used directly for micrococcal nuclease or DNase I digestion. Nuclei can be stored at −80°C indefinitely in buffer equivalent to storage buffer but containing only 20 mM KCl.

CELL DISRUPTION USING GLASS BEADS

ALTERNATE PROTOCOL 1

Additional Materials *(see APPENDIX 1 for items with ✓)*

✓ Glass bead disruption buffer
✓ Glass beads, 0.45- to 0.55-mm, chilled and acid-washed (Sigma)
Bead Beater and vessel (Biospec; optional)

1. Grow and harvest yeast cells (Basic Protocol, steps 1 to 3), stopping before resuspension in Zymolyase buffer. Determine packed cell volume. Carry out all subsequent steps at 4°C.

2. Resuspend cells in 1 vol glass bead disruption buffer.

   The cells can be frozen by slowly pouring this suspension into a plastic beaker filled with liquid nitrogen. Use enough liquid nitrogen to submerge frozen paste. This frozen "popcorn" can be stored at −80°C. Thaw overnight on ice before proceeding.

3. Mix cell paste with 2 vol glass bead disruption buffer. Add 4 vol chilled, acid-washed glass beads.

4a. *For packed cell volumes of* <10 ml: Transfer cell suspension to appropriately sized screw-cap centrifuge tube (suspension should occupy ≤60%-70% of capacity of tube). Vortex at maximum speed for 30 to 60 sec at 4°C; place tube on ice for 1 to 2 min. Repeat 3 to 5 times. Check amount of cell breakage under microscope. Proceed to step 5.

4b. *For packed cell volumes of* >10 to 20 ml: Transfer cell suspension to appropriately sized Bead Beater vessel (stainless steel is recommended for better heat transfer) and add glass bead disruption buffer to fill vessel almost to brim. The volume of buffer required to fill vessel should not exceed 1 cell-suspension volume. Attach blade and cap assembly, ensuring that all air is excluded from vessel (it is important to exclude air to prevent foaming and protein denaturation). Grind at high speed for 60 sec, then let sit 1 to 2 min on ice. Repeat 3 to 5 times.

5. Allow glass beads to settle out and decant supernatant. Add 2 to 4 vol glass bead disruption buffer to glass beads and invert tube 5 to 10 times. Allow beads to resettle and decant supernatant. Pool supernatants.

6. Centrifuge pooled supernatants 60 min at 12,000 × g, 4°C. Collect supernatant that represents crude extract. For long-term storage, aliquot into small tubes, quick-freeze tubes in liquid nitrogen, and store at −80°C.

ALTERNATE PROTOCOL 2

CELL DISRUPTION USING LIQUID NITROGEN

This protocol is designed for processing 200 ml to ~20 liters of cells. It can be easily scaled up for processing larger amounts (e.g., from a fermentor), in which case a larger blender and cup must be used.

Additional Materials *(also see Basic Protocol)*

Yeast cakes (optional; Red Star)
Liquid nitrogen

60-ml syringe
1-liter plastic beaker (Nalgene or equivalent)
1-liter stainless steel blender cup and blender (Waring or equivalent)

1. Grow cells to mid-log phase in YPD (or selective) medium with vigorous shaking. Centrifuge and discard supernatant. Commercial yeast cakes can be used in place of growing cells.

2. Vortex cell pellet to create a thick cell paste. If necessary, add a minimal amount of ice-cold water to allow paste to be poured or spooned. If using yeast cake, mix equal volumes of cells and water and blend into thick paste. Keep cell paste on ice in subsequent steps.

3. Remove plunger from 60-ml syringe. Seal bottom with plug or tightly wrapped parafilm. Add 50 ml of cell paste.

4. Place 400 ml liquid nitrogen into 1-liter Nalgene beaker. Hold syringe over liquid nitrogen; remove plug or Parafilm from bottom, insert plunger, and squeeze cell paste into liquid nitrogen. Long spaghetti-like aggregations of frozen cells form with thick cell paste; spherical popcorn-like clumps form with thinner paste. A maximum of ~200 ml cell paste can be processed at one time when using 1-liter blender cup. The frozen cells can be stored indefinitely at −80°C before disruption.

5. Carefully pour liquid nitrogen with frozen cells into blender cup that is already attached to blender and has been thoroughly dried prior to use (procedure can be done at room temperature if necessary).

   Only industrial-strength blenders and stainless steel blender cups can be used for this protocol (a Waring commercial blender is recommended). Care should be taken that all moving parts of both blender and blender cup are completely free of moisture to avoid freezing when adding liquid nitrogen.

6. Place lid on blender cup, making sure the lid vents are open so that pressure does not build up in cup as liquid nitrogen evaporates. Keeping lid held down tightly, grind at high speed in three successive 2-min bursts. Between bursts, mix frozen powder and, if necessary, add liquid nitrogen to just cover powder. Begin grinding as soon as possible after pouring in liquid nitrogen to avoid freezing moving parts.

7. After grinding, pour fine frozen yeast powder into beaker containing twice original cell-paste volume of ice-cold storage buffer. Mix and then pour suspension into a centrifuge bottle. Keep bottle on ice.

8. Repeat steps 3 to 7 until all cell paste has been processed.

9. Centrifuge 15 min at 5000 × *g*, 4°C. The supernatant contains a crude extract containing 10 to 20 mg protein/ml.

The supernatant can be used immediately for assays or as starting point for protein purification. For long-term storage, aliquot supernatant into plastic tubes, freeze quickly on dry ice, and store at –80°C. Alternatively, resuspend the frozen yeast powder in 1 vol lysis buffer and treat as in steps 11 to 14 of Basic Protocol.

Reference: Sorger et al., 1989.

Contributors: Barbara Dunn and C. Richard Wobbe

Interaction Trap/Two-Hybrid System to Identify Interacting Proteins

UNIT 13.14

To understand the function of a particular protein, it is often useful to identify other proteins with which it associates. This can be done by a selection or screen in which novel proteins that specifically interact with a target protein of interest are isolated from a library (Figs. 13.14.1 & 13.14.2).

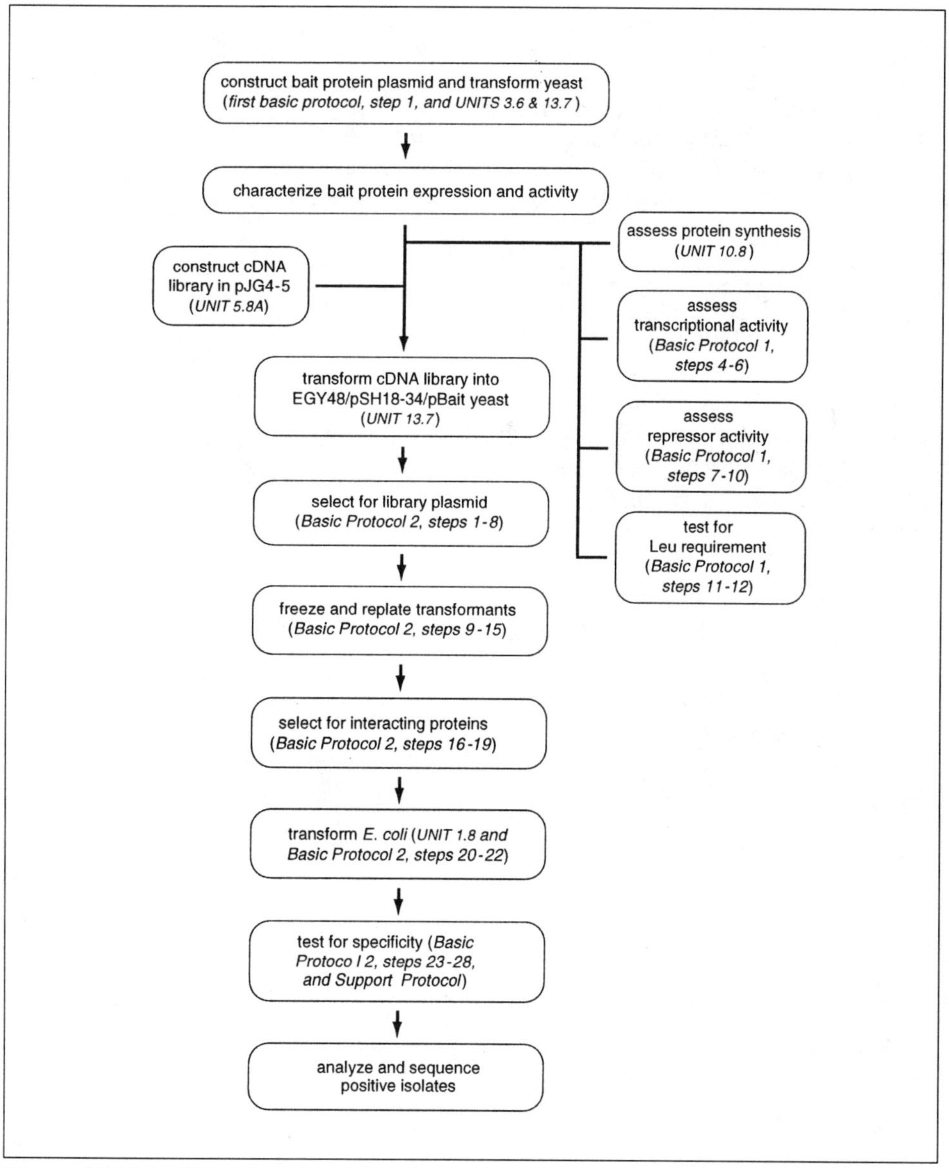

Figure 13.14.1 Flow chart for performing an interaction trap.

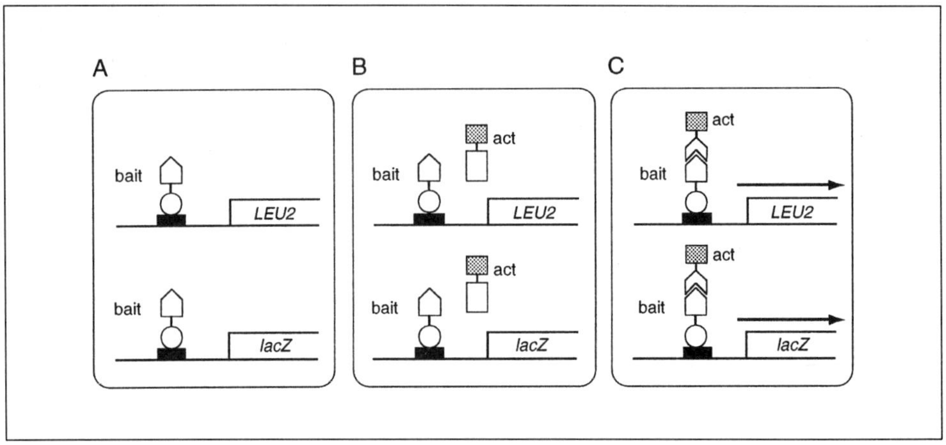

Figure 13.14.2 The interaction trap. (**A**) An EGY48 yeast cell containing two *lexA* operator–responsive reporters, one a chromosomally integrated copy of the *LEU2* gene (required for growth on –Leu media), the second a plasmid bearing a *GAL1* promoter–*lacZ* fusion gene (causing yeast to turn blue on media containing Xgal). The cell also contains a constitutively expressed chimeric protein, consisting of the DNA-binding domain of LexA fused to the probe or "bait" protein, shown as being unable to activate either of the two reporters. (**B**) and (**C**), EGY48/pSH18-34/pbait–containing yeast have been additionally transformed with an activation domain (act)–fused cDNA library in pJG4-5, and the library has been induced. In (**B**), the encoded protein does not interact specifically with the bait protein and the two reporters are not activated. In (**C**), a positive interaction is shown in which the library-encoded protein interacts with bait protein, resulting in activation of the two reporters (arrow), thus causing growth on media lacking Leu, and blue color on media containing Xgal. Symbols: black rectangle, *lexA* operator sequence; open circle, LexA protein; open pentagon, bait protein; open rectangle, library protein; shaded box, activator protein (acid blob in Fig. 13.14.4).

BASIC PROTOCOL 1

CHARACTERIZING A BAIT PROTEIN

The first step in an interactor hunt is to construct a protein in which LexA is fused to the protein of interest. A series of control experiments is performed to establish whether the construct is suitable as is or must be modified: a test for activation of the *lacZ* reporter (steps 1 to 6); a test for operator binding or a repression assay (steps 7 to 10); and a test for activation of the *Leu2* reporter (steps 11 and 12). The characterized bait protein plasmid (pBait) is used to screen a library for interacting proteins.

Materials *(see* APPENDIX 1 *for items with ✓)*

Plasmids (Table 13.14.1): pEG202 (Fig. 13.14.3), pSH18-34, pSH17-4, pRFHM1, pJK101, and pRS423 (available from Dr. R. Brent)

Yeast strain EGY48 *ura3 trp1 his3 lexA* operator–*LEU2* (available from Dr. R. Brent)

Complete minimal (CM) medium dropout plates (*UNIT 13.1*), supplemented with 2% (w/v) of the indicated sugars, in 100-mm plates:

- Glc/CM, –Ura –His
- Gal/CM, –Ura –His
- Gal/CM, –Ura –His –Leu

✓ 1× Z buffer with 1 mg/ml Xgal (Table 1.4.2)

Liquid Gal/CM –Ura –His dropout medium

LacZ activation assay

1. Using standard subcloning techniques, insert the DNA encoding the protein of interest into the polylinker of pEG202 (Fig. 13.14.3) to make an in-frame protein fusion.

Table 13.14.1 Plasmids Used for Interaction Traps

Plasmid[a]	Insert	Selectable marker	Comments
pEG202	*lexA* protein + polylinker	*HIS3 amp*^{r}	Used for cloning DNA for bait protein (pBait)
pJG4-5	*GAL1* promoter + expression cassette	*TRP1 amp*^{r}	Used to clone cDNA library; has nuclear localization and trancriptional activation sequences + HA epitope tag
pJK101	*GAL1* upstream activating sequences + 2 *lexA* operators + *lacZ* reporter	*URA3 amp*^{r}	Used as positive control for repression assay
pJK202	*lexA* protein + polylinker with nuclear localization motif	*HIS3 amp*^{r}	Used to ensure translocation of bait protein to the nucleus
pRFHM-1	*lexA* protein + gene for res 2-160 (PRD + domain + homeodomain) of the bicoid protein (N-terminal)	*HIS3 amp*^{r}	Negative control for activation or positive control for repression; control for specificity testing
pRS423	Vector encoding no protein	*HIS3 amp*^{r}	Negative control for repression
pSH17-4	*lexA* protein + *GAL4* activator protein-encoding sequence	*HIS3 amp*^{r}	Positive control for activation
pSH18-34	8 *lexA* operators + *lacZ* reporter	*URA3 amp*^{r}	Reporter gene plasmid

[a]All plasmids are described in Gyuris et al., 1993, and can be obtained from the R. Brent laboratory at Massachussetts General Hospital. Additional information is available online from the MGH molecular biology Gopher server.

2. Perform three separate lithium acetate transformations (UNIT 13.7) of EGY48 using the following combinations of plasmids:

   pBait + pSH18-34 (test)
   pSH17-4 + pSH18-34 (positive control for activation)
   pRFHM1 + pSH18-34 (negative control for activation).

3. Plate each transformation mixture on Glc/CM –Ura –His dropout plates. Incubate overnight at 30°C to select for yeast that contain both plasmids.

4. Streak a Glc/CM –Ura –His master dropout plate with at least five or six independent colonies obtained from each of the three transformations in step 3 (test, positive control, and negative control).

5. Lift colonies by gently placing a nylon filter on the yeast plate and allowing it to become wet through. Remove filter and air dry 5 min. Chill filter, colony side up, 10 min at –70°C.

6. Soak a piece of Whatman 3MM filter paper in 1× Z buffer containing 1 mg/ml Xgal. Place colony filter, colony side up, on Whatman 3MM paper, or float the filter in the lid of a petri dish containing ~2 ml Z buffer with 1 mg/ml Xgal. Incubate at 30°C and monitor for color changes.

Repression assay

7. Transform EGY48 yeast with the following combinations of plasmids (three transformations):

   pBait + pJK101 (test)
   pRS423 + pJK101 (negative control for repression)
   pRFHM1 + pJK101 (positive control for repression).

8. Plate each transformation mix on Glc/CM –Ura –His dropout plates to select yeast cells that contain the indicated pairs of plasmids. Incubate at 30°C until colonies appear; usually 2 to 3 days, but not longer than 4 days.

9. Streak colonies to a Gal/CM –Ura –His dropout plate and incubate overnight at 30°C.

10. Assay β-galactosidase activity of the three transformed strains (test, positive control, and negative control) by liquid assay (using liquid Gal/CM dropout medium), filter assay (steps 4 to 6), or both.

LEU2 assay

11. Disperse a colony of EGY48 containing pBait and pSH18-34 reporter plasmids into 500 µl sterile water. Dilute 100 µl of suspension into 1 ml sterile water. Make a series of 1/10 dilutions in sterile water to cover a 1000-fold concentration range.

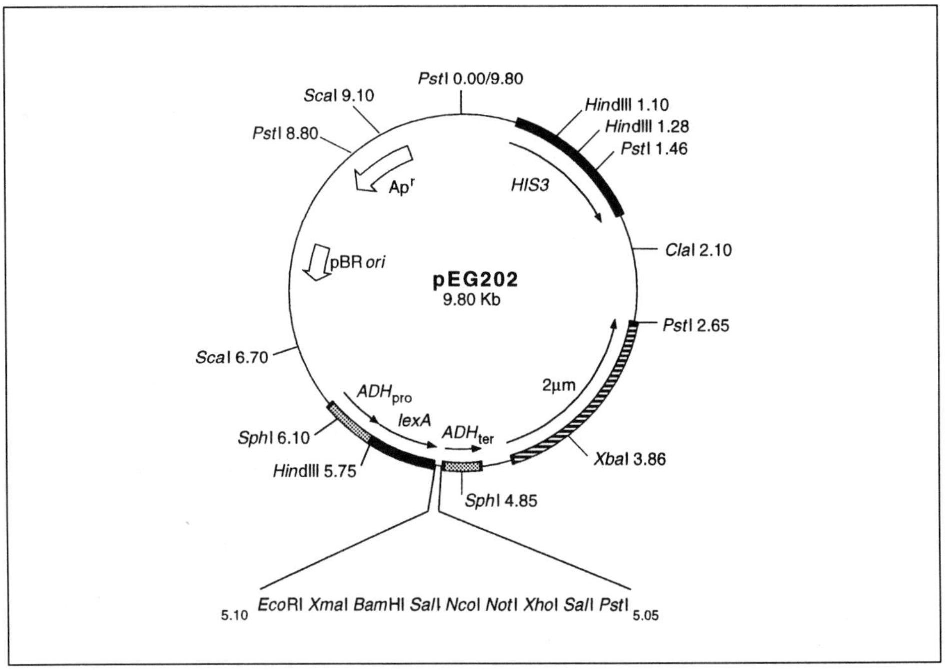

Figure 13.14.3 pEG202 LexA-fusion plasmid. pEG202, a derivative of Lex202 + PL (Ruden et al., 1991), uses the strong constitutive alcohol dehydrogenase promoter (ADH_{pro}) to express bait proteins as fusions to the DNA-binding protein lexA. A number of restriction sites immediately upstream of ADH_{ter} are available for insertion of coding sequences. The reading frame for insertion is GAA TTC CCG GGG ATC CGT CGA CCA TGG CGG CCG CTC GAG TCG ACC TGC AGC. The sequence CGT CAG CAG AGC TTC ACC ATT G can be used to design a primer to confirm correct reading frame for LexA fusions. The plasmid contains the *HIS3* selectable marker and the 2µm origin of replication to allow propagation in yeast, and the ampicillin resistance gene (amp^r) and the pBR origin (ori) of replication to allow propagation in *E. coli*. Endpoints of the *HIS3* and 2µm elements are drawn approximately with respect to the restriction data. Numbers indicate relative map positions.

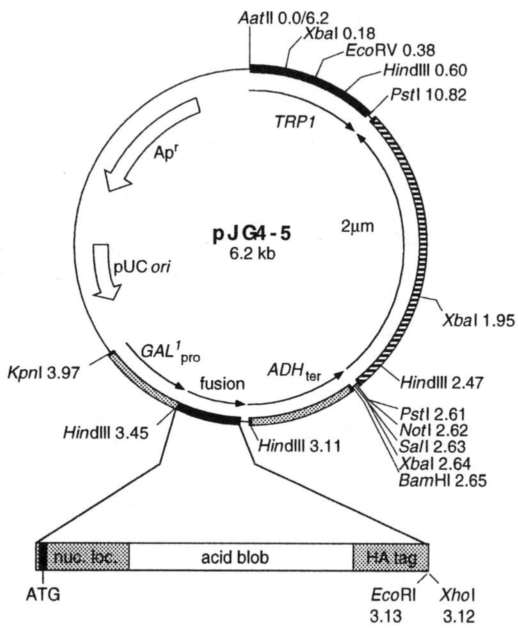

Figure 13.14.4 pJG4-5 library plasmid. pJG4-5 (Gyuris et al., 1993) expresses cDNAs or other coding sequences inserted into the unique *Eco*RI and *Xho*I sites as translational fusion to a cassette consisting of the SV40 nuclear localization (nuc. loc.) sequence (PPKKKRKVA), the acid blob B42, and the hemagglutinin (HA) epitope tag (YPYDVPDYA). Expression of sequences is under the control of the *GAL1* galactose–inducible promoter. The sequence CTG AGT GGA GAT GCC TCC can be used as a primer to identify inserts or confirm correct reading frame. The plasmid contains the *TRP1* selectable marker and the 2μm origin to allow propagation in yeast, and the ampicillin resistance (*amp*r) gene and the pUC origin (ori) to allow propagation in *E. coli.* Numbers indicate relative map positions.

12. Plate 100 μl from each dilution on Gal/CM –Ura, –His dropout plates and on Gal/CM –Ura, –His, –Leu dropout plates. Incubate overnight at 30°C.

13. Select clones for interaction trap library screening: pBaits that do not activate the *lacZ* and *LEU2* reporters and do produce levels of β-galactosidase activity less than that of JK101+ / 24523 and comparable to that of the JK101 + RFHMI combination are suitable; pBaits that activate slightly may also be suitable; and pBaits that activate strongly should be truncated.

PERFORMING AN INTERACTOR HUNT

BASIC PROTOCOL 2

An interactor hunt involves two successive large platings of yeast containing LexA-fused probes, reporters, and libraries in pJG4-5 (Fig. 13.14.4) with a cDNA expression cassette under control of the *GAL* promotor.

Materials *(see* APPENDIX 1 *for items with* ✓*)*

Yeast containing appropriate combinations of plasmids (Table 13.14.1):
- EGY48 containing pSH18-34 and pBait (Basic Protocol 1)
- EGY48 containing pSH18-34 and pRFHM-1
- EGY48 containing pSH18-34 and any nonspecific bait

Complete minimal (CM) dropout media (UNIT 13.1) supplemented with sugars as indicated [2% (w/v) Glc or 2% (w/v) Gal + 1% (w/v) Raff (raffinose)]:

Glc-CM –Ura –His
Gal/Raff/CM –Ura –His –Trp
TE buffer (pH 7.5)/0.1 M lithium acetate
Library DNA in pJG4-5 (Table 13.14.1 and Fig. 13.14.4)
High-quality sheared salmon sperm DNA (Support Protocol 1)
40% (w/v) PEG 4000/0.1 M lithium acetate/TE buffer, sterile
DMSO
Complete minimal (CM) medium dropout plates (*UNIT 13.1*) supplemented with sugars as indicated [2% (w/v) Glc or 2% (w/v) Gal + 1% (w/v) Raff]:
Glc/CM –Ura –His –Trp, 24 × 24–cm (Nunc) and 100-mm
Gal/Raff/CM –Ura –His –Trp, 100-mm
Gal/Raff/CM –Ura –His –Trp –Leu, 100-mm
Glc/CM –Ura –His, 100-mm
Glc/Xgal/CM –Ura –His –Trp, 100-mm
Gal/Raff/Xgal–CM –Ura –His –Trp, 100-mm
Glc/CM –Ura –His –Trp –Leu, 100-mm
Gal/CM –Ura –His –Trp –Leu, 100-mm
✓ TE buffer (pH 7.5), sterile (optional)
✓ Glycerol solution
E. coli KC8 (*pyrF*, *leuB600*, *trpC*, and *hisB463*; constructed by Dr. K. Struhl and available from Dr. R. Brent)
LB/ampicillin plates (*UNIT 1.1*)
Bacterial defined minimal plates: 1× A plates with vitamin B1 (*UNIT 1.1*), supplemented with 40 µg/ml each Ura, His, and Leu

1. To transform the library, grow an ~20-ml culture of EGY48 containing pSH18-34 and pBait in Glc/CM –Ura, –His liquid dropout medium overnight at 30°C.

2. Dilute culture into 300 ml Glc/CM –Ura –His liquid dropout medium to ~2×10^6 cell/ml (OD_{600} ~0.10). Incubate at 30°C until ~1×10^7 cells/ml (OD_{600} ~0.50).

3. Centrifuge 5 min at 1000–1500 × *g* in a low-speed centrifuge at room temperature to harvest cells. Resuspend in 30 ml sterile water and transfer to 50-ml conical tube.

4. Centrifuge 5 min at 1000 to 1500 × *g*. Decant supernatant and resuspend cells in 1.5 ml TE buffer/0.1 M lithium acetate.

5. Add 1 µg library DNA in pJG4-5 and 50 µg high-quality sheared salmon sperm carrier DNA in a total volume of <10 µl to each of 30 sterile 1.5-ml microcentrifuge tubes. Add 50 µl of the resuspended yeast solution from step 4.

6. Add 300 µl of sterile 40% PEG 4000/0.1 M lithium acetate/TE and invert to mix thoroughly. Incubate 30 min at 30°C.

7. Add DMSO to 10% (~40 µl per tube) and invert to mix. Heat shock 10 min in 42°C heating block.

8a. *For 28 tubes:* Plate the complete contents of one tube per 24 × 24–cm Glc-CM –Ura –His –Trp dropout plate and incubate at 30°C.

8b. *For two remaining tubes:* Plate 360 µl of each tube on 24 × 24–cm Glc/CM –Ura –His –Trp dropout plate. Use the remaining 40 µl from each tube to make a series of 1/10 dilutions in sterile water. Plate dilutions on 100-mm Glc/CM –Ura –His –Trp dropout plates. Incubate all plates at 30°C until colonies appear (2 to 3 days). Saturation screening of a mammalian library requires $\geq 10^4$ colonies. Use the dilution plates to establish transformation efficiency.

9. To harvest transformants, cool the thirty 24 × 24–cm plates containing transformants for several hours at 4°C to harden agar.

10. Wearing gloves and using a sterile glass microscope slide, gently scrape yeast cells off the plate. Pool cells from the 30 plates into one or two sterile 50-ml conical tubes.

11. Wash cells with one volume sterile TE or water. Centrifuge ~5 min at 1000 to 1500 × *g* room temperature, and discard supernatant. Repeat wash.

12. Resuspend pellet in 1 vol glycerol solution, mix well, and freeze in 1-ml aliquots at –70°C.

13. To establish viability, dilute an aliquot of frozen transformed yeast 1/10 with Gal/Raff/CM –Ura –His –Trp dropout medium. Incubate with shaking 4 hr at 30°C to induce the *GAL* promoter on the library.

14. Make serial dilutions of the yeast cells using Gal/Raff/CM –Ura –His –Trp dropout medium (1OD_{600} is ~2×10^7 cells). Plate on 100-mm Gal/Raff/CM –Ura –His –Trp dropout plates and incubate at 30°C until colonies are visible.

15. Count colonies and determine the number of cfu per aliquot of transformed yeast.

16. To select positive clones, thaw the appropriate quantity of transformed yeast, dilute, grow as in step 13, and plate on 100-mm Gal/Raff/CM –Ura –His –Trp –Leu dropout plates at a density of 10^6 cells/plate. Plate 3 to 10 cells for each primary transformant obtained. Perform dilutions in Gal/Raff/CM –Ura –His –Trp –Leu. Incubate 2 to 3 days at 30°C.

17. As they become visible, carefully pick colonies to a new Gal/Raff/CM –Ura –His –Trp –Leu master dropout plate. Colonies that arise earlier are generally more like to be physiological. Incubate at 30°C.

18. Restreak from the Gal/Raff/CM –Ura –His –Trp –Leu master dropout plate to a 100-mm Glc/CM –Ura –His –Trp master dropout plate. Incubate at 30°C until colonies form.

19. Restreak or replica plate from this plate to the following plates:

    Glc/Xgal/CM –Ura –His –Trp
    Gal/Raff/Xgal/CM –Ura –His –Trp
    Glc/CM –Ura –His –Trp –Leu
    Gal/Raff/CM –Ura –His –Trp –Leu.

20. For positive colonies (colonies that are blue on Xgal and grow on Gal/Raff/CM –Leu plates but not on Glc/CM –Leu plates), isolate plasmid DNA from yeast by the rapid miniprep protocol (UNIT 13.11). Bring total DNA from a 15-µl preparation to a final colume of 5 to 10 µl.

21. Use the isolated DNA to transform competent KC8 bacteria and plate on LB/ampicillin plates. Incubate overnight at 37°C.

22. Restreak or replica plate colonies arising on LB/ampicillin plates to bacterial defined minimal plates supplemented with Ura, His, and Leu but lacking Trp to select for JG4.5 plasmids. Incubate overnight at 37°C.

23. Purify library-containing plasmids using a bacterial miniprep procedure.

24. To conduct specificity testing in separate transformations, use purified plasmids from step 23 to transform yeast that already contain the following plasmids and are growing on Glc/CM –Ura –His plates:

    EGY48 containing pSH18-34 and pBait
    EGY48 containing pSH18-34 and pRFHM-1
    (opt) EGY48 containing pSH18-34 and a nonspecific bait.

25. Plate each transformation mix on Glc/CM –Ura –His –Trp dropout plates and incubate at 30°C until colonies grow.

26. Create a Glc/CM –Ura –His –Trp master dropout plate for each library plasmid being tested. Streak five or six independent colonies derived from each of the transformation plates adjacently. Incubate at 30°C until colonies form.

27. Restreak or replica plate from this master plate to the same series of test plates done in the actual screen:

    Glc/Xgal/CM –Ura –His –Trp
    Gal/Raff/Xgal/CM –Ura –His –Trp
    Glc/CM –Ura –His –Trp –Leu
    Gal/CM –Ura –His –Trp –Leu.

28. Conduct additional specificity tests (Support Protocol 2). Analyze and sequence positive isolates.

    "True" positives will give a positive phenotype (see step 20) with the lexA-pBait used to select them, but not with RFHMI or other nonspecific baits.

SUPPORT PROTOCOL 1

PREPARATION OF SHEARED SALMON SPERM CARRIER DNA

This protocol generates high-quality sheared salmon sperm DNA for use as carrier DNA in transformation and other applications where high-quality carrier DNA is needed (e.g., hybridization).

Materials *(see* APPENDIX 1 *for items with ✓)*

High-quality salmon sperm DNA (e.g., DNA Type III sodium salt from salmon testes, Sigma, or equivalent grade from Boehringer Mannheim)
✓ TE buffer, pH 7.5
TE saturated buffered phenol (*UNIT 2.1*)
1:1 (v/v) buffered phenol/chloroform
Chloroform
✓ 3 M sodium acetate, pH 5.2

1. Dissolve desiccated salmon sperm DNA in TE buffer at a concentration of 5 to 10 mg/ml by pipetting up and down in a 10-ml glass pipette. Place in a beaker with a stir-bar and stir overnight at 4°C to obtain a homogenous viscous solution.

2. Shear the DNA to 2 to 15 kb (7 kb average) by sonicating briefly using a large probe inserted into the beaker.

3. Extract the sssDNA solution by mixing with an equal volume of TE-saturated buffered phenol in a 50-ml conical tube. Centrifuge 5 to 10 min at 3000 × *g* (or until clear separation of phases is obtained). Transfer the upper phase containing the DNA to a fresh tube.

4. Repeat extraction using 1:1 (v/v) buffered phenol/chloroform, then chloroform alone. Transfer the DNA into a tube suitable for high-speed centrifugation.

5. Precipitate the DNA by adding 1/10 vol of 3 M sodium acetate and 2.5 vol of ice-cold 100% ethanol. Mix by inversion and centrifuge 15 min at ~12,000 × *g*.

6. Wash the pellet with 70% ethanol and dry. Resuspend the DNA in sterile TE at 5 to 10 mg/ml.

7. Denature the DNA 20 min at 100°C, then transfer to an ice-water bath. Aliquot the DNA into microcentrifuge tubes and store frozen at −20°C. Thaw as needed. Reboil briefly before use in transformation.

ADDITIONAL SPECIFICITY SCREENING

SUPPORT PROTOCOL 2

If other LexA-bait proteins that are related to the bait protein used in the initial library screen are available, substantial amounts of information can be gathered by additional specificity tests. For example, if the initial bait protein was LexA fused to the leucine zipper of c-fos, specificity screening of interactor-hunt positives against the leucine zippers of c-jun or GCN4 in addition to that of c-fos might allow discrimination between proteins that are specific for fos versus those that generically associate with leucine zippers.

Reference: Gyuris et al., 1993.

Contributors: Erica Golemis, Jeno Gyuris, and Roger Brent

In Situ Hybridization and Immunohistochemistry

14

The study of gene products using biochemical and molecular techniques often requires tissue samples containing a considerable amount of the target molecule. This presents a difficulty because many developmentally interesting genes are expressed either in a minority of cells in complex tissues or for only brief periods of time during the differentiation of an organism or tissue. This chapter presents some commonly used cytological techniques for determining the temporal and spatial expression patterns of both mRNA (in situ hybridization of tissue sections) and protein (immunohistochemistry) in individual cells.

The in situ hybridization techniques described here rely upon the hybridization of a specifically labeled nucleic acid probe to the cellular RNA in individual cells or tissue sections. Three areas of technical expertise are required. First, preparation of a suitable nucleic acid probe demands an understanding of the principles of molecular biology (e.g., subcloning, plasmid preparation, radiolabeling). Successful tissue preparation requires practical experience in the art of histology. Finally, as with all morphological techniques, the correct interpretation of the experimental results requires familiarity with cell biology, anatomy, or embryology.

Immunohistochemical localization of cellular molecules exploits the ability of antibodies to bind specific antigens (usually proteins) with high affinity. The technique may be used to localize antigens to subcellular compartments or individual cells within tissues. It is especially powerful when used together with in situ hybridization to localize both the mRNA and protein products of a particular gene.

Fixation and sectioning (*UNITS 14.1 & 14.2*), which critically affect the success of both in situ hybridization and immunohistochemistry experiments, must frequently be optimized for each experimental application. Pretreatment, in situ hybridization, and washing of cells and tissue sections are discussed in *UNIT 14.3*. In situ hybridization can be performed on either paraffin (*UNIT 14.1*) or frozen sections (cryosections; *UNIT 14.2*). The most important reason to choose one method over the other is equipment availability. If a microtome is available there is generally no reason to buy a cryostat and vice versa.

Finally, detection of hybridized probe and the counterstaining and mounting of treated cytological preparations are presented. Depending on the desired sensitivity and resolution, either film or emulsion autoradiography (*UNIT 14.4*) is used to detect the hybridized radioactive probe. A number of alternative staining and mounting procedures are presented in *UNIT 14.5*. *UNIT 14.6* surveys immunohistochemical techniques for protein localization and focuses on the use of indirect immunofluorescence as the primary detection method. Alternative methods (immunoperoxidase and immunogold labeling) suitable for bright-field microscopy are described. The protocols in *UNIT 14.7* present nonisotopic hybridization and detection approaches for determining the locations and relative levels of specific transcripts. Finally, *UNIT 14.8* describes in situ PCR (in situ amplification and hybridization) for detecting low abundance targets.

J.G. Seidman

Fixation, Embedding, and Sectioning of Tissues, Embryos, and Single Cells

Sections prepared according to the protocols in this unit can be used to examine cell and tissue morphology and in studies involving in situ hybridization, immunohistochemistry, and enzyme histochemistry.

BASIC PROTOCOL

PARAFORMALDEHYDE FIXATION AND PARAFFIN WAX EMBEDDING OF TISSUES AND EMBRYOS

Materials *(see APPENDIX 1 for items with ✓)*

✓ 4% paraformaldehyde (PFA) fixative, freshly prepared at 4°C
Paraffin wax (e.g., Paraplast)
50%, 70%, 95%, and 100% ethanol
Xylenes (Baker)
Silicone spray

20-ml snap-cap glass vials (silanized for small samples; *APPENDIX 3*)
60°C oven and heating block with holes to hold 20-ml glass vials
Embedding molds and rings (e.g., VWR Scientific)
Hot forceps or hot Pasteur pipet with end cut off
Hot Pasteur pipet with end drawn out and sealed

1. Place dissected organs or embryos in labeled 20-ml snap-cap glass vials (for very small samples use silanized glass vials). Fill vials with freshly prepared 4% PFA fixative, 4°C, and fix at 4°C for the desired time.

   Optimal fixation time is that which gives good morphology as well as good signal-to-noise ratio after in situ hybridization.

   For better and more homogenous fixation of individual organs, animals should be perfused with fixative prior to harvesting tissues. See support protocol for perfusion.

2. Begin melting paraffin wax in 60°C oven.

3. After fixation is completed, pour off fixative (be careful not to lose the sample), and replace with 50% ethanol. Immediately change 50% ethanol to fresh 50% ethanol. Incubate 20 min and change 50% ethanol a total of three times. Continue dehydration by incubating three times in 70% ethanol, 20 min each time, room temperature (samples can be stored in 70% ethanol for a few days at 4°C).

4. Incubate in 95% ethanol 20 min at room temperature. Repeat twice.

5. Incubate in 100% ethanol 20 min at room temperature. Repeat twice.

6. Replace 100% ethanol with xylenes. Immediately change to fresh xylenes and incubate 10 min. Follow with two more xylenes changes and 10-min incubations after each change.

   CAUTION: *Xylenes are toxic organic solvents—steps 6 and 7 should be carried out in a fume hood.*

7. Pour off xylenes and add 5 ml fresh xylenes to each vial. Add an equal amount of molten wax using a hot glass pipet (transfer quickly to avoid hardening). Mix and leave samples at overnight room temperature.

8. Transfer samples to 60°C oven to melt wax/xylenes mixture. Prepare a 60°C heating block (the 20-ml vials must fit its holes). When wax/xylenes mix is molten, transfer vials to heating block.

9. Pour off wax/xylenes mixture from one vial into a waste bottle. Immediately add fresh molten wax to vial with a hot glass pipet. Put vial back into heat block. Repeat procedure with all vials. Return vials to 60°C oven and incubate 1 hr.

   Work quickly to avoid hardening of wax. If hardening occurs, melt wax again at 60°C before initiating 1-hr incubation (step 9).

10. Remove vials from oven and place again into 60°C heating block. Repeat step 9 twice for a total of 3 hr of incubation.

11. Prepare embedding molds according to manufacturer's instructions (e.g., coated with silicone spray) and fill one of the molds with molten paraffin wax using a hot glass pipet. Immediately transfer sample to wax-filled mold using hot forceps or a hot cut-off Pasteur pipet.

12. Place an embedding ring on the mold and fill with paraffin wax. Label embedding ring to facilitate future identification of samples. Samples can be oriented within mold using a hot drawn-out and sealed Pasteur pipet. Leave cast blocks at room temperature to harden completely.

13. Remove cast blocks from embedding molds and store in a dry place at room temperature. Perform sectioning according to Support Protocol 2.

   Blocks stored this way are stable and can be used successfully for years. Moisture destroys the blocks; dry storage is essential.

PFA FIXATION OF SUSPENDED AND CULTURED CELLS

ALTERNATE PROTOCOL

Additional Materials *(also see Basic Protocol; see* APPENDIX 1 *for items with ✓)*

- 2×10^7 cells/ml (resuspended in serum-free medium if possible)
- ✓ 4% (w/v) PFA fixative, room temperature
- ✓ 3× and 1× PBS, pH 7.2
- Poly-L-lysine-coated glass slides (Support Protocol 3)
- Moist chamber (*UNIT 14.2*)
- Glass staining dishes

1. Pipet 10 µl of 2×10^7 cells/ml in a drop onto a poly-L-lysine-coated glass slide. Place in moist chamber and let cells settle 30 min.

2. Transfer slide into slide rack and immerse in glass staining dish filled with 4% PFA fixative. Fix for 20 min at room temperature.

   Alternatively, grow cells directly on poly-L-lysine-coated cover slips or slides, and fix with 4% PFA fixative for 20 min at room temperature.

3. Transfer slide rack to new dish filled with 3× PBS and incubate 2 min (this stops PFA fixation).

4. Transfer to new dish containing 1× PBS and incubate 2 min. Replace PBS with fresh 1× PBS and incubate another 2 min.

5. Dehydrate slides (in dishes) through a series of 5-min incubations in 50% ethanol, 70% ethanol, 95% ethanol, and 100% ethanol.

6. Air dry slides completely. Transfer to slide boxes containing desiccant and store at –70°C (can be stored several weeks).

Slides must be completely dry before storing in desiccant and freezing; otherwise, freezing artifacts can occur.

SUPPORT PROTOCOL 1

PERFUSION OF ADULT MICE

Perfusion of animals is essential for achieving good morphology and preservation of brains, kidneys, hearts, and many other organs. For the untrained person the first perfusion might not work—it is therefore advisable to learn this procedure using some control animals first, or to seek the advice of a trained animal pathologist or colleague familiar with this procedure. The protocol described here for mice is applicable to most other species.

Additional Materials *(also see Basic Protocol; see* APPENDIX 1 *for items with ✓)*

✓ 1× PBS
2 syringes (20- to 30-ml) equipped with 23-G needles
Container (bag) for mouse and CO_2 gas
Dissection instruments (scissors, forceps, etc.)

1. Fill one syringe with 1× PBS, and another with 4% PFA fixative, 4°C. Set aside.

2. Kill mouse with CO_2 gas in a bag. Immediately after respiratory arrest, lay mouse on its back and open the thorax carefully to avoid excessive bleeding. Cut carefully and quickly through rib cage and remove diaphragm for access to heart.

   If blood clots or main blood vessels are harmed, perfusion will not work.

3. Carefully insert syringe filled with 1× PBS into left ventricle. Cut open right ventricle for drainage, allowing the 1× PBS to be slowly but constantly perfused into heart (Fig. 14.1.1; if perfusion is working well, blood-rich organs such as liver, spleen, and kidneys turn grayish-white).

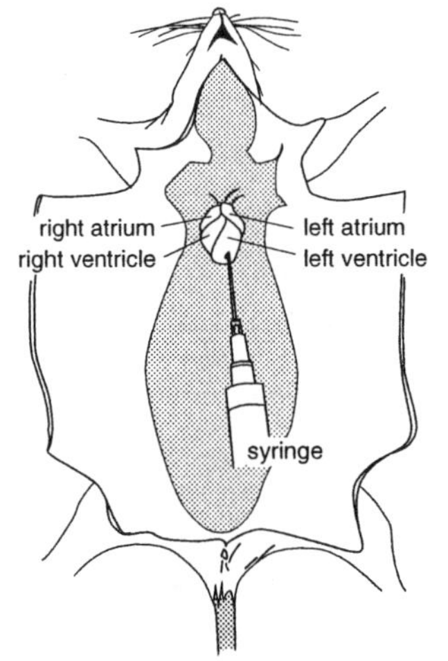

Figure 14.1.1 Perfusion of a mouse. This diagram illustrates the chambers of the heart and the correct positioning of the syringe in the left ventricle for perfusion.

4. After most of the blood has been flushed out, remove syringe with 1× PBS and insert syringe filled with 4% PFA fixative (4°C) into same puncture of left ventricle. Slowly perfuse mouse with ~20 ml fixative.

   A sign of a good perfusion is a muscle tremor best seen on the limbs and tail. By the end of this procedure the animal should be stiff. If perfusion is unsuccessful, it is not worthwhile continuing.

5. After perfusion, dissect out organs and tissues, place in well-labeled glass vials filled with 4% PFA fixative, and store on ice (4°C). Carry out further fixation and processing as in Basic Protocol or *UNIT 14.2*.

SECTIONING SAMPLES IN WAX BLOCKS

SUPPORT PROTOCOL 2

Sectioning paraffin blocks containing samples is a process requiring experience and should be learned from a trained individual if at all possible.

Additional Materials *(see APPENDIX 1 for items with ✓)*

Paraffin wax block(s) containing samples
✓ 0.2× gelatin subbing solution

Microtome, complete with knives
Gelatin-subbed glass slides (Support Protocol 3)
Fine brushes
Slide warmer or heating plate set between 45° and 50°C
42°C oven
Desiccant (e.g., Humicaps, United Desiccants-Gates)

1. Cut wax block containing samples into trapezoidal shape using razor blade. Carefully shave off extra wax and do not cut too close to embedded sample (sample might be destroyed or wax might shatter).

2. Attach trapezoid block (with wide edge facing knife) to holding clamp of microtome and begin sectioning (Fig. 14.2.1C-E). Cut 8-μm sections.

   CAUTION: *Microtome knives are very sharp. For proper and safe use of microtome, consult the manufacturer's manual.*

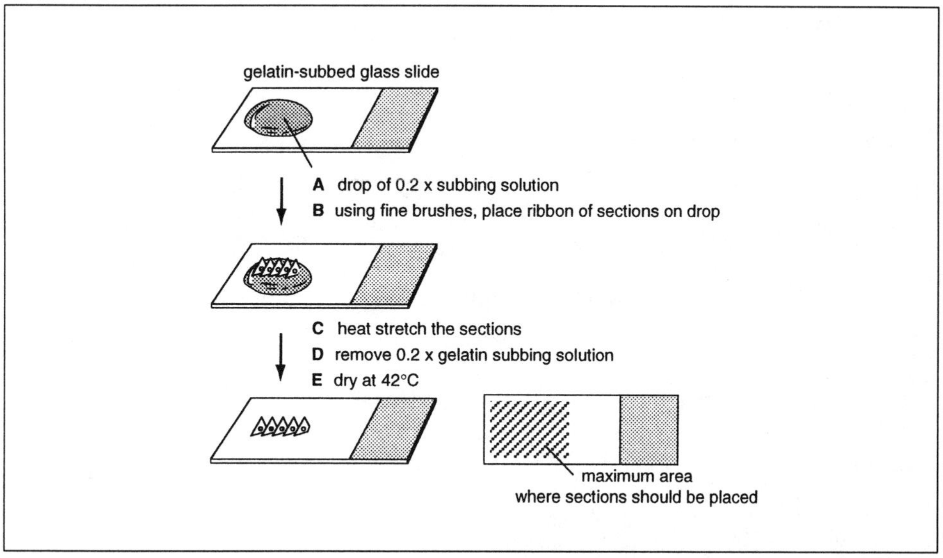

Figure 14.1.2 Placement of paraffin sections on gelatin-subbed glass slides. Steps **A** to **E** depicted here correspond to steps 3 to 8 in the Support Protocol.

3. Put a drop of 0.2× gelatin subbing solution on a gelatin-subbed glass slide (Fig. 14.1.2). Transfer ribbon of sections (up to 10 sections, depending on sample size) onto drop on subbed slide using fine brushes.

4. Transfer slide onto slide warmer or heating plate between 45° and 50°C.

   Heat stretches the sections and removes wrinkles; do not leave slides on slide warmer longer than necessary or sections may be destroyed.

5. After stretching is complete, remove slide from slide warmer and carefully remove remaining subbing solution using a Pasteur pipet. Dry slide at room temperature.

6. Incubate 1-2 days at 42°C to firmly attach sections to subbed slides.

SUPPORT PROTOCOL 3

PREPARATION OF COATED SLIDES

Poly-L-Lysine–Coated Glass Slides

Clean glass slides as described for gelatin-subbed slides (use slide racks and glass staining dishes). Prepare sufficient amount of 500 μg/ml poly-L-lysine (Sigma, mol. wt. >150,000) in water. Coat glass slides by dipping individually into solution, air dry, and store at 4°C. Use within a week. (A useful small dipping chamber is a plastic cytology slide mailer; *UNIT 14.4*.)

Gelatin-Subbed Glass Slides

1. Wash slides (25 × 75–mm precleaned Rite-on Microslides from Clay Adams) in water with detergent for a few minutes. Use any good soap.
2. Rinse slides in water 30 min. During this rinse, prepare 1× gelatin subbing solution (see *APPENDIX 1*).
3. Place slides in a slide rack and immerse 2 min in a glass staining dish filled with 1× gelatin subbing solution (4°C). Subbing must be done carefully to avoid air bubbles.
4. Remove slide rack from subbing solution and set on its side, with the frosted sides of the slides facing downward, to let excess subbing solution drip off. Dry overnight before use. Slides subbed with gelatin are stable at room temperature for several weeks.

Reference: Luna, 1968; Donovan and Brown, 1995.

Contributor: Rolf Zeller

UNIT 14.2

Cryosectioning

This unit describes sample preparation and sectioning methods for frozen tissue. A troubleshooting guide is provided in Table 14.2.1.

BASIC PROTOCOL

SPECIMEN PREPARATION AND SECTIONING

Materials

Liquid N_2
CryoKwik (Damon) or isopentane
OCT compound (Tissue Tek II, Miles)

Small Dewar flask or expanded polystyrene box
Filter paper cut into 1 × 7–cm strips (e.g., Whatman 50)
Cryostat and microtome
Cutting chuck (metal platform that supports specimen during sectioning)
Heat sink or CO_2 jet freezer
Fine brush and ¼-in. brush
Gelatin– or poly-L-lysine–coated slides (*UNIT 14.1*), prelabeled with specimen details in pencil

NOTE: All tools used in cryosectioning, including trimming razor blade, should be prechilled in cryostat chamber; however, *slides should not be chilled.*

1. Chill a 50-ml Pyrex beaker by immersing in a Dewar flask (or expanded polystyrene box) filled with liquid N_2. Fill beaker with CryoKwik (or isopentane).

2. Label one end of filter paper strip with pencil and place specimen on other end with forceps.

   Samples prepared as a slurry are best frozen as 200- to 300-µl drops applied to the filter paper strips with a Pasteur pipet. Most tissue samples will stick to filter paper naturally, but if they will not, it may be necessary to use a little OCT compound between sample and filter paper.

3. Ensure the CryoKwik is liquid (if necessary, melt it by touching with a warm metal rod). Immerse specimen in cold CryoKwik ~1 min.

   Some CryoKwik may solidify on sample, but this will evaporate in cryostat; specimens may be stored in either liquid N_2 or in a –70°C freezer.

4. Place filter paper strips with specimens in cryostat chamber and mount on cutting chucks with a thin layer of OCT compound (Fig. 14.2.1).

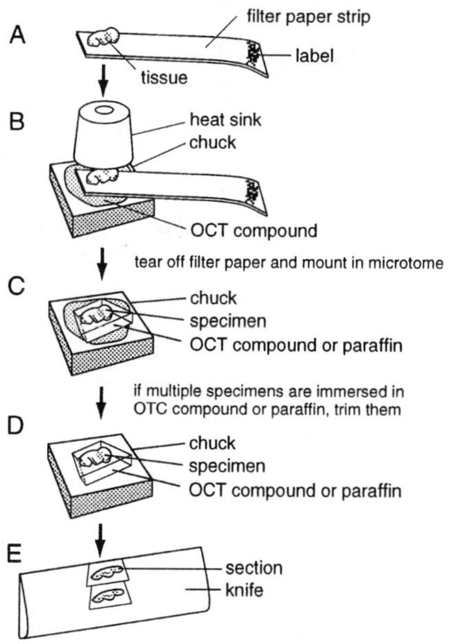

Figure 14.2.1 Sectioning procedure. **A** to **E** illustrate cryosectioning and correspond to steps 6 to 13 of Basic Protocol; **C** to **E** also illustrate paraffin wax sectioning and correspond to steps 1 to 4 of Support Protocol, *UNIT 14.1*.

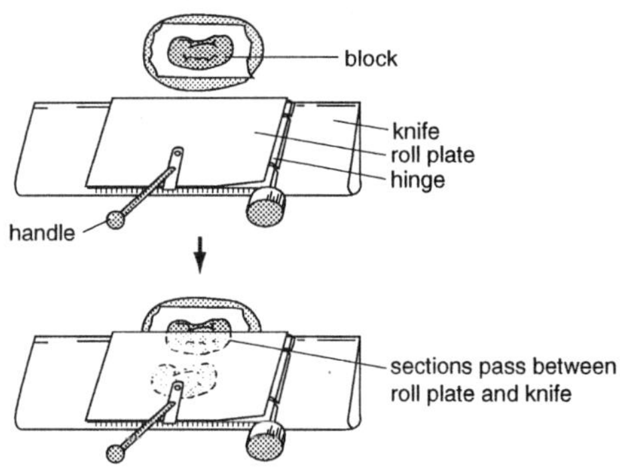

Figure 14.2.2 Plastic roll plate. A plastic roll plate is hinged at the back of the microtome's knife holder and placed parallel to the plane of the knife edge. The plate touches the knife at the leading edge and is tilted up from the knife at the rear so that sections may pass between the knife and the plate.

5. Cool specimen/chuck mount within cryostat with a precooled heat sink or a CO_2 jet freezer until the OCT compound solidifies. Leave within cryostat ≥10 min until temperature equilibrates.

6. Trim specimen block to trapezoid shape (if it is convenient and does not compromise morphology) using a prechilled razor blade. Tear off filter paper.

7. Mount chuck in microtome with parallel faces of trapezoid in line with knife and wide edge toward knife edge. Retract specimen block until it easily clears knife edge.

8. Produce smooth "face" on block using knife at rapid cutting speed (one section/sec; 5- to 10-μm thick). Once specimen block has been cut so that either the structure of interest can be seen within cut face or block dimensions are optimal (between 0.5 and 1 cm on a side), sections may be collected. Perform section collection as in step 9 (roll plate method).

9. The plastic roll-plate method (Fig. 14.2.2) is most commonly used as it allows easy serial sectioning, though it is less suitable for large block faces (for the latter, see roll-bar method; Watkins 1995):

   a. Prior to sectioning, flip roll plate down such that it touches knife and lies parallel to knife edge, though retracted slightly behind cutting edge of knife.

   b. Turn crank and cut sections, which under these conditions will roll up at cutting edge of knife.

   c. Lift roll plate and, with a small brush, brush cut sections from knife edge. Drop roll plate and advance it slightly.

   d. Repeat steps 9a to 9c until sections pass smoothly between roll plate and knife (generally when roll plate leads knife edge by ~0.5 mm), then proceed with step 10.

10. Lift roll plate and move cut sections to rear of knife with a fine brush. Collect sections by gently touching to a warm prelabeled gelatin– or poly-L-lysine–coated slide (see Table 14.2.1).

Mounted sections should be left in microtome until all sectioning is complete 1. If they will be used for in situ hybridization, treat immediately following fixation

Table 14.2.1 Troubleshooting Guide for Morphological Problems Encountered During Cryosectioning

Problem	Possible cause	Solution
Poor morphological preservation throughout tissue	Inadequate care during dissection or excessive delay between dissection and freezing	Take great care during dissection not to damage tissue by stretching or excessive bending; freeze tissue soon after dissection to prevent proteolysis by endogenous proteolytic enzymes; this is particularly important for delicate samples (e.g., small embryos or brain) or enzyme-rich tissues (e.g., liver); fixation and infusion with sucrose may improve morphology considerably (see Support Protocol).
Holes in tissue	Ice crystal formation in tissue during freezing, or thawing of the block and subsequent refreezing and ice crystal formation in the cryostat	Ensure that blocks are frozen in either isopentane or CryoKwik rather than a gas such as nitrogen. The latter will boil on contact with the warm tissue giving a "shell freezing" effect, where the boiling gas insulates the outer layers of the tissue from the cold liquid nitrogen, thereby slowing freezing and allowing ice crystal formation; ensure that when specimens are mounted in the microtome they are protected from warming by a large precooled heat sink; ensure that you do not touch the specimens with your fingers or with warm tools.
Morphology appears fuzzy	"Pressure artifact" due to excessive pressure on the back of the slide while picking up the section	Use less pressure when picking up the section; a mixture of warmth and static charge will normally ensure that sections stick to the slide; if this is not the case, the slide is too cold and may be locally warmed by touching a finger to the back of the slide.
Morphology appears smeared	Slide was moved while lifting section	Hold slide against the rear of the knife and use this as a fulcrum while gently rocking the slide down to touch the section, thereby ensuring no lateral motion of the slide.
Sections do not stick to slide, they wash away during labeling	Slides were not adequately coated	Ensure that sections stick to slides by dip-coating them as described in *UNIT 14.1*.
Sections blow away from knife edge during sectioning	A buildup of static electricity in the cryostat	There is no reliable cure; it is possible to buy polonium brushes which may help; antistatic guns (such as those designed for record players) have also been used successfully.
Sections have lines running up them and separate into ribbons on knife edge	Knife is dull	Sharpen knife; generally a knife will cut ~20 to 30 blocks of soft tissue before it needs to be sharpened; if starting with new equipment, a disposable blade system (e.g., Fisher) may be cheaper over the long term.

(Support Protocol 1). If they will be used for immunohistochemistry or enzyme histochemistry, sections may be stored at −70°C for a limited time (overnight) in an airtight container (see also UNIT 14.6 on air dried section for immunohistochemical study).

11. After lifting sections onto slide, clear condensed ice from the knife using thick (¼-in.) brush

    This brushing may need to be quite vigorous and, therefore, should always be toward knife edge; otherwise knife will be rapidly dulled and brush ruined.

SUPPORT PROTOCOL 1

FIXATION OF CRYOSECTIONS FOR IN SITU HYBRIDIZATION

Cryosections used for in situ hybridization must be fixed with paraformaldehyde and then dehydrated.

Additional Materials *(also see Basic Protocol; see APPENDIX 1 for items with ✓)*

✓ 4% (w/v) PFA fixative, freshly prepared at room temperature
✓ 3× and 1× PBS
30%, 60%, 80%, 95%, and 100% ethanol

Moist chamber (Fig. 14.2.3)
Desiccant (e.g., Humicaps, United Desiccants-Gates)

1. Put slides containing sections in moist chamber and cover sections with 4% PFA fixative. Cover chamber and incubate 20 min at room temperature for previously unfixed tissues or 5 min for fixed tissues.

2. Aspirate off fixative and flood section with 3× PBS from a pipet or squirt bottle. Incubate 5 min.

3. Repeat step 2 twice with 1× PBS. Remove PBS and flood with ethanol for 2 min at each of the following concentrations: 30%, 60%, 80%, 95%, and 100%.

4. Dry slides and store with desiccant up to several weeks at −70°C in an airtight box. Before opening box, bring to room temperature.

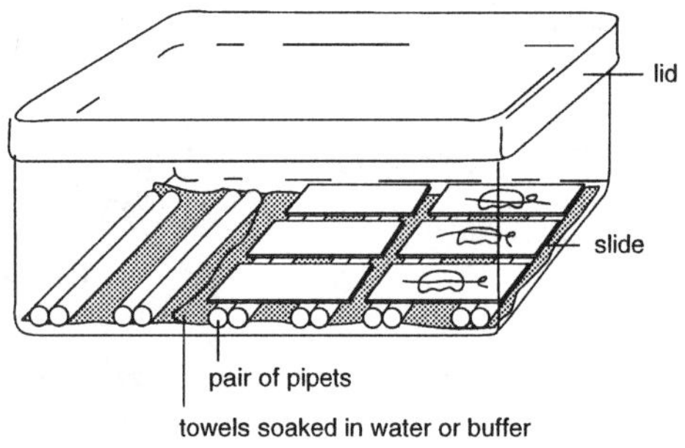

Figure 14.2.3 Moist chamber. Take a conveniently sized container with a tight lid (e.g., Tupperware or equivalent) and attach pairs of 5- to 10-ml plastic pipets to bottom so that they support slides at either end. Place pipets so that the maximum number of slides can be set on them. Alternatively, place a stainless steel cake rack in the bottom. Put absorbent paper on the bottom and drench with water or, for in situ hybridization, where incubation times are very long, use a solution of identical osmolarity and formamide concentration to your hybridization buffer (see *UNIT 14.3*).

TISSUE FIXATION AND SUCROSE INFUSION

SUPPORT PROTOCOL 2

Frequently for cryosectioning, morphology is much improved if the tissue is fixed and infused with sucrose prior to sectioning.

Additional Materials *(also see Basic Protocol; see* APPENDIX 1 *for items with ✓)*

✓ 1× PBS
✓ 2% *or* 4% (w/v) PFA fixative, freshly prepared at 4°C
30% sucrose in PBS

1. Fix small tissue samples or isolated embryos (e.g., day 7 or day 8 mouse embryos) in 2% (for immunohistochemistry) or 4% (for in situ hybridization) PFA fixative 30 min at 4°C.

   Larger embryos, and some tissues, may require fixation up to 4 hr. Larger organs should be fixed in situ within the animal by perfusion (UNIT 14.1) prior to dissection.

2. Wash tissue samples twice in PBS.

3. Infuse samples in 30% sucrose in PBS until tissue sinks (~1 to 3 hr to overnight).

   Paraformaldehyde fixation is a continuous fixation; therefore, fixation time should be minimized. Overfixation will result in loss of reactivity for both in situ hybridization or immunohistochemistry. This is particularly important for in situ hybridization (UNITS 14.1-14.3).

References: Hollands, 1962; Watkins, 1995; Zugibe, 1970.

Contributor: Simon Watkins

In Situ Hybridization to Cellular RNA

UNIT 14.3

In situ hybridization to cellular RNA is used to determine the cellular localization of specific messages within complex cell populations and tissues. Time parameters for in situ hybridization can be found in Table 14.3.1. Probes should be synthesized the day before hybridization.

HYBRIDIZATION USING PARAFFIN SECTIONS OR CELLS

BASIC PROTOCOL

Materials *(see* APPENDIX 1 *for items with ✓)*

Glass slides containing paraffin sections or cells (UNIT 14.1)
Dewaxing/rehydration (dehydration) series: 3 staining dishes of xylenes, 2 staining dishes of 100% ethanol, and 1 staining dish each of 95%, 70%, and 50% ethanol
✓ 0.2 M HCl
✓ 2× SSC, 70°C
✓ 1× and 3× PBS
✓ Pronase, predigested and lyophilized (optional)
2 mg/ml glycine in 1× PBS (optional)
4% (w/v) PFA fixative, freshly prepared at room temperature
10 mM DTT (high-quality; Sigma) in 1× PBS, freshly prepared at 45°C
✓ Blocking solution, prepared immediately before use at 45°C
✓ Triethanolamine buffer (0.1 M TEA), freshly prepared
Acetic anhydride

Table 14.3.1 Time Parameters for In situ Hybridization

Step	Paraffin	Cryosections
Fixation and embedding	Day 1—Fix or perfuse tissues in 4% PFA fixative; dehydrate.	Day 1—Fix or perfuse tissues in 4% PFA fixative. Infuse in sucrose; orient and embed in OCT compound; quick freeze and store at −70°C.
Sectioning	Days 3 to 4—Cut sections on microtome and place on slides; dry on slides for 1 to 2 days at 42°C; store desiccated ≤2 months at −20°C.	Day 2—Make cryostat sections and place on microscope slides; fix with 4% PFA fixative, dehydrate, and store desiccated ≤1 month at −70°C.
Pretreatment	Day 4—Dewax. HCl and heat treat; treat with blocking agents and acetic anhydride; dehydrate and store desiccated overnight at −70°C.	Day 3—Pronase digest; treat with acetic anhydride; dehydrate and use immediately.
Hybridization	Day 5—Overlay sections with probe and hybridize in moist chamber 1 to 4 hr.	Day 3—Overlay sections with probe and hybridize in moist chamber 1 to 4 hr.
Washes	Day 5—Wash, RNase-treat, and dehydrate.	Day 3—Wash, RNase-treat, and dehydrate.
Detection	Days 5-?—Expose to X-ray film, 1 to 3 days; estimate exposure time and dip multiple slides in autoradiographic emulsion; develop and analyze.	Days 3-?—Expose to X-ray film, 1 to 3 days; estimate exposure time and dip multiple slides in autoradiographic emulsion; develop and analyze.

[^{35}S]UTP-labeled riboprobes (Support Protocol 1)
S-riboprobe competitor (Support Protocol 1)
✓ 50 mM DTT, sterile
✓ Hybridization mix A
✓ Moist chamber solution A
✓ Wash solutions A, B, and C
✓ RNase digestion solution
50%, 70%, and 95% ethanol/0.3 M ammonium acetate solutions
100% ethanol

Two sets of slide racks (one clearly labeled for RNase use only)
≥10 glass staining dishes
45°, 55°, and 50°C water baths
Slide box with desiccant (e.g., Humicaps, United Desiccants-Gates)
100°C heating block or water bath
45°C incubator
Moist chambers (*UNIT 14.2*; one clearly labeled for RNase use only)
≥4 glass staining dishes, clearly labeled for RNase use only

NOTE: All of the following steps are performed by incubating slides containing specimens (held in a slide rack) in glass staining dishes in the indicated solutions. All solutions are made up fresh and are used only once unless indicated otherwise.

1. Prepare dewaxing/rehydration series (can be reused several times) and 0.2 N HCl while slides containing specimen sections (stored in slide boxes at −20° or −70°C; *UNIT 14.1*) are warmed to room temperature. Start preheating 2× SSC to 70°C (for use in step 5).

2. Dewax slides in dishes by three changes in xylenes, 2 min each (not necessary for slides containing unembedded single cells).

3. Rehydrate in dishes through the following regimen: 100% ethanol—twice, 2 min each; 95% ethanol—2 min; 70% ethanol—2 min; 50% ethanol—2 min.

4. Denature specimens 20 min at room temperature in 0.2 N HCl.

5. Heat denature 15 min at 70°C in 2× SSC. Rinse 2 min in 1× PBS.

   Optional step: Under certain circumstances, a pronase digestion step is included here. Specimens are digested 15 min at 37°C with 0.1 to 10 µg/ml predigested, lyophilized pronase (determined empirically; in general, the highest possible pronase concentration that still gives good cellular morphology is used). Pronase digestion is stopped by rinsing slides 30 sec in 2 mg/ml glycine prepared in 1× PBS.

6. Postfix specimens in freshly prepared 4% PFA fixative 5 min at room temperature. Block fixation 5 min in 3× PBS. Rinse twice, 30 sec each time, in 1× PBS.

7. Equilibrate specimens in 10 mM DTT prepared in 1× PBS 10 min at 45°C in a water bath.

8. Block with freshly prepared blocking solution 30 min at 45°C in water bath covered with aluminum foil (iodoacetamide is light sensitive).

   CAUTION: *Both iodoacetamide and N-ethylmaleimide are extremely toxic substances; handle very carefully.*

9. Rinse twice, 2 min each time, in 1× PBS at room temperature.

10. Equilibrate specimens 2 min in freshly prepared TEA buffer. Transfer slide rack to fresh TEA buffer and add acetic anhydride to 0.25% final. Mix quickly and incubate slides 5 min with agitation. Add additional acetic anhydride to 0.5% final and incubate 5 min.

11. Block specimens 5 min in 2× SSC.

12. Dehydrate specimens through 50% ethanol, 70% ethanol, 95% ethanol, and 100% ethanol (twice), 2 min each at room temperature (use same ethanol solutions as in step 3).

13. Air dry specimens (or dry in desiccator), making sure specimens are absolutely dry before proceeding. Store specimens in slide box with desiccant overnight at −70°C.

14. Centrifuge ethanol-precipitated antisense and sense ^{35}S-labeled riboprobes (Support Protocol 1), as well as S-riboprobe competitor. Dry pellets. Dissolve each pellet (corresponding to one reaction) in 5 µl sterile 50 mM DTT. Add 2.5 µl (half a reaction) of S-riboprobe competitor to both antisense and sense riboprobe.

15. Heat dissolved probes to 100°C for 3 min in heating block or water bath.

16. Immediately add enough hybridization mix A to obtain a 0.3 µg/ml final probe concentration (determination of probe mass synthesized is described in Support Protocol 1). Mix well and count 1 ìl (expected counts ≥1 × 10^5 cpm/µl). Place tubes in water bath at 45°C (hybridization temperature).

    If antisense and sense probes differ extensively (e.g., 5-fold) in their counts/µl, make sure it is not due to improper mixing and recount. If the difference is consistent, dilute

14.3

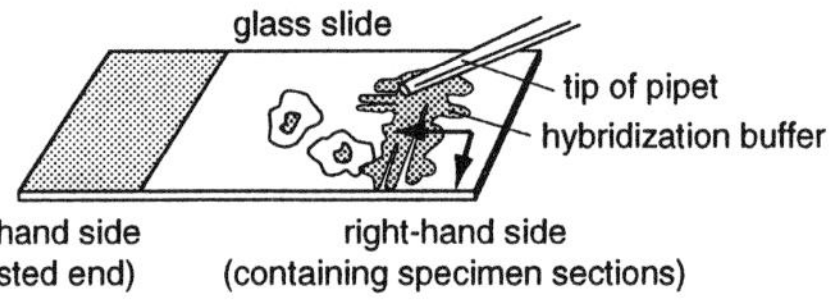

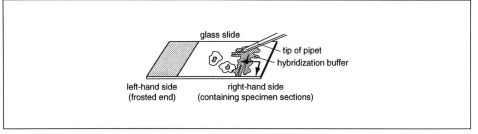

Figure 14.3.1 Application of hybridization mix to slides. Carefully spread hybridization mix on sections with the tip of a pipet (avoid touching sections). Arrows indicate directions of spreading of hybridization mix A (or B).

sample that has more counts to get a roughly equal counts/μl for both antisense and sense riboprobes. Labeled probes (substantially <10^5 cpm/ml) should not be used.

17. Set up hybridizations by carefully spreading an appropriate amount of probe on specimens (e.g., 20 μl/20 mm^3 using tip of a pipet; Fig. 14.3.1).

If background is a problem, a prehybridization step can be included prior to hybridization: take up one reaction of S-riboprobe competitor in 5 μl of 50 mM DTT, heat 3 min to 100°C, and add 500 μl hybridization mix A. Spread ≥20 μl of this prehybridization mix on specimens as in step 17. Incubate 1 to 2 hr at 45°C in humidified chamber. To remove mix, tip slides, causing mix to collect at one edge and carefully blot off buffer with Whatman 3MM paper. Proceed with step 17.

18. Place specimens (keep slides level) in moist chamber (Fig. 14.2.3) containing humidified chamber solution A and incubate at 45°C for appropriate hybridization time. Perform a series of hybridizations from 30 min to 4 hr (equilibrate and seal humidified chamber carefully because dried out specimens will result in high background; the osmolarity of the humidified chamber solution must equal that of hybridization mix to prevent its dilution or concentration).

19. During the last hour of hybridization, prepare and preheat wash solutions A, B, and C.

20. Start washing slides by dipping one at a time into 100 ml wash solution A at 55°C. Immediately place in a slide rack in a staining dish filled with wash solution A.

21. Incubate twice, 15 min each time, in wash solution A at 55°C. Incubate twice, 15 min each time, in wash solution B at 55°C. Incubate twice, 2 min each time, in wash solution C at room temperature.

The first dip into wash solution A removes much of the radioactivity, and should be disposed of into radioactive waste. For disposal of all other less radioactive wash solutions, follow local guidelines. Do not allow slides to dry during any hybridization and washing steps.

22. Add 500 μl RNase digestion solution per slide, covering all specimens, and place the slides in a moist chamber (containing water) as in Fig. 14.2.3 (but labeled "RNase"). Incubate 15 min, room temperature.

23. Wash slides twice, 30 min each time, in wash solution C at 50°C with gentle shaking. Wash slides twice, 30 min each time, in wash solution A at 50°C with gentle shaking. Wash slides twice, 5 min each time, in 2× SSC at room temperature.

24. Dehydrate through the following regimen (2 min each): 50% ethanol/0.3 M ammonium acetate; 70% ethanol/0.3 M ammonium acetate; 95% ethanol/0.3 M ammonium acetate; 100% ethanol.

25. Air dry slides. Expose slides at least overnight against film, and then perform emulsion autoradiography (*UNIT 14.4 & APPENDIX 3*).

HYBRIDIZATION USING CRYOSECTIONS

ALTERNATE PROTOCOL

Materials (also see Basic Protocol; see APPENDIX 1 for items with ✓)

Glass slides containing cryosections (*UNIT 14.2*)
50 mM Tris·Cl (pH 7.5)/5 mM EDTA
2 mg/ml glycine in 1× PBS
✓ 2× SSC
30%, 60%, 80%, 95%, and 100% ethanol
Labeled DNA or RNA probe (see Support Protocols)
✓ Hybridization mix B
✓ Deionized formamide
50% (w/v) dextran sulfate (Pharmacia)
✓ 3.3 M dithiothreitol (DTT), freshly prepared
✓ Moist chamber solution B
✓ DNA wash solution, prewarmed to 37°C
✓ RNA wash solutions I and II, prewarmed to 50°C
20 μg/ml boiled RNase A (*UNIT 3.13.*) in 0.5 M NaCl/10 mM Tris·Cl, pH 8.0
0.6 M NaCl in 30% ethanol *and* in 60% ethanol

2 sets of slide racks and jars (one set reserved for RNase use only)
50°C water bath, and 37° or 42°C incubator (or water bath)

1. Remove slides from freezer and allow to come to room temperature before opening box. Place slides in rack and immerse 10 min in predigested, lyophilized pronase in 50 mM Tris·Cl (pH 7.5)/5 mM EDTA.

   Digestion may sometimes be omitted for cryosections.

2. Rinse slides 30 sec at room temperature in PBS containing 2 mg/ml glycine, then twice in PBS for 30 sec each time.

3. Immerse slides 5 min in freshly prepared TEA buffer.

4. In a beaker, prepare enough TEA buffer to cover slides. Add acetic anhydride to 0.25% final concentration and pour on slides as quickly as possible. Jiggle rack to mix. Incubate 10 min.

5. Wash slides twice, 5 min each time, in 2× SSC.

6. Dehydrate 2 min each in 30%, 60%, 80%, 95%, and 100% ethanol. Dry and use immediately for hybridization.

7. Precipitate labeled DNA or RNA probes. Determine percent of label incorporated and estimate mass of probe synthesized. Calculate volume of final hybridization mix required to resuspend pellet at 0.2 μg/ml per kilobase final.

8. Resuspend pellet first in 2 parts hybridization mix B and 2 parts deionized formamide. Subsequently, add 1 part 50% dextran sulfate. Mix thoroughly.

   Probe may be prepared the previous day and stored frozen at −80°C.

9. Boil labeled DNA probes 2 min and chill in ice bath, or heat labeled RNA probes 30 sec at 80°C and hold at 50°C. After heating, add 3.3 M DTT to 50 mM final. Using pipet tip, distribute sufficient probe to cover sections, e.g., 20 µl/20 mm^2 (Fig. 14.3.1 and Support Protocols).

10. Incubate 4 hr in a well-sealed moist chamber containing moist chamber solution B. Incubate DNA probes at 37°C and RNA probes at 42°C.

11. *For DNA probes*: Wash 2 hr at 37°C with 4 to 5 changes of prewarmed DNA wash solution.

12. *For RNA probes:*

    a. Wash 15 min at least twice at 50°C in warm RNA wash solution I.

    b. Treat slides 30 min at 37°C with 20 µg/ml boiled RNase A in 0.5 M NaCl/10 mM Tris·Cl, pH 8.0.

    c. Wash 15 min at least twice at 50°C in warm RNA wash solution I.

    d. Wash 15 min twice at 50°C in prewarmed RNA wash solution II (wash temperature may be increased to 60°C, if desired, although morphology may suffer).

    NOTE: *Keep all glassware and racks used for washes and RNase treatment separate from those used for pretreatment and hybridizations.*

13. Dehydrate 2 min in each of the following solutions: 0.6 M NaCl in 30% ethanol, 0.6 M NaCl in 60% ethanol, 80% ethanol, 95% ethanol, and 100% ethanol.

14. Air dry slides and detect hybridized probe by autoradiography.

SUPPORT PROTOCOL 1

SYNTHESIS OF ^{35}S-LABELED RIBOPROBES AND S-RIBOPROBE COMPETITOR

Additional Materials *(also see Basic Protocol; see* APPENDIX 1 *for items with ✓)*

Gene of interest and appropriate vector—e.g., pSP64 or pSP65 (*UNIT 1.5*) or pT7 (*UNIT 16.2*)
✓ 5× transcription buffer
✓ 1 M dithiothreitol (DTT), freshly prepared
Ribonuclease inhibitor (e.g., Amersham placental ribonuclease inhibitor or Promega RNAsin)
10 mM CTP, ATP, and GTP (*UNIT 3.4*)
[^{35}S]UTP, specific activity 1000 to 1500 Ci/mmol (*UNIT 3.4*)
100 nM S-UTP (for S-riboprobe competitor; DuPont NEN)
SP6 or T7 RNA polymerase (*UNIT 3.8*)
10 mg/ml yeast tRNA or mouse poly(A) RNA for carrier
1 U/µl RNase-free DNase I (e.g., Promega RQ1 or *UNIT 3.12*)
✓ 3 M sodium acetate
✓ 7.5 M ammonium acetate
100% and 70% ethanol, ice-cold

NOTE: When ^{35}S-labeled probes are prepared, it is very important to add 10 mM DTT to all solutions containing [^{35}S]UTP, particularly after any step that may inactivate DTT (e.g., precipitation, column chromatography, and boiling). Extreme care should be taken to prevent RNase contamination.

1. Insert gene of interest into a vector containing an SP6, T3, or T7 promoter. Linearize (*UNIT 3.1*) the plasmid downstream of the coding sequence. Phenol/chloroform extract the DNA (*UNIT 2.1*), ethanol precipitate, wash pellet in 70% ethanol, dry, and redissolve at 1 µg/µl in sterile water.

   Alternatively, treat with 0.1% fresh diethylpyrocarbonate (DEPC) for 10 min at room temperature, heat to 65°C for 10 min, and ethanol precipitate as above. The DNA should be clean and free of salt after the restriction digestion in order to avoid terminating transcription.

2. Prepare following reaction mix at room temperature (20 µl total):

   4.0 µl 5× transcription buffer
   0.2 µl 1 M DTT
   60 U ribonuclease inhibitor
   1.0 µl each of the three 10 mM NTPs
   1.0 µl 1 µg/µl digested DNA (step 1)
   10.0 µl [^{35}S]UTP (use S-UTP for S-riboprobe competitor; see below)
   16 U SP6 or T7 RNA polymerase.

   Incubate 30 min at 37°C. Add 16 U more polymerase and incubate 40 min at 37°C.

   S-riboprobe competitor is synthesized using a vector without an insert as template. Replace the [^{35}S]UTP in the transcription reaction with 100 nM nonradioactive S-UTP. Half of a reaction of S-riboprobe competitor is required per ^{35}S-labeled riboprobe reaction.

3. Add the following to the reaction mix: 60 U ribonuclease inhibitor, 2.0 µl of 10 mg/ml carrier RNA, and 1.0 µl DNase I. Incubate 10 min at 37°C to remove template.

4. Add the following to the reaction mix: 0.8 µl 1 M DTT, 63.0 µl sterile H_2O, and 10.0 µl of 3 M sodium acetate. Remove 1 µl and determine cpm/µl.

5. Add 36.4 µl of 7.5 M ammonium acetate to reaction mix (2 M final). Add 50 to 100 µg of tRNA carrier and 272 µl of cold 100% ethanol. Precipitate 10 min on dry ice, pellet, wash with cold 70% ethanol, and dry. Repeat if desired. Resuspend in 100 µl of 10 mM DTT.

6. Determine cpm/µl and calculate percent incorporation. Riboprobes should be used within 2 to 3 days (70% to 90% of label is incorporated resulting in 70 to 90 ng labeled RNA.)

SYNTHESIS OF ^{35}S-LABELED DOUBLE-STRANDED DNA PROBES

SUPPORT PROTOCOL 2

Radioactive double-stranded DNA probe may be synthesized by nick translation or random oligonucleotide-primed synthesis using [^{35}S]dNTPs. Add 10 mM DTT to standard reaction mixes (*UNIT 3.5*) and replace [^{32}P]dNTP with two different [^{35}S]dNTPs (specific activity ≥1000 Ci/mmol) at as high a final molarity as possible, usually 2 to 4 µM. Note that because the rate of [^{35}S]dNTP incorporation is slower than that of [^{32}P]dNTPs, the reaction should be optimized for maximum incorporation and desired DNA length. Incorporation should be as high as 5×10^8 cpm/µg.

References: Hogan et al., 1986; Pardue, 1985.

Contributors: Rolf Zeller and Melissa Rogers

UNIT 14.4

Detection of Hybridized Probes

Autoradiographic film is used to detect ^{32}P- or ^{35}S-labeled probes that are hybridized to cytological preparations, and can be useful in experiments dealing with large organs or tissues. Emulsion autoradiography is required to obtain resolution at the level of a single cell.

BASIC PROTOCOL 1

FILM AUTORADIOGRAPHY

Tape slides to a backing such as cardboard or an old piece of film. Expose slides to Du Pont Cronex Video Imaging Film (MRF 34 Clear) at 4°C under light pressure. When using Du Pont Cronex Video Imaging Film, make sure that the emulsion side of the film is facing specimens.

BASIC PROTOCOL 2

EMULSION AUTORADIOGRAPHY

Materials

Diluted Kodak emulsion (Support Protocol)
Kodak D19 developer and fixer

Dipping chamber: plastic cytology slide mailer (Curtin Matheson)
Nonsparking fan or slide dryer (Oncor; optional)
Kodak safelight filter #2 (optional)
Black, light-tight slide boxes
Desiccant (e.g., Humicaps, United Desiccants-Gates)
42° to 45°C water bath
Slide racks and jars, for use in developing and fixing

15° to 20°C water bath (a styrofoam box works well)
Slide rack or wire test-tube rack for drying dipped slides

NOTE: Perform steps 1 to 7 in complete darkness or ~4 feet from safelight.

1. Melt aliquot of diluted emulsion 10 min in 42° to 45°C water bath. Pour or pipet emulsion into clean slide mailer (dipping chamber).
2. Dip slides slowly and smoothly into dipping chamber. Withdraw slowly and place vertically in a test-tube rack or a slide dryer 2 hr to dry (to prevent artifacts caused by running of emulsion with tissue sections, place dipped slides 10 min on a cold glass plate before drying).
3. Place thoroughly dry slides in a light-tight slide box with desiccant. Seal box with electrician's black tape, cover with foil, and expose at 4°C (do not expose in refrigerator used for storing ^{32}P or organic solvents).
4. Put developer, water, and fixer in slide jars and bring to 15° to 20°C in water bath. Take slides out of refrigerator and allow to warm to same temperature as developing solutions.
5. Transfer slides to slide racks and develop in light-tight darkroom as follows (monitor time carefully with timer): 2.5 min in developer; 30 sec in water; and 3 min in fixer (do not agitate vigorously).
6. Rinse slides 10 to 15 min under gently running cool tap water in light, then once in cool distilled water. While slides are still wet, scrape off emulsion on back of slide with razor, then dry in dust-free location.

7. Counterstain slides as desired and mount under clean cover slips (*UNIT 14.5*). Observe developed silver grains by dark-field microscopy, under which they appear as white dots on black background.

PREPARATION OF DILUTED EMULSION FOR AUTORADIOGRAPHY

SUPPORT PROTOCOL

Additional Materials *(also see Basic Protocol)*

Kodak NTB-2 autoradiographic emulsion

NOTE: Perform all steps in complete darkness or ~4 feet from safelight.

1. Heat 4-oz. bottle of Kodak NTB-2 autoradiographic emulsion in 42° to 45°C water bath 30 min in dark. At the same time, warm equal volume of water in 500-ml flask to same temperature.
2. After emulsion has melted, slowly pour down side of flask (held at an angle) and mix (swirl gently once or twice; avoid creating bubbles).
3. Aliquot emulsion into nylon scintillation vials or plastic slide mailers. Wrap each aliquot in aluminum foil or store in light-tight container in refrigerator that is never used for storing ^{32}P or organic chemicals.

   Remove cork inserts present in some scintillation vial caps. Emulsion stored in slide mailers must be placed vertically to prevent leakage before emulsion solidifies.
4. Test diluted emulsion before use by dipping and developing two clean, unused slides (Basic Protocol 2) and visualizing under a 100× lens. If >100 grains per microscope field are observed, emulsion should be returned to Kodak.

Reference: Pardue, 1985.

Contributor: Melissa Rogers

14.5

Counterstaining and Mounting of Autoradiographed In Situ Hybridization Slides

UNIT 14.5

The morphology of specimen sections and the identity of specific areas are defined by lightly counterstaining sections. After staining, slides are dehydrated, mounted with coverslips, hardened, and cleaned for examination under the microscope. In the protocols provided below, Giemsa stains predominantly the nuclei, hematoxylin/eosin stain differentiates both nuclei and cytoplasm, toluidine blue staining is a simpler procedure that lightly stains both nuclei and cytoplasm, and Hoechst staining of nuclei provides a fast, easy, and effective way to simultaneously view the entire tissue and the regions of hybridization.

GIEMSA STAINING

BASIC PROTOCOL

Materials *(see APPENDIX 1 for items with ✓)*

Hydrated, developed in situ hybridization slides (*UNITS 14.3 & 14.4*)
Giemsa stain (Fisher)
✓ 10 mM sodium phosphate buffer, pH 6.8 (1:50 dilution of 500 mM stock)
50%, 70%, 95%, and 100% ethanol
Xylenes

Mounting medium: Permount (Fisher) *or* Gelvatol (now called Airvol, Air Products and Chemicals; *APPENDIX 1*)

13 glass staining dishes
Whatman 3MM paper chips
42°C incubator

CAUTION: Xylenes are toxic organic solvents. All steps using xylenes must be carried out under a fume hood.

NOTE: Perform steps at room temperature except where noted otherwise.

1. Immerse developed slides in slide rack in staining dish filled with water.

2. Prepare the following regimen of staining dishes filled with solutions: 1 dish—25-fold dilution of Giemsa stain in 10 mM sodium phosphate buffer, pH 6.8; 3 dishes—H_2O; dehydration series (can be reused, or use the series prepared for dewaxing in *UNIT 14.3*)—3 dishes each filled with 50%, 70%, and 95% ethanol, respectively, and 2 dishes filled with 100% ethanol; and 3 dishes—xylenes.

3. Stain slides 20 sec in a 25-fold dilution of Giemsa stain (depending on staining time, a weak or intense stain can be achieved). Rinse slides in water three times, 2 min each time.

4. Dehydrate slides through ethanol series, 2 min in each dish. Transfer to xylenes and do three changes, 2 min each time.

   Perform dehydration and equilibration in xylenes carefully because Permount will not mix with water or ethanol.

5. In hood, mount slides as follows: With blunted forceps (in left hand), remove one slide from xylenes and hold it flat at its frosted end. Add ~4 drops Permount to the other end (where sections are located). With the right hand, grab a clean coverslip and place it very slowly on slide as in Figure 14.5.1 (do not let the slide start to dry before mounting medium and coverslip are added because microbubbles can cause artifacts).

   Use Gelvatol when the cytological preparation should not be dehydrated, e.g., in the case of immunohistochemistry. Mount slides with Gelvatol as follows: Place a small drop of Gelvatol on a coverslip. Carefully set slide containing the specimen on the drop, avoiding formation of bubbles. Avoid applying pressure as this may squash the specimen. Allow to harden 2 to 4 hr at room temperature.

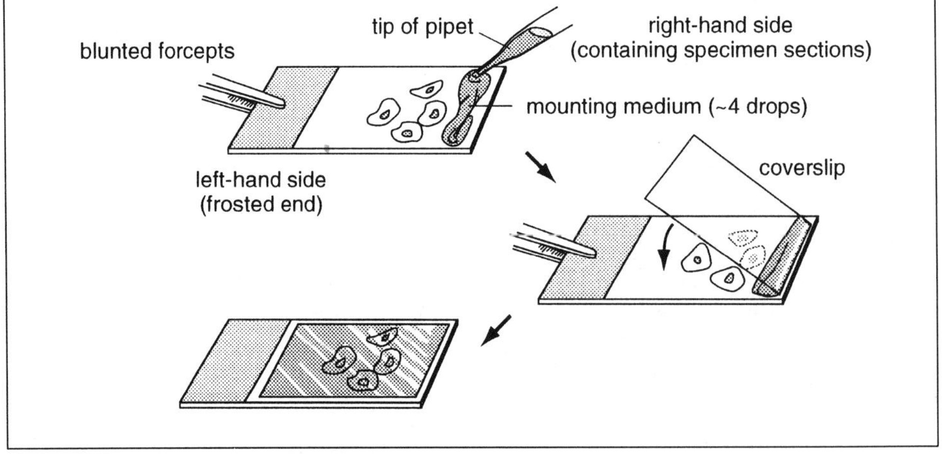

Figure 14.5.1 Mounting of slides with mounting medium.

6. Remove extra mounting medium/xylenes by blotting carefully around edges with 3MM paper and by wiping back of slide (but not coverslip).

   Do not store slides upright in slide boxes because mounting medium has not yet hardened.

7. Place slides flat on a cardboard tray and allow to harden ~2 days in a 42°C incubator.

8. Remove excess emulsion, stain, and mounting medium from backs of slides by carefully scraping with razor blade; remove dust with lens paper or plain tissue. Transfer to slide boxes. If necessary, relabel slides.

9. Examine slides microscopically. For weak signals, use dark-field illumination; for strong signals, use transillumination.

HEMATOXYLIN/EOSIN STAINING

ALTERNATE PROTOCOL 1

Additional Materials *(also see Basic Protocol; see* APPENDIX 1 *for items with ✓)*

Hematoxylin stain (Harris modified hematoxylin with acetic acid, mercury-free; Fisher)
0.1% ammonium hydroxide
✓ Eosin stain

1. Place developed slides (in slide rack) in staining dish filled with water.

2. Prepare the following regimen of solutions in staining dishes: 1 dish—hematoxylin stain; 2 dishes—H_2O; 1 dish—0.1% NH_4OH; 2 dishes—H_2O; 1 dish—eosin stain; 1 dish—95% ethanol; 2 dishes—100% ethanol; and 3 dishes—xylenes.

3. Stain slides 20 to 30 sec in hematoxylin. (Do not overstain.) Rinse two times in water, 2 min each time.

4. Quickly dip into 0.1% ammonium hydroxide, then rinse in water.

   If dip in 0.1% ammonium hydroxide is too long, autoradiographic emulsion may start to come off slides.

5. Wash 5 min in water, then eosin stain 20 to 30 sec.

   If overstaining occurs, destain by dipping slides in 50% ethanol for a short time (watch stain diffusing out). If stain is too weak, restain.

6. Dip slides eight times in the same 95% ethanol solution, then dip again eight times in the same 100% ethanol solution. Fully dehydrate 2 min in 100% ethanol. Equilibrate three times, 2 min each, in xylenes.

7. Mount and process as in steps 5 to 9 of Basic Protocol.

TOLUIDINE BLUE STAINING

ALTERNATE PROTOCOL 2

Additional Materials *(also see Basic Protocol; see* APPENDIX 1 *for items with ✓)*

✓ Toluidine blue stain

1. Dilute toluidine blue stain with water (determine dilution empirically according to degree of staining desired; start at ~1:100).

2. Dip slides briefly in diluted stain solution. Rinse several times with water and mount as desired.

ALTERNATE PROTOCOL 3

HOECHST STAINING

Additional Materials (also see Basic Protocol)

1 mg/ml Hoechst stain in dimethyl sulfoxide (Hoechst 33258 dye, bisbenzimide; store at −20°C)
Mounting medium: 0.5 g/ml Canada balsam in methyl salicylate (store at room temperature), *or* Gelvatol (*APPENDIX 1* and Basic Protocol)
Fluorescence microscope with Hoechst or DAPI filter and dark-field optics

CAUTION: Hoechst dye is a carcinogen.

1. Dilute 1 mg/ml Hoechst stain 1:500 with water to 2 µg/ml final. Cover sections with diluted stain or dip slides in diluted stain. Incubate 2 min at room temperature.
2. Wash 2 min in water at room temperature. Air dry at room temperature.
3. Bake slides 30 to 60 min in 42° to 55°C incubator. Remove and bring to room temperature.
4. Mount as in Figure 14.5.1 using 0.5 g/ml Canada balsam and air dry slides 2 days at room temperature (after mounting).

   Do not use Permount because autofluorescence occurs when nuclei are viewed. Canada balsam clears paraffin sections well and does not fluoresce under UV irradiation. Gelvatol may be used for cryosections but should not be used for paraffin sections, as it does not clear tissues.

5. View nuclei using Hoechst epifluorescence optics. Cover light source with red filter and view silver grains simultaneously with dark-field illumination. The silver grains will appear red and will contrast well with the fluorescent blue nuclei. To photograph, make a double exposure of nuclei and silver grains.

Reference: Luna, 1968.

Contributors: Rolf Zeller and Melissa Rogers

UNIT 14.6

Immunohistochemistry

This unit surveys immunohistochemical techniques for protein localization. The three protocols focus on indirect immunofluorescence, which is the easiest and most widely used method of optical detection. Alternate methods suitable for bright-field microscopy—such as immunoperoxidase and immunogold labeling—are then described; these protocols employ both primary and secondary antibodies (a fairly complete list is compiled annually in *Linscott's Directory*).

Today, commercially available secondary markers have very high specificity, purity, and signal strength, and are available directed against primary antibodies from most animals. If problems are encountered, refer to Table 14.6.1 for troubleshooting. Because of various problems of background and nonspecific labeling inherent in any immunohistochemical analysis of cells and tissues, the following controls are suggested (see also Table 14.6.1): (1) use of preimmune serum from the antibody-producing animal used at an equivalent dilution, (2) use of secondary antibody alone, (3) use of no antibodies at all, (4) use of an antibody, the antigen for which is not found in the tissue or cells to be studied, (5) use of a known positive antibody (e.g., actin), and (6) use of a limited dilution series for the primary antibody for each experiment.

IMMUNOFLUORESCENT LABELING OF CELLS GROWN AS MONOLAYERS

BASIC PROTOCOL 1

Materials *(see* APPENDIX 1 *for items with ✓)*

Confluent cells grown in a 3- to 5-cm dish or in Labtek multiwell slide (VWR; smaller vessels use less antibody; however, the vessel must fit under the microscope)

✓ PBS, 4°C

✓ 2% (w/v) paraformaldehyde (PFA) fixative (1:1 dilution with PBS of 4% PFA fixative) 4°C—for cell surface antigens

100% methanol, −10° to −20°C (solvent cooled in ice box of a refrigerator is ideal) *or* 2% PFA fixative containing 0.1% Triton X-100, 4°C—for cytoplasmic antigens

Primary antibody, ~5 to 10 µg/ml

Secondary antibody–fluorochrome conjugate specific to the source species of primary antibody

1. Cool confluent cells on ice. Pipet off culture medium and wash in 4°C PBS. Pipet off PBS.

2. If cell surface antigens are being studied, fix 30 min in 2% PFA fixative on ice. If cytoplasmic antigens are being studied, fix either 30 min in 2% PFA fixative/0.1% Triton X-100 on ice, or 15 min in 100% methanol in freezer compartment of a standard refrigerator (−10° to −20°C).

   All pipetting steps are done with Pasteur pipets attached to a pump. Both methods of fixation work well; however, the paraformaldehyde/Triton X-100 fix frequently gives a better final morphology. If methanol fixation is used, be sure that cells are washed thoroughly with methanol before putting into freezer or they will freeze.

3. Pipet off fixative and wash twice in 4°C PBS (5 min/wash). Microcentrifuge diluted primary antibody 2 min at 13,500 × g, 4°C.

4. Layer primary antibody into dish such that cells are just covered and incubate 1 hr at 4°C. Wash four times in 4°C PBS (5 min/wash).

5. Microcentrifuge diluted secondary antibody 2 min at 13,500 × g, 4°C.

6. Layer secondary antibody into dish and incubate 1 hr at 4°C. Wash four times in 4°C PBS.

7. Store cells in PBS. Unless cells are to be observed immediately, cover dishes, wrap in aluminum foil, and refrigerate. It is important to examine preparations of this type within 24 hr as fluorescence rapidly fades and/or dissociates from cells.

   *If Labtek slides are being used, strip chambers and mount in Gelvatol (*APPENDIX 1 *and Basic Protocol,* UNIT 14.5*) at this point.*

ALTERNATE PROTOCOL 1

IMMUNOFLUORESCENT LABELING OF SUSPENSION CELLS

Additional Materials

5×10^6 to 10^7 cells in media in 15-ml Falcon tube
Poly-L-lysine–coated slides (*UNIT 14.1* Support Protocol)

1. Cool cells on ice. Centrifuge 5 min at $800 \times g$, 4°C, in tabletop centrifuge (for lymphocytes). Pipet off medium and resuspend cells in 4°C PBS.

   Speed is adjusted for cell type but must be slow enough to prevent damage. Unless otherwise stated, all centrifuge steps are done under these conditions.

   All pipetting steps are performed with pipets attached to a pump.

2. Centrifuge, pipet off PBS, and fix by resuspending cells 30 min on ice in 1 to 2 ml of 2% PFA fixative or 2% PFA fixative/0.1% Triton X-100 (see Basic Protocol).
3. Centrifuge, pipet off fixative, and resuspend cells in 15 ml of 4°C PBS. Leave 5 min and repeat with a second PBS wash.
4. Microcentrifuge primary antibody 2 min at $13{,}500 \times g$, 4°C. An antibody volume of 250 µl is adequate for 5×10^6 cells.
5. Centrifuge cells, pipet off PBS, resuspend in primary antibody, and incubate 1 hr at 4°C.
6. Dilute cells and antibody to 15 ml with 4°C PBS.
7. Centrifuge, pipet off PBS, and resuspend cells in 4°C PBS. Repeat once.
8. Microcentrifuge secondary antibody 2 min at $13{,}500 \times g$, 4°C. An antibody volume of 250 µl is adequate for 5×10^6 cells.
9. Pipet off PBS, resuspend cells in secondary antibody, and incubate 1 hr at 4°C. Repeat steps 6 and 7.
10. Unless cells are to be observed immediately, wrap tube in aluminum foil and refrigerate (<24 hr)—prior to final concentration step below.
11. Centrifuge cells and resuspend in small volume of PBS (avoid water). Pipet onto poly-L-lysine–coated slide, cover with coverslip, and observe immediately if possible.

BASIC PROTOCOL 2

IMMUNOFLUORESCENT LABELING OF TISSUE SECTIONS

Materials (*see APPENDIX 1 for items with ✓*)

Specimen tissue cyrosections on glass slides (*UNIT 14.2*)
✓ PBS
Primary antibody, ~5 to 10 µg/ml
Secondary antibody–fluorochrome conjugate specific to the source species of primary antibody
Mounting medium (e.g., Gelvatol; *APPENDIX 1 & UNIT 14.5*)
Plastic slide box or moist chamber (*UNIT 14.2*)

1. Layer wet paper towels in base of slide box to make a moist chamber, or make a chamber as in *UNIT 14.2.* Remove slides with sections from cryostat or freezer,

place across slide box (~6 per side), or place in moist chamber (slides should not touch).

2. Once slides are at room temperature, and before they air dry, layer PBS over sections (do not flood slides).

3. Microcentrifuge diluted primary antibody 2 min at 13,500 × g, 4°C (40 to 50 μl antibody should cover sections on each slide).

4. Remove PBS from slides by aspirating at one end of sections with Pasteur pipet connected to pump and introduce antibody at other end. Close box and incubate 1 hr at room temperature.

5. Wash slides three times in PBS (5 min/wash).

   Introduce new buffer at one end and aspirate off old buffer from opposite end of sections.

6. Microcentrifuge diluted secondary antibody 2 min at 13,500 × g, 4°C, (allow 40 to 50 μl antibody/slide).

7. Layer secondary antibody over sections and incubate in moist chamber 1 hr at room temperature. Wash slides three times in PBS (5 min/wash).

8. Lay coverslips on paper towels and place a drop of Gelvatol in middle of coverslip. Invert slides on coverslip (do not apply pressure). Leave slides 30 min on bench under aluminum foil to keep out light to allow Gelvatol to harden.

   If Gelvatol is unavailable, glycerol or PBS may be used for temporary mounting.

9. Observe under microscope or store at 4°C in closed slide box.

IMMUNOFLUORESCENT LABELING USING STREPTAVIDIN-BIOTIN CONJUGATES

ALTERNATE PROTOCOL 2

This protocol describes a triple-step reaction technique that increases the sensitivity of the immunohistochemical reaction by using a streptavidin–secondary antibody conjugate and then a reaction with a biotin-fluorochrome conjugate (Fig. 14.6.1D)

Additional Materials *(also see Basic Protocol 2)*

Biotinylated secondary antibody (Vector Laboratories)
Fluorochrome-streptavidin conjugate (Vector Laboratories)

Follow steps 1 to 5 of Basic Protocol 2, then proceed as indicated.

5. Microcentrifuge diluted biotinylated secondary antibody 2 min at 13,500 × g, 4°C (~40 to 50 μl antibody/slide).

6. Lay secondary antibody over sections and incubate in moist chamber 1 hr at room temperature.

7. Wash slides three times in PBS (15 min/wash). Layer fluorochrome-streptavidin (~40 to 50 μl antibody/slide) over sections and incubate in moist chamber 1 hr at room temperature. Wash slides three times in PBS as in step 5 of basic protocol.

8. Follow Basic Protocol 2, steps 8 and 9.

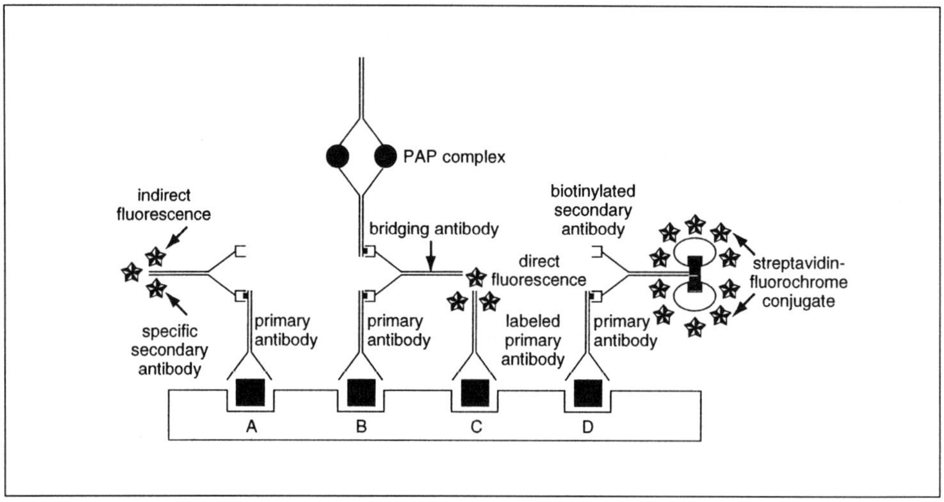

Figure 14.6.1 Various methods of immunohistochemical labeling. (**A**) The primary antibody bound to a specific epitope has been revealed by a secondary antibody conjugated to a fluorochrome. This may then be viewed by fluorescence microscopy. (**B**) The PAP–immunoperoxidase method. The primary antibody is bound to its epitope and has been coupled to PAP complex by a bridging antibody. Following development of the DAB substrate solutioin, a brown coloration at the epitopic site will be seen by bright-field microscopy. (**C**) The original method of labeling tissue or cells for immunofluorescence employed a primary antibody directly conjugated to a fluorochrome that could then be viewed by fluorescence microscopy. (**D**) The use of streptavidin-biotin conjugates allows considerable amplification of the section so that low-level binding may be clearly seen. The primary antibody has bound specifically to the substrate which is then labeled with biotinylated secondary antibody. This complex is then revealed with a streptavidin fluorochrome.

ALTERNATE PROTOCOL 3

IMMUNOGOLD LABELING OF TISSUE SECTIONS

Additional Materials (also see Basic Protocol 2)

Immunogold conjugate (Amersham)

Follow Basic Protocol 2, except as noted below.

7. Incubate 2 hr in immunogold-conjugated secondary antibody.

9. Observe under microscope or store at *room temperature* in closed slide box.

ALTERNATE PROTOCOL 4

IMMUNOPEROXIDASE LABELING OF TISSUE SECTIONS

This is a very sensitive method. The antibody constituting the HRPO-antiperoxidase (PAP) complex must be of the same type as the primary antibody to be recognized by the bridging antibody (Fig. 14.6.1B).

Additional Materials (also see Basic Protocol 2; see APPENDIX 1 for items with ✓)

0.25% hydrogen peroxide in PBS
Secondary antibody (a bridging antibody recognizing both primary antibody and PAP complex)
Horseradish peroxidase–antiperoxidase (PAP) complex (from same source species as the primary antibody)
✓ Diaminobenzidene (DAB) substrate solution

CAUTION: DAB is a carcinogen; handle with great care.

Follow steps 1 and 2 of Basic Protocol 2, then proceed as indicated below. All pipetting steps are performed with Pasteur pipets attached to a pump.

3. Incubate slides 30 min in 0.25% hydrogen peroxide in PBS at room temperature. Wash three times in PBS.

4. Follow steps 3 to 5 of Basic Protocol 2.

5. Incubate slides 1 hr at room temperature in specific secondary bridging antibody. Wash three times in PBS.

6. Incubate slides 1 hr at room temperature in PAP complex. Wash three times in PBS.

7. Develop 2 to 5 min (determined empirically) in DAB substrate solution at room temperature.

8. Wash and mount slides (Basic Protocol 2, steps 7 and 8). Keep slides in dark in slide box (it is unnecessary to refrigerate them).

IMMUNOFLUORESCENT DOUBLE-LABELING OF TISSUE SECTIONS

ALTERNATE PROTOCOL 5

Indirect immunofluorescence microscopy allows two or more antigens to be revealed on the same section at any one time. This is done by using fluorochromes that are excited by, and emit light of, different wavelengths. Generally this is limited to two fluorochromes, the most popular combination being rhodamine (excited in green range and emitting red) and fluorescein (excited in blue and emitting green), though more exotic combinations using fluorochromes such as phycoerythrin are possible. However, the following should be observed, regardless of the fluorochrome used:

1. When performing double-labeling experiments, the most important criterion is that the primary antibodies be of different types, such that secondary antibodies can recognize them independently. It is preferable that they originate in different animals; however, if monoclonals are being used, this may not be possible. When monoclonals are used, antibodies of different types such as IgG and IgM may be differentiated reliably, though antibodies of different classes (IgG1 and IgG2, for example) cannot generally be differentiated by secondary antibodies.

2. The protocol is the same as Basic Protocol 2 except that the primary and secondary antibodies are replaced by mixtures of two primary and two secondary antibodies.

3. Prior to attempting a double-labeling experiment, single-labeling experiments should be performed to find optimal dilutions for the different primary and secondary antibodies.

References: *Linscott's Directory of Immunological and Biological Reagents;* Sternberger, 1970.

Contributor: Simon Watkins

Table 14.6.1 Troubleshooting Guide for Immunofluorescent Labeling of Specimen Sections

Problem	Possible cause	Solution
Slides appear "faded"	Slides were not kept in the dark and the cold prior to observation.	It is extremely important that slides are stored in the cold in the dark.
Morphology appears grainy; labeling is grainy or weak	Slide dried out when removed from the cryostat or lyophilized in the cryostat or −80°C freezer.	Deal with slides promptly; *do not allow slides to dry out at any time,* unless the antigen is extremely robust and abundant.
Fluorescence is seen even in unlabeled material	Autofluorescence, frequently caused by the use of aldehyde fixatives	Block most autofluorescence by preincubating sections after the first PBS wash for 10 min in 0.5% sodium borohydride ($NaBH_4$) in water (prepare immediately before use); frequently, bubbles appear on the section; this is normal, but treatment is rigorous and may cause section loss.
	With monolayer labeling, autofluorescence may be caused by the culture dish; test by examining a new culture dish under the fluorescence microscope.	Grow cells on glass cover slips or in sterile culture chambers in glass slides.
Fluorescent labeling appears over section—even in the absence of a primary antibody	Secondary antibody is too concentrated	Dilute antibody further; normally, commercial immunoconjugates work at a 1:100 dilution, although this is a very general starting concentration; for best results, do a dilution series with each batch of secondary antibody.
Bright spots of fluorescence occur nonspecifically across section and slide	Inadequate washing during immunolabeling or a failure to spin unconjugated fluorochrome out of the secondary antibody	Make sure secondary antibodies are spun 5 min in microcentrifuge (13,500 × *g*) as soon as they are received; this is particularly important for antibodies shipped as lyophilized product; in most cases, a 10- to 20-µl pellet will result, which should be discarded.
Labeling appears specific, but background is excessive	Nonspecific binding of primary or secondary antibody	Preincubate sections 1 hr in 1% serum (eg., normal goat serum) prior to labeling with primary antibody; be sure that the blocking antibody is not recognized by the secondary antibody (as this will only compound labeling problems) by blocking with serum from same animal species as that used to make the secondary antibody; it is also possible to dilute all antibodies in 1% to 10% serum, thereby avoiding the blocking step.

continued

Table 14.6.1 Troubleshooting Guide for Immunofluorescent Labeling of Specimen Sections, continued

Problem	Possible cause	Solution
Labeling is specific, but not where expected, or mutually exclusive labeling patterns are seen on the same section	Antiserum contains tissue-specific antibodies other than those to the target antigen.	Confirm problem by labeling sections with preimmune serum, i.e., serum taken from the antibody-producing animal prior to injection of antigen; serum should always be taken prior to immunization and should be part of the antibody production process; avoid nonspecific labeling of this type by affinity purification of the antibody.
No labeling is seen on sections even though clear label is seen on western blots of the same material	The tissue section used does not contain the antigen.	Be sure to cut sufficient sections so that all possible structural features of the tissue are sampled.
	The antigen is not recognized by the antibody in its native form; this may occur if the antibody was raised against denatured rather than native antigen and happens more frequently with monoclonal than polyclonal antibodies, because the latter include antibodies to a variety of epitopes.	Make new antibody using native antigen.
	Tissue processing prior to labeling has rendered the antigen nonreactive.	While many antigens may be air dried in sections and still retain strong antigenic reactions (see Alternate Protocol for immunofluorescent staining using Coplin jars), if labeling is not seen using this protocol, use Basic Protocol.
	Overfixation leading to conformational changes in the antigen prior to freezing (*UNIT 14.2*)	Reduce the fixation time of the sample.
Labeling appears blurry and streaked across slide	On rare occasions, very soluble antigens may be found which dissolve out of the section during labeling; a more probable cause is smearing of the tissue during sectioning (see sectioning troubleshooting, Table 14.2.1).	Fixation of tissue prior to freezing or fixation of section itself.

In Situ Hybridization and Detection Using Nonisotopic Probes

Nonisotopic in situ hybridization can be used to determine the cellular location and the relative levels of expression of specific transcripts within cells and tissues. RNA in prepared specimens is hybridized with a probe labeled nonisotopically with biotin or digoxigenin. Nonisotopic probes are generally detected by fluorescence or enzymatic methods (Figs. 14.7.1 & 14.7.2).

NOTE: Water should be treated with DEPC and all solutions prepared with DEPC-treated water to inhibit RNase activity. See *APPENDIX 1* for instructions.

CAUTION: DEPC is a suspected carcinogen and should be handled carefully.

BASIC PROTOCOL 1

FLUORESCENCE IN SITU HYBRIDIZATION

Materials *(see APPENDIX 1 for items with ✓)*

Glass slide containing specimen (*UNIT 14.1*)
20 to 150 ng nonisotopically labeled DNA probe (*UNIT 3.18*)
✓ Deionized formamide (American Bioanalytical)
✓ 10 mg/ml sonicated salmon sperm DNA
✓ Master hybridization mix
50% (v/v) formamide (not deionized)/2× SSC
✓ 1×, 2×, and 4× SSC, pH 7.0
✓ Biotin detection solution *or* digoxigenin detection solution *or* biotin/digoxigenin detection solution
0.1% (v/v) Triton X-100/4× SSC
✓ DAPI *or* propidium iodide staining solution
✓ Appropriate antifade mounting medium

39°, 42°, and 72°C water baths
Phase-contrast microscope
22-mm^2 coverslips
Rubber cement
Moist chamber (Fig. 14.2.3)
Slide box with desiccant (Baxter Scientific)
Nail polish

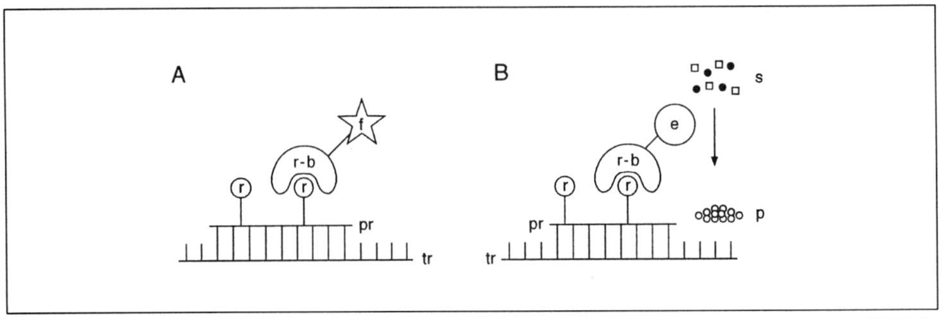

Figure 14.7.1 Detection of hybridized probe. (**A**) Fluorescence in situ hybridization (FISH). (**B**) Enzymatic detection. Abbreviations: e, enzyme (e.g., alkaline phosphatase or horseradish peroxidase); f, fluorochrome (e.g., fluorescein, rhodamine, or Texas red); p, colored precipitate product; pr, probe labeled with reporter molecule; r, reporter molecule (e.g., biotin, digoxigenin); r-b, reporter-binding molecule (e.g., avidin, streptavidin, or digoxigenin antibody); s, soluble substrate; tr, transcript. Arrow indicates reaction catalyzed by enzyme.

Fluorescence microscope with epillumination and filter set(s) appropriate for fluorochrome(s) used including dual-band pass filter (fluorescein/Texas red) or triple-band pass filter (fluorescein/Texas red/DAPI; Omega Optical or Chroma Technology)

Ektar-1000 *or* Ektachrome-400 color film *or* Technical Pan 2415 black-and-white film (Kodak)

CAUTION: Formamide, DAPI, and propidium iodide are hazardous; see manufacturer's information for guidelines on handling, storage, and disposal.

1a. *For hybridization to paraffin sections or cells:* Prepare slides by dewaxing, rehydrating, blocking, and dehydrating as in *UNIT 14.3*, Basic Protocol, steps 1 to 7, finishing with air drying the specimens (or drying in dessicator), followed by storage in a slide box with dessicant overnight at −70°C.

It is essential that the specimens be absolutely dry before they are stored.

1b. *For hybridization to cryosections:* Pretreat sections and acetylate as described in *UNIT 14.3*, Alternate Protocol, steps 1 to 7, finishing with dehydration of slides.

It is essential that the specimens be absolutely dry. Slides are now ready for hybridization (step 3), which should be done immediately.

2. For each hybridization precipitate 10 to 15 ng of probe in ethanol and dissolve in 10 µl deionized formamide. Add 5 µg (0.5 µl) sonicated salmon sperm DNA. Heat denature the probe 10 min at 70 to 80°C.

3. Add 10 µl master hybridization mix to denatured probe (0.1 to 0.5 µg/ml final probe concentration). Mix well, spin briefly at high speed in a microcentrifuge, and transfer onto preparation on slide. Cover with 22-mm^2 coverslip, remove any large air bubbles with gentle pressure, and incubate 2 to 4 hr in a moist chamber at 37°C.

Hybridizations can also be performed overnight. In this case, seal the coverslip using rubber cement.

4. During last 30 min of hybridization, warm 50 ml of 50% formamide/2× SSC wash solution and 50 ml of 2× SSC in Coplin jars in a 37°C water bath.

5. Remove slide from moist chamber. Peel off rubber cement (if present) and carefully remove coverslip. Wash hybridized slide 15 min in 37°C 50% formamide/2× SSC, 15 min in 37°C 2× SSC, and 15 min in room temperature 1× SSC.

6. Allow slide to equilibrate 5 min in 4× SSC at room temperature. Remove slide and drain excess buffer. Do not allow slide to dry at any point during the procedure.

7. Add 50 µl biotin detection solution, digoxigenin detection solution, or biotin/digoxigenin detection solution to hybridized preparation on slide and cover with 22-mm^2 square of Parafilm. Incubate 45 min in an aluminum foil–wrapped moist chamber at 37°C.

8. Soak slide sequentially in aluminum foil–wrapped Coplin jars containing room temperature 4× SSC, 0.1% Triton X-100/4× SSC, and 4× SSC—10 min in each solution.

9. Add 50 µl DAPI or propidium iodide staining solution to preparation on slide, cover with 22-mm^2 square of Parafilm, and allow to stain 5 min at room

temperature. Rinse briefly in 1× SSC in Coplin jar to remove excess stain. Blot slide, but do not dry.

10. Add 7 µl of appropriate antifade mounting medium to stained slide and add a coverslip. Gently squeeze out excess antifade medium, taking care not to damage tissue, and seal with nail polish. Store at –20°C in slide box with desiccant.

11. Examine slide using a fluorescence microscope with epillumination and filter set appropriate for the fluorochrome used.

12. Photograph using either Ektar-1000 (for prints) or Ektachrome-400 (for slides) color film.

    Exposure times vary with brightness of hybridization signal, but exposures for DAPI are ~2 sec and exposures through dual- or triple-band-pass filter sets are 30 to 90 sec and 3 to 8 sec, respectively. Probes with bright hybridization signals can be photographed with black-and-white film, e.g., Kodak Technical Pan 2415, exposed at ASA 200.

AMPLIFICATION OF HYBRIDIZATION SIGNALS

Signals of both biotin- and digoxigenin-labeled probes can be amplified if necessary (see Fig. 14.7.2). Amplification can be performed at any point after post-hybridization washes, including after viewing, and it may be repeated.

SUPPORT PROTOCOL 1

Amplification of Biotinylated Signals

Additional Materials *(also see Basic Protocol)*

1 to 3 µg/ml biotinylated anti-avidin antibodies (Vector Laboratories) in 4× SSC/1% (w/v) BSA (fraction V)

Biotin amplification solution: 2 to 5 µg/ml fluorescein-avidin DCS (Vector Laboratories) in 4× SSC/1% (w/v) BSA (fraction V)

NOTE: This procedure should be performed with minimum exposure to light.

1. Hybridize slide with biotinylated probe, wash, and perform first round of signal detection (Basic Protocol).

   Amplification of the signal from biotinylated probes can be carried out at any time after step 8 of the Basic Protocol.

2. If slide has been mounted and sealed, remove sealed coverslip by breaking the nail polish seal with a needle or scalpel, lifting off the nail polish, and soaking slide 15 min in 0.1% Triton X-100/4× SSC in an aluminum foil–wrapped Coplin jar with gentle agitation. If coverslip is not loose, carefully lift it off. Repeat wash twice.

3. Drain excess wash solution from slide and add 50 µl of 1 to 3 µg/ml biotinylated anti-avidin antibody to slide. Cover with 22-mm^2 square of Parafilm, place in moist chamber, wrap moist chamber in aluminum foil, and incubate 30 min at 37°C.

4. Remove Parafilm and wash 15 min in 0.1% Triton X-100/4× SSC with gentle agitation.

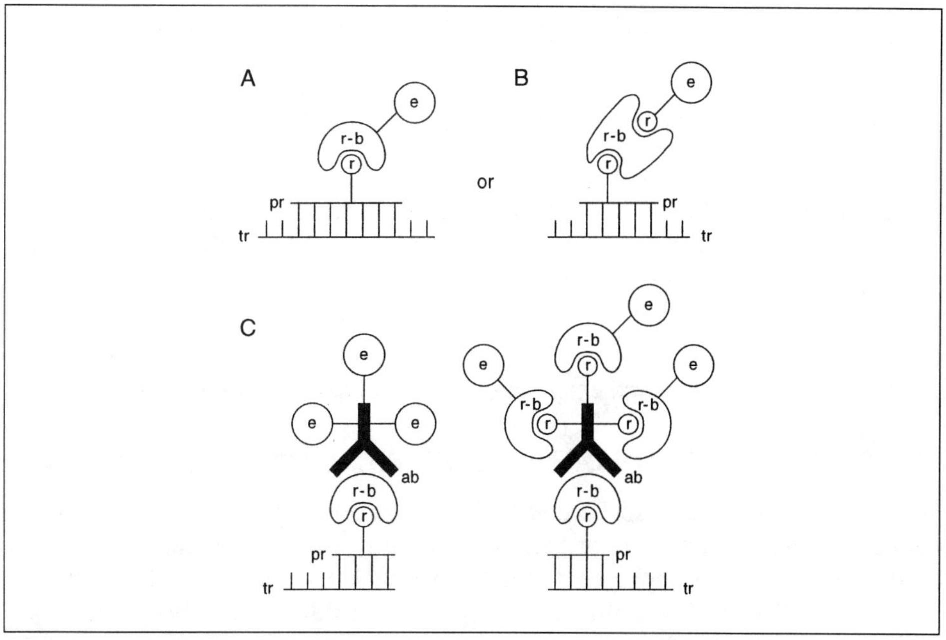

Figure 14.7.2 Enzyme-mediated detection of reporter molecules. (**A**) For direct detection, enzyme is conjugated to reporter-binding molecule. (**B**) For a two-step procedure, reporter-binding molecule is applied first, followed by an incubation with reporter-conjugated enzyme. (**C**) For signal amplification, incubation with reporter-binding molecule is followed by incubation with an antibody to reporter-binding molecule. The antibody may be conjugated with enzyme or reporter molecule. In the latter case an incubation with enzyme-conjugated reporter-binding molecule follows. The final step is addition of substrate (not shown). Fluorochrome can be substituted for enzyme molecule. Abbreviations: ab, antibody to reporter-binding molecule conjugated with enzyme or reporter molecule; e, enzyme (e.g, alkaline phosphatase or horseradish peroxidase); pr, probe labeled with reporter molecule; r, reporter molecule (e.g., biotin or digoxigenin); r-b, reporter-binding molecule (e.g., avidin, streptavidin, or anti-digoxigenin); tr, transcript.

5. Drain excess solution from slide and add 50 µl biotin amplification solution. Cover with Parafilm, then incubate 30 min at 37°C in a foil-wrapped moist chamber.

6. Remove Parafilm and wash 15 min in 0.1% Triton X-100/4× SSC, then 15 min in 4× SSC, both with gentle agitation.

   Multiple layers of avidin-fluorescein and biotinylated anti-avidin can be applied, but background will be significantly increased.

7. Counterstain, mount, and examine slide (Basic Protocol, steps 9 to 12).

Amplification of Signals from Digoxigenin-Labeled Probes

SUPPORT PROTOCOL 2

Additional Materials *(also see Basic Protocol)*

10 µg/ml Fab fragment of sheep anti-digoxigenin (Boehringer Mannheim) in 4× SSC/1% (w/v) BSA (fraction V)

Digoxigenin amplification solution: 3.5 to 7.0 µg/ml fluorescein-conjugated rabbit anti–sheep IgG (Sigma) in 4× SSC/1% (w/v) BSA (fraction V)

NOTE: This procedure should be performed with minimum exposure to light.

1. Hybridize slide with digoxigenin-labeled probe and wash (Basic Protocol, steps 1 through 6).

2. Add 50 μl of 10 μg/ml Fab fragment of sheep anti-digoxigenin to slide and cover with 22-mm^2 square of Parafilm. Incubate 30 min at 37°C.

3. Remove Parafilm. Wash in 4× SSC, 0.1% Triton X-100/4× SSC, then 4× SSC, 15 min each.

4. Drain excess 4× SSC and add 50 μl digoxigenin amplification solution to slide. Cover with 22-mm^2 square of Parafilm, place in moist chamber, wrap chamber in aluminum foil, and incubate 30 min at 37°C.

5. Repeat washes as in step 3.

6. Counterstain, mount, and examine slide (Basic Protocol, steps 9 to 12).

ENZYMATIC DETECTION OF NONISOTOPICALLY LABELED PROBES

Hybridized probes can also be detected by enzymatic reactions that produce a colored precipitate at the site of hybridization (Fig. 14.7.1). The most commonly used enzymes for this application are alkaline phosphatase (AP) and horseradish peroxidase (HRPO). To visualize the site of probe hybridization, a slide is incubated with the appropriate enzyme substrate. Table 14.7.1 summarizes the most commonly used combinations of enzyme and substrate for detection of in situ hybridization probes.

ALTERNATE PROTOCOL 1

Enzymatic Detection Using Horseradish Peroxidase

Additional Materials *(also see Basic Protocol; see* APPENDIX 1 *for items with ✓)*

Blocking solution: 1% (w/v) BSA in PBS
✓ Streptavidin solution
0.1% (v/v) Tween 20/PBS, 42°C
✓ Biotinylated horseradish peroxidase (HRPO) solution
DAB substrate solution: 500 μg/ml 3,3′-diaminobenzidine tetrahydrochloride (DAB) in PBS, prepared fresh

Table 14.7.1 Commonly Used Enzyme/Substrate Combinations for Detection of In Situ Hybridized Probes

Enzyme	Substrate[a]	Color[b]
Alkaline phosphatase	5-bromo-4-chloro-3-indolyl phosphate (BCIP) and nitroblue tetrazolium (NBT)	Bluish-purple precipitate
	Naphtol-AS-MX-phosphate and fast red TR	Red precipitate and red fluorescence
	Naphtol-AS-MX-phosphate and fast blue BN (Boehringer)	Blue precipitate
	Naphtol-AS-MX-phosphate and fast green BN (Boehringer)	Green precipitate
	Vector Red	Red precipitate
	Vector Black	Black precipitate
	Vector Blue	Blue precipitate
Horseradish peroxidase	3,3-diaminobenzidine tetrahydrochloride (DAB)	Brown precipitate[c]

[a]The main suppliers for these reagents are Boehringer Mannheim, Life Technologies, Promega, Sigma, and Vector Laboratories, but other suppliers deliver equivalent reagents.
[b]The detection systems listed result in hybridization signals that are generally analyzed by conventional bright-field microscopy. A fluorescence microscope is needed when utilizing the fluorescence signals produced by fast red. Optimization of the fluorescence detection procedure has been reported (Speel et al., 1992).
[c]Can be intensified by silver deposition.

3% H_2O_2
✓ PBS
✓ 90% (v/v) glycerol *or* appropriate antifade mounting medium

24 × 60–mm coverslips
42°C shaking water bath

CAUTION: DAB is hazardous; see manufacturer's information for guidelines on handling, storage, and disposal.

NOTE: This procedure must be performed in the dark with minimum exposure to ambient light.

1. Hybridize slide preparation with biotinylated probe and wash slide (Basic Protocol, steps 1 to 6).

2. Remove slide from 1× SSC. Drain as much buffer as possible, but do not allow slide to dry. Add 200 µl blocking solution to slide, and place 24 × 60–mm coverslip on top of applied solution. Place slide in an aluminum foil–wrapped moist chamber and incubate 30 min at 37°C.

3. Remove slide from chamber, tilt it to let coverslip slide off, and drain off as much blocking solution as possible without allowing slide to dry. Add 200 µl streptavidin solution to slide and place a 24 × 60–mm coverslip on top of applied solution. Place slide in an aluminum foil–wrapped moist chamber and incubate 30 min at 37°C.

4. Remove slide from chamber, tilt it to let the coverslip slide off, and place slide in Coplin jar containing 42°C 0.1% Tween 20/PBS. Place Coplin jar in a 42°C shaking water bath and agitate 5 min. Repeat wash twice with 42°C 0.1% Tween 20/PBS.

5. Remove slide and drain thoroughly without allowing to dry. Add 200 µl biotinylated HRPO solution to slide and place a 24 × 60–mm coverslip on top of applied solution. Incubate slide in an aluminum foil–wrapped moist chamber 30 min at 37°C.

6. Repeat wash as in step 4.

7. Remove slide from wash solution and drain thoroughly without allowing to dry. Add 0.015% H_2O_2 to DAB substrate solution. Immediately add 200 µl DAB substrate solution to slide and place a 24 × 60–mm coverslip on top of applied solution. Incubate slide 10 to 20 min in the dark at room temperature.

8. When the colored precipitate becomes visible to the eye, wash slide 5 min in room temperature PBS to stop the reaction.

9. If desired, apply a fluorescent counterstain to identify nuclei (Basic Protocol, step 9).

10. Mount slide in 90% glycerol *or* appropriate antifade mounting medium. View and photograph with a phase-contrast microscope.

ALTERNATE PROTOCOL 2

Enzymatic Detection Using Alkaline Phosphatase

Additional Materials (also see Basic Protocol; see APPENDIX 1 for items with ✓)

- Blocking solution: 1% (w/v) BSA/PBS
- ✓ Streptavidin solution
- 0.1% (v/v) Tween 20/PBS, 42°C
- ✓ Biotinylated alkaline phosphatase (AP) solution
- ✓ Alkaline phosphatase buffer, pH 9.5, 42°C
- ✓ NBT/BCIP substrate solution
- ✓ PBS
- ✓ 90% (v/v) glycerol *or* appropriate antifade mounting medium

- 24 × 60–mm coverslips
- 42°C shaking water bath

NOTE: This procedure must be performed in the dark with minimum exposure to ambient light.

1. Hybridize slide with biotinylated probe, wash, block, and incubate in streptavidin solution (Alternate Protocol 1, steps 1 to 4).
2. Remove slide from 0.1% Tween 20/PBS and drain without allowing it to dry. Add 200 µl biotinylated AP solution to slide and place a 24 × 60—mm coverslip on top of applied solution. Incubate slide in an aluminum foil–wrapped moist chamber 30 min at 37°C.
3. Remove slide from chamber, tilt it to let the coverslip slide off, and place slide in Coplin jar containing 42°C 0.1% Tween 20/PBS. Agitate 5 min at 42°C in shaking water bath. Repeat wash twice with 42°C 0.1% Tween 20/PBS.
4. Transfer slide to Coplin jar containing 42°C alkaline phosphatase buffer, pH 9.5. Agitate 5 min at 42°C. Replace buffer and agitate again 5 min at 42°C.
5. Place slide in aluminum foil–wrapped Coplin jar containing 50 ml freshly prepared NBT/BCIP substrate solution. Incubate in dark at 37°C or at room temperature (to slow down reaction) until color development is suitable (15 to 60 min).
6. Wash slide 5 min in PBS at room temperature to stop reaction.
7. If desired, apply a fluorescent counterstain to identify nuclei (Basic Protocol, step 9).
8. Mount in 90% glycerol or appropriate antifade mounting medium. View and photograph with a phase-contrast microscope.

References: Lichter et al., 1991; Speel et al., 1992.

Contributors: Omar Bagasra and Peter Lichter

In Situ Polymerase Chain Reaction and Hybridization to Detect Low-Abundance Nucleic Acid Targets

This unit presents a novel approach for detecting low-abundance nucleic acid targets in nuclear and cytoplasmic regions by amplification of specific target sequences using an in situ polymerase chain reaction (ISPCR). This is followed by in situ reverse transcription (if the target sequence is RNA) and in situ hybridization of the amplified sequences.

STRATEGIC PLANNING

Figure 14.8.1 outlines the major steps for in situ amplification and hybridization. It is essential that appropriate planning and consideration be given to each aspect of the project—including design of primers, amplification of parameters, detection and fixation conditions, and validation and controls—before proceeding with the actual protocols.

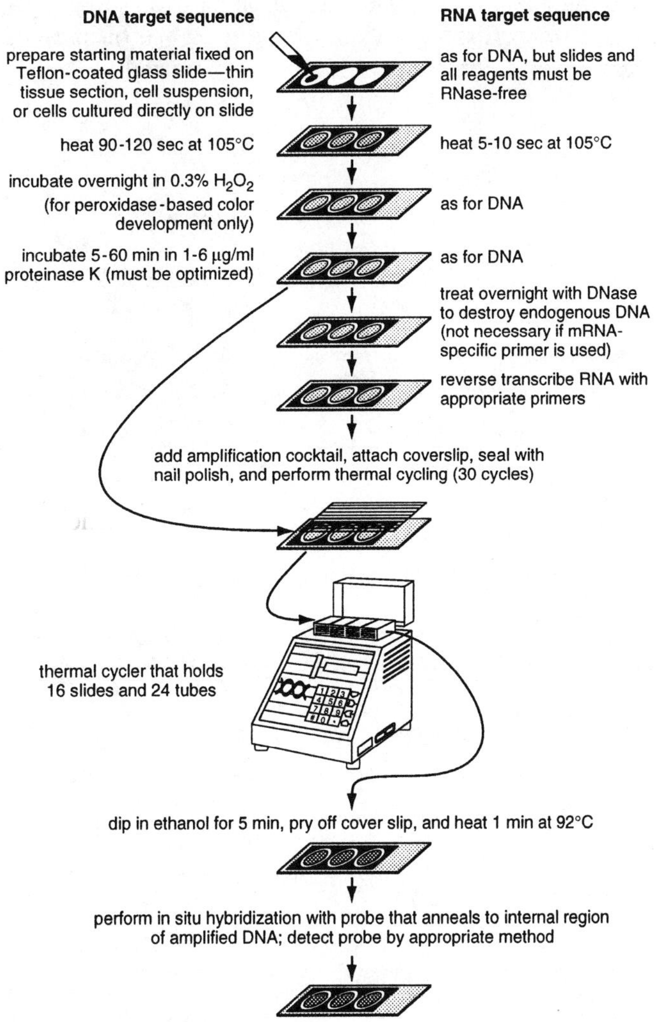

Figure 14.8.1 Flow chart for in situ amplification/hybridization.

Design of Primers to Reverse Transcribe and Amplify Target

The design of primers for amplification reactions in general requires careful planning (see *UNIT 15.1*); for in situ amplification, the design of the primer set is even more critical. Antisense downstream primers for the gene of interest are usually used. However, it is equally possible to use oligo(dT) primers to convert all mRNA populations into cDNA, and then perform the in situ amplification for a specific cDNA. This technique may be useful when several different gene transcripts are being amplified at the same time in a single cell.

There are two possible choices in designing a specific method to detect expression of a particular RNA. The more elegant method is to use primer pairs that are complementary to the sequence spanning spliced sequences of mRNA, as these particular sequences will be found only in spliced mRNA and will not fully hybridize to the encoding sequence in DNA (see Fig. 14.8.2). Thus, by using these primers, it is possible to omit the DNase treatment (to destroy endogenous DNA) and proceed directly to reverse transcription. The more brute-force, yet often necessary, approach is to treat the cells or tissue with DNase subsequent to the proteinase K digestion. This step destroys all of the endogenous DNA in the cells so that only RNA survives for amplification and subsequent detection.

In all reverse transcription reactions, it is advantageous to reverse transcribe only relatively small fragments of mRNA (<1500 bp). Larger fragments may not be completely reverse transcribed—a result of the presence of secondary structure. Furthermore, the reverse transcriptase enzymes—avian myeloblastosis (AMV)

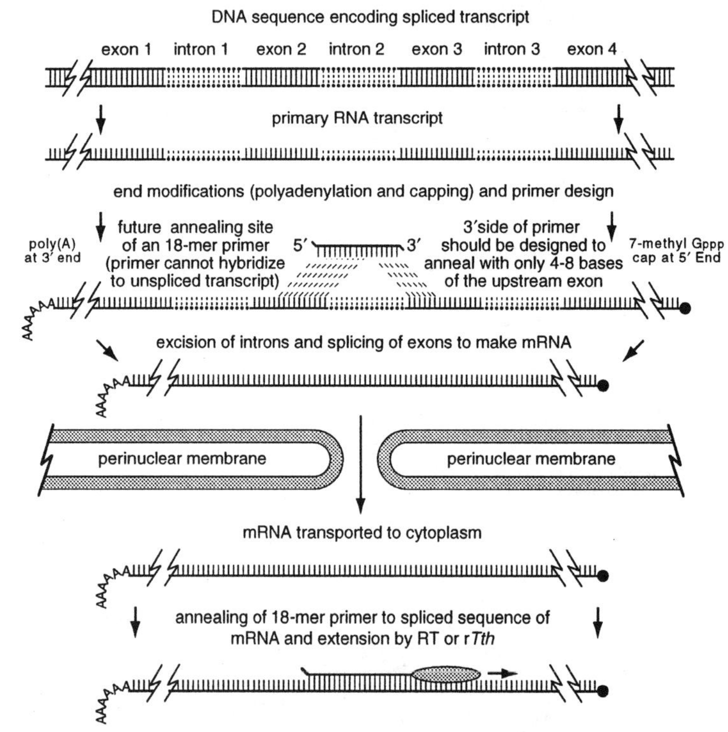

Figure 14.8.2 Design of primers to amplify mRNA without interference from DNA. Primers are designed so that their sequences span the splice junction where two exon sequences are joined. The resulting sequence thus exists only in mRNA, not in the encoding DNA. Use of such primers allows DNase treatment to be omitted and facilitates simultaneous amplification of RNA and DNA signals.

reverse transcriptase and Moloney murine leukemia virus (MoMuLV) reverse transcriptase, at least—are not very efficient in transcribing large mRNA fragments. However, this size restriction does not apply to DNA amplification; DNA target lengths up to 500 bp are routinely amplified in situ. Several new recombinant enzymes and buffer systems have been described that allow efficient amplifications of much larger fragments (up to 10 kb) of DNA or cDNA. An alternative approach is to create a multiple primer set, choosing primers positioned 200 to 300 nucleotides apart along the sequence to be amplified.

Several additional points should be kept in mind when designing primers for ISPCR and reverse transcription: (1) the length for both sense and antisense primers should be 18 to 22 bp; (2) at the 3′ ends, primers should contain at least one GC-type base pair (i.e., GG, CC, GC, or CG) to facilitate complementary-strand formation (as two GC-type base pairs will provide six hydrogen bonds as compared to four with two AT-type base pairs); (3) the preferred GC content of the primers is 45% to 55%; (4) primers should be designed so that they do not form intra- or interstrand base pairs; (5) the 3′ ends of the primers should not be complementary to each other, or they will form primer-dimers; and (6) reverse transcription primers should be designed so that they do not contain secondary structures.

Determining Optimal Annealing Temperature

The optimal primer annealing temperature for reverse transcription and DNA amplification is usually 2°C above the T_m of the primer, which is calculated according to the formula:

$$T_m \text{ of the primers} = 81.5°\text{C} + 16.6 (\text{Log } M) + 0.41 (\%\text{GC}) - 500/n$$

where T_m is the melting temperature of the primer, n is the length of the primer, and M is the molarity of the salt in the buffer (usually 0.047 M for DNA reactions and 0.070 M for reverse transcription reactions).

For AMV reverse transcriptase, the value of the melting temperature will be lower according to the following formula:

$$T_m \text{ of the primers} = 62.3°\text{C} + 0.41 (\%\text{GC}) - 500/n$$

These formulas provide only an approximate temperature for annealing, because base stacking, near-neighbor effect, and buffering capacity may play a significant role in determining the annealing temperature of a particular primer.

The logic of determining the correct annealing temperature for ISPCR is that during amplification, spurious products often appear in addition to those desired. Even if the cells do not contain DNA homologous to the primer sequences, many artifactual bands may appear as a result of false priming, which will occur if the determined melting temperature (T_m) between primer and template is not accurate. Annealing at a temperature too far above the T_m will yield no products, and annealing at a temperature too far below the T_m will often give unwanted products resulting from false priming. Many methods for eliminating high background resulting from false priming in PCR have appeared in the literature, including hot-start PCR, as well as use of DMSO, formamide, and anti-Taq DNA polymerase antibodies (e.g., TaqStart from Clontech).

The annealing temperature should be optimized with solution-based reactions before the corresponding in situ reactions are attempted, as in situ reactions are simply not as robust as solution-based ones. For ISPCR, determining the optimal annealing temperature is more critical because primers do not move as freely as a result of partial hindrance by intracellular organelles, multiple layers of membranes, and numerous subcompartments.

Validation and Controls

The validity of in situ amplification/hybridization should be examined in every run. Two or three sets of experiments can be run simultaneously in multiwell slides not only to validate the amplification, but also to confirm the subsequent hybridization/detection steps. Both positive and negative control samples should be included; dilutions of a positive control should be used to verify the proportionality of the response. An unrelated probe should be used as a control for nonspecific binding. RNAs for β-actin or HLA-DQα, or other abundant endogenous RNAs may be used as the positive markers. A reverse-transcription-negative control should always be used for reverse transcription/in situ amplification, as well as DNAse-treated and non-DNAse-treated controls. Controls without DNA polymerase, should always be included, with and without primers.

BASIC PROTOCOL 1

IN SITU PCR (ISPCR) AMPLIFICATION OF DNA AND RNA TARGETS WITH IN SITU REVERSE TRANSCRIPTION FOR RNA

Materials *(see APPENDIX 1 for items with ✓)*

Slides containing fixed specimens (Support Protocol 2)
0.3% H_2O_2 in PBS (prepare fresh)
✓ PBS
1 mg/ml proteinase K (Sigma; store in aliquots at −20°C)
✓ RNase-free DNase solution
Rinse buffer: RNase-free DNase solution without DNase
✓ DEPC-treated H_2O
✓ 10× AMV/MoMuLV reaction buffer
10 mM 4dNTP mix: 10 mM each dNTP in TE buffer, pH 7.5 (store at −20°C)
40 U/µl RNasin (Promega)
20 µM downstream primer (for reverse transcription; see Strategic Planning)
20 U/µl avian myeloblastosis virus (AMV) or Moloney murine leukemia virus (MoMuLV) reverse transcriptase *or* 20 U/µl SuperScript II (with 5× reaction buffer; Life Technologies)
✓ 0.1 M DTT
25 µM forward and reverse primers (for PCR; see Strategic Planning and *UNIT 15.1*)
✓ 1 M Tris·Cl, pH 8.3
✓ 1 M KCl
✓ 100 mM $MgCl_2$
0.01% (w/v) gelatin
5 U/µl *Taq* DNA polymerase
100% ethanol
✓ 2× SSC

55°C (optional), 92°C, 95°C, and 105°C heating blocks accommodating glass slides
Moist chamber (Fig. 14.2.3)
42°C incubator (optional)
20 × 60–mm glass coverslips
Clear nail polish or varnish
Thermal cycler accommodating glass slides (MJ Research)

NOTE: Where the target sequence is RNA, all reagents should be prepared with DEPC-treated water (*UNIT 4.1*). In addition, the AES-subbed glass slides (Support Protocol) and all glassware should be RNase free, e.g., by baking overnight in an oven usually set at 250° to 300°C.

1. Incubate slide with fixed specimens 5 to 120 sec on a 105°C heating block.

2. Incubate slide overnight at 37°C or room temperature in 0.3% H_2O_2 in PBS to inactivate endogenous peroxidase activity, then wash once with PBS.

3. Dilute 1 ml of 1 mg/ml proteinase K in 150 ml PBS (6 µg/ml final). Immerse slide in this solution and incubate at room temperature. After 5 min, examine cells under microscope at 400×; if the majority of cells of interest exhibit small round "bubbles," "blobs," or "peppery dots" (compare Figs. 14.8.3 and 14.8.4), proceed immediately to step 4. Otherwise, continue incubation up to 60 min, examining the slide at 400× at 5 min intervals and proceeding immediately to step 4 at the point where cell surface bubbles appear.

4. Heat slide 2 min in a 95°C heating block to inactivate the proteinase K, then rinse 10 sec in PBS and 10 sec in water. Allow slide to air dry.

   To reverse transcribe RNA targets, proceed with step 5 or step 7, if primer pairs are used that are complementary to the sequence joining spliced sequences of mRNA; to amplify DNA targets, proceed to step 10.

5. Add 10 µl RNase-free DNase solution to each well of slide. Incubate overnight at 37°C in a moist chamber.

6. Rinse slide once with rinse solution, then twice with DEPC-treated water and allow to air dry.

7a. *If using AMV or MoMuLV reverse transcriptase:* Make up the following reverse transcription cocktail (20 µl total volume):

    2 µl 10× AMV/MoMuLV RT buffer (1× final)
    2 µl 10 mM 4dNTP mix (1 mM final)
    0.5 µl 40 U/µl RNasin (1 U/µl final)
    1.0 µl 20 µM downstream primer (1 µM final
    0.5 µl 20 U/µl AMV or MoMuLV reverse transcriptase (0.5 U/µl final)
    8 µl DEPC-treated H_2O.

7b. *If using SuperScript II reverse transcriptase:* Make up the following reverse transcription cocktail (20 µl total volume):

    4 µl 5× reaction buffer (supplied with enzyme; 1× final)
    2 µl 10 mM 4dNTP mix (1 mM final)
    0.5 µl 4 U/µl RNasin (0.1 U/µl final)
    1.0 µl 20 µM downstream primer (1 µM final)
    0.5 µl 20 U/µl SuperScript II (0.5 U/µl final)
    1.2 µl 0.1 M dTT (6 mM final)
    4.8 µl DEPC-treated H_2O.

    SuperScript II (Life Technologies), has significantly reduced RNase H activity and is therefore suitable for reverse transcription of long mRNAs as well as routine reverse transcription.

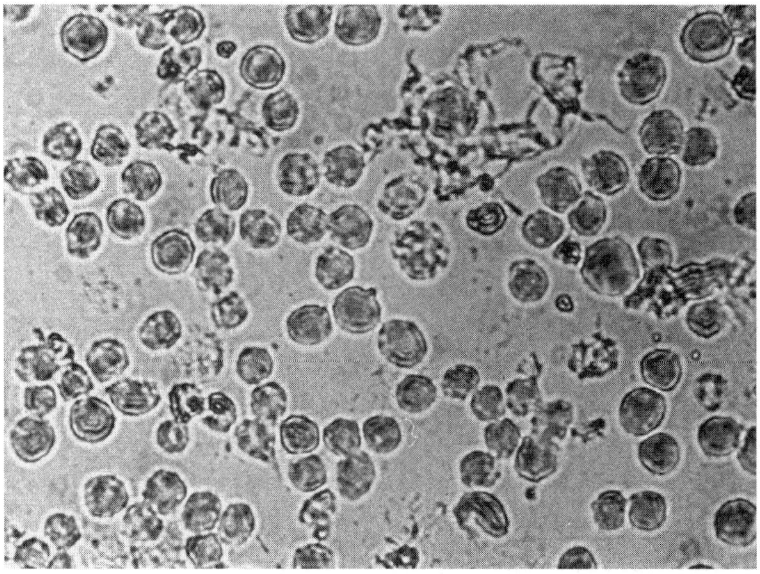

Figure 14.8.3 Lymphocytes before proteinase K treatment. Cells have been subjected to heat treatment and fixation. Note the smooth cytoplasmic membranes without "bubbles," "peppery dots," or "salt and pepper dots" on the cell surface.

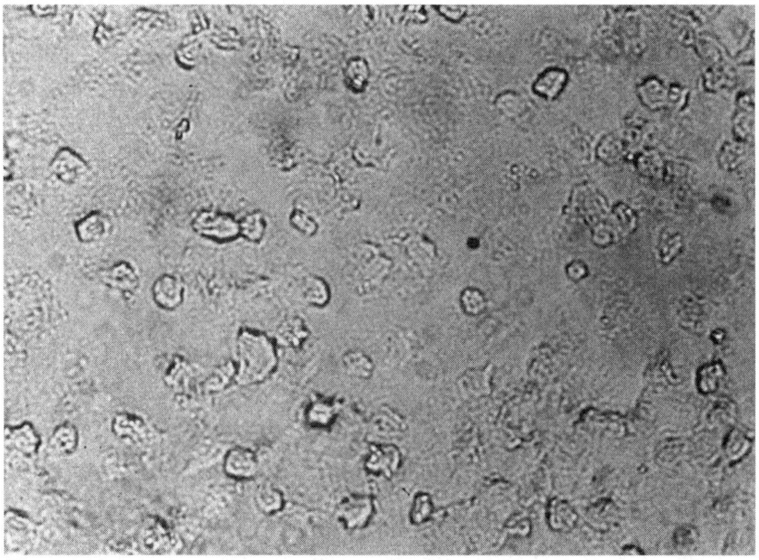

Figure 14.8.4 Lymphocytes after proteinase K treatment. Note the minute cell surface "bubbles," which are presumably the result of partial digestion of transmembrane proteins. The "bubbles" should appear uniform and be fairly evenly distributed.

8. Add 10 µl reverse transcription cocktail to each well of slide and carefully cover each well with a 20 × 60–mm glass coverslip. Incubate 1 hr at 42°C or 37°C in a moist chamber.

9. Incubate slides 2 min on a 92°C heating block. Remove coverslips and wash slides twice with water.

10. Make up the following amplification cocktail (100 µl total volume):

    5 µl 25 µM forward primer (1.25 µM final)
    5 µl 25 µM reverse primer (1.25 µM final)
    2.5 µl 10 mM 4dNTP mix (200 µM each dNTP final)

continued

1.0 μl 1.0 M Tris·Cl, pH 8.3 (10 mM final)
5.0 μl 1.0 M KCl (50 mM final)
2.5 μl 100 mM $MgCl_2$ (2.5 mM final)
10 μl 0.01% gelatin (0.001% final)
2 μl 5 U/μl *Taq* DNA polymerase (0.1 U/μl final)
66 μl H_2O.

11. Layer 8 μl (if using 3-well slide) or 12 to 20 μl (if using single-well slide) ISPCR amplification cocktail onto each well using a 20 μl micropipettor, so that the whole surface of the well is covered with the solution.

12. Place a 20 × 60–mm glass coverslip over each slide and carefully seal the edge of the coverslip to the slide with clear nail polish or varnish.

    If using tissue sections, use a second slide instead of a cover slip. Be certain to carefully paint the polish around the entire periphery of the coverslip or the edges of the dual slide to form a small reaction chamber that can contain the water vapor during thermal cycling.

13. Incubate slide 90 sec on a 92°C heating block, then transfer to thermal cycler.

14. Carry out PCR using the following amplification cycles or optimized conditions:

30 cycles:	30 sec	94°C	(denaturation)
	1 min	~45°C	(annealing)
	1 min	72°C	(extension)
Final step:	indefinitely	4°C	(hold).

15. Remove slide from thermal cycler and soak in 100% ethanol ≥5 min to dissolve nail polish. Pry off coverslip using a razor blade or other fine blade. Scratch off any remaining nail polish so that fresh coverslips can be placed evenly in the hybridization/detection steps (Basic Protocol 2).

16. Incubate slide 1 min on a 92°C heating block, then soak 5 min in 2× SSC at room temperature.

    Slides may be stored 2 to 3 weeks at 4°C.

ALTERNATE PROTOCOL

ONE-STEP REVERSE TRANSCRIPTION AND AMPLIFICATION

This method uses a single, recombinant enzyme—r*Tth* DNA polymerase—which performs both reverse transcription and DNA amplification at once in a single reaction, thus avoiding the need for two different buffer systems.

Additional Materials *(also see Basic Protocol 1; see* APPENDIX 1 *for items with ✓)*

3 mM 4dNTP mix: 3 mM each dNTP in TE buffer, pH 7.5 (store at –20°C)
10 mM $MnCl_2$
✓ 25 mM $MgCl_2$
10× r*Tth* transcription buffer: 100 mM Tris·Cl, pH 8.3/900 mM KCl
✓ 10× chelating buffer
✓ 1.7 mg/ml BSA
2.5 U/μl r*Tth* DNA polymerase (Perkin-Elmer or Life Technologies)

1. Pretreat slides with peroxidase, proteinase K, and DNase (Basic Protocol 1, steps 1 to 6).

2. Make up the following single-step reaction cocktail (100 μl total volume):

   0.5 μl 100 μM forward primer (0.5 μM final)
   0.5 μl 100 μM reverse primer (0.5 μM final)
   6 μl 3 mM 4dNTP mix (0.12 mM final)
   2 μl 10 mM $MnCl_2$ (0.2 mM final)
   10 μl 25 mM $MgCl_2$ (2.5 mM final)
   2 μl 10× r*Tth* transcription buffer (0.2× final)
   8 μl 10× chelating buffer (0.8× final)
   10 μl 1.7 mg/ml BSA (0.17 mg/ml final)
   2 μl 2.5 U/ml r*Tth* (9.05 U/ml final)
   59 μl DEPC-treated H_2O.

3. Carry out steps 11 to 13 of Basic Protocol 1, using the single-step reaction cocktail from step 2 in place of the ISPCR reaction cocktail.

4. Carry out the following thermal cycling program for reverse transcription:

1 cycle:	15 min	70°C
	3 min	92°C
	15 min	70°C
	3 min	92°C
	15 min	70°C

5. Carry out the following thermal cycling program for ISPCR:

29 cycles:	1 min	93°C	(denaturation)
	1 min	53°C	(annealing)
	1 min	72°C	(extension)
Final step:	indefinitely	4°C	(hold).

6. Carry out steps 15 and 16 of Basic Protocol 1.

BASIC PROTOCOL 2

HYBRIDIZATION AND DETECTION OF ISPCR-AMPLIFIED TARGET MATERIAL

Materials *(see* APPENDIX 1 *for items with* ✓*)*

200 pM probe, ^{33}P-labeled (see Support Protocol 3) or biotin- or digoxigenin-labeled (*UNIT 3.18*)
✓ Deionized formamide
✓ 20× and 2× SSC
✓ 100× Denhardt solution
✓ 10 mg/ml sonicated salmon sperm DNA (denature 10 min at 94°C before adding to hybridization mix)
✓ 10% (w/v) SDS
Slide containing ISPCR-amplified nucleic acids (Basic Protocol 1 or Alternate Protocol)
Diluted Kodak emulsion (*UNIT 14.4*)
Kodak D19 developer
Kodak Unifix fixer
2% (v/v) Gills hematoxylin (Sigma)
✓ PBS
✓ Streptavidin-peroxidase conjugate working solution
✓ AEC working solution

✓ Blocking solution
✓ Streptavidin–alkaline phosphatase conjugate working solution
100 mM Tris·Cl, pH 7.5/150 mM NaCl
✓ Alkaline phosphatase substrate buffer
75 mg/ml nitroblue tetrazolium (NBT) in 70% (v/v) dimethylformamide
50 mg/ml 5-bromo-4-chloro-3-indolyl phosphate (BCIP) in 100% dimethylformamide
Nuclear fast red stain (Sigma)
50%, 70%, 90%, and 100% (v/v) ethanol
50% (v/v) glycerol in PBS
Mounting medium, water-based (e.g., CrystalMount; Stephens Scientific or GelMount; Biomeda) or organic-based (e.g., Permount; Fisher)

20 × 60–mm glass coverslips
95°C heating block accommodating glass slides
48°C incubator
Moist chamber (Fig. 14.2.3)
Light-tight slide box
Dessicant
Coplin jars or glass staining dishes

NOTE: All washes and incubations are performed at room temperature unless otherwise noted.

1. Prepare the following hybridization mix:

   2 µl 200 pM probe (4 pM final)
   50 µl deionized formamide (50% final)
   10 µl 20× SSC (2× final)
   10 µl 100× Denhardt solution (10× final)
   10 µl 10 mg/ml sonicated salmon sperm DNA (1 mg/ml final)
   1 µl 10% SDS (1% final)
   7 µl H_2O.

2. Add 10 µl hybridization mix to each well of slide containing ISPCR-amplified nucleic acids. Cover wells with coverslips and heat slide 5 min on a 95°C heating block.

3. Incubate slides 2 to 4 hr at 48°C in a moist chamber.

   The optimal hybridization temperature is a function of the T_m (melting temperature) of the probe, which must be calculated for each probe. However, the hybridization temperatures used should not be too high. If it is necessary to use a very high hybridization temperature, 40% formamide should be substituted for 50% formamide in the hybidization mix in step 1.

For ^{33}P-labeled probe:

4a. Remove coverslips and wash slide in 2× SSC for 5 min.

5a. Dip slides in diluted Kodak emulsion.

6a. Air dry slides, then incubate 3 to 10 days in light-tight slide box with a dessicant.

7a. Develop slides in darkroom as follows:

   3 min in Kodak D19 developer
   30 sec rinse in H_2O
   3 min in Kodak Unifix fixer.

8a. Counterstain slide by incubating 2 to 3 min in 2% Gills hematoxylin in a Coplin jar at room temperature.

The slide is now ready for viewing; omit steps 9 to 11.

For peroxidase-based detection of biotin- or digoxigenin-labeled probe:

4b. Remove coverslips and wash slide twice in PBS, immersing for 5 min each time.

5b. Add 10 µl 100 µg/ml streptavidin-peroxidase to each well of slide. Gently cover wells with new coverslips and incubate 1 hr at 37°C.

6b. Remove coverslip and wash slides twice in PBS, immersing 5 min each time.

7b. In the dark, add 100 µl AEC working solution to each well of slide. Incubate 10 min at 37°C, then observe slides under microscope; if color is not strong, develop another 10 min.

8b. Rinse slide with tap water, then allow to dry.

Positive cells will appear brownish-red under the microscope.

For alkaline phosphatase–based detection of biotin- or digoxigenin-labeled probe:

4c. Remove coverslips and wash slides twice in 2× SSC, immersing 15 min each time at room temperature, then cover surface of each well with 100 µl blocking solution. Place slide flat in moist chamber and incubate 15 min at room temperature.

5c. For each well to be developed, mix 10 µl of 40 µg/ml streptavidin–alkaline phosphatase conjugate with 90 µl conjugate dilution buffer.

6c. Remove blocking solution by touching absorbent paper to edge of slide. Cover surface of each well with 100 µl of the diluted conjugate solution prepared in step 5c. Place slide flat in moist chamber and incubate in 15 min at room temperature.

Do not allow the tissue sample to dry after adding the conjugate.

7c. Wash slides twice in 100 mM Tris·Cl, pH 7.5/150 mM NaCl, 15 min each time at room temperature, then wash once more in alkaline phosphatase substrate buffer, 5 min at room temperature.

8c. Prewarm 50 ml alkaline phosphatase substrate buffer to 37°C in a Coplin jar. Add 200 µl 75 mg/ml NBT and 166 µl of 50 mg/ml BCIP. Mix well, then incubate slides in this solution at 37°C until desired level of signal is achieved (usually 10 min to 2 hr; check color development periodically by removing slide from solution and examining under 10× objective without allowing slide to dry), then stop reaction by rinsing slides in several changes of deionized water.

9. Stain slide for 5 min at room temperature in either 0.2% Gills hematoxylin (if peroxidase-based color development was used) or in 1% nuclear fast red stain (if alkaline phosphatase-based color development was used). Rinse slide in several changes of tap water.

10. Dehydrate slide by incubating 1 min at room temperature successively in 50%, 70%, 90%, and 100% ethanol. Air dry at room temperature.

11. Mount slide by applying one drop mounting medium per well and covering with a coverslip. View immediately, taking care not to disrupt the coverslip, or allow mounting medium to dry overnight at room temperature. Figures 14.8.5 and 14.8.6 show results obtained with FITC- and biotin-labeled probes.

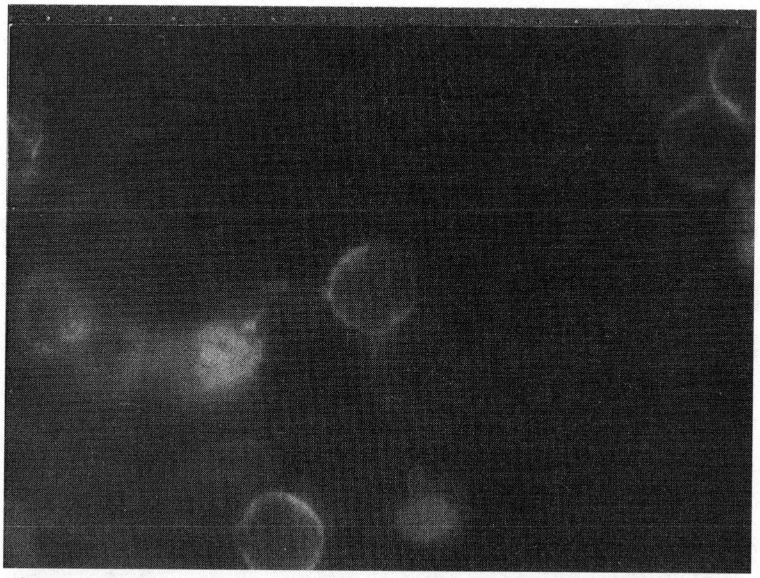

Figure 14.8.5 In situ reverse transcription/polymerase chain reaction using FITC-labeled (*tat-rev*) probes for HIV-1 expression in lymphocyte cell line chronically infected with HIV-1. Note cytoplasmic staining showing HIV-1 mRNA expression.

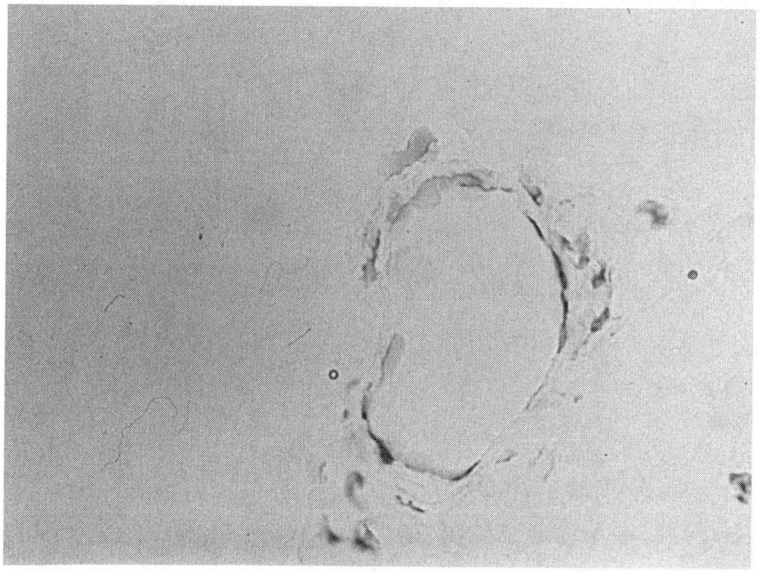

Figure 14.8.6 DNA-ISPCR of HIV-1-infected microvascular endothelial cell line. HIV-1 provirus gag sequence is amplified with an SK-38/39 primer pair and hybridized with a biotinylated SK-19 probe. Color was developed with AEC (see Basic Protocol 2) resulting in a red color.

PREPARATION OF AES-SUBBED SLIDES

SUPPORT PROTOCOL 1

Teflon-coated glass slides with wells are treated with 3-aminopropyltriethoxysilane (AES), rinsed extensively, and dried before use. The Teflon coating serves to form distinct wells, each of which serves as a small reaction chamber for in situ PCR.

Materials *(see APPENDIX 1 for items with ✓)*

Teflon-coated glass slides with three 10-, 12-, or 14-mm wells for cell suspensions or single oval well for tissue sections (Cel-Line Associates and Erie Scientific)

2% (v/v) 3-aminopropyltriethoxysilane (AES; Sigma) in acetone (prepare in Coplin jar or glass staining dish immediately before use)
✓ DEPC-treated H_2O

Coplin jars or glass staining dishes
Vessel accommodating 1000 ml of liquid

1. Dip Teflon-coated glass slides in 2% AES for 5 min, then allow to dry 10 to 15 min at room temperature.
2. Dip slides for five min in a separate vessel containing 1000 ml of DEPC-treated water.
3. Repeat step 2 three times using three changes of DEPC-treated water, then air dry in laminar-flow hood overnight.

   Slides should be used within 15 days of subbing.

SUPPORT PROTOCOL 2

PREPARATION OF SPECIMENS ON SLIDES FOR ISPCR

Paraffin-Fixed Tissue

Routinely prepared paraffin-fixed tissue sections (*UNIT 14.1*) on single-well 3-aminopropyltriethoxysilane-treated slides can be amplified quite successfully, permitting evaluation of individual cells in the tissue for the presence of a specific RNA or DNA sequence. Sections should be 3 to 5 µm thick unless the cells are large.

Plastic Sections

In situ amplification can be performed successfully on plastic sections. First, plastic-sectioned tissue must be deplasticized to remove methyl methacrylate by incubating in four successive baths of fresh methyl celluloacetate ether (MCA; Sigma) for 15 min each time, then in three successive baths of fresh acetone for 10 min each time, and then in multiple baths of xylene up to 4 hr each time. After this, the tissue is fixed in 4% paraformaldehyde (*APPENDIX 1*), and the in situ procedure is performed as with other tissue preparations.

Frozen Sections

It is possible to use frozen sections for in situ amplification, but the morphology of the tissue following the amplification process is generally not as good as with paraffin sections (see above). Certain immunohistochemical techniques require frozen sections.

To use frozen sections, the first step is to freeze the tissue properly (*UNIT 14.2*). For this purpose cut a 1 × 1–mm piece of stryrofoam from a sheet (or a cup). Cut a piece of tissue about the same size (1 × 1–mm; 0.2- to 0.3-mm thickness), attach it to the styrofoam with ~2 to 4 ml of OCT solution (*UNIT 14.2*) and immerse the whole specimen attached to the styrofoam in liquid nitrogen in an insulated vessel. The tissue will freeze in few seconds. The frozen tissue can then be mounted in a cryostat and cut with a microtome (*UNIT 14.2*) or frozen at –70°C for future use. This method prevents the formation of ice crystals in the tissues and preserves the morphology much better than simply storing the tissue in a deep freeze. If liquid nitrogen is not available, the tissue (with stryrofoam and OTC) may be wrapped in aluminum foil and placed on dry ice for 10 to 15 min before being stored in the deep freeze. As thin a slice as possible (3 to 4 µm) should be cut during sectioning. This is then applied to an AES-subbed slide (Support Protocol 1), dehydrated 10 min in 100% methanol, and air dried in a laminar-flow hood.

Cell Cultures

ISPCR has been successfully performed on cells cultured on either 4- or 8-well Nunc slides (VWR Scientific) or AES-subbed Teflon ISPCR slides (Support Protocol 1). If Nunc slides are used, 4- or 8-well slides with a glass base should be selected because plastic-base slides will melt during thermal cycling. This type of slide has a rubber gasket that should be left on, but the coverslip will not fit unless some protruding portions of the gasket are shaved off. The stability of these gaskets should be reinforced by applying a layer of nail polish at the junction of the gasket and the glass base. AES-subbed Teflon slides may also be used for tissue culture. They should be sterilized after silanization by incubating 30 min in 70% ethanol. Cells can be then placed on these slides and cultured overnight in a sterile humidified box. After this, the cells may be fixed as in *UNIT 14.1* and subjected to ISPCR.

LABELING OLIGONUCLEOTIDE PROBES USING ^{33}P

SUPPORT PROTOCOL 3

^{33}P-labeled probes are preferable for isotopic in situ hybridization (ISPCR) to avoid possible contamination of the thermal cycler with the more hazardous ^{32}P.

Materials *(see APPENDIX 1 for items with ✓)*

2 μM oligonucleotide probe (see Strategic Planning)
10× T4 polynucleotide kinase buffer (*UNIT 3.4*)
10 μCi/μl [γ-^{33}P]ATP (10 Ci/mmol; Amersham)
10 U/ul T4 polynucleotide kinase
0.8-ml Sephadex G-50 column (e.g., QuickSpin, Boehringer Mannheim)
✓ TE buffer, pH 7.4

1. Make up the following reaction mix in a microcentrifuge tube (20 μl total volume):

   1.0 μl 2 μM oligonucleotide probe (0.1 μM final)
   2 μl 10× polynucleotide kinase buffer (1× final)
   1.0 μl 10 μCi/μl [γ-^{33}P]ATP (0.5 μCi/μl final)
   15 μl H_2O
   1 μl 10 U/μl T4 polynucleotide kinase (0.5 U/μl final).

   Incubate 30 min at 37°C.

2. Apply reaction mix to an 0.8-ml Sephadex G-50 column. Elute with TE buffer, and collect fractions at the following volumes:

Fraction 1	300 μl
Fraction 2	100 μl
Fraction 3	100 μl
Fraction 4	100 μl
Fraction 5	100 μl
Fraction 6	100 μl.

3. Count radioactivity in 1.0 μl of each fraction and identify those containing the labeled probe.

   The labeled probe should be contained in fractions 2 through 4.

References: Bagasra et al., 1993; Hasse et al., 1990; Nuovo et al., 1991.

Contributors: Omar Bagasra, Thikkavarapu Seshamma, Roger Pomerantz, and John Hanson

The Polymerase Chain Reaction

15

The polymerase chain reaction (PCR) is a rapid procedure for in vitro enzymatic amplification of a specific segment of DNA. Like molecular cloning, PCR has spawned a multitude of experiments that were previously impossible. The number of applications of PCR seems infinite—and is still growing. They include direct cloning from genomic DNA or cDNA, in vitro mutagenesis and engineering of DNA, genetic fingerprinting of forensic samples, assays for the presence of infectious agents, prenatal diagnosis of genetic diseases, analysis of allelic sequence variations, analysis of RNA transcript structure, genomic footprinting, and direct nucleotide sequencing of genomic DNA and cDNA.

The theoretical basis of PCR is outlined in Figure 15.0.1. There are three nucleic acid segments: the segment of double-stranded DNA to be amplified and two single-stranded oligonucleotide primers flanking it. Additionally, there is a protein component (a DNA polymerase), appropriate deoxyribonucleoside triphosphates (dNTPs), a buffer, and salts.

The primers are added in vast excess compared to the DNA to be amplified. They hybridize to opposite strands of the DNA and are oriented with their 3′ ends facing each other so that synthesis by DNA polymerase (which catalyzes growth of new strands 5′→3′) extends across the segment of DNA between them. One round of

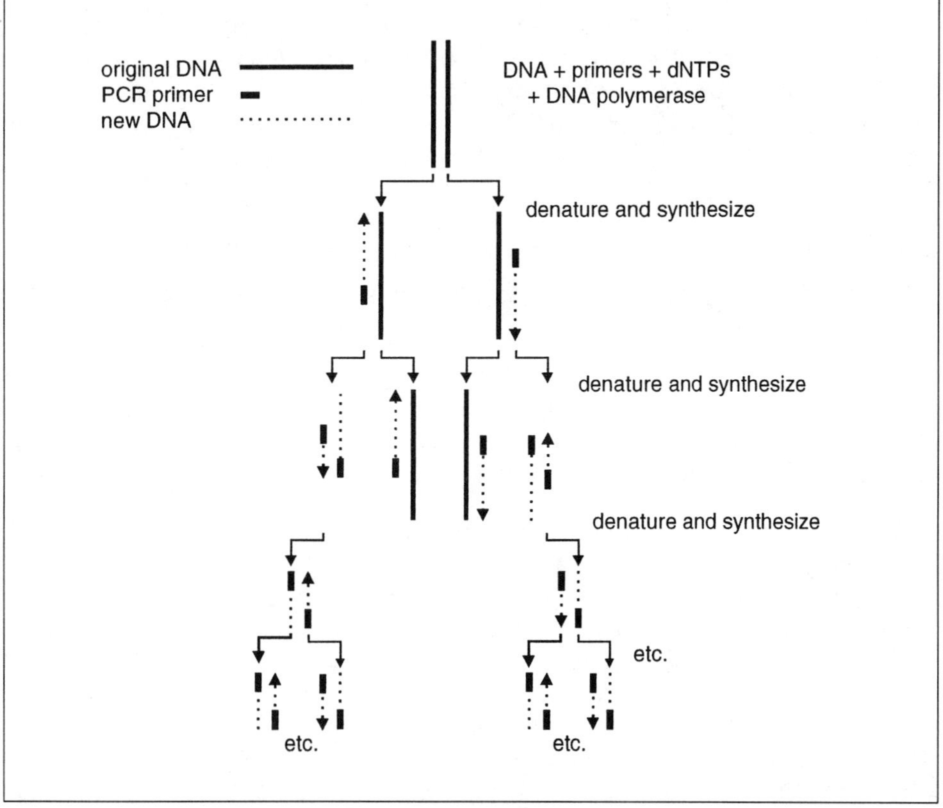

Figure 15.0.1 The polymerase chain reaction. DNA to be amplified is denatured by heating the sample. In the presence of DNA polymerase and excess dNTPs, oligonucleotides that hybridize specifically to the target sequence can prime new DNA synthesis. The first cycle is characterized by a product of indeterminate length; however, the second cycle produces the discrete "short product" which accumulates exponentially with each successive round of amplification. This can lead to the many million-fold amplification of the discrete fragment over the course of 20 to 30 cycles.

synthesis results in new strands of indeterminate length which, like the parental strands, can hybridize to the primers upon denaturation and annealing. These products accumulate only arithmetically with each subsequent cycle of denaturation, annealing to primers, and synthesis.

However, the second cycle of denaturation, annealing, and synthesis produces two single-stranded products that together compose a discrete double-stranded product which is exactly the length between the primer ends. Each strand of this discrete product is complementary to one of the two primers and can therefore participate as a template in subsequent cycles. The amount of this product doubles with every subsequent cycle of synthesis, denaturation, and annealing, accumulating exponentially so that 30 cycles should result in a 2^{28}-fold (270 million–fold) amplification of the discrete product.

This chapter consists of protocols that cover some of the more common applications of PCR. For many applications, the first step is simply to get PCR working with a known segment of DNA and a set of primers. Therefore, *UNIT 15.1* presents a basic PCR protocol and ways to optimize it for the sequence of interest.

PCR permits direct sequencing of nucleic acids without requiring cloning, thus avoiding cloning difficulties and artifacts. Several different protocols for preparing PCR products for sequencing using either dideoxy (Sanger) sequencing methods or chemical (Maxam-Gilbert) methods are presented in *UNIT 15.2*. This unit should permit the practitioner to choose a protocol best suited to the problem at hand and to his or her taste.

PCR is frequently used because it is the most sensitive assay for rare sequences. A protocol that not only detects rare DNAs but quantitates them as well is presented in *UNIT 15.3*. The downside of sensitivity is contamination by infinitesimal amounts of unwanted exogenous sequences. Procedures designed to avoid contamination with undesired DNA sequences are emphasized in this unit.

Another application is to detect transcripts, analyze their structure, and amplify their sequences to permit cloning and/or sequencing. *UNIT 15.4* presents procedures that adapt PCR to RNA templates, via production of a cDNA copy of the RNA by reverse transcriptase.

Several PCR methods have been developed that require knowledge of only a small stretch of sequence (30 to 40 bases). One of these, "anchored PCR," is applied in *UNIT 15.6* to the analysis of mRNAs. A second such method, ligation-mediated PCR (*UNIT 15.5*), has broad applications including genomic footprinting and sequencing.

PCR can be used to help clone and manipulate sequences. Various methods for generating suitable ends to facilitate the direct cloning of PCR products are detailed in *UNIT 15.7*. Protocols for relevant applications can be found elsewhere in this volume, including construction of recombinant DNA molecules with PCR (*UNIT 3.7*) and site-directed mutagenesis using PCR (*UNIT 8.5*).

A powerful application of PCR, called differential display, is described in *UNIT 15.8*. This technique allows the identification and subsequent isolation of differentially expressed genes through an approach that requires no knowledge of sequences, but rather involves reverse transcription of mRNAs using anchored oligo-dT primers, PCR amplification with an arbitrary decamer, and high-resolution PAGE analysis.

Donald M. Coen

Enzymatic Amplification of DNA by PCR: Standard Procedures and Optimization

UNIT 15.1

BASIC PROTOCOL

This unit describes a method for amplifying DNA enzymatically by the polymerase chain reaction (PCR) and for optimizing this reaction for the sequence and primer set of interest. Important variables that can influence the outcome of PCR include the $MgCl_2$ concentration and the cycling temperatures. Additives that promote polymerase stability and processivity or increase hybridization stringency, and strategies that reduce nonspecific primer-template interactions, especially prior to the critical first cycle, can greatly improve sensitivity, specificity, and yield. This protocol is designed to optimize the reaction components and conditions in one or two stages.

Materials *(see* APPENDIX 1 *for items with ✓)*

- Sterile H_2O
- 15 mM (L), 30 mM (M), and 45 mM (H) $MgCl_2$
- ✓ 10× amplification buffer, $MgCl_2$-free
- ✓ 25 mM 4dNTP mix
- 50 µM oligonucleotide primer 1: 50 pmol/µl in sterile H_2O (store at −20°C)
- 50 µM oligonucleotide primer 2: 50 pmol/µl in sterile H_2O (store at −20°C)
- Template DNA: 1 µg mammalian genomic DNA/10 µl or 0.1 ng plasmid DNA/10 µl
- 5 U/µl *Taq* DNA polymerase
- DMSO, cell culture grade
- Glycerol
- Perfect Match Polymerase Enhancer (PMPE, Stratagene)
- *Taq* pol + TaqStart: 1:1 mixture of *Taq* DNA polymerase and TaqStart antibody (Clontech; store at −20°C)
- Mineral oil
- Ficoll 400 (optional)
- Tartrazine dye (optional)
- Automated thermal cycler

NOTE: Do not use DEPC to treat water or reagents.

1. Prepare four reaction master mixes according to the recipes given in Table 15.1.1.
2. Aliquot 90 µl master mix I into each of three 0.5-ml microcentrifuge tubes labeled I-L, I-M, and I-H. Similarly, aliquot mixes II through IV into appropriately labeled tubes. Add 10 µl of 15 mM $MgCl_2$ into one tube of each master mix (labeled L; 1.5 mM final). Similarly, aliquot 10 µl of 30 mM and 45 mM $MgCl_2$ to separate tubes of each master mix (labeled M and H, respectively; 3.0 and 4.5 mM final concentrations respectively). Overlay the reaction mixture with 50 to 100 µl mineral oil (2 to 3 drops).
3. Program the automated thermal cycler according to manufacturers' instructions. Using the following guidelines, fill in the annealing temperature and extension time according to primer and product considerations. Run the program for 30 cycles.

    Denature 1 min at 94°C
    Anneal 1 min at _____°C: if GC content is ≤50%, anneal at 55°C; if >50%, anneal at 60°C.
    Extend _____ min at 72°C: if product length is ≤500 nucleotides, extend 1 min; if >500 nucleotides, extend 3 min.

Table 15.1.1 Master Mixes for Optimizing Reaction Components

Components	Final concentration	Per reaction	Master mix[a] (μl)			
			I	II	III	IV
10× PCR buffer	1×	10 μl	40.0	40.0	40.0	40.0
Primer 1	0.5 μM	1 μl	4.0	4.0	4.0	4.0
Primer 2	0.5 μM	1 μl	4.0	4.0	4.0	4.0
Template DNA	Undiluted	10 μl	40.0	40.0	40.0	40.0
25 mM 4dNTP mix[b]	0.2 mM	0.8 μl	3.2	3.2	3.2	3.2
Taq polymerase	2.5 U	0.5 μl	2.0	2.0	2.0	2.0
DMSO	5%	5 μl	—	20.0	—	—
Glycerol	10%	10 μl	—	—	40.0	—
PMPE	1%	1 μl	—	—	—	4.0
H_2O	—	To 90 μl	266.8	246.8	226.8	262.8

[a]Total volume = 360 μl (enough for *n* + 1 reactions).

[b]If 2 mM 4dNTP mix is preferred, use 10 μl per reaction, or 40 μl for each master mix; adjust the volume of water accordingly.

The number of cycles depends on both the efficiency of the reaction and the amount of template DNA in the reaction. Starting with as little as 100 ng of mammalian genomic DNA (~10^4 cell equivalents), after 30 cycles, 10% of the reaction should produce a band that is readily visible on an ethidium bromide–stained gel as a single predominant band. With more template, fewer cycles may suffice. With much less template, as many as 45 cycles may be necessary.

4. Electrophorese 10 μl from each reaction on an agarose, nondenaturing polyacrylamide, or sieving agarose gel appropriate for the PCR product size expected. Stain with ethidium bromide.

While checking the efficiency of the reaction on a gel (steps 5 and 6), samples can be stored at 4°C, and put through additional cycles if necessary.

5. Examine the stained gel to determine which condition resulted in the greatest amount of product.

Minor, nonspecific products may be present even under optimal conditions.

6. To ensure that the major product is the correct one, digest an aliquot of the reaction with a restriction endonuclease known to cut within the PCR product. Electrophorese the digestion product on a gel to verify that the resulting fragments have the expected sizes.

Alternatively, transfer the PCR products to a nitrocellulose or nylon membrane and hybridize with an oligonucleotide derived from the sequence internal to the primers.

These optional steps optimize initial hybridization and may improve efficiency and yield. They are used when primer-dimers or other nonspecific products are detected or when there is only a small amount of starting template.

7. Prepare four reaction mixtures using the optimal $MgCl_2$ concentration and additive requirement determined in step 5. Prepare the mixes according to the recipes in Table 15.1.2. Use the following variations for addition of *Taq* polymerase. Reactions A, C, and D are prepared at room temperature. All components of reaction B should be chilled in an ice slurry before they are combined.

To ensure that the reaction does not plateau and thereby obfuscate the results, use the smallest amount of template DNA necessary for visualization of the PCR product by ethidium bromide staining.

Table 15.1.2 Master Mixes for Optimizing First-Cycle Reactions

Components	Final concentration	Master mix (μl)[a]			
		A	B	C	D
10× PCR buffer	1×	10 μl	10	10	10
$MgCl_2$ (L, M, or H)	Optimal	10 μl	10	10	10
Primer 1	0.5 μM	1.0 μl	1.0	1.0	1.0
Primer 2	0.5 μM	1.0 μl	1.0	1.0	1.0
Additive	Optimal	V	V	V	V
Template DNA	—[b]	10 μl	10	10	10
25 mM 4dNTP mix[c]	0.2 mM	0.8 μl	0.8	0.8	0.8
Taq polymerase	2.5 U	0.5 μl	0.5	—	—
Taq pol + TaqStart	2.5 U	—	—	—	1.0
H_2O	To 100 μl	V	V	V	V
Preparation temperature		Room temperature	Ice slurry	Room temperature	Room temperature

[a]V, variable amount (total volume should be 100 μl).
[b]Use undiluted or diluted template DNA based on results obtained in step 6.
[c]If 2 mM 4dNTP mix is preferred, use 10 μl per reaction mix.

8. Overlay each reaction mixture with 50 to 100 μl mineral oil. Heat all reactions 5 min at 94°C.

   It is most convenient to use the automated thermal cycler for this step and then initiate the cycling program directly.

9. Cool the reactions to the appropriate annealing temperature as determined in step 4. Add 0.5 μl *Taq* DNA polymerase to reaction C, making sure the pipet tip is inserted through the layer of mineral oil into the reaction mix.

10. Begin amplification of all four reactions at once, using the same cycling parameters as before.

11. Analyze the PCR products on an agarose gel and evaluate the results as in steps 4 and 5.

12. Prepare a batch of the optimized reaction mixture, but omit *Taq* DNA polymerase, TaqStart antibody, PMPE, and 4dNTP mix—these ingredients should be added fresh just prior to use. If desired, add Ficoll 400 to a final concentration of 0.5% to 1% (v/v) and tartrazine to a final concentration of 1 mM.

    Adding Ficoll 400 and tartrazine dye to the reaction mix precludes the need for a gel loading buffer and permits direct application of PCR products to agarose or acrylamide gels.

    Ficoll 400 and tartrazine dye may be prepared as 10× stocks and stored indefinitely at room temperature.

Reference: Saiki et al., 1988

Contributors: Martha F. Kramer and Donald M. Coen

Direct DNA Sequencing of Polymerase Chain Reaction Products

PCR products can be sequenced using either the dideoxy (Sanger) approach or the chemical (Maxam-Gilbert) approach.

BASIC PROTOCOL 1

DIDEOXY SEQUENCING OF SINGLE-STRANDED PRODUCTS GENERATED BY ASYMMETRIC PCR

Materials *(see APPENDIX 1 for items with ✓)*

Oligonucleotide primers 1 and 2
^{32}P-labeled dNTPs (*UNIT 3.4*)
✓ 10 M ammonium acetate
100% and 70% ethanol, room temperature
✓ 0.1× TE buffer, pH 8.0
Centricon 30 or 100 column (optional; Amicon)

1. Assemble PCR with optimized components (*UNIT 15.1*) but use ~100:1 ratio of the two oligonucleotide primers (0.2-1 pmol for limiting primer and 10-30 pmol for primer present in excess). Carry out PCR using optimized times and temperatures for 40 to 45 cycles, ending with long extension step.

   Verify that there is just one predominant single-stranded product before proceeding. Add small amount (0.1% of cold dNTP concentration) of [^{32}P]dNTPs, or 0.001-0.1 pmol of [^{32}P] "nonlimiting" primer, during final 3 to 5 rounds of amplification. Visualize amplified products directly by autoradiography of agarose or nondenaturing polyacrylamide gel (UNITS 2.5A, 2.7, & APPENDIX 3).

2. Bring reaction slowly to room temperature and remove mineral oil with Pasteur pipet. Precipitate DNA by adding 10 M ammonium acetate to 2.5 M final, followed by 1 vol of room temperature 100% ethanol. Leave 5 min at room temperature.

3. Microcentrifuge 5 min at room temperature, wash pellet with 70% ethanol, and dry under vacuum. Resuspend pellet in 50 µl water or 50 µl of 0.1× TE buffer, pH 8.0.

4. Carry out dideoxy sequencing using asymmetric PCR–generated single-stranded templates (*UNIT 7.4*).

   The sequencing primer can be either a PCR-limiting oligonucleotide primer or any complementary sequence internal to 3′ end of single-stranded template.

ALTERNATE PROTOCOL 1

GENERATING SINGLE-STRANDED TEMPLATE FOR DIDEOXY SEQUENCING BY SINGLE-PRIMER REAMPLIFICATION

The use of asymmetric primer ratios does not always result in reproducible high yields of single-stranded products. This method of asymmetric reamplification supplies sufficient single-stranded template for a full set of dideoxy sequencing reactions.

1. Carry out 25 to 40 cycles of PCR under optimized conditions (*UNIT 15.1*).

2. Remove mineral oil with Pasteur pipet. Precipitate, wash, and dry amplified products (Basic Protocol 1, steps 2 and 3). Resuspend pellet in 30 µl of 0.1× TE buffer, pH 8.0.

3. Use 1 μl of resuspended double-stranded product as template for a new *asymmetric* PCR amplification. Add 20 to 40 pmol of *one* of two primers, and carry out 20 cycles of PCR under same conditions as in step 1.

4. Add 0.5-1 pmol of ^{32}P-labeled PCR primer (in addition to cold primer) to final three to five rounds of asymmetric reamplification. When PCR is completed, determine yield and quality of single-stranded product by autoradiography (APPENDIX 3).

   A single band corresponding to single-stranded product should be visible (a band corresponding to double-stranded product may occasionally appear, but should be much fainter than single-strand band).

5. Precipitate and dry amplified products as in step 2.

6. Carry out dideoxy sequencing using PCR-generated single-stranded template.

PREPARING DOUBLE-STRANDED PCR PRODUCTS FOR DIDEOXY SEQUENCING

ALTERNATE PROTOCOL 2

PCR-amplified double-stranded DNA can be purified and analyzed directly by dideoxy sequencing.

Additional Materials *(also see Basic Protocol 1; see* APPENDIX 1 *for items with ✓)*

Phenol and chloroform (UNIT 2.1)
PEG/NaCl solution: 20% (v/v) PEG 6000/2.5 M NaCl (store at room temperature)
70% ethanol, ice-cold
1 pmol (2 to 5 ng) sequencing primer
✓ 5× annealing buffer

1. Carry out PCR under optimized conditions (UNIT 15.1), using 50 pmol of each primer.

   Use enough template DNA and perform enough PCR cycles to yield products sufficient for several sequencing reactions.

2. Analyze a 5-μl aliquot of PCR product by electrophoresis in an agarose, nondenaturing polyacrylamide, or sieving agarose gel and stain with ethidium bromide.

   Identification and determination of purity of PCR product is important for deciding which method to use for purification. If desired PCR product is predominant species, proceed with step 3. If more than one PCR product is observed, desired product can be purified on low gelling/melting temperature agarose gel (UNIT 2.6) before denaturation (step 6).

3. Remove mineral oil with Pasteur pipet. Extract DNA with phenol and then chloroform. Transfer aqueous phase to fresh tube.

4. Add 0.6 vol PEG/NaCl solution and incubate 10 min at 37°C. Microcentrifuge 10 min at top speed and discard supernatant.

5. Wash pellet with cold 70% ethanol and dry under vacuum. Resuspend at ≥0.1 pmol/μl (~50 to 100 ng/μl of 1000- to 2000-bp fragments) in 0.1× TE buffer, pH 8.0.

6. Pipet ~0.1 pmol purified PCR product into microcentrifuge tube. Adjust to 9 μl with 0.1× TE buffer, pH 8.0.

7. Denature template by heating 2 min at 95°C, then quickly cool 1 min in dry ice/ethanol bath. Quickly add 1 µl sequencing primer and 2 µl of 5× annealing buffer to frozen DNA sample.

8. Microcentifuge tube briefly to thaw and mix reaction. Incubate template and primer 30 min at room temperature.

   A five-fold molar excess of primer over template is recommended.

9. Carry out dideoxy sequencing reactions (*UNIT 7.4*).

ALTERNATE PROTOCOL 3

DIDEOXY SEQUENCING OF SINGLE STRANDS GENERATED BY λ EXONUCLEASE DIGESTION OF DOUBLE-STRANDED PCR PRODUCTS

This approach does not require the use of unequal primer concentrations.

Additional Materials *(also see Basic Protocol 1; see APPENDIX 1 for items with ✓)*

✓ 3 M sodium acetate
λ exonuclease (*UNIT 3.11*) and 1× buffer (*UNIT 3.4*)

1. Phosphorylate 10 to 50 pmol of either PCR primer 1 or 2, using unlabeled ATP (*UNIT 3.10*).

2. Ethanol precipitate phosphorylated primer with 3 M sodium acetate (to 0.3 M final) and 2.5 vol cold 100% ethanol. Place 1 hr at −70°C, microcentrifuge 20 min at 4°C, and wash pellet once with 70% ethanol. Dry pellet briefly under vacuum.

3. Carry out PCR under optimized conditions (*UNIT 15.1*) with equimolar amounts (10-100 pmol) of both phosphorylated and nonphosphorylated primers.

4. Phenol extract PCR products, then ethanol precipitate, wash, and dry as in step 2. Resuspend pellet in 50 to 100 µl of 1× λ exonuclease buffer. Add 5 U λ exonuclease and incubate 15 to 30 min at 37°C.

5. Phenol extract twice, then ethanol precipitate, wash, dry, and resuspend single-stranded product as in step 3 or first basic protocol.

6. Carry out dideoxy sequencing with PCR- and λ exonuclease–generated single-stranded templates.

BASIC PROTOCOL 2

LABELING AND CHEMICAL SEQUENCING OF PCR PRODUCTS

Materials *(see APPENDIX 1 for items with ✓)*

5 to 10 pmol oligonucleotide primer 1 (to be end-labeled)
6000 Ci/mmol [γ-^{32}P]ATP
T4 polynucleotide kinase (*UNIT 3.10*)
20 to 40 pmol oligonucleotide primer 2
✓ 10× amplification buffer (optimized as in *UNIT 15.1*)
2 mM 4dNTP mix: 2 mM each dNTP in TE buffer, pH 7.5 (store at −20°C)
DNA template (20 to 1000 ng for eukaryotic DNA; 10 to 100 ng for bacterial DNA; 1 to 20 ng for cloned DNA inserts)
Taq DNA polymerase
Mineral oil

Disposable columns or dialysis cartridges for DNA purification (optional; NACS prepack cartridge, Life Technologies; Centricon 30 column, Amicon; or Select 6L column, 5 Prime→3 Prime)

1. In 10 µl, radioactively end label 5 to 10 pmol oligonucleotide primer 1 for 30 min with ≥6000 Ci/mmol [γ-^{32}P]ATP and T4 polynucleotide kinase (*UNIT 3.10*). Heat-inactivate kinase by incubating 10 min at 65°C.

2. Prepare microcentrifuge tube for PCR containing the following:

   10 µl 10× amplification buffer
   10 µl 2 mM 4dNTP mix
   20 to 40 pmol oligonucleotide primer 2 (not labeled in step 1)
   DNA template
   1.5 to 2.5 U *Taq* DNA polymerase
   10 µl kinase reaction (step 1)
   H_2O to 100 µl.

   Overlay with 100 µl mineral oil. Carry out PCR under optimized conditions for 5 to 10 cycles. Complete PCR with final long extension step of 5 to 7 min. Remove mineral oil with Pasteur pipet.

3. Separate labeled PCR product by electrophoresis on agarose or nondenaturing polyacrylamide gel (*UNITS 2.5 & 2.7*). Run gel until unincorporated primers and nucleotides migrate off gel and into lower buffer chamber. Autoradiograph PCR products 10 to 60 min (*APPENDIX 3*).

4. Cut and isolate desired labeled PCR product from gel (*UNITS 2.5, 2.6,* or *2.7*) and resuspend in 30 to 50 µl water.

   Amplified DNA can be also be purified using disposable columns. Chemical sequencing can also be carried out without gel or column purification. In such cases, PCR sample should be phenol extracted and ethanol precipitated. The columns and extraction procedures should not be used instead of a preparative gel unless it is known that only a single, homogeneous product has been amplified during PCR.

5. Carry out chemical sequencing using 8% to 10% of product for each sequencing reaction. After electrophoresis, wrap sequencing gel in plastic wrap and autoradiograph 2 to 48 hr at −20°C.

15.2

GENOMIC SEQUENCING OF PCR PRODUCTS

ALTERNATE PROTOCOL 4

This method is ideally suited to situations where large amounts of sequence are sought, or where more than one region is being amplified. Several amplified fragments can be mixed, simultaneously sequenced, run out on a single set of lanes, and the sequence of the different fragments successively visualized by the use of appropriate probes.

Additional Materials *(also see Basic Protocol 2)*

Filter paper, precut to gel size (Schleicher & Schuell)

1. Carry out PCR under optimized conditions, ending with final long extension step using 20-1000 ng for eukaryotic genomic DNA, 10-100 ng for bacterial DNA, and 1-20 ng for cloned DNA inserts. After final long extension step, remove mineral oil with Pasteur pipet.

2. Phenol extract product. Precipitate with 0.1 vol of 3 M sodium acetate, pH 5.2, and 2.5 vol of 100% ethanol by placing sample 20 min in dry ice/ethanol bath. Microcentrifuge 5 min, wash pellet twice with 70% ethanol, and dry under vacuum.

3. Resuspend pellet in 30-50 µl water. Electrophorese 10% of sample on an agarose or polyacrylamide gel to determine yield and purity.

   If amplification yields ≥2 bands of similar intensity, gel separation prior to sequencing is necessary, unless different products can be separately visualized using different oligonucleotide probes.

4. Carry out chemical sequencing using 8% to 10% of the amplified product for each sequencing reaction. Electrophorese on denaturing polyacrylamide sequencing gel.

5. Carefully remove one of glass gel plates. Place filter paper carefully on gel, making sure not to trap air bubbles between paper and gel. Lift gel (adhered to filter paper) onto appropriate transfer apparatus.

6. Wet gel with thin layer of TBE electrophoresis buffer and place nylon membrane over gel, being careful not to trap air bubbles between gel and membrane. Transfer DNA by electroblotting and UV-cross-link DNA onto filter (*UNIT 2.9*).

7. Hybridize membrane with appropriate probe. The membrane can now be probed, visualized, stripped, and reprobed as many as 40 times.

   Best results are obtained by probing with PCR oligonucleotide primers, ensuring that only completed PCR products are visualized. Degenerate probes or probes internal to the amplified fragment can also be used. A high-specific-activity probe can be obtained by "tailing" 3 to 6 pmol of probe using terminal deoxynucleotidyltranferase (UNIT 3.6).

15.3

Reference: Church and Kieffer-Higgins, 1988.

Contributors: Robert L. Dorit, Osamu Ohara, and Charles B.-C. Hwang

UNIT 15.3

Quantitation of Rare DNA by PCR

BASIC PROTOCOL

In this unit, PCR is used to quantitate the numbers of a particular DNA sequence from 1 to 20,000 molecules per sample. In addition, it helps assess the presence of contaminating sequences, the bane of this kind of procedure. For simplicity, this protocol is written in terms of quantitating viral DNA molecules relative to host cellular sequences; however, it can be readily adapted for other applications.

Materials *(see APPENDIX 1 for items with ✓)*

Cells or tissue sample (*UNIT 2.2*)
✓ Proteinase digestion buffer
20 mg/ml proteinase K (store at −20°C)
Phenol buffered with 50 mM Tris·Cl/10 mM EDTA, pH 7.4 (store at room temperature)
24:1 chloroform/isoamyl alcohol (*UNIT 2.1*)
✓ 10 M ammonium acetate
100% ethanol (ice cold) and 70% ethanol
✓ TE buffer, pH 7.5
✓ Reaction mix cocktail

Mineral oil
0.8 U/μl *Taq* DNA polymerase (*UNIT 15.1*)
✓ Oligonucleotide primers for hybridization (for step 16)

Screw-cap microcentrifuge tubes, autoclaved
Microcapillary pipets
94°, 72°, and 55°C water baths, or automated thermal cycler

NOTE: Use sterile, distilled water to prepare all reagents. Do NOT use diethylpyrocarbonate (DEPC) to treat reagents. To avoid contamination with unwanted nucleic acids, prepare reagents and solutions solely for use in this protocol. Wear disposable gloves and change them frequently.

1. Place cells or tissue sample in screw-cap microcentrifuge tube. Add ~100 μl proteinase digestion buffer per ~2×10^6 cells and 20 mg/ml proteinase K to 100 μg/ml. Incubate sample overnight at 50°C.

   Always include several negative control samples that contain no viral sequences. It is best to process samples in order of increasing likelihood of their containing sequences of interest. Perform extractions away from where PCR products or large quantities of plasmid DNA are handled.

2. Mix sample gently. Add 100 μl buffered phenol, mix gently, add 100 μl of 24:1 chloroform/isoamyl alcohol, and mix gently. Microcentrifuge 5 min and transfer aqueous phase to new microcentrifuge tube.

   Screw-cap microcentrifuge tubes and microcapillary pipets minimize aerosol contamination that commonly results from microcentrifuges and automatic pipettors. Dedicate a single bulb for each reagent used. Positive-displacement pipets with disposable tips and plungers are a more expensive, but easier-to-use alternative to microcapillary pipets. Avoid using a microcentrifuge that has been used for PCR products or large amounts of plasmid DNA.

3. Back-extract organic phase with proteinase digestion buffer (~½ original volume). Microcentrifuge 5 min and add aqueous phase to aqueous phase from step 2.

4. Extract aqueous phase twice with equal volume of 24:1 chloroform/isoamyl alcohol, centrifuging 5 min each time to separate phases.

5. To aqueous phase, add 10 M ammonium acetate to 2.5 M final and 2.5 vol cold 100% ethanol. Mix and place on dry ice 30 min. Microcentrifuge 15 min at 4°C, wash pellet with 70% ethanol, microcentrifuge again, and dry pellet.

   Ammonium acetate is critical during ethanol precipitation for subsequent efficient amplification. Avoid using a desiccator or evaporator that has been used for PCR products or large amounts of plasmid DNA.

6. Resuspend pellet in TE buffer, pH 7.5 (100 μl for sample prepared from 2×10^6 cells at 5-100 ng/μl). Store DNA at 4°C.

7. Estimate DNA concentration by running aliquot on agarose gel alongside standard DNA (*UNIT 2.5*) or by ethidium bromide dot quantitation (*UNIT 2.6*).

8. Prepare one tube each containing 110 and 90 ng DNA from uninfected cells or tissue, and several tubes containing 100 ng DNA. Use these tubes to make a set of 10-fold serial dilutions of sequence of interest as follows: Add a known amount (e.g., 20,000 molecules) of DNA containing sequence of interest to stock DNA and mix. Then add ⅒ of this material to stock DNA and mix. Add ⅒ of this material to new stock DNA and mix. Repeat several more times until

tube contains ≤10 molecules of sequence of interest. Add 1⁄10 of material from that tube to tube containing 90 ng of DNA. The final volume of each tube should be ≤71 μl.

9. For each amplification reaction, prepare screw-cap microcentrifuge tube containing 24 μl reaction mix cocktail and enough sterile distilled water for 100 μl final (after addition of DNA and *Taq* DNA polymerase). Mix and overlay each reaction with ~100 μl mineral oil.

10. Open only those tubes that will contain equivalent samples.Add 100 ng sample DNA to each appropriate tube and close tubes. Microcentrifuge briefly to mix.

    Begin with negative controls. Proceed in order of increasing likelihood that sample contains sequences of interest, adding to dilution tubes last.

11. Heat-denature samples 1 min at 94°C in water bath or in automated cycler.

12. Open tubes (containing equivalent samples) and add 5 μl of 0.8 U/μl *Taq* DNA polymerase. Close tubes and repeat steps 10 to 12 with next set of equivalent samples.

13. Microcentrifuge tubes briefly. Cycle tubes one time for 2 min at 55°C (reannealing) and 3 min at 72°C (extension).

14. Cycle tubes 29 times for 1 min at 94°C, 2 min at 55°C, and 3 min at 72°C. Extend for an additional 7 min at 72°C. Store completed reactions at 4°C.

15. Electrophorese aliquots (1⁄10 reaction) on appropriate agarose or polyacrylamide gel (*UNITS 2.5A & 2.7*). Include lanes with DNA molecular weight markers that will be visible both upon ethidium bromide staining and by autoradiography. Stain and photograph gel. The DNA fragment corresponding to PCR product from host DNA should be readily visible and measures the efficiency of each amplification reaction (don't be surprised to see other bands, especially of higher molecular weights, which are nonspecific PCR products).

16. Transfer to nitrocellulose/nylon membrane (*UNIT 2.9*) and UV-cross-link DNA to filter. Prehybridize and hybridize with end-labeled oligonucleotide specific for sequence of interest. Analyze by autoradiography.

    If each reaction is amplified with similar efficiency, the expected results are labeled bands of the appropriate size at monotonically decreasing intensities in the reconstructed dilution series, no bands in the negative controls at that size, and bands with varying intensities in the experimental samples. Depending on the probe and stringency of the hybridization and wash conditions, nonspecific sticking of the probe to the abundant internal control product may be seen. It is also common to see minor specific PCR products of slightly greater or lower mobility than the major specific product, especially at high copy number.

    A reasonable estimate for the number of molecules of sequence of interest in each sample can be obtained by comparison with dilution series, assuming that each reaction amplified with similar efficiency.

17. For more quantitative analysis, strip filter of previous probe by boiling in water 15 min (if necessary) and hybridize with probe specific for host single-copy sequence (*UNITS 2.10, 3.10, 4.6, 4.8, & 6.4*). Quantitate signals by densitometric scanning and compute standard curve from dilution series, normalizing to host sequence signals. Determine number of molecules in experimental samples by interpolation from the standard curve.

The standard curve will ideally give a linear relationship between log of autoradiographic signal and log of amount of DNA. However, it may not necessarily be completely linear, and it is unlikely to have a slope of 1. At the high end (0.1 to 1 copy per cell equivalent), it may plateau. If this is a problem, it may be necessary to reduce the number of amplification cycles or vary other parameters (UNIT 15.1).

Reference: Katz et al., 1990.

Contributor: Donald M. Coen

Enzymatic Amplification of RNA by PCR

UNIT 15.4

This unit describes methods for enzymatic amplification of RNA by PCR. The Basic Protocol is especially useful for rare RNAs because all steps (annealing, reverse transcription, and amplification) are performed under optimal conditions, maximizing efficiency and recovery. It is also recommended when amplifying heterogeneous RNA populations or large RNAs.

PCR AMPLIFICATION OF RNA UNDER OPTIMAL CONDITIONS

BASIC PROTOCOL

Materials *(see APPENDIX 1 for items with ✓)*

Poly(A)$^+$ RNA (*UNIT 4.5*) or crude RNA (Support Protocol)
25 µg/ml cDNA primer in H_2O
✓ 3 M sodium acetate, pH 5.5
100% and 70% ethanol
✓ 400 mM Tris·Cl, pH 8.3
✓ 400 mM KCl
✓ Reverse transcriptase buffer
32 U/µl AMV reverse transcriptase (*UNIT 3.7*; Boehringer-Mannheim)
10 mM Tris·Cl/10 mM EDTA, pH 7.5
Phenol buffered with 10 mM Tris·Cl/10 mM EDTA, pH 7.5 (store at room temperature)
Chloroform
~150 µg/ml amplification primers in H_2O (~20 µM each)
5 mM 4dNTP mix: 5 mM each dNTP in H_2O (*UNIT 3.4*)
✓ 10× amplification buffer
Taq DNA polymerase
Mineral oil
94°, 55°, and 72°C water baths or automated thermal cycler

NOTE: Reagents and solutions used for RNA procedures should be prepared using standard methods for handling RNA (*UNIT 4.1*).

1. Coprecipitate RNA and cDNA primer by adding the following to microcentrifuge tube: 2 µg poly(A)$^+$ RNA, 25 ng (3 pmol) cDNA primer, and H_2O to 90 µl. Mix and add 10 µl of 3 M sodium acetate, pH 5.5, and 200 µl of 100% ethanol. Incubate overnight at −20°C or 15 min at −70°C.

   The template source can also be total RNA or crude RNA (Support Protocol); <2 µg can be used, depending upon abundance of RNA within sample. The cDNA primer is frequently the same as one of the amplification primers.

2. Microcentrifuge 15 min and discard supernatant. Add 200 µl of 70% ethanol and mix gently. Microcentrifuge 5 min and dry pellet briefly.

3. Add the following to RNA pellet: 12 µl H_2O, 4 µl 400 mM Tris·Cl, pH 8.3, and 4 µl 400 mM KCl. Heat to 90°C, then cool slowly to 67°C.

4. Microcentrifuge sample briefly and incubate 3 hr at 52°C. Microcentrifuge 1 sec to collect condensate.

   This final annealing temperature can be adjusted according to base composition of primer and increased or decreased depending on specificity of annealing required (UNIT 6.4).

5. Add 29 µl reverse transcriptase buffer and 0.5 µl (16 U) AMV reverse transcriptase. Mix and incubate 1 hr at 42°C (adjust from 37° to 55°C depending on base composition of primer and RNA).

6. Add 150 µl 10 mM Tris·Cl/10 mM EDTA, pH 7.5, and mix. Add 200 µl buffered phenol and vortex briefly. Microcentrifuge 5 min.

7. Add 200 µl chloroform to aqueous phase and vortex briefly. Microcentrifuge 5 min.

8. Add 20 µl 3 M sodium acetate, pH 5.5, and 500 µl of 100% ethanol to aqueous phase. Mix and precipitate overnight at −20°C or 15 min at −70°C. Microcentrifuge 15 min, dry pellet briefly, and resuspend in 40 µl water.

9. Mix the following:

   5 µl cDNA (step 8)
   5 µl each amplification primer (~150 µg/ml or ~20 µM each)
   4 µl 5 mM 4dNTP mix
   10 µl 10× amplification buffer
   70.5 µl H_2O.

   Heat 2 min at 94°C. Microcentrifuge 1 sec to collect condensate.

   Usually one of the amplification primers is the same as cDNA primer. If a different amplification primer is used, the cDNA primer should be removed from cDNA reaction as in UNIT 15.2.

10. Add 0.5 µl (2.5 U) *Taq* DNA polymerase, mix, and microcentrifuge 1 sec. Overlay with 100 µl mineral oil.

11. Set up the following automated amplification cycles:

39 cycles:	2 min	55°C
	2 min	72°C
	1 min	94°C
1 cycle:	2 min	55°C
	7 min	72°C.

   The number of cycles can be varied depending upon abundance of RNA. Forty total cycles is sufficient for rare mRNAs in 2 µg poly(A)$^+$ RNA, but more cycles may be necessary if smaller amounts of template are used.

12. Analyze products by electrophoresis in agarose or nondenaturing polyacrylamide gels (UNITS 2.5A & 2.7).

AVOIDING LENGTHY COPRECIPITATION AND ANNEALING STEPS

ALTERNATE PROTOCOL 1

This protocol saves considerable time compared to the Basic Protocol because it eliminates the relatively lengthy coprecipitation and annealing. Instead, a quick heating and cooling is used. Although this reduces the specificity and quantitative efficiency of annealing, the losses incurred are frequently insignificant and should not affect most applications.

Replace steps 1 to 4 of Basic Protocol with the following:

1. Mix the following: 2 µg poly(A)$^+$ RNA, 25 ng (3 pmol) cDNA primer, and H_2O to 21 µl. Heat 3 to 15 min at 65°C.
2. Cool on ice and add to reverse transcriptase buffer as in step 5.

INTRODUCING cDNA DIRECTLY INTO THE AMPLIFICATION STEP

ALTERNATE PROTOCOL 2

This protocol saves time by eliminating the extraction and precipitation steps following cDNA synthesis in the Basic Protocol. The products of the cDNA reaction are diluted sufficiently to avoid interference when added to the enzymatic amplification step. This procedure may not be appropriate when the anticipated number of template molecules is small.

Replace steps 5 to 8 of Basic Protocol with the following:

1. Add 450 µl of 10 mM Tris·Cl/10 mM EDTA, pH 7.5, to 50 µl of reaction mix from step 4 of Basic Protocol.
2. Add 1 to 5 µl of this mixture to amplification reaction in step 9.

RAPID PREPARATION OF CRUDE RNA

SUPPORT PROTOCOL

In this protocol, RNA needn't be purified prior to enzymatic amplification.

Additional Materials *(also see Basic Protocol; see* APPENDIX 1 *for items with ✓)*

✓ PBS
✓ Diethylpyrocarbonate (DEPC) solution

CAUTION: DEPC is volatile and toxic and should be handled with precautions. DEPC is also rapidly inactivated in aqueous buffers and in the presence of Tris·Cl, so work quickly.

1. Suspend 10^6-10^7 cells (or equivalent amount of dispersed tissue) in PBS. Microcentrifuge 5 min.
2. Resuspend cell pellet in 200 to 400 µl DEPC solution and vortex briefly. Microcentrifuge nuclei 10 sec and transfer supernatant to new tube. Incubate 20 min at 37°C, then 10 min at 90°C.
3. Microcentrifuge 5 min at 4°C and transfer supernatant to a new tube. Use 5 to 10 µl directly in step 1 of Basic Protocol or ethanol precipitate and resuspend (*UNIT 2.1*). Store frozen.

Reference: Frohman et al., 1988.

Contributor: Stephen M. Beverley

Ligation-Mediated PCR for Genomic Sequencing and Footprinting

The polymerase chain reaction (PCR) can be used to exponentially amplify segments of DNA located between two specified primer hybridization sites in a single-sided PCR method that initially requires specification of only one primer hybridization site; the second is defined by the ligation-based addition of a unique DNA linker (Fig. 15.5.1). This linker, together with the flanking gene-specific primer, allows exponential amplification of any fragment of DNA. Because a defined, discrete-length sequence is added to every fragment, complex populations of DNA such as sequence ladders can be amplified intact with retention of single-base resolution.

NOTE: It is essential that fresh, high-quality reagents be used throughout this unit. In addition, all solutions should be prepared with glass-distilled water.

BASIC PROTOCOL

LIGATION-MEDIATED SINGLE-SIDED PCR

Materials *(see* APPENDIX 1 *for items with ✓)*

- 0.4 µg/µl cleaved genomic DNA in TE buffer, pH 7.5 (Support Protocol 1, 2, or 3)
- ✓ First-strand synthesis mix, containing oligonucleotide primer 1 (Figs. 15.5.2 and 15.5.3)
- ✓ 20 µM unidirectional linker mix
- ✓ Ligase dilution solution
- ✓ Ligase mix
- 2000 to 3000 "Weiss" U/ml T4 DNA ligase (*UNIT 3.14*; Promega or Pharmacia LKB)
- ✓ Precipitation salt mix
- 75% and 100% ethanol
- ✓ Amplification mix, containing linker primer and primer 2
- ✓ 2 U/µl Vent DNA polymerase mix
- Mineral oil
- ✓ End-labeling mix, containing end-labeled primer 3 (Fig. 15.5.3)
- ✓ Vent DNA polymerase stop solution
- 25:24:1 (v/v/v) phenol/chloroform/isoamyl alcohol (*UNIT 2.1*)
- ✓ Loading buffer

- 1.5-ml microcentrifuge tubes, silanized (*APPENDIX 3*) and with Lid-Loks (optional; Intermountain Scientific)
- 4° and 17°C water baths
- Automated thermal cycler *or* water baths at 60°, 76°, 95° and 60°–70°C

1. Transfer 5 µl (2 µg) cleaved genomic DNA to a silanized 1.5-ml microcentrifuge tube and chill several minutes in an ice-water bath.

   The cleaved DNA sample must be clean; contaminants left in the DNA (e.g., piperidine) will interfere with the reaction.

   High-quality, reproducible footprinting or sequencing reactions can be obtained starting from 6×10^5 genomes (3×10^5 diploid nuclei). This corresponds to 2 µg of mouse (mammalian) genomic DNA. For other species, the recommended minimum number of haploid genomes remains 6×10^5, but different genome sizes lead to corresponding differences in the absolute mass of DNA.

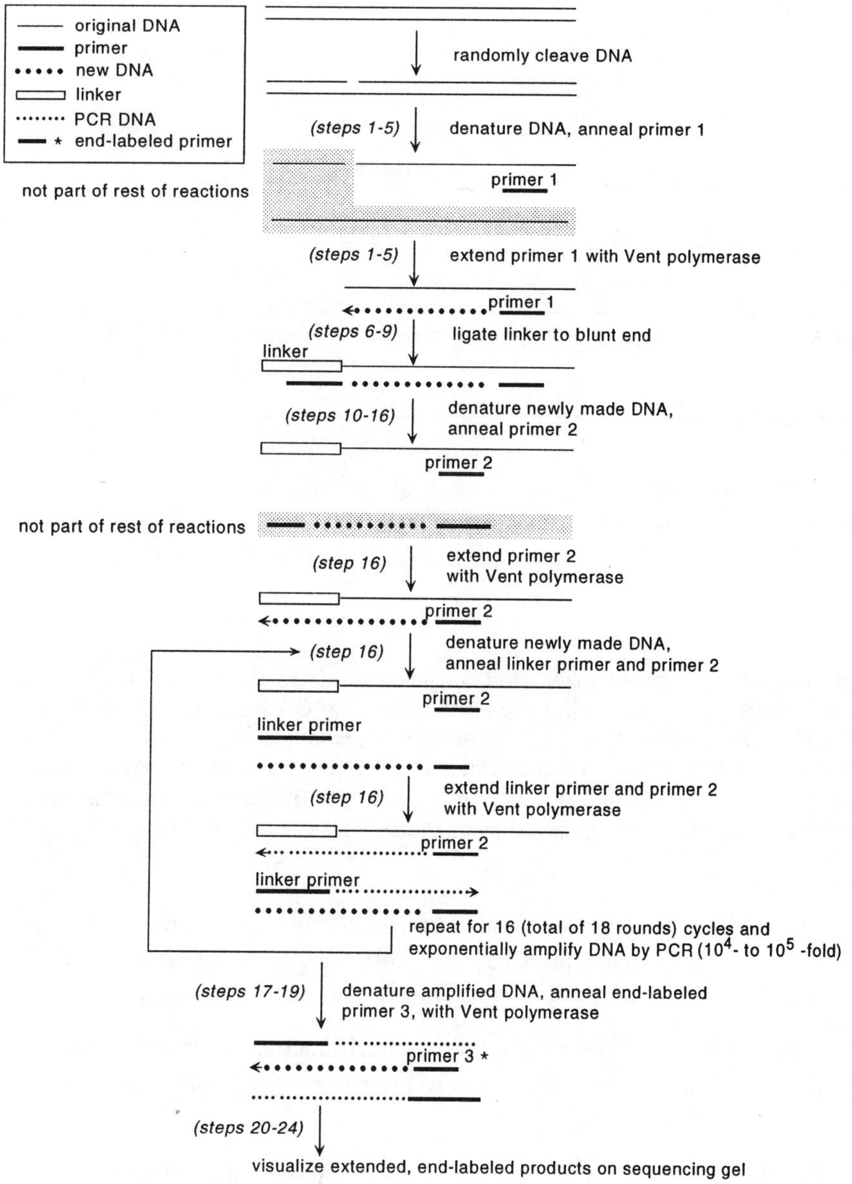

Figure 15.5.1 Flow chart of ligation-mediated PCR protocol (see text for details). The steps correspond to those listed in the Basic Protocol.

2. Prepare first-strand synthesis mix containing primer 1 and chill several minutes in an ice-water bath. Add 25 µl to DNA sample, gently mix with a pipettor, and return sample to the ice-water bath. Place Lid-Loks on sample to prevent tube from popping open during denaturation.

3. Denature DNA 5 min at 95°C, anneal primer 30 min at 60°C, and extend 10 min at 76°C. As first-strand synthesis proceeds, prepare solutions in step 4.

First-strand synthesis is most easily performed in an automated thermal cycler, but can be performed manually by transferring the tubes to different temperature water baths. Annealing should be carried out at 2°C above the T_m *and extension should be carried out at 76°C to minimize background. The extension step creates a blunt-end substrate for the subsequent ligation reaction (see Fig. 15.5.1).*

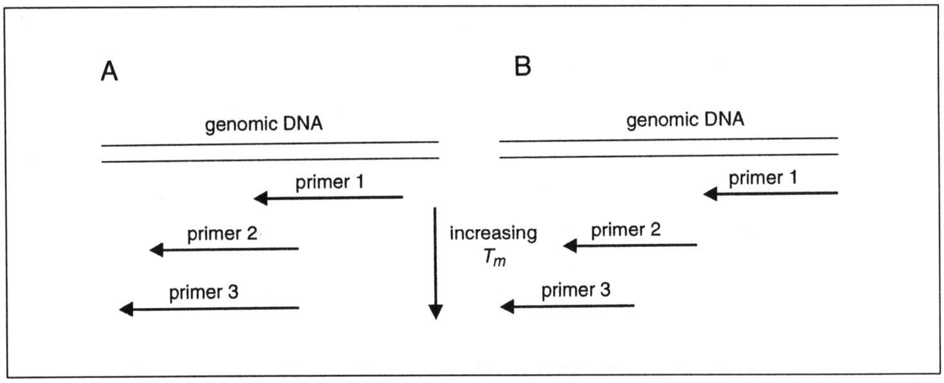

Figure 15.5.2 Two possible arrangements of gene-specific primers (see Critical Parameters).

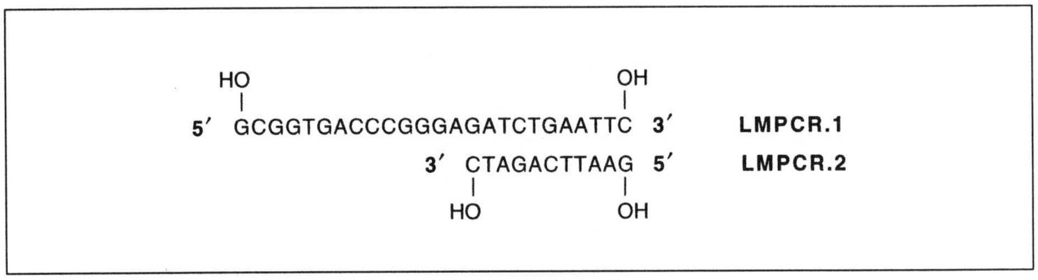

Figure 15.5.3 Staggered linker used in ligation-mediated PCR (LMPCR). This staggered linker is made by annealing the oligonucleotides LMPCR.1 and LMPCR.2 to each other. LMPCR.1 has 25 bases and a GC content of 60%, whereas LMPCR.2 has 11 bases and a GC content of 36%. Other oligonucleotides could be used in place of these as long as they meet the criteria defined in text. Note the lack of 5′ phosphates on oligonucleotides (i.e., 5′ OH instead) and the unconventional orientation of LMPCR.2 (3′ to 5′) in the figure.

4. Thaw 20 µM unidirectional linker mix in ice-water bath. Prepare ligase dilution solution and partially prepare ligase mix but do not add unidirectional linker and T4 DNA ligase until step 6. Chill both in ice-water bath.

5. When extension in step 3 is complete, immediately transfer sample to the ice-water bath. Microcentrifuge tube briefly at 4°C to collect condensation, then return to ice-water bath.

6. Finish preparing ice-cold ligase mix from step 4 by adding unidirectional linker mix, mixing, adding T4 DNA ligase, mixing, and keeping in ice-water bath.

7. Add 20 µl ice-cold ligase dilution solution from step 4 to sample, gently mix with a pipettor, and return sample to the ice-water bath. Add 25 µl ice-cold ligase mix from step 6 to sample, gently mix with a pipettor, and return to the ice-water bath. Microcentrifuge tube briefly at 4°C and incubate overnight in a 17°C water bath.

8. Prepare sample for precipitation by placing in the ice-water bath for several minutes. Microcentrifuge tube briefly at 4°C, then return to ice-water bath.

9. Prepare and chill precipitation salt mix. Add 9.4 µl ice-cold precipitation salt mix and 220 µl of ice-cold 100% ethanol to the sample. Mix thoroughly by inversion and chill ≥2 hr at –20°C.

10. Microcentrifuge precipitated ligation reaction 15 min at 4°C and discard supernatant.

11. Add 500 µl of room-temperature 75% ethanol and invert several times to wash pellet and walls of tube. Microcentrifuge sample ~5 min at room temperature and discard supernatant. Remove last traces of ethanol with a pipettor and allow any remaining ethanol to evaporate by air drying or by using a Speedvac evaporator.

12. Add 70 µl water and leave sample at room temperature to dissolve pellet. Vortex tube occasionally to assist dissolution, and after each vortexing, collect droplets by microcentrifuging 2 to 3 sec. When pellet is dissolved (usually ≤30 min), chill sample in ice-water bath. While pellet is dissolving, prepare and chill amplification mix containing linker primer and primer 2.

13. Add 30 µl ice-cold amplification mix to sample, gently mix with a pipettor, and return sample to the ice-water bath.

14. Add 3 µl (1 U) Vent DNA polymerase mix to sample and mix carefully with pipettor. Return sample to the ice-water bath.

    The reaction is very sensitive to the amount of Vent DNA polymerase used; excess polymerase results in high background.

15. Cover sample with 90 µl mineral oil, microcentrifuge briefly at 4°C, and return to the ice-water bath.

16. Carry out 18 cycles of PCR. The first denaturation should be 3 to 4 min at 95°C, and subsequent ones for 1 min. Anneal primers 2 min at a temperature 0° to 2°C above the calculated T_ms (if primer 2 and linker primer have different T_ms, use the lower T_m). Extend 3 min at 76°C; for every cycle, add an extra 5 sec to the extension step. Allow final extension to proceed 10 min. Transfer sample to ice-water bath, remove Lid-Lok (if applicable), and microcentrifuge briefly at 4°C to collect any condensation. Keep sample in ice-water bath.

    An important parameter in this step is the temperature of denaturation. If it is too low, no signal or shortened sequence ladders will be seen; if it is too high, the polymerase will be destroyed.

17. Prepare end-labeling mix containing labeled primer 3 and chill several minutes in the ice-water bath. Add 5 µl to the sample. Mix aqueous phase by gently pipetting up and down, keeping the sample on ice as much as possible. Microcentrifuge briefly at 4°C and return to ice-water bath.

    CAUTION: *The label mix contains a significant amount of ^{32}P. Appropriate care should be taken in handling and disposal of mix and samples to which it is added.*

18. Carry out two rounds of PCR to label the DNA. The first denaturation should be 3 to 4 min at 95°C; the second 1 min. Anneal end-labeled primer 3 for 2 min at a temperature 0° to 2°C above its calculated T_m. Extend 10 min at 76°C. When second extension is complete, transfer sample to ice-water bath.

19. Place sample at room temperature and immediately add 295 µl Vent DNA polymerase stop solution. Mix by vortexing. Microcentrifuge briefly to collect radioactive droplets from top and sides of tube. Add 500 µl phenol/chloroform/isoamyl alcohol and mix either by vigorous shaking or vortexing. Microcentrifuge 3 to 5 min at room temperature. Transfer upper aqueous layer (~400 µl; avoid interface if any) to a clean, silanized 1.5-ml microcentrifuge tube, and thoroughly mix. Microcentrifuge briefly to collect droplets from sides of tube.

20. Set up four clean, silanized 1.5-ml microcentrifuge tubes and add 235 µl of room-temperature 100% ethanol to each tube. Transfer 94 µl of the aqueous layer into each tube, thoroughly mix by vortexing, and chill ≥2 hr at −20°C. Discard any remaining aqueous layer.

21. Microcentrifuge precipitated samples 15 min at 4°C and discard supernatants.

22. Add 500 µl of room-temperature 75% ethanol, vortex, microcentrifuge samples 5 min at room temperature, and discard supernatants. Remove last traces of ethanol by using a pipettor and allow any remaining ethanol to evaporate by air drying or by using a Speedvac evaporator.

23. Add 7 µl loading buffer to each tube and leave at room temperature while pellets are dissolving. Vortex occasionally to assist pellets in dissolving and to recover any of the DNA pellet that is on the side of the tube. After each vortexing, collect droplets by microcentrifuging 2 to 3 sec. Samples usually resuspend rapidly (within 5 min). Check resuspension by removing the sample from the tube with a pipet and using a Geiger counter to ensure that the radioactivity is in the sample and not left behind in the tube. Return sample to same tube. If >10% of the total radioactivity has remained in the tube, vortex, microcentrifuge, and repeat the resuspension check.

24. Denature samples 5 min at 85° to 90°C. Load entire content of each tube on a 6% 0.35- to 0.56-mm thick sequencing gel. At the completion of the run, fix and dry gel, then autoradiograph 6 to 24 hr without an intensifying screen.

    The ladder will appear 25 bases longer than the original footprinting or sequencing products due to the addition of the 25-base linker.

SUPPORT PROTOCOL 1

PREPARATION OF GENOMIC DNA FROM MONOLAYER CELLS FOR DMS FOOTPRINTING

This protocol describes the preparation of in vivo and in vitro dimethyl sulfate (DMS)–treated genomic DNA from monolayer cells for use in conjunction with ligation-mediated PCR. DMS methylates guanine residues at the N7 position, rendering them susceptible to subsequent cleavage with piperidine.

This procedure works well with ~5 × 10^7 cells/set (i.e., one 15-cm plate at ~70% confluence for many fibroblasts). Extractions can be scaled up and <10^7 cells/set will work if extra precautions are taken in precipitating and handling the DNA.

Materials *(see* APPENDIX 1 *for items with ✓)*

✓ PBS
Tissue culture medium appropriate for sample cells
✓ Lysis solution
Duplicate 15-cm plates of cells in monolayer culture
100% dimethyl sulfate (DMS; Aldrich)
Equilibrated buffered phenol (*UNIT 2.1*)
✓ 25:24:1 (v:v:v) phenol/chloroform/isoamyl alcohol
24:1 (v/v) chloroform/isoamyl alcohol
Ethyl ether
Isopropanol
✓ TE buffer, pH 7.5
✓ 3 M sodium acetate, pH 7.0
100% and 75% ethanol

✓ DMS stop buffer, ice-cold
Piperidine (Aldrich)
8 M ammonium acetate

50- and 15-ml disposable polypropylene screw-cap tubes (e.g., Corning)
Aspirator attached to waste flask
Disposable cell scraper
Table top centrifuge (e.g., IEC Centra-7R)
1.5-ml microcentrifuge tubes, silanized (*APPENDIX 3*) and with Lid-Loks (Intermountain Scientific)
Sealed Pasteur pipet or thin glass rod

CAUTION: DMS and concentrated piperidine are extremely toxic. All manipulations that use DMS or piperidine should be performed in a properly functioning fume hood. Before starting this procedure, review precautions for working with and disposing of DMS and piperidine.

1. Prepare the following solutions before beginning the procedure: Prewarm PBS (~75 ml/15-cm plate) to 37°C in a water bath inside a fume hood. Aliquot 24 ml tissue culture medium into a 50-ml disposable polypropylene screw-cap tube and prewarm to 37°C as for PBS. Prepare lysis solution.

2. *For control cells and in vitro–treated DNA and cells:* Remove and discard medium from cells on one 15-cm tissue culture plate, and immediately add ~25 ml prewarmed PBS. Gently swirl plate a few times and remove and discard PBS. Immediately proceed to step 5.

3. *For in vivo DMS-treated cells:* Remove and discard medium from cells on the second 15-cm tissue culture plate. Add 24 µl of 100% DMS to aliquot of prewarmed medium (0.1% DMS final). Screw on cap, mix by inversion, and gently pour entire contents onto tissue culture plate. Leave this 0.1% DMS medium on cells exactly 2 min.

   Do not add DMS to the medium until immediately before it is needed; the water in the medium quickly inactivates it.

   It may be necessary to modify the incubation time and/or DMS concentration to get an optimal DMS footprint.

4. Remove and discard the 0.1% DMS medium using an aspirator attached to a flask to hold the DMS waste. Pour ~25 ml prewarmed PBS gently on the cells. Gently swirl plate a few times and aspirate PBS. Wash plate three times with prewarmed PBS, allowing PBS to sit on cells ~30 sec each wash.

5. *For both untreated and treated cells:* After removing as much PBS as possible, pipet 1.5 ml lysis solution onto cells. Gently tip and rock plate to spread solution evenly over cells, then let plate sit at room temperature ~5 min.

6. Tilt plate and scrape DNA/lysed-cell slurry to one side with a disposable cell scraper. Slowly remove as much of slurry as possible with a pipet and transfer to 15-ml polypropylene tube.

7. Incubate lysed-cell slurry 3 to 5 hr at 37°C. Mix by inversion every 30 to 60 min.

   During this incubation, proteinase K is digesting cellular proteins. After incubation, samples may be stored at −20°C indefinitely. They should be thawed and warmed to room temperature before proceeding with the protocol.

8. Add 1.25 vol buffered phenol, mix thoroughly by gently inverting ~30 times, and centrifuge in a tabletop centrifuge 10 min at 1300 × *g*, room temperature. Slowly remove phenol (bottom layer) from underneath by inserting a 9-in. Pasteur pipet through the aqueous layer and interface to the bottom of the tube. If the viscous DNA begins to trail along into the organic layer, it can be freed by blowing a few bubbles of air through the pipet using a pipet bulb. Do not remove the interface; much of the DNA may be trapped in it. Repeat phenol extraction once.

9. Add 1 vol phenol/chloroform/isoamyl alcohol, mix thoroughly by gently inverting ~30 times, and centrifuge 10 min at 500 × *g*, room temperature. Remove phenol/chloroform/isoamyl alcohol (bottom layer) from underneath with a 9-in. Pasteur pipet as in step 8. Leave the interface, if any, behind. Repeat this extraction once.

10. Add 1 vol of chloroform/isoamyl alcohol, mix thoroughly by gently inverting ~30 times, centrifuge 5 min at 200 × *g*, room temperature. Remove chloroform/isoamyl alcohol (bottom layer) from underneath with a 9-in. Pasteur pipet as in step 8.

11. Add 1 vol ethyl ether, mix thoroughly by gently inverting ~30 times, and allow phases to separate by gravity (~1 min). Remove ethyl ether (top layer) with a Pasteur pipet; evaporate any remaining ether in a fume hood (~5 min).

    CAUTION: *Ethyl ether is highly flammable; use appropriate caution and work in a fume hood.*

12. Add 1 vol isopropanol and mix thoroughly by gently inverting ~30 times.

13. Spool the stringy DNA precipitate around the end of a sealed 9-in. Pasteur pipet or thin glass rod. Carefully lift the spooled DNA out of the isopropanol solution, and gently touch the DNA to the inside of the tube to remove excess liquid.

14. Unspool the DNA in 3 ml TE buffer, pH 7.5. Redissolve by gently rocking the sample several hours to overnight at room temperature.

15. Add 330 μl of 3 M sodium acetate and 6.7 ml of ice-cold 100% ethanol; mix thoroughly by gently inverting ~30 times. Reprecipitate and spool DNA as in steps 12 and 13.

16. Unspool DNA in 200 to 500 μl TE buffer, pH 7.5. Redissolve DNA by storing it overnight at room temperature or for several days at 4°C, inverting tube gently a few times during the process. Quantitate DNA using a spectrophotometer and adjust concentration to 0.5 to 1 mg/ml. Store at 4°C.

    The control DNA (not treated with DMS) is of high quality, and at this point can also be used for genomic sequencing (Support Protocol 3) and genomic methyl cytosine analysis (Garrity and Wold, 1992).

17. Put ~75 to 175 μg control DNA in a silanized 1.5-ml microcentrifuge tube. Add TE buffer to bring the volume to 175 μl.

18. Make a 1% DMS solution by adding 5 μl of 100% DMS to 495 μl water. Mix thoroughly by vortexing 25 sec and microcentrifuge briefly to collect droplets.

    Several concentrations of DMS may have to be tested to find conditions that will perfectly match the in vivo conditions used in steps 3 and 4.

19. Add 25 μl of 1% DMS solution to the control DNA and mix thoroughly by vortexing gently ~25 sec. Avoid getting the DMS solution on the lid; microcentrifuge briefly if it does. Incubate exactly 2 min at room temperature, then add 50 μl ice-cold DMS stop buffer. Immediately add 750 μl of 100% ethanol prechilled on dry ice, mix by vigorously shaking, and plunge tube into powdered dry ice. Leave sample in dry ice for ~30 min.

20. Prepare in vivo–treated DNA for piperidine treatment in parallel with the in vitro–treated DMS samples. Mix 200 μl DNA (from step 16) with 50 μl ice-cold DMS stop buffer and vortex briefly (3 sec). Add 750 μl of ice-cold 100% ethanol, mix by vigorously shaking, and plunge tube into dry ice. Leave sample in dry ice ~30 min.

21. Microcentrifuge the two precipitated DNA samples (from steps 19 and 20) for 10 min at 4°C. Remove and discard supernatants, add 1 ml room-temperature 75% ethanol to the DNA pellets, and vortex briefly (~5 sec) until each pellet is dislodged. Microcentrifuge 10 min at 4°C, then remove and discard supernatants.

22. Dilute piperidine 1:10 in water (to 1 M final) and add 200 μl to each DNA pellet. Resuspend DNA by incubating at room temperature with intermittent vortexing; pellets usually dissolve in ~15 min. Carefully examine the samples to ensure that the DNA is dissolved; an undissolved pellet will appear as a clear, floating lens in the 1 M piperidine.

    CAUTION: *Aliquot piperidine in a fume hood.*

23. When pellets are completely dissolved, microcentrifuge briefly to collect droplets and place Lid-Loks on tubes. Heat 30 min at 90°C in the fume hood.

    Heat treatment with piperidine cleaves the genomic DNA at the methylated guanines, denatures the DNA, and destroys contaminating RNA.

24. Remove Lid-Loks and briefly microcentrifuge tubes to collect condensation. Chill samples on dry ice for 10 min, then evaporate piperidine in a Speedvac evaporator for 1 to 2 hr at room temperature.

25. Resuspend pellets in 360 μl TE buffer, pH 7.5, add 40 μl of 3 M sodium acetate, pH 7.0, and mix by vortexing. Add 1 ml of ice-cold 100% ethanol. Shake vigorously to mix and chill at −20°C for ≥2 hr.

26. Microcentrifuge samples 15 min at 4°C and discard supernatant. Resuspend pellets in 500 μl TE buffer, pH 7.5, and add 170 μl of 8 M ammonium acetate. Mix by vortexing. Add 670 μl isopropanol, shake vigorously to mix, and chill at −20°C for ≥2 hr.

27. Microcentrifuge samples 15 min at 4°C and discard supernatant. Add 500 μl room-temperature 75% ethanol, vortex, and microcentrifuge samples 15 min at 4°C. Remove and discard supernatants, and remove the last traces of ethanol with a pipettor.

28. Resuspend pellets in 50 μl water and dry ~1 hr in a Speedvac evaporator. Resuspend pellets in TE buffer such that the final DNA concentration is ~1 μg/ml.

29. Microcentrifuge samples 10 min at room temperature. Transfer supernatants to a new silanized 1.5-ml microcentrifuge tube; discard gelatinous pellet if present. Quantitate DNA with a spectrophotometer and adjust concentration to 0.4 µg/µl with TE buffer, pH 7.5. The samples are now ready for ligation-mediated PCR.

At this point most of the DNA will be single-stranded because of the piperidine treatment, so it can be assumed that an A_{260} of 1.0 = 40 µg/ml DNA. The DNA is stable for years at 4°C.

SUPPORT PROTOCOL 2

PREPARATION OF GENOMIC DNA FROM SUSPENSION CELLS FOR DMS FOOTPRINTING

Materials *(see* APPENDIX 1 *for items with ✓)*

Cells in suspension culture
✓ PBS
✓ Lysis solution
100% dimethyl sulfate (DMS; Aldrich)
100% ethanol, room-temperature

50-ml polypropylene screw-cap tubes (e.g., Corning)
Tabletop centrifuge (IEC Centra-7R or equivalent)

CAUTION: DMS is extremely toxic. All manipulations that use DMS should be performed in a properly functioning fume hood. Before starting this procedure, review precautions for working with and disposing of DMS.

1. Prepare three 49-ml aliquots of PBS in 50-ml polypropylene screw-cap tubes and chill at least 30 min on ice prior to use. Make lysis solution and prepare three 2.7-ml aliquots in 15-ml polypropylene screw-cap tubes.

2. Transfer duplicate aliquots of medium containing 0.5–1 × 10^8 cells to 50-ml polypropylene tubes. Centrifuge cells 5 min in a tabletop centrifuge at 500 × *g*, room temperature. Aspirate and discard supernatant, leaving behind sufficient tissue culture medium to resuspend cell pellet in a final volume of 1 ml. Resuspend cells by gently flicking the bottom of the tube with a finger or by gently pipetting up and down.

3. *For control and in vitro–treated suspension cells:* Transfer 1 ml resuspended cells to one ice-cold 49-ml PBS aliquot, mix by gentle inversion, and centrifuge 5 min at 500 ×*g*, 4°C. Proceed to step 8.

4. *For in vivo–treated suspension cells:* Transfer 1 ml resuspended cells to a 1.5-ml microcentrifuge tube and place in 37°C water bath inside a fume hood.

5. Make 10% DMS solution by adding 10 µl of room-temperature 100% DMS to 90 µl of 100% ethanol. Mix thoroughly by vortexing ~25 sec, and microcentrifuge briefly to collect droplets.

   The 10% DMS solution is made in ethanol because DMS is not soluble in water at this concentration.

6. Add 10 µl of 10% DMS solution to the warmed cells, mix by gentle inversion, and incubate 1 min at 37°C. Immediately transfer cells to the second ice-cold 49-ml PBS aliquot, mix by gentle inversion, and centrifuge 5 min at 500 × *g*, 4°C.

The incubation time and amount of DMS used in the treatment listed here are good starting conditions; it may be necessary to modify them to get an optimal DMS footprint.

7. Aspirate and discard supernatant. Quickly resuspend cells with gentle pipetting in 1 to 5 ml ice-cold PBS from the third aliquot. Immediately add ice-cold PBS to fill the 50-ml tube, mix by gentle inversion, and centrifuge 5 min at 500 × *g*, 4°C.

8. *For both control and treated cells:* Aspirate and discard supernatant. Resuspend cells in 300 µl ice-cold PBS and transfer to separate 2.7-ml aliquots of lysis solution. Mix by gentle inversion. Continue with DNA harvesting, in vitro DMS treatment of the control DNA, and piperidine cleavage (Support Protocol 1, steps 7 to 29).

PREPARATION OF GENOMIC DNA FOR CHEMICAL SEQUENCING

SUPPORT PROTOCOL 3

This protocol describes the preparation of genomic DNA for LMPCR-aided direct genomic sequencing (i.e., sequencing without intermediate cloning or amplification steps). This method also reveals in vivo cytosine methylation, as cytosines methylated at the C5 position do not participate in the C+T and C reactions.

Materials *(see* APPENDIX 1 *for items with ✓)*

0.5 to 1.0 µg/µl untreated genomic DNA in TE buffer, pH 7.5 (control DNA; Support Protocols 1 and 2, steps 1 to 16)
✓ TE buffer, pH 7.5
✓ 3 M sodium acetate, pH 7.0
100% and 75% ethanol
88% formic acid (Fischer Chemical)
✓ G+A stop solution
Hydrazine, 98% anhydrous (Aldrich)
✓ C+T/C stop solution
✓ 5 M NaCl

CAUTION: DMS, formic acid, and hydrazine are toxic. All manipulations using these chemicals should be performed in a properly functioning fume hood. Before starting this procedure, review precautions for working with and disposing of DMS and hydrazine.

1. Place ~50 to 175 µg control DNA in a 1.5-ml microcentrifuge tube. Add sufficient TE buffer, pH 7.5, to bring the total volume to 175 µl. Label this tube G>A.

2. Treat with DMS (Support Protocol 1, steps 18 and 19). Proceed to step 9.

3. Place 20 to 40 µg control DNA in each of three 1.5-ml microcentrifuge tubes. Add sufficient TE buffer to bring the total volume in each to 100 µl. Label one tube G+A, one tube C+T, and one tube C.

4. Add 11 µl of 3 M sodium acetate and 222 µl of room-temperature 100% ethanol to each tube. Mix thoroughly by gently inverting a few times.

5. Microcentrifuge precipitated samples 5 min at maximum speed, room temperature, then remove and discard supernatants. Add 500 µl of room-temperature 75% ethanol to each pellet and mix by inversion until pellets are dislodged.

Microcentrifuge 3 min at maximum speed, room temperature, then remove and discard supernatants.

6. Redissolve samples in water as follows: add 18 µl to G+A tube, 40 µl to C+T tube, and 10 µl to C tube. Incubate overnight at 4°C. Warm redissolved DNA to room temperature.

For the G+A sample:

7a. Add 54 µl of 88% formic acid, mix thoroughly by vortexing 25 sec, and microcentrifuge briefly to collect droplets. Incubate 7 min at room temperature.

8a. Add 164 µl ice-cold G+A stop solution and vortex briefly (3 sec). Add 750 µl of 100% ethanol prechilled on dry ice, mix by vigorously shaking, and plunge tube into powdered dry ice. Leave samples in dry ice for ~30 min, then proceed to step 9.

The extent of reaction can be changed by varying the incubation time; the average size of the final product after piperidine cleavage will decrease as incubation time is increased.

For the C+T sample:

7b. Add 60 µl hydrazine, mix thoroughly by vortexing 25 sec, and microcentrifuge briefly to collect droplets. Incubate 3 min at room temperature.

8b. Add 150 µl ice-cold C+T/C stop solution and vortex briefly (3 sec). Add 750 µl of 100% ethanol prechilled on dry ice, mix by vigorously shaking, and plunge tube into powdered dry ice. Leave samples in dry ice for ~30 min, then proceed to step 9.

For the C sample:

7c. Add 30 µl of 5 M NaCl, mix thoroughly by vortexing 25 sec, and microcentrifuge briefly to collect droplets. Add 60 µl hydrazine, mix thoroughly by vortexing 25 sec, and microcentrifuge briefly to collect droplets. Incubate 3 min at room temperature.

8c. Add 150 µl ice-cold C+T/C stop solution and vortex briefly (3 sec). Add 750 µl of 100% ethanol prechilled on dry ice, mix by vigorously shaking, and plunge tube into powdered dry ice. Leave samples in dry ice for ~30 min, then proceed to step 9.

9. Cleave DNA from each of the four reactions with piperidine (Support Protocol 1, steps 21 to 29). Proceed to ligation-mediated PCR (Basic Protocol).

References: Mueller and Wold, 1989, 1991.

Contributors: Paul R. Mueller, Barbara Wold, and Paul A. Garrity

cDNA Amplification Using One-Sided (Anchored) PCR

This protocol presents a modification of PCR called anchored PCR, which allows amplification of full-length mRNA when only a small amount of sequence information is available. Unlike the amplification of RNA or cDNA by conventional PCR (*UNIT 15.4*)—which requires prior knowledge of the sequences flanking the region of interest to design the PCR primers—anchored PCR can be employed when only a small region of sequence lying within the mRNA is known in advance.

In theory, only a single specific primer is required to perform anchored PCR. In practice, amplification of a single product requires a second round of PCR amplification, using a second sequence-specific primer. This scheme is depicted in Figures 15.6.1 and 15.6.2. Both the original and reamplifications use an oligo(dT) primer complementary either to the poly(A) tail of the mature mRNA (when amplifying downstream to the known sequence) or to an enzymatically synthesized homopolymer tail added to the cDNA following first-strand synthesis (when amplifying upstream to the known sequence). The two rounds of anchored PCR amplification result in a single product that can be directly sequenced or cloned into an appropriate vector for further analysis.

AMPLIFICATION OF REGIONS DOWNSTREAM (3′) OF KNOWN SEQUENCE

BASIC PROTOCOL 1

This method can be used to characterize messages of interest from any source of poly(A)$^+$ RNA (*UNITS 4.1-4.5*). Although the amount of poly(A)$^+$ RNA required will depend on the genome complexity of the organism being studied and on the relative abundance of the targeted mRNA, in general 100-300 ng of total poly(A)$^+$ RNA prepared from vertebrate tissue is sufficient. Three PCR primers are required: an oligo(dT) primer and two sequence-specific primers, one for the original amplification and the second for the reamplification (see Fig. 15.6.1). The internal sequence-specific primer can be immediately adjacent (3′) to or can partially overlap primer 1.

Materials *(see APPENDIX 1 for items with ✓)*

✓ 5× Moloney murine leukemia virus (MoMuLV) reverse transcriptase buffer
5 µg/µl bovine serum albumin (BSA)
10 mM 4dNTP mix: 10 mM each dNTP in TE buffer, pH 7.5 (store at −20°C)
500 ng/µl actinomycin D
200 U/µl MoMuLV reverse transcriptase (*UNIT 3.7*)
100 ng/µl oligo(dT)$_{20}$ primer
100 pmol/µl each of sequence-specific primers 1 and 2 (Fig. 15.6.1)
✓ TE buffer
2.5 mM 4dNTP mix: 2.5 mM each dNTP in TE buffer, pH 7.5 (store at −20°C)
✓ 10× PCR buffer
2.5 U/µl *Taq* DNA polymerase
Mineral oil

1. Place 100 ng poly(A)$^+$ RNA (≥100 ng/µl) in a 1.5-ml microcentrifuge tube and incubate 2 min at 65°C. Microcentrifuge briefly and place immediately on ice.

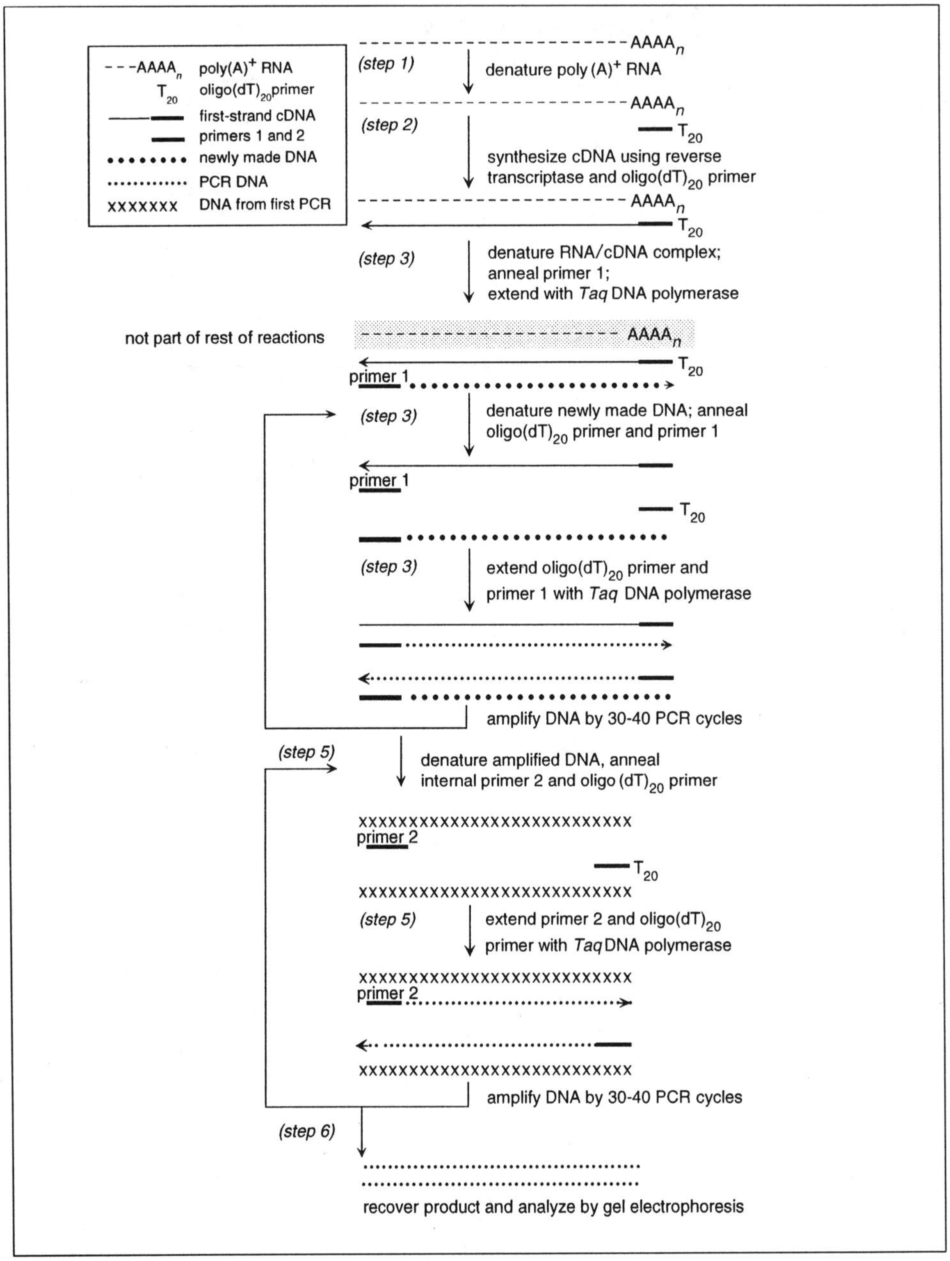

Figure 15.6.1 Flow chart of downstream (3′) anchored PCR protocol (see text for details). The steps correspond to those listed in Basic Protocol 1. Note that primer 2 can be immediately adjacent (3′) to, or can partially overlap with, primer 1.

2. In a separate 1.5-ml microcentrifuge tube, prepare on ice (10 µl total):

   2 µl 5× MoMuLV reverse transcriptase buffer
   1 µl (5 µg) BSA
   1 µl poly(A)$^+$ RNA (from step 1)
   1 µl 10 mM 4dNTP mix
   1 µl (500 ng) actinomycin D
   1 µl (200 U) MoMuLV reverse transcriptase
   1 µl (100 ng) oligo(dT)$_{20}$
   2 µl sterile H_2O.

   Mix gently, microcentrifuge briefly, and incubate 1 hr at 37°C. Add 40 µl TE buffer.

3. Prepare the following on ice (100 µl final):

 1 µl cDNA template (from step 2)
 10 µl 10× PCR buffer
 1 µl (100 pmol) oligo$(dT)_{20}$ primer
 1 µl (100 pmol) sequence-specific primer 1
 6 µl 2.5 mM 4dNTP mix
 1 µl (2.5 U) *Taq* DNA polymerase
 80 µl sterile H_2O.

 Overlay reaction mixture with mineral oil. Carry out 30 to 40 cycles of amplification (*UNIT 15.1*).

 Initially, amplifications should be carried out in the presence of 1.5 mM $MgCl_2$ (final concentration), with annealing temperatures not exceeding 42°C. These conditions can be subsequently modified and optimized (UNIT 15.1) for the specific template used. Primer concentrations can be reduced to 30 pmol of each primer.

4. Analyze an aliquot by agarose gel electrophoresis (*UNIT 2.5A*). The result of this first PCR is usually a smear around the expected size range.

5. Remove a 1-µl aliquot of product from step 3 to serve as the template for reamplification. Carry out a second round of PCR as in step 3, using 40 to 100 pmol each of the oligo$(dT)_{20}$ primer and internal sequence-specific primer 2.

6. Analyze an aliquot of second amplification by agarose gel electrophoresis. The amplified product should now appear as a single band. If desired, characterize product by cloning into an appropriate vector (*UNIT 3.16*) or by direct sequencing (*UNIT 15.2*).

AMPLIFICATION OF REGIONS UPSTREAM (5′) OF KNOWN SEQUENCE

BASIC PROTOCOL 2

In contrast to the first protocol, this uses one of the sequence-specific primers to initiate synthesis of the cDNA strand. This cDNA is modified by the addition of a poly(A) tail. PCR amplifications—mediated by two sequence-specific primers and a 20-mer oligo(dT) primer complementary to the newly synthesized tail—yield the desired unique product (Fig. 15.6.2).

Materials *(see APPENDIX 1 for items with ✓)*

 100 pmol sequence-specific primers 3 and 4 (Fig. 15.6.2)
 100 pmol oligo$(dT)_{20}$ primer
✓ 1 M NaCl
✓ 200 mM Tris·Cl, pH 7.5
✓ 25 mM EDTA
 100% and 70% ethanol, ice-cold
✓ 5× Moloney murine leukemia virus (MoMuLV) reverse transcriptase buffer
 5 µg/µl bovine serum albumin (BSA)
 10 mM 4dNTP mix: 2.5 mM each dNTP in TE buffer, pH 7.5 (store at −20°C)
 500 ng/µl actinomycin D
 200 U/µl MoMuLV reverse transcriptase (*UNIT 3.7*)
✓ 3 M sodium acetate
✓ TE buffer
 5× terminal deoxynucleotidyl transferase (TdT) buffer
 15 mM $CoCl_2$

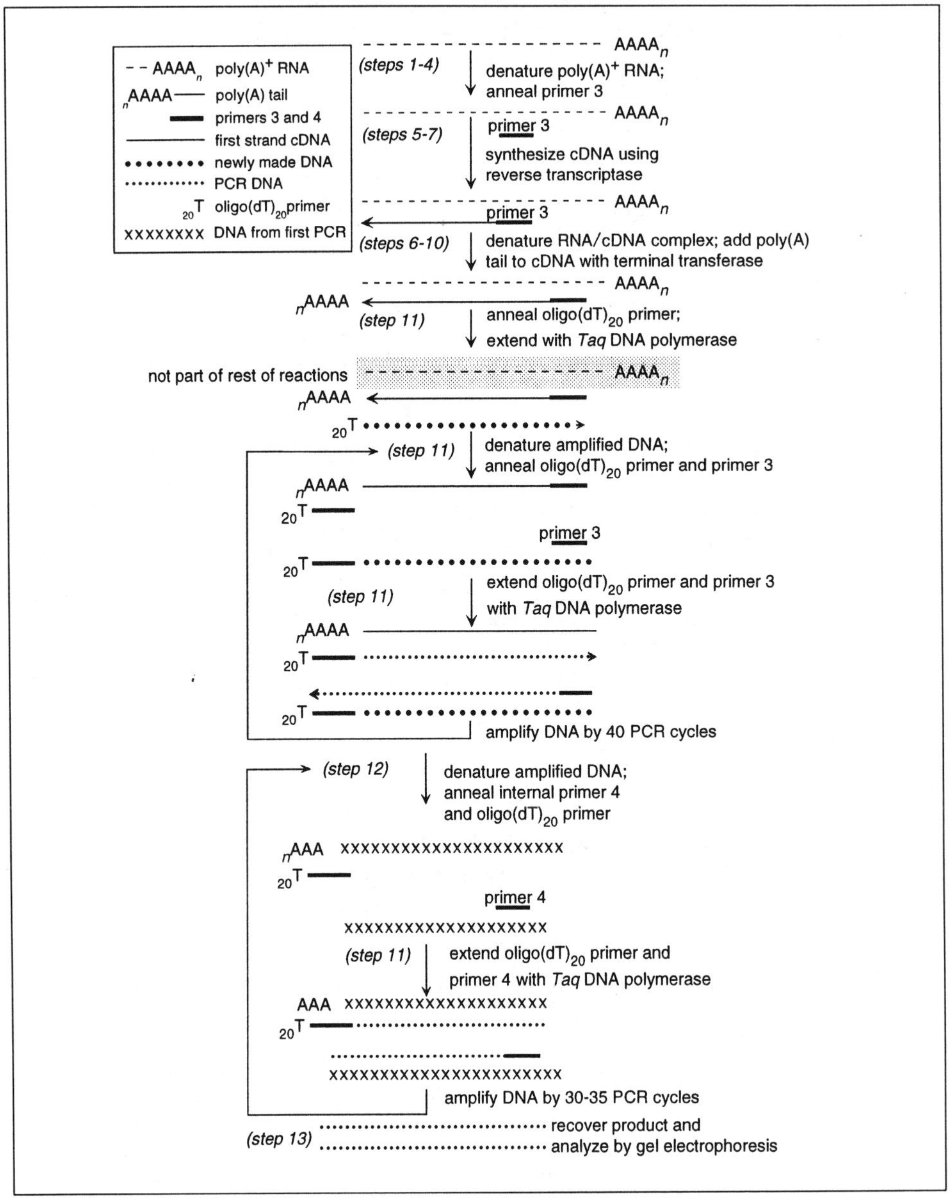

Figure 15.6.2 Flow chart of downstream (5′) anchored PCR protocol (see text for details). The steps correspond to those listed in Basic Protocol 2. Note that primer 4 can be immediately adjacent to, or can partially overlap with, primer 3.

1 mM dATP (*UNIT 3.4*)
Terminal transferase (*UNIT 3.6*)
40°C water bath

1. Prepare a 5-µl annealing mix containing:

   1 µl (1 fmol) sequence-specific primer 3
   1 µl 1 M NaCl
   1 µl 200 mM Tris·Cl, pH 7.5
   1 µl 25 mM EDTA
   1 µl (100 ng/µl) poly(A)+ RNA

   The amount of primer depends in part on the abundance of the target mRNA. As a general rule, a 5- to 10-fold molar excess of oligomer primer relative to target template yields the best results.

2. Incubate 3 min at 65°C, microcentrifuge briefly, and place immediately on ice. Incubate 3 to 4 hr at 40°C.

3. Add 15 µl ice-cold 100% ethanol, place 10 min in dry ice/ethanol bath, and microcentrifuge at high speed 10 min at 4°C.

4. Add 50 µl ice-cold 70% ethanol to pellet and gently invert tube several times. Microcentrifuge 2 min, dry pellet briefly under vacuum, and resuspend in 10 µl water.

5. Prepare mix on ice containing (25 µl total):

   5 µl 5× MoMuLV reverse transcriptase buffer
   2.5 µl (12.5 µg) BSA
   2.5 µl 10 mM 4dNTP mix
   2.5 µl (1.25 µg) actinomycin D
   10 µl annealed primer/template (from step 4)
   1.5 µl sterile H_2O.

   Add 1 µl (200 U) MoMuLV reverse transcriptase. Incubate 1 hr at 37°C.

6. Phenol extract (*UNIT 2.1*) and transfer supernatant to new microcentrifuge tube. Add 2.5 µl of 3 M sodium acetate (0.3 M final) and 75 µl ice-cold 100% ethanol, place 5 min in dry ice/ethanol bath, and microcentrifuge 20 min at high speed, 4°C.

   *In cases where small amounts of poly(A)$^+$ RNA are being used, 1 to 5 µg carrier tRNA (*UNIT 4.6*) may be added to facilitate precipitation.*

7. Resuspend pellet in 25 µl TE buffer. Add 2.5 µl of 3 M sodium acetate (0.3 M final) and repeat ethanol precipitation (step 6).

8. Add 100 µl ice-cold 70% ethanol to pellet and rinse by gently inverting tube. Microcentrifuge 5 min, dry pellet briefly under vacuum, and resuspend pellet in 5 µl water. Boil 2 min, microcentrifuge briefly, and place immediately on ice.

9. Prepare mix on ice containing (10 µl final):

   2 µl 5× TdT buffer
   1 µl 15 mM $CoCl_2$ (1.5 mM final)
   1 µl 1 mM dATP (100 µmol final)
   5 µl cDNA mix (from step 8).

   Add 1 ml (25 U) terminal transferase. Incubate 30 min at 37°C.

10. Inactivate enzyme by heating 2 min at 65°C. Add 1 µl of 3 M sodium acetate. Add 30 µl of ice-cold 100% ethanol and precipitate. Wash and dry pellet as in step 8.

11. Resuspend pellet in 10 µl water. Carry out 40 amplification cycles as in step 3 (and annotation therein) of Basic Protocol 1, using 40 to 100 pmol each of sequence-specific primer 3 and oligo$(dT)_{20}$.

12. Remove a 1-µl aliquot of the product to serve as template for a new round of amplification. Carry out 35 cycles of PCR (Basic Protocol 1, step 3), using 40 to 100 pmol each of primer 4 and the oligo(dT) primer.

   Primer 4 can be immediately adjacent to or even partially overlapping with primer 3. Because anchored PCR procedures are carried out independently in the 3′ and 5′

directions, sequence-specific primers 3 and 4 can be complements of primers 1 and 2 used in the previous protocol.

13. Analyze an aliquot by agarose gel electrophoresis. A single band is expected. If desired, characterize the PCR product by cloning into an appropriate vector (*UNIT 3.16*) and/or by direct sequencing (*UNIT 15.2*).

Reference: Ohara et al., 1989.

Contributor: Robert L. Dorit

UNIT 15.7

Molecular Cloning of PCR Products

Cloning PCR products is often desirable to establish a permanent source of cloned DNA for hybridization studies, to obtain high-quality DNA sequencing results, or to separate products when PCR amplification yields a complex mixture. The efficiency of direct cloning of PCR products can be improved by generating suitable ends on the amplified fragments.

BASIC PROTOCOL

GENERATION OF T-A OVERHANGS

Taq DNA polymerase normally adds a single nontemplated nucleotide (nearly always A) to the 3′ end of all duplex DNA strands, making direct blunt-ended cloning of PCR fragments inefficient. In the presence of dTTP alone, *Taq* DNA polymerase will add a single T to the blunt ends of a vector, generating the necessary complementary one-base overhang. This procedure is summarized in Figure 15.7.1.

Materials *(see APPENDIX 1 for items with ✓)*

5 µg vector DNA [e.g., pUC19 (*UNIT 1.5*) or M13mp18 (*UNIT 1.9*)]
✓ TE buffer, pH 8.0
✓ 5× amplification buffer (with optimized Mg^{2+} concentration)
5 mM dTTP (*UNIT 3.4*)
5 U/µl *Taq* DNA polymerase
Target DNA
Automated thermal cycler (e.g., MJ Research)

1. Digest 5 µg vector DNA with a restriction endonuclease that yields blunt ends. Check that digestion is complete by electrophoresing 50 ng on an agarose minigel. Extract the DNA with phenol/chloroform and precipitate with ethanol. Microcentrifuge 5 min at top speed, 4°C, and resuspend pellet in 25 µl TE buffer, pH 8.0.

2. Set up the following T-addition reaction:

 5 µg blunt-ended vector DNA
 20 µl 5× amplification buffer
 20 µl 5mM dTTP
 1 µl (5 U) *Taq* DNA polymerase
 H_2O to 100 µl.

 Incubate 2 hr at 75°C.

 For this reaction, the Mg^{2+} should be 2 to 5 mM. This yields sufficient T-tailed vector for many cloning procedures.

15.7

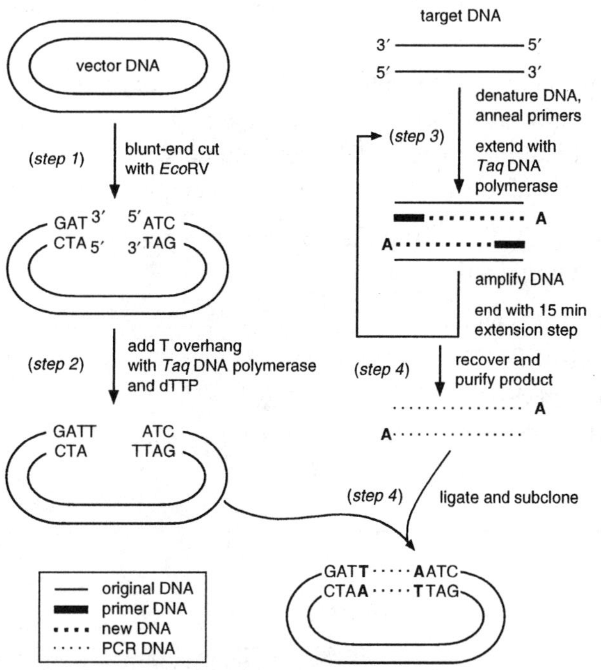

Figure 15.7.1 T-A overhang cloning. A single T is added to the 3′ ends of blunt-cut vector DNA using *Taq* DNA polymerase and dTTP. A normal PCR reaction produces a fragment with a single A added to each 3′ end. The As and Ts form complementary 1-base overhangs, facilitating ligation of insert to vector.

3. Carry out PCR to amplify the desired target DNA under optimized conditions (UNIT 15.1), ending the reaction with an extra 5- to 15-min extension at 70° to 75°C to be sure all fragments are A-tailed.

   The number of amplification cycles needed can vary from 15 to 40, depending on the input DNA. Overcycling can result in amplification artifacts.

4. Recover and subclone the PCR fragments, beginning with step 6 of the Basic Protocol in UNIT 3.16. Include a full set of controls the first time the preparation is used for cloning.

GENERATION OF HALF-SITES

ALTERNATE PROTOCOL 1

A second strategy for generating sticky ends on PCR products is to add to the 5′ end of each primer three bases, carefully chosen to create a full restriction site when the PCR products are concatemerized by blunt-end ligation. This procedure is summarized in Figure 15.7.2.

Additional Materials *(also see Basic Protocol; see* APPENDIX 1 *for items with ✓)*

T4 DNA ligase (UNIT 3.14) and 10× buffer (UNIT 3.4)
1 mM ATP (UNIT 3.4)
✓ 2 mM 4dNTP mix
Klenow fragment of *E. coli* DNA polymerase I (UNIT 3.5)
T4 polynucleotide kinase (UNIT 3.10)

1. Design and synthesize a pair of oligonucleotide primers with the three 3′ nucleotides of a palindromic six-base restriction site joined to the 5′ end of each specific primer.
2. Anneal primers to desired DNA sequences and carry out PCR under optimized conditions (*UNIT 15.1*). Analyze a 5-µl aliquot by agarose gel electrophoresis to verify amplification and quantitate yield. Remove the mineral oil with a pipettor. Extract DNA with phenol/chloroform and precipitate with ethanol.
3. Resuspend the DNA in TE buffer, pH 8.0, at a concentration of 0.2 µg/µl. Set up the following ligation reaction:

   5 µl (1 µg) DNA
   1 µl 10× T4 DNA ligase buffer
   1 µl 1 mM ATP
   1 µl 4dNTP mix

continued

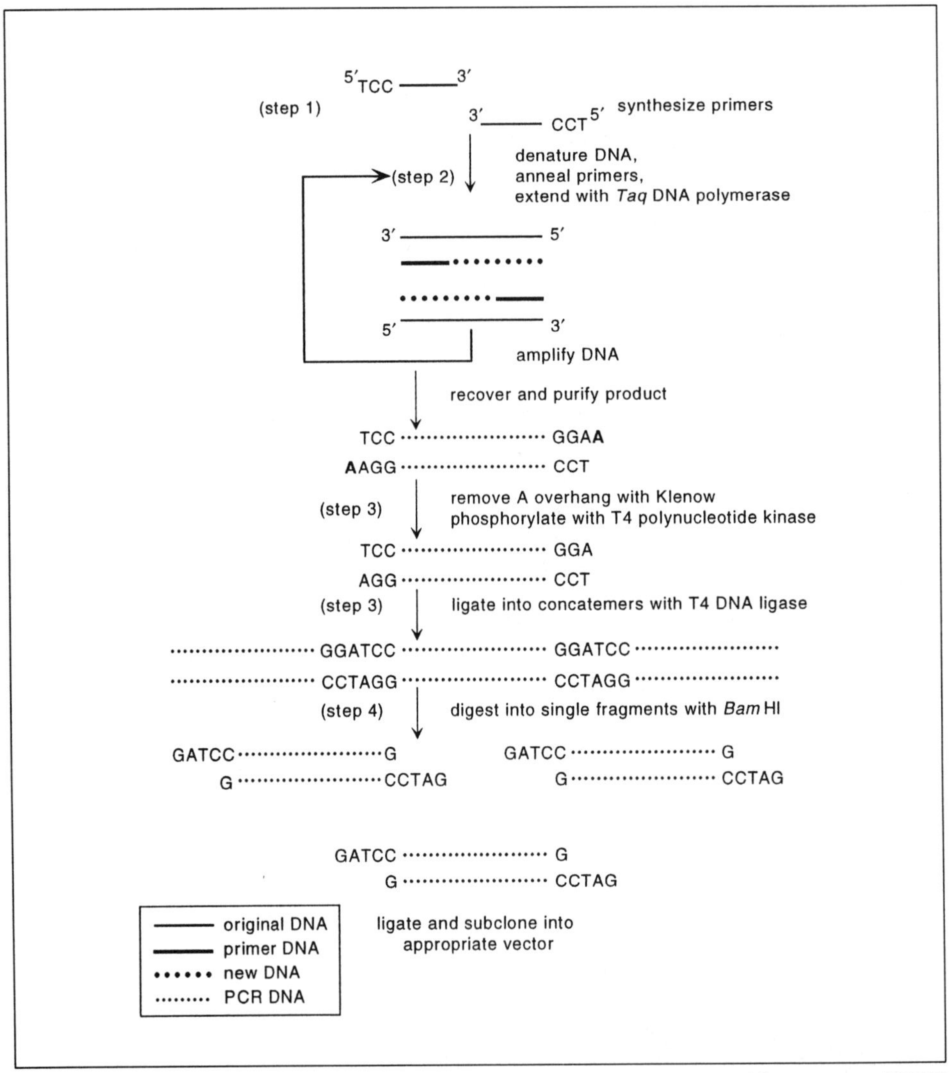

Figure 15.7.2 Restriction enzyme half-sites. Fragments are generated by a normal PCR using oligonucleotide primers containing the three 3′ nucleotides of a 6-base recognition site at their 5′ ends. In a single step, the nontemplated A is removed by Klenow fragment, and the fragments are phosphorylated at their 5′ ends by T4 polynucleotide kinase and ligated into concatemers. The ligation generates the 6-base recognition sequence, which can then be cleaved with the appropriate restriction enzyme. The resulting fragments can be easily cloned.

1 U Klenow fragment
1 U T4 polynucleotide kinase
500 U (cohesive end) T4 DNA ligase
H_2O to 10 µl.

Incubate ≥2 hr at 15°C. Analyze a 1-µl aliquot by agarose gel electrophoresis to verify ligation.

4. Subclone fragments into appropriate vectors, beginning with step 1 of the Basic Protocol in *UNIT 3.16*.

References: Marchuk et al., 1991.

Contributors: Michael Finney

Differential Display of mRNA by PCR

UNIT 15.8

BASIC PROTOCOL

This unit describes differential display to identify mRNA species for differentially expressed genes (Fig. 15.8.1). DNA sequences corresponding to these mRNAs can be recovered, cloned, sequenced, and used for hybridization or library screening probes. Specifically, an RNA sample is reverse transcribed with each of the four sets of degenerate anchored oligo(dT) primers (T_{12}MN), where M can be G, A, or C and N is G, A, T, and C. Each primer set is dictated by the 3′ base (N), with degeneracy in the penultimate (M) position. For example, the primer set where N = G consists of:

5′-TTTTTTTTTTTTGG-3′
5′-TTTTTTTTTTTTAG-3′
5′-TTTTTTTTTTTTCG-3′

Materials *(see APPENDIX 1 for items with ✓)*

Total cellular human RNA (*UNIT 4.2*) *or* poly(A)$^+$ RNA (*UNIT 4.5*)
1 U/µl human placental RNase inhibitor
10 U/µl DNase I (RNase-free)
✓ 0.1 M Tris·Cl, pH 8.3
✓ 0.5 M KCl
✓ 15 mM $MgCl_2$
3:1 (v/v) phenol/chloroform
✓ 3 M sodium acetate, pH 5.2
100%, 85%, and 70% ethanol
Diethylpyrocarbonate (DEPC)–treated H_2O
✓ 5× MoMuLV reverse transcriptase buffer
✓ 0.1 M dithiothreitol (DTT)
✓ 250 µM and 25 µM 4dNTP mixes
10 µM each degenerate anchored oligo(dT) primer set (e.g., GeneHunter): 5′-T_{12}MG-3′, 5′-T_{12}MA-3′, 5′-T_{12}MT-3′, and 5′-T_{12}MC-3′ (M represents G, A, or C)
200 U/µl Moloney murine leukemia virus (MoMuLV) reverse transcriptase
✓ 10× amplification buffer (prepare with 15 mM $MgCl_2$, but only 0.1 mg/ml gelatin; store at −20°C)
10 µCi/µl [α-^{35}S]dATP (>1200 Ci/mmol)

15.8

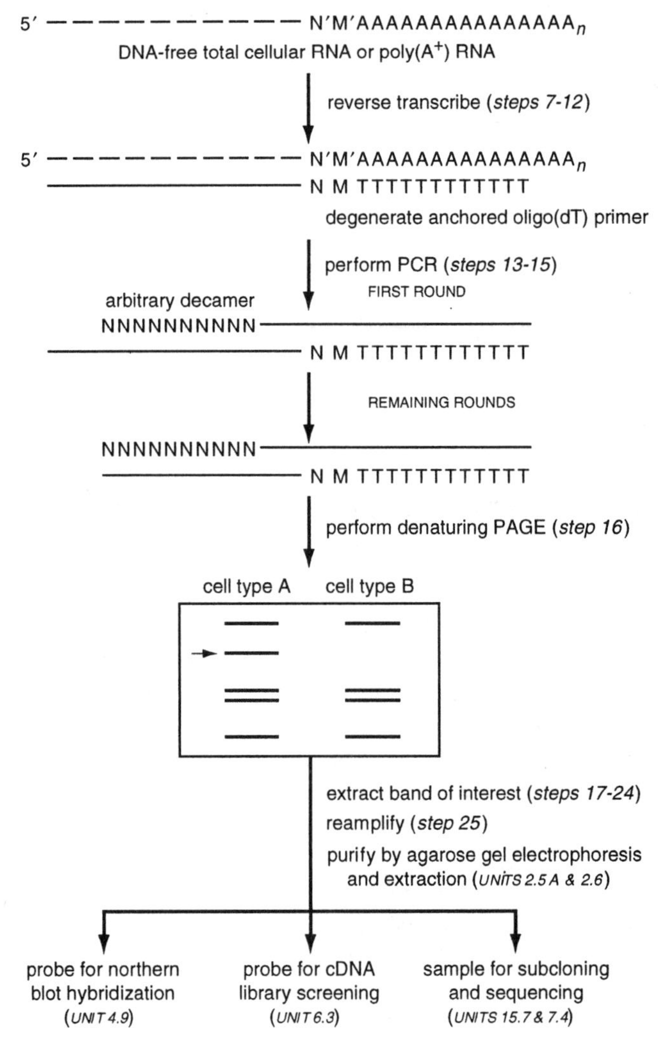

Figure 15.8.1 Schematic representation of differential display. Diagram of gel represents results with a single primer set for two cell types, A and B. Dashed line, RNA; solid line, DNA; $T_{12}MN$, degenerate oligo(dT) primer; M indicates A, C, or G (degenerate); N can be A, C, G or T.

2 µM arbitrary decamer (e.g., GeneHunter or Operon Technologies)
5 U/µl *Taq* DNA polymerase
Mineral oil
✓ Formamide loading buffer
10 mg/ml glycogen (DNA-free)

65°, 80°, 95°, and 100°C water baths
Thermal cycler
Whatman 3MM filter paper

CAUTION: This procedure should be performed only by personnel trained in the proper use of ^{35}S isotope and in NRC licensed sites. Standard precautions to prevent excessive exposure and radioactive contamination of personnel and equipment should be followed at all times.

NOTE: Experiments involving RNA require careful technique to prevent RNA degradation; see *UNIT 4.1*.

1. Digest DNA from total cellular RNA *or* poly(A)$^+$ RNA by mixing:

    50 µl 1 µg/µl RNA
    10 µl 1 U/µl human placental RNase inhibitor
    1 µl 10 U/µl RNase-free DNase I
    5 µl 0.1 M Tris·Cl, pH 8.3
    5 µl 0.5 M KCl
    5 µl 15 mM $MgCl_2$.

    Incubate 30 min at 37°C.

    It is essential that the RNA sample be free from any genomic DNA contamination. Amounts from 15 to 100 µg of total RNA can be cleaned with this procedure.

2. Add 50 µl phenol/chloroform (3:1), vortex, and microcentrifuge 2 min at maximum speed to separate phases.

    Vigorous mixing is important to allow complete extraction of DNase I.

3. Transfer upper phase to clean microcentrifuge tube and add 5 µl of 3 M sodium acetate and 200 ml of 100% ethanol. Incubate 30 min at −70°C to precipitate RNA.

4. Microcentrifuge 10 min at high speed. Remove supernatant and wash pellet (precipitated RNA) once with 500 µl of 70% ethanol.

5. Dissolve RNA pellet in 20 µl DEPC-treated water and quantitate the RNA concentration accurately by measuring the A_{260} with a spectrophotometer.

6. Electrophorese 3 µg of cleaned RNA on a denaturing agarose gel to check the integrity of the RNA to be used for differential display. Store DNA-free RNA at −80°C until used for differential display.

    DNA-free RNA should be stored at a concentration >1 µg/µl.

7. For each RNA sample, label four microcentrifuge tubes G, A, T, and C—one tube for each degenerate anchored oligo(dT) primer set.

8. Dilute 1 µg DNA-free RNA (step 5) to 0.1 µg/µl in DEPC-treated water and place on ice.

9. Set up reverse transcription of DNA-free total RNA *or* poly(A)$^+$ RNA with each of four different degenerate anchored oligo-dT primer sets ($T_{12}MN$: $T_{12}MG$, $T_{12}MA$, $T_{12}MT$, and $T_{12}MC$, where M is G, A or C) as follows:

    H_2O to give 19 µl total volume
    4 µl 5× MoMuLV reverse transcriptase buffer
    2 µl 0.1 M DTT
    1.6 µl 250 µM 4dNTP mix
    0.2 µg total RNA *or* 0.1 µg poly(A)$^+$ RNA
    2 µl of one 10 µM degenerate anchored oligo(dT) primer set ($T_{12}MN$).

    There will be four reactions for each RNA sample, each made with one degenerate primer set.

10. Incubate tube 5 min at 65°C to denature the mRNA secondary structure and incubate 10 min at 37°C to allow primer annealing.

11. Add 1 µl of 200 U/µl MoMuLV reverse transcriptase to each tube, mix well, and incubate 50 min at 37°C.

12. Incubate 5 min at 95°C to inactivate the reverse transcriptase and microcentrifuge briefly at high speed to collect condensation. Place tube on ice for immediate PCR amplification *or* store at −20°C for later use (stable at least 6 months).

13. Prepare a 20-μl reaction mix for each primer set as follows:

    9.2 μl H_2O
    2 μl 10× amplification buffer (1× final)
    1.6 μl 25 μM 4dNTP mix (2 μM final)
    1 μl 10 μCi/μl [α-^{35}S]dATP
    2 μl 10 μM degenerate anchored oligo(dT) primer set (T_{12}MN; 1 μM final)
    2 μl cDNA (step 10)
    0.2 μl 5 U/μl *Taq* DNA polymerase
    2 μl 2 μM arbitrary decamer (0.2 μM final).

    To avoid pipetting errors, prepare enough PCR reaction mix without the arbitrary decamer for 5 to 10 reactions and aliquot 18 μl to each tube. Then add the arbitrary decamer. Otherwise it is difficult to pipet accurately 0.2 μl of Taq *DNA polymerase.*

14. Pipet up and down to mix well and overlay with 25 μl mineral oil.

15. Carry out PCR in a thermal cycler using the following amplification cycles:

40 cycles:	30 sec	94°C	(denaturation)
	2 min	40°C	(annealing)
	30 sec	72°C	(extension)
1 cycle:	5 min	72°C	(extension)
Final step:	indefinitely	4°C	(hold).

    PCR products may be stored at 4°C until used.

16. Mix 3.5 μl PCR product with 2 μl formamide loading buffer and incubate 2 min at 80°C. Load sample onto a 6% denaturing polyacrylamide gel. Run the gel ~3 hr at 60 W until xylene cyanol runs to within 10 cm of the bottom.

17. Carefully remove one of the glass gel plates. Place a piece of Whatman 3MM filter paper over the gel without trapping air bubbles between filter paper and gel. Dry the gel ~1 hr at room temperature without fixing it in methanol/acetic acid.

18. Use either radioactive ink or needle punches to mark X-ray film and dried gel to orient them. Autoradiograph the gel 24 to 48 hr at room temperature.

19. Develop the film, align film with gel, and indicate DNA bands of interest (those differentially displayed in different lanes) either by marking beneath the film with a clean pencil or by cutting through the film.

    Typical results of differential display are shown in Figure 15.8.2.

20. Cut out gel slice and attached Whatman 3MM filter paper with a razor blade and place in a microcentrifuge tube. Add 100 μl H_2O and incubate 10 min at room temperature.

21. Cap tube tightly and boil 15 min.

    Place a Lid-Lok on the tube to prevent it from opening while boiling.

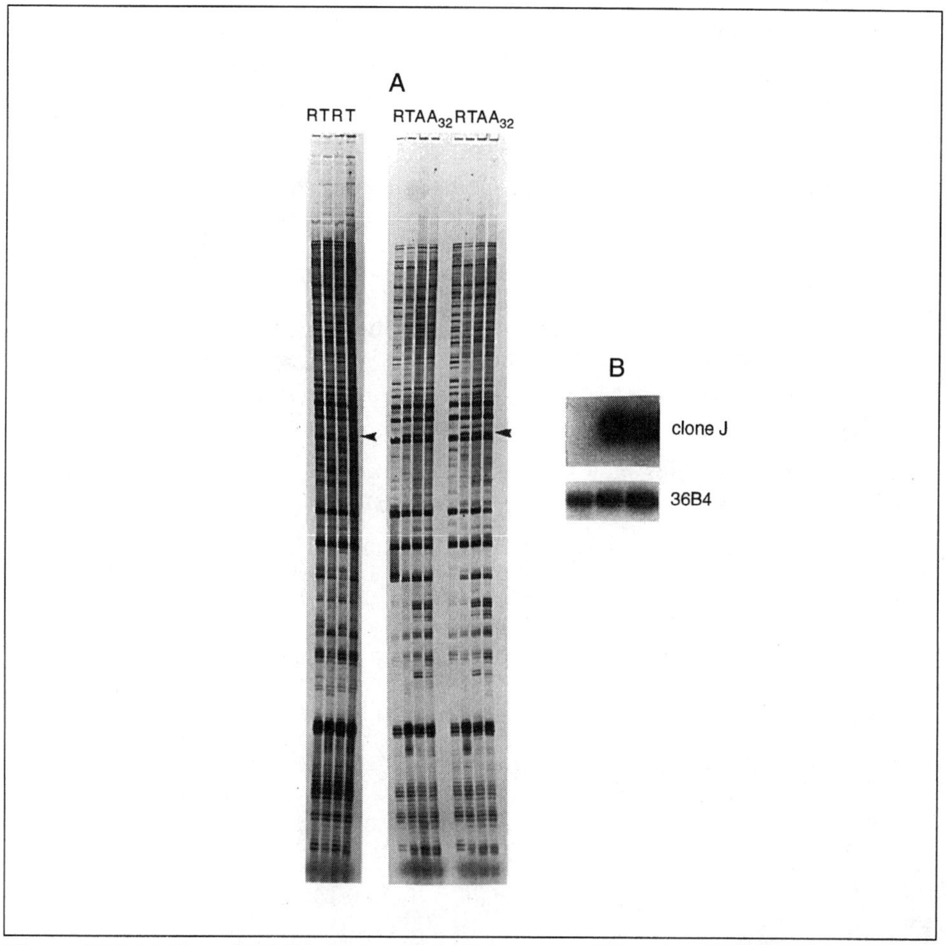

Figure 15.8.2 Reproducibility and multiple display of mRNAs from normal versus ras/p53 mutant transformed cells. (**A**) RNA samples from normal rat embryo fibroblasts REF (R) and its ras/p53 doubly transformed derivative T101-4 cells (T) were reverse transcribed and amplified in duplicate with T12MA and OPA17 primers (left four lanes). In a separate experiment, RNA samples from REF (R), T101-4 (T), and another ras/p53 temperature-sensitive mutant transformed cell line A1-5 grown at nonpermissive temperature (A) and shifted to permissive temperature for 24 hr (A_{32}) were reverse transcribed and amplified in duplicate with T12MA and OPA17 primers (right eight lanes). An arrowhead indicates a reproducible difference between normal and transformed cells. (**B**) Northern blot analysis of this reamplified cDNA probe (named as clone J). 20 mg of total RNA from REF, T101-4, and A1-5 cells were analyzed. 36B4 was used as a probe for RNA loading control.

22. Microcentrifuge 2 min at high speed to pellet gel slice and paper debris. Decant supernatant into clean tube.

23. Add 10 µl of 3 M sodium acetate (to give 0.3 M final) and 5 µl of 10 mg/ml glycogen (as a carrier) to supernatant. Add 400 µl of 100% ethanol and incubate 30 min at –70°C. Microcentrifuge 10 min at high speed, 4°C.

24. Rinse pellet with 500 µl of 85% ethanol, air-dry, and dissolve the DNA in 10 µl H_2O.

25. Reamplify 4 µl of the eluted DNA in a 40-µl reaction volume using the same degenerate anchored oligo(dT) primer set and PCR conditions as in steps 13 through 15, *except* add 1.6 µl of 250 µM 4dNTP mix (20 µM final) instead of 1.6 µl of 25 µM 4dNTP mix and omit isotope. Save the remaining recovered DNA at –20°C for future reamplification (stable indefinitely).

26. Electrophorese 30 µl of each PCR sample on a 1.5% agarose gel and stain with 0.5 µg/ml ethidium bromide. Store the remaining PCR samples at –20°C (stable for years).

Most amplified DNAs should be visible after the first reamplification. Fragment molecular weights should be checked after reamplification to ensure that they are consistent with those on the denaturing polyacrylamide gel. If a DNA is not visible after the first reamplification, 4 µl of 1/100 dilution (in water) of the first reamplification sample may be used for a second 40-cycle amplification.

27. Extract the desired reamplified DNA band from the agarose gel and use it as a probe for northern blot analysis and cDNA library screening.

Store extracted DNA at –20°C (stable for years) if it is not to be used immediately.

28. Characterize remaining PCR sample (from step 26) by subcloning and sequencing.

Contributors: Peng Liang and Arthur B. Pardee

Protein Expression

16

Protein expression, as used in this chapter, refers to the directed synthesis of large amounts of desired proteins. In early applications, molecular biologists interested in obtaining large amounts of prokaryotic regulatory proteins arranged their synthesis in large amounts, a process that came to be called overproduction, expression, or overexpression. These early techniques used genetic manipulations to select in vivo recombination events that inserted the desired gene into bacteriophages. Later as it was developed, recombinant DNA technology was used to create phages and plasmids in vitro to direct the synthesis of large amounts of the products of cloned genes.

This chapter describes methods to express proteins. In all these methods, a gene whose product is to be expressed is introduced into a plasmid or other vector. That vector is introduced into living cells. Typical expression vectors contain promoters that direct the synthesis of large amounts of mRNA corresponding to the gene. They may also include, for example, sequences that allow their autonomous replication within the host organism, sequences that encode genetic traits that allow cells containing the vectors to be selected, and sequences that increase the efficiency with which the mRNA is translated.

UNITS 16.1-16.8 describe techniques for expressing proteins in *E. coli*. *UNIT 16.1* contains an introduction to *E. coli* expression. *UNIT 16.2* describes the use of T7 vectors. In these vectors, synthesis of large amounts of foreign gene products is directed by the phage T7 gene 10 promoter, which uses T7 RNA polymerase. This polymerase transcribes the gene 10 promoter so efficiently that it uses up most of the ribonucleotide triphosphates in the cell and drastically inhibits transcription of genes by the host polymerase. *UNIT 16.3* describes the use of p_L-derived vectors and their appropriate host strains. These vectors carry the powerful bacteriophage p_L promoter and take advantage of a number of other useful aspects of phage lambda biology. The next units contain techniques for expression of fusion proteins in which the expressed protein carries an additional stretch of amino acids at its N terminus to aid its expression and purification. *UNIT 16.4* introduces the concept of fusions and provides methods for cleavage of fusion proteins. *UNITS 16.6*, *16.7*, and *16.8* describe techniques for expressing *lacZ* protein (β-galactosidase), *trpE* protein, glutathione-*S*-transferase, and thioredoxin fusions.

UNITS 16.9-16.11 describe the use of the baculovirus system. In this system, genes for proteins to be expressed are inserted into an insect virus in lieu of a highly expressed dispensable gene. The foreign protein is produced by growing the recombinant virus in cultured insect cells. *UNIT 16.9* introduces the system. *UNIT 16.10* describes how to grow the cultured insect cells and viral stocks. Finally, *UNIT 16.11* describes how to isolate recombinant baculoviruses and how to use them to produce the desired protein.

UNITS 16.12 and *16.13* describe techniques for expressing proteins in mammalian cells. *UNIT 16.12* introduces the general issues. *UNIT 16.13* describes expression using COS cell vectors. In this approach, vectors containing the gene to be expressed are transiently transfected into COS cells, which constitutively produce SV40 large T antigen. COS cell vectors contain an SV40 replication origin; when they are transfected into COS cells, they replicate, and protein is expressed from mRNA synthesized by hundreds of copies of the vectors.

All expression techniques have advantages and disadvantages that should be considered in choosing which one to use. *E. coli* expression techniques are probably the most popular: the organism is already used by most investigators, the techniques necessary to express usable amounts of protein are relatively simple, the amount of time necessary to generate an overexpressing strain is very short, and a familiarity with standard recombinant DNA techniques is all that is necessary to begin pilot expression experiments. *E. coli* has other advantages that have made it widely used for expression of commercially important proteins: it is cheap to grow, and the vast body of knowledge about it has made it possible to tinker intelligently with its genetics and physiology, so that strains producing 30% of their total protein as the expressed gene product can often be obtained. However, expression in *E. coli* does have some disadvantages. First, eukaryotic proteins expressed in *E. coli* are not properly modified. Second, proteins expressed in large amounts in *E. coli* often precipitate into insoluble aggregates called "inclusion bodies," from which they can only be recovered in an active form by solubilization in denaturing agents followed by careful renaturation. Third, it is relatively difficult to arrange the secretion of large amounts of expressed proteins from *E. coli*, although it has often been possible to secrete small amounts into the periplasmic space and to recover them by osmotic shock.

The baculoviral expression system has a number of advantages that have contributed to its recent popularity: proteins are almost always expressed at high levels; expressed proteins are usually expressed in the proper cellular compartment (that is, membrane proteins are usually localized to the membrane, nuclear proteins to the nucleus, and secreted proteins secreted into the medium); and the expressed protein is often properly modified. Expression using baculoviral vectors also has some drawbacks: the techniques to grow and work with the virus are still not very widely used and may be difficult for the beginner; the expressed proteins are not always properly modified; and, even for the sophisticated, generation of a recombinant baculovirus to express a given protein still takes a considerable amount of work.

Compared with the above systems, all mammalian expression techniques have certain advantages, particularly for the expression of higher eukaryotic proteins: expressed proteins are usually properly modified, and they almost always accumulate in the correct cellular compartment. Generally speaking, mammalian expression techniques are more difficult, time-consuming, and expensive than those used to express proteins in *E. coli*, and they are much more difficult to perform on a large scale; but they are quite practical for small- and medium-scale work by investigators already familiar with mammalian cell culture techniques. The COS cell and virus procedures are suitable for rapid small- and medium-scale protein production.

Roger Brent

Overview of Protein Expression in *E. coli*

The study of *Escherichia coli* during the 1960s and 1970s made it the best understood organism in nature (Chapter 1). Today's recombinant DNA technology is a direct extension of the genetic and biochemical analyses carried out at that time. Even before the advent of molecular cloning, genetically altered *E. coli* strains were used to produce quantities of proteins of scientific interest. When cloning techniques became available, most cloning vectors utilized *E. coli* as their host organism. Thus, it is not surprising that the first attempts to express large quantities of proteins encoded by cloned genes were carried out in *E. coli.*

E. coli has two characteristics that make it ideally suited as an expression system for many kinds of proteins: it is easy to manipulate and it grows quickly in inexpensive media. These characteristics, coupled with more than 10 years' experience with expression of foreign genes, have established *E. coli* as the leading host organism for most scientific applications of protein expression.

Despite a growing literature describing successful protein expression from cloned genes, each new gene still presents its own unique expression problems. No one, and certainly no laboratory manual, can provide a set of methods that will guarantee successful production of every protein in a useful form. Nevertheless, the vast body of accumulated knowledge has led to a general approach that often helps to solve specific expression problems. This unit introduces general considerations and strategies, while subsequent units (*16.2-16.8*) describe procedures that can be applied to specific expression problems.

GENERAL STRATEGY FOR GENE EXPRESSION IN *E. COLI*

The basic approach used to express all foreign genes in *E. coli* begins with insertion of the gene into an expression vector, usually a plasmid. This vector generally contains several elements: (1) sequences encoding a selectable marker that assure maintenance of the vector in the cell; (2) a controllable transcriptional promoter (e.g., *lac, trp,* or tac) which, upon induction, can produce large amounts of mRNA from the cloned gene; (3) translational control sequences, such as an appropriately positioned ribosome-binding site and initiator ATG; and (4) a polylinker to simplify the insertion of the gene in the correct orientation within the vector. Once constructed, the expression vector containing the gene to be expressed is introduced into an appropriate *E. coli* strain by transformation (*UNIT 1.8*).

16.1

SPECIFIC EXPRESSION SCENARIOS

Although this general approach—insertion of the gene of interest into an expression vector followed by transformation in *E. coli*—is common to all expression systems, specific procedures differ greatly. When choosing a procedure, it is helpful to consider the final application of the expressed protein, as this often dictates which expression strategy to use (*UNIT 16.4*).

Antigen Production

If the goal is to use the expressed protein as an antigen to make antibodies, several approaches are available to make protein reliably and to allow for rapid purification of the antigen. The two best approaches are synthesis of fusion proteins with specific "tag" sequences that can be retrieved by affinity chromatography (*UNITS 16.6, 16.7 & 16.8;* see also *UNIT 10.10*) and synthesis of the native protein, or a fragment of it, under conditions that cause it to precipitate into insoluble inclusion bodies (*UNIT 16.4*). These inclusion bodies can be purified sufficiently by differential centrifugation so preparative denaturing polyacrylamide gel electrophoresis (*UNIT 10.2*) will yield an isolated band that can be cut out and crushed, or electroeluted (*UNIT 10.4*), to provide antigenic material for injection into an animal.

Biochemical or Cell Biology Studies

If the goal is to use the expressed protein as a reagent in a series of biochemical or cell biology experiments, other considerations are relevant. In this case, the authenticity of the protein's function (e.g., high-specific-activity enzyme, binding protein, or growth factor) is very important, while the ease of preparing the protein matters less. For this application, it is possible to express the protein as a fusion protein containing a specific protease-sensitive cleavage site so the N-terminal peptide tail can be removed easily, leaving only the native amino acid sequence (*UNITS 16.4, 16.6, 16.7 & 16.8*). Alternatively, direct expression vectors of the type described in *UNITS 16.2 & 16.3* may be used to produce the authentic primary sequence. When expressed, the protein may be soluble and active, as is the case with many intracellular enzymes. If it is insoluble, as is the case for many secreted growth factors when they are made cytoplasmically in *E. coli*, it may be necessary to isolate inclusion bodies, solubilize the protein using denaturing agents, and refold the protein. Refolding is usually not too difficult when the protein is of moderate size. Whether the protein is expressed in a soluble form or whether it requires refolding, its integrity can usually be checked by specific enzyme assays or by bioassays.

Structural Studies

If the goal is to do structural studies of the expressed protein, the greatest constraints are imposed on the expression system. Because it is nearly impossible to show that a protein of unknown structure has been precisely refolded after denaturing, the protein must generally be made in a soluble form so its purification does not require a denaturation/renaturation step. Usually, the soluble form of the protein—either intracellular or secreted—must be made in strains and by induction protocols that minimize proteolytic degradation.

Soluble expression of most eukaryotic proteins is best achieved with systems that allow induction of synthesis without changing the temperature; for example, by inducing transcription from the *trp* or tac promoters. Maximum accumulation of soluble product is best achieved by testing expression in several strains and at several temperatures, and picking the combination that works best. This is an active area of research at present; the rules are not yet understood, so little more than trial and error can be recommended.

TROUBLESHOOTING GENE EXPRESSION

Once an expression strategy has been chosen and the gene is introduced into an appropriate expression vector, several strains of *E. coli* should be transformed with the vector and protein production should be monitored. Ideally, the protein of interest will be produced in an active form and in sufficient amounts to allow its isolation. Often, however, the protein will be made either in very small amounts or in an insoluble form, or both. If this happens, there are various approaches that may correct the problem.

If not enough protein is produced:

1. Reconstruct the 5′-end of the gene, maximizing its A+T content while preserving the protein sequence it encodes. This may reduce secondary structure within the mRNA, or it may alter an as yet undefined parameter of the reaction.

Regardless of the underlying cause, this procedure usually increases translation efficiency.

2. Determine if a transcriptional terminator is present. If the vector does not have a transcriptional terminator downstream from the site at which the gene is inserted, put one in. This often aids expression, probably by increasing mRNA stability and by decreasing nucleotide drain on the cell.

3. Examine the sequence of the cloned gene for codons used infrequently in *E. coli* genes. These so-called rare codons are usually not a rate-limiting problem, but if four or more happen to occur contiguously, they can reduce expression significantly, perhaps by causing ribosomes to pause. Ribosomal pausing can uncouple transcription from translation, leading to premature termination of the message. Even if transcription proceeds normally, the mRNA 3′ to the stalled ribosomes can be exposed to degradation by host ribonucleases, reducing its stability. Thus, if stretches of rare codons are found, they should be altered to codons more favorable to high expression in *E. coli.*

If enough protein is produced, but it is insoluble when the application requires it to be active and soluble:

1. Vary the growth temperature. As mentioned above, many proteins are more soluble at lower than at higher temperatures. On the other hand, some enzymes have a higher specific activity when made at temperatures >37°C. *E. coli* can synthesize proteins at temperatures ranging from 10° to 43°C, so trying expression at different temperatures is often worthwhile.

2. Change fermentation conditions. Many proteins contain metals as structural and catalytic cofactors. If the protein is being made faster than metals can be transported into the cell, the apoprotein without its metal cofactor will accumulate. This apoprotein will not fold correctly and will likely be insoluble. At the very least, the average specific activity of the expressed protein will be lower than expected. Different media and metal supplements can be tested and the best combination used. Clearly, if there is information about the metal content of the protein, these supplements can be designed more rationally. If no information is available, a more random approach must be tried.

3. Alter the rate of expression by using low-copy-number plasmids. This can be done by using the pACYC family or using single-copy chromosomal inserts of the cloned gene into a suitable target gene. Such reductions in gene dosage often reduce the final yield of protein, but the slower kinetics of synthesis they afford can sometimes result in production of soluble proteins.

To restate the obvious, protein expression is an inexact science at present. However, most proteins can be made in *E. coli* in a form that is useful for a variety of functions. The procedures employed are relatively quick and uncomplicated, and the rewards for success are great.

References: Hamilton et al, 1989; Schein, 1989.

Contributor: Paul F. Schendel

Expression Using the T7 RNA Polymerase/Promoter System

This unit describes expression of genes placed under control of the bacteriophage T7 RNA polymerase. This approach has a number of advantages compared to approaches that rely on *E. coli* RNA polymerase. First, T7 RNA polymerase synthesizes RNA at a rate several times that of *E. coli* RNA polymerase and terminates transcription less frequently. Second, T7 RNA polymerase is highly selective for initiation at its own promoter sequences and it does not initiate transcription from any sequences on *E. coli* DNA. Finally, T7 RNA polymerase is resistant to antibiotics such as rifampicin that inhibit *E. coli* RNA polymerase, resulting in the exclusive expression of genes under the control of a T7 RNA polymerase promoter (p_{T7}). Under optimal conditions, the gene product expressed by the T7 RNA polymerase/promoter system can accumulate to >25% of the total cellular protein; however, in most instances it is significantly less.

To use the two-plasmid p_{T7} system, it is necessary to clone the gene to be expressed into a plasmid containing a promoter recognized by the T7 RNA polymerase. The gene is then expressed by induction of T7 RNA polymerase. The gene for T7 RNA polymerase is present on a second DNA construction. This second construction can either permanently reside within the *E. coli* cell (Basic Protocol), or can be introduced into the cell at the time of induction by infection with a specialized phage, such as an M13 vector (mGP1-2; Alternate Protocol 2). A series of vectors (comparable to those described in this unit) have been developed by Studier et al. (1990) and are available from Novagen. Plasmid-encoded proteins can be selectively labeled (Alternate Protocol 1).

BASIC PROTOCOL

EXPRESSION USING THE TWO-PLASMID SYSTEM

Materials *(see APPENDIX 1 for items with ✓)*

- pT7-5, pT7-6, or pT7-7 vectors (available from author; Fig. 16.2.1)
- *E. coli* JM105, DH1, and K38 or equivalent (Table 1.4.5)
- LB plates and medium (*UNIT 1.1*) containing: 60 µg/ml ampicillin; 60 µg/ml kanamycin; and 60 µg/ml ampicillin + 60 µg/ml kanamycin
- pGP1-2 (available from author)
- ✓ Cracking buffer

1. Subclone a fragment containing the gene to be expressed into pT7-5, pT7-6, or pT7-7 (*UNITS 1.4 & 3.16*). Transform a standard *E. coli* strain (e.g., JM105 or DH1); this strain should *not* carry a plasmid that directs synthesis of T7 RNA polymerase (i.e., pGP1-2; *UNIT 1.8*). Plate transformants on LB/ampicillin plates and grow overnight at 37°C.

2. Grow individual transformants in LB/ampicillin medium at 37°C and obtain plasmid DNA by a miniprep procedure (*UNIT 1.6*). Confirm that the gene has been correctly inserted by restriction mapping (*UNITS 3.1-3.3*).

3. Transform *E. coli* K38 with pGP1-2, plate on LB/kanamycin plates, and grow overnight (~24 hr) at 30°C. Grow an individual *E. coli* K38/pGP1-2 transformant in LB/ kanamycin medium at 30°C.

4. Transform a vector containing gene to be expressed under control of p_{T7} into *E. coli* K38/pGP1-2 grown in LB/kanamycin medium (cells may be heat-shocked

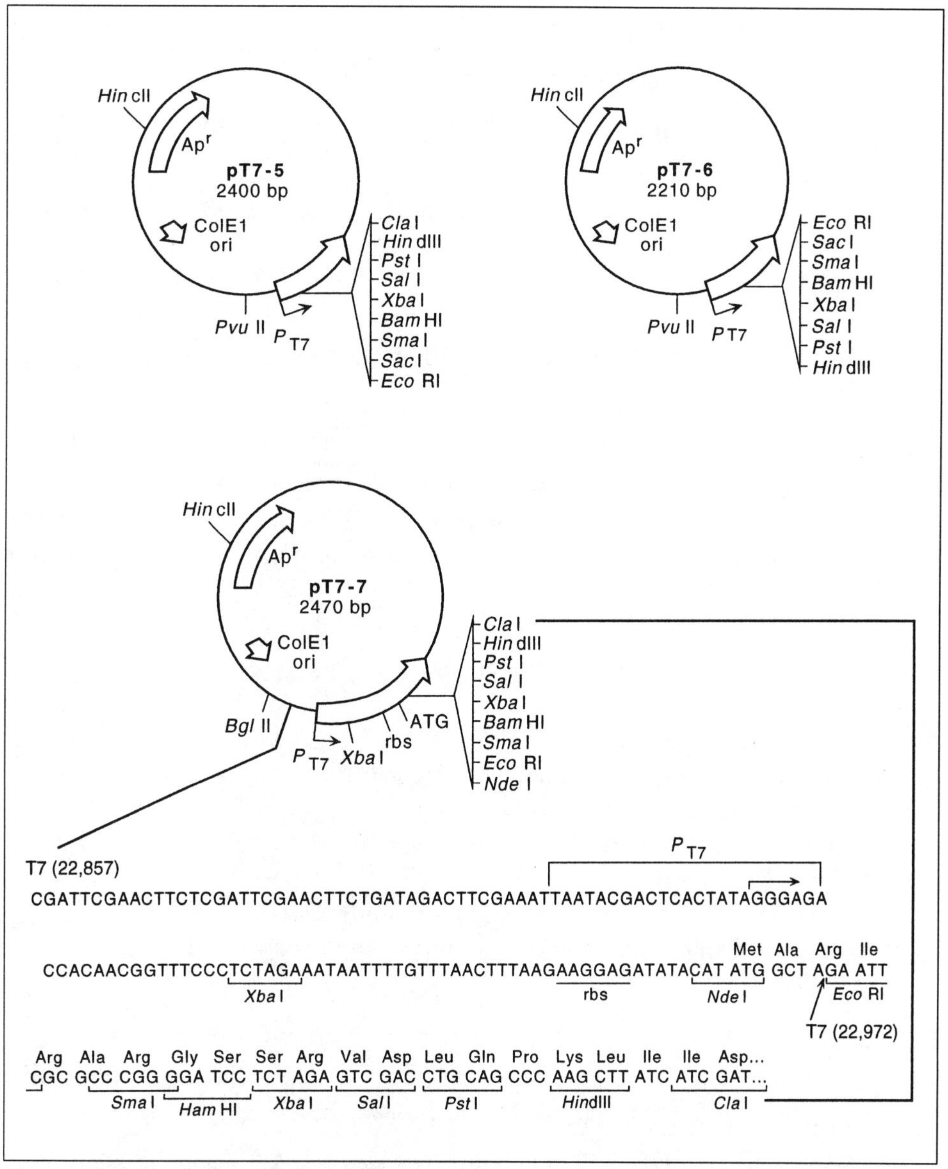

Figure 16.2.1 pT7-5, pT7-6, and pT7-7. pT7-5, pT7-6, and pT7-7 are cloning vectors that contain a T7 promoter and are used to express genes using T7 RNA polymerase. All three vectors contain a T7 RNA polymerase promoter, the gene encoding resistance to the antibiotic ampicillin and the ColE1 origin of replication. pT7-7 has a strong ribosome-binding site (rbs) and start codon (ATG) upstream of the polylinker sequence; the sequence of this region is shown below the map of pT7-7. pT7-5 and pT7-6 lack any ribosome-binding site upstream of the polylinker sequence and consequently are only useful when expressing genes that already contain the proper control sequences. pT7-5, pT7-6, and pT7-7 were constructed by S. Tabor and are derivatives of pT7-1 described in Tabor and Richardson (1985).

during transformation). Plate transformants (containing both plasmids) on LB/ampicillin/kanamycin plates and grow overnight at 30°C.

5. Pick a single *E. coli* colony that contains the two plasmids. Inoculate it into 5 ml LB/ampicillin/kanamycin medium and grow overnight at 30°C.

As a control, transform E. coli K38/pGP1-2 with the parent p_{T7} vector (without an insert). If the transformation efficiency of the vector containing the insert is significantly lower (>50-fold) than that of the parent vector, the gene product may be toxic to E. coli cells.

6. Dilute ~1 ml of overnight culture of cells 1:40 into LB/ampicillin/kanamycin medium and grow several hours at 30°C to an OD_{590} ≅0.4.

7. Induce gene for T7 RNA polymerase by quickly raising temperature to 42°C for 30 min.

   The E. coli RNA polymerase can be inhibited by adding rifampicin (after induction) to a final concentration of 200 µg/ml. Generally, add rifampicin to cells when the plasmid-encoded proteins are being labeled with [^{35}S]methionine (Alternate Protocol 1).

8. Reduce temperature to 37°C and grow cells an additional 90 min with shaking. Harvest cells by centrifugation.

9. To analyze induced proteins by SDS-PAGE, resuspend the equivalent of 1.0 ml of cells in 0.1 ml cracking buffer. Heat at 100°C for 5 min immediately prior to loading a 20-µl aliquot of each sample onto an SDS-polyacrylamide gel (*UNIT 10.2*). To analyze cells for induced enzymatic activity, prepare an appropriate cell extract from ~10 ml of cells.

ALTERNATE PROTOCOL 1

SELECTIVE LABELING OF PLASMID-ENCODED PROTEINS

Additional Materials *(also see Basic Protocol)*

M9 medium (*UNIT 1.1*) without and with 5% (v/v) 18 amino acid mix (*APPENDIX 1*)
20 mg/ml rifampicin in methanol (e.g., Sigma; store in dark at 4°C for 2 weeks; Table 1.4.1)
10 mCi/ml [^{35}S]methionine (800 Ci/mmol) diluted 1:10 in M9 medium
Fluorographic enhancing agent (e.g., Enlightning from Du Pont NEN or Amplify from Amersham)

1. Repeat steps 2 to 4 of Basic Protocol (using T7-promoter expression plasmid obtained from step 1 of Basic Protocol). When OD_{590} ≅0.4, remove 1 ml of cells, microcentrifuge 10 sec, and discard supernatant.

   An alternative to LB/ampicillin/kanamycin medium is M9 medium containing 25 µg/ml ampicillin, 25 µg/ml kanamycin, and any required nutrients. The addition of one part in twenty of the 18 amino acid mixture (0.1% stock, 0.005% final concentration) stimulates growth of cells in M9 medium without interfering with subsequent labeling of proteins with [^{35}S]methionine. The E. coli strain must be Cys^+ and Met^+.

2. Wash cell pellet with 1 ml M9 medium, microcentrifuge 10 sec at room temperature, and discard supernatant.

3. Resuspend cell pellet in 1 ml M9 medium containing 18 amino acid mixture. Grow cells 60 min at 30°C with shaking.

   The OD_{590} may not increase due to adaptation of cells to medium; however, labeling will occur.

4. Induce gene for T7 RNA polymerase by placing the cells in a 42°C water bath for 20 min. Add 20 mg/ml rifampicin to 200 µg/ml final. Keep cells at 42°C for an additional 10 min after adding rifampicin.

5. Move cells to a 30°C water bath for an additional 20 min. Remove 0.5 ml of cells for labeling with [^{35}S]methionine (the other 0.5 ml can be used to label the cells at a later time point).

6. Label newly synthesized proteins by adding 10 µl (10 µCi) diluted [^{35}S]methionine to 0.5 ml of cells and incubating for 5 min at 30°C.

7. Microcentrifuge cells 10 sec and discard supernatant. (*CAUTION*: the supernatant is radioactive; discard properly.) Resuspend cell pellet in 100 µl cracking buffer.

8. Heat samples to 100°C for 5 min. Load a 20-µl aliquot onto an SDS-polyacrylamide gel and electrophorese (*UNIT 10.2*).

9. Treat gel with a fluorographic-enhancing agent. Dry gel under vacuum 2 hr at 65°C and autoradiograph (*APPENDIX 3*; a 1-hr exposure should be adequate to visualize most proteins).

EXPRESSION BY INFECTION WITH M13 PHAGE mGP1-2

ALTERNATE PROTOCOL 2

Whenever the gene for T7 RNA polymerase is present in *E. coli* cells, low levels of T7 RNA polymerase are constitutively produced. This can be a problem when the gene products under control of p_{T7} are toxic. One strategy to avoid this is to keep the gene for T7 RNA polymerase out of the cell until the time of induction. In this protocol, T7 RNA polymerase is introduced into the cell by infection with the M13 phage mGP1-2, which contains the gene for T7 RNA polymerase under control of the *lac* promoter. Host cells for this phage (e.g., JM101 or K38) must carry the F factor.

Additional Materials *(also see Basic Protocol; see APPENDIX 1 for items with ✓)*

M13 phage mGP1-2 (available from author)
✓ PEG solution
100 mM IPTG (Table 1.4.2)

1. Prepare stock of M13 phage mGP1-2 and concentrate phage by precipitation with PEG solution (*UNIT 1.15*). (DO NOT proceed to add TE buffer or phenol.) Resuspend phage in M9 medium and titer (*UNIT 1.10*).

   If the cell proteins are to be labeled, it is important that the phage used to infect the cells are free of unlabeled methionine. Precipitate phage with PEG twice, each time resuspending the pellet in M9 medium.

2. Transform *E. coli* cells susceptible to M13 infection with T7-promoter expression plasmid from step 1 of Basic Protocol. Plate transformants on LB/ampicillin plates and grow overnight at 37°C. Pick a single colony and grow in LB/ampicillin medium overnight at 37°C.

3. Dilute overnight culture of cells 1:100 in LB/ampicillin medium and grow several hours at 37°C with gentle shaking to $OD_{590} \cong 0.5$ (avoid vigorous agitation).

4. Infect cells with M13 phage mGP1-2 (from step 1) at ~10 phage for each *E. coli* cell. Add 100 mM IPTG to 1 mM final to induce production of T7 RNA polymerase. Incubate cells 2 hr at 37°C.

   At $OD_{590} \cong 0.5$, the density of E. coli cells will be ~2×10^8 cells/ml. Add M13 mGP1-2 phage at a final concentration of 2×10^9 phage/ml to obtain a multiplicity of infection of 10.

5. Harvest cells and analyze induced proteins as in step 9 of Basic Protocol.

Reference: Studier et al., 1990.

Contributor: Stanley Tabor

Expression Using Vectors with Phage λ Regulatory Sequences

In the system described here, pBR222-based plasmids (pSKF) utilize regulatory signals—such as the powerful promoter p_L—from the bacteriophage λ. Transcription from p_L can be fully repressed and plasmids containing it are thus stabilized by the λ repressor, *c*I. The repressor is supplied by an *E. coli* host that contains an integrated copy of a portion of the λ genome. This so-called defective lysogen supplies the λ regulatory proteins *c*I and N but does not provide the lytic components that would normally lead to cell lysis. Thus, cells carrying these plasmids can be grown initially to high density without expression and subsequently induced to synthesize the product upon inactivation of the repressor. This system also ensures that p_L-directed transcription efficiently traverses any gene insert, which is accomplished by providing the phage l antitermination function, N, to the cell and by including on the p_L transcription unit a site necessary for N utilization (Nut site).

To express the coding sequence, efficient ribosome-recognition and translation-initiation sites have been engineered into the p_L transcription unit. Expression occurs after temperature or chemical induction inactivates the repressor (Basic Protocols). Restriction endonuclease sites for insertion of the desired gene have been introduced both upstream and downstream from an ATG initiation codon. Thus, the system allows either direct expression or indirect expression (via protein fusion) of any coding sequence, thereby potentially allowing expression of any gene insert. Direct expression generates "authentic" gene products (Support Protocol 1), while expression of heterologous genes fused to highly expressed gene partners generates chimeric proteins that differ from the native form. In the latter case, the fusion partner can be removed to obtain an unfused version of the gene product (Support Protocol 2).

Expression of most gene products as fusions with the first 81 amino acids of the NS1 protein (using pSKF301) can be achieved at levels of 5-30% of total cellular protein. Expression levels of nonfusion proteins (authentic) are less predictable and may vary from <1% to 30% of total cell protein.

BASIC PROTOCOL 1

TEMPERATURE INDUCTION OF GENE EXPRESSION

Materials *(see* APPENDIX 1 *for items with ✓)*

- Expression vector (e.g., pSKF series; Support Protocols)
- *E. coli* AR58 or equivalent (Table 1.4.5)
- LB plates and medium containing appropriate antibiotic (*UNIT 1.1*; medium at room temperature and prewarmed to 65°C)
- ✓ SDS sample buffer

1. Transform expression vector into an *E. coli* λ lysogen carrying a temperature-sensitive mutation in its repressor gene (λ *c*I857; *UNIT 1.8*). Plate on LB/antibiotic plates and incubate transformants at 32°C.

2. Grow transformed cells overnight at 32°C in LB/antibiotic medium.

3. Dilute overnight culture ≥1:20 into fresh LB/antibiotic medium. Grow culture at 32°C in gyrotory shaker at 250 to 300 rpm until OD_{650} = 0.6 to 0.8.

4. Add ⅓ vol of 65°C LB/antibiotic medium with swirling to elevate temperature rapidly to 42°C. Continue growing 2 to 3 hr at 42°C.

A rapid increase in temperature favors production. Small shake-flask cultures (≤25 ml) are more easily induced by transfer to a 42°C gyrotory water bath without addition of prewarmed media.

5. Remove 1-ml aliquot for analysis and harvest remainder of cells by centrifuging 15 min at 3000 × *g* (in a low-speed rotor), 4°C. Freeze cell pellet at −70°C until ready to isolate the gene product.

6. Microcentrifuge the 1-ml aliquot 1 min at top speed, then resuspend pellet in 50 µl SDS sample buffer. Boil 5 to 10 min and analyze gene product by SDS–polyacrylamide gel electrophoresis (*UNIT 10.2*).

CHEMICAL INDUCTION OF GENE EXPRESSION

BASIC PROTOCOL 2

Expression using the pSKF system can also be induced chemically in lysogens that carry a wild-type (*ind*$^{+}$) repressor gene (*c*I857 cannot be used as it is *ind*$^{-}$). In contrast to heat induction (product accumulates in 45 to 90 min), nalidixic acid–mediated induction of protein expression is comparatively slow (product accumulates in 5 to 6 hr).

Materials

Expression vector (e.g., pSKF series; Support Protocols)
E. coli AR120 or equivalent (Table 1.4.5)
LB plates and medium containing appropriate antibiotic (*UNIT 1.1*)
60 mg/ml nalidixic acid in 1 N NaOH (not necessary to filter sterilize; Table 1.4.1)

1. Transform expression vector into a replication-defective, *E. coli* *c*I^{+} lysogen. Plate on LB/antibiotic plates and incubate transformants at 37°C.

2. Grow transformed cells overnight at 37°C in LB/antibiotic medium.

3. Dilute overnight culture ≥1:20 into fresh LB/antibiotic medium. Grow culture at 37°C in gyrotory shaker at 250 to 300 rpm until OD_{650} = 0.4.

4. Add 1⁄1000 vol of 60 mg/ml nalidixic acid solution (60 µg/ml final). Continue growing culture 5 to 6 hr at 37°C.

5. Harvest cells and analyze gene product (Basic Protocol 1; steps 4 and 5).

AUTHENTIC GENE CLONING USING pSKF VECTORS

SUPPORT PROTOCOL 1

It is often most desirable to express a gene product in a form as similar to the native protein as possible. Such an "authentic" gene product will have the greatest chance of having a structure and activity identical to that of the native protein. Efficient translation of a coding sequence for an authentic gene product is typically accomplished by placing the inserted information immediately adjacent to a ribosome-binding site.

Strategic Planning

The translation-initiation signal utilized here is that of the phage λ *c*II gene. To make the translational information generally useful, the coding region of the gene has been removed from the vectors, leaving only their initiator *f*Met codon and upstream translational regulatory sequences. Additionally, these vectors have been engineered to provide restriction endonuclease sites on either side of the ATG, such that the initiation codon can be supplied by either the plasmid or the gene being inserted.

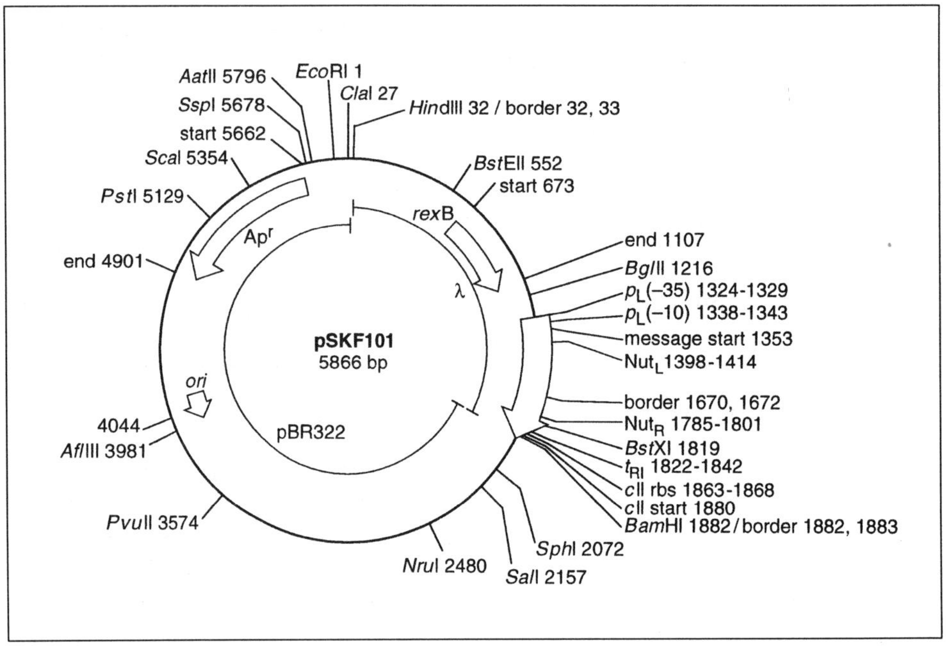

Figure 16.3.1 pSKF101. pSKF101 is a vector used for authentic gene cloning which allows direct expression of the inserted gene. It is a derivative of pBR322 (*UNIT 1.5*) containing sequences inserted between *Hin*dIII and *Bam*HI sites of pBR322. The inserted λ sequences contain the p_L promoter and *c*II ribosome-binding site (rbs); these are the transcriptional and translational requlatory sequences necessary to express heterologous genes in *E. coli.* Within this region are several unique restriction sites that permit insertion of the gene. The regions derived from pB322 and λ are indicated. This plasmid can be maintained stably in a λ-lysogenized *E. coli* strain. The selectable marker is ampicillin, encoded by β-lactamase.

An alternate name for pSKF101 is pASI (Rosenberg et al., 1983). Alternative names of related vectors are as follows: pSKF102 is pOTSV (Shatzman and Rosenberg, 1987); pSKF201 is pOTS-Nco (Shatzman and Rosenberg, 1987); and pSKF301 is pMG1.

Finally, restriction sites have also been engineered upstream of the translational regulatory region to permit insertion of other ribosome-binding sites. For example, pSKF101 (Fig. 16.3.1) and pSKF102 both have a *Bam*HI site adjacent to the initiation codon (ATGgatcc), while pSKF201 has a *Nco*I site (ccATGg) and pSKF301 (Fig. 16.3.2) has an *Nde*I site (catATG). The protocol presented below summarizes the steps to obtain an authentic gene clone using pSKF101 as an example.

Sample Protocol

Materials

Appropriate restriction endonucleases and buffers (*UNIT 3.1*)
pSKF101 vector (available from A. Shatzman; Fig. 16.3.1)
Competent *E. coli* AS1 (Table 1.4.5; also known as MM294*c*I$^+$)

1. Identify a unique restriction endonuclease site close to the 5′ end of the coding sequence of the gene to be expressed, as well as another unique site 3′ to this gene's termination codon (*UNIT 3.1*).

2. Synthesize two single-stranded DNA oligonucleotides that recreate the coding sequence immediately preceding unique restriction endonuclease site near the 5′ end of the gene to be expressed (*UNITS 2.11 & 2.12*). Purify and quantitate DNA, then anneal in order to obtain double-stranded DNA.

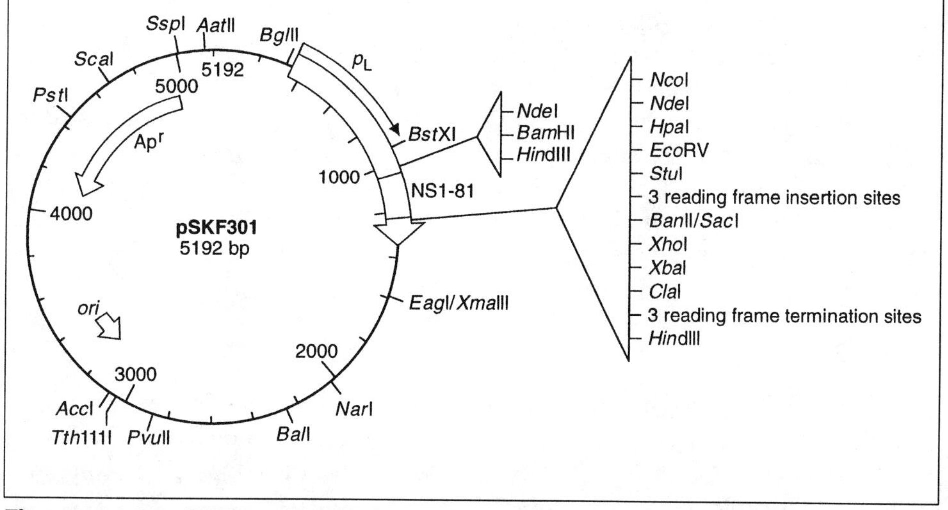

Figure 16.3.2 pSKF301. pSKF301 is a vector that can be used for both indirect and direct expression. it is similar to pSKF101 in that it contains the same transcriptional and translational regulatory sequences as well as selectable markers; it differs in that it contains a shorter segment of λ DNA than pSKF101. pSKF301 also contains the coding sequence of the first 81 amino acids of the influenza protein, NS1, shown as NS1-81. This region is adjacent to the *c*II ribosome-binding site (fbs) and contains restriction sites at the 3′ end of NS1-81 that allow construction of translational fusions in any of the three reading frames. Removal of NS1-81 permits direct expression of the cloned gene. (This vector is also known as pmG1.)

3. Digest 25 to 50 µg plasmid DNA containing gene to be expressed with restriction endonucleases identified in step 1 (*UNIT 3.1*). Electrophorese doubly digested plasmid DNA on a polyacrylamide gel (*UNIT 2.7*). Locate fragment of interest by staining with ethidium bromide and cut DNA fragment out of gel. Recover DNA by electroelution (*UNIT 2.6*) and quantitate amount of DNA (*APPENDIX 3*).

4. Digest 10 µg pSKF101 with *Bam*HI and a restriction endonuclease that generates ends compatible with 3′ end of coding sequence. Confirm that complete digestion has occurred by analysis of digested DNA on an agarose gel.

5. Prepare a ligation reaction (*UNIT 3.16*) by combining the following: 1 ng digested pSKF101 vector DNA, 10 ng of gene fragment to be expressed (step 3), 20 ng synthetic oligonucleotide (step 2), and T4 DNA ligase. Ligate 10 to 12 hr at 4°C.

6. Remove ⅓ of ligation reaction and transform 50 to 100 µl competent *E. coli* AS1 (*UNIT 1.8*). Plate on LB/ampicillin plates and incubate overnight at 37°C. Pick 12 to 24 colonies and transfer with sterile toothpick to 3 ml LB/ampicillin medium. Grow cells 5 to 18 hr and isolate DNA by miniprep method (cells may be harvested once broth appears turbid; *UNIT 1.6*).

7. Perform appropriate restriction digests to determine which clones contain desired construction of gene to be expressed. Transform an *E. coli* strain with DNA and express gene as in Basic Protocols.

CONSTRUCTION AND DISASSEMBLY OF FUSED GENES IN pSKF301

By fusing the gene to be expressed to a coding region of another gene (the fusion partner), a chimeric gene can be constructed in an appropriate vector. Most available vectors feature a fusion partner that is a highly expressed gene. When expression of the chimeric gene is induced, the resulting proteins carry additional peptide information at the N terminus. Although the fusion product may have physical and/or functional properties that differ from the "authentic" protein, advantages of the approach include highly efficient expression (up to 30% of total cell protein) without complicated alterations on the gene, and the presence of a "handle" on the expressed protein to help identify and purify it.

Strategic Planning

Plasmid pSKF301 (Fig. 16.3.2) has been constructed to permit initial expression of a gene as a fusion product, followed by removal of the DNA encoding the fusion portion by restriction digestion. Finally, the unfused version of the gene is expressed as an authentic protein.

pSKF301 contains a *Nde*I restriction site adjacent to the ATG following the *c*II ribosome-binding site (Fig. 16.3.3). This ATG also serves as the translational start of the *NS1* gene derived from the influenza nonstructural gene. This gene has been truncated to express only its first 81 amino acids. Just beyond the coding sequence for the 81st amino acid is a second *Nde*I site followed by three unique blunt-ended restriction sites, *Hpa*I, *Eco*RV, and *Stu*I, which allow for the insertion of genes into any of three reading frames. Immediately following the *Stu*I site are sequences coding for translational stops in any of the three reading frames.

The expression of a gene of interest as a fusion protein in pSKF301 may be achieved by utilizing *Nco*I, *Hpa*I, *Eco*RV, or *Stu*I. Choice of restriction site depends upon the reading frame necessary for translation of a specific protein sequence. First, a unique restriction site close to the 5′ end of the gene to be expressed must be identified. Second, the appropriate restriction endonuclease is selected for digesting pSKF301 such that the gene will be expressed. If the chosen restriction site is a blunt-end cutter, no further manipulation of that end is required. In the event the restriction site identified leaves a protruding end, further manipulation is required. "Filling in"

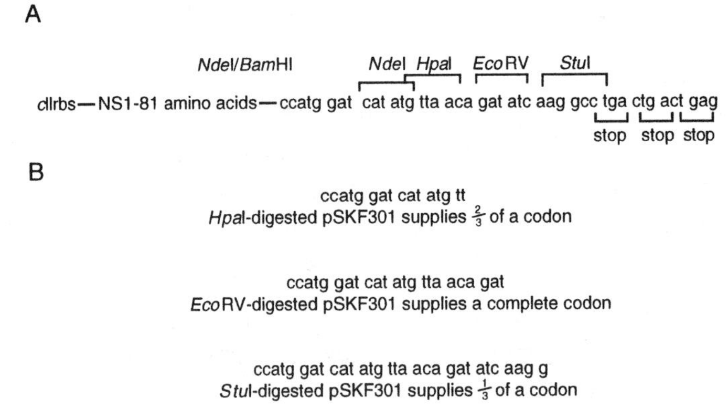

Figure 16.3.3 Sequence and restriction endonuclease sites (in the region used for cloning) of pSKF301 (**A**). Restriction endonuclease digestion shows the strategy utilized to obtain pSKF301 as a vehicle for expression in all three reading frames (**B**).

using the Klenow fragment of *E. coli* DNA polymerase for 5′ protrusions, or T4 DNA polymerase, S1 nuclease, or mung bean nuclease for 3′ protrusions, are methods of choice (*UNIT 3.16*).

Sample Protocol

Materials

Appropriate restriction endonucleases and buffers (*UNIT 3.1*)
Klenow fragment of *E. coli* DNA polymerase I (*UNIT 3.5*)
pSKF301 vector (available from A. Shatzman; Figs. 16.3.1 & 16.3.2)
T4 DNA ligase (*UNIT 3.14*)
Competent *E. coli* AS1 (Table 1.4.5; also known as MM294cI^{+})

1. Assume restriction site identified in gene is a *Bam*HI site (5′ protrusion). Digest with *Bam*HI (*UNIT 3.1*) to obtain:

```
GATCC   XXX   XXX   XXX   XXX
    G   YYY   YYY   YYY   YYY
```

2. Treat with Klenow fragment to fill in unpaired bases to obtain:

```
GATCC   XXX   XXX   XXX   XXX
CTAGG   YYY   YYY   YYY   YYY
```

3. Determine proper reading frame of gene. In this example assume XXX XXX XXX XXX is the proper reading frame; therefore, coding sequence of filled-in fragment should read:

```
GA   TCC   XXX   XXX   XXX   XXX
```

4. Determine which restriction endonuclease should be used to digest pSKF301 to allow expression of fusion protein. For this example, *Stu*I is required to yield:

```
ccatg gat cat atg tta aca gat atc aag g   GA TCC XXX XXX XXX XXX
|_____________________________________|   |____________________|
              pSKF301                           fusion gene
```

5. Prepare vector and gene fragment to be expressed (Support Protocol, steps 3 to 7; begin at recovery point; no synthetic DNA is required).

Once a gene has been expressed as a fusion protein, it may be desirable to obtain an unfused version as described below.

To convert a fusion protein to an unfused protein when using pSKF301, be certain that gene of interest does not contain an *Nde*I site. The following theoretical fusion construct will be used as an example in these steps:

```
 NdeI                            NdeI
CATATGGATCC---NS1-81---CCATGGATCATATGTT---fusion gene---tga
|______|                       |______|
```

6. Set up large-scale plasmid preparation of fusion construct to yield ~100 µg plasmid DNA (*UNIT 1.7*).

7. Digest 10 µg of construct with *Nde*I. Verify that all vector DNA has been completely digested by agarose gel electrophoresis (*UNIT 2.5*). A 280-bp fragment should be observed; this contains the NS1-81 gene sequence being liberated from the construct.

8. Purify digested construct by phenol/chloroform/isoamyl alcohol extraction followed by ethanol precipitation (*UNIT 2.1*).

9. Add T4 DNA ligase to 1 µg of *Nde*I-digested construct and incubate overnight at 4°C. Transform ligated DNA into competent *E. coli* AS1 cells (or any other suitable cI^+ lysogen; *UNIT 1.8*).

10. Determine that construct no longer contains NS1-81 gene sequence by restriction analysis (if *Nde*I digestion was complete, expect 95% to 100% of transformants to contain unfused construct).

11. Transform DNA and express gene by temperature or chemical induction as in Basic Protocols.

Reference: Shatzman and Rosenberg, 1987.

Contributors: Allan R. Shatzman, Mitchell S. Gross, and Martin Rosenberg

UNIT 16.4

Introduction to Expression by Fusion Protein Vectors

Expression—the directed synthesis of a foreign gene product—is often the logical next step for researchers who have isolated a gene and want to study the protein it encodes. During the early days of recombinant DNA technology, it was thought that a strong promoter and a start codon at the beginning of the gene would be sufficient for good expression in *Escherichia coli.* Since then it has been learned that the requirements for efficient translation are a good deal more complicated. In addition to a promoter and a start codon, good expression requires that the mRNA encoding the protein to be expressed contain a ribosome-binding site that is not blocked by mRNA secondary structure. The level of expression is also affected by codon preferences, especially in the second codon of the gene, and may be affected by the coding sequence in other ways that are not yet well understood (*UNIT 16.1*). In virtually all cases, these problems can be solved by altering the sequence preceding the start codon, and/or by making changes in the 5′ end of the coding sequence that do not change the protein sequence, taking advantage of the degeneracy of the genetic code.

However, it is often quicker to solve these problems by making fusions between genes. In this approach the cloned gene is introduced into an expression vector 3′ to a sequence (carrier sequence) coding for the amino terminus of a highly expressed protein (carrier protein). The carrier sequence is often from an *E. coli* gene, but it can be from any gene that is strongly expressed in *E. coli.* The carrier sequence provides the necessary signals for good expression, and the expressed fusion protein contains an N-terminal region encoded by the carrier. In such vectors, the portion of the fusion protein encoded by the carrier can be as small as one amino acid (*UNIT 16.3*), although expression from such vectors can still be subject to problems caused by the coding sequence of the expressed protein. Perhaps more typical examples of short carrier sequences are those contained in the *trpE* vectors (*UNIT 16.6*) or the λ cII vectors.

The carrier sequence can also code for an entire functional moiety or even for an entire protein. For example, *UNITS 16.6-16.8* describe the use of vectors that express β-galactosidase and *trpE* fusions, glutathione-S-transferase (GST) fusions, and thioredoxin (Trx) fusions. These carrier regions often can be exploited in purifying the protein, either with antibodies or with an affinity purification specific for that

16.4

carrier protein. Alternatively, unique physical properties of the carrier protein (e.g., heat stability) can be exploited to allow selective purification of the fusion protein. In addition, some carrier proteins such as Trx can be selectively released from intact cells by osmotic shock or freeze/thaw procedures, even though they reside in different cellular compartments. Often, proteins fused to these carriers can be separated from the bulk of intracellular contaminants by taking advantage of this attribute.

There are three problems often encountered when expressing fusion proteins: solubility of the expressed protein, stability of the expressed protein, and presence of the carrier protein. The first two problems are often encountered with both fusion and nonfusion expression systems (*UNIT 16.1*), while the third is unique to fusion systems.

SOLUBILITY OF THE EXPRESSED PROTEIN

The high-level expression of many proteins can lead to the formation of *inclusion bodies,* very dense aggregates of insoluble protein and RNA that contain most of the expressed protein. Precipitation of a protein into inclusion bodies sometimes can be an advantage, because inclusion bodies are insoluble and dense, and can be purified relatively easily by centrifugation (*UNIT 16.6*). In addition, some proteins that are degraded when expressed in the soluble fraction are quite stable as inclusion bodies. Once purified, protein in inclusion bodies can be solubilized by denaturation with guanidine·HCl or urea, and then can often be refolded by dialyzing away the denaturant. A problem, however, with denaturation/renaturation is that the yield of properly refolded protein is variable and sometimes quite low; some proteins, especially large ones, cannot be properly refolded at all (see *UNIT 16.6*).

If expression of a particular fusion protein produces insoluble aggregates and a soluble protein is required, there are several things to try. One important variable is temperature; for reasons not well understood, higher temperatures (37° and 42°C) promote inclusion-body formation and lower temperatures (30°C) inhibit it. Another variable is the level of expression; sometimes lowering the expression level can increase the proportion of protein that is soluble. A third variable is the strain background of the cells bearing the expression vector; large differences in the proportion of a particular expressed protein that is soluble are seen among different strains, but it is not known which of the genetic differences between the strains is responsible for the differences in solubility. Finally, it is worth noting that changes in the carrier protein can affect the solubility of an expressed fusion protein.

STABILITY OF THE EXPRESSED PROTEIN

Stability problems are often encountered when foreign proteins, especially eukaryotic proteins, are expressed in *E. coli.* The carrier protein can sometimes stabilize an expression fusion protein. Sometimes, however, the expressed protein is degraded but the carrier protein is not. Moreover, fusion proteins are sometimes cleaved in vivo at the fusion joint between the carrier and expressed portions of the fusion, which obviously creates problems if the carrier protein is to be used as an aid in purification. These facts about fusion proteins are consistent with a model in which the carrier and the rest of the protein form independent domains. In this view, it can be imagined that there are cases where the carrier domain folds correctly and the expressed protein does not (and is degraded). There are also cases where both domains fold correctly but the joint region between them is sensitive to one or more *E. coli* proteases.

Approaches that have been used to stabilize fusion proteins are generally the same as those used to stabilize nonfusion proteins. One method is to arrange for the fusion protein to be expressed as insoluble aggregates. Another method is to use *E. coli* strains deficient in known proteases. For example, a *lon htpR* double-mutant strain—which is deficient in several cytoplasmic proteases—shows reduced degradation of unstable proteins. Similarly, the *degP* mutant has been shown to stabilize fusion proteins in the periplasm, and *ompT* mutants have proven useful in preventing cleavage between exposed basic residues (e.g., Arg-Arg) in several nonfusion proteins during preparation of crude extracts. Finally, the stability of a particular fusion can vary even among different "wild-type" lab strains, perhaps due to uncharacterized differences in protease levels among the strains.

CLEAVAGE OF FUSION PROTEINS TO REMOVE THE CARRIER

The use of fusion proteins is growing rapidly for the many reasons described above. The various systems described in the following units have been used to produce many different kinds of proteins ranging from enzymes and growth factors to transmembrane receptors and DNA binding proteins. Often it is advantageous to remove the carrier protein moiety from the protein of interest to facilitate biochemical and functional analyses. Several methods for site-specific cleavage of fusion proteins have been developed (*UNIT 16.5*). The choice of method is usually determined by the composition, sequence, and physical characteristics of the particular protein. Chemical cleavage of fusion proteins can be accomplished with reagents such as cyanogen bromide (Met↓), 2-(2-nitrophenylsulphenyl)-3-methyl-3′-bromoindolenine (BNPS-skatole, Trp↓), hydroxylamine (Asn↓Gly), or low pH (Asp↓Pro). Chemical cleavage procedures tend to be inexpensive and efficient, and often can be accomplished under denaturing conditions to cleave otherwise insoluble fusion proteins. However, their use is hampered by the likely occurrence of cleavage sites in the protein of interest, along with the propensity for side reactions that result in unwanted modifications to the protein. As an alternative to chemical methods, enzymatic cleavage procedures are desirable for their relatively mild reaction conditions and, most importantly, for the high degree of specificity exhibited by some proteases commonly used for this purpose. Among the useful enzymes are factor Xa, thrombin, enterokinase, renin, and collagenase. All of these enzymes have extended substrate recognition sequences (up to 7 amino acids in the case of renin), which greatly reduces the likelihood of unwanted cleavages elsewhere in the protein. Of the above-mentioned proteases, factor Xa and enterokinase are most useful in this application because they cleave on the carboxy-terminal side of their respective recognition sequences, allowing the release of fusion partners containing their authentic amino-termini.

UNITS 16.6, 16.7 & 16.8 describe four different fusion protein vector systems; of these, only two include recognition sites for interdomain cleavage. The GST fusion system (*UNIT 16.7*) includes vectors that contain either a thrombin cleavage site, a factor Xa cleavage site, or an Asp-Pro acid cleavage site. The Trx fusion system (*UNIT 16.8*) uses an enterokinase cleavage site. *UNIT 16.5* describes fusion protein cleavages in detail, including specific protocols for cleaving fusion proteins produced with each of the aforementioned vector systems, along with methodologies for the site-specific cleavage of proteins using various chemical reagents.

References: Lee et al., 1984; Nagai and Thøgersen, 1987; Schein, 1989.

Contributors: Paul Riggs and Edward R. La Vallie

Enzymatic and Chemical Cleavage of Fusion Proteins

The use of gene fusion expression systems has become an increasingly popular method of producing foreign proteins in *Escherichia coli*. This popularity is due in large part to the development of fusion systems that are capable of producing large amounts of fusion protein in a soluble form. The glutathione-S-transferase (GST, *UNIT 16.7*), maltose binding protein (MED), and thioredoxin (Trx, *UNIT 16.8*) fusion systems have proven singularly successful in producing properly folded and biologically active proteins. Each of these systems also provides convenient methods for specific purification of the fusion protein from cellular contaminants. As a result, proteins produced using these systems are readily amenable to the study of their biological activities and/or interactions. As a consequence of the popularity of fusion protein expression strategies, the ability to cleave the N-terminal fusion "carrier" protein from the C-terminal protein of interest has become increasingly important.

This unit provides protocols for some commonly used methods of site-specific cleavage of fusion proteins. The first three protocols describe enzymatic cleavage of proteins using proteases that display highly restricted specificities, which greatly decrease the likelihood that unwanted secondary cuts will occur. Three additional protocols describe cleavage of fusion proteins with chemical reagents as an alternative to enzymatic cleavage. Chemical cleavage methods have the disadvantage of being less specific, and it is necessary to ensure that a susceptible peptide bond does not exist in the protein of interest.

ENZYMATIC CLEAVAGE OF FUSION PROTEINS WITH FACTOR Xa

BASIC PROTOCOL 1

Fusion proteins that have been produced with the MBP fusion vectors pMAL-c2, pMAL-p2, or the GST fusion vector pGEX3X contain a recognition sequence for coagulation factor Xa encoded in the DNA immediately preceding the polylinker cloning site, but factor Xa will not cleave if a proline residue follows the arginine of the recognition sequence.

Materials *(see APPENDIX 1 for items with ✓)*

1 mg/ml fusion protein
200 μg/ml factor Xa (New England Biolabs) in reaction buffer (see step 1)
✓ 2× SDS sample buffer
Boiling water bath

1. Prepare two small-scale trial reactions to determine optimum incubation time as follows:

   Reaction 1: 20 μl of 1 mg/ml fusion protein with 1 μl of 200 μg/ml factor Xa.

   Reaction 2: 5 μl of 1 mg/ml fusion protein and no factor Xa (mock digestion).

   Incubate at room temperature.

   Fusion protein in PBS from glutathione-agarose purification (UNIT 16.7) is suitable for factor Xa digestion; otherwise the protein should be prepared in 20 mM Tris·Cl (pH 8.0)/1 mM $CaCl_2$/100 mM NaCl. Although most fusion proteins could be kept at 4°C, any remaining fusion protein solution can be stored at −70°C, in 10% glycerol, until used in step 6.

16.5

2. At 2, 4, 8, and 24 hr, remove 5-μl aliquots of Reaction 1, add 5 μl of 2× SDS sample buffer, and freeze at −20°C.

3. At 24 hr mix 5 μl of Reaction 2 (mock digestion) with 5 μl of 2× SDS sample buffer.

4. Mix 5 μl of original fusion protein solution with 5 μl of 2× SDS sample buffer (uncut control).

5. Heat all samples 10 min in a boiling water bath and load onto an SDS-polyacrylamide gel. Evaluate extent of cleavage to determine correct incubation time.

   If only partial cleavage is evident, increase amount of enzyme and/or incubation time. If no cleavage is apparent, proceed to the Support Protocol.

6. Once satisfactory cleavage conditions have been determined, scale up the trial reaction for the remainder of the fusion protein sample, saving a small amount of uncleaved fusion protein for comparison purposes. Monitor the extent of cleavage by SDS-PAGE.

SUPPORT PROTOCOL

DENATURING A FUSION PROTEIN FOR FACTOR Xa CLEAVAGE

Some fusion proteins are resistant to cleavage with factor Xa. This problem can sometimes be alleviated by denaturing the fusion protein, renaturing it, and then incubating it with protease.

Additional Materials *(also see Basic Protocol 1)*

20 mM Tris·Cl (pH 7.4)/6 M guanidine·HCl
20 mM Tris·Cl (pH 8.0)/1 mM $CaCl_2$

1. Dialyze fusion protein for ≤4 hr against ≥10 vol of 20 mM Tris·Cl (pH 7.4)/6 M guanidine·HCl, or add guanidine·HCl to the fusion protein to give a final concentration of 6 M.

2. Dialyze the sample for 4 hr against 100 vol of 20 mM Tris·Cl (pH 8.0)/1 mM $CaCl_2$.

3. Repeat the second dialysis for an additional 4 hr against 100 vol fresh buffer.

   Rapid removal of denaturant sometimes results in precipitation of the protein; in these cases, gradual removal of denaturant by stepwise dialysis against 2-fold dilutions of the guanidine·HCl solution may keep the protein from precipitating.

4. Proceed with step 1 of Basic Protocol 1 for factor Xa cleavage.

ALTERNATE PROTOCOL 1

ENZYMATIC CLEAVAGE OF FUSION PROTEINS WITH THROMBIN

Thrombin is a mammalian serine protease that cleaves in a trypsin-like manner; that is, it cleaves after arginine and lysine residues. However, thrombin displays distinct subsite preferences, with optimum cleavage occurring at sites containing P4-P3-Pro-Arg↓P1′-P2′ (where P4 and P3 are hydrophobic amino acids and P1′ and P2′ are nonacidic amino acids).

Additional Materials *(also see Basic Protocol 1; see* APPENDIX 1 *for items with ✓)*

✓ Thrombin cleavage buffer
Heparin, sodium salt (with ≥140 U/mg activity, Sigma; optional)
Thrombin (human, with ~3000 U/mg activity; Sigma or Boehringer Mannheim)

1. Prepare two pilot cleavage reactions to determine optimal reaction conditions as follows:

   Reaction 1: 20 µl of 1 mg/ml fusion protein solution (in appropriate buffer) and 0.2 µg thrombin.

   Reaction 2: 5 µl of 1 mg/ml fusion protein solution only (mock digestion).

   Incubate at 25°C.

   GST fusion protein that has been eluted from glutathione-agarose in 50 mM Tris·Cl (pH 7.5)/5 mM reduced glutathione can be used after addition of NaCl to 150 mM and $CaCl_2$ to 2.5 mM and adjustment of the protein concentration to 1 mg/ml. Other fusion proteins can be resuspended or dialyzed in thrombin cleavage buffer (without glutathione) for subsequent cleavage.

   Addition of 10 µM heparin to the cleavage reaction is optional; it may increase the rate of some cleavages by 10% to 50%, apparently due to a direct interaction with the enzyme.

2. At 30 min, 1, 2, and 4 hr, remove 5 µl from Reaction 1 and mix with 5 µl of 2× SDS sample buffer. Freeze at −20°C.

3. At the 4-hr time point, add 5 µl of 2× SDS sample buffer to Reaction 2 (mock digestion).

4. Mix 5 µl of original fusion protein solution with 5 µl of 2× SDS sample buffer (untreated control).

5. Boil all samples 10 min and load on an SDS-polyacrylamide gel to analyze sample stability and efficiency of cleavage.

6. Use those conditions determined empirically to be best for cleaving the fusion protein to scale up the cleavage reaction for the desired quantity of protein.

ENZYMATIC CLEAVAGE OF MATRIX-BOUND GST FUSION PROTEINS

ALTERNATE PROTOCOL 2

Additional Materials *(also see Basic Protocol 1)*

GST fusion protein bound to glutathione-agarose beads (*UNIT 16.7*)
1% (v/v) Triton X-100 in PBS
GST wash buffer: 50 mM Tris·Cl (pH 7.5)/150 mM NaCl
GST elution buffer: 50 mM Tris·Cl (pH 8.0)/5 mM reduced glutathione
20- or 50- ml screw-cap tube

1. Wash GST fusion protein bound to glutathione-agarose beads with 20 vol of 1% Triton X-100 in PBS, using a 20- or 50-ml screw-cap tube. Centrifuge 10 sec in a tabletop centrifuge at 500 × *g*, room temperature, to pellet the beads. Carefully remove and discard the supernatant. Resuspend the beads in 20 vol Triton X-100 buffer and repeat wash.

2. After the second centrifugation, carefully remove and discard the supernatant. Resuspend the beads in 20 vol GST wash buffer.

3. Pellet the beads and discard the supernatant. Resuspend the beads in 20 vol thrombin cleavage buffer. Repeat the centrifugation and resuspend the beads in ≤1 ml thrombin cleavage buffer.

4. Remove a small aliquot of resuspended beads and add an equal volume of 2× SDS sample buffer. Store at −20°C until analyzed by SDS-PAGE (step 7).

5. Add thrombin to the remaining bead slurry at a ratio of 1% (w/w) thrombin to the estimated amount of bound fusion protein. Incubate 1 hr at 25°C.

6. Elute the cleaved and released protein by washing the beads with 1 bed volume of GST wash buffer. Centrifuge as in step 1 to pellet beads and collect supernatant. Repeat five times, but keep each wash fraction separate. Remove 20-μl aliquots from each wash fraction for SDS-PAGE.

7. Elute bound GST by repeating step 6 with GST elution buffer instead of GST wash buffer. Remove 20-μl aliquots from each fraction and analyze by SDS-PAGE to determine extent of cleavage. Include the aliquot of beads from step 4 on this gel.

ALTERNATE PROTOCOL 3

ENZYMATIC CLEAVAGE OF FUSION PROTEINS WITH ENTEROKINASE

Enterokinase (also called enteropeptidase) is a mammalian trypsin-like serine protease that displays a high degree of specificity for the sequence $(Asp)_4$-Lys, cleaving on the carboxy-terminal side of the lysine residue of the recognition sequence. Enterokinase is capable of cleaving fusion proteins under a wide range of reaction conditions, with pH ranging from 4.5 to 9.5 and temperatures ranging from 4° to 45°C.

The thioredoxin fusion vector pTRXFUS (*UNIT 16.8*) encodes an enterokinase cleavage site immediately preceding the polylinker cloning region. Proteins produced as Trx fusions using this system can be subsequently released by incubation with enterokinase, leaving their authentic amino-terminal sequence.

Additional Materials *(also see Basic Protocol 1)*

1 mg/ml thioredoxin fusion protein (*UNIT 16.8*) in 50 mM Tris·Cl (pH 8.0)/1 mM $CaCl_2$
Recombinant bovine enterokinase (rEK; EKMAX, Invitrogen) in 50 mM Tris·Cl (pH 8.0)/1 mM $CaCl_2$

NOTE: Many commercial preparations of enterokinase (bovine or porcine), with the exception of the source listed, are extremely impure and tend to be contaminated with, among other things, trypsin and chymotrypsin which can extensively degrade the fusion protein. It is recommended that only commercial enterokinase of the highest quality be used.

1. Perform a pilot experiment to monitor the efficiency of cleavage with various ratios of enterokinase to fusion protein. Prepare five reactions:

   Reactions 1 to 4: 20 μl of 1 mg/ml fusion protein, 0.2 U, 0.5 U, 1 U, and 2 U rEK, and 50 mM Tris·Cl (pH 8.0)/1 mM $CaCl_2$, to a total of 30 μl.
   Reaction 5: 20 μl of 1 mg/ml fusion protein and 10 μl of 50 mM Tris·Cl (pH 8.0)/1 mM $CaCl_2$ (mock digestion).

   Incubate samples ≥ 16 hr at 37°C.

   The fusion protein must be (at least) partially purified prior to digestion with enterokinase because the enzyme is inactive in crude bacterial lysates.

2. Stop the reaction by adding 30 μl of 2× SDS sample buffer to each reaction. Boil 10 min.

For larger-scale applications, the reaction can be stopped by adding p-aminobenzamidine (PABA) to 5 mM. PABA is a competitive inhibitor of most intestinal serine proteases.

3. Load 10 μl of each sample onto an SDS-polyacrylamide gel to analyze the extent of cleavage. Adjust enterokinase concentration and length of incubation accordingly to accomplish complete digestion.

4. Scale up the reaction components linearly to digest a larger amount of fusion protein.

If degradation of the cleaved fusion protein occurs, omit calcium and add 5 mM EDTA to the cleavage reaction to try to eliminate the problem.

CHEMICAL CLEAVAGE OF FUSION PROTEINS USING CYANOGEN BROMIDE

BASIC PROTOCOL 2

Cyanogen bromide (CNBr) has been used to cleave proteins at methionine residues for many years. The reaction is typically carried out at low pH in 70% formic acid, and cleavage occurs at the C-terminal side of methionine residues. Cleavage with CNBr is usually efficient, but side-chain modifications and nonspecific cleavages are common upon prolonged incubation at low pH. These problems, along with the potential for reduction of intramolecular disulfide bonds during treatment with 70% formic acid, can be minimized by replacing the formic acid with 6 M guanidine·HCl/0.2 M HCl.

CAUTION: Cyanogen bromide is extremely toxic. It should only be used in a properly ventilated fume hood. Exercise appropriate caution in its use and disposal.

Materials *(see* APPENDIX 1 *for items with ✓)*

1 mg/ml fusion protein
50 mg/ml cyanogen bromide (CNBr)/70% (v/v) formic acid
70% (v/v) formic acid
✓ 1× SDS sample buffer

1. Perform a pilot experiment to determine minimum incubation time. Lyophilize two 50-μl aliquots of fusion protein solution. Resuspend one aliquot in 50 μl of 50 mg/ml CNBr/70% formic acid. Resuspend the other in 50 μl of 70% formic acid *without* CNBr. Incubate at room temperature.

2. At 0, 8, 24, and 48 hr, remove a 5-μl aliquot and lyophilize.

3. Resuspend all aliquots in 20 μl of 1× SDS sample buffer, boil 10 min, and load onto an SDS-polyacrylamide gel.

4. Based on analysis of the gel, determine the minimum incubation time necessary to completely cleave the protein.

For proteins that are resistant to cleavage with cyanogen bromide or when the fusion protein to be cleaved is insoluble, guanidine·HCl can be added to the reaction at a final concentration of 6 M.

ALTERNATE PROTOCOL 4

CHEMICAL CLEAVAGE OF FUSION PROTEINS USING HYDROXYLAMINE

Hydroxylamine cleaves proteins at Asn-Gly bonds and can be used as a reagent for chemical cleavage of fusion proteins. This cleavage site is less common than that for cyanogen bromide, and therefore the presence of a susceptible bond in the protein of interest is less likely. Protein digestions by this technique are usually incomplete due to the nature of the cleavage mechanism, reducing yield and possibly complicating post-cleavage purification of the desired protein product. However, the technique does have advantages: speed, economy, and the ability to perform digestions under denaturing conditions (e.g., 6 M guanidine·HCl) for otherwise insoluble fusion proteins.

CAUTION: Hydroxylamine is potentially explosive if mishandled. Be sure to follow all precautions indicated by the manufacturer.

Additional Materials *(also see Basic Protocol 2; see* APPENDIX 1 *for items with* ✓*)*

1 mg/ml fusion protein in 10 mM Tris·Cl (pH 8.0)/150 mM NaCl
✓ 2× hydroxylamine cleavage solution
Guanidine·HCl (optional)
✓ 2× SDS sample buffer
Boiling water bath

1. Perform a pilot experiment to determine minimum incubation time. Mix 50 µl of 1 mg/ml fusion protein in 10 mM Tris·Cl (pH 8.0)/150 mM NaCl with 50 µl of 2× hydroxylamine cleavage solution in a 1.5-ml microcentrifuge tube. Incubate at 45°C.

   If the fusion protein is insoluble, guanidine·HCl can be added to the cleavage reaction at a final concentration of 6 M. This may also help in cases where a particular Asn-Gly bond appears to be resistant to cleavage.

2. At 0, 2, 4, 8, 16, and 24 hr, remove 10-µl aliquots from the cleavage reaction and mix with 10 µl of 2× SDS sample buffer. Freeze each tube on dry ice until all time points have been collected.

4. Heat samples 10 min in a boiling water bath. Load all samples onto an SDS-polyacrylamide gel to analyze the extent of cleavage.

5. Determine the minimum incubation time necessary for maximum cleavage.

   If cleavage after 24 hr is still poor, add guanidine·HCl to 6 M final, increase hydroxylamine concentration to 3 M final, or both.

ALTERNATE PROTOCOL 5

CHEMICAL CLEAVAGE OF FUSION PROTEINS BY HYDROLYSIS AT LOW pH

This method exploits the fact that the Asp-Pro bond is labile at low pH. Hydrolysis of this peptide bond occurs at elevated temperatures (37° to 40°C) under acidic conditions (pH 2.5). Reaction conditions are somewhat harsh and may result in denaturation or modification of the protein. The released protein will retain a proline residue at its amino-terminus.

Additional Materials *(also see Basic Protocol 2)*

Fusion protein containing an Asp-Pro bond between the component domains
70% (v/v) formic acid
13% (v/v) acetic acid

0.1 M Tris base
Guanidine·HCl

1. Perform a pilot experiment to determine optimal hydrolysis conditions. Prepare four reaction mixtures:

   Reaction 1: ~20 μg fusion protein in 70% formic acid
   Reaction 2: ~20 μg fusion protein in 70% formic acid/6 M guanidine·HCl
   Reaction 3: ~20 μg fusion protein in 13% acetic acid
   Reaction 4: ~20 μg fusion protein in 13% acetic acid/6 M guanidine·HCl.

   Incubate all samples at 37°C.

2. At 0, 24, 48, and 72 hr, remove a 5-μg aliquot of each reaction mixture and lyophilize to dryness.

3. Resuspend the hydrolyzed protein in 20 μl of 1× SDS sample buffer and neutralize by gradual addition of 0.1 M Tris base until the sample turns from yellow to blue. Analyze samples on a tricine SDS-polyacrylamide gel for extent of digestion.

4. Choose the mildest condition and shortest incubation time that give the desired extent of cleavage. Scale up to larger amounts of fusion protein accordingly.

References: Bornstein and Balian, 1977; Chang, 1985; Gross, 1967; Landon, 1977; LaVallie et al., 1993a; Maines et al., 1988; Nagai and Thøgersen, 1987; Spande et al., 1970.

Contributors: Edward R. LaVallie, Donald B. Smith, and Paul Riggs

Expression and Purification of *lacZ* and *trpE* Fusion Proteins

UNIT 16.6

16.6

BASIC PROTOCOL

Fusion proteins are commonly used as a source of antigen for producing antibodies and in many cases can be useful for biochemical analyses. Two widely used expression systems for *E. coli* are presented—*lacZ* fusions using the pUR series of vectors (UNIT 1.5) and *trpE* fusions using the pATH vectors.

Using these methods, a 400-ml culture of *E. coli* will yield 15 mg to 20 mg of total protein. About two-thirds of the total protein will be in the soluble fraction of the lysate, and one-third in the insoluble fraction (step 9). The induced protein usually represents ~1% of the soluble protein and 10% to 20% of the insoluble protein. Therefore, a typical yield from a 400-ml culture is between 0.5 mg and 1 mg of induced protein.

Materials *(see* APPENDIX 1 *for items with ✓)*

pUR (UNIT 1.5) or pATH (GenBank file name M32985) vectors
E. coli C600, HB101, RR1 or equivalent (Table 1.4.5)
LB plates and medium (UNIT 1.1) containing 50 μg/ml ampicillin
100 mM IPTG (Table 1.4.2; store at −20°C)

M9 plates and medium (*UNIT 1.1*) containing 50 µg/ml ampicillin, 0.5% Casamino acids, 10 µg/ml thiamine, and with/without 20 µg/ml tryptophan (supplemented M9)

2.5 mg/ml indoleacrylic acid (IAA) in 95% ethanol (store at –20°C)

✓ PBS, ice-cold

✓ HEMGN buffer, ice-cold

50 mg/ml lysozyme in 0.25 M Tris·Cl, pH 8.0 (prepare fresh and store at 4°C)

HEMGN buffer/8 M guanidine·HCl (prepare 100 ml and store at 4°C)

HEMGN buffer/1 M guanidine·HCl (prepare 500 ml and store at 4°C)

Sorvall GSA, Sorvall SS-34, and Beckman 60Ti rotors or equivalents

Sonicator with a microtip

Dialysis tubing, MWCO 12,000 to 14,000 (*APPENDIX 3*)

To express lacZ fusion proteins using pUR vectors:

1a. Subclone gene of interest into pUR vector in correct reading frame (*UNIT 3.16*), transform competent *E. coli* cells (*UNIT 1.8*), and select transformants on LB/ampicillin plates.

2a. Inoculate 2 to 5 ml LB/ampicillin medium with single colony containing expression vector. Grow overnight at 37°C with shaking.

3a. Add 1 ml overnight culture to 400 ml LB/ampicillin medium in 2-liter flask. Grow at 37°C with vigorous shaking until OD_{600} reaches 0.5.

4a. Add 1.6 ml of 100 mM IPTG (0.4 mM final). Grow cells 2 hr.

To express trpE fusion proteins using pATH vectors:

1b. Subclone gene of interest into pATH vector in correct reading frame (*UNIT 3.16*), transform competent *E. coli* cells (*UNIT 1.8*), and select transformants on supplemented M9/tryptophan plates.

2b. Inoculate 2 to 5 ml supplemented M9/tryptophan medium with single colony containing expression vector. Grow overnight as in step 2a.

3b. Add 1 ml of overnight culture to 400 ml supplemented M9 medium *without* tryptophan. Grow at 37°C with vigorous shaking until OD_{600} reaches 0.5.

4b. Add 1.6 ml of 2.5 mg/ml IAA (10 µg/ml final). Grow additional 2 hr.

Perform all subsequent steps on ice or at 4°C.

5. Split cell culture into two 200-ml centrifuge bottles. Centrifuge 10 min at 4000 × *g*.

6. Resuspend each pellet in 5 ml PBS by pipetting up and down. Transfer each to a 15-ml conical centrifuge tube. Centrifuge 10 min in clinical centrifuge at 3000 × *g*.

7. Resuspend cells in 2 ml HEMGN buffer with protease inhibitors. Add 20 µl of 50 mg/ml lysozyme. Incubate 15 to 30 min on ice.

8. Disrupt cells by sonicating two times for 15 sec each using a microtip. Place sample on ice between rounds of sonication. Pool lysate into one 50-ml centrifuge tube.

   Hold tube containing sample so tip of sonicator probe is at surface of solution. Adjust output level of sonicator to minimum setting required to achieve vigorous churning of the solution. Alternatively, a cup sonicator may be used. Sonicate two times for 30 sec each if using cup sonicator.

9. Centrifuge cell lysate 15 min at 27,000 × *g*. Pour off supernatant and save (check for presence of fusion protein in step 15). Save pellet, which usually contains almost all of induced protein in an insoluble form. To prepare soluble protein, proceed to next step.

   The insoluble material can be used for purification of fusion protein by gel electrophoresis.

10. Resuspend pellet in 2 ml HEMGN buffer with protease inhibitors.

    The pellet is usually very viscous and can be difficult to resuspend; scrape pellet off the side of tube using pipettor and pipet up and down a few times; resuspension can be aided by further sonication.

11. Add 2 ml HEMGN buffer/8 M guanidine·HCl (4 M final guanidine). Incubate with gentle shaking 30 min at 4°C.

12. Centrifuge 30 min in precooled ultracentrifuge at 87,000 × *g*. Alternatively centrifuge 20 min at 27,000 × *g*, 4°C.

13. Transfer supernatant to dialysis tubing and dialyze in three steps, each ≥3 hr to overnight: first, against 500 ml HEMGN buffer/1 M guanidine·HCl, and then twice against 1 liter HEMGN buffer excluding guanidine.

    During dialysis about half of the protein in the extract will come out of solution as the guanidine is removed, resulting in a large amount of white precipitate.

14. Transfer all material to centrifuge tube and centrifuge 5 min at 12,000 × *g*. Save supernatant (~4 ml), which should be a clear, colorless solution with a protein concentration of ~1 mg/ml. (Fusion protein typically constitutes between 1% and 10% of total protein.) Save insoluble pellet, which usually contains most of fusion protein and can be used for gel purifying the protein.

15. Determine protein concentration of supernatant and pellet from cell lysate (steps 8 and 9) and of final supernatant and pellet after guanidine extraction (step 14) by Bradford method (*UNIT 10.1*). Check results of induction by SDS-PAGE, loading 5 to 10 μg of protein/lane (*UNIT 10.2*).

    It is useful to compare extracts containing the fusion protein with control extracts prepared from uninduced cells or from cells containing the expression vector with no insert. The molecular weight of the β-gal protein is 116 kDa, and the trpE protein is 37 kDa.

Reference: Koerner et al., 1990.

Contributor: Timothy Hoey

UNIT 16.7 Expression and Purification of Glutathione-S-Transferase Fusion Proteins

BASIC PROTOCOL

pGEX vectors can be used for the expression and purification of individual polypeptides (including short peptides) for use as immunogens and as biochemical and biological reagents, and in the construction of cDNA expression libraries. Each pGEX vector contains an open reading frame encoding glutathione (GST), followed by unique restriction endonuclease sites for *Bam*HI, *Sma*I, and *Eco*RI, followed in turn by termination codons in all three frames (Fig. 16.7.1). The cloning sites are present in a different reading frame in each of the three vectors, so the vector in which the foreign polypeptide will be expressed in-frame with GST must be chosen

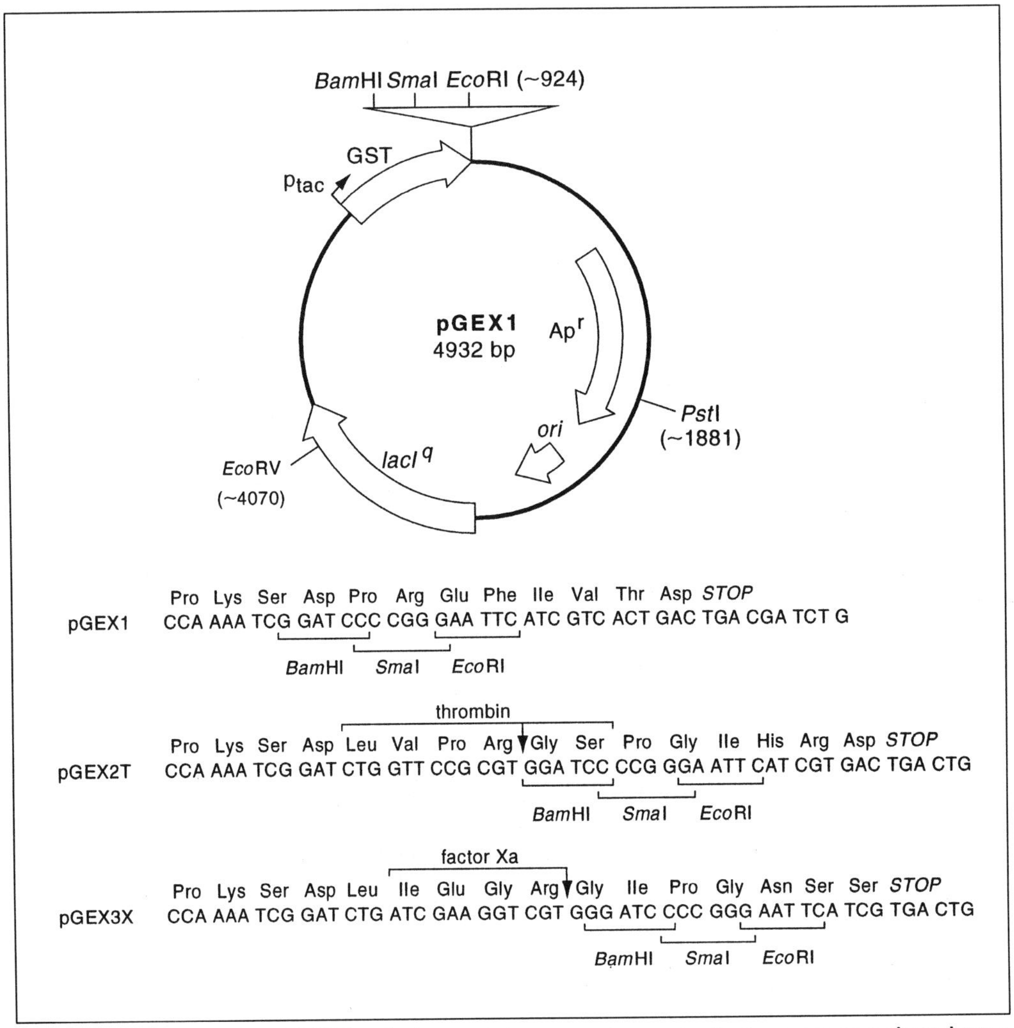

Figure 16.7.1 pGEX1. pGEX1 is a fusion-protein expression vector that expresses a cloned gene as a fusion protein to GST. The *lac* repressor (product of the *lacI* gene) binds to the p_{tac} promoter, repressing the expression of GST fusion protein. Upon induction with IPTG, derepression occurs and GST fusion protein is expressed. The gene of interest can be inserted into the polylinker located at the end of the GST gene. The polylinker sequences are shown below the map of pGEX1, where the restriction endonuclease cloning sites are bracketed. The polylinker of pGEX2T and pGEX3X contains protease cleavage sites so the cloned protein can be released from the GST moiety (or "carrier"). The recognition sequences for thrombin and factor Xa are bracketed above the polylinker sequences, with the actual cleavage site between Arg and Gly. The nucleotide sequence of pGEX1 is available under Genbank accession number M21676.

first. pGEX2T or pGEX3X should be used if the GST carrier is eventually to be removed by site-specific proteolysis (*UNIT 16.5*). Fusion proteins produced using pGEX1 can be cleaved by chemical hydrolysis at low pH (*UNIT 16.5*).

Materials *(see APPENDIX 1 for items with ✓)*

pGEX vector (pGEX1 from Amrad or pGEX2T and pGEX3X from Pharmacia LKB Biotech)
Transformation-competent *Escherichia coli* (*UNIT 1.8*)
LB plates containing 50 µg/ml ampicillin (*UNIT 1.1*)
LB medium containing 10 µg/ml ampicillin (*UNIT 1.1*)
100 mM isopropyl-1-thio-β-D-galactoside (IPTG), filter sterilized
✓ PBS, ice-cold
✓ Glutathione-agarose bead slurry
✓ 2× SDS sample buffer
10% (v/v) Triton X-100
50 mM Tris·Cl (pH 8.0)/5 mM reduced glutathione (freshly prepared; pH 7.5, final)
Glycerol

37°C shaking incubator
Beckman JA-10 and JA-20 rotors (or equivalents)
Probe sonicator (with 2- and 5-mm-diameter probes)

1. Subclone the chosen DNA fragment into the appropriate pGEX vector in the correct reading frame, transform competent *E. coli* cells, and select transformants on LB/ampicillin plates. Include a control of vector ligated to itself in the absence of insert DNA. Incubate plates 12 to 15 hr at 37°C.

2. Pick transformant colonies into 2 ml LB/ampicillin medium and streak out onto a master LB/ampicillin plate. Inoculate a control tube with bacteria transformed with the parental pGEX vector. Incubate the master plate 12 to 15 hr at 37°C. Grow liquid cultures with vigorous agitation in a 37°C shaking incubator until visibly turbid (3 to 5 hr).

3. Induce fusion protein expression by adding 100 mM IPTG to 0.1 mM. Continue incubation another 1 to 2 hr.

4. Transfer liquid cultures to labeled microcentrifuge tubes, microcentrifuge 5 sec at maximum speed, room temperature, and discard supernatants. Resuspend pellets in 300 µl ice-cold PBS. Remove 10 µl to labeled tubes (for use in step 7).

5. Lyse cells using a probe sonicator with a 2-mm-diameter probe. Microcentrifuge 5 min at maximum speed, 4°C, to remove insoluble material. Transfer supernatants to fresh tubes.

   Lysis is complete when the cloudy cell suspension becomes translucent. The frequency and intensity of sonication should be adjusted so no frothing occurs and lysis is complete in 10 sec.

6. Add 50 µl of 50% slurry of glutathione-agarose beads to each supernatant and mix gently ≥2 min at room temperature. Add 1 ml PBS, vortex briefly, microcentrifuge 5 sec at maximum speed, room temperature, to collect beads, and discard supernatants. Repeat the PBS wash twice.

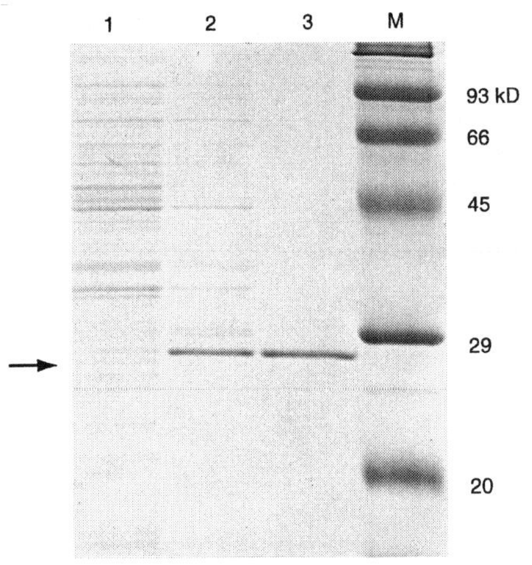

Figure 16.7.2 Stained protein gel showing expression of GST. The total cell lysate of a pGEX1 transformant is shown before IPTG induction (lane 1) and after IPTG induction (lane 2). The GST protein after elution from glutathione-agarose beads is shown at the position indicated by the arrow (lane 3). The lane marked M contains molecular-weight markers for sizes indicated.

7. Add an equal volume of 1× SDS sample buffer to the washed beads, and 30 µl to the 10-µl samples of resuspended whole cells (from step 4). Heat 3 min at 100°C, vortex briefly, and load onto a 10% SDS-polyacrylamide gel. Run the gel for the appropriate time and stain with Coomassie blue solution to visualize the parental GST (made in control cells carrying a pGEX vector) and the fusion protein (Fig. 16.7.2).

8. Inoculate a colony of the pGEX transformant into 100 ml LB/ampicillin medium and grow 12 to 15 hr at 37°C in a shaking incubator.

9. Dilute this culture 1:10 into 1 liter fresh LB/ampicillin medium, split between two 2-liter flasks, and grow 1 hr at 37°C.

10. Add 100 mM IPTG to 0.1 mM (final) and continue incubation an additional 3 to 7 hr.

11. Centrifuge 10 min at 5000 × *g*, room temperature, to collect cells. Discard supernatant and resuspend pellet in 10 to 20 ml ice-cold PBS.

12. Immerse the tube in ice and lyse cells using a probe sonicator with a 5-mm-diameter probe. Adjust the frequency and intensity of sonication so lysis occurs in ~30 sec, without frothing.

13. Add 10% Triton X-100 to 1% and mix. Centrifuge 5 min at 10,000 × *g*, 4°C, to remove insoluble material and intact cells. Alternatively, it may be convenient to microcentrifuge 1.5-ml aliquots for 5 min at top speed, 4°C. Collect supernatants (carefully avoiding the pellets) and pool them.

14. Add supernatant to 1 ml of 50% slurry of glutathione-agarose beads and mix gently ≥2 min at room temperature. Wash by adding 50 ml ice-cold PBS, mixing, and centrifuging 10 sec in a tabletop centrifuge at 500 × *g*, room temperature. Repeat the wash two more times. Resuspend the beads in a small volume (1 to 2 ml) of ice-cold PBS and transfer to a 1.5-ml microcentrifuge tube.

The capacity of glutathione-agarose is ≥8 mg protein/ml swollen beads.

15. Centrifuge 10 sec at 500 × *g*, room temperature, to collect beads; discard supernatant. Elute fusion protein by adding 1 ml of 50 mM Tris·Cl (pH 8.0)/5 mM reduced glutathione. Mix gently 2 min, centrifuge 10 sec at 500 × *g*, and collect supernatant. Repeat elution two to three times and analyze each fraction by SDS-PAGE. Store eluted protein in aliquots containing 10% glycerol at −70°C. Determine the yield of fusion protein by measuring the absorbance at 280 nm. For the GST carrier, $A_{280} = 1$ corresponds to a protein concentration of 0.5 mg/ml.

References: Smith, 1993; Smith and Johnson, 1988.

Contributors: Donald B. Smith and Lynn M. Corcoran

Expression and Purification of Thioredoxin Fusion Proteins

UNIT 16.8

The gene fusion expression system that uses thioredoxin, the product of the *Escherichia coli trxA* gene, as the fusion partner is particularly useful for high-level production of soluble fusion proteins in the *E. coli* cytoplasm; in many cases heterologous proteins produced as thioredoxin fusion proteins are correctly folded and display full biological activity. Although the thioredoxin gene fusion system is routinely used for protein production, high-level production of peptides—i.e., for use as antigens—is also possible because the prominent thioredoxin active-site loop is a very permissive site for the introduction of short amino acid sequences (10 to 30 residues in length). The inherent thermal stability of thioredoxin and its susceptibility to quantitative release from the *E. coli* cytoplasm by osmotic shock can also be exploited as useful tools for thioredoxin fusion protein purification.

CONSTRUCTION AND EXPRESSION OF A THIOREDOXIN FUSION PROTEIN

BASIC PROTOCOL

Materials *(see APPENDIX 1 for items with ✓)*

DNA fragment encoding desired sequence
Thioredoxin expression vectors (Fig. 16.8.1): pTRXFUS or pALtrxA-781 (Genetics Institute or Invitrogen) or hpTRXFUS (Genetics Institute)
E. coli strain GI724 (Genetics Institute or Invitrogen), grown in LB medium and made competent (Table 16.8.1, *UNIT 1.8*)
LB medium (*UNIT 1.1*)
✓ IMC plates containing 100 µg/ml ampicillin
✓ CAA/glycerol/ampicillin 100 medium
✓ IMC medium containing 100 µg/ml ampicillin
✓ 10 mg/ml tryptophan
✓ SDS-PAGE sample buffer

30°C convection incubator
18 × 50–mm culture tubes
Roller drum (New Brunswick Scientific)
250-ml culture flask
70°C water bath

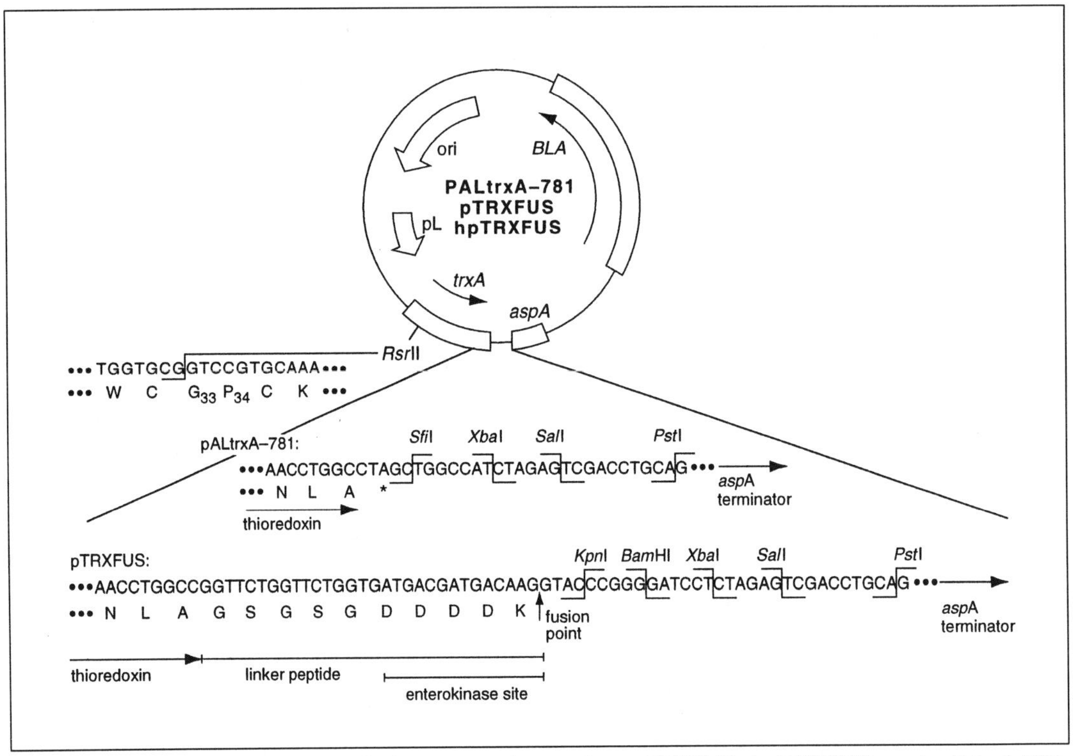

Figure 16.8.1 Thioredoxin gene fusion expression vectors pTRXFUS, hpTRXFUS, and pALtrxA-781. pALtrxA-781 contains a polylinker sequence at the 3′ end of the *trxA* gene. pTRXFUS and hpTRXFUS contain a linker region encoding a peptide that includes the enterokinase cleavage site between the *trxA* gene and the polylinker. The sequence surrounding the active site loop of thioredoxin has a single *Rsr*II site that can be used to insert peptide coding sequence. The asterisk indicates a translational stop codon. Abbreviations: *trxA*, *E. coli* thioredoxin gene; *BLA*, β-lactamase gene; ori, colE1 replication origin; p_L, bacteriophage λ major leftward promoter; *aspA* terminator, *E. coli* aspartate amino-transferase transcription terminator.

1. Use a DNA fragment encoding the desired sequence to construct either an in-frame fusion to the 3′-end of the *trxA* gene in pTRXFUS or hpTRXFUS, or a short peptide insertion into the unique *Rsr*II site of pALtrxA-781.

2. Transform the ligation mixture containing the new thioredoxin fusion plasmid into competent GI724 cells. Plate transformed cells onto IMC plates containing 100 μg/ml ampicillin to select transformants. Incubate plates in a 30°C convection incubator until colonies appear.

Strains GI698, GI723, and GI724 are all healthy prototrophs that can grow under a wide variety of growth conditions, including rich and minimal media and a broad range of growth temperatures (see Table 16.8.1). These strains can be prepared for transformation with p_L-containing vectors by growing them in LB medium at 37°C. LB medium may also be used for these strains during the short period of outgrowth immediately following transformation. This growth period of 30 min to 1 hr is often used to express drug resistance phenotypes before plating out plasmid transformations onto solid medium. Subsequently, however, these strains should be grown only on minimal or tryptophan-free rich media, such as IMC medium containing 100 μg/ml ampicillin (for expression of the fusion protein) or CAA/glycerol/ampicillin 100 medium (for plasmid DNA preparations). Except during transformation, LB medium should never be used with these three strains when they carry pL plasmids because LB contains tryptophan. The p_L promoter is extremely strong and should be maintained in an uninduced state until needed.

Table 16.8.1 *E. coli* Strains for Production of Thioredoxin Fusion Proteins at Varying Temperatures

Strain	Desired production temperature (°C)	Pre-induction growth temperature (°C)	Induction period (hr)
GI698	15	25	20
GI698	20	25	18
GI698	25	25	10
GI724	30	30	6
GI724	37	30	4
GI723	37	37	5

3. Grow candidate colonies in 5 ml CAA/glycerol/ampicillin 100 medium overnight at 30°C. Prepare minipreps of plasmid DNA and check for correct gene insertion into pTRXFUS by restriction mapping.

4. Sequence plasmid DNA of candidate clones to verify the junction region between thioredoxin and the gene or sequence of interest.

5. Streak out frozen stock culture of GI724 containing thioredoxin expression plasmid to single colonies on IMC plates containing 100 µg/ml ampicillin. Grow 20 hr at 30°C.

6. Pick a single fresh, well-isolated colony from the plate and use it to inoculate 5 ml IMC medium containing 100 mg/ml ampicillin in an 18 × 150–mm culture tube. Incubate overnight at 30°C on a roller drum.

7. Add 0.5 ml overnight culture to 50 ml fresh IMC medium containing 100 µg/ml ampicillin in a 250-ml culture flask (1:100 dilution). Grow at 30°C with vigorous aeration until absorbance at 550 nm reaches 0.4 to 0.6 OD/ml (~3.5 hr).

8. Remove a 1-ml aliquot of the culture (uninduced cells). Measure the absorbance at 550 nm and harvest the cells when they reach a density of 0.4 to 0.6 OD_{550}/ml by microcentrifuging 1 min at maximum speed, room temperature. Carefully remove all the spent medium with a pipet and store the cell pellet at –80°C.

9. Induce p_L by adding 0.5 ml of 10 mg/ml tryptophan (100 µg/ml final) to remaining cells immediately.

10. Incubate 4 hr at 37°C. At hourly intervals during this incubation, remove 1-ml aliquots of the culture and harvest cells as in step 8.

11. Harvest the remaining cells from the culture 4 hr post-induction by centrifuging 10 min at 3000 rpm (e.g., in a Beckman J6 rotor), 4°C. Store the cell pellet at –80°C.

12. Resuspend the pellets from the induction intervals (steps 8 and 10) in 200 µl of SDS-PAGE sample buffer/OD_{550} cells. Heat 5 min at 70°C to completely lyse the cells and denature the proteins.

13. Run the equivalent of 0.15 OD_{550} cells per lane (30 µl) on an SDS-polyacrylamide gel. Stain the gel 1 hr with Coomassie brilliant blue. Destain the gel and check for expression. Figure 16.8.2 shows typical results.

Most thioredoxin fusion proteins are produced at levels that vary from 5% to 20% of the total cell protein.

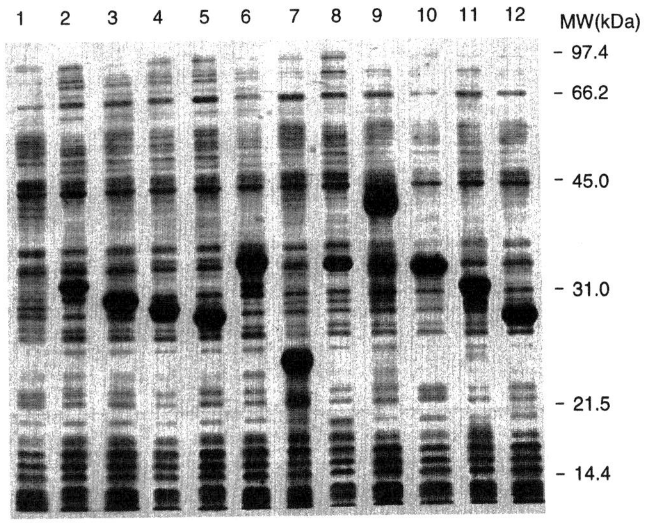

Figure 16.8.2 Expression of thioredoxin gene fusions. The gel shows proteins found in the soluble fractions derived from *E. coli* cells expressing eleven different thioredoxin gene fusions. Lane 1, host *E. coli* strain GI724 (negative control, 37°C); lane 2, murine interleukin-2 (IL-2; 15°C); lane 3, human IL-3 (15°C); lane 4, murine IL-4 (15°C); lane 5, murine IL-5 (15°C); lane 6, human IL-6 (25°C); lane 7, human MIP-1a (37°C); lane 8, human IL-11 (37°C); lane 9, human macrophage colony-stimulating factor (M-CSF; 37°C); lane 10, murine leukemia inhibitory factor (LIF; 25°C); lane 11, murine steel factor (SF; 37°C); and lane 12, human bone morphogenetic protein-2 (BMP-2; 25°C). Temperatures in parentheses are the production temperature chosen for expressing each fusion. This is a 10% SDS-polyacrylamide gel, stained with Coomassie brilliant blue.

SUPPORT PROTOCOL 1

E. COLI LYSIS USING A FRENCH PRESSURE CELL

A small 3.5-ml French pressure cell can be used as a convenient way to lyse *E. coli* cells. The whole-cell lysate can be fractionated into soluble and insoluble fractions by microcentrifugation. Other lysis procedures may be used—for example, sonication or treatment with lysozyme-EDTA (*UNIT 4.4*).

Additional Materials *(also see Basic Protocol; see APPENDIX 1 for items with ✓)*

Cell pellet from 4-hr post-induction culture (Basic Protocol)

✓ 20 mM Tris·Cl, pH 8.0, 4°C

Lysis buffer: 20 mM Tris·Cl (pH 8.0) with protease inhibitors (optional)—0.5 mM phenylmethylsulfonyl fluoride (PMSF), 1 mM *p*-aminobenzamidine (PABA), and 5 mM EDTA

French press and 3.5-ml mini-cell (Fig. 16.8.3; SLM Instruments), 4°C

1. Resuspend cell pellet from 4-hr post-induction culture in 20 mM Tri·Cl, pH 8.0, to a concentration of 5 OD_{550}/ml.

2. Place 1.5 ml resuspended cell pellet in the French pressure cell. Hold the cell upside down with the base removed, the piston fully extended downwards, and the outlet valve handle that holds the nylon ball seal in the open position (loose).

   Before filling the pressure cell, check that the nylon ball, which seals the outlet port and sits on the end of the outlet valve handle, is not deformed.

3. Bring the liquid in the pressure cell to the level of the outlet port by raising the piston slowly to expel excess air from the cell. With the outlet valve open and

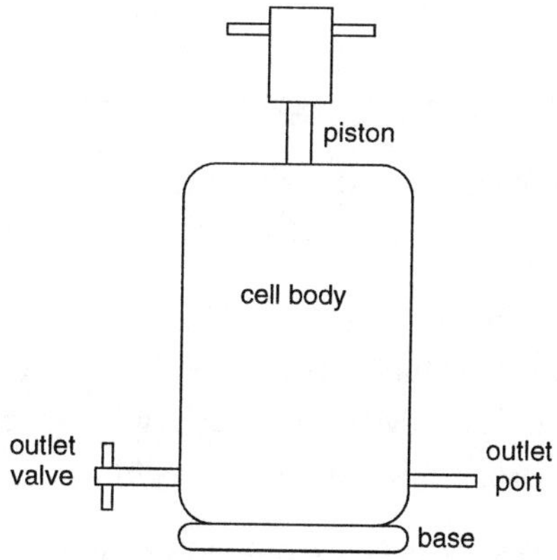

Figure 16.8.3 French pressure cell, equipped with 3.5-ml mini-cell.

at the same time maintaining the piston in position, install the pressure cell base. Gently close the outlet valve.

CAUTION: *Do not over-tighten the valve as this will deform the nylon ball and may irreparably damage its seat on the pressure cell body.*

4. Turn the sealed cell right side up and place it in the hydraulic press.

5. Turn the pressure regulator on the press fully counter-clockwise to reset it to zero pressure. Set the ratio selector to medium. Turn on the press.

   CAUTION: *The larger (50-ml) pressure cell is usually used with the selector set on high. The small (3.5-ml) cell is only used on medium ratio.*

6. Slowly turn the pressure regulator clockwise until the press just begins to move. Allow the press to compress the piston. It will stop moving after a few seconds.

7. Position a collection tube under the pressure cell outlet. Slowly increase the pressure in the cell by turning the pressure regulator clockwise. Monitor the reading on the gauge and increase the pressure to 1000 on the dial, corresponding to an internal cell pressure of 20,000 lb/in^2.

8. While continuously monitoring the gauge, very slowly open the outlet valve until lysate begins to trickle from the outlet.

   At 20,000 lb/in^2 and 5 OD_{550}/ml, cell lysis will be complete after one passage through the press. Lower pressures and/or higher cell densities may require a second passage.

9. Remove a 100-µl aliquot of the lysate and freeze at –80°C (whole-cell lysate). Fractionate the remainder of the lysate by microcentrifuging 10 min at maximum speed, 4°C.

10. Remove a 100-µl aliquot of the supernatant and freeze at –80°C (soluble fraction). Discard the remainder of the supernatant.

11. Resuspend the pellet in an equivalent volume of lysis buffer. Remove a 100-µl aliquot and freeze at –80°C (insoluble fraction).

12. Lyophilize the 100-µl aliquots to dryness in a Speedvac evaporator. Solubilize in 100 µl SDS-PAGE sample buffer. Analyze 30-µl samples by SDS-PAGE to

identify the fraction(s) that contains the protein of interest. Scale up the procedure for large-scale purifications.

SUPPORT PROTOCOL 2

OSMOTIC RELEASE OF THIOREDOXIN FUSION PROTEINS

Thioredoxin and some thioredoxin fusion proteins can be released with good yield from the *E. coli* cytoplasm by a simple osmotic shock procedure.

Additional Materials *(also see Basic Protocol)*

Cell pellet from 4-hr post-induction cultures (Basic Protocol)
20 mM Tris·Cl (pH 8.0)/2.5 mM EDTA/20% (w/v) sucrose, ice-cold
20 mM Tris·Cl (pH 8.0)/2.5 mM EDTA, ice-cold

1. Resuspend cell pellet from 4-hr post-induction cultures at a concentration of 5 OD_{550}/ml in ice-cold 20 mM Tris·Cl (pH 8.0)/2.5 mM EDTA/20% sucrose. Incubate 10 min on ice.

2. Microcentrifuge 30 sec at maximum speed, 4°C, to pellet the cells. Discard the supernatant and gently resuspend the cells in an equivalent volume of ice-cold 20 mM Tris·Cl (pH 8.0)/2.5 mM EDTA. Incubate 10 min on ice and mix occasionally by inverting the tube.

3. Microcentrifuge 30 sec at maximum speed, 4°C. Save the supernatant (osmotic shockate). Resuspend the cell pellet in an equivalent volume 20 mM Tris·Cl (pH 8.0)/2.5 mM EDTA (retentate).

4. Lyophilize 100-µl aliquots of osmotic shockate and retentate to dryness in a Speedvac evaporator.

5. Solubilize each in 100 µl SDS-PAGE sample buffer. Analyze 30-µl aliquots by SDS-PAGE.

   The osmotic shock procedure provides a substantial purification step for some thioredoxin fusion proteins.

SUPPORT PROTOCOL 3

PURIFICATION OF THIOREDOXIN FUSION PROTEINS BY HEAT TREATMENT

Wild-type thioredoxin is resistant to prolonged incubations at 80°C. A subset of thioredoxin fusion proteins also exhibit corresponding thermal stability.

Additional Materials *(also see Basic Protocol)*

Cell pellet from 4-hr post-induction cultures (Basic Protocol)
20 mM Tris·Cl (pH 8.0)/2.5 mM EDTA

80°C water bath
10-ml glass-walled tube

1. Resuspend cell pellet from 4-hr post-induction cultures at a concentration of 100 OD_{550}/ml in 20 mM Tris·Cl (pH 8.0)/2.5 mM EDTA.

   It is important to start off with a high protein concentration in the lysate to ensure efficient precipitation of denatured proteins.

2. Lyse the cells at 20,000 lb/in^2 in a French pressure cell as described (Support Protocol, steps 2 to 8). Collect whole-cell lysate in a 10-ml glass-walled tube.

3. Incubate whole-cell lysate 10 min at 80°C. Remove 100-µl aliquots after 30 sec, 1 min, 2 min and 5 min and plunge immediately into ice. At 10 min plunge the remaining heated lysate into ice.

4. Microcentrifuge the aliquots 10 min at maximum speed, 4°C to pellet heat-denatured, precipitated proteins.

5. Remove 2-µl aliquots of the supernatants and add 28 µl SDS-PAGE sample buffer. Analyze the samples by SDS-PAGE to determine the heat stability of the fusion protein and the minimum time of heat treatment required to obtain a good purification.

Reference: LaVallie et al., 1993.

Contributors: John McCoy and Edward R. LaVallie

Overview of the Baculovirus Expression System

UNIT 16.9

Baculoviruses have emerged as a popular system for overproducing recombinant proteins in eukaryotic cells. Several factors have contributed to this popularity. First, unlike bacterial expression systems, the baculovirus-based system is a eukaryotic expression system and thus uses many of the protein modification, processing, and transport systems present in higher eukaryotic cells. In addition, the baculovirus expression system uses a helper-independent virus that can be propagated to high titers in insect cells adapted for growth in suspension cultures, making it possible to obtain large amounts of recombinant protein with relative ease. The majority of this overproduced protein remains soluble in insect cells, in contrast to the insoluble proteins often obtained from bacteria. Furthermore, the viral genome is large (130 kbp) and thus can accommodate large segments of foreign DNA. Finally, baculoviruses are noninfectious to vertebrates, and their promoters have been shown to be inactive in mammalian cells, which gives them a possible advantage over other systems when expressing oncogenes or potentially toxic proteins.

Currently, the most widely used baculovirus expression system utilizes a lytic virus known as *Autographa californica* nuclear polyhedrosis virus (AcMNPV; hereafter called baculovirus). This virus is the prototype of the family *Baculoviridae*. It is a large, enveloped, double-stranded DNA virus that infects arthropods. The baculovirus expression system takes advantage of some unique features of the viral life cycle (Fig. 16.9.1). Two types of viral progeny are produced during the life cycle of the virus: extracellular virus particles (nonoccluded viruses) during the late phase and polyhedra-derived virus particles (occluded viruses) during the very late phase of infection. The polyhedrin protein of occluded viruses serves to sequester and thereby protect hundreds of virus particles from proteolytic inactivation during host cell lysis.

BACULOVIRUS EXPRESSION SYSTEM

The baculovirus expression system takes advantage of several facts about polyhedrin protein: (1) that it is expressed at very high levels in infected cells, constituting more than half of the total cellular protein late in the infectious cycle; (2) that it is nonessential for infection or replication of the virus, meaning that the recombinant virus does not require any helper function; and (3) that viruses lacking the polyhedrin gene have a plaque morphology that is distinct from that of viruses containing the

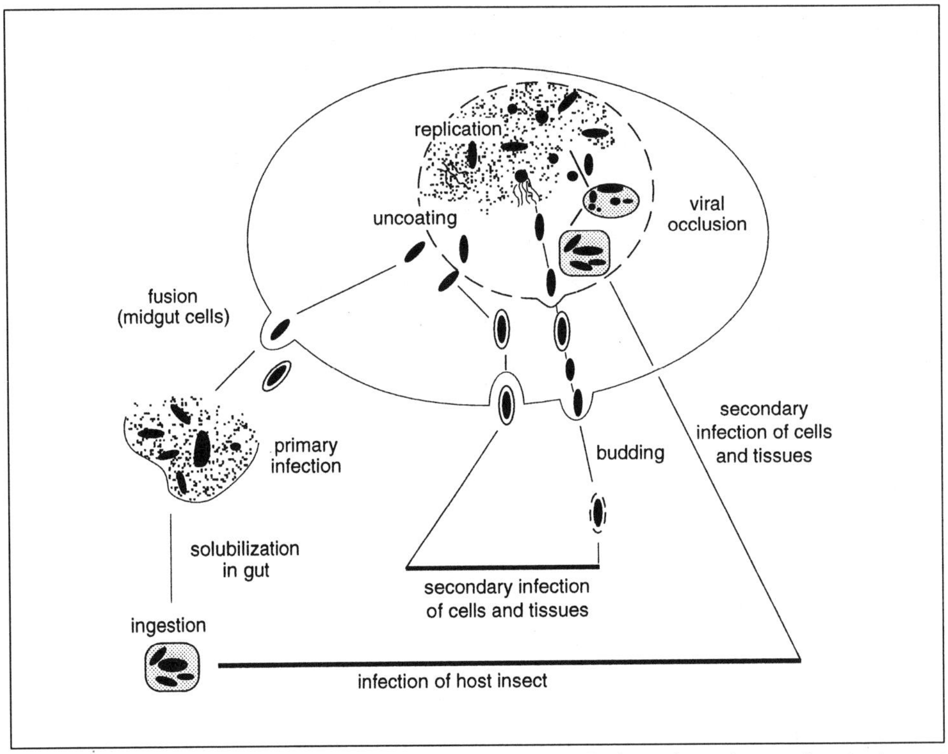

Figure 16.9.1 Baculovirus life cycle. Viruses enter cells by adsorptive endocytosis and move to the nucleus where their DNA is released. Both DNA replication and viral assembly take place in the nuclei of infected cells to generate two types of viral progeny. These include extracellular (nonoccluded) virus particles and polyhedra-derived (occluded) virus particles. Extracellular virus is released from the cell by budding, starting at ~12 hr postinfection and ending ~36 hr postinfection. Polyhedra-derived virus, on the other hand, appears later (~18 hr postinfection) and accumulates in the nuclei of infected cells ≤72 hr postinfection or until cellular lysis. Polyhedra-derived virus is embedded in proteinaceous viral occlusions, the major protein component of which is the viral polyhedrin protein. Secondary infection of cells and tissues occurs by two pathways. In the first, the extracellular virus, once budded from the site of primary infection, is free to infect neighboring cells by the pathway just described. Alternatively, polyhedra-derived virus is released from occlusion bodies after an infected food source is ingested by a new host. Reproduced from Summers and Smith (1987) with permission from the Texas Agricultural Experiment Station.

gene. Recombinant baculoviruses are generated by replacing the polyhedrin gene with a foreign gene through homologous recombination. In this system, the distinctive plaque morphology provides a simple visual screen for identifying the recombinants.

To produce a recombinant virus that expresses the gene of interest, the gene is first cloned into a transfer vector (described below). Most baculovirus transfer vectors contain the polyhedrin promoter followed by one or more restriction enzyme recognition sites for foreign gene insertion. Once cloned into the expression vector, the gene is flanked both 5′ and 3′ by viral-specific sequences. Next, the recombinant vector is transfected along with wild-type viral DNA into insect cells. In a homologous recombination event, the foreign gene is inserted into the viral genome and the polyhedrin gene is excised. Recombinant viruses lack the polyhedrin gene and in its place contain the inserted gene, whose expression is under the control of the polyhedrin promoter.

Homologous recombination between circular wild-type DNA and the recombinant plasmid DNA occurs at a low frequency (typically 0.2% to 5%). However, linearization of wild-type baculovirus DNA before cotransfection with plasmid DNA increases the proportion of recombinant virus to ~30%. If the DNA is linearized such that an essential portion of the 1629 open reading frame (ORF) downstream from the polyhedrin gene is deleted, 85% to 99% of the viruses obtained by cotransfection with a plasmid vector that complements the deletion express the heterologous gene. Two companies (Pharmingen and Clontech) market linear AcMNPV DNA containing such a deletion.

Once a virus stock is obtained after cotransfection, it is necessary to purify recombinant virus by plaque assay so the recombinant virus can be identified. Limiting dilution can also be used. One of the beauties of this expression system is a visual screen allowing recombinant viruses to be distinguished. Occlusion bodies up to 15 μm in size are present in nuclei of very late stage infected cells; they are highly refractile with a bright, shiny appearance that is readily visualized under the light microscope. Cells infected with recombinant viruses lack occlusion bodies. Thus, when the virus is plaqued onto *Sf*9 (*Spodoptera frugiperda*) cells, plaques can be screened for the presence (indicative of wild-type virus) or absence (indicative of recombinant virus) of occlusion bodies. Recombinant viruses can also be identified by DNA hybridization and polymerase chain reaction (PCR) amplification.

POST-TRANSLATIONAL MODIFICATION OF PROTEINS IN INSECT CELLS

Because baculoviruses infect invertebrate cells, it is possible that the processing of proteins produced by them is different from the processing of proteins produced by vertebrate cells. Although this seems to be the case for some post-translational modifications, it is not the case for others. In addition to myristylation, palmitylation has been shown to take place in insect cells. However, it has not been determined whether all or merely a subfraction of the total recombinant protein contains these modifications. Cleavage of signal sequences, removal of hormonal prosequences, and polyprotein cleavages have also been reported, although cleavage varies in its efficiency. Internal proteolytic cleavages at arginine- or lysine-rich sequences have been reported to be highly inefficient, and alpha-amidation, although it does not occur in cell culture, has been reported in larvae and pupae. In most of these cases a cell- or species-specific protease may be necessary for cleavage. Protein targeting seems conserved between insect and vertebrate cells. Thus, proteins can be secreted and localized faithfully to either the nucleus, cytoplasm, or plasma membrane. Although much remains to be learned about the nature of protein glycosylation in insect cells, proteins that are N-glycosylated in vertebrate cells will also generally be glycosylated in insect cells. However, with few exceptions the N-linked oligosaccharides in insect cell-derived glycoproteins are only high-mannose type and are not processed to complex-type oligosaccharides containing fucose, galactose, and sialic acid. O-linked glycosylations have been even less well characterized in *Sf*9 cells, but have been shown to occur.

STEPS FOR OVERPRODUCING PROTEINS USING THE BACULOVIRUS EXPRESSION SYSTEM

Overproduction of recombinant proteins using the baculovirus expression system is presented in detail in *UNITS 16.10-16.11* and outlined in Figure 16.9.2.

REAGENTS, SOLUTIONS, AND EQUIPMENT FOR THE BACULOVIRUS EXPRESSION SYSTEM

Commonly used reagents and solutions are summarized below.

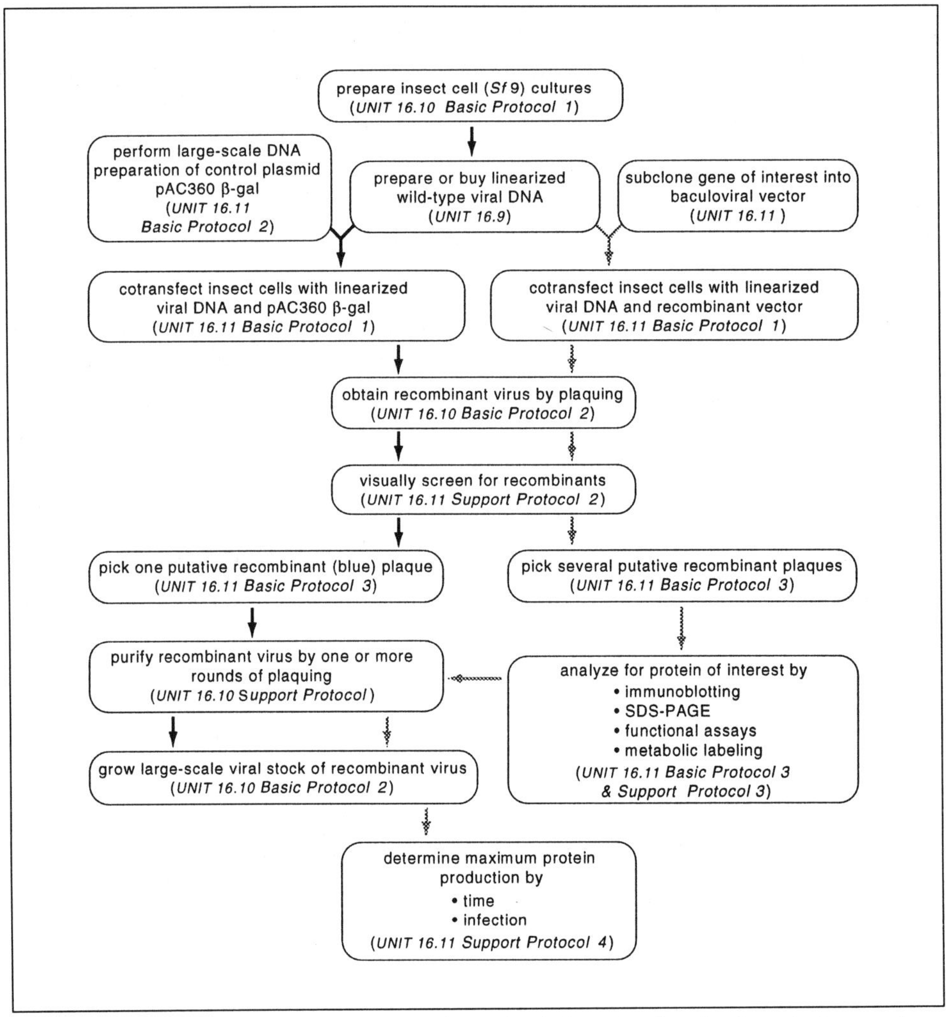

Figure 16.9.2 Flow chart for expression of proteins in insect cells using baculovirus vectors. The black arrows indicate the protocols used to optimize conditions and the light arrows indicate the protocols used to generate recombinant baculoviruses expressing the protein of interest.

1. *Spodoptera frugiperda* clone 9 (*Sf*9) cells from the American Type Culture Collection (#CRL 1711), Pharmingen, or Invitrogen; *Sf*21 cells from Clontech or Invitrogen. These cells are derived from fall armyworm ovaries. *Sf*9 is a clonal line derived from *Sf*21.

2. Graces insect cell culture medium, supplemented with lactalbumin hydrolysate and yeastolate and unsupplemented, 1× and 2× strength in powdered or liquid form, from Life Technologies. For instructions on media preparation from individual components, see O'Reilly et al. (1992).

3. Serum-free insect cell culture medium (Sf-900 II from Life Technologies or Ex-Cell 401 from JRH Biosciences).

4. Incubator at 27 ± 1°C; CO_2 is not required. The Biological Oxygen Demand (B.O.D.) low-temperature incubator (VWR Scientific) or the larger Isotemp (Fisher) are good examples.

5. Magnetic spinner flasks (Bellco; available in a variety of sizes).

6. Stir plate for multiple spinners (Bellco).

7. Fetal bovine serum (FBS) is available from many vendors. Obtain and test different lots of serum from a number of suppliers. The lot that promotes the best growth rate and cell viability should be purchased in bulk.

8. Seakem ME agarose (FMC Bioproducts).

9. 60-mm tissue culture plates (Falcon or Corning).

10. Antibiotics (optional): gentamicin and amphotericin B (Fungizone from Flow Laboratories).

11. Microscope, either an inverted light microscope or a dissecting microscope.

12. Appropriate cloning vectors, a manual of methods, and a wild-type baculovirus are available upon request from Dr. Max D. Summers, Department of Entomology, Texas Agricultural Experiment Station, Texas A & M University, College Station, Texas 77843 (409-845-9730). It is necessary to sign a licensing agreement before the material will be sent. Commercial kits are also available from Invitrogen, Pharmingen, and Clontech (see *APPENDIX 4*).

Reference: O'Reilly et al., 1992.

Contributors: Cheryl Isaac Murphy and Helen Piwnica-Worms

Preparation of Insect Cell Cultures and Baculovirus Stocks

UNIT 16.10

NOTE: All reagents and equipment coming into contact with live cells must be sterile.

MAINTENANCE AND CULTURE OF INSECT CELLS

BASIC PROTOCOL 1

Spodoptera frugiperda (*Sf*9) cells can be maintained and subcultured in both monolayer and suspension (spinner) cultures, in either serum-containing or serum-free medium.

Materials *(see APPENDIX 1 for items with ✓)*

✓ Complete insect cell medium with 10% fetal bovine serum (FBS)
Spodoptera frugiperda (*Sf*9) cells derived from fall armyworm ovaries (*UNIT 16.9*)
70% ethanol
0.4% (w/v) trypan blue (Life Technologies)
Serum-free insect cell culture medium (Sf-900 II from Life Technologies or ExCell 401 from JRH Biosciences)
✓ Complete insect cell medium with 10% FBS, with and without 20% (v/v) DMSO

25-cm^2 flasks
27°C incubator
Hemacytometer for tissue culture cells
Spinner culture flasks (for suspension cultures; Bellco)
Stir plate for multiple spinner flasks (Bellco)
Beckman GPR with GH-3.7 horizontal rotor (or equivalent)
Screw-top cryostat freezing vials

1. Place 4 ml complete insect cell medium/10% FBS in a 25-cm^2 flask. Thaw a frozen ampule of *Sf*9 cells rapidly in a 37°C water bath by shaking it back and forth vigorously by hand. When ampule contents are almost completely thawed, immerse ampule in 70% ethanol to sterilize the outside.

2. Break the neck of the ampule and transfer contents to 25-cm^2 flask. Rock flask gently by hand to distribute the cells evenly and incubate 2 to 3 hr at 27°C until cells have attached.

3. Remove old medium and replace with 5 ml fresh complete medium/10% FBS. Continue incubating, feeding culture every 3 days (by removing old medium and replacing with fresh), until cells reach confluency.

4. Prepare a new flask by placing 4 ml complete medium/10% FBS in a 25-cm^2 flask (or use a larger flask with more medium if preparing a larger culture of cells).

5. Remove the medium from a confluent flask of *Sf*9 cells. Resuspend the cells in fresh complete medium by gently spritzing them with medium/10% FBS from a pipet or by gently rapping the flask with the palm of your hand.

6. Count the cells using a hemacytometer designed for tissue culture cells. Seed $1–2 \times 10^6$ cells into the 25-cm^2 flask and rock it to evenly distribute the cells. Incubate at 27°C, feeding the culture every 3 days with complete medium/10% FBS, until the cells in the flask reach confluency (form a packed monolayer).

7. Remove medium and resuspend cells from confluent monolayer culture.

8. Count the cells using a hemacytometer. Seed the cells into a spinner culture flask at $\sim 4–5 \times 10^5$ cells/ml. Incubate at 27°C with constant stirring at 60 to 80 rpm. Leave the sidearm caps slightly open to ensure adequate aeration.

   The optional addition of Pluronic F-68 (1% v/v) to the culture medium helps to maintain growth of cells in suspension.

9. Count the cells every 2 to 3 days. Subculture when cells reach a concentration of $2–2.5 \times 10^6$ cells/ml by removing the appropriate number of cells to a new flask containing fresh complete medium/10% FBS for a final density of $4–5 \times 10^5$ cells/ml. Alternatively, pour out the appropriate volume of cell suspension and replace it with fresh medium.

10. Determine cell viability by adding 0.1 ml of 0.4% trypan blue to 1 ml log-phase cells. Examine the cells under a microscope at low power. Count the number of cells that take up trypan blue (dead cells) and count the total number of cells.

    To maintain a sufficient transfer of oxygen to the cells in suspension, a minimum ratio of surface area to volume of culture must be maintained.

11. Subculture the cells into medium composed of one part complete medium/10% FBS and one part serum-free medium (either Sf-900 II or ExCell 401). Allow cells to grow to confluency (monolayer cultures) or to a density of 2 to 3×10^6 cells/ml (suspension cultures).

12. Subculture the cells into a 1:4 mix of FBS-containing complete and serum-free media and allow cells to grow.

13. Repeat the subculture and growth procedure using a 1:8 to 1:10 mixture of FBS-containing complete and serum-free media.

14. Subculture the cells in serum-free medium.

15. Count cells to be frozen from an exponentially growing culture. Centrifuge cells 10 min at 1000 × *g*, room temperature, and discard supernatant.

16. Resuspend cell pellet to 1–2 × 10^7 cells/ml in complete medium/10% FBS. Add an equal volume of complete medium/10% FBS/20% DMSO and place cells on ice. Dispense 1-ml aliquots of cells into screw-top cryostat freezing vials and incubate 2 hr at –20°C, then overnight at –70°C.

    Use serum-free medium (with and without DMSO) if the cells have been adapted to serum-free medium.

17. Transfer frozen cells to a liquid nitrogen freezer for long-term storage.

PREPARATION OF BACULOVIRUS STOCKS

BASIC PROTOCOL 2

Materials *(see* APPENDIX 1 *for items with ✓)*

*Sf*9 cells in culture (Basic Protocol 1)
✓ Complete insect cell medium with and without 10% FBS
Autographa californica nuclear polyhedrosis virus (AcMNPV), wild-type and titered as pfu/ml [available from M.D. Summers (see *UNIT 16.9*), Invitrogen, or Clontech]
Recombinant virus (see *UNIT 16.11*)

150-cm² flasks
27°C incubator
Beckman GPR with GH-3.7 horizontal rotor (or equivalent), 4°C *or* room temperature
100-ml and 1-liter (optional) spinner culture flasks (Bellco)

From monolayer cultures:

1a. Seed two 150-cm² flasks with 1.8 × 10^7 *Sf*9 cells/flask in complete medium/10% FBS. Allow cells to attach for 3 hr at 27°C, then remove the complete medium. Add 5 ml serum-free complete medium supplemented with wild-type or recombinant virus at a multiplicity of infection (MOI) of 0.1. Incubate 1 hr at 27°C.

    The volume of viral inoculum needed to infect a given number of cells equals MOI (pfu/cell) × [number of cells/titer of viral stock (pfu/ml)].

2a. Remove the viral inoculum and replace it with 20 ml complete medium/10% FBS. Incubate cells at 27°C. Examine the cells daily under a light microscope for signs of occlusion bodies (for wild-type virus infection) or cytopathic effects (for recombinant virus infection).

    If the cells have been growing in serum-free medium, use serum-free insect cell culture medium instead of complete medium.

3a. Harvest the virus when the majority of cells contain occlusion bodies or show cytopathic effects (usually 4 to 5 days postinfection) by removing the infected culture medium to sterile tubes. Centrifuge 10 min at 1000 × *g*, 4°C, and transfer supernatant to new, sterile tubes. Dispense 1-ml aliquots into several screw-top cryostat freezing vials and place in a liquid nitrogen freezer for long-term storage. Store the remaining stock at 4°C in the dark.

    This should give ~40 ml of a virus stock at 10^8 to 10^9 pfu/ml.

    If the virus to be amplified is a single plaque isolate, suspend the agarose plug in 1 ml of serum-free complete medium. Infect Sf9 cells as in step 1a using 2 × 10^6 cells *in a 25-cm² flask and incubate 1 hr at 27°C. Add 3 ml of complete medium/10% FBS and incubate 4 days at 27°C. Harvest the virus as in step 3a. Determine the titer (see Support Protocol) and amplify the virus by repeating steps 1a to 3a to obtain a larger stock.*

From suspension cultures:

1b. Grow *Sf*9 cells in a 100-ml spinner culture flask with 50 ml complete medium/10% FBS to ~2×10^6 cells/ml with constant stirring. Centrifuge cells 10 min at 1000 × *g*, room temperature, and discard supernatant. Resuspend cell pellet in 10 to 20 ml serum-free complete medium supplemented with wild-type or recombinant virus at a multiplicity of infection (MOI) of 0.1. Incubate 1 hr at room temperature.

2b. Bring volume to 100 ml with complete medium/10% FBS and place the cells in a 100-ml spinner culture flask. Incubate 3 to 4 days at 27°C with constant stirring.

Virus stocks can also be prepared in suspension by growing Sf9 cells in serum-free insect cell culture medium in either a 100-ml or 1-liter spinner flask (filled to only 400 ml) until the cell density is between 1 and 2×10^6 cells/ml. Virus should be added at an MOI of 0.1 directly to the suspension culture, the cells incubated 4 to 5 days, and the virus harvested as in step 3b.

3b. Harvest the infected cells when the majority contain occlusion bodies or show cytopathic effects and store as described in step 3a.

SUPPORT PROTOCOL

TITERING BACULOVIRUS STOCKS USING PLAQUE ASSAYS

It is important to know the titer of a stock, expressed in plaque-forming units (pfu)/ml, when preparing new viral stocks or when carrying out infections for protein production and protein analyses.

Additional Materials *(also see Basic Protocol 2; see* APPENDIX 1 *for items with ✓)*

✓ Agarose overlay (prepare 30 min before use in step 5)
✓ Trypan blue overlay (optional)
60-mm tissue culture plates

1. Dilute a culture of exponentially growing *Sf*9 cells to ~5×10^5 cells/ml in complete medium/10% FBS. Seed cells into 60-mm tissue culture plates at two different densities—2×10^6 and 1.5×10^6 cells/plate—several hours before plaquing. Set up duplicate plates for each dilution of viral stock. Incubate at 27°C.

   The density of the cell monolayer is critical when plaquing virus. If the cells are too dense, plaques will not break through the monolayer to form a visible hole, and if they are too sparse, it will be difficult to distinguish plaques from areas where the cells have not filled in.

2. Make 5-ml serial dilutions of viral stocks in serum-free complete medium as follows:

   Pure viral stock—10^{-6}, 10^{-7}, and 10^{-8} dilutions
   Transfection supernatant—10^{-4}, 10^{-5}, and 10^{-6} dilutions (if using circular viral wild-type DNA) or 10^{-1} and 10^{-2} dilutions (if using linear wild-type viral DNA; *UNIT 16.11*)
   Single plaque—10^{-1}, 10^{-2}, and 10^{-3} dilutions (*UNIT 16.11*).

3. Carefully remove medium from cells with a sterile Pasteur pipet. Add 1 ml of each viral dilution to duplicate plates and incubate 1 hr at room temperature or 27°C, with periodic rocking to ensure even distribution of virus and medium over cells.

4. After 30 min of incubation, prepare the agarose overlay.

   The agarose overlay keeps virus released from infected cells from diffusing far from the site of initial infection. Thus, only cells immediately neighboring the cell of primary infection are subsequently infected by progeny virus.

5. Remove the viral supernatant from the cells with a sterile Pasteur pipet and add 4 ml agarose overlay. Allow the agarose to harden on the plates 10 to 20 min at room temperature (to allow condensation to escape). Wrap the plates individually with Parafilm (to avoid desiccation) and incubate 4 to 8 days at 27°C.

   Plaques continue to form for up to 2 weeks. If no plaques are visible by 1.5 to 2 weeks, the cells were probably plated at too high a density and should be replated at a lower density (between 1×10^6 and 1.3×10^6 cells/plate).

6. On plates containing plaques that are well formed and easily visualized, count the number of plaques at each dilution within a set. Calculate the viral titer (pfu/ml).

   Ten plaques at a 10^{-7} dilution or one plaque at at 10^{-8} dilution gives a titer of 10^8 pfu/ml.

7. If difficulties are encountered visualizing plaques, stain with trypan blue. Prepare the trypan blue overlay and dispense 1 ml on plates that have been incubating 3 to 5 days so plaques are well formed. Incubate the plates overnight at 27°C to allow the dye to diffuse into the dead cells. Count the number of blue plaques and determine the viral titer.

   Another procedure for determining virus titer is by endpoint dilution (O'Reilly et al., 1992). In this method, a series of viral dilutions are made and used to infect cells in microtiter wells. Each well is then scored for the presence or absence of viral infection and a 50% endpoint is determined. However, results from this method are often more difficult to interpret than plaque assays when titering recombinant virus stocks. Wild-type virus is very easy to score because of the accumulation of occlusion bodies, but recombinant-virus infection can sometimes be difficult to score because of the lack of occlusion bodies.

References: Maiorella et al., 1988; O'Reilly et al., 1992.

Contributors: Cheryl Isaac Murphy and Helen Piwnica-Worms

Generation of Recombinant Baculoviruses and Analysis of Recombinant Protein Expression

The procedure for cloning and screening of recombinant baculoviruses is outlined in Figure 16.11.1. The majority of available baculovirus vectors are pUC-based and confer ampicillin resistance (see Figure 16.11.2). Most contain the polyhedrin gene promoter, variable lengths of polyhedrin coding sequence, and insertion site(s) for cloning the foreign gene of interest flanked by viral sequences that lie 5′ to the promoter and 3′ to the foreign gene insert. These flanking sequences facilitate homologous recombination between the vector and wild-type baculovirus DNA.

A major consideration when choosing the appropriate baculovirus expression vector is whether to express the recombinant protein as a fusion or nonfusion protein in insect cells. The track record for expressing recombinant proteins is good; thus, if the gene contains its own initiation codon, it is preferable to express it as a nonfusion protein. For expressing the protein as a polyhedrin fusion protein, there are several available vectors. An improved expression vector that facilitates screening of recombinant baculoviruses has been developed: pBlueBacIII (Invitrogen), derived from pJV*Nhe*I. Because pBlueBacIII encodes β-galactosidase, the screening of recombinant baculoviruses is simplified. Recombinant virus generates plaques that are blue and lack occlusion bodies. Invitrogen also sells a set of three vectors, pBlueBacHis A, B, and C, which express proteins with an N-terminal peptide consisting of six histidine residues. These vectors allow easy purification of the recombinant protein using a nickel-chelating resin (*UNIT 10.11*).

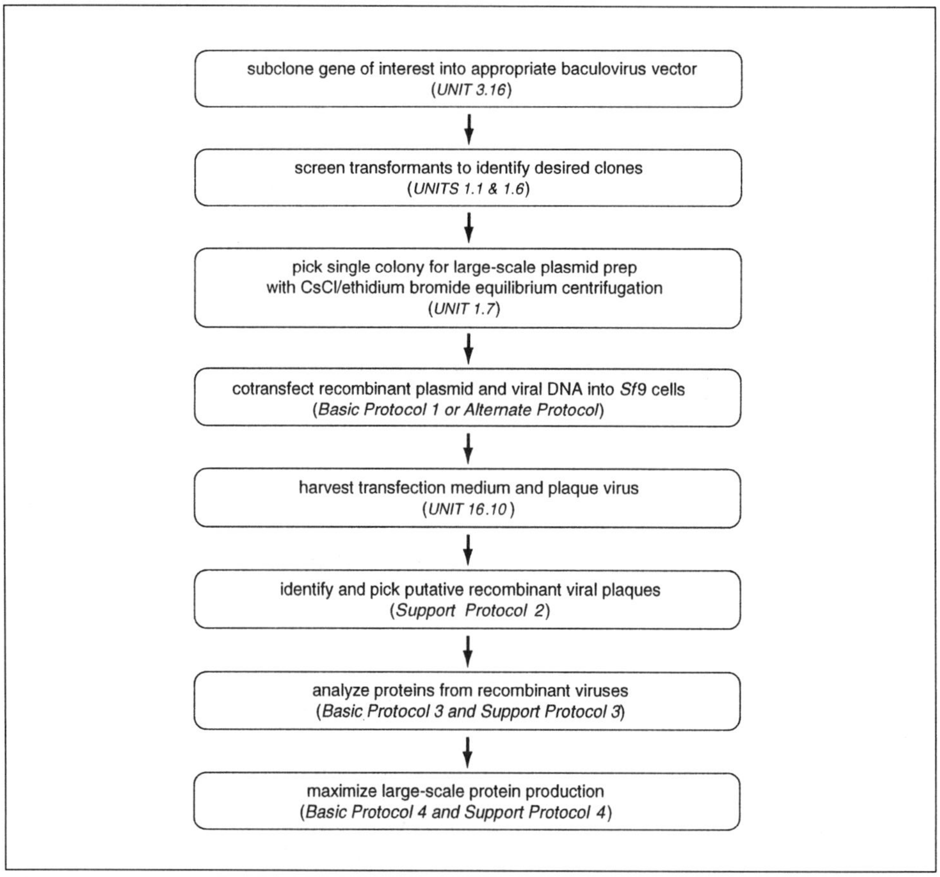

Figure 16.11.1 Flow chart showing cloning and screening of recombinant baculoviruses.

Nonfusion expression vector:	Cloning sites:
pVL941	*BamHI*
pVL1393	5′ *BamHI SmaI XbaI EcoRI NotI XmaIII BglII PstI* 3′ (polylinker)
pVL1392	5′ *BglII PstI NotI XmaIII EcoRI XbaI SmaI BamHI* 3′ (polylinker)
pBacPAK8	5′ *BamHI Sse8387I PstI StuI XhoI BstBI XbaI BglII Asp718I KpnI Ecl136II SacI EcoRI XmaI SmaI EagI NotI PacI* 3′
pBacPAK9	5′ *BamHI XmaI SmaI EcoRI Ecl136II SacI Asp718I KpnI BglII XbaI BstBI XhoI StuI PstI Sse8387I EagI NotI PacI* 3′

Fusion expression vector:	Cloning sites and sequence:
pAc700	ATG GAT CCG GTA CC (+1 at ATG; *BamHI*: GGATCC; *KpnI*: GGTACC)
pAc701	ATG GGA TCC GGT ACC GAG CTC (+1 at ATG; *BamHI*: GGATCC; *KpnI*: GGTACC)
pAc702	ATG CGG ATC CGG TAC C (+1 at ATG; *BamHI*: GGATCC; *KpnI*: GGTACC)
pAc360	ATG CCG GAT TAT TCA TAC CGT CCC ACC ATC GGG CCG GAT CC (+1 at ATG; *BamHI*: GGATCC)

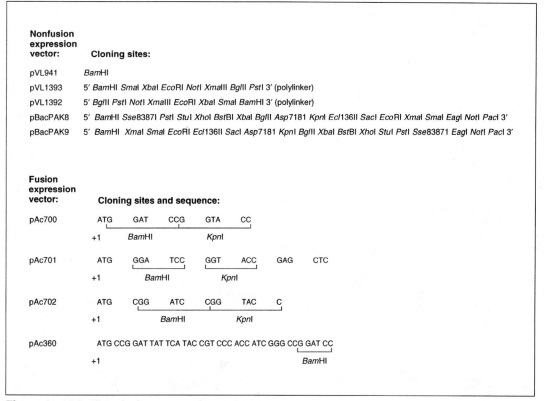

Figure 16.11.2 Baculovirus expression vectors.

NOTE: All reagents and equipment coming into contact with live cells must be sterile. All procedures should be performed with sterile technique unless otherwise noted.

COTRANSFECTION USING LINEAR WILD-TYPE VIRAL DNA

BASIC PROTOCOL 1

This method greatly increases the proportion of recombinant viruses obtained during cotransfection and thus makes screening for recombinants by plaque assay (Support Protocol 2) less difficult and time-consuming.

Materials *(see* APPENDIX 1 *for items with ✓)*

Spodoptera frugiperda (*Sf*9) cells (*UNIT 16.10*)
✓ Complete insect cell medium/10% FBS or serum-free insect cell culture medium
Sterile H_2O
Linearized AcMNPV (wild-type baculovirus) DNA (BacPAK6, Clontech; see Fig. 16.11.3)
Plasmid vector containing the gene of interest (100 ng DNA/μl)
✓ Complete insect cell medium without FBS
Lipofectin reagent (0.1 mg/ml, from Clontech or Life Technologies; see *UNIT 9.4*)
Serum-free insect cell culture medium containing 5% heat-inactivated (30 min, 56°C) fetal bovine serum (FBS)

60-mm tissue culture plates
Conical 15-ml polystyrene centrifuge tubes (Falcon or Corning)
27°C humidified incubator
Beckman GPR with GH-3.7 horizontal rotor (or equivalent), 4°C

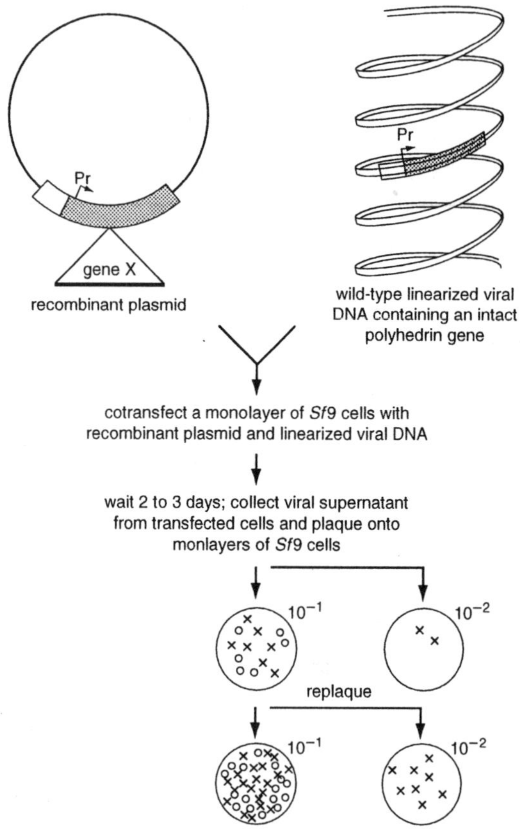

Figure 16.11.3 Generation and purification of recombinant baculoviruses. The foreign gene (X, depicted in black) is inserted downstream of the viral polyhedrin promoter (Pr) in the appropriate plasmid vector. The foreign gene is flanked both 5′ and 3′ by polyhedrin gene-specific sequences (hatched). The recombinant plasmid is then cotransfected into *Sf*9 cells with linearized viral DNA, purchased from a commercial vendor. After 2 to 3 days, the culture medium containing both wild-type (circles) and recombinant (X) virus is collected. Recombinant virus is purified from wild-type virus by one or more rounds of plaque purification.

1. Seed 60-mm plates with 2×10^6 *Sf*9 cells in complete medium/10% FBS or serum-free medium. Incubate 30 to 60 min at 27°C to allow cells to attach.

2. For each plate to be transfected pipet 40 µl sterile water into a sterile polystyrene tube. Add 5 µl of linear AcMNPV DNA and 5 µl of 100 ng/µl plasmid DNA. Thoroughly mix the Lipofectin and add 50 µl Lipofectin to the DNA solution. Mix gently. Incubate 15 min at room temperature.

3. During the 15-min incubation replace the medium in the plates with 1.5 ml complete medium without FBS.

4. Add the Lipofectin-DNA complexes dropwise to the medium while gently swirling the dish to mix. Incubate 4 to 5 hr at 27°C.

5. Add 1.5 ml complete medium/10% FBS (or serum-free medium containing 5% FBS) to each plate and incubate 60 to 72 hr in 27°C humidified incubator.

6. Transfer the transfection supernatants (comprised of culture medium and virus) from each plate to sterile conical 15-ml centrifuge tubes. Centrifuge 10 min at

1000 × *g*, 4°C. Transfer viral supernatants to new sterile tubes and store shielded from light at 4°C until needed.

Viral supernatants can be stored for several weeks until used for plaque assays (UNIT 16.10).

COTRANSFECTION USING CIRCULAR WILD-TYPE VIRAL DNA

ALTERNATE PROTOCOL

Some viral DNA preparations are more pure than others. In most instances, 1 µg viral DNA is sufficient for transfection; however, because contaminating host DNA and RNA may result in higher absorbance readings than expected, other concentrations of viral DNA should be tried to optimize the transfection procedure.

Additional Materials *(also see Basic Protocol 1; see APPENDIX 1 for items with ✓)*

Graces insect medium, unsupplemented (Life Technologies or JRH Biosciences), containing 10% fetal bovine serum (FBS) and 50 µg/ml gentamycin (optional)
Wild-type baculovirus DNA (Support Protocol 1)
pAC360β-gal vector (from M.D. Summers [see *UNIT 16.9*] or Invitrogen)
Plasmid vector containing the recombinant gene of interest
✓ Transfection buffer, room temperature
✓ 2.5 M $CaCl_2$, room temperature
✓ Complete insect cell medium/10% FBS with and without 150 µg/ml Xgal

25-cm^2 flasks
12 × 75–mm polystyrene tubes
Beckman GPR with GH-3.7 horizontal rotor (or equivalent), 4°C

1. Seed 25-cm^2 flasks with 2.5 × 10^6 *Sf*9 cells/flask. Incubate ~2 hr at 27°C to allow cells to attach. Remove medium and replace with 2 ml unsupplemented Graces insect medium/10% FBS. Leave flasks at room temperature.

2. For each transfection set up a sterile 12 × 75–mm polystyrene tube containing 1 µg (or other predetermined appropriate concentration) wild-type baculovirus DNA. Add 2 µg pAC360β-gal DNA or recombinant plasmid DNA and 950 µl transfection buffer, and mix.

   The plasmid pAC360β-gal provides an easy colorometric assay for determining transfection efficiencies and recombination frequencies. The plasmid contains the lacZ gene fused in-frame with the polyhedrin gene and promoter, flanked on both sides by polyhedrin gene sequences. A homologous recombination event produces recombinant viruses that express a polyhedrin–β-galactosidase fusion protein.

3. Add 50 µl of 2.5 M $CaCl_2$ to the DNA/transfection buffer mixture. Mix and aerate solution for 1 min by bubbling air through with a 1-ml pipet. Incubate tubes 25 min at room temperature (to allow a precipitate to form).

4. Add the solution containing the DNA precipitate to cells in the 25-cm^2 flasks. Incubate flasks 4 hr at 27°C.

5. Remove medium from each flask with a Pasteur pipet and rinse carefully with Graces insect medium/10% FBS. Add complete medium/10% FBS containing 150 µg/ml Xgal to the pAC360β-gal flasks and complete medium/10% FBS to the others. Incubate flasks 3 to 5 days at 27°C, and observe daily for occlusion bodies as well as for a blue tint in the medium (pAC360β-gal flasks).

6. When cells are well infected (4 to 5 days), transfer the transfection supernatant (comprised of culture medium and virus) from each flask to sterile conical 15-ml centrifuge tubes. Centrifuge 10 min at 1000 × *g*, 4°C. Transfer viral supernatants to fresh sterile tubes and store until needed at 4°C.

SUPPORT PROTOCOL 1

PURIFICATION OF WILD-TYPE BACULOVIRUS DNA

Additional Materials *(also see Basic Protocol 1; see* APPENDIX 1 *for items with ✓)*

✓ 0.1× and 1× TE buffer, pH 7.4
✓ Extraction buffer
10 mg/ml proteinase K (prepare fresh)
10% *N*-lauroylsarcosine (sodium salt; filter sterilize and store at 4°C)
25:24:1 phenol/chloroform/isoamyl alcohol (*UNIT 2.1*)
100% and 70% ethanol
✓ 3.0 M sodium acetate, pH 5.2

150-cm² flasks
50-ml conical tubes
Beckman GPR with GH-3.7 horizontal rotor (or equivalent), 4°C
Beckman SW-27 or SW-28 and SW-41 rotors and tubes (or equivalent), 4°C
15-ml polypropylene tubes
50°C water bath
5- or 10-ml wide-mouth pipets

1. Seed at least ten 150-cm² flasks with 1.8×10^7 *Sf*9 cells/flask. Incubate ~2 hr at 27°C to allow the cells to attach firmly, then infect them with AcMNPV wild-type virus at a multiplicity of infection (MOI) of 0.1 as in *UNIT 16.10*. Incubate 3 to 5 days at 27°C. When occlusion bodies are observed in most cells, transfer the viral supernatant (~200 ml) to four 50-ml conical tubes.

2. Centrifuge 10 min at 1000 × *g*, 4°C, and pour viral supernatant into four new tubes. Repeat centrifugation to remove any remaining cells.

3. Place viral supernatant in SW-27 ultracentrifuge tubes and balance the tubes. Centrifuge 30 min at 100,000 × *g*, 4°C, to pellet the virus. Pour off supernatant and invert tube on a Kimwipe to drain as much liquid as possible.

4. Examine the viral pellet carefully. If the viral pellet is pure (it should have an opaque, whitish appearance), proceed to step 10. In most instances, the viral pellet will appear yellowish because of the presence of cellular debris. If this is the case, separate the virus from cellular contaminants by one of the two methods below.

Purify viral pellet by sucrose-gradient fractionation:

5a. Add 1 ml of 0.1× TE buffer to the viral pellet and repeatedly pipet up and down to resuspend. If pellet is difficult to resuspend, incubate overnight at 4°C.

6a. Prepare two linear 25% to 56% sucrose gradients in SW-41 ultracentrifuge tubes using filter sterilized ultrapure sucrose in 0.1× TE buffer.

7a. Carefully layer 0.5 ml viral suspension on top of each sucrose gradient. Centrifuge 90 min at 100,000 × *g*, 4°C.

8a. Using a Pasteur pipet, transfer viral band to an SW-41 ultracentrifuge tube. Add 0.1× TE buffer to the top of the tube.

9a. Centrifuge 30 min at 100,000 × *g*, 4°C, to pellet the virus. Decant supernatant and invert tube on a Kimwipe to drain as much liquid as possible.

Purify viral pellet by microcentrifugation:

5b. Add 3 ml extraction buffer to pellet and repeatedly pipet up and down to resuspend. If pellet is difficult to resuspend, incubate overnight at 4°C.

6b. Transfer 1.5 ml viral suspension into each of two 1.5-ml microcentrifuge tubes. Microcentrifuge 5 min at maximum speed and transfer supernatants to one 15-ml polypropylene centrifuge tube.

7b. Wash the two pellets by resuspending each in 1 ml extraction buffer.

8b. Microcentrifuge 5 min at maximum speed and combine the two supernatants with the pooled supernatants in the 15-ml polypropylene tube.

9b. Bring volume to 9 ml with extraction buffer and transfer 4.5-ml aliquots to two 15-ml polypropylene centrifuge tubes. Proceed to step 11.

10. Resuspend the virus pellet in 9 ml extraction buffer and transfer 4.5-ml aliquots to two 15-ml polypropylene centrifuge tubes.

11. Add 200 µg of 10 mg/ml proteinase K to each tube and incubate 1 to 2 hr at 50°C.

12. Add 0.5 ml of 10% *N*-lauroylsarcosine to each tube and incubate 2 hr or overnight at 50°C.

13. Extract DNA twice with an equal volume of 25:24:1 phenol/chloroform/isoamyl alcohol.

14. Transfer the aqueous phase containing the DNA to another 15-ml tube using a wide-mouth (5- to 10-ml) pipet. Add 10 ml of 100% ethanol and mix gently by inverting the tubes several times. Incubate 10 min at −80°C.

15. Centrifuge 20 min at 1500 × *g* in a tabletop centrifuge, 4°C, and discard supernatant. Rinse the DNA pellet with 70% ethanol and air dry 30 to 60 min. Resuspend pellet in 800 µl of 1× TE buffer.

16. Transfer 400 µl to two microcentrifuge tubes and reprecipitate the DNA by adding 40 µl of 3.0 M sodium acetate and 2 vol of 100% ethanol to each tube. Incubate 10 min at −80°C.

17. Microcentrifuge 10 min and discard supernatant. Rinse DNA pellet with 70% ethanol and lyophilize. Resuspend DNA in 0.3 to 1.0 ml of 1× TE buffer. Read A_{260} and calculate yield. Store DNA at 4°C.

This method should yield 50 to 100 µg viral DNA.

BASIC PROTOCOL 2

PURIFICATION OF RECOMBINANT BACULOVIRUS ENCODING β-GALACTOSIDASE

To become familiar with the morphological differences between plaques produced by wild-type virus versus those produced by recombinant virus, it is helpful to first practice plaquing β-gal recombinant virus alongside wild-type virus.

Materials *(see APPENDIX 1 for items with ✓)*

✓ Agarose overlay containing 150 μg/ml Xgal (see recipe for Xgal stock)
✓ Complete insect cell medium with and without 10% fetal bovine serum (FBS)
Spodoptera frugiperda (*Sf*9) cells (*UNIT 16.10*)

25-cm² flask
1.5-ml screw-top cryostat tube

1. Prepare serial dilutions of the viral supernatants obtained from transfections using pAC360β-gal DNA (Alternate Protocol or Basic Protocol 1, step 6) in serum-free complete medium. Plaque the virus using an agarose overlay containing 150 μg/ml Xgal.

   If the cells are cultured in serum-free insect cell culture medium (Sf-900 II), the same medium can be used for preparing serial dilutions.

2. When plaques have formed, pick a well-isolated blue plaque with a sterile Pasteur pipet, and pipet the agarose plug into a sterile tube containing 1 ml serum-free complete medium. Vortex and prepare serial dilutions at 10^{-1}, 10^{-2}, and 10^{-3} in serum-free complete medium.

   A single plaque contains ~10,000 plaque-forming units (pfu).

3. Plaque the virus using agarose overlays containing 150 μg/ml Xgal (to aid in finding recombinants). Repeat plaquing until a pure stock of recombinant virus is obtained.

   It generally takes two to three rounds of plaquing from a transfection using circular DNA to obtain a recombinant virus free of contaminating wild-type virus. If linear DNA is used in the transfection, one round of plaquing is usually sufficient.

4. Using a sterile pipet, pick a pure recombinant plaque and pipet the agarose plug into a 25-cm² flask containing 2.5×10^6 *Sf*9 cells and 5 ml complete medium/10% FBS. Incubate 4 to 5 days at 27°C until the majority of the cells are infected.

5. Determine the virus titer (it should be between 5×10^7 and 1×10^8 pfu/ml).

6. Place 1 ml of this stock in a 1.5-ml screw-top cryostat tube at –80°C for long-term storage. Store the remainder at 4°C (marked as primary passage stock). Grow a large viral stock from the primary passage stock and titer.

SUPPORT PROTOCOL 2

VISUAL SCREENING FOR RECOMBINANT BACULOVIRUSES

To visualize the difference in morphology between wild-type and recombinant viral plaques, plaque the pure wild-type baculovirus stock (*UNIT 16.10*) and the pure pAC360β-gal recombinant virus stock (Basic Protocol 2, step 4) at dilutions of 10^{-6}, 10^{-7}, and 10^{-8}. Do not include Xgal in the agarose overlay because the blue color precludes a careful analysis of the plaque morphology produced by the β-gal recombinant virus.

Screening with a dissecting microscope. Invert plaquing dish on a black background beneath the microscope and shine a bright light at an acute angle on the plate. Scan cells infected with wild-type virus for plaques containing occlusion bodies. The plaques will look very refractile, with a slight yellowish color. Scan cells infected with the β-gal recombinant virus for plaques that lack occlusion bodies and appear grayish in color. Circle regions of the dish overlying the plaques. Examine the circled plaques under an inverted light microscope at 400× to train the eye to the subtleties of each plaque morphology.

Screening with an inverted light microscope. Place the plaquing dish right side up and scan for each type of plaque at 30× to 40×. Recombinant plaques have an orange tint, whereas wild-type plaques have a dark brown appearance due to the accumulation of occlusion bodies. While looking at the plaque under the microscope, place a dot on the region of the dish underlying the plaque. Remove the dish from the microscope and circle the plaque. Place the dish back on the microscope stage and view the plaque at 400× magnification.

Direct visual screening. If the plaques are well-formed, it is possible to distinguish a wild-type plaque from a recombinant plaque by holding the dish overhead and looking at the bottom of the dish directly. Regions surrounding wild-type plaques will look grayish-white whereas recombinant plaques will not.

After identifying putative recombinant viral plaques by one of these methods, pick several plaques and place the agarose plugs in 1 ml serum-free medium. Vortex and store up to several months at 4°C until needed.

ANALYSIS OF PROTEIN FROM PUTATIVE RECOMBINANT VIRUSES

BASIC PROTOCOL 3

Before proceeding with further plaque purification, it is recommended that the putative recombinants obtained in Support Protocol 2 be assayed for their ability to produce the protein of interest. Screening should be individually tailored to the properties of the protein being overproduced and the availability of detection reagents.

Materials *(see APPENDIX 1 for items with ✓)*

Spodoptera frugiperda (*Sf*9) cells (*UNIT 16.10*)
✓ Complete insect cell medium/10% FBS
Stock of recombinant plaque (Support Protocol 2)
✓ PBS
✓ 1× and 2× SDS sample buffer
Appropriate lysis buffer supplemented with protease inhibitors (optional)

25-cm² flasks
27°C incubator
Beckman GPR with GH-3.7 rotor (or equivalent), 4°C
15-ml polypropylene centrifuge tubes
Boiling water bath

1. Seed 2.5×10^6 *Sf*9 cells into 25-cm² flasks containing 5 ml complete medium/10% FBS. Prepare one flask for each putative recombinant plaque selected by the method described in Support Protocol 2. Incubate ≥2 hr at 27°C to allow cells to attach. Add 0.5 ml of the 1-ml stock containing serum-free complete medium and recombinant plaque (Support Protocol 2) to each flask. Store the remaining 0.5 ml of recombinant virus at 4°C for subsequent plaquings.

2. Incubate flasks 4 to 5 days at 27°C, monitoring daily for signs of infection. After 4 to 5 days, harvest the virus by transferring the culture medium to 15-ml centrifuge tubes.

3. Centrifuge 10 min at 1000 × *g*, 4°C. Transfer supernatant, which contains the expanded virus, to a sterile 15-ml polypropylene centrifuge tube.

4. Seed 2.5 x 10^6 *Sf*9 cells into 25-cm^2 flasks containing 5 ml complete medium/10% FBS. Prepare one flask for each viral stock to be tested and one flask as a noninfected control. Incubate ≥2 hr at 27°C to allow cells to attach. Remove the culture medium from the cells and add 1.5 ml of the expanded virus stock to the flask. Incubate 2 days at 27°C.

5. Harvest cells by gently dislodging them from flask and transfer cells and culture medium to 15-ml polypropylene centrifuge tubes.

6. Centrifuge 10 min at 1000 × *g*, 4°C. If the protein of interest is a secreted protein, transfer the culture supernatant to a new tube and proceed to step 8. If the protein of interest is an intracellular protein, discard the supernatant. Resuspend the cell pellet gently in PBS to rinse the cells, repeat centrifugation, and discard supernatant.

7. Boil the cell pellets directly in 500 µl of 1× SDS sample buffer and sonicate the sample if it is too viscous due to the presence of DNA. Continue to sonicate until viscosity clears, then proceed to step 9. Alternatively, lyse cell pellets in 0.5 ml of an appropriate lysis buffer supplemented with protease inhibitors. Microcentrifuge 10 min at 4°C to clarify the lysates and transfer supernatant to new tube. Add 0.1 ml of each lysate to 100 µl of 2× SDS sample buffer and boil 3 min in boiling water bath. Freeze remaining lysate up to several months at −80°C.

8. Determine the protein concentration in the culture supernatant (from step 6) for secreted proteins, or in the cell lysate (from step 7) for intracellular proteins using the Bradford method.

9. Analyze the proteins in each sample by one of the following methods:

   a. Immunoblotting (UNIT 10.7): Load 20 to 40 µg total cell protein per lane on a one-dimensional SDS-polyacrylamide gel. Remember to include the noninfected control.

   b. Coomassie brilliant blue staining (UNIT 10.5): Load 20 to 40 µg total cell protein per lane on a one-dimensional SDS-polyacrylamide gel. If the recombinant virus is not pure, recombinant protein will be detected only if it is produced at very high levels in the infected cells.

   c. Functional assays, such as mobility-shift DNA-binding assays (for a DNA-binding protein; UNIT 12.2), in vitro kinase assays (for a protein kinase), nucleotide-binding assays (for a protein that binds nucleotides), thymidine-incorporation assays (for a protein that is a growth factor), or any assay that is typically used to monitor the protein of interest.

   d. Metabolic labeling of recombinant proteins (Support Protocol 3).

   A simple dot hybridization technique to detect the presence of the gene is given in Summers and Smith (1987); alternatively PCR amplification can be used (O'Reilly et al., 1992). Several companies (e.g., Invitrogen and Clontech) sell PCR primers that will work for most baculovirus vectors.

10. Interpret results to identify which of the putative recombinant plaques is an actual recombinant that produces the desired protein. Plaque-purify it so it is free from any contaminating wild-type virus (Basic Protocol without Xgal). Prepare a large viral stock and titer the recombinant virus.

METABOLIC LABELING OF RECOMBINANT PROTEINS

SUPPORT PROTOCOL 3

Metabolic labeling in vivo is a sensitive way to detect recombinant proteins because at the time the recombinant protein is expressed, host protein synthesis is essentially terminated.

Additional Materials *(also see Basic Protocol 3)*

Autographa californica wild-type nuclear polyhedrosis virus (*UNIT 16.9*)
Methionine-free Graces insect cell culture medium or methionine-free/cysteine-free Sf-900 II (both from Life Technologies)
EXPRE^{35}S^{35}S, containing [^{35}S]methionine and [^{35}S]cysteine (>1000 Ci/mmol; Du Pont NEN)

1. Seed 2.5×10^6 cells into 60-mm tissue culture dishes containing 4 ml complete medium/10% FBS. Prepare one dish to be infected with each putative recombinant virus and one control dish to be infected with wild-type baculovirus. Incubate ~2 hr at 27°C to allow cells to attach. Aspirate medium, then add 1 ml recombinant virus (Basic Protocol 3, step 3) *or* 1 ml of serum-free complete medium containing wild-type virus at a multiplicity of infection (MOI) of 5 to 10. Incubate 1 hr at room temperature.

   If the cells are already adapted to Sf-900 II, this procedure can be done using Sf-900 II instead of complete medium. In this case use methione-free/cysteine-free Sf-900 II throughout instead of methionine-free Graces medium.

2. Remove the medium from each dish. Add 3 ml complete medium/10% FBS to the cells and incubate ~40 hr at 27°C.

3. Carefully remove culture medium, rinse cells once with methionine-free or methionine-free/cysteine-free medium, and add 1 ml of the same medium to which has been added 0.25 to 0.5 mCi/ml of EXPRE^{35}S^{35}S. Incubate 3 to 4 hr at 27°C.

4. Transfer cells and culture supernatant to a 15-ml polypropylene centrifuge tube, centrifuge 10 min at 1000 × *g*, 4°C, and follow steps 6 to 8 of Basic Protocol 3.

   If an antibody is available, it is recommended that the labeled lysates be immunoprecipitated (UNIT 10.15) prior to boiling in SDS sample buffer and resolution by SDS-PAGE. Immunoprecipitation will help detect the recombinant protein if it comigrates with a labeled host or late viral-specific protein.

5. Resolve the proteins in the lysate by SDS-PAGE, loading 20 to 40 µg total cell protein or the entire immunoprecipitate in each lane.

   Polyhedra are solubilized only under very alkaline conditions (0.1 M final NaOH concentration). Less than 10% of polyhedra will be solubilized in SDS sample buffer without prior disruption under alkaline conditions

6. Visualize proteins by autoradiography. Inspect the autoradiogram for protein of the expected molecular weight that appears in cells infected with recombinant virus but not with wild-type baculovirus.

SUPPORT PROTOCOL 4

DETERMINING TIME COURSE OF MAXIMUM PROTEIN PRODUCTION

Recombinant proteins are usually detected between 15 and 24 hr postinfection and accumulate until ~40 hr postinfection, at which time their accumulation levels off. Because individual proteins display differences in their stability in insect cells, it is recommended that the time course of protein accumulation be charted for each protein expressed using this system.

1. Seed 3×10^6 *Sf*9 cells into fifteen 60-mm tissue culture dishes containing 5 ml complete medium/10% FBS. Incubate ~2 hr at 27°C to allow cells to attach, then infect seven dishes with wild-type virus and seven dishes with recombinant virus, each at an MOI of 10. Leave one dish as a noninfected dish.

   Alternatively, Sf9 cells can be cultured in suspension in three 100-ml spinner flasks. When cells reach a density of 1.5×10^6 cells/ml, infect one with recombinant virus, one with wild-type virus, and leave one as an uninfected culture. Use a multiplicity of infection (MOI) of 1 to 2.

2. Harvest cells at various times after infection (from ~15 hr to 72 hr postinfection). Harvest noninfected cells 15 hr postinfection. To harvest, transfer cells and culture supernatants to centrifuge tubes and centrifuge 10 min at 1000 × *g*, 4°C.

   If cells are growing in suspension, remove 2-ml aliquots at each time point and process the cells and supernatants in the same way as the monolayer cultures.

3. Process and analyze the cells and supernatants (if the recombinant protein is secreted) according to steps 6 to 9 of Basic Protocol 3.

   When staining with Coomassie brilliant blue (Basic Protocol, step 9b), look for a protein that appears as a function of time post infection with the recombinant virus but not with the wild-type virus. The protein must be reasonably abundant for this method to be successful.

BASIC PROTOCOL 4

LARGE-SCALE PRODUCTION OF RECOMBINANT PROTEINS

Materials

Spodoptera frugiperda cells (*Sf*9; *UNIT 16.10*)
Serum-free insect cell culture medium (Sf-900 II from Life Technologies or ExCell 401 from JRH Biosciences)
High-titer recombinant virus (*UNIT 16.10*)

1- to 15-liter spinner flasks (Bellco; Fig. 16.11.4)
Two-port cap assemblies for spinner flasks (Bellco)
Silicone tubing (Cole-Palmer) with 3⁄16-in. (0.48-cm) i.d., 5⁄16-in. (0.8 cm) o.d., and 1⁄16-in. (0.16-cm) wall
0.2-µm filter units (Millipore)
4-in. (10.16-cm) cable ties (Cole-Palmer)
Tension tool (Cole-Palmer)
Stir plate for multiple spinner flasks (Bellco)
Air-supply pump (Bellco)

1. Grow *Sf*9 cells in suspension culture and adapt to serum-free medium.

2. Prepare spinner flasks to be used for scale-up of *Sf*9 cells by attaching the appropriate size two-port cap to one sidearm and a plain cap to the other sidearm (see Fig. 16.11.4). Put a short (~6-in. or 15-cm) piece of tubing on the air-vent port and attach a filter to the end. Secure the tubing to the port and to the filter with cable ties using the tension tool.

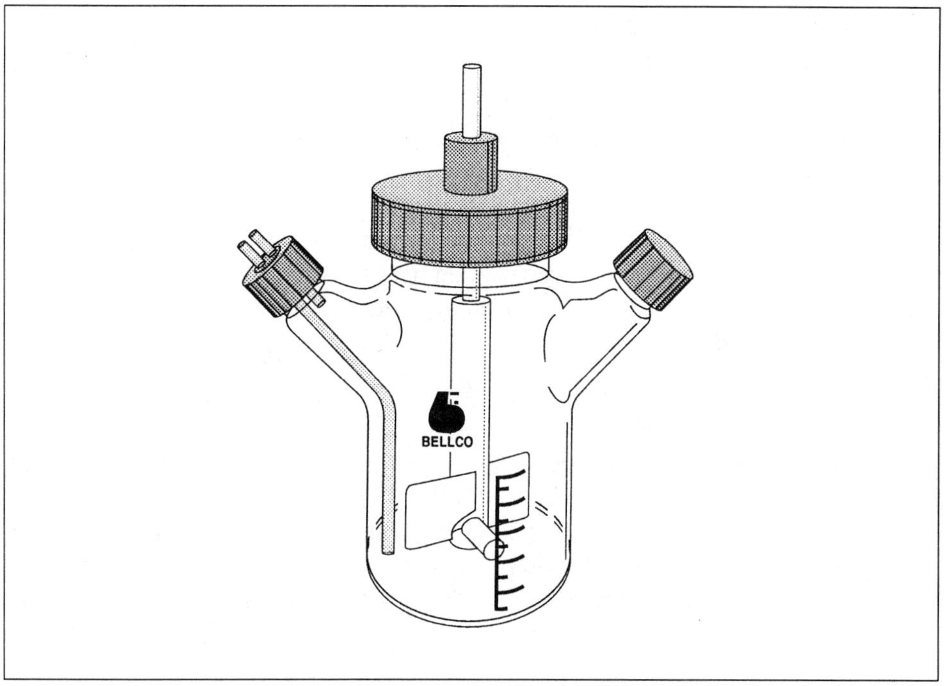

Figure 16.11.4 Bellco spinner flask with two-port cap assembly.

3. Put a longer piece of tubing (1 to 2 feet, 30 to 60 cm) on the air-supply port and attach a filter to the end. Secure with cable ties. Attach another piece of tubing to the other end of the filter with a cable tie. Cover the end of the tubing with aluminum foil.

4. Loosen the cap on the sidearm opposite to the two-port assembly a quarter turn and autoclave flask 1 hr.

5. Seed the autoclaved flask with *Sf*9 cells adapted to serum-free medium (from step 1). Fill the flask to between half full and full (e.g., 1.5 to 3 liters in a 3-liter flask), adding enough serum-free medium to make a final cell density of $5–6 \times 10^5$ cells/ml up to 1.5×10^6 cells/ml.

6. Place the flask on a magnetic stir plate in a 27°C incubator. Set the stir speed at 80 rpm. Remove the aluminum foil from the air intake tube and attach the end of the tube to the air-supply pump. Turn on the pump to a flow rate of 500 to 700 ml/min.

7. Grow the cells to a density of $\sim 1.5 \times 10^6$ cells/ml. In a laminar flow hood, add virus at a multiplicity of infection (MOI) of 1 to 2 directly to the flask through the sidearm.

8. Place the flask on the magnetic stir plate at 27°C and connect the air supply. Incubate the culture for the amount of time determined in Support Protocol 4.

9. Process the supernatant for secreted proteins or the cells for intracellular proteins.

References: Matsuura et al., 1987; Murphy et al., 1993; O'Reilly et al., 1992; Summers and Smith, 1987.

Contributors: Cheryl Isaac Murphy and Helen Piwnica-Worms

Overview of Protein Expression in Mammalian Cells

As described elsewhere in this manual (Chapter 9), mammalian cells are often used as hosts for expression of genes obtained from higher eukaryotes because the signals for synthesis, processing, and secretion of these proteins are usually recognized. The unit that follows describes vector systems or strategies for introducing foreign genes into mammalian cells (additional transfection methods can be found in *UNITS 9.1-9.4*). This method relies upon COS cells for rapid, transient expression of protein from specific vectors (*UNIT 16.13*).

The criteria for choosing a system include these considerations: whether DNA can be introduced directly by transfection methods or needs to be introduced by viral-mediated transfer, the identity of the control elements that can direct efficient mRNA expression and protein synthesis, and whether a particular host cell is appropriate for expression of the gene of interest. When transient expression is appropriate, the choice of which system to use depends upon the particular experiment. If a lower transfection efficiency is sufficient and if it is desirable that the cells continue to grow for several days, COS cells should be used.

VIRAL-MEDIATED GENE TRANSFER

Viral-mediated gene transfer provides a convenient, efficient means to introduce foreign DNA into most recipient cells. Representative expression levels obtained from SV40 recombinant viruses, retroviruses, and vaccinia viruses are shown in Table 16.12.1 in comparison to other expression strategies. A more detailed review of the different eukaryotic viral vectors can be found in Muzyczka (1989).

TRANSIENT EXPRESSION

The efficiency of expression from transient transfection depends on the number of cells that take up the transfected DNA, the gene copy number, and the expression level per gene. Most methods of DNA transfer allow 5% to 50% of the cells in the population to acquire DNA and express it transiently over a period of several days to several weeks. Transient DNA transfection is most frequently used to: (1) verify the identity of cloned genes based on their ability to express a particular activity, (2) rapidly study the effect of engineered mutations on either gene activity or protein function, and (3) isolate genes from cDNA libraries constructed in mammalian expression vectors based on their ability to express a particular activity in cells. The limitations of transient expression are that it is difficult to scale up for production of large quantities of protein (>1 mg), that it is difficult to study the consequences of gene expression only in the portion of the total population that has been transfected, and that the high copy number is eventually lethal; this lethality may significantly affect results.

UNIT 16.13 describes procedures and vectors used for transient expression in COS cells. This cell line is most frequently used for transient expression and is derived from African green monkey kidney cells by tranformation with an origin-defective simian virus 40 (SV40). COS cells express high levels of the SV40 large tumor (T) antigen which is required to initiate viral DNA replication at the origin of SV40. T antigen–mediated replication can amplify the copy number of plasmids containing the SV40 origin of replication to 100,000 per cell, which results in high expression levels from the transfected DNA.

Table 16.12.1 Expression Levels and Uses for Different Mammalian Cell Expression Systems[a]

Cell line	Expression method	Typical expression level (μg/ml)	Primary use
Monkey cells			
CV1	SV40 virus infection	1-10	Expression of wild-type and mutant proteins
COS	Transient DNA/ DEAE-dextran transfection	1	Cloning by expression in mammalian cells; rapid characterization of cDNA clones; expression of mutant proteins
CV1	Transient DNA/ DEAE-dextran transfection	0.05	
Murine fibroblasts			
C127	BPV stable transformant	1-5	High-level constitutive protein expression
3T3	Retrovirus infection	0.1-0.5	Gene transfer into animals; expression in different cell types
Other cells			
CHO(DHFR$^-$)	Stable DHFR$^+$ transformant	0.01-0.05	
	Amplified MTX^r	10	High-level constitutive protein expression
Primate	Vaccinia virus infection	1	Production of vaccines; expression of toxic proteins
	EBV vector	n.a.	Cloning by expression

[a]Abbreviations: BPV, bovine papilloma virus; EBV, Epstein-Barr virus.

STABLE DNA TRANSFECTION

If a selection procedure is applied after DNA transfection, it is possible to isolate cells that have stably integrated the foreign DNA into their genome (*UNIT 9.5*). Different cell lines exhibit different frequencies of stable transformation and vary in their capacity to incorporate foreign DNA. In most cases, the limiting factor for obtaining stable transformants is the frequency of DNA integration, not the frequency of DNA uptake. Cells selected for by incorporation and expression of one genetic marker will frequently incorporate a second gene provided on an independent plasmid during transfection; this ability to incorporate two separate plasmids into the chromosome has been termed cotransformation. Different cell lines and transfection methods yield varying frequencies of cotransformation.

Many recessive genetic selectable markers encode enzymes involved in the purine and pyrimidine biosynthetic pathways.

AMPLIFICATION OF TRANSFECTED DNA

Frequently, it is desirable to increase expression by selecting for increased copy number of the transfected DNA within the host chromosome. The ability to coamplify transfected DNA has permitted a 100- to 1000-fold increase in the expression of the proteins encoded by transfected DNA. Although there are over twenty selectable and amplifiable genes that have been described (Kaufman, 1990), the most experience and success has occurred when methotrexate selection has been used for amplification of transfected dihydrofolate reductase genes.

EXPRESSION VECTORS

Most mammalian cell expression vectors are designed to accomodate cDNAs rather than large genomic fragments because the small size of cDNA clones makes them more convenient to manipulate. Today most useful vectors contain multiple elements including: (1) an SV40 origin of replication for amplification to high copy number in COS monkey cells, (2) an efficient promoter element for high-level transcription initiation, (3) mRNA processing signals such as mRNA cleavage and polyadenylation sequences, and frequently intervening sequences as well, (4) polylinkers containing multiple restriction endonuclease sites for insertion of foreign DNA, (5) selectable markers that can be used to select cells that have stably integrated the plasmid DNA, and (6) plasmid backbone sequences to permint propagation in bacterial cells.

CHOICE OF EXPRESSION SYSTEM

In evaluating which approach to take in expressing a gene, it is most important to consider the goals of the expression work. If expression is required to demonstrate that a clone has some functional activity or to characterize this activity, then transient expression in COS cells is often the most convenient approach. If a large quantity of protein (>1 mg) is required, then stable coamplification in CHO cells is generally the most desirable approach. If the gene is potentially cytotoxic, high-level expression may be approached through vaccinia virus vectors or inducible promoter-vector systems. If there is a particular requirement for the host to produce the protein properly then that requirement will dictate the choice of the host.

TROUBLESHOOTING

If protein expression from the heterologous gene cannot be detected, it is important to examine the vector system in detail. In this sequence, each point should be satisfactorily addressed before proceeding to the next step.

1. Confirm the expected structure of the vector using restriction mapping (*UNITS 3.1 & 3.2*) and, if necessary, DNA sequencing (*UNIT 7.4*).
2. Determine transfection efficiency by including a positive control—e.g., the same vector with another insert.
3. Ensure that the RNA is of the expected size and amount compared to an appropriate control by preparing RNA (*UNITS 4.1, 4.2, & 9.8*) and analyzing it by northern hybridization (*UNIT 4.9*).
4. Use a completely different expression vector or system if the RNA transcript of the correct size cannot be detected in the transfected cells, as it is always possible that some unforeseen situation may result in aberrant splicing (Wise et al., 1989).
5. Determine if the coding region may contain a point mutation or other lesion that keeps it from encoding a full-length protein by carrying out in vitro translation to produce protein (*UNIT 10.16*) using mRNA isolated from transfected cells and using RNA transcribed by in vitro transcription (i.e., SP6; *UNIT 3.8*) of a vector containing the cDNA insert.

Reference: Kaufman, 1990.

Contributor: Randal J. Kaufman

Transient Expression of Proteins Using COS Cells

BASIC PROTOCOL

Three factors contribute to make COS cell expression systems appropriate for the high-level, short-term expression of proteins: (1) the high copy number achieved by SV40 origin–containing plasmids in COS cells 48 hr posttransfection, (2) the availability of good COS cell expression/shuttle vectors, and (3) the availability of simple methods for the efficient transfection of COS cells. Each COS cell transfected with DNA encoding a cell-surface antigen (in the appropriate vector) or cytoplasmic protein will express several thousand to several hundred thousand copies of the protein 72 hr posttransfection. If the transfected DNA encodes a secreted protein, up to 10 µg of protein can be recovered from the supernatant of the transfected COS cells 1 week posttransfection. COS cell transient expression systems have also been used to screen cDNA libraries, to isolate cDNAs encoding cell-surface proteins, secreted proteins, and DNA binding proteins, and to test protein expression vectors rapidly prior to the preparation of stable cell lines (*UNIT 9.5*).

This transfection protocol is a modification of that presented in *UNIT 9.2* and gives conditions for optimal transfection of COS cells (also *UNIT 9.9*). The main difference between this procedure and that in *UNIT 9.2* is the composition of the DEAE-dextran/chloroquine solution, which is prepared here in PBS, not TBS, and contains chloroquine to prevent acidification of endosomes presumed to carry DEAE-dextran/DNA into the cell. This acidification results in acid hydrolysis of DNA, giving rise to mutations and destruction of the DNA. With this protocol, 40-70% of the cells can be routinely transfected.

Materials *(see APPENDIX 1 for items with ✓)*

Appropriate vector (CDM8, pXM, or pDC201; Invitrogen)
COS-7 cells to be transfected (ATCC #CRL1651)
Dulbeccos minimum essential medium with 10% calf serum (DMEM-10 CS)
DMEM with 10% NuSerum (Collaborative Research) (DMEM-10 NS), 37°C
✓ PBS
DEAE-dextran/chloroquine solution: 10 mg/ml DEAE-dextran (Sigma) + 2.5 mM chloroquine (Sigma) in PBS
10% (v/v) dimethyl sulfoxide (DMSO; Sigma) in PBS
0.5 mM EDTA in PBS

100-mm tissue culture dishes
Humidified 37°C, 6% CO_2 incubator
Phase-contrast microscope
Sorvall RT-6000B rotor or equivalent

1. Subclone gene of interest into appropriate vector to get desired recombinant DNA (*UNIT 3.16*). Purify recombinant DNA by miniprep procedure (5-ml culture) or by CsCl/ethidium bromide centrifugation (*UNITS 1.6, 1.7 & 9.1*).

2. Split confluent COS-7 cells (~10^6 per 100-mm dish) in DMEM-10 CS at a 1:5 ratio the day prior to transfection so they will be ~50% confluent the next day. Grow cells overnight in a CO_2 incubator (6% CO_2) at 37°C to ~50% confluence.

3. Just before use (for each 100-mm dish of COS cells to be transfected), thoroughly mix 5 ml of 37°C DMEM-10 NS with 5 to 10 µg recombinant DNA and mix. Add 0.2 ml of DEAE-dextran/chloroquine solution.

It is important that DEAE-dextran be well mixed with medium before adding DNA; otherwise, the DNA will form large precipitates with DEAE-dextran. These large precipitates cannot be taken up by the cell, resulting in a reduced transfection efficiency.

4. Aspirate medium from COS cells and for each 100-mm dish, add DMEM-10 CS/DEAE-dextran/DNA prepared in step 2. Incubate cells 3 to 4 hr (may require optimization) in a CO_2 incubator at 37°C. Observe cells using phase-contrast microscope.

   The DEAE-dextran will cause cells to retract and become vacuolated. Efficiency of transfection increases with longer incubation periods; on the other hand, so does cell death.

5. Aspirate DMEM/DEAE-dextran/DNA and add 5 ml of 10% DMSO (prepared in PBS). Incubate cells 2 min at room temperature. Aspirate DMSO and add 10 ml DMEM-10 CS. Grow cells overnight (12 to 20 hr) in CO_2 incubator at 37°C.

6. Passage each 100-mm dish of transfected COS cells into two new 100-mm dishes. Grow cells at 37°C as in step 7a or 7b.

7a. When expressing secreted proteins, add 5 ml DMEM-10 CS 96 hr (4 days) after completing step 6 and incubate 4 days. Harvest medium, remove dead cells and debris by centrifuging 10 min at ~1000 × *g*, room temperature, and save supernatant. Detect secreted proteins by metabolic labeling (*UNIT 10.17*) and immunoprecipitation (*UNIT 10.15*), immunoaffinity chromatography (*UNIT 10.10*), radioimmunoassay, immunoblotting (*UNIT 10.7*), or bioassay (*UNIT 9.5*).

   Do not aspirate the old medium prior to addition of 5 ml DMEM-10 CS because this medium contains secreted protein. Addition of extra medium 96 hr posttransfection results in better yield of expressed protein; however, it also increases level of total protein because the medium contains 10% serum. To eliminate this problem, COS cells can be placed in serum-free medium 10 to 12 hr after they have been replated although this may result in a 10-fold lower yield of expressed protein than in the presence of serum.

7b. When expressing cell-surface or intracellular proteins, aspirate medium from cells 72 hr after transfection in step 5. Add 5 ml PBS, swirl, and aspirate PBS. Add 5 ml of 0.5 mM EDTA in PBS and incubate 15 min in CO_2 incubator at 37°C. Lift cells from dish by gently dislodging with a Pasteur pipet. Stain cell-surface proteins with appropriate fluorescent antibody and detect by microscopy or flow cytometry.

   Transfected COS cells will tend to clump when lifted from the dish. Pipetting the cells up and down will tend to disrupt these clumps. More effective dispersion of the clumps can be obtained by forcing the cells through a 100-μM nylon mesh.

Reference: Warren and Shields, 1984.

Contributor: Alejandro Aruffo

Analysis of Protein Phosphorylation

17

Most proteins undergo covalent modification of their amino acid side chains during or after synthesis. This modification, particularly if it is reversible, can provide an extraordinarily sensitive means by which the activity of a protein can be enhanced or diminished. Protein phosphorylation is probably the most common and important form of protein modification.

Fifteen years ago, only a few protein kinases had been identified and only a limited number of processes were suspected of being regulated by protein phosphorylation. The increasing power of molecular biology has facilitated the identification and cloning of the genes of literally hundreds of protein kinases and protein phosphatases, while advances in cell and biochemical techniques have allowed the detailed elucidation of complex cellular regulatory pathways. Today, many aspects of the regulation of cell function are known to be controlled by protein phosphorylation. Proliferation, differentiation, signal transduction, and metabolism are all regulated by the balance of activity of protein kinases and protein phosphatases upon critical target proteins. Both extracellular ligands, such as polypeptide growth factors, and key intracellular signaling elements such as the concentration of cyclic AMP or calcium closely regulate the activities of kinases and phosphatases, which are themselves subject to modification by phosphorylation.

This chapter presents an overview of protein phosphorylation and provides detailed methods designed to detect and identify phosphorylated proteins and analyze their specific sites of modification using both isotopic and nonisotopic approaches. The overview (*UNIT 17.1*) provides specific examples of regulatory processes controlled by phosphorylation and provides a description of and rationale for the various approaches to studying protein phosphorylation. Succeeding units describe the labeling of eukaryotic cells with inorganic phosphate and preparation of lysates for immunoprecipitation (*UNIT 17.2*), phosphoamino acid analysis (*UNIT 17.3*), and detection of protein phosphorylation using immunologic or enzymatic techniques (*UNIT 17.4*).

David Moore and
Bartholomew M. Sefton

Overview of Protein Phosphorylation

UNIT 17.1

17.1

Phosphorylation is the most common and important mechanism of acute and reversible regulation of protein function. It is fundamental to the regulation of cell proliferation. Most polypeptide growth factors (platelet-derived growth factor and epidermal growth factor are among the best studied) and cytokines (e.g., interleukin-2, colony stimulating factor 1, and γ-interferon) stimulate phosphorylation upon binding to their receptors. The induced phosphorylation in turn activates cytoplasmic protein kinases, such as raf, MEK, and MAP. Additionally, in all nucleated organisms, cell cycle progression is regulated at both the G1/S and the G2/M transitions by cyclin-dependent protein kinases. Differentiation and development are also controlled by phosphorylation. Development of the R7 cell in the *Drosophila* retina and of the vulva in *Caenorhabditis elegans* are both dependent on the function of receptor and cytoplasmic protein kinases. Finally, metabolism, in particular the interconversion of glucose and glycogen and the transport of glucose, is regulated by phosphorylation. Biologists of all stripes therefore find, often unexpectedly and occasionally reluctantly, that they must study protein phosphorylation in order to understand the regulation and function of their favorite gene and its product.

Protein phosphorylation is usually studied by biosynthetic labeling with ^{32}P-labeled inorganic phosphate ($^{32}P_i$). This is intrinsically quite simple—the label is just added to growth medium. A general protocol for biosynthetic labeling with $^{32}P_i$ that maximizes incorporation and minimizes radioactive exposure of workers in the lab and contamination of lab equipment is described in *UNIT 17.2*.

Most proteins are found to be phosphorylated at serine or threonine residues, and many proteins involved in signal transduction are also phosphorylated at tyrosine residues. These three hydroxyphosphoamino acids exhibit sufficient chemical stability at acidic pH that they can be recovered after acid hydrolysis and identified in a straightforward manner. A technique for the identification of phosphoserine, phosphothreonine, and phosphotyrosine by acid hydrolysis and two-dimensional thin-layer electrophoresis is described in *UNIT 17.3*. Proteins that contain covalently bound phosphate at histidine, cysteine, and aspartic acid residues, either as phosphoenzyme intermediates or as stable modifications, have been described. Each of these phosphoamino acids is chemically labile and impossible to study with the standard techniques used for the acid-stable phosphoamino acids. Indeed, they are often identified by inference or elimination. The analysis of acid-labile forms of protein phosphorylation is beyond the scope of this section, and the reader is referred to *Methods in Enzymology*, Volume 200, for techniques for identification of these novel phosphoamino acids.

Phosphotyrosine is not an abundant phosphoamino acid. Therefore, its detection in samples labeled with $^{32}P_i$ is often difficult, especially if the samples contain large quantities of proteins phosphorylated at serine residues or are contaminated with RNA. Detection of phosphotyrosine and phosphothreonine can be enhanced considerably by incubation of gel-fractionated samples in alkali. This hydrolyzes RNA and dephosphorylates phosphoserine, allowing visualization of minor tyrosine- and threonine-phosphorylated proteins. A simple procedure for alkaline treatment is described in *UNIT 17.3*.

If a protein is modified by phosphorylation, identification of the phosphoamino acid can often be accomplished without resorting to biosynthetic labeling. For example, tyrosine phosphorylation can be studied because proteins containing this rare phosphoamino acid can be detected with great specificity and sensitivity by antibodies to phosphotyrosine. More generally, because phosphorylation often alters the mobility of a protein during SDS–polyacrylamide gel electrophoresis and almost always alters its isoelectric point, the presence of phosphorylated residues in an unlabeled protein can be deduced from altered gel mobility after incubation of the protein with a phosphatase. These approaches are useful if the protein of interest is derived from tissues that are difficult to label biosynthetically or from in vitro translation, where the endogenous pool of ATP renders labeling with [γ-^{32}P]ATP very inefficient. *UNIT 17.4* describes protocols for detection of tyrosine phosphorylation by immunoblotting with antibodies to phosphotyrosine and for enzymatic dephosphorylation of proteins.

References: Cohen, 1985; Marshall, 1995.

Contributor: Bartholomew M. Sefton

Labeling Cultured Cells with $^{32}P_i$ and Preparing Cell Lysates for Immunoprecipitation

UNIT 17.2

This unit describes $^{32}P_i$ labeling and lysis of cultured cells to be used for subsequent immunoprecipitation of proteins. The approach is appropriate, however, for labeling any cellular constituent with ^{32}P. This procedure is suitable for insect, avian, and mammalian cells and can be used with both adherent and nonadherent cultures.

CAUTION: Unshielded ^{32}P will penetrate ~1 cm into flesh. Exposure to the skin and eyes is, therefore, of concern. Gloves and protective eyeware should always be worn when handling significant amounts of ^{32}P. A 1-in-thick (2.5-cm) Plexiglas shield, tall enough to look through when seated or standing comfortably, should be used when handling samples containing ^{32}P.

LABELING CULTURED CELLS WITH $^{32}P_i$ AND LYSIS USING MILD DETERGENT

BASIC PROTOCOL

Materials *(see APPENDIX 1 for items with ✓)*

Cell culture to be labeled
Labeling medium: phosphate-free tissue culture medium (e.g., DMEM) supplemented with the usual concentration of serum or serum dialyzed against phosphate-free saline, 37°C
1 Ci/ml $H_3{}^{32}PO_4$ in HCl (carrier free ICN)
✓ Tris-buffered saline (TBS) or PBS, cold
✓ Mild lysis buffer or RIPA lysis buffer

1-in-thick Plexiglas shield
Plugged, aerosol-resistant pipet tips
Plexiglas box, warmed to 37°C
Screw-cap microcentrifuge tubes
Plugged disposable pipet or disposable one-piece transfer pipet
Rubber policeman
Sorvall refrigerated centrifuge with SM 24 rotor and rubber adaptors, refrigerated microcentrifuguge, or equivalent, 4°C
Plexiglas sheet (10 × 10 × ¼–in.) or Plexiglas tube holder, 4°C

NOTE: All culture incubations are performed in a humidified 37°C, 10% CO_2 incubator unless otherwise specified.

1. Culture the cells to be labeled to an appropriate stage of growth.

   Phosphate transport is maximal in rapidly growing cells. Therefore, except in those cases where phosphorylation of a protein in quiescent cells is to be examined, cells to be labeled should be subconfluent (adherent cells) or at less than maximal density (nonadherent cells). It is useful to change the medium to fresh growth medium 3 to 18 hr prior to labeling.

For adherent cells:

2a. Remove growth medium by aspiration. Wash away any residual phosphate-containing medium by adding 37°C labeling medium supplemented with serum, but lacking the label, and removing the wash medium by aspiration.

17.2

3a. Add prewarmed labeling medium to cultures, using 0.5 to 1 ml per 35-mm dish, 1 to 2 ml per 50-mm dish, or 2 to 4 ml per 100-mm dish of adherent cells.

For nonadherent cells:

2b. Gently centrifuge the culture 1 min at 1800 × *g* and aspirate the medium away from the cell pellet. Resuspend the cells in labeling medium supplemented with serum, but lacking the label. Centrifuge and remove the medium.

3b. Add 2 ml medium per 10^7 cells and transfer to an appropriate size petri dish.

4. Working behind a Plexiglas shield, use a micropipettor with plugged, aerosol-resistant pipet tips to add $^{32}P_i$ to a final concentration of 0.1 to 2 mCi/ml.

   Labeling for 1 to 2 hr is usually sufficient, but cells will tolerate as much as 2 mCi/ml for 6 hr and lower concentrations (0.1 to 0.5 mCi/ml) for 18 hr.

5. Place dishes in a warmed Plexiglas box and put box in the incubator.

6. At the end of the labeling period, carry the labeled cells, still in the Plexiglas box, into the cold room. Place the cells behind a Plexiglas shield.

For adherent cells:

7a. Take the dish out of the box and remove the labeling medium manually, using either a plugged disposable pipet or a disposable one-piece plastic Pasteur transfer pipet. Discard the medium and pipet as radioactive wastes.

   Labeling medium should be removed without using a vacuum aspirator. Vacuum aspiration generates radioactive aerosols and leaves a radioactive film on the equipment.

   Continue to use shielding with cells and lysate.

8a. Wash cells once with 2 to 10 ml cold TBS. Remove the wash buffer manually, as in step 7, and discard as radioactive waste.

9a. Add lysis buffer to cells, using 0.3 ml per 35-mm dish, 0.6 ml per 50-mm dish, or 1.0 ml per 100-mm dish of adherent cells. Dislodge adherent cells by scraping with a rubber policeman, but leave the lysate in the dish. Incubate 20 min at 4°C. With a rubber policeman, scrape the lysate of adherent cells to the side of the dish and transfer lysate to a screw-cap microcentrifuge tube.

   If a low background of nonspecific contaminants is critical, use RIPA lysis buffer. If maintenance of enzymatic activity or the structure of protein complexes is critical, use a milder lysis buffer containing either 3[(3-cholamidopropyl)-dimethylammonio]-1-propane-sulfonate (CHAPS) or Nonidet P-40 (NP-40) as the only detergent. If complete solubilization of the cells and denaturation of the protein is desired, use SDS for lysis (see Alternate Protocol).

For nonadherent cells:

7b. Take the dish out of the box, transfer the cells to a screw-cap centrifuge tube, pellet them by centrifugation (1 min at 1800 × *g*), and remove the medium.

8b. Resuspend pelleted cells gently in a small volume of cold TBS, transfer to a screw-cap microcentrifuge tube, and pellet the cells by microcentrifuging 1 min at 1800 × *g*.

9b. Add 0.5 to 1 ml of the appropriate lysis buffer per 10^7 cells and resuspend the pellet by gentle agitation with a disposable plastic Pasteur pipet. Incubate 20 min at 4°C.

10. Cap the tube, and clarify the lysate by centrifuging 30 min at 26,000 × *g*, 4°C.

    It is best to half-fill the tube to prevent spilling.

    Lysates prepared with RIPA buffer often become viscous due to lysis of nuclei. If this occurs, increase the time of centrifugation to 90 min, or add 50 μl of fixed Staphylococcus aureus bacteria (Pansorbin, Calbiochem) in RIPA buffer to the lysate prior to centrifugation. Either modification will cause the solubilized DNA to pellet.

11. After centrifugation, transfer the supernatant (lysate) to a new tube and discard the tube and pellet as radioactive waste.

12. Analyze the labeled lysate using gel electrophoresis, immunoprecipitation, or protein purification. Carry out all analytical procedures at 4°C using adequate shielding.

LYSIS OF CELLS BY BOILING IN SDS

ALTERNATE PROTOCOL

Some proteins, such as eukaryotic RNA polymerase II, are difficult to solubilize with mild lysis buffer or RIPA lysis buffer, and some analytical procedures use antibodies that recognize epitopes exposed only in denatured proteins. In these cases, it is useful to solubilize labeled cells completely in SDS and then adjust the composition of the lysate solution to match that of RIPA buffer for immunoprecipitation. To avoid the formation of spurious disulfide bonds, lysis and washing during immunoprecipitation are carried out in the presence of fresh 1 mM dithiothreitol (DTT).

Additional Materials *(also see Basic Protocol; see* APPENDIX 1 *for items with ✓)*

✓ SDS lysis buffer
✓ RIPA correction buffer
✓ Immunoprecipitate wash buffer
Fixed *Staphylococcus aureus* bacteria (Pansorbin, Calbiochem; optional)
Boiling water bath

1. Label and wash cells (Basic Protocol, steps 1 to 7).

2a. *For adherent cells:* Add SDS lysis buffer using 0.1 ml for a 35-mm dish, 0.25 ml for a 50-mm dish, or 0.5 ml for an 100-mm dish. Immediately scrape the dish with a rubber policeman and transfer the cell lysate to a screw-cap microcentrifuge tube.

2b. *For nonadherent cells:* Vortex briefly to loosen the cell pellet, add 1 ml SDS lysis buffer per 5×10^7 cells and vortex again.

3. Boil the samples 2 to 5 min, then add 4 vol RIPA correction buffer and mix well.

4. Clarify the cell lysate by centrifuging 90 min at 26,000 × *g*, 4°C or at maximum speed, 4°C in a refrigerated microcentrifuge.

    Lysate may also be clarified by adding 50 μl fixed Staphylococcus aureus and centrifuging 30 min at 26,000 × g, 4°C.

5. Carry out immunoprecipitation (UNIT 10.15) as usual using immunoprecipitate wash buffer for washes.

Contributor: Bartholomew M. Sefton

Phosphoamino Acid Analysis

It is often valuable to identify the phosphorylated residue in a protein. In the case of proteins phosphorylated at serine, threonine, or tyrosine, this is readily accomplished by partial acid hydrolysis in HCl followed by two-dimensional thin-layer electrophoresis of the labeled phosphoamino acid.

NOTE: Wear gloves and use blunt-end forceps to handle membrane.

BASIC PROTOCOL

ACID HYDROLYSIS AND TWO-DIMENSIONAL ELECTROPHORETIC ANALYSIS OF PHOSPHOAMINO ACIDS

The protein to be acid hydrolyzed is transferred to a PVDF membrane using the same technique used for immunoblotting (*UNIT 10.7*) or for microsequencing (*UNIT 10.18*).

Materials *(see APPENDIX 1 for items with ✓)*

^{32}P-labeled phosphoprotein (*UNIT 18.2*)
India ink solution: 1 µl/ml India ink in TBS/0.02% (v/v) Tween 20, pH 6.5 (prepare fresh or store indefinitely at room temperature); *or* radioactive or phosphorescent alignment markers
✓ 6 M HCl
✓ Phosphoamino acid standards mixture
✓ pH 1.9 electrophoresis buffer
✓ pH 3.5 electrophoresis buffer
0.25% (w/v) ninhydrin in acetone in a freon (aerosol, gas-driven) atomizer/sprayer

PVDF membrane (Immobilon-P, Millipore)
110°C oven
Screw-cap microcentrifuge tubes
20 cm × 20 cm × 100 µm glass-backed cellulose thin-layer chromatography plate (EM Sciences)
Large blotter: two 25 × 25–cm layers of Whatman 3MM paper sewn together at the edges, with four 2-cm holes that align with the origins on the TLC plate
Glass tray or plastic box
Whatman 3MM paper
Thin-layer electrophoresis apparatus (e.g., HTLE 7000, CBS Scientific)
Fan
Small blotters: 4 × 25–cm, 5 × 25–cm, and 10 × 25–cm pieces of Whatman 3MM paper
50° to 80°C drying oven
Sheets of transparency film for overhead projector

1. Run radiolabeled phosphoprotein on a preparative SDS-polyacrylamide gel. Transfer proteins electrophoretically to a PVDF membrane. Wash the membrane several times with water. Do not let the membrane dry.
2. Locate the band of interest by staining the membrane 5 to 10 min in 30 to 50 ml India ink solution with shaking until bands are detectable, *or* by wrapping the membrane in plastic wrap, applying radioactive or phosphorescent alignment markers, and performing autoradiography.
3. Excise the piece of membrane containing the band of interest with a clean razor blade. Rewet the piece of membrane with methanol for 1 min and then rewet it

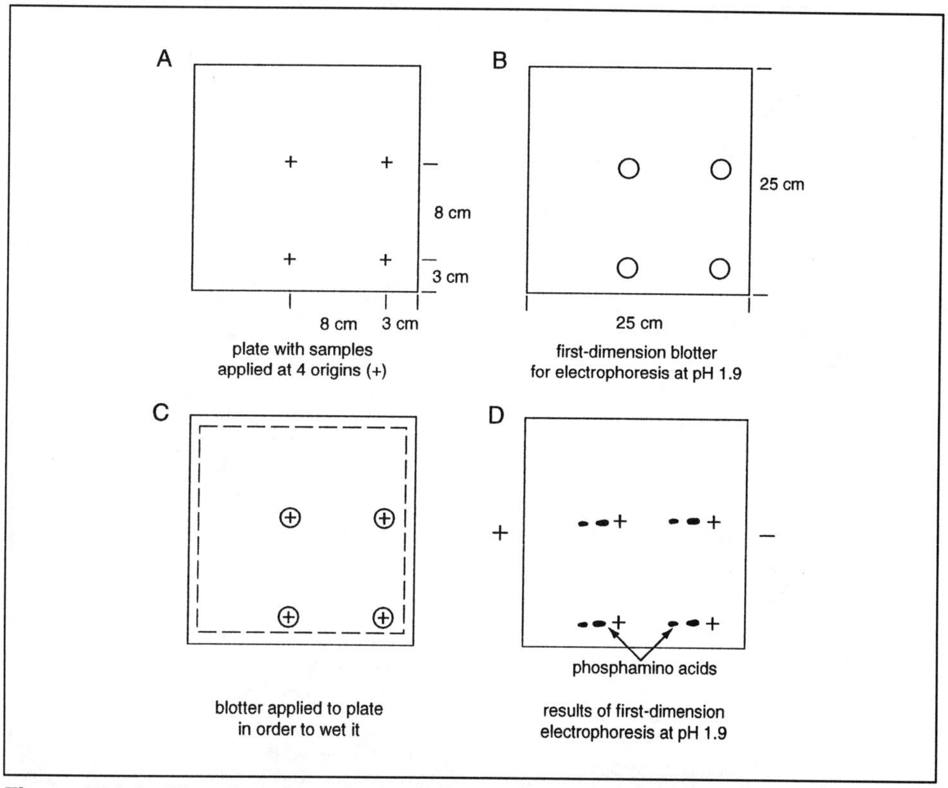

Figure 17.3.1 First-dimension electrophoretic separation of phosphoamino acids at pH 1.9. (**A**) Positions of the four origins on a single 20 × 20–cm plate; (**B**) blotter used for wetting the plate with pH 1.9 electrophoresis buffer; (**C**) placement of the blotter on the plate (underneath; indicated by dashed outline); and (**D**) orientation of the plate between the + and – electrodes with the positions of the phosphoamino acids after electrophoresis.

in >0.5 ml water. Place the piece of membrane in a screw-cap microcentrifuge tube.

4. Add enough 6 M HCl to submerge the membrane. Screw the cap on the tube tightly and incubate 60 min in 110°C oven.

5. Let cool. Microcentrifuge 2 min at maximum speed. Transfer the liquid hydrolysate to a fresh microcentrifuge tube and dry with a Speedvac evaporator (2 hr).

6. Dissolve the sample in 6 to 10 µl water by vortexing vigorously. Microcentrifuge 5 min at maximum speed.

7. Spot 25% to 50% of the sample, in 0.25- to 0.50-µl aliquots, on one origin of a 20 cm × 20 cm × 100 µm glass-backed cellulose thin-layer chromatography plate (see Fig. 17.3.1 for arrangement of samples). Between each application, dry the sample spot with compressed air delivered through a Pasteur pipet plugged with cotton.

8. Spot 1 µl nonradioactive phosphoamino acid standards mixture (containing phosphoserine, phosphothreonine, and phosphotyrosine) on top of each sample in 0.25- to 0.50-µl aliquots as above.

9. Wet the large blotter (with four holes) by submerging it in pH 1.9 electrophoresis buffer in a large glass tray or plastic box. Briefly allow the excess buffer to drain off. Lower the wet blotter onto the prespotted plate with the origins on the plate

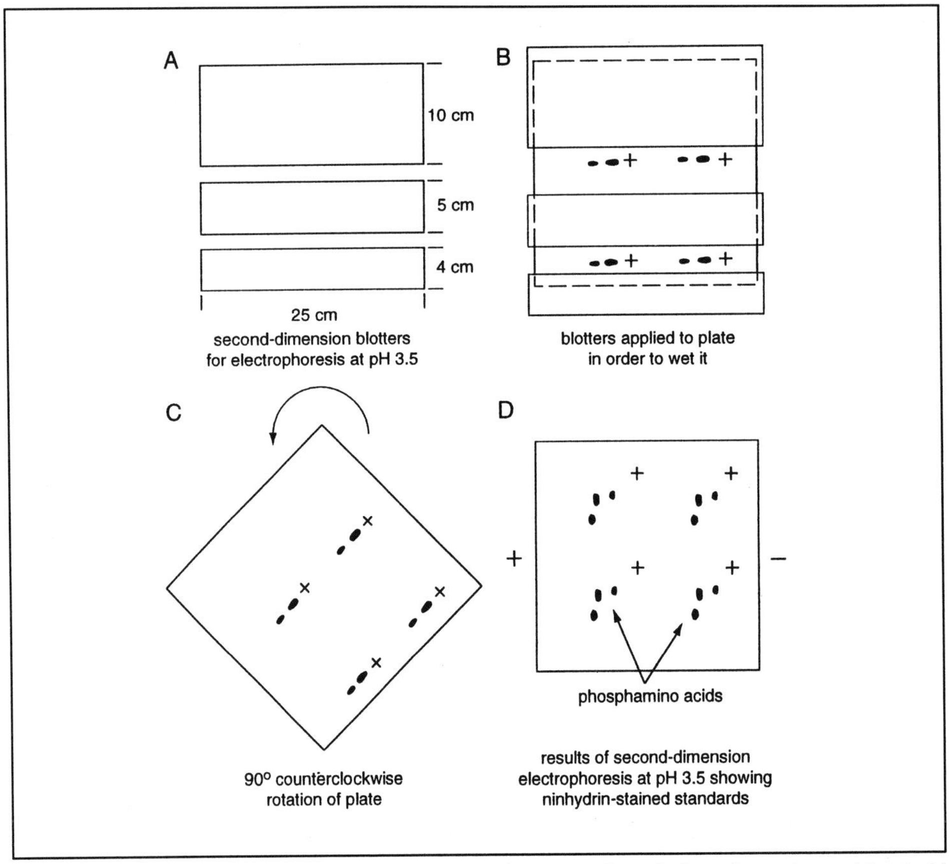

Figure 17.3.2 Second-dimension electrophoretic separation of phosphoamino acids at pH 3.5. (**A**) The three pieces of Whatman 3MM paper used for wetting the plate with pH 3.5 electrophoresis buffer; (**B**) proper placement of the blotters on the plate (underneath; indicated by dashed outline); (**C**) reorientation of the plate for electrophoresis in the second dimension; and (**D**) orientation of the plate between the + and – electrodes with the position of the phosphoamino acids after electrophoresis.

in the centers of the four holes in the blotter (Fig. 17.3.1). Press on the blotter gently to achieve even wetting of the cellulose and concentration of the samples. When the plate is uniformly wet, remove the blotter.

10. Place the thin-layer plate in the electrophoresis apparatus and overlap 0.5 cm of the right and left sides of the plate with wicks made of Whatman 3MM paper. If the apparatus has an air bag, be sure to inflate it. Close the cover and start electrophoresis. With an HTLE 7000, double-thickness Whatman 3MM wicks, and a plate with four samples, electrophorese 20 min at 1.5 kV.

    For other electrophoresis apparatuses the appropriate duration of electrophoresis can be determined empirically by examining the rate of migration of the phosphoamino acid standards.

11. Following electrophoresis, remove the plate and quickly air dry with a fan without heating (20 min).

12. Wet the small blotters in pH 3.5 electrophoresis buffer and use them to wet the plate using the method described in step 10 to achieve even wetting without puddling (Fig. 17.3.2). Avoid placing the blotters over the sample.

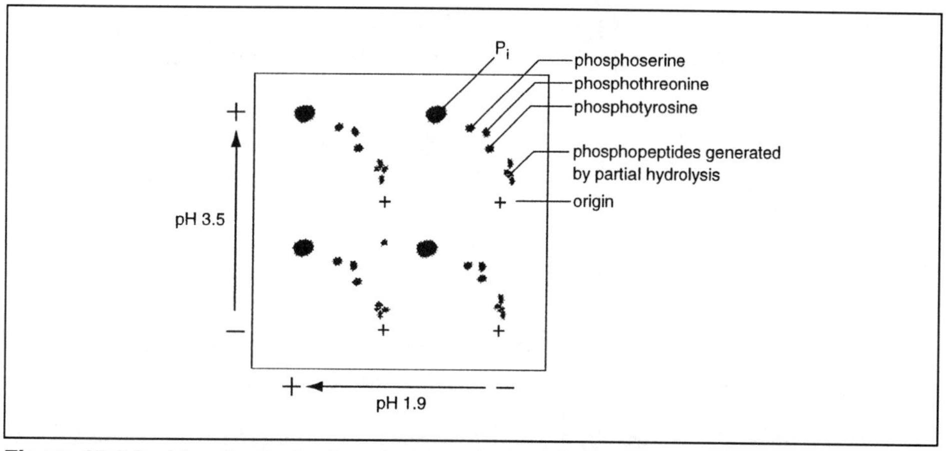

Figure 17.3.3 Hypothetical autoradiogram of a two-dimensional separation. Four samples of acid-hydrolyzed, ^{32}P-labeled proteins are applied at the origins, one in each of the four quandrants. This diagram shows the origins, the directions of electrophoresis, the positions of phosphoserine, phosphothreonine, and phosphotyrosine, the position of P$_i$, and the position of partially hydrolyzed fragments of the proteins for the upper right-hand sample. Every protein generates different partial hydrolysis peptide fragments.

13. Remove the blotters, rotate the plate 90° counterclockwise, and electrophorese 16 min at 1.3 kV in pH 3.5 electrophoresis buffer if using the HTLE 7000 apparatus.

14. At the end of the electrophoresis run, remove the plate and dry 20 to 30 min in an oven at 50° to 80°C. When dry, spray with 0.25% ninhydrin in acetone, then reheat in the oven 5 to 10 min to visualize the phosphoamino acid standards.

15. Place radioactive or phosphorescent alignment marks on the plate and autoradiograph with an intensifying screen overnight to 10 days at –70°C.

16. Following autoradiography, trace the alignment markers and the stained phosphoamino acid markers onto a transparent sheet used for overhead projectors. Save this template. Align the film with the plate and identify radioactive phosphoamino acids (Fig. 17.3.3).

ALKALI TREATMENT TO ENHANCE DETECTION OF TYR- AND THR-PHOSPHORYLATED PROTEINS BLOTTED ONTO FILTERS

ALTERNATE PROTOCOL

Because phosphothreonine and phosphotyrosine are much more stable to hydrolysis in alkali than RNA or phosphoserine, detection of proteins containing phosphothreonine and phosphotyrosine can often be enhanced by mild alkaline hydrolysis of gel-fractionated samples. Alkaline hydrolysis does not preclude subsequent phosphoamino acid analysis. A band from a blot that has been treated with alkali can be excised and subjected to acid hydrolysis.

Additional Materials *(also see Basic Protocol; see* APPENDIX 1 *for items with ✓)*

1 M KOH
TN buffer: 10 mM Tris·Cl (pH 7.4 at room temperature)/0.15 M NaCl
✓ 1 M Tris·Cl, pH 7.0 at room temperature

Covered plastic container (e.g., Tupperware box)
55°C oven or water bath

1. Run radiolabeled phosphoprotein on a preparative SDS–polyacrylamide gel and transfer proteins electrophoretically to a PVDF membrane (Basic Protocol, steps 1 and 2).

   A nylon membrane may be used in place of a PVDF membrane, but in that case, the bands cannot subsequently be analyzed by acid hydrolysis, as nylon membrane will dissolve in 6 M HCl.

2. Wash membrane thoroughly with water: three 2-min incubations in 1 liter water are sufficient.

3. Incubate membrane 120 min at 55°C in an oven or water bath in sufficient 1 M KOH to cover the filter in a covered Tupperware container.

4. Discard KOH. Wash membrane and neutralize remaining KOH by rinsing once for 5 min in 500 ml TN buffer, once for 5 min in 500 ml of 1 M Tris·Cl (pH 7.0), and twice for 5 min in 500 ml water. Wrap the membrane in plastic wrap and autoradiograph overnight with flashed film and an intensifying screen at –70°C.

Reference: Kamps and Sefton, 1989.

Contributor: Bartholomew M. Sefton

UNIT 17.4

Analysis of Phosphorylation of Unlabeled Proteins

BASIC PROTOCOL 1

IMMUNOBLOTTING WITH ANTI-PHOSPHOTYROSINE ANTIBODIES AND DETECTION USING [^{125}I]PROTEIN A

Materials *(see* APPENDIX 1 *for items with ✓)*

Protein sample: cultured cells, tissue, lysate, or immunoprecipitate
Transfer buffer containing 100 µM sodium vanadate
✓ Blocking buffer
Anti-phosphotyrosine antibody: 2 µg/ml rabbit polyclonal anti-phosphotyrosine (UBI) *or* mouse monoclonal anti-phosphotyrosine—e.g., py20 (Leinco, ICN Biomedicals, Zymed, Transduction Laboratories) or 4G10 (UBI)—in blocking buffer
TN buffer: 10 mM Tris·Cl (pH 7.4 at room temperature)/0.15 M NaCl
TNA solution: TN buffer (recipe above)/0.01% (w/v) sodium azide (store at room temperature)
NP-40 wash solution: 0.05% (v/v) Nonidet P-40 (NP-40)/TNA solution (store at room temperature)
0.5 µCi/ml ^{125}I-labeled protein A (30 mCi/mg, ICN) in blocking buffer
India ink solution: 1 µl India ink in TBS/0.02% (w/v) Tween 20, pH 6.5 (prepare fresh or store indefinitely at room temperature; optional)

Plastic container with lid (e.g., Tupperware box)
Blotting paper

NOTE: Wear gloves and use blunt-end forceps to handle membrane.

17.4

1. Dissolve cultured cells, tissue, or lysate immunoprecipitate in an equal volume of 2× SDS sample buffer and boil 5 min.

   Prevent phosphorylation or dephosphorylation from occurring following cell lysis by using 50 mM sodium fluoride to inhibit serine and threonine phosphatases, 0.2 mM sodium vanadate to inhibit tyrosine phosphatases, and 2 mM EDTA to inhibit kinases.

2. Electrophorese sample on an SDS-polyacrylamide gel. Transfer proteins to a membrane suitable for immunoblotting using transfer buffer that contains 100 µM sodium vanadate.

3. Incubate membrane in 30 to 50 ml blocking buffer ≥30 min at room temperature.

   Blotto and other blocking mixtures containing dry milk are unsuitable because anti-phosphotyrosine antibodies bind to a number of the protein constituents of milk. Blocking buffer can be reused several times.

4. Incubate the membrane in 30 to 50 ml anti-phosphotyrosine antibody solution in a covered plastic box, 60 to 120 min at room temperature with occasional agitation.

   The antibody solution can be saved in the plastic box, stored at 4°C for at least a month, and reused up to five to ten times.

5. Remove the membrane from the staining box and wick off excess liquid with blotting paper. In a new plastic box, wash the membrane twice for 10 min with 50 ml TBSA, twice for 10 min with 50 ml NP-40 wash solution, and twice for 5 min with 50 ml TBSA. Discard washes.

6. Incubate membrane in 30 to 50 ml of 0.5 µCi/ml ^{125}I-labeled protein A in blocking buffer 60 min at room temperature with gentle agitation in a covered plastic box.

7. Remove the membrane and wick off excess liquid with blotting paper. Wash the membrane as described in step 5 in a clean plastic box.

   Protein A solution can be stored at least 1 month at 4°C and reused ~3 to 5 times.

8. If desired, stain the membrane 5 to 10 min with 30 to 50 ml India ink solution until bands are detectable. Wash the membrane with water to remove excess ink.

9. Remove excess moisture with blotting paper, wrap membrane in plastic wrap, attach radioactive or fluorescent alignment markers, and expose by fluorography using an intensifying screen and preflashed film for 18 hr to 10 days.

ALTERNATE PROTOCOL

DETECTION OF BOUND ANTIBODIES BY ENHANCED CHEMILUMINESCENCE (ECL)

Bound anti-phosphotyrosine antibodies can also be detected using enhanced chemiluminescence (ECL) which requires autoradiographic exposure of only minutes rather than days. It also avoids the use of radioactivity. However, results obtained with the two techniques are not always identical and exhibit quantitative differences. The technique appears to work somewhat better with nitrocellulose membranes than with Immobilon-P. Therefore, handling of the membranes requires somewhat more delicacy.

Additional Materials *(also see Basic Protocol 1; see* APPENDIX 1 *for items with ✓)*

✓ Blocking buffer *without* azide
0.05% (v/v) Nonidet P-40 (NP-40)/TN buffer (Basic Protocol 1)
Horseradish peroxidase–conjugated secondary antibody: anti-rabbit or anti-mouse antibody diluted 1/1000 to 1/2000 in blocking buffer *without* azide
ECL detection reagents A and B (Amersham)
Luminol reagent for ECL detection (Amersham)
Oxidizing reagent for ECL detection (Amersham)

Nitrocellulose membrane
Plastic container slightly larger than the membrane
Plastic sheet protector

NOTE: The reagents for ECL may be purchased as a kit, the Enhanced Chemiluminescence Western Blotting Detection System, from Amersham.

NOTE: Wear gloves and use blunt-end forceps to handle membrane.

1. Electrophorese sample prepared as described in Basic Protocol step 1 on an SDS-polyacrylamide gel. Transfer proteins to a nitrocellulose membrane using a transfer buffer that contains 100 µM sodium vanadate.
2. In a plastic container, incubate membrane in blocking buffer without azide ≥30 min at room temperature.
3. Incubate membrane in 30 to 50 ml anti-phosphotyrosine antibody 60 to 120 min at room temperature with occasional agitation.
4. In a clean plastic container, wash the membrane twice for 10 min with TN buffer, twice for 10 min with 0.05% NP-40/TN buffer, and twice for 5 min with TN buffer. Discard the washes.
5. In a plastic container, incubate the membrane in 30 to 50 ml of horseradish peroxidase–conjugated secondary antibody appropriate for the primary antibody 60 min at room temperature with gentle agitation.
6. Remove membrane and wash it in a clean plastic container as described in step 4.
7. In a plastic container only slightly larger than the membrane, mix equal volumes of the Luminol reagent and the oxidizing reagent for ECL detection.
8. Put the membrane, with the side containing protein facing up, into the mixed reagents. Agitate gently 60 sec.
9. Remove the blot, wick off excess moisture, and place it, protein-side up, in a sheet protector.
10. In the darkroom, expose the blot to X-ray film 15 sec to 30 min.

Table 17.4.1 Phosphatase Reaction Conditions[a]

Phosphatase (supplier[b])	Amino acid specificity	Buffer	Amount	Incubation conditions	Inhibitor
Bacterial alkaline phosphatase (Pharmacia Biotech)	Phosphoserine, phosphothreonine, and phosphotyrosine	20 mM HEPES, pH 7.0 150 mM NaCl 0.1% (v/v) Triton X-100 10% (v/v) glycerol	2-4 U per reaction	60 min at 30°C	100 mM sodium phosphate
Potato acid phosphatase (Sigma)	Phosphoserine, phosphothreonine, and phosphotyrosine	40 mM PIPES, pH 6.0 1 mM DTT 20 μg/ml aprotinin 20 μM leupeptin	100 μg/ml	10 min at 30°C or 30 min at 4°C	100 mM sodium phosphate
Protein phosphatase 2A (UBI or Calbiochem)	Phosphoserine and phosphothreonine	20 mM HEPES, pH 7.0 1 mM DTT 1 mM $MnCl_2$ 100 μg/ml BSA 50 μM leupeptin	10 U/ml	30 to 60 min at 30°C	20 μM okadaic acid
PTP-1B (UBI) or *Yersinia* PTP (Calbiochem)	Phosphotyrosine	25 mM imidazole·HCl, pH 7.0 1 mg/ml BSA 0.1% (v/v) 2-ME	0.5 mg per reaction	60 min at 30°C	200 μM sodium vanadate

[a]Abbreviations: BSA, bovine serum albumin; DTT, dithiothreitol; HEPES, *N*-[2-hydroxyethyl]piperazine-*N*′-[2-ethanesulfonic acid]; 2-ME, 2-mercaptoethanol; PIPES, piperazine-*N,N*′-bis[2-ethanesulfonic acid]; PTP, protein tyrosine phosphatase.

[b]See *APPENDIX 4* for supplier contact information.

IDENTIFICATION OF PHOSPHORYLATED PROTEINS BY PHOSPHATASE DIGESTION

BASIC PROTOCOL 2

Phosphorylation of a protein often alters its mobility during SDS-polyacrylamide gel electrophoresis. An effect of phosphatase digestion on the gel mobility is therefore diagnostic of phosphorylation and is a useful means to determine whether an unlabeled protein is phosphorylated. Additionally, the effect of phosphorylation on the activity of a protein can be assessed by measuring changes in activity following enzymatic dephosphorylation.

Enzymatic protein dephosphorylation has traditionally been accomplished with alkaline phosphatase or potato acid phosphatase. Both enzymes dephosphorylate phosphoserine, phosphothreonine, and phosphotyrosine. However, phosphoamino acid–specific phosphatases are now commercially available and offer the ability to dephosphorylate a protein specifically at only serine and threonine residues, or at only tyrosine residues. Some prepartions of phosphatases are contaminated with proteases, so it is advisable to include protease inhibitor(s)—e.g., 10,000 IU/ml Trasylol, 40 μg/ml leupeptin, 400 μM phenylmethylsulfonyl chloride (PMSF), or 40 μg/ml soybean trypsin inhibitor—in the reaction mix.

Enzymatic dephosphorylation is intrinsically quite simple. The protein is dissolved in, or in the case of an immunoprecipitated protein suspended in, a buffer suitable for the phosphatase of interest and incubated with the phosphatase for 30 to 60 min. Table 17.4.1 describes the reaction conditions for a number of phosphatases.

Dephosphorylated protein present in an immunoprecipitate can be recovered by centrifugation and analyzed by gel electrophoresis or an enzymatic assay. Alternatively, soluble protein can be analyzed by adding an equal volume of 2× SDS sample buffer (*UNIT 10.2*) to the reaction mix and subjecting the digested protein to gel electrophoresis. Soluble protein can be assayed enzymatically after dephosphorylation if an inhibitor of the phosphatase is added to the reaction mixture. Okadaic acid (20 μM) is a potent and very specific inhibitor of the serine/threonine–specific phosphatase protein phosphatase 2A (PP2A). Sodium vanadate (200 μM) is an efficient inhibitor of all known tyrosine phosphatases. Sodium phosphate (100 mM) is an inhibitor of all protein phosphatases.

References: Kamps and Sefton, 1988; Morla and Wang, 1986.

Contributor: Bartholomew M. Sefton

Appendices

REAGENTS AND SOLUTIONS

This appendix includes all recipes for the reagents and solutions used in *Short Protocols* except for media and enzyme buffers which are cross-referenced parenthetically in the materials lists. Solutions are listed alphabetically, with the unit(s) in which the recipes are used listed in parentheses. These cross-references are critical because reagents from different protocols may have similar names, but the recipes differ. No cross-references are provided for commonly used buffers and solutions (e.g., TE buffer and 1 M $MgCl_2$, respecively). Ingredients in some recipes include reagents for which recipes exist elsewhere in APPENDIX 1; these items are indicated by ✓. Many commonly used ingredients are abbreviated throughout this appendix; see APPENDIX 2 for a full listing of abbreviations and their definitions.

NOTE: All reagents are prepared using distilled, deionized water. It is important to follow laboratory safety guidelines and heed manufacturers' precautions when working with noxious or carcinogenic chemicals.

Acetate buffer, pH 4.3 *(UNIT 11.11)*

To 900 ml H_2O, add:
6.80 g sodium acetate (trihydrate; 0.05 M final)
8.77 g NaCl (0.15 M final)
Titrate with acetic acid to pH 4.3
Add H_2O to 1 liter

Acetate buffer, 50 mM (pH 5.0) *(UNIT 14.8)*

Add 74 ml of 0.2 M acetic acid (11.55 ml glacial acetic acid/liter) and 176 ml of 0.2 M sodium acetate (27.2 g sodium acetate trihydrate in 1 liter) to 1 liter of deionized water and mix.

Acid phenol *(UNIT 13.12)*

Add sufficient water to a bottle of solid phenol such that phenol is water-saturated; pH will be ~5.0. Do not buffer phenol. Store at 4°C, protected from light.

Acid precipitation solution

✓ 1 M HCl
0.1 M sodium pyrophosphate

Nucleic acids also can be precipitated with a 10% (v/v) solution of trichloroacetic acid; however, this recipe is cheap, easy to prepare, and efficient.

Acrylamide/bisacrylamide solutions *(UNITS 2.7, 2.11, 10.2 & 10.3)*

Proportion (w/w)	Acrylamide	Bisacrylamide	H_2O
29%/1% *(UNIT 2.7)*	29 g	1 g	to 100 ml
30%/0.8% *(UNITS 10.2 & 10.3)*	300 g	8 g	to 1000 ml
30%/1.8% *(UNIT 10.3)*	30 g	1.8 g	to 100 ml
40%/2% *(UNIT 2.11)*	100 g	5 g	to 250 ml

Prepare by adding approximately half the final volume of water to acrylamide (it is convenient to use the original container of acrylamide for preparing and storing the solution). Add *N,N′*-methylene-bisacrylamide (Bio-Rad) and mix until dissolved. If desired, add Amberlite MB-1 (Sigma) to ~2% and stir again. Bring to final volume in a graduated cylinder and stir. Filter through Whatman No. 5 filter

continued

Table A.1.1 Molarities and Specific Gravities of Concentrated Acids and Bases

Acid/base	Molecular weight	% by weight	Molarity (approx.)	1 M solution (ml/l)	Specific gravity
Acetic acid (glacial)	60.05	99.6	17.4	57.5	1.05
Ammonium hydroxide	35.0	28	14.8	67.6	0.90
Formic acid	46.03	90	23.6	42.4	1.205
		98	25.9	38.5	1.22
Hydrochloric acid	36.46	36	11.6	85.9	1.18
Nitric acid	63.01	70	15.7	63.7	1.42
Perchloric acid	100.46	60	9.2	108.8	1.54
		72	12.2	82.1	1.70
Phosphoric acid	98.00	85	14.7	67.8	1.70
Sulfuric acid	98.07	98	18.3	54.5	1.835

paper or through 0.2- to 0.45-µm filter. Store at 4°C in tightly capped amber bottle ≤30 days; acrylamide gradually hydrolyzes to acrylic acid and ammonia.

CAUTION: *Acrylamide monomer is a neurotoxic. Wear gloves and proceed carefully when working with the unpolymerized monomer.*

AEC solution *(UNIT 14.8)*

Stock solution: Dissolve one tablet (20 mg) of 3-amino-9-ethylcarbazole (AEC; Sigma) in 2.5 ml *N,N* dimethylformamide. Store at 4°C in the dark up to 6 months.

Working solution: Add 250 µl stock solution to 5 ml of 50 mM acetate buffer, pH 5 (see recipe), then add 25 µl 30% H_2O_2. Make fresh before each use; keep solution in the dark.

Agarose, 1% (w/v) agarose *(UNIT 16.10)*

Place 0.5 g (1%) SeaKem agarose (FMC Bioproducts) into each of ten 100-ml Wheaton bottles. Add 50 ml H_2O to each bottle, close loosely, and autoclave 20 min. When the solutions have cooled, tighten lids. Store indefinitely at room temperature.

Agarose, 1% or 3% *(UNITS 2.5B & 16.10)*

Place 0.5 g (1%) or 1.5 g (3%) SeaPlaque agarose (FMC Bioproducts) into each of ten 100-ml Wheaton bottles. Add 50 ml H_2O to each bottle, loosen lids, and autoclave 45 min. When solutions have cooled, tighten lids. Store indefinitely at room temperature.

SeaPlaque is a low-melting-temperature agarose and is recommended in all tissue culture procedures because of its high level of purity.

Agarose gel *(UNIT 2.5A)*

Gels typically contain ~1% agarose in 1× TAE or TBE. Electrophoresis- grade agarose powder is added to 1× gel buffer and melted by boiling for several minutes. Be sure all agarose particles are completely melted. To facilitate visualization of DNA fragments during the run, ethidium bromide can be added to 0.5 µg/ml in the gel.

SeaKem agarose is recommended over SeaPlaque agarose for plaque assays because it can be used at a lower concentration. High concentrations of agarose tend to obscure the plaques.

Agarose overlay *(UNIT 16.10)*

40 ml 2× Graces insect medium, supplemented (Life Technologies)
10 ml heat-inactivated (30 min at 56°C) fetal bovine serum (FBS)
0.25 ml gentamicin (10 mg/ml stock; optional)
✓ 50 ml 1% (w/v) agarose (sterile)

Mix the first three ingredients together, filter sterilize, and equilibrate to 42°C. Microwave the 1% agarose ~2 min until liquid and equilibrate to 42°C. Mix the two solutions just prior to the end of the 1-hr viral incubation. Use immediately (this solution cannot be stored).

This is enough for 24 plates, using 4 ml/plate. 42°C is the optimum temperature for equilibration because if the overlay is too hot the cells will lyse, and if it is too cool the agarose will solidify into clumps, making visual screening difficult.

Agarose solution, 0.5%, hot *(UNIT 10.3)*

0.25 g agarose (standard low-Mr; Bio-Rad)
✓ 50 ml reservoir buffer

Heat in a boiling water bath to dissolve the agarose and keep in a boiling water bath. Prepare fresh each time.

Agarose solution, 1%, hot *(UNIT 10.3)*

Prepare as for hot 0.5% agarose solution, substituting 0.5 g agarose for 0.25 g agarose.

AHC medium and plates (–Ura –Trp; *UNIT 6.10)*

1.7 g yeast nitrogen base *without* amino acids and *without* ammonium sulfate (Difco)
5 g ammonium sulfate
10 g casein hydrolysate-acid, salt-free and vitamin-free (U.S. Biochemical)
50 ml (for medium) or 10 ml (for plates) of 2 mg/ml adenine hemisulfate (Sigma)
Dissolve in a final volume of 900 ml H_2O
Adjust pH to 5.8

Autoclave 30 min, then add 100 ml sterile 20% (w/v) glucose. For AHC plates, add 20 g agar prior to autoclaving. Store at 4°C for 6 weeks.

Airvol, see Gelvatol

Alkaline loading buffer *(UNIT 4.6)*

✓ 30 mM NaOH
✓ 1 mM EDTA, pH 8.0
10% (w/v) Ficoll
0.025% (w/v) bromcresol green

Alkaline phosphatase pH 7.5 (AP 7.5) buffer *(UNIT 3.18)*

✓ 0.1 M TrisCl, pH 7.5
✓ 0.1 M NaCl
✓ 2 mM $MgCl_2$
Autoclave or filter sterilize and store at room temperature

Alkaline phosphatase pH 9.5 (AP 9.5) buffer *(UNIT 3.18)*

✓ 0.1 M Tris·Cl, pH 9.5
✓ 0.1 M NaCl
✓ 50 mM $MgCl_2$
Autoclave or filter sterilize and store at room temperature

Alkaline phosphatase buffer, pH 9.5 *(UNIT 14.7)*

✓ 0.1 M Tris·Cl, pH 9.5
✓ 0.1 M NaCl
✓ 50 mM $MgCl_2$ (add immediately before use)
Store up to 1 year (without $MgCl_2$) at room temperature

Alkaline phosphatase substrate buffer *(UNIT 14.8)*

✓ 100 mM Tris·Cl, pH 9.5
✓ 150 mM NaCl
✓ 50 mM $MgCl_2$
Store at 4°C up to 3 months

Alkaline phosphate substrate buffer *(UNIT 10.7)*

✓ 100 mM Tris·Cl, pH 9.5
✓ 100 mM NaCl
✓ 5 mM $MgCl_2$

Alkaline pour buffer, 50× *(UNIT 4.6)*

✓ 2.5 M NaCl
✓ 50 mM EDTA, pH 8.0
Dilute to 1× for working solution

Alkaline running buffer, 50× *(UNIT 4.6)*

✓ 1.5 M NaOH
✓ 50 mM EDTA, pH 8.0
Dilute to 1× for working solution

18 amino acid mix *(UNIT 16.2)*

Prepare solution containing 0.1% (w/v) of each amino acid *except* cysteine and *except* methionine. Filter sterilize through a 0.2-μm filter. Store at −20°C for several years.

Ammonium acetate, 10 M

Dissolve 385.4 g ammonium acetate in 150 ml H_2O
Add H_2O to 500 ml

Ammonium persulfate, 10% *(UNIT 10.3)*

10 g ammonium persulfate (Bio-Rad)
H_2O to 100 ml total volume
Store refrigerated ~2 weeks

Amplification buffer, 5× *(UNIT 15.5)*

✓ 200 mM NaCl
✓ 100 mM Tris·Cl, pH 8.9
✓ 25 mM $MgSO_4$
0.05% (w/v) gelatin (Sigma #1393)
0.5% (v/v) Triton X-100
Store at −20°C

Amplification buffer, 10× *(UNITS 8.5, 15.3, 15.8)*

✓ 500 mM KCl
✓ 100 mM Tris·Cl, pH 8.4
Autoclave (not necessary for some applications)
1 mg/ml gelatin
Store at −20°C

The concentration (x) of $MgCl_2$ depends upon the sequence and primer set of interest. The optimal concentration is first determined empirically (steps 1 to 5, UNIT 15.1) using a 10× $MgCl_2$-free PCR amplification buffer. Based upon these results, a 10× amplification buffer containing $MgCl_2$ at the optimal concentration is prepared.

Amplification buffer, 10×, $MgCl_2$-free *(UNIT 15.1)*

✓ 500 mM KCl
✓ 100 mM Tris·Cl, pH 9.0 (at 25°C)
0.1% (v/v) Triton X-100
Store indefinitely at −20°C

$MgCl_2$ must be added at the optimized concentration for the primers being used for the reaction; see the note to the preceding recipe.

This buffer can be obtained from Promega; it is supplied with Taq DNA polymerase.

Amplification mix *(UNIT 15.7)*

Per reaction (30 μl):

✓ 20.0 μl 5× amplification buffer

1.0 μl 10 pmol/μl oligonucleotide LMPCR.1 (linker primer)

1.0 μl 10 pmol/μl oligonucleotide primer 2

✓ 0.8 μl 25 mM 4dNTP mix, pH 7.0

7.2 μl H_2O

Prepare immediately before use and chill on ice

AMV/MoMuLV reaction buffer, 10× *(UNIT 14.8)*

✓ 100 mM Tris·Cl, pH 8.3

✓ 500 mM KCl

✓ 15 mM $MgCl_2$

Prepare fresh

Annealing buffer, 5× *(UNIT 15.2)*

✓ 200 mM Tris HCl, pH 7.5

✓ 100 mM $MgCl_2$

✓ 250 mM NaCl

Anode buffer *(UNIT 10.2)*

121.1 g Tris base

500 ml H_2O

Adjust to pH 8.9 with concentrated HCl

Dilute to 5 liters with H_2O

Store at 4°C up to 1 month

Final concentration is 0.2 M Tris·Cl, pH 8.9.

Antibody-conjugate solution *(UNIT 7.5)*

Prepare antibody-conjugate (e.g., anti-digoxigenin, anti-fluorescein, or anti-DNP alkaline phosphatase; Table 7.5.1) dilutions in blocking buffer II (see recipe). Follow conjugate manufacturer's recommendations for the appropriate dilutions; these are typically 1/2500 to 1/10,000. Prepare fresh.

Antifade mounting media *(UNIT 14.7)*

DABCO mounting medium: Dissolve 0.233 g 1,4-diazobicyclo-[2.2.2]octane (Sigma; 0.21 M final) in 800 μl H_2O. Add 200 μl 1 M Tris·Cl, pH 8.0 (0.02 M final) and 9 ml glycerol (90% final). Mix. Store 100-μl aliquots, wrapped in foil, at −20°C. Thaw and use once.

CAUTION: *DABCO is hazardous; see manufacturer's information for guidelines on handling, storage, and disposal.*

Phenylenediamine dihydrochloride: Dissolve 50 mg *p*-phenylenediamine dihydrochloride in 5 ml PBS (9 mM final). Adjust to pH 8 with 0.5 M carbonate/bicarbonate buffer, pH 9.0 (see recipe). Add to 45 ml glycerol (90% final), mix, and filter through 0.22 μm filter. Store in small aliquots in dark at −20°C. Thaw and use at once.

Vectashield: Purchase from Vector Laboratories; follow manufacturer's instructions.

Aqueous prehybridization/hybridization (APH) solution *(UNIT 2.10)*

✓ 5× SSC

✓ 5× Denhardt solution

1% (w/v) SDS

Add 100 μg/ml denatured salmon sperm DNA (see recipe below) just before use

Alternatives to Denhardt solution and denatured salmon sperm DNA as blocking agents are listed in Table 2.10.5.

Aqueous hybridization solution, 3× *(UNIT 4.6)*

✓ 3 M NaCl
0.5 M HEPES, pH 7.5
✓ 1 mM EDTA, pH 8.0

Assay buffer *(UNIT 7.5)*

Dissolve DEA (99% purity) in H_2O to 0.1 M (final), adjust pH to 10.0 with concentrated HCl, and add $MgCl_2$ to 1 mM (final).

Add sodium azide to 0.02% (w/v) to store solution for 2 to 7 days at 4°C.

Assay buffer *(UNIT 12.4)*

✓ 10 mM Tris·Cl
✓ 5 mM $MgCl_2$
✓ 1 mM $CaCl_2$
✓ 2 mM DTT
✓ 50 μg/ml BSA
✓ 2 μg/ml calf thymus DNA
✓ 100 mM KCl
Titrate to pH 8.0

β-agarase buffer, 10× *(UNIT 2.6)*

100 mM bis-Tris, pH 6.5
✓ 10 mM EDTA
Filter sterilize and store indefinitely at room temperature

β-galactosidase reaction buffer *(UNIT 9.7B)*

✓ 100 mM sodium phosphate, pH 8.0
✓ 1 mM magnesium chloride
1× 3-(4-methoxyspiro[1,2-dioxetane-3,2′-(5′-chloro)-tricyclo[3.3.3.$1^{3,7}$]decan]-4-yl)phenyl β-D-galactopyranoside (Galacton) chemiluminescent substrate (Tropix)
Store up to one month at 4°C

This solution is available as part of the Galacto-Light assay kit (Tropix).

Bal 31 nuclease buffer, 10× *(UNIT 7.2)*

✓ 2.0 M NaCl
✓ 0.2 M Tris·Cl, pH 8.0
✓ 0.12 M $MgCl_2$
✓ 0.12 M $CaCl_2$
✓ 0.02 M EDTA

BCIP/NBT visualization solution *(UNIT 10.7)*

Mix 33 μl NBT stock (100 mg NBT in 2 ml 70% DMF, stored at 4°C, <1 year) and 5 ml alkaline phosphate substrate buffer (see recipe). Add 17 μl BCIP stock (100 mg BCIP in 2 ml 100% DMF, stored at 4°C, <1 year) and mix. Stable 1 hr at room temperature.

Alternatively, BCIP/NBT substrates may be purchased from Sigma, Kirkegaard & Perry, Moss, and Vector Labs (APPENDIX 4).

BES-buffered solution (BBS), 2× *(UNIT 9.1)*

50 mM *N,N-bis*(2-hydroxyethyl)-2-aminoethanesulfonic acid (BES; Calbiochem)
✓ 280 mM NaCl
1.5 mM Na_2HPO_4, pH 6.95
800 ml H_2O
Adjust to pH 6.95 with 1 N NaOH, room temperature
H_2O to 1 liter
Filter sterilize through 0.45-μm nitrocellulose filter (Nalgene)
Store in aliquots at −20°C (can be frozen and thawed repeatedly)

The pH of this solution is critical (pH 6.95 to 6.98). When a new batch of 2× BES buffer is prepared, its pH should be checked against a reference stock prepared (and tested) earlier.

Binding buffer (*UNIT 12.7*)

10 mM Tris·Cl, pH 7.5
50 mM NaCl
1 mM EDTA
5 mM DTT

Binding buffer, 5× (*UNIT 12.8*)

✓ 100 mM Tris·Cl, pH 7.4
✓ 250 mM KCl
✓ 15 mM $MgCl_2$
✓ 5 mM EDTA
500 μg/ml gelatin

Biotin-11-dUTP, 0.5 mM (*UNIT 3.18*)

Prepare 0.5 mM stock of lyophilized biotin-11-dUTP (e.g., Sigma) in 20 mM Tris·Cl, pH 7.5 (at 289 nm, molar extinction coefficient ε = 7100; Table A.3.2). Check pH of stock; if <7.5, add a few microliters of 1 M Tris·Cl, pH 7.5, to adjust pH to 7.5. Store at −20°C.

Biotin-cellulose binding buffer (*UNIT 12.6*)

12% (v/v) glycerol
12 mM HEPES-NaOH, pH 7.9
✓ 4 mM Tris·Cl, pH 7.9
✓ 60 mM KCl
✓ 1 mM EDTA
✓ 1 mM DTT
Store at −20°C upon addition of DTT

This is a typical binding buffer. The composition of binding buffer—especially with respect to pH, ionic strength, and the presence or absence of $MgCl_2$—should be determined by those conditions which optimize the binding of the protein of interest to its recognition site.

Biotin-cellulose elution buffer (*UNIT 12.6*)

12% (v/v) glycerol
✓ 20 Tris·Cl, pH 6.8
✓ 1 M KCl
✓ 5 mM $MgCl_2$
✓ 1 mM EDTA
✓ 1 mM DTT
✓ 200 μg/ml BSA
Store at −20°C upon addition of DTT

This is a typical elution buffer; see annotation to biotin-cellulose binding buffer recipe regarding optimization. Another carrier protein, such as insulin or hemoglobin, may be substituted for BSA. The small size of insulin can be useful if protein gels will be run to determine the size of the regulatory factor.

Biotin detection solution (*UNIT 14.7*)

Dilute fluorescein-avidin DCS *or* rhodamine-avidin D (Vector Laboratories) to 2 μg/ml in 4× SSC/1% (w/v) BSA (fraction V). Prepare fresh daily.

Biotin/digoxigenin detection solution (*UNIT 14.7*)

Dilute fluorescein-avidin DCS (Vector Laboratories) and rhodamine-conjugated Fab fragment of sheep anti-digoxigenin (Boehringer Mannheim) to 2 μg/ml each in 4× SSC/1% (w/v) BSA (fraction V). Prepare fresh daily.

Biotinylated alkaline phosphatase (*UNIT 3.19*)

0.38 mg/ml biotin conjugate
✓ 3M NaCl
✓ 1 mM $MgCl_2$
0.1 mM $ZnCl_2$
30 mM triethanolamine acetate (TEA), pH 7.5
Store at 4°C

Biotinylated alkaline phosphatase solution (*UNIT 7.5*)

Dilute biotinylated alkaline phosphatase to 0.5 μg/ml in two-step blocking solution (see recipe). Prepare fresh.

Biotinylated alkaline phosphatase that has been tested for chemiluminescent detection, available as part of kits from Millipore and New England Biolabs, is highly recommended.

Biotinylated alkaline phosphatase solution (*UNIT 14.7*)

✓ *PBS containing:*
✓ 1% (w/v) BSA
0.1% (v/v) Tween 20
2.5 μg/ml biotinylated alkaline phosphatase
Prepare fresh

Biotinylated horseradish peroxidase solution (*UNIT 14.7*)

✓ *PBS containing:*
✓ 1% (w/v) BSA
0.1% (v/v) Tween 20
3 μg/ml biotinylated horseradish peroxidase
Prepare fresh

Biotinylated random octamers (*UNIT 3.18*)

Obtain commercially (e.g., Millipore or New England Biolabs) or prepare on a DNA synthesizer. Prepare a 5× stock (60 O.D./ml) in 250 mM Tris·Cl, pH 8.0, 1 M HEPES, pH 6.6, 25 mM $MgCl_2$, and 10 mM DTT. Store at −20°C.

Blocking buffer (*UNIT 17.4*)

✓ 5% (w/v) BSA
1% (w/v) hen ovalbumin
✓ 10 mM Tris·Cl, pH 7.4, at room temperature
✓ 0.15 M NaCl
0.01% (w/v) sodium azide
Store ≤6 months at 4°C

For enhanced chemiluminescence, omit the sodium azide.

Blocking buffer (*UNIT 10.7*)

Colorimetric detection

For nitrocellulose and PVDF: 0.1% (v/v) Tween 20 in TBS (TTBS).

For neutral and positively charged nylon: Tris-buffered saline (TBS) containing 10% (w/v) nonfat dry milk.

TTBS can be stored ≤1 week at 4°C. Prepare blocking buffer containing nonfat dry milk immediately prior to use as the milk blocking solution is not stable.

Luminescence detection

For nitrocellulose, PVDF, and neutral nylon (e.g., Pall Biodyne A): 0.2% casein (e.g., Hammarsten grade or I-Block; Tropix) in TTBS. Prepare just before use.

For positively charged nylon: 6% (w/v) casein/1% (v/v) polyvinyl pyrrolidone (PVP) in TTBS. Prepare just before use.

With constant mixing, add casein and PVP to warm (65°C) TTBS. Stir for 5 min. Cool before use.

Blocking buffer (*UNIT 11.2*)

Borate-buffered saline (see recipe) containing:
0.05% (v/v) Tween 20
✓ 1 mM EDTA
✓ 0.25% (w/v) BSA
0.05% (w/v) NaN_3
Store at 4°C

Gelatin or 5% instant milk may be substituted for BSA, although the latter may interfere nonspecifically with antibody binding.

continued

Blocking buffer I *(UNIT 8.5)*

✓ 1× PBS
0.2% (w/v) purified casein or membrane-blocking reagent (Table 7.5.1)
0.5% (w/v) SDS

Prepare 1× PBS solution from 10× stock. Add casein or membrane-blocking reagent, heat to 70°C with constant stirring, then add SDS. Do not boil.

Add sodium azide to 0.02% (w/v) to store solution for 2 to 7 days at 4°C.

Blocking buffer II *(UNIT 7.5)*

✓ 1× PBS
0.2% (w/v) purified casein or membrane-blocking reagent (Table 7.5.1)
0.1% (v/v) Tween 20

Prepare 1× PBS solution from 10× stock. Add casein or membrane-blocking reagent, heat to 70°C with constant stirring, then add Tween 20. Do not boil.

Add sodium azide to 0.02% (w/v) to store solution for 2 to 7 days at 4°C.

Blocking solution *(UNIT 3.19)*

5% (w/v) SDS
17 mM Na_2HPO_4
8 mM NaH_2PO_4
Store at room temperature
Filter sterilize with a 0.45-µm filter if necessary

Blocking solution *(UNIT 14.3)*

Preheat 400 ml of 1× PBS to 45°C. Immediately before use, add the following: 0.617 g DTT, 0.74 g iodoacetamide, and 0.5 g *N*-ethylmaleimide. Mix well and use immediately. Cover with aluminum foil.

CAUTION: *These substances are very toxic—exercise proper precautions during preparation and handle very carefully.*

Blocking solution *(UNIT 14.8)*

✓ 100 mM Tris·Cl, pH 7.8
✓ 150 mM NaCl
✓ 50 mg/ml BSA
0.2 mg/ml sodium azide
Prepare fresh

BLOTTO *(UNIT 12.7)*

5% (w/v) Carnation nonfat milk powder
✓ 50 mM Tris·Cl, pH 7.5
✓ 50 mM NaCl
✓ 1 mM EDTA
✓ 1 mM DTT
Store 1 to 2 weeks at 4°C

Thoroughly dissolve milk in sterile water containing other components.

Borate buffers *(UNIT 11.8)*

0.1 M, pH 8.5:
18.55 g boric acid
2850 ml H_2O
✓ 10 M NaOH to pH 8.5
H_20 to 3000 ml

0.1 M, pH 10:
0.618 g boric acid
95 ml H_2O
10 M NaOH to pH 10
H_2O to 100 ml

Bovine serum albumin (BSA), 0.5 mg/ml *(UNIT 10.1)*

The concentration of BSA is determined using A_{280} = 6.6 for a 10 mg/ml solution of BSA measured in a 1-cm pathlength cuvette (e.g., 0.5 mg/ml solution will have an A_{280} of 0.33).

Breaking buffer (*UNIT 13.11*)

2% (v/v) Triton X-100
1% (w/v) SDS
✓ 100 mM NaCl
✓ 10 mM Tris·Cl, pH 8.0
✓ 1 mM EDTA
Store at room temperature up to 1 year

Buffer-gradient gel solutions (*UNIT 7.6*)

0.5× *TBE (clear solution):*

Reagent	Acrylamide concentration (%)		
	4	6	8
Urea (ultrapure) (g)	21	21	21
38% acrylamide/2% bisacrylamide (ml)	5.0	7.5	10.0
10× TBE (ml)	2.5	2.5	2.5
H_2O (ml)	26.5	24	21.5
Total volume (ml)	*50.0*	*50.0*	*50.0*

2.5× *TBE (blue solution):*

Reagent	Acrylamide concentration (%)		
	4	6	8
Urea (ultrapure) (g)	10.5	10.5	10.5
38% acrylamide/2% bisacrylamide (ml)	2.5	3.75	5.0
10× TBE (ml)	6.25	6.25	6.25
Sucrose (g)	2.5	2.5	2.5
1% (wt/vol) bromphenol blue (μl)	250	250	250
H_2O (ml)	6.25	5.0	3.75
Total volume (ml)	*25.0*	*25.0*	*25.0*

Buffer preparation (*UNITS 10.12 & 10.13*)

All buffer salts, organic solvents, and water must be of high purity, preferably HPLC grade. Buffers should be filtered through a 0.45-μm nylon filter into a thick-walled 4-liter Erlenmeyer flask and degassed using a trapped vacuum pump prior to the addition of a volatile buffer solute, or else filtered and degassed after the addition of a nonvolatile buffer solute (see Support Protocol, *UNIT 10.12*). Buffer pH must be carefully adjusted to within 0.1 U to ensure reproducibility of chromatography; adjust after all ingredients have been dissolved and the solutions have been filtered, as described above. To minimize the growth of bacteria in IEX buffers, 2% to 5% (v/v) methanol could be added to each buffer. When not in use, columns should be stored in a refrigerator. As a matter of practice, buffers should be made fresh daily. The UV cutoff (i.e., the wavelength at which the absorbance becomes high) of any organic solvent (e.g., acetonitrile or propanol) should be the lowest obtainable.

CAUTION: *The chemicals used in HPLC are toxic, especially TFA, heptafluorobutyric acid, formic acid, and acetonitrile, and should be used in a well ventilated laboratory.*

Buffer TM *(UNIT 12.9)*

✓ 50 mM Tris·Cl, pH 7.9
✓ 0 M *or* 1 M KCl
✓ 12.5 mM $MgCl_2$
✓ 1 mM dithiothreitol (DTT; add fresh just before use)
20% (v/v) glycerol
0.1% (v/v) Nonidet P-40 (NP-40)

Do not make a 10× buffer. To generate aliquots of buffer with the range of KCl concentrations described in the protocol, make two 500-ml batches of 1× buffer containing no KCl and 1 M KCl respectively and mix together appropriate quantities. Store at 4°C. Just prior to use, place 1.5-ml aliquots of each concentration in separate 1.5-ml microcentrifuge tubes, and add DTT.

Buffer Z *(UNIT 12.9)*

25 mM HEPES (K^+ salt), pH 7.6
✓ 0 M *or* 1 M KCl
✓ 12.5 mM $MgCl_2$
✓ 1 mM DTT (add fresh just before use)
20% (v/v) glycerol
0.1% (v/v) NP-40
Adjust pH to 7.6 with KOH

Do not make a 10× buffer. To generate aliquots of buffer with the range of KCl concentrations described in the protocol, make two 500-ml batches of 1× buffer containing no KCl and 1 M KCl respectively and mix together appropriate quantities. Store at 4°C. Just prior to use, place 1.5-ml aliquots of each concentration in separate 1.5-ml microcentrifuge tubes, and add DTT.

Buffer Z^e *(UNIT 12.9)*

25 mM HEPES (K^+ salt), pH 7.6
✓ 0 M *or* 1 M KCl
✓ 1 mM DTT (add fresh just before use)
20% (v/v) glycerol
0.1% (v/v) NP-40
Adjust the pH to 7.6 with KOH

Do not make a 10× buffer. To generate aliquots of buffer with the range of KCl concentrations described in the protocol, make two 500-ml batches of 1× buffer containing no KCl and 1 M KCl respectively and mix together appropriate quantities. Store at 4°C. Just prior to use, place 1.5-ml aliquots of each concentration in separate 1.5-ml microcentrifuge tubes, and add DTT.

C+T/C stop solution *(UNIT 15.5)*

✓ 400 mM sodium acetate, pH 7.5
✓ 0.14 mM EDTA, pH 8.0

Prepare fresh and chill on ice before use

CAA/glycerol/ampicillin 100 medium *(UNIT 16.8)*

✓ 800 ml 2% (w/v) Casamino Acids (1.6% final)
✓ 100 ml 10× M9 salts (1× final)
100 ml 10% (v/v) glycerol (sterile; 1% final)
✓ 1 ml 1 M $MgCl_2$ (sterile; 1 mM final)
✓ 0.1 ml 1 M $CaCl_2$ (sterile; 0.1 mM final)
1 ml 2% (w/v) vitamin B1 (sterile; 0.002% final)
10 ml 10 mg/ml ampicillin (sterile; 100 µg/ml final)
Prepare fresh

$CaCl_2$, 1 M

147 g $CaCl_2 \cdot 2H_2O$
H_2O to 1 liter

A1

$CaCl_2$, 2.5 M (UNIT 9.1)

183.7 g $CaCl_2$ dihydride (Sigma; tissue culture grade)
H_2O to 500 ml
Filter sterilize through a 0.45-µm nitrocellulose filter (Nalgene)
Store at −20°C in 10-ml aliquots (*UNIT 9.1*; can be frozen and thawed repeatedly) or in 0.5-ml aliquots (*UNIT 16.10*; use fresh tube for each set of transfections)

$CaCl_2$, 2.5 M (UNIT 16.11)

Prepare using tissue culture grade $CaCl_2$. Filter sterilize and dispense in 0.5-ml aliquots. Store ≤1 year at −20°C. Use a fresh tube for each set of transfections.

$CaCl_2$ solution (UNIT 1.8)

✓ 60 mM $CaCl_2$
15% (v/v) glycerol
10 mM PIPES, pH 7.0
Filter sterilize using a disposable filter unit, or autoclave

$CaCl_2$ solution (UNIT 13.7)

✓ 0.1 M Tris·Cl, pH 7.4
✓ 0.1 M $CaCl_2$
Autoclave or filter sterilize and store at room temperature

Calf thymus DNA standard solutions (APPENDIX 3D)

Kits containing calf thymus DNA standard for fluorometry are available (Fluorometry Reference Standard Kits, Hoefer). Premeasured, CsCl-gradient-purified DNA of defined GC content, for use in absorption and fluorometric spectroscopy, is available from Sigma (e.g., calf thymus DNA, 42% GC; *Clostridium perfringens* DNA, 26.5% GC).

Carbonate/bicarbonate buffer (pH 9.0), 0.5 M (UNIT 14.7)

0.42 g $NaHCO_3$
10 ml H_2O
Adjust pH to 9 with NaOH
Store up to 1 year at room temperature

Carbonate buffer, pH 9.2, 0.1 M (UNIT 11.1)

1.36 g sodium carbonate
7.35 g sodium bicarbonate
950 ml H_2O
Adjust pH to 9.2 with 1 M HCl or 1 M NaOH, if necessary
Add H_2O to 1 liter

Carbonate developing solution (UNIT 10.5)

0.5 ml 37% formaldehyde per liter solution
3% (w/v) sodium carbonate
Prepare in distilled, deionized water

Casamino Acids (CAA), 2% (w/v; UNIT 16.8)

20 g Casamino Acids (Difco certified)
H_2O to 1 liter
Autoclave or filter sterilize through a 0.45-µm filter
Store ≤2 months at room temperature

Do not use technical-grade Casamino Acids because it has a higher NaCl content.

Cathode buffer (UNIT 10.2)

12.11 g Tris base (0.1 M)
17.92 g tricine (0.1 M)
1 g SDS [1% (w/v)]
Dilute to 1 liter with H_2O
Do not adjust pH
Store at 4°C up to 1 month

Final concentrations are 0.1 M Tris, 0.1 M tricine, and 0.1% (w/v) SDS.

Chase medium *(UNIT 10.17)*

Culture medium containing 15 mg/liter of nonradioactive methionine or equivalent excess of other amino acid that was used for radiolabeling.

Chelating buffer, 10× *(UNIT 14.8)*

✓ 100 mM Tris·Cl, pH 8.3
✓ 1 M KCl
7.5 mM EGTA
0.5% (w/v) Tween 20
50% (v/v) glycerol

[³H]Chloramphenicol solution for phase-extraction assay *(UNIT 9.7A)*

Prepare a 0.2 µCi/µl [^{3}H]chloramphenicol stock by adding 960 µl of 100% ethanol and 40 µl of 100 mg/ml unlabeled chloramphenicol to 250 µl of [^{3}H]chloramphenicol (250 µCi/250 µl in ethanol; 42.0 to 58.2 Ci/mmol). Preextract the stock by first diluting it 20-fold in water, and then extracting this mixture with an equal volume of xylenes by vigorous shaking. Centrifuge 2 min in microcentrifuge at room temperature to separate phases, and discard the top xylenes phase. Extract aqueous phase one more time, centrifuge, and discard the top xylenes phase. This creates a working solution of 0.01 µCi/µl [^{3}H]chloramphenicol.

When using ^{3}H-labeled instead of ^{14}C-labeled chloramphenicol, preextraction with xylenes is necessary to reduce the background. Preextraction of [^{14}C]chloramphenicol reduces background, but not to the same extent as with [^{3}H]chloramphenicol.

Chromic acid cleaning solution *(UNIT 10.3)*

Add a 25 ml bottle of Chromerge (Fisher) to a 9-lb bottle of concentrated sulfuric acid. Add ~5 ml at a time and stir.

CAUTION: *This is very corrosive and toxic. Carefully read and observe package instructions.*

Citrate buffer, pH 5.5 *(UNIT 11.4)*

2.45 g citric acid (anhydrous) } (0.05 M citrate final)
10.96 g trisodium citrate dihydrate }
8.77 g NaCl (0.15 M final)
Add H_2O to 1 liter

4CN visualization solution *(UNIT 10.7)*

Mix 20 ml ice-cold methanol with 60 mg 4CN. Separately mix 60 µl of 30% H_2O_2 with 100 ml TBS (see recipe) at room temperature. Rapidly mix the two solutions and use immediately.

CNBr/acetonitrile, *see Cyanogen bromide/acetonitrile*

Column regeneration buffer *(UNIT 12.9)*

✓ 10 mM Tris·Cl, pH 7.8
✓ 1 mM EDTA, pH 8.0
✓ 2.5 M NaCl
1% (v/v) NP-40
Store at room temperature

The solution will be cloudy and separate into two phases (NP-40 and aqueous) upon storage. Mix by swirling and shaking just before use.

Column storage buffer *(UNIT 12.9)*

✓ 10 mM Tris·Cl, pH 7.8
✓ 1 mM EDTA, pH 8.0
✓ 0.3 M NaCl
0.04% (w/v) sodium azide

Store at room temperature without sodium azide. Make a 4% (w/v) sodium azide stock solution and add just before use.

Column storage solutions *(UNIT 10.10)*

✓ *TSA solution containing:*

1 mM EDTA + 20 µg/ml gentamycin *or* 0.01% thimerosal (Aldrich)

Complete DMEM-5, -10, or -20, *(UNITS 9.4, 9.12 & 9.13)*

Dulbeccos minimum essential medium, high-glucose formulation, containing:

5%, 10%, or 20% (v/v) FBS, heat-inactivated 1 hr at 56°C (optional; see below)

1% (w/v) nonessential amino acids

2 mM L-glutamine

50 µM 2-ME

100 U/ml penicillin

100 µg/ml streptomycin sulfate

DMEM containing 4500 mg/liter D-glucose can be obtained from Life Technologies. DMEM is also known as Dulbeccos modified Eagle medium.

Complete DMEM-10 *(UNIT 11.3)*

Dulbeccos minimum essential medium, high glucose formulation, containing:

33.3 mM sodium bicarbonate

20 mM HEPES

2 mM L-glutamine

1 mM sodium pyruvate

50 U/ml penicillin

50 µg/ml streptomycin

See annotations to complete DMEM recipe (UNIT 9.4).

Complete DMEM-10 NS *(UNIT 9.2)*

Prepare medium as for complete DMEM (*UNIT 9.2*), but use 10% (v/v) NuSerum (Collaborative Research) instead of FBS.

Complete insect cell medium *(UNIT 16.10)*

To 500 ml 1× Graces insect medium supplemented with yeastolate and lactalbumin hydrolysate (Life Technologies), add 10 mg/ml gentamicin (optional), and (if needed) 50 ml heat-inactivated fetal bovine serum (FBS; 10% final; see annotation). Filter sterilize. Incubate a 5-ml sample 2 days at 37°C to check sterility, and store remainder at 4°C until manufacturer's expiration date.

For medium with 20% DMSO, add sterilized (autoclaved) DMSO to 20% (v/v).

To prepare 25× stock yeastolate or 25× stock lactalbumin hydrosylate (both obtained from VWR Scientific), dissolve 41.25 grams yeastolate or lactalbumin hydrolysate in 500 ml deionized distilled H_2O. Dispense into 50-ml aliquots, autoclave to sterilize, and store at 4°C.

To heat-inactivate FBS, incubate 30 min at 56°C and store in 50-ml aliquots ≤1 year at −20°C.

Complete serum-free RPMI medium *(UNIT 10.18)*

RPMI 1640 medium (Life Technologies) containing:

2 mM L-glutamine

100 U/ml penicillin

100 µg/ml streptomycin sulfate

Concentrated bromphenol blue *(UNIT 10.3)*

50% (v/v) aqueous glycerol

0.01 mg/ml bromphenol blue

Conjugate dilution buffer *(UNIT 14.8)*

✓ 100 mM Tris·Cl, pH 7.3

✓ 150 mM $MgCl_2$

✓ 10 mg/ml BSA

0.2 mg/ml sodium azide

Prepare fresh

Conjugate solution *(UNIT 7.5)*

Prepare streptavidin–alkaline phosphatase (Table 7.4.6) in blocking buffer I (see recipe). Follow manufacturer's recommendations for appropriate dilution; 1/5000 is typically used. Prepare fresh.

Coomassie blue G-250 staining solution *(UNIT 10.2)*

200 ml acetic acid [10% (v/v)]
1800 ml H_2O
0.5 g Coomassie blue G-250 [0.025% (w/v)]
Mix 1 hr and filter (Whatman no. 1 paper)
Store indefinitely at room temperature

Coomassie brilliant blue solution *(UNIT 10.1)*

In a 1-liter volumetric flask, dissolve 100 mg Coomassie brilliant blue G-250 in 50 ml of 95% ethanol. Add 100 ml of 85% phosphoric acid. Bring to volume with water. Filter through Whatman No. 1 filter paper. Store at 4°C.

Kits are available from Pierce and Bio-Rad.

Coomassie staining solution *(UNIT 10.5)*

50% (v/v) methanol
0.05% (v/v) Coomassie brilliant blue R-250 (Bio-Rad or Pierce)
10% (v/v) acetic acid
40% H_2O
Prepare in distilled, deionized water

Dissolve the Coomassie brilliant blue R-250 in methanol before adding acetic acid and water (filtration is not generally needed). Solution can be stored for 6 months. If precipitate is observed following prolonged storage, filter to obtain a homogeneous solution.

Coupling buffer, pH 8.3 *(0.125 M phosphate; UNIT 11.4)*

Solution A: 17.75 g Na_2HPO_4 (anhydrous)/liter H_2O
Solution B: 1.95 g $NaH_2PO_4 \cdot 2H_2O$/100 ml H_2O
Titrate solution A with solution B to pH 8.3

Cracking buffer *(UNIT 16.2)*

60 mM Tris·Cl, pH 6.8
1% (w/v) 2-ME
1% (w/v) SDS
10% (v/v) glycerol
0.01% (w/v) bromphenol blue

CsCl, 5.7 M, DEPC-treated, *(UNIT 4.2)*

Dissolve CsCl in 0.1 M EDTA. Add 0.002 vol DEPC, shake 20 to 30 min, and autoclave. Weigh the bottle of solution before and after autoclaving and make up the weight lost to evaporation during autoclaving with DEPC-treated H_2O. This ensures that the solution is actually 5.7 M.

CTAB extraction solution *(UNIT 2.3)*

2% (w/v) CTAB
✓ 100 mM Tris·Cl, pH 8.0
✓ 20 mM EDTA, pH 8.0
✓ 1.4 M NaCl
Store at room temperature (stable several years)

CTAB/NaCl solution *(10% CTAB/0.7 M NaCl; UNIT 2.3)*

Dissolve 4.1 g NaCl in 80 ml H_2O and slowly add 10 g CTAB (hexadecyltrimethyl ammonium bromide) while heating and stirring. If necessary, heat to 65°C to dissolve. Adjust final volume to 100 ml.

CTAB precipitation solution *(UNIT 2.3)*

1% (w/v) CTAB
✓ 50 mM Tris·Cl, pH 8.0
✓ 10 mM EDTA, pH 8.0
Store at room temperature (stable several years)

Cyanogen bromide (CNBr)/acetonitrile *(UNIT 10.15)*

To 25 g of cyanogen bromide *(NOTE: CNBr should be white, not yellow, crystals)*, add 50 ml acetonitrile to make a 62.5% (w/v) solution. This may be stored indefinitely at −20°C in a desiccator over silica. Allow to warm before opening.

CAUTION: *CNBr is a highly toxic lachrymator; use in fume hood.*

Cytoplasmic extract buffer, 10× *(UNIT 12.1)*

0.3 M HEPES, pH 7.9 at 4°C
✓ 1.4 M KCl
✓ 0.03 M $MgCl_2$

DAB/NiCl₂ visualization solution *(UNIT 10.7)*

✓ 5 ml 100 mM Tris·Cl, pH 7.5
100 µl DAB stock (40 mg/ml in H_2O, stored in 100-µl aliquots at −20°C)
25 µl $NiCl_2$ stock (80 mg/ml in H_2O, stored in 100-µl aliquots at −20°C)
15 µl 3% H_2O_2
Mix just before use

CAUTION: *Handle DAB carefully, wearing gloves and mask; it is a carcinogen.*

Suppliers of peroxidase substrates are Sigma, Kirkegaard & Perry, Moss, and Vector Laboratories (APPENDIX 4).

DAB substrate solution, *see Diaminobenzidine (DAB) substrate solution*

DAPI staining solution *(UNIT 14.7)*

Stock solution: Dissolve 1 mg 4′,6-diamidino-2-phenylindole (DAPI; 0.3 mM final) in 10 ml H_2O. (Add a few drops of methanol before adding H_2O to help dissolve DAPI). Aliquot into aluminum foil–wrapped tubes and store a year or more at −20°C.

Working solution: Dilute stock solution 1/1000 in PBS. Store in aluminum foil–wrapped tubes several weeks at 4°C.

CAUTION: *DAPI is hazardous; see manufacturer's information for guidelines on handling, storage, and disposal.*

DEAE elution buffer *(UNIT 2.7)*

✓ 10 mM Tris·Cl, pH 8.0
✓ 1 mM EDTA, pH 8.0
✓ 1 M NaCl

DEAE elution solution *(UNIT 12.2)*

✓ 0.4 ml 1 M Tris·Cl, pH 7.9
✓ 0.08 ml 0.5 M EDTA, pH 8.0
✓ 8.0 ml 5 M NaCl
H_2O to 40 ml
Filter through 0.2-µm filter

DEAE-dextran in TBS or STBS, 10 mg/ml *(UNIT 9.2)*

Use ~2 × 10^6 molecular weight dextran. Make up a 10 mg/ml stock in TBS or STBS (see recipes). Filter sterilize and store at 4°C.

Make sure the stock DEAE-dextran solution is mixed well just before use.

Deionized formamide *(UNITS 14.3, 14.7 & 14.8)*

Mix 50 ml of good quality formamide (e.g., Fluka) and ~5 g of mixed-bed, ion-exchange resin (e.g., Bio-Rad AG 501-X8, 20 to 50 mesh) and stir 30 min at room temperature. Filter through Whatman filter paper, dispense into 1-ml aliquots, and store at −20°C.

Denatured salmon sperm DNA *(UNIT 2.10)*

Dissolve 10 mg Sigma type III salmon sperm DNA (sodium salt) in 1 ml water. Pass vigorously through a 17-G needle 20 times to shear the DNA. Place in a boiling water bath for 10 min, then chill. Use immediately or store at −20°C in small aliquots. If stored, reheat to 100°C for 5 min and chill on ice immediately before using.

Denaturing acrylamide gel solution *(UNIT 7.6)*

Reagent	Acrylamide concentration (%)		
	4	6	8
Urea (ultrapure; g)	25.2	25.2	25.2
✓38% acrylamide/2% bisacrylamide (ml)	6.0	9.0	12.0
✓10× TBE electrophoresis buffer (ml)	6.0	6.0	6.0
H_2O (ml)	27	24	21
Total volume (ml)	60	*60*	*60*

Filter solution through Whatman No. 1 filter paper. Quantities are for a single sequencing gel. If gels are poured daily, make gel solutions in quantity—e.g., make 1 liter of gel solution by multiplying the above quantities by 16.7. Store 2 to 4 weeks at 4°C. Solutions of acrylamide deteriorate quickly, especially when exposed to light or left at room temperature.

Denaturing solution *(UNIT 4.2)*

4 M guanidinium thiosulfate
25 mM sodium citrate, pH 7.0
0.1 M 2-ME (added as noted below)
0.5% (w/v) *N*-laurylsarkosine (Sarkosyl)

Prepare a stock solution by dissolving 250 g guanidinium thiosulfate in a solution of 293 ml H_2O, 17.6 ml of 0.75 M sodium citrate, pH 7.0, and 26.4 ml of 10% Sarkosyl at 60 to 65°C with stirring. The stock solution can be stored up to 3 months at room temperature.

Prepare a working solution by adding 0.35 ml of 2-ME per 50 ml stock solution. The working solution can be stored 1 month at room temperature.

Denaturing solution *(UNIT 4.9)*

500 µl formamide
162 µl 12.3 M (37%) formaldehyde
✓ 100 µl MOPS buffer

If formamide has a yellow color, deionize as follows: add 5 g of mixed-bed ion-exchange resin [e.g., Bio-Rad AG 501-X8 or X8(D) resins] per 100 ml formamide, stir 1 hr at room temperature, and filter through Whatman #1 filter paper.

CAUTION: *Formamide is a teratogen. Handle with care.*

Denhardt solution, 100×

10 g Ficoll 400 [2% (w/v)]
10 g polyvinylpyrrolidone [2% (w/v)]
10 g BSA (Pentax Fraction V; 20 mg/ml)
H_2O to 500 ml
Filter and store at −20°C in 25-ml aliquots

DEPC (diethylpyrocarbonate) treatment of solutions

Add 0.2 ml DEPC to 100 ml of the solution to be treated. Shake vigorously to get the DEPC into solution. Autoclave the solution to inactivate the remaining DEPC. Many investigators keep the solutions they use for RNA work separate to ensure that "dirty" pipets do not go into them. See also separate entry entitled CsCl, 5.7 M, DEPC-treated.

CAUTION: *Wear gloves and use a fume hood when using DEPC, as it is a suspected carcinogen.*

DEPC solution *(UNIT 15.4)*

Dilute DEPC 1:10 in 100% ethanol. *Immediately* before use, further dilute 1:10 DEPC 1:1000 in lysis buffer (see recipe).

DEPC-treated CsCl, 5.7 M, *see CsCl, 5.7 M, DEPC-treated*

Desalting solution *(UNIT 10.8)*

0.05 M NH_4OH or acetic acid are convenient, and since they are volatile can be completely removed during lyophilization.

Destaining solution *(UNIT 10.4)*

5% (v/v) acetic acid
16.5% (v/v) methanol
78.5% H_2O

Destaining solution *(UNIT 10.5)*

7% (v/v) acetic acid
5% (v/v) methanol
88% H_2O

Detergent stock solutions *(UNIT 10.10)*

10% (v/v) Triton X-100 (store in the dark to prevent photooxidation) *or* 5% (w/v) sodium deoxycholate

Sterilize each solution separately by Millipore filtration. Both solutions are stable 5 years at room temperature.

Developing solution *(UNIT 10.5)*

0.5 g sodium citrate [0.5% (w/v)]
0.5 ml 37% formaldehyde solution (Kodak)
H_2O to 100 ml

Dialysis buffer *(UNIT 12.1)*

20 mM HEPES, pH 7.9, 4°C
20% (v/v) glycerol
✓ 100 mM KCl
✓ 0.2 mM EDTA
0.2 mM PMSF (add dropwise just before use; prepare 0.2 M stock in anhydrous isopropanol; stable for 9 months)
0.5 mM DTT (add just before use)

Dialysis buffer *(0.1 M NH_4HCO_3; UNIT 10.4)*

Dissolve 4.0 g NH_4HCO_3 in 500 ml H_2O containing 0.02% (w/v) SDS.

Dialyzed fetal bovine serum *(UNIT 9.9)*

Purchase from commercial supplier or prepare as follows:

1. Heat inactivate fetal bovine serum at 56°C for 60 min.
2. Soak Spectrapor dialysis tubing (MWCO 6000-8000) in PBS. Remove, rinse tubing, clip one end closed, and fill with FBS.
3. Dialyze 6 to 8 hr in cold room against PBS. Change dialysis solution at least once.
4. Filter sterilize using a 0.02-mm filter and store frozen (–20°C) in 50-ml aliquots.

Diaminobenzidene (DAB) substrate solution *(UNIT 14.6)*

Dissolve 60 mg 3,3′ diaminobenzidene in 200 ml PBS [0.03% (w/v)]. Store frozen in 5- to 10-ml aliquots until use. Defrost DAB aliquot and add hydrogen peroxide to a final concentration of 0.1%. (NOTE: concentrated hydrogen peroxide is generally sold as a 30% solution.) Filter through syringe (Millipore) and use immediately.

CAUTION: *DAB is a carcinogen and must be used with great care. Exposure to dry DAB may be minimized by preparing large batches of DAB solution in the fume hood and storing them as frozen aliquots.*

Diethylpyrocarbonate, *see DEPC*

Digestion buffer *(UNIT 2.2)*

✓ 100 mM NaCl
✓ 10 mM Tris·Cl, pH 8.0
✓ 25 mM EDTA, pH 8.0
0.5% (w/v) SDS
0.1 mg/ml proteinase K

The proteinase K is labile and must be added fresh with each use.

Digestion buffer *(UNIT 10.18)*

✓ 0.1 M Tris·Cl, pH 8
✓ 1 mM $CaCl_2$
10% (v/v) acetonitrile
Store frozen

Digoxigenin-11-dUTP/dTTP, 10× stock *(UNIT 3.18)*

0.375 mM dTTP
0.125 mM digoxigenin-11-dUTP
✓ 20 mM Tris·Cl, pH 7.8
Store at −20°C

Digoxigenin detection solution *(UNIT 14.7)*

2 µg/ml fluorescein *or* rhodamine-conjugated Fab fragment of sheep anti-digoxigenin (Boehringer Mannhein) in 4× SSC/1% (w/v) BSA (fraction V). Prepare fresh daily.

Dilution buffer *(UNIT 10.15)*

0.1% (v/v) Triton X-100 (store at room temperature in dark)
0.1% (w/v) bovine hemoglobin (store frozen)
Prepare in TSA solution (see recipe)

Dioxetane detection solution *(UNIT 7.5)*

Prepare a 1/100 dilution of concentrated (100×) dioxetane (Table 7.4.6) in assay buffer (see recipe). Make fresh and do not reuse solution.

Dioxetane phosphate substrate buffer *(UNIT 10.7)*

✓ 1 mM $MgCl_2$
0.1 M diethanolamine
0.02% (w/v) sodium azide (optional)
Adjust to pH 10 with HCl and use fresh

Traditionally, the AMPPD substrate buffer has been a solution containing 1 mM $MgCl_2$ and 50 mM sodium carbonate/bicarbonate, pH 9.6 (Gillespie and Hudspeth, 1991). The use of diethanolamine results in better light output (Tropix Western Light instructions).

Alternatively, 100 mM Tris·Cl (pH 9.5)/100 mM NaCl/5 mM $MgCl_2$ can be used.

Dioxetane phosphate visualization solution *(UNIT 10.7)*

Prepare 0.1 mg/ml AMPPD or CSPD (Tropix) *or* Lumigen-PPD (Lumigen; see Table 10.7.1) substrate in dioxetane phosphate substrate buffer (see recipe). Prepare just before use. Lumi-Phos 530 (Boehringer Mannheim or Lumigen) is a ready-to-use solution and can be applied directly to the membrane.

This concentration (240 μM) of AMPPD substrate is the minimum recommended by Tropix Western Light. Ten-fold lower concentrations can be used but require longer exposures.

Dithiothreitol (DTT), 1 M

Dissolve 15.45 g DTT in 100 ml H_2O
Store at −20°C

DMS reaction buffer *(UNIT 12.3)*

50 mM sodium cacodylate, pH 8.0
✓ 1 mM EDTA, pH 8.0
Store at 4°C

DMS stop buffer *(UNITS 12.3 & 15.5)*

✓ 1.5 M sodium acetate, pH 7.0
1.0 M 2-ME
Filter sterilize, then add tRNA to 100 μg/ml final
Store indefinitely at −20°C

DMS stop buffer *(UNIT 12.3)*

✓ 1.5 M sodium acetate, pH 7.0
1 M 2-ME
Store at 4°C

DNA dilution buffer *(UNIT 3.18)*

✓ 0.1 μg/l sheared salmon sperm DNA prepared in 6× SSC (see recipe).
Store at 4°C (for long-term storage, aliquot and store at −20°C).

DNA wash solution *(UNIT 14.3)*

✓ 0.6 M NaCl
✓ 10 mM Tris·Cl, pH 7.5
✓ 1 mM EDTA
50% (v/v) formamide (use inexpensive grade, e.g. Fluka, and do not deionize)
0.1% (v/v) 2-ME

DNase I, RNase-free *(UNITS 2.10, 4.1, 4.4 & 4.7)*

Commercially prepared enzymes such as Worthington grade DPRF are satisfactory. If supplied as a powder, redissolve in TE buffer containing 50% (v/v) glycerol and store at −20°C.

DNase I stop solution *(UNIT 12.4)*

For each 200-μl binding-reaction mixture:
645 μl 100% ethanol
5 μl tRNA stock solution (1 mg/ml)
50 μl saturated ammonium acetate
Store at −20°C for up to 1 to 2 weeks

DNase I storage buffer *(UNIT 12.4)*

✓ 50 mM Tris·Cl, pH 7.2
✓ 10 mM $MgSO_4$
✓ 1 mM DTT
50% (v/v) glycerol
Store DNase I in small aliquots at −70°C

DNase solution, RNase-free *(UNIT 14.8)*

✓ 40 mM Tris·Cl, pH 7.4
✓ 6 mM $MgCl_2$
✓ 2 mM $CaCl_2$
1 U/µl RNase-free DNase (e.g., RQ1, Promega)
Prepare fresh

For use with cells that are particularly rich in RNase, add 1000 U/ml placental ribonuclease inhibitor (e.g., RNasin, Promega) and 1 mM DTT to the above solution.

DNase stop mix *(UNIT 4.1)*

✓ 50 mM EDTA
✓ 1.5 M sodium acetate
1% (w/v) SDS

The SDS may come out of solution at room temperature. Heat briefly to redissolve.

DNase-free RNase, *see RNase, DNase-free*

dNTP/BrdU solution, 50× *(UNIT 12.5)*

2.5 mM dATP
2.5 mM dGTP
250 µM dCTP
2.5 mM BrdU (5-bromo-2′-deoxyuridine triphosphate; Pharmacia LKB)

4dNTP mix *(UNIT 15.1)*

For 2 mM 4dNTP mix: Prepare 2 mM each dNTP in TE buffer, pH 7.5. Store up to 1 year at −20°C in 1-ml aliquots.

For 25 mM 4dNTP mix: Combine equal volumes of 100 mM dNTPs (Promega). Store indefinitely at −20°C in 1-ml aliquots.

4dNTP mix, 25 mM, pH 7.0 *(UNIT 15.5)*

Prepare as in *UNIT 3.4*. Alternatively, purchase as individual 100-mM stock solutions from Pharmacia LKB and combine equal amounts of dATP, dCTP, dGTP, and dTTP to obtain the 25 mM 4dNTP mix. Store at −20°C.

dNTP/biotin mix *(UNIT 3.18)*

Prepare a mixture of 1 mM dATP, 1 mM dCTP, 1 mM dGTP, and 0.65 mM dTTP. Add biotin-16-dUTP (Enzo Biochem) to 0.35 mM final. Store at −20°C.

dNTP/TTP solution, 50× *(UNIT 12.5)*

2.5 mM dATP
2.5 mM dGTP
250 µM dCTP
2.5 mM TTP

Dowex AG50W-X8 resin, 100-200 mesh *(UNIT 1.7)*

Prepare resin in large batches (200 to 400 ml packed resin) by the following series of washing steps. These washes can be conveniently performed using a large Buchner funnel. Use filter paper to collect the resin between changes of wash solution.

1. Wash resin in ≥10 vol 0.5 N NaOH until no color is observed in wash solution. (Resin will retain its buff color.)
2. Wash with 5 to 10 vol 0.5 N HCl.
3. Wash with 5 to 10 vol 0.5 M NaCl.
4. Wash with 5 to 10 vol H_2O.
5. Wash with 5 to 10 vol 0.5 N NaOH.
6. Wash with distilled water until pH = 9.
7. Store prepared resin in 0.5 M NaCl/0.1 M Tris, pH 7.5, at 4°C indefinitely.

Drying solution *(UNIT 10.5)*

10% (v/v) ethanol
4% (v/v) glycerol
Can be stored ~1 month at room temperature

Dulbeccos minimum essential (or modified Eagle) medium, *see Complete DMEM*

EDTA (ethylenediamine tetraacetic acid), 0.5 M

Dissolve 186.1 g $Na_2EDTA \cdot 2H_2O$ in 700 ml H_2O
Adjust pH to 8.0 with 10 M NaOH (~50 ml)
Add H_2O to 1 liter

Electroporation buffers *(UNIT 9.3)*

Choice of electroporation buffer depends on the cells being used in the experiment. The following buffers (stored at 4°C) can be used:

✓ 1. PBS *without* Ca^{++} or Mg^{++}
✓ 2. HEPES-buffered saline (HeBS)
3. Tissue culture medium *without* FBS (introduction to Chapter 9)
4. Phosphate-buffered sucrose: 272 mM sucrose/7 mM K_2HPO_4 (adjusted to pH 7.4 with phosphoric acid)/1 mM $MgCl_2$

Elution buffer *(UNITS 2.7, 8.2A & 8.3)*

✓ 0.5 M ammonium acetate
✓ 1 mM EDTA
Final pH should be 8.0

This stock solution should be protected from light, and is stable for several months at room temperature.

Elution buffer *(UNIT 4.7)*

✓ 2 M ammonium acetate
1% (w/v) SDS
25 μg/ml tRNA

Elution buffer *(UNIT 10.4)*

Dissolve 1.98 g NH_4HCO_3 in 500 ml H_2O containing 0.1% (w/v) SDS
Final buffer is 0.05 M NH_4HCO_3

Elution buffers *(UNIT 11.4)*

To 900 ml H_2O, add:
11.76 g trisodium citrate dihydrate (0.04 M final)
1.17 g NaCl (0.02 M final)
Titrate with HCl to pH 6.0, 5.0, 4.0, or 3.2
Add H_2O to 1 liter

Elution buffer *(UNIT 12.7)*

✓ 20 mM Tris·Cl, pH 7.5
✓ 200 mM NaCl
✓ 1 mM EDTA
Store at room temperature

Elutip high-salt solution *(UNIT 2.6)*

✓ 1 M NaCl
✓ 20 mM Tris·Cl, pH 7.5
✓ 1 mM EDTA
Filter sterilize and store at room temperature

Elutip low-salt solution *(UNIT 2.6)*

✓ 0.2 M NaCl
✓ 20 mM Tris·Cl, pH 7.5
✓ 1 mM EDTA
Filter sterilize and store at room temperature

End-labeling mix, 5 μl *(UNIT 15.5)*

✓ 1.0 μl 5× amplification buffer
2.3 μl 1 pmol/μl end-labeled primer 3 (Fig. 15.5.3)
✓ 0.4 μl 25 mM 4dNTP mix, pH 7.0
0.8 μl H_2O
0.5 μl 2 U/μl *Thermococcus litoralis* DNA polymerase (Vent; New England Biolabs)

Prepare immediately before use. Mix first four components and chill on ice before adding Vent DNA polymerase. Then add polymerase and keep on ice.

5′ end-labeled primer 3 can be prepared as in UNIT 3.10 (forward reaction) with the following modifications: end-label 20 to 100 pmol of primer 3 with labeling-grade (i.e., less expensive) [γ-^{32}]P ATP (~6000 Ci/mmol); incubate at 37°C for 30 min instead of 60 min. Remove unincorporated ^{32}P by gel purification (UNIT 2.11) or by using a Nensorb-20 nucleic acid purification cartridge (Du Pont NEN)—elute primer with 50% ethanol instead of 50% methanol. Resuspend primer 3 in TE buffer, pH 7.5, so that its concentration is 1 pmol/μl; the specific activity should be 4–9 × 10^6 cpm/pmol.

Eosin stain *(UNIT 14.5)*

12 g eosin Y [Fisher; 2.4% (w/v)]
3 g phloxine B [Fisher; 0.6% (w/v)]
500 ml 70% ethanol

Equilibration buffer *(UNIT 10.3)*

3.75 g Tris base (0.123 M)
25 ml glycerol [10% (v/v)]
5.25 g SDS [2% (v/v)]
333 mg dithiothreitol (DTT; 7 mM)

Dissolve Tris base in H_2O and adjust pH to 6.8 with 6 M HCl. Add other ingredients. Add H_2O to 250 ml final volume. This buffer can be stored for several weeks at room temperature.

Ethanolamine, pH 8.0, 1.0 M *(UNIT 11.4)*

61.1 ml ethanolamine
Titrate with HCl to pH 8.0
Add H_2O to 1 liter

1 M ethanolamine hydrochloride, pH 8.0 *(UNIT 12.9)*

1 M ethanolamine
Adjust pH to 8.0 with HCl
Filter sterilize
Store at room temperature

Ethidium bromide, 10 mg/ml

Dissolve 0.2 g ethidium bromide in 20 ml H_2O
Mix well and store at 4°C in dark

CAUTION: *Ethidium bromide is a mutagen and must be handled carefully.*

Ethidium bromide assay solution *(APPENDIX 3D)*

Add 10 ml of 10× TNE buffer (see recipe) to 89.5 ml H_2O. Filter through a 0.45-μm filter, then add 0.5 ml of 1 mg/ml ethidium bromide.

Add the dye after filtering, as ethidium bromide will bind to most filtration membranes.

CAUTION: *Ethidium bromide is hazardous; wear gloves and use appropriate care in handling, storage, and disposal.*

Ethidium bromide solution *(UNITS 2.5A, 2.6 & 12.2)*

1000× stock solution, 0.5 mg/ml:
50 mg ethidium bromide
100 ml H_2O
Working solution, 0.5 μg/ml:
Dilute stock 1:1000 for gels or stain solution

Protect solutions from light. CAUTION: *Ethidium bromide is a mutagen and must be handled carefully.*

Exo III ligation buffer *(UNIT 7.2)*

✓ 80 mM Tris·Cl, pH 7.5
✓ 30 mM DTT
✓ 20 mM $MgCl_2$
Store at −20°C

Extract buffer *(UNIT 12.7)*

✓ 50 mM Tris·Cl, pH 7.5
✓ 1 mM EDTA
✓ 1 mM DTT
1 mM phenylmethylsulfonylfluoride (PMSF)
Store at −20°C

Extraction buffer *(UNIT 2.3)*

✓ 100 mM Tris·Cl, pH 8.0
✓ 100 mM EDTA
✓ 250 mM NaCl
100 μg/ml proteinase K

Extraction buffer *(UNIT 9.7B)*

✓ 0.1 M potassium phosphate, pH 7.8
✓ 1 mM DTT added just before use
Store at room temperature

Extraction buffer *(UNIT 13.13)*

✓ Lysis buffer
0.8 M ammonium sulfate
20% (v/v) glycerol

Extraction buffer *(UNIT 16.11)*

✓ 100 mM Tris·Cl, pH 7.5
✓ 90 mM EDTA
✓ 200 mM KCl
Filter sterilize and store ≤1 year at 4°C

Fetal bovine serum (FBS), dialyzed, *see Dialyzed bovine serum*

Ficoll buffer *(UNIT 13.13)*

18% (w/v) Ficoll-400 (Pharmacia LKB)
20 mM PIPES, pH 6.3
✓ 0.5 mM $CaCl_2$
✓ 1 mM DTT
✓ 0.1 mM EDTA
✓ 1× protease inhibitor mix
1 mM PMSF

First-strand buffer, 5× *(UNIT 15.5)*

✓ 200 mM NaCl
✓ 50 mM Tris·Cl, pH 8.9
✓ 25 mM $MgSO_4$
0.05% (w/v) gelatin (Sigma)
Store at −20°C

First-strand synthesis mix, 25 µl *(UNIT 15.5):*

✓ 6.0 µl 5× first-strand buffer
0.3 µl 1 pmol/µl oligonucleotide primer 1
✓ 0.24 µl 25 mM 4dNTP mix, pH 7.0
18.21 µl H_2O
0.25 µl 2 U/µl *Thermococcus litoralis* DNA polymerase (Vent; New England Biolabs)

Prepare immediately before use. Mix all components except Vent DNA polymerase and chill on ice. Then add polymerase and keep on ice.

Primer 1 should be stored in a silanized tube at a concentration ≥10 pmol, then diluted in TE buffer, pH 7.5, immediately before use. This will minimize primer loss due to sticking to the tube walls.

Fixative solution *(0.05% glutaraldehyde; UNIT 9.11)*

Glutaraldehyde (Sigma) is typically supplied as a 25% solution and can be stored frozen at –20°C in aliquots. Aliquots of the 25% solution can be frozen and thawed many times. Immediately before use, thaw an aliquot and dilute 500-fold in PBS to 0.05%.

CAUTION: *Glutaraldehyde is a carcinogen and causes allergic reactions. Use in fume hood and discard as directed by your institution.*

Fixative solution *(2% paraformaldehyde; UNIT 9.11)*

Dissolve 2 g paraformaldehyde (e.g., from BDH) in 100 ml of 0.1 M PIPES buffer, pH 6.9 [30.24 g piperazine-*N,N′*-bis(2-ethanesulfonic acid) per liter] containing 2 mM $MgCl_2$ (200 µl of a 1 M solution) and 1.25 mM EGTA (250 µl of a 0.5 M solution, pH 8.0). Heat in the fume hood with stirring 5 to 10 min (do not boil). Cool to 4°C. Always use paraformaldehyde solution that is well buffered and <1 week old.

CAUTION: *Formaldehyde is a carcinogen and may cause allergic reaction. Use in fume hood and discard as directed by your institution.*

Fixing solution *(UNIT 10.5)*

50% (v/v) methanol
10% (v/v) acetic acid
40% H_2O
Can be stored ~1 month at room temperature

Formaldehyde fixing solution *(UNIT 10.5)*

40% (v/v) methanol
0.5 ml of 37% formaldehyde per liter solution
Can be stored ~1 month at room temperature

Formaldehyde loading buffer *(UNIT 4.9)*

✓ 1 mM EDTA, pH 8.0
0.25% (w/v) bromphenol blue
0.25% (w/v) xylene cyanol
50% (v/v) glycerol

Formamide, deionized, *see Deionized formamide*

Formamide gel solution *(UNIT 7.6)*

Reagent[a]	Acrylamide concentration (%)		
	4	6	8
Urea (ultrapure) (g)	42	42	42
✓38% acrylamide/2% bisacrylamide (ml)	10	15	20
✓10× TBE electrophoresis buffer (ml)	10	10	10
Deionized formamide (ml)[b]	40	40	40
H_2O (ml)	10	5	0
Total volume (ml)	*100*	*100*	*100*

[a]Heat the ingredients in a beaker while stirring. When dissolved, add H_2O to 100 ml total volume.
[b]Deionize formamide by stirring with Amberlite MB-1 (Sigma) or equivalent mixed-bed resin 30 min at 4°C. Filter through Whatman No. 1 filter paper and store at −20°C. Formamide that has been deionized is available from American Bioanalytical.

Formamide loading buffer, 2× *(UNIT 2.11)*

Mix 1 part 10× TBE electrophoresis buffer (see recipe) with 9 parts deionized formamide (see recipe). Add bromphenol blue to 0.5%.

High-quality formamide that does not require deionization is available from Fluka, IBI, and American Laboratory.

Formamide loading buffer *(UNITS 4.6, 4.8 & 12.4)*

✓ 0.2 ml 0.5 M EDTA, pH 8.0
10 mg bromphenol blue [0.1% (w/v)]
10 mg xylene cyanol [0.1% (w/v)]
10 ml formamide

High-quality formamide that does not require deionization is available from Fluka, IBI, and American Laboratory.

Formamide loading buffer *(UNIT 12.9)*

✓ 90 ml deionized formamide
✓ 10 ml 10× TBE electrophoresis buffer
40 mg xylene cyanol [0.04% (w/v)]
40 mg bromphenol blue [0/04% (w/v)]
Store at −20°C

Formamide loading buffer *(UNIT 15.8)*

95% (v/v) formamide
0.09% (w/v) bromphenol blue
0.09% (w/v) xylene cyanol FF
Store at 4°C

Formamide prehybridization/hybridization (FPH) solution *(UNITS 2.10 & 4.9)*

✓ 5× SSC
✓ 5× Denhardt solution
50% (w/v) formamide
1% (w/v) SDS

Add 100 µg/ml denatured salmon sperm DNA (see recipe) just before use

Alternatives to Denhardt solution and denatured salmon sperm DNA as blocking agents are listed in Table 2.10.2.

Commercial formamide is usually satisfactory for use. If the liquid has a yellow color, deionize as follows: add 5 g of mixed-bed ion-exchange resin [e.g., Bio-Rad AG 501-X8 or 501-X8(D) resins] per 100 ml formamide, stir at room temperature for 1 hr, and filter through Whatman no. 1 paper.

CAUTION: *Formamide is a teratogen. Handle with care.*

G+A stop solution *(UNIT 15.5)*

✓ 360 mM sodium acetate, pH 7.5
✓ 0.14 mM EDTA, pH 8.0
Prepare fresh and chill on ice before use

Gel buffer *(UNIT 10.3)*

Dissolve 90.8 g Tris base (1.5 M) in 300 ml H_2O. Adjust to pH 8.6 with 6 M HCl. Add H_2O to 500 ml. This buffer can be stored for several weeks in the refrigerator.

4× gel buffer *(UNIT 10.18)*

41.24 g *bis*-Tris (0.493 M)
7.08 ml 37.5% HCl (reagent grade; 0.007%)
H_2O to 400 ml

Final pH should be 6.61; do not adjust the pH with acid or base. This buffer may be stored indefinitely at −20°C or ≤2 months at 4°C.

Gelatin subbing solution, 1× and 0.2× *(UNIT 14.1)*

Prepare 1× solution by dissolving gelatin (type Bloom 275) in water at 70°C to 0.5%. Cool solution to room temperature, and add 0.05% (w/v) $CrK(SO_4)_2 \cdot 12H_2O$ (chromium potassium sulfate). Cool solution to 4°C on ice. Use immediately, as gelling occurs after prolonged storage of the solution on ice.

Prepare 0.2× gelatin solution by diluting 1× solution 5-fold with water. It can be used for several days if prepared using sterile water and stored at room temperature. If cloudiness or bacterial growth occurs, discard.

Gelvatol *(UNIT 14.5)*

Gelvatol (available as Airvol from Air Products and Chemicals) is a water-soluble mounting medium made from polyvinyl alcohol (PVA) and glycerol. The following recipe is a simplification of the original method: Add 5 g PVA 2000 to 100 ml 1× PBS every hour for 4 hr (total 20 g) while stirring constantly. Keep covered and stir overnight at 4°C. Add 3 more grams PVA 2000 and stir until dissolved. Add a single sodium azide crystal and 50 ml glycerol. Mix thoroughly, aliquot, and store at 4°C in sealed storage vials.

GF (gel-filtration) buffer *(UNIT 10.8)*

Tris, phosphate, and acetate buffers at various pHs are most commonly used, although almost any buffer may be used. An ionic strength of at least 0.05 M is recommended to reduce nonspecific interactions between the proteins being separated and the chromatographic matrix. If required, NaCl, urea, and guanidine·Cl may be added to GF buffer and the sample solution in order to completely solubilize the proteins in a sample mixture.

Glass bead disruption buffer *(UNIT 13.13)*

✓ 20 mM Tris·Cl, pH 7.9
✓ 10 mM $MgCl_2$
✓ 1 mM EDTA
5% (v/v) glycerol
✓ 1 mM DTT
✓ 0.3 M ammonium sulfate
✓ 1× protease inhibitor mix
1 mM PMSF

The concentration of ammonium sulfate in the buffer can be varied between 0.1 and 1.0 M. Final concentrations above 0.25 M strip specific DNA-binding proteins and histones off chromatin and are therefore useful in obtaining factors that interact with nucleic acids. KCl and NaCl can also be added to final concentrations between 0.1 and 2.0 M.

A1

Glass beads, chilled and acid-washed, *0.45- to 0.55-mm (UNITS 13.11, 13.12 & 13.13)*

Wash the beads by soaking 1 hr in concentrated nitric acid. Rinse thoroughly with water. Dry the beads in a baking oven, cool to room temperature, and store at 4°C until needed.

Glass beads suspension *(UNITS 2.1 & 2.6)*

Transfer a volume of ~200 to 300 µl of 200-µm glass beads (National Scientific Supply) into a 1.5-ml microcentrifuge tube; add an equal volume of water. Vortex briefly to suspend just before using.

If glass beads do not come acid-washed, prepare as follows: Wash by soaking 1 hr in concentrated nitric acid. Rinse thoroughly with water. Dry in a baking oven, cool to room temperature, and store at 4°C until needed.

Glucose/Tris/EDTA solution *(UNITS 1.6 & 1.7)*

50 mM glucose
✓ 25 mM Tris·Cl, pH 8.0
✓ 10 mM EDTA

Glutaraldehyde fixative, 0.05%, *see Fixative solution (0.05% glutaraldehyde)*

Glutathione-agarose bead slurry *(UNIT 16.7)*

Preswell *S*-linkage glutathione-agarose beads (Sigma or Pharmacia Biotech) 1 hr in 10 vol PBS. Wash twice with PBS and store as a 50% (v/v) slurry ≤1 month at 4°C.

The beads can be recycled by boiling 5 min in PBS containing 1% SDS (Frangioni and Neel, 1993), but should then only be used to purify the same fusion protein to prevent cross-contamination.

Glycerol solution *(UNIT 1.3)*

65% (v/v) glycerol
✓ 0.1 M $MgSO_4$
✓ 0.025 M Tris·Cl, pH 8.0
Store at room temperature

Glycerol solution *(UNIT 13.14)*

65% (v/v) glycerol, sterile
0.1 M $MgSO_4$
25 mM Tris·Cl, pH 8.0
Store at room temperature (stable at least 1 year)

Glycine buffer *(UNIT 10.10)*

50 mM glycine·HCl, pH 2.5
0.1% (v/v) Triton X-100 (see recipe for detergent stock solutions)
✓ 0.15 M NaCl

Glycine·Cl buffer, pH 2.3 *(UNIT 11.4)*

To 900 ml H_2O, add:
3.75 g glycine (0.05 M final)
8.77 g NaCl (0.15 M final)
Titrate with HCl to pH 2.3
Add H_2O to 1 liter

Glycine·OH buffer, pH 8.9 *(UNIT 11.4)*

To 700 ml H_2O, add:
108.9 g glycine (1.45 M final)
175.3 g NaCl (3 M final)
Titrate with NaOH to pH 8.9
Add H_2O to 1 liter

Glycine buffer *(UNIT 10.10)*

50 mM glycine·Cl, pH 2.5
✓ 0.1% (v/v) Triton X-100
✓ 0.15 M NaCl

6 M glyoxal, deionized (*UNIT 4.9*)

Immediately before use, deionize glyoxal by passing through a small column of mixed-bed ion-exchange resin (e.g., Bio-Rad AG 501-X8 or X8(D) resins) until the pH is >5.0.

Glyoxal loading buffer (*UNIT 4.9*)

10 mM $NaPO_4$, pH 7.0
0.25% (w/v) bromphenol blue
0.25% (w/v) xylene cyanol
50% (v/v) glycerol

Gram$^-$ lysing buffer (*UNIT 4.4*)

10 mM Tris·Cl, pH 8.0
10 mM NaCl
1 mM sodium citrate
1.5% (w/v) SDS

Grinding buffer (*UNIT 4.3*)

✓ 0.18 M Tris
0.09 M LiCl
✓ 4.5 mM EDTA
1% (w/v) SDS
pH to 8.2 with HCl

This buffer is equivalent to TLE solution with $^1/_{10}$ vol 10% SDS added.

GTBE buffer (*UNITS 2.5B & 6.10*)

✓ 50 ml 10× TBE electrophoresis buffer
50 ml 2 M glycine
900 ml H_20

Guanidinium solution (*UNIT 4.2*)

4 M guanidinium isothiocyanate
✓ 20 mM sodium acetate, pH 5.2
✓ 0.1 mM DTT
0.5% (w/v) Sarkosyl

Dissolve the guanidinium isothiocyanate in water and the appropriate amount of sodium acetate. Heating the solution slightly (65°C) may be necessary to get the guanidinium into solution. Add the DTT and Sarkosyl. Check the pH—it should be ~5.5. If not, adjust with acetic acid. Bring to volume and filter the solution through a Nalgene filter. Store at room temperature.

GuMCAC buffers (*UNIT 10.11*)

GuMCAC-0 buffer:
✓ 20 mM Tris·Cl, pH 7.9
✓ 0.5 M NaCl
6 M guanidine·HCl
10% (v/v) glycerol

GuMCAC-500 buffer:
✓ 20 mM Tris·Cl, pH 7.9
✓ 0.5 M NaCl
6 M guanidine·HCl
0.5 M imidazole
10% (v/v) glycerol

GuMCAC-20 buffer and other GuMCAC buffers containing different concentrations of imidazole are made by mixing GuMCAC-0 and GuMCAC-500 in the appropriate ratios: e.g., for GuMCAC-20, use 96:4 (v/v) GuMCAC-0/GuMCAC-500.

Store GuMCAC buffers ≤6 months at 4°C.

GuMCAC-EDTA buffer (*UNIT 10.11*)

✓ 20 mM Tris·Cl, pH 7.9
✓ 0.5 M NaCl
6 M guanidine·HCl
✓ 0.1 M EDTA, pH 8.0
10% (v/v) glycerol
Store ≤6 months at 4°C

Hanks balanced salt solution (HBSS)

5.4 mM KCl
0.3 mM Na_2HPO_4
0.4 mM KH_2PO_4
4.2 mM $NaHCO_3$
1.3 mM $CaCl_2$
0.5 mM $MgCl_2$
0.6 mM $MgSO_4$
137 mM NaCl
5.6 mM D-glucose
0.02% (w/v) phenol red (optional)
Add H_2O to 1 liter and adjust pH to 7.4

HBSS can be purchased from Biofluids or Whittaker.

HBSS may be made or purchased without $CaCl_2$ and $MgCl_2$. These are optional components that usually have no effect on an experiment. In some cases, however, their presence may be detrimental to a procedure. Consult the individual protocol to see if the presence or absence of these components is recommended in the materials list.

HCl, 1 M

Mix in the following order:
913.8 ml H_2O
86.2 ml concentrated HCl

HEMGN buffer *(UNIT 16.6)*

✓ 100 mM KCl
25 mM HEPES, pH 7.6
✓ 0.1 mM EDTA, pH 8.0
✓ 12.5 mM $MgCl_2$
10% (v/v) glycerol
0.1% (v/v) Nonidet P-40

Prepare 2 liters of above solution. For resuspending cells and pellet, take 10 ml and add 10 µl of each of the following just before using (save remainder of the 2 liters for dialysis):

1 M DTT (1 mM final)
2 mg/ml aprotinin (2 µg/ml final; optional)
1 mg/ml leupeptin (1 µg/ml final; optional)
1 mg/ml pepstatin in methanol (1 µg/ml final; optional)
100 mM PMSF in ethanol (0.1 mM final)
100 mM sodium *meta*-bisulfite (0.1 mM final)

The last five ingredients are protease inhibitors. Store solutions of DTT, aprotinin, leupeptin, and pepstatin at –20°C and PMSF solution at 4°C. Make sodium meta-bisulfite fresh and store on ice during use. For dialysis, only PMSF and sodium meta-bisulfite are necessary for protease inhibition. DTT should be included in all dialysis buffers.

HEPES binding buffers *(UNIT 12.7)*

25 mM HEPES, pH 7.9
✓ 25 mM NaCl
✓ 5 mM $MgCl_2$
✓ 0.5 mM DTT

Prepare separate solutions *without* and *with* 6 M guanidine·HCl. Prepare fresh for each use and keep briefly at 4°C.

HEPES-buffered saline (HeBS), 2× *(UNIT 9.1)*

16.4 g NaCl (0.283 M)
11.9 g HEPES acid (0.023 M)
0.21 g Na_2HPO_4 (1.5 mM)
H_2O to 1 liter

Titrate to pH 7.05 with 5 M NaOH (an exact pH is extremely important for efficient transfection). Filter sterilize. Many researchers make up large batches of 2× HeBS, test for transfection efficiency, and store the solution frozen in 50-ml aliquots.

There can be wide variability in the efficiency of transfection obtained between batches of 2× HeBS. Efficiency should be checked with each new batch. The 2× HeBS solution can be rapidly tested by mixing 0.5 ml 2× HeBS with 0.5 ml 250 mM $CaCl_2$ and vortexing. A fine precipitate should develop that is readily visible in the microscope. Transfection efficiency must still be checked, but if the solution does not form a precipitate in this test, there is something wrong.

HEPES-buffered saline (HeBS) *(UNIT 9.11)*

✓ 137 mM NaCl
✓ 5 mM KCl
0.7 mM Na_2HPO_4
6 mM dextrose
21 mM HEPES
Adjust pH to 7.05; check carefully as this is a critical parameter

HEPES-buffered saline (HeBS), 10× *(UNIT 16.11)*

✓ 1.37 M NaCl
0.06 M D^+ glucose
✓ 0.05 M KCl
0.007 M $Na_2HPO_4 \cdot 7H_20$
0.2 M HEPES
Dissolve in H_2O to specified concentration. Adjust to pH 7.1 with freshly prepared 5 M NaOH. Filter sterilize and store at −20°C.

Herring sperm DNA, sonicated, see Sonicated herring sperm DNA

High-ionic-strength electrophoresis buffer *(UNIT 12.2)*

✓ Prepare by diluting 5× Tris/glycine stock solution (see recipe) 5-fold.

High-ionic-strength gel mix *(UNIT 12.2)*

✓ 8.0 ml Tris/glycine stock solution
5.33 ml 30% (w/v) acrylamide
1.0 ml 2% (w/v) bisacrylamide
2.0 ml 50% (v/v) glycerol
23.7 ml H_2O
Filter through 0.2-µm filter
Use witin 24 hr of preparation

High-salt buffer *(UNIT 12.1)*

20 mM HEPES, pH 7.9, 4°C
25% (v/v) glycerol
✓ 1.5 mM $MgCl_2$
✓ 1.2 M KCl
✓ 0.2 mM EDTA
0.2 mM PMSF (add just before use)
✓ 0.5 mM DTT (add just before use)

High-salt immunoprecipitation buffer *(UNIT 6.8)*

Same recipe as immunoprecipitation buffer, except 0.5 M NaCl.

High-salt TE buffer *(UNIT 2.3)*

✓ 10 mM Tris·Cl, pH 8.0
✓ 0.1 mM EDTA, pH 8.0
✓ 1 M NaCl
Store at room temperature (stable for several years)

A1

High-stringency wash buffer I *(UNIT 6.3)*

✓ 0.2× SSC (dilute 20× recipe in this appendix)
0.1% (w/v) SDS

High-stringency wash buffer II *(UNIT 6.3)*

✓ 1 mM Na_2EDTA
40 mM $NaHPO_4$, pH 7.2
1% (w/v) SDS

Hoechst 33258 assay solutions *(APPENDIX 3D)*

Stock solution: Dissolve in H_2O at 1 mg/ml. Stable for ~6 months at 4°C.

Working solution: Add 10 ml of 10× TNE buffer (see recipe) to 90 ml H_2O. Filter through a 0.45-μm filter, then add 10 μl of 1 mg/ml Hoechst 33258.

Hoechst 33258 is a fluorochrome dye with a molecular weight of 624 and a molar extinction coefficient of 4.2×10^4 $M^{-1}cm^{-1}$ *at 338 nm.*

The dye is added after filtering because it will bind to most filtration membranes.

CAUTION: *Hoechst 33258 is hazardous; use appropriate care in handling, storage, and disposal.*

Hybridization buffer *(UNIT 4.7)*

5× stock solution:
200 mM PIPES, pH 6.4
✓ 2 M NaCl
✓ 5 mM EDTA

Working solution:
4 parts formamide
1 part 5× stock solution

Prepare hybridization buffer fresh as needed from frozen 5× stock and formamide freshly deionized by vortexing with mixed-bed resin beads. Alternatively, store in small aliquots at –70°C.

10× hybridization buffer *(UNIT 4.8)*

✓ 1.5 M KCl
✓ 0.1 M Tris·Cl, pH 8.3
✓ 10 mM EDTA

Hybridization mix A *(UNIT 14.3)*

✓ 50% (v/v) deionized formamide
✓ 0.3 M NaCl, sterile
✓ 10 mM Tris·Cl, pH 8.0
✓ 1 mM EDTA
✓ 1× Denhardt solution
✓ 500 μg/ml yeast tRNA
500 μg/ml poly(A) (Pharmacia)
✓ 50 mM DTT
10% polyethylene glycol (MW 6000; EM Science)
Store in aliquots at –70°C

Hybridization mix B *(UNIT 14.3)*

✓ 1.2 M NaCl
✓ 20 mM Tris·Cl, pH 7.5
✓ 4 mM EDTA
✓ 2× Denhardt solution
✓ 1 mg/ml yeast tRNA
200 μg/ml poly(A) (Pharmacia)
Store in aliquots at –20°C

Hybridization solution I *(UNIT 6.3)*

Mix following ingredients for range of volumes indicated (in milliliters). Solution may be stored for prolonged periods at room temperature.

Formamide	24	48	72	120	240	480
✓20× SSC	12	24	36	60	120	240
✓2 M Tris·Cl, pH 7.6	0.5	1.0	1.5	2.5	5.0	10
✓100× Denhardt solution	0.5	1.0	1.5	2.5	5.0	10
Deionized H_2O	2.5	5.0	7.5	12.5	25	50
50% dextran sulfate[a]	10	20	30	50	100	200
10% (w/v) SDS[b]	0.5	1	1.5	2.5	5	10
Total volume	50	*100*	*150*	*250*	*500*	*1000*

[a]Use high-quality dextran sulfate (e.g., Pharmacia).

[b]In place of SDS, Sarkosyl may be used. Add SDS (or Sarkosyl) last.

Hybridization solution II *(UNIT 6.3)*

1% (w/v) crystalline BSA (fraction V)
✓ 1 mM EDTA
✓ 0.5 M $NaHPO_4$ buffer, pH 7.2
7% (w/v) SDS

1 M $NaHPO_4$ = 134 g $Na_2HPO_4 \cdot 7H_2O$ plus 4 ml 85% H_3PO_4 per liter.

Hybridization solution IV *(UNIT 6.8)*

65% (v/v) deionized formamide
✓ 0.4 M NaCl
0.2% (w/v) SDS
30 mM PIPES, pH 6.5
50 µg yeast tRNA
50 to 500 µg poly(A)$^+$mRNA

This solution should be made fresh prior to use.

Hydroxylamine cleavage solution, 2× *(UNIT 16.5)*

4 M hydroxylamine
0.4 M CHES buffer
Adjust pH to 9.5 with NaOH
Prepare fresh

Hypotonic buffer *(UNIT 9.7A)*

✓ 25 mM Tris·Cl, pH 7.5
✓ 2 mM $MgCl_2$
Store up to 6 months at room temperature

Hypotonic buffer *(UNIT 12.1)*

10 mM HEPES, pH 7.9, 4°C
✓ 1.5 mM $MgCl_2$
✓ 10 mM KCl
0.2 mM PMSF (add just before use)
✓ 0.5 mM DTT (add just before use)

IEX protein standards *(UNIT 10.13)*

Weigh 5 mg ovalbumin, α-chymotrypsinogen, ribonuclease A, and lysozyme into a 1.5-ml microcentrifuge tube. Add 2.5 mg α-lactalbumin, and dissolve the protein mixture with 1 ml distilled H_2O. Incubate at room temperature for 1 hr. Dilute the resulting solution 10-fold with 0.02 M Tris·Cl, pH 7.5, or 0.02 M sodium phosphate, pH 6.0, for anion and cation exchange, respectively.

IMC medium (UNIT 16.8)
✓ 200 ml 2% (w/v) Casamino Acids (0.4% final)
✓ 100 ml 10× M9 salts (1× final)
40 ml 20% (w/v) glucose (sterile; 0.5% final)
✓ 1 ml 1 M $MgCl_2$ (sterile; 1 mM final)
✓ 0.1 ml 1 M $CaCl_2$ (sterile; 0.1 mM final)
1 ml 2% (w/v) vitamin B1 (sterile; 0.002% final)
658 ml glass-distilled H_2O (sterile)
10 ml 10 mg/ml ampicillin (sterile; optional; 100 µg/ml final)
Use fresh

IMC plates (UNIT 16.8)
15 g agar [Difco; 1.5% (w/v)]
4 g Casamino Acids [Difco-certified; 0.4% (w/v)]
858 ml glass-distilled H_2O (sterile)
Autoclave 30 min
Cool in a 50°C water bath
✓ 100 ml 10× M9 salts (1× final)
40 ml 20% (w/v) glucose (sterile; 0.5% final)
✓ 1 ml 1 M $MgCl_2$ (sterile; 1 mM final)
✓ 0.1 ml 1 M $CaCl_2$ (sterile; 0.1 mM final)
1 ml 2% (w/v) vitamin B1 (sterile; 0.002% final)
10 ml 10 mg/ml ampicillin (sterile; optional; 100 µg/ml final)
Mix well and pour into Petri plates
Store ≤1 month at 4°C

Immunoprecipitate wash buffer for boiled sample (UNIT 17.2)
✓ RIPA lysis buffer
✓ 1 mM DTT, added fresh

DTT is added from a 1 M stock solution stored at −20°C.

Immunoprecipitation buffer (UNIT 6.8)
10 mM Tris·Cl, pH 7.4
2 mM EDTA
0.15 M NaCl
10% (v/v) Nonidet P-40

Immunoscreening buffer (UNIT 6.7)
Prepare in PBS (see recipe):
5% (w/v) nonfat dry milk
0.1% (v/v) Nonidet P-40
0.05% (w/v) sodium azide (from 5% stock)

KCl, 1 M
74.6 g KCl
H_2O to 1 liter

Klenow fragment buffer (UNIT 7.4)
✓ 10 mM Tris·Cl, pH 7.5
✓ 50 mM NaCl
✓ 0.1 mM EDTA, pH 8.0
✓ 5 mM DTT
Store frozen

Labeling buffer (UNIT 2.10)
✓ 200 mM Tris·Cl, pH 7.5
✓ 30 mM $MgCl_2$
10 mM spermidine

Labeling mixes *(UNIT 7.4)*

To obtain sequencing reaction products that are relatively short (<350 bases), use the short mix. Prepare the long mix when relatively long (>150 bases) products are desired. Prepare labeling mixes for Sequenase and Klenow fragment in 10 mM Tris·Cl (pH 7.5)/0.1 mM EDTA using nucleotide stock solutions (see recipes). Prepare labeling mixes for *Taq* DNA polymerase in 0.1 mM EDTA. Store mixes at −20°C.

Short mix: 1.5 μM each dCTP, dGTP, and dTTP
Long mix: 7.5 μM each dCTP, dGTP, and dTTP

To substitute 7-deaza-dGTP for dGTP, replace dGTP with an equimolar concentration of 7-deaza-dGTP in each mix.

To substitute dITP for dGTP in Sequenase labeling mixes, replace dGTP with 3 μM dITP in the short mix and 15 μM dITP in the long mix. Substitution of dITP is not recommended when using Taq DNA polymerase.

Lambda PEG solution *(UNIT 7.3)*

20% (w/v) PEG 6000
2.5 M sodium acetate, pH 6.0

Light-emission accelerator solution *(UNIT 9.7B)*

10% (v/v) Emerald luminescence amplifier (Tropix)
✓ 0.2 N sodium hydroxide
Store up to 1 year at 4°C

A modified version of this solution that produces lower, nonenzymatic background is available as part of the Galacto-Light assay kit (Tropix).

Ligase dilution solution *(UNIT 15.5)*

Per reaction (20 μl):
✓ 2.2 μl 1 M Tris·Cl, pH 7.5 (110 mM final)
✓ 0.35 μl 1 M $MgCl_2$ (17.5 mM final)
✓ 1.0 μl 1 M DTT (50 mM final)
✓ 0.25 μl 10 mg/ml BSA (DNase-free; Pharmacia LKB; 125 μg/ml final)
16.2 μl H_2O
Prepare immediately before use and chill on ice

Ligase mix *(UNIT 15.5)*

Per reaction, (25.0 μl):
✓ 0.25 μl 1 M $MgCl_2$ (10 mM final)
✓ 0.50 μl 1 M DTT (20 mM final)
0.75 μl 100 mM ATP, pH 7.0 (Pharmacia LKB; 3 mM final)
✓ 0.125 μl 10 mg/ml BSA (DNase-free; Pharmacia LKB; 50 μg/ml final)
17.375 μl H_2O
✓ 5.0 μl 20 μM unidirectional linker mix (4 μM linker and 50 mM Tris·Cl final)
1.0 μl 3 "Weiss" U/μl T4 DNA ligase (*UNIT 3.14*; 3 U final)

Prepare immediately before use. First, mix $MgCl_2$, DTT, rATP, BSA, and H_2O and chill on ice. Next, add ice-cold unidirectional linker mix and T4 DNA ligase.

Tris·Cl added with the linker (step 6) is 50 mM, pH 7.7, at final concentration.

Linker-kinase buffer, 10× *(UNIT 12.9)*

✓ 660 mM Tris·Cl, pH 7.6
✓ 100 mM $MgCl_2$
✓ 100 mM DTT
10 mM spermidine
Store at −20°C
Add an extra 10 mM DTT just before use

Lithium lysis solution *(UNIT 6.10)*

1% (w/v) lithium dodecyl sulfate (Sigma)
✓ 100 mM EDTA
✓ 10 mM Tris·Cl, pH 8.0
Filter sterilize and store indefinitely at room temperature

Loading buffer, 10× *(UNITS 2.5A, 2.7, 3.1, 7.2 & 12.2)*

20% (w/v) Ficoll 400
✓ 0.1 M Na_2EDTA, pH 8.0
1.0% (w/v) SDS
0.25% (w/v) bromphenol blue
0.25% (w/v) xylene cyanol (optional; runs ~50% as fast as bromphenol blue and can interfere with visualization of bands of moderate molecular weight, but helpful for monitoring very long runs)

Loading buffer *(UNIT 12.8)*

✓ 1× binding buffer
20% (w/v) glycerol
1 mg/ml (w/v) bromophenol blue
1 mg/ml (w/v) xylene cyanol

Loading buffer *(UNIT 15.5)*

✓ 80% (v/v) deionized formamide
45 mM Tris base
45 mM boric acid
✓ 1 mM EDTA
0.05% (w/v) bromphenol blue
0.05% (w/v) xylene cyanol
Store in aliquots at −20°C and discard after 3 months

Long-term labeling medium *(90% methionine-free RPMI 1640 medium; UNIT 10.17)*

Mix 9 vol methionine-free RPMI with 1 vol complete serum-free RPMI medium (see recipe) and add fetal bovine serum to 10%. Alternatively, substitute 90% leucine- or cysteine-free RPMI 1640 for methionine-free medium, according to amino acid used for labeling cells. Store at 4°C up to 2 weeks.

The amino acid used to label cells is omitted from this medium (e.g., if [^{35}S]methionine is employed, methionine-free RPMI 1640 must be used). Specific amino acid–free media can be obtained from several tissue culture media suppliers. Alternatively, prepare medium lacking leucine, cysteine, or methionine from amino acid–free medium by adding the individual amino acid components, except the one that will be used to label. A kit for the preparation of such media is available from Life Technologies (Select-Amine).

Medium other than RPMI 1640 (e.g., DMEM) can be substituted, as long as it lacks the amino acid used for labeling. Special cell growth requirements (e.g., 60 μM 2-ME for T cells) should be considered when selecting the medium.

If lengthy handling of the cells out of the CO_2 incubator is anticipated, buffer the medium by adding HEPES, pH 7.4, to 25 mM. The pH should remain 7.4.

Low-ionic-strength electrophoresis buffer *(UNIT 12.2)*

✓ 26.9 ml 1 M Tris·Cl, pH 7.9
✓ 13.2 ml 1 M sodium acetate, pH 7.9
✓ 8.0 ml 0.5 M EDTA, pH 8.0
H_2O to 4 liters

Low-ionic-strength gel mix *(UNIT 12.2)*

✓ 270 µl 1 M Tris·Cl, pH 7.9
✓ 80 µl 0.5 M EDTA, pH 7.9
5.33 ml 30% (w/v) acrylamide
1.0 ml 2% (w/v) bisacrylamide
2.0 ml 50% (v/v) glycerol
31.0 ml H_2O
Filter through 0.2-µm filter
Store at 4°C for several months

A1

Low-salt buffer (UNIT 12.1)

Prepare high-salt buffer (see recipe), substituting 0.02 M KCl for 1.2 M KCl.

Low-stringency wash buffer I (UNIT 6.3)

✓ 2× SSC (dilute 20×)
0.1% (w/v) SDS

Low-stringency wash buffer II (UNIT 6.3)

✓ 0.5% (w/v) BSA (fraction V)
✓ 1 mM Na_2EDTA
✓ 40 mM $NaHPO_4$, pH 7.2
5% (w/v) SDS

10× lower reservoir buffer (UNIT 10.18)

52.4 g *bis*-Tris (0.626M)
12.2 ml 37.5% HCl (1.1%)
H_2O to 400 ml

Final pH should be 5.90; do not adjust the pH with acid or base. This buffer may be stored indefinitely at −20°C or ≤2 months at 4°C.

Luciferase assay buffer (UNIT 9.7B)

25 mM glycylglycine, pH 7.8
✓ 15 mM potassium phosphate, pH 7.8
✓ 15 mM $MgSO_4$
4 mM EGTA
2 mM ATP (Sigma)
✓ 1 mM DTT added just before use
Store at 4°C

Luciferin stock solution (UNIT 9.7B)

1 mM D-luciferin, synthetic crystalline (Sigma)
25 mM glycylglycine, pH 7.8
✓ 10 mM DTT
Store 1-ml aliquots in a light-tight box at −70°C for several months

Luminol visualization solution (UNIT 10.7)

0.5 ml 10× luminol stock [40 mg luminol (Sigma) in 10 ml DMSO]
0.5 ml 10× *p*-iodophenol stock [optional; 10 mg (Aldrich) in 10 ml DMSO]
✓ 2.5 ml 100 mM Tris·Cl, pH 7.5 (50 mM final)
25 µl 3% H_2O_2 (0.00015%)
H_2O to 5 ml
Prepare just before use

Premixed luminol substrate mix (Mast Immunosystems, Amersham, Du Pont NEN, Kirkegaard & Perry) may also be used. p-Iodophenol is an optional enhancing agent that increases light output. Luminol and p-iodophenol stocks can be stored for ≤6 months at −20°C.

Lysis buffer (UNIT 2.5B)

✓ 100 mM EDTA, pH 8.0
✓ 10 mM Tris·Cl, pH 8.0
1% (w/v) *N*-lauroylsarcosine sodium salt (Sarkosyl)
100 µg/ml proteinase K (add just before use from a 20 mg/ml stock)
Store at room temperature, *without proteinase K*

Lysis buffer *(UNIT 4.1)*

✓ *Prepare with DEPC-treated H_2O*

✓ 50 mM Tris·Cl, pH 8.0

✓ 100 mM NaCl

✓ 5 mM $MgCl_2$

0.5% (v/v) Nonidet P-40

Filter sterilize

If the RNA is to be used for northern blot analysis or the cells are particularly rich in ribonuclease, add ribonuclease inhibitors to the lysis buffer: 1000 U/ml placental ribonuclease inhibitor (e.g., RNAsin) plus 1 mM DTT or 10 mM vanadyl-ribonucleoside complex.

Lysis buffer *(UNIT 4.4)*

✓ 30 mM Tris·Cl, pH 7.4

✓ 100 mM NaCl

✓ 5 mM EDTA

1% (w/v) SDS

Add proteinase K to 100 µg/ml just before use

Lysis buffer *(UNIT 6.10)*

✓ 0.5 M Tris·Cl, pH 8.0

3% (v/v) *N*-lauroylsarcosine (Sarkosyl)

✓ 0.2 M EDTA, pH 8.0

Store indefinitely at room temperature. Add 1 mg/ml proteinase K just before use.

Lysis buffer *(UNIT 10.10)*

✓ *TSA solution containing:*

2% (v/v) Triton X-100 (see recipe for detergent stock solutions)

5 mM iodoacetamide

Aprotinin (0.2 trypsin inhibitor U/ml)

1 mM phenylmethylsulfonyl fluoride (PMSF), added fresh from 100 mM stock

Solution prepared in absolute ethanol

NOTE: Iodoacetomide is a protease inhibitor and prevents oxidation of free cysteines to disulfide-bonded cysteines. It should be omitted for enzymes that require cysteines for activity.

Lysis buffer *(UNIT 10.15)*

1% (v/v) Triton X-100

1% (v/v) bovine hemoglobin

1 mM iodoacetamide (freshly prepared)

Aprotinin (0.2 trypsin inhibitor U/ml)

1 mM phenylmethylsulfonyl fluoride (PMSF; add fresh from 100 mM stock in absolute ethanol)

✓ Prepare in TSA solution

Lysis buffer *(UNIT 13.13)*

✓ 50 mM Tris·Cl, pH 7.5

✓ 10 mM $MgSO_4$

✓ 1 mM EDTA

✓ 10 mM potassium acetate

✓ 1 mM DTT

✓ 1× protease inhibitor mix

1 mM PMSF

Lysis buffer *(UNIT 15.4)*

✓ 140 mM NaCl

✓ 10 mM Tris·Cl, pH 8.0

✓ 15 mM $MgCl_2$

0.5% (v/v) Nonidet P-40 (NP-40)

Lysis solution *(UNIT 15.5)*

✓ 300 mM NaCl
✓ 50 mM Tris·Cl, pH 8.0, room temperature
✓ 25 mM EDTA, pH 8.0
0.2% (v/v) SDS (prepare fresh; add just before use)
0.2 mg/ml proteinase K (prepare fresh; add just before use)

Lysis solution without SDS and proteinase K can be stored indefinitely at room temperature. Immediately before use, add 10 µl of 20% SDS and 10 µl of 20 mg/ml proteinase K per milliliter lysis solution.

Lysozyme solution *(UNIT 11.2)*

5 mg chicken egg white lysozyme (Sigma Grade VI)
✓ 1 ml Tris/EDTA/NaCl (TEN) buffer
Make fresh immediately before use

M9 salts, 10× *(UNIT 16.8)*

60 g Na_2HPO_4 (0.42 M)
30 g KH_2PO_4 (0.24 M)
5 g NaCl (0.09 M)
10 g NH_4Cl (0.19 M)
H_2O to 1 liter
Adjust pH to 7.4 with NaOH
Autoclave or filter sterilize through a 0.45-µm filter
Store ≤6 months at room temperature

M9ZB medium *(UNIT 10.11)*

Dissolve 10 g N-Z-Amine A (Sigma) and 5 g NaCl in 889 ml water. Autoclave, cool, and add 100 ml of 10× M9 medium (*UNIT 1.1*), 1 ml of 1 M sterile $MgSO_4$, and 10 ml of 40% (w/v) glucose (filter sterilized). Store at room temperature for ≤1 year.

M13 PEG solution *(UNIT 7.3)*

20% (w/v) PEG 800
✓ 2.5 M NaCl

Master hybridization mix *(UNIT 14.7)*

✓ 1 ml 20× SSC (4× final)
0.5 ml 20 mg/ml nuclease-free BSA (2 mg/ml final)
1.5 ml sterile H_2O
2 ml 50% (w/v) dextran sulfate (Pharmacia Biotech, mol. wt. 500,000; autoclaved; 20% final)
Store at 4°C and use up to 6 weeks

MCAC buffers *(UNIT 10.11)*

MCAC-0 buffer:
✓ 20 mM Tris·Cl, pH 7.9
✓ 0.5 M NaCl
10% (v/v) glycerol
1 mM PMSF

MCAC-1000 buffer:
✓ 20 mM Tris·Cl, pH 7.9
✓ 0.5 M NaCl
1 M imidazole
10% (v/v) glycerol
1 mM PMSF

Add PMSF immediately before use from a 0.2 M stock in ethanol stored at room temperature.

All buffers used with Ni^{2+}-NTA resin contain high salt concentrations to reduce nonspecific electrostatic interactions between proteins and resin. Lower salt concentrations can be used but may lead to nonspecific binding of unwanted proteins to the resin.

continued

MCAC-20, MCAC-40, MCAC-60, MCAC-80, MCAC-100, and MCAC-200 buffers:

These buffers (containing different concentrations of imidazole) are made by mixing MCAC-0 buffer and MCAC-1000 buffer in the appropriate ratios: e.g., for MCAC-60 buffer, 94:6 (v/v) MCAC-0/MCAC-1000.

Store MCAC buffers ≤6 months at 4°C.

MCAC-EDTA buffer *(UNIT 10.11)*

✓ 20 mM Tris·Cl, pH 7.9
✓ 0.5 M NaCl
✓ 0.1 M EDTA, pH 8.0
10% (v/v) glycerol
1 mM PMSF
Store ≤6 months at 4°C

Methylase buffer, 10× *(UNIT 3.1)*

NaCl
✓ Tris·Cl, pH 7.5
✓ EDTA
2-mercaptoethanol (2-ME) or dithiothreitol (DTT)
S-adenosylmethionine (SAM)

The concentrations of buffer components depend upon the methylase (Table 3.1.3).

$MgCl_2$, 1 M

20.3 g $MgCl_2 \cdot 6H_2O$
H_2O to 100 ml

$MgSO_4$, 1 M

24.6 g $MgSO_4 \cdot 7H_2O$
H_2O to 100 ml

Middle wash buffer *(UNIT 4.5)*

0.15 M LiCl
10 mM Tris·Cl, pH 7.5
1 mM EDTA
0.1% (v/v) SDS

Mild lysis buffer *(UNIT 17.2)*

10 mM 3[(3-cholamidopropyl)-dimethylammonio]-1-propane-sulfonate (CHAPS) or 1% (w/w) Nonidet P-40 (NP-40)
✓ 0.15 M NaCl
✓ 0.01 M sodium phosphate, pH 7.2
✓ 2 mM EDTA
50 mM sodium fluoride
0.2 mM sodium vanadate added fresh from 0.2 M stock solution
100 U/ml aprotinin (Trasylol, Pentex/Miles)
Store buffer without vanadate at 4°C up to 1 year

CHAPS is a milder detergent than NP-40, but yields precipitates with a higher background and may solubilize some proteins less efficiently.

Sodium vanadate stock solution can be stored in plastic at room temperature.

Mild stripping solution *(UNIT 2.10)*

✓ 5 mM Tris·Cl, pH 8.0
✓ 2 mM EDTA
✓ 0.1× Denhardt solution

Mn buffer *(UNIT 7.4)*

0.15 M sodium isocitrate
✓ 0.1 M $MnCl_2$

Moderate stripping solution *(UNIT 2.10)*

✓ 200 mM Tris·Cl, pH 7.0
✓ 0.1× SSC
0.1% (w/v) SDS

Modified lysis buffer *(UNIT 10.15)*

Prepare in Tris/saline/azide (TSA) solution (see recipe):
0.5% (v/v) Triton X-100
1 mM PMSF
5 mM iodoacetamide
Aprotinin (0.2 trypsin inhibitor U/ml)

Moist chamber solution A *(UNIT 14.3)*

50% (v/v) formamide (inexpensive grade)
✓ 0.3 M NaCl
✓ 10 mM Tris·Cl, pH 8.0
✓ 1 mM EDTA

Moist chamber solution B *(UNIT 14.3)*

50% (v/v) formamide (inexpensive grade)
✓ 0.6 M NaCl
✓ 10 mM Tris·Cl, pH 7.5
✓ 2 mM EDTA

MoMuLV reverse transcriptase buffer, 5× *(UNITS 15.6 & 15.8)*

✓ 250 mM Tris·Cl, pH 8.3
✓ 375 mM KCl
✓ 50 mM DTT
✓ 15 mM $MgCl_2$
Store at −20°C for ≤3 months

MOPS buffer *(UNIT 4.9)*

0.2 M MOPS [3-(*N*-morpholino)-propanesulfonic acid], pH 7.0
0.5 M sodium acetate
✓ 0.01 M EDTA

Store in the dark and discard if it turns yellow.

10× MOPS running buffer *(UNIT 4.9)*

0.4 M MOPS, pH 7.0
✓ 0.1 M sodium acetate
✓ 0.01 M EDTA

MUP substrate solution *(UNIT 11.2)*

0.2 mM 4-methylumbelliferyl phosphate (MUP; Sigma)
0.05 M Na_2CO_3
✓ 0.05 mM $MgCl_2$
Store at room temperature

NA-45 elution buffer *(UNIT 2.6)*

✓ 1 M NaCl
0.05 M arginine (free base)
Filter sterilize and store indefinitely at room temperature

NaCl, 5 M

292 g NaCl
H_2O to 1 liter

Na_2EDTA, see EDTA

NaOH, 0.02 M *(UNIT 10.3)*

Just before using, add 0.5 ml of 10 M NaOH to 250 ml deaerated water.

It is especially important to make 0.02 M NaOH with deaerated water.

NaOH, 10 M

Dissolve 400 g NaOH in 450 ml H_2O
Add H_2O to 1 liter

NaOH/SDS solution *(UNITS 1.6 & 1.7)*

✓ 0.2 M NaOH
1% (w/v) SDS
Prepare fresh from stock solutions of 10 M NaOH and 10% SDS

$NaPO_4$ buffer, *see Sodium phosphate buffer*

NBT/BCIP substrate solution *(UNIT 14.7)*

Add 220 µl nitroblue tetrazolium (NBT) solution (75 mg/ml in dimethylformamide; 330 µg/ml final) to 50 ml alkaline phosphatase buffer, pH 9.5 (see recipe) and mix gently (do not vortex). Add 170 µl 5-bromo-4-chloro-3-indolyl phosphate (BCIP) solution (50 mg/ml in dimethylformamide; 170 µg/ml final) and mix gently again. Prepare fresh each time.

CAUTION: *Dimethylformamide is hazardous; see manufacturer's information for guidelines on handling, storage, and disposal.*

NDS solution, 100% *UNIT 6.10)*

Mix 350 ml H_2O, 93 g EDTA (0.5 M), and 0.6 g Tris base (0.1M). Adjust pH to ~8.0 with 100 to 200 pellets of solid NaOH. Add 5 g *N*-lauroylsarcosine [predissolved in 50 ml water; 1% (w/v)] and adjust to pH 9.0 with concentrated NaOH. Bring volume to 500 ml with water. Filter sterilize and store indefinitely at 4°C. Dilute 1:5 with H_2O (20% final) just before use.

Neutralization solution *(UNIT 6.8)*

✓ 200 ml 20× SSC (10×)
✓ 100 ml 1 N HCl (0.25 N)
✓ 100 ml 1 M Tris·Cl, pH 8.0 (0.25 M)

Neutralizing buffer, pH 7.7 *(0.5 M sodium phosphate; UNIT 11.11)*

Solution A: 70.98 g Na_2HPO_4 (anhydrous)/liter H_20
Solution B: 7.80 g $NaH_2PO_4·2H_2O$/liter H_20
Titrate solution A with solution B to pH 7.7

NPP substrate solution *(UNIT 11.2)*

3 mM *p*-nitrophenyl phosphate (NPP; Sigma)
0.05 M $NaCO_3$
✓ 0.05 mM $MgCl_2$
Store at 4°C

NTE buffer *(UNIT 9.12)*

✓ 100 mM NaCl
✓ 10 mM Tris·Cl, pH 7.4
✓ 1 mM EDTA
Filter sterilize

Nuclei buffer *(UNIT 13.13)*

Prepare as for Ficoll buffer (see recipe) except substitute 1 M sorbitol for 18% Ficoll.

Nucleotide mix *(UNIT 2.10)*

2.5 mM ATP
2.5 mM CTP
2.5 mM GTP
✓ 20 mM Tris·Cl, pH 7.5
Store at −20°C

Nucleotide stock solutions *(UNIT 7.4)*

Purchase (Table 7.4.1) or prepare sequencing grade dNTPs (including dITP and 7-deaza-dGTP) and ddNTPs as concentrated 20 mM stock solutions in 0.1 mM EDTA. Determine the actual concentrations spectrophotometrically using the extinction coefficients in Table 3.4.3; the extinction coefficient for each ddNTP is the same as that of the corresponding dNTP. Store at −20°C up to 1 year.

Oligonucleotide primers for amplification and hybridization *(UNIT 15.3)*

One pair each for amplification and one primer each for detection of each target sequence. Prepare each primer at 50 pmol/μl in sterile, distilled water. Store at −20°C.

Paraformaldehyde fixative, 2%, *see Fixative solution (2% paraformaldehyde)*

Paraformaldehyde (PFA) fixative, 4% *(UNITS 14.1, 14.2 & 14.8)*

Heat to 60°C a volume of water equal to slightly less than ⅔ the desired final volume of fixative. Weigh out a quantity of paraformaldehyde (Baker) that will make a 4% solution and add it with a stir bar to the water. Cover. Transfer to fume hood and maintain on heating plate at 60°C with stirring. Add 1 drop 2 N NaOH with a Pasteur pipet. The solution should become almost clear fairly rapidly, but will still have some fine particles that will not go away. Be careful not to overheat solution. Remove from heat and add ⅓ vol 3× PBS. Bring pH of solution to 7.2 with HCl, add water to final volume, and filter using a Millipore or Nalgene filter. Cool to room temperature, or to 4°C on ice.

CAUTION: *Formaldehyde is a carcinogen and may cause allergic reactions.*

PBS, *see Phosphate-buffered saline*

PCR amplification buffer, *see Amplification buffer*

PCR buffer, 10× *(UNIT 15.6)*

✓ 500 mM KCl
✓ 100 mM Tris·Cl, pH 8.8
✓ 15 mM $MgCl_2$
✓ 30 mM DTT
✓ 1 mg/ml BSA
Store at −20°C for ≤3 months

PCR reaction mix *(UNIT 6.10)*

✓ 1.5 mM $MgCl_2$
✓ 50 mM KCl
✓ 10 mM Tris·Cl, pH 8.3
0.2 mM each dATP, dCTP, dGTP, and dTTP
0.05 U AmpliTaq polymerase (Perkin-Elmer/Cetus)/μl reaction mixture
0.03 μl Perfect Match Enhancer (Stratagene)/μl reaction mixture
Store all components at −20°C and mix just before use

PEG/$CaCl_2$ solution *(UNIT 6.2)*

45% (w/v) PEG 3350
✓ 10 mM Tris·Cl, pH 7.4 *(APPENDIX 2)*
✓ 10 mM $CaCl_2$
Filter sterilize and store at room temperature

PEG solution *(UNIT 16.2)*

30% (w/v) PEG 8000
✓ 1.6 M NaCl
Store at 4°C

PFA fixative, *see Paraformaldehyde fixative*

pH 1.9 electrophoresis buffer *(UNIT 17.3)*

50 ml 88% formic acid (0.58 M final concentration)
156 ml glacial acetic acid (1.36 M final concentration)
1794 ml H_2O
Store indefinitely in a sealed bottle at room temperature

pH 3.5 electrophoresis buffer *(UNIT 17.3)*

100 ml glacial acetic acid (0.87 M final concentration)
10 ml pyridine [0.5% (v/v) final concentration]
✓ 10 ml 100 mM EDTA (0.5 mM final concentration)
1880 ml H_2O
Store indefinitely in a sealed bottle at room temperature

Phage loading buffer *(UNIT 7.2)*

0.25% (w/v) bromphenol blue
15.0% (v/v) Ficoll 400
2.0% (w/v) SDS
✓ 10 mM EDTA
Store at room temperature

Phenol/chloroform/isoamyl alcohol *(UNIT 15.5)*

Mix 25 parts phenol (equilibrated in 150 mM NaCl/50 mM Tris·Cl (pH 7.5)/1 mM EDTA) with 24 parts chloroform and 1 part isoamyl alcohol. Add 8-hydroxyquinoline to 0.1%. Store in aliquots at −20°C and discard after 6 months.

Phenol equilibrated with TLE *(UNIT 4.3)*

Equilibrate freshly liquefied phenol (250 ml for a 15-g prep) with TLE solution (see recipe) on the day of preparation. First, extract with an equal volume of TLE solution plus 0.5 ml of 15 M NaOH (this should bring the pH close to 8.0), then extract two more times with TLE.

Phosphate buffer, *see Potassium or sodium phosphate buffer*

Phosphate-buffered saline (PBS), 10× and 1× *(all chapters except 7 & 14)*

10× stock solution, 1 liter:	*1× working solution, pH ~ 7.3:*
80 g NaCl	137 mM NaCl
2 g KCl	2.7 mM KCl
11.5 g $Na_2HPO_4·7H_2O$	4.3 mM $Na_2HPO_4·7H_2O$
2 g KH_2PO_4	1.4 mM KH_2PO_4

Phosphate-buffered saline (PBS), pH 7.2, 3× and 1× *(CHAPTER 14)*

Prepare following solutions: 390 mM NaCl/30 mM Na_2HPO_4 and 390 mM NaCl/30 mM NaH_2PO_4. Mix to obtain 3× PBS, pH 7.2, and autoclave or filter sterilize. Prepare 1× PBS by diluting 3× PBS 3-fold with water.

4× phosphate/SDS, pH 7.2 *(UNIT 10.2)*

46.8 g NaH_2PO_4 } (0.4 M sodium phosphate final)
231.6 g $Na_2HPO_4·7H_2O$ }
12 g SDS [0.4% (w/v) final]
3 liters H_2O
Store up to 3 months at 4°C

Phosphate/SDS electrophoresis buffer *(UNIT 10.2)*

Dilute 500 ml of 4× phosphate/SDS, pH 7.2 with H_2O to 2 liters. Store at 4°C up to 1 month.

Final concentrations are 0.1 M sodium phosphate (pH 7.2)/0.1% (w/v) SDS.

Phosphate/SDS sample buffer, 2× (for continuous systems; UNIT 10.2)

✓ 0.5 ml 4× phosphate/SDS, pH 7.2 (20 mM)
0.2 g SDS 2%
0.1 mg bromphenol blue [0.01% (w/v)]
0.31 g DTT (0.2 M)
2.0 ml glycerol [20% (v/v)]
Add H_2O to 10 ml and mix

Phosphoamino acid standards mixture (UNIT 17.3)

Prepare a solution of phosphoserine, phosphothreonine, and phosphotyrosine (Sigma) in water at a final concentration of 0.3 µg/ml each. Store in 1-ml aliquots indefinitely at −20°C.

Phosphoric acid, 0.085% (UNIT 10.3)

Just before using, dilute 300 ml of 0.85% phosphoric acid to 3.0 liters with deaerated water.

Plant electroporation buffer (UNIT 9.3)

✓ *Prepare in PBS:*
0.4 M mannitol
✓ 5 mM $CaCl_2$
Store at 4°C

Plug solution (UNIT 10.2)

✓ 0.125 M Tris·Cl, pH 8.8
50% (w/v) sucrose
0.001% (w/v) bromphenol blue
Store at 4°C up to 1 month

Poly(A) loading buffer (UNIT 4.5)

0.5 M LiCl
✓ 10 mM Tris·Cl, pH 7.5
✓ 1 mM EDTA
0.1% (w/v) SDS

Polyacrylamide/urea gel, 16% (UNIT 12.9)

50 ml 40% (w/v) 19:1 acrylamide/bisacrylamide
✓ 12.5 ml 10× TBE electrophoresis buffer
62.5 g urea
17 ml H_2O
Mix, filter, briefly degas, and then add:
✓ 750 µl 10% (w/v) ammonium persulfate
20 µl TEMED

Pour immediately into prepared gel plates.

Polyethylene glycol solution, see PEG solution

Polymerase mix, 5× (UNIT 8.1)

✓ 100 mM Tris·Cl, pH 8.8
✓ 10 mM DTT
✓ 50 mM $MgCl_2$
2.5 mM each of dATP, dTTP, dGTP, dCTP
5 mM ATP

Use of highly purified (e.g., by HPLC) dNTPs is critical.

Polynucleotide kinase buffer, 10× (UNITS 4.6 & 4.8)

✓ 700 mM Tris·Cl, pH 7.5
✓ 100 mM $MgCl_2$
✓ 50 mM DTT
1 mM spermidine·Cl
✓ 1 mM EDTA

Ponceau S solution *(UNIT 10.7)*

Dissolve 0.5 g Ponceau S in 1 ml glacial acetic acid. Bring to 100 ml with H_2O. Prepare just before use.

Potassium acetate buffer, 0.1 M

Solution A: 11.55 ml glacial acetic acid/liter (0.2 M).

Solution B: 19.6 g potassium acetate ($KC_2H_3O_2$)/liter (0.2 M).

Referring to Table A.1.2 for desired pH, mix the indicated volumes of solutions A and B, then dilute with H_2O to 200 ml.

This may be made as a 5- or 10-fold concentrate by scaling up the amount of potassium acetate in the same volume. Acetate buffers show concentration-dependent pH changes, so check concentrate pH by diluting an aliquot to the final concentration.

To prepare buffers with pH intermediate between the points listed in Table A.1.2, prepare closest higher pH, then titrate with solution A.

Table A.1.2 Preparation of 0.1 M Sodium and Potassium Acetate Buffers[a]

Desired pH	Solution A (ml)	Solution B (ml)
3.6	46.3	3.7
3.8	44.0	6.0
4.0	41.0	9.0
4.2	36.8	13.2
4.4	30.5	19.5
4.6	25.5	24.5
4.8	20.0	30.0
5.0	14.8	35.2
5.2	10.5	39.5
5.4	8.8	41.2
5.6	4.8	45.2

[a]Adapted by permission from CRC, 1975.

Potassium acetate solution, pH 4.8, 5 M *(UNIT 1.6)*

29.5 ml glacial acetic acid
KOH pellets to pH 4.8 (several)
H_2O to 100 ml
Store at room temperature (do not autoclave)

Potassium acetate solution, pH ~5.5, 3 M *(UNIT 1.7)*

294 g potassium acetate
50 ml 90% formic acid (1.18 M final)
H_2O to 1 liter

Potassium phosphate buffer, 0.1 M

Solution A: 27.2 g KH_2PO_4 per liter (0.2 M).

Solution B: 45.6 g K_2HPO_4 per liter (0.2 M).

Referring to Table A.1.3 for desired pH, mix the indicated volumes of solutions A and B, then dilute with H_2O to 200 ml.

This may be made as a 5- or 10-fold concentrate by scaling up the amount of potassium phosphate in the same volume. Phosphate buffers show concentration-dependent pH changes, so check concentrate pH by diluting an aliquot to the final concentration.

Table A.1.3 Preparation of 0.1 M Sodium and Potassium Phosphate Buffers[a]

Desired pH	Solution A (ml)	Solution B (ml)	Desired pH	Solution A (ml)	Solution B (ml)
5.7	93.5	6.5	6.9	45.0	55.0
5.8	92.0	8.0	7.0	39.0	61.0
5.9	90.0	10.0	7.1	33.0	67.0
6.0	87.7	12.3	7.2	28.0	72.0
6.1	85.0	15.0	7.3	23.0	77.0
6.2	81.5	18.5	7.4	19.0	81.0
6.3	77.5	22.5	7.5	16.0	84.0
6.4	73.5	26.5	7.6	13.0	87.0
6.5	68.5	31.5	7.7	10.5	90.5
6.6	62.5	37.5	7.8	8.5	91.5
6.7	56.5	43.5	7.9	7.0	93.0
6.8	51.0	49.0	8.0	5.3	94.7

[a]Adapted by permission from CRC, 1975.

1 M potassium phosphate buffer, pH 8.0

A: 1 M K_2HPO_4
B: 1 M KH_2PO_4
Add A to B until pH = 8.0

Precipitation salt mix (*UNIT 15.5*)

Per reaction (9.4 µl):
✓ 8.4 µl 3 M sodium acetate, pH 7.0 (2.7 M final)
✓ 1.0 µl 10 mg/ml yeast tRNA (~1 mg/ml final)
Prepare immediately before use

Prehybridization solution (*UNIT 6.4*)

✓ 6× SSC (dilute 20×)
✓ 5× Denhardt solution (dilute 100×)
0.05% (w/v) sodium pyrophosphate
100 µg/ml boiled herring sperm DNA
0.5% (w/v) SDS

Primer termination mixes (*UNIT 7.5*)

Concentrations of dNTPs and ddNTPs in the A, C, G, and T mixes are listed below in mM. Prepare all mixes in highly purified H_2O. Store indefinitely at –20°C.

	A	C	G	T
dATP	4	80	80	80
dCTP	80	4	80	80
dGTP	80	80	4	80
dTTP	80	80	80	4
ddATP	200	—	—	—
ddCTP	—	100	—	—
ddGTP	—	—	60	—
ddTTP	—	—	—	225

Pronase, predigested and lyophilized *(UNIT 14.3)*

Predigest pronase by incubating a 40 mg/ml pronase solution in water for 4 hr at 37°C. Lyophilize in aliquots and store in a nondefrosting freezer at −20°C. Determine optimal pronase concentration by digesting a series of sections with sequential dilutions of predigested pronase (0.1 to 10 μg/ml) resuspended in 50 mM Tris·Cl (pH 7.5)/5 mM EDTA. Pretreat experimental sections with the highest possible pronase concentration that allows the retention of adequate cellular morphology.

Propidium iodide staining solution *(UNIT 14.7)*

Stock solution: Dissolve 1 mg propidium iodide in 10 ml H_2O (0.15 mM final). Aliquot into aluminum foil–wrapped tubes and store up to 1 year at −20°C.

Working solution: Dilute stock solution 1/1000 in PBS and store in aluminum foil–wrapped tube up to 6 months at 4°C.

CAUTION: *Propidium iodide is hazardous; see manufacturer's information for guidelines on handling, storage, and disposal.*

Protease inhibitor cocktail, 150× *(UNIT 10.11)*

Stock solutions:
2 mg/ml aprotinin in H_2O
1 mg/ml leupeptin in H_2O
1 mg/ml pepstatin A in methanol

Combine 1.5 vol of each stock solution with 5.5 vol sterile water.

Protease inhibitors can be obtained from Sigma or U.S. Biochemical. The stock solutions and cocktail are stable for ≥6 months at −20°C.

Protease inhibitor mix, 100× *(UNIT 13.13)*

Listed below are representative protease inhibitors; different combinations may be more appropriate for individual applications.

10 μg/ml chymostatin
200 μg/ml aprotinin
100 μg/ml pepstatin A
110 μg/ml phosphoramidon
720 μg/ml E-64
50 μg/ml leupeptin
250 μg/ml antipain
10 mM benzamidine
10 mM sodium metabisulfite

Protein A–Sepharose suspension *(UNIT 6.8)*

1.5 g of Sepharose–protein A (Pharmacia LKB) is resuspended in :

✓ 10 mM Tris·Cl, pH 7.5
✓ 0.15 M NaCl
0.4% (v/v) Triton X-100
0.5% (v/v) aprotinin (Sigma)

Shake for 5 min. Spin down beads, wash 3 times with same buffer, then resuspend in 11 ml of the buffer.

Protein and peptide sample preparation *(UNITS 10.12 & 10.13)*

For reversed-phased chromatography *(UNIT 10.12)*, proteins and peptides must be fully solubilized. If visual inspection indicates turbidity or particulates, add TFA/guanidine buffer to solubilize most mixtures of protein fragments. For ion exchange chromatography *(UNIT 10.13)*, proteins must be fully solubilized in a buffer of identical pH and lower ionic strength than the starting buffer of the gradient separation. If visual inspection indicates insolubility, add urea, organic solvent (e.g., methanol or propanol), or detergent (SDS or CHAPS) to solubilize the protein. If a small amount of sample (low-microgram level) is being solubilized, it may be necessary to magnify the sample 10 to 20 times with a magnifying lens. All samples should be centrifuged prior to injection in order to remove insoluble particulates.

Proteinase digestion buffer *(UNIT 15.3)*

20 mM Tris·Cl, pH 7.4 (prepared from autoclaved, 1 M stock)
✓ 20 mM EDTA, pH 8.0 (prepared from autoclaved, 0.5 M stock)
0.5% (w/v) SDS
Store at room temperature

Protoplast solution *(UNIT 9.3)*

2% (w/v) cellulase (Yakult Biochemical)
1% (w/v) macerozyme (Yakult Biochemical)
0.01% (w/v) pectylase
0.4 M mannitol
✓ 40 mM $CaCl_2$
10 mM 2-[*N*-morpholino]ethanesulfonic acid (MES), pH 5.5
Prepare fresh

Protoplasting buffer *(UNIT 4.4)*

✓ 15 mM Tris·Cl, pH 8.0
0.45 M sucrose
✓ 8 mM EDTA

Reaction mix cocktail (per amplification reaction) *(UNIT 15.3)*

10 µl 10× amplification buffer
10 µl 2 mM 4dNTP mix (2 mM each dNTP in TE buffer, pH 7.5)
1 µl each oligo primer for amplification (4 µl total)

Reservoir buffer *(UNIT 10.3)*

15.0 g Tris base (0.025 M)
72.0 g glycine (0.2 M)
5.0 g SDS [1% (w/v)]
H_2O to 5 liters

For convenience, make up as a 10× stock or store the preweighed dry ingredients in packets for future use. The 10× stock can be stored for several weeks in the refrigerator.

Restriction endonuclease buffers *(UNIT 3.1)*

10× sodium chloride–based buffers
✓ 100 mM Tris·Cl, pH 7.5
✓ 100 mM $MgCl_2$
✓ 10 mM dithiothreitol (DTT)
✓ 1 mg/ml bovine serum albumin (BSA)
✓ 0, 0.5, 1.0, *or* 1.5 M NaCl

The concentration of NaCl depends upon the restriction endonuclease (Table 3.1.2). The four different NaCl concentrations listed above are sufficient to cover the range for essentially all commercially available enzymes except those requiring a buffer containing KCl instead of NaCl. Autoclaved gelatin (at 1 mg/ml) can be used instead of BSA.

Note that most restriction enzymes are provided by the suppliers with the appropriate buffers.

10× KCl-based buffer
✓ 60 mM Tris·Cl, pH 8.0
✓ 60 mM $MgCl_2$
✓ 200 mM KCl
60 mM 2-ME
✓ 1 mg/ml BSA

10× potassium acetate–based buffer
660 mM potassium acetate
330 mM Tris·acetate, pH 7.9
100 mM magnesium acetate
✓ 5 mM DTT
✓ 1 mg/ml BSA (optional)

2× potassium glutamate–based buffer
200 mM potassium glutamate
50 mM Tris·acetate, pH 7.5
✓ 100 µg/ml BSA
1 mM 2-ME

Reverse transcriptase buffer *(UNIT 15.4)*

Per sample:

✓ 2.5 µl 400 mM Tris·Cl, pH 8.3

✓ 2.5 µl 400 mM KCl

✓ 1 µl 300 mM $MgCl_2$

✓ 5 µl 100 mM DTT

✓ 5 µl 5 mM 4dNTP mix

2 µl actinomycin D

11 µl H_2O

Optional: 10 U RNasin (Promega; reduce volume of H_2O accordingly)

Reverse transcriptase (RT) reaction cocktail *(UNIT 9.13)*

Reagent[a]	Number of assays			
	20	40	80	100
✓1 M Tris·Cl, pH 8.3, *µl*	50	100	200	250
✓0.2 M DTT, *µl*	100	200	400	500
✓0.02 M $MnCl_2$, *µl*	30	60	120	150
✓3 M NaCl, *µl*	20	40	80	100
100 µg/ml oligo(dT), *µl*	50	100	200	250
100 µg/ml poly(rA), *µl*	100	200	400	500
0.2 mM dTTP, *µl*	50	100	200	250
1% (v/v) NP-40, *µl*	50	100	200	250
[α-^{32}P]dTTP, *µCi*[b]	100	200	400	500
Final volume (add dH_2O), *ml*	1	2	4	5

[a]Cocktail components are for assay of murine RT (Goff et al., 1981). For avian RT, substitute 1% 2-ME (final) for DTT; 10 mM $MgCl_2$ (final) for $MnCl_2$, and 100 µM dTTP (final) for 10 µM dTTP (Omer and Faras, 1982).

[b]Specific activity >500 mCi/mmol. Since volume will vary with different lots, the volume of dH_20 to add can be varied to compensate.

Ribonuclease digestion buffer *(UNIT 4.7)*

✓ 10 mM Tris·Cl, pH 7.5

✓ 300 mM NaCl

✓ 5 mM EDTA

Add 1/50 vol of 50× ribonuclease mix:

2 mg/ml ribonuclease A *(UNIT 3.13)*

0.1 mg/ml ribonuclease T1 *(UNIT 3.13)*

RNase digestion buffer may be stored at room temperature. Add RNases from frozen stocks as needed. Be sure to use disposable tubes to make this reagent.

Ribonucleoside triphosphate mix, 5× *(UNIT 10.16)*

5 mM each ATP, UTP, CTP

5 mM diguanosine triphosphate (G-5′ppp5′-G)TP

0.5 mM GTP

RIPA buffer *(UNIT 10.15)*

1% (w/v) sodium deoxycholate

0.1% (w/v) SDS

✓ Prepare in lysis buffer

RIPA (RadioImmunoProtection Assay) correction buffer for boiled sample *(UNIT 17.2)*

1.25% (w/w) Nonidet P-40 (NP-40)

1.25% (w/v) sodium deoxycholate

✓ 0.0125 M sodium phosphate, pH 7.2

✓ 2 mM EDTA

0.2 mM sodium vanadate added fresh from 0.2 M stock solution

50 mM sodium fluoride

100 U/ml aprotinin (Trasylol, Pentex/Miles)

Store buffer without vanadate at 4°C up to 1 year

Sodium vanadate stock solution can be stored in plastic at room temperature.

RIPA (RadioImmunoPrecipitation Assay) lysis buffer *(UNIT 17.2)*

1% (w/w) Nonidet P-40 (NP-40)
1% (w/v) sodium deoxycholate
0.1% (w/v) SDS
✓ 0.15 M NaCl
✓ 0.01 M sodium phosphate, pH 7.2
✓ 2 mM EDTA
50 mM sodium fluoride
0.2 mM sodium vanadate added fresh from 0.2 M stock solution
100 U/ml aprotinin (Trasylol, Pentex/Miles)
Store buffer without vanadate at 4°C up to 1 year

Sodium vanadate stock solution can be stored in plastic at room temperature.

RNA buffer *(UNIT 13.12)*

✓ 0.5 M NaCl
✓ 200 mM Tris·Cl, pH 7.5
✓ 10 mM EDTA

RNA loading buffer *(UNIT 4.7)*

80% (v/v) formamide
1 mM EDTA, pH 8.0
0.1% (w/v) bromphenol blue
0.1% (w/v) xylene cyanol

Do not use formamide loading buffer from UNIT 7.5 or any buffer containing NaOH.

RNA wash solution I *(UNIT 14.3)*

✓ 2× SSC (dilute 20×)
50% formamide (use inexpensive grade, e.g., Fluka, and do not deionize)
0.1% (v/v) 2-ME

RNA wash solution II *(UNIT 14.3)*

✓ 0.1× SSC (dilute 20×)
1% (v/v) 2-ME

RNase, DNase-free *(UNIT 6.10)*

Prepare RNase A (derived from bovine pancreas) free of DNase by dissolving RNase A in TE buffer at 1 mg/ml, boiling 10 to 30 min, and allowing to cool to room temperature. Store aliquots at −20°C to prevent microbial growth.

RNase A, heat-inactivated *(UNIT 7.3)*

Dissolve RNase A at 10 mg/ml in 10 mM Tris·Cl (pH 7.5)/15 mM NaCl. Heat 10 min at 80°C, slowly cool to room temperature, and store in aliquots at −20°C. Dilute to 1 mg/ml or 10 μg/ml as required.

RNase digestion solution *(UNIT 14.3)*

40 μg/ml RNase A (Sigma; *UNIT 3.13)*
2 μg/ml RNase T1 (Sigma; *UNIT 3.13)*
10 mM Tris·Cl (pH 7.5)/5 mM EDTA
✓ 0.3 M NaCl

RNase-free DNase I*, see DNase I, RNase-free*

RNase reaction mix *(UNIT 4.8)*

100 μg/ml salmon sperm DNA
✓ 20 μg/ml RNase A (DNase-free) in TEN 100 buffer

A1

RP peptide standards *(UNIT 10.12)*

Dissolve 1 mg α-lactoglobulin A in 1 ml of 0.1 M ammonium bicarbonate and add 20 µl of 1 mg/ml trypsin (tosylphenyl-alanine chloromethyl ketone treated). Incubate 1 to 24 hr at room temperature. Stop the enzymic digestion by adding 10 µl of 0.1% (v/v) trifluoroacetic acid. Following cessation of CO_2 formation as evidenced by tiny bubble formation, check that the pH is <3 by spotting a tiny drop onto pH paper.

RP protein standards *(UNIT 10.12)*

Weigh into the same vial 1 mg amounts of insulin, cytochrome *c*, α-lactalbumin, carbonic anhydrase, and ovalbumin. Add 1 ml of 0.1% (v/v) trifluoroacetic acid and gently mix.

S1 hybridization solution *(UNITS 4.6 & 4.8)*

80% (v/v) deionized formamide
40 mM PIPES, pH 6.4
400 mM NaCl
1 mM EDTA, pH 8.0
Store in 1-ml aliquots at −70°C

S1 nuclease buffer, 2× *(UNIT 4.6)*

✓ 0.56 M NaCl
✓ 0.1 M sodium acetate, pH 4.5
9 mM $ZnSO_4$
Filter sterilize and store at 4°C

S1 nuclease buffer *(UNIT 7.2)*

✓ 16 mM sodium acetate, pH 4.6
✓ 400 mM NaCl
1.6 mM $ZnSO_4$
8% (v/v) glycerol

S1 nuclease stop buffer *(UNIT 7.2)*

✓ 0.8 M Tris·Cl, pH 8.0
✓ 20 mM EDTA, pH 8.0
✓ 80 mM $MgCl_2$

S1 stop buffer *(UNIT 4.6)*

✓ 4 M ammonium acetate
✓ 20 mM EDTA, pH 8.0
40 µg/ml tRNA
Store at 4°C

Sample buffer *(UNIT 10.18)*

✓ 50 mM Tris·Cl, pH 6.8
5% (v/v) 2-ME
10% (v/v) glycerol
1% (w/v) SDS
Store frozen in small aliquots for ≤2 months

Saturated ammonium sulfate (SAS) solution *(UNITS 11.1 & 11.6)*

Prepare 0.01 M Tris solution by adding 1.21 g Tris base to 990 ml water, adjust to pH 7.0, and bring to a final volume of 1 liter. Weigh 767 g $(NH_4)_2SO_4$ and dissolve in 1 liter 0.01 M Tris by stirring and gently warming. Adjust the pH to 7.0 and store at 4°C. $(NH_4)_2SO_4$ crystals should be seen at the bottom of the solution at 4°C.

SCE buffer *(UNIT 6.10)*

0.9 M sorbitol (Fisher, molecular biology grade)
0.1 M sodium citrate
✓ 0.06 M EDTA, pH 8.0
Adjust pH to 7.0
Store at room temperature 3 months

SCEM buffer *(UNIT 6.10)*

✓ 4.9 ml SCE buffer
0.1 ml 2-ME

Add 1 to 2 mg Lyticase (Sigma or ICN Biomedicals) just before use.

SDS, recrystallized *(UNIT 10.2)*

High-purity SDS is available from several suppliers, but for some sensitive applications (e.g., protein sequencing) recrystallization is useful. Commercially available electrophoresis-grade SDS is usually of sufficient purity for most applications.

Add 100 g SDS to 450 ml ethanol and heat to 55°C. While stirring, gradually add 50 to 75 ml hot H_2O until all SDS dissolves. Add 10 g activated charcoal (Norit 1, Sigma) to solution. After 10 min, filter solution through Whatman no. 5 paper on a Buchner funnel to remove charcoal. Chill filtrate 24 hr at 4°C and 24 hr at –20°C. Collect crystalline SDS on a coarse-frit (porosity A) sintered-glass funnel and wash with 800 ml –20°C ethanol (reagent grade). Repeat crystallization without adding activated charcoal. Dry recrystallized SDS under vacuum overnight at room temperature. Store in a desiccator over P_2O_5 in a dark bottle.

If proteins will be electroeluted or electroblotted for protein sequence analysis, it may be desirable to crystallize the SDS twice from ethanol/H_2O (Hunkapiller et al., 1983).

SDS column buffer *(UNIT 3.18)*

✓ 10 mM Tris·Cl, pH 8.0
✓ 1 mM EDTA
0.1% (w/v) SDS
Autoclave or filter sterilize

SDS gel solution *(UNIT 12.5)*

✓ 0.3 M Tris·Cl, pH 6.8
6% (w/v) SDS
15% (v/v) glycerol
✓ 70 mM DTT
Store at –20°C

SDS electrophoresis buffer, 5× *(UNIT 10.2)*

15.1 g Tris base (0.125 M)
72.0 g glycine (0.96)
5.0 g SDS [0.5 (w/v)]
H_2O to 1000 ml
Prepare in distilled, deionized water
Dilute to 1× before use

Do not adjust the pH of the stock solution, as the pH is 8.3 when diluted. Store at 0° to 4°C until use. Use purified SDS if appropriate. Prepare without SDS for nondenaturing protocol in UNIT 10.2.

SDS lysis buffer *(UNIT 17.2)*

0.5% (w/v) SDS
✓ 0.05 M Tris·Cl, pH 8.0
✓ 1 mM DTT, added fresh

SDS and Tris·Cl solutions can be made in advance and stored at room temperature. DTT is added from a 1 M stock solution stored at –20°C.

SDS sample buffer, 2× *(UNITS 6.8, 10.2, 10.11, 10.15, 12.5, 16.3, 16.5, 16.7 & 16.11)*

✓ 25 ml 4× Tris·Cl/SDS, pH 6.8 (0.1 M)
20 ml glycerol [20% (w/v)]
4 g SDS [4% (w/v)]
2 ml 2-ME *or* 3.1 g DTT
1 mg bromphenol blue [0.001% (w/v)]
Add H_2O to 100 ml and mix
Prepare in distilled, deionized water
Store in 1-ml aliquots at –70°C

Use purified SDS if appropriate. Omit 2-ME or DTT (reducing agent) to avoid reducing proteins to subunits and make 0.01 M iodoacetamide to prevent disulfide interchange. Prepare without SDS and reducing agent for nondenaturing protocol in UNIT 10.2.

SDS sample buffer, 6× *(UNIT 10.2)*

✓ 7 ml 4× Tris·Cl/SDS, pH 6.8(0.28 M)
3.0 ml 3.8 g glycerol [30% (v/v)]
1 g SDS [1% (w/v)]
0.93 g DTT (0.5 M)
1.2 mg bromphenol blue [0.0012% (w/v)]
Add H_2O to 10 ml if needed
Prepare in distilled, deionized water
Store in 0.5 ml aliquots at –70°C

Prepare without SDS and reducing agent for the nondenaturing protocol in UNIT 10.2.

SDS solubilization buffer *(UNIT 10.3)*

0.1 g 2-(*N*-cyclohexylamino)ethanesulfonic acid (CHES; 0.05 M)
0.2 g SDS [2% (w/v)]
0.1 g DTT (0.55 M)
1.0 ml glycerol [10% (v/v)]
H_2O to 10 ml total volume
Store aliquots at –70°C

SDS-PAGE sample buffer *(UNIT 16.8)*

15% (v/v) glycerol
✓ 0.125 M Tris·Cl, pH 6.8
✓ 5 mM Na_2EDTA
2% (w/v) SDS
0.1% (w/v) bromphenol blue
1% (v/v) 2-mercaptoethanol (2-ME; add immediately before use)
Store indefinitely at room temperature

SE buffer *(UNIT 10.14)*

✓ 20 mM sodium acetate, pH 5.6
✓ 150 mM NaCl

All water and salts used for buffer preparation must be HPLC grade [see entry in this appendix for "Buffer preparation (UNITS 10.12 & 10.13)"], and the glassware used to prepare and to store the buffer must be thoroughly cleaned. Any contaminants add to background at low-wavelength detection.

SE protein standards *(UNIT 10.14)*

Weigh into a plastic scintillation vial:
2.0 mg thyroglobulin
4.0 mg catalase
3.0 mg BSA
3.0 mg ovalbumin
4.0 mg ribonuclease A
Add 6.0 ml SE buffer (see recipe) and gently mix

Proteins must be fully solubilized. Visual inspection will indicate the presence of turbidity or particulates. Magnification 10 to 20 times is often necessary to determine

continued

that a sample at the microgram level is solubilized. All samples should be centrifuged prior to injection.

Selective regeneration agar *(UNIT 13.7)*

5 g ammonium sulfate
1.7 g yeast nitrogen base (without amino acids or ammonium sulfate)
20 g dextrose
30 g agar
1 pellet NaOH
1.3 g appropriate amino acid dropout powder (Table 13.1.1)

To above ingredients, add 500 ml of 2 M sorbitol, 20 ml YPD medium *(UNIT 13.1)*, and 480 ml water. Mix well and autoclave 15 min in 2-liter flask. Aliquot 250-ml portions into 500-ml sterile bottles, using sterile technique. Store at room temperature or 4°C.

SEM buffer *(UNIT 6.10)*

1 M sorbitol
✓ 20 mM EDTA, pH 8.0
14 mM 2-ME
Filter sterilize
Store at 4°C for ~6 weeks

SEMT buffer *(UNIT 6.10)*

1 M sorbitol
✓ 20 mM EDTA
14 mM 2-ME
✓ 10 mM Tris·Cl, pH 8.0
Filter sterilize

Add 1 mg/ml Lyticase (Sigma or ICN Biomedicals) just before use.

Sepharose–protein A suspension, *see Protein A–Sepharose suspension*

Sequenase buffer, 10× *(UNIT 7.4)*

✓ 400 mM Tris·Cl, pH 7.5
✓ 200 mM $MgCl_2$
✓ 500 mM NaCl
✓ 50 mM DTT

Sequenase diluent *(UNIT 7.4)*

✓ 10 mM Tris·Cl, pH 7.5
✓ 5 mM DTT
✓ 0.5 mg/ml BSA

Sequenase/pyrophosphatase mix *(UNIT 7.4)*

Pyrophosphorolysis, the reversal of the polymerase reaction, is a problem when sequencing with Sequenase under some conditions, particularly when dITP is used. Inorganic pyrophosphatase prevents pyrophosphorolysis, and thus should be included in all sequencing reactions using Sequenase, particularly when dITP is present.

Pyrophosphatase is included with U.S. Biochemical's Sequenase sequencing kit. A mixture of the two enzymes can be prepared and stored in 20 mM KPO_4 buffer (pH 7.4)/0.1 mM DTT/0.1 mM EDTA/50% glycerol (v/v) at −20°C. Mix 10 ng of sequencing grade yeast inorganic pyrophosphatase (0.005 U) per 2 to 4 U of Sequenase (i.e., per set of reactions), keeping the Sequenase stock as concentrated as possible. These enzymes are stable together at least 1 year.

Sequenase termination mixes (*UNIT 7.4*)

Concentrations of dNTPs and ddNTPs in the A, C, G, and T mixes are listed below in μM. Prepare all mixes in 10 mM Tris·Cl (pH 7.5)/0.1 mM EDTA using nucleotide stock solutions (see recipe).

	A	C	G	T
dATP	80	80	80	80
dCTP	80	80	80	80
dGTP	80	80	80	80
dTTP	80	80	80	80
ddATP	8	—	—	—
ddCTP	—	8	—	—
ddGTP	—	—	8	—
ddTTP	—	—	—	8
NaCl (mM)	50	50	50	50

To substitute dITP for dGTP, replace dGTP with 160 μM dITP in each termination mix and use 1.6 μM ddGTP in the G termination mix. Inorganic pyrophosphatase should be used in reactions with dITP.

To substitute 7-deaza-dGTP for dGTP, replace dGTP with 80 μM 7-deaza-dGTP in each termination mix; do not alter the concentration of ddGTP in G termination mix.

Short-term labeling medium (*UNIT 10.17*)

Complete serum-free RPMI medium (see recipe) lacking specific amino acid, but containing:

5% (v/v) FBS (dialyze overnight against a saline solution to remove unlabeled amino acids that would decrease labeling efficiency; see entry above for dialyzed fetal bovine serum)

Store at 4°C up to 2 weeks

See annotations to recipe for long-term labeling medium concerning amino acid–free media, substitution of DMEM for RPMI, and use of HEPES buffer.

Silanized glass wool (*UNIT 12.6*)

Submerge glass wool in 1:100 dilution of a silanizing agent such as Prosil 28 (VWR) for 15 sec with shaking. Rinse glass wool extensively with distilled H_2O; autoclave for 10 min and store at room temperature.

Silver nitrate solution, ammoniacal (*UNIT 10.5*)

Add 3.5 ml concentrated NH_4OH (~30%) to 42 ml of 0.36% NaOH and bring the volume to 200 ml with H_2O. Mix with a magnetic stirrer and slowly add 8 ml of 19.4% (1.6 g/8 ml) silver nitrate.

If the solution is cloudy, carefully add NH_4OH until it clears. The solution should be used within 20 min.

CAUTION: *This solution is potentially explosive when dry and therefore should be precipitated by the addition of an equal volume of 1 M HCl. The resultant silver chloride can be washed down a drain with a large volume of cold water.*

SM medium (*UNIT 7.3*)

Per liter:

5.8 g NaCl (0.1 M)
2 g $MgSo_4 \cdot 7H_2O$ (0.01 M)
50 ml 1 M Tris·Cl, pH 7.5 (0.05 M)
0.01% (w/v) gelatin (Difco)

Soaking buffer *(0.4 M NH_4HCO_3)* (*UNIT 10.4*)

Dissolve 3.16 g NH_4HCO_3 in 100 ml H_2O containing 2% (w/v) SDS.

SOC medium *(UNIT 1.8)*

0.5% (w/v) yeast extract
2% (w/v) tryptone
✓ 10 mM NaCl
✓ 2.5 mM KCl
✓ 10 mM $MgCl_2$
✓ 20 mM $MgSO_4$
20 mM glucose

Sodium acetate, 3 M

Dissolve 408 g sodium acetate·$3H_2O$ in H_2O
Adjust pH to 5.2 with 3 M acetic acid
Add H_2O to 1 liter

Sodium acetate buffer, 0.1 M

Solution A: 11.55 ml glacial acetic acid/liter (0.2 M)

Solution B: 27.2 g sodium acetate ($NaC_2H_3O_2 \cdot 3H_2O$)/liter (0.2 M)

Referring to Table A.1.2 for desired pH, mix the indicated volumes of solutions A and B, then dilute with H_2O to 200 ml. (See Potassium acetate buffer recipe for further details.)

Sodium iodide (NaI) solution, 6 M *(UNITS 2.1 & 2.6)*

Dissolve 0.75 g Na_2SO_3 in 40 ml H_2O. Add 45 g NaI (Sigma) and stir until dissolved (~30 min). Filter through Whatman paper or nitrocellulose and store 3 to 4 months in the dark (in aluminum foil). Discard if precipitate is observed.

Sodium phosphate, 100 mM and 10 mM (pH 7.0) *(UNIT 4.9)*

100 mM stock solution:
5.77 ml 1 M Na_2HPO_4
4.23 ml 1 M NaH_2PO_4
H_2O to 100 ml

10 mM solution:
Dilute 100 mM stock 1/10 with H_2O

Sodium phosphate buffer, 0.1 M

Solution A: 27.6 g $NaH_2PO_4 \cdot H_2O$ per liter (0.2 M)

Solution B: 53.65 g $Na_2HPO_4 \cdot 7H_2O$ per liter (0.2 M)

Referring to Table A.1.3 for desired pH, mix the indicated volumes of solutions A and B, then dilute with H_2O to 200 ml. (See Potassium phosphate buffer recipe for further details.)

Sodium phosphate buffer, pH 6.3 *(UNIT 10.10)*

✓ 50 mM sodium phosphate, pH 6.3
0.1% (v/v) Triton X-100
✓ 0.5 M NaCl

Sodium phosphate buffer (PB), pH 6.8 *(UNIT 2.10)*

Mix equal volumes of 1 M Na_2HPO_4 and 1 M NaH_2PO_4 to make 1 M PB, pH 6.8.
Dilute 12 ml of 1 M PB with H_2O to final volume of 100 ml to make 0.12 M PB.
Dilute 40 ml of 1 M PB with H_2O to final volume of 100 ml to make 0.4 M PB.

Sodium phosphate buffer, pH 6.8, 0.5 M stock solution *(UNIT 14.5)*

35.5 g Na_2HPO_4
34.5 g NaH_2PO_4
H_2O to 1 liter
Adjust pH to 6.8 if necessary

Sonicated herring sperm DNA, 2 mg/ml *(UNITS 3.4 & 6.3)*

Resuspend 1 g herring sperm DNA (Boehringer Mannheim) in a convenient volume (50 ml of water) by sonicating briefly. The DNA is now ready to be sheared into short molecules by sonication. Place tube containing the herring sperm DNA solution in an ice bath (the tube must be stable even if ice begins to melt). The sonicator probe is placed in the DNA solution (without touching bottom of vessel). The sonicator is turned on to 50% power 20 min, or until there is a uniform and obvious decrease in viscosity. At no time should tube containing the DNA become hot to the touch. After sonication, the DNA is diluted to a final concentration of 2 mg/ml, frozen in 50-ml aliquots, and thawed as needed.

Sonicated salmon sperm DNA, 10 mg/ml *(UNITS 3.4, 14.7 & 14.8)*

Dissolve 10 mg salmon sperm DNA (Worthington) in 1 ml sterile water in a polycarbonate tube. Sonicate five times, 30 sec each time, at maximum power, chilling tube on ice between bursts. Check molecular size of DNA by gel electrophoresis *(UNIT 2.5A)*; it should be 200 to 400 bp. Store in 50-µl aliquots up to 1 year at −20°C.

Sorbitol/$CaCl_2$ solution *(UNIT 13.7)*

1 M sorbitol
✓ 10 mM Tris·Cl, pH 7.4
✓ 10 mM $CaCl_2$
Filter sterilize or autoclave and store at room temperature

Sorbitol selection plates *(UNIT 13.7)*

Supplement CM dropout plates *(UNIT 13.1)* with sorbitol to 1 M final concentration. Sorbitol can be added as a powder prior to autoclaving. *NOTE*: Addition of sorbitol makes the agar more viscous and more prone to boiling over upon removal from the autoclave.

Spectrapor 6 dialysis membrane *(UNIT 10.4)*

Available in molecular mass cutoff values of 1000, 2000, 3500, 8000, 10,000, 25,000, and 50,000. Soak tubing in 1% (w/v) ammonium bicarbonate at 60°C for 1 hr, wash extensively with H_2O, soak in 0.1% (w/v) SDS at 60°C for 1 hr, and wash extensively with H_2O. Cut discs of dialysis membrane from the tubular dialysis membrane with a sharpened cork borer (No. 8 size). Discs should be handled with Teflon-coated forceps and stored in 0.1% (w/v) SDS/0.1% (w/v) sodium azide at room temperature.

SSC, 20×

✓ 3 M NaCl (175 g/liter)
0.3 M trisodium citrate $2H_2O$ (88 g/liter)
Adjust pH to 7.0 with 1 M HCl

SSC hybridization solution *(UNIT 6.4)*

✓ 6× SSC
✓ 1× Denhardt solution
100 µg/ml yeast tRNA
0.05% sodium pyrophosphate

Stacking gel buffer *(UNIT 10.3)*

15.0 g Tris base
1.0 g SDS
Dissolve Tris base and SDS in 200 ml H_2O. Adjust to pH 6.8 with 6 M HCl. Add H_2O to 250 ml. The 5× stock can be stored for several weeks in the refrigerator.

Staining solution *(UNIT 10.4)*

1 part acetic acid
3 parts isopropanol
6 parts H_2O
0.5% (w/v) Coomassie brilliant blue

Standard enzyme diluent (SED)

✓ 20 mM Tris·Cl, pH 7.5
500 µg/ml BSA (Pentax Fraction V)
10 mM 2-ME
Store at 4°C for up to 1 month

STE buffer

✓ 10 mM Tris·Cl, pH 7.5
✓ 10 mM NaCl
✓ 1 mM EDTA

STET lysing solution *(UNIT 4.4)*

8% (w/v) sucrose
5% (v/v) Triton X-100
✓ 50 mM EDTA
✓ 50 mM Tris·Cl, pH 7.0
Prepare in DEPC-treated stock solutions and store at 4°C

STET solution *(UNIT 1.6)*

8% (w/v) sucrose
0.5% (v/v) Triton X-100
✓ 50 mM EDTA
✓ 50 mM Tris·Cl, pH 8.0
Filter sterilize and store at 4°C

Stop buffer *(UNIT 4.4)*

✓ 200 mM Tris·Cl, pH 8.0
✓ 20 mM EDTA
20 mM sodium azide
20 mM aurintricarboxylic acid (ATA; Sigma)

Do not include ATA if RNA is needed for primer extension of S1 nuclease mapping; store in brown bottle at room temperature.

Stop/loading dye *(UNITS 4.8, 7.4 & 12.3)*

Prepare in deionized formamide (see recipe):
0.05% (w/v) bromphenol blue
0.05% (w/v) xylene cyanol
✓ 20 mM EDTA

A high-quality grade of formamide that does not require deionization is available from Fluka.

Stop solution *(UNIT 7.5)*

✓ 95% (v/v) deionized formamide
✓ 10 mM EDTA
0.1% (w/v) xylene cyanol
0.1% (w/v) bromphenol blue
Store indefinitely at –20°C

Storage buffer *(UNIT 13.13)*

✓ 20 mm Tris·Cl, pH 7.5
✓ 0.1 mM EDTA
10% (v/v) glycerol
✓ 100 mM KCl
✓ 1 mM DTT
✓ 1× protease inhibitor mix
1 mM PMSF

Streptavidin–alkaline phosphatase conjugate solution *(UNIT 14.8)*

Stock solution: 40 µg/ml streptavidin–alkaline phosphatase conjugate (e.g., Life Technologies). Store at 4°C up to 8 months.

Working solution: For each well to be developed, dilute 10 µl stock solution with 90 µl conjugate dilution buffer (see recipe) immediately before use.

Streptavidin-peroxidase conjugate solution *(UNIT 14.8)*

Stock solution: 1 mg/ml streptavidin–horseradish peroxidase conjugate (e.g., Life Technologies) in PBS, pH 7.2 *(APPENDIX2)*. Store at 4°C.

Working solution: When stock solution is fresh (i.e., stored <1 month), dilute 1:100 in PBS, pH 7.2, to prepare working solution. Because the activity of the solution declines with prolonged storage, dilute 1:80 after 1 month of storage, 1:50 after 2 months of storage, and 1:30 after 3 months of storage. Discard after 4 months of storage.

Streptavidin solution, pH 7.2 *(UNIT 7.5)*

1 mg/ml streptavidin
6.8 mM Na_2HPO_4
3.2 mM NaH_2PO_4
✓ 150 mM NaCl
0.05% (w/v) sodium azide
Stable ≤6 months at 4°C

Streptavidin that has been tested for chemiluminescent detection, available from Millipore and New England Biolabs, is highly recommended.

Streptavidin solution *(UNIT 14.7)*

✓ *PBS containing:*
1% (w/v) BSA
0.1% (v/v) Tween 20
3 µg/ml streptavidin
Prepare fresh

Sucrose/Tris/EDTA solution *(UNIT 1.7)*

25% (w/v) sucrose
✓ 50 mM Tris·Cl, pH 8.0
✓ 100 mM EDTA, pH 8.0

Sucrose/Tris/EDTA solution *(UNIT 9.1)*

25% (w/v) sucrose
✓ 50 mM Tris·Cl, pH 8.0
✓ 40 mM EDTA, pH 8.0

Suspension medium (SM) *(UNIT 5.3)*

5.8 g NaCl (0.1 M)
2 g $MgSO_4 \cdot 7H_2O$ (8 mM)
✓ 50 ml 1 M Tris·Cl, pH 7.5 (0.05 M)
H_2O to 1 liter
Solid gelatin (Difco) to 0.01% (w/v)
Heat to just below boiling to dissolve

Suspension TBS (STBS) solution *(UNIT 9.2)*

✓ 25 mM Tris·Cl, pH 7.4
✓ 137 mM NaCl
✓ 5 mM KCl
0.6 mM Na_2HPO_4
✓ 0.7 mM $CaCl_2$
✓ 0.5 mM $MgCl_2$

Make up in distilled H_2O and filter sterilize. A 10× stock of this solution can be made that is diluted to 1× with sterile distilled H_2O prior to use.

T4 DNA ligase buffer, 2× *(UNIT 3.16)*

✓ 100 mM Tris·Cl, pH 7.5
✓ 20 mM $MgCl_2$
✓ 20 mM DTT

T4 polynucleotide kinase buffer, 10× *(UNIT 12.9)*

✓ 500 mM Tris·Cl, pH 7.6
✓ 100 mM $MgCl_2$
✓ 50 mM DTT
1 mM spermidine
✓ 1 mM EDTA, pH 8.0
Store at −20°C
Add an extra 50 mM DTT just before use

TAE electrophoresis buffer

50× stock solution, pH ~8.5:
242 g Tris base
57.1 ml glacial acetic acid
37.2 g $Na_2EDTA \cdot 2H_2O$
H_2O to 1 liter

1× working solution:
40 mM Tris acetate
2 mM EDTA

Taq sequencing buffer, 10× *(UNIT 7.4)*

✓ 500 mM Tris·Cl, pH 9.0
✓ 150 mM $MgCl_2$

Taq Sanger mixes *(UNIT 7.4)*

Concentrations of dNTPs and ddNTPs in the A, C, G, and T mixes are listed below in µM. Prepare all mixes in 0.1 mM EDTA using nucleotide stock solutions (see recipe).

	A	C	G	T
dATP	2	2	2	2
dCTP	20	20	20	20
dGTP	20	20	20	20
dTTP	20	20	20	20
ddATP	150	—	—	—
ddCTP	—	500	—	—
ddGTP	—	—	222	—
ddTTP	—	—	—	520

To substitute 7-deaza-dGTP for dGTP, replace dGTP with an equimolar concentration of 7-deaza-dGTP in each termination mix; do not alter the concentration of ddGTP in the G mix.

TBE electrophoresis buffer

10× stock solution:
108 g Tris base
55 g boric acid
✓ 40 ml 0.5 M EDTA, pH 8.0
H_2O to 1 liter

1× working solution:
89 mM Tris base
89 mM boric acid
2 mM EDTA

TBS, *see Tris-buffered saline*

TCS stop/loading dye (*UNIT 7.4*)

Prepare in deionized formamide:

0.3 (w/v) bromphenol blue
0.3 (w/v) xylene cyanol
✓ 12 mM EDTA

A high-quality grade of formamide that does not require deionization is available from Fluka.

TE buffer, pH 7.4, 7.5, or 8.0

✓ 10 mM Tris·Cl, pH 7.4, 7.5, or 8.0
✓ 1 mM EDTA, pH 8.0

TEN (Tris/EDTA/NaCl) solution (*UNIT 9.7A*)

✓ 40 mM Tris·Cl, pH 7.5
✓ 1 mM EDTA, pH 8.0
✓ 150 mM NaCl
Store up to 6 months at room temperature

Terminal deoxynucleotidyl transferase (TdT) buffer, 5× (*UNIT 15.6*)

1 M potassium cacodylate
✓ 125 mM Tris·Cl, pH 7.4
✓ 1.25 µg/µl BSA
Store at −20°C for ≤6 weeks

TES buffer (*UNIT 6.8*)

✓ 10 mM Tris·Cl, pH 7.6
✓ 1 mM EDTA
✓ 0.15 M NaCl

TES solution (*UNIT 4.2*)

✓ 10 mM Tris·Cl, pH 7.4
✓ 5 mM EDTA
1% (w/v) SDS

TES solution (*UNIT 13.12*)

✓ 10 mM Tris·Cl, pH 7.5
✓ 10 mM EDTA
✓ 0.5% (w/v) SDS
Store indefinitely at room temperature

Tetramethylammonium chloride, *see TMAC*

Thiosulfate developing solution (*UNIT 10.5*)

3% (w/v) sodium carbonate
0.0004% (w/v) sodium thiosulfate
0.5 ml of 37% formaldehyde per liter solution (add immediately before use)
Store indefinitely without formaldehyde at room temperature

Thrombin cleavage buffer (*UNIT 16.5*)

✓ 50 mM Tris·Cl, pH 7.5
✓ 150 mM NaCl
✓ 2.5 mM $CaCl_2$
Store indefinitely at −20°C

Tissue guanidinium solution, *1 liter* (*UNIT 4.2*)

Dissolve 590.8 g guanidinium isothiocyanate in ~400 ml DEPC-treated H_2O (see recipe). Add 25 ml of 2 M Tris·Cl, pH 7.5 (0.05 M final) and 20 ml of 0.5 M Na_2EDTA, pH 8.0 (0.01 M final). Stir overnight, adjust the volume to 950 ml, and filter. Finally, add 50 ml 2-ME.

Tissue resuspension buffer *(UNIT 4.2)*

✓ 5 mM EDTA
0.5% (v/v) Sarkosyl
5% (v/v) 2-ME

TLE solution *(UNIT 4.3)*

✓ 0.2 M Tris
0.1 M LiCl
✓ 5 mM EDTA
pH to 8.2 with HCl

See annotation to recipe for grinding buffer.

TM buffer, 10×

✓ 100 mM Tris·Cl, pH 8.0
✓ 100 mM $MgCl_2$

TMAC (tetramethylammonium chloride), 6 M stock solution *(UNIT 6.4)*

Dissolve 657.6 g TMAC (mol. wt. = 109.6) in H_2O and bring to 1 liter. Filter the solution through Whatman No. 1 filter paper and determine the precise concentration of the solution by measuring the refractive index (n) of a 3-fold diluted solution. The molarity (M) of the diluted solution = $55.6(n - 1.331)$ and the molarity of the stock solution = $3 \times M$. TMAC can be stored at room temperature in brown bottles.

CAUTION: *TMAC can irritate eyes, skin, and mucous membranes. Use with adequate ventilation in a fume hood and collect/discard used TMAC solutions as hazardous and/or radioactive waste. Small amounts (<10 ml) can be flushed down the drain with a large quantity of tap water.*

TMAC hybridization solution *(UNIT 6.4)*

✓ 3 M TMAC stock solution
0.1 M $NaPO_4$, pH 6.8
✓ 1 mM EDTA, pH 8.0
✓ 5× Denhardt solution
0.6% (w/v) SDS
✓ 100 µg/ml denatured salmon sperm DNA

TMAC wash solution *(UNIT 6.4)*

✓ 3 M TMAC stock solution
✓ 50 mM Tris·Cl, pH 8.0
0.2% (w/v) SDS

TNE buffer, 10× *(APPENDIX 3D)*

0.1 M Tris base
✓ 10 mM EDTA
✓ 2.0 M NaCl
Adjust pH to 7.4 with concentrated HCl

Toluidine blue stain *(UNIT 14.5)*

1.0 g sodium borate (3 mM)
0.5 g toluidine blue [0.5% (w/v)]
H_2O to 100 ml; filter before use

Transcription buffers, 5× *(UNIT 4.7)*

T7, T3 RNA polymerases:

✓ 200 mM Tris·Cl, pH 8.0
✓ 40 mM $MgCl_2$
10 mM spermidine
✓ 250 mM NaCl

SP6 RNA polymerase:

✓ 200 mM Tris·Cl, pH 7.5
✓ 30 mM $MgCl_2$
10 mM spermidine

These buffers are best made up fresh from frozen dedicated stock solutions. Use DEPC-treated water (see recipe).

Transcription buffer, 5× *(UNIT 14.3)*

✓ 200 mM Tris·Cl, pH 8.3
✓ 30 mM $MgCl_2$
10 mM spermidine·Cl
0.1% (v/v) Triton X-100
Store at −70°C in aliquots

Transfection buffer *(UNIT 16.11)*

✓ 10 ml 10× HeBS
✓ 1.5 ml 1 mg/ml carrier DNA (sonicated calf thymus or salmon sperm DNA)
88.5 ml sterile H_2O
Combine ingredients and adjust pH to 7.05-7.10 with 5 M NaOH
Refilter if necessary and store at 4°C

Transfer buffer *(UNIT 10.7)*

Add 18.2 g Tris base and 86.5 g glycine to 4 liters of H_2O. Add 1200 ml methanol and bring to 6 liters with H_2O. The pH of the solution is ~8.3 to 8.4. For use with PVDF filters, decrease methanol concentration to 15%; for nylon filters, omit methanol.

CAPS transfer buffer can also be used. Add 2.21 g cyclohexylaminopropane sulfonic acid (CAPS; free acid), 0.5 g DTT, 150 ml methanol, and H_2O to 1 liter. Adjust to pH 10.5 with NaOH and chill to 4°C. For proteins >60 kDa, reduce methanol content to 1% (Moos, 1992).

Transfer buffer *(UNIT 10.18)*

2.21 g cyclohexylaminopropane sulfonic acid (CAPS), free acid (0.01 M)
0.5 g dithiothreitol (3 mM)
150 ml methanol [15% (v/v)]
H_2O to 1 liter
Adjust pH to 10.5 with NaOH and chill to 4°C
Prepare just before use

For proteins >60 kDa, reduce amount of methanol to 1%.

Transformation and storage solution (TSS), 2× *(UNIT 1.8)*

Diluting sterile (autoclaved) 40% (w/v) polyethylene glycol (PEG) 3350 to 20% PEG in sterile LB medium containing 100 mM $MgCl_2$. Add dimethyl sulfoxide (DMSO) to 10% (v/v) and adjust to pH 6.5.

Tricine sample buffer, 2× *(UNIT 10.2)*

✓ 2 ml 4× Tris·Cl/SDS, pH 6.8 (0.08 M)
2.4 ml (3.0 g) glycerol [24% (v/v)]
0.8 g SDS [8% (w/v)]
0.31 g DTT (0.2 M)
2 mg Coomassie blue G-250 [0.02% (w/v)]
Add H_2O to 10 ml and mix

Final concentrations are 0.1 M Tris, 24% (w/v) glycerol, 8% (w/v) SDS, 0.2 M DTT, and 0.02% (w/v) Coomassie blue G-250.

Triethanolamine solution *(UNIT 10.10)*

50 mM triethanolamine, pH ~11.5
0.1% (w/v) Triton X-100
✓ 0.15 M NaCl

Triethanolamine (TEA) buffer *(0.1 M TEA; UNIT 14.3)*

Add 18.57 g triethanolamine·Cl to 900 ml water. Dissolve and adjust pH to 8.0 with NaOH. Adjust to 1 liter with water for 0.1 M solution. Use same day.

Tris buffer, pH 8.6 (*UNIT 11.11*)

To 900 ml H_2O, add:

6.06 g Tris·Cl (0.05 M final)
8.77 g NaCl (0.15 M final)
0.2 g NaN_3 (0.02% final)
Titrate with HCl to pH 8.6
Add H_2O to 1 liter

Tris buffers, pH 8.0 and pH 9.0 (*UNIT 10.10*)

✓ 50 mM Tris·Cl, pH 8.0 *or* pH 9.0
0.1% (v/v) Triton X-100
✓ 0.5 M NaCl

Tris-buffered saline (TBS) (*UNIT 9.2*)

Solution A:
80 g/liter NaCl (1.38 M)
3.8 g/liter KCl (0.05 M)
2 g/liter Na_2HPO_4 (0.014 M)
30 g/liter Tris base (0.248 M)
Adjust pH to 7.5
Filter sterilize
Store at −20°C

Solution B:
15 g/liter $CaCl_2$ (0.135 M)
10 g/liter $MgCl_2$ (0.11 M)
Filter sterilize
Store at −20°C

For 100 ml, add 10 ml solution A to 89 ml H_2O. While stirring rapidly, add 1 ml solution B slowly, drop by drop. Filter sterilize and store at 4°C.

Make certain that the stock TBS solution is mixed well just before use by pipeting up and down with a sterile 10-ml pipet several times.

Tris-buffered saline (*TBS; UNIT 10.7*)

✓ 100 mM Tris·Cl, pH 7.5
✓ 0.9% (w/v) NaCl
Store at 4°C for several months

Tris·Cl [tris(hydroxymethyl)aminomethane], 1 M

Dissolve 121 g Tris base in 800 ml H_2O
Adjust to desired pH with concentrated HCl
Mix and add H_2O to 1 liter

Desired pH values can also be obtained by mixing the indicated amounts of 0.1 M HCl with 100 ml of 0.1 M Tris base using the following table:

pH, 25°C	0.1 M HCl (ml)	pH, 25°C	0.1 M HCl (ml)	pH, 25°C	0.1 M HCl (ml)
7.2	89.4	7.8	69.0	8.4	34.4
7.3	86.8	7.9	64.0	8.5	29.4
7.4	84.0	8.0	58.4	8.6	24.8
7.5	80.6	8.1	52.4	8.7	20.6
7.6	77.0	8.2	45.8	8.8	17.0
7.7	73.2	8.3	39.8	8.9	14.0

NOTE: *The pH of Tris buffers changes significantly with temperature, decreasing approximately 0.028 pH units per 1°C. Tris-buffered solutions should be adjusted to the desired pH at the temperature at which they will be used. Since the pK_a of Tris is 8.08, Tris should not be used as a buffer below pH ~7.2 or above pH ~9.0.*

4× *Tris·Cl, pH 6.8* (*UNIT 10.2*)

Dissolve 6.05 g Tris base (0.5 M) in 40 ml H_2O. Adjust to pH 6.8 with 1 N HCl. Add H_2O to 100 ml total volume. Filter the solution through a 0.45-μm filter and store at 4°C up to 1 month.

8× ***Tris·Cl, pH 8.8*** *(UNIT 10.2)*

Dissolve 182 g Tris base (3.0 M) in 300 ml H_2O. Adjust to pH 8.8 with 1 N HCl. Add H_2O to 500 ml total volume. Filter the solution through a 0.45-µm filter and store at 4°C up to 1 month.

4× ***Tris·Cl/SDS, pH 8.8*** *(UNIT 10.2)*

Dissolve 91 g Tris base (1.5 M) in 300 ml H_2O. Adjust pH to 8.8 with 1 N HCl. Add H_2O to 500 ml total volume. Filter the solution through a 0.45-µm filter, add 2 g SDS [0.4% (w/v)], and store at 4°C up to 1 month.

8× ***Tris·Cl/SDS, pH 6.8*** *(UNIT 10.2)*

Dissolve 6.05 g Tris base (0.5 M) in 40 ml H_2O. Adjust to pH 6.8 with 1 N HCl. Add H_2O to 100 ml total volume. Filter the solution through a 0.45-µm filter, add 0.4 g SDS [0.4% (w/v)], and store at 4°C up to 1 month.

Tris·Cl/SDS, pH 8.45 *(UNIT 10.2)*

Dissolve 182 g Tris base (3.0 M) in 300 ml H_2O. Adjust pH to 8.45 with 1 N HCl. Add H_2O to 500 ml total volume. Filter the solution through a 0.45-µm filter, add 1.5 g SDS [0.3% (w/v)], and store at 4°C up to 1 month.

Tris/EDTA/glucose buffer *(UNIT 7.3)*

✓ 25 mM Tris·Cl, pH 8.0
✓ 10 mM EDTA
50 mM glucose

Tris/EDTA/NaCl (TEN) buffer, pH 7.2 *(UNITS 11.1 & 11.2)*

To 930 ml H_2O, add:
6.06 g Tris base (0.05 M)
0.37 g Na_2EDTA (1 M)
8.77 g NaCl (0.151 M)
Adjust pH to 7.2 with HCl
Add H_2O to 1 liter

Tris/EDTA/proteinase K *(UNIT 7.3)*

✓ 10 mM Tris·Cl, pH 7.5
✓ 1 mM EDTA
50 µg/ml proteinase K

Prepare fresh for each use at the final concentrations listed above using stock solutions of 1 mg/ml proteinase K and TE buffer, pH 7.5 (see recipe).

Tris/glycine stock solution, 5× *(UNITS 12.2 & 12.3)*

30.28 g Tris base (0.25 M)
142.7 g glycine (2 M)
3.92 g EDTA (10 mM)
H_2O to 1 liter
pH should be ~8.5
Can be stored for several months

Tris/ovalbumin buffer *(UNIT 11.1)*

✓ 0.05 M Tris·Cl, pH 8.0
5% (w/v) ovalbumin
✓ 5 mM $MgCl_2$
0.5% (w/v) NaN_3
0.5% (w/v) merthiolate

Tris/saline/azide (TSA) solution *(UNIT 10.10)*

✓ 0.002 M Tris·Cl, pH 8.0 (at 4°C)
✓ 0.14 M NaCl
0.025% (w/v) NaN_3

CAUTION: *Sodium azide (NaN_3) is poisonous; wear gloves.*

Tris/saline/azide (TSA) solution *(UNIT 10.16)*

✓ 10 mM Tris·Cl, pH 8.0
✓ 0.14 M NaCl
0.025% (w/v) NaN_3

CAUTION: *Sodium azide (NaN_3) is poisonous; wear gloves.*

Triton/glycylglycine lysis buffer *(UNIT 9.7A)*

1% (v/v) Triton X-100
25 mM glycylglycine, pH 7.8
✓ 15 mM $MgSO_4$
4 mM EGTA
✓ 1 mM DTT added just before each use
Store at 4°C

Triton lysis solution *(UNIT 9.7B)*

✓ 100 mM potassium phosphate, pH 7.8
0.2% (v/v) Triton X-100
✓ 1 mM DTT (add fresh immediately prior to use)
Store at 4°C; stable several months without DTT

This solution is available as part of the Galacto-Light assay kit (Tropix).

Triton lytic mix *(UNIT 9.1)*

2% (v/v) Triton X-100
✓ 40 mM EDTA, pH 8.0
✓ 50 mM Tris·Cl, pH 8.0

tRNA, 10 mg/ml

Dissolve in H_2O at 10 to 20 mg/ml and extract repeatedly with buffered phenol.

Trypan blue overlay *(UNIT 16.10)*

Prepare a 1% (w/v) trypan blue solution in H_2O and filter sterilize. Microwave 1% (w/v) agarose (sterile; see recipe) 2 min until liquid and equilibrate to 42°C. Equilibrate the 1% trypan blue solution to 42°C. Add 4 ml trypan blue solution to 50 ml of 1% agarose, and mix well. Use immediately (this solution cannot be stored).

Tryptophan, 10 mg/ml *(UNIT 16.8)*

Heat 500 ml glass-distilled H_2O to 80°C. Stir in 5 g L-tryptophan until dissolved. Filter sterilize the solution through a 0.45 μm filter and store ≤6 months in the dark at 4°C.

TSA, *see Tris/saline/azide (TSA) solution*

Tween 20/TBS (TTBS) *(UNIT 10.7)*

0.1% (v/v) Tween 20 in Tris-buffered saline (see recipe)
Store at 4°C for several months

Two-step blocking solution, pH 7.2 *(UNIT 7.5)*

5% (w/v) SDS
17 mM Na_2HPO_4
8 mM NaH_2PO_4

Add sodium azide to 0.02% (w/v) to store solution for 2 to 7 days at 4°C.

Two-step wash solution I *(UNIT 7.5)*

Prepare a 1/10 dilution of two-step blocking solution (see recipe) in H_2O. Make fresh and do not reuse.

Two-step wash solution II *(UNIT 7.5)*

✓ 10 mM Tris·Cl, pH 9.5
✓ 10 mM NaCl
✓ 1 mM $MgCl_2$
Autoclave to sterilize, then store indefinitely at room temperature.

Unidirectional linker mix, 20 μM (*UNIT 15.5*)

20 μM oligonucleotide LMPCR.1 (Fig. 15.5.3)
20 μM oligonucleotide LMPCR.2 (Fig. 15.5.3)
✓ 250 mM Tris·Cl, pH 7.7

Prepare this mix in advance as follows: (1) gel-purify the oligonucleotides on denaturing polyacrylamide gels (*UNIT 2.11*), (2) combine the two oligonucleotides and Tris·Cl and heat 5 min at 95°C, (3) transfer to 70°C and gradually cool ~1 hr to room temperature, (4) leave ~1 hr at room temperature and then gradually cool ~1 hr to 4°C, and (5) leave ~12 hr at 4°C and store in aliquots at −20°C. Thaw linker on ice when needed.

Hybridization of the linker undoubtedly takes place more rapidly than allowed for in this procedure but this has reproducibly worked well. Because Tris·Cl is used as a salt in this hybridization reaction, it is difficult to calculate the kinetics of oligonucleotide annealing and this has not been studied in detail. Tris·Cl is used in place of more traditional monovalent salts (e.g., NaCl) because these inhibit T4 DNA ligase. In addition, Tris·Cl is required by T4 DNA ligase.

10× upper reservoir buffer (*UNIT 10.18*)

40.28 g *N*-tris-(hydroxymethyl)-methyl-2-aminoethanesulfonic acid (TES; 0.439 M)
94.64 g *bis*-Tris (1.131 M)
4.0 g SDS [1% (w/v)]
H_2O to 400 ml

Final pH should be 7.25; do not adjust the pH with acid or base. This buffer may be stored indefinitely at −20°C or ≤2 months at 4°C.

Urea solubilization buffer (*UNIT 10.3*)

54 g urea (9 M)
4 ml NP-40 [4% (w/v)]
10 ml 20% (w/v) stock ampholyte (pH 9 to 11)
2 ml 2-ME (Kodak)
H_2O to 100 ml total volume
If pH is <9.5, add NaOH
Store aliquots at −70°C

Vent DNA polymerase mix, 3 μl (*UNIT 15.5*)

Mix 0.6 μl of 5× amplification buffer with 1.9 μl H_2O and chill several minutes in an ice-water bath. Add 0.5 μl (10 U) *Thermococcus litoralis* DNA polymerase (Vent; New England Biolabs), gently mix, and return to ice-water bath. Prepare just before use.

Vent DNA polymerase stop solution, 295 μl (*UNIT 15.5*)

25 μl 3 M sodium acetate, pH 7.0 (260 mM final)
✓ 266 μl TE buffer, pH 7.5 (10 mM Tris·Cl, pH 7.5, final)
✓ 2 μl 0.5 M EDTA, pH 8.0 (~4 mM final)
2 μl 10 mg/ml yeast tRNA (68 μg/ml final)
Prepare immediately before use

10× Vent$_R$ (exo$^-$) sequencing buffer (*UNIT 7.4*)

✓ 100 mM KCl
100 mM $(NH_4)_2SO_4$
✓ 200 mM Tris·Cl, pH 8.8
50 mM $MgSO_4$

Vent$_R$ (exo$^-$) sequencing mixes *(UNIT 7.4)*

Concentrations of dNTPs and ddNTPs in the A, C, G, and T mixes are listed below in μM. Prepare all mixes in 1× Vent$_R$ (exo$^-$) sequencing buffer using nucleotide stock solutions (see recipe).

	A	C	G	T
dATP	30	30	30	30
dCTP	100	37	100	100
dGTP	100	100	37	100
dTTP	100	100	100	33
ddATP	900	—	—	—
ddCTP	—	480	—	—
ddGTP	—	—	400	—
ddTTP	—	—	—	720

To substitute 7-deaza-dGTP for dGTP, replace dGTP with an equimolar concentration of 7-deaza-dGTP in each of the Vent$_R$ (exo$^-$) sequencing mix (Sears et al., 1992); do not alter the ddGTP concentration.

dITP is not recommended for use with Vent$_R$ (exo$^-$) DNA polymerase.

Wash buffer *(UNIT 10.10)*

✓ 0.01 M Tris·Cl, pH 8.0 (at 4°C)
✓ 0.14 M NaCl
0.025% (w/v) NaN_3 *(handle cautiously!)*
0.5% (v/v) Triton X-100
0.5% (v/v) sodium deoxycholate

Wash buffer *(UNIT 11.2)*

✓ Hanks balanced salt solution (HBSS)
1% (v/v) fetal bovine serum (FBS; heat-inactivated 60 min, 56°C)
0.05% (w/v) NaN_3
Store at 4°C
CAUTION: *Sodium azide is poisonous; handle with care.*

Wash buffer I *(UNIT 3.19)*

Dilute blocking solution 1:10 with water just prior to use

Wash buffer I *(UNIT 7.5)*

Prepare 0.5% (w/v) SDS in 1× PBS (see recipe). Make fresh if possible and do not reuse.

If necessary, add sodium azide to 0.02% (w/v) to store solution 2 to 7 days at 4°C.

Wash buffer II, 10× stock *(UNIT 3.19)*

✓ 100 mM Tris·Cl, pH 9.5
✓ 100 mM NaCl
✓ 10 mM $MgCl_2$

Prepare stock and store at 4°C
Dilute 1:10 with water just prior to use

Wash buffer II *(UNIT 7.5)*

Prepare 0.1% (v/v) Tween-20 in 1× PBS (see recipe). Make fresh if possible and do not reuse.

If necessary, add sodium azide to 0.02% (w/v) to store solution 2 to 7 days at 4°C.

Washing buffer *(UNIT 11.4)*

To 900 ml H_2O, add:
13.6 g sodium acetate (trihydrate; 0.1 M final)
29.2 g NaCl (0.5 M final)
Titrate with acetic acid to pH 4.0
Add H_2O to 1 liter

Wash solution *(UNIT 2.1)*

✓ 20 mM Tris·Cl, pH 7.4
✓ 1 mM EDTA
✓ 100 mM NaCl

Add an equal volume of 100% ethanol and store 3 to 4 months at 0°C.

Wash solution A *(UNIT 14.3)*

50% (v/v) formamide (inexpensive grade, e.g., Fluka)
✓ 2× SSC (dilute 20×)
20 mM 2-ME
Preheat half to 55°C and half to 50°C

Wash solution B *(UNIT 14.3)*

50% (v/v) formamide (inexpensive grade, e.g., Fluka)
✓ 2× SSC (dilute 20×)
20 mM 2-ME
0.5% (v/v) Triton X-100
Preheat to 55°C

Wash solution C *(UNIT 14.3)*

✓ 2× SSC (dilute 20×)
20 mM 2-ME
Preheat to 50°C

Water, high-quality and sterile *(UNIT 13.7)*

Water used in these protocols for washes and in solution preparation must be of the highest-possible quality (e.g., Millipore Milli-Q); the resistance must be at least as high as 10 MΩ/cm. Sterilize by autoclaving and store at room temperature.

Xgal, 20 mg/ml *(UNIT 16.11)*

Prepare in sterile dimethylformamide and store in the dark in a glass container up to several months at −20°C.

Xgal solution *(UNIT 9.11)*

Prepare in PBS:
5 to 35 mM $K_3Fe(CN)_6$ (potassium ferricyanide)
5 to 35 mM $K_4Fe(CN)_6 \cdot 3H_2O$ (potassium ferrocyanide)
✓ 1 to 2 mM $MgCl_2$ *or* $MgSO_4$
Just before use, add 40× Xgal in *N,N*-dimethyl formamide (Table 1.4.2) to 1 mg/ml final

CAUTION: *Avoid contact and inhalation of cyanide. Discard waste as directed by your institution.*

The amount of ferric and ferrous cyanide to use is a matter of preference. The higher amount causes precipitation of indole to occur more quickly and thus reduces diffusion. However, it may cause a greenish background upon prolonged incubation (overnight or longer) in some tissues, although not usually on cell lines.

The first three ingredients (i.e., excluding Xgal) can be stored at least a few months at room temperature. 40× Xgal can be stored in a glass container covered with foil at −20°C.

Z buffer *(UNITS 13.4 & 13.6)*

16.1 g $Na_2HPO_4 \cdot 7H_2O$ (60 mM final)
5.5 g $NaH_2PO_4 \cdot H_2O$ (40 mM final)
0.75 g KC1 (10 mM final)
0.246 g $MgSO_4 \cdot 7H_2O$ (1 mM final)
2.7 ml 2-ME (50 mM final)
Adjust to pH 7.0 and bring to 1 liter with H_2O (do not autoclave)

Zymolyase buffer (*UNIT 13.13*)

✓ 50 mM Tris·Cl, pH 7.5
✓ 10 mM $MgCl_2$
1 M sorbitol
✓ 1 mM *or* 30 mM DTT

For step 3 of Basic Protocol use 30 mM DTT in this buffer. For all other applications, use 1 mM DTT.

STANDARD MEASUREMENTS, DATA, AND ABBREVIATIONS

Common Abbreviations

*A*260 absorbance at 260 nm

A adenine or adenosine; one-letter code for alanine

Ab antibody

ABTS [2,2′-azino-di(3-ethylbenzothiazoline sulfonate)]

acetyl CoA acetyl coenzyme A

AcMNPV *Autographica californica* multiply-enveloped nuclear polyhedrosis virus

ADA adenosine deaminase

ADP adenosine 5′-diphosphate

AEX anion exchange

Ag antigen

AMP adenosine 5′-monophosphate

AMV avian myeloblastosis virus

AP sites apyrimidinic sites

APH aminoglycoside phosphotransferase

ARS autonomous replication sequences

ATP adenosine 5′-triphosphate

AUFS absorbance units, full scale

β-gal β-galactosidase

BAP bacterial alkaline phosphatase

BBS BES-buffered solution or borate-buffered saline

BHI brain heart infusion (medium)

biotin-11-dUTP 8-(2,4-dinitrophenyl-2,6-aminohexyl)aminoadenosine-5′-triphosphate or 2′-deoxyuridine-5′triphosphate-5′allylamin biotin

bis, **bisacrylamide** *N,N′*-methylene-bisacrylamide

bis-**Tris** 2-*bis*[2-hydroxyethyl]amino-2-[hydroxymethyl]-1,3-propanediol

bp base pair

BrdU 5-bromodeoxyuridine

BSA bovine serum albumin

C cytosine or cytidine; one-letter code for cysteine

cAMP adenosine 3′,5′-cyclic-monophosphate

CAPS [cyclohexylamino]-1-propanesulphonic acid

CAT chloramphenicol acetyltransferase

cDNA complementary deoxyribonucleic acid

CDP cytidine 5′-diphosphate

CEX cation exchange

CHAPS 3-[(3-cholamidopropyl)-dimethylammonio]-1-propane-sulfonate

Ci curie

CIP calf intestine alkaline phosphatase

cM centimorgans

CM complete minimal (medium); carboxymethyl

CMP cytidine 5′-monophosphate

cpm counts per minute

CTAB cetyltrimethylammonium bromide

CTP cytidine 5′-triphosphate

CWS cell wall skeleton

dA deoxyadenosine

Da Dalton

DAB 3,3′-diaminobenzidene

dAMP deoxyadenosine monophosphate

DAPI 4′6-diamidino-2-phenylindole

d(A-T) deoxyadenylate-deoxythymidylate

dATP deoxyadenosine triphosphate

DBM diazobenzyloxymethyl

dC deoxycytosine

DCA dichloroacetic acid

DCC dextran-coated charcoal

dCF 2′-deoxycoformycin

dCMP deoxycytidine monophosphate

dCTP deoxycytidine triphosphate

ddATP dideoxyadenosine triphosphate

ddCTP dideoxycytidine triphosphate

ddGTP dideoxyguanosine triphosphate

ddNTP dideoxynucleoside triphosphate

ddTTP dideoxythymidine triphosphate

DEA diethylamine

DEAE diethylaminoethyl

DEPC diethylpyrocarbonate

DES diethylstilbestrol

dG deoxyguanosine

dGTP deoxyguanosine triphosphate

DHFR dihydrofolate reductase

dITP deoxyinosine 5′-triphosphate

DMEM Dulbeccos minimum essential (or modified Eagle) medium

DMF dimethylformamide

DMS dimethyl sulfate

DMSO dimethyl sulfoxide

DMT dimethoxytrityl

DNA deoxyribonucleic acid

DNase deoxyribonuclease

dNTP deoxynucleoside triphosphate

ds double stranded

dT deoxythymidine

DTT dithiothreitol

dTTP deoxythymidine triphosphate

dUMP deoxyuridine monophosphate

dUTP deoxyuridine triphosphate

ECTEOLA epichlorohydrin triethanolamine

EDTA ethylenediaminetetraacetic acid

EGTA ethylene glycol-*bis*(β-aminoethyl ether)-*N,N,N′,N′*-tetraacetic acid

ELISA enzyme-linked immunosorbent assay

EMBL European Molecular Biology Laboratory

EMS ethyl methanesulfonate

exo exonuclease

F Farad

FBS fetal bovine serum

FCS fetal calf serum

FITC fluorescein isothiocyanate

FOA fluoroorotic acid

FPLC fast protein, peptide, and polynucleotide liquid chromatography

g gravity

G gauge; guanine or guanosine; one-letter code for glycine

GDP guanosine 5′-diphosphate

GF gel filtration

GMP guanosine monophosphate

GST glutathione *S*-transferase

GTP guanosine 5′-triphosphate

HAT hypoxanthine/aminopterin/thymidine (medium)

HeBS HEPES-buffered saline

HEPES *N*-2-hydroxyethylpiperazine-*N′*-2-ethanesulfonic acid

HFBA heptafluorobutyric acid

hGH human growth hormone

HGPRT hypoxathine-guanine phosphoribosyltransferase

HIC hydrophobic-interaction chromatography

HPCF high-performance chromatofocusing

HPH hygromycin-B-phosphotransferase

HPHIC high-performance hydrophobic-interaction chromatography

HPLC high-performance liquid chromatography

HRPO horseradish peroxidase

HSV herpes simplex virus

IAA 3-β indoleacrylic acid; indole-3-acetic acid

IEF isoelectric focusing

IEX ion exchange

Ig immunoglobulin

imm immunity region

IODOGEN (1,3,4,6-tetrachloro-3α,6α-diphenylglycouril)

IPTG isopropyl-1-thio-β-D-galactoside

K_m Michaelis constant

KLH keyhole limpet hemocyanin

mAb monoclonal antibody

MBP maltose-binding protein

MBS *m*-maleimidobenzoic acid *N*-hydroxysuccinimide ester

MES 2-(*N*-morpholino)ethanesulfonic acid

MMT monomethoxytrityl

MOI multiplicity of infection

MOMuLV Moloney murine leukemia virus

MOPS 3-(*N*-morpholino)propane sulfonic acid

mp melting point

M_r relative molecular weight

mRNA messenger ribonucleic acid

MTX methotrexate

MWCO molecular weight cutoff

NAD nicotinamide adenine dinucleotide

NIH National Institutes of Health

NTP nucleoside triphosphate

OD_{260} optical density at 260 nm

oligo oligonucleotide

oligo(dT) oligodeoxythymidylic acid

ONPG *o*-nitrophenyl-β-D-galactosidase

ORF open reading frame

ori origin of replication

PAGE polyacrylamide gel electrophoresis

PAP peroxidase–anti-peroxidase

par partition loci on plasmid DNA

PB phosphate buffer

PBS phosphate-buffered saline

PCR polymerase chain reaction
PEG polyethylene glycol
PEI polyethylenimine
PFA paraformaldehyde
pfu plaque-forming units
pI isoelectric point
PIPES piperazine-*N,N′*-bis(2-ethane-sulfonic acid)
PMSF phenylmethylsulfonyl fluoride
poly(A) polyadenylic acid or polyadenylate
poly(A)$^+$ polyadenylated (mRNA)
poly(A)$^-$ nonpolyadenylated (mRNA)
poly(dA-dT) poly(deoxyadenylic acid-deoxythymidylic acid)
poly(U) polyuridylic acid or polyuridylate
PPO 2,5-diphenyloxazole
Pristane 2,6,10,14-tetramethylpentadecane
Pth 3-phenyl-2-thiohydantoin
PVA polyvinyl alcohol
PVC polyvinyl chloride
PVDF polyvinylidene difluoride
rA riboadenylate
RE restriction endonuclease
RF replicative form
RFLP restriction-fragment-length polymorphisms
RIA radioimmunoassay
RNA ribonucleic acid
RNase ribonuclease
RP reversed phase (HPLC)
rRNA ribosomal ribonucleic acid
RT reverse transcriptase
SAM *S*-adenosylmethionine
Sarkosyl *N*-lauroylsarcosine
SAX strong anion exchange
SCX strong cation-exchange
SDS sodium dodecyl sulfate
SE size exclusion (HPLC)
SED standard enzyme diluent
SM suspension medium
ss single stranded
SSC sodium chloride/sodium citrate (buffer)
STBS suspension Tris-buffered saline
T thymine or thymidine; one-letter code for threonine
TAE Tris/acetate (buffer)
Taq *Thermus aquaticus* DNA (polymerase)
TBE Tris/borate electrophoresis (buffer)
TBS Tris-buffered saline
TCA trichloracetic acid
TDM trehalose dimycolate
TE Tris/EDTA (buffer)
TEAE triethylaminoethyl
TEMED *N,N,N′,N′*-tetramethylethylenediamine
TES *N*-tris(hydroxymethyl)methyl-2-aminoethanesulfonic acid
TFA trifluoroacetic acid
THF tetrahydrofuran
TK thymidine kinase
TLC thin-layer chromatography
T_m melting (or midpoint) temperature; thermal denaturation
TMAC tetramethylammonium chloride
TMP trimethylphosphate; thymidine monophosphate
TONPG orthonitrophenyl-β-D-thiogalacoside
Tris tris(hydroxymethyl)aminomethane
Tris·Cl Tris hydrochloride
TRITC tetramethylrhodamine isothiocyanate
tRNA transfer ribonucleic acid
TTP thymidine 5′-triphosphate
U unit; uracil or uridine
UDP uridine 5′-diphosphate
UMP uridine 5′-monophosphate
UTP uridine 5′-triphosphate
UV ultraviolet
UWGCG University of Wisconsin Genetics Computer Group
WAX weak anion exchange
WCX weak cation exchange
Xgal 5-bromo-4-chloro-3-indolyl-β-D-galactoside
XGPRT xanthine-guanine phosphoribosyl transferase
Xyl-A 9-β-D-xylofuranosyl adenine
YCp yeast centromeric plasmid
YEp yeast episomal plasmid
YIp yeast integrating plasmid
YNB–AA/AS yeast nitrogen base without amino acids or ammonium sulfate
YPD yeast/peptone/dextrose (medium)
YRp yeast replicating plasmid

Useful Measurements and Data

Figure A.2.1 A physical chemist's view of the cell. The data in this figure were assembled from Alberts et al. (1983), and represent the approximate concentrations of a variety of intracellular components.

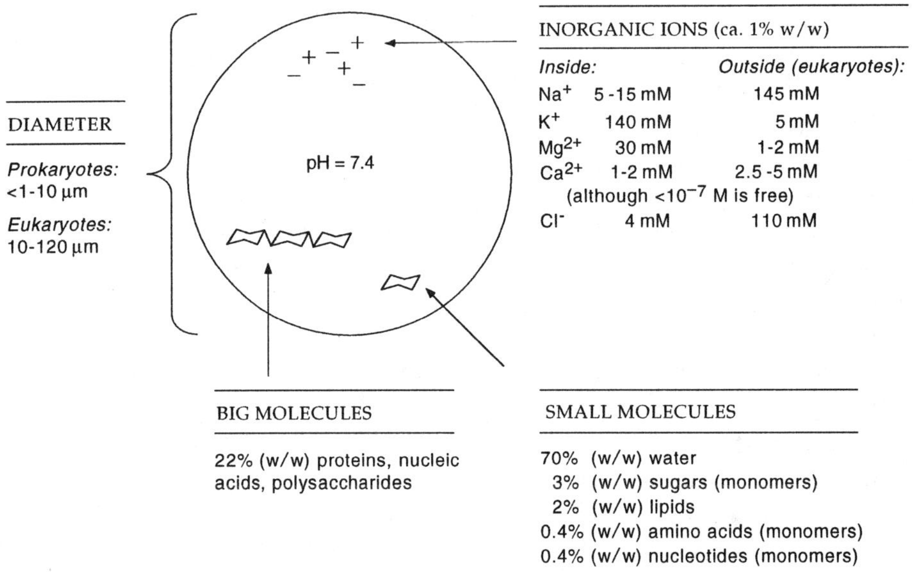

Table A.2.1 Conversion Factors

Molecular weight (ave.) of DNA base pair: 649 Da	1 kb DNA: 333 amino acids of coding capacity
Molecular weight (ave.) of amino acid: 110 Da	≈ 36,000 Da
1 μg/ml DNA: 3.08 μM phosphate	6.5×10^5 Da of double-stranded DNA (sodium salt)
1 μg/ml of 1 kb DNA: 3.08 nM 5′ ends	3.3×10^5 Da of single-stranded DNA (sodium salt)
1 μmol pBR322 (4363 bp): 2.83 g	3.4×10^5 Da of single-stranded RNA (sodium salt)
1 pmol linear pBR322 5′ ends: 1.4 μg	
1 A_{260} double-stranded DNA: 50 μg/ml	10 kDa protein ≈ 91 amino acids
1 A_{260} single-stranded DNA: 37 μg/ml	≈ 273 nucleotides

Table A.2.2 Genome Size of Various Organisms[a]

Organism	Base pairs/ haploid genome	Organism	Base pairs/ haploid genome
SV40	5,243	*Drosophila melanogaster*	1.4×10^8
ΦX174	5,386	*Gallus domesticus* (chicken)	1.2×10^9
Adenovirus 2	35,937	*Mus musculus* (mouse)	2.7×10^9
Lambda	48,502	*Rattus norvigeticus* (rat)	3.0×10^9
Escherichia coli	4.7×10^6	*Xenopus laevis*	3.1×10^9
Saccharomyces cerevisiae	1.5×10^7	*Homo sapiens*	3.3×10^9
Dictyostelium discoideum	5.4×10^7	*Zea mays*	3.9×10^9
Arabidopsis thaliana	7.0×10^7	*Nicotiana tabacum*	4.8×10^9
Caenorhabditis elegans	8.0×10^7		

[a]Genome size determined either by direct sequence analysis (viruses), electrophoretic analysis (*E. coli, S. cerevisiae*), or a combination of DNA content per cell and hybridization kinetics. Some data are from Lewin (1980).

Table A.2.3 Physical Characteristics of the Amino Acids

Amino acid	3-letter code	1-letter code	Mol. wt. (g/mol)	Accessible surface area[a]	Hydrophobicity[b]	Relative mutability[c]	Surface probability[d]
Alanine	Ala	A	89.1	115	–0.40	100	62
Arginine	Arg	R	174.2	225	–0.59	65	99
Asparagine	Asn	N	132.1	160	–0.92	134	88
Aspartate	Asp	D	133.1	150	–1.31	106	85
Cysteine	Cys	C	121.2	135	0.17	20	55
Glutamate	Glu	E	147.1	190	–1.22	102	82
Glutamine	Gln	Q	146.2	180	–0.91	93	93
Glycine	Gly	G	75.1	75	–0.67	49	64
Histidine	His	H	155.2	195	–0.64	66	83
Isoleucine	Ile	I	131.2	175	1.25	96	40
Leucine	Leu	L	131.2	170	1.22	40	55
Lysine	Lys	K	146.2	200	–0.67	56	97
Methionine	Met	M	149.2	185	1.02	94	60
Phenylalanine	Phe	F	165.2	210	1.92	41	50
Proline	Pro	P	115.1	145	–0.49	56	82
Serine	Ser	S	105.1	115	–0.55	120	78
Threonine	Thr	T	119.1	140	–0.28	97	77
Tryptophan	Trp	W	204.2	255	0.50	18	73
Tyrosine	Tyr	Y	181.2	230	1.67	41	85
Valine	Val	V	117.1	155	0.91	74	46

[a]Accessible surface area is in $Å^2$ and is for the amino acid as part of a polypeptide backbone (Chothia, 1976).

[b]Hydrophobicity is in arbitrary units and is based on the OMH scale of Sweet and Eisenberg (1983), which emphasizes the ability of amino acids to replace one another during the course of evolution.

[c]Relative mutability is also in arbitrary units (with alanine set to 100) and represents the probability that an amino acid will mutate within a given time. Thus, as two closely related proteins diverge, a given tryptophan residue is only 18% as likely as a given alanine residue to mutate (Dayhoff et al., 1978).

[d]Surface probability is the likelihood that 5% or more of the surface area of an amino acid will be exposed to the solution surrounding a protein (Chothia, 1976). Thus, while some portion of almost all the arginines will help make up the surface of a protein, less than half of the valines will be exposed to solution. To understand in more detail how amino acids are buried, see Rose et al. (1985; for example, although tyrosine is often found exposed to the surface of a protein, a substantial proportion of its surface area is typically buried).

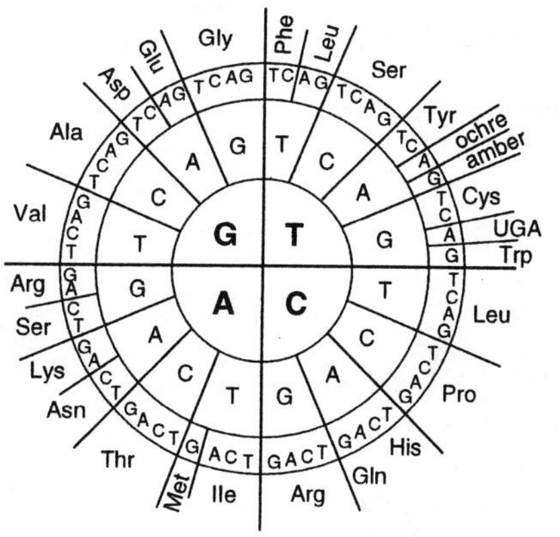

Figure A.2.2 The genetic code. Names of amino acids and chain termination codons are on the periphery of the circle. The first base of the codon is identified in the center ring; the second base of the codon is in the middle ring; and the third base(s) of the codon is in the outer ring of the circle.

A2

A

Amino acids with dissociable protons

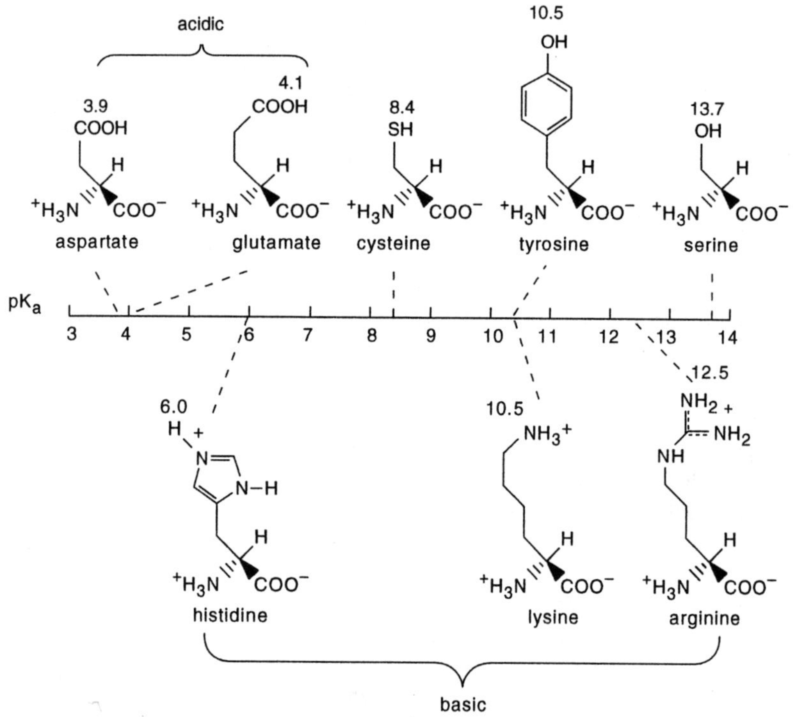

Other amino acids with polar side chains

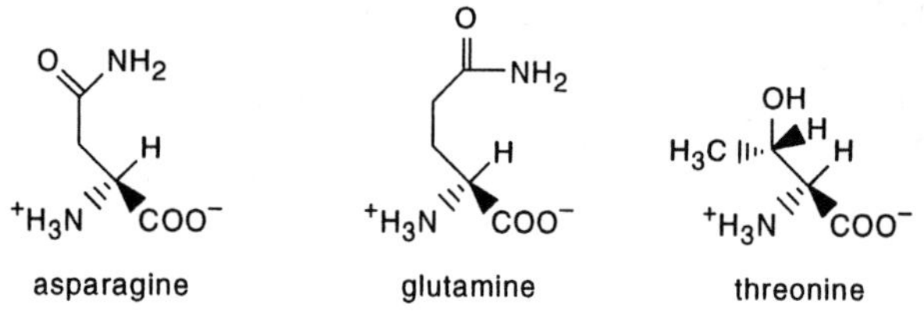

B

Nonpolar amino acids

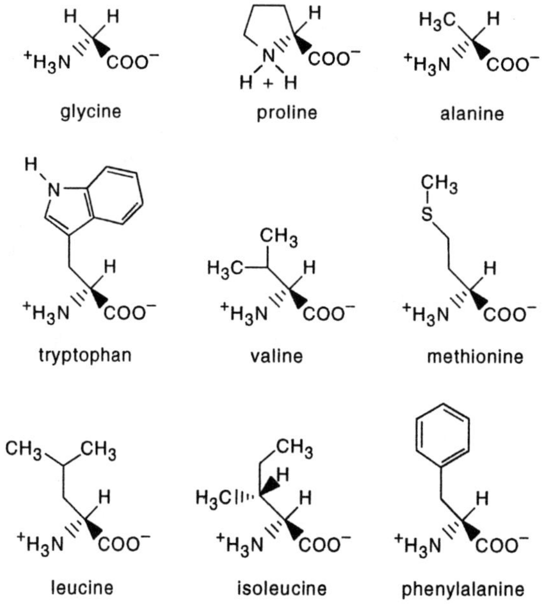

A2

Figure A.2.3 (at left) Line drawings of the amino acids. The amino acids are roughly divided into three groups: amino acids with dissociable protons, other amino acids with polar side chains, and nonpolar amino acids. These groupings are designed to facilitate an understanding of enzymology and the thermodynamics of protein folding.

In this representation, hydrogens are omitted except in showing ionization or stereochemistry. In the case of arginine the delocalized positive charge is indicated by dashed double bonds. At stereocenters, bold lines indicate a group is coming out of the page toward the viewer, while hashed lines indicate that the group goes into the page away from the viewer.

Amino acids with dissociable protons are generally intimately involved in the chemistry of enzymes. Acidic and basic groups can form salt bridges to substrates or to each other. They can also act as proton donors/acceptors in mechanisms that rely on acid/base catalysis. The polar side chains of some of these amino acids (notably cysteine, serine and histidine) can act as nucleophiles. The pK_a values for the free amino acids are shown, but these values can markedly change when these groups are buried in proteins. The pK_as of the α-amino groups range from 8.7 to 10.7, while the pK_as of the α-carboxylates range from 1.8 to 2.4.

Amino acids with polar side chains can form hydrogen bonds to substrates or to each other. Cysteine, serine, and tyrosine could also be included in this group, since the ionized forms of these amino acids do not generally perform structural roles in proteins. In general, these amino acids (and the amino acids with dissociable protons) will be found on the surfaces of proteins. Cysteine is an exception, since it is slightly hydrophobic and can often be buried as a disulfide bond.

The nonpolar amino acids are often found in the interiors of proteins or in hydrophobic substrate-binding pockets. They interact with one another like jigsaw pieces, forming tight-fitting associations that have a density similar to that of an amino acid crystal. Proline is buried less frequently than might be expected because of its predominance in turns, which are often found on the periphery of a protein.

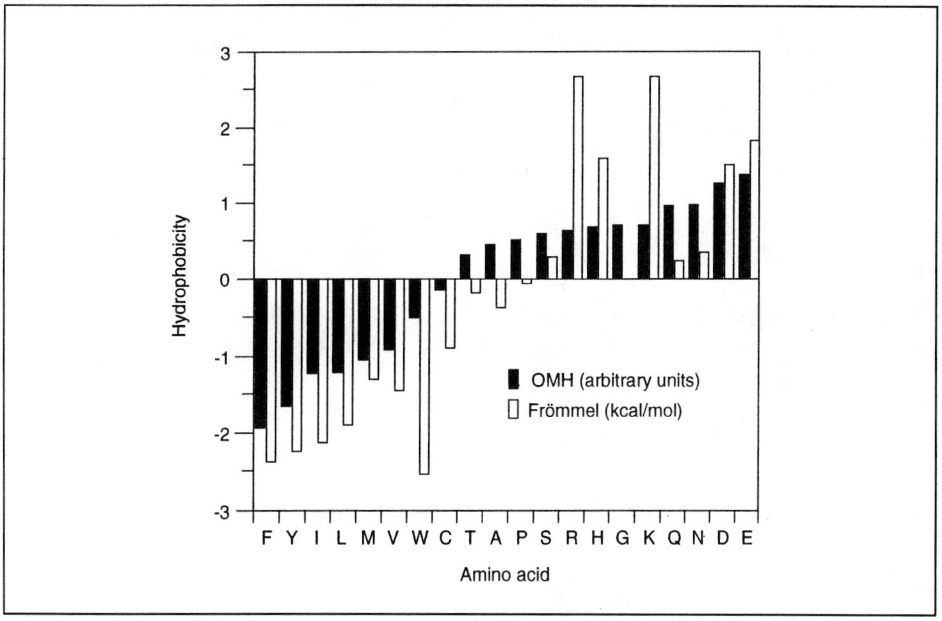

Figure A.2.4 Amino acid hydrophobicity. The hydrophobicity of an amino acid is the degree to which it prefers a nonpolar medium, such as ethanol or the interior of a protein, to a polar medium, such as water. In this graph, the more hydrophobic amino acids "sink" below zero, while the more hydrophilic amino acids "float" above the surface.

Two scales are used: the Frömmel scale (Frömmel, 1984) represents the free energy of transfer from a hydrophobic medium to water. This value is an intrinsic property of an amino acid, separate from its role in a protein. In contrast, the OMH scale (Sweet and Eisenberg, 1983) is a measure of how likely a given amino acid will be replaced by a different hydrophobic or "buried" amino acid in a protein. In effect, this scale is how evolution views the hydrophobicity of an amino acid.

The distinction between physical and evolutionary properties is important. For example, while arginine is definitely a charged, polar amino acid (Sambrook et al., 1989), it can substitute more freely for nonpolar amino acids in the interior of a protein than glutamate (also a charged, polar amino acid) because of its long aliphatic side chain.

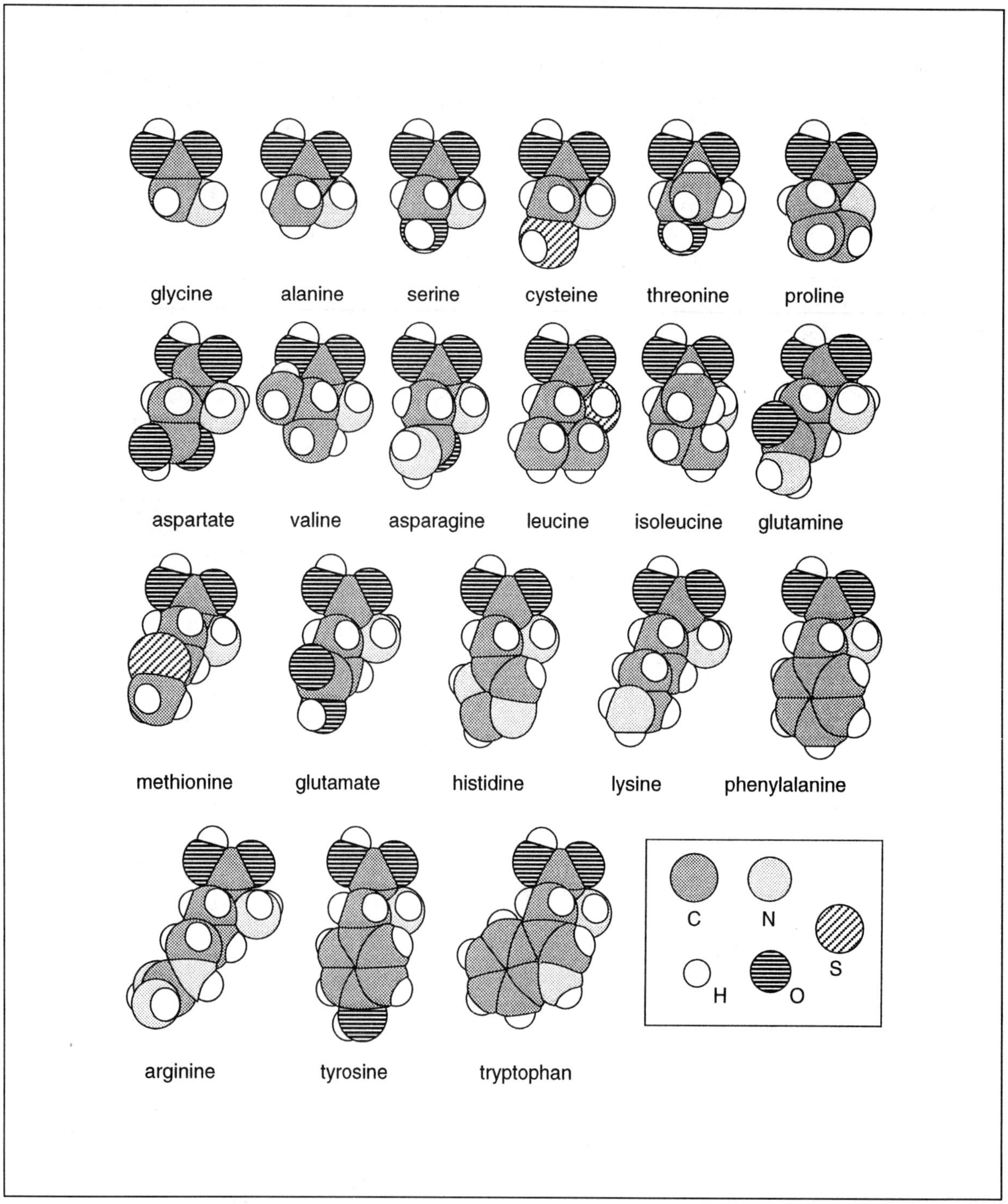

Figure A.2.5 Space-filling representations of the amino acids. The amino acids are arranged in order of size. The conformations shown maximize the two-dimensional area but are not necessarily the most stable geometries.

A2

Characteristics of Nucleic Acids

Table A.2.4 Physical Characteristics of the Nucleotides[a]

Nucleotide	Mol. wt. (g/mol)	λ_{max} (nm)	λ_{min} (nm)	ε_{max} (mM^{-1} cm^{-1})	A_{280}/A_{260}	TLC mobility[b]		
						A	B	C
ATP	507.2	259	227	15.4	0.15	0	6	34
ADP	427.2	259	227	15.4	0.16	0	26	54
AMP	347.2	259	227	15.4	0.16	11	52	65
Adenosine[c]	267.2	260	227	14.9	0.14	—	—	—
dATP[d]	491.2	259	226	15.4	0.15	0	—	35
dAMP[d]	331.2	259	226	15.2	0.15	11	52	—
dA	251.2	260	225	15.2	0.15	—	—	—
CTP	483.2	271	249	9.0	0.97	0	11	41
CDP	403.2	271	249	9.1	0.98	0	33	64
CMP	323.2	271	249	9.1	0.98	15	64	75
Cytidine	243.2	271	250	9.1	0.93	—	—	—
dCTP[d]	467.2	272	—	9.1	0.98	0	—	43
dCMP	307.2	271	249	9.3	0.99	18	65	—
dC	227.2	271	250	9.0	0.97	—	—	—
GTP	523.2	253	223	13.7	0.66	0	5	25
GDP	443.2	253	224	13.7	0.66	0	17	45
GMP	363.2	252	224	13.7	0.66	6	40	51
Guanosine[c]	283.2	253	223	13.6	0.67	—	—	—
dGTP[d]	507.2	252	222	13.7	0.66	0	—	26
dGMP[d]	347.2	253	222	13.7	0.67	6	41	—
dG	267.2	254	223	13.0	0.68	—	—	—
UTP	484.2	262	230	10.0	0.38	0	14	49
UDP	404.2	262	230	10.0	0.39	0	41	71
UMP	324.2	262	230	10.0	0.39	20	75	80
Uridine	244.2	262	230	10.1	0.35	—	—	—
TTP[d]	482.2	267	—	9.6	0.73	0	—	52
TMP[d]	322.2	267	234	9.6	0.73	24	74	—
Thymidine[d]	242.2	267	235	9.7	0.70	—	—	—

[a] Spectral data are assembled from Fasman (1975) at pH 7.0 except where footnoted otherwise.

[b] TLC mobility is expressed as the percent distance a given spot migrates relative to the solvent front (*Rf*) in three different TLC systems using 0.5-mm polyethylenimine cellulose plates: "A" is 0.25 M LiCl, "B" is 1.0 M LiCl, and "C" is 1.6 M LiCl.

[c] Spectral measurements taken at pH 6.0.

[d] Spectral data assembled from Dawson et al. (1987).

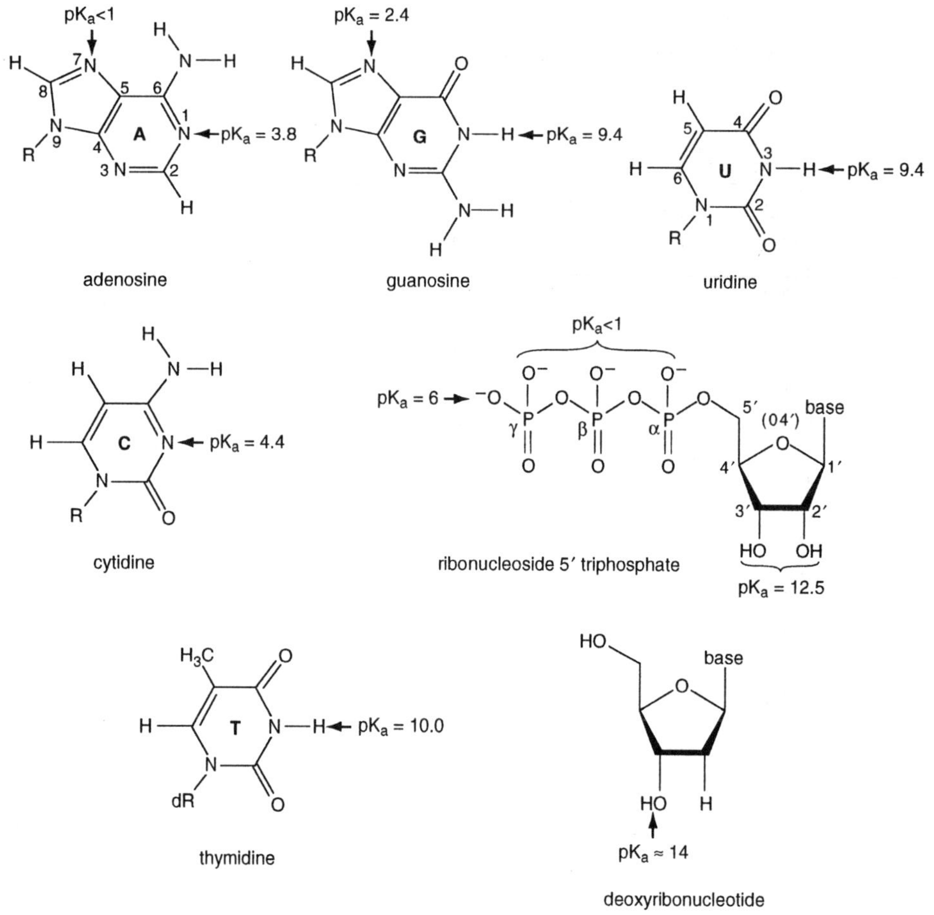

Figure A.2.6 Line drawings of the nucleotides. The chemical structure that predominates at neutral pH is shown. Drawings of the nucleotide bases and their associated sugars, either ribose (R) or deoxyribose (dR), are shown separately. In the representations of ribose (as a nucleoside triphosphate) and deoxyribose (as a nucleotide), the bold lines indicate that this portion of the sugar is coming out of the page toward the reader. In this view, the base is found above the plane of the sugar, while the 3′ hydroxyl group is found below the plane of the sugar.

The pK_a values for all groups are shown; pK_as above 7 imply proton dissociation from the pictured structure, while pK_as below 7 imply proton association to the pictured structure. The tautomeric form of a given base may change at different pH values. The pK_a values given are for nucleotide monophosphates and were taken from Dawson et al. (1987); a fuller discussion of the chemical basis for these values can be found in a review by T'so (1974).

The small numbers adjacent to adenosine, uridine, and ribose indicate the nomenclature of the purines, pyrimidines, and sugars, respectively. Groups appended to a ring have the same numbering as the position to which they are linked; thus, the "O6" moiety of guanosine is the carbonyl oxygen bonded to C6 in the ring. Similarly, "O3′" on ribose or deoxyribose indicates the oxygen of the hydroxyl group bonded to C3′ in the ring. The α, β, and γ phosphates in a nucleoside triphosphate are also indicated.

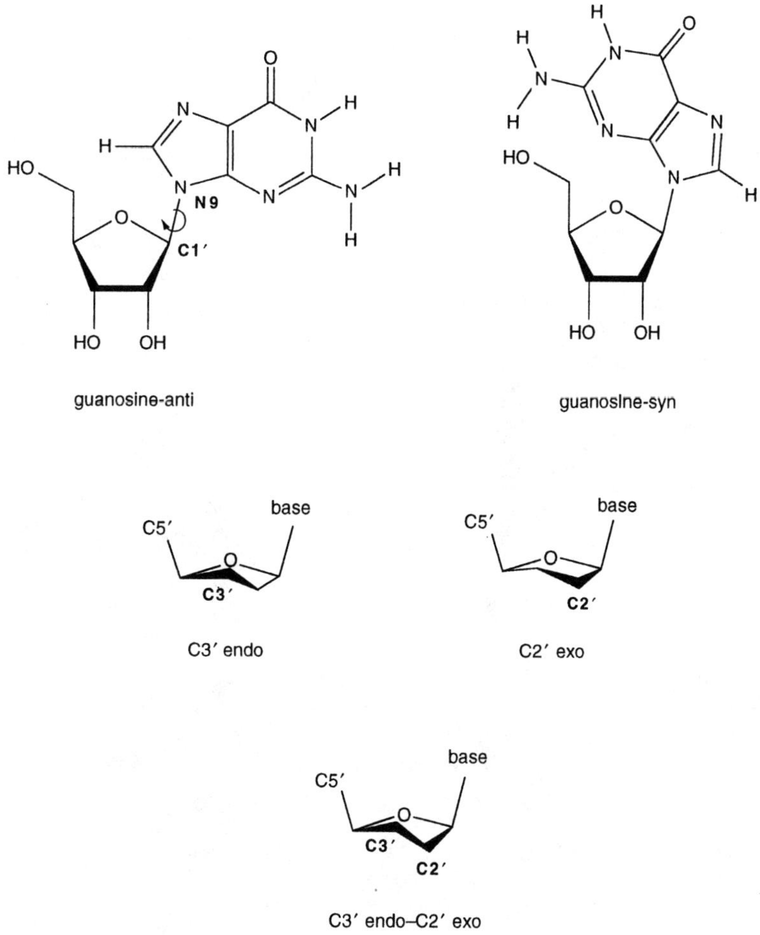

Figure A.2.7 Nucleotide stereochemistry. Depending on the rotation about the bond between C1′ of the sugar and either N1 (for pyrimidines) or N9 (for purines), a nucleotide can be described as either "anti" or "syn." Because of steric constraints, nucleotides are generally found in the "anti" configuration, with their Watson-Crick hydrogen bond donor-acceptors swung outward away from the plane of the sugar ring. However, guanosine is sometimes found in a "syn" configuration, both in polynucleotides and in solution. In this form, the bulk of the purine ring is positioned directly over the plane of the sugar. The sugar ring can also adopt different stereochemistries. These are labeled according to which group is bent out of the plane of the ring, and in which direction. If a portion of the ring is bent "upward" toward the base, this is known as "endo," while if it is bent "downward" away from the base, this is known as "exo." In the figure, plain lines represent bonds that are within the plane of the sugar, while bold lines indicate that the bond is bent out of the plane. Hence, "C3′ endo–C2′ exo" describes a furanose ring in which the 2′ and 3′ carbons have been twisted in opposite directions and the bond connecting them crosses the plane of the ring.

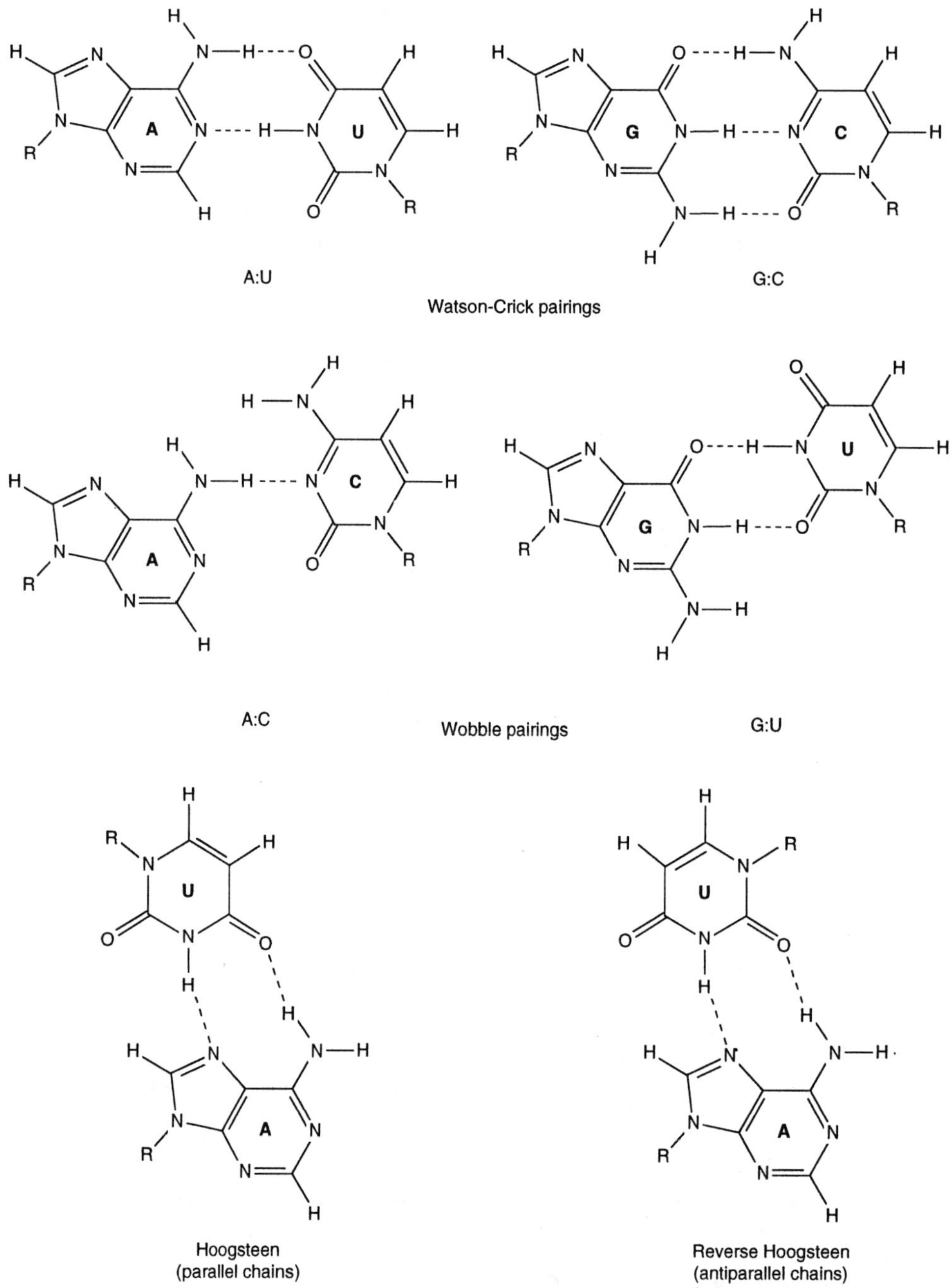

Figure A.2.8 Base pairing schemas. The chemical structures of the nucleotide bases determine the formation of secondary and tertiary structures in nucleic acids. A wide variety of hydrogen bonding schemas (indicated by dashed lines) are possible between different bases. Watson-Crick pairings are perhaps the most widely known and are the basis of the double helical structure of complementary, anti-parallel DNA strands. Other base pairs can also be accommodated within the double helix, such as "wobble pairings," in which the bases are slightly off-center with respect to each other. By using the N7 hydrogen bond acceptor of the purine bases adenosine and guanosine, an even wider variety of structures becomes possible, including Hoogsteen base pairs and a G-G pairing in which one of the guanosine residues assumes a "syn" conformation. Bonds involving N7 of the purine bases allow tertiary structural interactions to occur in nucleic acids, including triple base pairs (such as those found in tRNA) and the recently described "G quartet" (Sen and Gilbert, 1988). A discussion of the structural possibilities of base pairing can be found in Saenger's superlative book, *Principles of Nucleic Acid Structure* (1984).

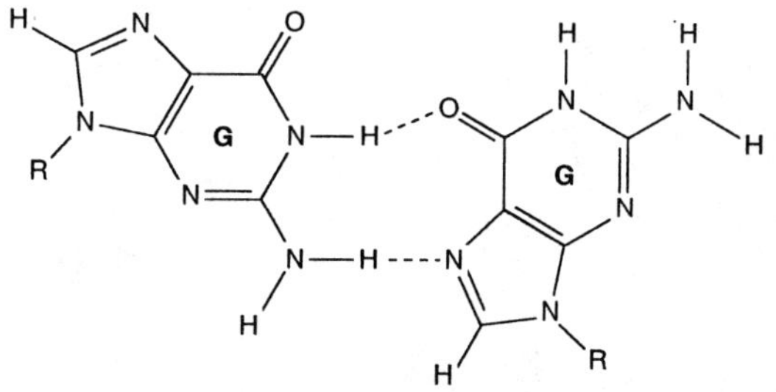

G-anti:G-syn (antiparallel chains)

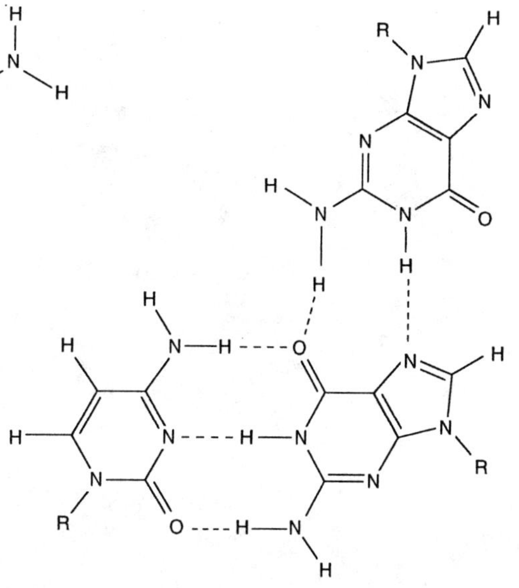

C:G—G (as in tRNA Phe)

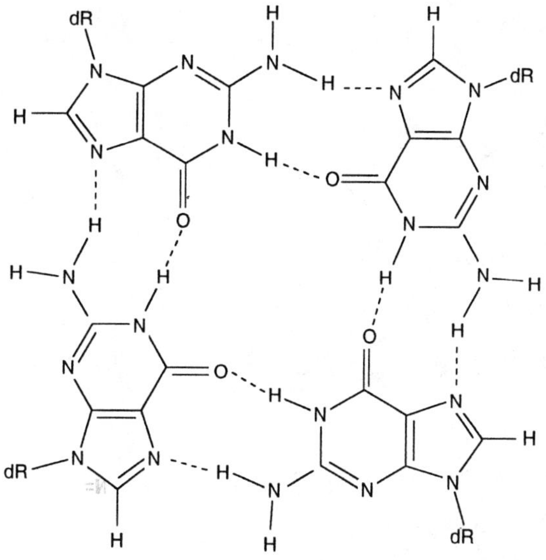

"G quartet"

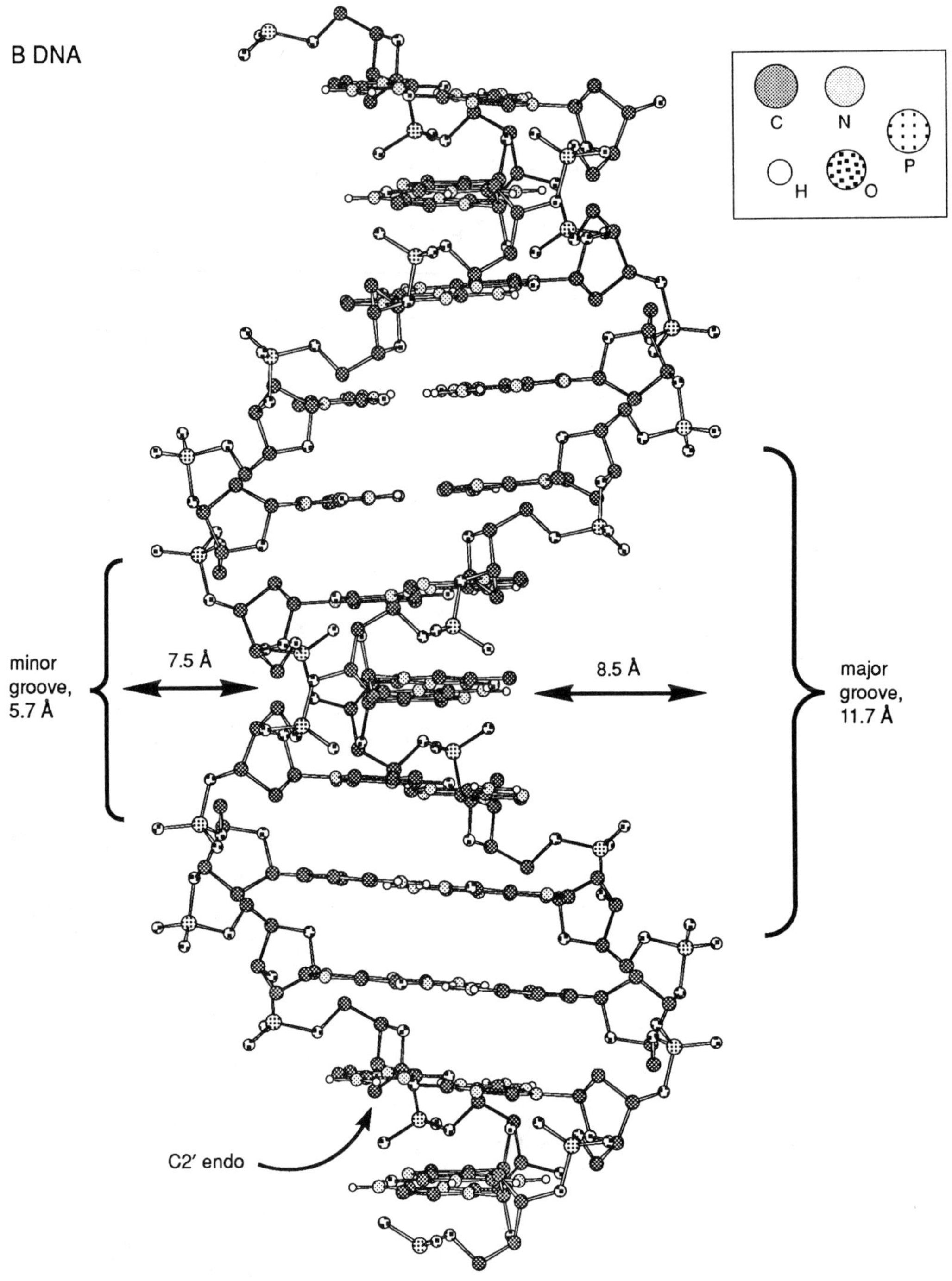

Figure A.2.9 Nucleic acid secondary structures. The structural consequence of the ability of nucleotides to form Watson-Crick base pairs is nucleic acid double helices. In this figure, the self-complementary 12-mer CGCGAATTCGCG is shown as both A- and B-form helices. Two representations of the A helix have been shown in order to emphasize the depth of the major groove. The arrows and brackets in these figures are not drawn to scale.

While both of these helices are right-handed (in terms of anthropomorphic referents, if you were to point your thumb along a strand in a 5′ to 3′ manner, the twist of the helix would be the same as the curl of your right hand), their structural details are very different: B DNA has roughly 10 bases per full turn, while A DNA and A RNA have 11 to 12; the major groove of B-form helices is wide and the minor groove is narrow, while for A-form helices this is reversed; in B-forms the base pairs are located close to the helix axis (as can be seen in end-on views), while in A-forms the base pairs are pushed out away from the long helical axis, leaving a "hole" in the middle of the polynucleotide coil (if one imagines DNA as a flat ribbon, then B DNA is twisted from its ends, while A DNA is coiled on itself).

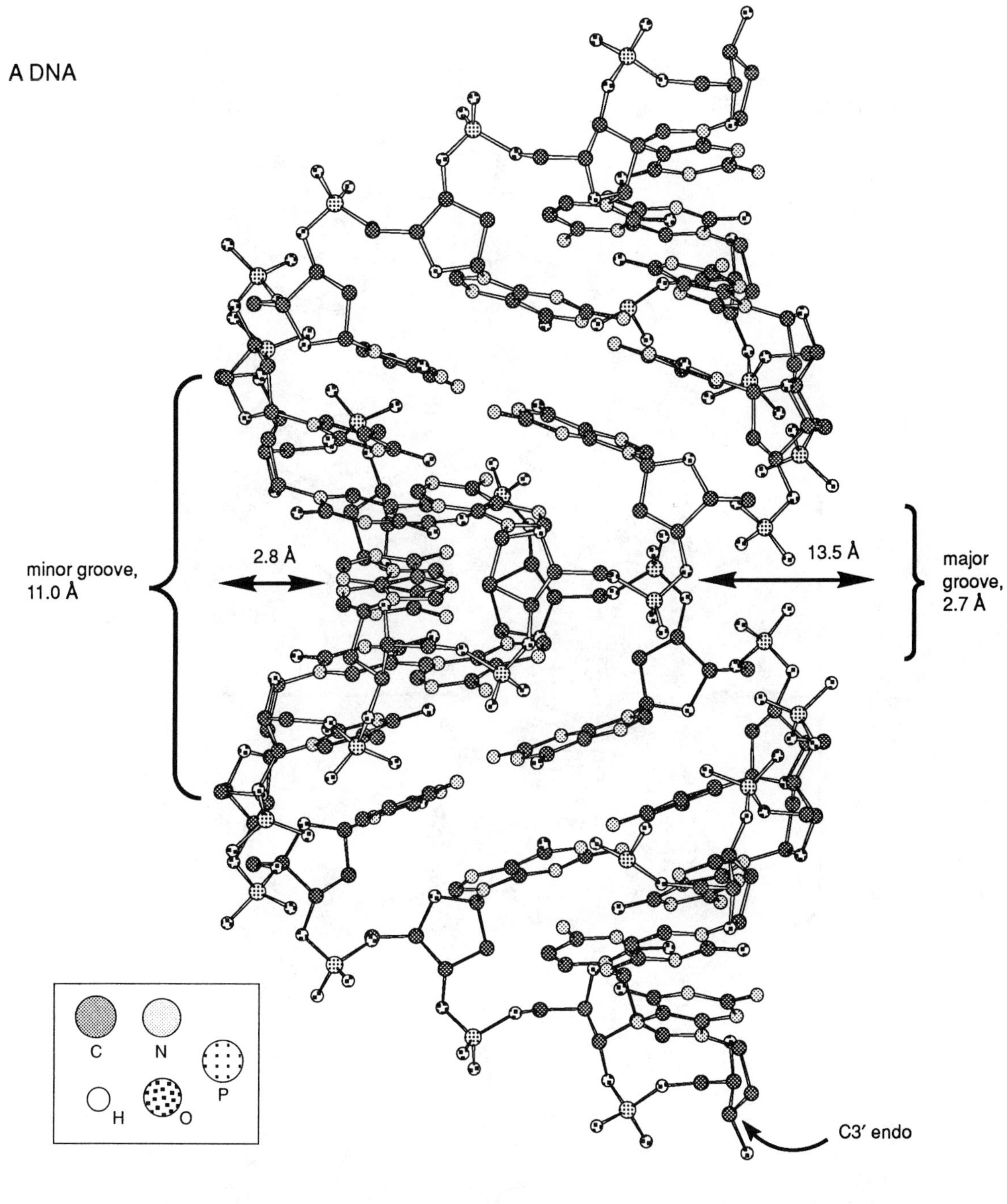

Different helical forms are largely due to differences in sugar stereochemistry. Examples of a 2′ endo deoxyribose (found in B DNA) and a 3′ endo deoxyribose (found in A DNA) are indicated.

While there are a variety of other helical forms, the most striking is that found in Z DNA. The Z DNA coil is left- rather than right-handed and contains G:C base pairs where the G is in the "syn" conformation (shown in the inset).

The uneven progression, or zigzag, of Z DNA can be more easily seen when the polynucleotide backbone is shown in isolation; the inset shows the connectivity between phosphates by 5′ to 3′ vector arrows. Because of its odd shape, base pairs actually protrude from what would be a cavity in A or B DNA; thus, Z DNA has a minor but no major groove. This diagram is based on the original structure of alternating C:G/G:C base pairs (Wang et al., 1979).

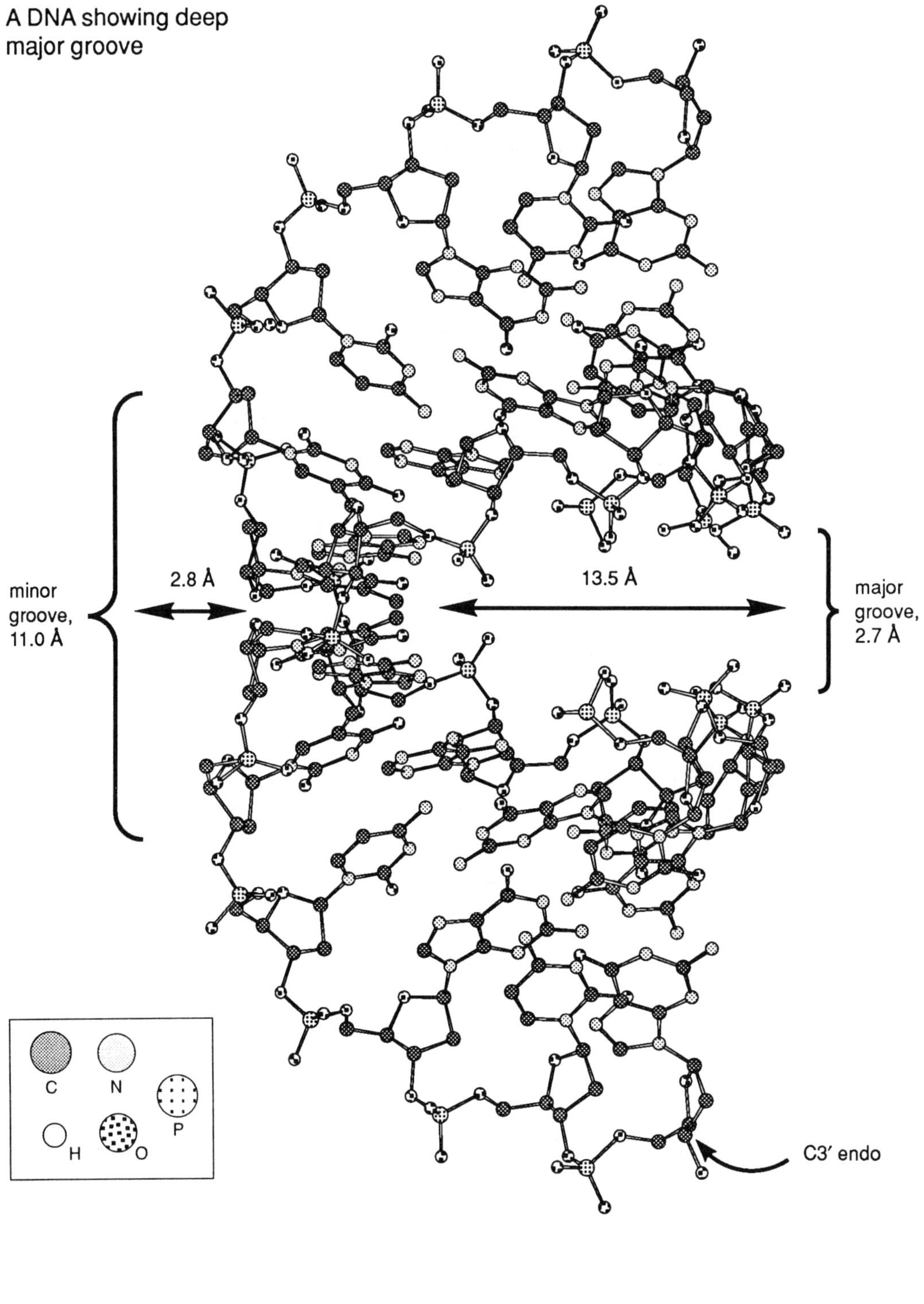
A DNA showing deep
major groove
2.8 Å
13.5 Å
minor
groove,
11.0 Å
major
groove,
2.7 Å
C
N
P
H
O
C3′ endo

A2

B form

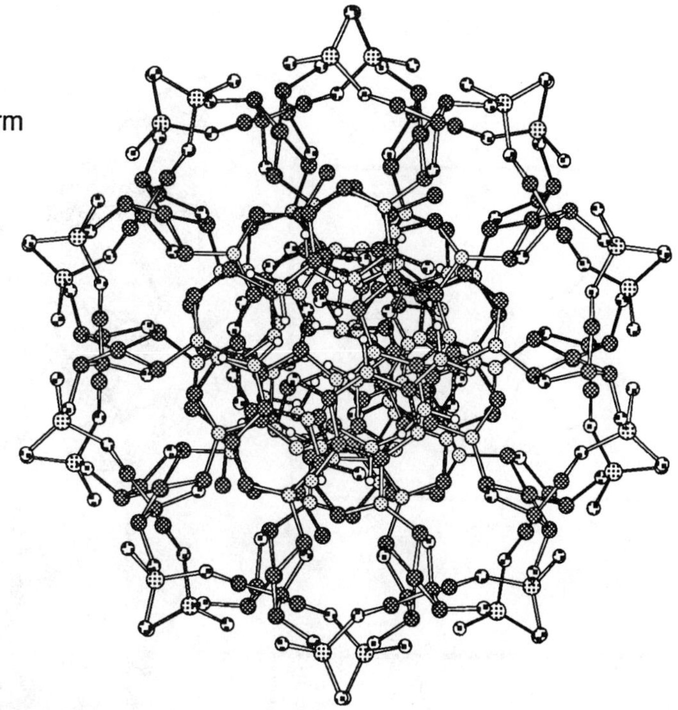

A form

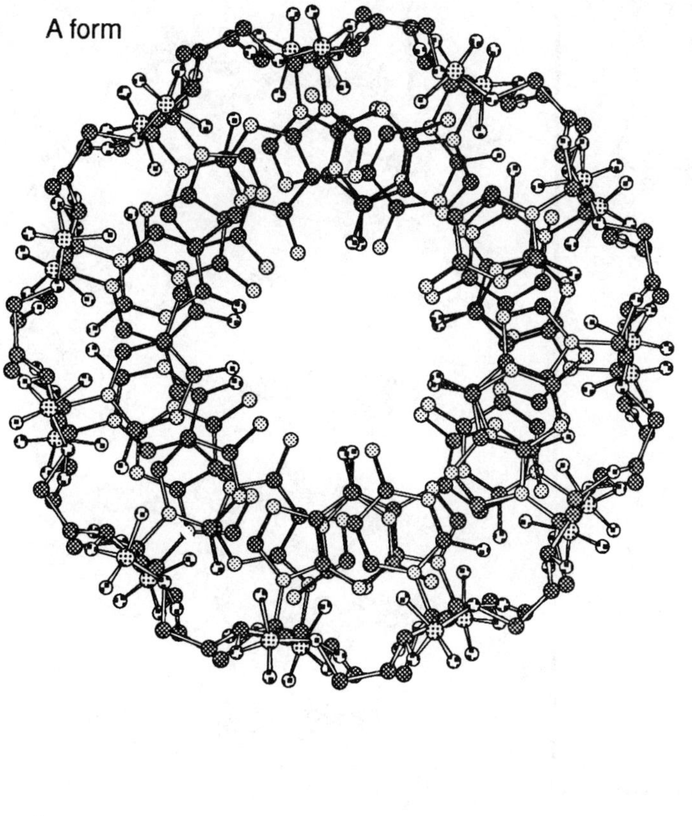

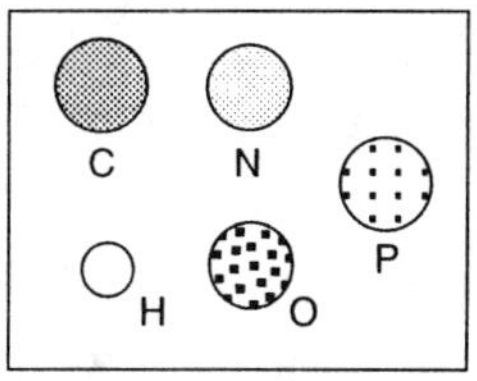

Z DNA

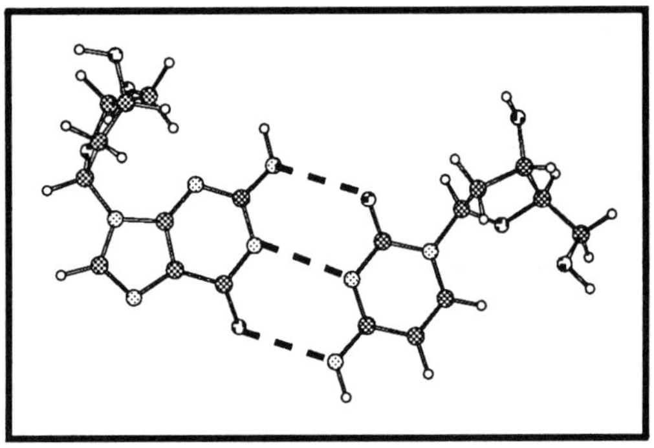

G-syn:C-anti

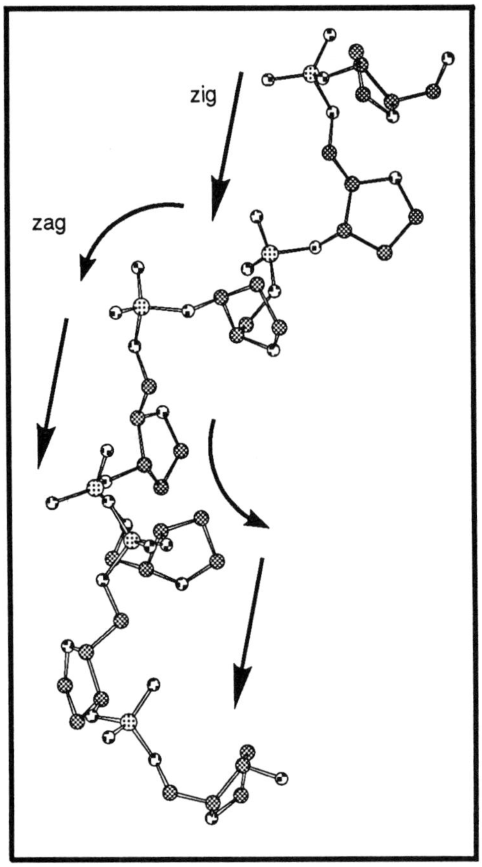

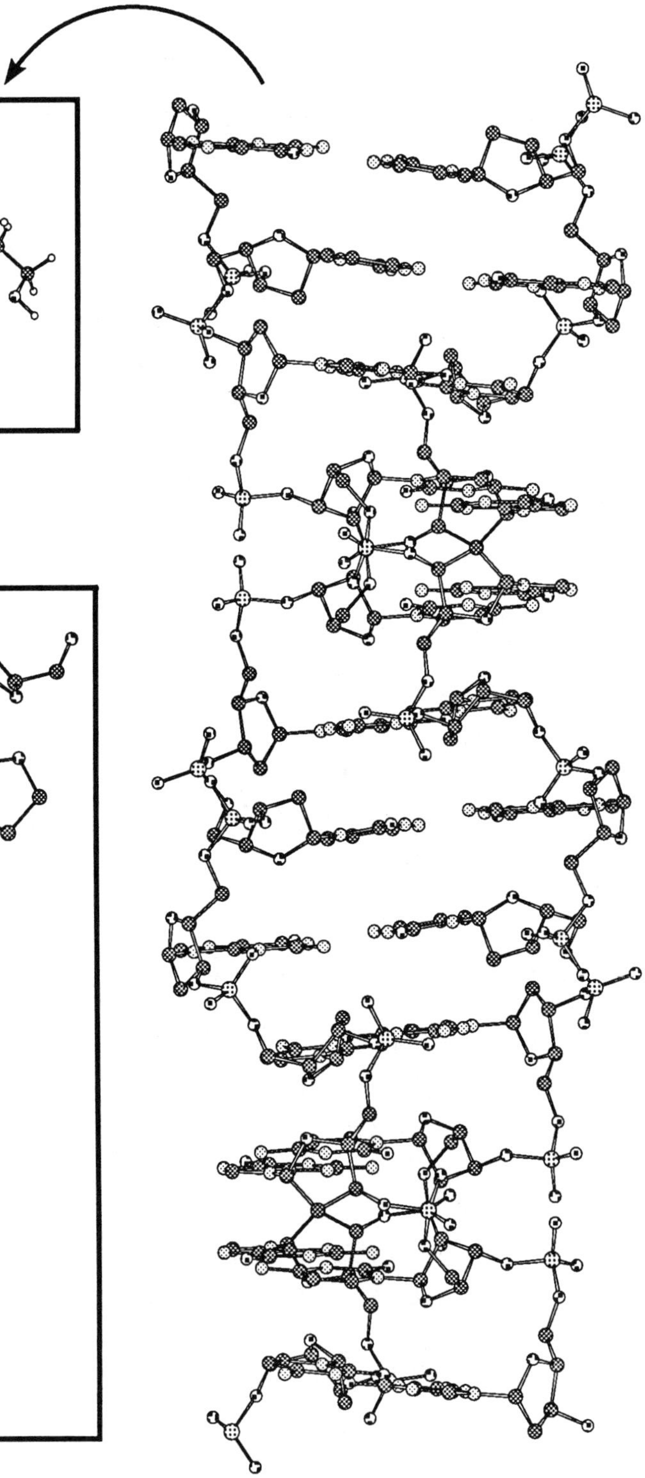

Radioactivity

Table A.2.5 Physical Characteristics of Commonly Used Radionuclides[a]

Nuclide	Half-life	Emission	Energy, max (MeV)	Range of emission, max	Approx. specific activity at 100% enrichment (Ci/mg)	Atom resulting from decay	Target organ
^{3}H	12.43 years	β	0.0186	0.42 cm (air)	9.6	$^{3}_{2}He$	Whole body
^{14}C	5370 years	β	0.156	21.8 cm (air)	4.4 mCi/mg	$^{14}_{7}N$	Bone, fat
^{32}P[b]	14.3 days	β	1.71	610 cm (air) 0.8 cm (water) 0.76 cm (Plexiglas)	285	$^{33}_{16}S$	Bone
^{33}P[b]	25.4 days	β	0.249	49 cm	156	$^{33}_{16}S$	Bone
^{35}S	87.4 days	β	0.167	24.4 cm (air)	43	$^{35}_{17}Cl$	Testes
^{125}I[c]	60 days	γ	0.27–0.035	0.2 mm (lead)	14.2	$^{125}_{52}Te$	Thyroid
^{131}I[c]	8.04 days	β	0.606	165 cm (air)	123	$^{130}_{54}Xe$	Thyroid
		γ	0.364	2.4 cm (lead)			

[a]Table compiled based on information in Lederer et al. (1967) and Shleien (1987).
[b]Recommended shielding is Plexiglas; half-value layer measurement is 1 cm.
[c]Recommended shielding is lead; half-value layer measurement is 0.02 mm.

Table A.2.6 Shielding Radioactive Emission[a]

β emitters

Energy (MeV)	Mass (mg)/cm^2 to reduce intensity by 50%	Thickness (mm) to reduce intensity by 50%			
		Water	Glass	Lead	Plexiglas
0.1	1.3	0.013	0.005	0.0011	0.0125
1.0	48	0.48	0.192	0.042	0.38
2.0	130	1.3	0.52	0.115	1.1
5.0	400	4.0	1.6	0.35	4.2

γ emitters

Energy (MeV)	Thickness of material (cm) to attenuate a broad beam of γ-rays by a factor of 10			
	Water	Aluminum	Iron	Lead
0.5	54.6	20.3	6.1	1.8
1.0	70.0	24.4	8.2	3.8
2.0	76.0	32.0	11.0	5.9
3.0	89.0	37.0	12.0	6.4

[a]From Dawson et al. (1986). Reprinted with permission.

Table A.2.7 Decay factors for calculating the amount of radioactivity present at a given time after a reference date. For example, a vial containing 1.85 MBq (50 μCi) of an ^{35}S-labeled compound on the reference date will have the following activity 33 days later: 1.85 × 0.770 = 1.42 MBq; 50 × 0.770 = 38.5 μCi.

^{125}I Half-life: 60.0 days

Days ↓ / Days →	0	2	4	6	8	10	12	14	16	18
0	1.000	0.977	0.955	0.933	0.912	0.891	0.871	0.851	0.831	0.812
20	0.794	0.776	0.758	0.741	0.724	0.707	0.691	0.675	0.660	0.645
40	0.630	0.616	0.602	0.588	0.574	0.561	0.548	0.536	0.524	0.512
60	0.500	0.489	0.477	0.467	0.456	0.445	0.435	0.425	0.416	0.406
80	0.397	0.388	0.379	0.370	0.362	0.354	0.345	0.338	0.330	0.322
100	0.315	0.308	0.301	0.294	0.287	0.281	0.274	0.268	0.262	0.256
120	0.250	0.244	0.239	0.233	0.228	0.223	0.218	0.213	0.208	0.203
140	0.198	0.194	0.189	0.185	0.181	0.177	0.173	0.169	0.165	0.161
160	0.157	0.154	0.150	0.147	0.144	0.140	0.137	0.134	0.131	0.128
180	0.125	0.122	0.119	0.117	0.114	0.111	0.109	0.106	0.104	0.102
200	0.099	0.097	0.095	0.093	0.090	0.088	0.086	0.084	0.082	0.081
220	0.079	0.077	0.075	0.073	0.072	0.070	0.069	0.067	0.065	0.064
240	0.063	0.061	0.060	0.058	0.057	0.056	0.054	0.053	0.052	0.051

^{32}P Half-life: 14.3 days

Days ↓ / Hours →	0	12	24	36	48	60	72	84
0	1.000	0.976	0.953	0.930	0.908	0.886	0.865	0.844
4	0.824	0.804	0.785	0.766	0.748	0.730	0.712	0.695
8	0.679	0.662	0.646	0.631	0.616	0.601	0.587	0.573
12	0.559	0.546	0.533	0.520	0.507	0.495	0.483	0.472
16	0.460	0.449	0.439	0.428	0.418	0.408	0.398	0.389
20	0.379	0.370	0.361	0.353	0.344	0.336	0.328	0.320
24	0.312	0.305	0.298	0.291	0.284	0.277	0.270	0.264
28	0.257	0.251	0.245	0.239	0.234	0.228	0.223	0.217
32	0.212	0.207	0.202	0.197	0.192	0.188	0.183	0.179
36	0.175	0.170	0.166	0.162	0.159	0.155	0.151	0.147
40	0.144	0.140	0.137	0.134	0.131	0.127	0.124	0.121
44	0.119	0.116	0.113	0.110	0.108	0.105	0.102	0.100
48	0.098	0.095	0.093	0.091	0.089	0.086	0.084	0.082
52	0.080	0.078	0.077	0.075	0.073	0.071	0.070	0.068

^{131}I Half-life: 8.04 days

Days ↓ / Hours →	0	6	12	18	24	30	36	42	48	54	60	66
0	1.000	0.979	0.958	0.937	0.917	0.898	0.879	0.860	0.842	0.824	0.806	0.789
3	0.772	0.756	0.740	0.724	0.708	0.693	0.678	0.664	0.650	0.636	0.622	0.609
6	0.596	0.583	0.571	0.559	0.547	0.533	0.524	0.513	0.502	0.491	0.481	0.470
9	0.460	0.450	0.441	0.431	0.422	0.413	0.405	0.396	0.387	0.379	0.371	0.363
12	0.355	0.348	0.340	0.333	0.326	0.319	0.312	0.306	0.299	0.293	0.286	0.280
15	0.274	0.269	0.263	0.257	0.252	0.246	0.241	0.236	0.231	0.226	0.221	0.216
18	0.212	0.207	0.203	0.199	0.194	0.190	0.186	0.182	0.178	0.175	0.171	0.167
21	0.164	0.160	0.157	0.153	0.150	0.147	0.144	0.141	0.138	0.135	0.132	0.129
24	0.126	0.124	0.121	0.118	0.116	0.113	0.111	0.109	0.106	0.104	0.102	0.100
27	0.098	0.095	0.093	0.091	0.089	0.088	0.086	0.084	0.082	0.080	0.079	0.077
30	0.075	0.074	0.072	0.071	0.069	0.068	0.066	0.065	0.064	0.063	0.061	0.059
33	0.058	0.057	0.056	0.054	0.053	0.052	0.051	0.050	0.049	0.048	0.047	0.046
36	0.045	0.044	0.043	0.042	0.041	0.040	0.039	0.039	0.038	0.037	0.036	0.035

^{35}S Half-life: 87.4 days

Weeks ↓ / Days →	0	1	2	3	4	5	6
0	1.000	0.992	0.984	0.976	0.969	0.961	0.954
1	0.946	0.939	0.931	0.924	0.916	0.909	0.902
2	0.895	0.888	0.881	0.874	0.867	0.860	0.853
3	0.847	0.840	0.833	0.827	0.820	0.814	0.807
4	0.801	0.795	0.788	0.782	0.776	0.770	0.764
5	0.758	0.752	0.746	0.740	0.734	0.728	0.722
6	0.717	0.711	0.705	0.700	0.694	0.689	0.683
7	0.678	0.673	0.667	0.662	0.657	0.652	0.646
8	0.641	0.636	0.631	0.626	0.621	0.616	0.612
9	0.607	0.602	0.597	0.592	0.588	0.583	0.579
10	0.574	0.569	0.565	0.560	0.556	0.552	0.547
11	0.543	0.539	0.534	0.530	0.526	0.522	0.518
12	0.514	0.510	0.506	0.502	0.498	0.494	0.490

^{33}P Half-life: 25.4 days

Days ↓ / Days →	0	1	2	3	4	5	6	7	8	9
0	1.000	0.973	0.947	0.921	0.897	0.872	0.849	0.826	0.804	0.782
10	0.761	0.741	0.721	0.701	0.683	0.664	0.646	0.629	0.612	0.595
20	0.579	0.564	0.549	0.534	0.520	0.506	0.492	0.479	0.466	0.453
30	0.441	0.429	0.418	0.406	0.395	0.385	0.374	0.364	0.355	0.345
40	0.336	0.327	0.318	0.309	0.301	0.293	0.285	0.277	0.270	0.263
50	0.256	0.249	0.242	0.236	0.229	0.223	0.217	0.211	0.205	0.200
60	0.195	0.189	0.184	0.179	0.174	0.170	0.165	0.161	0.156	0.152
70	0.148	0.144	0.140	0.136	0.133	0.129	0.126	0.122	0.119	0.116
80	0.113	0.110	0.107	0.104	0.101	0.098	0.096	0.093	0.091	0.088
90	0.086	0.084	0.081	0.079	0.077	0.075	0.073	0.071	0.069	0.067
100	0.065	0.064	0.062	0.060	0.059	0.057	0.055	0.054	0.053	0.051
110	0.050	0.048	0.047	0.046	0.045	0.043	0.042	0.041	0.040	0.039
120	0.038	0.037	0.036	0.035	0.034	0.033	0.032	0.031	0.030	0.030

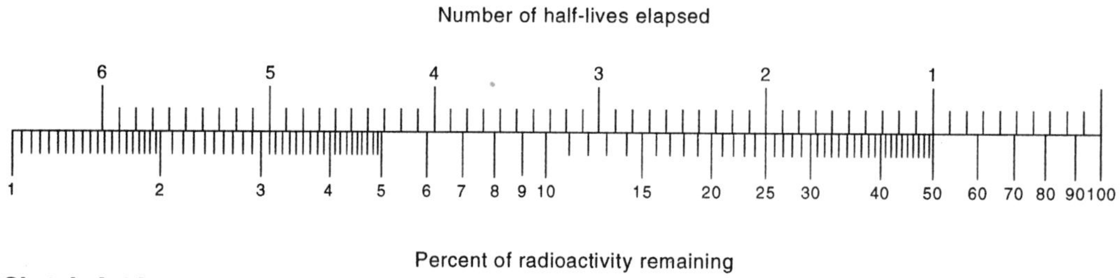

Sketch A.1A

Centrifuges and Rotors

Table A.2.8 Low-Speed Rotor Conversion Values[a]

Rotor type[a]	Relative centrifugal field (× g) at r_{max}[b]	Radial distance (mm)		
		r_{max}	r_{av}	r_{min}
JA-21	50,400	102	73	45
JA-20.1	51,500 (outer row)	115	89	64
	44,000 (inner row)	98	73	47
JA-20	48,400	108	70	32
JA-17	39,700	123	90	56
JA-14	30,100	137	86	35
JA-10	17,700	158	98	38
JS-13	26,750	141	92	42
JS-4.2	5,010	254	184	114
Microtest plate carrier for JSA-4.2 (Beckman #341803)[c]	1,500 (r_{max}) 1,400 (r_{av}) 1,300 (r_{min})	214	200	186

[a]All rotors from Beckman Instruments of SmithKline Beckman.

[b]Relative centrifugal field (RCF) values represent centrifugal acceleration relative to standard acceleration of gravity ($g = 9807$ mm/sec^2) at maximum rotor speed. The strength of the centrifugal field is proportional to the square of the rotor speed, in rpm. At lower speeds, RCF can be determined using accompanying nomograms, or calculated as: RCF = $1.12\ r(\text{rpm}/1000)^2$, where r = radius in mm, rpm = revolutions per minute.

CAUTION: *Do not exceed maximum rotor speed!* The maximum speed for each rotor is denoted by its name, i.e., the maximum speed of the JA-21 is 21,000 rpm. This speed refers only to centrifugation of solutions below a particular allowed density, which differs among rotors (see user manual).

[c]The microtest plates can be centrifuged in stacks of three, with slightly different RCFs.

A2

Table A.2.9 High-Speed Rotor Conversion Values[a]

Rotor type[a]	Relative centrifugal field (× g) at r_{max}[b]	Radial distance (mm)		
		r_{max}	r_{av}	r_{min}
Vertical tube				
VTi80	510,000	71.1	64.5	57.9
VTi65.2	416,000	87.9	81.2	74.7
VTi65	404,000	85.4	78.7	72.1
VTi50	242,000	86.6	73.7	60.8
VTi45	202,000	88.9	64.9	50.8
Swinging bucket				
SW65Ti	421,000	89.0	65.1	41.2
SW60Ti	485,000	120.3	91.7	63.1
SW55Ti	368,000	108.5	84.6	60.8
SW50.1	300,000	107.3	83.5	59.7
SW41Ti	288,000	153.1	110.2	67.4
SW40Ti	285,000	158.8	112.7	66.7
SW30/30.1	124,000	123.0	99.2	75.3
SW28.1	150,000	171.3	122.1	72.9
SW28	141,000	161	118.2	75.3
SW25.1	90,400	129.2	92.7	56.2
Fixed angle				
80Ti	602,000	84.0	62.5	41.0
75Ti	502,000	79.7	58.3	36.9
70.1Ti	450,000	82.0	61.2	40.5
70Ti	504,000	91.9	65.7	39.5
65	368,000	77.8	57.3	36.8
60Ti	362,000	89.9	63.4	36.9
55.2Ti	340,000	100.3	73.5	46.8
50.3Ti	223,000	79.5	64.2	48.9
50.2Ti	302,000	107.9	81.2	54.4
50Ti	226,000	80.8	59.1	37.4
50	196,000	70.1	53.6	37.0
45Ti	235,000	103.8	69.8	35.9
42.1	195,000	98.6	68.8	39.1
42.2Ti	223,000	113	108.5	104
40.3	142,000	79.5	64.2	48.9
40.2	146,000	81.4	58.8	36.2
40	145,000	80.8	59.1	37.4
35	143,000	104	69.5	35
30.2	95,000	94.2	78.3	62.4
30	106,000	104.8	77	49.1
25	92,500	132.1	122.8	113.4
21	60,000	121.5	90.9	60.3
19	53,900	133.5	83.9	34.4
15	35,800	142.1	91.2	40.7

[a]All rotors from Beckman Instruments of SmithKline Beckman.

[b]CAUTION: *Do not exceed maximum rotor speed!* The maximum speed for each rotor is denoted by its name, i.e., the maximum speed of the VTi80 is 80,000 rpm. This speed refers only to centrifugation of solutions below a particular allowed density, which differs among rotors (see user manual). For centrifugation of high-density solutions, rotor maximum speed can be determined as: reduced rpm = $rpm_{max}(A/B)^{1/2}$, where A = allowed density and B = density of solution. A = 1.7 g/ml for several vertical rotors (including VTi80 and VTi50), and 1.2 g/ml for several swinging-bucket rotors (including SW55Ti, 28, 28.1, 40Ti, 50.1). For gradients using heavy salts such as CsCl, particularly at low temperatures, maximum rpm should be reduced to prevent precipitation (see user manual).

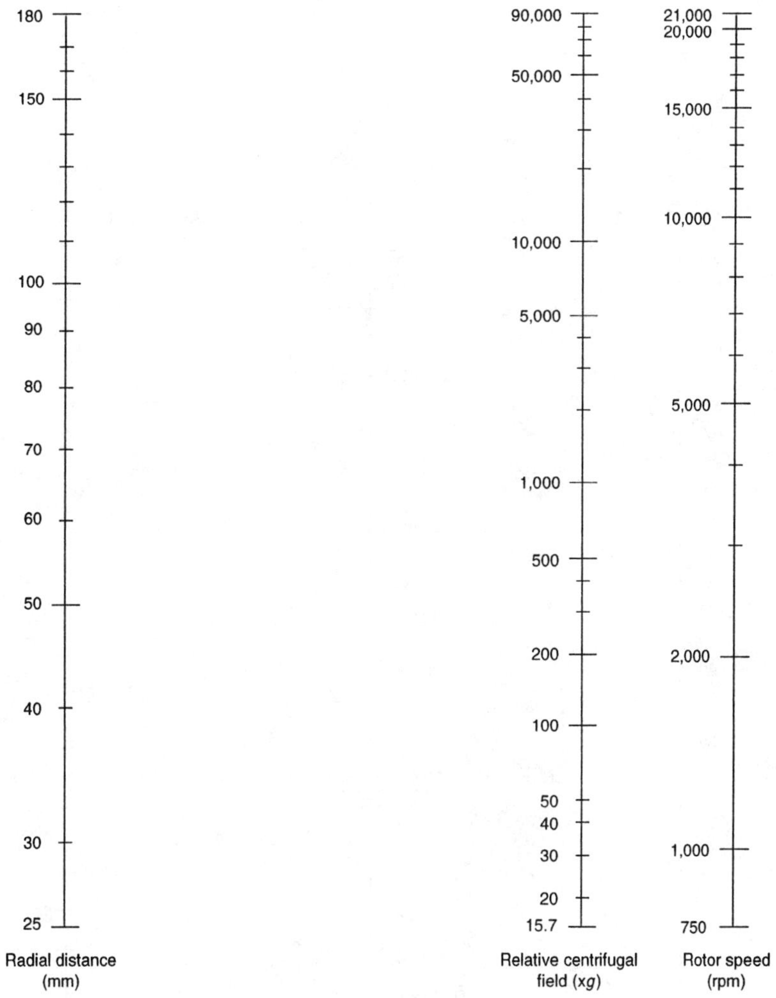

Figure A.2.10 Relative centrifugal field nomogram for low-speed rotors. To determine an unknown value in a given column, align ruler through known values in the other two columns. Desired value is found at the intersection of the ruler with the column of interest. See Table A.2.8 for low-speed rotor conversion values and accompanying equation for more precise calculations of RCF conversion.

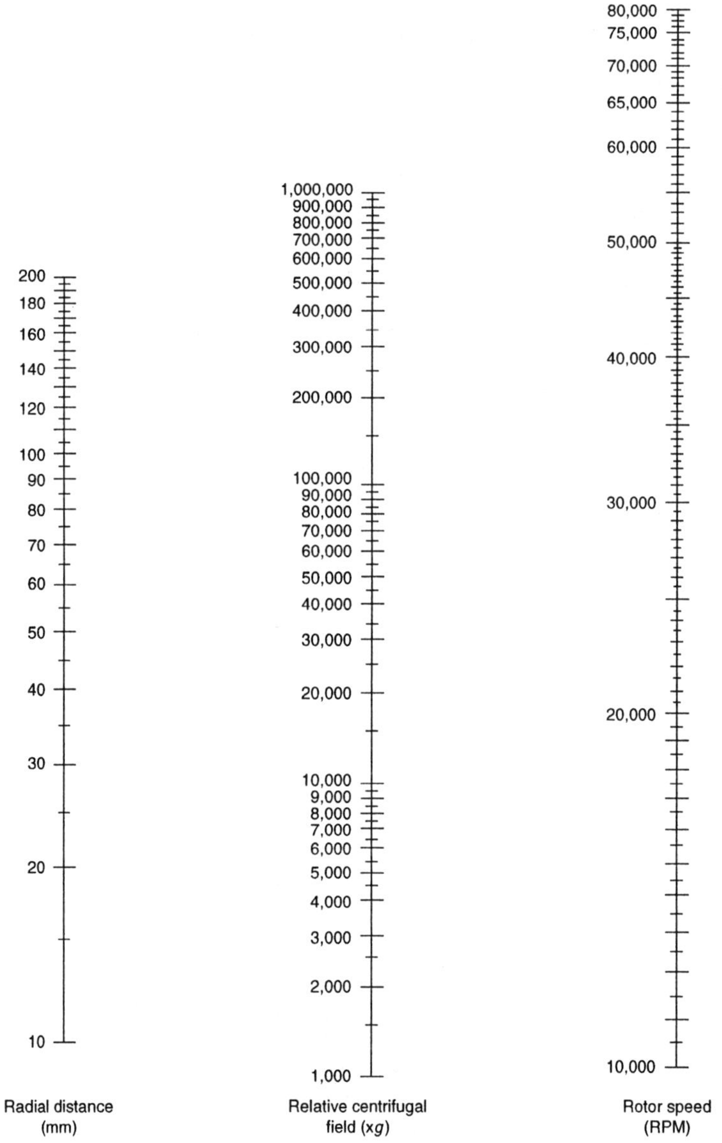

Figure A.2.11 Relative centrifugal field nomogram for high-speed rotors. To determine an unknown value in a given column, align ruler through known values in the other two columns. Desired value is found at the intersection of the ruler with the column of interest. See Table A.2.9 for high-speed rotor conversion values and accompanying equation for more precise calculations of RCF conversion.

A2

COMMONLY USED TECHNIQUES IN BIOCHEMISTRY AND MOLECULAR BIOLOGY

APPENDIX 3

Autoradiography

APPENDIX 3A

BASIC PROTOCOL

Autoradiography is used to visualize and quantitate on film radioactive molecules hybridized to membranes (e.g., Southern, northern, western blots), electrophoresed through agarose or polyacrylamide gels, or chromatographed through paper or thin-layer plates. For enhanced film images, solid-state scintillation is frequently employed to convert energy released by radioactive molecules to visible light: (1) organic scintillants are incorporated into the sample to increase the proportion of emitted energy detected from low-energy β particles (^{3}H, ^{14}C, ^{35}S; fluorography), and (2) high density, fluorescent "intensifying screens" are placed next to the sample and used to capture the excess energy of γ rays (^{125}I) and high-energy β particles (^{32}P).

FILM PREPARATION AND CHOICE

A photon of light, β particle, or γ ray can "activate" a silver bromide crystal on a film emulsion, rendering it capable of being reduced through the developing process to form silver metal (a "grain"). An activated silver bromide crystal, however, is highly unstable, and quickly reverts back to its stable form ($t_{½}$ = ~1 sec at room temperature). The absorbance of several photons increases the stability of activated silver bromide but does not ensure development; approximately 5 photons of light are required to obtain a 50% probability that any single silver bromide crystal will become developed during film processing. This inefficiency means that film images produced by very low levels of exposure will be disproportionately faint. However, two measures can be taken to maximize efficiency and linearity of exposure at low levels commonly encountered in ordinary use. First, the film should be preexposed to a hypersensitizing flash of light, which provides several photons per silver bromide crystal and stably activates them, without providing enough exposure to cause them to become developed. This allows a linear relationship to be drawn between the film image and the amount of radioactivity in the sample, even for low levels of radioactivity. Second, film exposure should be conducted at low temperatures (−70°C) to slow the reversal of activated silver bromide crystals to their stable form.

Film can be hypersensitized by exposure to a flash of light (≤1 msec) provided by a photographic flash unit or a stroboscope. The intensity of the light flash should be adjusted to increase the absorbance of developed, preflashed film to 0.15(A_{540}) over that of developed, unflashed film (this can be measured on a spectrophotometer). As the optimal light intensity required for preexposure varies with the type of film and the flash unit being used, adjustments can be made by adding a neutral density filter or an orange filter to the flash unit (Kodak Wratten #21 or 22), which serves to decrease the intensity of emitted light (X-ray films are most sensitive to blue light), and/or adjusting the distance between the flash unit and the film (a distance of at least 50 cm is usually required to obtain uniform illumination). An uneven fog level on the preexposed film can be remedied by including a porous paper diffuser such as Whatman No. 1 filter paper. The hypersensitized side of the film is then placed directly onto the radioactive sample for exposure of the autoradiogram.

The choice of film is critical for autoradiography, especially if intensifying screens or organic scintillants are to be used. In order to record the visible light emitted by these image enhancers, "screen-type" X-ray films are required. Kodak X-OMAT R and Fuji RX are two commercially available screen-type films ideal for autoradiography. Films such as No Screen or Kodirex, which are not sensitive to visible light, should only be used for samples not requiring scintillants or screens.

INTENSIFYING SCREENS

Intensifying screens are used to enhance the film image of radioactive molecules in either wet or dried gels, nylon or nitrocellulose membranes, or chromatograms of paper or thin-layer plates. They are used strictly in conjunction with strong-β-emitting isotopes such as ^{32}P or γ-emitting isotopes such as ^{125}I. Emissions from these forms of radiation will frequently pass completely through a film but can be absorbed by an intensifying screen, which fluoresces and returns them to the film as multiple photons of light. While an intensifying screen will substantially enhance the film image over direct exposure (Table A.3A.1), some loss of image resolution will occur due to light scatter.

The best intensifying screens for autoradiography are those made from calcium tungstate such as Du Pont Cronex Lightening Plus and Fuji Mach 2. To obtain an autoradiogram, a preflashed film is placed between the sample and the intensifying screen (Fig. A.3A.1). If the sample is wet (e.g., undried gel or nylon membrane) it must be covered with plastic wrap. A second screen, placed on the other side of the radioactive sample, can increase image enhancement with ^{32}P (screen, sample, film, screen), but this causes further loss in resolution due to light scatter (a second screen will not effectively enhance images from samples containing ^{125}I). The sandwich is clamped into a film cassette or placed between glass plates and allowed to expose at –70°C. Table A.3A.1 indicates the level of enhancement that can be expected with intensifying screens for both ^{32}P and ^{125}I.

Table A.3A.1 Sensitivities of Fluorography and Intensifying Screens for Isotope Detection[a]

Isotope	Method	dpm/cm² required for detectable image (A_{540} = 0.02) in 24 hr	Enhancement over direct autoradiography
^{125}I	Screen	100	16
^{32}P	Screen	50	10.5
^{14}C	PPO	400	15
^{35}S	PPO	400	15
^{3}H	PPO	8000	>1000

[a]Exposures conducted at –70°C using preexposed film. Direct autoradiography for comparison was performed on Kodirex film (Laskey, 1980).

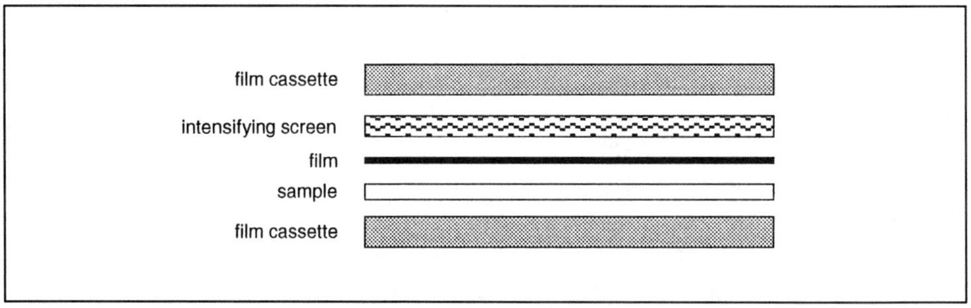

Figure A.3A.1 Autoradiography setup: intensifying screen, film, and sample in film cassette.

A3

FLUOROGRAPHY

Organic scintillants can be included in radioactive samples to obtain autoradiograms of weak β-emitting isotopes such as ^{3}H, ^{14}C, and ^{35}S. The scintillant fluoresces upon absorption of β particles from these isotopes, facilitating film exposure. Fluorographs of radioactive molecules in polyacrylamide and agarose gels have been conventionally accomplished using the scintillant PPO (2,5-diphenyloxazole). PPO, however, has been largely replaced with commercial scintillation formulations which reduce the amount of preparation time and are considerably safer to use (Enlightening and Enhance, New England Nuclear). These scintillants come with complete instructions for their use with both polyacrylamide and agarose gels. Spray applicators are also available (New England Nuclear) which can be used on membranes or thin-layer plates before exposure to film. The expected levels of image enhancement obtained through fluorography are listed in Table A.3A.1.

QUANTITATION OF FILM IMAGES

Film images obtained by any of the autoradiographic methods described can be quantitated by microdensitometry. The absorbance of the film image will be proportional to the amount of radioactivity in the gel providing that the film has been properly preexposed. If the preexposure is excessive, however, [>0.2(A_{540}) treated film/untreated film], smaller amounts of radioactivity will produce disproportionately dense images. Autoradiograms that exceed an absorbance of 1.4(A_{540}) have saturated all available silver bromide crystals and also cannot be evaluated quantitatively. Quantitation by microdensitometry can be confirmed by counting the radioactivity in the sample in a liquid scintillation spectrometer.

References: Laskey, 1980; Laskey and Mills, 1975, 1977.

Contributor: Daniel Voytas

Silanizing Glassware

APPENDIX 3B

BASIC PROTOCOL

Glassware is silanized (siliconized) to prevent adsorption of solute to the glass surface or to increase its hydrophobicity. This is particularly important when dealing with low concentrations of particularly "sticky" solutes such as single-stranded nucleic acids or proteins.

Materials

Chlorotrimethylsilane *or* dichlorodimethylsilane
Vacuum pump
Desiccator, equipped with a valve

1. In a fume hood, place glassware or equipment to be silanized into desiccator along with a beaker containing 1 to 3 ml of chlorotrimethylsilane or dichlorodimethylsilane.

   CAUTION: *Chlorotrimethylsilane and dichlorodimethylsilane vapors are toxic and highly flammable.*

   Items too large to fit in a desiccator can be silanized by briefly rinsing with or soaking in a solution of approximately 5% (v/v) dichlorodimethylsilane in various volatile organic solvents such as chloroform or heptane. The organic solvent is removed by evaporation, depositing the dichlorodimethylsilane on the surface. This approach is particularly useful for treating glass plates for denaturing polyacrylamide sequencing gels.

2. Connect desiccator to vacuum pump until silane starts to boil, and close connection to pump (maintaining vacuum in desiccator). Disconnect the vacuum pump. Leave the desiccator evacuated and closed until liquid silane is gone (~1 to 3 hr).

   During the incubation the silane will evaporate, be deposited on the surface of the glassware, and polymerize. Do not leave the desiccator attached to the vacuum pump. This will suck away the silane, minimizing deposition and damaging the pump.

3. Open desiccator in a fume hood, and leave open for several minutes to disperse silane vapors.

4. If desired, bake or autoclave the glassware or apparatus.

   Autoclaving or rinsing with water removes the reactive chlorosilane end of the dimethylsiloxane polymer generated by dichlorodimethylsilane.

CAUTION: If a flammable solvent is used, do not bake the glassware until the solvent is completely evaporated.

Contributor: Brian Seed

APPENDIX 3C

Dialysis and Ultrafiltration

DIALYSIS

Conventional dialysis separates small molecules from large molecules by allowing diffusion of only the small molecules through selectively permeable membranes. Dialysis is usually used to change the salt (small-molecule) composition of a macromolecule-containing solution. Concomitant with the movement of small solutes across the membrane, however, is the movement of solvent in the opposite direction. This can result in some sample dilution (usually <50%).

BASIC PROTOCOL

Large-Volume Dialysis

This protocol describes the use of membranes, prepared using the Support Protocol, for dialysis of samples in large, easily handled volumes, typically 0.1 to 500 ml.

Materials

Macromolecule-containing sample to be dialyzed
Appropriate dialysis buffer

Dialysis membrane (Support Protocol)
Clamps (Spectrapor Closures, Spectrum, or equivalent)

1. Remove dialysis membrane from ethanol storage solution and rinse with distilled water. Secure clamp to one end of the membrane or knot one end.

   Always use gloves to handle dialysis membrane because the membrane is susceptible to cellulolytic microorganisms.

2. Fill membrane with water or buffer, hold the unclamped end closed, and squeeze membrane. A fine spray of liquid indicates a pinhole in the membrane; discard and try a new membrane.

3. Replace the water or buffer in dialysis membrane with the macromolecule-containing sample and clamp the open end. Again, squeeze to check the integrity of the membrane and clamps.

   If dialyzing a concentrated or high-salt sample, leave some space in the clamped membrane; there will be a net flow of water into the sample, and if sufficient pressure builds up the membrane can burst.

4. Immerse dialysis membrane in a beaker or flask containing a large volume (relative to the sample) of the desired buffer. Dialyze for several hours at the desired temperature with gentle stirring of the buffer.

   Dialysis rates are dependent on membrane pore size, sample viscosity, and ratio of membrane surface to sample volume. Temperature has little effect on dialysis rate, but low temperatures are usually chosen to enhance macromolecule stability. Common salts will equilibrate across a 15,000-MWCO membrane in ~3 hr with stirring.

5. Change dialysis buffer two or more times as necessary.

6. Remove dialysis membrane from the buffer. Hold the membrane vertically and remove excess buffer trapped in end of membrane outside upper clamp. Release upper clamp and remove the sample with a Pasteur pipet.

Small-Volume Dialysis

ALTERNATE PROTOCOL

The method described below can easily dialyze volumes of 10 to 100 μl.

Additional Materials *(also see Basic Protocol)*

0.5- or 1.5-ml microcentrifuge tube
Cork borer

1. Cut a hole in the cap of a 0.5- or 1.5-ml microcentrifuge tube using a cork borer. Be sure that there are no rough edges on the inside of the cap (which will be in contact with the dialysis membrane).

2. Place sample in the microcentrifuge tube, cover tube opening with dialysis membrane, and snap on the tube cap in which a hole has been cut.

3. To get the sample in contact with the dialysis membrane, *gently* centrifuge the microcentrifuge tube in an inverted position in a tabletop centrifuge.

4. Keep the tube inverted, immerse it in dialysis buffer, and anchor it so that it will remain inverted and immersed.

5. Use a bent Pasteur pipet to blow out any air bubbles trapped under the cap to allow dialysis buffer to contact the membrane.

6. Stir the dialysis buffer and dialyze at an appropriate temperature for at least several hours (Basic Protocol, step 4).

7. To recover the sample, remove microcentrifuge tube from the buffer and centrifuge briefly right-side-up.

   Commercially available alternatives to this method include use of individual hollow-fiber filters from Spectrum (with sample capacities of 1 to 140 μl; see Ultrafiltration), and many different microdialysis machines, both single-sample and multisample (Spectrum, Cole-Parmer, Hoefer, and others).

SUPPORT PROTOCOL

Selection and Preparation of Dialysis Membrane

Dialysis membranes are available in a number of thicknesses and pore sizes. Thicker membranes are tougher, but restrict solute flow and reach equilibrium more slowly. Pore size is defined by "molecular weight cutoff" (MWCO)—i.e., the size of the smallest particle that cannot penetrate the membrane. Use a membrane with a pore size that is much smaller than the macromolecule of interest. For most plasmid and protein dialyses, a MWCO of 12,000 to 14,000 is appropriate.

Most dialysis membranes are made of derivatives of cellulose. They come in a wide variety of MWCO values, ranging from 500 to 500,000, and also vary in cleanliness, sterility, and cost. Spectrum has the most impressive inventory. The least expensive membranes come dry on rolls; these contain glycerol to keep them flexible as well as residual sulfides and traces of heavy metals from their manufacture. For most DNA dialysis applications, such as removing CsCl from gradient-purified samples (*UNIT 1.7*) and electroelution (*UNIT 2.6*), these dialysis membranes can be used directly from the roll after wetting and a brief rinse in water. If glycerol, sulfur compounds, or small amounts of heavy metals are problematic, clean membranes should be purchased or membranes should be prepared as described below. Protein dialysis should only be done with clean membranes.

Additional Materials *(also see Basic Protocol; see APPENDIX 1 for items with ✓)*

10 mM sodium bicarbonate
✓ 10 mM Na_2EDTA, pH 8.0
20% to 50% (v/v) ethanol

1. Remove membrane from the roll and cut into usable lengths (usually 8 to 12 in.).

   Always use gloves to handle dialysis membrane, as it is susceptible to a number of cellulolytic microorganisms. Membrane is available as sheets or preformed tubing.

2. Wet membrane and boil it several minutes in a large excess of 10 mM sodium bicarbonate.

3. Boil several minutes in 10 mM Na_2EDTA. Repeat.

   Boiling speeds up the treatment process but is not necessary. A 30-min soak with some agitation can substitute for the boiling steps.

4. Wash several times in distilled water.

5. Store at 4°C in 20% to 50% ethanol to prevent growth of cellulolytic microorganisms.

   Alternatively, noxious buffers (e.g., sodium azide or sodium cacodylate) may be used for storage; however, ethanol is preferred for ease and convenience.

References: Craig, 1967; McPhie, 1971.

ULTRAFILTRATION

Ultrafiltration (UF) is a type of pressure-driven membrane filtration in which a pressure differential across a semipermeable membrane is used to separate solvent and small solute molecules from a solution containing larger molecules. UF usually uses membranes with MWCOs of 1000 to 1,000,000. The pressure differential is usually achieved by vacuum, pressurized inert gas, or centrifugal force. UF is frequently used to concentrate protein solutions; because buffer concentration does not change, this is one of the most gentle methods for concentrating proteins. UF is also used to separate molecules of grossly different molecular weight (e.g., plasmids or PCR products from dNTPs or primers).

The major technical difficulty with UF is concentration polarization, in which rejected large-molecule solute builds up as a gel layer that obstructs the membrane. Various stirring and pumping regimes are used to keep the solution flowing tangentially to the membrane, and this serves to dissipate the gel layer.

UF can be broadly divided into vacuum-dialysis, centrifugal-concentration, stirred-cell dialysis, and tangential-flow systems. Choice of system will depend on the nature of the solution (i.e., volume, viscosity, presence of particulates, and shear lability), concentration rate needed, and cost. In general, vacuum dialysis is cheapest, but is also slowest and has the lowest volume capacity. Centrifuge concentrators are fairly inexpensive (on a per-use basis), moderately slow, and restricted to volumes ≤20 ml. Stirred cells come next on the scale of increasing volume capacity, concentration rate, and cost. Tangential-flow systems can concentrate large volumes rapidly, but at substantial cost.

Vacuum dialysis can be done with common laboratory equipment (see Basic Protocol). Centrifuge concentrators are single-use disposable units. The other types of UF generally require a commercial apparatus. There is a plethora of such apparatuses; because their design and availability can change rapidly, the following discussion is meant as a general guideline to the available methods. It should be noted that most UF apparatus can also be used for large-volume dialysis.

NOTE: For a more complete listing of suppliers of ultrafiltration equipment, see the *Biotech Buyer's Guide* (American Chemical Society), published annually.

Ultrafiltration Using Vacuum Dialysis

BASIC PROTOCOL

In vacuum dialysis, a vacuum is placed on the dialysis buffer while the sample remains at atmospheric pressure. This is a very gentle method used with dilute samples ≤20 ml.

Materials

Sample
Appropriate dialysis buffer

Vacuum flask
One-hole cork stopper
Pasteur pipet
Dialysis membrane of desired MWCO

1. Connect outlet of vacuum flask to vacuum line or aspirator (see Fig. A.3C.1).

2. Prepare or purchase a one-hole cork stopper into which the large end of a Pasteur pipet fits snugly, even with lubrication. Cut off the thin end of a Pasteur pipet and smooth the end using the flame from a Bunsen burner.

3. Prepare the dialysis membrane, clamp one end, and slip the other end through the hole in the stopper. Insert the Pasteur pipet into the open end of the dialysis tubing, using glycerol as a lubricant if necessary.

   A 6.4-mm-diameter membrane fits snugly over a Pasteur pipet.

4. Insert the Pasteur pipet with dialysis membrane into the stopper until firmly seated. Keep the membrane wet at all times.

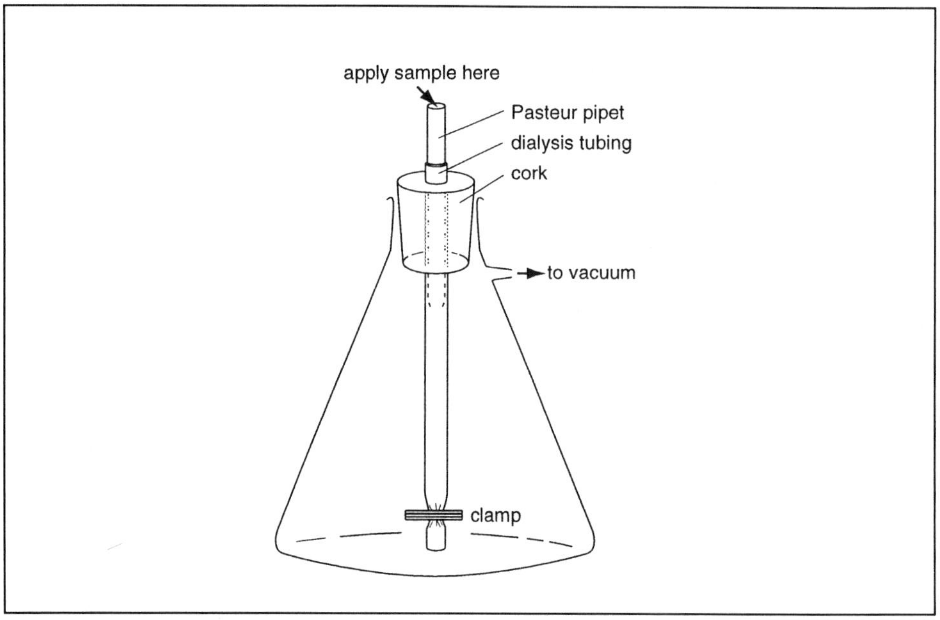

Figure A.3C.1 Vacuum dialysis apparatus (from Craig, 1967).

5. Put the stopper/Pasteur pipet/dialysis membrane assembly into the mouth of a buffer-filled vacuum flask and apply a mild vacuum. The dialysis membrane will inflate.

6. If no leaks are detected, put the sample into the membrane via the Pasteur pipet and dialyze under vacuum until the desired sample volume is achieved. Add sample continuously or at intervals throughout dialysis.

   Because vacuum pumps can rupture the membrane, the use of an aspirator with overnight stirring at 4°C is recommended. This reduces 10 to 15 ml of fairly dilute protein solution to <0.5 ml. Vacuum ≤3 lb/in^2 (150 mm Hg) will cause little membrane distortion, but 5 lb/in^2 will cause some change in pore size.

ALTERNATE PROTOCOL 1

Ultrafiltration Using Centrifuge Concentrators

Centrifuge concentrators use centrifugal force imparted by a centrifuge to force solvent through a membrane. They are more expensive than home-made vacuum dialyzers, but are faster and more amenable to small samples and multiple samples. Centrifuge concentrators are currently available to handle sample volumes of 0.05 to 20 ml, with a huge variety of MWCOs ranging from 100 to 500,000. Polarization is avoided by spinning in fixed-angle rotors so the membrane is not at a right angle to the force; any gel that forms is pushed off to one side of the membrane. This design also prevents complete drying of the sample. Centrifuge concentrators are made by Amicon, Corning, Millipore, and Spectrum, among others.

ALTERNATE PROTOCOL 2

Ultrafiltration Using Stirred-Cell Apparatus

Stirred-cell units (Amicon or Spectrum) are a relatively inexpensive means for concentrating moderate volumes of dilute, nonparticulate samples. They use pressurized inert gas (usually N_2) to drive the solvent through membrane disks. A stir bar suspended above the disk keeps the solution stirred in an attempt to dissipate any gel layer. The disks are available in various MWCOs and compositions. It is possible to concentrate 3 to 400 ml with stirred-cell units. They are cheaper than a tangential-flow apparatus (see below), but have a lower concentration rate and are more prone to concentration polarization than tangential-flow systems. However, they are less prone to concentration polarization than vacuum dialysis.

Ultrafiltration Using Tangential-Flow Systems

ALTERNATE PROTOCOL 3

Like stirred-cell apparatuses, tangential-flow ultrafiltration systems are pressure-driven. However, instead of stirring the solution, these units pump the pressurized solution tangentially with respect to the membrane, greatly reducing the concentration-polarization problem. Tangential-flow systems are more expensive than stirred-cell units. Most tangential-flow systems can also be adapted for large-volume dialysis.

One design, known as thin-channel or flat-plate, uses a flat ultrafiltration membrane. The Thin Channel System (Amicon) and Minitan System (Millipore) are examples of this design. The solution is pumped across one face of the membrane through either a spiral channel or a series of linear channels. This design has a surface area similar to that of stirred cells, but can handle larger volumes (~200 ml to 2 liters). The technique is much less susceptible to concentration polarization and can thus handle more concentrated solutions. The concentration rate is greater than that of stirred cells, as greater pressure can be achieved before concentration polarization becomes a problem. Tangential-flow systems are somewhat cheaper and simpler than the high-capacity systems described below.

Stacked-flat-plate, spiral-wound, and hollow-fiber systems (e.g., Amicon, Millipore, Spectrum) are designed for large-volume samples (1 to >1000 liters) and high concentration rates, requirements usually encountered in an industrial setting. The choice between them will normally depend more on sample characteristics than on cost (see below). Also, the filter cartridges are not cheap, so preventing sample cross-contamination usually involves extensive cleaning, rather than simply changing a membrane as with stirred-cell or lower-capacity tangential-flow systems.

The principle behind the stacked-flat-plate system (e.g., Pellicon System; Millipore) is self-explanatory. The surface area of such systems is limited, and the apparatus may be difficult to clean.

Spiral-wound systems consist of a large membrane wound on a core with a mesh separating the two faces of the membrane, thereby creating two noncontiguous spiral channels. Pressure is applied to the channel containing the input solution, and filtrate is collected from the other. This design achieves a very high surface-to-volume ratio, and thus a high concentration rate. These systems are susceptible to clogging, and are recommended only for particle-free solutions.

Hollow-fiber systems are collections of hollow fibers of ultrafiltration membrane; pressure is applied to the sample as it is pumped through the fibers. This design has the advantage of handling particulate samples (i.e., cellular material) and precipitates. They are also easy to clean, but cannot achieve high pressures, and so have lower concentration rates than spiral-channel systems.

References: American Chemical Society, 1995; McGregor, 1986; Pohl, 1990.

CONCENTRATION USING SOLID-PHASE OR SLURRY ABSORBENTS

A cheap, low-tech alternative to dialysis and ultrafiltration is concentration with solid-phase or slurry absorbents. In this method, the sample is placed in dialysis membrane and surrounded with dry matrix or immersed in an absorbent slurry, whereupon water and small solute molecules are drawn out of the sample. This technique is slow and requires constant monitoring; otherwise, complete drying can occur, to the potential detriment of the sample. Common absorbents are polyethylene glycol (PEG) and carboxymethyl cellulose (Aquacides; Calbiochem), Sephadex beads, and polyacrylamide derivatives (Spectra/Gel, Spectrum).

Contributor: Louis Zumstein

APPENDIX 3D

Quantitation of DNA and RNA with Absorption and Fluorescence Spectroscopy

Reliable quantitation of nanogram and microgram amounts of DNA and RNA in solution is essential to researchers in molecular biology. In addition to the traditional absorbance measurements at 260 nm (A_{260}), two more sensitive fluorescence techniques are presented below. These three procedures cover a range from 5 to 10 ng/ml DNA to 50 µg/ml DNA (see Table A.3D.1).

Absorbance measurements are straightforward as long as any contribution from contaminants and the buffer components are taken into account. Fluorescence assays are less prone to interference than A_{260} measurements and are also simple to perform. As with absorbance measurement, a reading from the reagent blank is taken prior to adding the DNA. In instruments where the readout can be set to indicate concentration, a known concentration is used for calibration and subsequent readings are taken in ng/ml or µg/ml DNA.

Table A.3D.1 Properties of Absorbance and Fluorescence Spectrophotometric Assays for DNA and RNA

Property	Absorbance (A_{260})	Fluorescence	
		H33258	EtBr
Sensitivity (µg/ml)			
DNA	1-50	0.01-15	0.1-10
RNA	1-40	n.a.	0.2-10
Ratio of signal (DNA/RNA)	0.8	400	2.2

A3

DETECTION OF NUCLEIC ACIDS USING ABSORPTION SPECTROSCOPY

BASIC PROTOCOL

Absorption of the sample is measured at several different wavelengths to assess purity and concentration of nucleic acids. A_{260} measurements are quantitative for relatively pure nucleic acid preparations in microgram quantities. Absorbance readings cannot discriminate between DNA and RNA; however, the ratio of A at 260 and 280 nm can be used as an indicator of nucleic acid purity. Proteins, for example, have a peak absorption at 280 nm that will reduce the A_{260}/A_{280} ratio. Absorbance at 325 nm indicates particulates in the solution or dirty cuvettes; contaminants containing peptide bonds or aromatic moieties such as protein and phenol absorb at 230 nm.

This protocol is designed for a single-beam ultraviolet to visible range (UV-VIS) spectrophotometer. If available, a double-beam spectrophotometer will simplify the measurements, as it will automatically compare the cuvette holding the sample solution to a reference cuvette that contains the blank. In addition, more sophisticated double-beam instruments will scan various wavelengths and report the results automatically.

Materials *(see* APPENDIX 1 *for items with ✓)*

✓ 1× TNE buffer
DNA sample to be quantitated
✓ Calf thymus DNA standard solutions

Matched quartz semi-micro spectrophotometer cuvettes (1-cm pathlength)
Single- or dual-beam spectrophotometer (ultraviolet to visible)

1. Pipet 1.0 ml of 1× TNE buffer into a quartz cuvette. Place the cuvette in a single- or dual-beam spectrophotometer, read at 325 nm (note contribution of the blank relative to distilled water if necessary), and zero the instrument. Use this blank solution as the reference in double-beam instruments. For single-beam spectrophotometers, remove blank cuvette and insert cuvette containing DNA sample or standard suspended in the same solution as the blank. Take reading. Repeat this process at 280, 260, and 230 nm.

   It is important that the DNA be suspended in the same solution as the blank.

2. To determine the concentration (C) of DNA present, use the A_{260} reading in conjunction with one of the following equations:

Single-stranded DNA:
$$C\ (\text{pmol}/\mu\text{l}) = \frac{A260}{10 \times S}$$
$$C\ (\mu\text{g/ml}) = \frac{A260}{0.027}$$

Double-stranded DNA:
$$C\ (\text{pmol}/\mu\text{l}) = \frac{A260}{13.2 \times S}$$
$$C\ (\mu\text{g/ml}) = \frac{A260}{0.020}$$

Single-stranded RNA:
$$C\ (\mu\text{g/ml}) = \frac{A260}{0.025}$$

Oligonucleotide:
$$C\ (\text{pmol}/\mu\text{l}) = A260 \times \frac{100}{1.5\,N_A + 0.71\,N_C + 1.20\,N_G + 0.84\,N_T}$$

A3

Table A.3D.2 Molar Extinction Coefficients of DNA Bases[a]

Base	$\varepsilon_{260\text{ nm}}^{1M}$
Adenine	15,200
Cytosine	7,050
Guanosine	12,010
Thymine	8,400

[a]Measured at 260 nm; see Wallace and Miyada, 1987. Detailed spectrophotometric properties of nucleoside triphosphates are listed in UNIT 3.4.

Table A.3D.3 Spectrophotometric Measurements of Purified DNA[a]

Wavelength (nm)	Absorbance	A_{260}/A_{280}	Conc. (μg/ml)
325	0.01	—	—
280	0.28	—	—
260	0.56	2.0	28
230	0.30	—	—

[a]Typical absorbancy readings of highly purified calf thymus DNA suspended in 1× TNE buffer. The concentration of DNA was nominally 25 μg/ml.

where *S* represents the size of the DNA in kilobases and *N* is the number or residues of base A, G, C, or T.

For double- or single-stranded DNA and single-stranded RNA: These equations assume a 1-cm-pathlength spectrophotometer cuvette and neutral pH. The calculations are based on the Lambert-Beer law, A = ECl, where A is the absorbance at a particular wavelength, C is the concentration of DNA, l is the pathlength of the spectrophotometer cuvette (typically 1 cm), and E is the extinction coefficient. For solution concentrations given in mol/liter and a cuvette of 1-cm pathlength, E is the molar extinction coefficient and has units of $M^{-1}cm^{-1}$*. If concentration units of μg/ml are used, then E is the specific absorption coefficient and has units of* $(\mu g/ml)^{-1}cm^{-1}$*. The values of E used here are as follows: ssDNA, 0.027* $(\mu g/ml)^{-1}cm^{-1}$*; dsDNA, 0.020* $(\mu g/ml)^{-1}cm^{-1}$*; ssRNA, 0.025* $(\mu g/ml)^{-1}cm^{-1}$*. Using these calculations, an* A_{260} *of 1.0 indicates 50 μg/ml double-stranded DNA, ~37 μg/ml single-stranded DNA, or ~40 μg/ml single- stranded RNA (adapted from Applied Biosystems, 1989).*

For oligonucleotides: Concentrations are calculated in the more convenient units of pmol/μl. The base composition of the oligonucleotide has significant effects on absorbance, because the total absorbance is the sum of the individual contributions of each base (Table A.3D.2).

3. Use the A_{260}/A_{280} ratio and readings at A_{230} and A_{325} to estimate the purity of the nucleic acid sample.

Ratios of 1.8 to 1.9 and 1.9 to 2.0 indicate highly purified preparations of DNA and RNA, respectively. Contaminants that absorb at 280 nm (e.g., protein) will lower this ratio.

Absorbance at 230 nm reflects contamination of the sample by phenol or urea, whereas absorbance at 325 nm suggests contamination by particulates and dirty cuvettes. Light scatter at 325 nm can be magnified 5-fold at 260 nm (K. Hardy, pers. comm.).

Typical values at the four wavelengths for a highly purified preparation are shown in Table A.3D.3.

DNA DETECTION USING THE DNA-BINDING FLUOROCHROME HOECHST 33258

Use of fluorometry to measure DNA concentration has gained popularity because it is simple and much more sensitive than spectrophotometric measurements. Specific for nanogram amounts of DNA, the Hoechst 33258 fluorochrome has little affinity for RNA and works equally well with either whole-cell homogenates or purified preparations of DNA. The fluorochrome is, however, sensitive to changes in DNA composition, with preferential binding to AT-rich regions. A fluorometer capable of an excitation wavelength of 365 nm and an emission wavelength of 460 nm is required for this assay.

Additional Materials *(also see Basic Protocol; see* APPENDIX 1 *for items with ✓)*

✓ Hoechst 33258 assay solution (working solution)
Dedicated filter fluorometer (Hoefer TKO100) *or* scanning fluorescence spectrophotometer (Shimadzu model RF-5000 or Perkin-Elmer model LS-5B or LS-3B)
Fluorometric square glass cuvettes *or* disposable acrylic cuvettes (Sarstedt)
Teflon stir rod

1. Prepare the scanning fluorescence spectrophotometer by setting the excitation wavelength to 365 nm and the emission wavelength to 460 nm.

2. Pipet 2.0 ml Hoechst 33258 assay solution into cuvette and place in sample chamber. Take a reading without DNA and use as background.

   If the fluorometer has a concentration readout mode or is capable of creating a standard curve, set instrument to read 0 with the blank solution. Otherwise note the readings in relative fluorescence units. Be sure to take a blank reading for each cuvette used, as slight variations can cause changes in the background reading.

3. With the cuvette still in the sample chamber, add 2 µl DNA standard to the blank Hoechst 33258 assay solution. Mix in the cuvette with a Teflon stir rod or by

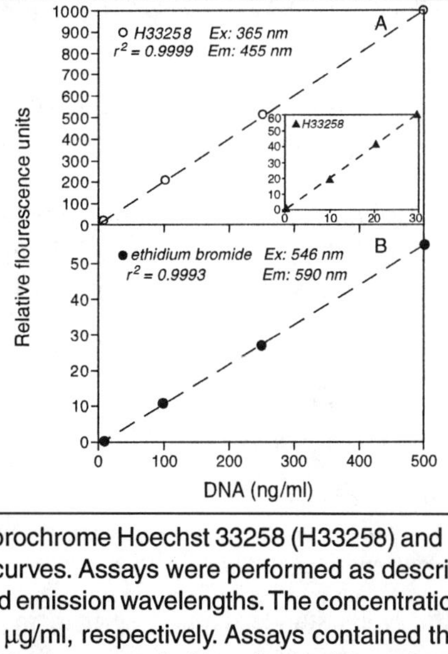

Figure A.3D.1 **(A)** Fluorochrome Hoechst 33258 (H33258) and **(B)** ethidium bromide DNA concentration standard curves. Assays were performed as described in Alternate Protocols, at indicated excitation and emission wavelengths. The concentrations of H33258 and ethidium bromide were 0.1 and 5 µg/ml, respectively. Assays contained the indicated concentrations of calf thymus DNA standards suspended in a final volume of 2.0 ml. Inset shows low DNA concentration curve for the H33258 assay. Note that, under these conditions, H33258 produces ~20 times more relative fluorescence units than ethidium bromide. A Shimadzu RF-5000 scanning fluorescence spectrometer was used for both assays.

capping and inverting the cuvette. Read emission in relative fluorescence units or set the concentration readout equal to the final DNA concentration. Repeat measurements with remaining DNA standards using fresh assay solution (take background zero reading and zero instrument if needed). Read samples in duplicate or triplicate, with a blank reading taken each time. Construct a standard curve (see Fig. A.3D.1).

The GC content of the standard should be similar to that of the sample. Eukaryotic cells vary somewhat in GC content but are generally in the range of 39% to 46%. In contrast, the GC content of prokaryotes can vary from 26% to 77%, causing considerable variation in the fluorescence signal. Calf thymus DNA may be used with prokaryotic DNA samples with an appropriate correction factor determined by quantitating standard and sample DNAs by transmission spectrophotometry.

Unusual or unstable blank readings indicate a dirty cuvette or particulate material in the solution, respectively.

4. Repeat step 3 with unknown samples and determine the concentration from the standard curve.

A dye concentration of 0.1 μg/ml is adequate for final DNA concentrations up to ~500 ng/ml. Increasing the working dye concentration to 1 μg/ml Hoechst 33258 will extend the assay's range to 15 μg/ml DNA, but will limit sensitivity at low concentrations (5 to 10 ng/ml). Sample volumes of ≤10 μl can be added to the 2.0-ml aliquot of Hoechst 33258 assay solution.

ALTERNATE PROTOCOL 2

DNA AND RNA DETECTION WITH ETHIDIUM BROMIDE FLUORESCENCE

In contrast to the fluorochrome Hoechst 33258, ethidium bromide is relatively unaffected by differences in the base composition of DNA. Ethidium bromide is not as sensitive as Hoechst 33258 and, although capable of detecting nanogram levels of DNA, it will also bind to RNA. In preparations of DNA with minimal RNA contamination or with DNA samples having an unusually high guanine and cytosine (GC) content where the Hoechst 33258 signal can be quite low, ethidium bromide offers a relatively sensitive alternative to the more popular Hoechst 33258 DNA assay. A fluorometer capable of an excitation wavelength of 302 or 546 nm and an emission wavelength of 590 nm is required for this assay.

Additional Materials *(also see Basic Protocol; see* APPENDIX 1 *for items with ✓)*

✓ Ethidium bromide assay solution

1. Pipet 2.0 ml ethidium bromide assay solution into cuvette and place in sample chamber. Set excitation wavelength to 302 nm or 546 nm and emission wavelength to 590 nm. Take an emission reading without DNA and use as background.

If the instrument has a concentration readout mode or is capable of creating a standard curve, set instrument to read 0 with the blank solution. Otherwise note the readings in relative fluorescence units.

The excitation wavelength of this assay can be either in the UV range (~302 nm) using a quartz cuvette or in the visible range (546 nm) using a glass cuvette. In both cases the emission wavelength is 590 nm.

2. Read and calibrate the DNA standards as described in step 3 of the Hoechst 33258 assay. Construct a standard curve (see Fig. A.3D.1)

3. Read emissions of the unknown samples as in step 4 of the Hoechst 33258 assay.

A dye concentration of 5 µg/ml in the ethidium bromide assay solution is appropriate for final DNA concentrations up to 1000 ng/ml. 10 µg/ml ethidium bromide in the ethidium bromide assay solution will extend the assay's range to 10 µg/ml DNA, but is only used for DNA concentrations >1 µg/ml. Sample volumes of up to 10 µl can be added to the 2.0-ml aliquot of ethidium bromide assay solution.

References: Applied Biosystems, 1987; Labarca and Paigen, 1980; Le Pecq, 1971.

Contributor: Sean R. Gallagher

Introduction of Restriction Enzyme Recognition Sequences by Silent Mutation

APPENDIX 3E

Site-directed mutagenesis has been an invaluable tool in understanding structure-function relationships. A number of techniques have been developed for introducing site-directed mutations. These techniques exploit in vitro and in vivo features of enzyme systems that act on nucleic acids (see Chapter 8) and have been immensely aided by development of the polymerase chain reaction (PCR; *UNIT 15.7*). Any changes in primer sequence can be easily incorporated into the DNA product of PCR; this facilitates subsequent incorporation of the changes into the gene sequence.

Sequence analysis and other strategies have identified biologically important motifs within proteins. This knowledge together with the tools of recombinant DNA technology can be used to design, construct, and express hybrid sequences that have a particular function. For example, development of multivalent recombinant vaccines requires design of a protein with immune-neutralizing domains from several pathogens expressed as a fusion protein. Alternatively, very small regulatory sequences can be introduced as part of the coding sequence to develop novel products that exert multiple effects as products of transcription and translation. Thus it is often necessary to generate gene cassette structures that encompass these motifs to facilitate manipulation of each motif for subsequent studies. Restriction endonuclease recognition sequences are rarely found flanking such biologically important domains and motifs, and it is often desirable to introduce restriction sites at flanking regions. Because the restriction endonuclease recognition sequence will form part of the coding sequence, it is imperative that the introduction of restriction sites does not alter the amino acid sequence of the encoded protein. The following strategy enables identification of a large number of potential sites where restriction endonuclease recognition sequences can be introduced via silent mutations.

The first step for introducing a restriction enzyme recognition sequence by silent mutagenesis involves translating the sequence in all the three reading frames (Tables A.3E.1 and A.3E.2). Due to degeneracy of the genetic code, translation of a six-base recognition sequence yields a large number of di- and tripeptides. Next, the sequence of the protein to be modified is analyzed for the presence of any of these peptides. The presence of a peptide indicates that it is possible to introduce the corresponding restriction endonuclease recognition sequence at that site by silent mutagenesis. For example, translation of the recognition sequence for *Pst*I (CTGCAG) yields a dipeptide Leu-Gln. However, translation of the same sequence in the second and third reading frames yields an additional 20 and 30 tripeptide sequences respectively. Hence, it is now possible to introduce *Pst*I recognition sequence at any of the 51 di- or tripeptide motifs by silent mutagenesis. Similarly, an analysis of the recognition sequences for *Not*I, an eight-base restriction enzyme, identifies 75 different tri- and

Table A.3E.1 Translation Sequences for Silent Mutagenesis Based on Six-Base Restriction Enzymes

Restriction enzyme[a]	Recognition sequence	First reading frame		Second reading frame			Third reading frame		
		aa 1[b,c]	aa 2	aa 1	aa 2	aa 3	aa 1	aa 2	aa 3
*Hind*III	AAGCTT	K	L	XQKE	A	FLSYXCW	LSXPQRITKVAEG	S	FL
*Mlu*I	ACGCGT	T	R	YHND	A	FLSYXCW	LSXPQRITKVAEG	R	V
*Spe*I	ACTAGT	T	S	YHND	X	FLSYXCW	LSXPQRITKVAEG	L	V
*Bgl*II	AGATCT	R	S	XQKE	I	FLSYXCW	LSXPQRITKVAEG	D	L
*Stu*I	AGGCCT	R	P	XQKE	A	FLSYXCW	LSXPQRITKVAEG	G	L
*Bsp*DI/*Cla*I	ATCGAT	I	D	YHND	R	FLSYXCW	LSXPQRITKVAEG	S	IM
*Pvu*II	CAGCTG	Q	L	SPTA	A	VADEG	FSYCLPHRITNVADG	S	CXW
*Nde*I	CATATG	H	M	SPTA	Y	VADEG	FSYCLPHRITNVADG	I	CXW
*Nco*I	CCATGG	P	W	SPTA	M	VADEG	FSYCLPHRITNVADG	H	G
*Sma*I/*Xma*I	CCCGGG	P	G	SPTA	R	VADEG	FSYCLPHRITNVADG	P	G
*Sac*II	CCGCGG	P	R	SPTA	A	VADEG	FSYCLPHRITNVADG	R	G
*Pvu*I	CGATCG	R	S	SPTA	I	VADEG	FSYCLPHRITNVADG	D	R
*Eag*I/*Xma*III	CGGCCG	R	P	SPTA	A	VADEG	FSYCLPHRITNVADG	G	R
*Pae*R7I/*Xho*I	CTCGAG	L	E	SPTA	R	VADEG	FSYCLPHRITNVADG	S	SR
*Pst*I	CTGCAG	L	Q	SPTA	A	VADEG	FSYCLPHRITNVADG	C	SR
*Eco*RI	GAATTC	E	F	XRG	I	LPHQR	LSXWPQRMTKVAEG	N	S
*Sac*I	GAGCTC	E	L	XRG	A	LPHQR	LSXWPQRMTKVAEG	S	S
*Eco*RV	GATATC	D	I	XRG	Y	LPHQR	LSXWPQRMTKVAEG	I	S
*Sph*I	GCATGC	A	C	CRSG	M	LPHQR	LSXWPQRMTKVAEG	H	A
*Nae*I	GCCGGC	A	G	CRSG	R	LPHQR	LSXWPQRMTKVAEG	P	A
*Nhe*I	GCTAGC	A	S	CRSG	X	LPHQR	LSXWPQRMTKVAEG	L	A
*Bam*HI	GGATCC	G	S	WRG	I	LPHQR	LSXWPQRMTKVAEG	D	P
*Nar*I	GGCGCC	G	A	WRG	R	LPHQR	LSXWPQRMTKVAEG	A	P
*Apa*I	GGGCCC	G	P	WRG	A	LPHQR	LSXWPQRMTKVAEG	G	P
*Acc*65I/*Kpn*I	GGTACC	G	T	WRG	Y	LPHQR	LSXWPQRMTKVAEG	V	P
*Sal*I	GTCGAC	V	D	CRSG	R	LPHQR	LSXWPQRMTKVAEG	S	T
*Apa*LI	GTGCAC	V	H	CRSG	A	LPHQR	LSXWPQRMTKVAEG	C	T
*Hpa*I	GTTAAC	V	N	CRSG	X	LPHQR	LSXWPQRMTKVAEG	L	T
*Bsp*EI	TCCGGA	S	G	FLIV	R	IMTNKSR	FSYCLPHRITNVADG	P	DE
*Nru*I	TCGCGA	S	R	FLIV	A	IMTNKSR	FSYCLPHRITNVADG	R	DE
*Xba*I	TCTAGA	S	R	FLIV	X	IMTNKSR	FSYCLPHRITNVADG	L	DE
*Bcl*I	TGATCA	X	S	LMV	I	IMTNKSR	FSYCLPHRITNVADG	D	HQ
*Bal*I	TGGCCA	W	P	LMV	A	IMTNKSR	FSYCLPHRITNVADG	G	HQ

[a]Sequences recognized by six-base restriction enzymes that do not contain degeneracy in the recognition sequence.
[b]Identity of amino acid at position 1, 2, or 3 in the translation sequence.
[c]Single-letter amino acid code (see Table A.2.3); X indicates a stop codon.

tetrapeptide sequences that have the potential to be changed by silent mutations to incorporate the recognition sequence.

Introduction of restriction endonuclease recognition sequences by silent mutations also provides a simple strategy to generate unique cohesive sites at the ends of DNA fragments (see also UNIT 3.17). It is possible to ligate several fragments containing unique overhangs at their ends, thus generating a large fragment containing all the desired mutations. Introduction of silent mutation can be coupled with introduction of a second mutation downstream of the restriction site. Digestion of the DNA

Table A.3E.2 Translation Sequences for Silent Mutagenesis Based on Eight-Base Restriction Enzymes

Restriction enzyme[a]	Recognition sequence	First reading frame			Second reading frame			Third reading frame			
		aa 1[b,c]	aa 2	aa 3	aa 1	aa 2	aa 3	aa 1	aa 2	aa 3	aa 4
*Swa*I	ATTTAAAT	I	X	IM	YHND	L	N	LSXPQRITKVAEG	F	K	FLSYXCW
*Sse*8387I	CCTGCAGG	P	A	G	SPTA	C	R	FSYCLPHRITNVADG	L	Q	VADEG
*Srf*I	GCCCGGGC	A	R	A	CRSG	P	G	LSXWPQRMTKVAEG	P	G	LPHQR
*Not*I	GCGGCCGC	A	A	A	CRSG	G	R	LSXWPQRMTKVAEG	R	P	LPHQR
*Asc*I	GGCGCGCC	G	A	P	WRG	R	A	LSXWPQRMTKVAEG	A	R	LPHQR
*Pme*I	GTTTAAAC	V	X	T	CRSG	L	N	LSXWPQRMTKVAEG	F	K	LPHQR
*Pac*I	TTAATTAA	L	I	NK	FLIV	N	X	FSYCLPHRITNVADG	X	F	IMTNKSR

[a]Sequences recognized by eight-base restriction enzymes that do not contain degeneracy in the recognition sequence.

[b]Identity of amino acid at position 1, 2, 3, or 4 in the translation sequence.

[c]Single-letter amino acid code (see Table A.2.3); X indicates a stop codon.

fragment generated by PCR using such a primer will generate unique cohesive sites at the ends to facilitate subsequent ligation.

Tables A.3E.1 and A.3E.2 list di-, tri-, and tetrapeptide translation sequences that can be used to introduce sequences recognized by common restriction endonucleases that do not contain any degeneracy. A computer program, SILMUT, uses either amino acid or nucleic acid sequence to identify potential sites for silent mutations. The program is available from the author and at many anonymous FTP sites. When using this mutagenesis strategy, due consideration should be given to codon usage of the host organism.

Reference: Shankarappa et al., 1992.

Contributor: Raj Shankarappa

Selected Suppliers of Reagents and Equipment

Listed below are addresses and phone numbers of commercial suppliers who have been recommended for particular items used in this manual because: (1) the particular brand has actually been found to be of superior quality, or (2) the item is difficult to find in the marketplace. Consequently, this compilation does not include many important vendors of biological supplies. For comprehensive listings of worldwide suppliers, see *Linscott's Directory of Immunological and Biological Reagents* (Santa Rosa, Calif.; updated regularly), *The Biotechnology Directory 1994* (Stockton Press, New York), as well as the annual Buyers' Guide supplement to the journal *Bio/Technology*.

Accurate Chemical and Scientific
300 Shames Drive
Westbury, NY 11590
(800) 645-6264
FAX: (516) 997-4948
(516) 333-2221

Air Products and Chemicals
7201 Hamilton Boulevard
Allentown, PA 18195
(800) 345-3148
FAX: (215) 481-4381
(610) 481-3864

Aldrich Chemical
P.O. Box 2060
Milwaukee, WI 53201
(800) 558-9160
FAX: (800) 962-9591
(414) 273-3850
FAX: (414) 273-4979

Allelix
6850 Gorway Drive
Mississauga, Ontario
Canada L4V 1P1
(416) 677-0831
FAX: (416) 677-9595

Altec Plastics
116 B Street
Boston, MA 02127
(617) 269-1400
FAX: (617) 269-8484

American Bioanalytical
10 Huron Drive
Natick, MA 01760
(800) 443-0600
FAX: (508) 655-2754
(508) 655-4336

American Laboratory Supply
See American Bioanalytical

American Scientific Products
See Baxter Scientific Products

American Type Culture Collection
(ATCC)
12301 Parklawn Drive
Rockville, MD 20852
(800) 638-6597
FAX: (301) 231-5826
(301) 881-2600

Amersham
2636 South Clearbrook Drive
Arlington Heights, IL 60005
(800) 323-9750
FAX: (800) 228-8735
(708) 593-6300
FAX: (708) 593-8010

Amicon
Scientific Systems Division
72 Cherry Hill Drive
Beverly, MA 01915
(800) 4AMICON
FAX: (508) 777-6204
(508) 777-3622

Amrad
17-27 Cotham Road
Kew, Victoria 3101, Australia
(03) 853-0022
FAX: (03) 853-0202

Anglian Biotec
Whitehall Road
Colchester, Essex CO2 8HA UK
(44) 0206 866007
FAX: (44) 0206 860516

Ann Arbor Plastics
2289 S. State Road
Ann Arbor, MI 48104
(800) 783-3674
FAX: (313) 994-8622
(313) 994-3674

Applied Biosystems
850 Lincoln Centre Drive
Foster City, CA 94404
(800) 327-3002
FAX: (415) 572-2743
(415) 570-6667

Ariad Pharmaceuticals
26 Landsdowne Street
Cambridge, MA 02139
(617) 494-0400
FAX: (617) 494-8144

ATCC
See American Type Culture Collection

ATR
P.O. Box 460
Laurel, MD 20725-0460
(800) 827-5931
FAX: (410) 792-2837
(301) 470-2799

Baekon
18866 Allendale Avenue
Saratoga, CA 95070
(408) 972-8779
FAX: (408) 741-0944

Barnstead/Thermolyne
P.O. Box 797
2555 Keeper Boulevard
Dubuque, IA 52004-0797
(800) 553-0039
FAX: (319) 589-0530
(319) 589-0538

Baxter Scientific Products
100 Raritan Road
Edison, NJ 08818
(800) 234-5227
FAX: (908) 417-4601
(908) 225-4700

BDH Chemicals
Broom Road
Parkstone, Poole
Dorset, UK BH124NN
(44) 020-2745520
FAX: (44) 020-2738299

Beckman Instruments
8920 Route 108
Columbia, MD 21045
(800) 742-2345
FAX: (800) 643-4366
(714) 871-4848
FAX: (714) 773-8283

Becton Dickinson Labware
2 Bridgewater Lane
Lincoln Park, NJ 07035
(800) 235-5953
FAX: (201) 628-1533
(201) 847-4222
Distributes Falcon (= BD labware) and some Damon products

Becton Dickinson Primary Care Diagnostics
7 Loveton Circle
Sparks, MD 21152
(410) 316-3300
FAX: (410) 316-4723

Bellco Biotechnology
340 Edrudo Road
Vineland, NJ 08360
(800) 257-7043
FAX: (609) 691-3247
(609) 691-1075

Beral Enterprises
9400 Lurline Avenue, Unit F
Chatsworth, CA 91311
(800) 342-3725
FAX: (818) 882-3229
(818) 882-6544

Bethesda Research Laboratories (BRL)
See Life Technologies

BIO 101
P.O. Box 2284
La Jolla, CA 92038-2284
(800) 424-6101
FAX: (619) 598-0116
(619) 598-7299

Biocomp Instruments
Fredericton, New Brunswick
Canada E3B 1P6
(800) 561-4221
FAX: (506) 453-3583
(506) 453-4812

Bio Image
777 Eisenhower Parkway S/950
Ann Arbor, MI 48108
(800) 246-4624
FAX: (313) 930-0990
(313) 930-9900

Biomeda
1155 Triton Drive, Suite E
P.O. Box 8045
Foster City, CA 94404
(800) 341-8787
FAX: (415) 341-2299
(415) 341-8787

Bio-Rad Laboratories
1000 Alfred Nobel Drive
Hercules, CA 94547
(800) 424-6723
FAX: (516) 756-2594
(516) 756-2575

Biospec Products
P.O. Box 722
Bartlesville, OK 74005
(918) 336-3363
FAX: (918) 336-3363

Biozyme Laboratories
9939 Hibert Street, Suite 101
San Diego, CA 92131-1030
(800) 423-8199
FAX: (619) 549-0138
(619) 549-4484

Boehringer Mannheim
Biochemicals Division
P.O. Box 50414
Indianapolis, IN 46250
(800) 262-1640
FAX: (317) 576-2754
(317) 849-9350

Branson Ultrasonic
41 Eagle Road
Danbury, CT 06810
(203) 796-0400
FAX: (203) 796-0381

Brinkmann Instruments
Subsidiary of Sybron Corp.
1 Cantiague Road
Westbury, NY 11590
(800) 645-3050
FAX: (516) 334-7506
(516) 334-7500

Brownlee Labs
See Applied Biosystems

BTX
11199-A Sorrento Valley Road
San Diego, CA 92121-1334
(800) 289-2465
FAX: (619) 597-9594
(619) 597-6006

Burdick and Jackson
A Division of Baxter Scientific Products
1953 S. Harvey Street
Muskegon, MI 49442
(800) 368-0050
FAX: (616) 728-8226
(616) 726-3171

Calbiochem-Novabiochem
P.O. Box 12087
La Jolla, CA 92039-2087
(800) 854-3417
FAX: (800) 776-0999
(619) 450-9600
FAX: (619) 453-3552

Cappel Laboratories
See Organon Teknika Cappel

CBS Scientific
P.O. Box 856
Del Mar, CA 92014
(800) 243-4959
FAX: (619) 755-0733
(619) 755-4959

Cel-Line Associates
33 Gorgo Lane
P.O. Box 648
Newfield, NJ 08344
(800) 662-0973
FAX: (609) 697-9728
(609) 697-4590

Celltech
250 Bath Road
Slough, Berkshire SL1 4EN, UK
(44) 0753 34655
FAX: (44) 0753 36632

ChemGenes
145 Newton Street
Waltham, MA 02154
(800) 762-9323
FAX: (617) 647-3063
(617) 899-2959

Clay Adams
See Becton Dickinson Primary Care Diagnostics

Clontech Laboratories
4030 Fabian Way
Palo Alto, CA 94303
(800) 662-2566
FAX: (415) 424-1352
(415) 424-1352

Cole-Parmer Instrument
7425 North Oak Park Avenue
Miles, IL 60714
(800) 323-4340
FAX: (708) 647-9860
(708) 647-7600

Collaborative Biomedical Products
Becton Dickinson Labware
2 Oak Park
Bedford, MA 01730
(800) 343-2035
FAX: (617) 275-0043
(617) 275-0004

Corning
HP-AB-03
Corning, NY 14831
(800) 222-7740
FAX: (607) 974-7919
(607) 974-0353

Costar
1 Alewife Center
Cambridge, MA 02140
(800) 492-1110
FAX: (617) 868-2076
(617) 868-6200

Coy Corp.
14500 Coy Drive
Grass Lake, MI 49240
(313) 475-2200
FAX: (313) 475-1846

Curtin Matheson Scientific
9999 Veterans Memorial Drive
Houston, TX 77038
(800) 392-3353
FAX: (713) 878-3598
(713) 878-3500

Dako A/S
42 Produktionsvej
P.O. Box 1359
DK-2600 Glostrup
Denmark
(045) 4492-0044
FAX: (045) 4284-1822

Damon
115 Fourth Avenue
Needham Heights, MA 02194
(800) 225-8856
FAX: (617) 461-0915
(617) 329-4800

Damon, IEC
See International Equipment Co.

Diagen GmbH
Max-Volmer Strasse 4
4010 Hilden, Germany
(049) 2103-892-230
FAX: (049) 2103-892-222

Difco Laboratories
P.O. Box 331058
Detroit, MI 48232-7058
(800) 521-0851
FAX: (313) 462-8517
(313) 462-8500

Diversified Biotech
1208 VFW Parkway
Boston, MA 02132
(617) 965-8557
FAX: (617) 323-5641

DNA ProScan
P.O. Box 121585
Nashville, TN 37212
(800) 841-4362
FAX: (615) 292-1436
(615) 298-3524

Dow Chemical
Customer Service Center
2040 Willard H. Dow Center
Midland, MI 48674
(800) 232-2436
FAX: (517) 832-1190
(409) 238-9321

Drummond Scientific
P.O. Box 700
Broomall, PA 19008
(800) 523-7480
FAX: (215) 353-6204
(215) 353-0200

DuPont Merck Pharmaceuticals
331 Treble Cove Road
Billerica, MA 01862
(800) 225-1572
FAX: (508) 436-7501
(508) 667-9531

DuPont Biotechnology Systems
and
DuPont NEN
549 Albany Street
Boston, MA 02118
(800) 551-2121
FAX: (617) 426-3038
(617) 350-9338

Dynalab
P.O. Box 112
Rochester, NY 14692
(800) 828-6595
FAX: (716) 334-9496
(716) 334-2060

Dynatech Laboratories
14340 Sully Field Circle
Chantilly, VA 22021
(800) 336-4543
FAX: (703) 631-7816
(703) 631-7800

Eastman Kodak
1001 Lee Road
Rochester, NY 14650
(800) 225-5352
FAX: (800) 879-4979
(716) 722-5780
FAX: (716) 477-8040

EC Apparatus
3831 Tyrone Boulevard North
St. Petersburg, FL 33709
(800) 327-2643
FAX: (813) 343-5730
(813) 344-1644

E. Merck
Frankfurterstrasse 250
D-1600 Darmstadt 1, Germany
(049) 6151-720

EM Science
480 Democrat Road
Gibbstown, NJ 08027
(800) 222-0342
FAX: (609) 423-4389
(609) 423-6300

Enzo Biochem
60 Executive Boulevard
Farmingdale, NY 11735
(800) 221-7705
FAX: (516) 694-7501
(516) 496-8080

Epicentre Technologies
1402 Emil Street
Madison, WI 53713
(800) 284-8474
FAX: (608) 258-3088
(608) 258-3080

Erie Scientific
Portsmouth Industrial Park
20 Post Road
Portsmouth, NH 03801
(800) 258-0834
FAX: (603) 431-2918
(603) 431-8410

E-Y Laboratories
107 N. Amphlett Boulevard
San Mateo, CA 94401
(800) 821-0044
FAX: (415) 342-2648
(415) 342-3296

Falcon
See Becton Dickinson Labware

Fisher Scientific
711 Forbes Avenue
Pittsburgh, PA 15219
(800) 766-7000
FAX: (412) 963-3328
(412) 562-8300

5 Prime → 3 Prime
5603 Arapahoe Road
Boulder, CO 80303
(800) 533-5703
FAX: (303) 440-0835
(303) 440-3705

Fluka Chemical
980 South 2nd Street
Ronkonkoma, NY 11779
(800) 358-5287
FAX: (516) 467-0663
(516) 467-0980

FMC BioProducts
191 Thomaston Street
Rockland, ME 04841
(800) 341-1574
FAX: (800) 362-5552
(207) 594-3400
FAX: (207) 594-3491

Fotodyne
950 Walnut Ridge Drive
Hartland, WI 53029-9388
(800) 362-3686
FAX: (800) 362-3642
(414) 369-7000
FAX: (414) 369-7013

Gelman Sciences
600 South Wagner Road
Ann Arbor, MI 48106
(800) 521-1520
FAX: (313) 668-2495
(313) 665-0651

Gene Codes
2901 Hubbar Road
Ann Arbor, MI 48105
(800) 497-4939
FAX: (313) 930-0145
(313) 769-7249

Genehunter
50 Boylston Street
Brookline, MA 02146
(617) 739-6771
FAX: (617) 734-5482

Genetics Computer Group
575 Science Drive
Madison, WI 53711
(608) 231-5200
FAX: (608) 231-0388

Genetics Institute
87 Cambridge Park Drive
Cambridge, MA 02140
(617) 876-1170
FAX: (617) 876-0388

Genzyme
1 Kendall Square
Cambridge, MA 02139
(800) 332-1042
FAX: (617) 252-7759
(617) 252-7500

GIBCO/BRL
See Life Technologies

Gilson
P.O. Box 620027
Middletown, WI 53562-0027
(800) 445-7661
FAX: (608) 831-4451
(608) 836-1551

Glas-Col Apparatus
P.O. Box 2128
Terre Haute, IN 47802
(812) 235-6167
FAX: (812) 234-6975

Hamilton
P.O. Box 10030
Reno, NV 89520
(800) 648-5950
FAX: (702) 856-7259
(702) 786-7077

Hazelton Biologics
See J.R.H. Biosciences

Health Products
See Pierce Chemical

Hoefer Pharmacia Biotech
654 Minnesota Street
P.O. Box 77387
San Francisco, CA 94107
(800) 227-4750
FAX: (415) 821-1081
(415) 282-2307

Hybaid
111-113 Waldegrave Road
Teddington, Middx TW11 8LL
United Kingdom
(081) 614-1000
FAX: (081) 977-0170

IBI
(International Biotechnologies)
See Eastman Kodak
For technical service (800) 243-2555
(203) 786-5600

ICN Biomedicals
3300 Hyland Avenue
Costa Mesa, CA 92626
(800) 854-0530
FAX: (800) 334-6999
(714) 545-0100
FAX: (714) 641-7275

ICN Flow
See ICN Biomedicals

IEC
See International Equipment Co.

Imclone Systems
180 Varick Street
New York, NY 10014
(212) 645-1405
FAX: (212) 645-2054

Inotech Biosystems
P.O. Box 21064
Lansing, MI 48909
(800) 635-4070
FAX: (517) 487-1013
(517) 487-1800

Integrated Separation Systems (ISS)
21 Strathmore Road
Natick, MA 01760
(800) 433-6433
FAX: (508) 655-8501
(508) 655-1500

IntelliGenetics/Betagen
7000 East El Camino Real
Mountain View, CA 94040
(800) 947-3113
FAX: (415) 962-7302
(415) 962-7300

Intermountain Scientific
420 N. Keys Drive
Kaysville, UT 84037
(800) 999-2901
FAX: (800) 574-7892
(801) 547-5047
FAX: (801) 547-5051

International Equipment Co. (IEC)
300 Second Avenue
Needham Heights, MA 02194
(800) 843-1113
FAX: (617) 444-6743
(617) 449-0800

Invitrogen
3985B Sorrento Valley Road
San Diego, CA 92121
(800) 955-6288
FAX: (619) 259-8683
(619) 597-6200

ISS
See Integrated Separation Systems

Janssen Life Sciences Products
See Amersham

John's Scientific
175 Hanson Street
Toronto, Ontario
Canada M4C1A7
(416) 699-5555
FAX: (416) 699-6536

Jordan Scientific
4315 S. State Road 446
Bloomington, IN 47401
(800) 222-2092
FAX: (812) 334-1509
(812) 334-1509

JR Scientific and
J.R.H. Biosciences
P.O. Box 14848
Lenexa, KS 66215
(800) 231-3735
FAX: (913) 469-5584
(913) 469-5580

J.T. Baker
222 Red School Lane
Phillipsburg, NJ 08865
(800) 582-2537
FAX: (908) 859-9318
(908) 859-2151

Kirkegaard & Perry
2 Cessna Court
Gaithersburg, MD 20879
(800) 638-3167
FAX: (301) 948-5815
(301) 948-7755

Kodak
See Eastman Kodak

Kontes Glass
1022 Spruce Street
Vineland, NJ 08360
(800) 223-7150
FAX: (609) 692-3242
(609) 692-8500

Kraft Apparatus
See Glas-Col Apparatus

Labconco
8811 Prospect Avenue
Kansas City, MO 64132
(800) 821-5525
FAX: (816) 363-0130
(816) 333-8811

LI-COR
P.O. Box 4425
4421 Superior Street
Lincoln, NE 68504
(402) 467-3576
FAX: (402) 467-2819

Life Sciences
2900 72nd Street North
St. Petersburg, FL 33710
(800) 237-4323
FAX: (813) 347-2957
(813) 345-9371

Life Technologies
1 Kendall Square
Grand Island, NY 14072-0068
(800) 828-6686
FAX: (800) 331-2286
(716) 774-6700

LKB Instruments
See Pharmacia Biotech

Lumigen
24485 W. Ten Mile Road
Southfield, MI 48034
(313) 351-5600
FAX: (313) 351-0518

Marsh Biomedical Products
565 Blossom Road
Rochester, NY 14610
(800) 445-2812
FAX: (716) 445-4810
(716) 654-4800

Merck
See EM Sciences

Micron Separations
See MSI

Miles
Diagnostics Division
(Order Services)
P.O. Box 3100
Elkhart, IN 46515-3100
(800) 248-2637
FAX: (219) 262-6704
(914) 524-3341

Miles Laboratories
195 W. Birch Street
Kankakee, IL 60901
(800) 227-9412
FAX: (815) 937-8285
(815) 937-8270

MilliGen/Biosearch
See Millipore

Millipore
397 Williams Street
Marlborough, MA 01752
(800) 645-5476
FAX: (508) 624-8873
(508) 624-8400

MJ Research
149 Grove Street
Watertown, MA 02172
(800) 729-2165
FAX: (617) 924-2148
(617) 923-8000

Molecular Biosystems
10030 Barnes Canyon Road
San Diego, CA 92121
(619) 452-0681
FAX: (619) 452-6187

Molecular Probes
4849 Pitchford Avenue
Eugene, OR 97402
(503) 344-3007
FAX: (503) 344-6504

Monsanto Chemical
800 North Lindbergh Boulevard
St. Louis, MO 63167
(314) 694-1000
FAX: (314) 694-4105

Moss
P.O. Box 189
Pasadena, MD 21122
(800) 932-6677
FAX: (410) 768-3971
(410) 768-3442

MSI (Micron Separations)
135 Flanders Road
P.O. Box 1046
Westborough, MA 01581
(800) 444-8212
FAX: (508) 366-5840
(508) 366-8212

Nalge
Nalgene Labware
75 Panorama Creek Drive
Rochester, NY 14602
(716) 586-8800
FAX: (716) 586-8987

National Bag Company
2233 Old Mill Road
Hudson, OH 44236
(800) 247-6000
FAX: (216) 425-9800
(216) 425-2600

National Biosciences
3650 Annapolis Lane, Suite 140
Plymouth, MN 55447
(800) 747-4362
FAX: (800) 369-5118
(612) 550-2012
FAX: (612) 550-9625

National Scientific Supply
2505 Kerner Boulevard
San Rafael, CA 94901
(800) 525-1779
FAX: (415) 459-2954
(415) 459-6070

NEB
See New England Biolabs

The Nest Group
45 Valley Road
Southboro, MA 01772-1306
(800) 347-6378
FAX: (508) 458-5736
(508) 481-6223

New Brunswick Scientific
44 Talmadge Road
Edison, NJ 08818-4005
(800) 631-5417
FAX: (908) 287-4222
(908) 287-1200

New England Biolabs
32 Tozer Road
Beverly, MA 01915
(800) 632-5227
FAX: (508) 921-1350
(508) 927-5054

New England Nuclear (NEN)
See Du Pont NEN Products

Nichols Institute Diagnostics
33608 Ortega Highway
San Juan Capistrano, CA 92690-6130
(800) 642-4657
FAX: (714) 728-4972
(714) 728-4000

Novagen
597 Science Drive
Madison, WI 53711
(800) 526-7319
FAX: (608) 238-1388
(608) 238-6110

Oncor
209 Perry Parkway
Gaithersburg, MD 20877
(800) 776-6267
FAX: (301) 926-6129
(301) 963-3500

Organon Teknika Cappel
100 Akzo Avenue
Durham, NC 27702
(800) 523-7620
FAX: (800) 432-9682
(919) 620-2000
FAX: (919) 620-2100

Operon Technologies
1000 Atlantic Avenue
Alameda, CA 94501
(800) 688-2248
FAX: (510) 865-5225
(510) 865-8644

Owl Scientific Plastics
P.O. Box 566
Cambridge, MA 02139
(800) 242-5560
FAX: (617) 935-8499
(617) 935-9499

Oxford GlycoSystems
Cross Island Plaza
133-33 Brookville Boulevard
Rosedale, NY 11422
(800) 722-2597
FAX: (718) 712-3364
(718) 712-2693

PALL
2200 Northern Boulevard
East Hills, NY 11548
(800) 645-6262
FAX: (516) 484-5228
(516) 671-4000

PCR
P.O. Box 1466
Gainesville, FL 32602
(800) 331-6313
FAX: (904) 371-6246
(904) 376-8246

Peninsula Laboratories
611 Taylor Way
Belmont, CA 94002
(800) 922-1516
FAX: (415) 595-4071
(415) 592-5392

Perkin-Elmer
850 Lincoln Center Drive
Foster, CA 94404
(800) 327-3002
FAX: (415) 572-2743
(415) 570-6667

Pharmacia ENI
See Pharmacia Diagnostics

Pharmacia Biotech
800 Centennial Avenue
Piscataway, NJ 08855
(800) 526-3593
FAX: (908) 457-8100
(908) 457-8000

Pharmacia LKB Biotechnology
See Pharmacia Biotech

Pierce Chemical
P.O. Box 117
Rockford, IL 61105
(800) 874-3723
FAX: (800) 842-5007
(815) 968-0747
FAX: (815) 968-7316

PolyLC
9052 Bellwart Way
Columbia, MD 21045
(301) 776-2410
FAX: (301) 604-4395

Polytech Products
285 Washington Street
Somerville, MA 02143
(617) 666-5064
FAX: (617) 625-0975

Primary Care Diagnostics
See Becton Dickinson Primary Care Diagnostics

Promega
2800 Woods Hollow Road
Madison, WI 53711
(800) 356-9526
FAX: (800) 356-1970
(608) 274-4330
FAX: (608) 277-2516

Protein Databases (PDI)
405 Oakwood Road
Huntington Station, NY 11746
(800) 777-6834
FAX: (516) 673-4502
(516) 673-3939

Qiagen
9600 DeSoto Avenue
Chatsworth, CA 91311
(800) 426-8157
FAX: (818) 718-2056
(818) 718-9870

Quality Biological
7581 Lindbergh Drive
Gaithersburg, MD 20879
(800) 443-9331
FAX: (301) 840-5450
(301) 840-9331

Radiometer AS
Emdrupvej 72, DK-2400
Copenhagen NV Denmark
(01) 69 63 11 Telex: 15411

Rainin Instrument
Mack Road
P.O. Box 4626
Woburn, MA 01888-4026
(800) 225-5392
FAX: (617) 938-1152
(617) 935-3050

Reagents International
Inland Farm Drive
South Windham, ME 04062
(207) 892-3266
FAX: (207) 892-6774

Research Plus
P.O. Box 324
Bayonne, NJ 07002
(800) 341-2296
FAX: (201) 823-9590
(201) 823-3592

Research Products International
410 N. Business Center Drive
Mount Prospect, IL 60056
(708) 635-7330
FAX: (708) 635-1177

Ribi ImmunoChem Research
563 Old Corvalis Road
Hamilton, MT 59840-3131
(800) 548-7424
FAX: (406) 363-6129
(406) 363-6214

Sarstedt
P.O. Box 468
Newton, NC 28658-0468
(800) 257-5101
FAX: (704) 465-4003
(704) 465-4000

Sartorius
131 Heartsland Boulevard
Edgewood, NY 11717
(800) 368-7178
FAX: (516) 254-4253
(516) 254-4261

Savant Instruments
110-103 Bi-County Boulevard
Farmingdale, NY 11735
(516) 249-4600
FAX: (516) 249-4639

Schleicher & Schuell
10 Optical Avenue
Keene, NH 03431
(800) 245-4024
FAX: (603) 357-3627
(603) 352-3810

Scripps Clinic and Research Foundation
Instrumentation and Design Lab, IMM 20
10666 N. Torrey Pines Road
La Jolla, CA 92037
(619) 554-8170
FAX: (619) 554-6705

Seikagaku America
30 West Guide Drive, Suite 260
Rockville, MD 20850
(800) 237-4512
FAX: (301) 424-6961
(301) 424-0546

Sepracor
33 Locke Drive
Marlborough, MA 01752
(800) 752-5277
FAX: (508) 481-7683
(508) 481-6802

Serva Biochemicals
1324 Motor Parkway
Hauppauge, NY 11788
(800) 645-3412
FAX: (516) 348-0913
(516) 348-0333

Shimadzu Scientific Instruments
7102 Riverwood Drive
Columbia, MD 21046
(410) 381-1227
FAX: (410) 381-1222

Sigma Chemical
P.O. Box 14508
St. Louis, MO 63178
(800) 325-3010
FAX: (800) 325-5025
(314) 771-5750
FAX: (314) 771-5757

SLM Instruments
810 W. Antony Drive
Urbana, IL 61801
(800) 637-7689
FAX: (217) 384-7744
(217) 384-7730

SmithKline Beecham
200 North 16th Street
P.O. Box 7929
Philadelphia, PA 19101
(215) 751-4000
FAX: (215) 751-4992

Sonics & Materials
W. Kenosia Avenue
Danbury, CT 06810
(800) 745-1105
FAX: (203) 798-8350
(203) 744-4400

Sorvall
P.O. Box 5509
Newtown, CT 06470-5509
(800) 551-2121
FAX: (203) 270-2166
(203) 270-2080

Southern Biotechnology Associates
P.O. Box 26221
Birmingham, AL 35226
(800) 722-2255
FAX: (205) 945-8768
(205) 945-1774

Spectrum Medical Industries
1100 Rankin Road
Houston, TX 77073
(800) 634-3300
FAX: (800) 445-7330
(713) 443-2900
FAX: (713) 443-3100

Stephens Scientific
107 Riverdale Road
Riverdale, NJ 07457
(800) 831-8099
FAX: (201) 831-8009
(201) 831-9800

Stratagene
11099 N. Torrey Pines Road
La Jolla, CA 92037
(800) 424-5444
FAX: (619) 535-0045
(619) 535-5400

SynChrom
P.O. Box 5868
Lafayette, IN 47903
(800) 283-4752
FAX: (317) 447-5703
(317) 447-6178

TAO Biomedical
73 Manassas Court
Laurel Springs, NJ 08021
(609) 782-8622
FAX: (609) 782-8622

Tago
950 Flynn Road, Suite A
Camarillo, CA 93012
(800) 242-0607
FAX: (805) 987-3385
(805) 987-0086

Techne
3700 Brunswick Pike
Princeton, NJ 08540-6192
(800) 225-9243
FAX: (609) 987-8177
(609) 452-9275

Tekmar
P.O. Box 429576
Cincinnati, OH 45242-9576
(800) 543-4461
FAX: (800) 841-5262
(513) 247-7000
FAX: (513) 247-7050

Thomas Scientific
99 High Hill Road at I-295
Swedesboro, NJ 08085
(800) 345-2100
FAX: (609) 467-3087
(609) 467-2000

TosoHaas
156 Keystone Drive
Montgomeryville, PA 18036
(800) 456-4502
FAX: (215) 283-5035
(215) 283-5020

Toyo Soda
Now TosoHaas

Tropix
47 Wiggins Avenue
Bedford, MA 01730
(800) 542-2369
FAX: (617) 275-8581
(617) 271-0045

TSI Center for Diagnostic Products
25 Birch Street
Milford, MA 01757
(800) 282-7879
FAX: (508) 473-9701
(508) 478-4030

United Desiccants-Gates
1010 Haddonfield-Berlin Road S/307
Voorhees, NJ 08043
(609) 782-0500
FAX: (609) 782-1105

United States Biochemical
P.O. Box 22400
Cleveland, OH 44122
(800) 321-9322
FAX: (800) 535-0898
(708) 593-6300
FAX: (708) 437-1640

USA/Scientific Plastics
346 SW 57th Avenue
Ocala, FL 34474
(800) 522-8477
FAX: (904) 351-2057
(904) 237-6288

UVP (Ultraviolet Products)
2066 W. 11th Street
Upland, CA 91786
(800) 452-6788
FAX: (909) 946-3597
(909) 946-3197

Value Plastics
3350 Eastbrook Drive
Ft. Collins, CO 80525
(303) 223-8306
FAX: (303) 223-0953

Vangard International
P.O. Box 308
Neptune, NJ 07754-0308
(800) 922-0784
FAX: (908) 922-0557
(908) 922-4900

Vector Laboratories
30 Ingold Road
Burlingame, CA 94010
(800) 227-6666
FAX: (415) 697-0339
(415) 697-3600

VWR Scientific
P.O. Box 1380
Piscataway, NJ 08855
(800) 932-5000
FAX: (908) 756-5098
(908) 756-8030

Waring Products
283 Main Street
New Hartford, CT 06057
(203) 379-0731
FAX: (203) 738-0249

Waters Chromatography
Division of Millipore
397 Williams Street
P.O. Box 9162
Marlboro, MA 01752-9162
(800) 252-HPLC
FAX: (508) 624-8873
(508) 624-8400

Whatman
9 Bridewell Place
Clifton, NJ 07014
(800) 631-7290
FAX: (201) 773-6138
(201) 773-5800

Wheaton
1501 North 10th Street
Millville, NJ 08332
(800) 225-1437
FAX: (800) 368-3108
(609) 825-1100
FAX: (609) 825-1368

Worthington Biochemical
Halls Mill Road
Freehold, NJ 07728
(800) 445-9603
FAX: (800) 368-3108
(908) 462-3838
FAX: (908) 308-4453

Yakult Biochemical
8-21 Jingikan Machi
Nishinomiya-Shi, Hyogo, Japan
(079) 805-8960

REFERENCES

Aboud, M., Wolfson, M., Hassan, Y., and Huleihel, M. 1982. Rapid purification of extracellular and intracellular Moloney murine leukemia virus. *Arch. Virol.* 71:185-195.

Alam, J. and Cook, J.L. 1990. Reporter genes: Application to the study of mammalian gene transcription. *Anal. Biochem.* 188:245-254.

Alberts, B., Bray, D. Lewis, J., Raff, M., Roberts, K., and Watson, J.D. 1983. The Molecular Biology of the Cell. Garland Publishing, New York.

Alexander, M., Heppel, L.A., and Hurwitz, J. 1961. The purification and properties of micrococcal nuclease. *J. Biol. Chem.* 236:3014-3019.

Amann, E., Brosius, J., and Ptashne, M. 1983. Vectors bearing a hybrid *trp-lac* promoter useful for regulated expression of cloned genes in *Escherichia coli*. *Gene* 25:167-178.

Amann, E., Ochs, B., and Abel, K.-J. 1988.Tightly regulated tac promotor vectors useful for the expression of unfused and fused proteins in *Escherichia coli*. *Gene* 69:301-314.

American Chemical Society. 1995. Biotech Buyers' Guide 1995. ACS, Washington, D.C.

Anderson, N.L. 1988. Two-Dimensional Electrophoresis. Operation of the ISO-DALT (R) System. Large Scale Biology Press, Washington, D.C.

Andreadis, A., Hsu, Y.-P., Koldhaw, G.B., and Schimmel, P. 1982. Nucleotide sequence of yeast *LEU2* shows 5′ noncoding region has sequences cognate to leucine. *Cell* 31:319-325.

Andrew, S.M. and Titus, J.A. 1995. Purification of immunoglobulin G. *In* Current Protocols in Immunology (J.E. Coligan, A.M. Kruisbeek, D.H. Margulies, E.M. Shevach, and W. Strober, eds.) pp. 2.7.1-2.7.12. John Wiley & Sons, New York.

Applied Biosystems, 1984. Evaluation and Purification of Synthetic Oligonucleotides. *User Bulletin,* Issue #13. Foster City, Calif.

Applied Biosystems. 1987. *User Bulletin,* Issue #11, Model No. 370. Applied Biosystems, Foster City, Calif.

Arber, W., Enquist, L., Hohn, B., Murray, N., and Murray, K. 1983. Experimental methods for use with lambda. *In* Lambda II (R.W., Hendrix, J.W. Roberts, F.W. Stahl, and R.A. Weisberg, eds.) pp. 433-466. CSH Laboratory, Cold Spring Harbor, N.Y.

Arnold, W. and Puhler, A. 1988. A family of high-copy-number plasmid vectors with single end-labeled sites in rapid nucleotide sequencing. *Gene* 70:171-179.

Austin, C.P. and Cepko, C.L. 1990. Cellular migration patterns in the developing mouse cerebral cortex. *Development* 110:713-732.

Ausubel, F.M., Brent, R., Kingston, R.E., Moore, D.D., Seidman, J.G., Smith, J.A., Struhl, K. 1992. Current Protocols in Molecular Biology. John Wiley & Sons, New York.

Aviv, H. and Leder, P. 1972. Purification of biologically active globin messenger RNA by chromatography on oligothymidylic acid–cellulose. *Proc. Natl. Acad. Sci. U.S.A.* 69:1408-1412.

Bach, M.I., LaCroute, F., and Botstein D. 1979. Evidence for transcriptional regulation of orotidine-5′-phosphate decarboxylase in yeast by hybridization of mRNA to the structural gene cloned in *E. coli*. *Proc. Natl. Acad. Sci. U.S.A.* 76:386-390.

Bachmann, B.J. 1983. Linkage map of *Escherichia coli* K-12, edition 7. *Microbiol. Rev.* 47:180-230.

Bagasra, O., Seshamma, T., and Pomerantz, R.J. 1993. Polymerase chain reaction in situ: Intracellular amplification and detection of HIV-1 proviral DNA and other specific genes. *J. Immunol. Methods* 158:131-145.

Barkley, M.D. and Bourgeois, S. 1978. Repressor recognition of operator and effectors. *In* The Operon (J. Miller, ed.) pp. 177-220. CSH Laboratory, Cold Spring Harbor, N.Y.

Barnes, D.A. and Thorner, J. 1986. Genetic manipulation of *Saccharomyces cerevisiae* by use of the *LYS2* gene. *Molec. Cell Biol.* 6:2828-2838.

Bebenek, K. and Kunkel, T.A. 1989. The use of T7 DNA polymerase for site-directed mutagnesis. *Nucl. Acids. Res.* 17:5408.

Beck, S. and Koster, H. 1990. Applications of dioxetane chemiluminescent probes to molecular biology. *Anal. Chem.* 62:2258-2270.

Becker, D.M. and Guarente, L. 1991. High-efficiency transformation of yeast by electroporation. *Methods Enzymol.* 194:182-187.

Becker, D.M. and Lundblad, V. 1995. Introduction of DNA into Yeast Cells. *In* Current Protocols in Molecular Biology (F. Ausubel, D. Moore, J.G. Seidman, J. Smith, and K. Struhl, eds.) pp. 13.7.1-13.7.10. John Wiley & Sons, New York.

Berger, J., Hauber, J., Hauber, R., Geiger, R., and Cullen, B.R. 1988. Secreted placental alkaline phosphatase: A powerful new quantitative indicator of gene expression in eukaryotic cells. *Gene* 66:1-10.

Bio-Rad Price List L. Bio-Rad, Richmond, Calif.

Birnboim, H.C. 1983. A rapid alkaline extraction method for the isolation of plasmid DNA. *Methods Enzymol.* 100:243-255.

Bjerrum, O.J. and Schafer-Nielsen, C. 1986. Buffer systems and transfer parameters for semidry electroblotting with a horizontal apparatus. *In* Electrophoresis '86 (M.J. Dunn, ed.) pp. 315-327. VCH Publishers, Deerfield Beach, Fla.

Bloom, H., Beier, H., and Gross, H.S. 1987. Improved silver staining of plant proteins, RNA and DNA in polyacrylamide gels. *Electrophoresis* 8:93-99.

Boeke, J.D., LaCroute, F., and Fink, G.R. 1984. A positive selection for mutants lacking orotidine-5′-phosphate decarboxylase activity in yeast: 5-fluoroorotic acid resistance. *Mol. Gen. Genet.* 197:345-346.

Boeke, J.D., Trueheart, J., Natsoulis, G., and Fink, G.R. 1987. 5-

A5

fluoroorotic acid as a selective agent in yeast molecular genetics. *Methods Enzymol.* 154:164-175.

Bolivar, F., Rodriguez, R.L., Greene, P.J., Betlach, M.C., Heynecker, H.L, and Boyer, H.W. 1977. Construction of useful cloning vectors. *Gene* 2:95-113.

Bornstein, P. and Balian, G. 1970. The specific nonenzymatic cleavage of bovine ribonuclease with hydroxylamine. *J. Biol. Chem.* 245:4854-4856.

Botstein, D., Falco, S.C., Stewart, S.E., Brennan, M., Scherer, S., Stinchcomb, D.T., Struhl, K., and Davis, R.W. 1979. Sterile host yeast (SHY). A eukaryotic system of biological containment for recombinant DNA experiments. *Gene* 8:17-24.

Brenowitz, M., Senear, D.F., Shea, M.A., and Ackers, G.K. 1986a. Footprint titrations yield valid thermodynamic isotherms. *Proc. Natl. Acad. Sci. U.S.A.* 83:8462-8466.

Brenowitz, M., Senear, D.F., Shea, M.A., and Ackers, G.K. 1986b. Quantitative DNase I footprint titration: A method for studying protein-DNA interactions. *Methods Enzymol.* 130:132-181.

Broach, J.R., Atkin, J.F., McGill C., and Chow, L. 1979. Identification and mapping of the transcriptional and translational products of the yeast plasmid 2μm circle. *Cell* 16:827-839.

Bronner-Fraser, M. 1985. Alterations in neural crest migration by a monoclonal antibody that affects cell adhesion. *J. Cell Biol.* 101:610-617.

Bronstein, I., Fortin, J., Stanley, P.E., Stewart, G.S., and Kricka, L.J. 1994. Chemiluminescent and bioluminescent reporter gene assays. *Anal. Biochem.* 219:169-181.

Brown, P.A. and Szostak, J.W. 1983. Yeast vectors with negative selection. *Methods Enzymol.* 101:278-290.

Brown, T.A., ed. 1991. Molecular Biology Labfax. BIOS Scientific Publishers, Oxford.

Burke, D.T., Carle, G.F., and Olson, M.V. 1987. Cloning of large segments of exogenous DNA into yeast by means of artificial chromosome vectors. *Science* 236:806-812.

Burnet, F.M. 1959. The Clonal Selection Theory of Acquired Immunity. Vanderbilt University Press, Nashville.

Carlson, M. and Botstein, D. 1982. Two differentially regulated mRNAs with different 5′ ends encode secreted and intracellular forms of yeast invertase. *Cell* 28:145-154.

Celis, J.E. and Bravo, R. (eds.) 1984. Two-Dimensional Gel Electrophoresis of Proteins. Academic Press, San Diego.

Celis, J.E. and Smith, J.D. 1979. Nonsense Mutations and tRNA Suppressors. Academic Press, London.

Cepko, C.L. 1989. Lineage analysis and immortalization of neural cells via retrovirus vectors. *In* Neuromethods, Vol. 16: Molecular Neurobiological Techniques (A.A. Boulton, G.B. Baker, and A.T. Campagnoni, eds.) pp. 367-392. Humana Press, Clifton, N.J.

Chaconas, G. and van de Sande, J.H. 1980. 5′-^{32}P labeling of RNA and DNA restriction fragments. *Methods Enzymol.* 65:75-88.

Challberg, M.D. and Englund, P.T. 1980. Specific labeling of 3′ termini with T4 DNA polymerase. *Methods Enzymol.* 65:39-43.

Chamberlin, M.J. and Ryan, T. 1982. Bacteriophage DNA-dependent RNA polymerases. *In* The Enzymes, Vol. 15B (P.D. Boyer, ed.) pp. 87-109. Academic Press, San Diego.

Chang, J.-Y. 1985. Thrombin specificity. *Eur. J. Biochem.* 151:217-224.

Chang, A.C.Y. and Cohen, S.N. 1978. Construction and characterization of amplifiable multicopy DNA cloning vehicles derived from the P15A cryptic miniplasmid. *J. Bacteriol.* 134:1141-1156.

Chemical Rubber Company. 1975. CRC Handbook of Biochemistry and Molecular Biology, Physical and Chemical Data, 3rd ed., Vol. 1. CRC Press, Boca Raton, Fla.

Chen, C. and Okayama, H. 1987. High efficiency transformation of mammalian cells by plasmid DNA. *Mol. Cell. Biol.* 7:2745-2752.

Chen, C. and Okayama, H. 1988. Calcium phosphate–mediated gene transfer: A highly efficient system for stably transforming cells with plasmid DNA. *BioTechniques* 6:632-638.

Chirgwin, J.J., Przbyla, A.E., MacDonald, R.J., and Rutter, W.J. 1979. Isolation of biologically active ribonucleic acid from sources enriched in ribonuclease. *Biochemistry* 18:5294.

Chodosh, L.A., Carthew, R.W., and Sharp, P.A. 1986. A single polypeptide possesses the binding and activities of the adenovirus major late transcription factor. *Mol. Cell. Biol.* 6:4723-4733.

Chomczynski, P. 1989. Product and process for isolating RNA. U.S. Patent #4,843,155.

Chomczynski, P. and Sacchi, N. 1987. Single-step method of RNA isolation by acid guanidinium thiocyanate-phenol-chloroform extraction. *Anal. Biochem.* 162:156-159.

Chothia, C. 1976. The nature of the accessible and buried surfaces in proteins. *J. Mol. Biol.* 105:1-14.

Chou, P.T. and Fasman, G.D. 1978.Empirical prediction of protein conformation. *Annu. Rev. Biochem.* 47:251-276.

Chrambach, A. and Rodbard, D. 1971. Polyacrylamide gel electrophoresis. *Science* 172:440-451.

Chu, G. 1989. Pulsed-field electrophoresis in contour-clamped homogeneous electric fields for the resolution of DNA by size or topology. *Electrophoresis* 10:290-295.

Church, G. and Gilbert, W. 1984. Genomic sequencing. *Proc. Natl. Acad. Sci. U.S.A.* 81:1991-1995.

Church, G.M. and Kieffer-Higgins, S. 1988. Multiplex DNA sequencing. *Science* 240:185-188.

Clarke, L. and Carbon, J. 1978. Functional expression of cloned yeast DNA in *Escherichia coli*: Specific complementation of argininosuccinate lyase (*argH*) mutations. *J. Mol. Biol.* 120:517-532.

Clewell, D.B. and Helinski, D.R. 1970. Properties of a desoxyribonucleic acid-protein relaxation complex and strand specificity of the relaxation event. *Biochemistry* 9:4428-4440.

Clewell, D.B. and Helinski, D.R. 1972. Nature of ColE1 plasmid replication in *Escherichia coli* in the presence of chloramphenicol. *J. Bacteriol.* 110:667-676.

A5

Cohen, P. 1985. The role of protein phosphorylation in the hormonal control of enzyme activity. *Eur. J. Biochem.* 15:439-448.

Coligan, J.E., Gates III, F.T., Kimball, E.S., and Maloy, W.L. 1983. Radiochemical sequence analysis of biosynthetically labeled proteins. *Methods Enzymol.* 91:413-434.

Coligan, J.E., Kruisbeek, A.M., Margulies, D.H., Shevach, E.M., and Strober, W., eds. 1995. Current Protocols in Immunology, Chapter 5: Immunofluorescence and cell sorting. John Wiley & Sons, New York.

Cooper, T.G. 1977. Tools in Biochemistry. John Wiley & Sons, New York.

Corrigan, A.J. and Huang, P.C. 1982. A basic microcomputer program for plotting the secondary structure of proteins. *Comput. Programs Biomed.* 3:163-168.

Cossett, F-L., Legras, C., Chebloune, Y., Savatier, P., Thoraval, P., Thomas, J.L., Samarut, J., Nigon, V.M., and Verdier, G. 1990. A new avian leukosis virus (ALV)-based packaging cell line using two separate transcomplementing helper genomes. *J. Virol.* 64:1070-1078.

Craig, L.C. 1967. Techniques for the study of peptides and proteins by dialysis and diffusion. *Methods Enzymol.* 11:870-905.

Creasey, A., D'Angio, L.M., Dunne, T., Kissinger, C., O'Keefe, T., Perry-O'Keefe, H., Moran, L., Roskey, M., Shildkraut, I., Sears, L., and Slatko, B. 1991. Application of a novel chemiluminescent-based DNA detection method to single-vector and multiplex DNA sequencing. *Biotechniques* 11:102-109.

Cullen, B.R. and Malim, M.H. 1992. Secreted placental alkaline phosphatase as a eukaryotic reporter gene. *Methods Enzymol.* 216:362-368

Dabrowiak, J.C. and Goodisman, J. 1989. Quantitative footprinting analysis of drug-DNA interactions. *In* Chemistry and Physics of DNA-Ligand Interactions (N.R. Kallenback, ed.). Adenine Press.

Danna, A.J. 1980. Determination of fragment order through partial digests and multiple enzyme digests. *Methods Enzymol.* 65:449-467.

Danos, O. and Mulligan, R.C. 1988. Safe and efficient generation of recombinant retroviruses with amphotropic and ecotropic host ranges. *Proc. Natl. Acad. Sci. U.S.A.* 85:6460-6464.

Darbre, A. 1986. Analytical methods. *In* Practical Protein Chemistry: A Handbook (A. Darbre, ed.) pp. 227-335. John Wiley and Sons, New York.

Davis, R.W., Botstein, D., and Roth, J.R. 1980. A Manual for Genetic Engineering: Advanced Bacterial Genetics, pp. 70-113. CSH Laboratory, Cold Spring Harbor, N.Y.

Dawson, M.C., Elliott, D.C., Elliott, W.H., and Jones, K.M. (eds.). 1987. Data for Biochemical Research, 3rd ed. Clarendon Press, Oxford.

Dawson, R.M.C., Elliot. D.C., Elliott, W.H., and Jones, K.M. (eds.) 1986. Data for Biochemical Research, Alden Press, London.

Dayhoff, M.O., Schwartz, R.M., and Orcutt, B.C. 1978. A model of evolutionary change in proteins, *In* Atlas of Protein Sequence and Structure, (M. Dayhoff ed.) Vol. 5, pp. 345-352. National Biomedical Research Foundation, Washington, D.C.

Dellaporta, S.L., Wood, J., and Hicks, J.B. 1983. A plant DNA minipreparation: Version II. *Plant Mol. Biol. Rep.* 1(4):19.

Denhardt, D. 1966. A membrane filter technique for the detection of complementary DNA. *Biochem. Biophys. Res. Commun.* 23:641-646.

Dente, L., Cesarini, G., and Cortese, R. 1983. pEMBL: A new family of single-stranded plasmids. *Nucl. Acids Res.* 11:1645-1655.

Derbyshire, V. Freemont, P.S., Sanderson, M.R., Beese, L., Friedman, J.M., Joyce, C.M., and Steitz, T. 1988. Genetic and crystallographic studies of the 3′, 5′-exonucleolytic site of DNA polymerase I, *Science* 240:199-201.

Dignam, J.D., Lebovitz, R.M., and Roeder, R.G. 1983. Accurate transcription initiation by RNA polymerase II in a soluble extract from isolated mammalian nuclei. *Nucl. Acids Res.* 11:1475-1489.

Donovan, J. and Brown, P. 1995. Care and handling of laboratory animals. *In* Current Protocols in Immunology (J.E. Coligan, A.M. Kruisbeek, D.H. Margulies, E.M. Shevach, and W. Strober, eds.) pp. 1.0.1-1.8.4. John Wiley & Sons, New York.

Doolittle, R.F. 1986. Of URFS and ORFS: A Primer on How to Analyze Derived Amino Acid Sequences. University Science Books, Mill Valley, Calif.

Dower, W.J., Miller, J.F., and Ragdale, C.W. 1988. High efficiency transformation of *E. coli* by high voltage electroporation. *Nucl. Acids Res.* 16:6127-6145.

Dunbar, B.S. 1987. Two-dimensional electrophoresis and immunological techniques. Plenum, New York.

Dustin, M.L., Rothlein, R., Bhan, A.K., Dinarello, D.A., and Springer, T.A. 1986. Induction by IL-1 and interferon, tissue distribution, biochemistry, and function of a natural adherence molecule (ICAM-1). *J. Immunol.* 137:245-254.

Dyson, N.J. 1991. Immobilization of nucleic acids and hybridization analysis. *In* Essential Molecular Biology: A Practical Approach, Vol. 2 (T.A. Brown, ed.) pp 111-156. IRL Press, Oxford.

Eckert, R. 1987. New vectors for rapid sequencing of DNA fragments by chemical degradation. *Gene* 51:245-252.

Edmonds, M. 1982. Poly(A) adding enzymes. *In* The Enzymes, Vol. 15B (P.D. Boyer, ed.) pp. 218-245. Academic Press, San Diego.

Eibel, H. aned Phillippsen, P. 1983. Identification of the cloned *S. cerevisiae LYS2* gene by an integrative transformation approach. *Mol. Gen. Genet.* 191:66-73.

Eickbush, T.H. and Moudrianakis, E.N. 1978. The compaction of DNA helices into either continuous supercoils or folded-fiber rods and toroids. *Cell* 13:295-306.

Elledge, S.J. and Davis, R.W. 1988. A family of versatile centromeric vectors designed for use in the sectoring-shuffle mutagenesis assay in *Saccharomyces cerevisiae. Proc. Natl. Acad. Sci. U.S.A.* In press.

Engler, M.J. and Richardson, D.C. 1982. DNA ligases. *In* The Enzymes, Vol. 15B (P.D. Boyer, ed.) pp. 3-30. Academic Press, San Diego.

Engvall, E. and Perlman, P. 1971. Enzyme-linked immunosorbent assay (ELISA): Quantitative assay of immunoglobulin G. *Immunochemistry* 8:871-879.

Ey, P.L, Prowse, S.J., and Jenkin, S.R. 1978. Isolation of pure IgG_1, IgG_{2a} and IgG_{2b} immunoglobulins from mouse serum using protein A-Sepharose. *Immunochemistry* 15:429-436.

Fasman, G. (ed.). 1975. Handbook of Biochemistry and Molecular Biology, Vol. 1: Nucleic Acids, 3rd ed. CRC Press, Boca Raton, Fla.

Favoloro, J., Treisman, R., and Kamen, R. 1980. Transcription maps of polyoma virus-specific RNA: Analysis by two-dimensional nuclease S1 gel mapping. *Methods Enzymol.* 65:718-749.

Feinberg, A.P. and Vogelstein, B. 1983. A technique for radiolabeling DNA restriction endonuclease fragments to high specific activity. *Anal. Biochem.* 132:6-13.

Felgner, P.L., Gadek, T.R., Holm, M., Roman, R., Chan, H.W., Wenz, M., Northrop, J.P., Ringold, G.M., and Danielson, M. 1987. Lipofectin: A highly efficient, lipid-mediated DNA/transfection procedure. *Proc. Natl. Acad. Sci. U.S.A.* 84:7413-7417.

Fischer, L. 1980. Gel-filtration Chromatography. Elsevier, Amsterdam.

Fitzgerald-Hayes, M., Clarke, L., and Carbon, J. 1982. Nucleotide sequence comparisons and functional analysis of yeast centromere DNAs. *Cell* 29:235-244.

Foster, T.J. 1983. Plasmid-determined resistance to antimicrobial drugs and toxic metal ions in bacteria. *Microbiol. Rev.* 47:361-409.

Frangioni, J.V. and Neel, B.G. 1993. Solubilization and purification of enzymatically active glutathione-S-tranwferase (pGEX) fusion proteins. *Anal. Biochem.* 210:179-187.

Frischauf, A.-M., Lehrach, H., Polstka, A., and Murray, N.M. 1983. Lambda replacement vectors carrying polylinker sequences. *J. Mol. Biol.* 170:827-842.

Frohman, M.A., Dush, M.K., and Martin, G.R. 1988. Rapid production of full length cDNAs from rare transcripts: Amplification using a single gene-specific oligonucleotide primer. *Proc. Natl. Acad. Sci. U.S.A.* 85: 8998-9002.

Frömmel, C. 1984. The apolar surface area of amino acids and its empirical correlation with hydrophobic free energy. *J. Theor. Biol.* 111:247-260.

Fuchs, R., and Blakesley, R. 1983. Guide to the use of type II restriction endonucleases. *Methods Enzymol.* 100:3-38.

Garrity, P.A. and Wold, B.J. 1992. Effects of different DNA polymerases in ligation-mediated PCR: Enhanced genomic sequencing and in vivo footprinting. *Proc. Natl. Acad. Sci. U.S.A.* In press.

Geitz, R.D. and Sugino, A. 1988. New yeast. *E. coli* shuttle vectors contructed with in vitro mutagenized yeast genes lacking six-base-pair restriction sites. *Gene* 74:527-534.

Gillespie, P.G. and Hudspeth, A.J. 1991. Chemiluminescence detection of proteins from single cells. *Proc. Natl. Acad. Sci. U.S.A.* 88:2563-2567.

Gilman, C.M., Wilson, R.N., and Weinberg, R.A. 1986. Multiple binding sites in the 5′-flanking region regulates c-fos expression. *Mol. Cell Biol.* 6:4305-4316.

Goff, S., Trakman, P., and Baltimore, D. 1981. Isolation and properties of murine leukemia virus mutants: Use of a rapid assay for release of virion reverse transciptase. *J. Virol.* 38:239-248.

Gorman, C.M., Moffat, L.F., and Howard, B.H. 1982. Recombinant genomes which express chloramphenicol acetyltransferase in mammalian cells. *Mol. Cell. Biol.* 2:1044-1051.

Gottlieb, D. and Shaw, P.D. 1967. Antibiotics. I. Mechanism of Action. Springer-Verlag, New York.

Gould, S.J. and Subramani, S. 1988. Firefly luciferase as a tool in molecular and cell biology. *Anal. Biochem.* 7:5-13.

Gritz, L. and Davies, J. 1983. Plasmid-encoded hygromycin-β resistance: The sequence of hygromycin-β-phosphotransferase gene and its expression in *E. coli* and *S. cerevisiae*. *Gene* 25:179-188.

Gross, E. 1967. The cyanogen bromide reaction. *Methods Enzymol.* 11:238-255.

Gross-Bellard, M., Oudet, P., and Chambon, P. 1972. Isolation of high-molecular-weight DNA from mammalian cells. *Eur. J. Biochem.* 36:32.

Guarente, L. 1983. Yeast promoters and *lacZ* fusions designed to study expression of cloned genes in yeast. *Methods Enzymol.* 101:181-191.

Guarente, L., Lauer, G., Roberts, T.M., and Ptashne, M. 1980. Improved methods for maximizing expression of a cloned gene: a bacterium that synthesizes rabbit β-globin. *Cell* 20:543-553.

Gyuris, J., Golemis, E.A., Chertkov, H., and Brent, R. 1993. Cdi1, a human G1 and S phase protein phosphatase that associates with Cdk2. *Cell* 75:791-803.

Haase, A.T., Retzel, E.F., and Staskus, K.A. 1990. Amplification and detection of lentiviral DNA inside cells. *Proc. Natl. Acad. Sci. U.S.A.* 37:4971-4975.

Hames, B.D. and Rickwood, D. (eds.) 1981. Gel Electrophoresis of Proteins. IRL Press, Washington, D.C.

Hames, B.D. and Rickwood, D. (eds.) 1990. Gel Electrophoresis of Proteins: A Practical Approach, 2nd ed. Oxford University Press, New York.

Hamilton, C.M. Aldea, M., Washburn, B.K., Babitzke, P., and Kushner, S.R. 1989. New methods for generating deletions and gene replacements in *Escherichia coli*. *J. Bacteriol.* 171:4617-4622.

Hanahan, D. 1983. Studies on transformation of *Escherichia coli* with plasmids. *J. Mol. Biol.* 166:557-580.

Hanahan, D. and Meselson, M. 1983. Plasmid screening at high density. *Methods Enzymol.* 100:333-342.

Hancock, K. and Tsang, V.C.M. 1983. India ink staining of proteins on nitrocellulose paper. *Anal. Biochem.* 133:157-162.

Hancock, W.S. (ed.) 1984. Handbook of HPLC for the Separation of Amino Acids, Peptides, and Proteins, Vols. I and II. CRC Press, Boca Raton, Fla.

Harlow, E., and Lane, D. 1988. Antibodies: A Laboratory Manual.

CSH Laboratory. Cold Spring Harbor, N.Y.

Hartley, J.L. and Donelson, J.E. 1980. Nucleotide sequence of the yeast plasmid. *Nature* 286:860-865.

He, M., Kaderbhai, M.A., Adcock, L., and Austen, B.M. 1991. An improved rapid procedure for isolating RNA-free *Escherichia coli* plasmid DNA. *Gene Anal. Tech.* 8:107-110.

Hendrickson, W. and Schleif, R. 1985. A dimer of AraC protein contacts three adjacent major groove regions at the Ara I DNA site. *Proc. Natl. Acad. Sci. U.S.A.* 82:3129-3133.

Henikoff, S. 1984. Unidirectional digestion with exonuclease III creates targeted breakpoints for DNA sequencing. *Gene* 28:351-359.

Hill, D.E., Oliphant, A.R., and Struhl, K. 1987. Mutagenesis with degenerate oligonucleotides: An efficient method for saturating a defined DNA region with base pair substitutions. *Methods Enzymol.* 155:558-568.

Hinnen, A., Hicks, J.B., and Fink, G.R. 1978. Transformation of yeast. *Proc. Natl. Acad. Sci. U.S.A.* 75:1929-1933.

Hjelmeland, J.M. and Chrambach, A. 1984. Solubilization of functional membrane proteins. *Methods Enzymol.* 104:305-318.

Hochuli, E. 1990. Purification of recombinant proteins with metal chelate adsorbent. *In* Genetic Engineering, Principles and Practice, Vol. 12 (J. Setlow, ed.) pp. 87-98. Plenum, New York.

Hoffman, C.S. and Winston, F. 1987. A ten-minute DNA preparation from yeast efficiently releases autonomous plasmids for transformation of *Escherichia coli. Gene* 57:267-272.

Hogan, B., Constantini, F., and Lacy, E. 1986. Manipulating the Mouse Embryo. CSH Laboratory, Cold Spring Harbor, N.Y.

Hoheisel, J. and Pohl, F.M. 1986. Simplified preparation of unidirectional deletion clones. *Nucl. Acids Res.* 14:3605.

Hollands, B. 1962. Histochemistry and microtomy of fresh-frozen tissue. *In* Progress in Medical Laboratory Technique (F.J. Baker, ed.) pp. 112-135. Butterworth, London.

Holmes, D.S. and Quigley, M. 1981. A rapid boiling method for the preparation of bacterial plasmids. *Anal. Biochem.* 114:193-197.

Holmes, K. and Fowlkes, B.J. 1995. Preparation of cells and reagents for flow cytometry. *In* Current Protocols in Immunology (J.E. Coligan, A.M. Kruisbeek, D.H. Margulies, E.M. Shevach, and W. Strober, eds.) pp. 5.3.1-5.3.11. John Wiley & Sons, New York.

Hope, I.A. and Struhl, K. 1985. GCN4 protein, synthesized in vitro, binds *HIS3* regulatory sequences: Implications for general control of amino acid biosynthetic genes in yeast. *Cell* 43:177-188.

Hope, I.A. and Struhl, K. 1987. GCN4, a eukaryotic transcriptional activator protein, binds DNA as a dimer. *EMBO J.* 6:2781-2784.

Hopp, T.P. and Woods, K.R 1983. A computer program for predicting protein antigenic determinants. *Mol. Immunol.* 20:483-489.

Hughes, S.H., Greenhouse, J.J., Petropoulos, C.J., and Sutrave, P. 1987. Adapter plasmids simplify the insertion of foreign DNA into helper-independent retroviral vectors. *J. Virol.* 61:3001-3012.

Hultman, T., Beregh, S., Moks, T., and Uhlen, M. 1991. Bidirectional solid-phase sequencing of in vitro-amplified plasmid DNA. *BioTechniques* 10:84-93.

Hunkapiller, M.W., Lujan, E., Ostrander, F., and Hood, L.E. 1983. Isolation of microgram quantities of proteins from polyacrylamide gels for amino acid sequence analysis. *Methods Enzymol.* 91:227-236.

Hunkapiller, M.W. and Lujan, E. 1986. Purification of microgram quantities of proteins by polyacrylamide gel electrophoresis. *In* Methods of Protein Microcharacterization (J. Shively, ed.) pp. 89-101. Humana Press, Clifton, N.J.

Hunter, E. 1979. Biological techniques for avian sarcoma viruses. *Methods Enzymol.* 58:379-393.

Huynh, T.V., Young, R.A., and Davis, R.W. 1984. Construction and screening of cDNA libraries in gt10 and gt11. *In* DNA Cloning: A Practical Approach, Vol. 1 (D.M. Glover, ed.) pp 49-78. IRL Press, Oxford.

Innis, M.A., Myambo, K.B., Gelfand, D.H., and Brow, M.D. 1988. DNA sequencing with *Thermus aquaticus* DNA polymerase, and direct sequencing of PCR-amplified DNA. *Proc. Natl. Acad. Sci. U.S.A.* 85:9436-9440.

Innis, M.A., Gelfand, D.H., Sninsky, J.J., and White, T.J. (eds.) 1990. PCR Protocols. Academic Press, San Diego.

Ish-Horowitz, D. and Burke, J.F. 1981. Rapid and efficient cosmid cloning. *Nucl. Acids Res.* 9:2989-2998.

Ishiura, M., Hirose, S., Uchida, T., Hamada, Y., Suzuki, Y., and Okada, Y., 1982. Phage particle–mediated gene transfer to cultured mammalian cells. *Mol. Cell. Biol.* 2:607-616.

Jacobs, K.A., Rudersdorf, R., Neill, S.D., Dougherty, J.P., Brown, E.L., and Fritsch, E.F. 1988. The thermal stability of oligonucleotide duplexes is sequence independent in tetraalkylammonium salt solutions: Application to identifying recombinant DNA clones. *Nucl. Acids Res.* 16:4637-4650.

Jain, V. and Magrath, I. 1991. A chemiluminescent assay for quantitation of β-galactosidase in the femtogram range: Application to quantitation of β-galactosidase in lacZ-transfected cells. *Anal. Biochem.* 199:119-124.

Johnson, M.L. and Frasier, S.G. 1985. Nonlinear least-squares analysis. *Meth. Enzymol.* 117:301-342.

Kadonaga, J.T. 1991. Purification of sequence-specific DNA binding proteins by DNA affinity chromatography. *Methods Enzymol.* 208:10-23.

Kadonaga, J.T. and Tjian, R. 1986. Affinity purification of sequence-specific DNA binding proteins. *Proc. Natl. Acad. Sci. U.S.A.* 83:5889-5893.

Kahn, M., Kolter, R., Thomas, C., Figurski, D., Meyer, R., Remaut, E., and Helinski, D.R. 1979. Plasmid cloning vehicles derived from plasmids ColE1, F, R6K, and RK2. *Methods Enzymol.* 68:268-280.

Kaiser, K. and Murray, N.E. 1984. The use of phage lambda replacement vectors in the construction of representative genomic DNA libraries. *In* DNA Cloning: A Practical Approach, Vol. 1 (D.M. Glover, ed.) pp. 1-47. IRL Press, Oxford.

Kamps, M.P. and Sefton, B.M. 1988. Identification of multiple novel polypeptide substrates of the v-src, v-yes, v-fps, v-ros, and v-erb-B oncogenic tyrosine protein kinases utilizing antisera against phosphotyrosine. *Oncogene* 2:305-315.

Kamps, M.P. and Sefton, B.M. 1989. Acid and base hydrolysis of phosphoproteins bound to Immobilon facilitates the analysis of phosphoamino acids in gel-fractionated proteins. *Anal. Biochem.* 176:22-27.

Kasher, M.S., Pintel, D., and Ward, D.C. 1986. Rapid enrichment of HeLa transcription factors IIIB and IIIC by using affinity chromatography based on avidin-biotin interactions. *Mol. Cell. Biol.* 6:3117-3127.

Kato, Y. 1984. Toyo Soda high-performance gel filtration columns. *In* Handbook of HPLC for the Separation of Amino Acids, Peptides, and Proteins, Vol. II (W.S. Hancock, ed.) pp. 363-369. CRC Press, Boca Raton, Fla.

Katz, J.P., Bodin, E.T., and Coen, D.M. 1990. Quantitative polymerase chain reaction analysis of herpes simplex virus DNA in ganglia of mice infected with replication-incompetent mutants. *J. Virol.* 64:4288-4295.

Kaufman, R.J. 1990. Overview of vectors used for expression in mammalian cells. *Methods Enzymol.* 185:487-511.

Kaufman, R.J., Murtha, P., Ingolia, D.E., Yeung, C-Y., and Kellems, R.E. 1986. Selection and amplification of heterologous genes encoding adenosine deaminase in mammalian cells. *Proc. Natl. Acad. Sci. U.S.A.* 83:3136-3140.

Kingston, R.E., Cowie, A., Morimoto, R.I., and Gwinn, K.A. 1986. Binding of polyomavirus large T antigen to the human hsp70 promoter is not required for trans-activation. *Mol. Cell. Biol.* 6:3180-3190.

Klinman, N.R. and Press, J. 1975. The B cell specificity repertoire: Its relationship to definable subpopulations. *Transplant. Rev.* 24:41-83.

Koerner, T.J., Hill, J.E., Myers, A.M., and Tzagoloff, A. 1990. High-expression vectors with multiple cloning sites for construction of *trpE*-fusion genes: pATH vectors. *Methods Enzymol.* 194:477-490.

Kohler, G. and Milstein, C. 1975. Continuous cultures of fused cells secreting antibody of predefined specificity. *Nature* 256:495-497.

Kornberg, A. 1980. DNA Replication. W.H. Freeman, New York.

Kroeker, W.D., Kowalski, D., and Laskowski, M. 1976. Mung bean nuclease I. Terminally directed hydrolysis of native DNA. *Biochemistry* 15:4463-4467.

Kunkel, T.A. 1985. Rapid and efficient site-specific mutagenesis without phenotypic selection. *Proc. Natl. Acad. Sci. U.S.A.* 82:488-492.

Kunkel, T.A., Roberts, J.D., and Zakour, R.A. 1987. Rapid and efficient site-specific mutagenesis without phenotypic selection. *Methods Enzymol.* 154:367-382.

Kyte, J. and Doolittle, R.F. 1982. A simple method for displaying the hydropathic character of a protein. *J. Molec. Biol.* 157:112-122.

Labarca, C. and Paigen, K. 1980. A simple, rapid, and sensitive DNA assay procedure. *Anal. Biochem. 102:344-352.*

Laemmli, U.K. 1970. Cleavage of structural proteins during the assembly of the head of bacteriophage T4. *Nature* 227:680-685.

Laiminis, L.A., Gruss, P., Pozzatti, R., and Khoury, G. 1984. Characterization of enhancer elements in the long terminal repeat of Moloney murine sarcoma virus. *J. Virol.* 49:183-189.

Landon, M. 1977. Cleavage at aspartyl-prolyl bonds. *Methods Enzymol.* 47(E):145-149.

Langer, P.R., Waldrop, A.A., and Ward, D.C. 1981. Enzymatic synthesis of biotinylated polynucleotides: Novel nucleic acid affinity probes. *Proc. Natl. Acad. Sci. U.S.A.* 78:6633-6637.

Langone, J.J. and Van Vunakis, H., eds. 1986. Immunological techniques, Part I: Hybridoma technology and monoclonal antibodies. *Methods Enzymol.* 121:1-947.

Laskey, R.A. 1980. The use of intensifying screens or organic scintillators for visualizing radioactive molecules resolved by gel electrophoresis. *Methods Enzymol.* 65:363-371.

Laskey, R.A. and Mills, A.D. 1975. Quantitative film detection of ^{3}H and ^{14}C in polyacrylamide gels by fluorography. *Eur. J. Biochem.* 56:335-341.

Laskey, R.A. and Mills. A.D. 1977. Enhanced autoradioactive detection of ^{32}P and ^{125}I using intensifying screens and hypersensitized film. *FEBS Lett.* 82:314-316.

Lathe, R. 1985. Synthetic oligonucleotide probes deduced from amino acid sequence data. Theoretical and practical considerations. *J. Mol. Biol.* 183:1-12.

Lau, P.P. and Gray, H.B. 1979. Extracellular nucleases of *Alteromonas espejiana Bal* 31. IV. The single strand-specific deoxyriboendonuclease activity as a probe for regions of altered secondary structure in negatively and positively supercoiled closed circular DNA. *Nucl. Acids Res.* 6:331-357.

LaVallie, E.R., Rehemtulla, A., Racie., L.A., DeBlasio, E.A., Ferenz, C., Grant, K.L., Light, A., and McCoy, J.M. 1993a. Cloning and functional expression of a cDNA encoding the catalytic subunit of bovine enterokinase. *J. Biol. Chem.* 268:23311-233217.

LaVallie, E.R., DiBlasio, E.A., Kovacic, S., Grant, K.L., Schendel, P.F., and McCoy, J.M. 1993b. A thioredoxin gene fusion expression system that circumvents inclusion body formation in the *E. coli* cytoplasm. *Bio/Technology* 11:187-193.

Lederberg, J. and Tatum, E.L. 1953. Novel genotypes in mixed cultures of biochemical mutants of bacteria. *Cold Spring Harbor Symp. Quant. Biol.* 18:75.

Lederer, C.M., Hollander, J.M., and Perlman, J., eds. 1967. Table of Radioisotopes, 6th edition. John Wiley & Sons, New York.

Lee, N., Cozzikorto, J., Wainwright, N., and Testa, D. 1984. Cloning with tandem gene systems for high level gene expression. *Nucl. Acids Res.* 12:

Lemischka, I.R., Raulet, D.H., and Mulligan, R.C. 1986. Developmental potential and dynamic behavior of hematopoietic stem cells. *Cell* 45:917-927.

Le Pecq, J-B. 1971. Use of ethidium bromide for separation and determination of nucleic acids of various conformational forms and measurement of their associated enzymes. *In* Methods of Biochemical Analysis, Vol. 20 (D. Glick, ed.) pp. 41-86. John Wiley & Sons, New York.

Lewin, B. 1980. *In* Gene Expression 2, p. 962. John Wiley & Sons, New York.

Lichter, P., Boyle, A.L., Cremer, T., and Ward, D.C. 1991. Analysis of genes and chromosomes by non-isotopic in situ hybridization. *Genet. Anal. Techn. Appl.* 8:24-35.

Linscott's Directory of Immunological and Biological Reagents. Mill Valley, Calif.

Lis, J.T. 1980. Fractionation of DNA fragments by polyethylene glycol-induced precipitation. *Methods Enzymol.* 65:347-353.

Littlefield, J.W. 1964. Selection of hybrids from matings of fibroblasts in vitro and their presumed recombinants. *Science* 145:709-710.

Lopata, M.A., Cleveland, D.W., and Sollner-Webb, B. 1984. High-level expression of a chloramphenicol acetyltransferase gene by DEAE-dextran-mediated DNA transfection coupled with a dimethyl sulfoxide or glycerol shock treatment. *Nucl. Acids Res.* 12:5707.

Luna, L.G. 1968. Manual of Histologic Staining: Methods of the Armed Forces Institute of Pathology (3rd ed.). McGraw-Hill, New York.

Ma, H., Kunes, S., Schatz, P.J., and Botstein, D. 1987. Plasmid construction by homologous recombination in yeast. *Gene* 58:201-216.

Maina, C.V., Riggs, P.D., Grandea, A.G., Slatko, B.E., Moran, L.S., Tagliamonte, J.A., McReynolds, L.A., and Guan, C. 1988. An *Escherichia coli* vector to express and purify foreign proteins by fusion to and separation from maltose-binding protein. *Gene* 74:365-373.

Maiorella, B., Inlow, D., Shauger, A., and Harano, D. 1988. Large-scale insect cell culture for recombinant protein production. *Bio/Technology* 6:1406-1410.

Maniatis, T., Jeffrey, A., and van deSande, H. 1975. Chain length determination of small double- and single-stranded DNA molecules by polyacrylamide gel electrophoresis. *Biochemistry* 14:3787-3794.

Maniatis, T. and Ptashne, M. 1973a. Multiple repressor binding at the operators in bacteriophage. *Proc. Natl. Acad. Sci. U.S.A.* 70:1531-1535.

Maniatis, T. and Ptashne, M. 1973b. Structure of the operators. *Nature* 246:133-136.

Mann, C. and Davis, R.W. 1986. Structure and sequence of the centromeric DNA of chromosome 4 in *Saccharomyces cerevisiae. Mol. Cell Biol. 6:241-245.*

Mann, R., Mulligan, R.C., and Baltimore, D. 1983. Construction of a retrovirus package mutant and its use to produce helper-free defective retrovirus. *Cell* 33:153-159.

Marchuk, D., Drumm, M., Saulino, A., and Collins, F.S. 1991. Construction of T-vectors, a rapid and general system for direct cloning of unmodified PCR products. *Nucl. Acids Res.* 19:1154.

Mardis, E.R. and Roe, B.A. 1989. Automated methods for single-stranded DNA isolation and dideoxynucleotide DNA sequencing reactions on a robotic workstation. *BioTechniques* 7:840-850.

Markowitz, D., Goff, S., and Bank, A. 1988. A safe packaging line for gene transfer: separating viral genes on two different plasmids. *J. Virol.* 62:1120-1124.

Marmur, J. 1961. A procedure for the isolation of desoxyribonucleic acids from microorganisms. *J. Mol. Biol.* 3:208-218.

Marshall, C.J. 1995. Specificity of receptor tyrosine kinase signalling: Transient versus sustained extracellular signal-regulated kinase activation. *Cell* 80:179-185.

Martin, C., Bresnick, L., Juo, R.-R., Voyta, J.C., and Bronstein, I. 1991. Improved chemiluminescent DNA sequencing. *BioTechniques* 11:102-109.

Matsuura, Y., Possee, R.D., and Bishop, D.H.L. 1987. Baculovirus expression vectors: The requirements for high level expression of proteins, including glycoproteins. *J. Gen. Virol.* 68:1233-1250.

Maxam, A.M. and Gilbert, W. 1977. A new method for sequencing DNA. *Proc. Natl. Acad. Sci. U.S.A.* 74:560-564.

Maxam, A.M. and Gilbert, W. 1980. Sequencing end-labeled DNA with base-specific chemical cleavages. *Methods Enzymol.* 65:499-559.

McClelland, M., Hanish, J., Nelson, M., and Patel, Y. 1988. KGB: A single buffer for all restriction endonucleases. *Nucl. Acids Res.* 16:364.

McGregor, W.C. 1986. Selection and use of ultrafiltration membranes. *In* Membrane Separations in Biotechnology (W.C. McGregor, ed.) pp. 1-36. Marcel Dekker, New York.

McKimm-Breschkin, J.L. 1990. The use of tetramethylbenzidine for solid phase immunoassays. *J. Immunol. Methods* 135:277-280.

McKnight, S.L. and Kingsbury, R. 1982. Transcription control signals of a eukaryotic protein-coding gene. *Science* 217:316-324.

McPhie, P. 1971. Dialysis. *Methods Enzymol.* 22:23-32.

Melton, D.A., Krieg, P.A., Rebagliati, M.R., Maniatis, T., Zinn, K., and Green, M.R. 1984. Efficient in vitro synthesis of biologically active RNA and RNA hybridization probes from plasmids containing a bacteriophage SP6 promoter. *Nucl. Acids Res.* 12:7035-7056.

Merril, C.R., Goldman, D., and Van Keuren, M.L. 1984. Gel protein stains: Silver stain. *Methods Enzymol.* 104:441-447.

Messing, J. 1983. New M13 vectors for cloning. *Methods Enzymol.* 101:20-78.

Messing, J. 1988. M13, the universal primer and the polylinker. *Focus (BRL)* 10(2)21-26.

Mierendorf, R.C. and Pfeffer, D. 1987. Sequencing of RNA transcripts synthesized in vitro from plasmids containing bacteriophage promoters. *Methods Enzymol.* 152:563-566.

Miller, A.D. and Buttimore, C. 1986. Redesign of retrovirus packaging cell lines to avoid recombination leading to helper virus production. *Mol. Cell. Biol.* 6:2895-2902.

Miller, A.D., Law, M-F., and Verma, I.M. 1985. Generation of helper-free amphotropic retroviruses that transduce a dominant-acting, methotrexate-resistant dihydrofolate reductase gene. *Mol. Cell. Biol.* 5:431-437.

Miller, J. 1972. Experiments in Molecular Genetics. CSH Laboratory, Cold Spring Harbor, N.Y.

Miller, J.H. 1978. The *lacI* gene: Its role in *lac* operon control and its uses as a genetic system. *In* The Operon (J. Miller, ed.) pp. 31-88. CSH Laboratory, Cold Spring Harbor, N.Y.

Miller, J.H., Lebrowski, J.S., Griese, K.S., and Calos, M.P. 1984. Specificity of mutations induced in transferred DNA by mammalian cells. *EMBO J.* 3:3117-3121.

Miller, J.J., Schultz, G.S., and Levy, R.S. 1984. Rapid purification of radioiodinated peptides with Sep-Pak™ reverse-phase cartridges and HPLC. *Int. J. Pept. Protein Res.* 24:112-122.

Moazed, D. and Noller, H.F. 1987. Interaction of antibiotics with functional sites in 16S ribosomal RNA. *Nature* 327:389-394.

Moeremans, M., Daneels, G., and De Mey, J. 1985. Sensitive colloidal metal (gold or silver) staining of protein blots on nitrocellulose membranes. *Anal. Biochem.* 145:315-321.

Moore, S. 1981. Pancreatic DNase. *In* The Enzymes, Vol. 14A (P.D. Boyer, ed.) pp. 281-298. Academic Press, San Diego.

Moos, M. 1995. Isolation of proteins for microsequemce analysis. *In* Current Protocols in Immunology (J.E. Coligan, A.M. Kruisbeek, D.H. Margulies, E.M. Shevach, and W. Strober, eds.) pp. 8.7.1-8.7.12. John Wiley & Sons, New York.

Moos, M., Nguyen, N.Y., and Liu, T-Y. 1988. Reproducible, high-yield sequencing of proteins electrophoretically separated and transferred to an inert support. *J. Biol. Chem.* 263:6005-6008.

Morgenstern, J.P. and Land, H. 1990. Advanced mammalian gene transfer: High-titer retroviral vectors with multiple drug selection markers and a complementary helper-free packaging cell line. *Nucl. Acids Res.* 18:3587-3596.

Morla, A. and Wang, J.Y.J. 1986. Protein tyrosine phosphorylation in the cell cycle of BALB/c 3T3 fibroblasts. *Proc. Natl. Acad. Sci. U.S.A.* 83:8191-8195.

Morrissey, J.H. 1981. Silver stain for proteins in polyacrylamide gels: A modified procedure with enhanced uniform sensitivity. *Anal. Biochem.* 117:307-310.

Moscovici, C., Moscovici, M.G., Jiminez, H., Lai, M.M.C., Hayman, M.J., and Vogt, P.K. 1977. Continuous tissue culture cell lines derived from chemically induced tumors of Japanese quail. *Cell.* 11:95-103.

Mueller, P.R., and Wold, B. 1989. In vivo footprinting of a muscle specific enhancer by ligation mediated PCR. *Science* 246:780-786.

Mueller, P.R. and Wold, B. 1991. Ligation-mediated PCR: Applications to genomic footprinting. *Methods* 2:20-31.

Mueller, P.R., Salser, S.J., and Wold, B. 1988. Constitutive and metal inducible protein:DNA interactions at the mouse metallothionein-I promoter examined by in vivo and in vitro footprinting. *Genes & Dev.* 2:412-427.

Mulligan, R.C. and Berg, P. 1981. Selection for animal cells that express the *E. coli* gene coding for xanthine–guanine phosphoribosyltransferase. *Proc. Natl. Acad. Sci. U.S.A.* 78:2072-2076.

Muneoka, K., Wanek, N., and Bryant, S.V. 1986. Mouse embryos develop normally exo utero. *J. Exp. Zool.* 239:289-293.

Murphy, C.I., McIntire, J.R., Davis, D.vR., Hodgdon, H., Seals, J.R., and Young, E. 1993. Enhanced expression, secretion, and large-scale purification of recombinant HIV-1 gp120 in insect cells using the baculovirus egt and p67 signal peptides. *Prot. Expr. Purif.* 4:349-357.

Murray, A.W. and Szostak, J.W. 1983. Pedigree analysis of plasmid segregation in yeast. *Cell* 34:961-970.

Murray, A.W., Schultes, N.P., and Szostak, J.W. 1986. Chromosome length controls meiotic chromosome segregation in yeast. *Cell* 45:529-536.

Murray, M.G. and Thompson, W.F. 1980. Rapid isolation of high-molecular-weight plant DNA. *Nucl. Acids Res.* 8:4321-4325.

Muzyczka, N. ed. 1989. Eukaryotic Viral Expression Vectors: Current Topics in Microbiology and Immunology. Springer-Verlag, Berlin.

Myers, R.M., Lerman, L.S., and Maniatis, T. 1985. A general method for saturation mutagenesis of cloned DNA fragments. *Science* 229:242-246.

Myers, R.M., Maniatis, T., and Lerman, L.S. 1987. Detection and localization of single-base changes by denaturing gradient gel electrophoresis. *Methods Enzymol.* 155:501-527.

Nagai, K. and Thøgersen, H.C. 1987. Synthesis and sequence-specific proteolysis of hybrid proteins produced in *Escherichia coli*. *Methods Enzymol.* 153:461-481.

Nasmyth, K.A. and Reed, S.I. 1980. Isolation of genes by complementation in yeast. Molecular cloning of a cell cycle gene. *Proc. Natl. Acad. Sci. U.S.A.* 77:2119-2123.

Needleman, S.B. and Wunsch, C.D. 1970. A general method applicable to the search for similarities in the amino acid sequence of two proteins. *J. Mol. Biol.* 48:443-453.

Neihardt, F.C., Ingraham, J.L., Low, K.B., Magasanik, B., Schaechter, M., and Umbarger, H.E., eds. 1987. *Escherichia coli* and *Salmonella typhinurium:* Cellular and Molecular Biology. American Scoiety for Microbiology, Washington, D.C.

Nordeen, S.K. 1988. Luciferase reporter gene vectors for analysis of promoters and enhancers. *BioTechniques* 6:454-457.

Norrander, J., Kempe, T., and Messing, J. 1983. Construction of improved M13 vectors using oligonucleotide-directed mutagenesis. *Gene* 26:101-106.

Nuovo, G.J., Gallery, F., MacConnell, P., Becker, J., and Bloch, W. 1991. An improved technique for the in situ detection of DNA after polymerase chain reaction amplification. *Am. J. Path.* 139:1239-1244.

O'Farrell, P.H. 1975. High-resolution two-dimensional electrophoresis of proteins. *J. Biol. Chem.* 250:4007-4021.

O'Farrell, P.H., Kutter, E., and Nakanishe, M. 1980. A restriction map of bacteriophage T4 genome. *Mol. Gen. Genet.* 179:411-435.

Ohara, O., Dorit, R.L., and Gilbert, W. 1989. One-sided polymerase chain reaction: The amplification of cDNA. *Proc. Natl. Acad. Sci. U.S.A.* 86:5673-5677.

Okajima, T., Tanabe, T., and Yasuda, T. 1993. Nonurea sodium dodecyl sulfate-polyacrylamide gel electrophoresis with high-molarity buffers for the separation of proteins and peptides. *Anal. Biochem.* 211:293-300.

Olesen, C.E.M., Martin, C.S., and Bronstein, I. 1993. Chemiluminescent DNA sequencing with multiplex labeling. *BioTechniques* 15:480-485.

Omer, C.A. and Faras, A.J., 1982. Mechanism of release of the avian retrovirus tRNAtrp primer molecule from viral DNA by ribonuclease H during reverse transcription. *Cell* 30:797-805.

O'Reilly, D.R., Miller, L.K., and Luckow, V.A. 1992. Baculovirus Expression Vectors. W.H. Freeman and Company, New York.

Orr-Weaver, T.L., Szostak, J.W., and Rothstein, R.J. 1983. Genetic applications of yeast transformation with linear and gapped plasmids. *Methods Enzymol.* 101:228-245.

Palmer, T.D., Hock, R.A., Osborne, W.R.A., and Miller, A.D. 1987. Efficient retrovirus-mediated transfer and expression of a human adenosine deaminase gene in diploid skin fibroblasts from an adenosine-deficient human. *Proc. Natl. Acad. Sci. U.S.A.* 84:1055-1059.

Palmiter, R.D. 1974. Magnesium precipitation of ribonucleoprotein complexes: Expedient techniques for the isolation of undegraded polysomes and messenger ribonucleic acid. *Biochemistry* 13:3606.

Pardue, M.L. 1985. In situ Hybridization. *In* Nucleic Acid Hybridization: A Practical Approach (B.D. Hames and S.J. Higgins, eds.) pp. 179-202. IRL Press, Oxford.

Parent, S.A., Fenimore, C.M., and Bostian, K.A. 1985. Vector systems for the expression, analysis and cloning of DNA sequences in *S. cerevisiae. Yeast* 1:83-138.

Parnes, J.R., Velan, B., Felsenfeld, A., Ramanathan, L., Ferrini, U., Appella, E., and Seidman, J.G. 1981. Mouse β2-microglobulin cDNA clones: A screening procedure for cDNA clones corresponding to rare mRNAs. *Proc. Natl. Acad. Sci. U.S.A.* 78:2253.

Perucho, M., Hanahan, D., and Wigler, M. 1980. Genetic and physical linkage of exogenous sequences in transformed cells. *Cell* 22:309-317.

Pharmacia. Gel Filtration: Theory and Practice. Uppsala, Sweden.

Pharmacia. Ion-Exchange Chromatography: Principles and Methods. Uppsala, Sweden.

Pohl, T. 1990. Concentration of proteins and removal of solutes. *Methods Enzymol.* 182:68-83.

Poncz, M., Solowiejczyk, D., Ballantine, M., Schwartz, E., and Surrey, S. 1982. "Nonrandom" DNA sequence analysis in bacteriophage M13 by the dideoxy chain-termination method. *Proc. Natl. Acad. Sci. U.S.A.* 79:4298-4302.

Potter, H. 1988. Electroporation in biology: Methods, applications, and instrumentation. *Anal. Biochem.* 74:361-373.

Potter, H., Weir, L., and Leder, P. 1984. Enhancer-dependent expression of human *k* immunoglobulin genes introduced into mouse pre-B lymphocytes by electroporation. *Proc. Natl. Acad. Sci. U.S.A.* 81:7161.

Pouwels, P.H., Enger-Valk, B.E., and Brammar, W.J. 1985. Cloning Vectors: A Laboratory Manual. Elsevier Science Publishing, Amsterdam.

Pu, W.T. and Struhl, K. 1992. Uracil interference, a rapid and general method for defining protein-DNA interactions involving the 5-methyl group of thymines: The GCN4-DNA complex. *Nucl. Acids Res.* 20:771-775.

Radloff, R., Bauer, W., and Vinograd, J. 1967. A dye-buoyant-density method for the detection and isolation of closed circular duplex DNA: The closed circular DNA in HeLa cells. *Proc. Natl. Acad. Sci. U.S.A.* 57:1514-1521.

Rasched, I. and Oberer, E. 1986. Ff coliphages: Structural and functional relationships. *Microbiol. Rev.* 50:401-427.

Ratliff, R.L. 1981. Terminal deoxynucleotidyltransferase. *In* The Enzymes, Vol. 14A (P.D. Boyer, ed.) pp. 105-118. Academic Press, San Diego.

Reddy, K.J., Webb, R., and Sherman, L.A. 1990. Bacterial RNA isolation with one-hour centrifugation in a tabletop centrifuge. *BioTechniques* 8:250-251.

Regnier, F. 1984. High-performance ion-exchange chromatography. *Methods Enzymol.* 104:170-189.

Roberts, R.J. and Macelis, D. 1992. Restriction enzymes and their isoschizomers. *Nucl. Acids Res.* 20 (Supp. 1):2167-2180.

Robins, D.M., Ripley, S., Henderson, A.S., and Axel, R. 1981. Transforming DNA integrates into the host chromosome. *Cell* 23:29-39.

Rose, G.D., Geselowitz, A.R., Lesser, G.J., Lee, R.H., and Zehfus, M.H. 1985. Hydrophobicity of amino acid residues in globular proteins. *Science* 229:834-838.

Rose, M.D., Grisaffi, P., and Botstein D. 1984. Structure and function of the yeast *URA3* gene: Expression in *Escherichia coli. Gene* 29:113-124.

Rose, M.D. 1987. Isolation of genes by complementation in yeast. *Methods Enzymol.* 152:481-504.

Rose, M.D., Novick, P., Thomas, J.H., Botstein, D., and Fink, G.R. 1987. A *Saccharomyces cerevisiae* genomic plasmid bank based on a centromere-containing shuttle vector. *Gene* 60;237-243.

Rose, M.D., Winston, F., and Hieter, P. 1990. Laboratory Course Manual for Methods in Yeast Genetics. CSH Laboratory, Cold Spring Harbor, N.Y.

Rosenberg, M., Ho, Y.S., and Shatzman, A.R. 1983. The use of pKC30 and its derivatives for controlled expression of genes. *Methods Enzymol.* 101:123-138.

Rothstein, R. 1985. Cloning in yeast. *In* DNA Cloning, Vol. 2: A Practical Approach (D.M. Glover, ed.) 4th ed., pp. 45-67. IRL Press, Oxford.

Rothstein, R.J. 1983. One-step gene disruption in yeast. *Methods Enzymol.* 101:202-210.

Ruden, D.M., Ma, J., Li,Y., Wood, K., and Ptashne, M. 1991. Generating yeast transcriptional activators containing no yeast protein sequences. *Nature* 350:426-430.

Saenger, W. 1984. Principles of Nucleic Acid Structure. Springer-Verlag, New York.

Saiki, R.K., Gelfand, D.H., Stoffel, S., Scharf, S.J., Higuchi, R., Horn, G.T., Mullis, K.B., and Erlich,

H.A. 1988. Primer-directed enzymatic amplification of DNA with a thermostable DNA polymerase. *Science* 239:487-491.

Sambrook, J., Fritsch, E.F., and Maniatis, T.M. (eds.) 1989. Molecular Cloning: A Laboratory Manual, 2nd ed. CSH Laboratory Press, Cold Spring Harbor, New York.

Sanger, F., Nicklen, S., and Coulson, A.R. 1977. DNA sequencing with chain-terminating inhibitors. *Proc. Natl. Acad. Sci. U.S.A.* 74:5463-5467.

Sanger, F., Coulson, A.R., Barrell, B.G., Smith, A.J.M., and Roe, B.A. 1980. Cloning in single-stranded bacteriophage as an aid to rapid DNA sequencing. *J. Mol. Biol.* 143:161-178.

Schagger, H. and von Jagow, G. 1987. Tricine-sodium dodecyl sulfate-polyacrylamide gel electrophoresis for the separation of proteins in the range from 1 to 100 kDa. *Anal. Biochem.* 166:368-379.

Schein, C.H. 1989. Production of soluble recombinant proteins in bacteria. *Bio/Technology* 7:1141-1149.

Scherer, S. and Davis, R.W. 1979. Replacement of chromosome segments with altered DNA sequences constructed in vitro. *Proc. Natl. Acad. Sci. U.S.A.* 76:4951-4955.

Schiestl, R.H. and Gietz, R.D. 1989. High-efficiency transformation of intact yeast cells using single-stranded nucleic acids as a carrier. *Curr. Genet.* 16:339-346.

Schneider, J.C. and Guarente, L. 1991. Vectors for expression of cloned genes in yeast. *Methods Enzymol.* 194:382-383.

Schneppenheim, R., Budde, U., Dahlmann, N., and Rautenberg, P. 1991. Luminography—a new, highly sensitive visualization method for electrophoresis. *Electrophoresis* 12:367-372.

Schwartz, D.C. and Cantor, C.R. 1984. Separation of yeast chromosome-sized DNAs by pulsed-field gradient electrophoresis. *Cell* 37:67-75.

Sears, L.E., Moran, L.S., Kissinger, C., Creasey, T., Perry-O'Keefe, H., Roskey, M., Sutherland, E., and Slatko, B.S. 1992. Circum-Vent thermal cycle sequencing and alternative manual and automated DNA sequencing protocols using the highly thermostable Vent_R (exo$^-$) DNA polymerase. *BioTechniques.* 13:626-633.

Seed, B. and Sheen, J.-Y. 1988. A simple phase-extraction assay for chloramphenicol acetyltransferase activity. *Gene* 67:271-277.

Seed, B., Parker, R.C., and Davidson, N. 1982. Representation of DNA sequences in recombinant DNA libraries prepared by restriction enzyme partial digestion. *Gene* 19:201-209.

Selden, R.F, Burke-Howie, K., Rowe, M.E., Goodman, H.M., and Moore, D.D. 1986. Human growth hormone as a reporter gene in regulation studies employing transient gene expression. *Mol. Cell. Biol.* 6:3173-3179.

Sen, D. and Gilbert, W. 1988. Formation of parallel four-stranded complexes by guanine-rich motifs in DNA and its implications for meiosis. *Nature* 334:364-366.

Shankarappa, B., Balachandran, R., Gupta, P., and Ehrlich, G.D. 1992. Introduction of multiple restriction enzyme sites by in vitro mutagenesis using the polymerase chain reaction. *PCR Method. Appl.* 1:277-278.

Sharp, P.A., Berk, A.J., and Berget, S.M. 1980. Transcription maps of adenovirus. *Methods Enzymol.* 65:750-768.

Shatzman, A.R. and Rosenberg, M. 1987. Expression, identification and characterization of recombinant gene products in *Escherichia coli. Methods. Enzymol.* 152:661-673.

Sherman, F. 1987. Micromanipulators for yeast genetic studies. *Applied Microbiol.* 26:829.

Sherman, F., Fink, G.R., and Lawrence, C.W. 1979. Methods in Yeast Genetics. CSH Laboratory, Cold Spring Harbor, N.Y.

Shleien, B. (ed.) 1987. Radiation Safety Manual for Users of Radioisotopes in Research and Academic Institutions. Nucleon Lectern Associates, Olney, Md.

Shortle, D. Haber, J.E., and Botstein, D. 1982. Lethal disruption of the yeast actin gene by integrative DNA transformation. *Science* 217:373-373.

Sigma. Molecular weight markers for proteins kit (Technical Bulletin MWS-877L). Sigma Chemical Company, St. Louis, Mo.

Sikorski, R.S. and Hieter, P. 1989. A system of shuttle vectors and yeast host strains designed for efficient manipulation of DNA in *Saccaromyces cerevisae. Genetics* 122:19-27.

Simonsen, C.C. and Levinson, A.D. 1983. Isolation and expression of an altered mouse dihydrofolate reductase cDNA. *Proc. Natl. Acad. Sci. U.S.A.* 80:2495-2499.

Singh, H., Clerc, R.G., LeBowitz, J.H. 1989. Molecular cloning of sequence-specific DNA-binding proteins using recognition-site probes. *BioTechniques* 7:252-261.

Slatko, B. 1991a. Protocols for manual dideoxy DNA sequencing. *In* Methods in Nucleic Acids Research (J. Karam, L. Chao, and G. Warr, eds.) pp. 83-129. CRC Press, Boca Raton, Fla.

Slatko, B.E. 1991b. Sources of reagnets and supplies for dideoxy DNA sequencing and other applications. *In* Methods in Nucleic Acids Research (J. Karam, L. Chao, and G. Warr, eds.) pp. 379-392. CRC Press, Boca Raton, Fla.

Smith, D.B. 1993. Purification of glutathione-S-transferase fusion proteins. *Methods Mol. Cell Biol.* 4:220-229.

Smith, D.B. and Johnson, K.S. 1988. Single-step purification of polypeptides expressed in *Escherichia coli* as fusions with glutathione *S*-transferase. *Gene* 67:31-40.

Smith, T.F. and Waterman, M.S. 1981. Identification of common molecular subsequences. *J. Mol. Biol.* 147:195-197.

Sorger, P.K., Ammerer, G., and Shore, D. 1989. Identification and purification of sequence-specific DNA-binding proteins. *In* Protein Function: A Practical Approach (T.E. Creighton, ed.) pp. 199-223. IRL Press, Oxford.

Southern, E.M. 1975. Detection of specific sequences among DNA fragments separated by gel electrophoresis. *J. Mol. Biol.* 98:503-517.

Southern, E. 1979. Gel electrophoresis of restriction fragments. *Methods Enzymol.* 68:152-176.

Southern, P.J. and Berg, P. 1982. Transformation of mammalian cells to antibiotic resistance with a bacterial gene under control of the SV40 early region promoter. *J. Mol. Appl. Gen.* 1:327-341.

Spande, T.F., Witkop, B., Degani, Y., and Patchornik, A. 1970. Selective cleavage and modification of peptides and proteins. *Adv. Prot. Chem.* 24:97-260.

Speel, E.J.M., Schutte, B., Wiegant, J., Ramaekers, F.C., and Hopman, A.H.N. 1992. A novel fluorescence detection method for in situ hybridization, based on the alkaline phosphatase–fast red reaction. *J. Histochem. Cytochem.* 40:1299-1308.

Springer, T.A. 1981. Monoclonal antibody analysis of complex biological systems: Combination of cell hybridization and immunoadsorbents in a novel cascade procedure and its application to the macrophage cell surface. *J. Biol. Chem.* 256:3833-3839.

Sternberger, L.A., Hardy, P.H. Jr., Cuculis, J.J., and Meyer, H.G. 1970. The unlabeled antibody enzyme method of immunohistochemistry. Preparation of soluble antigen-antibody complex (horseradish peroxidase anti-horseradish peroxidase) and its use in the identification of spirochetes. *J. Histochem. Cytochem.* 18:315-333.

Stinchcomb, D.T., Mann, C., and Davis, R.W. 1982. Centromeric DNA from *Saccharomyces cerevisiae. J. Mol. Biol.* 158:157-190.

Stoker, A. and Bissell, M.J. 1988. Development of avian sarcoma and leukosis virus-based vector-packaging cell lines. *J. Virol.* 62:1008-1015.

Stoker, A.W. and Bissell, M.J. 1987. Quantitative immunocytochemical assay for infectious avian retroviruses. *J. Gen. Virol.* 68:22481-2485.

Stoker, A.W., Hatier, C., and Bissell, M.J. 1990. The embryonic environment strongly attenuates v-src oncogenesis in mesenchymal and epithelial tissues, but not endothelia. *J. Cell Biol.* 111:217-228.

Strathern, J.N., Jones, E.W., and Broach, J.R. (eds.) 1981. The Molecular Biology of the Yeast *Saccharomyces*: Life Cycle and Inheritance. CSH Laboratory, Cold Spring Harbor, N.Y.

Strathern, J.N., Jones, E.W., and Broach, J.R.(eds.) 1982. The Molecular Biology of the Yeast *Saccharomyces*: Metabolism and Gene Expression. CSH Laboratory, Cold Spring Harbor, N.Y.

Struhl, K. 1985. A rapid method for creating recombinant DNA molecules. *BioTechniques* 3:452-453.

Struhl, K., Stinchcomb, D.T., Scherer, S., and Davis, R.W. 1979. High-frequency transformation of yeast: Autonomous replication of hybrid DNA molecules. *Proc. Natl. Acad. Sci. U.S.A.* 76:1035-1039.

Studier, F.W., Rosenberg, A.H., Dunn, J.J., and Dubendorff, J.W. 1990. Use of T7 RNA polymerase to direct the expression of cloned genes. *Methods Enzymol.* 185:60-89.

Summers, M.D. and Smith, G.E. 1987. A manual of methods for baculovirus vectors and insect cell culture procedures. Texas Agricultural Experiment Station Bulletin No. 1555. College Station, Tex.

Summers, W.C. 1970. A simple method for extraction of RNA from *E. coli* utilizing diethylpyrocarbonate. *Anal. Biochem.* 33:459-463.

Sussman, D.J. and Milman, G. 1984. Short-term, high-efficiency expression of transfected DNA. *Mol. Cell. Biol.* 4:1641.

Sutcliffe, J.G. 1978. Complete nucleotide sequence of the *Escherichia coli* plasmid pBR322. *Cold Spring Harbor Symp. Quant. Biol.* 43:77-90.

Sutherland, M.W. and Skerritt, J.H. 1986. Alkali enhancement of protein staining on nitrocellulose. *Electrophoresis* 7:401-406.

Sweet, R.M. and Eisenberg, D. 1983. Correlation of sequence hydrophobicities measures similarity in three-dimensional protein structure. *J. Mol. Biol.* 171:479-488.

Tabor, S. and Richardon, C.C. 1985. A bacteriophage T7 RNA polymerase/promotor system for controlled exclusive expression of specific genes. *Proc. Natl. Acad. Sci. U.S.A.* 82:1074-1078.

Tabor, S. and Richardson, C.C. 1987a. DNA sequence analysis with a modified bacteriophage T7 DNA polymerase. *Proc. Natl. Acad. Sci. U.S.A.* 84:4767-4771.

Tabor, S. and Richardson, C.C. 1987b. Selective oxidation of the exonuclease domain of bacteriophage T7 DNA polymerase. *J. Biol. Chem.* 262:15330-15333.

Tabor, S. and Richardson, C.C. 1989. Selective inactivation of the exonuclease activity of bacteriophage T7 DNA polymerase by in vitro mutagenesis. *J. Biol. Chem.* 264:6447-6458.

Tabor, S. and Richardson, C.C. 1990. DNA sequence analysis with a modified bacteriophage T7 DNA polymerase: Effect of pyrophorolysis and metal ions. *J. Biol. Chem.* 265:8322-8328.

Takagi, H., Morinaga, Y., Tsuchiya, M., Ikemura, G., and Inouye, M. 1988. Control of folding of proteins secreted by a high expression secretion vector, pIN-III-*ompA*: 16-fold increase in production of active subtilisin E in *Escherichia coli. Bio/Technology* 6:948-950.

Tempst, P., Link, A.J., Riviere, L.R., Fleming, M., and Elicone, C. 1990. Internal sequence analysis of proteins separated on polyacrylamide gels at the picomole level: Improved methods, applications and gene cloning strategies. *Electrophoresis* 11:537-553.

Thomas, K.R. and Olivera, B.M. 1978. Processivity of DNA exonucleases. *J. Biol. Chem.* 253:424-429.

Thomas, P. S. 1980. Hybridization of denatured RNA and small DNA fragments transferred to nitrocellulose. *Proc. Natl. Acad. Sci. U.S.A.* 77:5201.

Toh-e, A. and Wickner, R.B. 1981. Curing of the 2μ DNA plasmid from *Saccharomyces cerevisiae. J. Bacteriol.* 145:1421-1424.

Tonegawa, S. 1985. The molecules of the immune system. *Sci. Am.* 253:122-131.

Treisman, R. 1986. Transient accumulation of c-fos RNA following serum stimulation requires a conserved 5′ element and c-fos 3′ sequences. *Cell* 42:889-902.

Tschumper, G. and Carbon, J. 1980. Sequence of a yeast DNA fragment containing a chromosomal replicator and the *TRP1* gene. *Gene* 10:157-166.

T'so, P.O.P. 1974. Bases, nucleosides, and nucleotides. *In* Principles in Nuclei Acid Chemistry, Vol 1

(P.O.P. T'so, ed.) pp. 453-584. Academic Press, San Diego.

Uchida, T. and Egami, F. 1971. Microbial ribonucleases with special reference to RNases T1, T2, N1, and U2. *In* The Enzymes, Vol. IV (P.D. Boyer, ed.) pp. 205-250. Academic Press, San Diego.

Uhlenbeck, O.C. and Gumport, R.I. 1982. T4 RNA ligase. *In* The Enzymes, Vol. 15B (P.D. Boyer, ed.) pp. 31-60. Academic Press, San Diego.

Uhlmann, E., 1988. An alternate approach in gene synthesis: Use of long self-priming oligodeoxynucleotides for construction of double-stranded DNA. *Gene* 71:29-40.

Unger, K. 1984. High-performance size exclusion chromatography. *Methods Enzymol.* 104:154-169.

Van Vunakis, H. and Langone, J.J., eds. 1980. Immunochemical techniques. *Methods Enzymol.* 70:1-525.

Verma, I.M. 1977. Reverse transcriptase. *In* The Enzymes, Vol. 14A (P.D. Boyer, ed.) pp. 87-104. Academic Press, San Diego.

Vinson, C.R., LaMarco, K.L., Johnson, P.F., Landschulz, W.H., and McKnight, S.L. 1988. *In situ* detection of sequence-specific DNA binding activity specified by a recombinant bacteriophage. *Genes & Dev.* 2:801-806.

Vogt, V.M. 1980. Purification and properties of S1 nuclease from *Aspergillus. Methods Enzymol.* 65:248-254.

Wallace, R.B. and Miyada, C.G. 1987. Oligonucelotide probes for the screening of recombinant DNA libraries. *Methods Enzymol.* 152:432-442.

Wang, A.R., Quigley, G.J., Kolpak, F.J., Crawford, J.L., Van Boom, J.H., Van der Marel, G., and Rich, A. 1979. Molecular structure of a left-handed double helical DNA fragment at atomic resolution. *Nature* 282:680-686.

Warren, T.G. and Shields, D. 1984. Expression of preprosomatostatin in heterologous cells: Biosynthesis, posttranslational processing, and secretion of mature somatostatin. *Cell* 39:547-555.

Watkins, S. 1995. Cryosectioning. *In* Current Protocols in Molecular Biology (F. Ausubel, D. Moore, J.G. Seidman, J. Smith, and K. Struhl, eds.) pp. 14.2.1-14.2.8. John Wiley & Sons, New York.

Watson, J.D., Hopkins, N.H., Roberts, J.W., Steitz, J.A., and Weiner, A.M. 1987. Yeasts as the *E. coli of eukaryotic cells. In* Molecular Biology of the Gene, Vol. 1, pp. 550-594. Benjamin/ Cummings, Menlo Park, Calif.

Weber, K., Pringle, J.R., and Osborn, M. 1972. Measurement of molecular weights by electrophoresis on SDS-acrylamide gel. *Methods Enzymol.* 26:3-27.

Weinstock, G.M., Rhys, C. Berman, M.L., Hampar, B., Jackson, D., Silhavy, T.L., Weisemann, J., and Zweig, M. 1983. Open reading frame expression vectors: A general method for antigen productin in *Escherichia coli* using protein fusions to β-galactosidase. *Proc. Natl. Acad. Sci. U.S.A.* 80:4432-4436.

Weiss, R., Teich, N., Varmus, H., and Coffin, J. 1984 and 1985. RNA Tumor Viruses. CSH Laboratory, Cold Spring Harbor, N.Y.

Wienand, U., Schwarz, Z., and Felix, G. 1978. Electrophoretic elution of nucleic acids from gels adapted for subsequent biological tests: Application for analysis of mRNAs from maize endosperm. *FEBS Lett.* 98:319-323. (2.6)

Wigler, M., Silverstein, S., Lee, L-S., Pellicer, A., Cheng, Y-C., and Axel, R. 1977. Transfer of purified herpes virus thymidine kinase gene to cultured mouse cells. *Cell* 11:223-232.

Wilbur, W.J. and Lipman, D.J. 1983. Rapid similarity searches of nucleic acid and protein data banks. *Proc. Natl. Acad. Sci. U.S.A.* 80:726-730.

Wilchek, M., Miron, T., and Kohn, J. 1984. Affinity chromatography. *Methods Enzymol.* 104:3-55.

Wise, R.J., Orkin, S.H., and Collins, T. 1989. Aberrant expression of platelet-derived growth factor A-chain cDNAs due to cryptic splicing of RNA transcripts in COS-1 cells. *Nucl. Acids Res.*17:6591-6601.

Wood, W.I., Gitschier, J., Lasky, L.A., and Lawn, R.M. 1985. Base composition-independent hybridization in tetramethylammonium chloride: A method for oligonucleotide screening of highly complex gene libraries. *Proc. Natl. Acad. Sci. U.S.A.* 82:1585-1588.

Woods, D.E., Miarkham, A.F., Ricker, A.T., Goldberger, G., and Colten, H.R. 1982. Isolation of cDNA clones for the human complement protein factor B, a class III major histocompatibility complex gene product. *Proc. Natl. Acad. Sci. U.S.A.* 79:5661-5665.

Yanisch-Perron, C., Vieira, J., and Messing, J. 1985. Improved M13 phage cloning vectors and host strains: Nucleotide sequences of the M13mp18 and pUC19 vectors. *Gene* 33:103-119.

Yokoyama, W.M. 1995. Cryopreservation of cells. *In* Current Protocols in Immunology (J.E. Coligan, A.M. Kruisbeek, D.H. Margulies, E.M. Shevach, and W. Strober, eds.) pp. A.3.15-A.3.17. John Wiley and Sons, New York.

Zabin, L. and Fowler, A.V. 1978. β-galactosidase, the lactose permease protein, and thiogalactosdide transacetylase. *In* The Operon (J. Miller, ed.) pp. 89-122. CSH Laboratory, Cold Spring Harbor, N.Y.

Zagursky, R.J., Conway, P.S., and Kashdan, M.A. 1991. Use of ^{33}P for Sanger DNA sequencing. *BioTechniques* 11:36-38.

Zinn, K., DiMaio, D., and Maniatis, T. 1983. Identification of two distinct regulatory regions adjacent to the human β-interferon gene. *Cell* 34:865-879.

Zugibe, F. 1970. Diagnostic Histochemistry. C.V. Mosby, St. Louis, Mo.

INDEX

Page numbers in this book are hyphenated: the number before the hyphen refers to the chapter and the number after the hyphen refers to the page within the chapter (e.g., 12-3 is page 3 of Chapter 12). A range of pages is indicated by an arrow connecting the page numbers (e.g., 12-3→12-5 refers to pages 3 through 5 of Chapter 12). Whole units are indicated by numbers with decimals (e.g. UNIT 2.7). Within the body of the book, units are distinguished by black margin tabs on every page.

A

B

C

I

D

E

F

G

H

M

N

O

P

R

S

T

U

V

W

X

Y

Z